EQUATION SHEET

Ideal-gas law: $p = \rho RT$, $R_{\text{air}} = 287$ J/kg-K	Surface tension: $\Delta p = Y(R$
Hydrostatics, constant density: $$p_2 - p_1 = -\gamma(z_2 - z_1), \quad \gamma = \rho g$$	Hydrostatic panel force: $F = \gamma h_{\text{CG}}A$, $$y_{\text{CP}} = -I_{xx}\sin\theta/(h_{\text{CG}}A), \; x_{\text{CP}} = -I_{xy}\sin\theta/(h_{\text{CG}}A)$$
Buoyant force: $$F_B = \gamma_{\text{fluid}}(\text{displaced volume})$$	CV mass: $d/dt(\int_{\text{cv}}\rho d\upsilon) + \sum(\rho AV)_{\text{out}}$ $$- \sum(\rho AV)_{\text{in}} = 0$$
CV momentum: $d/dt(\int_{\text{cv}}\rho\mathbf{V}d\upsilon)$ $$+ \sum[(\rho AV)\mathbf{V}]_{\text{out}} - \sum[(\rho AV)\mathbf{V}]_{\text{in}} = \sum\mathbf{F}$$	CV angular momentum: $d/dt(\int_{\text{cv}}\rho(\mathbf{r_0}\times\mathbf{V})d\upsilon)$ $$+ \sum\rho AV(\mathbf{r_0}\times\mathbf{V})_{\text{out}} - \sum\rho AV(\mathbf{r_0}\times\mathbf{V})_{\text{in}} = \sum\mathbf{M_0}$$
Steady flow energy: $(p/\gamma + \alpha V^2/2g + z)_{\text{in}} =$ $(p/\gamma + \alpha V^2/2g + z)_{\text{out}} + h_{\text{friction}} - h_{\text{pump}} + h_{\text{turbine}}$	Acceleration: $d\mathbf{V}/dt = \partial\mathbf{V}/\partial t$ $$+ u(\partial\mathbf{V}/\partial x) + v(\partial\mathbf{V}/\partial y) + w(\partial\mathbf{V}/\partial z)$$
Incompressible continuity: $\nabla \cdot \mathbf{V} = 0$	Navier-Stokes: $\rho(d\mathbf{V}/dt) = \rho\mathbf{g} - \nabla p + \mu\nabla^2\mathbf{V}$
Incompressible stream function $\psi(x,y)$: $$u = \partial\psi/\partial y; \quad v = -\partial\psi/\partial x$$	Velocity potential $\phi(x,y,z)$: $$u = \partial\phi/\partial x; \; v = \partial\phi/\partial y; \; w = \partial\phi/\partial z$$
Bernoulli unsteady irrotational flow: $$\partial\phi/\partial t + \int dp/\rho + V^2/2 + gz = \text{Const}$$	Turbulent friction factor: $1/\sqrt{f} =$ $$-2.0\log_{10}[\varepsilon/(3.7d) + 2.51/(\text{Re}_d\sqrt{f})]$$
Pipe head loss: $h_f = f(L/d)V^2/(2g)$ where f = Moody chart friction factor	Orifice, nozzle, venturi flow: $$Q = C_d A_{\text{throat}}[2\Delta p/\{\rho(1-\beta^4)\}]^{1/2}, \quad \beta = d/D$$
Laminar flat plate flow: $\delta/x = 5.0/\text{Re}_x^{1/2}$, $$c_f = 0.664/\text{Re}_x^{1/2}, \quad C_D = 1.328/\text{Re}_L^{1/2}$$	Turbulent flat plate flow: $\delta/x = 0.16/\text{Re}_x^{1/7}$, $$c_f = 0.027/\text{Re}_x^{1/7}, \; C_D = 0.031/\text{Re}_L^{1/7}$$
$C_D = \text{Drag}/(\frac{1}{2}\rho V^2 A)$; $C_L = \text{Lift}/(\frac{1}{2}\rho V^2 A)$	2-D potential flow: $\nabla^2\phi = \nabla^2\psi = 0$
Isentropic flow: $T_0/T = 1 + \{(k-1)/2\}\text{Ma}^2$, $$\rho_0/\rho = (T_0/T)^{1/(k-1)}, \quad p_0/p = (T_0/T)^{k(k-1)}$$	One-dimensional isentropic area change: $$A/A^* = (1/\text{Ma})[1 + \{(k-1)/2\}\text{Ma}^2]^{(1/2)(k+1)/(k-1)}$$
Prandtl-Meyer expansion: $K = (k+1)/(k-1)$, $$\omega = K^{1/2}\tan^{-1}[(\text{Ma}^2-1)/K]^{1/2} - \tan^{-1}(\text{Ma}^2-1)^{1/2}$$	Uniform flow, Manning's n, SI units: $$V_0(\text{m/s}) = (1.0/n)[R_h(m)]^{2/3}S_0^{1/2}$$
Gradually varied channel flow: $$dy/dx = (S_0 - S)/(1 - \text{Fr}^2), \text{Fr} = V/V_{\text{crit}}$$	Euler turbine formula: $$\text{Power} = \rho Q(u_2 V_{t2} - u_1 V_{t1}), \; u = r\omega$$

Fluid Mechanics

McGraw-Hill Series in Mechanical Engineering

Alciatore/Histand
Introduction to Mechatronics and Measurement Systems

Anderson
Computational Fluid Dynamics: The Basics with Applications

Anderson
Fundamentals of Aerodynamics

Anderson
Introduction to Flight

Anderson
Modern Compressible Flow

Barber
Intermediate Mechanics of Materials

Beer/Johnston
Vector Mechanics for Engineers: Statics and Dynamics

Beer/Johnston
Mechanics of Materials

Budynas
Advanced Strength and Applied Stress Analysis

Çengel
Heat and Mass Transfer: A Practical Approach

Çengel
Introduction to Thermodynamics & Heat Transfer

Çengel/Boles
Thermodynamics: An Engineering Approach

Çengel/Cimbala
Fluid Mechanics: Fundamentals and Applications

Çengel/Turner
Fundamentals of Thermal-Fluid Sciences

Crespo da Silva
Intermediate Dynamics

Dieter
Engineering Design: A Materials & Processing Approach

Dieter
Mechanical Metallurgy

Doebelin
Measurement Systems: Application & Design

Dorf/Byers
Technology Ventures: From Idea to Enterprise

Dunn
Measurement & Data Analysis for Engineering & Science

EDS, Inc.
I-DEAS Student Guide

Finnemore/Franzini
Fluid Mechanics with Engineering Applications

Hamrock/Schmid/Jacobson
Fundamentals of Machine Elements

Heywood
Internal Combustion Engine Fundamentals

Holman
Experimental Methods for Engineers

Holman
Heat Transfer

Hutton
Fundamentals of Finite Element Analysis

Kays/Crawford/Weigand
Convective Heat and Mass Transfer

Meirovitch
Fundamentals of Vibrations

Norton
Design of Machinery

Palm
System Dynamics

Reddy
An Introduction to Finite Element Method

Schaffer et al.
The Science and Design of Engineering Materials

Schey
Introduction to Manufacturing Processes

Budynas/Nisbett
Shigley's Mechanical Engineering Design

Smith/Hashemi
Foundations of Materials Science and Engineering

Turns
An Introduction to Combustion: Concepts and Applications

Ugural
Mechanical Design: An Integrated Approach

Ullman
The Mechanical Design Process

White
Fluid Mechanics

White
Viscous Fluid Flow

Zeid
CAD/CAM Theory and Practice

Zeid
Mastering CAD/CAM

Chapter 1
Introduction

1.1 Preliminary Remarks

Fluid mechanics is the study of fluids either in motion (fluid *dynamics*) or at rest (fluid *statics*). Both gases and liquids are classified as fluids, and the number of fluid engineering applications is enormous: breathing, blood flow, swimming, pumps, fans, turbines, airplanes, ships, rivers, windmills, pipes, missiles, icebergs, engines, filters, jets, and sprinklers, to name a few. When you think about it, almost everything on this planet either is a fluid or moves within or near a fluid.

The essence of the subject of fluid flow is a judicious compromise between theory and experiment. Since fluid flow is a branch of mechanics, it satisfies a set of well-documented basic laws, and thus a great deal of theoretical treatment is available. However, the theory is often frustrating because it applies mainly to idealized situations, which may be invalid in practical problems. The two chief obstacles to a workable theory are geometry and viscosity. The basic equations of fluid motion (Chap. 4) are too difficult to enable the analyst to attack arbitrary geometric configurations. Thus most textbooks concentrate on flat plates, circular pipes, and other easy geometries. It is possible to apply numerical computer techniques to complex geometries, and specialized textbooks are now available to explain the new *computational fluid dynamics* (CFD) approximations and methods [1–4].[1] This book will present many theoretical results while keeping their limitations in mind.

The second obstacle to a workable theory is the action of viscosity, which can be neglected only in certain idealized flows (Chap. 8). First, viscosity increases the difficulty of the basic equations, although the boundary-layer approximation found by Ludwig Prandtl in 1904 (Chap. 7) has greatly simplified viscous-flow analyses. Second, viscosity has a destabilizing effect on all fluids, giving rise, at frustratingly small velocities, to a disorderly, random phenomenon called *turbulence*. The theory of turbulent flow is crude and heavily backed up by experiment (Chap. 6), yet it can be quite serviceable as an engineering estimate. This textbook only introduces the standard experimental correlations for turbulent time-mean flow. Meanwhile, there are advanced texts on both time-mean *turbulence and turbulence modeling* [5, 6] and on the newer, computer-intensive *direct numerical simulation* (DNS) of fluctuating turbulence [7, 8].

---

[1]Numbered references appear at the end of each chapter.

Thus there is theory available for fluid flow problems, but in all cases it should be backed up by experiment. Often the experimental data provide the main source of information about specific flows, such as the drag and lift of immersed bodies (Chap. 7). Fortunately, fluid mechanics is a highly visual subject, with good instrumentation [9–11], and the use of dimensional analysis and modeling concepts (Chap. 5) is widespread. Thus experimentation provides a natural and easy complement to the theory. You should keep in mind that theory and experiment should go hand in hand in all studies of fluid mechanics.

1.2 History and Scope of Fluid Mechanics

Like most scientific disciplines, fluid mechanics has a history of erratically occurring early achievements, then an intermediate era of steady fundamental discoveries in the eighteenth and nineteenth centuries, leading to the twentieth-century era of "modern practice," as we self-centeredly term our limited but up-to-date knowledge. Ancient civilizations had enough knowledge to solve certain flow problems. Sailing ships with oars and irrigation systems were both known in prehistoric times. The Greeks produced quantitative information. Archimedes and Hero of Alexandria both postulated the parallelogram law for addition of vectors in the third century B.C. Archimedes (285–212 B.C.) formulated the laws of buoyancy and applied them to floating and submerged bodies, actually deriving a form of the differential calculus as part of the analysis. The Romans built extensive aqueduct systems in the fourth century B.C. but left no records showing any quantitative knowledge of design principles.

From the birth of Christ to the Renaissance there was a steady improvement in the design of such flow systems as ships and canals and water conduits but no recorded evidence of fundamental improvements in flow analysis. Then Leonardo da Vinci (1452–1519) stated the equation of conservation of mass in one-dimensional steady flow. Leonardo was an excellent experimentalist, and his notes contain accurate descriptions of waves, jets, hydraulic jumps, eddy formation, and both low-drag (streamlined) and high-drag (parachute) designs. A Frenchman, Edme Mariotte (1620–1684), built the first wind tunnel and tested models in it.

Problems involving the momentum of fluids could finally be analyzed after Isaac Newton (1642–1727) postulated his laws of motion and the law of viscosity of the linear fluids now called newtonian. The theory first yielded to the assumption of a "perfect" or frictionless fluid, and eighteenth-century mathematicians (Daniel Bernoulli, Leonhard Euler, Jean d'Alembert, Joseph-Louis Lagrange, and Pierre-Simon Laplace) produced many beautiful solutions of frictionless-flow problems. Euler, Fig. 1.1, developed both the differential equations of motion and their integrated form, now called the Bernoulli equation. D'Alembert used them to show his famous paradox: that a body immersed in a frictionless fluid has zero drag. These beautiful results amounted to overkill, since perfect-fluid assumptions have very limited application in practice and most engineering flows are dominated by the effects of viscosity. Engineers began to reject what they regarded as a totally unrealistic theory and developed the science of *hydraulics,* relying almost entirely on experiment. Such experimentalists as Chézy, Pitot, Borda, Weber, Francis, Hagen, Poiseuille, Darcy, Manning, Bazin, and Weisbach produced data on a variety of flows such as open channels, ship resistance, pipe flows, waves, and turbines. All too often the data were used in raw form without regard to the fundamental physics of flow.

Fig. 1.1 Leonhard Euler (1707–1783) was the greatest mathematician of the eighteenth century and used Newton's calculus to develop and solve the equations of motion of inviscid flow. He published over 800 books and papers. [*Courtesy of the School of Mathematics and Statistics, University of St Andrew, Scotland.*]

Fig. 1.2 Ludwig Prandtl (1875–1953), often called the "father of modern fluid mechanics" [15], developed boundary layer theory and many other innovative analyses. He and his students were pioneers in flow visualization techniques. [*Aufnahme von Fr. Struckmeyer, Gottingen, courtesy AIP Emilio Segre Visual Archives, Lande Collection.*]

At the end of the nineteenth century, unification between experimental *hydraulics* and theoretical *hydrodynamics* finally began. William Froude (1810–1879) and his son Robert (1846–1924) developed laws of model testing; Lord Rayleigh (1842–1919) proposed the technique of dimensional analysis; and Osborne Reynolds (1842–1912) published the classic pipe experiment in 1883, which showed the importance of the dimensionless Reynolds number named after him. Meanwhile, viscous-flow theory was available but unexploited, since Navier (1785–1836) and Stokes (1819–1903) had successfully added newtonian viscous terms to the equations of motion. The resulting Navier-Stokes equations were too difficult to analyze for arbitrary flows. Then, in 1904, a German engineer, Ludwig Prandtl (1875–1953), Fig. 1.2, published perhaps the most important paper ever written on fluid mechanics. Prandtl pointed out that fluid flows with small viscosity, such as water flows and airflows, can be divided into a thin viscous layer, or *boundary layer*, near solid surfaces and interfaces, patched onto a nearly inviscid outer layer, where the Euler and Bernoulli equations apply. Boundary-layer theory has proved to be a very important tool in modern flow analysis. The twentieth-century foundations for the present state of the art in fluid mechanics were laid in a series of broad-based experiments and theories by Prandtl and his two chief friendly competitors, Theodore von Kármán (1881–1963) and Sir Geoffrey I. Taylor (1886–1975). Many of the results sketched here from a historical point of view will, of course, be discussed in this textbook. More historical details can be found in Refs. 12 to 14.

Since the earth is 75 percent covered with water and 100 percent covered with air, the scope of fluid mechanics is vast and touches nearly every human endeavor. The sciences of meteorology, physical oceanography, and hydrology are concerned with naturally occurring fluid flows, as are medical studies of breathing and blood circulation. All transportation problems involve fluid motion, with well-developed specialties in aerodynamics of aircraft and rockets and in naval hydrodynamics of ships and submarines. Almost all our electric energy is developed either from water flow or from steam flow through turbine generators. All combustion problems involve fluid motion as do the more classic problems of irrigation, flood control, water supply, sewage disposal, projectile motion, and oil and gas pipelines. The aim of this book is to present enough fundamental concepts and practical applications in fluid mechanics to prepare you to move smoothly into any of these specialized fields of the science of flow—and then be prepared to move out again as new technologies develop.

1.3 Problem-Solving Techniques

Fluid flow analysis is packed with problems to be solved. The present text has more than 1600 problem assignments. Solving a large number of these is a key to learning the subject. One must deal with equations, data, tables, assumptions, unit systems, and solution schemes. The degree of difficulty will vary, and we urge you to sample the whole spectrum of assignments, with or without the Answers in the Appendix. Here are the recommended steps for problem solution:

1. Read the problem and restate it with your summary of the results desired.
2. From tables or charts, gather the needed property data: density, viscosity, etc.
3. Make sure you understand what is *asked*. Students are apt to answer the wrong question—for example, pressure instead of pressure gradient, lift force instead of drag force, or mass flow instead of volume flow. Read the problem carefully.

4. Make a detailed, *labeled* sketch of the system or control volume needed.

5. Think carefully and list your *assumptions*. You must decide if the flow is steady or unsteady, compressible or incompressible, viscous or inviscid, and whether a control volume or partial differential equations are needed.

6. Find an algebraic solution if possible. Then, if a numerical value is needed, use either the SI or BG unit systems, to be reviewed in Sec. 1.6.

7. Report your solution, *labeled*, with the proper units and the proper number of significant figures (usually two or three) that the data uncertainty allows.

We shall follow these steps, where appropriate, in our example problems.

1.4 The Concept of a Fluid

From the point of view of fluid mechanics, all matter consists of only two states, fluid and solid. The difference between the two is perfectly obvious to the layperson, and it is an interesting exercise to ask a layperson to put this difference into words. The technical distinction lies with the reaction of the two to an applied shear or tangential stress. *A solid can resist a shear stress by a static deflection; a fluid cannot.* Any shear stress applied to a fluid, no matter how small, will result in motion of that fluid. The fluid moves and deforms continuously as long as the shear stress is applied. As a corollary, we can say that a fluid at rest must be in a state of zero shear stress, a state often called the hydrostatic stress condition in structural analysis. In this condition, Mohr's circle for stress reduces to a point, and there is no shear stress on any plane cut through the element under stress.

Given this definition of a fluid, every layperson also knows that there are two classes of fluids, *liquids* and *gases.* Again the distinction is a technical one concerning the effect of cohesive forces. A liquid, being composed of relatively close-packed molecules with strong cohesive forces, tends to retain its volume and will form a free surface in a gravitational field if unconfined from above. Free-surface flows are dominated by gravitational effects and are studied in Chaps. 5 and 10. Since gas molecules are widely spaced with negligible cohesive forces, a gas is free to expand until it encounters confining walls. A gas has no definite volume, and when left to itself without confinement, a gas forms an atmosphere that is essentially hydrostatic. The hydrostatic behavior of liquids and gases is taken up in Chap. 2. Gases cannot form a free surface, and thus gas flows are rarely concerned with gravitational effects other than buoyancy.

Figure 1.3 illustrates a solid block resting on a rigid plane and stressed by its own weight. The solid sags into a static deflection, shown as a highly exaggerated dashed line, resisting shear without flow. A free-body diagram of element A on the side of the block shows that there is shear in the block along a plane cut at an angle θ through A. Since the block sides are unsupported, element A has zero stress on the left and right sides and compression stress $\sigma = -p$ on the top and bottom. Mohr's circle does not reduce to a point, and there is nonzero shear stress in the block.

By contrast, the liquid and gas at rest in Fig. 1.3 require the supporting walls in order to eliminate shear stress. The walls exert a compression stress of $-p$ and reduce Mohr's circle to a point with zero shear everywhere—that is, the hydrostatic condition. The liquid retains its volume and forms a free surface in the container. If the walls are removed, shear develops in the liquid and a big splash results. If the container is

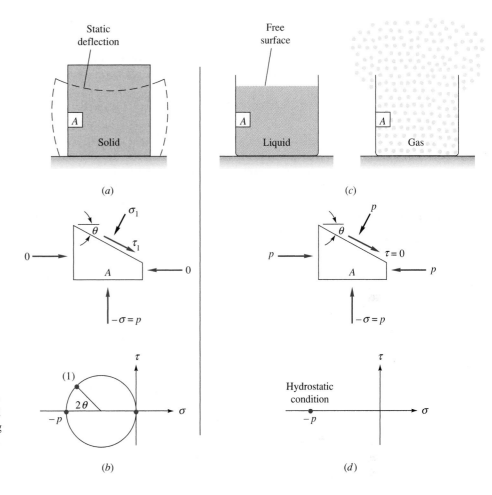

Fig. 1.3 A solid at rest can resist shear. (*a*) Static deflection of the solid; (*b*) equilibrium and Mohr's circle for solid element *A*. A fluid cannot resist shear. (*c*) Containing walls are needed; (*d*) equilibrium and Mohr's circle for fluid element *A*.

tilted, shear again develops, waves form, and the free surface seeks a horizontal configuration, pouring out over the lip if necessary. Meanwhile, the gas is unrestrained and expands out of the container, filling all available space. Element *A* in the gas is also hydrostatic and exerts a compression stress $-p$ on the walls.

In the previous discussion, clear decisions could be made about solids, liquids, and gases. Most engineering fluid mechanics problems deal with these clear cases—that is, the common liquids, such as water, oil, mercury, gasoline, and alcohol, and the common gases, such as air, helium, hydrogen, and steam, in their common temperature and pressure ranges. There are many borderline cases, however, of which you should be aware. Some apparently "solid" substances such as asphalt and lead resist shear stress for short periods but actually deform slowly and exhibit definite fluid behavior over long periods. Other substances, notably colloid and slurry mixtures, resist small shear stresses but "yield" at large stress and begin to flow as fluids do. Specialized textbooks are devoted to this study of more general deformation and flow, a field called *rheology* [16]. Also, liquids and gases can coexist in two-phase mixtures, such as steam–water mixtures or water with entrapped air bubbles.

Specialized textbooks present the analysis of such *multiphase flows* [17]. Finally, in some situations the distinction between a liquid and a gas blurs. This is the case at temperatures and pressures above the so-called *critical point* of a substance, where only a single phase exists, primarily resembling a gas. As pressure increases far above the critical point, the gaslike substance becomes so dense that there is some resemblance to a liquid and the usual thermodynamic approximations like the perfect-gas law become inaccurate. The critical temperature and pressure of water are $T_c = 647$ K and $p_c = 219$ atm (atmosphere[2]) so that typical problems involving water and steam are below the critical point. Air, being a mixture of gases, has no distinct critical point, but its principal component, nitrogen, has $T_c = 126$ K and $p_c = 34$ atm. Thus typical problems involving air are in the range of high temperature and low pressure where air is distinctly and definitely a gas. This text will be concerned solely with clearly identifiable liquids and gases, and the borderline cases just discussed will be beyond our scope.

1.5 The Fluid as a Continuum

We have already used technical terms such as *fluid pressure* and *density* without a rigorous discussion of their definition. As far as we know, fluids are aggregations of molecules, widely spaced for a gas, closely spaced for a liquid. The distance between molecules is very large compared with the molecular diameter. The molecules are not fixed in a lattice but move about freely relative to each other. Thus fluid density, or mass per unit volume, has no precise meaning because the number of molecules occupying a given volume continually changes. This effect becomes unimportant if the unit volume is large compared with, say, the cube of the molecular spacing, when the number of molecules within the volume will remain nearly constant in spite of the enormous interchange of particles across the boundaries. If, however, the chosen unit volume is too large, there could be a noticeable variation in the bulk aggregation of the particles. This situation is illustrated in Fig. 1.4, where the "density" as calculated from molecular mass δm within a given volume $\delta \mathcal{V}$ is plotted versus the size of the unit volume. There is a limiting volume $\delta \mathcal{V}^*$ below which molecular variations

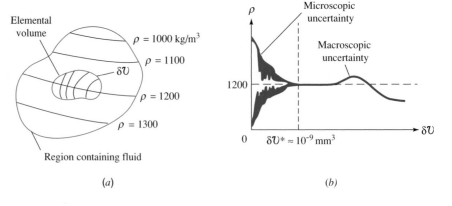

Fig. 1.4 The limit definition of continuum fluid density: (*a*) an elemental volume in a fluid region of variable continuum density; (*b*) calculated density versus size of the elemental volume.

[2]One atmosphere equals 2116 lbf/ft^2 = 101,300 Pa.

may be important and above which aggregate variations may be important. The *density* ρ of a fluid is best defined as

$$\rho = \lim_{\delta V \to \delta V^*} \frac{\delta m}{\delta V} \tag{1.1}$$

The limiting volume δV^* is about 10^{-9} mm^3 for all liquids and for gases at atmospheric pressure. For example, 10^{-9} mm^3 of air at standard conditions contains approximately 3×10^7 molecules, which is sufficient to define a nearly constant density according to Eq. (1.1). Most engineering problems are concerned with physical dimensions much larger than this limiting volume, so that density is essentially a point function and fluid properties can be thought of as varying continually in space, as sketched in Fig. 1.4a. Such a fluid is called a *continuum,* which simply means that its variation in properties is so smooth that differential calculus can be used to analyze the substance. We shall assume that continuum calculus is valid for all the analyses in this book. Again there are borderline cases for gases at such low pressures that molecular spacing and mean free path[3] are comparable to, or larger than, the physical size of the system. This requires that the continuum approximation be dropped in favor of a molecular theory of rarefied gas flow [18]. In principle, all fluid mechanics problems can be attacked from the molecular viewpoint, but no such attempt will be made here. Note that the use of continuum calculus does not preclude the possibility of discontinuous jumps in fluid properties across a free surface or fluid interface or across a shock wave in a compressible fluid (Chap. 9). Our calculus in analyzing fluid flow must be flexible enough to handle discontinuous boundary conditions.

1.6 Dimensions and Units

A *dimension* is the measure by which a physical variable is expressed quantitatively. A *unit* is a particular way of attaching a number to the quantitative dimension. Thus length is a dimension associated with such variables as distance, displacement, width, deflection, and height, while centimeters and inches are both numerical units for expressing length. Dimension is a powerful concept about which a splendid tool called *dimensional analysis* has been developed (Chap. 5), while units are the numerical quantity that the customer wants as the final answer.

In 1872 an international meeting in France proposed a treaty called the Metric Convention, which was signed in 1875 by 17 countries including the United States. It was an improvement over British systems because its use of base 10 is the foundation of our number system, learned from childhood by all. Problems still remained because even the metric countries differed in their use of kiloponds instead of dynes or newtons, kilograms instead of grams, or calories instead of joules. To standardize the metric system, a General Conference of Weights and Measures, attended in 1960 by 40 countries, proposed the *International System of Units* (SI). We are now undergoing a painful period of transition to SI, an adjustment that may take many more years to complete. The professional societies have led the way. Since July 1, 1974, SI units have been required by all papers published by the American Society of Mechanical

---

[3]The mean distance traveled by molecules between collisions (see Prob. P1.5).

Table 1.1 Primary Dimensions in SI and BG Systems

Primary dimension	SI unit	BG unit	Conversion factor
Mass $\{M\}$	Kilogram (kg)	Slug	1 slug = 14.5939 kg
Length $\{L\}$	Meter (m)	Foot (ft)	1 ft = 0.3048 m
Time $\{T\}$	Second (s)	Second (s)	1 s = 1 s
Temperature $\{\Theta\}$	Kelvin (K)	Rankine (°R)	1 K = 1.8°R

Engineers, and there is a textbook explaining the SI [19]. The present text will use SI units together with British gravitational (BG) units.

Primary Dimensions

In fluid mechanics there are only four *primary dimensions* from which all other dimensions can be derived: mass, length, time, and temperature.[4] These dimensions and their units in both systems are given in Table 1.1. Note that the kelvin unit uses no degree symbol. The braces around a symbol like $\{M\}$ mean "the dimension" of mass. All other variables in fluid mechanics can be expressed in terms of $\{M\}$, $\{L\}$, $\{T\}$, and $\{\Theta\}$. For example, acceleration has the dimensions $\{LT^{-2}\}$. The most crucial of these secondary dimensions is force, which is directly related to mass, length, and time by Newton's second law. Force equals the time rate of change of momentum or, for constant mass,

$$\mathbf{F} = m\mathbf{a} \tag{1.2}$$

From this we see that, dimensionally, $\{F\} = \{MLT^{-2}\}$.

The International System (SI)

The use of a constant of proportionality in Newton's law, Eq. (1.2), is avoided by defining the force unit exactly in terms of the other basic units. In the SI system, the basic units are newtons $\{F\}$, kilograms $\{M\}$, meters $\{L\}$, and seconds $\{T\}$. We define

$$1 \text{ newton of force} = 1 \text{ N} = 1 \text{ kg} \cdot 1 \text{ m/s}^2$$

The newton is a relatively small force, about the weight of an apple (0.225 lbf). In addition, the basic unit of temperature $\{\Theta\}$ in the SI system is the degree Kelvin, K. Use of these SI units (N, kg, m, s, K) will require no conversion factors in our equations.

The British Gravitational (BG) System

In the BG system also, a constant of proportionality in Eq. (1.2) is avoided by defining the force unit exactly in terms of the other basic units. In the BG system, the basic units are pound-force $\{F\}$, slugs $\{M\}$, feet $\{L\}$, and seconds $\{T\}$. We define

$$1 \text{ pound of force} = 1 \text{ lbf} = 1 \text{ slug} \cdot 1 \text{ ft/s}^2$$

One lbf $\approx$ 4.4482 N and approximates the weight of four apples. We will use the abbreviation *lbf* for pound-force and *lbm* for pound-mass. The slug is a rather hefty

---

[4]If electromagnetic effects are important, a fifth primary dimension must be included, electric current $\{I\}$, whose SI unit is the ampere (A).

mass, equal to 32.174 lbm. The basic unit of temperature $\{\Theta\}$ in the BG system is the degree Rankine, °R. Recall that a temperature difference 1 K = 1.8°R. Use of these BG units (lbf, slug, ft, s, °R) will require no conversion factors in our equations.

Other Unit Systems

There are other unit systems still in use. At least one needs no proportionality constant: the CGS system (dyne, gram, cm, s, K). However, CGS units are too small for most applications (1 dyne = 10^{-5} N) and will not be used here.

In the USA, some still use the English Engineering system, (lbf, lbm, ft, s, °R), where the basic mass unit is the *pound of mass*. Newton's law (1.2) must be rewritten:

$$\mathbf{F} = \frac{m\mathbf{a}}{g_c}, \quad \text{where} \quad g_c = 32.174 \frac{\text{ft} \cdot \text{lbm}}{\text{lbf} \cdot \text{s}^2} \tag{1.3}$$

The constant of proportionality, g_c, has both dimensions and a numerical value not equal to 1.0. The present text uses only the SI and BG systems and will not solve problems or examples in the English Engineering system. Because Americans still use them, a few problems in the text will be stated in truly awkward units: acres, gallons, ounces, or miles. Your assignment will be to convert these and solve in the SI or BG systems.

The Principle of Dimensional Homogeneity

In engineering and science, *all* equations must be *dimensionally homogeneous,* that is, each additive term in an equation must have the same dimensions. For example, take Bernoulli's incompressible equation, to be studied and used throughout this text:

$$p + \frac{1}{2}\rho V^2 + \rho g Z = \text{constant}$$

Each and every term in this equation *must* have dimensions of pressure $\{ML^{-1}T^{-2}\}$. We will examine the dimensional homogeneity of this equation in detail in Ex. 1.3.

A list of some important secondary variables in fluid mechanics, with dimensions derived as combinations of the four primary dimensions, is given in Table 1.2. A more complete list of conversion factors is given in App. C.

Table 1.2 Secondary Dimensions in Fluid Mechanics

Secondary dimension	SI unit	BG unit	Conversion factor
Area $\{L^2\}$	m²	ft²	1 m² = 10.764 ft²
Volume $\{L^3\}$	m³	ft³	1 m³ = 35.315 ft³
Velocity $\{LT^{-1}\}$	m/s	ft/s	1 ft/s = 0.3048 m/s
Acceleration $\{LT^{-2}\}$	m/s²	ft/s²	1 ft/s² = 0.3048 m/s²
Pressure or stress $\{ML^{-1}T^{-2}\}$	Pa = N/m²	lbf/ft²	1 lbf/ft² = 47.88 Pa
Angular velocity $\{T^{-1}\}$	s⁻¹	s⁻¹	1 s⁻¹ = 1 s⁻¹
Energy, heat, work $\{ML^2T^{-2}\}$	J = N · m	ft · lbf	1 ft · lbf = 1.3558 J
Power $\{ML^2T^{-3}\}$	W = J/s	ft · lbf/s	1 ft · lbf/s = 1.3558 W
Density $\{ML^{-3}\}$	kg/m³	slugs/ft³	1 slug/ft³ = 515.4 kg/m³
Viscosity $\{ML^{-1}T^{-1}\}$	kg/(m · s)	slugs/(ft · s)	1 slug/(ft · s) = 47.88 kg/(m · s)
Specific heat $\{L^2T^{-2}\Theta^{-1}\}$	m²/(s² · K)	ft²/(s² · °R)	1 m²/(s² · K) = 5.980 ft²/(s² · °R)

EXAMPLE 1.1

A body weighs 1000 lbf when exposed to a standard earth gravity $g = 32.174$ ft/s^2. (*a*) What is its mass in kg? (*b*) What will the weight of this body be in N if it is exposed to the moon's standard acceleration $g_{moon} = 1.62$ m/s^2? (*c*) How fast will the body accelerate if a net force of 400 lbf is applied to it on the moon or on the earth?

Solution

We need to find the (*a*) mass; (*b*) weight on the moon; and (*c*) acceleration of this body. This is a fairly simple example of conversion factors for differing unit systems. No property data is needed. The example is too low-level for a sketch.

Part (a)

Newton's law (1.2) holds with known weight and gravitational acceleration. Solve for *m:*

$$F = W = 1000 \text{ lbf} = mg = (m)(32.174 \text{ ft/s}^2), \quad or \quad m = \frac{1000 \text{ lbf}}{32.174 \text{ ft/s}^2} = 31.08 \text{ slugs}$$

Convert this to kilograms:

$$m = 31.08 \text{ slugs} = (31.08 \text{ slugs})(14.5939 \text{ kg/slug}) = 454 \text{ kg} \qquad \textit{Ans. (a)}$$

Part (b)

The mass of the body remains 454 kg regardless of its location. Equation (1.2) applies with a new gravitational acceleration and hence a new weight:

$$F = W_{moon} = mg_{moon} = (454 \text{ kg})(1.62 \text{ m/s}^2) = 735 \text{ N} \qquad \textit{Ans. (b)}$$

Part (c)

This part does not involve weight or gravity or location. It is simply an application of Newton's law with a known mass and known force:

$$F = 400 \text{ lbf} = ma = (31.08 \text{ slugs}) \, a$$

Solve for

$$a = \frac{400 \text{ lbf}}{31.08 \text{ slugs}} = 12.87 \frac{\text{ft}}{\text{s}^2}\left(0.3048\frac{\text{m}}{\text{ft}}\right) = 3.92\frac{\text{m}}{\text{s}^2} \qquad \textit{Ans. (c)}$$

Comment (c):   This acceleration would be the same on the earth or moon or anywhere.

Many data in the literature are reported in inconvenient or arcane units suitable only to some industry or specialty or country. The engineer should convert these data to the SI or BG system before using them. This requires the systematic application of conversion factors, as in the following example.

EXAMPLE 1.2

Industries involved in viscosity measurement [27, 36] continue to use the CGS system of units, since centimeters and grams yield convenient numbers for many fluids. The absolute viscosity (μ) unit is the *poise,* named after J. L. M. Poiseuille, a French physician who in 1840 performed pioneering experiments on water flow in pipes; 1 poise = 1 g/(cm-s). The kinematic viscosity (ν) unit is the *stokes,* named after G. G. Stokes, a British physicist who in 1845 helped develop the basic partial differential equations of fluid momentum; 1 stokes $= 1$ cm^2/s. Water at 20°C has $\mu \approx 0.01$ poise and also $\nu \approx 0.01$ stokes. Express these results in (a) SI and (b) BG units.

Solution

Part (a)

- *Approach:* Systematically change grams to kg or slugs and change centimeters to meters or feet.
- *Property values:* Given $\mu = 0.01$ g/(cm-s) and $\nu = 0.01$ cm^2/s.
- *Solution steps:* (a) For conversion to SI units,

$$\mu = 0.01 \frac{g}{cm \cdot s} = 0.01 \frac{g(1 \text{ kg}/1000g)}{cm(0.01 \text{ m}/cm)s} = 0.001 \frac{kg}{m \cdot s}$$

$$\nu = 0.01 \frac{cm^2}{s} = 0.01 \frac{cm^2(0.01 \text{ m}/cm)^2}{s} = 0.000001 \frac{m^2}{s} \qquad \textit{Ans. (a)}$$

Part (b)

- For conversion to BG units

$$\mu = 0.01 \frac{g}{cm \cdot s} = 0.01 \frac{g(1 \text{ kg}/1000 \text{ g})(1 \text{ slug}/14.5939 \text{ kg})}{(0.01 \text{ m}/cm)(1 \text{ ft}/0.3048 \text{ m})s} = 0.0000209 \frac{slug}{ft \cdot s}$$

$$\nu = 0.01 \frac{cm^2}{s} = 0.01 \frac{cm^2(0.01 \text{ m}/cm)^2(1 \text{ ft}/0.3048 \text{ m})^2}{s} = 0.0000108 \frac{ft^2}{s} \qquad \textit{Ans. (b)}$$

- *Comments:* This was a laborious conversion that could have been shortened by using the direct viscosity conversion factors in App. C. For example, $\mu_{BG} = \mu_{SI}/47.88$.

We repeat our advice: Faced with data in unusual units, convert them immediately to either SI or BG units because (1) it is more professional and (2) theoretical equations in fluid mechanics are *dimensionally consistent* and require no further conversion factors when these two fundamental unit systems are used, as the following example shows.

EXAMPLE 1.3

A useful theoretical equation for computing the relation between pressure, velocity, and altitude in a steady flow of a nearly inviscid, nearly incompressible fluid with negligible heat transfer and shaft work[5] is the *Bernoulli relation,* named after Daniel Bernoulli, who published a hydrodynamics textbook in 1738:

$$p_0 = p + \tfrac{1}{2}\rho V^2 + \rho g Z \qquad (1)$$

where p_0 = stagnation pressure
p = pressure in moving fluid
V = velocity
ρ = density
Z = altitude
g = gravitational acceleration

---

[5]That's an awful lot of assumptions, which need further study in Chap. 3.

(*a*) Show that Eq. (1) satisfies the principle of dimensional homogeneity, which states that all additive terms in a physical equation must have the same dimensions. (*b*) Show that consistent units result without additional conversion factors in SI units. (*c*) Repeat (*b*) for BG units.

Solution

Part (a)

We can express Eq. (1) dimensionally, using braces, by entering the dimensions of each term from Table 1.2:

$$\{ML^{-1}T^{-2}\} = \{ML^{-1}T^{-2}\} + \{ML^{-3}\}\{L^2T^{-2}\} + \{ML^{-3}\}\{LT^{-2}\}\{L\}$$

$$= \{ML^{-1}T^{-2}\} \text{ for all terms} \qquad Ans.\ (a)$$

Part (b)

Enter the SI units for each quantity from Table 1.2:

$$\{N/m^2\} = \{N/m^2\} + \{kg/m^3\}\{m^2/s^2\} + \{kg/m^3\}\{m/s^2\}\{m\}$$

$$= \{N/m^2\} + \{kg/(m \cdot s^2)\}$$

The right-hand side looks bad until we remember from Eq. (1.3) that $1\ kg = 1\ N \cdot s^2/m$.

$$\{kg/(m \cdot s^2)\} = \frac{\{N \cdot s^2/m\}}{\{m \cdot s^2\}} = \{N/m^2\} \qquad Ans.\ (b)$$

Thus all terms in Bernoulli's equation will have units of pascals, or newtons per square meter, when SI units are used. No conversion factors are needed, which is true of all theoretical equations in fluid mechanics.

Part (c)

Introducing BG units for each term, we have

$$\{lbf/ft^2\} = \{lbf/ft^2\} + \{slugs/ft^3\}\{ft^2/s^2\} + \{slugs/ft^3\}\{ft/s^2\}\{ft\}$$

$$= \{lbf/ft^2\} + \{slugs/(ft \cdot s^2)\}$$

But, from Eq. (1.3), $1\ slug = 1\ lbf \cdot s^2/ft$, so that

$$\{slugs/(ft \cdot s^2)\} = \frac{\{lbf \cdot s^2/ft\}}{\{ft \cdot s^2\}} = \{lbf/ft^2\} \qquad Ans.\ (c)$$

All terms have the unit of pounds-force per square foot. No conversion factors are needed in the BG system either.

There is still a tendency in English-speaking countries to use pound-force per square inch as a pressure unit because the numbers are more manageable. For example, standard atmospheric pressure is $14.7\ lbf/in^2 = 2116\ lbf/ft^2 = 101{,}300\ Pa$. The pascal is a small unit because the newton is less than $\frac{1}{4}$ lbf and a square meter is a very large area.

Consistent Units

Note that not only must all (fluid) mechanics equations be dimensionally homogeneous, one must also use *consistent units;* that is, each additive term must have the same units. There is no trouble doing this with the SI and BG systems, as in Example 1.3, but woe unto those who try to mix colloquial English units. For example, in Chap. 9, we often use the assumption of steady adiabatic compressible gas flow:

$$h + \tfrac{1}{2}V^2 = \text{constant}$$

where h is the fluid enthalpy and $V^2/2$ is its kinetic energy per unit mass. Colloquial thermodynamic tables might list h in units of British thermal units per pound mass (Btu/lb), whereas V is likely used in ft/s. It is completely erroneous to add Btu/lb to ft^2/s^2. The proper unit for h in this case is ft · lbf/slug, which is identical to ft^2/s^2. The conversion factor is 1 Btu/lb $\approx$ 25,040 ft^2/s^2 = 25,040 ft · lbf/slug.

Homogeneous versus Dimensionally Inconsistent Equations

All theoretical equations in mechanics (and in other physical sciences) are *dimensionally homogeneous;* that is, each additive term in the equation has the same dimensions. However, the reader should be warned that many empirical formulas in the engineering literature, arising primarily from correlations of data, are dimensionally inconsistent. Their units cannot be reconciled simply, and some terms may contain hidden variables. An example is the formula that pipe valve manufacturers cite for liquid volume flow rate Q (m^3/s) through a partially open valve:

$$Q = C_V \left(\frac{\Delta p}{\text{SG}} \right)^{1/2}$$

where Δp is the pressure drop across the valve and SG is the specific gravity of the liquid (the ratio of its density to that of water). The quantity C_V is the *valve flow coefficient,* which manufacturers tabulate in their valve brochures. Since SG is dimensionless $\{1\}$, we see that this formula is totally inconsistent, with one side being a flow rate $\{L^3/T\}$ and the other being the square root of a pressure drop $\{M^{1/2}/L^{1/2}T\}$. It follows that C_V must have dimensions, and rather odd ones at that: $\{L^{7/2}/M^{1/2}\}$. Nor is the resolution of this discrepancy clear, although one hint is that the values of C_V in the literature increase nearly as the square of the size of the valve. The presentation of experimental data in homogeneous form is the subject of *dimensional analysis* (Chap. 5). There we shall learn that a homogeneous form for the valve flow relation is

Table 1.3 Convenient Prefixes for Engineering Units

Multiplicative factor	Prefix	Symbol
10^{12}	tera	T
10^9	giga	G
10^6	mega	M
10^3	kilo	k
10^2	hecto	h
10	deka	da
10^{-1}	deci	d
10^{-2}	centi	c
10^{-3}	milli	m
10^{-6}	micro	μ
10^{-9}	nano	n
10^{-12}	pico	p
10^{-15}	femto	f
10^{-18}	atto	a

$$Q = C_d A_{\text{opening}} \left(\frac{\Delta p}{\rho} \right)^{1/2}$$

where ρ is the liquid density and A the area of the valve opening. The *discharge coefficient C_d* is dimensionless and changes only moderately with valve size. Please believe—until we establish the fact in Chap. 5—that this latter is a *much* better formulation of the data.

Meanwhile, we conclude that dimensionally inconsistent equations, though they occur in engineering practice, are misleading and vague and even dangerous, in the sense that they are often misused outside their range of applicability.

Convenient Prefixes in Powers of 10

Engineering results often are too small or too large for the common units, with too many zeros one way or the other. For example, to write p = 114,000,000 Pa is long and awkward. Using the prefix "M" to mean 10^6, we convert this to a concise p = 114 MPa (megapascals). Similarly, t = 0.000000003 s is a proofreader's nightmare compared to the equivalent t = 3 ns (nanoseconds). Such prefixes are common and convenient, in both the SI and BG systems. A complete list is given in Table 1.3.

EXAMPLE 1.4

In 1890 Robert Manning, an Irish engineer, proposed the following empirical formula for the average velocity V in uniform flow due to gravity down an open channel (BG units):

$$V = \frac{1.49}{n}R^{2/3}S^{1/2} \qquad (1)$$

where R = hydraulic radius of channel (Chaps. 6 and 10)
$\quad\quad\ S$ = channel slope (tangent of angle that bottom makes with horizontal)
$\quad\quad\ n$ = Manning's roughness factor (Chap. 10)

and n is a constant for a given surface condition for the walls and bottom of the channel. (*a*) Is Manning's formula dimensionally consistent? (*b*) Equation (1) is commonly taken to be valid in BG units with n taken as dimensionless. Rewrite it in SI form.

Solution

- *Assumption:*   The channel slope S is the tangent of an angle and is thus a dimensionless ratio with the dimensional notation $\{1\}$—that is, not containing M, L, or T.
- *Approach (a):*   Rewrite the dimensions of each term in Manning's equation, using brackets $\{\ \}$:

$$\{V\} = \left\{\frac{1.49}{n}\right\}\{R^{2/3}\}\{S^{1/2}\} \quad \text{or} \quad \left\{\frac{L}{T}\right\} = \left\{\frac{1.49}{n}\right\}\{L^{2/3}\}\{1\}$$

This formula is incompatible unless $\{1.49/n\} = \{L^{1/3}/T\}$. If n is dimensionless (and it is never listed with units in textbooks), the number 1.49 must carry the dimensions of $\{L^{1/3}/T\}$. *Ans.* (*a*)

- *Comment (a):*   Formulas whose numerical coefficients have units can be disastrous for engineers working in a different system or another fluid. Manning's formula, though popular, is inconsistent both dimensionally and physically and is valid only for water flow with certain wall roughnesses. The effects of water viscosity and density are hidden in the numerical value 1.49.
- *Approach (b):*   Part (*a*) showed that 1.49 has dimensions. If the formula is valid in BG units, then it must equal 1.49 $\text{ft}^{1/3}$/s. By using the SI conversion for length, we obtain

$$(1.49\ \text{ft}^{1/3}/\text{s})(0.3048\ \text{m/ft})^{1/3} = 1.00\ \text{m}^{1/3}/\text{s}$$

Therefore Manning's inconsistent formula changes form when converted to the SI system:

$$\text{SI units:} \quad V = \frac{1.0}{n}R^{2/3}S^{1/2} \qquad\qquad \textit{Ans. (b)}$$

with R in meters and V in meters per second.
- *Comment (b):*   Actually, we misled you: This is the way Manning, a metric user, first proposed the formula. It was later converted to BG units. Such dimensionally inconsistent formulas are dangerous and should either be reanalyzed or treated as having very limited application.

1.7 Properties of the Velocity Field

In a given flow situation, the determination, by experiment or theory, of the properties of the fluid as a function of position and time is considered to be the *solution* to the problem. In almost all cases, the emphasis is on the space–time distribution of the fluid properties. One rarely keeps track of the actual fate of the specific fluid particles.[6] This treatment of properties as continuum-field functions distinguishes fluid mechanics from solid mechanics, where we are more likely to be interested in the trajectories of individual particles or systems.

Eulerian and Lagrangian Desciptions

There are two different points of view in analyzing problems in mechanics. The first view, appropriate to fluid mechanics, is concerned with the field of flow and is called the *eulerian* method of description. In the eulerian method we compute the pressure field $p(x, y, z, t)$ of the flow pattern, not the pressure changes $p(t)$ that a particle experiences as it moves through the field.

The second method, which follows an individual particle moving through the flow, is called the *lagrangian* description. The lagrangian approach, which is more appropriate to solid mechanics, will not be treated in this book. However, certain numerical analyses of sharply bounded fluid flows, such as the motion of isolated fluid droplets, are very conveniently computed in lagrangian coordinates [1].

Fluid dynamic measurements are also suited to the eulerian system. For example, when a pressure probe is introduced into a laboratory flow, it is fixed at a specific position (x, y, z). Its output thus contributes to the description of the eulerian pressure field $p(x, y, z, t)$. To simulate a lagrangian measurement, the probe would have to move downstream at the fluid particle speeds; this is sometimes done in oceanographic measurements, where flowmeters drift along with the prevailing currents.

The two different descriptions can be contrasted in the analysis of traffic flow along a freeway. A certain length of freeway may be selected for study and called the field of flow. Obviously, as time passes, various cars will enter and leave the field, and the identity of the specific cars within the field will constantly be changing. The traffic engineer ignores specific cars and concentrates on their average velocity as a function of time and position within the field, plus the flow rate or number of cars per hour passing a given section of the freeway. This engineer is using an eulerian description of the traffic flow. Other investigators, such as the police or social scientists, may be interested in the path or speed or destination of specific cars in the field. By following a specific car as a function of time, they are using a lagrangian description of the flow.

The Velocity Field

Foremost among the properties of a flow is the velocity field $\mathbf{V}(x, y, z, t)$. In fact, determining the velocity is often tantamount to solving a flow problem, since other properties follow directly from the velocity field. Chapter 2 is devoted to the calculation of the pressure field once the velocity field is known. Books on heat transfer (for example, Ref. 20) are largely devoted to finding the temperature field from known velocity fields.

---

[6]One example where fluid particle paths are important is in water quality analysis of the fate of contaminant discharges.

In general, velocity is a vector function of position and time and thus has three components u, v, and w, each a scalar field in itself:

$$V(x, y, z, t) = \mathbf{i}u(x, y, z, t) + \mathbf{j}v(x, y, z, t) + \mathbf{k}w(x, y, z, t) \qquad (1.4)$$

The use of u, v, and w instead of the more logical component notation V_x, V_y, and V_z is the result of an almost unbreakable custom in fluid mechanics. Much of this textbook, especially Chaps. 4, 7, 8, and 9, is concerned with finding the distribution of the velocity vector $\mathbf{V}$ for a variety of practical flows.

1.8 Thermodynamic Properties of a Fluid

While the velocity field $\mathbf{V}$ is the most important fluid property, it interacts closely with the thermodynamic properties of the fluid. We have already introduced into the discussion the three most common such properties:

1. Pressure p
2. Density ρ
3. Temperature T

These three are constant companions of the velocity vector in flow analyses. Four other intensive thermodynamic properties become important when work, heat, and energy balances are treated (Chaps. 3 and 4):

4. Internal energy $\hat{u}$
5. Enthalpy $h = \hat{u} + p/\rho$
6. Entropy s
7. Specific heats c_p and c_v

In addition, friction and heat conduction effects are governed by the two so-called *transport properties:*

8. Coefficient of viscosity μ
9. Thermal conductivity k

All nine of these quantities are true thermodynamic properties that are determined by the thermodynamic condition or *state* of the fluid. For example, for a single-phase substance such as water or oxygen, two basic properties such as pressure and temperature are sufficient to fix the value of all the others:

$$\rho = \rho(p, T) \qquad h = h(p, T) \qquad \mu = \mu(p, T) \qquad (1.5)$$

and so on for every quantity in the list. Note that the specific volume, so important in thermodynamic analyses, is omitted here in favor of its inverse, the density ρ.

Recall that thermodynamic properties describe the state of a *system*—that is, a collection of matter of fixed identity that interacts with its surroundings. In most cases here the system will be a small fluid element, and all properties will be assumed to be continuum properties of the flow field: $\rho = \rho(x, y, z, t)$, and so on.

Recall also that thermodynamics is normally concerned with *static* systems, whereas fluids are usually in variable motion with constantly changing properties.

Do the properties retain their meaning in a fluid flow that is technically not in equilibrium? The answer is yes, from a statistical argument. In gases at normal pressure (and even more so for liquids), an enormous number of molecular collisions occur over a very short distance of the order of 1 μm, so that a fluid subjected to sudden changes rapidly adjusts itself toward equilibrium. We therefore assume that all the thermodynamic properties just listed exist as point functions in a flowing fluid and follow all the laws and state relations of ordinary equilibrium thermodynamics. There are, of course, important nonequilibrium effects such as chemical and nuclear reactions in flowing fluids, which are not treated in this text.

Pressure

Pressure is the (compression) stress at a point in a static fluid (Fig. 1.3). Next to velocity, the pressure p is the most dynamic variable in fluid mechanics. Differences or *gradients* in pressure often drive a fluid flow, especially in ducts. In low-speed flows, the actual magnitude of the pressure is often not important, unless it drops so low as to cause vapor bubbles to form in a liquid. For convenience, we set many such problem assignments at the level of 1 atm = 2116 lbf/ft^2 = 101,300 Pa. High-speed (compressible) gas flows (Chap. 9), however, are indeed sensitive to the magnitude of pressure.

Temperature

Temperature T is related to the internal energy level of a fluid. It may vary considerably during high-speed flow of a gas (Chap. 9). Although engineers often use Celsius or Fahrenheit scales for convenience, many applications in this text require *absolute* (Kelvin or Rankine) temperature scales:

$$°R = °F + 459.69$$
$$K = °C + 273.16$$

If temperature differences are strong, *heat transfer* may be important [20], but our concern here is mainly with dynamic effects.

Density

The density of a fluid, denoted by ρ (lowercase Greek rho), is its mass per unit volume. Density is highly variable in gases and increases nearly proportionally to the pressure level. Density in liquids is nearly constant; the density of water (about 1000 kg/m^3) increases only 1 percent if the pressure is increased by a factor of 220. Thus most liquid flows are treated analytically as nearly "incompressible."

In general, liquids are about three orders of magnitude more dense than gases at atmospheric pressure. The heaviest common liquid is mercury, and the lightest gas is hydrogen. Compare their densities at 20°C and 1 atm:

$$\text{Mercury: } \rho = 13{,}580 \text{ kg/m}^3 \qquad \text{Hydrogen: } \rho = 0.0838 \text{ kg/m}^3$$

They differ by a factor of 162,000! Thus the physical parameters in various liquid and gas flows might vary considerably. The differences are often resolved by the use of *dimensional analysis* (Chap. 5). Other fluid densities are listed in Tables A.3 and A.4 (in App. A) and in Ref. 21.

Specific Weight

The *specific weight* of a fluid, denoted by γ (lowercase Greek gamma), is its weight per unit volume. Just as a mass has a weight $W = mg$, density and specific weight are simply related by gravity:

$$\gamma = \rho g \qquad (1.6)$$

The units of γ are weight per unit volume, in lbf/ft^3 or N/m^3. In standard earth gravity, $g = 32.174 \ ft/s^2 = 9.807 \ m/s^2$. Thus, for example, the specific weights of air and water at 20°C and 1 atm are approximately

$$\gamma_{air} = (1.205 \ kg/m^3)(9.807 \ m/s^2) = 11.8 \ N/m^3 = 0.0752 \ lbf/ft^3$$
$$\gamma_{water} = (998 \ kg/m^3)(9.807 \ m/s^2) = 9790 \ N/m^3 = 62.4 \ lbf/ft^3$$

Specific weight is very useful in the hydrostatic pressure applications of Chap. 2. Specific weights of other fluids are given in Tables A.3 and A.4.

Specific Gravity

Specific gravity, denoted by SG, is the ratio of a fluid density to a standard reference fluid, usually water at 4°C (for liquids) and air (for gases):

$$SG_{gas} = \frac{\rho_{gas}}{\rho_{air}} = \frac{\rho_{gas}}{1.205 \ kg/m^3} \qquad (1.7)$$

$$SG_{liquid} = \frac{\rho_{liquid}}{\rho_{water}} = \frac{\rho_{liquid}}{1000 \ kg/m^3}$$

For example, the specific gravity of mercury (Hg) is $SG_{Hg} = 13{,}580/1000 \approx 13.6$. Engineers find these dimensionless ratios easier to remember than the actual numerical values of density of a variety of fluids.

Potential and Kinetic Energies

In thermostatics the only energy in a substance is that stored in a system by molecular activity and molecular bonding forces. This is commonly denoted as *internal energy* $\hat{u}$. A commonly accepted adjustment to this static situation for fluid flow is to add two more energy terms that arise from newtonian mechanics: potential energy and kinetic energy.

The potential energy equals the work required to move the system of mass m from the origin to a position vector $\mathbf{r} = \mathbf{i}x + \mathbf{j}y + \mathbf{k}z$ against a gravity field $\mathbf{g}$. Its value is $-m\mathbf{g} \cdot \mathbf{r}$, or $-\mathbf{g} \cdot \mathbf{r}$ per unit mass. The kinetic energy equals the work required to change the speed of the mass from zero to velocity V. Its value is $\frac{1}{2}mV^2$ or $\frac{1}{2}V^2$ per unit mass. Then by common convention the total stored energy e per unit mass in fluid mechanics is the sum of three terms:

$$e = \hat{u} + \tfrac{1}{2}V^2 + (-\mathbf{g} \cdot \mathbf{r}) \qquad (1.8)$$

Also, throughout this book we shall define z as upward, so that $\mathbf{g} = -g\mathbf{k}$ and $\mathbf{g} \cdot \mathbf{r} = -gz$. Then Eq. (1.8) becomes

$$e = \hat{u} + \tfrac{1}{2}V^2 + gz \qquad (1.9)$$

The molecular internal energy $\hat{u}$ is a function of T and p for the single-phase pure substance, whereas the potential and kinetic energies are kinematic quantities.

State Relations for Gases

Thermodynamic properties are found both theoretically and experimentally to be related to each other by state relations that differ for each substance. As mentioned, we shall confine ourselves here to single-phase pure substances, such as water in its liquid phase. The second most common fluid, air, is a mixture of gases, but since the mixture ratios remain nearly constant between 160 and 2200 K, in this temperature range air can be considered to be a pure substance.

All gases at high temperatures and low pressures (relative to their critical point) are in good agreement with the *perfect-gas law*

$$p = \rho RT \qquad R = c_p - c_v = \text{gas constant} \qquad (1.10)$$

where the specific heats c_p and c_v are defined in Eqs. (1.14) and (1.15).

Since Eq. (1.10) is dimensionally consistent, R has the same dimensions as specific heat, $\{L^2T^{-2}\Theta^{-1}\}$, or velocity squared per temperature unit (kelvin or degree Rankine). Each gas has its own constant R, equal to a universal constant Λ divided by the molecular weight

$$R_{\text{gas}} = \frac{\Lambda}{M_{\text{gas}}} \qquad (1.11)$$

where $\Lambda = 49{,}700$ ft-lbf/(slugmol · °R) = 8314 kJ/(kmol · K). Most applications in this book are for air, whose molecular weight is $M = 28.97/\text{mol}$:

$$R_{\text{air}} = \frac{49{,}700 \text{ ft} \cdot \text{lbf/(slugmol} \cdot \text{°R)}}{28.97/\text{mol}} = 1716 \frac{\text{ft} \cdot \text{lbf}}{\text{slug} \cdot \text{°R}} = 1716 \frac{\text{ft}^2}{\text{s}^2\text{°R}} = 287 \frac{\text{m}^2}{\text{s}^2 \cdot \text{K}} \quad (1.12)$$

Standard atmospheric pressure is 2116 lbf/ft² = 2116 slug/(ft · s²), and standard temperature is 60°F = 520°R. Thus standard air density is

$$\rho_{\text{air}} = \frac{2116 \text{ slug/(ft} \cdot \text{s}^2)}{[1716 \text{ ft}^2/(\text{s}^2 \cdot \text{°R})](520\text{°R})} = 0.00237 \text{ slug/ft}^3 = 1.22 \text{ kg/m}^3 \quad (1.13)$$

This is a nominal value suitable for problems. For other gases, see Table A.4.

One proves in thermodynamics that Eq. (1.10) requires that the internal molecular energy $\hat{u}$ of a perfect gas vary only with temperature: $\hat{u} = \hat{u}(T)$. Therefore the specific heat c_v also varies only with temperature:

$$c_v = \left(\frac{\partial \hat{u}}{\partial T}\right)_\rho = \frac{d\hat{u}}{dT} = c_v(T)$$

or

$$d\hat{u} = c_v(T)dT \qquad (1.14)$$

In like manner h and c_p of a perfect gas also vary only with temperature:

$$h = \hat{u} + \frac{p}{\rho} = \hat{u} + RT = h(T)$$

$$c_p = \left(\frac{\partial h}{\partial T}\right)_p = \frac{dh}{dT} = c_p(T) \qquad (1.15)$$

$$dh = c_p(T)dT$$

The ratio of specific heats of a perfect gas is an important dimensionless parameter in compressible flow analysis (Chap. 9)

$$k = \frac{c_p}{c_v} = k(T) \geq 1 \qquad (1.16)$$

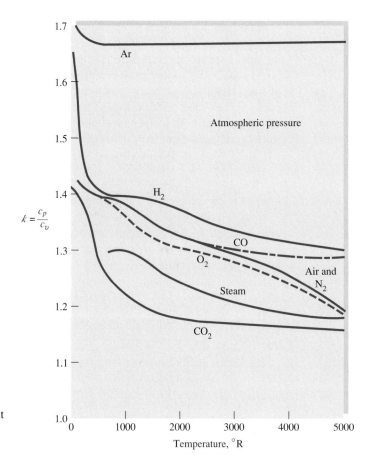

Fig. 1.5 Specific-heat ratio of eight common gases as a function of temperature. *(Data from Ref. 22.)*

As a first approximation in airflow analysis we commonly take c_p, c_v, and k to be constant:

$$k_{\text{air}} \approx 1.4$$

$$c_v = \frac{R}{k-1} \approx 4293 \text{ ft}^2/(\text{s}^2 \cdot {}^\circ\text{R}) = 718 \text{ m}^2/(\text{s}^2 \cdot \text{K}) \qquad (1.17)$$

$$c_p = \frac{kR}{k-1} \approx 6010 \text{ ft}^2/(\text{s}^2 \cdot {}^\circ\text{R}) = 1005 \text{ m}^2/(\text{s}^2 \cdot \text{K})$$

Actually, for all gases, c_p and c_v increase gradually with temperature, and k decreases gradually. Experimental values of the specific-heat ratio for eight common gases are shown in Fig. 1.5.

Many flow problems involve steam. Typical steam operating conditions are relatively close to the critical point, so that the perfect-gas approximation is inaccurate. Since no simple formulas apply accurately, steam properties are available both in EES (see Sec. 1.12) and on a CD-ROM [23] and even on the Internet, as a MathPad Corp. applet [24]. Meanwhile, the error of using the perfect-gas law can be moderate, as the following example shows.

EXAMPLE 1.5

Estimate ρ and c_p of steam at 100 lbf/in^2 and 400°F, in English units, (a) by the perfect-gas approximation and (b) by the ASME Steam Tables [23] or by EES.

Solution

- *Approach (a)—the perfect-gas law:*   Although steam is not an ideal gas, we can estimate these properties with moderate accuracy from Eqs. (1.10) and (1.17). First convert pressure from 100 lbf/in^2 to 14,400 lbf/ft^2, and use absolute temperature, (400°F + 460) = 860°R. Then we need the gas constant for steam, in English units. From Table A.4, the molecular weight of H$_2$O is 18.02, whence

$$R_{steam} = \frac{\Lambda_{English}}{M_{H_2O}} = \frac{49{,}700 \text{ ft} \cdot \text{lbf/(slugmol °R)}}{18.02/\text{mol}} = 2758 \frac{\text{ft} \cdot \text{lbf}}{\text{slug °R}}$$

Then the density estimate follows from the perfect-gas law, Eq. (1.10):

$$\rho \approx \frac{p}{RT} = \frac{14{,}400 \text{ lbf/ft}^2}{\left[2758 \text{ ft} \cdot \text{lbf/(slug} \cdot \text{°R)}\right](860 \text{ °R})} \approx 0.00607 \frac{\text{slug}}{\text{ft}^3} \qquad \textit{Ans. (a)}$$

At 860°R, from Fig. 1.5, $k_{steam} = c_p/c_v \approx 1.30$. Then, from Eq. (1.17),

$$c_p \approx \frac{kR}{k-1} = \frac{(1.3)(2758 \text{ ft} \cdot \text{lbf/(slug °R)})}{(1.3-1)} \approx 12{,}000 \frac{\text{ft} \cdot \text{lbf}}{\text{slug °R}} \qquad \textit{Ans. (a)}$$

- *Approach (b)—tables or software:*   One can either read the steam tables or program a few lines in EES. In either case, the English units (psi, Btu, lbm) are awkward when applied to fluid mechanics formulas. Even so, when using EES, make sure that the *Variable Information* menu specifies English units: psia and °F. EES statements for evaluating density and specific heat of steam are, for these conditions,

$$Rho = DENSITY(steam, P = 100, T = 400)$$

$$Cp = SPECHEAT(steam, P = 100, T = 400)$$

Note that the software is set up for psia and °F, without converting. EES returns the curve-fit values

$$Rho \approx 0.2027 \text{ lbm/ft}^3 \quad ; \quad Cp \approx 0.5289 \text{ Btu/(lbm-F)}$$

As just stated, Btu and lbm are extremely unwieldy when applied to mass, momentum, and energy problems in fluid mechanics. Therefore, either convert to ft-lbf and slugs using your own resources, or use the "Convert" function in EES, placing the old and new units in single quote marks:

$$Rho2 = Rho*CONVERT('lbm/ft^3', 'slug/ft^3')$$

$$Cp2 = Cp*CONVERT('Btu/lbm-F', 'ft^2/s^2-R')$$

Note that (1) you multiply the old Rho and Cp by the CONVERT function; and (2) units to the right of the division sign "/" in CONVERT are assumed to be in the denominator. EES returns these results:

$$Rho2 = 0.00630 \text{ slug/ft}^3 \quad Cp2 = 13{,}200 \text{ ft}^2/(s^2\text{-R}) \qquad \textit{Ans. (b)}$$

- *Comments:*   The steam tables would yield results quite close to EES. The perfect-gas estimate of ρ is 4 percent low, and the estimate of c_p is 9 percent low. The chief reason for the discrepancy is that this temperature and pressure are rather close to the critical point and saturation line of steam. At higher temperatures and lower pressures, say, 800°F and 50 lbf/in², the perfect-gas law yields properties with an accuracy of about ±1 percent.

Once again let us warn that English units (psia, lbm Btu) are awkward and must be converted in most fluid mechanics formulas. EES handles SI units nicely, with no conversion factors needed.

State Relations for Liquids

The writer knows of no "perfect-liquid law" comparable to that for gases. Liquids are nearly incompressible and have a single, reasonably constant specific heat. Thus an idealized state relation for a liquid is

$$\rho \approx \text{const} \qquad c_p \approx c_v \approx \text{const} \qquad dh \approx c_p \, dT \tag{1.18}$$

Most of the flow problems in this book can be attacked with these simple assumptions. Water is normally taken to have a density of 998 kg/m³ and a specific heat $c_p = 4210$ m²/(s² · K). The steam tables may be used if more accuracy is required.

The density of a liquid usually decreases slightly with temperature and increases moderately with pressure. If we neglect the temperature effect, an empirical pressure–density relation for a liquid is

$$\frac{p}{p_a} \approx (B + 1)\left(\frac{\rho}{\rho_a}\right)^n - B \tag{1.19}$$

where B and n are dimensionless parameters that vary slightly with temperature and p_a and ρ_a are standard atmospheric values. Water can be fitted approximately to the values $B \approx 3000$ and $n \approx 7$.

Seawater is a variable mixture of water and salt and thus requires three thermodynamic properties to define its state. These are normally taken as pressure, temperature, and the *salinity* $\hat{S}$, defined as the weight of the dissolved salt divided by the weight of the mixture. The average salinity of seawater is 0.035, usually written as 35 parts per 1000, or 35 ‰. The average density of seawater is 2.00 slugs/ft³ ≈ 1030 kg/m³. Strictly speaking, seawater has three specific heats, all approximately equal to the value for pure water of 25,200 ft²/(s² · °R) = 4210 m²/(s² · K).

EXAMPLE 1.6

The pressure at the deepest part of the ocean is approximately 1100 atm. Estimate the density of seawater in slug/ft³ at this pressure.

Solution

Equation (1.19) holds for either water or seawater. The ratio p/p_a is given as 1100:

$$1100 \approx (3001)\left(\frac{\rho}{\rho_a}\right)^7 - 3000$$

or

$$\frac{\rho}{\rho_a} = \left(\frac{4100}{3001}\right)^{1/7} = 1.046$$

Assuming an average surface seawater density $\rho_a = 2.00$ slugs/ft^3, we compute

$$\rho \approx 1.046(2.00) = 2.09 \text{ slugs/ft}^3 \qquad Ans.$$

Even at these immense pressures, the density increase is less than 5 percent, which justifies the treatment of a liquid flow as essentially incompressible.

1.9 Viscosity and Other Secondary Properties

The quantities such as pressure, temperature, and density discussed in the previous section are *primary* thermodynamic variables characteristic of any system. Certain secondary variables also characterize specific fluid mechanical behavior. The most important of these is viscosity, which relates the local stresses in a moving fluid to the strain rate of the fluid element.

Viscosity

Viscosity is a quantitative measure of a fluid's resistance to flow. More specifically, it determines the fluid strain rate that is generated by a given applied shear stress. We can easily move through air, which has very low viscosity. Movement is more difficult in water, which has 50 times higher viscosity. Still more resistance is found in SAE 30 oil, which is 300 times more viscous than water. Try to slide your hand through glycerin, which is five times more viscous than SAE 30 oil, or blackstrap molasses, another factor of five higher than glycerin. Fluids may have a vast range of viscosities.

Consider a fluid element sheared in one plane by a single shear stress τ, as in Fig. 1.6a. The shear strain angle $\delta\theta$ will continuously grow with time as long as the stress τ is maintained, the upper surface moving at speed δu larger than the lower. Such common fluids as water, oil, and air show a linear relation between applied shear and resulting strain rate:

$$\tau \propto \frac{\delta\theta}{\delta t} \qquad (1.20)$$

Fig. 1.6 Shear stress causes continuous shear deformation in a fluid: (*a*) a fluid element straining at a rate $\delta\theta/\delta t$; (*b*) newtonian shear distribution in a shear layer near a wall.

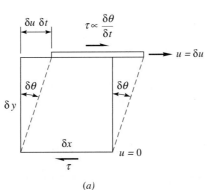
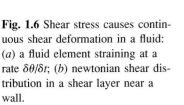

(*a*)

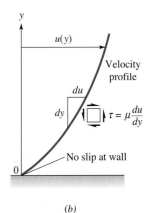

(*b*)

From the geometry of Fig. 1.4*a*, we see that

$$\tan \delta\theta = \frac{\delta u \, \delta t}{\delta y} \tag{1.21}$$

In the limit of infinitesimal changes, this becomes a relation between shear strain rate and velocity gradient:

$$\frac{d\theta}{dt} = \frac{du}{dy} \tag{1.22}$$

From Eq. (1.20), then, the applied shear is also proportional to the velocity gradient for the common linear fluids. The constant of proportionality is the viscosity coefficient μ:

$$\tau = \mu \frac{d\theta}{dt} = \mu \frac{du}{dy} \tag{1.23}$$

Equation (1.23) is dimensionally consistent; therefore μ has dimensions of stress–time: $\{FT/L^2\}$ or $\{M/(LT)\}$. The BG unit is slugs per foot-second, and the SI unit is kilograms per meter-second. The linear fluids that follow Eq. (1.23) are called *newtonian fluids,* after Sir Isaac Newton, who first postulated this resistance law in 1687.

We do not really care about the strain angle $\theta(t)$ in fluid mechanics, concentrating instead on the velocity distribution $u(y)$, as in Fig. 1.6*b*. We shall use Eq. (1.23) in Chap. 4 to derive a differential equation for finding the velocity distribution $u(y)$—and, more generally, $\mathbf{V}(x, y, z, t)$—in a viscous fluid. Figure 1.6*b* illustrates a shear layer, or *boundary layer,* near a solid wall. The shear stress is proportional to the slope of the velocity profile and is greatest at the wall. Further, at the wall, the velocity u is zero relative to the wall: This is called the *no-slip condition* and is characteristic of all viscous fluid flows.

The viscosity of newtonian fluids is a true thermodynamic property and varies with temperature and pressure. At a given state (p, T) there is a vast range of values among the common fluids. Table 1.4 lists the viscosity of eight fluids at standard pressure and temperature. There is a variation of six orders of magnitude from hydrogen up to glycerin. Thus there will be wide differences between fluids subjected to the same applied stresses.

Generally speaking, the viscosity of a fluid increases only weakly with pressure. For example, increasing p from 1 to 50 atm will increase μ of air only 10 percent.

Table 1.4 Viscosity and Kinematic Viscosity of Eight Fluids at 1 atm and 20°C

Fluid	μ, kg/(m · s)[†]	Ratio $\mu/\mu(H_2)$	ρ, kg/m³	ν m²/s[†]	Ratio $\nu/\nu(Hg)$
Hydrogen	9.0 E–6	1.0	0.084	1.05 E–4	910
Air	1.8 E–5	2.1	1.20	1.50 E–5	130
Gasoline	2.9 E–4	33	680	4.22 E–7	3.7
Water	1.0 E–3	114	998	1.01 E–6	8.7
Ethyl alcohol	1.2 E–3	135	789	1.52 E–6	13
Mercury	1.5 E–3	170	13,550	1.16 E–7	1.0
SAE 30 oil	0.29	33,000	891	3.25 E–4	2,850
Glycerin	1.5	170,000	1,260	1.18 E–3	10,300

[†] 1 kg/(m · s) = 0.0209 slug/(ft · s); 1 m²/s = 10.76 ft²/s.

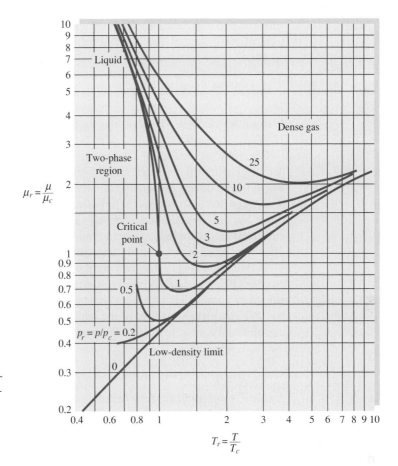

$$\mu_r = \frac{\mu}{\mu_c}$$

$$T_r = \frac{T}{T_c}$$

Fig. 1.7 Fluid viscosity nondimensionalized by critical-point properties. This generalized chart is characteristic of all fluids but is accurate only to ±20 percent. *(From Ref. 25.)*

Temperature, however, has a strong effect, with μ increasing with T for gases and decreasing for liquids. Figure A.1 (in App. A) shows this temperature variation for various common fluids. It is customary in most engineering work to neglect the pressure variation.

The variation $\mu(p, T)$ for a typical fluid is nicely shown by Fig. 1.7, from Ref. 25, which normalizes the data with the *critical-point state* (μ_c, p_c, T_c). This behavior, called the *principle of corresponding states,* is characteristic of all fluids, but the actual numerical values are uncertain to ±20 percent for any given fluid. For example, values of $\mu(T)$ for air at 1atm, from Table A.2, fall about 8 percent low compared to the "low-density limit" in Fig. 1.7.

Note in Fig. 1.7 that changes with temperature occur very rapidly near the critical point. In general, critical-point measurements are extremely difficult and uncertain.

The Reynolds Number

The primary parameter correlating the viscous behavior of all newtonian fluids is the dimensionless *Reynolds number:*

$$\mathrm{Re} = \frac{\rho V L}{\mu} = \frac{V L}{\nu} \tag{1.24}$$

where V and L are characteristic velocity and length scales of the flow. The second form of Re illustrates that the ratio of μ to ρ has its own name, the *kinematic viscosity:*

$$\nu = \frac{\mu}{\rho} \tag{1.25}$$

It is called kinematic because the mass units cancel, leaving only the dimensions $\{L^2/T\}$.

Generally, the first thing a fluids engineer should do is estimate the Reynolds number range of the flow under study. Very low Re indicates viscous *creeping* motion, where inertia effects are negligible. Moderate Re implies a smoothly varying *laminar* flow. High Re probably spells *turbulent* flow, which is slowly varying in the time-mean but has superimposed strong random high-frequency fluctuations. Explicit numerical values for low, moderate, and high Reynolds numbers cannot be stated here. They depend on flow geometry and will be discussed in Chaps. 5 through 7.

Table 1.4 also lists values of ν for the same eight fluids. The pecking order changes considerably, and mercury, the heaviest, has the smallest viscosity relative to its own weight. All gases have high ν relative to thin liquids such as gasoline, water, and alcohol. Oil and glycerin still have the highest ν, but the ratio is smaller. For given values of V and L in a flow, these fluids exhibit a spread of four orders of magnitude in the Reynolds number.

Flow between Plates

A classic problem is the flow induced between a fixed lower plate and an upper plate moving steadily at velocity V, as shown in Fig. 1.8. The clearance between plates is h, and the fluid is newtonian and does not slip at either plate. If the plates are large, this steady shearing motion will set up a velocity distribution $u(y)$, as shown, with $v = w = 0$. The fluid acceleration is zero everywhere.

With zero acceleration and assuming no pressure variation in the flow direction, you should show that a force balance on a small fluid element leads to the result that the shear stress is constant throughout the fluid. Then Eq. (1.23) becomes

$$\frac{du}{dy} = \frac{\tau}{\mu} = \text{const}$$

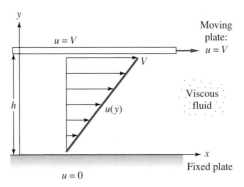

Fig. 1.8 Viscous flow induced by relative motion between two parallel plates.

which we can integrate to obtain

$$u = a + by$$

The velocity distribution is linear, as shown in Fig. 1.8, and the constants a and b can be evaluated from the no-slip condition at the upper and lower walls:

$$u = \begin{cases} 0 = a + b(0) & \text{at } y = 0 \\ V = a + b(h) & \text{at } y = h \end{cases}$$

Hence $a = 0$ and $b = V/h$. Then the velocity profile between the plates is given by

$$u = V\frac{y}{h} \tag{1.26}$$

as indicated in Fig. 1.8. Turbulent flow (Chap. 6) does not have this shape.

Although viscosity has a profound effect on fluid motion, the actual viscous stresses are quite small in magnitude even for oils, as shown in the following example.

EXAMPLE 1.7

Suppose that the fluid being sheared in Fig. 1.8 is SAE 30 oil at 20°C. Compute the shear stress in the oil if $V = 3$ m/s and $h = 2$ cm.

Solution

- *System sketch:*   This is shown earlier in Fig. 1.8.
- *Assumptions:*   Linear velocity profile, laminar newtonian fluid, no slip at either plate surface.
- *Approach:*   The analysis of Fig. 1.8 leads to Eq. (1.26) for laminar flow.
- *Property values:*   From Table 1.4 for SAE 30 oil, the oil viscosity $\mu = 0.29$ kg/(m-s).
- *Solution steps:*   In Eq. (1.26), the only unknown is the fluid shear stress:

$$\tau = \mu\frac{V}{h} = \left(0.29 \frac{\text{kg}}{\text{m} \cdot \text{s}}\right)\frac{(3 \text{ m/s})}{(0.02 \text{ m})} = 43.5 \frac{\text{kg} \cdot \text{m/s}^2}{\text{m}^2} = 43.5 \frac{\text{N}}{\text{m}^2} \approx 44 \text{ Pa} \qquad Ans.$$

- *Comments:*   Note the unit identities, 1 kg-m/s$^2 \equiv$ 1 N and 1 N/m$^2 \equiv$ 1 Pa. Although oil is very viscous, this shear stress is modest, about 2400 times less than atmospheric pressure. Viscous stresses in gases and thin (watery) liquids are even smaller.

Variation of Viscosity with Temperature

Temperature has a strong effect and pressure a moderate effect on viscosity. The viscosity of gases and most liquids increases slowly with pressure. Water is anomalous in showing a very slight decrease below 30°C. Since the change in viscosity is only a few percent up to 100 atm, we shall neglect pressure effects in this book.

Gas viscosity increases with temperature. Two common approximations are the power law and the Sutherland law:

$$\frac{\mu}{\mu_0} \approx \begin{cases} \left(\dfrac{T}{T_0}\right)^n & \text{power law} \\[2mm] \dfrac{(T/T_0)^{3/2}(T_0 + S)}{T + S} & \text{Sutherland law} \end{cases} \tag{1.27}$$

where μ_0 is a known viscosity at a known absolute temperature T_0 (usually 273 K). The constants n and S are fit to the data, and both formulas are adequate over a wide range of temperatures. For air, $n \approx 0.7$ and $S \approx 110$ K $= 199°$R. Other values are given in Ref. 26.

Liquid viscosity decreases with temperature and is roughly exponential, $\mu \approx ae^{-bT}$; but a better fit is the empirical result that ln μ is quadratic in $1/T$, where T is absolute temperature:

$$\ln\frac{\mu}{\mu_0} \approx a + b\left(\frac{T_0}{T}\right) + c\left(\frac{T_0}{T}\right)^2 \tag{1.28}$$

For water, with $T_0 = 273.16$ K, $\mu_0 = 0.001792$ kg/(m · s), suggested values are $a = -1.94$, $b = -4.80$, and $c = 6.74$, with accuracy about ± 1 percent. The viscosity of water is tabulated in Table A.1. Curve-fit viscosity formulas for 355 organic liquids are given by Yaws et al. [27]. For further viscosity data, see Refs. 28 and 29.

Thermal Conductivity

Just as viscosity relates applied stress to resulting strain rate, there is a property called *thermal conductivity* k that relates the vector rate of heat flow per unit area **q** to the vector gradient of temperature ∇T. This proportionality, observed experimentally for fluids and solids, is known as *Fourier's law of heat conduction:*

$$\mathbf{q} = -k\nabla T \tag{1.29a}$$

This can also be written as three scalar equations:

$$q_x = -k\frac{\partial T}{\partial x} \qquad q_y = -k\frac{\partial T}{\partial y} \qquad q_z = -k\frac{\partial T}{\partial z} \tag{1.29b}$$

The minus sign satisfies the convention that heat flux is positive in the direction of decreasing temperature. Fourier's law is dimensionally consistent, and k has SI units of joules per second-meter-kelvin. Thermal conductivity k is a thermodynamic property and varies with temperature and pressure in much the same way as viscosity. The ratio k/k_0 can be correlated with T/T_0 in the same manner as Eqs. (1.27) and (1.28) for gases and liquids, respectively.

Further data on viscosity and thermal conductivity variations can be found in Ref. 21.

Nonnewtonian Fluids

Fluids that do not follow the linear law of Eq. (1.23) are called *nonnewtonian* and are treated in books on *rheology* [16]. Figure 1.9a compares some examples to a newtonian fluid.

Dilatant.   This fluid is *shear-thickening,* increasing its resistance with increasing strain rate. Examples are suspensions of corn starch or sand in water. The classic case is *quicksand,* which stiffens up if one thrashes about.

Pseudoplastic.   A *shear-thinning* fluid is less resistant at higher strain rates. A very strong thinning is called *plastic.* Some of the many examples are polymer solutions, colloidal suspensions, paper pulp in water, latex paint, blood plasma, syrup, and

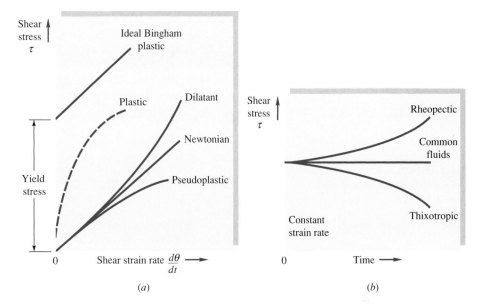

Fig. 1.9 Rheological behavior of various viscous materials: (*a*) stress versus strain rate; (*b*) effect of time on applied stress.

molasses. The classic case is *paint,* which is thick when poured but thin when brushed at a high strain rate.

Bingham plastic.   The limiting case of a plastic substance is one that requires a finite yield stress before it begins to flow. Figure 1.9*a* shows yielding followed by linear behavior, but nonlinear flow can also occur. Some examples are clay suspensions, drilling mud, toothpaste, mayonnaise, chocolate, and mustard. The classic case is *catsup,* which will not come out of the bottle until you stress it by shaking.

A further complication of nonnewtonian behavior is the transient effect shown in Fig. 1.9*b.* Some fluids require a gradually increasing shear stress to maintain a constant strain rate and are called *rheopectic.* The opposite case of a fluid that thins out with time and requires decreasing stress is termed *thixotropic.* We neglect nonnewtonian effects in this book; see Ref. 16 for further study.

Surface Tension

A liquid, being unable to expand freely, will form an *interface* with a second liquid or gas. The physical chemistry of such interfacial surfaces is quite complex, and whole textbooks are devoted to this specialty [30]. Molecules deep within the liquid repel each other because of their close packing. Molecules at the surface are less dense and attract each other. Since half of their neighbors are missing, the mechanical effect is that the surface is in tension. We can account adequately for surface effects in fluid mechanics with the concept of surface tension.

If a cut of length dL is made in an interfacial surface, equal and opposite forces of magnitude $Y\,dL$ are exposed normal to the cut and parallel to the surface, where Y is called the *coefficient of surface tension.* The dimensions of Y are $\{F/L\}$, with SI units of newtons per meter and BG units of pounds-force per foot. An alternate concept is to open up the cut to an area dA; this requires work to be done of amount $Y\,dA$. Thus the coefficient Y can also be regarded as the surface energy per unit area of the interface, in $N \cdot m/m^2$ or $ft \cdot lbf/ft^2$.

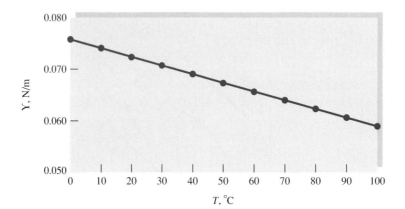

Fig. 1.10 Surface tension of a clean air–water interface. Data from Table A.5.

The two most common interfaces are water–air and mercury–air. For a clean surface at 20°C = 68°F, the measured surface tension is

$$Y = \begin{cases} 0.0050 \text{ lbf/ft} = 0.073 \text{ N/m} & \text{air–water} \\ 0.033 \text{ lbf/ft} = 0.48 \text{ N/m} & \text{air–mercury} \end{cases} \tag{1.30}$$

These are design values and can change considerably if the surface contains contaminants like detergents or slicks. Generally Y decreases with liquid temperature and is zero at the critical point. Values of Y for water are given in Fig. 1.10 and Table A.5.

If the interface is curved, a mechanical balance shows that there is a pressure difference across the interface, the pressure being higher on the concave side, as illustrated in Fig. 1.11. In Fig. 1.11a, the pressure increase in the interior of a liquid cylinder is balanced by two surface-tension forces:

$$2RL \, \Delta p = 2YL$$

or

$$\Delta p = \frac{Y}{R} \tag{1.31}$$

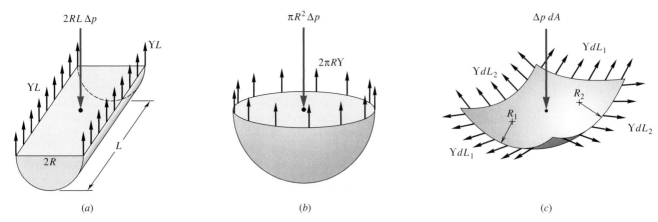

(a)   (b)   (c)

Fig. 1.11 Pressure change across a curved interface due to surface tension: (*a*) interior of a liquid cylinder; (*b*) interior of a spherical droplet; (*c*) general curved interface.

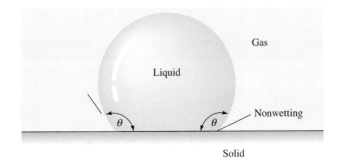

Gas

Liquid

Nonwetting

θ θ

Solid

Fig. 1.12 Contact-angle effects at liquid–gas–solid interface. If $\theta < 90°$, the liquid "wets" the solid; if $\theta > 90°$, the liquid is nonwetting.

We are not considering the weight of the liquid in this calculation. In Fig. 1.11*b*, the pressure increase in the interior of a spherical droplet balances a ring of surface-tension force:

$$\pi R^2 \, \Delta p = 2\pi R Y$$

or

$$\Delta p = \frac{2Y}{R} \qquad (1.32)$$

We can use this result to predict the pressure increase inside a soap bubble, which has *two* interfaces with air, an inner and outer surface of nearly the same radius R:

$$\Delta p_{\text{bubble}} \approx 2 \, \Delta p_{\text{droplet}} = \frac{4Y}{R} \qquad (1.33)$$

Figure 1.11*c* shows the general case of an arbitrarily curved interface whose principal radii of curvature are R_1 and R_2. A force balance normal to the surface will show that the pressure increase on the concave side is

$$\boxed{\Delta p = Y(R_1^{-1} + R_2^{-1})} \qquad (1.34)$$

Equations (1.31) to (1.33) can all be derived from this general relation; for example, in Eq. (1.31), $R_1 = R$ and $R_2 = \infty$.

A second important surface effect is the *contact angle* θ, which appears when a liquid interface intersects with a solid surface, as in Fig. 1.12. The force balance would then involve both Y and θ. If the contact angle is less than 90°, the liquid is said to *wet* the solid; if $\theta > 90°$, the liquid is termed *nonwetting*. For example, water wets soap but does not wet wax. Water is extremely wetting to a clean glass surface, with $\theta \approx 0°$. Like Y, the contact angle θ is sensitive to the actual physicochemical conditions of the solid–liquid interface. For a clean mercury–air–glass interface, $\theta = 130°$.

Example 1.8 illustrates how surface tension causes a fluid interface to rise or fall in a capillary tube.

EXAMPLE 1.8

Derive an expression for the change in height h in a circular tube of a liquid with surface tension Y and contact angle θ, as in Fig. E1.8.

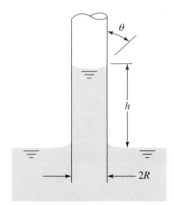

E1.8

Solution

The vertical component of the ring surface-tension force at the interface in the tube must balance the weight of the column of fluid of height h:

$$2\pi R Y \cos \theta = \gamma \pi R^2 h$$

Solving for h, we have the desired result:

$$h = \frac{2Y \cos \theta}{\gamma R} \qquad\qquad Ans.$$

Thus the capillary height increases inversely with tube radius R and is positive if $\theta < 90°$ (wetting liquid) and negative (capillary depression) if $\theta > 90°$.

Suppose that $R = 1$ mm. Then the capillary rise for a water–air–glass interface, $\theta \approx 0°$, $Y = 0.073$ N/m, and $\rho = 1000$ kg/m^3 is

$$h = \frac{2(0.073 \text{ N/m})(\cos 0°)}{(1000 \text{ kg/m}^3)(9.81 \text{ m/s}^2)(0.001 \text{ m})} = 0.015 \text{ (N} \cdot \text{s}^2)/\text{kg} = 0.015 \text{ m} = 1.5 \text{ cm}$$

For a mercury–air–glass interface, with $\theta = 130°$, $Y = 0.48$ N/m, and $\rho = 13{,}600$ kg/m^3, the capillary rise is

$$h = \frac{2(0.48)(\cos 130°)}{13{,}600(9.81)(0.001)} = -0.0046 \text{ m} = -0.46 \text{ cm}$$

When a small-diameter tube is used to make pressure measurements (Chap. 2), these capillary effects must be corrected for.

Vapor Pressure

Vapor pressure is the pressure at which a liquid boils and is in equilibrium with its own vapor. For example, the vapor pressure of water at 68°F is 49 lbf/ft^2, while that of mercury is only 0.0035 lbf/ft^2. If the liquid pressure is greater than the vapor pressure, the only exchange between liquid and vapor is evaporation at the interface. If, however, the liquid pressure falls below the vapor pressure, vapor bubbles begin to appear in the liquid. If water is heated to 212°F, its vapor pressure rises to 2116 lbf/ft^2, and thus water at normal atmospheric pressure will boil. When the liquid pressure is dropped below the vapor pressure due to a flow phenomenon, we call the process *cavitation*. If water is accelerated from rest to about 50 ft/s, its pressure drops by about 15 lbf/in^2, or 1 atm. This can cause cavitation [31].

The dimensionless parameter describing flow-induced boiling is the *cavitation number*

$$\text{Ca} = \frac{p_a - p_v}{\frac{1}{2}\rho V^2} \qquad\qquad (1.35)$$

where p_a = ambient pressure
p_v = vapor pressure
V = characteristic flow velocity
ρ = fluid density

Depending on the geometry, a given flow has a critical value of Ca below which the flow will begin to cavitate. Values of surface tension and vapor pressure of water are given in Table A.5. The vapor pressure of water is plotted in Fig. 1.13.

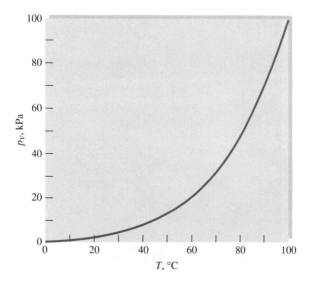

Fig. 1.13 Vapor pressure of water. Data from Table A.5.

Figure 1.14*a* shows cavitation bubbles being formed on the low-pressure surfaces of a marine propeller. When these bubbles move into a higher-pressure region, they collapse implosively. Cavitation collapse can rapidly spall and erode metallic surfaces and eventually destroy them, as shown in Fig. 1.14*b*.

EXAMPLE 1.9

A certain torpedo, moving in fresh water at 10°C, has a minimum-pressure point given by the formula

$$p_{min} = p_0 - 0.35\, \rho V^2 \tag{1}$$

where $p_0 = 115$ kPa, ρ is the water density, and V is the torpedo velocity. Estimate the velocity at which cavitation bubbles will form on the torpedo. The constant 0.35 is dimensionless.

Solution

- *Assumption:* Cavitation bubbles form when the minimum pressure equals the vapor pressure p_v.
- *Approach:* Solve Eq. (1) above, which is related to the Bernoulli equation from Example 1.3, for the velocity when $p_{min} = p_v$. Use SI units (m, N, kg, s).
- *Property values:* At 10°C, read Table A.1 for $\rho = 1000$ kg/m^3 and Table A.5 for $p_v = 1.227$ kPa.
- *Solution steps:* Insert the known data into Eq. (1) and solve for the velocity, using SI units:

$$p_{min} = p_v = 1227 \text{ Pa} = 115{,}000 \text{ Pa} - 0.35\left(1000\,\frac{\text{kg}}{\text{m}^3}\right)V^2, \text{ with } V \text{ in m/s}$$

$$\text{Solve } V^2 = \frac{(115{,}000 - 1227)}{0.35(1000)} = 325\frac{\text{m}^2}{\text{s}^2} \text{ or } V = \sqrt{325} \approx 18.0 \text{ m/s} \qquad Ans.$$

- *Comments:* Note that the use of SI units requires no conversion factors, as discussed in Example 1.3*b*. Pressures must be entered in pascals, not kilopascals.

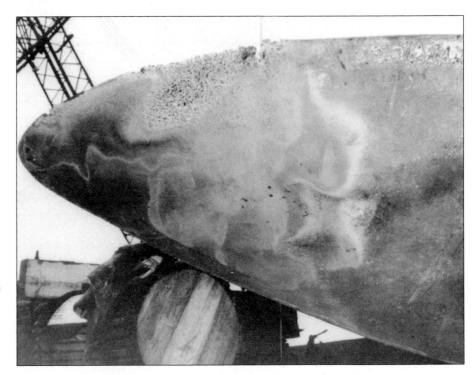

Fig. 1.14 Two aspects of cavitation bubble formation in liquid flows: (*a*) Beauty: spiral bubble sheets form from the surface of a marine propeller *(courtesy of the Garfield Thomas Water Tunnel, Pennsylvania State University);* (*b*) ugliness: collapsing bubbles erode a propeller surface *(courtesy of Thomas T. Huang, David Taylor Research Center).*

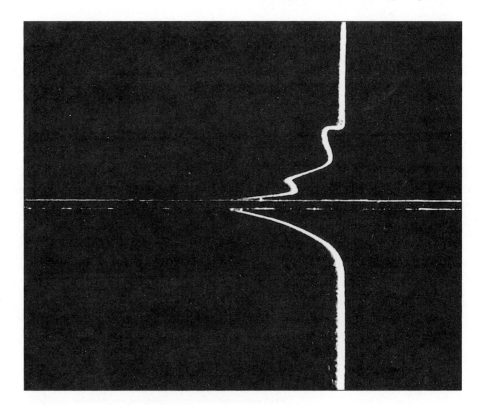

Fig. 1.15 The no-slip condition in water flow past a thin fixed plate. The upper flow is turbulent; the lower flow is laminar. The velocity profile is made visible by a line of hydrogen bubbles discharged from the wire across the flow. *(National Committee for Fluid Mechanics Films, Education Development Center, Inc, © 1972)*

No-Slip and No-Temperature-Jump Conditions

When a fluid flow is bounded by a solid surface, molecular interactions cause the fluid in contact with the surface to seek momentum and energy equilibrium with that surface. All liquids essentially are in equilibrium with the surfaces they contact. All gases are, too, except under the most rarefied conditions [18]. Excluding rarefied gases, then, all fluids at a point of contact with a solid take on the velocity and temperature of that surface:

$$V_{fluid} \equiv V_{wall} \qquad T_{fluid} \equiv T_{wall} \tag{1.36}$$

These are called the *no-slip* and *no-temperature-jump conditions,* respectively. They serve as *boundary conditions* for analysis of fluid flow past a solid surface. Figure 1.15 illustrates the no-slip condition for water flow past the top and bottom surfaces of a fixed thin plate. The flow past the upper surface is disorderly, or turbulent, while the lower surface flow is smooth, or laminar.[7] In both cases there is clearly no slip at the wall, where the water takes on the zero velocity of the fixed plate. The velocity profile is made visible by the discharge of a line of hydrogen bubbles from the wire shown stretched across the flow.

To decrease the mathematical difficulty, the no-slip condition is partially relaxed in the analysis of inviscid flow (Chap. 8). The flow is allowed to "slip" past the surface but not to permeate through the surface

$$V_{normal}(fluid) \equiv V_{normal}(solid) \tag{1.37}$$

---

[7]Laminar and turbulent flows are studied in Chaps. 6 and 7.

while the tangential velocity V_t is allowed to be independent of the wall. The analysis is much simpler, but the flow patterns are highly idealized.

For high-viscosity newtonian fluids, the linear velocity assumption and the no-slip conditions can yield some sophisticated approximate analyses for two- and three-dimensional viscous flows. The following example, for a type of rotating-disk viscometer, will illustrate.

EXAMPLE 1.10

A oil film of viscosity μ and thickness $h \ll R$ lies between a solid wall and a circular disk, as in Fig. E1.10. The disk is rotated steadily at angular velocity Ω. Noting that both velocity and shear stress vary with radius r, derive a formula for the torque M required to rotate the disk. Neglect air drag.

Solution

• *System sketch:* Figure E1.10 shows a side view (*a*) and a top view (*b*) of the system.

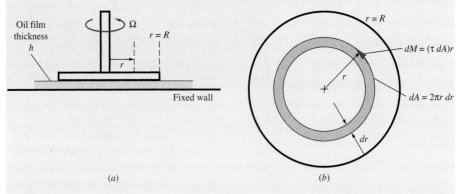

E1.10

• *Assumptions:* Linear velocity profile, laminar flow, no-slip, local shear stress given by Eq. (1.23).
• *Approach:* Estimate the shear stress on a circular strip of width dr and area $dA = 2\pi r\, dr$ in Fig. E1.10*b*, then find the moment dM about the origin caused by this shear stress. Integrate over the entire disk to find the total moment M.
• *Property values:* Constant oil viscosity μ. In this steady flow, oil density is not relevant.
• *Solution steps:* At radius r, the velocity in the oil is tangential, varying from zero at the fixed wall (no-slip) to $u = \Omega r$ at the disk surface (also no-slip). The shear stress at this position is thus

$$\tau = \mu \frac{du}{dy} \approx \mu \frac{\Omega r}{h}$$

This shear stress is everywhere perpendicular to the radius from the origin (see Fig. E1.10*b*). Then the total moment about the disk origin, caused by shearing this circular strip, can be found and integrated:

$$dM = (\tau)(dA)r = \left(\frac{\mu\Omega r}{h}\right)(2\pi r\,dr)r, \quad M = \int dM = \frac{2\pi\mu\Omega}{h}\int_0^R r^3\,dr = \frac{\pi\mu\Omega R^4}{2h} \quad Ans.$$

- *Comments:* This is a simplified engineering analysis, which neglects possible edge effects, air drag on the top of the disk, and the turbulence that might ensue if the disk rotates too fast.

Speed of Sound

In gas flow, one must be aware of *compressibility* effects (significant density changes caused by the flow). We shall see in Sec. 4.2 and in Chap. 9 that compressibility becomes important when the flow velocity reaches a significant fraction of the speed of sound of the fluid. The *speed of sound a* of a fluid is the rate of propagation of small-disturbance pressure pulses ("sound waves") through the fluid. In Chap. 9 we shall show, from momentum and thermodynamic arguments, that the speed of sound is defined by

$$a^2 = \left(\frac{\partial p}{\partial \rho}\right)_s = k\left(\frac{\partial p}{\partial \rho}\right)_T \qquad k = \frac{c_p}{c_v} \tag{1.38}$$

This is true for either a liquid or a gas, but it is for *gases* that the problem of compressibility occurs. For an ideal gas, Eq. (1.10), we obtain the simple formula

$$a_{\text{ideal gas}} = (kRT)^{1/2} \tag{1.39}$$

where R is the gas constant, Eq. (1.11), and T the absolute temperature. For example, for air at 20°C, $a = \{(1.40)[287\ \text{m}^2/(\text{s}^2\cdot\text{K})](293\ \text{K})\}^{1/2} \approx 343$ m/s (1126 ft/s = 768 mi/h). If, in this case, the air velocity reaches a significant fraction of a, say, 100 m/s, then we must account for compressibility effects (Chap. 9). Another way to state this is to account for compressibility when the *Mach number* Ma = V/a of the flow reaches about 0.3.

The speed of sound of water is tabulated in Table A.5. The speed of sound of air (or any approximately perfect gas) is simply calculated from Eq. (1.39).

EXAMPLE 1.11

A commercial airplane flies at 540 mi/h at a standard altitude of 30,000 ft. What is its Mach number?

Solution

- *Approach:* Find the "standard" speed of sound; divide it into the velocity, using proper units.
- *Property values:* From Table A.6, at 30,000 ft (9144 m), $a \approx 303$ m/s. Check this against the standard temperature, estimated from the table to be 229 K. From Eq. (1.39) for air,

$$a = [kR_{\text{air}}T]^{1/2} = [1.4(287)(229)]^{1/2} \approx 303\ \text{m/s}.$$

- *Solution steps:* Convert the airplane velocity to m/s:

$$V = (540 \text{ mi/h})[0.44704 \text{ m/s/(mi/h)}] \approx 241 \text{ m/s}.$$

Then the Mach number is given by

$$\text{Ma} = V/a = (241 \text{ m/s})/(303 \text{ m/s}) = 0.80 \qquad\qquad Ans.$$

- *Comments:* This value, Ma = 0.80, is typical of present-day commercial airliners.

1.10 Basic Flow Analysis Techniques

There are three basic ways to attack a fluid flow problem. They are equally important for a student learning the subject, and this book tries to give adequate coverage to each method:

1. Control-volume, or *integral* analysis (Chap. 3).
2. Infinitesimal system, or *differential* analysis (Chap. 4).
3. Experimental study, or *dimensional* analysis (Chap. 5).

In all cases, the flow must satisfy the three basic laws of mechanics plus a thermodynamic state relation and associated boundary conditions:

1. Conservation of mass (continuity).
2. Linear momentum (Newton's second law).
3. First law of thermodynamics (conservation of energy).
4. A state relation like $\rho = \rho(p, T)$.
5. Appropriate boundary conditions at solid surfaces, interfaces, inlets, and exits.

In integral and differential analyses, these five relations are modeled mathematically and solved by computational methods. In an experimental study, the fluid itself performs this task without the use of any mathematics. In other words, these laws are believed to be fundamental to physics, and no fluid flow is known to violate them.

1.11 Flow Patterns: Streamlines, Streaklines, and Pathlines

Fluid mechanics is a highly visual subject. The patterns of flow can be visualized in a dozen different ways, and you can view these sketches or photographs and learn a great deal qualitatively and often quantitatively about the flow.

Four basic types of line patterns are used to visualize flows:

1. A *streamline* is a line everywhere tangent to the velocity vector at a given instant.
2. A *pathline* is the actual path traversed by a given fluid particle.
3. A *streakline* is the locus of particles that have earlier passed through a prescribed point.
4. A *timeline* is a set of fluid particles that form a line at a given instant.

The streamline is convenient to calculate mathematically, while the other three are easier to generate experimentally. Note that a streamline and a timeline are instantaneous lines, while the pathline and the streakline are generated by the passage of time. The velocity profile shown in Fig. 1.15 is really a timeline generated earlier by a single discharge of bubbles from the wire. A pathline can be found by a time exposure

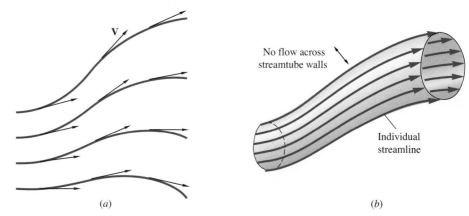

Fig. 1.16 The most common method of flow-pattern presentation: (*a*) Streamlines are everywhere tangent to the local velocity vector; (*b*) a streamtube is formed by a closed collection of streamlines.

(*a*)   (*b*)

of a single marked particle moving through the flow. Streamlines are difficult to generate experimentally in unsteady flow unless one marks a great many particles and notes their direction of motion during a very short time interval [32]. In steady flow, where velocity varies only with position, the situation simplifies greatly:

Streamlines, pathlines, and streaklines are identical in steady flow.

In fluid mechanics the most common mathematical result for visualization purposes is the streamline pattern. Figure 1.16*a* shows a typical set of streamlines, and Fig. 1.16*b* shows a closed pattern called a *streamtube*. By definition the fluid within a streamtube is confined there because it cannot cross the streamlines; thus the streamtube walls need not be solid but may be fluid surfaces.

Figure 1.17 shows an arbitrary velocity vector. If the elemental arc length dr of a streamline is to be parallel to **V**, their respective components must be in proportion:

Streamline:
$$\frac{dx}{u} = \frac{dy}{v} = \frac{dz}{w} = \frac{dr}{V} \tag{1.41}$$

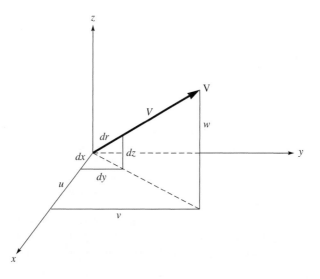

Fig. 1.17 Geometric relations for defining a streamline.

If the velocities (u, v, w) are known functions of position and time, Eq. (1.41) can be integrated to find the streamline passing through the initial point (x_0, y_0, z_0, t_0). The method is straightforward for steady flows (Example 1.12) but may be laborious for unsteady flow.

The pathline, or displacement of a particle, is defined by integration of the velocity components:

Pathline: $$x = \int u \, dt \qquad y = \int v \, dt \qquad z = \int w \, dt \qquad (1.42)$$

Given (u, v, w) as known functions of position and time, the integration is begun at a specified initial position (x_0, y_0, z_0, t_0). Again the integration may be laborious.

Streaklines, easily generated experimentally with smoke, dye, or bubble releases, are very difficult to compute analytically. See Ref. 33 for mathematical details.

EXAMPLE 1.12

Given the steady two-dimensional velocity distribution

$$u = Kx \qquad v = -Ky \qquad w = 0 \qquad (1)$$

where K is a positive constant, compute and plot the streamlines of the flow, including directions, and give some possible interpretations of the pattern.

Solution

Since time does not appear explicitly in Eq. (1), the motion is steady, so that streamlines, pathlines, and streaklines will coincide. Since $w = 0$ everywhere, the motion is two-dimensional, in the xy plane. The streamlines can be computed by substituting the expressions for u and v into Eq. (1.41):

$$\frac{dx}{Kx} = -\frac{dy}{Ky}$$

or $$\int \frac{dx}{x} = -\int \frac{dy}{y}$$

Integrating, we obtain $\ln x = -\ln y + \ln C$, or

$$xy = C \qquad\qquad Ans. (2)$$

This is the general expression for the streamlines, which are hyperbolas. The complete pattern is plotted in Fig. E1.12 by assigning various values to the constant C. The arrowheads can be determined only by returning to Eq. (1) to ascertain the velocity component directions, assuming K is positive. For example, in the upper right quadrant ($x > 0$, $y > 0$), u is positive and v is negative; hence the flow moves down and to the right, establishing the arrowheads as shown.

Note that the streamline pattern is entirely independent of constant K. It could represent the impingement of two opposing streams, or the upper half could simulate the flow of a single downward stream against a flat wall. Taken in isolation, the upper right quadrant is

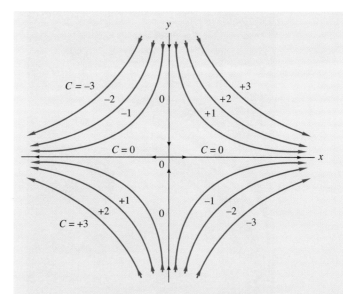

Fig. E1.12 Streamlines for the velocity distribution given by Eq. (1), for $K > 0$.

similar to the flow in a 90° corner. This is definitely a realistic flow pattern and is discussed again in Chap. 8.

Finally note the peculiarity that the two streamlines ($C = 0$) have opposite directions and intersect. This is possible only at a point where $u = v = w = 0$, which occurs at the origin in this case. Such a point of zero velocity is called a *stagnation point*.

Flow Visualization

Clever experimentation can produce revealing images of a fluid flow pattern, as shown earlier in Figs. 1.14a and 1.15. For example, streaklines are produced by the continuous release of marked particles (dye, smoke, or bubbles) from a given point. If the flow is steady, the streaklines will be identical to the streamlines and pathlines of the flow.

Some methods of flow visualization include the following [34–36]:

1. Dye, smoke, or bubble discharges.
2. Surface powder or flakes on liquid flows.
3. Floating or neutral-density particles.
4. Optical techniques that detect density changes in gas flows: shadowgraph, schlieren, and interferometer.
5. Tufts of yarn attached to boundary surfaces.
6. Evaporative coatings on boundary surfaces.
7. Luminescent fluids, additives, or bioluminescence.
8. Particle image velocimetry (PIV).

Figures 1.14a and 1.15 were both visualized by bubble releases. Another example is the use of particles in Fig. 1.18 to visualize a flow negotiating a 180° turn in a serpentine channel [42].

(a)

Fig. 1.18. Two visualizations of flow making a 180° turn in a serpentine channel: (a) particle streaklines at a Reynolds number of 1000; (b) time-mean particle image velocimetry (PIV) at a turbulent Reynolds number of 30,000 [*From Ref. 42, by permission of the American Society of Mechanical Engineers.*]

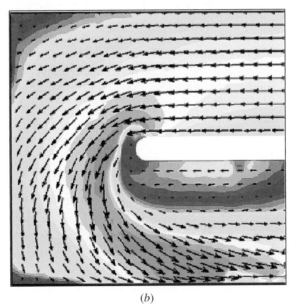

(b)

Figure 1.18*a* is at a low, laminar Reynolds number of 1000. The flow is steady, and the particles form streaklines showing that the flow cannot make the sharp turn without separating away from the bottom wall.

Figure 1.18*b* is at a higher, turbulent Reynolds number of 30,000. The flow is unsteady, and the streaklines would be chaotic and smeared, unsuitable for visualization. The image is thus produced by the new technique of particle image velocimetry [37]. In PIV, hundreds of particles are tagged and photographed at two closely spaced times. Particle movements thus indicate local velocity vectors. These hundreds of vectors are then smoothed by repeated computer operations until the time-mean flow pattern in Fig. 1.18*b* is achieved. Modern flow experiments and numerical models use computers extensively to create their visualizations, as described in the text by Yang [38].

Mathematical details of streamline/streakline/pathline analysis are given in Ref. 33. References 39–41 are beautiful albums of flow photographs. References 34–36 are monographs on flow visualization techniques.

Fluid mechanics is a marvelous subject for visualization, not just for still (steady) patterns, but also for moving (unsteady) motion studies. An outstanding list of available flow movies and videotapes is given by Carr and Young [43].

1.12 The Engineering Equation Solver

Most of the examples and exercises in this text are amenable to direct calculation without guessing or iteration or looping. Until recently, only such direct problem assignments, whether "plug-and-chug" or more subtle, were appropriate for undergraduate engineering courses. However, the introduction of computer software *solvers* makes almost any set of algebraic relations viable for analysis and solution. The solver recommended here is the *Engineering Equation Solver* (EES) developed by Klein and Beckman [44] and described in App. E.

Any software solver should handle a purely mathematical set of relations, such as the one posed in Ref. 44: $X \ln (X) = Y^3$, $X^{1/2} = 1/Y$. Submit that pair to any commercial solver and you will no doubt receive the answer: $X = 1.467$, $Y = 0.826$. However, for engineers, in the author's opinion, EES is superior to most solvers because (1) equations can be entered in any order; (2) scores of mathematical formulas are built-in, such as the Bessel functions; and (3) thermophysical properties of many fluids are built-in, such as the steam tables [23]. Both metric and English units are allowed. Equations need not be written in the traditional BASIC or FORTRAN style. For example, $X - Y + 1 = 0$ is perfectly satisfactory; there is no need to retype this as $X = Y - 1$.

For example, reconsider Example 1.7 as an EES exercise. One would first enter the reference properties p_0 and ρ_0 plus the curve-fit constants B and n:

$$Pz = 1.0$$

$$Rhoz = 2.0$$

$$B = 3000$$

$$n = 7$$

Then specify the given pressure ratio and the curve-fit relation, Eq. (1.19), for the equation of state of water:

$$P = 1100*Pz$$

$$P/Pz = (B + 1)*(Rho/Rhoz)^n - B$$

If you request an initial opinion from the CHECK/FORMAT menu, EES states that there are six equations in six unknowns and there are no obvious difficulties. Then request SOLVE from the menu and EES quickly prints out Rho = 2.091, the correct answer as seen already in Example 1.6. It also prints out values of the other five variables. Occasionally EES reports "unable to converge" and states what went wrong (division by zero, square root of a negative number, etc.). One needs only to improve the guesses and ranges of the unknowns in Variable Information to assist EES to the solution.

In subsequent chapters we will illustrate some implicit (iterative) examples by using EES and will also assign some advanced problem exercises for which EES is an ideal approach. The use of an engineering solver, notably EES, is recommended to all engineers in this era of the personal computer. If EES is not available, the writer recommends using an Excel spreadsheet.

1.13 Uncertainty in Experimental Data

Uncertainty is a fact of life in engineering. We rarely know any engineering properties or variables to an extreme degree of accuracy. The *uncertainty* of data is normally defined as the band within which one is 95 percent confident that the true value lies. Recall from Fig. 1.7 that the uncertainty of the ratio μ/μ_c was estimated as ± 20 percent. There are whole monographs devoted to the subject of experimental uncertainty [45–47], so we give only a brief summary here.

All experimental data have uncertainty, separated into two causes: (1) a *systematic* error due to the instrument or its environment and (2) a *random* error due to scatter in repeated readings. We minimize the systematic error by careful calibration and then estimate the random error statistically. The judgment of the experimenter is of crucial importance.

Here is the accepted mathematical estimate. Suppose a desired result P depends upon a single experimental variable x. If x has an uncertainty δx, then the uncertainty δP is estimated from the calculus:

$$\delta P \approx \frac{\partial P}{\partial x} \delta x$$

If there are multiple variables, $P = P(x_1, x_2, x_3, \ldots x_N)$, the overall uncertainty δP is calculated as a root-mean-square estimate [48]:

$$\delta P = \left[\left(\frac{\partial P}{\partial x_1} \delta x_1 \right)^2 + \left(\frac{\partial P}{\partial x_2} \delta x_2 \right)^2 + \cdots + \left(\frac{\partial P}{\partial x_N} \delta x_N \right)^2 \right]^{1/2} \tag{1.43}$$

This calculation is statistically much more probable than simply adding linearly the various uncertainties δx_i, thereby making the unlikely assumption that all variables simultaneously attain maximum error. Note that it is the responsibility of the experimenter to establish and report accurate estimates of all the relevant uncertainties δx_i.

If the quantity P is a simple power-law expression of the other variables, for example, $P = \text{Const } x_1^{n_1} x_2^{n_2} x_3^{n_3} \ldots$, then each derivative in Eq. (1.43) is proportional to P and the relevant power-law exponent and is inversely proportional to that variable.

If $P = \text{Const } x_1^{n_1} x_2^{n_2} x_3^{n_3} \ldots$, then

$$\frac{\partial P}{\partial x_1} = \frac{n_1 P}{x_1}, \quad \frac{\partial P}{\partial x_2} = \frac{n_2 P}{x_2}, \quad \frac{\partial P}{\partial x_3} = \frac{n_3 P}{x_3}, \ldots$$

Thus, from Eq. (1.43),

$$\frac{\delta P}{P} = \left[\left(n_1 \frac{\delta x_1}{x_1}\right)^2 + \left(n_2 \frac{\delta x_2}{x_2}\right)^2 + \left(n_3 \frac{\delta x_3}{x_3}\right)^2 + \cdots\right]^{1/2} \tag{1.44}$$

Evaluation of δP is then a straightforward procedure, as in the following example.

EXAMPLE 1.13

The so-called dimensionless Moody pipe friction factor f, plotted in Fig. 6.13, is calculated in experiments from the following formula involving pipe diameter D, pressure drop Δp, density ρ, volume flow rate Q, and pipe length L:

$$f = \frac{\pi^2}{8} \frac{D^5 \Delta p}{\rho Q^2 L}$$

Measurement uncertainties are given for a certain experiment: $D = 0.5$ percent, $\Delta p = 2.0$ percent, $\rho = 1.0$ percent, $Q = 3.5$ percent, and $L = 0.4$ percent. Estimate the overall uncertainty of the friction factor f.

Solution

The coefficient $\pi^2/8$ is assumed to be a pure theoretical number, with no uncertainty. The other variables may be collected using Eqs. (1.43) and (1.44):

$$U = \frac{\delta f}{f} = \left[\left(5\frac{\delta D}{D}\right)^2 + \left(1\frac{\delta \Delta p}{\Delta p}\right)^2 + \left(1\frac{\delta \rho}{\rho}\right)^2 + \left(2\frac{\delta Q}{Q}\right)^2 + \left(1\frac{\delta L}{L}\right)^2\right]^{1/2}$$

$$= [\{5(0.5\%)\}^2 + (2.0\%)^2 + (1.0\%)^2 + \{2(3.5\%)\}^2 + (0.4\%)^2]^{1/2} \approx 7.8\% \quad Ans.$$

By far the dominant effect in this particular calculation is the 3.5 percent error in Q, which is amplified by doubling, due to the power of 2 on flow rate. The diameter uncertainty, which is quintupled, would have contributed more had δD been larger than 0.5 percent.

1.14 The Fundamentals of Engineering (FE) Examination

The road toward a professional engineer's license has a first stop, the Fundamentals of Engineering Examination, known as the FE exam. It was formerly known as the Engineer-in-Training (E-I-T) Examination. This eight-hour national test will probably soon be required of all engineering graduates, not just for licensure, but as a student assessment tool. The 120-problem four-hour morning session covers many general studies:

Mathematics—15%	Ethics and business practices—7%	Material properties—7%
Engineering probability and statistics—7%	Engineering economics—8%	**Fluid mechanics—7%**
Chemistry—9%	Engineering mechanics—10%	Electricity and magnetism—9%
Computers—7%	Strength of materials—7%	Thermodynamics—7%

For the 60-problem, four-hour afternoon session you may choose one of seven modules: chemical, civil, electrical, environmental, industrial, mechanical, and other/general engineering. Note that fluid mechanics is an integral topic of the examination. Therefore, for practice, this text includes a number of end-of-chapter FE problems where appropriate.

The format for the FE exam questions is multiple-choice, usually with five selections, chosen carefully to tempt you with plausible answers if you used incorrect units, forgot to double or halve something, are missing a factor of π, or the like. In some cases, the selections are unintentionally ambiguous, such as the following example from a previous exam:

> Transition from laminar to turbulent flow occurs at a Reynolds number of
> (A) 900   (B) 1200   (C) 1500   (D) 2100   (E) 3000

The "correct" answer was graded as (D), Re = 2100. Clearly the examiner was thinking, but forgot to specify, Re_d for *flow in a smooth circular pipe,* since (see Chaps. 6 and 7) transition is highly dependent on geometry, surface roughness, and the length scale used in the definition of Re. The moral is not to get peevish about the exam but simply to go with the flow (pun intended) and decide which answer best fits an undergraduate training situation. Every effort has been made to keep the FE exam questions in this text unambiguous.

Problems

Most of the problems herein are fairly straightforward. More difficult or open-ended assignments are labeled with an asterisk as in Prob. 1.18. Problems labeled with an EES icon (for example, Prob. 1.61) will benefit from the use of the Engineering Equation Solver (EES), while problems labeled with a computer disk may require the use of a computer. The standard end-of-chapter problems 1.1 to 1.90 (categorized in the problem list below) are followed by fundamentals of engineering (FE) exam problems FE1.1 to FE1.10 and comprehensive problems C1.1 to C1.11.

Problem Distribution

Section	Topic	Problems
1.1, 1.4, 1.5	Fluid continuum concept	1.1–1.3
1.6	Dimensions and units	1.4–1.23
1.8	Thermodynamic properties	1.24–1.37
1.9	Viscosity, no-slip condition	1.38–1.61
1.9	Surface tension	1.62–1.71
1.9	Vapor pressure; cavitation	1.72–1.74
1.9	Speed of sound, Mach number	1.75–1.79
1.11	Streamlines and pathlines	1.80–1.84
1.2	History of fluid mechanics	1.85a–n
1.13	Experimental uncertainty	1.86–1.90

P1.1 A gas at 20°C may be considered *rarefied,* deviating from the continuum concept, when it contains less than

10^{12} molecules per cubic millimeter. If Avogadro's number is 6.023 E23 molecules per mole, what absolute pressure (in Pa) for air does this represent?

P1.2 Table A.6 lists the density of the standard atmosphere as a function of altitude. Use these values to estimate, crudely—say, within a factor of 2—the number of molecules of air in the entire atmosphere of the earth.

P1.3 For the triangular element in Fig. P1.3, show that a *tilted free liquid surface,* in contact with an atmosphere at pressure p_a, must undergo shear stress and hence begin to flow. *Hint:* Account for the weight of the fluid and show that a no-shear condition will cause horizontal forces to be out of balance.

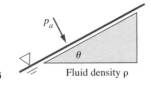

P1.3

P1.4 The Saybolt Universal Viscometer, now outdated but still sold in scientific catalogs, measures the kinematic viscosity of lubricants [Ref. 49, p. 40]. A specialized container, held at constant temperature, is filled with the test fluid. Measure the time t for 60 ml of the fluid to drain from a small hole or short tube in the bottom. This time unit, called *Saybolt universal seconds,* or SUS, is correlated with kinematic viscosity ν, in centistokes (1 stoke = 1 cm^2/s), by the following curve-fit formula:

$$\nu = 0.215t - \frac{145}{t} \qquad \text{for} \quad 40 < t < 100 \text{ SUS}$$

(a) Comment on the dimensionality of this equation. (b) Is the formula physically correct? (c) Since ν varies strongly with temperature, how does temperature enter into the formula? (d) Can we easily convert ν from centistokes to mm^2/s?

P1.5   The *mean free path* of a gas, l, is defined as the average distance traveled by molecules between collisions. A proposed formula for estimating l of an ideal gas is

$$l = 1.26 \frac{\mu}{\rho \sqrt{RT}}$$

What are the dimensions of the constant 1.26? Use the formula to estimate the mean free path of air at 20°C and 7 kPa. Would you consider air rarefied at this condition?

P1.6   From the (correct) theory of Prob. P1.5, estimate the pressure, in pascals, of carbon dioxide at 20°C for which the mean free path is (a) 1 micron (μm) and (b) 43.3 nm.

P1.7   To determine the flow rate of water at 20°C through a hose, a student finds that the hose fills a 55-gallon drum in 2 minutes and 37 seconds. Estimate (a) the volume flow rate in m^3/s and (b) the weight flow in N/s.

P1.8   Suppose we know little about the strength of materials but are told that the bending stress σ in a beam is *proportional* to the beam half-thickness y and also depends on the bending moment M and the beam area moment of inertia I. We also learn that, for the particular case $M = 2900$ in · lbf, $y = 1.5$ in, and $I = 0.4$ in^4, the predicted stress is 75 MPa. Using this information and dimensional reasoning only, find, to three significant figures, the only possible dimensionally homogeneous formula $\sigma = y\, f(M, I)$.

P1.9   A fluid is weighed in a laboratory. It is found that 1.5 U.S. gallons of the fluid weigh 136.2 ounces. (a) What is the fluid's density, in kg/m^3? (b) What fluid could this be? Assume standard gravity, $g = 9.807$ m/s^2.

P1.10   The Stokes-Oseen formula [33] for drag force F on a sphere of diameter D in a fluid stream of low velocity

V, density ρ, and viscosity μ is

$$F = 3\pi\mu DV + \frac{9\pi}{16}\rho V^2 D^2$$

Is this formula dimensionally homogeneous?

P1.11   In 1851, Sir George Stokes theorized that the drag force F on a particle in high-viscosity (low Reynolds number) flow depended only upon viscosity μ, particle velocity V, and particle size D. Use the concept of dimensional homogeneity to deduce a possible formula for the force.

P1.12   For low-speed (laminar) steady flow through a circular pipe, as shown in Fig. P1.12, the velocity u varies with radius and takes the form

$$u = B\frac{\Delta p}{\mu}(r_0^2 - r^2)$$

where μ is the fluid viscosity and Δp is the pressure drop from entrance to exit. What are the dimensions of the constant B?

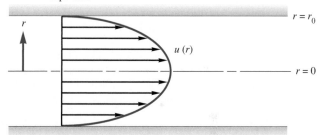

P1.12

P1.13   The efficiency η of a pump is defined as the (dimensionless) ratio of the power developed by the flow to the power required to drive the pump:

$$\eta = \frac{Q\Delta p}{\text{input power}}$$

where Q is the volume rate of flow and Δp is the pressure rise produced by the pump. Suppose that a certain pump develops a pressure rise of 35 lbf/in^2 when its flow rate is 40 L/s. If the input power is 16 hp, what is the efficiency?

*P1.14   Figure P1.14 shows the flow of water over a dam. The volume flow Q is known to depend only on crest width B, acceleration of gravity g, and upstream water height H above the dam crest. It is further known that Q is proportional to B. What is the form of the only possible dimensionally homogeneous relation for this flow rate?

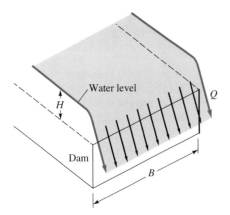

P1.14

P1.15 Mott [49] recommends the following formula for the friction head loss h_f, in ft, for flow through a pipe of length L and diameter D (both must be in ft):

$$h_f = L\left(\frac{Q}{0.551AC_hD^{0.63}}\right)^{1.852}$$

where Q is the volume flow rate in ft^3/s, A is the pipe cross-section area in ft^2, and C_h is a dimensionless coefficient whose value is approximately 100. Determine the dimensions of the constant 0.551.

P1.16 Algebraic equations such as Bernoulli's relation, Eq. (1) of Example 1.3, are dimensionally consistent, but what about differential equations? Consider, for example, the boundary-layer x-momentum equation, first derived by Ludwig Prandtl in 1904:

$$\rho u\frac{\partial u}{\partial x} + \rho v\frac{\partial u}{\partial y} = -\frac{\partial p}{\partial x} + \rho g_x + \frac{\partial \tau}{\partial y}$$

where τ is the boundary-layer shear stress and g_x is the component of gravity in the x direction. Is this equation dimensionally consistent? Can you draw a general conclusion?

P1.17 The Hazen-Williams hydraulics formula for volume rate of flow Q through a pipe of diameter D and length L is given by

$$Q \approx 61.9D^{2.63}\left(\frac{\Delta p}{L}\right)^{0.54}$$

where Δp is the pressure drop required to drive the flow. What are the dimensions of the constant 61.9? Can this formula be used with confidence for various liquids and gases?

*P1.18 For small particles at low velocities, the first term in the Stokes-Oseen drag law, Prob. 1.10, is dominant; hence, $F \approx KV$, where K is a constant. Suppose a particle of

mass m is constrained to move horizontally from the initial position $x = 0$ with initial velocity V_0. Show (a) that its velocity will decrease exponentially with time and (b) that it will stop after traveling a distance $x = mV_0/K$.

P1.19 *Marangoni convection* arises when a surface has a difference in surface tension along its length. The dimensionless Marangoni number M is a combination of thermal diffusivity $\alpha = k/(\rho c_p)$ (where k is the thermal conductivity), length scale L, viscosity μ, and surface tension difference δY. If M is proportional to L, find its form.

P1.20 A baseball, with $m = 145$ g, is thrown directly upward from the initial position $z = 0$ and $V_0 = 45$ m/s. The air drag on the ball is CV^2, where $C \approx 0.0013$ N $\cdot$ s^2/m^2. Set up a differential equation for the ball motion, and solve for the instantaneous velocity $V(t)$ and position $z(t)$. Find the maximum height z_{max} reached by the ball, and compare your results with the classical case of zero air drag.

P1.21 In 1908, Prandtl's student, Heinrich Blasius, proposed the following formula for the wall shear stress τ_w at a position x in viscous flow at velocity V past a flat surface:

$$\tau_w = 0.332\,\rho^{1/2}\mu^{1/2}V^{3/2}x^{-1/2}$$

Determine the dimensions of the constant 0.332.

P1.22 The *Richardson number,* Ri, which correlates the production of turbulence by buoyancy, is a dimensionless combination of the acceleration of gravity g, the fluid temperature T_o, the local temperature gradient $\partial T/\partial z$, and the local velocity gradient $\partial u/\partial z$. Determine an acceptable form for the Richardson number (most workers put $\partial T/\partial z$ in the numerator).

P1.23 During World War II, Sir Geoffrey Taylor, a British fluid dynamicist, used dimensional analysis to estimate the energy released by an atomic bomb explosion. He assumed that the energy released E, was a function of blast wave radius R, air density ρ, and time t. Arrange these variables into a single dimensionless group, which we may term the *blast wave number*.

P1.24 Consider carbon dioxide at 10 atm and 400°C. Calculate ρ and c_p at this state and then estimate the new pressure when the gas is cooled isentropically to 100°C. Use two methods: (a) an ideal gas and (b) the gas tables or EES.

P1.25 A tank contains 0.9 m^3 of helium at 200 kPa and 20°C. Estimate the total mass of this gas, in kg, (a) on earth and (b) on the moon. Also, (c) how much heat transfer, in MJ, is required to expand this gas at constant temperature to a new volume of 1.5 m^3?

P1.26 When we in the United States say a car's tire is filled "to 32 lb," we mean that its internal pressure is 32 lbf/in^2 above the ambient atmosphere. If the tire is at sea level, has a volume of 3.0 ft^3, and is at 75°F, estimate the total weight of air, in lbf, inside the tire.

P1.27   For steam at 40 lbf/in², some values of temperature and specific volume are as follows, from Ref. 23:

T, °F	400	500	600	700	800
v, ft³/lbm	12.624	14.165	15.685	17.195	18.699

Is steam, for these conditions, nearly a perfect gas, or is it wildly nonideal? If reasonably perfect, find a least-squares[†] value for the gas constant R, in m²/(s² · K); estimate the percentage error in this approximation; and compare with Table A.4.

P1.28   Wet atmospheric air at 100 percent relative humidity contains saturated water vapor and, by Dalton's law of partial pressures,

$$p_{atm} = p_{dry\ air} + p_{water\ vapor}$$

Suppose this wet atmosphere is at 40°C and 1 atm. Calculate the density of this 100 percent humid air, and compare it with the density of dry air at the same conditions.

P1.29   A compressed-air tank holds 5 ft³ of air at 120 lbf/in² "gage," that is, above atmospheric pressure. Estimate the energy, in ft-lbf, required to compress this air from the atmosphere, assuming an ideal isothermal process.

P1.30   Repeat Prob. 1.29 if the tank is filled with compressed *water* instead of air. Why is the result thousands of times less than the result of 215,000 ft · lbf in Prob. 1.29?

P1.31   One cubic foot of argon gas at 10°C and 1 atm is compressed isentropically to a pressure of 600 kPa. (*a*) What will be its new pressure and temperature? (*b*) If it is allowed to cool at this new volume back to 10°C, what will be the final pressure?

P1.32   A blimp is approximated by a prolate spheroid 90 m long and 30 m in diameter. Estimate the weight of 20°C gas within the blimp for (*a*) helium at 1.1 atm and (*b*) air at 1.0 atm. What might the *difference* between these two values represent (see Chap. 2)?

*P1.33   Experimental data for the density of mercury versus pressure at 20°C are as follows:

p, atm	1	500	1,000	1,500	2,000
ρ, kg/m³	13,545	13,573	13,600	13,625	13,653

Fit these data to the empirical state relation for liquids, Eq. (1.22), to find the best values of B and n for mercury. Then, assuming the data are nearly isentropic, use these values to estimate the speed of sound of mercury at 1 atm and compare with Table 9.1.

P1.34   Consider steam at the following state near the saturation line: (p_1, T_1) = (1.31 MPa, 290°C). Calculate and com-

[†]The concept of "least-squares" error is very important and should be learned by everyone.

pare, for an ideal gas (Table A.4) and the steam tables (or the EES software), (*a*) the density ρ_1 and (*b*) the density ρ_2 if the steam expands isentropically to a new pressure of 414 kPa. Discuss your results.

P1.35   In Table A.4, most common gases (air, nitrogen, oxygen, hydrogen) have a specific heat ratio $k \approx 1.40$. Why do argon and helium have such high values? Why does NH_3 have such a low value? What is the lowest k for any gas that you know of?

P1.36   The isentropic bulk modulus B of a fluid is defined as the isentropic change in pressure per fractional change in density:

$$B = \rho\left(\frac{\partial p}{\partial \rho}\right)_s$$

What are the dimensions of B? Using theoretical $p(\rho)$ relations, estimate the bulk modulus of (*a*) N_2O, assumed to be an ideal gas, and (*b*) water, at 20°C and 1 atm.

P1.37   A near-ideal gas has a molecular weight of 44 and a specific heat c_v = 610 J/(kg · K). What are (*a*) its specific heat ratio, k, and (*b*) its speed of sound at 100°C?

P1.38   In Fig. 1.8, if the fluid is glycerin at 20°C and the width between plates is 6 mm, what shear stress (in Pa) is required to move the upper plate at 5.5 m/s? What is the Reynolds number if L is taken to be the distance between plates?

P1.39   Knowing μ for air at 20°C from Table 1.4, estimate its viscosity at 500°C by (*a*) the power law and (*b*) the Sutherland law. Also make an estimate from (*c*) Fig. 1.7. Compare with the accepted value of $\mu \approx$ 3.58 E-5 kg/m · s.

*P1.40   For liquid viscosity as a function of temperature, a simplification of the log-quadratic law of Eq. (1.31) is *Andrade's equation* [21], $\mu \approx A \exp (B/T)$, where (A, B) are curve-fit constants and T is absolute temperature. Fit this relation to the data for water in Table A.1 and estimate the percentage error of the approximation.

P1.41   Some experimental values of the viscosity of argon gas at 1 atm are as follows:

T, K	300	400	500	600	700	800
μ, kg/(m·s)	2.27 E-5	2.85 E-5	3.37 E-5	3.83 E-5	4.25 E-5	4.64 E-5

Fit these value to either (*a*) a power law or (*b*) the Sutherland law, Eq. (1.30).

P1.42   Experimental values for the viscosity of helium at 1 atm are as follows:

T, K	200	400	600	800	1000	1200
μ, kg/(m·s)	1.50 E-5	2.43 E-5	3.20 E-5	3.88 E-5	4.50 E-5	5.08 E-5

Fit these values to either (*a*) a power law or (*b*) the Sutherland law, Eq. (1.30).

P1.43 According to rarefied gas theory [18], the no-slip condition begins to fail in tube flow when the mean free path of the gas is as large as 0.005 times the tube diameter. Consider helium at 20°C (Table A.4) flowing in a tube of diameter 1 cm. Using the theory of Prob. P1.5 (which is "correct", not "proposed"), find the helium pressure for which this no-slip failure begins.

P1.44 The values for SAE 30 oil in Table 1.4 are strictly "representative," not exact, because lubricating oils vary considerably according to the type of crude oil from which they are refined. The Society of Automotive Engineers [26] allows certain kinematic viscosity *ranges* for all lubricating oils: for SAE 30, $9.3 < \nu < 12.5$ mm^2/s at 100°C. SAE 30 oil density can also vary ± 2 percent from the tabulated value of 891 kg/m^3. Consider the following data for an acceptable grade of SAE 30 oil:

T, °C	0	20	40	60	80	100
μ, kg/(m·s)	2.00	0.40	0.11	0.042	0.017	0.0095

How does this oil compare with the plot in Appendix Fig. A.1? How well does the data fit Andrade's equation in Prob. 1.40?

P1.45 A block of weight W slides down an inclined plane while lubricated by a thin film of oil, as in Fig. P1.45. The film contact area is A and its thickness is h. Assuming a linear velocity distribution in the film, derive an expression for the "terminal" (zero-acceleration) velocity V of the block. Find the terminal velocity of the block if the block mass is 6 kg, $A = 35$ cm^2, $\theta = 15°$, and the film is 1-mm-thick SAE 30 oil at 20°C.

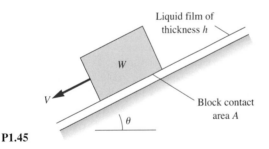

Liquid film of thickness h

W

V

Block contact area A

θ

P1.45

P1.46 A simple and popular model for two nonnewtonian fluids in Fig. 1.9a is the *power-law*:

$$\tau \approx C \left(\frac{du}{dy} \right)^n$$

where C and n are constants fit to the fluid [16]. From Fig. 1.9a, deduce the values of the exponent n for which the fluid is (a) newtonian, (b) dilatant, and (c) pseudoplastic.

Consider the specific model constant $C = 0.4$ N · s^n/m^2, with the fluid being sheared between two parallel plates as in Fig. 1.8. If the shear stress in the fluid is 1200 Pa, find the velocity V of the upper plate for the cases (d) $n = 1.0$, (e) $n = 1.2$, and (f) $n = 0.8$.

P1.47 A shaft 6.00 cm in diameter is being pushed axially through a bearing sleeve 6.02 cm in diameter and 40 cm long. The clearance, assumed uniform, is filled with oil whose properties are $\nu = 0.003$ m^2/s and SG = 0.88. Estimate the force required to pull the shaft at a steady velocity of 0.4 m/s.

P1.48 A thin plate is separated from two fixed plates by very viscous liquids μ_1 and μ_2, respectively, as in Fig. P1.48. The plate spacings h_1 and h_2 are unequal, as shown. The contact area is A between the center plate and each fluid. (a) Assuming a linear velocity distribution in each fluid, derive the force F required to pull the plate at velocity V. (b) Is there a necessary *relation* between the two viscosities, μ_1 and μ_2?

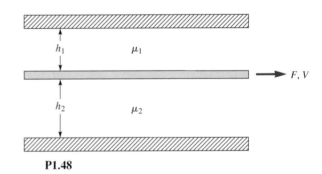

h_1 μ_1

h_2 μ_2

F, V

P1.48

P1.49 An amazing number of commercial and laboratory devices have been developed to measure fluid viscosity, as described in Refs. 29 and 49. Consider a concentric shaft, as in Prob. 1.47, but now fixed axially and rotated inside the sleeve. Let the inner and outer cylinders have radii r_i and r_o, respectively, with total sleeve length L. Let the rotational rate be Ω (rad/s) and the applied torque be M. Using these parameters, derive a theoretical relation for the viscosity μ of the fluid between the cylinders.

P1.50 A simple viscometer measures the time t for a solid sphere to fall a distance L through a test fluid of density ρ. The fluid viscosity μ is then given by

$$\mu \approx \frac{W_{net} t}{3 \pi D L} \quad \text{if} \quad t \geq \frac{2 \rho D L}{\mu}$$

where D is the sphere diameter and W_{net} is the sphere net weight in the fluid. (a) Prove that both of these formulas are dimensionally homogeneous. (*b*) Suppose that a 2.5 mm diameter aluminum sphere (density 2700 kg/m³) falls in an oil of density 875 kg/m³. If the time to fall 50 cm is 32 s, estimate the oil viscosity and verify that the inequality is valid.

P1.51 An approximation for the boundary-layer shape in Figs. 1.6*b* and P1.51 is the formula

$$u(y) \approx U \sin\left(\frac{\pi y}{2\delta}\right), \qquad 0 \le y \le \delta$$

where U is the stream velocity far from the wall and δ is the boundary layer thickness, as in Fig. P.151. If the fluid is helium at 20°C and 1 atm, and if $U = 10.8$ m/s and $\delta = 3$ cm, use the formula to (*a*) estimate the wall shear stress τ_w in Pa, and (*b*) find the position in the boundary layer where τ is one-half of τ_w.

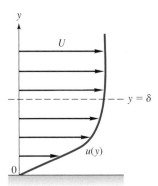

P1.51

P1.52 The belt in Fig. P1.52 moves at a steady velocity V and skims the top of a tank of oil of viscosity μ, as shown. Assuming a linear velocity profile in the oil, develop a simple formula for the required belt-drive power P as a function of (h, L, V, b, μ). What belt-drive power P, in watts, is required if the belt moves at 2.5 m/s over SAE 30W oil at 20°C, with $L = 2$ m, $b = 60$ cm, and $h = 3$ cm?

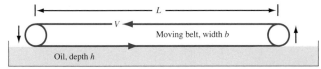

P1.52

*P1.53 A solid cone of angle 2θ, base r_0, and density ρ_c is rotating with initial angular velocity ω_0 inside a conical seat, as shown in Fig. P1.53. The clearance h is filled with oil of viscosity μ. Neglecting air drag, derive an analytical expression for the cone's angular velocity $\omega(t)$ if there is no applied torque.

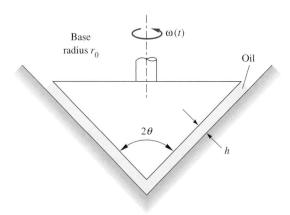

P1.53

*P1.54 A disk of radius R rotates at an angular velocity Ω inside a disk-shaped container filled with oil of viscosity μ, as shown in Fig. P1.54. Assuming a linear velocity profile and neglecting shear stress on the outer disk edges, derive a formula for the viscous torque on the disk.

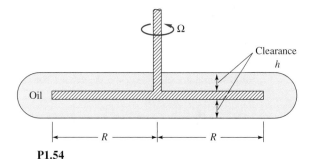

P1.54

P1.55 A block of weight W is being pulled over a table by another weight W_o, as shown in Fig. P1.55. Find an algebraic formula for the steady velocity U of the block if it slides on an oil film of thickness h and viscosity μ. The block bottom area A is in contact with the oil. Neglect the cord weight and the pulley friction. Assume a linear velocity profile in the oil film.

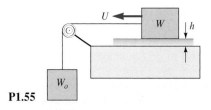

P1.55

*P1.56 The device in Fig. P1.56 is called a *cone-plate viscometer* [29]. The angle of the cone is very small, so that $\sin \theta \approx \theta$, and the gap is filled with the test liquid. The torque M to rotate the cone at a rate Ω is measured. Assuming a linear velocity profile in the fluid film, derive an expression for fluid viscosity μ as a function of (M, R, Ω, θ).

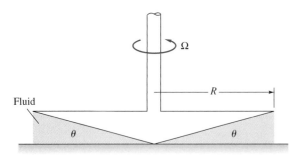

P1.56

P1.57   For the geometry of Prob. P1.55, (a) solve the *unsteady* problem $U(t)$ where the block starts from rest and accelerates toward the final steady velocity U_o of Prob. P1.55. (b) As a separate issue, if the table were instead *sloped* at an angle θ toward the pulley, state the criterion for whether the block moves up or down the table.

*P1.58   The laminar pipe flow example of Prob. 1.12 can be used to design a *capillary viscometer* [29]. If Q is the volume flow rate, L is the pipe length, and Δp is the pressure drop from entrance to exit, the theory of Chap. 6 yields a formula for viscosity:

$$\mu = \frac{\pi r_0^4 \Delta p}{8LQ}$$

Pipe end effects are neglected [29]. Suppose our capillary has $r_0 = 2$ mm and $L = 25$ cm. The following flow rate and pressure drop data are obtained for a certain fluid:

Q, m³/h	0.36	0.72	1.08	1.44	1.80
Δp, kPa	159	318	477	1274	1851

What is the viscosity of the fluid? *Note:* Only the first three points give the proper viscosity. What is peculiar about the last two points, which were measured accurately?

P1.59   A solid cylinder of diameter D, length L, and density ρ_s falls due to gravity inside a tube of diameter D_0. The clearance, $D_0 - D \ll D$, is filled with fluid of density ρ and viscosity μ. Neglect the air above and below the cylinder. Derive a formula for the terminal fall velocity of the cylinder. Apply your formula to the case of a steel cylinder, $D = 2$ cm, $D_0 = 2.04$ cm, $L = 15$ cm, with a film of SAE 30 oil at 20°C.

P1.60   Pipelines are cleaned by pushing through them a close-fitting cylinder called a *pig*. The name comes from the squealing noise it makes sliding along. Reference 50 describes a new nontoxic pig, driven by compressed air, for cleaning cosmetic and beverage pipes. Suppose the pig diameter is 5-15/16 in and its length 26 in. It cleans a 6-in-diameter pipe at a speed of 1.2 m/s. If the clearance is filled with glycerin at 20°C, what pressure difference, in pascals, is needed to drive the pig? Assume a linear velocity profile in the oil and neglect air drag.

*P1.61   An air-hockey puck has a mass of 50 g and is 9 cm in diameter. When placed on the air table, a 20°C air film, of 0.12-mm thickness, forms under the puck. The puck is struck with an initial velocity of 10 m/s. Assuming a linear velocity distribution in the air film, how long will it take the puck to (a) slow down to 1 m/s and (b) stop completely? Also, (c) how far along this extremely long table will the puck have traveled for condition (a)?

P1.62   The hydrogen bubbles that produced the velocity profiles in Fig. 1.15 are quite small, $D \approx 0.01$ mm. If the hydrogen–water interface is comparable to air–water and the water temperature is 30°C, estimate the excess pressure within the bubble.

P1.63   Derive Eq. (1.34) by making a force balance on the fluid interface in Fig. 1.11c.

P1.64   A shower head emits a cylindrical jet of clean 20°C water into air. The pressure inside the jet is approximately 200 Pa greater than the air pressure. Estimate the diameter of the jet in mm.

P1.65   The system in Fig. P1.65 is used to calculate the pressure p_1 in the tank by measuring the 15-cm height of liquid in the 1-mm-diameter tube. The fluid is at 60°C. Calculate the true fluid height in the tube and the percentage error due to capillarity if the fluid is (a) water or (b) mercury.

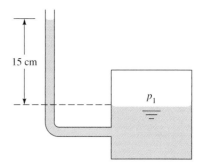

P1.65

P1.66   A thin wire ring, 3 cm in diameter, is lifted from a water surface at 20°C. Neglecting the wire weight, what is the force required to lift the ring? Is this a good way to measure surface tension? Should the wire be made of any particular material?

P1.67 A vertical concentric annulus, with outer radius r_o and inner radius r_i, is lowered into a fluid of surface tension Y and contact angle $\theta < 90°$. Derive an expression for the capillary rise h in the annular gap if the gap is very narrow.

*P1.68 Make an analysis of the shape $\eta(x)$ of the water–air interface near a plane wall, as in Fig. P1.68, assuming that the slope is small, $R^{-1} \approx d^2\eta/dx^2$. Also assume that the pressure difference across the interface is balanced by the specific weight and the interface height, $\Delta p \approx \rho g \eta$. The boundary conditions are a wetting contact angle θ at $x = 0$ and a horizontal surface $\eta = 0$ as $x \to \infty$. What is the maximum height h at the wall?

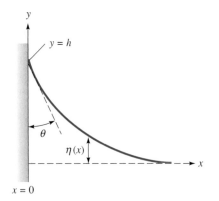

P1.68 $x = 0$

P1.69 A solid cylindrical needle of diameter d, length L, and density ρ_n may float in liquid of surface tension Y. Neglect buoyancy and assume a contact angle of $0°$. Derive a formula for the maximum diameter $d_{\max}$ able to float in the liquid. Calculate $d_{\max}$ for a steel needle (SG $= 7.84$) in water at $20°C$.

P1.70 Derive an expression for the capillary height change h for a fluid of surface tension Y and contact angle θ between two vertical parallel plates a distance W apart, as in Fig. P1.70. What will h be for water at $20°C$ if $W = 0.5$ mm?

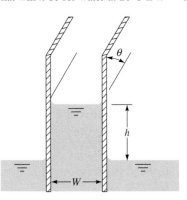

P1.70

*P1.71 A soap bubble of diameter D_1 coalesces with another bubble of diameter D_2 to form a single bubble D_3 with the same amount of air. Assuming an isothermal process, derive an expression for finding D_3 as a function of D_1, D_2, p_{atm}, and Y.

P1.72 Early mountaineers boiled water to estimate their altitude. If they reach the top and find that water boils at $84°C$, approximately how high is the mountain?

P1.73 A small submersible moves at velocity V, in fresh water at $20°C$, at a 2-m depth, where ambient pressure is 131 kPa. Its critical cavitation number is known to be $C_a = 0.25$. At what velocity will cavitation bubbles begin to form on the body? Will the body cavitate if $V = 30$ m/s and the water is cold ($5°C$)?

P1.74 Oil, with a vapor pressure of 20 kPa, is delivered through a pipeline by equally spaced pumps, each of which increases the oil pressure by 1.3 MPa. Friction losses in the pipe are 150 Pa per meter of pipe. What is the maximum possible pump spacing to avoid cavitation of the oil?

P1.75 An airplane flies at 555 mi/h. At what altitude in the standard atmosphere will the airplane's Mach number be exactly 0.8?

P1.76 Estimate the speed of sound of steam at $200°C$ and 400 kPa (*a*) by an ideal-gas approximation (Table A.4) and (*b*) using EES (or the steam tables) and making small isentropic changes in pressure and density and approximating Eq. (1.38).

*P1.77 The density of $20°C$ gasoline varies with pressure approximately as follows:

p, atm	1	500	1000	1500
ρ, lbm/ft^3	42.45	44.85	46.60	47.98

Use these data to estimate (*a*) the speed of sound (m/s) and (*b*) the bulk modulus (MPa) of gasoline at 1 atm.

P1.78 Sir Isaac Newton measured the speed of sound by timing the difference between seeing a cannon's puff of smoke and hearing its boom. If the cannon is on a mountain 5.2 mi away, estimate the air temperature in degrees Celsius if the time difference is (*a*) 24.2 s and (*b*) 25.1 s.

P1.79 Even a tiny amount of dissolved gas can drastically change the speed of sound of a gas–liquid mixture. By estimating the pressure–volume change of the mixture, Olson [51] gives the following approximate formula:

$$a_{\text{mixture}} \approx \sqrt{\frac{p_g K_l}{[x\rho_g + (1 - x)\rho_l][xK_l + (1 - x)p_g]}}$$

Where x is the volume fraction of gas, K is the bulk modulus, and subscripts l and g denote the liquid and gas, respectively. (a) Show that the formula is dimensionally homogeneous. (b) For the special case of air bubbles (density 1.7 kg/m^3 and pressure 150 kPa) in water (density 998 kg/m^3 and bulk modulus 2.2 (GPa), plot the mixture speed of sound in the range $0 \leq x \leq 0.002$ and discuss.

*P1.80 A two-dimensional steady velocity field is given by $u = x^2 - y^2$, $v = -2xy$. Derive the streamline pattern and sketch a few streamlines in the upper half plane. *Hint:* The differential equation is exact.

P1.81 Repeat Example 1.12 by letting the velocity components increase linearly with time:

$$V = Kxt\mathbf{i} - Kyt\mathbf{j} + 0\mathbf{k}$$

Find and sketch, for a few representative times, the instantaneous streamlines. How do they differ from the steady flow lines in Example 1.12?

P1.82 A velocity field is given by $u = V\cos\theta$, $v = V\sin\theta$, and $w = 0$, where V and θ are constants. Derive a formula for the streamlines of this flow.

*P1.83 A two-dimensional unsteady velocity field is given by $u = x(1 + 2t)$, $v = y$. Find the equation of the time-varying streamlines that all pass through the point (x_0, y_0) at some time t. Sketch a few of these.

*P1.84 Repeat Prob. 1.83 to find and sketch the equation of the *pathline* that passes through (x_0, y_0) at time $t = 0$.

P1.85 Do some reading and report to the class on the life and achievements, especially vis-à-vis fluid mechanics, of

(a) Evangelista Torricelli (1608–1647)
(b) Henri de Pitot (1695–1771)
(c) Antoine Chézy (1718–1798)
(d) Gotthilf Heinrich Ludwig Hagen (1797–1884)
(e) Julius Weisbach (1806–1871)
(f) George Gabriel Stokes (1819–1903)
(g) Moritz Weber (1871–1951)
(h) Theodor von Kármán (1881–1963)
(i) Paul Richard Heinrich Blasius (1883–1970)
(j) Ludwig Prandtl (1875–1953)
(k) Osborne Reynolds (1842–1912)
(l) John William Strutt, Lord Rayleigh (1842–1919)
(m) Daniel Bernoulli (1700–1782)
(n) Leonhard Euler (1707–1783)

P1.86 A right circular cylinder volume v is to be calculated from the measured base radius R and height H. If the uncertainty in R is 2 percent and the uncertainty in H is 3 percent, estimate the overall uncertainty in the calculated volume.

P1.87 Use the theory of Prob. 1.49 for a shaft 8 cm long, rotating at 1200 r/min, with $r_i = 2.00$ cm and $r_o = 2.05$ cm. (a) If the measured torque is 0.293 N · m, what is the fluid viscosity? (b) Suppose that the uncertainties in the experiment are as follows: L (±0.5 mm), M (±0.003 N · m), Ω (±1 percent), r_i and r_o (±0.02 mm). Estimate the overall uncertainty of the measured viscosity.

P1.88 The device in Fig. P1.54 is called a *rotating disk viscometer* [29]. Suppose that $R = 5$ cm and $h = 1$ mm. (a) If the torque required to rotate the disk at 900 r/min is 0.537 N · m, what is the viscosity of the fluid? (b) If the uncertainty in each parameter (M, R, h, Ω) is ±1 percent, what is the overall uncertainty in the viscosity?

P1.89 For the cone-plate viscometer of Fig. P1.56, suppose $R = 6$ cm and $\theta = 3°$. (a) If the torque required to rotate the cone is 0.157 N · m, what is the viscosity of the fluid? (b) If the uncertainty in each parameter (M, R, θ, Ω) is ±2 percent, what is the overall uncertainty in the viscosity?

P1.90 The dimensionless *drag coefficient* C_D of a sphere, to be studied in Chaps. 5 and 7, is

$$C_D = \frac{F}{(1/2)\rho V^2 (\pi/4)D^2}$$

where F is the drag force, ρ the fluid density, V the fluid velocity, and D the sphere diameter. If the uncertainties of these variables are F (±3 percent), ρ (±1.5 percent), V (±2 percent), and D (±1 percent), what is the overall uncertainty in the measured drag coefficient?

Fundamentals of Engineering Exam Problems

FE1.1 The absolute viscosity μ of a fluid is primarily a function of
(a) Density, (b) Temperature, (c) Pressure, (d) Velocity, (e) Surface tension

FE1.2 If a uniform solid body weighs 50 N in air and 30 N in water, its specific gravity is
(a) 1.5, (b) 1.67, (c) 2.5, (d) 3.0, (e) 5.0

FE1.3 Helium has a molecular weight of 4.003. What is the weight of 2 m^3 of helium at 1 atm and 20°C?
(a) 3.3 N, (b) 6.5 N, (c) 11.8 N, (d) 23.5 N, (e) 94.2 N

FE1.4 An oil has a kinematic viscosity of 1.25 E-4 m^2/s and a specific gravity of 0.80. What is its dynamic (absolute) viscosity in kg/(m · s)?
(a) 0.08, (b) 0.10, (c) 0.125, (d) 1.0, (e) 1.25

FE1.5 Consider a soap bubble of diameter 3 mm. If the surface tension coefficient is 0.072 N/m and external pressure is 0 Pa gage, what is the bubble's internal gage pressure?
(a) −24 Pa, (b) +48 Pa, (c) +96 Pa, (d) +192 Pa, (e) −192 Pa

FE1.6 The only possible dimensionless group that combines velocity V, body size L, fluid density ρ, and surface tension coefficient σ is
(a) $L\rho\sigma/V$, (b) $\rho VL^2/\sigma$, (c) $\rho\sigma V^2/L$, (d) $\sigma LV^2/\rho$, (e) $\rho LV^2/\sigma$

FE1.7 Two parallel plates, one moving at 4 m/s and the other fixed, are separated by a 5-mm-thick layer of oil of specific gravity 0.80 and kinematic viscosity 1.25 E-4 m^2/s. What is the average shear stress in the oil?
(a) 80 Pa, (b) 100 Pa, (c) 125 Pa, (d) 160 Pa, (e) 200 Pa

FE1.8 Carbon dioxide has a specific heat ratio of 1.30 and a gas constant of 189 J/(kg · °C). If its temperature rises from 20 to 45°C, what is its internal energy rise?
(a) 12.6 kJ/kg, (b) 15.8 kJ/kg, (c) 17.6 kJ/kg, (d) 20.5 kJ/kg, (e) 25.1 kJ/kg

FE1.9 A certain water flow at 20°C has a critical cavitation number, where bubbles form, $Ca \approx 0.25$, where $Ca = 2(p_a - p_{vap})/\rho V^2$. If $p_a = 1$ atm and the vapor pressure is 0.34 pounds per square inch absolute (psia), for what water velocity will bubbles form?
(a) 12 mi/h, (b) 28 mi/h, (c) 36 mi/h, (d) 55 mi/h, (e) 63 mi/h

FE1.10 Example 1.10 gave an analysis that predicted that the viscous moment on a rotating disk $M = \pi\mu\Omega R^4/(2h)$. If the uncertainty of each of the four variables (μ, Ω, R, h) is 1.0 percent, what is the estimated overall uncertainty of the moment M?
(a) 4.0 percent (b) 4.4 percent (c) 5.0 percent (d) 6.0 percent (e) 7.0 percent

Comprehensive Problems

C1.1 Sometimes we can develop equations and solve practical problems by knowing nothing more than the dimensions of the key parameters in the problem. For example, consider the heat loss through a window in a building. Window efficiency is rated in terms of "R value," which has units of (ft^2 · h · °F)/Btu. A certain manufacturer advertises a double-pane window with an R value of 2.5. The same company produces a triple-pane window with an R value of 3.4. In either case the window dimensions are 3 ft by 5 ft. On a given winter day, the temperature difference between the inside and outside of the building is 45°F.
(a) Develop an equation for the amount of heat lost in a given time period Δt, through a window of area A, with R value R, and temperature difference ΔT. How much heat (in Btu) is lost through the double-pane window in one 24-h period?
(b) How much heat (in Btu) is lost through the triple-pane window in one 24-h period?
(c) Suppose the building is heated with propane gas, which costs $3.25 per gallon. The propane burner is 80 percent efficient. Propane has approximately 90,000 Btu of available energy per gallon. In that same 24-h period, how much money would a homeowner save per window by installing triple-pane rather than double-pane windows?
(d) Finally, suppose the homeowner buys 20 such triple-pane windows for the house. A typical winter has the equivalent of about 120 heating days at a temperature difference of 45°F. Each triple-pane window costs $85 more than the double-pane

window. Ignoring interest and inflation, how many years will it take the homeowner to make up the additional cost of the triple-pane windows from heating bill savings?

C1.2 When a person ice skates, the surface of the ice actually melts beneath the blades, so that he or she skates on a thin sheet of water between the blade and the ice.
(a) Find an expression for total friction force on the bottom of the blade as a function of skater velocity V, blade length L, water thickness (between the blade and the ice) h, water viscosity μ, and blade width W.
(b) Suppose an ice skater of total mass m is skating along at a constant speed of V_0 when she suddenly stands stiff with her skates pointed directly forward, allowing herself to coast to a stop. Neglecting friction due to air resistance, how far will she travel before she comes to a stop? (Remember, she is coasting on *two* skate blades.) Give your answer for the total distance traveled, x, as a function of V_0, m, L, h, μ, and W.
(c) Find x for the case where $V_0 = 4.0$ m/s, $m = 100$ kg, $L = 30$ cm, $W = 5.0$ mm, and $h = 0.10$ mm. Do you think our assumption of negligible air resistance is a good one?

C1.3 Two thin flat plates, tilted at an angle α, are placed in a tank of liquid of known surface tension Υ and contact angle θ, as shown in Fig. C1.3. At the free surface of the liquid in the tank, the two plates are a distance L apart and have width b into the page. The liquid rises a distance h between the plates, as shown.

(a) What is the total upward (z-directed) force, due to surface tension, acting on the liquid column between the plates?

(b) If the liquid density is ρ, find an expression for surface tension Y in terms of the other variables.

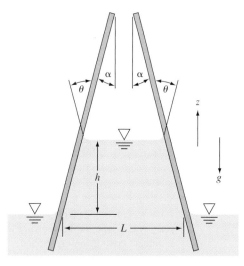

C1.3

C1.4   Oil of viscosity μ and density ρ drains steadily down the side of a tall, wide vertical plate, as shown in Fig. C1.4. In the region shown, *fully developed* conditions exist; that is, the velocity profile shape and the film thickness δ are independent of distance z along the plate. The vertical velocity w becomes a function only of x, and the shear resistance from the atmosphere is negligible.

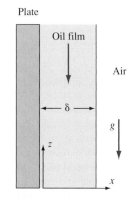

C1.4

(a) Sketch the approximate shape of the velocity profile $w(x)$, considering the boundary conditions at the wall and at the film surface.

(b) Suppose film thickness δ, and the slope of the velocity profile at the wall, $(dw/dx)_{wall}$, are measured by a laser Doppler anemometer (to be discussed in Chap. 6). Find an expression for the viscosity of the oil as a function of ρ, δ, $(dw/dx)_{wall}$, and the gravitational acceleration g. Note that, for the coordinate system given, both w and $(dw/dx)_{wall}$ are negative.

C1.5   Viscosity can be measured by flow through a thin-bore or *capillary* tube if the flow rate is low. For length L, (small) diameter $D \ll L$, pressure drop Δp, and (low) volume flow rate Q, the formula for viscosity is $\mu = D^4\Delta p/(CLQ)$, where C is a constant. (a) Verify that C is dimensionless. The following data are for water flowing through a 2-mm-diameter tube which is 1 meter long. The pressure drop is held constant at $\Delta p = 5$ kPa.

T, °C	10.0	40.0	70.0
Q, L/min	0.091	0.179	0.292

(b) Using proper SI units, determine an average value of C by accounting for the variation with temperature of the viscosity of water.

C1.6   The *rotating-cylinder viscometer* in Fig. C1.6 shears the fluid in a narrow clearance Δr, as shown. Assuming a linear velocity distribution in the gaps, if the driving torque M is measured, find an expression for μ by (a) neglecting and (b) including the bottom friction.

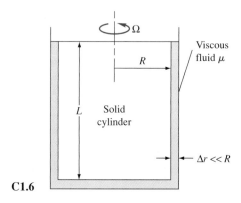

C1.6

C1.7   SAE 10W oil at 20°C flows past a flat surface, as in Fig. 1.6b. The velocity profile $u(y)$ is measured, with the following results:

y, m	0.0	0.003	0.006	0.009	0.012	0.015
u, m/s	0.0	1.99	3.94	5.75	7.29	8.46

Using your best interpolating skills, estimate the shear stress in the oil (a) at the wall and (b) at $y = 15$ mm.

C1.8 A mechanical device that uses the rotating cylinder of Fig. C1.6 is the *Stormer viscometer* [29]. Instead of being driven at constant Ω, a cord is wrapped around the shaft and attached to a falling weight W. The time t to turn the shaft a given number of revolutions (usually five) is measured and correlated with viscosity. The formula is

$$t \approx \frac{A\mu}{W - B}$$

where A and B are constants that are determined by calibrating the device with a known fluid. Here are calibration data for a Stormer viscometer tested in glycerol, using a weight of 50 N:

μ, kg/m-s	0.23	0.34	0.57	0.84	1.15
t, sec	15	23	38	56	77

(*a*) Find reasonable values of A and B to fit this calibration data. [*Hint:* The data are not very sensitive to the value of B.] (*b*) A more viscous fluid is tested with a 100 N weight and the measured time is 44 s. Estimate the viscosity of this fluid.

C1.9 The lever in Fig. C1.9 has a weight W at one end and is tied to a cylinder at the left end. The cylinder has negligible weight and buoyancy and slides upward through a film of heavy oil of viscosity μ. (*a*) If there is no acceleration (uniform lever rotation), derive a formula for the rate of fall V_2 of the weight. Neglect the lever weight. Assume a linear velocity profile in the oil film. (*b*) Estimate the fall velocity of the weight if $W = 20$ N, $L_1 = 75$ cm, $L_2 = 50$ cm, $D = 10$ cm, $L = 22$ cm, $\Delta R = 1$ mm, and the oil is glycerin at 20°C.

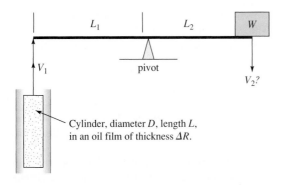

C1.9

C1.10 A popular gravity-driven instrument is the *Cannon-Ubbelohde viscometer*, shown in Fig. C1.10. The test liquid is drawn up above the bulb on the right side and allowed to drain by gravity through the capillary tube below the bulb. The time t for the meniscus to pass from upper to lower timing marks is recorded. The kinematic viscosity is computed by the simple formula:

$$\nu = Ct$$

where C is a calibration constant. For ν in the range of 100–500 mm²/s, the recommended constant is $C = 0.50$ mm²/s², with an accuracy less than 0.5 percent.

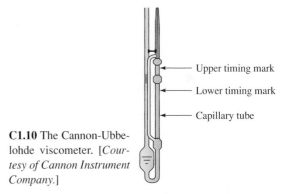

C1.10 The Cannon-Ubbelohde viscometer. [*Courtesy of Cannon Instrument Company.*]

(*a*) What liquids from Table A.3 are in this viscosity range? (*b*) Is the calibration formula dimensionally consistent? (*c*) What system properties might the constant C depend upon? (*d*) What problem in this chapter hints at a formula for estimating the viscosity?

C1.11 Mott [Ref. 49, p. 38] discusses a simple falling-ball viscometer, which we can analyze later in Chapter 7. A small ball of diameter D and density ρ_b falls though a tube of test liquid (ρ, μ). The fall velocity V is calculated by the time to fall a measured distance. The formula for calculating the viscosity of the fluid is

$$\mu = \frac{(\rho_b - \rho)gD^2}{18\,V}$$

This result is limited by the requirement that the Reynolds number ($\rho VD/\mu$) be less than 1.0. Suppose a steel ball (SG = 7.87) of diameter 2.2 mm falls in SAE 25W oil (SG = 0.88) at 20°C. The measured fall velocity is 8.4 cm/s. (*a*) What is the viscosity of the oil, in kg/m-s? (*b*) Is the Reynolds number small enough for a valid estimate?

References

1. J. C. Tannehill, D. A. Anderson, and R. H. Pletcher, *Computational Fluid Mechanics and Heat Transfer,* 2d ed., Taylor and Francis, Bristol, PA, 1997.
2. J. Blazek, *Computational Fluid Dynamics Principles,* 2d ed., Elsevier Science and Technology Books, Burlington, 2005.
3. J. D. Anderson, *Computational Fluid Dynamics,* McGraw-Hill, New York, 1995.
4. T. Cebeci, *Computational Fluid Dynamics for Engineers,* Springer-Verlag, New York, 2005.
5. C. J. Chen and S. Y. Jaw, *Fundamentals of Turbulence Modeling,* Taylor and Francis, Bristol, PA, 1997.
6. S. B. Pope, *Turbulent Flows,* Cambridge University Press, New York, 2000.
7. T. Baritaud, *Direct Numerical Simulation for Turbulent Reacting Flows,* Editions Technip, Paris, 1996.
8. B. Geurts, *Elements of Direct and Large Eddy Simulation,* R. T. Edwards Inc., Flourtown, PA, 2003.
9. R. J. Goldstein (ed.), *Fluid Mechanics Measurements,* 2d ed., Taylor and Francis, Bristol, PA, 1997.
10. R. C. Baker, *Introductory Guide to Flow Measurement,* Wiley, New York, 2002.
11. R. W. Miller, *Flow Measurement Engineering Handbook,* 3d ed., McGraw-Hill, New York, 1996.
12. H. Rouse and S. Ince, *History of Hydraulics,* Iowa Institute of Hydraulic Research, Univ. of Iowa, Iowa City, IA, 1957; reprinted by Dover, New York, 1963.
13. H. Rouse, *Hydraulics in the United States 1776–1976,* Iowa Institute of Hydraulic Research, Univ. of Iowa, Iowa City, IA, 1976.
14. G. Garbrecht, *Hydraulics and Hydraulic Research: An Historical Review,* Gower Pub., Aldershot, UK, 1987.
15. Cambridge University Press, "Ludwig Prandtl—Father of Modern Fluid Mechanics," URL <www.fluidmech.net/msc/prandtl.htm>.
16. R. I. Tanner, *Engineering Rheology,* 2d ed., Oxford University Press, New York, 2000.
17. C. E. Brennen, *Fundamentals of Multiphase Flow,* Cambridge University Press, New York, 2005. See also URL <http://caltechbook.library.caltech.edu/51/01/multiph.htm>.
18. C. Shen, *Rarefied Gas Dynamics: Fundamentals, Simulations, and Microflows,* Springer-Verlag, New York, 2005.
19. F. Carderelli and M. J. Shields, *Scientific Unit Conversion: A Practical Guide to Metrification,* 2d ed., Springer-Verlag, New York, 1999.
20. J. P. Holman, *Heat Transfer,* 9th ed., McGraw-Hill, New York, 2001.
21. R. C. Reid, J. M. Prausnitz, and T. K. Sherwood, *The Properties of Gases and Liquids,* 4th ed., McGraw-Hill, New York, 1987.
22. J. Hilsenrath et al., "Tables of Thermodynamic and Transport Properties," *U.S. Nat. Bur. Standards Circular 564,* 1955; reprinted by Pergamon, New York, 1960.
23. W. T. Parry et al., *ASME International Steam Tables for Industrial Use,* ASME Press, New York, 2000 (software also available).
24. MathPad Corp., *Internet Steam Tables Calculator,* URL <http://www.mathpad.com/public/htmls/main/left/StInfo.html>.
25. O. A. Hougen and K. M. Watson, *Chemical Process Principles Charts,* Wiley, New York, 1960.
26. F. M. White, *Viscous Fluid Flow,* 3d ed., McGraw-Hill, New York, 2005.
27. C. L. Yaws, X. Lin, and L. Bu, "Calculate Viscosities for 355 Compounds: An Equation Can Be Used to Calculate Liquid Viscosity as a Function of Temperature," *Chemical Engineering,* vol. 101, no. 4, April 1994, pp. 119–128.
28. *SAE Fuels and Lubricants Standards Manual,* Society of Automotive Engineers, Warrendale, PA, 2001.
29. C. L. Yaws, *Handbook of Viscosity,* 3 vols., Elsevier Science and Technology, New York, 1994.
30. A. W. Adamson and A. P. Gast, *Physical Chemistry of Surfaces,* Wiley, New York, 1999.
31. C. E. Brennen, *Cavitation and Bubble Dynamics,* Oxford University Press, New York, 1994.
32. National Committee for Fluid Mechanics Films, *Illustrated Experiments in Fluid Mechanics,* M.I.T. Press, Cambridge, MA, 1972.
33. I. G. Currie, *Fundamental Mechanics of Fluids,* 3d ed., Marcel Dekker, New York, 2003.
34. W.-J. Yang (ed.), *Handbook of Flow Visualization,* 2d ed., Taylor and Francis, New York, 2001.
35. W. Merzkirch, *Flow Visualization,* 2d ed., Elsevier, New York, 1987.
36. T. T. Lim and A. J. Smits (eds.), *Flow Visualization: Techniques and Examples,* Imperial College Press, London, 2000.
37. M. Raffel, C. Willert, and J. Kompenhaus, *Particle Image Velocimetry: A Practical Guide,* Springer-Verlag, New York, 1998.
38. Wen-Jai Yang, *Computer-Assisted Flow Visualization,* Begell House, New York, 1994.
39. M. van Dyke, *An Album of Fluid Motion,* Parabolic Press, Stanford, CA, 1982.
40. Y. Nakayama and Y. Tanida (eds.), *Visualized Flow,* vol. 1, Elsevier, New York, 1993; vols. 2 and 3, CRC Press, Boca Raton, FL, 1996.
41. M. Samimy, K. S. Breuer, L. G. Leal, and P. H. Steen, *A Gallery of Fluid Motion,* Cambridge University Press, New York, 2003.
42. S. Y. Son et al., "Coolant Flow Field Measurements in a Two-Pass Channel Using Particle Image Velocimetry," 1999 Heat Transfer Gallery, *Journal of Heat Transfer,* vol. 121, August, 1999.

43. B. Carr and V. E. Young, "Videotapes and Movies on Fluid Dynamics and Fluid Machines," in *Handbook of Fluid Dynamics and Fluid Machinery,* vol. II, J. A. Schetz and A. E. Fuhs (eds.), Wiley, New York, 1996, pp. 1171–1189.

44. Sanford Klein and William Beckman, *Engineering Equation Solver (EES),* F-Chart Software, Middleton, WI, 2005.

45. H. W. Coleman and W. G. Steele, *Experimentation and Uncertainty Analysis for Engineers,* 2d ed., Wiley, New York, 1998.

46. R. Cooke, *Uncertainty Analysis,* Wiley, New York, 2006.

47. J. R. Taylor, *An Introduction to Error Analysis*, 2d ed., University Science Books, Herndon, VA, 1997.

48. S. J. Kline and F. A. McClintock, "Describing Uncertainties in Single-Sample Experiments," *Mechanical Engineering,* January, 1953, pp. 3–9.

49. R. L. Mott, *Applied Fluid Mechanics,* Pearson Prentice-Hall, Upper Saddle River, NJ, 2006.

50. "Putting Porky to Work," Technology Focus, *Mechanical Engineering,* August 2002, p. 24.

51. R. M. Olson and S. J. Wright, *Essentials of Engineering Fluid Mechanics,* 5th ed., HarperCollins, New York, 1990.

Roosevelt Dam in Arizona. Hydrostatic pressure, due to the weight of a standing fluid, can cause enormous forces and moments on large-scale structures such as a dam. Hydrostatic fluid analysis is the subject of the present chapter. *(Courtesy of Dr. E. R. Degginger/Color-Pic Inc.)*

Chapter 2
Pressure Distribution in a Fluid

Motivation. Many fluid problems do not involve motion. They concern the pressure distribution in a static fluid and its effect on solid surfaces and on floating and submerged bodies.

When the fluid velocity is zero, denoted as the *hydrostatic condition,* the pressure variation is due only to the weight of the fluid. Assuming a known fluid in a given gravity field, the pressure may easily be calculated by integration. Important applications in this chapter are (1) pressure distribution in the atmosphere and the oceans, (2) the design of manometer pressure instruments, (3) forces on submerged flat and curved surfaces, (4) buoyancy on a submerged body, and (5) the behavior of floating bodies. The last two result in Archimedes' principles.

If the fluid is moving in *rigid-body motion,* such as a tank of liquid that has been spinning for a long time, the pressure also can be easily calculated because the fluid is free of shear stress. We apply this idea here to simple rigid-body accelerations in Sec. 2.9. Pressure measurement instruments are discussed in Sec. 2.10. As a matter of fact, pressure also can be analyzed in arbitrary (nonrigid-body) motions $V(x, y, z, t)$, but we defer that subject to Chap. 4.

2.1 Pressure and Pressure Gradient

In Fig. 1.1 we saw that a fluid at rest cannot support shear stress and thus Mohr's circle reduces to a point. In other words, the normal stress on any plane through a fluid element at rest is a point property called the *fluid pressure p,* taken positive for compression by common convention. This is such an important concept that we shall review it with another approach.

Figure 2.1 shows a small wedge of fluid at rest of size Δx by Δz by Δs and depth b into the paper. There is no shear by definition, but we postulate that the pressures p_x, p_z, and p_n may be different on each face. The weight of the element also may be important. The element is assumed small, so the pressure is constant on each

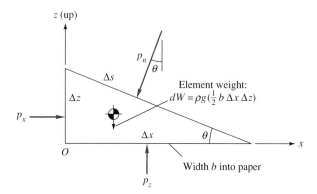

Fig. 2.1 Equilibrium of a small wedge of fluid at rest.

face. Summation of forces must equal zero (no acceleration) in both the x and z directions.

$$\sum F_x = 0 = p_x b\, \Delta z - p_n b\, \Delta s \sin \theta$$
$$\sum F_z = 0 = p_z b\, \Delta x - p_n b\, \Delta s \cos \theta - \tfrac{1}{2}\rho g b\, \Delta x\, \Delta z \tag{2.1}$$

But the geometry of the wedge is such that

$$\Delta s \sin \theta = \Delta z \qquad \Delta s \cos \theta = \Delta x \tag{2.2}$$

Substitution into Eq. (2.1) and rearrangement give

$$p_x = p_n \qquad p_z = p_n + \tfrac{1}{2}\rho g\, \Delta z \tag{2.3}$$

These relations illustrate two important principles of the hydrostatic, or shear-free, condition: (1) There is no pressure change in the horizontal direction, and (2) there is a vertical change in pressure proportional to the density, gravity, and depth change. We shall exploit these results to the fullest, starting in Sec. 2.3.

In the limit as the fluid wedge shrinks to a "point," $\Delta z \to 0$ and Eqs. (2.3) become

$$p_x = p_z = p_n = p \tag{2.4}$$

Since θ is arbitrary, we conclude that the pressure p in a static fluid is a point property, independent of orientation.

Pressure Force on a Fluid Element

Pressure (or any other stress, for that matter) causes a net force on a fluid element when it varies *spatially*.[1] To see this, consider the pressure acting on the two x faces in Fig. 2.2. Let the pressure vary arbitrarily

$$p = p(x, y, z, t)$$

The net force in the x direction on the element in Fig. 2.2 is given by

$$dF_x = p\, dy\, dz - \left(p + \frac{\partial p}{\partial x}\, dx\right) dy\, dz = -\frac{\partial p}{\partial x}\, dx\, dy\, dz$$

[1]An interesting application for a large element is seen in Fig. 3.7.

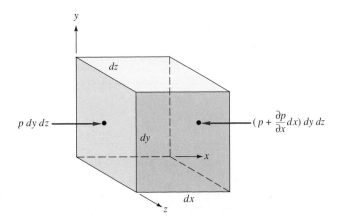

Fig. 2.2 Net *x* force on an element due to pressure variation.

In like manner the net force dF_y involves $-\partial p/\partial y$, and the net force dF_z concerns $-\partial p/\partial z$. The total net-force vector on the element due to pressure is

$$d\mathbf{F}_{\text{press}} = \left(-\mathbf{i}\frac{\partial p}{\partial x} - \mathbf{j}\frac{\partial p}{\partial y} - \mathbf{k}\frac{\partial p}{\partial z}\right) dx\, dy\, dz \tag{2.5}$$

We recognize the term in parentheses as the negative vector gradient of *p*. Denoting **f** as the net force per unit element volume, we rewrite Eq. (2.5) as

$$\mathbf{f}_{\text{press}} = -\nabla p \tag{2.6}$$

Thus it is not the pressure but the pressure *gradient* causing a net force that must be balanced by gravity or acceleration or some other effect in the fluid.

2.2 Equilibrium of a Fluid Element

The pressure gradient is a *surface* force that acts on the sides of the element. There may also be a *body* force, due to electromagnetic or gravitational potentials, acting on the entire mass of the element. Here we consider only the gravity force, or weight of the element:

$$d\mathbf{F}_{\text{grav}} = \rho\mathbf{g}\, dx\, dy\, dz$$

$$\tag{2.7}$$

or
$$\mathbf{f}_{\text{grav}} = \rho\mathbf{g}$$

In addition to gravity, a fluid in motion will have *surface* forces due to viscous stresses. By Newton's law, Eq. (1.2), the sum of these per-unit-volume forces equals the mass per unit volume (density) times the acceleration **a** of the fluid element:

$$\sum \mathbf{f} = \mathbf{f}_{\text{press}} + \mathbf{f}_{\text{grav}} + \mathbf{f}_{\text{visc}} = -\nabla p + \rho\mathbf{g} + \mathbf{f}_{\text{visc}} = \rho\mathbf{a} \tag{2.8}$$

This general equation will be studied in detail in Chap. 4. Note that Eq. (2.8) is a *vector* relation, and the acceleration may not be in the same vector direction as the velocity. For our present topic, *hydrostatics,* the viscous stresses and the acceleration are zero.

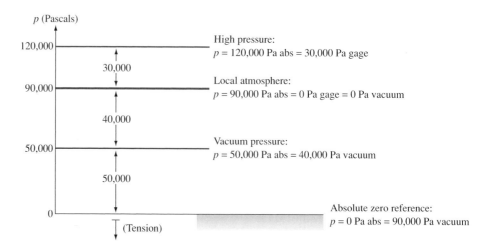

Fig. 2.3 Illustration of absolute, gage, and vacuum pressure readings.

Gage Pressure and Vacuum Pressure: Relative Terms

Before embarking on examples, we should note that engineers are apt to specify pressures as (1) the *absolute* or total magnitude or (2) the value *relative* to the local ambient atmosphere. The second case occurs because many pressure instruments are of *differential* type and record, not an absolute magnitude, but the difference between the fluid pressure and the atmosphere. The measured pressure may be either higher or lower than the local atmosphere, and each case is given a name:

1. $p > p_a$ *Gage* pressure: $p(\text{gage}) = p - p_a$
2. $p < p_a$ *Vacuum* pressure: $p(\text{vacuum}) = p_a - p$

This is a convenient shorthand, and one later adds (or subtracts) atmospheric pressure to determine the absolute fluid pressure.

A typical situation is shown in Fig. 2.3. The local atmosphere is at, say, 90,000 Pa, which might reflect a storm condition in a sea-level location or normal conditions at an altitude of 1000 m. Thus, on this day, p_a = 90,000 Pa absolute = 0 Pa gage = 0 Pa vacuum. Suppose gage 1 in a laboratory reads p_1 = 120,000 Pa absolute. This value may be reported as a *gage* pressure, p_1 = 120,000 − 90,000 = 30,000 Pa *gage*. (One must also record the atmospheric pressure in the laboratory, since p_a changes gradually.) Suppose gage 2 reads p_2 = 50,000 Pa absolute. Locally, this is a *vacuum* pressure and might be reported as p_2 = 90,000 − 50,000 = 40,000 Pa *vacuum*. Occasionally, in the problems section, we will specify gage or vacuum pressure to keep you alert to this common engineering practice. If a pressure is listed without the modifier gage or vacuum, we assume it is absolute pressure.

2.3 Hydrostatic Pressure Distributions

If the fluid is at rest or at constant velocity, $\mathbf{a} = 0$ and $\mathbf{f}_{\text{visc}} = 0$. Equation (2.8) for the pressure distribution reduces to

$$\nabla p = \rho \mathbf{g} \qquad (2.9)$$

This is a *hydrostatic* distribution and is correct for all fluids at rest, regardless of their viscosity, because the viscous term vanishes identically.

Recall from vector analysis that the vector ∇p expresses the magnitude and direction of the maximum spatial rate of increase of the scalar property p. As a result, ∇p is perpendicular everywhere to surfaces of constant p. Thus Eq. (2.9) states that a fluid in hydrostatic equilibrium will align its constant-pressure surfaces everywhere normal to the local-gravity vector. The maximum pressure increase will be in the direction of gravity—that is, "down." If the fluid is a liquid, its free surface, being at atmospheric pressure, will be normal to local gravity, or "horizontal." You probably knew all this before, but Eq. (2.9) is the proof of it.

In our customary coordinate system z is "up." Thus the local-gravity vector for small-scale problems is

$$\mathbf{g} = -g\mathbf{k} \qquad (2.10)$$

where g is the magnitude of local gravity, for example, 9.807 m/s^2. For these coordinates Eq. (2.9) has the components

$$\frac{\partial p}{\partial x} = 0 \qquad \frac{\partial p}{\partial y} = 0 \qquad \frac{\partial p}{\partial z} = -\rho g = -\gamma \qquad (2.11)$$

the first two of which tell us that p is independent of x and y. Hence $\partial p/\partial z$ can be replaced by the total derivative dp/dz, and the hydrostatic condition reduces to

$$\frac{dp}{dz} = -\gamma$$

or

$$p_2 - p_1 = -\int_1^2 \gamma \, dz \qquad (2.12)$$

Equation (2.12) is the solution to the hydrostatic problem. The integration requires an assumption about the density and gravity distribution. Gases and liquids are usually treated differently.

We state the following conclusions about a hydrostatic condition:

> Pressure in a continuously distributed uniform static fluid varies only with vertical distance and is independent of the shape of the container. The pressure is the same at all points on a given horizontal plane in the fluid. The pressure increases with depth in the fluid.

An illustration of this is shown in Fig. 2.4. The free surface of the container is atmospheric and forms a horizontal plane. Points a, b, c, and d are at equal depth in a horizontal plane and are interconnected by the same fluid, water; therefore all points have the same pressure. The same is true of points A, B, and C on the bottom, which all have the same higher pressure than at a, b, c, and d. However, point D, although at the same depth as A, B, and C, has a different pressure because it lies beneath a different fluid, mercury.

Effect of Variable Gravity

For a spherical planet of uniform density, the acceleration of gravity varies inversely as the square of the radius from its center

$$g = g_0\left(\frac{r_0}{r}\right)^2 \qquad (2.13)$$

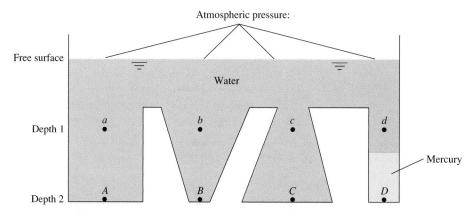

Fig. 2.4 Hydrostatic-pressure distribution. Points a, b, c, and d are at equal depths in water and therefore have identical pressures. Points A, B, and C are also at equal depths in water and have identical pressures higher than a, b, c, and d. Point D has a different pressure from A, B, and C because it is not connected to them by a water path.

where r_0 is the planet radius and g_0 is the surface value of g. For earth, $r_0 \approx 3960$ statute mi ≈ 6400 km. In typical engineering problems the deviation from r_0 extends from the deepest ocean, about 11 km, to the atmospheric height of supersonic transport operation, about 20 km. This gives a maximum variation in g of $(6400/6420)^2$, or 0.6 percent. We therefore neglect the variation of g in most problems.

Hydrostatic Pressure in Liquids

Liquids are so nearly incompressible that we can neglect their density variation in hydrostatics. In Example 1.6 we saw that water density increases only 4.6 percent at the deepest part of the ocean. Its effect on hydrostatics would be about half of this, or 2.3 percent. Thus we assume constant density in liquid hydrostatic calculations, for which Eq. (2.12) integrates to

$$\text{Liquids:} \qquad p_2 - p_1 = -\gamma(z_2 - z_1) \qquad (2.14)$$

$$\text{or} \qquad z_1 - z_2 = \frac{p_2}{\gamma} - \frac{p_1}{\gamma}$$

We use the first form in most problems. The quantity γ is called the *specific weight* of the fluid, with dimensions of weight per unit volume; some values are tabulated in Table 2.1. The quantity p/γ is a length called the *pressure head* of the fluid.

Table 2.1 Specific Weight of Some Common Fluids

Fluid	Specific weight γ at 68°F = 20°C	
	lbf/ft³	N/m³
Air (at 1 atm)	0.0752	11.8
Ethyl alcohol	49.2	7,733
SAE 30 oil	55.5	8,720
Water	62.4	9,790
Seawater	64.0	10,050
Glycerin	78.7	12,360
Carbon tetrachloride	99.1	15,570
Mercury	846	133,100

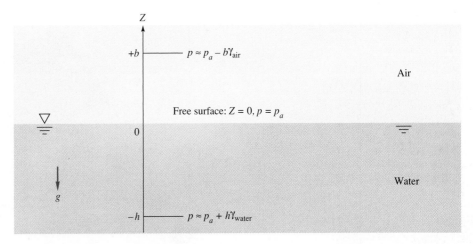

Fig. 2.5 Hydrostatic-pressure distribution in oceans and atmospheres.

For lakes and oceans, the coordinate system is usually chosen as in Fig. 2.5, with $z = 0$ at the free surface, where p equals the surface atmospheric pressure p_a. When we introduce the reference value $(p_1, z_1) = (p_a, 0)$, Eq. (2.14) becomes, for p at any (negative) depth z,

Lakes and oceans: $$p = p_a - \gamma z \qquad (2.15)$$

where γ is the average specific weight of the lake or ocean. As we shall see, Eq. (2.15) holds in the atmosphere also with an accuracy of 2 percent for heights z up to 1000 m.

EXAMPLE 2.1

Newfound Lake, a freshwater lake near Bristol, New Hampshire, has a maximum depth of 60 m, and the mean atmospheric pressure is 91 kPa. Estimate the absolute pressure in kPa at this maximum depth.

Solution

- *System sketch:* Imagine that Fig. 2.5 is Newfound Lake, with $h = 60$ m and $z = 0$ at the surface.
- *Property values:* From Table 2.1, $\gamma_{water} = 9790$ N/m³. We are given that $p_{atmos} = 91$ kPa.
- *Solution steps:* Apply Eq. (2.15) to the deepest point. Use SI units, pascals, not kilopascals:

$$p_{max} = p_a - \gamma z = 91{,}000 \text{ Pa} - (9790 \frac{\text{N}}{\text{m}^3})(-60\,\text{m}) = 678{,}400 \text{ Pa} \approx 678 \text{ kPa} \qquad Ans.$$

- *Comments:* Kilopascals are awkward. Use pascals in the formula, then convert the answer.

The Mercury Barometer

The simplest practical application of the hydrostatic formula (2.14) is the barometer (Fig. 2.6), which measures atmospheric pressure. A tube is filled with mercury and inverted while submerged in a reservoir. This causes a near vacuum in the closed upper end because mercury has an extremely small vapor pressure at room temperatures

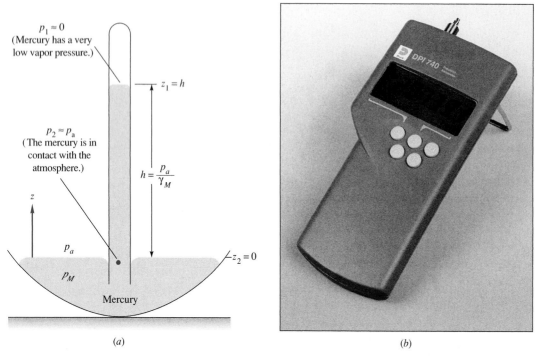

$p_1 \approx 0$
(Mercury has a very low vapor pressure.)

$z_1 = h$

$p_2 \approx p_a$
(The mercury is in contact with the atmosphere.)

$h = \dfrac{p_a}{\gamma_M}$

z

p_a

p_M

$z_2 = 0$

Mercury

(a)

(b)

Fig. 2.6 A barometer measures local absolute atmospheric pressure: (*a*) the height of a mercury column is proportional to p_{atm}; (*b*) a modern portable barometer, with digital readout, uses the resonating silicon element of Fig. 2.28c. *(Courtesy of Paul Lupke, Druck Inc.)*

(0.16 Pa at 20°C). Since atmospheric pressure forces a mercury column to rise a distance h into the tube, the upper mercury surface is at zero pressure.

From Fig. 2.6, Eq. (2.20) applies with $p_1 = 0$ at $z_1 = h$ and $p_2 = p_a$ at $z_2 = 0$:

$$p_a - 0 = -\gamma_M(0 - h)$$

or

$$h = \frac{p_a}{\gamma_M} \tag{2.16}$$

At sea-level standard, with $p_a = 101{,}350$ Pa and $\gamma_M = 133{,}100$ N/m^3 from Table 2.1, the barometric height is $h = 101{,}350/133{,}100 = 0.761$ m or 761 mm. In the United States the weather service reports this as an atmospheric "pressure" of 29.96 inHg (inches of mercury). Mercury is used because it is the heaviest common liquid. A water barometer would be 34 ft high.

Hydrostatic Pressure in Gases

Gases are compressible, with density nearly proportional to pressure. Thus density must be considered as a variable in Eq. (2.12) if the integration carries over large pressure changes. It is sufficiently accurate to introduce the perfect-gas law $p = \rho RT$ in Eq. (2.12):

$$\frac{dp}{dz} = -\rho g = -\frac{p}{RT}g$$

Separate the variables and integrate between points 1 and 2:

$$\int_1^2 \frac{dp}{p} = \ln\frac{p_2}{p_1} = -\frac{g}{R}\int_1^2 \frac{dz}{T} \tag{2.17}$$

The integral over z requires an assumption about the temperature variation $T(z)$. One common approximation is the *isothermal atmosphere*, where $T = T_0$:

$$p_2 = p_1 \exp\left[-\frac{g(z_2 - z_1)}{RT_0}\right] \tag{2.18}$$

The quantity in brackets is dimensionless. (Think that over; it must be dimensionless, right?) Equation (2.18) is a fair approximation for earth, but actually the earth's mean atmospheric temperature drops off nearly linearly with z up to an altitude of about 36,000 ft (11,000 m):

$$T \approx T_0 - Bz \tag{2.19}$$

Here T_0 is sea-level temperature (absolute) and B is the *lapse rate*, both of which vary somewhat from day to day. By international agreement [1] the following standard values are assumed to apply from 0 to 36,000 ft:

$$T_0 = 518.69°R = 288.16 \text{ K} = 15°C$$
$$B = 0.003566°R/ft = 0.00650 \text{ K/m}$$

This lower portion of the atmosphere is called the *troposphere*. Introducing Eq. (2.19) into (2.17) and integrating, we obtain the more accurate relation

$$p = p_a\left(1 - \frac{Bz}{T_0}\right)^{g/(RB)} \quad \text{where } \frac{g}{RB} = 5.26 \text{ (air)} \tag{2.20}$$

in the troposphere, with $z = 0$ at sea level. The exponent $g/(RB)$ is dimensionless (again it must be) and has the standard value of 5.26 for air, with $R = 287 \text{ m}^2/(\text{s}^2 \cdot \text{K})$.

The U.S. standard atmosphere [1] is sketched in Fig. 2.7. The pressure is seen to be nearly zero at $z = 30$ km. For tabulated properties see Table A.6.

EXAMPLE 2.2

If sea-level pressure is 101,350 Pa, compute the standard pressure at an altitude of 5000 m, using (*a*) the exact formula and (*b*) an isothermal assumption at a standard sea-level temperature of 15°C. Is the isothermal approximation adequate?

Solution

Part (a) Use absolute temperature in the exact formula, Eq. (2.20):

$$p = p_a\left[1 - \frac{(0.00650 \text{ K/m})(5000 \text{ m})}{288.16 \text{ K}}\right]^{5.26} = (101,350 \text{ Pa})(0.8872)^{5.26}$$
$$= 101,350(0.5328) = 54,000 \text{ Pa} \qquad\qquad \textit{Ans. (a)}$$

This is the standard-pressure result given at $z = 5000$ m in Table A.6.

Part (b)   If the atmosphere were isothermal at 288.16 K, Eq. (2.18) would apply:

$$p \approx p_a \exp\left(-\frac{gz}{RT}\right) = (101,350 \text{ Pa}) \exp\left\{-\frac{(9.807 \text{ m/s}^2)(5000 \text{ m})}{[287 \text{ m}^2/(\text{s}^2 \cdot \text{K})](288.16 \text{ K})}\right\}$$

$$= (101,350 \text{ Pa}) \exp(-0.5929) \approx 56,000 \text{ Pa} \qquad\qquad Ans.\ (b)$$

This is 4 percent higher than the exact result. The isothermal formula is inaccurate in the troposphere.

Is the Linear Formula Adequate for Gases?

The linear approximation from Eq. (2.14), $\delta p \approx -\rho g\, \delta z$, is satisfactory for liquids, which are nearly incompressible. For gases, it is inaccurate unless δz is rather small. Problem P2.4 asks you to show, by binomial expansion of Eq. (2.20), that the error in using constant gas density to estimate δp from Eq. (2.14) is small if

$$\delta z \ll \frac{2T_0}{(n-1)B} \tag{2.21}$$

where T_o is the local absolute temperature, B is the lapse rate from Eq. (2.19), and $n = g/(RB)$ is the exponent in Eq. (2.20). The error is less than 1 percent if $\delta z < 200$ m.

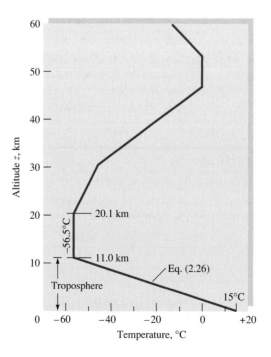
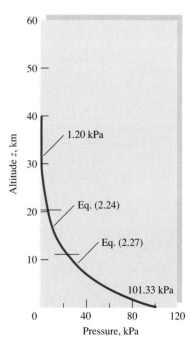

Fig. 2.7 Temperature and pressure distribution in the U.S. standard atmosphere. *(From Ref. 1.)*

2.4 Application to Manometry

From the hydrostatic formula (2.14), a change in elevation $z_2 - z_1$ of a liquid is equivalent to a change in pressure $(p_2 - p_1)/\gamma$. Thus a static column of one or more liquids or gases can be used to measure pressure differences between two points. Such a device is called a *manometer*. If multiple fluids are used, we must change the density in the formula as we move from one fluid to another. Figure 2.8 illustrates the use of the formula with a column of multiple fluids. The pressure change through each fluid is calculated separately. If we wish to know the total change $p_5 - p_1$, we add the successive changes $p_2 - p_1$, $p_3 - p_2$, $p_4 - p_3$, and $p_5 - p_4$. The intermediate values of p cancel, and we have, for the example of Fig. 2.8,

$$p_5 - p_1 = -\gamma_0(z_2 - z_1) - \gamma_w(z_3 - z_2) - \gamma_G(z_4 - z_3) - \gamma_M(z_5 - z_4) \quad (2.22)$$

No additional simplification is possible on the right-hand side because of the different densities. Notice that we have placed the fluids in order from the lightest on top to the heaviest at bottom. This is the only stable configuration. If we attempt to layer them in any other manner, the fluids will overturn and seek the stable arrangement.

Pressure Increases Downward

The basic hydrostatic relation, Eq. (2.14), is mathematically correct but vexing to engineers because it combines two negative signs to have the pressure increase downward. When calculating hydrostatic pressure changes, engineers work instinctively by simply having the pressure increase downward and decrease upward. If point 2 is a distance h below point 1 in a uniform liquid, then $p_2 = p_1 + \rho g h$. In the meantime, Eq. (2.14) remains accurate and safe if used properly. For example, Eq. (2.22) is correct as shown, or it could be rewritten in the following "multiple downward increase" mode:

$$p_5 = p_1 + \gamma_0 |z_1 - z_2| + \gamma_w |z_2 - z_3| + \gamma_G |z_3 - z_4| + \gamma_M |z_4 - z_5|$$

That is, keep adding on pressure increments as you move down through the layered fluid. A different application is a manometer, which involves both "up" and "down" calculations.

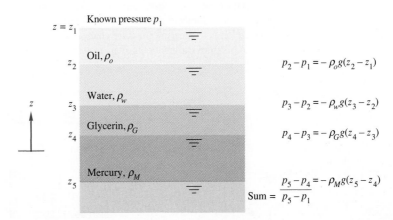

Fig. 2.8 Evaluating pressure changes through a column of multiple fluids.

$z = z_1$ — Known pressure p_1

z_2 — Oil, ρ_o — $p_2 - p_1 = -\rho_o g(z_2 - z_1)$

z_3 — Water, ρ_w — $p_3 - p_2 = -\rho_w g(z_3 - z_2)$

z_4 — Glycerin, ρ_G — $p_4 - p_3 = -\rho_G g(z_4 - z_3)$

z_5 — Mercury, ρ_M — $p_5 - p_4 = -\rho_M g(z_5 - z_4)$

Sum = $p_5 - p_1$

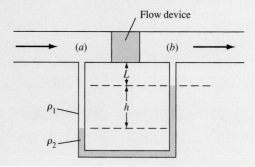

Open, p_a

$z_2, p_2 \approx p_a$

z_A, p_A —

A

p_1

Jump across

z_1, p_1 —

$p = p_1$ at $z = z_1$ in fluid 2

p_2

Fig. 2.9 Simple open manometer for measuring p_A relative to atmospheric pressure.

Application: A Simple Manometer

Figure 2.9 shows a simple U-tube open manometer that measures the *gage* pressure p_A relative to the atmosphere, p_a. The chamber fluid ρ_1 is separated from the atmosphere by a second, heavier fluid ρ_2, perhaps because fluid A is corrosive, or more likely because a heavier fluid ρ_2 will keep z_2 small and the open tube can be shorter.

We first apply the hydrostatic formula (2.14) from A down to z_1. Note that we can then go down to the bottom of the U-tube and back up on the right side to z_1, and the pressure will be the same, $p = p_1$. Thus we can "jump across" and then up to level z_2:

$$p_A + \gamma_1 |z_A - z_1| - \gamma_2 |z_1 - z_2| = p_2 \approx p_{atm} \qquad (2.23)$$

Another physical reason that we can "jump across" at section 1 is that a continuous length of the same fluid connects these two equal elevations. The hydrostatic relation (2.14) requires this equality as a form of Pascal's law:

Any two points at the same elevation in a continuous mass of the same static fluid will be at the same pressure.

This idea of jumping across to equal pressures facilitates multiple-fluid problems. It will be inaccurate however if there are bubbles in the fluid.

EXAMPLE 2.3

The classic use of a manometer is when two U-tube legs are of equal length, as in Fig. E2.3, and the measurement involves a pressure difference across two horizontal points. The typical application is to measure pressure change across a flow device, as shown. Derive a formula for the pressure difference $p_a - p_b$ in terms of the system parameters in Fig. E2.3.

Flow device

(a)

(b)

L

h

ρ_1

ρ_2

E2.3

Solution

Using Eq. (2.14), start at (a), evaluate pressure changes around the U-tube, and end up at (b):

$$p_a + \rho_1 gL + \rho_1 gh - \rho_2 gh - \rho_1 gL = p_b$$

or
$$p_a - p_b = (\rho_2 - \rho_1)gh \qquad\qquad Ans.$$

The measurement only includes h, the manometer reading. Terms involving L drop out. Note the appearance of the *difference* in densities between manometer fluid and working fluid. It is a common student error to fail to subtract out the working fluid density ρ_1—a serious error if both fluids are liquids and less disastrous numerically if fluid 1 is a gas. Academically, of course, such an error is always considered serious by fluid mechanics instructors.

Although Example 2.3, because of its popularity in engineering experiments, is sometimes considered to be the "manometer formula," it is best *not* to memorize it but rather to adapt Eq. (2.14) to each new multiple-fluid hydrostatics problem. For example, Fig. 2.10 illustrates a multiple-fluid manometer problem for finding the difference in pressure between two chambers A and B. We repeatedly apply Eq. (2.14), jumping across at equal pressures when we come to a continuous mass of the same fluid. Thus, in Fig. 2.10, we compute four pressure differences while making three jumps:

$$
\begin{aligned}
p_A - p_B &= (p_A - p_1) + (p_1 - p_2) + (p_2 - p_3) + (p_3 - p_B) \\
&= -\gamma_1(z_A - z_1) - \gamma_2(z_1 - z_2) - \gamma_3(z_2 - z_3) - \gamma_4(z_3 - z_B)
\end{aligned}
\tag{2.24}
$$

The intermediate pressures $p_{1,2,3}$ cancel. It looks complicated, but really it is merely *sequential*. One starts at A, goes down to 1, jumps across, goes up to 2, jumps across, goes down to 3, jumps across, and finally goes up to B.

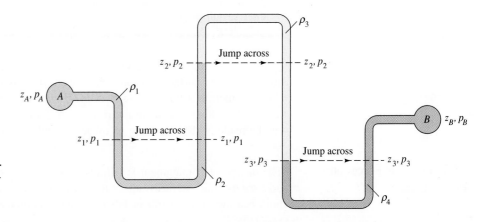

Fig. 2.10 A complicated multiple-fluid manometer to relate p_A to p_B. This system is not especially practical but makes a good homework or examination problem.

EXAMPLE 2.4

Pressure gage B is to measure the pressure at point A in a water flow. If the pressure at B is 87 kPa, estimate the pressure at A in kPa. Assume all fluids are at 20°C. See Fig. E2.4.

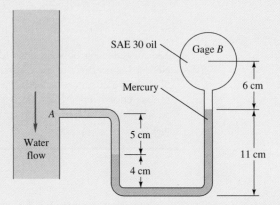

E2.4

Solution

- *System sketch:* The system is shown in Fig. E2.4.
- *Assumptions:* Hydrostatic fluids, no mixing, vertical "up" in Fig. E2.4.
- *Approach:* Sequential use of Eq. (2.14) to go from A to B.
- *Property values:* From Table 2.1 or Table A.3:

$$\gamma_{\text{water}} = 9790 \text{ N/m}^3; \qquad \gamma_{\text{mercury}} = 133{,}100 \text{ N/m}^3; \qquad \gamma_{\text{oil}} = 8720 \text{ N/m}^3$$

- *Solution steps:* Proceed from A to B, "down" then "up," jumping across at the left mercury meniscus:

$$p_A + \gamma_w \, |\Delta z|_w - \gamma_m \, |\Delta z_m| - \gamma_o \, |\Delta z|_o = p_B$$

or $p_A + (9790 \text{ N/m}^3)(0.05 \text{ m}) - (133{,}100 \text{ N/m}^3)(0.07 \text{ m}) - (8720 \text{ N/m}^3)(0.06 \text{ m}) = 87{,}000$

or $p_A + 490 - 9317 - 523 = 87{,}000$ Solve for $p_A = 96{,}350 \text{ N/m}^2 \approx 96.4 \text{ kPa}$ *Ans.*

- *Comments:* Note that we abbreviated the units N/m^2 to pascals, or Pa. The intermediate five-figure result, $p_A = 96{,}350$ Pa, is unrealistic, since the data are known to only about three significant figures.

In making these manometer calculations we have neglected the capillary height changes due to surface tension, which were discussed in Example 1.8. These effects cancel if there is a fluid interface, or *meniscus,* between similar fluids on both sides of the U-tube. Otherwise, as in the right-hand U-tube of Fig. 2.10, a capillary correction can be made or the effect can be made negligible by using large-bore (≥ 1 cm) tubes.

2.5 Hydrostatic Forces on Plane Surfaces

The design of containment structures requires computation of the hydrostatic forces on various solid surfaces adjacent to the fluid. These forces relate to the weight of fluid bearing on the surface. For example, a container with a flat, horizontal bottom

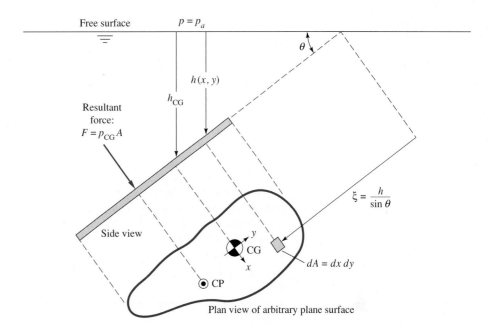

2.11 Hydrostatic force and center of pressure on an arbitrary plane surface of area A inclined at an angle θ below the free surface.

of area A_b and water depth H will experience a downward bottom force $F_b = \gamma H A_b$. If the surface is not horizontal, additional computations are needed to find the horizontal components of the hydrostatic force.

If we neglect density changes in the fluid, Eq. (2.14) applies and the pressure on any submerged surface varies linearly with depth. For a plane surface, the linear stress distribution is exactly analogous to combined bending and compression of a beam in strength-of-materials theory. The hydrostatic problem thus reduces to simple formulas involving the centroid and moments of inertia of the plate cross-sectional area.

Figure 2.11 shows a plane panel of arbitrary shape completely submerged in a liquid. The panel plane makes an arbitrary angle θ with the horizontal free surface, so that the depth varies over the panel surface. If h is the depth to any element area dA of the plate, from Eq. (2.14) the pressure there is $p = p_a + \gamma h$.

To derive formulas involving the plate shape, establish an xy coordinate system in the plane of the plate with the origin at its centroid, plus a dummy coordinate ξ down from the surface in the plane of the plate. Then the total hydrostatic force on one side of the plate is given by

$$F = \int p \, dA = \int (p_a + \gamma h) \, dA = p_a A + \gamma \int h \, dA \qquad (2.25)$$

The remaining integral is evaluated by noticing from Fig. 2.11 that $h = \xi \sin \theta$ and, by definition, the centroidal slant distance from the surface to the plate is

$$\xi_{CG} = \frac{1}{A} \int \xi \, dA$$

Therefore, since θ is constant along the plate, Eq. (2.35) becomes

$$F = p_a A + \gamma \sin \theta \int \xi \, dA = p_a A + \gamma \sin \theta \, \xi_{CG} A$$

Finally, unravel this by noticing that $\xi_{CG} \sin \theta = h_{CG}$, the depth straight down from the surface to the plate centroid. Thus

$$F = p_a A + \gamma h_{CG} A = (p_a + \gamma h_{CG})A = p_{CG} A \qquad (2.26)$$

The force on one side of any plane submerged surface in a uniform fluid equals the pressure at the plate centroid times the plate area, independent of the shape of the plate or the angle θ at which it is slanted.

Equation (2.26) can be visualized physically in Fig. 2.12 as the resultant of a linear stress distribution over the plate area. This simulates combined compression and bending of a beam of the same cross section. It follows that the "bending" portion of the stress causes no force if its "neutral axis" passes through the plate centroid of area. Thus the remaining "compression" part must equal the centroid stress times the plate area. This is the result of Eq. (2.26).

However, to balance the bending-moment portion of the stress, the resultant force F acts not through the centroid but below it toward the high-pressure side. Its line of action passes through the *center of pressure* CP of the plate, as sketched in Fig. 2.11. To find the coordinates (x_{CP}, y_{CP}), we sum moments of the elemental force $p \, dA$ about the centroid and equate to the moment of the resultant F. To compute y_{CP}, we equate

$$F y_{CP} = \int y p \, dA = \int y(p_a + \gamma \xi \sin \theta) \, dA = \gamma \sin \theta \int y \xi \, dA$$

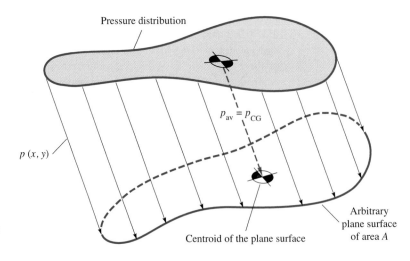

Pressure distribution

$p_{av} = p_{CG}$

$p(x, y)$

Arbitrary plane surface of area A

Centroid of the plane surface

Fig. 2.12 The hydrostatic pressure force on a plane surface is equal, regardless of its shape, to the resultant of the three-dimensional linear pressure distribution on that surface $F = p_{CG} A$.

The term $\int p_a y\, dA$ vanishes by definition of centroidal axes. Introducing $\xi = \xi_{CG} - y$, we obtain

$$Fy_{CP} = \gamma \sin \theta \left(\xi_{CG} \int y\, dA - \int y^2\, dA \right) = -\gamma \sin \theta\, I_{xx}$$

where again $\int y\, dA = 0$ and I_{xx} is the area moment of inertia of the plate area about its centroidal x axis, computed in the plane of the plate. Substituting for F gives the result

$$\boxed{y_{CP} = -\gamma \sin \theta \, \frac{I_{xx}}{p_{CG}A}} \qquad (2.27)$$

The negative sign in Eq. (2.27) shows that y_{CP} is below the centroid at a deeper level and, unlike F, depends on angle θ. If we move the plate deeper, y_{CP} approaches the centroid because every term in Eq. (2.27) remains constant except p_{CG}, which increases.

The determination of x_{CP} is exactly similar:

$$Fx_{CP} = \int xp\, dA = \int x[p_a + \gamma(\xi_{CG} - y)\sin \theta]\, dA$$

$$= -\gamma \sin \theta \int xy\, dA = -\gamma \sin \theta\, I_{xy}$$

where I_{xy} is the product of inertia of the plate, again computed in the plane of the plate. Substituting for F gives

$$\boxed{x_{CP} = -\gamma \sin \theta \, \frac{I_{xy}}{p_{CG}A}} \qquad (2.28)$$

For positive I_{xy}, x_{CP} is negative because the dominant pressure force acts in the third, or lower left, quadrant of the panel. If $I_{xy} = 0$, usually implying symmetry, $x_{CP} = 0$ and the center of pressure lies directly below the centroid on the y axis.

Gage Pressure Formulas

In most cases the ambient pressure p_a is neglected because it acts on both sides of the plate; for example, the other side of the plate is inside a ship or on the dry side of a gate or dam. In this case $p_{CG} = \gamma h_{CG}$, and the center of pressure becomes independent of specific weight:

$$\boxed{F = \gamma h_{CG}A \qquad y_{CP} = -\frac{I_{xx} \sin \theta}{h_{CG}A} \qquad x_{CP} = -\frac{I_{xy} \sin \theta}{h_{CG}A}} \qquad (2.29)$$

Figure 2.13 gives the area and moments of inertia of several common cross sections for use with these formulas. Note that θ is the angle between the plate and the horizon.

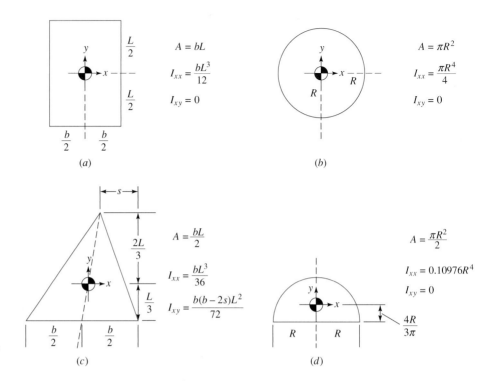

Fig. 2.13 Centroidal moments of inertia for various cross sections: (*a*) rectangle, (*b*) circle, (*c*) triangle, and (*d*) semicircle.

EXAMPLE 2.5

The gate in Fig. E2.5*a* is 5 ft wide, is hinged at point *B*, and rests against a smooth wall at point *A*. Compute (*a*) the force on the gate due to seawater pressure, (*b*) the horizontal force *P* exerted by the wall at point *A*, and (*c*) the reactions at the hinge *B*.

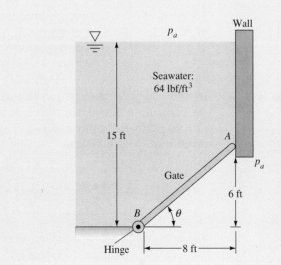

E2.5a

Solution

Part (a)

By geometry the gate is 10 ft long from A to B, and its centroid is halfway between, or at elevation 3 ft above point B. The depth h_{CG} is thus $15 - 3 = 12$ ft. The gate area is $5(10) = 50$ ft^2. Neglect p_a as acting on both sides of the gate. From Eq. (2.26) the hydrostatic force on the gate is

$$F = p_{CG}A = \gamma h_{CG}A = (64 \text{ lbf/ft}^3)(12 \text{ ft})(50 \text{ ft}^2) = 38,400 \text{ lbf} \qquad \textit{Ans. (a)}$$

Part (b)

First we must find the center of pressure of F. A free-body diagram of the gate is shown in Fig. E2.5b. The gate is a rectangle, hence

$$I_{xy} = 0 \quad \text{and} \quad I_{xx} = \frac{bL^3}{12} = \frac{(5 \text{ ft})(10 \text{ ft})^3}{12} = 417 \text{ ft}^4$$

The distance l from the CG to the CP is given by Eq. (2.29) since p_a is neglected.

$$l = -y_{CP} = +\frac{I_{xx} \sin \theta}{h_{CG}A} = \frac{(417 \text{ ft}^4)(\frac{6}{10})}{(12 \text{ ft})(50 \text{ ft}^2)} = 0.417 \text{ ft}$$

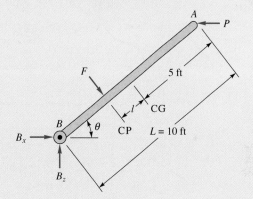

E2.5b

The distance from point B to force F is thus $10 - l - 5 = 4.583$ ft. Summing the moments counterclockwise about B gives

$$PL \sin \theta - F(5 - l) = P(6 \text{ ft}) - (38,400 \text{ lbf})(4.583 \text{ ft}) = 0$$

or
$$P = 29,300 \text{ lbf} \qquad \textit{Ans. (b)}$$

Part (c)

With F and P known, the reactions B_x and B_z are found by summing forces on the gate:

$$\sum F_x = 0 = B_x + F \sin \theta - P = B_x + 38,400 \text{ lbf } (0.6) - 29,300 \text{ lbf}$$

or
$$B_x = 6300 \text{ lbf}$$

$$\sum F_z = 0 = B_z - F \cos \theta = B_z - 38,400 \text{ lbf } (0.8)$$

or
$$B_z = 30,700 \text{ lbf} \qquad \textit{Ans. (c)}$$

This example should have reviewed your knowledge of statics.

The solution of Example 2.5 was achieved with the moment of inertia formulas, Eqs. (2.29). They simplify the calculations, but one loses a physical feeling for the forces. Let us repeat Parts (a) and (b) of Example 2.5 using a more visual approach.

EXAMPLE 2.6

Repeat Example 2.5 to sketch the pressure distribution on plate AB, and break this distribution into rectangular and triangular parts to solve for (a) the force on the plate and (b) the center of pressure.

Solution

Part (a)

Point A is 9 ft deep, hence $p_A = \gamma h_A = (64\ \text{lbf/ft}^3)(9\ \text{ft}) = 576\ \text{lbf/ft}^2$. Similarly, Point B is 15 ft deep, hence $p_B = \gamma h_B = (64\ \text{lbf/ft}^3)(15\ \text{ft}) = 960\ \text{lbf/ft}^2$. This defines the linear pressure distribution in Fig. E2.6. The rectangle is 576 by 10 ft by 5 ft into the paper. The triangle is $(960 - 576) = 384\ \text{lbf/ft}^2 \times 10$ ft by 5 ft. The centroid of the rectangle is 5 ft down the plate from A. The centroid of the triangle is 6.67 ft down from A. The total force is the rectangle force plus the triangle force:

$$F = \left(576\frac{\text{lbf}}{\text{ft}^2}\right)(10\ \text{ft})(5\ \text{ft}) + \left(\frac{384}{2}\frac{\text{lbf}}{\text{ft}^2}\right)(10\ \text{ft})(5\ \text{ft})$$

$$= 28{,}800\ \text{lbf} + 9600\ \text{lbf} = 38{,}400\ \text{lbf} \qquad\qquad Ans.\ (a)$$

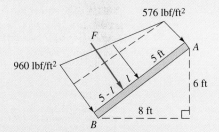

E2.6

Part (b)

The moments of these forces about point A are

$$\Sigma M_A = (28{,}800\ \text{lbf})(5\ \text{ft}) + (9600\ \text{lbf})(6.67\ \text{ft}) = 144{,}000 + 64{,}000 = 208{,}000\ \text{ft}\cdot\text{lbf}$$

Then $\quad 5\ \text{ft} + l = \dfrac{M_A}{F} = \dfrac{208{,}000\ \text{ft}\cdot\text{lbf}}{38{,}400\ \text{lbf}} = 5.417\ \text{ft}\quad$ hence $l = 0.417\ \text{ft}\qquad Ans.\ (b)$

Comment: We obtain the same force and center of pressure as in Example 2.5 but with more understanding. However, this approach is awkward and laborious if the plate is not a rectangle. It would be difficult to solve Example 2.7 with the pressure distribution alone because the plate is a triangle. Thus moments of inertia can be a useful simplification.

EXAMPLE 2.7

A tank of oil has a right-triangular panel near the bottom, as in Fig. E2.7. Omitting p_a, find the (a) hydrostatic force and (b) CP on the panel.

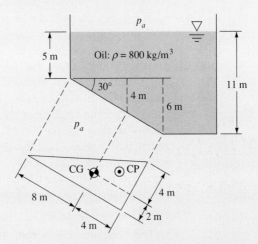

E2.7

Solution

Part (a)

The triangle has properties given in Fig. 2.13c. The centroid is one-third up (4 m) and one-third over (2 m) from the lower left corner, as shown. The area is

$$\tfrac{1}{2}(6 \text{ m})(12 \text{ m}) = 36 \text{ m}^2$$

The moments of inertia are

$$I_{xx} = \frac{bL^3}{36} = \frac{(6 \text{ m})(12 \text{ m})^3}{36} = 288 \text{ m}^4$$

and

$$I_{xy} = \frac{b(b - 2s)L^2}{72} = \frac{(6 \text{ m})[6 \text{ m} - 2(6 \text{ m})](12 \text{ m})^2}{72} = -72 \text{ m}^4$$

The depth to the centroid is $h_{CG} = 5 + 4 = 9$ m; thus the hydrostatic force from Eq. (2.26) is

$$F = \rho g h_{CG} A = (800 \text{ kg/m}^3)(9.807 \text{ m/s}^2)(9 \text{ m})(36 \text{ m}^2)$$
$$= 2.54 \times 10^6 \text{ (kg} \cdot \text{m)/s}^2 = 2.54 \times 10^6 \text{ N} = 2.54 \text{ MN} \qquad \textit{Ans. (a)}$$

Part (b)

The CP position is given by Eqs. (2.29):

$$y_{CP} = -\frac{I_{xx} \sin \theta}{h_{CG} A} = -\frac{(288 \text{ m}^4)(\sin 30°)}{(9 \text{ m})(36 \text{ m}^2)} = -0.444 \text{ m}$$

$$x_{CP} = -\frac{I_{xy} \sin \theta}{h_{CG} A} = -\frac{(-72 \text{ m}^4)(\sin 30°)}{(9 \text{ m})(36 \text{ m}^2)} = +0.111 \text{ m} \qquad \textit{Ans. (b)}$$

The resultant force $F = 2.54$ MN acts through this point, which is down and to the right of the centroid, as shown in Fig. E2.7.

2.6 Hydrostatic Forces on Curved Surfaces

The resultant pressure force on a curved surface is most easily computed by separating it into horizontal and vertical components. Consider the arbitrary curved surface sketched in Fig. 2.14a. The incremental pressure forces, being normal to the local area element, vary in direction along the surface and thus cannot be added numerically. We could sum the separate three components of these elemental pressure forces, but it turns out that we need not perform a laborious three-way integration.

Figure 2.14b shows a free-body diagram of the column of fluid contained in the vertical projection above the curved surface. The desired forces F_H and F_V are exerted by the surface on the fluid column. Other forces are shown due to fluid weight and horizontal pressure on the vertical sides of this column. The column of fluid must be in static equilibrium. On the upper part of the column $bcde$, the horizontal components F_1 exactly balance and are not relevant to the discussion. On the lower, irregular portion of fluid abc adjoining the surface, summation of horizontal forces shows that the desired force F_H due to the curved surface is exactly equal to the force F_H on the vertical left side of the fluid column. This left-side force can be computed by the plane surface formula, Eq. (2.26), based on a vertical projection of the area of the curved surface. This is a general rule and simplifies the analysis:

> The horizontal component of force on a curved surface equals the force on the plane area formed by the projection of the curved surface onto a vertical plane normal to the component.

If there are two horizontal components, both can be computed by this scheme. Summation of vertical forces on the fluid free body then shows that

$$F_V = W_1 + W_2 + W_{air} \tag{2.30}$$

We can state this in words as our second general rule:

> The vertical component of pressure force on a curved surface equals in magnitude and direction the weight of the entire column of fluid, both liquid and atmosphere, above the curved surface.

Fig. 2.14 Computation of hydrostatic force on a curved surface: (a) submerged curved surface; (b) free-body diagram of fluid above the curved surface.

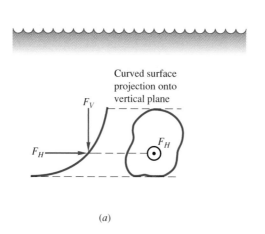

(a)

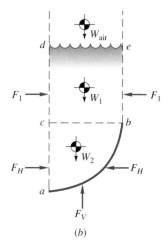

(b)

Thus the calculation of F_V involves little more than finding centers of mass of a column of fluid—perhaps a little integration if the lower portion *abc* in Fig. 2.14*b* has a particularly vexing shape.

EXAMPLE 2.8

A dam has a parabolic shape $z/z_0 = (x/x_0)^2$ as shown in Fig. E2.8*a*, with $x_0 = 10$ ft and $z_0 = 24$ ft. The fluid is water, $\gamma = 62.4$ lbf/ft^3, and atmospheric pressure may be omitted. Compute the forces F_H and F_V on the dam and their line of action. The width of the dam is 50 ft.

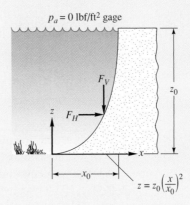

$p_a = 0$ lbf/ft^2 gage

z_0

z

F_V

F_H

x

x_0

$z = z_0 \left(\dfrac{x}{x_0}\right)^2$

E2.8a

Solution

- *System sketch:* Figure E2.8*b* shows the various dimensions. The dam width is $b = 50$ ft.
- *Approach:* Calculate F_H and its line of action from Eqs. (2.26) and (2.29). Calculate F_V and its line of action by finding the weight of fluid above the parabola and the centroid of this weight.
- *Solution steps for the horizontal component:* The vertical projection of the parabola lies along the z axis in Fig. E2.8*b* and is a rectangle 24 ft high and 50 ft wide. Its centroid is halfway down, or $h_{CG} = 24/2 = 12$ ft. Its area is $A_{proj} = (24\text{ ft})(50\text{ ft}) = 1200$ ft^2. Then, from Eq. (2.26),

$$F_H = \gamma h_{CG} A_{proj} = \left(62.4\frac{\text{lbf}}{\text{ft}^3}\right)(12\text{ ft})(1200\text{ ft}^2) = 898{,}560\text{ lbf} \approx 899 \times 10^3\text{ lbf}$$

The line of action of F_H is below the centroid of A_{proj}, as given by Eq. (2.29):

$$y_{CP,\,proj} = -\frac{I_{xx}\sin\theta}{h_{CG}A_{proj}} = -\frac{(1/12)(50\text{ ft})(24\text{ ft})^3\sin 90^\circ}{(12\text{ ft})(1200\text{ ft}^2)} = -4\text{ ft}$$

Thus F_H is $12 + 4 = 16$ ft, or two-thirds of the way down from the surface (8 ft up from the bottom).

- *Comments:* Note that you calculate F_H and its line of action from the *vertical projection* of the parabola, not from the parabola itself. Since this projection is *vertical*, its angle $\theta = 90^\circ$.
- *Solution steps for the vertical component:* The vertical force F_V equals the weight of water above the parabola. Alas, a parabolic section is not in Fig. 2.13, so we had to look

it up in another book. The area and centroid are shown in Fig. E2.8b. The weight of this parabolic amount of water is

$$F_V = \gamma A_{section} b = \left(62.4\frac{\text{lbf}}{\text{ft}^3}\right)\left[\frac{2}{3}(24\text{ ft})(10\text{ ft})\right](50\text{ ft}) = 499,200\text{ lbf} \approx 499 \times 10^3\text{ lbf}$$

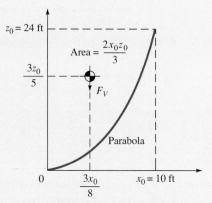

E2.8b

This force acts downward, through the centroid of the parabolic section, or at a distance $3x_0/8 = 3.75$ ft over from the origin, as shown in Figs. E2.8b,c. The resultant hydrostatic force on the dam is

$$F = (F_H^2 + F_V^2)^{1/2} = [(899\text{E3 lbf})^2 + (499\text{E3 lbf})^2]^{1/2} = 1028 \times 10^3\text{ lbf at } 29° \quad Ans.$$

This resultant is shown in Fig. E2.8c and passes through a point 8 ft up and 3.75 ft over from the origin. It strikes the dam at a point 5.43 ft over and 7.07 ft up, as shown.

- *Comments:* Note that entirely different formulas are used to calculate F_H and F_V. The concept of center of pressure CP is, in the writer's opinion, stretched too far when applied to curved surfaces.

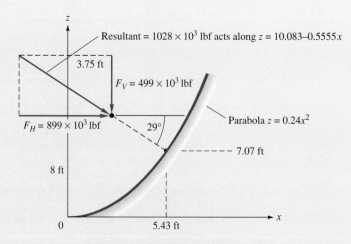

E2.8c

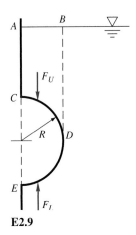

E2.9

EXAMPLE 2.9

Find an algebraic formula for the net vertical force F on the submerged semicircular projecting structure CDE in Fig. E2.9. The structure has uniform width b into the paper. The liquid has specific weight γ.

Solution

The net force is the difference between the upward force F_L on the lower surface DE and the downward force F_U on the upper surface CD, as shown in Fig. E2.9. The force F_U equals γ times the volume $ABDC$ above surface CD. The force F_L equals γ times the volume $ABDEC$ above surface DE. The latter is clearly larger. The difference is γ times the volume of the structure itself. Thus the net upward fluid force on the semicylinder is

$$F = \gamma_{\text{fluid}} (\text{volume } CDE) = \gamma_{\text{fluid}} \frac{\pi}{2} R^2 b \qquad \qquad Ans.$$

This is the principle upon which the laws of buoyancy, Sec. 2.8, are founded. Note that the result is independent of the depth of the structure and depends upon the specific weight of the *fluid*, not the material within the structure.

2.7 Hydrostatic Forces in Layered Fluids

The formulas for plane and curved surfaces in Secs. 2.5 and 2.6 are valid only for a fluid of uniform density. If the fluid is layered with different densities, as in Fig. 2.15, a single formula cannot solve the problem because the slope of the linear pressure

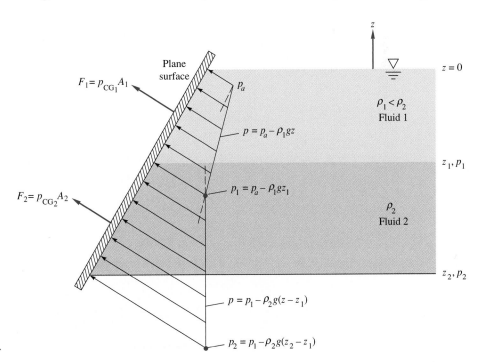

Fig. 2.15 Hydrostatic forces on a surface immersed in a layered fluid must be summed in separate pieces.

distribution changes between layers. However, the formulas apply separately to each layer, and thus the appropriate remedy is to compute and sum the separate layer forces and moments.

Consider the slanted plane surface immersed in a two-layer fluid in Fig. 2.15. The slope of the pressure distribution becomes steeper as we move down into the denser second layer. The total force on the plate does *not* equal the pressure at the centroid times the plate area, but the plate portion in each layer does satisfy the formula, so that we can sum forces to find the total:

$$F = \Sigma F_i = \Sigma p_{CG_i} A_i \tag{2.31}$$

Similarly, the centroid of the plate portion in each layer can be used to locate the center of pressure on that portion:

$$y_{CP_i} = -\frac{\rho_i g \sin \theta_i I_{xx_i}}{p_{CG_i} A_i} \qquad x_{CP_i} = -\frac{\rho_i g \sin \theta_i I_{xy_i}}{p_{CG_i} A_i} \tag{2.32}$$

These formulas locate the center of pressure of that particular F_i with respect to the centroid of that particular portion of plate in the layer, not with respect to the centroid of the entire plate. The center of pressure of the total force $F = \Sigma F_i$ can then be found by summing moments about some convenient point such as the surface. The following example will illustrate this.

EXAMPLE 2.10

A tank 20 ft deep and 7 ft wide is layered with 8 ft of oil, 6 ft of water, and 4 ft of mercury. Compute (*a*) the total hydrostatic force and (*b*) the resultant center of pressure of the fluid on the right-hand side of the tank.

Solution

Part (a)

Divide the end panel into three parts as sketched in Fig. E2.10, and find the hydrostatic pressure at the centroid of each part, using the relation (2.26) in steps as in Fig. E2.10:

$$p_{CG_1} = (55.0 \text{ lbf/ft}^3)(4 \text{ ft}) = 220 \text{ lbf/ft}^2$$
$$p_{CG_2} = (55.0)(8) + 62.4(3) = 627 \text{ lbf/ft}^2$$
$$p_{CG_3} = (55.0)(8) + 62.4(6) + 846(2) = 2506 \text{ lbf/ft}^2$$

These pressures are then multiplied by the respective panel areas to find the force on each portion:

$$F_1 = p_{CG_1} A_1 = (220 \text{ lbf/ft}^2)(8 \text{ ft})(7 \text{ ft}) = 12,300 \text{ lbf}$$
$$F_2 = p_{CG_2} A_2 = 627(6)(7) = 26,300 \text{ lbf}$$
$$F_3 = p_{CG_3} A_3 = 2506(4)(7) = \underline{70,200 \text{ lbf}}$$
$$F = \Sigma F_i = 108,800 \text{ lbf} \qquad \qquad Ans. \ (a)$$

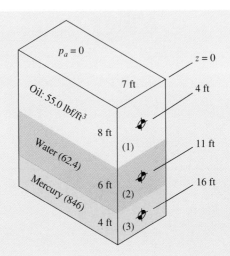

E2.10

Part (b)

Equations (2.32) can be used to locate the CP of each force F_i, noting that $\theta = 90°$ and $\sin \theta = 1$ for all parts. The moments of inertia are $I_{xx_1} = (7 \text{ ft})(8 \text{ ft})^3/12 = 298.7 \text{ ft}^4$, $I_{xx_2} = 7(6)^3/12 = 126.0 \text{ ft}^4$, and $I_{xx_3} = 7(4)^3/12 = 37.3 \text{ ft}^4$. The centers of pressure are thus at

$$y_{CP_1} = -\frac{\rho_1 g I_{xx_1}}{F_1} = -\frac{(55.0 \text{ lbf/ft}^3)(298.7 \text{ ft}^4)}{12{,}300 \text{ lbf}} = -1.33 \text{ ft}$$

$$y_{CP_2} = -\frac{62.4(126.0)}{26{,}300} = -0.30 \text{ ft} \qquad y_{CP_3} = -\frac{846(37.3)}{70{,}200} = -0.45 \text{ ft}$$

This locates $z_{CP_1} = -4 - 1.33 = -5.33 \text{ ft}$, $z_{CP_2} = -11 - 0.30 = -11.30 \text{ ft}$, and $z_{CP_3} = -16 - 0.45 = -16.45 \text{ ft}$. Summing moments about the surface then gives

$$\sum F_i z_{CP_i} = F z_{CP}$$

or $\qquad 12{,}300(-5.33) + 26{,}300(-11.30) + 70{,}200(-16.45) = 108{,}800 z_{CP}$

or $\qquad\qquad\qquad z_{CP} = -\dfrac{1{,}518{,}000}{108{,}800} = -13.95 \text{ ft}$ $\qquad\qquad$ *Ans.* (b)

The center of pressure of the total resultant force on the right side of the tank lies 13.95 ft below the surface.

2.8 Buoyancy and Stability

The same principles used to compute hydrostatic forces on surfaces can be applied to the net pressure force on a completely submerged or floating body. The results are the two laws of buoyancy discovered by Archimedes in the third century B.C.:

1. A body immersed in a fluid experiences a vertical buoyant force equal to the weight of the fluid it displaces.
2. A floating body displaces its own weight in the fluid in which it floats.

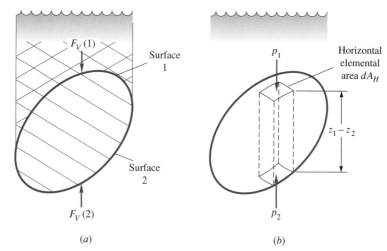

Fig. 2.16 Two different approaches to the buoyant force on an arbitrary immersed body: (*a*) forces on upper and lower curved surfaces; (*b*) summation of elemental vertical-pressure forces.

These two laws are easily derived by referring to Fig. 2.16. In Fig. 2.16*a*, the body lies between an upper curved surface 1 and a lower curved surface 2. From Eq. (2.30) for vertical force, the body experiences a net upward force

$$F_B = F_V(2) - F_V(1)$$
$$= \text{(fluid weight above 2)} - \text{(fluid weight above 1)}$$
$$= \text{weight of fluid equivalent to body volume} \qquad (2.33)$$

Alternatively, from Fig. 2.16*b*, we can sum the vertical forces on elemental vertical slices through the immersed body:

$$F_B = \int_{\text{body}} (p_2 - p_1)\, dA_H = -\gamma \int (z_2 - z_1)\, dA_H = (\gamma)(\text{body volume}) \quad (2.34)$$

These are identical results and equivalent to Archimedes' law 1.

Equation (2.34) assumes that the fluid has uniform specific weight. The line of action of the buoyant force passes through the center of volume of the displaced body; that is, its center of mass is computed as if it had uniform density. This point through which F_B acts is called the *center of buoyancy,* commonly labeled *B* or CB on a drawing. Of course, the point *B* may or may not correspond to the actual center of mass of the body's own material, which may have variable density.

Equation (2.34) can be generalized to a layered fluid (LF) by summing the weights of each layer of density ρ_i displaced by the immersed body:

$$(F_B)_{\text{LF}} = \sum \rho_i g (\text{displaced volume})_i \qquad (2.35)$$

Neglect the displaced air up here.

CG

W

F_B

B

(Displaced volume) × (γ of fluid) = body weight

Fig. 2.17 Static equilibrium of a floating body.

Each displaced layer would have its own center of volume, and one would have to sum moments of the incremental buoyant forces to find the center of buoyancy of the immersed body.

Since liquids are relatively heavy, we are conscious of their buoyant forces, but gases also exert buoyancy on any body immersed in them. For example, human beings have an average specific weight of about 60 lbf/ft^3. We may record the weight of a person as 180 lbf and thus estimate the person's total volume as 3.0 ft^3. However, in so doing we are neglecting the buoyant force of the air surrounding the person. At standard conditions, the specific weight of air is 0.0763 lbf/ft^3; hence the buoyant force is approximately 0.23 lbf. If measured in a vacuum, the person would weigh about 0.23 lbf more. For balloons and blimps the buoyant force of air, instead of being negligible, is the controlling factor in the design. Also, many flow phenomena, such as natural convection of heat and vertical mixing in the ocean, are strongly dependent on seemingly small buoyant forces.

Floating bodies are a special case; only a portion of the body is submerged, with the remainder poking up out of the free surface. This is illustrated in Fig. 2.17, where the shaded portion is the displaced volume. Equation (2.34) is modified to apply to this smaller volume:

$$F_B = (\gamma)(\text{displaced volume}) = \text{floating-body weight} \qquad (2.36)$$

Not only does the buoyant force equal the body weight, but also they are *collinear* since there can be no net moments for static equilibrium. Equation (2.36) is the mathematical equivalent of Archimedes' law 2, previously stated.

EXAMPLE 2.11

A block of concrete weighs 100 lbf in air and "weighs" only 60 lbf when immersed in fresh water (62.4 lbf/ft^3). What is the average specific weight of the block?

Solution

A free-body diagram of the submerged block (see Fig. E2.11) shows a balance between the apparent weight, the buoyant force, and the actual weight:

$$\sum F_z = 0 = 60 + F_B - 100$$

or

$$F_B = 40 \text{ lbf} = (62.4 \text{ lbf/ft}^3)(\text{block volume, ft}^3)$$

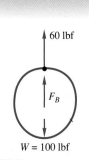

60 lbf

F_B

$W = 100$ lbf

E2.11

Solving gives the volume of the block as $40/62.4 = 0.641$ ft^3. Therefore the specific weight of the block is

$$\gamma_{block} = \frac{100 \text{ lbf}}{0.641 \text{ ft}^3} = 156 \text{ lbf/ft}^3 \qquad Ans.$$

Occasionally, a body will have exactly the right weight and volume for its ratio to equal the specific weight of the fluid. If so, the body will be *neutrally buoyant* and will remain at rest at any point where it is immersed in the fluid. Small neutrally buoyant particles are sometimes used in flow visualization, and a neutrally buoyant body called a *Swallow float* [2] is used to track oceanographic currents. A submarine can achieve positive, neutral, or negative buoyancy by pumping water in or out of its ballast tanks.

Stability

A floating body as in Fig. 2.17 may not approve of the position in which it is floating. If so, it will overturn at the first opportunity and is said to be statically *unstable*, like a pencil balanced on its point. The least disturbance will cause it to seek another equilibrium position that is stable. Engineers must design to avoid floating instability. The only way to tell for sure whether a floating position is stable is to "disturb" the body a slight amount mathematically and see whether it develops a restoring moment that will return it to its original position. If so, it is stable; if not, unstable. Such calculations for arbitrary floating bodies have been honed to a fine art by naval architects [3], but we can at least outline the basic principle of the static stability calculation. Figure 2.18 illustrates the computation for the usual case of a symmetric floating body. The steps are as follows:

1. The basic floating position is calculated from Eq. (2.36). The body's center of mass G and center of buoyancy B are computed.
2. The body is tilted a small angle $\Delta\theta$, and a new waterline is established for the body to float at this angle. The new position B' of the center of buoyancy is calculated. A vertical line drawn upward from B' intersects the line of symmetry at a point M, called the *metacenter,* which is independent of $\Delta\theta$ for small angles.
3. If point M is above G (that is, if the *metacentric height* $\overline{MG}$ is positive), a restoring moment is present and the original position is stable. If M is below G (negative $\overline{MG}$), the body is unstable and will overturn if disturbed. Stability increases with increasing $\overline{MG}$.

Thus the metacentric height is a property of the cross section for the given weight, and its value gives an indication of the stability of the body. For a body of varying cross section and draft, such as a ship, the computation of the metacenter can be very involved.

Stability Related to Waterline Area

Naval architects [3] have developed the general stability concepts from Fig. 2.18 into a simple computation involving the area moment of inertia of the *waterline area* about

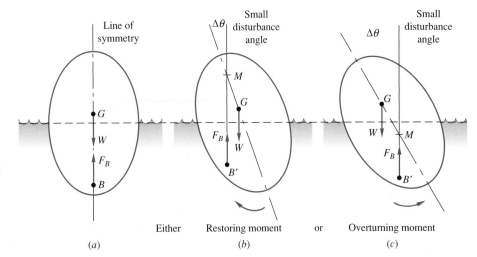

Fig. 2.18 Calculation of the meta-center M of the floating body shown in (a). Tilt the body a small angle $\Delta\theta$. Either (b) B' moves far out (point M above G denotes stability); or (c) B' moves slightly (point M below G denotes instability).

the axis of tilt. The derivation assumes that the body has a smooth shape variation (no discontinuities) near the waterline and is derived from Fig. 2.19.

The y axis of the body is assumed to be a line of symmetry. Tilting the body a small angle θ then submerges the small wedge Obd and uncovers an equal wedge cOa, as shown. The new position B' of the center of buoyancy is calculated as the centroid of the submerged portion $aObde$ of the body:

$$\bar{x}\,\upsilon_{abOde} = \underbrace{\int x\,d\upsilon}_{cOdea} + \underbrace{\int x\,d\upsilon}_{Obd} - \underbrace{\int x\,d\upsilon}_{cOa} = 0 + \underbrace{\int x\,(L\,dA)}_{Obd} - \underbrace{\int x\,(L\,dA)}_{cOa}$$

$$= 0 + \underbrace{\int x\,L\,(x\tan\theta\,dx)}_{Obd} - \underbrace{\int xL\,(-x\tan\theta\,dx)}_{cOa} = \tan\theta\underbrace{\int x^2\,dA_{\text{waterline}}}_{\text{waterline}} = I_O\tan\theta$$

where I_O is the area moment of inertia of the *waterline footprint* of the body about its tilt axis O. The first integral vanishes because of the symmetry of the original

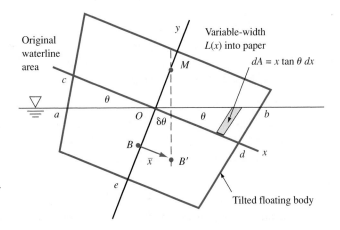

Fig. 2.19 A floating body tilted through a small angle θ. The movement $\bar{x}$ of the center of buoyancy B is related to the waterline area moment of inertia.

submerged portion $cOdea$. The remaining two "wedge" integrals combine into I_O when we notice that $L\,dx$ equals an element of *waterline area*. Thus we determine the desired distance from M to B:

$$\frac{\bar{x}}{\tan \theta} = \overline{MB} = \frac{I_O}{v_{\text{submerged}}} = \overline{MG} + \overline{GB} \quad \text{or} \quad \overline{MG} = \frac{I_O}{v_{\text{sub}}} - \overline{GB} \qquad (2.37)$$

The engineer would determine the distance from G to B from the basic shape and design of the floating body and then make the calculation of I_O and the submerged volume v_{sub}. If the metacentric height $\overline{MG}$ is positive, the body is stable for small disturbances. Note that if $\overline{GB}$ is negative, that is, B is *above* G, the body is always stable.

EXAMPLE 2.12

A barge has a uniform rectangular cross section of width $2L$ and vertical draft of height H, as in Fig. E2.12. Determine (a) the metacentric height for a small tilt angle and (b) the range of ratio L/H for which the barge is statically stable if G is exactly at the waterline as shown.

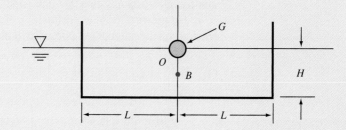

E2.12

Solution

If the barge has length b into the paper, the waterline area, relative to tilt axis O, has a base b and a height $2L$; therefore, $I_O = b(2L)^3/12$. Meanwhile, $v_{\text{sub}} = 2LbH$. Equation (2.37) predicts

$$\overline{MG} = \frac{I_o}{v_{\text{sub}}} - \overline{GB} = \frac{8bL^3/12}{2LbH} - \frac{H}{2} = \frac{L^2}{3H} - \frac{H}{2} \qquad \text{Ans. (a)}$$

The barge can thus be stable only if

$$L^2 > 3H^2/2 \quad \text{or} \quad 2L > 2.45H \qquad \text{Ans. (b)}$$

The wider the barge relative to its draft, the more stable it is. Lowering G would help also.

Even an expert will have difficulty determining the floating stability of a buoyant body of irregular shape. Such bodies may have two or more stable positions. For

Fig. 2.20 A North Atlantic iceberg formed by calving from a Greenland glacier. These, and their even larger Antarctic sisters, are the largest floating bodies in the world. Note the evidence of further calving fractures on the front surface. (© *Corbis.*)

example, a ship may float the way we like it, so that we can sit on the deck, or it may float upside down (capsized). An interesting mathematical approach to floating stability is given in Ref. 11. The author of this reference points out that even simple shapes, such as a cube of uniform density, may have a great many stable floating orientations, not necessarily symmetric. Homogeneous circular cylinders can float with the axis of symmetry tilted from the vertical.

Floating instability occurs in nature. Fish generally swim with their planes of symmetry vertical. After death, this position is unstable and they float with their flat sides up. Giant icebergs may overturn after becoming unstable when their shapes change due to underwater melting. Iceberg overturning is a dramatic, rarely seen event.

Figure 2.20 shows a typical North Atlantic iceberg formed by calving from a Greenland glacier that protruded into the ocean. The exposed surface is rough, indicating that it has undergone further calving. Icebergs are frozen fresh, bubbly, glacial water of average density 900 kg/m^3. Thus, when an iceberg is floating in seawater, whose average density is 1025 kg/m^3, approximately 900/1025, or seven-eighths, of its volume lies below the water.

2.9 Pressure Distribution in Rigid-Body Motion

In rigid-body motion, all particles are in combined translation and rotation, and there is no relative motion between particles. With no relative motion, there are no strains or strain rates, so that the viscous term in Eq. (2.8) vanishes, leaving a balance between pressure, gravity, and particle acceleration:

$$\nabla p = \rho(\mathbf{g} - \mathbf{a}) \tag{2.38}$$

The pressure gradient acts in the direction $\mathbf{g} - \mathbf{a}$, and lines of constant pressure (including the free surface, if any) are perpendicular to this direction. The general case of combined translation and rotation of a rigid body is discussed in Chap. 3, Fig. 3.11.

Fluids can rarely move in rigid-body motion unless restrained by confining walls for a long time. For example, suppose a tank of water is in a car that starts a constant acceleration. The water in the tank would begin to slosh about, and that sloshing

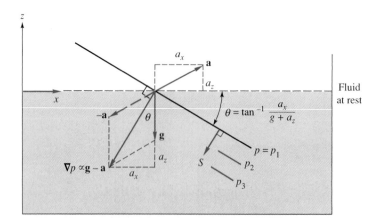

Fig. 2.21 Tilting of constant-pressure surfaces in a tank of liquid in rigid-body acceleration.

would damp out very slowly until finally the particles of water would be in approximately rigid-body acceleration. This would take so long that the car would have reached hypersonic speeds. Nevertheless, we can at least discuss the pressure distribution in a tank of rigidly accelerating water.

Uniform Linear Acceleration

In the case of uniform rigid-body acceleration, Eq. (2.38) applies, **a** having the same magnitude and direction for all particles. With reference to Fig. 2.21, the parallelogram sum of **g** and $-\mathbf{a}$ gives the direction of the pressure gradient or greatest rate of increase of p. The surfaces of constant pressure must be perpendicular to this and are thus tilted at a downward angle θ such that

$$\theta = \tan^{-1}\frac{a_x}{g + a_z} \tag{2.39}$$

One of these tilted lines is the free surface, which is found by the requirement that the fluid retain its volume unless it spills out. The rate of increase of pressure in the direction $\mathbf{g} - \mathbf{a}$ is greater than in ordinary hydrostatics and is given by

$$\frac{dp}{ds} = \rho G \quad \text{where } G = [a_x^2 + (g + a_z)^2]^{1/2} \tag{2.40}$$

These results are independent of the size or shape of the container as long as the fluid is continuously connected throughout the container.

EXAMPLE 2.13

A drag racer rests her coffee mug on a horizontal tray while she accelerates at 7 m/s². The mug is 10 cm deep and 6 cm in diameter and contains coffee 7 cm deep at rest. (*a*) Assuming rigid-body acceleration of the coffee, determine whether it will spill out of the mug. (*b*) Calculate the gage pressure in the corner at point A if the density of coffee is 1010 kg/m³.

Solution

- *System sketch:* Figure E2.13 shows the coffee tilted during the acceleration.

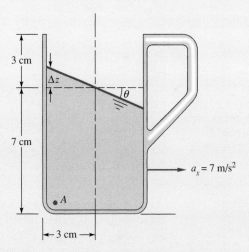

E2.13

- *Assumptions:* Rigid-body horizontal acceleration, $a_x = 7$ m/s². Symmetric coffee cup.
- *Property values:* Density of coffee given as 1010 kg/m³.
- *Approach (a):* Determine the angle of tilt from the known acceleration, then find the height rise.
- *Solution steps:* From Eq. (2.39), the angle of tilt is given by

$$\theta = \tan^{-1}\frac{a_x}{g} = \tan^{-1}\frac{7.0 \text{ m/s}^2}{9.81 \text{ m/s}^2} = 35.5°$$

If the mug is symmetric, the tilted surface will pass through the center point of the rest position, as shown in Fig. E2.13. Then the rear side of the coffee free surface will rise an amount Δz given by

$$\Delta z = (3 \text{ cm})(\tan 35.5°) = 2.14 \text{ cm} < 3 \text{ cm} \qquad \text{therefore no spilling} \qquad Ans. (a)$$

- *Comment (a):* This solution neglects sloshing, which might occur if the start-up is uneven.
- *Approach (b):* The pressure at A can be computed from Eq. (2.40), using the perpendicular distance Δs from the surface to A. When at rest, $p_A = \rho g h_{\text{rest}} = (1010 \text{ kg/m}^3)$ $(9.81 \text{ m/s}^2)(0.07 \text{ m}) = 694$ Pa. When accelerating,

$$p_A = \rho G \, \Delta s = \left(1010\frac{\text{kg}}{\text{m}^3}\right)\left[\sqrt{(9.81)^2 + (7.0)^2}\right][(0.07 + 0.0214)\cos 35.5°] \approx 906 \text{ Pa} \; Ans. (b)$$

- *Comment (b):* The acceleration has increased the pressure at A by 31 percent. Think about this alternative: why does it work? Since $a_z = 0$, we may proceed vertically down the left side to compute

$$p_A = \rho g(z_{\text{surf}} - z_A) = (1010 \text{ kg/m}^3)(9.81 \text{ m/s}^2)(0.0214 + 0.07 \text{ m}) = 906 \text{ Pa}$$

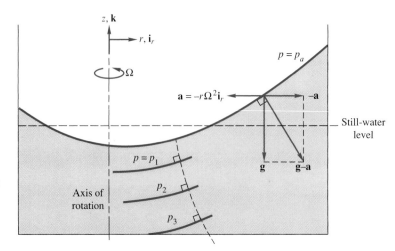

Fig. 2.22 Development of paraboloid constant-pressure surfaces in a fluid in rigid-body rotation. The dashed line along the direction of maximum pressure increase is an exponential curve.

Rigid-Body Rotation

As a second special case, consider rotation of the fluid about the z axis without any translation, as sketched in Fig. 2.22. We assume that the container has been rotating long enough at constant Ω for the fluid to have attained rigid-body rotation. The fluid acceleration will then be a centripetal term. In the coordinates of Fig. 2.22, the angular-velocity and position vectors are given by

$$\mathbf{\Omega} = \mathbf{k}\Omega \quad \mathbf{r_0} = \mathbf{i}_r r \tag{2.41}$$

Then the acceleration is given by

$$\mathbf{\Omega} \times (\mathbf{\Omega} \times \mathbf{r_0}) = -r\Omega^2 \mathbf{i}_r \tag{2.42}$$

as marked in the figure, and Eq. (2.38) for the force balance becomes

$$\nabla p = \mathbf{i}_r \frac{\partial p}{\partial r} + \mathbf{k}\frac{\partial p}{\partial z} = \rho(\mathbf{g} - \mathbf{a}) = \rho(-g\mathbf{k} + r\Omega^2 \mathbf{i}_r)$$

Equating like components, we find the pressure field by solving two first-order partial differential equations:

$$\frac{\partial p}{\partial r} = \rho r \Omega^2 \quad \frac{\partial p}{\partial z} = -\gamma \tag{2.43}$$

The right-hand sides of (2.43) are known functions of r and z. One can proceed as follows: Integrate the first equation "partially," holding z constant, with respect to r. The result is

$$p = \tfrac{1}{2}\rho r^2 \Omega^2 + f(z) \tag{2.44}$$

where the "constant" of integration is actually a function $f(z)$.[2] Now differentiate this with respect to z and compare with the second relation of (2.43):

$$\frac{\partial p}{\partial z} = 0 + f'(z) = -\gamma$$

[2] This is because $f(z)$ vanishes when differentiated with respect to r. If you don't see this, you should review your calculus.

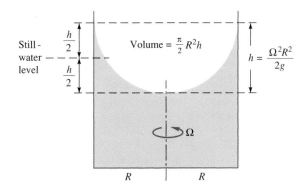

Fig. 2.23 Determining the free-surface position for rotation of a cylinder of fluid about its central axis.

or

$$f(z) = -\gamma z + C$$

where C is a constant. Thus Eq. (2.44) now becomes

$$p = \text{const} - \gamma z + \tfrac{1}{2}\rho r^2 \Omega^2 \qquad (2.45)$$

This is the pressure distribution in the fluid. The value of C is found by specifying the pressure at one point. If $p = p_0$ at $(r, z) = (0, 0)$, then $C = p_0$. The final desired distribution is

$$\boxed{p = p_0 - \gamma z + \tfrac{1}{2}\rho r^2 \Omega^2} \qquad (2.46)$$

The pressure is linear in z and parabolic in r. If we wish to plot a constant-pressure surface, say, $p = p_1$, Eq. (2.45) becomes

$$z = \frac{p_0 - p_1}{\gamma} + \frac{r^2 \Omega^2}{2g} = a + br^2 \qquad (2.47)$$

Thus the surfaces are paraboloids of revolution, concave upward, with their minimum points on the axis of rotation. Some examples are sketched in Fig. 2.22.

As in the previous example of linear acceleration, the position of the free surface is found by conserving the volume of fluid. For a noncircular container with the axis of rotation off-center, as in Fig. 2.22, a lot of laborious mensuration is required, and a single problem will take you all weekend. However, the calculation is easy for a cylinder rotating about its central axis, as in Fig. 2.23. Since the volume of a paraboloid is one-half the base area times its height, the still-water level is exactly halfway between the high and low points of the free surface. The center of the fluid drops an amount $h/2 = \Omega^2 R^2/(4g)$, and the edges rise an equal amount.

EXAMPLE 2.14

The coffee cup in Example 2.13 is removed from the drag racer, placed on a turntable, and rotated about its central axis until a rigid-body mode occurs. Find (a) the angular velocity that will cause the coffee to just reach the lip of the cup and (b) the gage pressure at point A for this condition.

Solution

Part (a)

The cup contains 7 cm of coffee. The remaining distance of 3 cm up to the lip must equal the distance $h/2$ in Fig. 2.23. Thus

$$\frac{h}{2} = 0.03 \text{ m} = \frac{\Omega^2 R^2}{4g} = \frac{\Omega^2 (0.03 \text{ m})^2}{4(9.81 \text{ m/s}^2)}$$

Solving, we obtain

$$\Omega^2 = 1308 \qquad \text{or} \qquad \Omega = 36.2 \text{ rad/s} = 345 \text{ r/min} \qquad \textit{Ans. (a)}$$

Part (b)

To compute the pressure, it is convenient to put the origin of coordinates r and z at the bottom of the free-surface depression, as shown in Fig. E2.14. The gage pressure here is $p_0 = 0$, and point A is at $(r, z) = (3 \text{ cm}, -4 \text{ cm})$. Equation (2.46) can then be evaluated:

$$p_A = 0 - (1010 \text{ kg/m}^3)(9.81 \text{ m/s}^2)(-0.04 \text{ m})$$
$$+ \tfrac{1}{2}(1010 \text{ kg/m}^3)(0.03 \text{ m})^2(1308 \text{ rad}^2/\text{s}^2)$$
$$= 396 \text{ N/m}^2 + 594 \text{ N/m}^2 = 990 \text{ Pa} \qquad \textit{Ans. (b)}$$

This is about 43 percent greater than the still-water pressure $p_A = 694$ Pa.

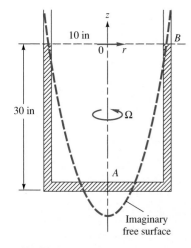

E2.14

Here, as in the linear acceleration case, it should be emphasized that the paraboloid pressure distribution (2.46) sets up in *any* fluid under rigid-body rotation, regardless of the shape or size of the container. The container may even be closed and filled with fluid. It is only necessary that the fluid be continuously interconnected throughout the container. The following example will illustrate a peculiar case in which one can visualize an imaginary free surface extending outside the walls of the container.

EXAMPLE 2.15

A U-tube with a radius of 10 in and containing mercury to a height of 30 in is rotated about its center at 180 r/min until a rigid-body mode is achieved. The diameter of the tubing is negligible. Atmospheric pressure is 2116 lbf/ft². Find the pressure at point A in the rotating condition. See Fig. E2.15.

Solution

Convert the angular velocity to radians per second:

$$\Omega = (180 \text{ r/min}) \frac{2\pi \text{ rad/r}}{60 \text{ s/min}} = 18.85 \text{ rad/s}$$

From Table 2.1 we find for mercury that $\gamma = 846$ lbf/ft³ and hence $\rho = 846/32.2 = 26.3$ slugs/ft³. At this high rotation rate, the free surface will slant upward at a fierce angle [about 84°; check this from Eq. (2.47)], but the tubing is so thin that the free surface will remain at approximately the same 30-in height, point B. Placing our origin of coordinates at this

E2.15

height, we can calculate the constant C in Eq. (2.45) from the condition $p_B = 2116$ lbf/ft^2 at $(r, z) = (10$ in, $0)$:

$$p_B = 2116 \text{ lbf/ft}^2 = C - 0 + \tfrac{1}{2}(26.3 \text{ slugs/ft}^3)(\tfrac{10}{12} \text{ ft})^2(18.85 \text{ rad/s})^2$$

or $$C = 2116 - 3245 = -1129 \text{ lbf/ft}^2$$

We then obtain p_A by evaluating Eq. (2.46) at $(r, z) = (0, -30$ in$)$:

$$p_A = -1129 - (846 \text{ lbf/ft}^3)(-\tfrac{30}{12} \text{ ft}) = -1129 + 2115 = 986 \text{ lbf/ft}^2 \qquad Ans.$$

This is less than atmospheric pressure, and we can see why if we follow the free-surface paraboloid down from point B along the dashed line in the figure. It will cross the horizontal portion of the U-tube (where p will be atmospheric) and fall *below* point A. From Fig. 2.23 the actual drop from point B will be

$$h = \frac{\Omega^2 R^2}{2g} = \frac{(18.85)^2(\tfrac{10}{12})^2}{2(32.2)} = 3.83 \text{ ft} = 46 \text{ in}$$

Thus p_A is about 16 inHg below atmospheric pressure, or about $\tfrac{16}{12}(846) = 1128$ lbf/ft^2 below $p_a = 2116$ lbf/ft^2, which checks with the answer above. When the tube is at rest,

$$p_A = 2116 - 846(-\tfrac{30}{12}) = 4231 \text{ lbf/ft}^2$$

Hence rotation has reduced the pressure at point A by 77 percent. Further rotation can reduce p_A to near-zero pressure, and cavitation can occur.

An interesting by-product of this analysis for rigid-body rotation is that the lines everywhere parallel to the pressure gradient form a family of curved surfaces, as sketched in Fig. 2.22. They are everywhere orthogonal to the constant-pressure surfaces, and hence their slope is the negative inverse of the slope computed from Eq. (2.47):

$$\left.\frac{dz}{dr}\right|_{\text{GL}} = -\frac{1}{(dz/dr)_{p=\text{const}}} = -\frac{1}{r\Omega^2/g}$$

where GL stands for gradient line

or $$\frac{dz}{dr} = -\frac{g}{r\Omega^2} \qquad (2.48)$$

Separating the variables and integrating, we find the equation of the pressure-gradient surfaces:

$$r = C_1 \exp\left(-\frac{\Omega^2 z}{g}\right) \qquad (2.49)$$

Notice that this result and Eq. (2.47) are independent of the density of the fluid. In the absence of friction and Coriolis effects, Eq. (2.49) defines the lines along which

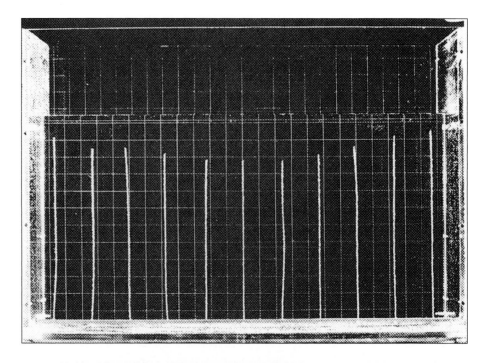

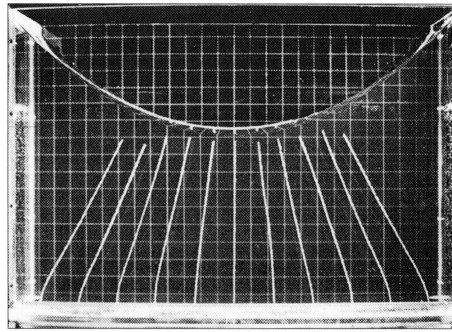

Fig. 2.24 Experimental demonstration with buoyant streamers of the fluid force field in rigid-body rotation: (*top*) fluid at rest (streamers hang vertically upward); (*bottom*) rigid-body rotation (streamers are aligned with the direction of maximum pressure gradient). *(© The American Association of Physics Teachers. Reprinted with permission from 'The Apparent Field of Gravity in a Rotating Fluid System' by R. Ian Fletcher. American Journal of Physics vol. 40, pp. 959–965, July 1972.)*

the apparent net gravitational field would act on a particle. Depending on its density, a small particle or bubble would tend to rise or fall in the fluid along these exponential lines, as demonstrated experimentally in Ref. 5. Also, buoyant streamers would align themselves with these exponential lines, thus avoiding any stress other than pure tension. Figure 2.24 shows the configuration of such streamers before and during rotation.

2.10 Pressure Measurement

Pressure is a derived property. It is the force per unit area as related to fluid molecular bombardment of a surface. Thus most pressure instruments only *infer* the pressure by calibration with a primary device such as a deadweight piston tester. There are many such instruments, for both a static fluid and a moving stream. The instrumentation texts in Refs. 7 to 10, 12, 13, and 16–17 list over 20 designs for pressure measurement instruments. These instruments may be grouped into four categories:

1. *Gravity-based:* barometer, manometer, deadweight piston.
2. *Elastic deformation:* bourdon tube (metal and quartz), diaphragm, bellows, strain-gage, optical beam displacement.
3. *Gas behavior:* gas compression (McLeod gage), thermal conductance (Pirani gage), molecular impact (Knudsen gage), ionization, thermal conductivity, air piston.
4. *Electric output:* resistance (Bridgman wire gage), diffused strain gage, capacitative, piezoelectric, potentiometric, magnetic inductance, magnetic reluctance, linear variable differential transformer (LVDT), resonant frequency.
5. *Luminescent coatings* for surface pressures [15].

The gas-behavior gages are mostly special-purpose instruments used for certain scientific experiments. The deadweight tester is the instrument used most often for calibrations; for example, it is used by the U.S. National Institute for Standards and Technology (NIST). The barometer is described in Fig. 2.6.

The manometer, analyzed in Sec. 2.4, is a simple and inexpensive hydrostatic-principle device with no moving parts except the liquid column itself. Manometer measurements must not disturb the flow. The best way to do this is to take the measurement through a *static hole* in the wall of the flow, as illustrated in Fig. 2.25a. The hole should be normal to the wall, and burrs should be avoided. If the hole is small enough (typically 1-mm diameter), there will be no flow into the measuring tube once the pressure has adjusted to a steady value. Thus the flow is almost undisturbed. An oscillating flow pressure, however, can cause a large error due to possible dynamic response of the tubing. Other devices of smaller dimensions are used for dynamic-pressure measurements. The manometer in Fig. 2.25a measures the gage pressure p_1. The instrument in Fig. 2.25b is a digital *differential* manometer, which can measure the difference between two different points in the flow, with stated accuracy of 0.1 percent of full scale. The world of instrumentation is moving quickly toward digital readings.

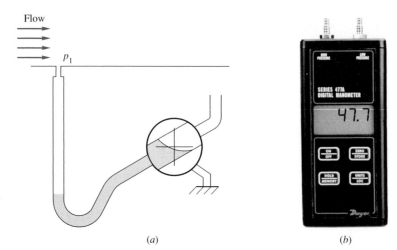

(a)   (b)

Fig. 2.25 Two types of accurate manometers for precise measurements: (*a*) tilted tube with eyepiece; (*b*) a capacitive-type digital manometer of rated accuracy ±0.1 percent. (*Courtesy of Dwyer Instruments, Inc.*)

In category 2, elastic-deformation instruments, a popular, inexpensive, and reliable device is the *bourdon tube,* sketched in Fig. 2.26. When pressurized internally, a curved tube with flattened cross section will deflect outward. The deflection can be measured by a linkage attached to a calibrated dial pointer, as shown. Or the deflection can be used to drive electric-output sensors, such as a variable transformer. Similarly, a membrane or *diaphragm* will deflect under pressure and can either be sensed directly or used to drive another sensor.

An interesting variation of Fig. 2.26 is the *fused-quartz, force-balanced bourdon tube,* shown in Fig. 2.27, whose spiral-tube deflection is sensed optically and returned to a zero reference state by a magnetic element whose output is proportional to the

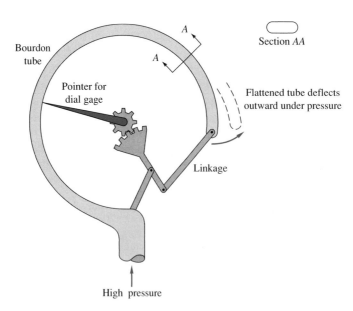

Fig. 2.26 Schematic of a bourdon-tube device for mechanical measurement of high pressures.

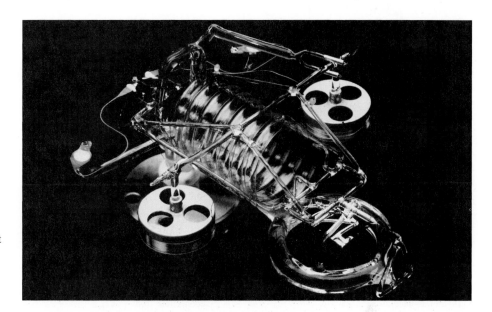

Fig. 2.27 The fused-quartz, force-balanced bourdon tube is the most accurate pressure sensor used in commercial applications today. *(Courtesy of Ruska Instrument Corporation, Houston, TX.)*

fluid pressure. The fused-quartz, force-balanced bourdon tube is reported to be one of the most accurate pressure sensors ever devised, with uncertainty on the order of ±0.003 percent.

The quartz gages, both the bourdon type and the resonant type, are expensive but extremely accurate, stable, and reliable [14]. They are often used for deep-ocean pressure measurements, which detect long waves and tsunami activity over extensive time periods.

The last category, *electric-output sensors,* is extremely important in engineering because the data can be stored on computers and freely manipulated, plotted, and analyzed. Three examples are shown in Fig. 2.28, the first being the *capacitive* sensor in Fig. 2.28*a.* The differential pressure deflects the silicon diaphragm and changes the capacitance of the liquid in the cavity. Note that the cavity has spherical end caps to prevent overpressure damage. In the second type, Fig. 2.28*b,* strain gages and other sensors are chemically diffused or etched onto a chip, which is stressed by the applied pressure. Finally, in Fig. 2.28*c,* a micromachined silicon sensor is arranged to deform under pressure such that its natural vibration frequency is proportional to the pressure. An oscillator excites the element's resonant frequency and converts it into appropriate pressure units.

Another kind of dynamic electric-output sensor is the *piezoelectric transducer,* shown in Fig. 2.29. The sensing elements are thin layers of quartz, which generate an electric charge when subjected to stress. The design in Fig. 2.29 is flush-mounted on a solid surface and can sense rapidly varying pressures, such as blast waves. Other designs are of the cavity type. This type of sensor primarily detects transient pressures, not steady stress, but if highly insulated can also be used for short-term static events. Note also that it measures *gage* pressure—that is, it detects only a change from ambient conditions.

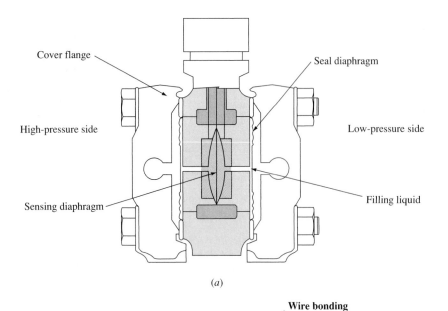

Cover flange

Seal diaphragm

High-pressure side

Low-pressure side

Sensing diaphragm

Filling liquid

(a)

Strain gages
Diffused into integrated
silicon chip

Wire bonding
Stitch bonded
connections from
chip to body plug

Etched cavity
Micromachined
silicon sensor

(b)

Temperature sensor
On-chip diode for
optimum temperature
performance

Fig. 2.28 Pressure sensors with
electric output: (*a*) a silicon
diaphragm whose deflection
changes the cavity capacitance
*(Courtesy of Johnson-Yokogawa
Inc.);* (*b*) a silicon strain gage that
is stressed by applied pressure;
(*c*) a micromachined silicon
element that resonates at a
frequency proportional to applied
pressure. *[(b) and (c) are courtesy
of Druck, Inc., Fairfield, CT.]*

(c)

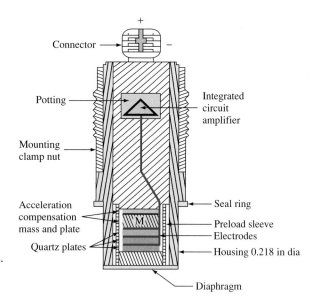

Fig. 2.29 A piezoelectric transducer measures rapidly changing pressures. *(Courtesy of PCB Piezotronics, Inc. Depew, New York.)*

Connector

Potting

Integrated circuit amplifier

Mounting clamp nut

Acceleration compensation mass and plate

Quartz plates

Seal ring

Preload sleeve

Electrodes

Housing 0.218 in dia

Diaphragm

Summary

This chapter has been devoted entirely to the computation of pressure distributions and the resulting forces and moments in a static fluid or a fluid with a known velocity field. All hydrostatic (Secs. 2.3 to 2.8) and rigid-body (Sec. 2.9) problems are solved in this manner and are classic cases that every student should understand. In arbitrary viscous flows, both pressure and velocity are unknowns and are solved together as a system of equations in the chapters that follow.

Problems

Most of the problems herein are fairly straightforward. More difficult or open-ended assignments are indicated with an asterisk, as in Prob. 2.9. Problems labeled with an EES icon (for example, Prob. 2.62) will benefit from the use of the Engineering Equation Solver (EES), while problems labeled with a disk icon may require the use of a computer. The standard end-of-chapter problems 2.1 to 2.159 (categorized in the problem distribution) are followed by word problems W2.1 to W2.8, fundamentals of engineering exam problems FE2.1 to FE2.10, comprehensive problems C2.1 to C2.8, and design projects D2.1 to D2.3.

Problem Distribution

Section	Topic	Problems
2.1, 2.2	Stresses; pressure gradient; gage pressure	2.1–2.6
2.3	Hydrostatic pressure; barometers	2.7–2.23
2.3	The atmosphere	2.24–2.29
2.4	Manometers; multiple fluids	2.30–2.47
2.5	Forces on plane surfaces	2.48–2.80
2.6	Forces on curved surfaces	2.81–2.100
2.7	Forces in layered fluids	2.101–2.102
2.8	Buoyancy; Archimedes' principles	2.103–2.126
2.8	Stability of floating bodies	2.127–2.136
2.9	Uniform acceleration	2.137–2.151
2.9	Rigid-body rotation	2.152–2.159
2.10	Pressure measurements	None

P2.1 For the two-dimensional stress field shown in Fig. P2.1 it is found that

$$\sigma_{xx} = 3000 \ \text{lbf/ft}^2 \quad \sigma_{yy} = 2000 \ \text{lbf/ft}^2 \quad \sigma_{xy} = 500 \ \text{lbf/ft}^2$$

Find the shear and normal stresses (in lbf/ft^2) acting on plane *AA* cutting through the element at a 30° angle as shown.

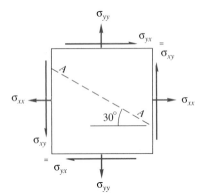

P2.1

P2.2   For the two-dimensional stress field shown in Fig. P2.1 suppose that

$$\sigma_{xx} = 2000 \text{ lbf/ft}^2 \quad \sigma_{yy} = 3000 \text{ lbf/ft}^2 \quad \sigma_n(AA) = 2500 \text{ lbf/ft}^2$$

Compute (a) the shear stress σ_{xy} and (b) the shear stress on plane AA.

P2.3   A vertical, clean, glass piezometer tube has an inside diameter of 1 mm. When pressure is applied, water at 20°C rises into the tube to a height of 25 cm. After correcting for surface tension, estimate the applied pressure in Pa.

P2.4   For gases that undergo large changes in height, the linear approximation, Eq. (2.14), is inaccurate. Expand the troposphere power-law, Eq. (2.20), into a power series, and show that the linear approximation $p \approx p_a - \rho_a\, gz$ is adequate when

$$\delta z \ll \frac{2T_0}{(n-1)B} \quad \text{where } n = \frac{g}{RB}$$

P2.5   Denver, Colorado, has an average altitude of 5300 ft. On a standard day (Table A.6), pressure gage A in a laboratory experiment reads 83 kPa and gage B reads 105 kPa. Express these readings in gage pressure or vacuum pressure (Pa), whichever is appropriate.

P2.6   Any pressure reading can be expressed as a length or head, $h = p/\rho g$. What is standard sea-level pressure expressed in (a) ft of glycerin, (b) inHg, (c) m of water, and (d) mm of ethanol? Assume all fluids are at 20°C.

P2.7   The deepest known point in the ocean is 11,034 m in the Mariana Trench in the Pacific. At this depth the specific weight of seawater is approximately 10,520 N/m³. At the surface, $\gamma \approx 10,050$ N/m³. Estimate the absolute pressure at this depth, in atm.

P2.8   A diamond mine is two miles below sea level. (a) Estimate the air pressure at this depth. (b) If a barometer,

accurate to 1 mm of mercury, is carried into this mine, how accurately can it estimate the depth of the mine? List your assumptions carefully.

*P2.9   For a liquid, integrate the hydrostatic relation, Eq. (2.12), by assuming that the *isentropic bulk modulus*, $B = \rho(\partial p/\partial \rho)_s$, is constant—see Eq. (9.18). Find an expression for $p(z)$ and apply the Mariana Trench data as in Prob. P2.7, using B_{seawater} from Table A.3.

P2.10   A closed tank contains 1.5 m of SAE 30 oil, 1 m of water, 20 cm of mercury, and an air space on top, all at 20°C. The absolute pressure at the bottom of the tank is 60 kPa. What is the pressure in the air space?

P2.11   In Fig. P2.11, pressure gage A reads 1.5 kPa (gage). The fluids are at 20°C. Determine the elevations z, in meters, of the liquid levels in the open piezometer tubes B and C.

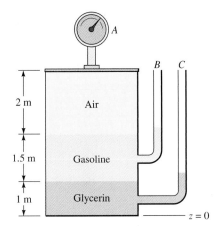

P2.11

P2.12   In Fig. P2.12 the tank contains water and immiscible oil at 20°C. What is h in cm if the density of the oil is 898 kg/m³?

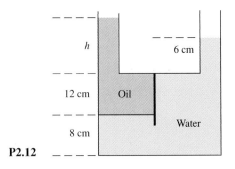

P2.12

P2.13   In Fig. P2.13 the 20°C water and gasoline surfaces are open to the atmosphere and at the same elevation. What is the height h of the third liquid in the right leg?

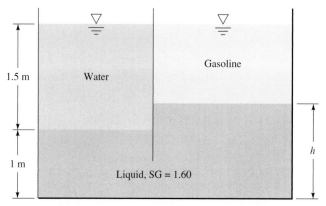

P2.13

P2.14 The symmetric vee-shaped tube in Fig. P2.14 contains static water and air at 20°C. What is the pressure of the air in the closed section at point *B*?

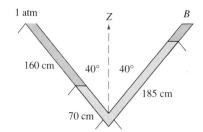

P2.14

P2.15 The air–oil–water system in Fig. P2.15 is at 20°C. Knowing that gage *A* reads 15 lbf/in² absolute and gage *B* reads 1.25 lbf/in² less than gage *C*, compute (*a*) the specific weight of the oil in lbf/ft³ and (*b*) the actual reading of gage *C* in lbf/in² absolute.

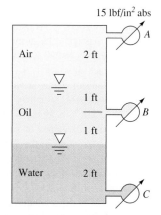

P2.15

P2.16 Suppose that a barometer, using carbon tetrachloride as the working fluid (not recommended), is installed on a

standard day in Denver, Colorado. (*a*) How high would the fluid rise in the barometer tube? (*Note:* Don't forget the vapor pressure.) (*b*) Compare this result with a mercury barometer.

P2.17 The system in Fig. P2.17 is at 20°C. If the pressure at point *A* is 1900 lbf/ft², determine the pressures at points *B*, *C*, and *D* in lbf/ft².

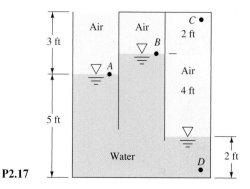

P2.17

P2.18 The system in Fig. P2.18 is at 20°C. If atmospheric pressure is 101.33 kPa and the pressure at the bottom of the tank is 242 kPa, what is the specific gravity of fluid *X*?

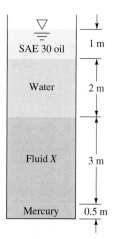

P2.18

P2.19 The U-tube in Fig. P2.19 has a 1-cm ID and contains mercury as shown. If 20 cm³ of water is poured into the right-hand leg, what will the free-surface height in each leg be after the sloshing has died down?

P2.20 The hydraulic jack in Fig. P2.20 is filled with oil at 56 lbf/ft³. Neglecting the weight of the two pistons, what force *F* on the handle is required to support the 2000-lbf weight for this design?

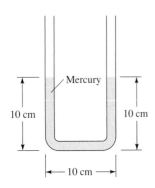

P2.19

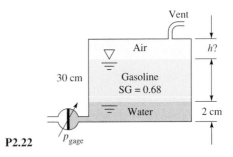

P2.22

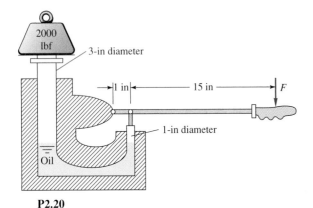

P2.20

P2.21 At 20°C gage A reads 350 kPa absolute. What is the height h of the water in cm? What should gage B read in kPa absolute? See Fig. P2.21.

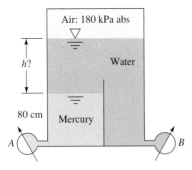

P2.21

P2.22 The fuel gage for a gasoline tank in a car reads proportional to the bottom gage pressure as in Fig. P2.22. If the tank is 30 cm deep and accidentally contains 2 cm of water plus gasoline, how many centimeters of air remain at the top when the gage erroneously reads "full"?

P2.23 In Fig. P2.23 both fluids are at 20°C. If surface tension effects are negligible, what is the density of the oil, in kg/m^3?

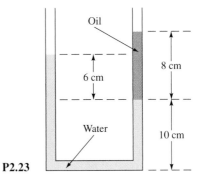

P2.23

P2.24 In Prob. 1.2 we made a crude integration of the density distribution $\rho(z)$ in Table A.6 and estimated the mass of the earth's atmosphere to be $m \approx$ 6 E18 kg. Can this result be used to estimate sea-level pressure on the earth? Conversely, can the actual sea-level pressure of 101.35 kPa be used to make a more accurate estimate of the atmospheric mass?

P2.25 Venus has a mass of 4.90 E24 kg and a radius of 6050 km. Its atmosphere is 96 percent CO_2, but let us assume it to be 100 percent. Its surface temperature averages 730 K, decreasing to 250 K at an altitude of 70 km. The average surface pressure is 9.1 MPa. Estimate the atmospheric pressure of Venus at an altitude of 5 km.

*P2.26 A *polytropic atmosphere* is defined by the power law $p/p_0 = (\rho/\rho_0)^m$, where m is an exponent of order 1.3 and p_0 and ρ_0 are sea-level values of pressure and density. (a) Integrate this expression in the static atmosphere and find a distribution $p(z)$. (b) Assuming an ideal gas, $p = \rho RT$, show that your result in (a) implies a linear temperature distribution as in Eq. (2.19). (c) Show that the standard value $B = 0.0065$ K/m is equivalent to $m = 1.235$.

P2.27 Conduct an experiment to illustrate atmospheric pressure. *Note:* Do this over a sink or you may get wet! Find a drinking glass with a very smooth, uniform rim at the top. Fill the glass nearly full with water. Place a smooth, light, flat plate on top of the glass such that the entire rim of the glass is covered. A glossy postcard works best. A small index card or one flap of a greeting card will also work. See Fig. P2.27a.

(a) Hold the card against the rim of the glass and turn the glass upside down. Slowly release pressure on the card. Does the water fall out of the glass? Record your experimental observations. (b) Find an expression for the pressure at points 1 and 2 in Fig. P2.27b. Note that the glass is now inverted, so the original top rim of the glass is at the bottom of the picture, and the original bottom of the glass is at the top of the picture. The weight of the card can be neglected. (c) Estimate the theoretical maximum glass height at which this experiment could still work, such that the water would not fall out of the glass.

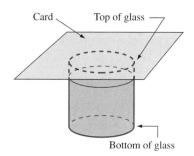

P2.27a Card   Top of glass   Bottom of glass

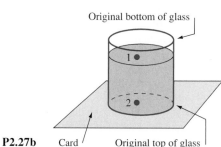

P2.27b Card   Original bottom of glass   Original top of glass

P2.28 A correlation of numerical calculations indicates that, all other things being equal, the distance traveled by a well-hit baseball varies inversely as the cube root of the air density. If a home-run ball hit in New York City travels 400 ft, estimate the distance it would travel in (a) Denver, Colorado, and (b) La Paz, Bolivia.

*P2.29 Under some conditions the atmosphere is *adiabatic*, $p \approx (\text{const})(\rho^k)$, where k is the specific heat ratio. Show that, for an adiabatic atmosphere, the pressure variation is given by

$$p = p_0\left[1 - \frac{(k-1)gz}{kRT_0}\right]^{k/(k-1)}$$

Compare this formula for air at $z = 5000$ m with the standard atmosphere in Table A.6.

P2.30 A mercury manometer is connected at two points to a horizontal 20°C water-pipe flow. If the manometer reading is $h = 55$ cm, what is the pressure drop between the two points?

P2.31 In Fig. P2.31 all fluids are at 20°C. Determine the pressure difference (Pa) between points A and B.

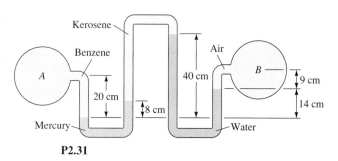

P2.31

P2.32 For the inverted manometer of Fig. P2.32, all fluids are at 20°C. If $p_B - p_A = 97$ kPa, what must the height H be in cm?

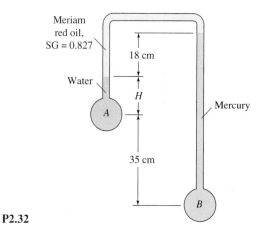

P2.32

P2.33 In Fig. P2.33 the pressure at point A is 25 lbf/in². All fluids are at 20°C. What is the air pressure in the closed chamber B, in Pa?

*P2.34 Sometimes manometer dimensions have a significant effect. In Fig. P2.34 containers (a) and (b) are cylindrical and conditions are such that $p_a = p_b$. Derive a formula for the pressure difference $p_a - p_b$ when the oil–water interface on the right rises a distance $\Delta h < h$, for (a) $d \ll D$ and (b) $d = 0.15D$. What is the percentage change in the value of Δp?

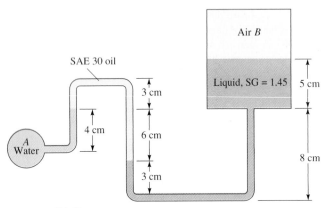

P2.33

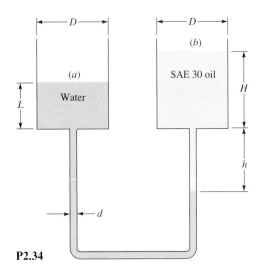

P2.34

P2.35 Water flows upward in a pipe slanted at 30°, as in Fig. P2.35. The mercury manometer reads $h = 12$ cm. Both fluids are at 20°C. What is the pressure difference $p_1 - p_2$ in the pipe?

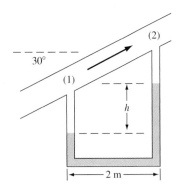

P2.35

P2.36 In Fig. P2.36 both the tank and the tube are open to the atmosphere. If $L = 2.13$ m, what is the angle of tilt θ of the tube?

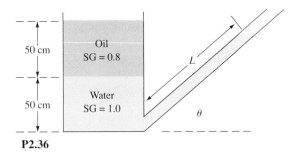

P2.36

P2.37 The inclined manometer in Fig. P2.37 contains Meriam red manometer oil, SG = 0.827. Assume that the reservoir is very large. If the inclined arm is fitted with graduations 1 in apart, what should the angle θ be if each graduation corresponds to 1 lbf/ft^2 gage pressure for p_A?

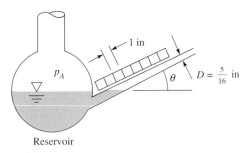

P2.37

P2.38 An interesting article appeared in the *AIAA Journal* (vol. 30, no. 1, January 1992, pp. 279–280). The authors explain that the air inside fresh plastic tubing can be up to 25 percent more dense than that of the surroundings, due to outgassing or other contaminants introduced at the time of manufacture. Most researchers, however, assume that the tubing is filled with room air at standard air density, which can lead to significant errors when using this kind of tubing to measure pressures. To illustrate this, consider a U-tube manometer with manometer fluid ρ_m. One side of the manometer is open to the air, while the other is connected to new tubing that extends to pressure measurement location 1, some height H higher in elevation than the surface of the manometer liquid. For consistency, let ρ_a be the density of the air in the room, ρ_t be the density of the gas inside the tube, ρ_m be the density of the manometer liquid, and h be the height difference between the two sides of the manometer. See Fig. P2.38. (*a*) Find an expression for the gage pressure at the measurement point. *Note:* When calculating gage pressure, use the local atmospheric pressure at the elevation of the measurement point. You may assume that $h \ll H$; that is, assume the gas in the entire left side of the manometer is of density ρ_t. (*b*) Write an expression for the error caused by assuming that the gas inside the tubing has the same density as that of the surrounding air. (*c*) How much error (in Pa) is caused by ignoring this density difference for the following conditions: $\rho_m = 860$ kg/m^3, $\rho_a = 1.20$ kg/m^3, $\rho_t = 1.50$ kg/m^3, $H = 1.32$ m, and $h = 0.58$ cm? (*d*) Can you think of a simple way to avoid this error?

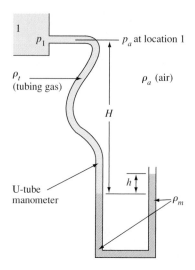

P2.38

P2.39　In Fig. P2.39 the right leg of the manometer is open to the atmosphere. Find the gage pressure, in Pa, in the air gap in the tank.

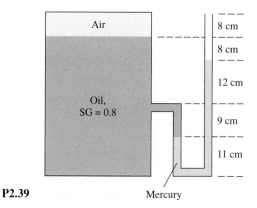

P2.39

P2.40　In Fig. P2.40 the pressures at A and B are the same, 100 kPa. If water is introduced at A to increase p_A to 130 kPa, find and sketch the new positions of the mercury menisci. The connecting tube is a uniform 1-cm diameter. Assume no change in the liquid densities.

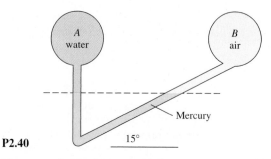

P2.40

P2.41　The system in Fig. P2.41 is at 20°C. Compute the pressure at point A in lbf/ft^2 absolute.

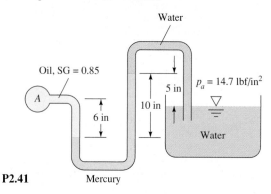

P2.41

P2.42　Very small pressure differences $p_A - p_B$ can be measured accurately by the two-fluid differential manometer in Fig. P2.42. Density ρ_2 is only slightly larger than that of the upper fluid ρ_1. Derive an expression for the proportionality between h and $p_A - p_B$ if the reservoirs are very large.

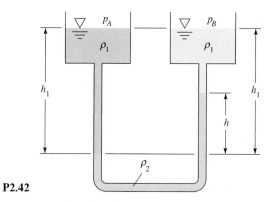

P2.42

P2.43   The traditional method of measuring blood pressure uses a *sphygmomanometer*, first recording the highest (*systolic*) and then the lowest (*diastolic*) pressure from which flowing "Korotkoff" sounds can be heard. Patients with dangerous hypertension can exhibit systolic pressures as high as 5 lbf/in². Normal levels, however, are 2.7 and 1.7 lbf/in², respectively, for systolic and diastolic pressures. The manometer uses mercury and air as fluids. (*a*) How high in cm should the manometer tube be? (*b*) Express normal systolic and diastolic blood pressure in millimeters of mercury.

P2.44   Water flows downward in a pipe at 45°, as shown in Fig. P2.44. The pressure drop $p_1 - p_2$ is partly due to gravity and partly due to friction. The mercury manometer reads a 6-in height difference. What is the total pressure drop $p_1 - p_2$ in lbf/in²? What is the pressure drop due to friction only between 1 and 2 in lbf/in²? Does the manometer reading correspond only to friction drop? Why?

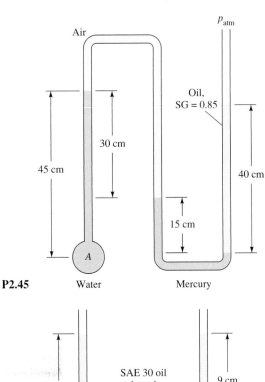

P2.45

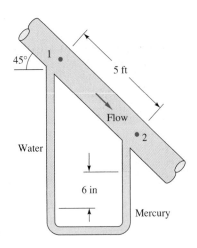

P2.44

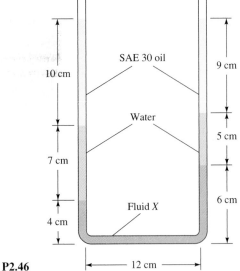

P2.46

P2.45   In Fig. P2.45, determine the gage pressure at point A in Pa. Is it higher or lower than atmospheric?

P2.46   In Fig. P2.46 both ends of the manometer are open to the atmosphere. Estimate the specific gravity of fluid X.

P2.47   The cylindrical tank in Fig. P2.47 is being filled with water at 20°C by a pump developing an exit pressure of 175 kPa. At the instant shown, the air pressure is 110 kPa and $H = 35$ cm. The pump stops when it can no longer raise the water pressure. For isothermal air compression, estimate H at that time.

P2.48   Conduct the following experiment to illustrate air pressure. Find a thin wooden ruler (approximately 1 ft in length) or a thin wooden paint stirrer. Place it on the edge of a desk or table with a little less than half of it hanging over the edge lengthwise. Get two full-size sheets of newspaper; open them up and place them on top of the ruler, covering only the portion of the ruler resting on the desk as illustrated in Fig. P2.48. (*a*) Estimate the total force on top of the newspaper due to air pressure in the room. (*b*) *Careful!* To avoid potential injury, make sure nobody is standing directly in front of

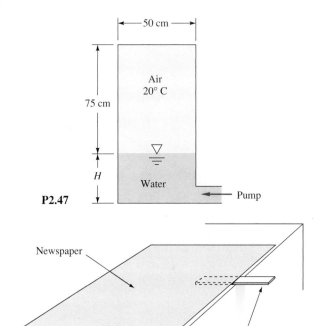

P2.47

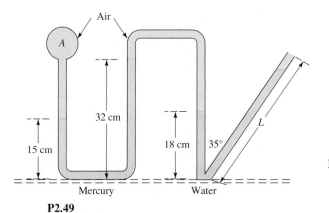

P2.48

the desk. Perform a karate chop on the portion of the ruler sticking out over the edge of the desk. Record your results. (*c*) Explain your results.

P2.49 The system in Fig. P2.49 is open to 1 atm on the right side. (*a*) If *L* = 120 cm, what is the air pressure in container *A*? (*b*) Conversely, if p_A = 135 kPa, what is the length *L*?

P2.49

P2.50 A vat filled with oil (SG = 0.85) is 7 m long and 3 m deep and has a trapezoidal cross section 2 m wide at the bottom and 4 m wide at the top. Compute (*a*) the weight of oil in the vat, (*b*) the force on the vat bottom, and (*c*) the force on the trapezoidal end panel.

P2.51 Gate *AB* in Fig. P2.51 is 1.2 m long and 0.8 m into the paper. Neglecting atmospheric pressure, compute the force *F* on the gate and its center-of-pressure position *X*.

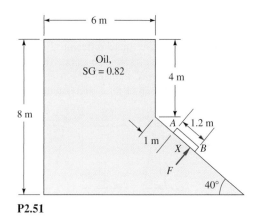

P2.51

P2.52 Example 2.5 calculated the force on plate *AB* and its line of action, using the   moment-of-inertia approach. Some teachers say it is more instructive to calculate these by *direct integration* of the pressure forces. Using Figs. P2.52 and E2.5*a*, (*a*) find an expression for the pressure variation $p(\xi)$ along the plate; (*b*) integrate this expression to find the total force *F*; (*c*) integrate the moments about point *A* to find the position of the center of pressure.

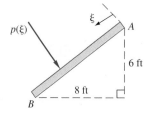

P2.52

P2.53 Panel *ABC* in the slanted side of a water tank is an isosceles triangle with the vertex at *A* and the base *BC* = 2 m, as in Fig. P2.53. Find the water force on the panel and its line of action.

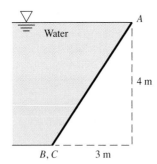

P2.53

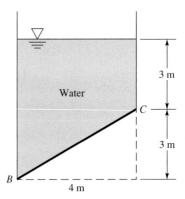

P2.57

P2.54 In Fig. P2.54, the hydrostatic force F is the same on the bottom of all three containers, even though the weights of liquid above are quite different. The three bottom shapes and the fluids are the same. This is called the *hydrostatic paradox*. Explain why it is true and sketch a free body of each of the liquid columns.

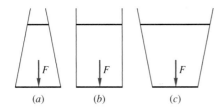

P2.54   (*a*)   (*b*)   (*c*)

P2.55 Gate AB in Fig. P2.55 is 5 ft wide into the paper, hinged at A, and restrained by a stop at B. The water is at 20°C. Compute (*a*) the force on stop B and (*b*) the reactions at A if the water depth h = 9.5 ft.

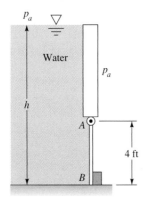

P2.55

P2.56 In Fig. P2.55, gate AB is 5 ft wide into the paper, and stop B will break if the water force on it equals 9200 lbf. For what water depth h is this condition reached?

P2.57 The tank in Fig. P2.57 is 2 m wide into the paper. Neglecting atmospheric pressure, find the resultant hydrostatic force on panel BC (*a*) from a single formula and (*b*) by computing horizontal and vertical forces separately, in the spirit of Section 2.6.

P2.58 In Fig. P2.58, the cover gate AB closes a circular opening 80 cm in diameter. The gate is held closed by a 200-kg mass as shown. Assume standard gravity at 20°C. At what water level h will the gate be dislodged? Neglect the weight of the gate.

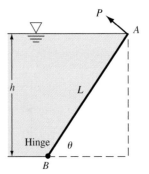

P2.58

*P2.59 Gate AB has length L and width b into the paper, is hinged at B, and has negligible weight. The liquid level h remains at the top of the gate for any angle θ. Find an analytic expression for the force P, perpendicular to AB, required to keep the gate in equilibrium in Fig. P2.59.

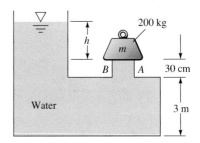

P2.59

P2.60 In 1960, Auguste and Jacques Picard's self-propelled bathyscaphe *Trieste* set a record by descending to a depth of 35,800 feet in the Pacific Ocean, near Guam.

The passenger sphere was 7 ft in diameter, 6 in. thick, and had a window diameter of 16 in. (*a*) Estimate the hydrostatic force on the window at that depth. (*b*) If the window is vertical, how far below its center is the center of pressure?

*P2.61 Gate *AB* in Fig. P2.61 is a homogeneous mass of 180 kg, 1.2 m wide into the paper, hinged at *A*, and resting on a smooth bottom at *B*. All fluids are at 20°C. For what water depth *h* will the force at point *B* be zero?

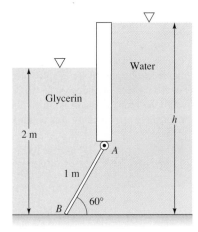

P2.61

P2.62 Gate *AB* in Fig. P2.62 is 15 ft long and 8 ft wide into the paper and is hinged at *B* with a stop at *A*. The water is at 20°C. The gate is 1-in-thick steel, SG = 7.85. Compute the water level *h* for which the gate will start to fall.

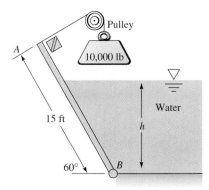

P2.62

P2.63 The tank in Fig. P2.63 has a 4-cm-diameter plug at the bottom on the right. All fluids are at 20°C. The plug will pop out if the hydrostatic force on it is 25 N. For this condition, what will be the reading *h* on the mercury manometer on the left side?

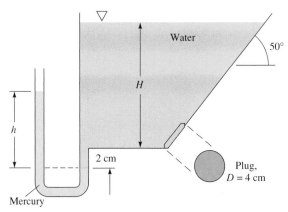

P2.63

*P2.64 Gate *ABC* in Fig. P2.64 has a fixed hinge line at *B* and is 2 m wide into the paper. The gate will open at *A* to release water if the water depth is high enough. Compute the depth *h* for which the gate will begin to open.

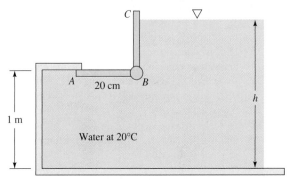

P2.64

*P2.65 Gate *AB* in Fig. P2.65 is semicircular, hinged at *B*, and held by a horizontal force *P* at *A*. What force *P* is required for equilibrium?

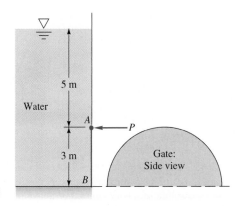

P2.65

P2.66 Dam *ABC* in Fig. P2.66 is 30 m wide into the paper and made of concrete (SG = 2.4). Find the hydrostatic force on surface *AB* and its moment about *C*. Assuming no seepage of water under the dam, could this force tip the dam over? How does your argument change if there is seepage under the dam?

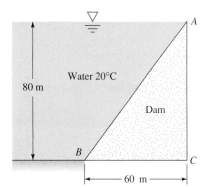

P2.66

***P2.67** Generalize Prob. P2.66 as follows. Denote length *AB* as *H*, length *BC* as *L*, and angle *ABC* as *θ*. Let the dam material have specific gravity SG. The width of the dam is *b*. Assume no seepage of water under the dam. Find an analytic relation between SG and the critical angle $θ_c$ for which the dam will just tip over to the right. Use your relation to compute $θ_c$ for the special case SG = 2.4 (concrete).

P2.68 Isosceles triangle gate *AB* in Fig. P2.68 is hinged at *A* and weighs 1500 N. What horizontal force *P* is required at point *B* for equilibrium?

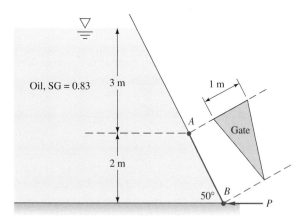

P2.68

P2.69 Consider the slanted plate *AB* of length *L* in Fig. P2.69. (*a*) Is the hydrostatic force *F* on the plate equal to the weight of the *missing water* above the plate? If not, cor-

rect this hypothesis. Neglect the atmosphere. (*b*) Can a "missing water" theory be generalized to *curved* surfaces of this type?

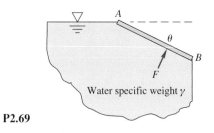

P2.69

P2.70 The swing-check valve in Fig. P2.70 covers a 22.86-cm diameter opening in the slanted wall. The hinge is 15 cm from the centerline, as shown. The valve will open when the hinge moment is 50 N · m. Find the value of *h* for the water to cause this condition.

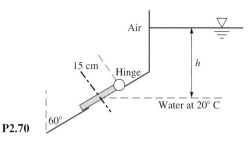

P2.70

***P2.71** In Fig. P2.71 gate *AB* is 3 m wide into the paper and is connected by a rod and pulley to a concrete sphere (SG = 2.40). What diameter of the sphere is just sufficient to keep the gate closed?

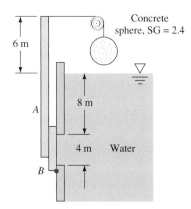

P2.71

P2.72 Gate *B* in Fig. P2.72 is 30 cm high, 60 cm wide into the paper, and hinged at the top. What water depth *h* will first cause the gate to open?

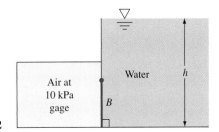

P2.72

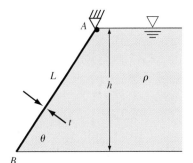

P2.73 Gate *AB* is 5 ft wide into the paper and opens to let fresh water out when the ocean tide is dropping. The hinge at *A* is 2 ft above the freshwater level. At what ocean level *h* will the gate first open? Neglect the gate weight.

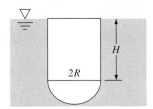

P2.73

P2.74 Find the height *H* in Fig. P2.74 for which the hydrostatic force on the rectangular panel is the same as the force on the semicircular panel below.

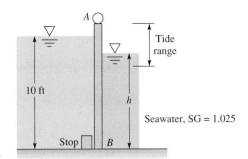

P2.74

P2.75 B

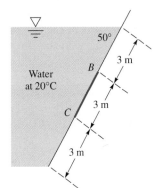

P2.76

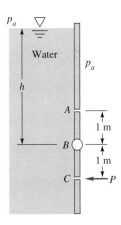

P2.77

*P2.75** Gate *AB* in Fig. P2.75 is hinged at *A*, has width *b* into the paper, and makes smooth contact at *B*. The gate has density ρ_s and uniform thickness *t*. For what gate density ρ_s, expressed as a function of (h, t, ρ, θ), will the gate just begin to lift off the bottom? Why is your answer independent of gate length *L* and width *b*?

P2.76 Panel *BC* in Fig. P2.76 is circular. Compute (*a*) the hydrostatic force of the water on the panel, (*b*) its center of pressure, and (*c*) the moment of this force about point *B*.

P2.77 The circular gate *ABC* in Fig. P2.77 has a 1-m radius and is hinged at *B*. Compute the force *P* just sufficient to keep the gate from opening when $h = 8$ m. Neglect atmospheric pressure.

P2.78 Repeat Prob. P2.77 to derive an analytic expression for *P* as a function of *h*. Is there anything unusual about your solution?

P2.79 Gate *ABC* in Fig. P2.79 is 1 m square and is hinged at *B*. It will open automatically when the water level *h* becomes high enough. Determine the lowest height for which the gate will open. Neglect atmospheric pressure. Is this result independent of the liquid density?

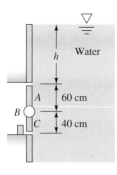

P2.79

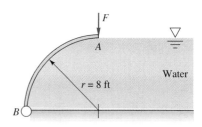

P2.83

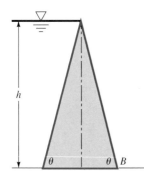

*P2.80   A concrete dam (SG = 2.5) is made in the shape of an isosceles triangle, as in Fig. P2.80. Analyze this geometry to find the range of angles θ for which the hydrostatic force will tend to tip the dam over at point B. The width into the paper is b.

P2.80

P2.81   For the semicircular cylinder *CDE* in Example 2.9, find the vertical hydrostatic force by integrating the vertical component of pressure around the surface from $\theta = 0$ to $\theta = \pi$.

*P2.82   The dam in Fig. P2.82 is a quarter circle 50 m wide into the paper. Determine the horizontal and vertical components of the hydrostatic force against the dam and the point CP where the resultant strikes the dam.

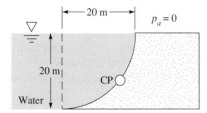

P2.82

*P2.83   Gate *AB* in Fig. P2.83 is a quarter circle 10 ft wide into the paper and hinged at *B*. Find the force F just sufficient to keep the gate from opening. The gate is uniform and weighs 3000 lbf.

P2.84   Determine (*a*) the total hydrostatic force on the curved surface *AB* in Fig. P2.84 and (*b*) its line of action. Neglect atmospheric pressure, and let the surface have unit width.

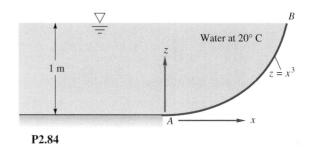

P2.84

P2.85   Compute the horizontal and vertical components of the hydrostatic force on the quarter-circle panel at the bottom of the water tank in Fig. P2.85.

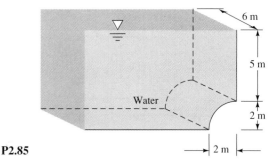

P2.85

P2.86   The quarter circle gate *BC* in Fig. P2.86 in hinged at *C*. Find the horizontal force P required to hold the gate stationary. Neglect the weight of the gate.

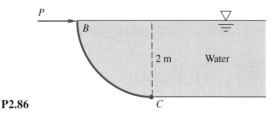

P2.86

P2.87   The bottle of champagne (SG = 0.96) in Fig. P2.87 is under pressure, as shown by the mercury-manometer reading. Compute the net force on the 2-in-radius hemispherical end cap at the bottom of the bottle.

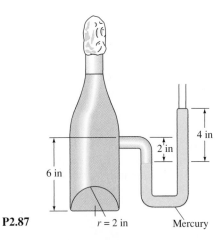

P2.87

*P2.88   Gate *ABC* is a circular arc, sometimes called a *Tainter gate*, which can be raised and lowered by pivoting about point *O*. See Fig. P2.88. For the position shown, determine (*a*) the hydrostatic force of the water on the gate and (*b*) its line of action. Does the force pass through point *O*?

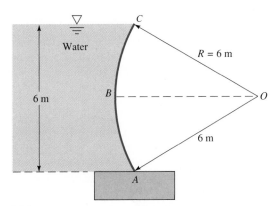

P2.88

P2.89   The tank in Fig. P2.89 contains benzene and is pressurized to 200 kPa (gage) in the air gap. Determine the vertical hydrostatic force on circular-arc section *AB* and its line of action.

P2.90   The tank in Fig. P2.90 is 120 cm long into the paper. Determine the horizontal and vertical hydrostatic forces on the quarter-circle panel *AB*. The fluid is water at 20°C. Neglect atmospheric pressure.

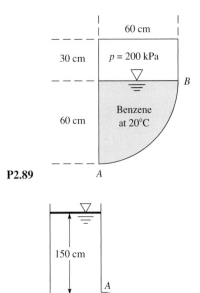

P2.89

P2.90

P2.91   The hemispherical dome in Fig. P2.91 weighs 30 kN and is filled with water and attached to the floor by six equally spaced bolts. What is the force in each bolt required to hold down the dome?

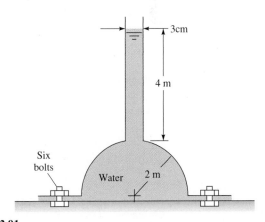

P2.91

P2.92   A 4-m-diameter water tank consists of two half cylinders, each weighing 4.5 kN/m, bolted together as shown in Fig. P2.92. If the support of the end caps is neglected, determine the force induced in each bolt.

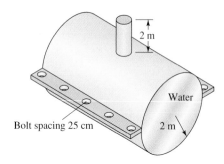

P2.92

*P2.93   In Fig. P2.93, a one-quadrant spherical shell of radius R is submerged in liquid of specific weight γ and depth $h > R$. Find an analytic expression for the resultant hydrostatic force, and its line of action, on the shell surface.

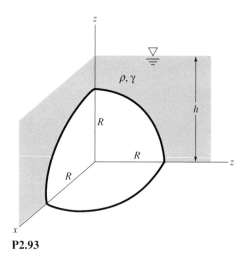

P2.93

P2.94   The 4-ft-diameter log (SG = 0.80) in Fig. P2.94 is 8 ft long into the paper and dams water as shown. Compute the net vertical and horizontal reactions at point C.

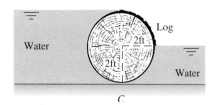

P2.94

*P2.95   The uniform body A in Fig. P2.95 has width b into the paper and is in static equilibrium when pivoted about hinge O. What is the specific gravity of this body if (a) $h = 0$ and (b) $h = R$?

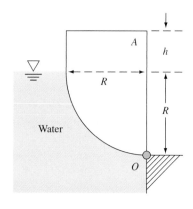

P2.95

P2.96   Curved panel BC in Fig. P2.96 is a 60° arc, perpendicular to the bottom at C. If the panel is 4 m wide into the paper, estimate the resultant hydrostatic force of the water on the panel.

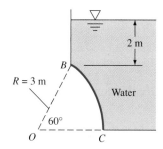

P2.96

P2.97   Gate AB in Fig. P2.97 is a three-eighths circle, 3 m wide into the paper, hinged at B, and resting against a smooth wall at A. Compute the reaction forces at points A and B.

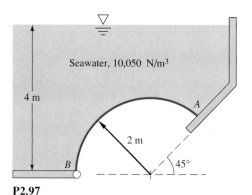

P2.97

P2.98   Gate ABC in Fig. P2.98 is a quarter circle 8 ft wide into the paper. Compute the horizontal and vertical hydrostatic forces on the gate and the line of action of the resultant force.

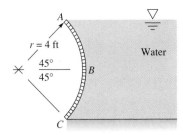

P2.98

P2.99 A 2-ft-diameter sphere weighing 400 lbf closes a 1-ft-diameter hole in the bottom of the tank in Fig. P2.99. Compute the force F required to dislodge the sphere from the hole.

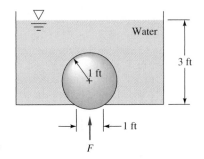

P2.99

P2.100 Pressurized water fills the tank in Fig. P2.100. Compute the net hydrostatic force on the conical surface ABC.

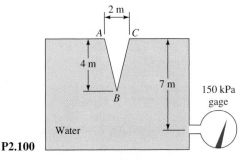

P2.100

P2.101 The closed layered box in Fig. P2.101 has square horizontal cross sections everywhere. All fluids are at 20°C. Estimate the gage pressure of the air if (a) the hydrostatic force on panel AB is 48 kN or (b) the hydrostatic force on the bottom panel BC is 97 kN.

P2.102 A cubical tank is $3 \times 3 \times 3$ m and is layered with 1 meter of fluid of specific gravity 1.0, 1 meter of fluid with SG = 0.9, and 1 meter of fluid with SG = 0.8. Neglect atmospheric pressure. Find (a) the hydrostatic force on the bottom and (b) the force on a side panel.

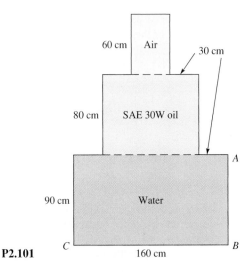

P2.101

P2.103 A solid block, of specific gravity 0.9, floats such that 75 percent of its volume is in water and 25 percent of its volume is in fluid X, which is layered above the water. What is the specific gravity of fluid X?

P2.104 The can in Fig. P2.104 floats in the position shown. What is its weight in N?

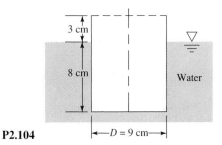

P2.104

P2.105 It is said that Archimedes discovered the buoyancy laws when asked by King Hiero of Syracuse to determine whether his new crown was pure gold (SG = 19.3). Archimedes measured the weight of the crown in air to be 11.8 N and its weight in water to be 10.9 N. Was it pure gold?

P2.106 A spherical helium balloon is 2.5 m in diameter and has a total mass of 6.7 kg. When released into the U.S. standard atmosphere, at what altitude will it settle?

P2.107 Repeat Prob. 2.62, assuming that the 10,000-lbf weight is aluminum (SG = 2.71) and is hanging submerged in the water.

P2.108 A 7-cm-diameter solid aluminum ball (SG = 2.7) and a solid brass ball (SG = 8.5) balance nicely when submerged in a liquid, as in Fig. P2.108. (a) If the fluid is water at 20°C, what is the diameter of the brass ball?

(*b*) If the brass ball has a diameter of 3.8 cm, what is the density of the fluid?

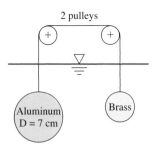

P2.108

P2.109   A *hydrometer* floats at a level that is a measure of the specific gravity of the liquid. The stem is of constant diameter *D*, and a weight in the bottom stabilizes the body to float vertically, as shown in Fig. P2.109. If the position $h = 0$ is pure water (SG = 1.0), derive a formula for *h* as a function of total weight *W*, *D*, SG, and the specific weight γ_0 of water.

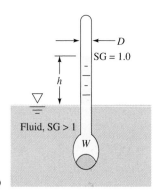

P2.109

P2.110   An average table tennis ball has a diameter of 3.81 cm and a mass of 2.6 g. Estimate the (small) depth at which this ball will float in water at 20°C and sea-level standard air if air buoyancy is (*a*) neglected and (*b*) included.

P2.111   A hot-air balloon must be designed to support basket, cords, and one person for a total weight of 1300 N. The balloon material has a mass of 60 g/m². Ambient air is at 25°C and 1 atm. The hot air inside the balloon is at 70°C and 1 atm. What diameter spherical balloon will just support the total weight? Neglect the size of the hot-air inlet vent.

P2.112   The uniform 5-m-long round wooden rod in Fig. P2.112 is tied to the bottom by a string. Determine (*a*) the tension in the string and (*b*) the specific gravity of the wood. Is it possible for the given information to determine the inclination angle θ? Explain.

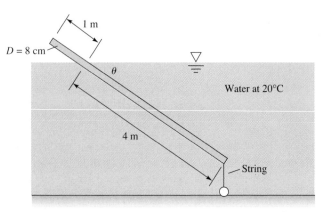

P2.112

P2.113   A *spar buoy* is a buoyant rod weighted to float and protrude vertically, as in Fig. P2.113. It can be used for measurements or markers. Suppose that the buoy is maple wood (SG = 0.6), 2 in by 2 in by 12 ft, floating in seawater (SG = 1.025). How many pounds of steel (SG = 7.85) should be added to the bottom end so that $h = 18$ in?

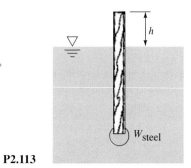

P2.113

P2.114   The uniform rod in Fig. P2.114 is hinged at point *B* on the waterline and is in static equilibrium as shown when 2 kg of lead (SG = 11.4) are attached to its end. What is the specific gravity of the rod material? What is peculiar about the rest angle $\theta = 30°$?

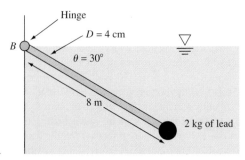

P2.114

P2.115 The 2-in by 2-in by 12-ft spar buoy from Fig. P2.113 has 5 lbm of steel attached and has gone aground on a rock, as in Fig. P2.115. Compute the angle θ at which the buoy will lean, assuming that the rock exerts no moments on the spar.

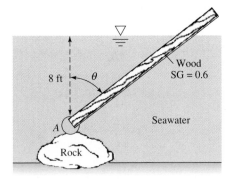

P2.115

P2.116 Ocean currents can be tracked by *Swallow floats* [2], named after Dr. John Swallow, of the United Kingdom, who designed them in 1955. There have been many design changes, but the original float was an aluminum tube, 6-cm-outside diameter and about 3 m long, sealed at the ends and slightly pressurized. The tubes had to be etched to obtain the right thickness. Estimate the tube thickness to cause neutral buoyancy at a seawater density of 1030 kg/m³.

P2.117 The balloon in Fig. P2.117 is filled with helium and pressurized to 135 kPa and 20°C. The balloon material has a mass of 85 g/m². Estimate (*a*) the tension in the mooring line and (*b*) the height in the standard atmosphere to which the balloon will rise if the mooring line is cut.

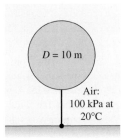

P2.117

P2.118 An intrepid treasure-salvage group has discovered a steel box, containing gold doubloons and other valuables, resting in 80 ft of seawater. They estimate the weight of the box and treasure (in air) at 7000 lbf. Their plan is to attach the box to a sturdy balloon, inflated with air to 3 atm pressure. The empty balloon weighs 250 lbf. The box is 2 ft wide, 5 ft long, and 18 in high. What is the proper diameter of the balloon to ensure an upward lift force on the box that is 20 percent more than required?

P2.119 When a 5-lbf weight is placed on the end of the uniform floating wooden beam in Fig. P2.119, the beam tilts at an angle θ with its upper right corner at the surface, as shown. Determine (*a*) the angle θ and (*b*) the specific gravity of the wood. (*Hint:* Both the vertical forces and the moments about the beam centroid must be balanced.)

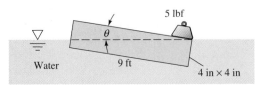

P2.119

P2.120 A uniform wooden beam (SG = 0.65) is 10 cm by 10 cm by 3 m and is hinged at *A*, as in Fig. P2.120. At what angle θ will the beam float in the 20°C water?

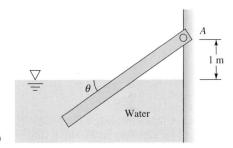

P2.120

P2.121 The uniform beam in Fig. P2.121, of size *L* by *h* by *b* and with specific weight γ_b, floats exactly on its diagonal when a heavy uniform sphere is tied to the left corner, as

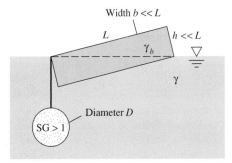

P2.121

shown. Show that this can happen only (*a*) when $\gamma_b = \gamma/3$ and (*b*) when the sphere has size

$$D = \left[\frac{Lhb}{\pi(SG - 1)} \right]^{1/3}$$

P2.122 A uniform block of steel (SG = 7.85) will "float" at a mercury–water interface as in Fig. P2.122. What is the ratio of the distances *a* and *b* for this condition?

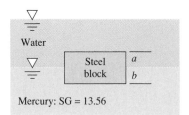

Water

Steel block — *a*, *b*

Mercury: SG = 13.56

P2.122

P2.123 A barge has the trapezoidal shape shown in Fig. P2.123 and is 22 m long into the paper. If the total weight of barge and cargo is 350 tons, what is the draft *H* of the barge when floating in seawater?

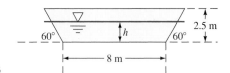

60°   *h*   60°   2.5 m

8 m

P2.123

P2.124 A balloon weighing 3.5 lbf is 6 ft in diameter. It is filled with hydrogen at 18 lbf/in² absolute and 60°F and is released. At what altitude in the U.S. standard atmosphere will this balloon be neutrally buoyant?

P2.125 Suppose that the balloon in Prob. P2.111 is constructed to have a diameter of 14 m, is filled at sea level with hot air at 70°C and 1 atm, and is released. If the air inside the balloon remains constant and the heater maintains it at 70°C, at what altitude in the U.S. standard atmosphere will this balloon be neutrally buoyant?

P2.126 A block of wood (SG = 0.6) floats in fluid *X* in Fig. P2.126 such that 75 percent of its volume is submerged in fluid *X*. Estimate the vacuum pressure of the air in the tank.

*P2.127 Consider a cylinder of specific gravity *S* < 1 floating vertically in water (*S* = 1), as in Fig. P2.127. Derive a formula for the stable values of *D/L* as a function of *S* and apply it to the case *D/L* = 1.2.

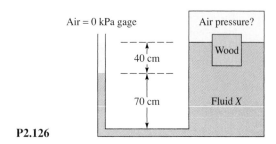

Air = 0 kPa gage   Air pressure?

40 cm   Wood

70 cm   Fluid *X*

P2.126

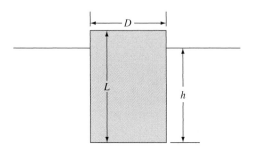

D

L   *h*

P2.127

P2.128 An iceberg can be idealized as a cube of side length *L*, as in Fig. P2.128. If seawater is denoted by *S* = 1.0, then glacier ice (which forms icebergs) has *S* = 0.88. Determine if this "cubic" iceberg is stable for the position shown in Fig. P2.128.

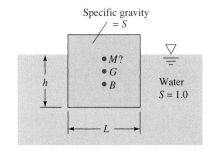

Specific gravity = *S*

M?
G
B

h

Water
S = 1.0

L

P2.128

P2.129 The iceberg idealization in Prob. P2.128 may become unstable if its sides melt and its height exceeds its width. In Fig. P2.128 suppose that the height is *L* and the depth into the paper is *L*, but the width in the plane of the paper is *H* < *L*. Assuming *S* = 0.88 for the iceberg, find the ratio *H/L* for which it becomes neutrally stable (about to overturn).

P2.130 Consider a wooden cylinder (SG = 0.6) 1 m in diameter and 0.8 m long. Would this cylinder be stable if placed to float with its axis vertical in oil (SG = 0.8)?

P2.131 A barge is 15 ft wide and 40 ft long and floats with a draft of 4 ft. It is piled so high with gravel that its center of gravity is 2 ft above the waterline. Is it stable?

P2.132   A solid right circular cone has SG = 0.99 and floats vertically as in Fig. P2.132. Is this a stable position for the cone?

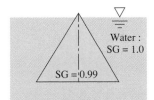

P2.132

P2.133   Consider a uniform right circular cone of specific gravity $S < 1$, floating with its vertex *down* in water ($S = 1$). The base radius is R *and* the cone height is H. Calculate and plot the stability MG of this cone, in dimensionless form, versus H/R for a range of $S < 1$.

P2.134   When floating in water (SG = 1.0), an equilateral triangular body (SG = 0.9) might take one of the two positions shown in Fig. P2.134. Which is the more stable position? Assume large width into the paper.

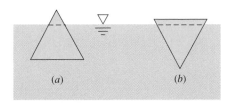

P2.134

P2.135   Consider a homogeneous right circular cylinder of length L, radius R, and specific gravity SG, floating in water (SG = 1). Show that the body will be stable with its axis vertical if

$$\frac{R}{L} > [2SG(1 - SG)]^{1/2}$$

P2.136   Consider a homogeneous right circular cylinder of length L, radius R, and specific gravity SG = 0.5, floating in water (SG = 1). Show that the body will be stable with its axis horizontal if $L/R > 2.0$.

P2.137   A tank of water 4 m deep receives a constant upward acceleration a_z. Determine (a) the gage pressure at the tank bottom if $a_z = 5$ m²/s and (b) the value of a_z that causes the gage pressure at the tank bottom to be 1 atm.

P2.138   A 12-fl-oz glass, of 3-in diameter, partly full of water, is attached to the edge of an 8-ft-diameter merry-go-round, which is rotated at 12 r/min. How full can the glass be before water spills? (*Hint:* Assume that the glass is much smaller than the radius of the merry-go-round.)

P2.139   The tank of liquid in Fig. P2.139 accelerates to the right with the fluid in rigid-body motion. (a) Compute a_x in m/s². (b) Why doesn't the solution to part (a) depend on the density of the fluid? (c) Determine the gage pressure at point A if the fluid is glycerin at 20°C.

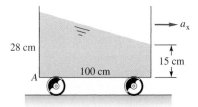

P2.139

P2.140   Suppose an elliptical-end fuel tank that is 10 m long and has a 3-m horizontal major axis and 2-m vertical minor axis is filled completely with fuel oil ($\rho = 890$ kg/m³). Let the tank be pulled along a horizontal road. For rigid-body motion, find the acceleration, and its direction, for which (a) a constant-pressure surface extends from the top of the front end wall to the bottom of the back end and (b) the top of the back end is at a pressure 0.5 atm lower than the top of the front end.

P2.141   The same tank from Prob. P2.139 is now moving with constant acceleration up a 30° inclined plane, as in Fig. P2.141. Assuming rigid-body motion, compute (a) the value of the acceleration a, (b) whether the acceleration is up or down, and (c) the gage pressure at point A if the fluid is mercury at 20°C.

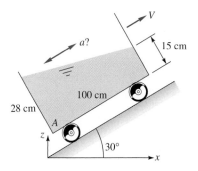

P2.141

P2.142   The tank of water in Fig. P2.142 is 12 cm wide into the paper. If the tank is accelerated to the right in rigid-body motion at 6.0 m/s², compute (a) the water depth on side AB and (b) the water-pressure force on panel AB. Assume no spilling.

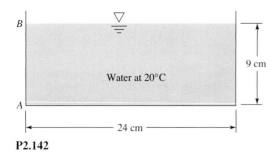

P2.142

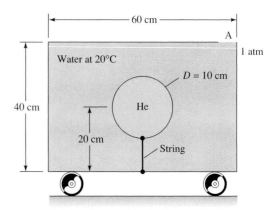

accelerates to the right at 5 m/s² in rigid-body motion, at what angle will the balloon lean? Will it lean to the right or to the left?

P2.143 The tank of water in Fig. P2.143 is full and open to the atmosphere at point A. For what acceleration a_x in ft/s² will the pressure at point B be (a) atmospheric and (b) zero absolute?

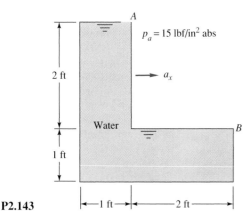

P2.143

P2.146

P2.147 The tank of water in Fig. P2.147 accelerates uniformly by freely rolling down a 30° incline. If the wheels are frictionless, what is the angle θ? Can you explain this interesting result?

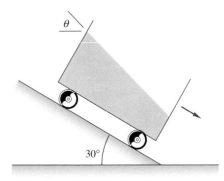

P2.147

P2.144 Consider a hollow cube of side length 22 cm, filled completely with water at 20°C. The top surface of the cube is horizontal. One top corner, point A, is open through a small hole to a pressure of 1 atm. Diagonally opposite to point A is top corner B. Determine and discuss the various rigid-body accelerations for which the water at point B begins to cavitate, for (a) horizontal motion and (b) vertical motion.

P2.145 A fish tank 14 in deep by 16 by 27 in is to be carried in a car that may experience accelerations as high as 6 m/s². What is the maximum water depth that will avoid spilling in rigid-body motion? What is the proper alignment of the tank with respect to the car motion?

P2.146 The tank in Fig. P2.146 is filled with water and has a vent hole at point A. The tank is 1 m wide into the paper. Inside the tank, a 10-cm balloon, filled with helium at 130 kPa, is tethered centrally by a string. If the tank

P2.148 A child is holding a string onto which is attached a helium-filled balloon. (a) The child is standing still and suddenly accelerates forward. In a frame of reference moving with the child, which way will the balloon tilt, forward or backward? Explain. (b) The child is now sitting in a car that is stopped at a red light. The helium-filled balloon is not in contact with any part of the car (seats, ceiling, etc.) but is held in place by the string, which is in turn held by the child. All the windows in the car are closed. When the traffic light turns green, the car accelerates forward. In a frame of reference moving with the car and child, which way will the balloon tilt, forward or backward? Explain. (c) Purchase or borrow a helium-filled balloon. Conduct a scientific experiment to see if your predictions in parts (a) and (b) above are correct. If not, explain.

P2.149   The 6-ft-radius waterwheel in Fig. P2.149 is being used to lift water with its 1-ft-diameter half-cylinder blades. If the wheel rotates at 10 r/min and rigid-body motion is assumed, what is the water surface angle θ at position A?

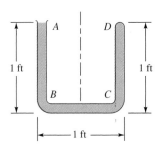

P2.151

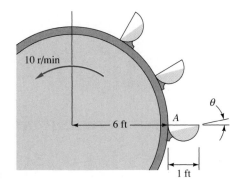

P2.149

P2.150   A cheap accelerometer, probably worth the price, can be made from a U-tube as in Fig. P2.150. If $L = 18$ cm and $D = 5$ mm, what will h be if $a_x = 6$ m/s²? Can the scale markings on the tube be linear multiples of a_x?

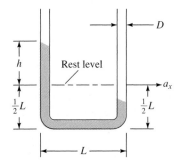

P2.150

P2.151   The U-tube in Fig. P2.151 is open at A and closed at D. If accelerated to the right at uniform a_x, what acceleration will cause the pressure at point C to be atmospheric? The fluid is water (SG = 1.0).

P2.152   A 16-cm-diameter open cylinder 27 cm high is full of water. Compute the rigid-body rotation rate about its central axis, in r/min, (a) for which one-third of the water will spill out and (b) for which the bottom will be barely exposed.

P2.153   Suppose the U-tube in Fig. P2.150 is not translated but rather rotated about its right leg at 95 r/min. What will be the level h in the left leg if $L = 18$ cm and $D = 5$ mm?

P2.154   A very tall 10-cm-diameter vase contains 1178 cm³ of water. When spun steadily to achieve rigid-body rotation, a 4-cm-diameter dry spot appears at the bottom of the vase. What is the rotation rate, in r/min, for this condition?

P2.155   For what uniform rotation rate in r/min about axis C will the U-tube in Fig. P2.155 take the configuration shown? The fluid is mercury at 20°C.

P2.155

P2.156   Suppose that the U-tube of Fig. P2.151 is rotated about axis DC. If the fluid is water at 122°F and atmospheric pressure is 2116 lbf/ft² absolute, at what rotation rate will the fluid within the tube begin to vaporize? At what point will this occur?

P2.157   The 45° V-tube in Fig. P2.157 contains water and is open at A and closed at C. What uniform rotation rate in r/min about axis AB will cause the pressure to be equal at points B and C? For this condition, at what point in leg BC will the pressure be a minimum?

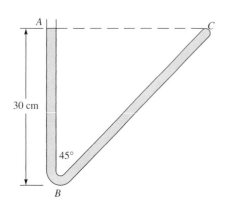

P2.157

*P2.158 It is desired to make a 3-m-diameter parabolic tele-
scope mirror by rotating molten glass in rigid-body
motion until the desired shape is achieved and then
cooling the glass to a solid. The focus of the mirror is
to be 4 m from the mirror, measured along the cen-

terline. What is the proper mirror rotation rate, in
r/min, for this task?

P2.159 The three-legged manometer in Fig. P2.159 is filled with
water to a depth of 20 cm. All tubes are long and have
equal small diameters. If the system spins at angular
velocity Ω about the central tube, (a) derive a formula
to find the change of height in the tubes; (b) find the
height in cm in each tube if Ω = 120 r/min. (*Hint:* The
central tube must supply water to *both* the outer legs.)

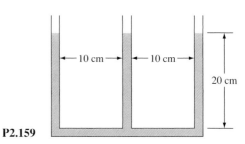

P2.159

Word Problems

W2.1 Consider a hollow cone with a vent hole in the vertex at
the top, along with a hollow cylinder, open at the top, with
the same base area as the cone. Fill both with water to the
top. The *hydrostatic paradox* is that both containers have
the same force on the bottom due to the water pressure,
although the cone contains 67 percent less water. Can you
explain the paradox?

W2.2 Can the temperature ever *rise* with altitude in the real
atmosphere? Wouldn't this cause the air pressure to
increase upward? Explain the physics of this situation.

W2.3 Consider a submerged curved surface that consists of a
two-dimensional circular arc of arbitrary angle, arbitrary
depth, and arbitrary orientation. Show that the resultant
hydrostatic pressure force on this surface must pass
through the center of curvature of the arc.

W2.4 Fill a glass approximately 80 percent with water, and add a
large ice cube. Mark the water level. The ice cube, having
SG ≈ 0.9, sticks up out of the water. Let the ice cube melt
with negligible evaporation from the water surface. Will the
water level be higher than, lower than, or the same as before?

W2.5 A ship, carrying a load of steel, is trapped while floating
in a small closed lock. Members of the crew want to get
out, but they can't quite reach the top wall of the lock.
A crew member suggests throwing the steel overboard in
the lock, claiming the ship will then rise and they can
climb out. Will this plan work?

W2.6 Consider a balloon of mass *m* floating neutrally in the
atmosphere, carrying a person/basket of mass *M* > *m*.
Discuss the stability of this system to disturbances.

W2.7 Consider a helium balloon on a string tied to the seat of
your stationary car. The windows are closed, so there is
no air motion within the car. The car begins to accelerate forward. Which way will the balloon lean, forward or
backward? (*Hint:* The acceleration sets up a horizontal
pressure gradient in the air within the car.)

W2.8 Repeat your analysis of Prob. W2.7 to let the car move
at constant velocity and go around a curve. Will the balloon lean in, toward the center of curvature, or out?

Fundamentals of Engineering Exam Problems

FE2.1 A gage attached to a pressurized nitrogen tank reads a
gage pressure of 28 in of mercury. If atmospheric pressure is 14.4 psia, what is the absolute pressure in the
tank?
(*a*) 95 kPa, (*b*) 99 kPa, (*c*) 101 kPa, (*d*) 194 kPa,
(*e*) 203 kPa

FE2.2 On a sea-level standard day, a pressure gage, moored
below the surface of the ocean (SG = 1.025), reads
an absolute pressure of 1.4 MPa. How deep is the
instrument?
(*a*) 4 m, (*b*) 129 m, (*c*) 133 m, (*d*) 140 m,
(*e*) 2080 m

FE2.3 In Fig. FE2.3, if the oil in region B has SG = 0.8 and the absolute pressure at point A is 1 atm, what is the absolute pressure at point B?
(a) 5.6 kPa, (b) 10.9 kPa, (c) 107 kPa, (d) 112 kPa, (e) 157 kPa

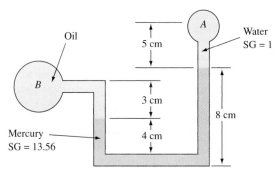

FE2.3

FE2.4 In Fig. FE2.3, if the oil in region B has SG = 0.8 and the absolute pressure at point B is 14 psia, what is the absolute pressure at point A?
(a) 11 kPa, (b) 41 kPa, (c) 86 kPa, (d) 91 kPa, (e) 101 kPa

FE2.5 A tank of water (SG = 1.0) has a gate in its vertical wall 5 m high and 3 m wide. The top edge of the gate is 2 m below the surface. What is the hydrostatic force on the gate?

(a) 147 kN, (b) 367 kN, (c) 490 kN, (d) 661 kN, (e) 1028 kN

FE2.6 In Prob. FE2.5, how far below the surface is the center of pressure of the hydrostatic force?
(a) 4.50 m, (b) 5.46 m, (c) 6.35 m, (d) 5.33 m, (e) 4.96 m

FE2.7 A solid 1-m-diameter sphere floats at the interface between water (SG = 1.0) and mercury (SG = 13.56) such that 40 percent is in the water. What is the specific gravity of the sphere?
(a) 6.02, (b) 7.28, (c) 7.78, (d) 8.54, (e) 12.56

FE2.8 A 5-m-diameter balloon contains helium at 125 kPa absolute and 15°C, moored in sea-level standard air. If the gas constant of helium is 2077 $m^2/(s^2 \cdot K)$ and balloon material weight is neglected, what is the net lifting force of the balloon?
(a) 67 N, (b) 134 N, (c) 522 N, (d) 653 N, (e) 787 N

FE2.9 A square wooden (SG = 0.6) rod, 5 cm by 5 cm by 10 m long, floats vertically in water at 20°C when 6 kg of steel (SG = 7.84) are attached to one end. How high above the water surface does the wooden end of the rod protrude?
(a) 0.6 m, (b) 1.6 m, (c) 1.9 m, (d) 2.4 m, (e) 4.0 m

FE2.10 A floating body will be stable when its
(a) center of gravity is above its center of buoyancy, (b) center of buoyancy is below the waterline, (c) center of buoyancy is above its metacenter, (d) metacenter is above its center of buoyancy, (e) metacenter is above its center of gravity

Comprehensive Problems

C2.1 Some manometers are constructed as in Fig. C2.1, where one side is a large reservoir (diameter D) and the other side is a small tube of diameter d, open to the atmosphere. In such a case, the height of manometer liquid on the reservoir side does not change appreciably. This has the advantage that only one height needs to be measured rather than two. The manometer liquid has density ρ_m while the air has density ρ_a. Ignore the effects of surface tension. When there is no pressure difference across the manometer, the elevations on both sides are the same, as indicated by the dashed line. Height h is measured from the zero pressure level as shown. (a) When a high pressure is applied to the left side, the manometer liquid in the large reservoir goes down, while that in the tube at the right goes up to conserve mass. Write an exact expression for $p_{1\text{gage}}$, taking into account the movement of the surface of the reservoir. Your equation should give $p_{1\text{gage}}$ as a function of h, ρ_m, and the physical parameters in the problem, h, d, D, and gravity constant g. (b) Write an approximate expression for $p_{1\text{gage}}$, neglecting the change in elevation of the surface of the reservoir liquid. (c) Suppose $h =$

0.26 m in a certain application. If $p_a = 101,000$ Pa and the manometer liquid has a density of 820 kg/m^3, estimate the ratio D/d required to keep the error of the approximation of part (b) within 1 percent of the exact measurement of part (a). Repeat for an error within 0.1 percent.

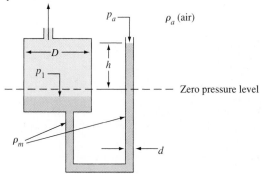

C2.1

C2.2 A prankster has added oil, of specific gravity SG_0, to the left leg of the manometer in Fig. C2.2. Nevertheless, the U-tube is still useful as a pressure-measuring device. It is attached to a pressurized tank as shown in the figure. (a) Find an expression for h as a function of H and other parameters in the problem. (b) Find the special case of your result in (a) when $p_{tank} = p_a$. (c) Suppose $H = 5.0$ cm, p_a is 101.2kPa, p_{tank} is 1.82 kPa higher than p_a, and $SG_0 = 0.85$. Calculate h in cm, ignoring surface tension effects and neglecting air density effects.

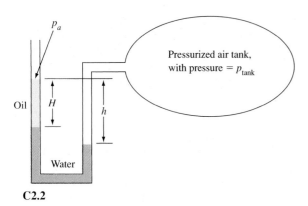

C2.2

C2.3 Professor F. Dynamics, riding the merry-go-round with his son, has brought along his U-tube manometer. (You never know when a manometer might come in handy.) As shown in Fig. C2.3, the merry-go-round spins at constant angular velocity and the manometer legs are 7 cm apart. The manometer center is 5.8 m from the axis of rotation. Determine the height difference h in two ways: (a) approximately, by assuming rigid-body translation with **a** equal to the average manometer acceleration; and (b) exactly, using rigid-body rotation theory. How good is the approximation?

C2.4 A student sneaks a glass of cola onto a roller coaster ride. The glass is cylindrical, twice as tall as it is wide, and filled to the brim. He wants to know what percent of the cola he should drink before the ride begins, so that none of it spills during the big drop, in which the roller coaster achieves 0.55-g acceleration at a 45° angle below the horizontal. Make the calculation for him, neglecting sloshing and assuming that the glass is vertical at all times.

C2.5 *Dry adiabatic lapse rate* (DALR) is defined as the negative value of atmospheric temperature gradient, dT/dz, when temperature and pressure vary in an isentropic fashion. Assuming air is an ideal gas, DALR $= -dT/dz$ when $T = T_0(p/p_0)^a$, where exponent $a = (k - 1)/k$, $k = c_p/c_v$ is the ratio of specific heats, and T_0 and p_0 are the temperature and pressure at sea level, respectively. (a) Assuming that hydrostatic conditions exist in the atmosphere, show that the dry adiabatic lapse rate is constant and is given by DALR $= g(k - 1)/(kR)$, where R is the ideal gas constant for air. (b) Calculate the numerical value of DALR for air in units of °C/km.

C2.6 In "soft" liquids (low bulk modulus β), it may be necessary to account for liquid compressibility in hydrostatic calculations. An approximate density relation would be

$$dp \approx \frac{\beta}{\rho} d\rho = a^2 d\rho \qquad \text{or} \qquad p \approx p_0 + a^2(\rho - \rho_0)$$

where a is the speed of sound and (p_0, ρ_0) are the conditions at the liquid surface $z = 0$. Use this approximation to show that the density variation with depth in a soft liquid is $\rho = \rho_0 e^{-gz/a^2}$ where g is the acceleration of gravity and z is positive upward. Then consider a vertical wall of width b, extending from the surface ($z = 0$) down to depth $z = -h$. Find an analytic expression for the hydrostatic force F on this wall, and compare it with the incompressible result $F = \rho_0 g h^2 b/2$. Would the center of pressure be below the incompressible position $z = -2h/3$?

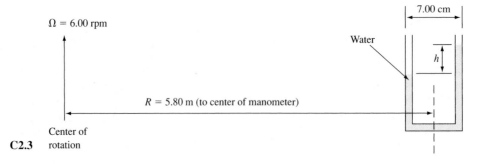

C2.3   Center of rotation

C2.7 Venice, Italy, is slowly sinking, so now, especially in winter, plazas and walkways are flooded during storms. The proposed solution is the floating levee of Fig. C2.7. When filled with air, it rises to block off the sea. The levee is 30 m high, 5 m wide, and 20 m deep. Assume a uniform density of 300 kg/m^3 when floating. For the 1-m sea–lagoon difference shown, estimate the angle at which the levee floats.

C2.8 What is the uncertainty in using pressure measurement as an altimeter? A gage on the side of an airplane measures a local pressure of 54 kPa, with an uncertainty of 3 kPa. The estimated lapse rate that day is 0.007 K/m, with an uncertainty of 0.001 K/m. Effective sea-level temperature is 10°C, with an uncertainty of 4°C. Effective sea-level pressure is 100 kPa, with an uncertainty of 3 kPa. Estimate the airplane's altitude and its uncertainty.

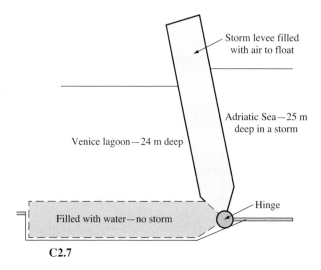

C2.7

Design Projects

D2.1 It is desired to have a bottom-moored, floating system that creates a nonlinear force in the mooring line as the water level rises. The design force F need only be accurate in the range of seawater depths h between 6 and 8 m, as shown in the accompanying table. Design a buoyant system that will provide this force distribution. The system should be practical (of inexpensive materials and simple construction).

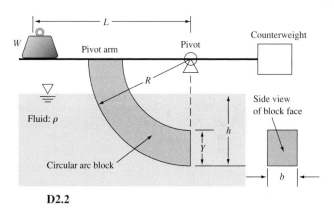

D2.2

h, m	F, N	h, m	F, N
6.00	400	7.25	554
6.25	437	7.50	573
6.50	471	7.75	589
6.75	502	8.00	600
7.00	530		

D2.2 A laboratory apparatus used in some universities is shown in Fig. D2.2. The purpose is to measure the hydrostatic force on the flat face of the circular-arc block and compare it with the theoretical value for given depth h. The counterweight is arranged so that the pivot arm is horizontal when the block is not submerged, whence the weight W can be correlated with the hydrostatic force when the submerged arm is again brought to horizontal. First show that the apparatus concept is valid in principle; then derive a formula for W as a function of h in terms of the system parameters. Finally, suggest some appropriate values of Y, L, and so on for a suitable apparatus and plot theoretical W versus h for these values.

D2.3 The Leary Engineering Company (see *Popular Science*, November 2000, p. 14) has proposed a ship hull with hinges that allow it to open into a flatter shape when entering shallow water. A simplified version is shown in Fig. D2.3. In deep water, the hull cross section would be triangular, with large draft. In shallow water, the hinges would open to an angle as high as $\theta = 45°$. The dashed line indicates that the bow and stern would be closed. Make a parametric study of this configuration for various θ, assuming a reasonable weight and center of gravity location. Show how the draft, the metacentric height, and the ship's stability vary as the hinges are opened. Comment on the effectiveness of this concept.

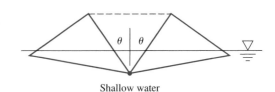

D2.3

Deep water   Shallow water

References

1. *U.S. Standard Atmosphere,* 1976, Government Printing Office, Washington, DC, 1976.
2. G. L. Pickard, *Descriptive Physical Oceanography,* Butterworth-Heinemann, Woburn, MA, 1990.
3. E. C. Tupper, *Introduction to Naval Architecture,* 4th ed., Elsevier, New York, 2004.
4. D. T. Greenwood, *Principles of Dynamics,* 2d ed., Prentice-Hall, Englewood Cliffs, NJ, 1988.
5. R. I. Fletcher, "The Apparent Field of Gravity in a Rotating Fluid System," *Am. J. Phys.,* vol. 40, July 1972, pp. 959–965.
6. National Committee for Fluid Mechanics Films, *Illustrated Experiments in Fluid Mechanics,* M.I.T. Press, Cambridge, MA, 1972.
7. J. P. Holman, *Experimental Methods for Engineers,* 7th ed., McGraw-Hill, New York, 2000.
8. R. P. Benedict, *Fundamentals of Temperature, Pressure, and Flow Measurement,* 3d ed., Wiley, New York, 1984.
9. T. G. Beckwith and R. G. Marangoni, *Mechanical Measurements,* 5th ed., Addison-Wesley, Reading, MA, 1993.
10. J. W. Dally, W. F. Riley, and K. G. McConnell, *Instrumentation for Engineering Measurements,* 2d ed., Wiley, New York, 1993.
11. E. N. Gilbert, "How Things Float," *Am. Math. Monthly,* vol. 98, no. 3, 1991, pp. 201–216.
12. R. J. Figliola and D. E. Beasley, *Theory and Design for Mechanical Measurements,* 3d ed., Wiley, New York, 2000.
13. R. W. Miller, *Flow Measurement Engineering Handbook,* 3d ed., McGraw-Hill, New York, 1996.
14. L. D. Clayton, E. P. EerNisse, R. W. Ward, and R. B. Wiggins, "Miniature Crystalline Quartz Electromechanical Structures," *Sensors and Actuators,* vol. 20, Nov. 15, 1989, pp. 171–177.
15. J. H. Bell et al., "Surface Pressure Measurements Using Luminescent Coatings," *Annual Review of Fluid Mechanics,* vol. 33, 2001, pp. 155–206.
16. B. G. Liptak (ed.), *Instrument Engineer's Handbook: Process Measurement and Analysis,* 4th ed., vol. 1, CRC Press, Boca Raton, FL, 2003.
17. D. R. Gillum, *Industrial Pressure, Level and Density Measurement,* Insrument Society of America, Research Triangle Park, NC, 1995.

Table tennis ball suspended by an air jet. The control volume momentum principle, studied in this chapter, requires a force to change the direction of a flow. The jet flow deflects around the ball, and the force is the ball's weight. *(Courtesy of Paul Silverman/Fundamental Photographs.)*

Chapter 3
Integral Relations for a Control Volume

Motivation. In analyzing fluid motion, we might take one of two paths: (1) seeking to describe the detailed flow pattern at every point (x, y, z) in the field or (2) working with a finite region, making a balance of flow in versus flow out, and determining gross flow effects such as the force or torque on a body or the total energy exchange. The second is the "control volume" method and is the subject of this chapter. The first is the "differential" approach and is developed in Chap. 4.

We first develop the concept of the control volume, in nearly the same manner as one does in a thermodynamics course, and we find the rate of change of an arbitrary gross fluid property, a result called the *Reynolds transport theorem*. We then apply this theorem, in sequence, to mass, linear momentum, angular momentum, and energy, thus deriving the four basic control volume relations of fluid mechanics. There are many applications, of course. The chapter then ends with a special case of frictionless, shaft-work-free momentum and energy: the *Bernoulli equation*. The Bernoulli equation is a wonderful, historic relation, but it is extremely restrictive and should always be viewed with skepticism and care in applying it to a real (viscous) fluid motion.

3.1 Basic Physical Laws of Fluid Mechanics

It is time now to really get serious about flow problems. The fluid statics applications of Chap. 2 were more like fun than work, at least in this writer's opinion. Statics problems basically require only the density of the fluid and knowledge of the position of the free surface, but most flow problems require the analysis of an arbitrary state of variable fluid motion defined by the geometry, the boundary conditions, and the laws of mechanics. This chapter and the next two outline the three basic approaches to the analysis of arbitrary flow problems:

1. Control volume, or large-scale, analysis (Chap. 3).
2. Differential, or small-scale, analysis (Chap. 4).
3. Experimental, or dimensional, analysis (Chap. 5).

The three approaches are roughly equal in importance. Control volume analysis, the present topic, is accurate for any flow distribution but is often based on average or "one-dimensional" property values at the boundaries. It always gives useful "engineering" estimates. In principle, the differential equation approach of Chap. 4 can be applied to any problem. Only a few problems, such as straight pipe flow, yield to exact analytical solutions. But the differential equations can be modeled numerically, and the flourishing field of computational fluid dynamics (CFD)[8] can now be used to give good estimates for almost any geometry. Finally, the dimensional analysis of Chap. 5 applies to any problem, whether analytical, numerical, or experimental. It is particularly useful to reduce the cost of experimentation. Differential analysis began with Euler and Lagrange in the eighteenth century, and dimensional analysis was pioneered by Lord Rayleigh in the late nineteenth century. The control volume, although proposed by Euler and used again by Osborne Reynolds in the late nineteenth century, was not developed as a general analytical tool until the 1940s.

Systems versus Control Volumes

All the laws of mechanics are written for a *system,* which is defined as an arbitrary quantity of mass of fixed identity. Everything external to this system is denoted by the term *surroundings,* and the system is separated from its surroundings by its *boundaries.* The laws of mechanics then state what happens when there is an interaction between the system and its surroundings.

First, the system is a fixed quantity of mass, denoted by m. Thus the mass of the system is conserved and does not change.[1] This is a law of mechanics and has a very simple mathematical form, called *conservation of mass:*

$$m_{\text{syst}} = \text{const}$$

or
$$\frac{dm}{dt} = 0 \tag{3.1}$$

This is so obvious in solid mechanics problems that we often forget about it. In fluid mechanics, we must pay a lot of attention to mass conservation, and it takes a little analysis to make it hold.

Second, if the surroundings exert a net force $\mathbf{F}$ on the system, Newton's second law states that the mass in the system will begin to accelerate:[2]

$$\mathbf{F} = m\mathbf{a} = m\frac{d\mathbf{V}}{dt} = \frac{d}{dt}(m\mathbf{V}) \tag{3.2}$$

In Eq. (2.8) we saw this relation applied to a differential element of viscous incompressible fluid. In fluid mechanics Newton's second law is called the linear momentum relation. Note that it is a vector law that implies the three scalar equations $F_x = ma_x$, $F_y = ma_y$, and $F_z = ma_z$.

Third, if the surroundings exert a net moment $\mathbf{M}$ about the center of mass of the system, there will be a rotation effect

$$\mathbf{M} = \frac{d\mathbf{H}}{dt} \tag{3.3}$$

---

[1]We are neglecting nuclear reactions, where mass can be changed to energy.
[2]We are neglecting relativistic effects, where Newton's law must be modified.

where $\mathbf{H} = \Sigma(\mathbf{r} \times \mathbf{V})\delta m$ is the angular momentum of the system about its center of mass. Here we call Eq. (3.3) the angular momentum relation. Note that it is also a vector equation implying three scalar equations such as $M_x = dH_x/dt$.

For an arbitrary mass and arbitrary moment, $\mathbf{H}$ is quite complicated and contains nine terms (see, for example, Ref. 1). In elementary dynamics we commonly treat only a rigid body rotating about a fixed x axis, for which Eq. (3.3) reduces to

$$M_x = I_x \frac{d}{dt}(\omega_x) \tag{3.4}$$

where ω_x is the angular velocity of the body and I_x is its mass moment of inertia about the x axis. Unfortunately, fluid systems are not rigid and rarely reduce to such a simple relation, as we shall see in Sec. 3.5.

Fourth, if heat δQ is added to the system or work δW is done by the system, the system energy dE must change according to the energy relation, or first law of thermodynamics:

$$\delta Q - \delta W = dE$$

or

$$\dot{Q} - \dot{W} = \frac{dE}{dt} \tag{3.5}$$

Like mass conservation, Eq. (3.1), this is a scalar relation having only a single component.

Finally, the second law of thermodynamics relates entropy change dS to heat added dQ and absolute temperature T:

$$dS \geq \frac{\delta Q}{T} \tag{3.6}$$

This is valid for a system and can be written in control volume form, but there are almost no practical applications in fluid mechanics except to analyze flow loss details (see Sec. 9.5).

All these laws involve thermodynamic properties, and thus we must supplement them with state relations $p = p(\rho, T)$ and $e = e(\rho, T)$ for the particular fluid being studied, as in Sec. 1.8. Although thermodynamics is not the main topic of this book, it is very important to the general study of fluid mechanics. Thermodynamics is crucial to compressible flow, Chap. 9. The student should review the first law and the state relations, as discussed in Refs. 6 and 7.

The purpose of this chapter is to put our four basic laws into the control volume form suitable for arbitrary regions in a flow:

1. Conservation of mass (Sec. 3.3).
2. The linear momentum relation (Sec. 3.4).
3. The angular momentum relation (Sec. 3.5).
4. The energy equation (Sec. 3.6).

Wherever necessary to complete the analysis we also introduce a state relation such as the perfect-gas law.

Equations (3.1) to (3.6) apply to either fluid or solid systems. They are ideal for solid mechanics, where we follow the same system forever because it represents the product we are designing and building. For example, we follow a beam as it deflects under load. We follow a piston as it oscillates. We follow a rocket system all the way to Mars.

But fluid systems do not demand this concentrated attention. It is rare that we wish to follow the ultimate path of a specific particle of fluid. Instead it is likely that the fluid forms the environment whose effect on our product we wish to know. For the three examples just cited, we wish to know the wind loads on the beam, the fluid pressures on the piston, and the drag and lift loads on the rocket. This requires that the basic laws be rewritten to apply to a specific *region* in the neighborhood of our product. In other words, where the fluid particles in the wind go after they leave the beam is of little interest to a beam designer. The user's point of view underlies the need for the control volume analysis of this chapter.

In analyzing a control volume, we convert the system laws to apply to a specific region, which the system may occupy for only an instant. The system passes on, and other systems come along, but no matter. The basic laws are reformulated to apply to this local region called a control volume. All we need to know is the flow field in this region, and often simple assumptions will be accurate enough (such as uniform inlet and/or outlet flows). The flow conditions away from the control volume are then irrelevant. The technique for making such localized analyses is the subject of this chapter.

Volume and Mass Rate of Flow   All the analyses in this chapter involve evaluation of the volume flow Q or mass flow $\dot{m}$ passing through a surface (imaginary) defined in the flow.

Suppose that the surface S in Fig. 3.1a is a sort of (imaginary) wire mesh through which the fluid passes without resistance. How much volume of fluid passes through S in unit time? If, typically, $\mathbf{V}$ varies with position, we must integrate over the elemental surface dA in Fig. 3.1a. Also, typically $\mathbf{V}$ may pass through dA at an angle θ off the normal. Let $\mathbf{n}$ be defined as the unit vector normal to dA. Then the amount of fluid swept through dA in time dt is the volume of the slanted parallelepiped in Fig. 3.1b:

$$d\mathcal{V} = V \, dt \, dA \cos \theta = (\mathbf{V} \cdot \mathbf{n}) \, dA \, dt$$

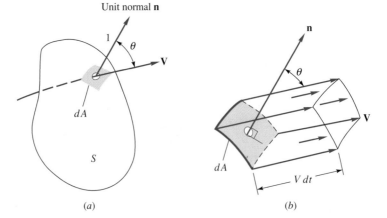

Fig. 3.1 Volume rate of flow through an arbitrary surface: (*a*) an elemental area dA on the surface; (*b*) the incremental volume swept through dA equals $V \, dt \, dA \cos \theta$.

The integral of $d\mathcal{V}/dt$ is the total volume rate of flow Q through the surface S:

$$Q = \int_s (\mathbf{V} \cdot \mathbf{n})\, dA = \int_s V_n\, dA \qquad (3.7)$$

We could replace $\mathbf{V} \cdot \mathbf{n}$ by its equivalent, V_n, the component of $\mathbf{V}$ normal to dA, but the use of the dot product allows Q to have a sign to distinguish between inflow and outflow. By convention throughout this book we consider $\mathbf{n}$ to be the *outward* normal unit vector. Therefore $\mathbf{V} \cdot \mathbf{n}$ denotes outflow if it is positive and inflow if negative. This will be an extremely useful housekeeping device when we are computing volume and mass flow in the basic control volume relations.

Volume flow can be multiplied by density to obtain the mass flow $\dot{m}$. If density varies over the surface, it must be part of the surface integral:

$$\dot{m} = \int_s \rho(\mathbf{V} \cdot \mathbf{n})\, dA = \int_s \rho V_n\, dA$$

If density and velocity are constant over the surface S, a simple expression results:

One-dimensional approximation: $\qquad \dot{m} = \rho Q = \rho A V$

3.2 The Reynolds Transport Theorem

To convert a system analysis to a control volume analysis, we must convert our mathematics to apply to a specific region rather than to individual masses. This conversion, called the *Reynolds transport theorem,* can be applied to all the basic laws. Examining the basic laws (3.1) to (3.3) and (3.5), we see that they are all concerned with the time derivative of fluid properties m, $\mathbf{V}$, $\mathbf{H}$, and E. Therefore what we need is to relate the time derivative of a system property to the rate of change of that property within a certain region.

The desired conversion formula differs slightly according to whether the control volume is fixed, moving, or deformable. Figure 3.2 illustrates these three cases. The fixed control volume in Fig. 3.2a encloses a stationary region of interest to a nozzle designer. The control surface is an abstract concept and does not hinder the flow in any way. It slices through the jet leaving the nozzle, encloses the surrounding atmosphere, and slices through the flange bolts and the fluid within the nozzle. This particular control volume exposes the stresses in the flange bolts,

Fig. 3.2 Fixed, moving, and deformable control volumes: (*a*) fixed control volume for nozzle stress analysis; (*b*) control volume moving at ship speed for drag force analysis; (*c*) control volume deforming within cylinder for transient pressure variation analysis.

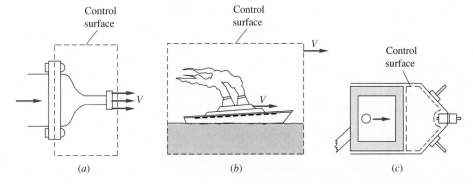

which contribute to applied forces in the momentum analysis. In this sense the control volume resembles the *free-body* concept, which is applied to systems in solid mechanics analyses.

Figure 3.2*b* illustrates a moving control volume. Here the ship is of interest, not the ocean, so that the control surface chases the ship at ship speed *V*. The control volume is of fixed volume, but the relative motion between water and ship must be considered. If *V* is constant, this relative motion is a steady flow pattern, which simplifies the analysis.[3] If *V* is variable, the relative motion is unsteady, so that the computed results are time-variable and certain terms enter the momentum analysis to reflect the noninertial (accelerating) frame of reference.

Figure 3.2*c* shows a deforming control volume. Varying relative motion at the boundaries becomes a factor, and the rate of change of shape of the control volume enters the analysis. We begin by deriving the fixed control volume case, and we consider the other cases as advanced topics.

Arbitrary Fixed Control Volume

Figure 3.3 shows a fixed control volume with an arbitrary flow pattern passing through. There are variable slivers of inflow and outflow of fluid all about the control surface. In general, each differential area *dA* of surface will have a different velocity **V** making a different angle θ with the local normal to *dA*. Some elemental areas will have inflow volume $(VA \cos \theta)_{\text{in}} \, dt$, and others will have outflow volume $(VA \cos \theta)_{\text{out}} \, dt$,

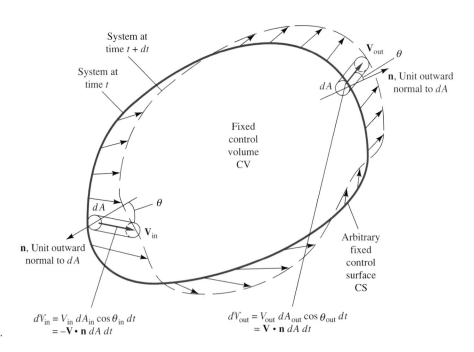

Fig. 3.3 An arbitrary control volume with an arbitrary flow pattern.

$$d\mathcal{V}_{\text{in}} = V_{\text{in}} \, dA_{\text{in}} \cos \theta_{\text{in}} \, dt$$
$$= -\mathbf{V} \cdot \mathbf{n} \, dA \, dt$$

$$d\mathcal{V}_{\text{out}} = V_{\text{out}} \, dA_{\text{out}} \cos \theta_{\text{out}} \, dt$$
$$= \mathbf{V} \cdot \mathbf{n} \, dA \, dt$$

[3]A *wind tunnel* uses a fixed model to simulate flow over a body moving through a fluid. A *tow tank* uses a moving model to simulate the same situation.

as seen in Fig. 3.3. Some surfaces might correspond to streamlines ($\theta = 90°$) or solid walls ($\mathbf{V} = 0$) with neither inflow nor outflow.

Let B be any property of the fluid (energy, momentum, enthalpy, etc.) and let $\beta = dB/dm$ be the *intensive* value, or the amount of B per unit mass in any small element of the fluid. The total amount of B in the control volume (the solid curve in Fig. 3.3) is thus

$$B_{\text{CV}} = \int_{\text{CV}} \beta \, dm = \int_{\text{CV}} \beta \rho \, d\mathcal{V} \qquad \beta = \frac{dB}{dm} \tag{3.8}$$

Examining Fig. 3.3, we see three sources of changes in B relating to the control volume:

$$\text{A change within the control volume } \frac{d}{dt}\left(\int_{\text{CV}} \beta \rho \, d\mathcal{V} \right)$$

$$\text{Outflow of } \beta \text{ from the control volume } \int_{\text{CS}} \beta \rho V \cos \theta \, dA_{\text{out}} \tag{3.9}$$

$$\text{Inflow of } \beta \text{ to the control volume } \int_{\text{CS}} \beta \rho V \cos \theta \, dA_{\text{in}}$$

The notations CV and CS refer to the control volume and control surface, respectively. Note, in Fig. 3.3, that the *system* has moved a bit, gaining the outflow sliver and losing the inflow sliver. In the limit as $dt \to 0$, the instantaneous change of B in the system is the sum of the change within, plus the outflow, minus the inflow:

$$\frac{d}{dt}(B_{\text{syst}}) = \frac{d}{dt}\left(\int_{\text{CV}} \beta \rho \, d\mathcal{V} \right) + \int_{\text{CS}} \beta \rho V \cos \theta \, dA_{\text{out}} - \int_{\text{CS}} \beta \rho V \cos \theta \, dA_{\text{in}} \tag{3.10}$$

This is the *Reynolds transport theorem* for an arbitrary fixed control volume. By letting the property B be mass, momentum, angular momentum, or energy, we can rewrite all the basic laws in control volume form. Note that all three of the integrals are concerned with the intensive property β. Since the control volume is fixed in space, the elemental volumes $d\mathcal{V}$ do not vary with time, so that the time derivative of the volume integral vanishes unless either β or ρ varies with time (unsteady flow).

Equation (3.10) expresses the basic formula that a system derivative equals the rate of change of B within the control volume plus the flux of B out of the control surface minus the flux of B into the control surface. The quantity B (or β) may be any vector or scalar property of the fluid. Two alternate forms are possible for the flux terms. First we may notice that $V \cos \theta$ is the component of V normal to the area element of the control surface. Thus we can write

$$\text{Flux terms} = \int_{\text{CS}} \beta \rho V_n \, dA_{\text{out}} - \int_{\text{CS}} \beta \rho V_n \, dA_{\text{in}} = \int_{\text{CS}} \beta \, d\dot{m}_{\text{out}} - \int_{\text{CS}} \beta \, d\dot{m}_{\text{in}} \tag{3.10a}$$

where $d\dot{m} = \rho V_n \, dA$ is the differential mass flux through the surface. Form (3.10a) helps us visualize what is being calculated.

A second, alternative form offers elegance and compactness as advantages. If **n** is defined as the *outward* normal unit vector everywhere on the control surface, then $\mathbf{V} \cdot \mathbf{n} = V_n$ for outflow and $\mathbf{V} \cdot \mathbf{n} = -V_n$ for inflow. Therefore the flux terms can be represented by a single integral involving $\mathbf{V} \cdot \mathbf{n}$ that accounts for both positive outflow and negative inflow:

$$\text{Flux terms} = \int_{\text{CS}} \beta \rho (\mathbf{V} \cdot \mathbf{n}) \, dA \tag{3.11}$$

The compact form of the Reynolds transport theorem is thus

$$\frac{d}{dt}(B_{\text{syst}}) = \frac{d}{dt}\left(\int_{\text{CV}} \beta \rho \, d\mathcal{V} \right) + \int_{\text{CS}} \beta \rho (\mathbf{V} \cdot \mathbf{n}) \, dA \tag{3.12}$$

This is beautiful but only occasionally useful, when the coordinate system is ideally suited to the control volume selected. Otherwise the computations are easier when the flux of B out is added and the flux of B in is subtracted, according to (3.10) or (3.11).

The time derivative term can be written in the equivalent form

$$\frac{d}{dt}\left(\int_{\text{CV}} \beta \rho \, d\mathcal{V} \right) = \int_{\text{CV}} \frac{\partial}{\partial t}(\beta \rho) \, d\mathcal{V} \tag{3.13}$$

for the fixed control volume since the volume elements do not vary.

Control Volume Moving at Constant Velocity

If the control volume is moving uniformly at velocity $\mathbf{V}_s$, as in Fig. 3.2b, an observer fixed to the control volume will see a relative velocity $\mathbf{V}_r$ of fluid crossing the control surface, defined by

$$\mathbf{V}_r = \mathbf{V} - \mathbf{V}_s \tag{3.14}$$

where **V** is the fluid velocity relative to the same coordinate system in which the control volume motion $\mathbf{V}_s$ is observed. Note that Eq. (3.14) is a vector subtraction. The flux terms will be proportional to $\mathbf{V}_r$, but the volume integral of Eq. (3.12) is unchanged because the control volume moves as a fixed shape without deforming. The Reynolds transport theorem for this case of a uniformly moving control volume is

$$\frac{d}{dt}(B_{\text{syst}}) = \frac{d}{dt}\left(\int_{\text{CV}} \beta \rho \, d\mathcal{V} \right) + \int_{\text{CS}} \beta \rho (\mathbf{V}_r \cdot \mathbf{n}) \, dA \tag{3.15}$$

which reduces to Eq. (3.12) if $\mathbf{V}_s \equiv 0$.

Control Volume of Constant Shape but Variable Velocity[4]

If the control volume moves with a velocity $\mathbf{V}_s(t)$ that retains its shape, then the volume elements do not change with time, but the boundary relative velocity $\mathbf{V}_r = \mathbf{V}(\mathbf{r}, t) - \mathbf{V}_s(t)$ becomes a somewhat more complicated function. Equation (3.15) is unchanged in form, but the area integral may be more laborious to evaluate.

---

[4]This section may be omitted without loss of continuity.

Arbitrarily Moving and Deformable Control Volume[5]

The most general situation is when the control volume is both moving and deforming arbitrarily, as illustrated in Fig. 3.4. The flux of volume across the control surface is again proportional to the relative normal velocity component $V_r \cdot n$, as in Eq. (3.15). However, since the control surface has a deformation, its velocity $V_s = V_s(r, t)$, so that the relative velocity $V_r = V(r, t) - V_s(r, t)$ is or can be a complicated function, even though the flux integral is the same as in Eq. (3.15). Meanwhile, the volume integral in Eq. (3.15) must allow the volume elements to distort with time. Thus the time derivative must be applied *after* integration. For the deforming control volume, then, the transport theorem takes the form

$$\frac{d}{dt}(B_{\text{syst}}) = \frac{d}{dt}\left(\int_{\text{CV}} \beta\rho \, d\mathcal{V}\right) + \int_{\text{CS}} \beta\rho(V_r \cdot n) \, dA \qquad (3.16)$$

This is the most general case, which we can compare with the equivalent form for a fixed control volume:

$$\frac{d}{dt}(B_{\text{syst}}) = \int_{\text{CV}} \frac{\partial}{\partial t}(\beta\rho) \, d\mathcal{V} + \int_{\text{CS}} \beta\rho(V \cdot n) \, dA \qquad (3.17)$$

The moving and deforming control volume, Eq. (3.16), contains only two complications: (1) The time derivative of the first integral on the right must be taken outside, and (2) the second integral involves the *relative* velocity V_r between the fluid system and the control surface. These differences and mathematical subtleties are best shown by examples.

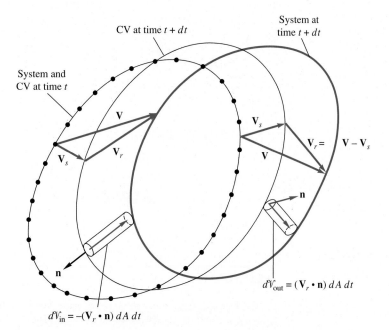

Fig. 3.4 Relative velocity effects between a system and a control volume when both move and deform. The system boundaries move at velocity V, and the control surface moves at velocity V_s.

[5]This section may be omitted without loss of continuity.

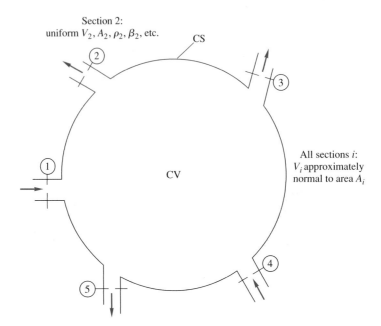

Fig. 3.5 A control volume with simplified one-dimensional inlets and exits.

One-Dimensional Flux Term Approximations

In many situations, the flow crosses the boundaries of the control surface only at simplified inlets and exits that are approximately *one-dimensional;* that is, flow properties are nearly uniform over the cross section. For a fixed control volume, the surface integral in Eq. (3.12) reduces to a sum of positive (outlet) and negative (inlet) product terms for each cross section:

$$\frac{d}{dt}(B_{\text{syst}}) = \frac{d}{dt}\left(\int_{\text{CV}} \beta \, dm\right) + \sum_{\text{outlets}} \beta_i \dot{m}_i \big|_{\text{out}} - \sum_{\text{inlets}} \beta_i \dot{m}_i \big|_{\text{in}} \quad \text{where } \dot{m}_i = \rho_i A_i V_i \quad (3.18)$$

To the writer, this is an attractive way to set up a control volume analysis without using the dot product notation. An example of multiple one-dimensional fluxes is shown in Fig. 3.5. There are inlet flows at sections 1 and 4 and outflows at sections 2, 3, and 5. Equation (3.18) becomes

$$\frac{d}{dt}(B_{\text{syst}}) = \frac{d}{dt}\left(\int_{\text{CV}} \beta \, dm\right) + \beta_2(\rho AV)_2 + \beta_3(\rho AV)_3 + \beta_5(\rho AV)_5$$
$$- \beta_1(\rho AV)_1 - \beta_4(\rho AV)_4 \qquad (3.19)$$

with no contribution from any other portion of the control surface because there is no flow across the boundary.

EXAMPLE 3.1

A fixed control volume has three one-dimensional boundary sections, as shown in Fig. E3.1. The flow within the control volume is steady. The flow properties at each section are tabulated below. Find the rate of change of energy of the system that occupies the control volume at this instant.

E3.1

Section	Type	ρ, kg/m^3	V, m/s	A, m^2	e, J/kg
1	Inlet	800	5.0	2.0	300
2	Inlet	800	8.0	3.0	100
3	Outlet	800	17.0	2.0	150

Solution

- *System sketch:*   Figure E3.1 shows two inlet flows, 1 and 2, and a single outlet flow, 3.
- *Assumptions:*   Steady flow, fixed control volume, one-dimensional inlet and exit flows.
- *Approach:*   Apply Eq. (3.17) with *energy* as the property, where $B = E$ and $\beta = dE/dm = e$. Use the one-dimensional flux approximation and then insert the data from the table.
- *Solution steps:*   Outlet 3 contributes a positive term, and inlets 1 and 2 are negative. The appropriate form of Eq. (3.12) is

$$\left(\frac{dE}{dt}\right)_{\text{syst}} = \frac{d}{dt}\left(\int_{\text{CV}} e\,\rho\,d\upsilon\right) + e_3 \dot{m}_3 - e_1 \dot{m}_1 - e_2 \dot{m}_2$$

Since the flow is steady, the time-derivative volume integral term is zero. Introducing $(\rho A V)_i$ as the mass flow grouping, we obtain

$$\left(\frac{dE}{dt}\right)_{\text{syst}} = -e_1\rho_1 A_1 V_1 - e_2\rho_2 A_2 V_2 + e_3\rho_3 A_3 V_3$$

Introducing the numerical values from the table, we have

$$\left(\frac{dE}{dt}\right)_{\text{syst}} = -(300 \text{ J/kg})(800 \text{ kg/m}^3)(2 \text{ m}^2)(5 \text{ m/s}) - 100(800)(3)(8) + 150(800)(2)(17)$$

$$= (-2,400,000 - 1,920,000 + 4,080,000) \text{ J/s}$$

$$= -240,000 \text{ J/s} = -0.24 \text{ MJ/s} \qquad\qquad Ans.$$

Thus the system is losing energy at the rate of 0.24 MJ/s = 0.24 MW. Since we have accounted for all fluid energy crossing the boundary, we conclude from the first law that there must be heat loss through the control surface, or the system must be doing work on the environment through some device not shown. Notice that the use of SI units leads to a consistent result in joules per second without any conversion factors. We promised in Chap. 1 that this would be the case.

- *Comments:*   This problem involves energy, but suppose we check the balance of mass also. Then $B = $ mass m, and $B = dm/dm = $ unity. Again the volume integral vanishes for steady flow, and Eq. (3.17) reduces to

$$\left(\frac{dm}{dt}\right)_{\text{syst}} = \int_{\text{CS}} \rho(\mathbf{V} \cdot \mathbf{n})\, dA = -\rho_1 A_1 V_1 - \rho_2 A_2 V_2 + \rho_3 A_3 V_3$$

$$= -(800 \text{ kg/m}^3)(2 \text{ m}^2)(5 \text{ m/s}) - 800(3)(8) + 800(17)(2)$$

$$= (-8000 - 19,200 + 27,200) \text{ kg/s} = 0 \text{ kg/s}$$

Thus the system mass does not change, which correctly expresses the law of conservation of system mass, Eq. (3.1).

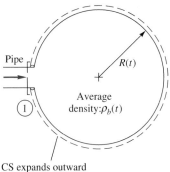

Pipe

$R(t)$

Average density:$\rho_b(t)$

①

CS expands outward
with balloon radius $R(t)$

E3.2

EXAMPLE 3.2

The balloon in Fig. E3.2 is being filled through section 1, where the area is A_1, velocity is V_1, and fluid density is ρ_1. The average density within the balloon is $\rho_b(t)$. Find an expression for the rate of change of system mass within the balloon at this instant.

Solution

- *System sketch:* Figure E3.2 shows one inlet, no exits. The control volume and system expand together, hence the relative velocity $V_r = 0$ on the balloon surface.
- *Assumptions:* Unsteady flow (the control volume mass increases), deformable control surface, one-dimensional inlet conditions.
- *Approach:* Apply Eq. (3.16) with $V_r = 0$ on the balloon surface and $V_r = V_1$ at the inlet.
- *Solution steps:* The property being studied is *mass*, $B = m$ and $\beta = dm/dm =$ unity. Apply Eq. (3.16). The volume integral is evaluated based on average density ρ_b, and the surface integral term is negative (for an inlet):

$$\left(\frac{dm}{dt}\right)_{\text{syst}} = \frac{d}{dt}\left(\int_{\text{CV}} \rho \, d\mathcal{V}\right) + \int_{\text{CS}} \rho(\mathbf{V}_r \cdot \mathbf{n})dA = \frac{d}{dt}\left(\rho_b \frac{4\pi}{3} R^3\right) - \rho_1 A_1 V_1 \qquad Ans.$$

- *Comments:* The relation given is the answer to the question that was asked. Actually, by the conservation law for mass, Eq. (3.1), $(dm/dt)_{\text{syst}} = 0$, and the answer could be rewritten as

$$\frac{d}{dt}(\rho_b R^3) = \frac{3}{4\pi}\rho_1 A_1 V_1$$

This is a first-order ordinary differential equation relating gas density and balloon radius. It could form part of an engineering analysis of balloon inflation. It cannot be solved without further use of mechanics and thermodynamics to relate the four unknowns ρ_b, ρ_1, V_1, and R. The pressure and temperature and the elastic properties of the balloon would also have to be brought into the analysis.

For advanced study, many more details of the analysis of deformable control volumes can be found in Hansen [4] and Potter et al. [5].

3.3 Conservation of Mass

The Reynolds transport theorem, Eq. (3.16) or (3.17), establishes a relation between system rates of change and control volume surface and volume integrals. But system derivatives are related to the basic laws of mechanics, Eqs. (3.1) to (3.5). Eliminating system derivatives between the two gives the control volume, or *integral,* forms of the laws of mechanics of fluids. The dummy variable B becomes, respectively, mass, linear momentum, angular momentum, and energy.

For conservation of mass, as discussed in Examples 3.1 and 3.2, $B = m$ and $\beta = dm/dm = 1$. Equation (3.1) becomes

$$\left(\frac{dm}{dt}\right)_{\text{syst}} = 0 = \frac{d}{dt}\left(\int_{\text{CV}} \rho \, d\mathcal{V}\right) + \int_{\text{CS}} \rho(\mathbf{V}_r \cdot \mathbf{n}) \, dA \qquad (3.20)$$

This is the integral mass conservation law for a deformable control volume. For a fixed control volume, we have

$$\int_{CV} \frac{\partial \rho}{\partial t}\, d\mathcal{V} + \int_{CS} \rho(\mathbf{V} \cdot \mathbf{n})\, dA = 0 \tag{3.21}$$

If the control volume has only a number of one-dimensional inlets and outlets, we can write

$$\boxed{\int_{CV} \frac{\partial \rho}{\partial t}\, d\mathcal{V} + \sum_i (\rho_i A_i V_i)_{\text{out}} - \sum_i (\rho_i A_i V_i)_{\text{in}} = 0} \tag{3.22}$$

Other special cases occur. Suppose that the flow within the control volume is steady; then $\partial \rho / \partial t \equiv 0$, and Eq. (3.21) reduces to

$$\int_{CS} \rho(\mathbf{V} \cdot \mathbf{n})\, dA = 0 \tag{3.23}$$

This states that in steady flow the mass flows entering and leaving the control volume must balance exactly.[6] If, further, the inlets and outlets are one-dimensional, we have for steady flow

$$\sum_i (\rho_i A_i V_i)_{\text{in}} = \sum_i (\rho_i A_i V_i)_{\text{out}} \tag{3.24}$$

This simple approximation is widely used in engineering analyses. For example, referring to Fig. 3.5, we see that if the flow in that control volume is steady, the three outlet mass fluxes balance the two inlets:

$$\text{Outflow} = \text{inflow}$$
$$\rho_2 A_2 V_2 + \rho_3 A_3 V_3 + \rho_5 A_5 V_5 = \rho_1 A_1 V_1 + \rho_4 A_4 V_4 \tag{3.25}$$

The quantity ρAV is called the *mass flow* $\dot{m}$ passing through the one-dimensional cross section and has consistent units of kilograms per second (or slugs per second) for SI (or BG) units. Equation (3.25) can be rewritten in the short form

$$\dot{m}_2 + \dot{m}_3 + \dot{m}_5 = \dot{m}_1 + \dot{m}_4 \tag{3.26}$$

and, in general, the steady-flow–mass-conservation relation (3.23) can be written as

$$\sum_i (\dot{m}_i)_{\text{out}} = \sum_i (\dot{m}_i)_{\text{in}} \tag{3.27}$$

If the inlets and outlets are not one-dimensional, one has to compute $\dot{m}$ by integration over the section

$$\dot{m}_{\text{cs}} = \int_{cs} \rho(\mathbf{V} \cdot \mathbf{n})\, dA \tag{3.28}$$

where "cs" stands for cross section. An illustration of this is given in Example 3.4.

---

[6]Throughout this section we are neglecting *sources* or *sinks* of mass that might be embedded in the control volume. Equations (3.20) and (3.21) can readily be modified to add source and sink terms, but this is rarely necessary.

Incompressible Flow

Still further simplification is possible if the fluid is incompressible, which we may define as having density variations that are negligible in the mass conservation requirement.[7] As we saw in Chap. 1, all liquids are nearly incompressible, and gas flows can *behave* as if they were incompressible, particularly if the gas velocity is less than about 30 percent of the speed of sound of the gas.

Again consider the fixed control volume. For nearly incompressible flow, the term $\partial\rho/\partial t$ is small, so the time-derivative volume integral in Eq. (3.21) can be neglected. The constant density can then be removed from the surface integral for a nice simplification:

$$\frac{d}{dt}\left(\int_{CV} \frac{\partial\rho}{\partial t}\, dv\right) + \int_{CS} \rho(\mathbf{V}\cdot\mathbf{n})\, dA = 0 = \int_{CS} \rho(\mathbf{V}\cdot\mathbf{n})\, dA = \rho\int_{CS}(\mathbf{V}\cdot\mathbf{n})\, dA$$

or

$$\int_{CS}(\mathbf{V}\cdot\mathbf{n})\, dA = 0 \qquad (3.29)$$

If the inlets and outlets are one-dimensional, we have

$$\sum_i (V_i A_i)_{\text{out}} = \sum_i (V_i A_i)_{\text{in}} \qquad (3.30)$$

or

$$\sum Q_{\text{out}} = \sum Q_{\text{in}}$$

where $Q_i = V_i A_i$ is called the *volume flow* passing through the given cross section.

Again, if consistent units are used, $Q = VA$ will have units of cubic meters per second (SI) or cubic feet per second (BG). If the cross section is not one-dimensional, we have to integrate

$$Q_{CS} = \int_{CS}(\mathbf{V}\cdot\mathbf{n})\, dA \qquad (3.31)$$

Equation (3.31) allows us to define an *average velocity* V_{av} that, when multiplied by the section area, gives the correct volume flow:

$$V_{av} = \frac{Q}{A} = \frac{1}{A}\int(\mathbf{V}\cdot\mathbf{n})\, dA \qquad (3.32)$$

This could be called the *volume-average velocity*. If the density varies across the section, we can define an average density in the same manner:

$$\rho_{av} = \frac{1}{A}\int \rho\, dA \qquad (3.33)$$

But the mass flow would contain the product of density and velocity, and the average product $(\rho V)_{av}$ would in general have a different value from the product of the averages:

$$(\rho V)_{av} = \frac{1}{A}\int \rho(\mathbf{V}\cdot\mathbf{n})\, dA \approx \rho_{av} V_{av} \qquad (3.34)$$

---

[7]Be warned that there is subjectivity in specifying incompressibility. Oceanographers consider a 0.1 percent density variation very significant, while aerodynamicists may neglect density variations in highly compressible, even hypersonic, gas flows. Your task is to justify the incompressible approximation when you make it.

We illustrate average velocity in Example 3.4. We can often neglect the difference or, if necessary, use a correction factor between mass average and volume average.

EXAMPLE 3.3

Write the conservation-of-mass relation for steady flow through a streamtube (flow everywhere parallel to the walls) with a single one-dimensional inlet 1 and exit 2 (Fig. E3.3).

Solution

For steady flow Eq. (3.24) applies with the single inlet and exit:

$$\dot{m} = \rho_1 A_1 V_1 = \rho_2 A_2 V_2 = \text{const}$$

Thus, in a streamtube in steady flow, the mass flow is constant across every section of the tube. If the density is constant, then

$$Q = A_1 V_1 = A_2 V_2 = \text{const} \qquad \text{or} \qquad V_2 = \frac{A_1}{A_2} V_1$$

The volume flow is constant in the tube in steady incompressible flow, and the velocity increases as the section area decreases. This relation was derived by Leonardo da Vinci in 1500.

EXAMPLE 3.4

For steady viscous flow through a circular tube (Fig. E3.4), the axial velocity profile is given approximately by

$$u = U_0 \left(1 - \frac{r}{R} \right)^m$$

so that u varies from zero at the wall ($r = R$), or no slip, up to a maximum $u = U_0$ at the centerline $r = 0$. For highly viscous (laminar) flow $m \approx \frac{1}{2}$, while for less viscous (turbulent) flow $m \approx \frac{1}{7}$. Compute the average velocity if the density is constant.

Solution

The average velocity is defined by Eq. (3.32). Here $\mathbf{V} = \mathbf{i}u$ and $\mathbf{n} = \mathbf{i}$, and thus $\mathbf{V} \cdot \mathbf{n} = u$. Since the flow is symmetric, the differential area can be taken as a circular strip $dA = 2\pi r\, dr$. Equation (3.32) becomes

$$V_{av} = \frac{1}{A} \int u\, dA = \frac{1}{\pi R^2} \int_0^R U_0 \left(1 - \frac{r}{R} \right)^m 2\pi r\, dr$$

or

$$V_{av} = U_0 \frac{2}{(1 + m)(2 + m)} \qquad\qquad\qquad Ans.$$

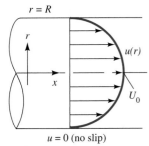

E3.4

E3.3

For the laminar flow approximation, $m \approx \frac{1}{2}$ and $V_{av} \approx 0.53U_0$. (The exact laminar theory in Chap. 6 gives $V_{av} = 0.50U_0$.) For turbulent flow, $m \approx \frac{1}{7}$ and $V_{av} \approx 0.82U_0$. (There is no exact turbulent theory, and so we accept this approximation.) The turbulent velocity profile is more uniform across the section, and thus the average velocity is only slightly less than maximum.

EXAMPLE 3.5

The tank in Fig. E3.5 is being filled with water by two one-dimensional inlets. Air is trapped at the top of the tank. The water height is h. (a) Find an expression for the change in water height dh/dt. (b) Compute dh/dt if $D_1 = 1$ in, $D_2 = 3$ in, $V_1 = 3$ ft/s, $V_2 = 2$ ft/s, and $A_t = 2$ ft^2, assuming water at 20°C.

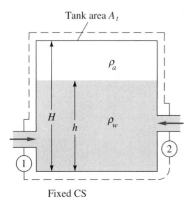

Tank area A_t

p_a

H h p_w

(2)

(1)

Fixed CS

E3.5

Solution

Part (a)

A suggested control volume encircles the tank and cuts through the two inlets. The flow within is unsteady, and Eq. (3.22) applies with no outlets and two inlets:

$$\frac{d}{dt}\left(\int_{CV} \rho \, d\mathcal{V}\right) - \rho_1 A_1 V_1 - \rho_2 A_2 V_2 = 0 \tag{1}$$

Now if A_t is the tank cross-sectional area, the unsteady term can be evaluated as follows:

$$\frac{d}{dt}\left(\int_{CV} \rho \, d\mathcal{V}\right) = \frac{d}{dt}(\rho_w A_t h) + \frac{d}{dt}[\rho_a A_t(H - h)] = \rho_w A_t \frac{dh}{dt} \tag{2}$$

The ρ_a term vanishes because it is the rate of change of air mass and is zero because the air is trapped at the top. Substituting (2) into (1), we find the change of water height

$$\frac{dh}{dt} = \frac{\rho_1 A_1 V_1 + \rho_2 A_2 V_2}{\rho_w A_t} \qquad \qquad Ans. \ (a)$$

For water, $\rho_1 = \rho_2 = \rho_w$, and this result reduces to

$$\frac{dh}{dt} = \frac{A_1 V_1 + A_2 V_2}{A_t} = \frac{Q_1 + Q_2}{A_t} \tag{3}$$

Part (b)

The two inlet volume flows are

$$Q_1 = A_1 V_1 = \tfrac{1}{4}\pi(\tfrac{1}{12} \text{ ft})^2(3 \text{ ft/s}) = 0.016 \text{ ft}^3/\text{s}$$

$$Q_2 = A_2 V_2 = \tfrac{1}{4}\pi(\tfrac{3}{12} \text{ ft})^2(2 \text{ ft/s}) = 0.098 \text{ ft}^3/\text{s}$$

Then, from Eq. (3),

$$\frac{dh}{dt} = \frac{(0.016 + 0.098) \text{ ft}^3/\text{s}}{2 \text{ ft}^2} = 0.057 \text{ ft/s} \qquad Ans. \ (b)$$

Suggestion: Repeat this problem with the top of the tank open.

An illustration of a mass balance with a deforming control volume has already been given in Example 3.2.

The control volume mass relations, Eq. (3.20) or (3.21), are fundamental to all fluid flow analyses. They involve only velocity and density. Vector directions are of no consequence except to determine the normal velocity at the surface and hence whether the flow is *in* or *out*. Although your specific analysis may concern forces or moments or energy, you must always make sure that mass is balanced as part of the analysis; otherwise the results will be unrealistic and probably incorrect. We shall see in the examples that follow how mass conservation is constantly checked in performing an analysis of other fluid properties.

3.4 The Linear Momentum Equation

In Newton's second law, Eq. (3.2), the property being differentiated is the linear momentum $m\mathbf{V}$. Therefore our dummy variable is $\mathbf{B} = m\mathbf{V}$ and $\beta = d\mathbf{B}/dm = \mathbf{V}$, and application of the Reynolds transport theorem gives the linear momentum relation for a deformable control volume:

$$\frac{d}{dt}(m\mathbf{V})_{\text{syst}} = \sum \mathbf{F} = \frac{d}{dt}\left(\int_{\text{CV}} \mathbf{V}\rho\, d\mathcal{V}\right) + \int_{\text{CS}} \mathbf{V}\rho(\mathbf{V}_r \cdot \mathbf{n})\, dA \qquad (3.35)$$

The following points concerning this relation should be strongly emphasized:

1. The term $\mathbf{V}$ is the fluid velocity relative to an *inertial* (nonaccelerating) coordinate system; otherwise Newton's second law must be modified to include noninertial relative acceleration terms (see the end of this section).

2. The term $\sum \mathbf{F}$ is the *vector* sum of all forces acting on the system material considered as a free body; that is, it includes surface forces on all fluids and solids cut by the control surface plus all body forces (gravity and electromagnetic) acting on the masses within the control volume.

3. The entire equation is a vector relation; both the integrals are vectors due to the term $\mathbf{V}$ in the integrands. The equation thus has three components. If we want only, say, the x component, the equation reduces to

$$\sum F_x = \frac{d}{dt}\left(\int_{\text{CV}} u\rho\, d\mathcal{V}\right) + \int_{\text{CS}} u\rho(\mathbf{V}_r \cdot \mathbf{n})\, dA \qquad (3.36)$$

and similarly, $\sum F_y$ and $\sum F_z$ would involve v and w, respectively. Failure to account for the vector nature of the linear momentum relation (3.35) is probably the greatest source of student error in control volume analyses.

For a fixed control volume, the relative velocity $\mathbf{V}_r \equiv \mathbf{V}$, and Eq. (3.35) becomes

$$\sum \mathbf{F} = \frac{d}{dt}\left(\int_{\text{CV}} \mathbf{V}\rho\, d\mathcal{V}\right) + \int_{\text{CS}} \mathbf{V}\rho(\mathbf{V} \cdot \mathbf{n})\, dA \qquad (3.37)$$

Again we stress that this is a vector relation and that $\mathbf{V}$ must be an inertial-frame velocity. Most of the momentum analyses in this text are concerned with Eq. (3.37).

One-Dimensional Momentum Flux

By analogy with the term *mass flow* used in Eq. (3.28), the surface integral in Eq. (3.37) is called the *momentum flux term*. If we denote momentum by $\mathbf{M}$, then

$$\dot{\mathbf{M}}_{CS} = \int_{sec} \mathbf{V}\rho(\mathbf{V} \cdot \mathbf{n})\,dA \tag{3.38}$$

Because of the dot product, the result will be negative for inlet momentum flux and positive for outlet flux. If the cross section is one-dimensional, $\mathbf{V}$ and ρ are uniform over the area and the integrated result is

$$\dot{\mathbf{M}}_{seci} = \mathbf{V}_i(\rho_i V_{ni} A_i) = \dot{m}_i \mathbf{V}_i \tag{3.39}$$

for outlet flux and $-\dot{m}_i \mathbf{V}_i$ for inlet flux. Thus if the control volume has only one-dimensional inlets and outlets, Eq. (3.37) reduces to

$$\sum \mathbf{F} = \frac{d}{dt}\left(\int_{CV} \mathbf{V}\rho\,d\mathcal{V}\right) + \sum (\dot{m}_i \mathbf{V}_i)_{out} - \sum (\dot{m}_i \mathbf{V}_i)_{in} \tag{3.40}$$

This is a commonly used approximation in engineering analyses. It is crucial to realize that we are dealing with vector sums. Equation (3.40) states that the net vector force on a fixed control volume equals the rate of change of vector momentum within the control volume plus the vector sum of outlet momentum fluxes minus the vector sum of inlet fluxes.

Net Pressure Force on a Closed Control Surface

Generally speaking, the surface forces on a control volume are due to (1) forces exposed by cutting through solid bodies that protrude through the surface and (2) forces due to pressure and viscous stresses of the surrounding fluid. The computation of pressure force is relatively simple, as shown in Fig. 3.6. Recall from Chap. 2 that

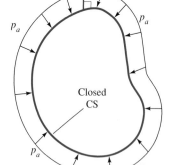

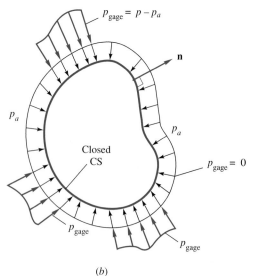

Fig. 3.6 Pressure force computation by subtracting a uniform distribution: (*a*) uniform pressure, $\mathbf{F} = -p_a \int \mathbf{n}\,dA \equiv 0$; (*b*) nonuniform pressure, $\mathbf{F} = -\int (p - p_a)\mathbf{n}\,dA$.

(*a*)

(*b*)

the external pressure force on a surface is normal to the surface and *inward*. Since the unit vector **n** is defined as *outward*, one way to write the pressure force is

$$\mathbf{F}_{\text{press}} = \int_{\text{CS}} p(-\mathbf{n}) \, dA \qquad (3.41)$$

Now if the pressure has a uniform value p_a all around the surface, as in Fig. 3.7a, the net pressure force is zero:

$$\mathbf{F}_{\text{UP}} = \int p_a(-\mathbf{n}) \, dA = -p_a \int \mathbf{n} \, dA \equiv 0 \qquad (3.42)$$

where the subscript UP stands for uniform pressure. This result is *independent of the shape of the surface*[8] as long as the surface is closed and all our control volumes are closed. Thus a seemingly complicated pressure force problem can be simplified by subtracting any convenient uniform pressure p_a and working only with the pieces of gage pressure that remain, as illustrated in Fig. 3.6b. So Eq. (3.41) is entirely equivalent to

$$\mathbf{F}_{\text{press}} = \int_{\text{CS}} (p - p_a)(-\mathbf{n}) \, dA = \int_{\text{CS}} p_{\text{gage}}(-\mathbf{n}) \, dA$$

This trick can mean quite a savings in computation.

EXAMPLE 3.6

A control volume of a nozzle section has surface pressures of 40 lbf/in² absolute at section 1 and atmospheric pressure of 15 lbf/in² absolute at section 2 and on the external rounded part of the nozzle, as in Fig. E3.6a. Compute the net pressure force if $D_1 = 3$ in and $D_2 = 1$ in.

Solution

- *System sketch:*   The control volume is the *outside* of the nozzle, plus the cut sections (1) and (2). There would also be *stresses* in the cut nozzle wall at section 1, which we are neglecting here. The pressures acting on the control volume are shown in Fig. E3.6a. Figure E3.6b shows the pressures after 15 lbf/in² has been subtracted from all sides. Here we compute the net pressure force only.

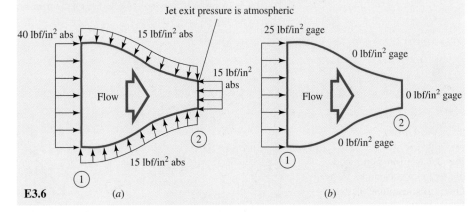

E3.6   (a)   (b)

---

[8]Can you prove this? It is a consequence of Gauss's theorem from vector analysis.

- *Assumptions:* Known pressures, as shown, on all surfaces of the control volume.
- *Approach:* Since three surfaces have $p = 15$ lbf/in^2, subtract this amount everywhere so that these three sides reduce to zero "gage pressure" for convenience. This is allowable because of Eq. (3.42).
- *Solution steps:* For the modified pressure distribution, Fig. E3.6b, only section 1 is needed:

$$\mathbf{F}_{press} = p_{gage,1}\,(-\mathbf{n})_1\,A_1 = \left(25\frac{\text{lbf}}{\text{in}^2}\right)\left[-(-\mathbf{i})\right]\left[\frac{\pi}{4}(3\text{ in})^2\right] = 177\mathbf{i}\text{ lbf} \qquad Ans.$$

- *Comments:* This "uniform subtraction" artifice, which is entirely legal, has greatly simplified the calculation of pressure force. *Note:* We were a bit too informal when multiplying pressure in lbf/in^2 times area in square inches. We achieved lbf correctly, but it would be better practice to convert all data to standard BG units. *Further note:* In addition to $\mathbf{F}_{press}$, there are other forces involved in this flow, due to tension stresses in the cut nozzle wall and the fluid weight inside the control volume.

Pressure Condition at a Jet Exit

Figure E3.6 illustrates a pressure boundary condition commonly used for jet exit flow problems. When a fluid flow leaves a confined internal duct and exits into an ambient "atmosphere," its free surface is exposed to that atmosphere. Therefore the jet itself will essentially be at atmospheric pressure also. This condition was used at section 2 in Fig. E3.6.

Only two effects could maintain a pressure difference between the atmosphere and a free exit jet. The first is surface tension, Eq. (1.31), which is usually negligible. The second effect is a *supersonic* jet, which can separate itself from an atmosphere with expansion or compression waves (Chap. 9). For the majority of applications, therefore, we shall set the pressure in an exit jet as atmospheric.

EXAMPLE 3.7

A fixed control volume of a streamtube in steady flow has a uniform inlet flow (ρ_1, A_1, V_1) and a uniform exit flow (ρ_2, A_2, V_2), as shown in Fig. 3.7. Find an expression for the net force on the control volume.

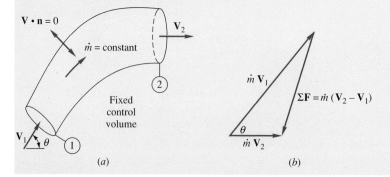

Fig. 3.7 Net force on a one-dimensional streamtube in steady flow: (*a*) streamtube in steady flow; (*b*) vector diagram for computing net force.

Solution

Equation (3.40) applies with one inlet and exit:

$$\sum \mathbf{F} = \dot{m}_2 \mathbf{V}_2 - \dot{m}_1 \mathbf{V}_1 = (\rho_2 A_2 V_2)\mathbf{V}_2 - (\rho_1 A_1 V_1)\mathbf{V}_1$$

The volume integral term vanishes for steady flow, but from conservation of mass in Example 3.3 we saw that

$$\dot{m}_1 = \dot{m}_2 = \dot{m} = \text{const}$$

Therefore a simple form for the desired result is

$$\sum \mathbf{F} = \dot{m}(\mathbf{V}_2 - \mathbf{V}_1) \qquad\qquad Ans.$$

This is a *vector* relation and is sketched in Fig. 3.7*b*. The term $\Sigma\,\mathbf{F}$ represents the net force acting on the control volume due to all causes; it is needed to balance the change in momentum of the fluid as it turns and decelerates while passing through the control volume.

EXAMPLE 3.8

As shown in Fig. 3.8*a*, a fixed vane turns a water jet of area A through an angle θ without changing its velocity magnitude. The flow is steady, pressure is p_a everywhere, and friction on the vane is negligible. (*a*) Find the components F_x and F_y of the applied vane force. (*b*) Find expressions for the force magnitude F and the angle ϕ between F and the horizontal; plot them versus θ.

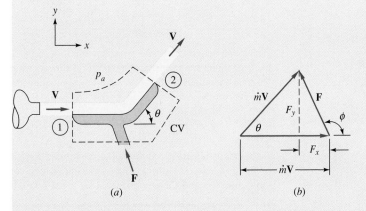

Fig. 3.8 Net applied force on a fixed jet-turning vane: (*a*) geometry of the vane turning the water jet; (*b*) vector diagram for the net force.

(*a*)   (*b*)

Solution

Part (a)

The control volume selected in Fig. 3.8*a* cuts through the inlet and exit of the jet and through the vane support, exposing the vane force **F**. Since there is no cut along the vane–jet interface, vane friction is internally self-canceling. The pressure force is zero in the uniform atmosphere. We neglect the weight of fluid and the vane weight within the control volume. Then Eq. (3.40) reduces to

$$\mathbf{F}_{\text{vane}} = \dot{m}_2 \mathbf{V}_2 - \dot{m}_1 \mathbf{V}_1$$

But the magnitude $V_1 = V_2 = V$ as given, and conservation of mass for the streamtube requires $\dot{m}_1 = \dot{m}_2 = \dot{m} = \rho A V$. The vector diagram for force and momentum change becomes an isosceles triangle with legs $\dot{m}\mathbf{V}$ and base $\mathbf{F}$, as in Fig. 3.8b. We can readily find the force components from this diagram:

$$F_x = \dot{m}V(\cos\theta - 1) \qquad F_y = \dot{m}V\sin\theta \qquad \qquad \textit{Ans. (a)}$$

where $\dot{m}V = \rho A V^2$ for this case. This is the desired result.

Part (b)   The force magnitude is obtained from part (a):

$$F = (F_x^2 + F_y^2)^{1/2} = \dot{m}V[\sin^2\theta + (\cos\theta - 1)^2]^{1/2} = 2\dot{m}V\sin\frac{\theta}{2} \qquad \textit{Ans. (b)}$$

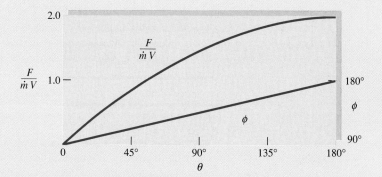

E3.8

From the geometry of Fig. 3.8b we obtain

$$\phi = 180° - \tan^{-1}\frac{F_y}{F_x} = 90° + \frac{\theta}{2} \qquad \qquad \textit{Ans. (b)}$$

These can be plotted versus θ as shown in Fig. E3.8. Two special cases are of interest. First, the maximum force occurs at $\theta = 180°$—that is, when the jet is turned around and thrown back in the opposite direction with its momentum completely reversed. This force is $2\dot{m}V$ and acts to the *left*; that is, $\phi = 180°$. Second, at very small turning angles ($\theta < 10°$) we obtain approximately

$$F \approx \dot{m}V\theta \qquad \phi \approx 90°$$

The force is linearly proportional to the turning angle and acts nearly normal to the jet. This is the principle of a lifting vane, or airfoil, which causes a slight change in the oncoming flow direction and thereby creates a lift force normal to the basic flow.

EXAMPLE 3.9

A water jet of velocity V_j impinges normal to a flat plate that moves to the right at velocity V_c, as shown in Fig. 3.9a. Find the force required to keep the plate moving at constant velocity if the jet density is 1000 kg/m^3, the jet area is 3 cm^2, and V_j and V_c are 20 and 15 m/s, respectively. Neglect the weight of the jet and plate, and assume steady flow with respect to the moving plate with the jet splitting into an equal upward and downward half-jet.

Solution

The suggested control volume in Fig. 3.9a cuts through the plate support to expose the desired forces R_x and R_y. This control volume moves at speed V_c and thus is fixed relative to the plate, as in Fig. 3.9b. We must satisfy both mass and momentum conservation for the assumed steady flow pattern in Fig. 3.9b. There are two outlets and one inlet, and Eq. (3.30) applies for mass conservation:

$$\dot{m}_{out} = \dot{m}_{in}$$

or

$$\rho_1 A_1 V_1 + \rho_2 A_2 V_2 = \rho_j A_j (V_j - V_c) \tag{1}$$

We assume that the water is incompressible $\rho_1 = \rho_2 = \rho_j$, and we are given that $A_1 = A_2 = \frac{1}{2} A_j$. Therefore Eq. (1) reduces to

$$V_1 + V_2 = 2(V_j - V_c) \tag{2}$$

Strictly speaking, this is all that mass conservation tells us. However, from the symmetry of the jet deflection and the neglect of gravity on the fluid trajectory, we conclude that the two velocities V_1 and V_2 must be equal, and hence Eq. (2) becomes

$$V_1 = V_2 = V_j - V_c \tag{3}$$

This equality can also be predicted by Bernoulli's equation in Sect 3.7. For the given numerical values, we have

$$V_1 = V_2 = 20 - 15 = 5 \text{ m/s}$$

Now we can compute R_x and R_y from the two components of momentum conservation. Equation (3.40) applies with the unsteady term zero:

$$\sum F_x = R_x = \dot{m}_1 u_1 + \dot{m}_2 u_2 - \dot{m}_j u_j \tag{4}$$

where from the mass analysis, $\dot{m}_1 = \dot{m}_2 = \frac{1}{2}\dot{m}_j = \frac{1}{2}\rho_j A_j (V_j - V_c)$. Now check the flow directions at each section: $u_1 = u_2 = 0$, and $u_j = V_j - V_c = 5$ m/s. Thus Eq. (4) becomes

$$R_x = -\dot{m}_j u_j = -[\rho_j A_j (V_j - V_c)](V_j - V_c) \tag{5}$$

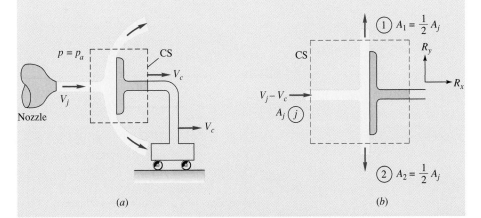

Fig. 3.9 Force on a plate moving at constant velocity: (a) jet striking a moving plate normally; (b) control volume fixed relative to the plate.

For the given numerical values we have

$$R_x = -(1000 \text{ kg/m}^3)(0.0003 \text{ m}^2)(5 \text{ m/s})^2 = -7.5 \text{ (kg} \cdot \text{m)/s}^2 = -7.5 \text{ N} \qquad Ans.$$

This acts to the *left*; that is, it requires a restraining force to keep the plate from accelerating to the right due to the continuous impact of the jet. The vertical force is

$$F_y = R_y = \dot{m}_1 v_1 + \dot{m}_2 v_2 - \dot{m}_j v_j$$

Check directions again: $v_1 = V_1$, $v_2 = -V_2$, $v_j = 0$. Thus

$$R_y = \dot{m}_1(V_1) + \dot{m}_2(-V_2) = \tfrac{1}{2}\dot{m}_j(V_1 - V_2) \qquad (6)$$

But since we found earlier that $V_1 = V_2$, this means that $R_y = 0$, as we could expect from the symmetry of the jet deflection.[9] Two other results are of interest. First, the relative velocity at section 1 was found to be 5 m/s up, from Eq. (3). If we convert this to absolute motion by adding on the control-volume speed $V_c = 15$ m/s to the right, we find that the absolute velocity $\mathbf{V}_1 = 15\mathbf{i} + 5\mathbf{j}$ m/s, or 15.8 m/s at an angle of 18.4° upward, as indicated in Fig. 3.9a. Thus the absolute jet speed changes after hitting the plate. Second, the computed force R_x does not change if we assume the jet deflects in all radial directions along the plate surface rather than just up and down. Since the plate is normal to the x axis, there would still be zero outlet x-momentum flux when Eq. (4) was rewritten for a radial deflection condition.

EXAMPLE 3.10

The sluice gate in Fig. E3.10a controls flow in open channels. At sections 1 and 2, the flow is uniform and the pressure is hydrostatic. Neglecting bottom friction and atmospheric pressure, derive a formula for the horizontal force F required to hold the gate. Express your final formula in terms of the inlet velocity V_1, eliminating V_2.

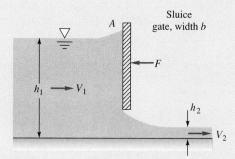

E3.10a

Solution

Choose a control volume, Fig. E3.10b, that cuts through known regions (section 1 and section 2, the bottom and the atmosphere) and that cuts along regions where unknown information is desired (the gate, with its force F).

---

[9]Symmetry can be a powerful tool if used properly. Try to learn more about the uses and misuses of symmetry conditions.

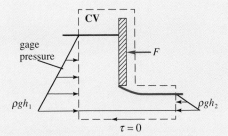

E3.10b

Assume steady incompressible flow with no variation across the width b. The inlet and outlet mass flows balance:

$$\dot{m} = \rho V_1 h_1 b = \rho V_2 h_2 b \quad \text{or} \quad V_2 = V_1(h_1/h_2)$$

We may use gage pressures for convenience because a uniform atmospheric pressure causes no force, as shown earlier in Fig. 3.6. With x positive to the right, equate the net horizontal force to the x-directed momentum change:

$$\Sigma F_x = -F_{\text{gate}} + \frac{\rho}{2}gh_1(h_1 b) - \frac{\rho}{2}gh_2(h_2 b) = \dot{m}(V_2 - V_1)$$

$$\dot{m} = \rho h_1 b V_1$$

Solve for F_{gate}, and eliminate V_2 using the mass flow relation. The desired result is:

$$F_{\text{gate}} = \frac{\rho}{2}gbh_1^2\left[1 - \left(\frac{h_2}{h_1}\right)^2\right] - \rho h_1 b V_1^2\left(\frac{h_1}{h_2} - 1\right) \qquad Ans.$$

This is a powerful result from a relatively simple analysis. Later, in Sec. 10.4, we will be able to calculate the actual flow rate from the water depths and the gate opening height.

EXAMPLE 3.11

Example 3.9 treated a plate at normal incidence to an oncoming flow. In Fig. 3.10 the plate is parallel to the flow. The stream is not a jet but a broad river, or *free stream,* of uniform velocity $\mathbf{V} = U_0\mathbf{i}$. The pressure is assumed uniform, and so it has no net force on the plate. The plate does not block the flow as in Fig. 3.9, so the only effect is due to boundary shear, which was neglected in the previous example. The no-slip condition at the wall brings the fluid there to a halt, and these slowly moving particles retard their neighbors above, so that at the end of the plate there is a significant retarded shear layer, or *boundary layer,* of thickness $y = \delta$. The viscous stresses along the wall can sum to a finite drag force on the plate. These effects are illustrated in Fig. 3.10. The problem is to make an integral analysis and find the drag force D in terms of the flow properties ρ, U_0, and δ and the plate dimensions L and b.[10]

Solution

Like most practical cases, this problem requires a combined mass and momentum balance. A proper selection of control volume is essential, and we select the four-sided region from

---

[10]The general analysis of such wall shear problems, called *boundary-layer theory,* is treated in Sec. 7.3.

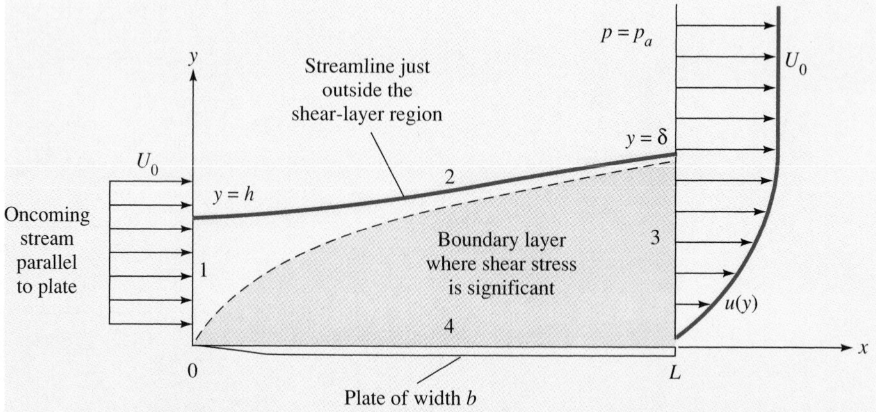

Fig. 3.10 Control volume analysis of drag force on a flat plate due to boundary shear. The control volume is bounded by sections 1, 2, 3, and 4.

0 to h to δ to L and back to the origin 0, as shown in Fig. 3.10. Had we chosen to cut across horizontally from left to right along the height $y = h$, we would have cut through the shear layer and exposed unknown shear stresses. Instead we follow the streamline passing through $(x, y) = (0, h)$, which is outside the shear layer and also has no mass flow across it. The four control volume sides are thus

1. From $(0, 0)$ to $(0, h)$: a one-dimensional inlet, $\mathbf{V} \cdot \mathbf{n} = -U_0$.
2. From $(0, h)$ to (L, δ): a streamline, no shear, $\mathbf{V} \cdot \mathbf{n} \equiv 0$.
3. From (L, δ) to $(L, 0)$: a two-dimensional outlet, $\mathbf{V} \cdot \mathbf{n} = +u(y)$.
4. From $(L, 0)$ to $(0, 0)$: a streamline just above the plate surface, $\mathbf{V} \cdot \mathbf{n} = 0$, shear forces summing to the drag force $-D\mathbf{i}$ acting from the plate onto the retarded fluid.

The pressure is uniform, and so there is no net pressure force. Since the flow is assumed incompressible and steady, Eq. (3.37) applies with no unsteady term and fluxes only across sections 1 and 3:

$$\sum F_x = -D = \rho \int_1 u(0, y)(\mathbf{V} \cdot \mathbf{n})\, dA + \rho \int_3 u(L, y)\, (\mathbf{V} \cdot \mathbf{n})\, dA$$

$$= \rho \int_0^h U_0(-U_0)b\, dy + \rho \int_0^\delta u(L, y)[+u(L, y)]b\, dy$$

Evaluating the first integral and rearranging give

$$D = \rho U_0^2 bh - \rho b \int_0^\delta u^2 dy \,|_{x=L} \tag{1}$$

This could be considered the answer to the problem, but it is not useful because the height h is not known with respect to the shear layer thickness δ. This is found by applying mass conservation, since the control volume forms a streamtube:

$$\rho \int_{CS} (\mathbf{V} \cdot \mathbf{n})\, dA = 0 = \rho \int_0^h (-U_0)b\, dy + \rho \int_0^\delta ub\, dy \,|_{x=L}$$

or

$$U_0 h = \int_0^\delta u\, dy \,|_{x=L} \tag{2}$$

after canceling b and ρ and evaluating the first integral. Introduce this value of h into Eq. (1) for a much cleaner result:

$$D = \rho b \int_0^\delta u(U_0 - u)\, dy \,|_{x=L} \qquad\qquad Ans. \text{ (3)}$$

This result was first derived by Theodore von Kármán in 1921.[11] It relates the friction drag on one side of a flat plate to the integral of the *momentum deficit* $\rho u(U_0 - u)$ across the trailing cross section of the flow past the plate. Since $U_0 - u$ vanishes as y increases, the integral has a finite value. Equation (3) is an example of *momentum integral theory* for boundary layers, which is treated in Chap. 7.

Momentum Flux Correction Factor

For flow in a duct, the axial velocity is usually nonuniform, as in Example 3.4. For this case the simple momentum flux calculation $\int u\rho(\mathbf{V} \cdot \mathbf{n})\, dA = \dot{m}V = \rho A V^2$ is somewhat in error and should be corrected to $\beta \rho A V^2$, where β is the dimensionless momentum flux correction factor, $\beta \geq 1$.

The factor β accounts for the variation of u^2 across the duct section. That is, we compute the exact flux and set it equal to a flux based on average velocity in the duct:

$$\rho \int u^2 dA = \beta \dot{m} V_{av} = \beta \rho A V_{av}^2$$

or

$$\beta = \frac{1}{A} \int \left(\frac{u}{V_{av}}\right)^2 dA \qquad\qquad (3.43a)$$

Values of β can be computed based on typical duct velocity profiles similar to those in Example 3.4. The results are as follows:

Laminar flow:
$$u = U_0\left(1 - \frac{r^2}{R^2}\right) \qquad \beta = \frac{4}{3} \qquad\qquad (3.43b)$$

Turbulent flow:
$$u \approx U_0\left(1 - \frac{r}{R}\right)^m \qquad \frac{1}{9} \leq m \leq \frac{1}{5}$$

$$\beta = \frac{(1+m)^2(2+m)^2}{2(1+2m)(2+2m)} \qquad\qquad (3.43c)$$

The turbulent correction factors have the following range of values:

Turbulent flow:	m	$\frac{1}{5}$	$\frac{1}{6}$	$\frac{1}{7}$	$\frac{1}{8}$	$\frac{1}{9}$
	β	1.037	1.027	1.020	1.016	1.013

These are so close to unity that they are normally neglected. The laminar correction may be important.

---

[11]The autobiography of this great twentieth-century engineer and teacher [2] is recommended for its historical and scientific insight.

To illustrate a typical use of these correction factors, the solution to Example 3.8 for nonuniform velocities at sections 1 and 2 would be modified as

$$\sum \mathbf{F} = \dot{m}(\beta_2 \mathbf{V}_2 - \beta_1 \mathbf{V}_1) \tag{3.43d}$$

Note that the basic parameters and vector character of the result are not changed at all by this correction.

Linear Momentum Tips

The previous examples make it clear that the vector momentum equation is more difficult to handle than the scalar mass and energy equations. Here are some momentum tips to remember:

- The momentum relation is a *vector* equation. The forces and the momentum terms are directional and can have three components. A *sketch* of these vectors will be indispensable for the analysis.
- The momentum flux terms, such as $\int \mathbf{V}(\rho \mathbf{V} \cdot \mathbf{n})dA$, link *two* different sign conventions, so special care is needed. First, the vector coefficient $\mathbf{V}$ will have a sign depending on its direction. Second, the mass flow term $(\rho \mathbf{V} \cdot \mathbf{n})$ will have a sign $(+, -)$ depending on whether it is (out, in). For example, in Fig. 3.8, the x-components of $\mathbf{V}_2$ and $\mathbf{V}_1$, u_2 and u_1, are both positive; that is, they both act to the right. Meanwhile, the mass flow at (2) is positive (out) and at (1) is negative (in).
- The *one-dimensional approximation,* Eq. (3.40), is glorious, because non-uniform velocity distributions require laborious integration, as in Eq. 3.11. Thus the momentum flux correction factors β are very useful in avoiding this integration, especially for pipe flow.
- The applied forces $\Sigma\mathbf{F}$ act on *all the material in the control volume*—that is, the surfaces (pressure and shear stresses), the solid supports that are cut through, and the weight of the interior masses. Stresses on non-control-surface parts of the interior are self-canceling and should be ignored.
- If the fluid exits subsonically to an atmosphere, the fluid pressure there is *atmospheric*.
- Where possible, choose inlet and outlet surfaces *normal to the flow,* so that pressure is the dominant force and the normal velocity equals the actual velocity.

Clearly, with that many helpful tips, substantial practice is needed to achieve momentum skills.

Noninertial Reference Frame[12]

All previous derivations and examples in this section have assumed that the coordinate system is inertial—that is, at rest or moving at constant velocity. In this case the rate of change of velocity equals the absolute acceleration of the system, and Newton's law applies directly in the form of Eqs. (3.2) and (3.35).

In many cases it is convenient to use a *noninertial,* or accelerating, coordinate system. An example would be coordinates fixed to a rocket during takeoff. A second example is any flow on the earth's surface, which is accelerating relative to the fixed

---

[12]This section may be omitted without loss of continuity.

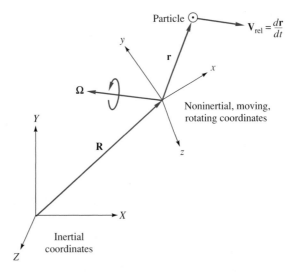

Fig. 3.11 Geometry of fixed versus accelerating coordinates.

stars because of the rotation of the earth. Atmospheric and oceanographic flows experience the so-called *Coriolis acceleration,* outlined next. It is typically less than $10^{-5}g$, where g is the acceleration of gravity, but its accumulated effect over distances of many kilometers can be dominant in geophysical flows. By contrast, the Coriolis acceleration is negligible in small-scale problems like pipe or airfoil flows.

Suppose that the fluid flow has velocity $\mathbf{V}$ relative to a noninertial *xyz* coordinate system, as shown in Fig. 3.11. Then $d\mathbf{V}/dt$ will represent a noninertial acceleration that must be added vectorially to a relative acceleration $\mathbf{a}_{rel}$ to give the absolute acceleration $\mathbf{a}_i$ relative to some inertial coordinate system *XYZ*, as in Fig. 3.11. Thus

$$\mathbf{a}_i = \frac{d\mathbf{V}}{dt} + \mathbf{a}_{rel} \tag{3.44}$$

Since Newton's law applies to the absolute acceleration,

$$\sum \mathbf{F} = m\mathbf{a}_i = m\left(\frac{d\mathbf{V}}{dt} + \mathbf{a}_{rel}\right)$$

or

$$\sum \mathbf{F} - m\mathbf{a}_{rel} = m\frac{d\mathbf{V}}{dt} \tag{3.45}$$

Thus Newton's law in noninertial coordinates *xyz* is analogous to adding more "force" terms $-m\mathbf{a}_{rel}$ to account for noninertial effects. In the most general case, sketched in Fig. 3.11, the term $\mathbf{a}_{rel}$ contains four parts, three of which account for the angular velocity $\boldsymbol{\Omega}(t)$ of the inertial coordinates. By inspection of Fig. 3.11, the absolute displacement of a particle is

$$\mathbf{S}_i = \mathbf{r} + \mathbf{R} \tag{3.46}$$

Differentiation gives the absolute velocity

$$\mathbf{V}_i = \mathbf{V} + \frac{d\mathbf{R}}{dt} + \boldsymbol{\Omega} \times \mathbf{r} \tag{3.47}$$

A second differentiation gives the absolute acceleration:

$$\mathbf{a}_i = \frac{d\mathbf{V}}{dt} + \frac{d^2\mathbf{R}}{dt^2} + \frac{d\mathbf{\Omega}}{dt} \times \mathbf{r} + 2\mathbf{\Omega} \times \mathbf{V} + \mathbf{\Omega} \times (\mathbf{\Omega} \times \mathbf{r}) \tag{3.48}$$

By comparison with Eq. (3.44), we see that the last four terms on the right represent the additional relative acceleration:

1. $d^2\mathbf{R}/dt^2$ is the acceleration of the noninertial origin of coordinates xyz.
2. $(d\mathbf{\Omega}/dt) \times \mathbf{r}$ is the angular acceleration effect.
3. $2\mathbf{\Omega} \times \mathbf{V}$ is the Coriolis acceleration.
4. $\mathbf{\Omega} \times (\mathbf{\Omega} \times \mathbf{r})$ is the centripetal acceleration, directed from the particle normal to the axis of rotation with magnitude $\Omega^2 L$, where L is the normal distance to the axis.[13]

Equation (3.45) differs from Eq. (3.2) only in the added inertial forces on the left-hand side. Thus the control volume formulation of linear momentum in noninertial coordinates merely adds inertial terms by integrating the added relative acceleration over each differential mass in the control volume:

$$\sum \mathbf{F} - \int_{CV} \mathbf{a}_{\text{rel}}\, dm = \frac{d}{dt}\left(\int_{CV} \mathbf{V}\rho\, d\mathcal{V} \right) + \int_{CS} \mathbf{V}\rho(\mathbf{V}_r \cdot \mathbf{n})\, dA \tag{3.49}$$

where

$$\mathbf{a}_{\text{rel}} = \frac{d^2\mathbf{R}}{dt^2} + \frac{d\mathbf{\Omega}}{dt} \times \mathbf{r} + 2\mathbf{\Omega} \times \mathbf{V} + \mathbf{\Omega} \times (\mathbf{\Omega} \times \mathbf{r})$$

This is the noninertial analog of the inertial form given in Eq. (3.35). To analyze such problems, one must know the displacement $\mathbf{R}$ and angular velocity $\mathbf{\Omega}$ of the noninertial coordinates.

If the control volume is fixed in a moving frame, Eq. (3.49) reduces to

$$\sum \mathbf{F} - \int_{CV} \mathbf{a}_{\text{rel}}\, dm = \frac{d}{dt}\left(\int_{CV} \mathbf{V}\rho\, d\mathcal{V} \right) + \int_{CS} \mathbf{V}\rho(\mathbf{V} \cdot \mathbf{n})\, dA \tag{3.50}$$

In other words, the right-hand side reduces to that of Eq. (3.37).

EXAMPLE 3.12

A classic example of an accelerating control volume is a rocket moving straight up, as in Fig. E3.12. Let the initial mass be M_0, and assume a steady exhaust mass flow $\dot{m}$ and exhaust velocity V_e relative to the rocket, as shown. If the flow pattern within the rocket motor is steady and air drag is neglected, derive the differential equation of vertical rocket motion $V(t)$ and integrate using the initial condition $V = 0$ at $t = 0$.

---

[13]A complete discussion of these noninertial coordinate terms is given, for example, in Ref. 4, pp. 49–51.

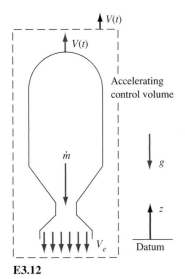

E3.12

Solution

The appropriate control volume in Fig. E3.12 encloses the rocket, cuts through the exit jet, and accelerates upward at rocket speed $V(t)$. The z-momentum equation (3.49) becomes

$$\sum F_z - \int a_{\text{rel}}\, dm = \frac{d}{dt}\left(\int_{\text{CV}} w\, dm\right) + (\dot{m}w)_e$$

or $-mg - m\dfrac{dV}{dt} = 0 + \dot{m}(-V_e)$ with $m = m(t) = M_0 - \dot{m}t$

The term $a_{\text{rel}} = dV/dt$ of the rocket. The control volume integral vanishes because of the steady rocket flow conditions. Separate the variables and integrate, assuming $V = 0$ at $t = 0$:

$$\int_0^V dV = \dot{m}\, V_e \int_0^t \frac{dt}{M_0 - \dot{m}t} - g \int_0^t dt \quad \text{or} \quad V(t) = -V_e \ln\left(1 - \frac{\dot{m}t}{M_0}\right) - gt \qquad Ans.$$

This is a classic approximate formula in rocket dynamics. The first term is positive and, if the fuel mass burned is a large fraction of initial mass, the final rocket velocity can exceed V_e.

3.5 The Angular Momentum Theorem[14]

A control volume analysis can be applied to the angular momentum relation, Eq. (3.3), by letting our dummy variable **B** be the angular-momentum vector **H**. However, since the system considered here is typically a group of nonrigid fluid particles of variable velocity, the concept of mass moment of inertia is of no help, and we have to calculate the instantaneous angular momentum by integration over the elemental masses dm. If O is the point about which moments are desired, the angular momentum about O is given by

$$\mathbf{H}_o = \int_{\text{syst}} (\mathbf{r} \times \mathbf{V})\, dm \qquad (3.51)$$

where **r** is the position vector from 0 to the elemental mass dm and **V** is the velocity of that element. The amount of angular momentum per unit mass is thus seen to be

$$\beta = \frac{d\mathbf{H}_o}{dm} = \mathbf{r} \times \mathbf{V}$$

The Reynolds transport theorem (3.16) then tells us that

$$\left.\frac{d\mathbf{H}_o}{dt}\right|_{\text{syst}} = \frac{d}{dt}\left[\int_{\text{CV}} (\mathbf{r} \times \mathbf{V})\rho\, d\mathcal{V}\right] + \int_{\text{CS}} (\mathbf{r} \times \mathbf{V})\rho(\mathbf{V}_r \cdot \mathbf{n})\, dA \qquad (3.52)$$

for the most general case of a deformable control volume. But from the angular momentum theorem (3.3), this must equal the sum of all the moments about point O applied to the control volume

$$\frac{d\mathbf{H}_o}{dt} = \sum \mathbf{M}_o = \sum (\mathbf{r} \times \mathbf{F})_o$$

Note that the total moment equals the summation of moments of all applied forces about point O. Recall, however, that this law, like Newton's law (3.2), assumes that the particle velocity **V** is relative to an *inertial* coordinate system. If not, the moments about

---

[14]This section may be omitted without loss of continuity.

point O of the relative acceleration terms $\mathbf{a}_{rel}$ in Eq. (3.49) must also be included:

$$\sum \mathbf{M}_o = \sum (\mathbf{r} \times \mathbf{F})_o - \int_{CV} (\mathbf{r} \times \mathbf{a}_{rel})\, dm \qquad (3.53)$$

where the four terms constituting $\mathbf{a}_{rel}$ are given in Eq. (3.49). Thus the most general case of the angular momentum theorem is for a deformable control volume associated with a noninertial coordinate system. We combine Eqs. (3.52) and (3.53) to obtain

$$\sum (\mathbf{r} \times \mathbf{F})_o - \int_{CV} (\mathbf{r} \times \mathbf{a}_{rel})\, dm = \frac{d}{dt}\left[\int_{CV} (\mathbf{r} \times \mathbf{V})\rho\, d\mathcal{V} \right] + \int_{CS} (\mathbf{r} \times \mathbf{V})\rho(\mathbf{V}_r \cdot \mathbf{n})\, dA \qquad (3.54)$$

For a nondeformable inertial control volume, this reduces to

$$\boxed{\sum \mathbf{M}_0 = \frac{\partial}{\partial t}\left[\int_{CV} (\mathbf{r} \times \mathbf{V})\rho\, d\mathcal{V} \right] + \int_{CS} (\mathbf{r} \times \mathbf{V})\rho(\mathbf{V} \cdot \mathbf{n})\, dA} \qquad (3.55)$$

Further, if there are only one-dimensional inlets and exits, the angular momentum flux terms evaluated on the control surface become

$$\int_{CS} (\mathbf{r} \times \mathbf{V})\rho(\mathbf{V} \cdot \mathbf{n})\, dA = \sum (\mathbf{r} \times \mathbf{V})_{out}\, \dot{m}_{out} - \sum (\mathbf{r} \times \mathbf{V})_{in}\, \dot{m}_{in} \qquad (3.56)$$

Although at this stage the angular momentum theorem can be considered a supplementary topic, it has direct application to many important fluid flow problems involving torques or moments. A particularly important case is the analysis of rotating fluid flow devices, usually called *turbomachines* (Chap. 11).

EXAMPLE 3.13

As shown in Fig. E3.13a, a pipe bend is supported at point A and connected to a flow system by flexible couplings at sections 1 and 2. The fluid is incompressible, and ambient pressure p_a is zero. (a) Find an expression for the torque T that must be resisted by the support at A, in terms of the flow properties at sections 1 and 2 and the distances h_1 and h_2. (b) Compute this torque if $D_1 = D_2 = 3$ in, $p_1 = 100$ lbf/in^2 gage, $p_2 = 80$ lbf/in^2 gage, $V_1 = 40$ ft/s, $h_1 = 2$ in, $h_2 = 10$ in, and $\rho = 1.94$ slugs/ft^3.

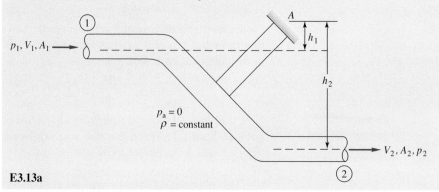

E3.13a

Solution

Part (a)

The control volume chosen in Fig. E3.13b cuts through sections 1 and 2 and through the support at A, where the torque T_A is desired. The flexible couplings description specifies that there is no torque at either section 1 or 2, and so the cuts there expose no moments. For the angular momentum terms $\mathbf{r} \times \mathbf{V}$, $\mathbf{r}$ should be taken from point A to sections 1 and 2. Note that the gage pressure forces p_1A_1 and p_2A_2 both have moments about A. Equation (3.55) with one-dimensional flux terms becomes

$$\sum \mathbf{M}_A = \mathbf{T}_A + \mathbf{r}_1 \times (-p_1A_1\mathbf{n}_1) + \mathbf{r}_2 \times (-p_2A_2\mathbf{n}_2)$$
$$= (\mathbf{r}_2 \times \mathbf{V}_2)(+\dot{m}_{out}) + (\mathbf{r}_1 \times \mathbf{V}_1)(-\dot{m}_{in}) \tag{1}$$

Figure E3.13c shows that all the cross products are associated with either $r_1 \sin \theta_1 = h_1$ or $r_2 \sin \theta_2 = h_2$, the perpendicular distances from point A to the pipe axes at 1 and 2. Remember that $\dot{m}_{in} = \dot{m}_{out}$ from the steady flow continuity relation. In terms of counterclockwise moments, Eq. (1) then becomes

$$T_A + p_1A_1h_1 - p_2A_2h_2 = \dot{m}(h_2V_2 - h_1V_1) \tag{2}$$

Rewriting this, we find the desired torque to be

$$T_A = h_2(p_2A_2 + \dot{m}V_2) - h_1(p_1A_1 + \dot{m}V_1) \qquad Ans.\ (a)\ (3)$$

counterclockwise. The quantities p_1 and p_2 are gage pressures. Note that this result is independent of the shape of the pipe bend and varies only with the properties at sections 1 and 2 and the distances h_1 and h_2.[15]

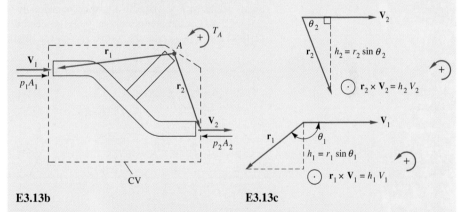

E3.13b   **E3.13c**

Part (b)

For the numerical example, convert all data to BG units:

$$D_1 = D_2 = 3 \text{ in} = 0.25 \text{ ft} \quad p_1 = 100 \frac{\text{lbf}}{\text{in}^2} = 14{,}400 \frac{\text{lbf}}{\text{ft}^2} \quad p_2 = 80 \frac{\text{lbf}}{\text{in}^2} = 11{,}520 \frac{\text{lbf}}{\text{ft}^2}$$

$$h_1 = 2 \text{ in} = \frac{2}{12} \text{ ft} \quad h_2 = 10 \text{ in} = \frac{10}{12} \text{ ft} \quad \rho = 1.94 \frac{\text{slug}}{\text{ft}^3}$$

The inlet and exit areas are the same, $A_1 = A_2 = (\pi/4)(0.25 \text{ ft})^2 = 0.0491 \text{ ft}^2$. Since the density is constant, we conclude from mass conservation, $\rho A_1 V_1 = \rho A_2 V_2$, that $V_1 = V_2 = 40$ ft/s.

---

[15]Indirectly, the pipe bend shape probably affects the pressure change from p_1 to p_2.

The mass flow is

$$\dot{m} = \rho A_1 V_1 = \left(1.94 \, \frac{\text{slug}}{\text{ft}^3}\right)(0.0491 \, \text{ft}^2)\left(40 \, \frac{\text{ft}}{\text{s}}\right) = 3.81 \, \frac{\text{slug}}{\text{s}}$$

- *Evaluation of the torque:* The data can now be substituted into Eq. (3):

$$T_A = \left(\frac{10}{12} \, \text{ft}\right)\left[\left(11{,}520 \, \frac{\text{lbf}}{\text{ft}^2}\right)(0.0491 \, \text{ft}^2) + \left(3.81 \, \frac{\text{slug}}{\text{s}}\right)\left(40 \, \frac{\text{ft}}{\text{s}}\right)\right]$$

$$- \left(\frac{2}{12} \, \text{ft}\right)\left[\left(14{,}400 \, \frac{\text{lbf}}{\text{ft}^2}\right)(0.0491 \, \text{ft}^2) + \left(3.81 \, \frac{\text{slug}}{\text{s}}\right)\left(40 \, \frac{\text{ft}}{\text{s}}\right)\right]$$

$$= 598 \, \text{ft} \cdot \text{lbf} - 143 \, \text{ft} \cdot \text{lbf} = 455 \, \text{ft} \cdot \text{lbf counterclockwise} \qquad \textit{Ans. (b)}$$

- *Comments:* The use of standard BG units is crucial when combining dissimilar terms, such as pressure times area and mass flow times velocity, into proper additive units for a numerical solution.

EXAMPLE 3.14

Figure 3.12 shows a schematic of a centrifugal pump. The fluid enters axially and passes through the pump blades, which rotate at angular velocity ω; the velocity of the fluid is changed from V_1 to V_2 and its pressure from p_1 to p_2. (a) Find an expression for the torque T_o that must be applied to these blades to maintain this flow. (b) The power supplied to the pump would be $P = \omega T_o$. To illustrate numerically, suppose $r_1 = 0.2$ m, $r_2 = 0.5$ m, and $b = 0.15$ m. Let the pump rotate at 600 r/min and deliver water at 2.5 m³/s with a density of 1000 kg/m³. Compute the torque and power supplied.

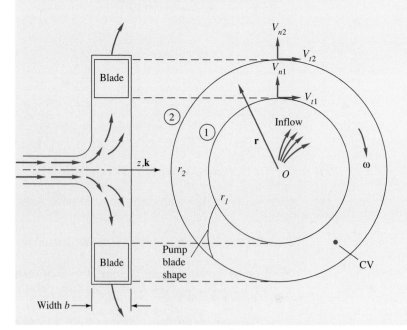

Fig. 3.12 Schematic of a simplified centrifugal pump.

Solution

Part (a)

The control volume is chosen to be the annular region between sections 1 and 2 where the flow passes through the pump blades (see Fig. 3.12). The flow is steady and assumed incompressible. The contribution of pressure to the torque about axis O is zero since the pressure forces at 1 and 2 act radially through O. Equation (3.55) becomes

$$\sum \mathbf{M}_o = \mathbf{T}_o = (\mathbf{r}_2 \times \mathbf{V}_2)\dot{m}_{out} - (\mathbf{r}_1 \times \mathbf{V}_1)\dot{m}_{in} \tag{1}$$

where steady flow continuity tells us that

$$\dot{m}_{in} = \rho V_{n1} 2\pi r_1 b = \dot{m}_{out} = \rho V_{n2} 2\pi r_2 b = \rho Q$$

The cross product $\mathbf{r} \times \mathbf{V}$ is found to be clockwise about O at both sections:

$$\mathbf{r}_2 \times \mathbf{V}_2 = r_2 V_{t2} \sin 90° \, \mathbf{k} = r_2 V_{t2}\mathbf{k} \quad \text{clockwise}$$
$$\mathbf{r}_1 \times \mathbf{V}_1 = r_1 V_{t1}\mathbf{k} \quad \text{clockwise}$$

Equation (1) thus becomes the desired formula for torque:

$$T_o = \rho Q (r_2 V_{t2} - r_1 V_{t1})\mathbf{k} \quad \text{clockwise} \qquad \textit{Ans. (a)} \;(2a)$$

This relation is called *Euler's turbine formula*. In an idealized pump, the inlet and outlet tangential velocities would match the blade rotational speeds $V_{t1} = \omega r_1$ and $V_{t2} = \omega r_2$. Then the formula for torque supplied becomes

$$T_o = \rho Q \omega (r_2^2 - r_1^2) \quad \text{clockwise} \tag{2b}$$

Part (b)

Convert ω to $600(2\pi/60) = 62.8$ rad/s. The normal velocities are not needed here but follow from the flow rate

$$V_{n1} = \frac{Q}{2\pi r_1 b} = \frac{2.5 \text{ m}^3/\text{s}}{2\pi(0.2 \text{ m})(0.15 \text{ m})} = 13.3 \text{ m/s}$$

$$V_{n2} = \frac{Q}{2\pi r_2 b} = \frac{2.5}{2\pi(0.5)(0.15)} = 5.3 \text{ m/s}$$

For the idealized inlet and outlet, tangential velocity equals tip speed:

$$V_{t1} = \omega r_1 = (62.8 \text{ rad/s})(0.2 \text{ m}) = 12.6 \text{ m/s}$$
$$V_{t2} = \omega r_2 = 62.8(0.5) = 31.4 \text{ m/s}$$

Equation (2a) predicts the required torque to be

$$T_o = (1000 \text{ kg/m}^3)(2.5 \text{ m}^3/\text{s})[(0.5 \text{ m})(31.4 \text{ m/s}) - (0.2 \text{ m})(12.6 \text{ m/s})]$$
$$= 33,000 \; (\text{kg} \cdot \text{m}^2)/\text{s}^2 = 33,000 \text{ N} \cdot \text{m} \qquad \textit{Ans.}$$

The power required is

$$P = \omega T_o = (62.8 \text{ rad/s})(33,000 \text{ N} \cdot \text{m}) = 2,070,000 \; (\text{N} \cdot \text{m})/\text{s}$$
$$= 2.07 \text{ MW} \;(2780 \text{ hp}) \qquad \textit{Ans.}$$

In actual practice the tangential velocities are considerably less than the impeller-tip speeds, and the design power requirements for this pump may be only 1 MW or less.

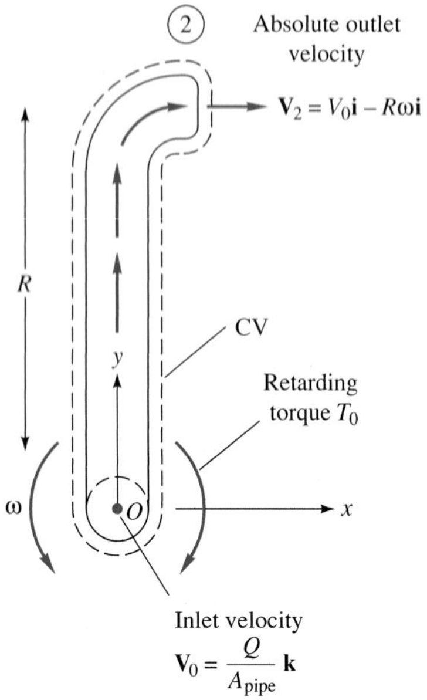

Fig. 3.13 View from above of a single arm of a rotating lawn sprinkler.

EXAMPLE 3.15

Figure 3.13 shows a lawn sprinkler arm viewed from above. The arm rotates about O at constant angular velocity ω. The volume flux entering the arm at O is Q, and the fluid is incompressible. There is a retarding torque at O, due to bearing friction, of amount $-T_o\mathbf{k}$. Find an expression for the rotation ω in terms of the arm and flow properties.

Solution

The entering velocity is $V_0\mathbf{k}$, where $V_0 = Q/A_{\text{pipe}}$. Equation (3.55) applies to the control volume sketched in Fig. 3.14 only if $\mathbf{V}$ is the absolute velocity relative to an inertial frame. Thus the exit velocity at section 2 is

$$\mathbf{V}_2 = V_0\mathbf{i} - R\omega\mathbf{i}$$

Equation (3.55) then predicts that, for steady flow,

$$\sum \mathbf{M}_o = -T_o\mathbf{k} = (\mathbf{r}_2 \times \mathbf{V}_2)\dot{m}_{\text{out}} - (\mathbf{r}_1 \times \mathbf{V}_1)\dot{m}_{\text{in}} \tag{1}$$

where, from continuity, $\dot{m}_{\text{out}} = \dot{m}_{\text{in}} = \rho Q$. The cross products with reference to point O are

$$\mathbf{r}_2 \times \mathbf{V}_2 = R\mathbf{j} \times (V_0 - R\omega)\mathbf{i} = (R^2\omega - RV_0)\mathbf{k}$$
$$\mathbf{r}_1 \times \mathbf{V}_1 = 0\mathbf{j} \times V_0\mathbf{k} = 0$$

Equation (1) thus becomes

$$-T_o\mathbf{k} = \rho Q(R^2\omega - RV_0)\mathbf{k}$$

$$\omega = \frac{V_o}{R} - \frac{T_o}{\rho Q R^2} \qquad\qquad Ans.$$

The result may surprise you: Even if the retarding torque T_o is negligible, the arm rotational speed is limited to the value V_0/R imposed by the outlet speed and the arm length.

3.6 The Energy Equation[16]

As our fourth and final basic law, we apply the Reynolds transport theorem (3.12) to the first law of thermodynamics, Eq. (3.5). The dummy variable B becomes energy E, and the energy per unit mass is $\beta = dE/dm = e$. Equation (3.5) can then be written for a fixed control volume as follows:[17]

$$\frac{dQ}{dt} - \frac{dW}{dt} = \frac{dE}{dt} = \frac{d}{dt}\left(\int_{\text{CV}} e\rho \, d\mathcal{V}\right) + \int_{\text{CS}} e\rho(\mathbf{V} \cdot \mathbf{n}) \, dA \tag{3.57}$$

Recall that positive Q denotes heat added to the system and positive W denotes work done by the system.

The system energy per unit mass e may be of several types:

$$e = e_{\text{internal}} + e_{\text{kinetic}} + e_{\text{potential}} + e_{\text{other}}$$

---

[16]This section should be read for information and enrichment even if you lack formal background in thermodynamics.

[17]The energy equation for a deformable control volume is rather complicated and is not discussed here. See Refs. 4 and 5 for further details.

where e_{other} could encompass chemical reactions, nuclear reactions, and electrostatic or magnetic field effects. We neglect e_{other} here and consider only the first three terms as discussed in Eq. (1.9), with z defined as "up":

$$e = \hat{u} + \tfrac{1}{2}V^2 + gz \tag{3.58}$$

The heat and work terms could be examined in detail. If this were a heat transfer book, dQ/dt would be broken down into conduction, convection, and radiation effects and whole chapters written on each (see, for example, Ref. 3). Here we leave the term untouched and consider it only occasionally.

Using for convenience the overdot to denote the time derivative, we divide the work term into three parts:

$$\dot{W} = \dot{W}_{shaft} + \dot{W}_{press} + \dot{W}_{viscous\ stresses} = \dot{W}_s + \dot{W}_p + \dot{W}_v$$

The work of gravitational forces has already been included as potential energy in Eq. (3.58). Other types of work, such as those due to electromagnetic forces, are excluded here.

The shaft work isolates the portion of the work that is deliberately done by a machine (pump impeller, fan blade, piston, or the like) protruding through the control surface into the control volume. No further specification other than $\dot{W}_s$ is desired at this point, but calculations of the work done by turbomachines will be performed in Chap. 11.

The rate of work $\dot{W}_p$ done by pressure forces occurs at the surface only; all work on internal portions of the material in the control volume is by equal and opposite forces and is self-canceling. The pressure work equals the pressure force on a small surface element dA times the normal velocity component into the control volume:

$$d\dot{W}_p = -(p\ dA)V_{n,\ in} = -p(-\mathbf{V} \cdot \mathbf{n})\ dA$$

The total pressure work is the integral over the control surface:

$$\dot{W}_p = \int_{CS} p(\mathbf{V} \cdot \mathbf{n})\ dA \tag{3.59}$$

A cautionary remark: If part of the control surface is the surface of a machine part, we prefer to delegate that portion of the pressure to the *shaft work* term $\dot{W}_s$, not to $\dot{W}_p$, which is primarily meant to isolate the fluid flow pressure work terms.

Finally, the shear work due to viscous stresses occurs at the control surface and consists of the product of each viscous stress (one normal and two tangential) and the respective velocity component:

$$d\dot{W}_v = -\tau \cdot \mathbf{V}\ dA$$

or

$$\dot{W}_v = -\int_{CS} \tau \cdot \mathbf{V}\ dA \tag{3.60}$$

where τ is the stress vector on the elemental surface dA. This term may vanish or be negligible according to the particular type of surface at that part of the control volume:

Solid surface. For all parts of the control surface that are solid confining walls, $\mathbf{V} = 0$ from the viscous no-slip condition; hence $\dot{W}_v =$ zero identically.

Surface of a machine. Here the viscous work is contributed by the machine, and so we absorb this work in the term $\dot{W}_s$.

An inlet or outlet. At an inlet or outlet, the flow is approximately normal to the element dA; hence the only viscous work term comes from the normal stress $\tau_{nn}V_n\, dA$. Since viscous normal stresses are extremely small in all but rare cases, such as the interior of a shock wave, it is customary to neglect viscous work at inlets and outlets of the control volume.

Streamline surface. If the control surface is a streamline such as the upper curve in the boundary layer analysis of Fig. 3.11, the viscous work term must be evaluated and retained if shear stresses are significant along this line. In the particular case of Fig. 3.11, the streamline is outside the boundary layer, and viscous work is negligible.

The net result of this discussion is that the rate-of-work term in Eq. (3.57) consists essentially of

$$\dot{W} = \dot{W}_s + \int_{CS} p(\mathbf{V} \cdot \mathbf{n})\, dA - \int_{CS} (\tau \cdot \mathbf{V})_{ss}\, dA \tag{3.61}$$

where the subscript SS stands for stream surface. When we introduce (3.61) and (3.58) into (3.57), we find that the pressure work term can be combined with the energy flux term since both involve surface integrals of $\mathbf{V} \cdot \mathbf{n}$. The control volume energy equation thus becomes

$$\dot{Q} - \dot{W}_s - \dot{W}_v = \frac{\partial}{\partial t}\left(\int_{CV} e\rho\, d\mathcal{V}\right) + \int_{CS}\left(e + \frac{p}{\rho}\right)\rho(\mathbf{V} \cdot \mathbf{n})\, dA \tag{3.62}$$

Using e from (3.58), we see that the enthalpy $\hat{h} = \hat{u} + p/\rho$ occurs in the control surface integral. The final general form for the energy equation for a fixed control volume becomes

$$\boxed{\dot{Q} - \dot{W}_s - \dot{W}_v = \frac{\partial}{\partial t}\left[\int_{CV}\left(\hat{u} + \tfrac{1}{2}V^2 + gz\right)\rho\, d\mathcal{V}\right] + \int_{CS}\left(\hat{h} + \tfrac{1}{2}V^2 + gz\right)\rho(\mathbf{V} \cdot \mathbf{n})\, dA}$$

$$\tag{3.63}$$

As mentioned, the shear work term $\dot{W}_v$ is rarely important.

One-Dimensional Energy-Flux Terms

If the control volume has a series of one-dimensional inlets and outlets, as in Fig. 3.5, the surface integral in (3.63) reduces to a summation of outlet fluxes minus inlet fluxes:

$$\int_{CS}(\hat{h} + \tfrac{1}{2}V^2 + gz)\rho(\mathbf{V} \cdot \mathbf{n})\, dA$$

$$= \sum(\hat{h} + \tfrac{1}{2}V^2 + gz)_{out}\dot{m}_{out} - \sum(\hat{h} + \tfrac{1}{2}V^2 + gz)_{in}\dot{m}_{in} \tag{3.64}$$

where the values of $\hat{h}$, $\tfrac{1}{2}V^2$, and gz are taken to be averages over each cross section.

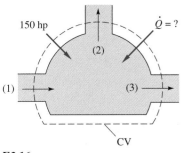

E3.16

EXAMPLE 3.16

A steady flow machine (Fig. E3.16) takes in air at section 1 and discharges it at sections 2 and 3. The properties at each section are as follows:

Section	A, ft^2	Q, ft^3/s	T, °F	p, lbf/in^2 abs	z, ft
1	0.4	100	70	20	1.0
2	1.0	40	100	30	4.0
3	0.25	50	200	?	1.5

Work is provided to the machine at the rate of 150 hp. Find the pressure p_3 in lbf/in^2 absolute and the heat transfer $\dot{Q}$ in Btu/s. Assume that air is a perfect gas with $R = 1716$ and $c_p = 6003$ ft · lbf/(slug · °R).

Solution

- *System sketch:* Figure E3.16 shows inlet 1 (negative flux) and outlets 2 and 3 (positive fluxes).
- *Assumptions:*   Steady flow, one-dimensional inlets and outlets, ideal gas, negligible shear work. The flow is *not* incompressible. Note that $Q_1 \neq Q_2 + Q_3$ because the densities are different.
- *Approach:*   Evaluate the velocities and densities and enthalpies and substitute into Eq. (3.63). Use BG units for all properties, including the pressures. With Q_i given, we evaluate $V_i = Q_i/A_i$:

$$V_1 = \frac{Q_1}{A_1} = \frac{100 \text{ ft}^3/\text{s}}{0.4 \text{ ft}^2} = 250 \frac{\text{ft}}{\text{s}} \qquad V_2 = \frac{40 \text{ ft}^3/\text{s}}{1.0 \text{ ft}^2} = 40 \frac{\text{ft}}{\text{s}} \qquad V_3 = \frac{50 \text{ ft}^3/\text{s}}{0.25 \text{ ft}^2} = 200 \frac{\text{ft}}{\text{s}}$$

The densities at sections 1 and 2 follow from the ideal gas law:

$$\rho_1 = \frac{p_1}{RT_1} = \frac{(20 \times 144) \text{ lbf/ft}^2}{[1716 \text{ ft} - \text{lbf/(slug°R)}][(70 + 460)°\text{R}]} = 0.00317 \frac{\text{slug}}{\text{ft}^3}$$

$$\rho_2 = \frac{(30 \times 144)}{(1716)(100 + 460)} = 0.00450 \frac{\text{slug}}{\text{ft}^3}$$

However, p_3 is unknown, so how do we find ρ_3? Use the steady flow continuity relation:

$$\dot{m}_1 = \dot{m}_2 + \dot{m}_3 \qquad \text{or} \qquad \rho_1 Q_1 = \rho_2 Q_2 + \rho_3 Q_3 \qquad (1)$$

$$\left(0.00317 \frac{\text{slug}}{\text{ft}^3}\right)\left(100 \frac{\text{ft}^3}{\text{s}}\right) = 0.00450(40) + \rho_3(50) \quad \text{solve for } \rho_3 = 0.00274 \frac{\text{slug}}{\text{ft}^3}$$

Knowing ρ_3 enables us to find p_3 from the ideal-gas law:

$$p_3 = \rho_3 R T_3 = \left(0.00274 \frac{\text{slug}}{\text{ft}^3}\right)\left(1716 \frac{\text{ft} - \text{lbf}}{\text{slug °R}}\right)(200 + 460°\text{R}) = 3100 \frac{\text{lbf}}{\text{ft}^2} = 21.5 \frac{\text{lbf}}{\text{in}^2} \quad Ans.$$

- *Final solution steps:*   For an ideal gas, simply approximate enthalpies as $h_i = c_p T_i$. The shaft work is *negative* (into the control volume) and viscous work is neglected for this solid-wall machine:

$$\dot{W}_v \approx 0 \qquad \dot{W}_s = (-150 \text{ hp})\left(550 \frac{\text{ft} - \text{lbf}}{\text{s} - \text{hp}}\right) = -82{,}500 \frac{\text{ft} - \text{lbf}}{\text{s}} \quad \text{(work } on \text{ the system)}$$

For steady flow, the volume integral in Eq. (3.63) vanishes, and the energy equation becomes

$$\dot{Q} - \dot{W}_s = -\dot{m}_1\left(c_p T_1 + \tfrac{1}{2}V_1^2 + gz_1\right) + \dot{m}_2\left(c_p T_2 + \tfrac{1}{2}V_2^2 + gz_2\right) + \dot{m}_3\left(c_p T_3 + \tfrac{1}{2}V_3^2 + gz_3\right) \qquad (2)$$

From our continuity calculations in Eq. (1) above, the mass flows are

$$\dot{m}_1 = \rho_1 Q_1 = (0.00317)(100) = 0.317 \frac{\text{slug}}{\text{s}} \qquad \dot{m}_2 = \rho_2 Q_2 = 0.180 \frac{\text{slug}}{\text{s}}$$

$$\dot{m}_3 = \rho_3 Q_3 = 0.137 \frac{\text{slug}}{\text{s}}$$

It is instructive to separate the flux terms in the energy equation (2) for examination:

$$
\begin{aligned}
\text{Enthalpy flux} &= c_p(-\dot{m}_1 T_1 + \dot{m}_2 T_2 + \dot{m}_3 T_3) \\
&= (6003)[(-0.317)(530) + (0.180)(560) + (0.137)(660)] \\
&= -1{,}009{,}000 + 605{,}000 + 543{,}000 \approx +139{,}000 \text{ ft} - \text{lbf/s}
\end{aligned}
$$

$$
\begin{aligned}
\text{Kinetic energy flux} &= \tfrac{1}{2}(-\dot{m}_1 V_1^2 + \dot{m}_2 V_2^2 + \dot{m}_3 V_3^2) \\
&= \tfrac{1}{2}[-0.317(250)^2 + (0.180)(40)^2 + (0.137)(200)^2] \\
&= -9900 + 140 + 2740 \approx -7000 \text{ ft} - \text{lbf/s}
\end{aligned}
$$

$$
\begin{aligned}
\text{Potential energy flux} &= g(-\dot{m}_1 z_1 + \dot{m}_2 z_2 + \dot{m}_3 z_3) \\
&= (32.2)[-0.317(1.0) + 0.180(4.0) + 0.137(1.5)] \\
&= -10 + 23 + 7 \approx +20 \text{ ft} - \text{lbf/s}
\end{aligned}
$$

Equation (2) may now be evaluated for the heat transfer:

$$\dot{Q} - (-82{,}500) = 139{,}000 - 7{,}000 + 20$$

or
$$\dot{Q} \approx \left(+49{,}520 \, \frac{\text{ft} - \text{lbf}}{\text{s}}\right)\left(\frac{1 \text{ Btu}}{778.2 \text{ ft} - \text{lbf}}\right) = +64 \, \frac{\text{Btu}}{\text{s}} \qquad \textit{Ans.}$$

- *Comments:*   The heat transfer is positive, which means *into* the control volume. It is typical of gas flows that potential energy flux is negligible, enthalpy flux is dominant, and kinetic energy flux is small unless the velocities are very high (that is, high subsonic or supersonic).

The Steady Flow Energy Equation

For steady flow with one inlet and one outlet, both assumed one-dimensional, Eq. (3.63) reduces to a celebrated relation used in many engineering analyses. Let section 1 be the inlet and section 2 the outlet. Then

$$\dot{Q} - \dot{W}_s - \dot{W}_v = \dot{m}_1(\hat{h}_1 + \tfrac{1}{2}V_1^2 + gz_1) + \dot{m}_2(\hat{h}_2 + \tfrac{1}{2}V_2^2 + gz_2) \qquad (3.65)$$

But, from continuity, $\dot{m}_1 = \dot{m}_2 = \dot{m}$, we can rearrange (3.65) as follows:

$$\hat{h}_1 + \tfrac{1}{2}V_1^2 + gz_1 = (\hat{h}_2 + \tfrac{1}{2}V_2^2 + gz_2) - q + w_s + w_v \qquad (3.66)$$

where $q = \dot{Q}/\dot{m} = dQ/dm$, the heat transferred to the fluid per unit mass. Similarly, $w_s = W_s/\dot{m} = dW_s/dm$ and $w_v = \dot{W}_v/\dot{m} = dW_v/dm$. Equation (3.66) is a general form of the *steady flow energy equation*, which states that the upstream *stagnation enthalpy* $H_1 = (h + \tfrac{1}{2}V^2 + gz)_1$ differs from the downstream value H_2 only if there is heat transfer, shaft work, or viscous work as the fluid passes between sections 1 and 2. Recall that q is positive if heat is added to the control volume and that w_s and w_v are positive if work is done by the fluid on the surroundings.

Each term in Eq. (3.66) has the dimensions of energy per unit mass, or velocity squared, which is a form commonly used by mechanical engineers. If we divide through by g, each term becomes a length, or head, which is a form preferred by civil engineers. The traditional symbol for head is h, which we do not wish to confuse with enthalpy. Therefore we use internal energy in rewriting the head form of the energy relation:

$$\frac{p_1}{\gamma} + \frac{\hat{u}_1}{g} + \frac{V_1^2}{2g} + z_1 = \frac{p_2}{\gamma} + \frac{\hat{u}_2}{g} + \frac{V_1^2}{2g} + z_2 - h_q + h_s + h_v \tag{3.67}$$

where $h_q = q/g$, $h_s = w_s/g$, and $h_v = w_u/g$ are the head forms of the heat added, shaft work done, and viscous work done, respectively. The term p/γ is called *pressure head*, and the term $V^2/2g$ is denoted as *velocity head*.

Friction and Shaft Work in Low-Speed Flow

A common application of the steady flow energy equation is for low-speed (incompressible) flow through a pipe or duct. A pump or turbine may be included in the pipe system. The pipe and machine walls are solid, so the viscous work is zero. Equation (3.67) may be written as

$$\left(\frac{p_1}{\gamma} + \frac{V_1^2}{2g} + z_1\right) = \left(\frac{p_2}{\gamma} + \frac{V_2^2}{2g} + z_2\right) + \frac{\hat{u}_2 - \hat{u}_1 - q}{g} \tag{3.68}$$

Every term in this equation is a length, or *head*. The terms in parentheses are the upstream (1) and downstream (2) values of the useful or *available head* or *total head* or the flow, denoted by h_0. The last term on the right is the difference $(h_{01} - h_{02})$, which can include pump head input, turbine head extraction, and the friction head loss h_f, always *positive*. Thus, in incompressible flow with one inlet and one outlet, we may write

$$\left(\frac{p}{\gamma} + \frac{V^2}{2g} + z\right)_{in} = \left(\frac{p}{\gamma} + \frac{V^2}{2g} + z\right)_{out} + h_{friction} - h_{pump} + h_{turbine} \tag{3.69}$$

Most of our internal flow problems will be solved with the aid of Eq. (3.69). The h terms are all positive; that is, friction loss is always positive in real (viscous) flows, a pump adds energy (increases the left-hand side), and a turbine extracts energy from the flow. If h_p and/or h_t are included, the pump and/or turbine must lie *between* points 1 and 2. In Chaps. 5 and 6 we shall develop methods of correlating h_f losses with flow parameters in pipes, valves, fittings, and other internal flow devices.

EXAMPLE 3.17

Gasoline at 20°C is pumped through a smooth 12-cm-diameter pipe 10 km long, at a flow rate of 75 m³/h (330 gal/min). The inlet is fed by a pump at an absolute pressure of 24 atm. The exit is at standard atmospheric pressure and is 150 m higher. Estimate the frictional head loss h_f, and compare it to the velocity head of the flow $V^2/(2g)$. (These numbers are quite realistic for liquid flow through long pipelines.)

Solution

- *Property values:* From Table A.3 for gasoline at 20°C, $\rho = 680$ kg/m³, or $\gamma = (680)(9.81) = 6670$ N/m³.

- *Assumptions:* Steady flow. No shaft work, thus $h_p = h_t = 0$. If $z_1 = 0$, then $z_2 = 150$ m.
- *Approach:* Find the velocity and the velocity head. These are needed for comparison. Then evaluate the friction loss from Eq. (3.69).
- *Solution steps:* Since the pipe diameter is constant, the average velocity is the same everywhere:

$$V_{in} = V_{out} = \frac{Q}{A} = \frac{Q}{(\pi/4)D^2} = \frac{(75 \text{ m}^3/\text{h})/(3600 \text{ s/h})}{(\pi/4)(0.12 \text{ m})^2} \approx 1.84 \frac{\text{m}}{\text{s}}$$

$$\text{Velocity head} = \frac{V^2}{2g} = \frac{(1.84 \text{ m/s})^2}{2(9.81 \text{ m/s}^2)} \approx 0.173 \text{ m}$$

Substitute into Eq. (3.69) and solve for the friction head loss. Use pascals for the pressures and note that the velocity heads cancel because of the constant-area pipe.

$$\frac{p_{in}}{\gamma} + \frac{V_{in}^2}{2g} + z_{in} = \frac{p_{out}}{\gamma} + \frac{V_{out}^2}{2g} + z_{out} + h_f$$

$$\frac{(24)(101,350 \text{ N/m}^2)}{6670 \text{ N/m}^3} + 0.173 \text{ m} + 0 \text{ m} = \frac{101,350 \text{ N/m}^2}{6670 \text{ N/m}^3} + 0.173 \text{ m} + 150 \text{ m} + h_f$$

or
$$h_f = 364.7 - 15.2 - 150 \approx 199 \text{ m} \qquad \textit{Ans.}$$

The friction head is larger than the elevation change Δz, and the pump must drive the flow against both changes, hence the high inlet pressure. The ratio of friction to velocity head is

$$\frac{h_f}{V^2/(2g)} \approx \frac{199 \text{ m}}{0.173 \text{ m}} \approx 1150 \qquad \textit{Ans.}$$

- *Comments:* This high ratio is typical of long pipelines. (Note that we did not make direct use of the 10,000-m pipe length, whose effect is hidden within h_f.) In Chap. 6 we can state this problem in a more direct fashion: Given the flow rate, fluid, and pipe size, what inlet pressure is needed? Our correlations for h_f will lead to the estimate $p_{inlet} \approx 24$ atm, as stated here.

EXAMPLE 3.18

Air [$R = 1716$, $c_p = 6003$ ft · lbf/(slug · °R)] flows steadily, as shown in Fig. E3.18, through a turbine that produces 700 hp. For the inlet and exit conditions shown, estimate (*a*) the exit velocity V_2 and (*b*) the heat transferred Q in Btu/h.

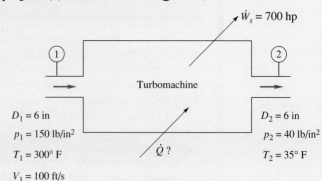

$\dot{W}_s = 700$ hp

Turbomachine

$D_1 = 6$ in
$p_1 = 150$ lb/in²
$T_1 = 300°$ F
$V_1 = 100$ ft/s

$\dot{Q}$?

$D_2 = 6$ in
$p_2 = 40$ lb/in²
$T_2 = 35°$ F

E3.18

Solution

Part (a)

The inlet and exit densities can be computed from the perfect-gas law:

$$\rho_1 = \frac{p_1}{RT_1} = \frac{150(144)}{1716(460 + 300)} = 0.0166 \text{ slug/ft}^3$$

$$\rho_2 = \frac{p_2}{RT_2} = \frac{40(144)}{1716(460 + 35)} = 0.00679 \text{ slug/ft}^3$$

The mass flow is determined by the inlet conditions

$$\dot{m} = \rho_1 A_1 V_1 = (0.0166)\frac{\pi}{4}\left(\frac{6}{12}\right)^2(100) = 0.325 \text{ slug/s}$$

Knowing mass flow, we compute the exit velocity

$$\dot{m} = 0.325 = \rho_2 A_2 V_2 = (0.00679)\frac{\pi}{4}\left(\frac{6}{12}\right)^2 V_2$$

or $V_2 = 244$ ft/s   *Ans. (a)*

Part (b)

The steady flow energy equation (3.65) applies with $\dot{W}_v = 0$, $z_1 = z_2$, and $\hat{h} = c_p T$:

$$\dot{Q} - \dot{W}_s = \dot{m}(c_p T_2 + \tfrac{1}{2}V_2^2 - c_p T_1 - \tfrac{1}{2}V_1^2)$$

Convert the turbine work to foot-pounds-force per second with the conversion factor 1 hp = 550 ft · lbf/s. The turbine work $\dot{W}_s$ is positive

$$\dot{Q} - 700(550) = 0.325[6003(495) + \tfrac{1}{2}(244)^2 - 6003(760) - \tfrac{1}{2}(100)^2]$$

$$= -510,000 \text{ ft} \cdot \text{lbf/s}$$

or $\dot{Q} = -125,000$ ft · lbf/s

Convert this to British thermal units as follows:

$$\dot{Q} = (-125,000 \text{ ft} \cdot \text{lbf/s})\frac{3600 \text{ s/h}}{778.2 \text{ ft} \cdot \text{lbf/Btu}}$$

$$= -578,000 \text{ Btu/h}$$ *Ans. (b)*

The negative sign indicates that this heat transfer is a *loss* from the control volume.

Kinetic Energy Correction Factor

Often the flow entering or leaving a port is not strictly one-dimensional. In particular, the velocity may vary over the cross section, as in Fig. E3.4. In this case the kinetic energy term in Eq. (3.64) for a given port should be modified by a dimensionless correction factor α so that the integral can be proportional to the square of the average velocity through the port:

$$\int_{port} (\tfrac{1}{2}V^2)\rho(\mathbf{V} \cdot \mathbf{n})\, dA \equiv \alpha(\tfrac{1}{2}V_{av}^2)\dot{m}$$

where $V_{av} = \dfrac{1}{A}\displaystyle\int u\, dA$ for incompressible flow

If the density is also variable, the integration is very cumbersome; we shall not treat this complication. By letting u be the velocity normal to the port, the first equation above becomes, for incompressible flow,

$$\tfrac{1}{2}\rho \int u^3 dA = \tfrac{1}{2}\rho \alpha V_{av}^3 A$$

or

$$\alpha = \frac{1}{A}\int \left(\frac{u}{V_{av}}\right)^3 dA \tag{3.70}$$

The term α is the kinetic energy correction factor, having a value of about 2.0 for fully developed laminar pipe flow and from 1.04 to 1.11 for turbulent pipe flow. The complete incompressible steady flow energy equation (3.69), including pumps, turbines, and losses, would generalize to

$$\left(\frac{p}{\rho g} + \frac{\alpha}{2g}V^2 + z\right)_{in} = \left(\frac{p}{\rho g} + \frac{\alpha}{2g}V^2 + z\right)_{out} + h_{turbine} - h_{pump} + h_{friction} \tag{3.71}$$

where the head terms on the right (h_t, h_p, h_f) are all numerically positive. All additive terms in Eq. (3.71) have dimensions of length $\{L\}$. In problems involving turbulent pipe flow, it is common to assume that $\alpha \approx 1.0$. To compute numerical values, we can use these approximations to be discussed in Chap. 6:

Laminar flow:

$$u = U_0\left[1 - \left(\frac{r}{R}\right)^2\right]$$

from which

$$V_{av} = 0.5U_0$$

and

$$\alpha = 2.0 \tag{3.72}$$

Turbulent flow:

$$u \approx U_0\left(1 - \frac{r}{R}\right)^m \qquad m \approx \frac{1}{7}$$

from which, in Example 3.4,

$$V_{av} = \frac{2U_0}{(1 + m)(2 + m)}$$

Substituting into Eq. (3.70) gives

$$\alpha = \frac{(1 + m)^3(2 + m)^3}{4(1 + 3m)(2 + 3m)} \tag{3.73}$$

and numerical values are as follows:

Turbulent flow:	m	$\frac{1}{5}$	$\frac{1}{6}$	$\frac{1}{7}$	$\frac{1}{8}$	$\frac{1}{9}$
	α	1.106	1.077	1.058	1.046	1.037

These values are only slightly different from unity and are often neglected in elementary turbulent flow analyses. However, α should never be neglected in laminar flow.

EXAMPLE 3.19

A hydroelectric power plant (Fig. E3.19) takes in 30 m³/s of water through its turbine and discharges it to the atmosphere at $V_2 = 2$ m/s. The head loss in the turbine and penstock system is $h_f = 20$ m. Assuming turbulent flow, $\alpha \approx 1.06$, estimate the power in MW extracted by the turbine.

Solution

We neglect viscous work and heat transfer and take section 1 at the reservoir surface (Fig. E3.19), where $V_1 \approx 0$, $p_1 = p_{atm}$, and $z_1 = 100$ m. Section 2 is at the turbine outlet.

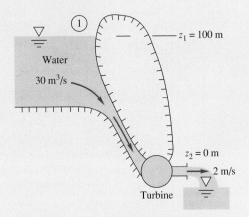

E3.19

The steady flow energy equation (3.71) becomes, in head form,

$$\frac{p_1}{\gamma} + \frac{\alpha_1 V_1^2}{2g} + z_1 = \frac{p_2}{\gamma} + \frac{\alpha_2 V_2^2}{2g} + z_2 + h_t + h_f$$

$$\frac{p_a}{\gamma} + \frac{1.06(0)^2}{2(9.81)} + 100 \text{ m} = \frac{p_a}{\gamma} + \frac{1.06(2.0 \text{ m/s})^2}{2(9.81 \text{ m/s}^2)} + 0 \text{ m} + h_t + 20 \text{ m}$$

The pressure terms cancel, and we may solve for the turbine head (which is positive):

$$h_t = 100 - 20 - 0.2 \approx 79.8 \text{ m}$$

The turbine extracts about 79.8 percent of the 100-m head available from the dam. The total power extracted may be evaluated from the water mass flow:

$$P = \dot{m} w_s = (\rho Q)(g h_t) = (998 \text{ kg/m}^3)(30 \text{ m}^3/\text{s})(9.81 \text{ m/s}^2)(79.8 \text{ m})$$

$$= 23.4 \text{ E6 kg} \cdot \text{m}^2/\text{s}^3 = 23.4 \text{ E6 N} \cdot \text{m/s} = 23.4 \text{ MW} \qquad \textit{Ans.}$$

The turbine drives an electric generator that probably has losses of about 15 percent, so the net power generated by this hydroelectric plant is about 20 MW.

EXAMPLE 3.20

The pump in Fig. E3.20 delivers water (62.4 lbf/ft³) at 1.5 ft³/s to a machine at section 2, which is 20 ft higher than the reservoir surface. The losses between 1 and 2 are given by

$h_f = KV_2^2/(2g)$, where $K \approx 7.5$ is a dimensionless loss coefficient (see Sec. 6.7). Take $\alpha \approx 1.07$. Find the horsepower required for the pump if it is 80 percent efficient.

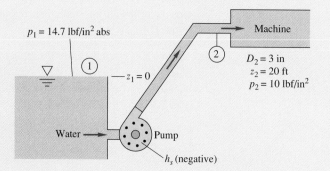

E3.20

Solution

- *System sketch:*   Figure E3.20 shows the proper selection for sections 1 and 2.
- *Assumptions:*   Steady flow, negligible viscous work, large reservoir ($V_1 \approx 0$).
- *Approach:*   First find the velocity V_2 at the exit, then apply the steady flow energy equation.
- *Solution steps:*   Use BG units, $p_1 = 14.7(144) = 2117$ lbf/ft² and $p_2 = 10(144) = 1440$ lbf/ft². Find V_2 from the known flow rate and the pipe diameter:

$$V_2 = \frac{Q}{A_2} = \frac{1.5 \text{ ft}^3/\text{s}}{(\pi/4)(3/12 \text{ ft})^2} = 30.6 \text{ ft/s}$$

The steady flow energy equation (3.71), with a pump (no turbine) plus $z_1 \approx 0$ and $V_1 \approx 0$, becomes

$$\frac{p_1}{\gamma} + \frac{\alpha_1 V_1^2}{2g} + z_1 = \frac{p_2}{\gamma} + \frac{\alpha_2 V_2^2}{2g} + z_2 - h_p + h_f, \quad h_f = K\frac{V_2^2}{2g}$$

or

$$h_p = \frac{p_2 - p_1}{\gamma} + z_2 + (\alpha_2 + K)\frac{V_2^2}{2g}$$

- *Comment:*   The pump must balance four different effects: the pressure change, the elevation change, the exit jet kinetic energy, and the friction losses.
- *Final solution:*   For the given data, we can evaluate the required pump head:

$$h_p = \frac{(1440 - 2117)\,\text{lbf/ft}^2}{62.4 \text{ lbf/ft}^3} + 20 + (1.07 + 7.5)\frac{(30.6 \text{ ft/s})^2}{2(32.2 \text{ ft/s}^2)} = -11 + 20 + 124 = 133 \text{ ft}$$

With the pump head known, the delivered pump power is computed similar to the turbine in Example 3.19:

$$P_{\text{pump}} = \dot{m}w_s = \gamma Q h_p = \left(62.4 \frac{\text{lbf}}{\text{ft}^3}\right)\left(1.5 \frac{\text{ft}^3}{\text{s}}\right)(133 \text{ ft})$$

$$= 12450 \frac{\text{ft} - \text{lbf}}{\text{s}} = \frac{12{,}450 \text{ ft} - \text{lbf/s}}{550 \text{ ft} - \text{lbf}/(\text{s} - \text{hp})} = 22.6 \text{ hp}$$

If the pump is 80 percent efficient, then we divide by the efficiency to find the input power required:

$$P_{\text{input}} = \frac{P_{\text{pump}}}{\text{efficiency}} = \frac{22.6 \text{ hp}}{0.80} = 28.3 \text{ hp} \qquad \textit{Ans.}$$

• *Comment:* The inclusion of the kinetic energy correction factor α in this case made a difference of about 1 percent in the result. The friction loss, not the exit jet, was the dominant parameter.

3.7 Frictionless Flow: The Bernoulli Equation

Closely related to the steady flow energy equation is a relation between pressure, velocity, and elevation in a frictionless flow, now called the *Bernoulli equation*. It was stated (vaguely) in words in 1738 in a textbook by Daniel Bernoulli. A complete derivation of the equation was given in 1755 by Leonhard Euler. The Bernoulli equation is very famous and very widely used, but one should be wary of its restrictions—all fluids are viscous and thus all flows have friction to some extent. To use the Bernoulli equation correctly, one must confine it to regions of the flow that are nearly frictionless. This section (and, in more detail, Chap. 8) will address the proper use of the Bernoulli relation.

Consider Fig. 3.14, which is an elemental fixed streamtube control volume of variable area $A(s)$ and length ds, where s is the streamline direction. The properties (ρ, V, p) may vary with s and time but are assumed to be uniform over the cross section A. The streamtube orientation θ is arbitrary, with an elevation change $dz = ds \sin \theta$. Friction on the streamtube walls is shown and then neglected—a very restrictive assumption. Note that the limit of a vanishingly small area means that the streamtube is equivalent to a *streamline* of the flow. Bernoulli's equation is valid for both and is usually stated as holding "along a streamline" in frictionless flow.

Conservation of mass [Eq. (3.20)] for this elemental control volume yields

$$\frac{d}{dt}\left(\int_{\text{CV}} \rho \, d\mathcal{V} \right) + \dot{m}_{\text{out}} - \dot{m}_{\text{in}} = 0 \approx \frac{\partial \rho}{\partial t} d\mathcal{V} + d\dot{m}$$

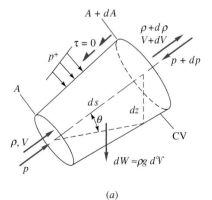

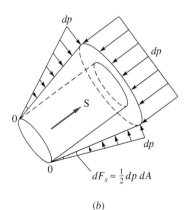

Fig. 3.14 The Bernoulli equation for frictionless flow along a streamline: (*a*) forces and fluxes; (*b*) net pressure force after uniform subtraction of *p*.

(*a*)   (*b*)

where $\dot{m} = \rho A V$ and $dV \approx A\ ds$. Then our desired form of mass conservation is

$$dm = d(\rho A V) = -\frac{\partial \rho}{\partial t} A\ ds \tag{3.74}$$

This relation does not require an assumption of frictionless flow.

Now write the linear momentum relation [Eq. (3.37)] in the streamwise direction:

$$\sum dF_s = \frac{d}{dt}\left(\int_{CV} V\rho\ dV\right) + (\dot{m}V)_{\text{out}} - (\dot{m}V)_{\text{in}} \approx \frac{\partial}{\partial t}(\rho V) A\ ds + d(\dot{m}V)$$

where $V_s = V$ itself because s is the streamline direction. If we neglect the shear force on the walls (frictionless flow), the forces are due to pressure and gravity. The streamwise gravity force is due to the weight component of the fluid within the control volume:

$$dF_{s,\text{grav}} = -dW \sin\theta = -\gamma A\ ds \sin\theta = -\gamma A\ dz$$

The pressure force is more easily visualized, in Fig. 3.14b, by first subtracting a uniform value p from all surfaces, remembering from Fig. 3.6 that the net force is not changed. The pressure along the slanted side of the streamtube has a streamwise component that acts not on A itself but on the outer ring of area increase dA. The net pressure force is thus

$$dF_{s,\text{press}} = \tfrac{1}{2} dp\ dA - dp(A + dA) \approx -A\ dp$$

to first order. Substitute these two force terms into the linear momentum relation:

$$\sum dF_s = -\gamma A\ dz - A\ dp = \frac{\partial}{\partial t}(\rho V) A\ ds + d(\dot{m}V)$$

$$= \frac{\partial \rho}{\partial t} VA\ ds + \frac{\partial V}{\partial t}\rho A\ ds + \dot{m}\ dV + V\ d\dot{m}$$

The first and last terms on the right cancel by virtue of the continuity relation [Eq. (3.74)]. Divide what remains by ρA and rearrange into the final desired relation:

$$\frac{\partial V}{\partial t} ds + \frac{dp}{\rho} + V\ dV + g\ dz = 0 \tag{3.75}$$

This is Bernoulli's equation for *unsteady frictionless flow along a streamline*. It is in differential form and can be integrated between any two points 1 and 2 on the streamline:

$$\int_1^2 \frac{\partial V}{\partial t} ds + \int_1^2 \frac{dp}{\rho} + \frac{1}{2}(V_2^2 - V_1^2) + g(z_2 - z_1) = 0 \tag{3.76}$$

Steady Incompressible Flow

To evaluate the two remaining integrals, one must estimate the unsteady effect $\partial V/\partial t$ and the variation of density with pressure. At this time we consider only steady ($\partial V/\partial t = 0$) incompressible (constant-density) flow, for which Eq. (3.76) becomes

$$\frac{p_2 - p_1}{\rho} + \frac{1}{2}(V_2^2 - V_1^2) + g(z_2 - z_1) = 0$$

or
$$\frac{p_1}{\rho} + \frac{1}{2}V_1^2 + gz_1 = \frac{p_2}{\rho} + \frac{1}{2}V_2^2 + gz_2 = \text{const} \qquad (3.77)$$

This is the Bernoulli equation for steady frictionless incompressible flow along a streamline.

Relation between the Bernoulli and Steady Flow Energy Equations

Equation (3.77) is a widely used form of the Bernoulli equation for incompressible steady frictionless streamline flow. It is clearly related to the steady flow energy equation for a streamtube (flow with one inlet and one outlet), from Eq. (3.66), which we state as follows:

$$\frac{p_1}{\rho} + \frac{\alpha_1 V_1^2}{2} + gz_1 = \frac{p_2}{\rho} + \frac{\alpha_2 V_2^2}{2} + gz_2 + (\hat{u}_2 - \hat{u}_1 - q) + w_s + w_v \qquad (3.78)$$

This relation is much more general than the Bernoulli equation because it allows for (1) friction, (2) heat transfer, (3) shaft work, and (4) viscous work (another frictional effect).

The Bernoulli equation is a momentum-based force relation and was derived using the following restrictive assumptions:

1. *Steady flow:* a common situation, application to most flows in this text.
2. *Incompressible flow:* appropriate if the flow Mach number is less than 0.3. This restriction is removed in Chap. 9 by allowing for compressibility.
3. *Frictionless flow:* restrictive—solid walls and mixing introduce friction effects.
4. *Flow along a single streamline:* but different streamlines may have different "Bernoulli constants" $w_o = p/\rho + V^2/2 + gz$, depending upon the flow conditions.

The Bernoulli derivation does not account for possible energy exchange due to heat or work. These thermodynamic effects are accounted for in the steady flow energy Equation (3.66). We are thus warned that the Bernoulli equation may be modified by such an energy exchange.

Figure 3.15 illustrates some practical limitations on the use of Bernoulli's equation (3.77). For the wind tunnel model test of Fig. 3.15a, the Bernoulli equation is valid in the core flow of the tunnel but not in the tunnel wall boundary layers, the model surface boundary layers, or the wake of the model, all of which are regions with high friction.

In the propeller flow of Fig. 3.15b, Bernoulli's equation is valid both upstream and downstream, but with a different constant $w_0 = p/\rho + V^2/2 + gz$, caused by the work addition of the propeller. The Bernoulli relation (3.77) is not valid near the propeller blades or in the helical vortices (not shown, see Fig. 1.14) shed downstream of the blade edges. Also, the Bernoulli constants are higher in the flowing "slipstream" than in the ambient atmosphere because of the slipstream kinetic energy.

For the chimney flow of Fig. 3.15c, Eq. (3.77) is valid before and after the fire, but with a change in Bernoulli constant that is caused by heat addition. The Bernoulli equation is not valid within the fire itself or in the chimney wall boundary layers.

Hydraulic and Energy Grade Lines

A useful visual interpretation of Bernoulli's equation is to sketch two grade lines of a flow. The *energy grade line* (EGL) shows the height of the total Bernoulli constant $h_0 = z + p/\gamma + V^2/(2g)$. In frictionless flow with no work or heat transfer [Eq. (3.77)]

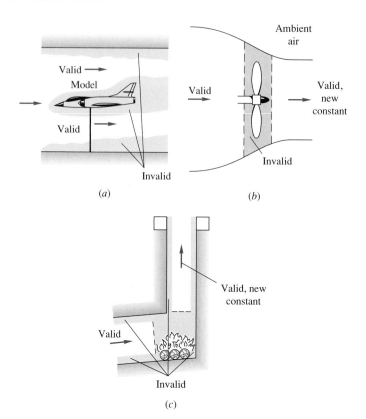

Fig. 3.15 Illustration of regions of validity and invalidity of the Bernoulli equation: (*a*) tunnel model, (*b*) propeller, (*c*) chimney.

the EGL has constant height. The *hydraulic grade line* (HGL) shows the height corresponding to elevation and pressure head $z + p/\gamma$—that is, the EGL minus the velocity head $V^2/(2g)$. The HGL is the height to which liquid would rise in a piezometer tube (see Prob. 2.11) attached to the flow. In an open-channel flow the HGL is identical to the free surface of the water.

Figure 3.16 illustrates the EGL and HGL for frictionless flow at sections 1 and 2 of a duct. The piezometer tubes measure the static pressure head $z + p/\gamma$ and thus outline the HGL. The pitot stagnation-velocity tubes measure the total head $z + p/\gamma + V^2/(2g)$, which corresponds to the EGL. In this particular case the EGL is constant, and the HGL rises due to a drop in velocity.

In more general flow conditions, the EGL will drop slowly due to friction losses and will drop sharply due to a substantial loss (a valve or obstruction) or due to work extraction (to a turbine). The EGL can rise only if there is work addition (as from a pump or propeller). The HGL generally follows the behavior of the EGL with respect to losses or work transfer, and it rises and/or falls if the velocity decreases and/or increases.

As mentioned before, no conversion factors are needed in computations with the Bernoulli equation if consistent SI or BG units are used, as the following examples will show.

In all Bernoulli-type problems in this text, we consistently take point 1 upstream and point 2 downstream.

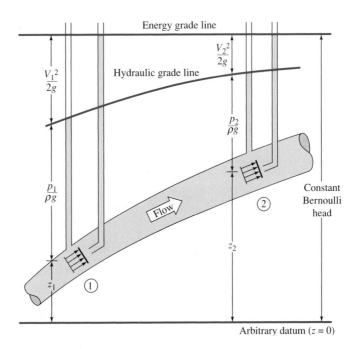

Fig. 3.16 Hydraulic and energy grade lines for frictionless flow in a duct.

EXAMPLE 3.21

Find a relation between nozzle discharge velocity V_2 and tank free surface height h as in Fig. E3.21. Assume steady frictionless flow.

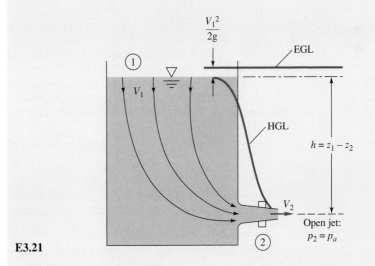

E3.21

Solution

As mentioned, we always choose point 1 upstream and point 2 downstream. Try to choose points 1 and 2 where maximum information is known or desired. Here we select point 1 as the tank free surface, where elevation and pressure are known, and point 2 as the nozzle exit, where again pressure and elevation are known. The two unknowns are V_1 and V_2.

Mass conservation is usually a vital part of Bernoulli analyses. If A_1 is the tank cross section and A_2 the nozzle area, this is approximately a one-dimensional flow with constant density, Eq. (3.30):

$$A_1 V_1 = A_2 V_2 \tag{1}$$

Bernoulli's equation (3.77) gives

$$\frac{p_1}{\rho} + \tfrac{1}{2}V_1^2 + gz_1 = \frac{p_2}{\rho} + \tfrac{1}{2}V_2^2 + gz_2$$

But since sections 1 and 2 are both exposed to atmospheric pressure $p_1 = p_2 = p_a$, the pressure terms cancel, leaving

$$V_2^2 - V_1^2 = 2g(z_1 - z_2) = 2gh \tag{2}$$

Eliminating V_1 between Eqs. (1) and (2), we obtain the desired result:

$$V_2^2 = \frac{2gh}{1 - A_2^2/A_1^2} \qquad\qquad Ans.\ (3)$$

Generally the nozzle area A_2 is very much smaller than the tank area A_1, so that the ratio A_2^2/A_1^2 is doubly negligible, and an accurate approximation for the outlet velocity is

$$V_2 \approx (2gh)^{1/2} \qquad\qquad Ans.\ (4)$$

This formula, discovered by Evangelista Torricelli in 1644, states that the discharge velocity equals the speed that a frictionless particle would attain if it fell freely from point 1 to point 2. In other words, the potential energy of the surface fluid is entirely converted to kinetic energy of efflux, which is consistent with the neglect of friction and the fact that no net pressure work is done. Note that Eq. (4) is independent of the fluid density, a characteristic of gravity-driven flows.

Except for the wall boundary layers, the streamlines from 1 to 2 all behave in the same way, and we can assume that the Bernoulli constant h_0 is the same for all the core flow. However, the outlet flow is likely to be nonuniform, not one-dimensional, so that the average velocity is only approximately equal to Torricelli's result. The engineer will then adjust the formula to include a dimensionless *discharge coefficient* c_d:

$$(V_2)_{av} = \frac{Q}{A_2} = c_d (2gh)^{1/2} \tag{5}$$

As discussed in Sec. 6.12, the discharge coefficient of a nozzle varies from about 0.6 to 1.0 as a function of (dimensionless) flow conditions and nozzle shape.

Before proceeding with more examples, we should note carefully that a solution by Bernoulli's equation (3.77) does *not* require a control volume analysis, only a selection of two points 1 and 2 along a given streamline. The control volume was used to derive the differential relation (3.75), but the integrated form (3.77) is valid all along the streamline for frictionless flow with no heat transfer or shaft work, and a control volume is not necessary.

A classical Bernoulli application is the familiar process of siphoning a fluid from one container to another. No pump is involved; a hydrostatic pressure difference provides the motive force. We analyze this in the following example.

EXAMPLE 3.22

Consider the water siphon shown in Fig. E3.22. Assuming that Bernoulli's equation is valid, (*a*) find an expression for the velocity V_2 exiting the siphon tube. (*b*) If the tube is 1 cm in diameter and $z_1 = 60$ cm, $z_2 = -25$ cm, $z_3 = 90$ cm, and $z_4 = 35$ cm, estimate the flow rate in cm^3/s.

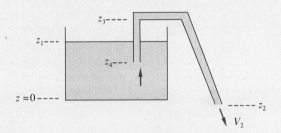

E3.22

Solution

- *Assumptions:* Frictionless, steady, incompressible flow. Write Bernoulli's equation starting from where information is known (the surface, z_1) and proceeding to where information is desired (the tube exit, z_2).

$$\frac{p_1}{\rho} + \frac{V_1^2}{2} + gz_1 = \frac{p_2}{\rho} + \frac{V_2^2}{2} + gz_2$$

Note that the velocity is approximately zero at z_1, and a streamline goes from z_1 to z_2. Note further that p_1 and p_2 are both atmospheric, $p = p_{atm}$, and therefore cancel. (*a*) Solve for the exit velocity from the tube:

$$V_2 = \sqrt{2g(z_1 - z_2)} \qquad\qquad Ans. (a)$$

The velocity exiting the siphon increases as the tube exit is lowered below the tank surface. There is no siphon effect if the exit is at or above the tank surface. Note that z_3 and z_4 do not directly enter the analysis. However, z_3 should not be too high because the pressure there will be lower than atmospheric, and the liquid might vaporize. (*b*) For the given numerical information, we need only z_1 and z_2 and calculate, in SI units,

$$V_2 = \sqrt{2(9.81 \text{ m/s}^2)[0.6 \text{ m} - (-0.25) \text{ m}]} = 4.08 \text{ m/s}$$

$$Q = V_2 A_2 = (4.08 \text{ m/s})(\pi/4)(0.01 \text{ m})^2 = 321 \text{ E} - 6 \text{ m}^3/\text{s} = 321 \text{ cm}^3/\text{s} \qquad Ans. (b)$$

- *Comments:* Note that this result is independent of the density of the fluid. As an exercise, you may check that, for water (998 kg/m^3), p_3 is 11,300 Pa *below* atmospheric pressure. In Chap. 6 we will modify this example to include friction effects.

EXAMPLE 3.23

A constriction in a pipe will cause the velocity to rise and the pressure to fall at section 2 in the throat. The pressure difference is a measure of the flow rate through the pipe. The smoothly necked-down system shown in Fig. E3.23 is called a *venturi tube*. Find an expression for the mass flux in the tube as a function of the pressure change.

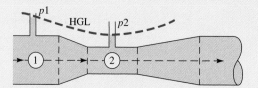

E3.23

Solution

Bernoulli's equation is assumed to hold along the center streamline:

$$\frac{p_1}{\rho} + \tfrac{1}{2} V_1^2 + g z_1 = \frac{p_2}{\rho} + \tfrac{1}{2} V_2^2 + g z_2$$

If the tube is horizontal, $z_1 = z_2$ and we can solve for V_2:

$$V_2^2 - V_1^2 = \frac{2 \, \Delta p}{\rho} \qquad \Delta p = p_1 - p_2 \tag{1}$$

We relate the velocities from the incompressible continuity relation:

$$A_1 V_1 = A_2 V_2$$

or $\qquad\qquad V_1 = \beta^2 V_2 \qquad \beta = \dfrac{D_2}{D_1} \tag{2}$

Combining (1) and (2), we obtain a formula for the velocity in the throat:

$$V_2 = \left[\frac{2 \, \Delta p}{\rho(1 - \beta^4)} \right]^{1/2} \tag{3}$$

The mass flux is given by

$$\dot{m} = \rho A_2 V_2 = A_2 \left(\frac{2 \rho \, \Delta p}{1 - \beta^4} \right)^{1/2} \tag{4}$$

This is the ideal frictionless mass flux. In practice, we measure $\dot{m}_{\text{actual}} = c_d \dot{m}_{\text{ideal}}$ and correlate the discharge coefficient c_d.

EXAMPLE 3.24

A 10-cm fire hose with a 3-cm nozzle discharges 1.5 m^3/min to the atmosphere. Assuming frictionless flow, find the force F_B exerted by the flange bolts to hold the nozzle on the hose.

Solution

We use Bernoulli's equation and continuity to find the pressure p_1 upstream of the nozzle, and then we use a control volume momentum analysis to compute the bolt force, as in Fig. E3.24.

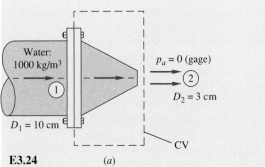

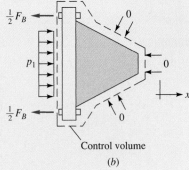

E3.24   (a)   (b)

The flow from 1 to 2 is a constriction exactly similar in effect to the venturi in Example 3.23, for which Eq. (1) gave

$$p_1 = p_2 + \tfrac{1}{2}\rho(V_2^2 - V_1^2) \tag{1}$$

The velocities are found from the known flow rate $Q = 1.5$ m^3/min or 0.025 m^3/s:

$$V_2 = \frac{Q}{A_2} = \frac{0.025 \text{ m}^3/\text{s}}{(\pi/4)(0.03 \text{ m})^2} = 35.4 \text{ m/s}$$

$$V_1 = \frac{Q}{A_1} = \frac{0.025 \text{ m}^3/\text{s}}{(\pi/4)(0.1 \text{ m})^2} = 3.2 \text{ m/s}$$

We are given $p_2 = p_a = 0$ gage pressure. Then Eq. (1) becomes

$$p_1 = \tfrac{1}{2}(1000 \text{ kg/m}^3)[(35.4^2 - 3.2^2)\text{m}^2/\text{s}^2]$$
$$= 620,000 \text{ kg/(m} \cdot \text{s}^2) = 620,000 \text{ Pa gage}$$

The control volume force balance is shown in Fig. E3.24b:

$$\sum F_x = -F_B + p_1 A_1$$

and the zero gage pressure on all other surfaces contributes no force. The x-momentum flux is $+\dot{m}V_2$ at the outlet and $-\dot{m}V_1$ at the inlet. The steady flow momentum relation (3.40) thus gives

$$-F_B + p_1 A_1 = \dot{m}(V_2 - V_1)$$

or
$$F_B = p_1 A_1 - \dot{m}(V_2 - V_1) \tag{2}$$

Substituting the given numerical values, we find

$$\dot{m} = \rho Q = (1000 \text{ kg/m}^3)(0.025 \text{ m}^3/\text{s}) = 25 \text{ kg/s}$$

$$A_1 = \frac{\pi}{4}D_1^2 = \frac{\pi}{4}(0.1 \text{ m})^2 = 0.00785 \text{ m}^2$$

$$F_B = (620,000 \text{ N/m}^2)(0.00785 \text{ m}^2) - (25 \text{ kg/s})[(35.4 - 3.2)\text{m/s}]$$
$$= 4872 \text{ N} - 805 \text{ (kg} \cdot \text{m)/s}^2 = 4067 \text{ N (915 lbf)} \qquad \textit{Ans.}$$

Notice from these examples that the solution of a typical problem involving Bernoulli's equation almost always leads to a consideration of the continuity equation as an equal partner in the analysis. The only exception is when the complete velocity distribution is already known from a previous or given analysis, but that means the continuity relation has already been used to obtain the given information. The point is that the continuity relation is always an important element in a flow analysis.

Summary

This chapter has analyzed the four basic equations of fluid mechanics: conservation of (1) mass, (2) linear momentum, (3) angular momentum, and (4) energy. The equations were attacked "in the large"—that is, applied to whole regions of a flow. As such, the typical analysis will involve an approximation of the flow field within the region, giving somewhat crude but always instructive quantitative results. However, the basic control volume relations are rigorous and correct and will give exact results if applied to the exact flow field.

There are two main points to a control volume analysis. The first is the selection of a proper, clever, workable control volume. There is no substitute for experience, but the following guidelines apply. The control volume should cut through the place where the information or solution is desired. It should cut through places where maximum information is already known. If the momentum equation is to be used, it should *not* cut through solid walls unless absolutely necessary, since this will expose possible unknown stresses and forces and moments that make the solution for the desired force difficult or impossible. Finally, every attempt should be made to place the control volume in a frame of reference where the flow is steady or quasi-steady, since the steady formulation is much simpler to evaluate.

The second main point to a control volume analysis is the reduction of the analysis to a case that applies to the problem at hand. The 24 examples in this chapter give only an introduction to the search for appropriate simplifying assumptions. You will need to solve 24 or 124 more examples to become truly experienced in simplifying the problem just enough and no more. In the meantime, it would be wise for the beginner to adopt a very general form of the control volume conservation laws and then make a series of simplifications to achieve the final analysis. Starting with the general form, one can ask a series of questions:

1. Is the control volume nondeforming or nonaccelerating?
2. Is the flow field steady? Can we change to a steady flow frame?
3. Can friction be neglected?
4. Is the fluid incompressible? If not, is the perfect-gas law applicable?
5. Are gravity or other body forces negligible?
6. Is there heat transfer, shaft work, or viscous work?
7. Are the inlet and outlet flows approximately one-dimensional?
8. Is atmospheric pressure important to the analysis? Is the pressure hydrostatic on any portions of the control surface?
9. Are there reservoir conditions that change so slowly that the velocity and time rates of change can be neglected?

In this way, by approving or rejecting a list of basic simplifications like these, one can avoid pulling Bernoulli's equation off the shelf when it does not apply.

Problems

Most of the problems herein are fairly straightforward. More difficult or open-ended assignments are labeled with an asterisk. Problems labeled with an EES icon (for example, Prob. P3.5) will benefit from the use of the Engineering Equation Solver (EES), while figures labeled with a computer disk may require the use of a computer. The standard end-of-chapter problems P3.1 to P3.185 (categorized in the problem list here) are followed by word problems W3.1 to W3.7, fundamentals of engineering (FE) exam problems FE3.1 to FE3.10, comprehensive problems C3.1 to C3.5, and design project D3.1.

Problem Distribution

Section	Topic	Problems
3.1	Basic physical laws; volume flow	P3.1–P3.6
3.2	The Reynolds transport theorem	P3.7–P3.11
3.3	Conservation of mass	P3.12–P3.38
3.4	The linear momentum equation	P3.39–P3.109
3.5	The angular momentum theorem	P3.110–P3.125
3.6	The energy equation	P3.126–P3.146
3.7	The Bernoulli equation	P3.147–P3.185

P3.1   Discuss Newton's second law (the linear momentum relation) in these three forms:

$$\sum \mathbf{F} = m\mathbf{a} \qquad \sum \mathbf{F} = \frac{d}{dt}(m\mathbf{V})$$

$$\sum \mathbf{F} = \frac{d}{dt}\left(\int_{\text{system}} \mathbf{V}\rho \, d\mathcal{V} \right)$$

Are they all equally valid? Are they equivalent? Are some forms better for fluid mechanics as opposed to solid mechanics?

P3.2   Consider the angular momentum relation in the form

$$\sum \mathbf{M}_O = \frac{d}{dt}\left[\int_{\text{system}} (\mathbf{r} \times \mathbf{V})\rho \, d\mathcal{V} \right]$$

What does $\mathbf{r}$ mean in this relation? Is this relation valid in both solid and fluid mechanics? Is it related to the *linear* momentum equation (Prob. 3.1)? In what manner?

P3.3   For steady low-Reynolds-number (laminar) flow through a long tube (see Prob. 1.12), the axial velocity distribution is given by $u = C(R^2 - r^2)$, where R is the tube radius and $r \leq R$. Integrate $u(r)$ to find the total volume flow Q through the tube.

P3.4   A fire hose has a 5-in inside diameter and water is flowing at 600 gal/min. The flow exits through a nozzle contraction with a diameter D_n. For steady flow, what should D_n be, in inches, to create an average exit velocity of 25 m/s?

*P3.5   A theory proposed by S. I. Pai in 1953 gives the following velocity values $u(r)$ for turbulent (high-Reynolds-number) airflow in a 4-cm-diameter tube:

r, cm	0	0.25	0.5	0.75	1.0	1.25	1.5	1.75	2.0
u, m/s	6.00	5.97	5.88	5.72	5.51	5.23	4.89	4.43	0.00

Comment on these data vis-à-vis laminar flow, Prob. P3.3. Estimate, as best you can, the total volume flow Q through the tube, in m^3/s.

P3.6   When a gravity-driven liquid jet issues from a slot in a tank, as in Fig. P3.6, an approximation for the exit velocity distribution is $u \approx \sqrt{2g(h - z)}$, where h is the depth of the jet centerline. Near the slot, the jet is horizontal, two-dimensional, and of thickness $2L$, as shown. Find a general expression for the total volume flow Q issuing from the slot; then take the limit of your result if $L \ll h$.

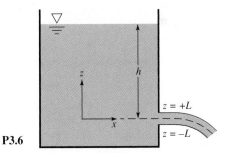

P3.6

P3.7   A spherical tank, of diameter 35 cm, is leaking air through a 5-mm-diameter hole in its side. The air exits the hole at 360 m/s and a density of 2.5 kg/m^3. Assuming uniform mixing, (a) find a formula for the rate of change of average density in the tank and (b) calculate a numerical value for ($d\rho/dt$) in the tank for the given data.

P3.8   Three pipes steadily deliver water at 20°C to a large exit pipe in Fig. P3.8. The velocity $V_2 = 5$ m/s, and the exit flow rate $Q_4 = 120 \text{ m}^3/\text{h}$. Find (a) V_1, (b) V_3, and (c) V_4 if it is known that increasing Q_3 by 20 percent would increase Q_4 by 10 percent.

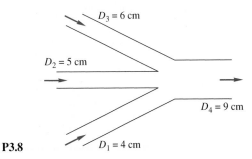

P3.8

P3.9   A laboratory test tank contains seawater of salinity S and density ρ. Water enters the tank at conditions (S_1, ρ_1, A_1, V_1) and is assumed to mix immediately in the tank. Tank water leaves through an outlet A_2 at velocity V_2. If salt is a "conservative" property (neither created nor destroyed), use the Reynolds transport theorem to find an expression for the rate of change of salt mass M_{salt} within the tank.

P3.10   Water flowing through an 8-cm-diameter pipe enters a porous section, as in Fig. P3.10, which allows a uniform radial velocity v_w through the wall surfaces for a distance of 1.2 m. If the entrance average velocity V_1 is 12 m/s, find the exit velocity V_2 if (a) v_w = 15 cm/s out of the pipe walls or (b) v_w = 10 cm/s into the pipe. (c) What value of v_w will make V_2 = 9 m/s?

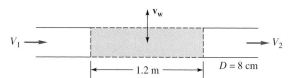

P3.10

P3.11   A room contains dust of uniform concentration $C = \rho_{\text{dust}}/\rho$. It is to be cleaned up by introducing fresh air at velocity V_i through a duct of area A_i on one wall and exhausting the room air at velocity V_o through a duct A_o on the opposite wall. Find an expression for the instantaneous rate of change of dust mass within the room.

P3.12   The pipe flow in Fig. P3.12 fills a cylindrical surge tank as shown. At time $t = 0$, the water depth in the tank is 30 cm. Estimate the time required to fill the remainder of the tank.

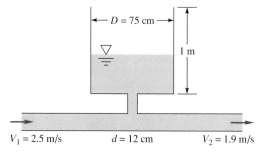

P3.12

P3.13   The cylindrical container in Fig. P3.13 is 20 cm in diameter and has a conical contraction at the bottom with an exit hole 3 cm in diameter. The tank contains fresh water at standard sea-level conditions. If the water surface is falling at the nearly steady rate $dh/dt \approx -0.072$ m/s, estimate the average velocity V out of the bottom exit.

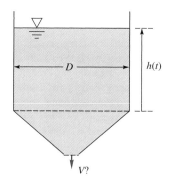

P3.13

P3.14   The open tank in Fig. P3.14 contains water at 20°C and is being filled through section 1. Assume incompressible flow. First derive an analytic expression for the water-level change dh/dt in terms of arbitrary volume flows (Q_1, Q_2, Q_3) and tank diameter d. Then, if the water level h is constant, determine the exit velocity V_2 for the given data V_1 = 3 m/s and Q_3 = 0.01 m³/s.

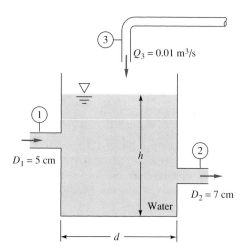

P3.14

P3.15   Water, assumed incompressible, flows steadily through the round pipe in Fig. P3.15. The entrance velocity is constant, $u = U_0$, and the exit velocity approximates turbulent flow, $u = u_{\text{max}}(1 - r/R)^{1/7}$. Determine the ratio U_0/u_{max} for this flow.

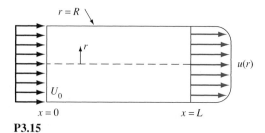

P3.15

P3.16 An incompressible fluid flows past an impermeable flat plate, as in Fig. P3.16, with a uniform inlet profile $u = U_0$ and a cubic polynomial exit profile

$$u \approx U_0\left(\frac{3\eta - \eta^3}{2}\right) \text{ where } \eta = \frac{y}{\delta}$$

Compute the volume flow Q across the top surface of the control volume.

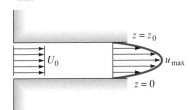

Solid plate, width b into paper

P3.16

P3.17 Incompressible steady flow in the inlet between parallel plates in Fig. P3.17 is uniform, $u = U_0 = 8$ cm/s, while downstream the flow develops into the parabolic laminar profile $u = az(z_0 - z)$, where a is a constant. If $z_0 = 4$ cm and the fluid is SAE 30 oil at 20°C, what is the value of u_{max} in cm/s?

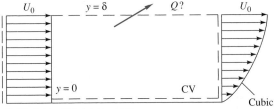

P3.17

P3.18 An incompressible fluid flows steadily through the rectangular duct in Fig. P3.18. The exit velocity profile is given approximately by

$$u = u_{max}\left(1 - \frac{y^2}{b^2}\right)\left(1 - \frac{z^2}{h^2}\right)$$

(*a*) Does this profile satisfy the correct boundary conditions for viscous fluid flow? (*b*) Find an analytical expression for the volume flow Q at the exit. (*c*) If the inlet flow is 300 ft³/min, estimate u_{max} in m/s for $b = h = 10$ cm.

P3.19 Water from a storm drain flows over an outfall onto a porous bed that absorbs the water at a uniform vertical velocity of 8 mm/s, as shown in Fig. P3.19. The system is 5 m deep into the paper. Find the length L of the bed that will completely absorb the storm water.

P3.20 Oil (SG = 0.89) enters at section 1 in Fig. P3.20 at a weight flow of 250 N/h to lubricate a thrust bearing. The steady oil flow exits radially through the narrow clearance between thrust plates. Compute (*a*) the outlet volume flux in mL/s and (*b*) the average outlet velocity in cm/s.

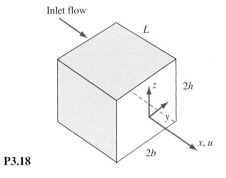

Inlet flow

P3.18

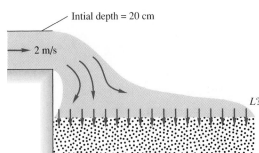

Intial depth = 20 cm

2 m/s

L?

P3.19

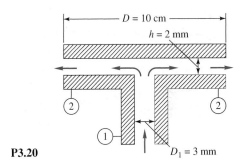

$D = 10$ cm
$h = 2$ mm
$D_1 = 3$ mm

P3.20

P3.21 Modify Prob. P3.16 as follows. Let the plate be $L = 125\delta$ long from inlet to exit. The plate is *porous* and is drawing in fluid from the boundary layer at a uniform suction velocity v_w. (*a*) Calculate Q across the top if $v_w = 0.002U_0$. (*b*) Find the ratio v_w/U_0 for which Q across the top is zero.

P3.22 The converging–diverging nozzle shown in Fig. P3.22 expands and accelerates dry air to supersonic speeds at the exit, where $p_2 = 8$ kPa and $T_2 = 240$ K. At the throat, $p_1 = 284$ kPa, $T_1 = 665$ K, and $V_1 = 517$ m/s. For steady compressible flow of an ideal gas, estimate (*a*) the mass flow in kg/h, (*b*) the velocity V_2, and (*c*) the Mach number Ma_2.

P3.23 The hypodermic needle in Fig. P3.23 contains a liquid serum (SG = 1.05). If the serum is to be injected steadily at 6 cm³/s, how fast in in/s should the plunger be

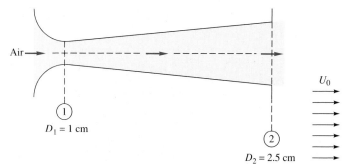

P3.22

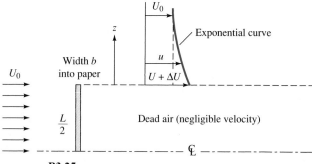

P3.25

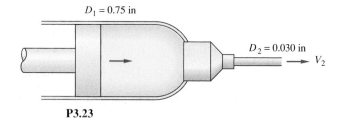

P3.23

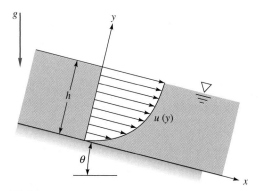

P3.26

advanced (*a*) if leakage in the plunger clearance is neglected and (*b*) if leakage is 10 percent of the needle flow?

*P3.24   Water enters the bottom of the cone in Fig. P3.24 at a uniformly increasing average velocity $V = Kt$. If d is very small, derive an analytic formula for the water surface rise $h(t)$ for the condition $h = 0$ at $t = 0$. Assume incompressible flow.

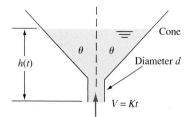

P3.24

P3.25   As will be discussed in Chaps. 7 and 8, the flow of a stream U_0 past a blunt flat plate creates a broad low-velocity *wake* behind the plate. A simple model is given in Fig. P3.25, with only half of the flow shown due to symmetry. The velocity profile behind the plate is idealized as "dead air" (near-zero velocity) behind the plate, plus a higher velocity, decaying vertically above the wake according to the variation $u \approx U_0 + \Delta U\, e^{-z/L}$, where L is the plate height and $z = 0$ is the top of the wake. Find ΔU as a function of stream speed U_0.

P3.26   A thin layer of liquid, draining from an inclined plane, as in Fig. P3.26, will have a laminar velocity profile $u \approx U_0(2y/h - y^2/h^2)$, where U_0 is the surface velocity. If the

plane has width b into the paper, determine the volume rate of flow in the film. Suppose that $h = 0.5$ in and the flow rate per foot of channel width is 1.25 gal/min. Estimate U_0 in ft/s.

P3.27   Consider a highly pressurized air tank at conditions (p_0, ρ_0, T_0) and volume v_0. In Chap. 9 we will learn that, if the tank is allowed to exhaust to the atmosphere through a well-designed converging nozzle of exit area A, the outgoing mass flow rate will be

$$\dot{m} = \frac{\alpha\, p_0 A}{\sqrt{RT_0}} \quad \text{where } \alpha \approx 0.685 \text{ for air}$$

This rate persists as long as p_0 is at least twice as large as the atmospheric pressure. Assuming constant T_0 and an ideal gas, (*a*) derive a formula for the change of density $\rho_0(t)$ within the tank. (*b*) Analyze the time Δt required for the density to decrease by 25 percent.

P3.28   According to Torricelli's theorem, the velocity of a fluid draining from a hole in a tank is $V \approx (2gh)^{1/2}$, where h is the depth of water above the hole, as in Fig. P3.28. Let the hole have area A_o and the cylindrical tank have cross-section area $A_b \gg A_o$. Derive a formula for the time to drain the tank completely from an initial depth h_o.

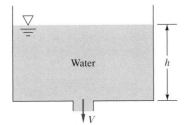

P3.28

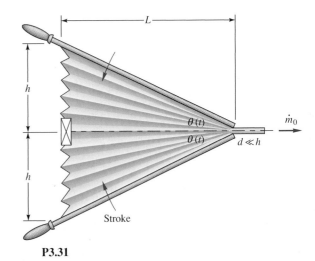

P3.31

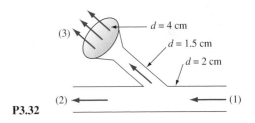

P3.32

P3.29 In elementary compressible flow theory (Chap. 9), compressed air will exhaust from a small hole in a tank at the mass flow rate $\dot{m} \approx C\rho$, where ρ is the air density in the tank and C is a constant. If ρ_0 is the initial density in a tank of volume $\mathcal{V}$, derive a formula for the density change $\rho(t)$ after the hole is opened. Apply your formula to the following case: a spherical tank of diameter 50 cm, with initial pressure 300 kPa and temperature 100°C, and a hole whose initial exhaust rate is 0.01 kg/s. Find the time required for the tank density to drop by 50 percent.

P3.30 A steady two-dimensional water jet, 4 cm thick with a weight flow rate of 1960 N/s, strikes an angled barrier as in Fig. P3.30. Pressure and water velocity are constant everywhere. Thirty percent of the jet passes through the slot. The rest splits symmetrically along the barrier. Calculate the horizontal force F needed to hold the barrier per unit thickness into the paper.

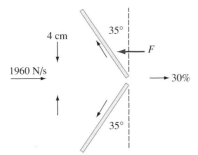

P3.30

P3.31 A bellows may be modeled as a deforming wedge-shaped volume as in Fig. P3.31. The check valve on the left (pleated) end is closed during the stroke. If b is the bellows width into the paper, derive an expression for outlet mass flow $\dot{m}_0$ as a function of stroke $\theta(t)$.

P3.32 Water at 20°C flows steadily through the piping junction in Fig. P3.32, entering section 1 at 20 gal/min. The average velocity at section 2 is 2.5 m/s. A portion of the flow is diverted through the showerhead, which contains 100 holes of 1-mm diameter. Assuming uniform shower flow, estimate the exit velocity from the showerhead jets.

P3.33 In some wind tunnels the test section is perforated to suck out fluid and provide a thin viscous boundary layer. The test section wall in Fig. P3.33 contains 1200 holes of 5-mm diameter each per square meter of wall area. The suction velocity through each hole is $V_s = 8$ m/s, and the test-section entrance velocity is $V_1 = 35$ m/s. Assuming incompressible steady flow of air at 20°C, compute (*a*) V_0, (*b*) V_2, and (*c*) V_f, in m/s.

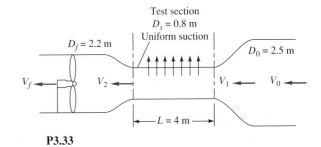

P3.33

P3.34 A rocket motor is operating steadily, as shown in Fig. P3.34. The products of combustion flowing out the exhaust nozzle approximate a perfect gas with a molecular weight of 28. For the given conditions calculate V_2 in ft/s.

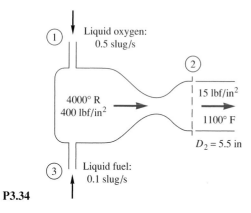

P3.34

P3.35 In contrast to the liquid rocket in Fig. P3.34, the solid-propellant rocket in Fig. P3.35 is self-contained and has no entrance ducts. Using a control volume analysis for the conditions shown in Fig. P3.35, compute the rate of mass loss of the propellant, assuming that the exit gas has a molecular weight of 28.

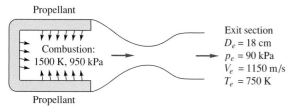

P3.35

P3.36 The jet pump in Fig. P3.36 injects water at $U_1 = 40$ m/s through a 3-in-pipe and entrains a secondary flow of water $U_2 = 3$ m/s in the annular region around the small pipe. The two flows become fully mixed downstream, where U_3 is approximately constant. For steady incompressible flow, compute U_3 in m/s.

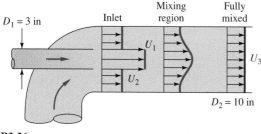

P3.36

P3.37 A solid steel cylinder, 4.5 cm in diameter and 12 cm long, with a mass of 1500 g, falls concentrically through a 5-cm-diameter vertical container filled with oil (SG =

0.89). Assuming the oil is incompressible, estimate the oil average velocity in the annular clearance between cylinder and container (*a*) relative to the container and (*b*) relative to the cylinder.

P3.38 An incompressible fluid in Fig. P3.38 is being squeezed outward between two large circular disks by the uniform downward motion V_0 of the upper disk. Assuming one-dimensional radial outflow, use the control volume shown to derive an expression for $V(r)$.

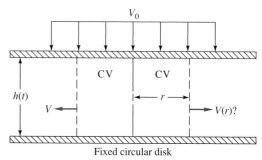

P3.38

P3.39 A wedge splits a sheet of 20°C water, as shown in Fig. P3.39. Both wedge and sheet are very long into the paper. If the force required to hold the wedge stationary is $F = 124$ N per meter of depth into the paper, what is the angle θ of the wedge?

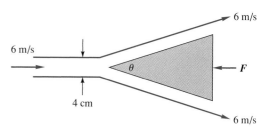

P3.39

P3.40 The water jet in Fig. P3.40 strikes normal to a fixed plate. Neglect gravity and friction, and compute the force F in newtons required to hold the plate fixed.

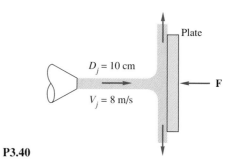

P3.40

P3.41 In Fig. P3.41 the vane turns the water jet completely around. Find an expression for the maximum jet velocity V_0 if the maximum possible support force is F_0.

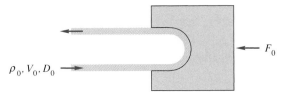

ρ_0, V_0, D_0

P3.41

P3.42 A liquid of density ρ flows through the sudden contraction in Fig. P3.42 and exits to the atmosphere. Assume uniform conditions (p_1, V_1, D_1) at section 1 and (p_2, V_2, D_2) at section 2. Find an expression for the force F exerted by the fluid on the contraction.

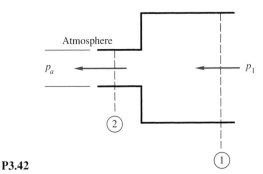

P3.42

P3.43 Water at 20°C flows through a 5-cm-diameter pipe that has a 180° vertical bend, as in Fig. P3.43. The total length of pipe between flanges 1 and 2 is 75 cm. When the weight flow rate is 230 N/s, $p_1 = 165$ kPa and $p_2 = 134$ kPa. Neglecting pipe weight, determine the total force that the flanges must withstand for this flow.

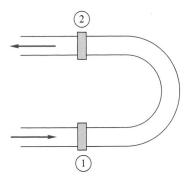

P3.43

*P3.44 When a uniform stream flows past an immersed thick cylinder, a broad low-velocity wake is created downstream, idealized as a V shape in Fig. P3.44. Pressures p_1 and p_2 are approximately equal. If the flow is two-dimensional and incompressible, with width b into the paper, derive a formula for the drag force F on the cylinder. Rewrite your result in the form of a dimensionless *drag coefficient* based on body length $C_D = F/(\rho U^2 bL)$.

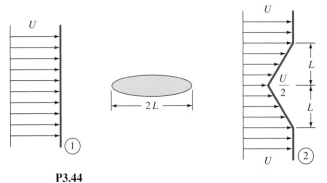

P3.44

P3.45 A 12-cm-diameter pipe, containing water flowing at 200 N/s, is capped by an orifice plate, as in Fig. P3.45. The exit jet is 25 mm in diameter. The pressure in the pipe at section 1 is 800 kPa (gage). Calculate the force F required to hold the orifice plate.

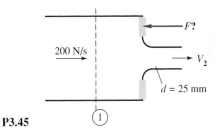

P3.45

P3.46 When a jet strikes an inclined fixed plate, as in Fig. P3.46, it breaks into two jets at 2 and 3 of equal velocity $V = V_{\text{jet}}$ but unequal fluxes αQ at 2 and $(1 - \alpha)Q$ at section 3, α being a fraction. The reason is that for frictionless flow the fluid can exert no tangential force F_t on the plate. The condition $F_t = 0$ enables us to solve for α. Perform this analysis, and find α as a function of the plate angle θ. Why doesn't the answer depend on the properties of the jet?

P3.47 A liquid jet of velocity V_j and diameter D_j strikes a fixed hollow cone, as in Fig. P3.47, and deflects back as a conical sheet at the same velocity. Find the cone angle θ for which the restraining force $F = \frac{3}{2}\rho A_j V_j^2$.

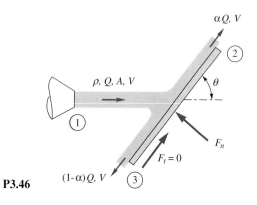

P3.46

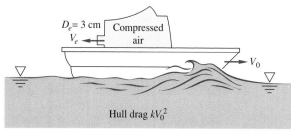

P3.47

P3.48 The small boat in Fig. P3.48 is driven at a steady speed V_0 by a jet of compressed air issuing from a 3-cm-diameter hole at $V_e = 343$ m/s. Jet exit conditions are $p_e = 1$ atm and $T_e = 30°C$. Air drag is negligible, and the hull drag is kV_0^2, where $k \approx 19$ N $\cdot$ s^2/m^2. Estimate the boat speed V_0 in m/s.

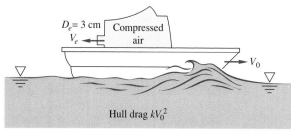

Hull drag kV_0^2

P3.48

P3.49 The horizontal nozzle in Fig. P3.49 has $D_1 = 12$ in and $D_2 = 6$ in, with inlet pressure $p_1 = 38$ lbf/in²absolute and $V_2 = 56$ ft/s. For water at 20°C, compute the horizontal force provided by the flange bolts to hold the nozzle fixed.

P3.50 The jet engine on a test stand in Fig. P3.50 admits air at 20°C and 1 atm at section 1, where $A_1 = 0.5$ m² and $V_1 = 250$ m/s. The fuel-to-air ratio is 1:30. The air

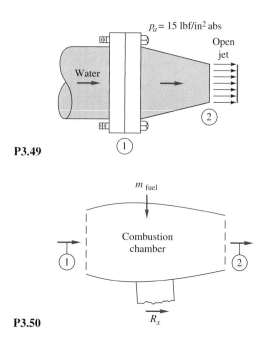

P3.49

P3.50

leaves section 2 at atmospheric pressure and higher temperature, where $V_2 = 900$ m/s and $A_2 = 0.4$ m². Compute the horizontal test stand reaction R_x needed to hold this engine fixed.

P3.51 A liquid jet of velocity V_j and area A_j strikes a single 180° bucket on a turbine wheel rotating at angular velocity Ω, as in Fig. P3.51. Derive an expression for the power P delivered to this wheel at this instant as a function of the system parameters. At what angular velocity is the maximum power delivered? How would your analysis differ if there were many, many buckets on the wheel, so that the jet was continually striking at least one bucket?

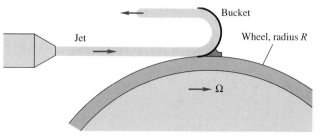

P3.51

P3.52 The vertical gate in a water channel is partially open, as in Fig. P3.52. Assuming no change in water level and a hydrostatic pressure distribution, derive an expression for the streamwise force F_x on one-half of the gate as a

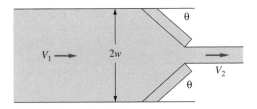

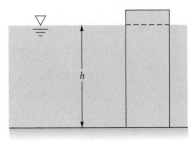

Top view

Side view

P3.52

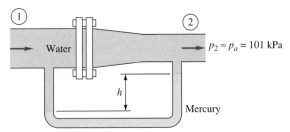

P3.54

P3.55 In Fig. P3.55 the jet strikes a vane that moves to the right at constant velocity V_c on a frictionless cart. Compute (*a*) the force F_x required to restrain the cart and (*b*) the power P delivered to the cart. Also find the cart velocity for which (*c*) the force F_x is a maximum and (*d*) the power P is a maximum.

function of $(\rho, h, w, \theta, V_1)$. Apply your result to the case of water at 20°C, $V_1 = 0.8$ m/s, $h = 2$ m, $w = 1.5$ m, and $\theta = 50°$.

P3.53 Consider incompressible flow in the entrance of a circular tube, as in Fig. P3.53. The inlet flow is uniform, $u_1 = U_0$. The flow at section 2 is developed pipe flow. Find the wall drag force F as a function of (p_1, p_2, ρ, U_0, R) if the flow at section 2 is

(*a*) Laminar: $u_2 = u_{max}\left(1 - \dfrac{r^2}{R^2}\right)$

(*b*) Turbulent: $u_2 \approx u_{max}\left(1 - \dfrac{r}{R}\right)^{1/7}$

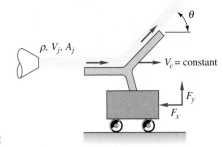

P3.55

P3.56 Water at 20°C flows steadily through the box in Fig. P3.56, entering station (1) at 2 m/s. Calculate the (*a*) horizontal and (*b*) vertical forces required to hold the box stationary against the flow momentum.

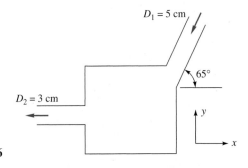

P3.56

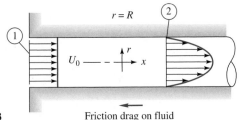

P3.53

Friction drag on fluid

P3.54 For the pipe-flow-reducing section of Fig. P3.54, $D_1 = 8$ cm, $D_2 = 5$ cm, and $p_2 = 1$ atm. All fluids are at 20°C. If $V_1 = 5$ m/s and the manometer reading is $h = 58$ cm, estimate the total force resisted by the flange bolts.

P3.57 Water flows through the duct in Fig. P3.57, which is 50 cm wide and 1 m deep into the paper. Gate *BC* completely closes the duct when $\beta = 90°$. Assuming one-dimensional flow, for what angle β will the force of the exit jet on the plate be 3 kN?

P3.58 The water tank in Fig. P3.58 stands on a frictionless cart and feeds a jet of diameter 4 cm and velocity 8 m/s,

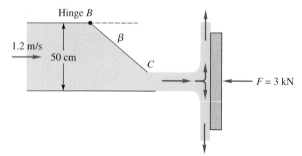

P3.57

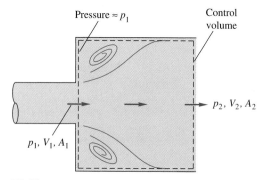

P3.58

which is deflected 60° by a vane. Compute the tension in the supporting cable.

P3.59 When a pipe flow suddenly expands from A_1 to A_2, as in Fig. P3.59, low-speed, low-friction eddies appear in the corners and the flow gradually expands to A_2 downstream. Using the suggested control volume for incompressible steady flow and assuming that $p \approx p_1$ on the corner annular ring as shown, show that the downstream pressure is given by

$$p_2 = p_1 + \rho V_1^2 \frac{A_1}{A_2}\left(1 - \frac{A_1}{A_2}\right)$$

Neglect wall friction.

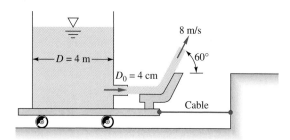

P3.59

P3.60 Water at 20°C flows through the elbow in Fig. P3.60 and exits to the atmosphere. The pipe diameter is $D_1 = 10$ cm, while $D_2 = 3$ cm. At a weight flow rate of 150 N/s, the pressure $p_1 = 2.3$ atm (gage). Neglecting the weight of water and elbow, estimate the force on the flange bolts at section 1.

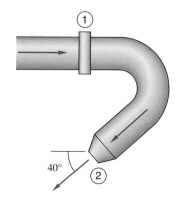

P3.60

P3.61 A 20°C water jet strikes a vane mounted on a tank with frictionless wheels, as in Fig. P3.61. The jet turns and falls into the tank without spilling out. If $\theta = 30°$, evaluate the horizontal force F required to hold the tank stationary.

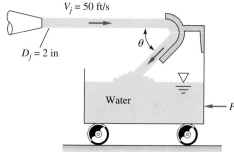

P3.61

P3.62 Water at 20°C exits to the standard sea-level atmosphere through the split nozzle in Fig. P3.62. Duct areas are

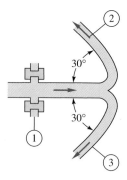

P3.62

$A_1 = 0.02$ m^2 and $A_2 = A_3 = 0.008$ m^2. If $p_1 = 135$ kPa (absolute) and the flow rate is $Q_2 = Q_3 = 275$ m^3/h, compute the force on the flange bolts at section 1.

P3.63 In Example 3.10, the gate force F is a function of both water depth and velocity. (*a*) Nondimensionalize the force by dividing by ($\rho g b h_1^2$), and plot this force versus $h_2/h_1 \leq 1.0$. (*b*) The plot involves a second dimensionless parameter involving V_1. Do you know its name? (*c*) For what condition h_2/h_1 is the force largest? (*d*) For small values of V_1, the force becomes negative (to the right), which is totally unrealistic. Can you explain why?

P3.64 The 6-cm-diameter 20°C water jet in Fig. P3.64 strikes a plate containing a hole of 4-cm diameter. Part of the jet passes through the hole, and part is deflected. Determine the horizontal force required to hold the plate.

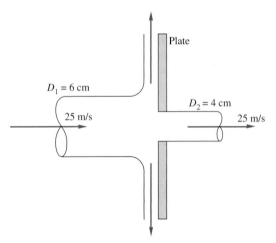

Plate

$D_1 = 6$ cm

25 m/s

$D_2 = 4$ cm

25 m/s

P3.64

P3.65 The box in Fig. P3.65 has three 0.5-in holes on the right side. The volume flows of 20°C water shown are steady, but the details of the interior are not known. Compute the force, if any, that this water flow causes on the box.

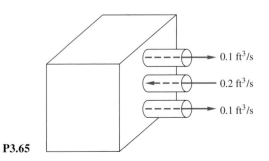

0.1 ft^3/s

0.2 ft^3/s

0.1 ft^3/s

P3.65

P3.66 The tank in Fig. P3.66 weighs 500 N empty and contains 600 L of water at 20°C. Pipes 1 and 2 have equal diameters of 6 cm and equal steady volume flows of 300 m^3/h. What should the scale reading W be in N?

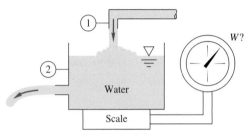

W?

Water

Scale

P3.66

P3.67 Gravel is dumped from a hopper, at a rate of 650 N/s, onto a moving belt, as in Fig. P3.67. The gravel then passes off the end of the belt. The drive wheels are 80 cm in diameter and rotate clockwise at 150 r/min. Neglecting system friction and air drag, estimate the power required to drive this belt.

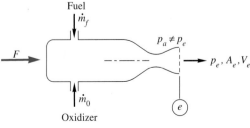

P3.67

P3.68 The rocket in Fig. P3.68 has a supersonic exhaust, and the exit pressure p_e is not necessarily equal to p_a. Show that the force F required to hold this rocket on the test stand is $F = \rho_e A_e V_e^2 + A_e(p_e - p_a)$. Is this force F what we term the *thrust* of the rocket?

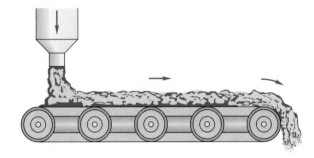

Fuel

$\dot{m}_f$

$p_a \neq p_e$

F

p_e, A_e, V_e

$\dot{m}_0$

Oxidizer

e

P3.68

P3.69   A uniform rectangular plate, 40 cm long and 30 cm deep into the paper, hangs in air from a hinge at its top (the 30-cm side). It is struck in its center by a horizontal 3-cm-diameter jet of water moving at 8 m/s. If the gate has a mass of 16 kg, estimate the angle at which the plate will hang from the vertical.

P3.70   The dredger in Fig. P3.70 is loading sand (SG = 2.6) onto a barge. The sand leaves the dredger pipe at 4 ft/s with a weight flux of 850 lbf/s. Estimate the tension on the mooring line caused by this loading process.

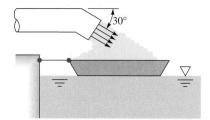

P3.70

P3.71   Suppose that a deflector is deployed at the exit of the jet engine of Prob. P3.50, as shown in Fig. P3.71. What will the reaction R_x on the test stand be now? Is this reaction sufficient to serve as a braking force during airplane landing?

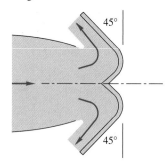

P3.71

P3.72   When immersed in a uniform stream, a thick elliptical cylinder creates a broad downstream wake, as idealized in Fig. P3.72. The pressure at the upstream and downstream sections are approximately equal, and the fluid is water at 20°C. If U_0 = 4 m/s and L = 80 cm, estimate the drag force on the cylinder per unit width into the paper. Also compute the dimensionless drag coefficient $C_D = 2F/(\rho U_0^2 bL)$.

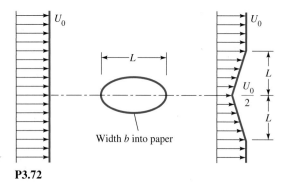

P3.72

P3.73   A pump in a tank of water at 20°C directs a jet at 45 ft/s and 200 gal/min against a vane, as shown in Fig. P3.73. Compute the force F to hold the cart stationary if the jet follows (a) path A or (b) path B. The tank holds 550 gal of water at this instant.

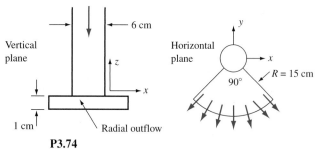

P3.73

P3.74   Water at 20°C flows down through a vertical, 6-cm-diameter tube at 300 gal/min, as in Fig. P3.74. The flow then turns horizontally and exits through a 90° radial duct segment 1 cm thick, as shown. If the radial outflow is uniform and steady, estimate the forces (F_x, F_y, F_z) required to support this system against fluid momentum changes.

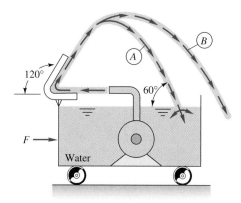

P3.74

P3.75   A jet of liquid of density ρ and area A strikes a block and splits into two jets, as in Fig. P3.75. Assume the same velocity V for all three jets. The upper jet exits at an angle θ and area αA. The lower jet is turned 90° downward. Neglecting fluid weight, (a) derive a formula for the forces (F_x, F_y) required to support the block against fluid momentum changes. (b) Show that $F_y = 0$ only if $\alpha \geq 0.5$. (c) Find the values of α and θ for which both F_x and F_y are zero.

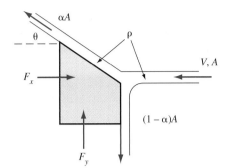

P3.75

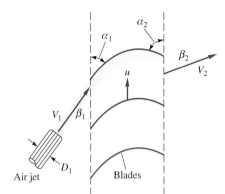

P3.78

P3.76 A two-dimensional sheet of water, 10 cm thick and moving at 7 m/s, strikes a fixed wall inclined at 20° with respect to the jet direction. Assuming frictionless flow, find (*a*) the normal force on the wall per meter of depth, and find the widths of the sheet deflected (*b*) upstream and (*c*) downstream along the wall.

P3.77 Water at 20°C flows steadily through a reducing pipe bend, as in Fig. P3.77. Known conditions are $p_1 = 350$ kPa, $D_1 = 25$ cm, $V_1 = 2.2$ m/s, $p_2 = 120$ kPa, and $D_2 = 8$ cm. Neglecting bend and water weight, estimate the total force that must be resisted by the flange bolts.

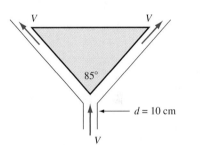

P3.79

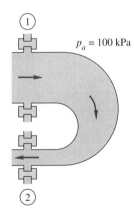

P3.77

P3.78 A fluid jet of diameter D_1 enters a cascade of moving blades at absolute velocity V_1 and angle β_1, and it leaves at absolute velocity V_2 and angle β_2, as in Fig. P3.78. The blades move at velocity u. Derive a formula for the power P delivered to the blades as a function of these parameters.

P3.79 Air at 20°C and 1 atm enters the bottom of an 85° conical flowmeter duct at a mass flow of 0.3 kg/s, as shown in Fig. P3.79. It is able to support a centered conical body by steady annular flow around the cone, as shown. The air velocity at the upper edge of the body equals the entering velocity. Estimate the weight of the body, in newtons.

P3.80 A river of width b and depth h_1 passes over a submerged obstacle, or "drowned weir," in Fig. P3.80, emerging at a new flow condition (V_2, h_2). Neglect atmospheric pressure, and assume that the water pressure is hydrostatic at both sections 1 and 2. Derive an expression for the force exerted by the river on the obstacle in terms of V_1, h_1, h_2, b, ρ, and g. Neglect water friction on the river bottom.

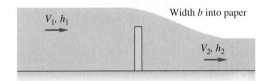

P3.80

P3.81 Torricelli's idealization of efflux from a hole in the side of a tank is $V = \sqrt{2gh}$, as shown in Fig. P3.81. The cylindrical tank weighs 150 N when empty and contains water at 20°C. The tank bottom is on very smooth ice (static friction coefficient $\zeta \approx 0.01$). The hole diameter is 9 cm. For what water depth h will the tank just begin to move to the right?

*P3.82 The model car in Fig. P3.82 weighs 17 N and is to be accelerated from rest by a 1-cm-diameter water jet moving at 75 m/s. Neglecting air drag and wheel friction, estimate the velocity of the car after it has moved forward 1 m.

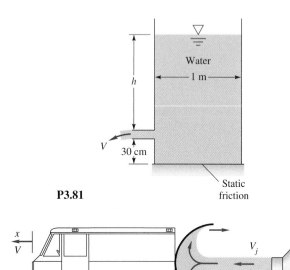

P3.81

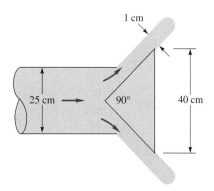

P3.82

P3.83 Gasoline at 20°C is flowing at $V_1 = 12$ m/s in a 5-cm-diameter pipe when it encounters a 1-m length of uniform radial wall suction. At the end of this suction region, the average fluid velocity has dropped to $V_2 = 10$ m/s. If $p_1 = 120$ kPa, estimate p_2 if the wall friction losses are neglected.

P3.84 Air at 20°C and 1 atm flows in a 25-cm-diameter duct at 15 m/s, as in Fig. P3.84. The exit is choked by a 90° cone, as shown. Estimate the force of the airflow on the cone.

1 cm
25 cm ⟶ 90° 40 cm

P3.84

P3.85 The thin-plate orifice in Fig. P3.85 causes a large pressure drop. For 20°C water flow at 500 gal/min, with pipe $D = 10$ cm and orifice $d = 6$ cm, $p_1 - p_2 \approx 145$ kPa. If the wall friction is negligible, estimate the force of the water on the orifice plate.

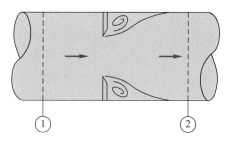

P3.85 (1) (2)

P3.86 For the water jet pump of Prob. P3.36, add the following data: $p_1 = p_2 = 25$ lbf/in^2, and the distance between sections 1 and 3 is 80 in. If the average wall shear stress between sections 1 and 3 is 7 lbf/ft^2, estimate the pressure p_3. Why is it higher than p_1?

P3.87 Figure P3.87 simulates a *manifold* flow, with fluid removed from a porous wall or perforated section of pipe. Assume incompressible flow with negligible wall friction and small suction $V_w \ll V_1$. If (p_1, V_1, V_w, ρ, D) are known, derive expressions for (a) V_2 and (b) p_2.

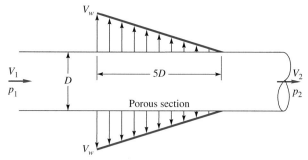

P3.87

P3.88 The boat in Fig. P3.88 is jet-propelled by a pump that develops a volume flow rate Q and ejects water out the stern at velocity V_j. If the boat drag force is $F = kV^2$, where k is a constant, develop a formula for the steady forward speed V of the boat.

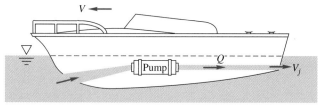

P3.88

P3.89 Consider Fig. P3.36 as a general problem for analysis of a mixing ejector pump. If all conditions (p, ρ, V) are known at sections 1 and 2 and if the wall friction is negligible, derive formulas for estimating (a) V_3 and (b) p_3.

P3.90   As shown in Fig. P3.90, a liquid column of height h is confined in a vertical tube of cross-sectional area A by a stopper. At $t = 0$ the stopper is suddenly removed, exposing the bottom of the liquid to atmospheric pressure. Using a control volume analysis of mass and vertical momentum, derive the differential equation for the downward motion $V(t)$ of the liquid. Assume one-dimensional, incompressible, frictionless flow.

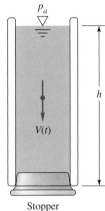

P3.90   Stopper

P3.91   Extend Prob. P3.90 to include a linear (laminar) average wall shear stress resistance of the form $\tau \approx cV$, where c is a constant. Find the differential equation for dV/dt and then solve for $V(t)$, assuming for simplicity that the wall area remains constant.

*P3.92   A more involved version of Prob. P3.90 is the elbow-shaped tube in Fig. P3.92, with constant cross-sectional area A and diameter $D \ll h, L$. Assume incompressible flow, neglect friction, and derive a differential equation for dV/dt when the stopper is opened. *Hint:* Combine two control volumes, one for each leg of the tube.

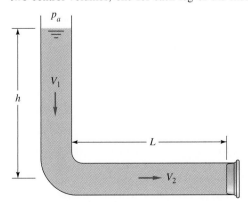

P3.92

P3.93   Extend Prob. P3.92 to include a linear (laminar) average wall shear stress resistance of the form $\tau \approx cV$, where c is a constant. Find the differential equation for dV/dt and then solve for $V(t)$, assuming for simplicity that the wall area remains constant.

P3.94   Attempt a numerical solution of Prob. P3.93 for SAE 30 oil at 20°C. Let $h = 20$ cm, $L = 15$ cm, and $D = 4$ mm. Use the laminar shear approximation from Sec. 6.4: $\tau \approx 8\mu V/D$, where μ is the fluid viscosity. Account for the decrease in wall area wetted by the fluid. Solve for the time required to empty (a) the vertical leg and (b) the horizontal leg.

P3.95   A tall water tank discharges through a well-rounded orifice, as in Fig. P3.95. Use the Torricelli formula of Prob. P3.81 to estimate the exit velocity. (a) If, at this instant, the force F required to hold the plate is 40 N, what is the depth h? (b) If the tank surface is dropping at the rate of 2.5 cm/s, what is the tank diameter D?

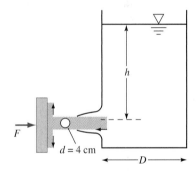

P3.95

P3.96   Extend Prob. P3.90 to the case of the liquid motion in a frictionless U-tube whose liquid column is displaced a distance Z upward and then released, as in Fig. P3.96.

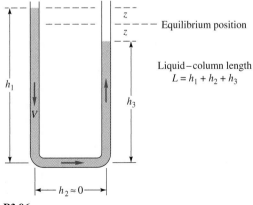

P3.96

Neglect the short horizontal leg, and combine control volume analyses for the left and right legs to derive a single differential equation for $V(t)$ of the liquid column.

*P3.97   Extend Prob. P3.96 to include a linear (laminar) average wall shear stress resistance of the form $\tau \approx 8\mu V/D$, where μ is the fluid viscosity. Find the differential equation for dV/dt and then solve for $V(t)$, assuming an initial displacement $z = z_0$, $V = 0$ at $t = 0$. The result should be a damped oscillation tending toward $z = 0$.

*P3.98   As an extension of Example 3.10, let the plate and its cart (see Fig. 3.10a) be unrestrained horizontally, with frictionless wheels. Derive (a) the equation of motion for cart velocity $V_c(t)$ and (b) a formula for the time required for the cart to accelerate from rest to 90 percent of the jet velocity (assuming the jet continues to strike the plate horizontally). (c) Compute numerical values for part (b) using the conditions of Example 3.10 and a cart mass of 2 kg.

P3.99   Let the rocket of Fig. E3.12 start at $z = 0$, with constant exit velocity and exit mass flow, and rise vertically with zero drag. (a) Show that, as long as fuel burning continues, the vertical height $S(t)$ reached is given by

$$S = \frac{V_e M_o}{\dot{m}} [\zeta \ln \zeta - \zeta + 1], \text{ where } \zeta = 1 - \frac{\dot{m}t}{M_o}$$

(b) Apply this to the case $V_e = 1500$ m/s and $M_o = 1000$ kg to find the height reached after a burn of 30 seconds, when the final rocket mass is 400 kg.

P3.100   Suppose that the solid-propellant rocket of Prob. P3.35 is built into a missile of diameter 70 cm and length 4 m. The system weighs 1800 N, which includes 700 N of propellant. Neglect air drag. If the missile is fired vertically from rest at sea level, estimate (a) its velocity and height at fuel burnout and (b) the maximum height it will attain.

P3.101   Modify Prob. P3.100 by accounting for air drag on the missile $F \approx C\rho D^2 V^2$, where $C \approx 0.02$, ρ is the air density, D is the missile diameter, and V is the missile velocity. Solve numerically for (a) the velocity and altitude at burnout and (b) the maximum altitude attained.

P3.102   As can often be seen in a kitchen sink when the faucet is running, a high-speed channel flow (V_1, h_1) may "jump" to a low-speed, low-energy condition (V_2, h_2) as in Fig. P3.102. The pressure at sections 1 and 2 is approximately hydrostatic, and wall friction is negligible. Use the continuity and momentum relations to find h_2 and V_2 in terms of (h_1, V_1).

*P3.103   Suppose that the solid-propellant rocket of Prob. P3.35 is mounted on a 1000-kg car to propel it up a long slope of 15°. The rocket motor weighs 900 N, which includes 500 N of propellant. If the car starts from rest when

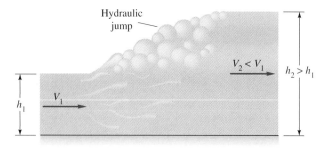

P3.102

the rocket is fired, and if air drag and wheel friction are neglected, estimate the maximum distance that the car will travel up the hill.

P3.104   A rocket is attached to a rigid horizontal rod hinged at the origin as in Fig. P3.104. Its initial mass is M_0, and its exit properties are $\dot{m}$ and V_e relative to the rocket. Set up the differential equation for rocket motion, and solve for the angular velocity $\omega(t)$ of the rod. Neglect gravity, air drag, and the rod mass.

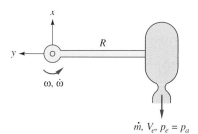

P3.104

P3.105   Extend Prob. P3.104 to the case where the rocket has a linear air drag force $F = cV$, where c is a constant. Assuming no burnout, solve for $\omega(t)$ and find the *terminal* angular velocity—that is, the final motion when the angular acceleration is zero. Apply to the case $M_0 = 6$ kg, $R = 3$ m, $\dot{m} = 0.05$ kg/s, $V_e = 1100$ m/s, and $c = 0.075$ N · s/m to find the angular velocity after 12 s of burning.

P3.106   Extend Prob. P3.104 to the case where the rocket has a quadratic air drag force $F = kV^2$, where k is a constant. Assuming no burnout, solve for $\omega(t)$ and find the *terminal* angular velocity—that is, the final motion when the angular acceleration is zero. Apply to the case $M_0 = 6$ kg, $R = 3$ m, $\dot{m} = 0.05$ kg/s, $V_e = 1100$ m/s, and $k = 0.0011$ N · s^2/m^2 to find the angular velocity after 12 s of burning.

P3.107   The cart in Fig. P3.107 moves at constant velocity $V_0 = 12$ m/s and takes on water with a scoop 80 cm wide that dips $h = 2.5$ cm into a pond. Neglect air drag and wheel friction. Estimate the force required to keep the cart moving.

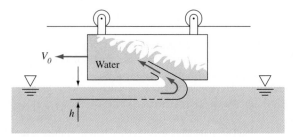

P3.107

*P3.108 A rocket sled of mass M is to be decelerated by a scoop, as in Fig. P3.108, which has width b into the paper and dips into the water a depth h, creating an upward jet at $60°$. The rocket thrust is T to the left. Let the initial velocity be V_0, and neglect air drag and wheel friction. Find an expression for $V(t)$ of the sled for (a) $T = 0$ and (b) finite $T \neq 0$.

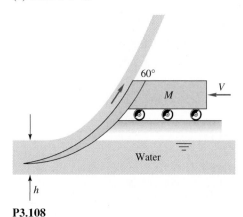

P3.108

P3.109 Apply Prob. P3.108 to the following case: $M_{total} = 900$ kg, $b = 60$ cm, $h = 2$ cm, and $V_0 = 120$ m/s, with the rocket of Prob. P3.35 attached and burning. Estimate V after 3 s.

P3.110 The horizontal lawn sprinkler in Fig. P3.110 has a water flow rate of 4.0 gal/min introduced vertically through the center. Estimate (a) the retarding torque required to keep the arms from rotating and (b) the rotation rate (r/min) if there is no retarding torque.

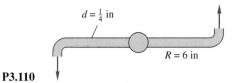

P3.110

P3.111 In Prob. P3.60 find the torque caused around flange 1 if the center point of exit 2 is 1.2 m directly below the flange center.

P3.112 The wye joint in Fig. P3.112 splits the pipe flow into equal amounts $Q/2$, which exit, as shown, a distance R_0 from the axis. Neglect gravity and friction. Find an expression for the torque T about the x axis required to keep the system rotating at angular velocity Ω.

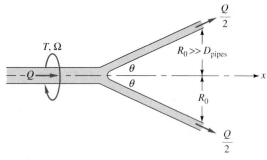

P3.112

P3.113 Modify Example 3.14 so that the arm starts from rest and spins up to its final rotation speed. The moment of inertia of the arm about O is I_0. Neglecting air drag, find $d\omega/dt$ and integrate to determine the angular velocity $\omega(t)$, assuming $\omega = 0$ at $t = 0$.

P3.114 The three-arm lawn sprinkler of Fig. P3.114 receives $20°C$ water through the center at 2.7 m³/h. If collar friction is negligible, what is the steady rotation rate in r/min for (a) $\theta = 0°$ and (b) $\theta = 40°$?

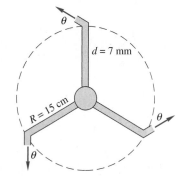

P3.114

P3.115 Water at $20°C$ flows at 30 gal/min through the 0.75-in-diameter double pipe bend of Fig. P3.115. The pressures are $p_1 = 30$ lbf/in² and $p_2 = 24$ lbf/in². Compute the torque T at point B necessary to keep the pipe from rotating.

P3.116 The centrifugal pump of Fig. P3.116 has a flow rate Q and exits the impeller at an angle θ_2 relative to the blades, as shown. The fluid enters axially at section 1. Assuming incompressible flow at shaft angular velocity ω, derive a formula for the power P required to drive the impeller.

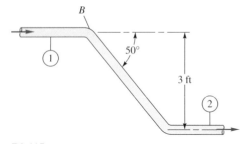

P3.115

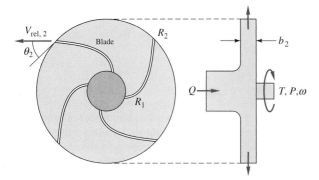

P3.116

P3.117 A simple turbomachine is constructed from a disk with two internal ducts that exit tangentially through square holes, as in Fig. P3.117. Water at 20°C enters normal to the disk at the center, as shown. The disk must drive, at 250 r/min, a small device whose retarding torque is 1.5 N · m. What is the proper mass flow of water, in kg/s?

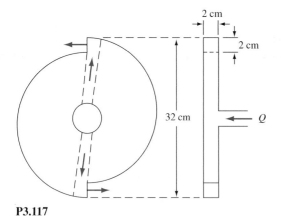

P3.117

P3.118 Reverse the flow in Fig. P3.116, so that the system operates as a radial-inflow *turbine*. Assuming that the outflow

into section 1 has no tangential velocity, derive an expression for the power P extracted by the turbine.

P3.119 Revisit the turbine cascade system of Prob. P3.78, and derive a formula for the power P delivered, using the *angular* momentum theorem of Eq. (3.55).

P3.120 A centrifugal pump impeller delivers 4000 gal/min of water at 20°C with a shaft rotation rate of 1750 r/min. Neglect losses. If $r_1 = 6$ in, $r_2 = 14$ in, $b_1 = b_2 = 1.75$ in, $V_{t1} = 10$ ft/s, and $V_{t2} = 110$ ft/s, compute the absolute velocities (*a*) V_1 and (*b*) V_2 and (*c*) the horsepower required. (*d*) Compare with the ideal horsepower required.

P3.121 The pipe bend of Fig. P3.121 has $D_1 = 27$ cm and $D_2 = 13$ cm. When water at 20°C flows through the pipe at 4000 gal/min, $p_1 = 194$ kPa (gage). Compute the torque required at point B to hold the bend stationary.

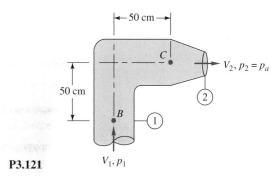

P3.121

*P3.122 Extend Prob. P3.46 to the problem of computing the center of pressure L of the normal face F_n, as in Fig. P3.122. (At the center of pressure, no moments are required to hold the plate at rest.) Neglect friction. Express your result in terms of the sheet thickness h_1 and the angle θ between the plate and the oncoming jet 1.

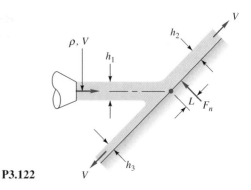

P3.122

P3.123 The waterwheel in Fig. P3.123 is being driven at 200 r/min by a 150-ft/s jet of water at 20°C. The jet diameter is 2.5 in. Assuming no losses, what is the horsepower developed by the wheel? For what speed Ω r/min will the horsepower developed be a maximum? Assume that there are many buckets on the waterwheel.

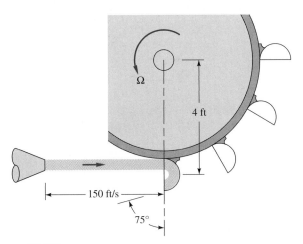

P3.123

P3.124 A rotating dishwasher arm delivers at 60°C to six nozzles, as in Fig. P3.124. The total flow rate is 3.0 gal/min. Each nozzle has a diameter of $\frac{3}{16}$ in. If the nozzle flows are equal and friction is neglected, estimate the steady rotation rate of the arm, in r/min.

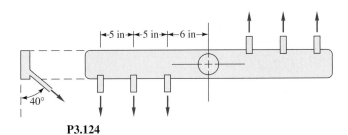

P3.124

*P3.125 A liquid of density ρ flows through a 90° bend as shown in Fig. P3.125 and issues vertically from a uniformly porous section of length L. Neglecting pipe and liquid weight, derive an expression for the torque M at point 0 required to hold the pipe stationary.

P3.126 There is a steady isothermal flow of water at 20°C through the device in Fig. P3.126. Heat-transfer, gravity, and temperature effects are negligible. Known data are $D_1 = 9$ cm, $Q_1 = 220$ m³/h, $p_1 = 150$ kPa, $D_2 = 7$ cm, $Q_2 = 100$ m³/h, $p_2 = 225$ kPa, $D_3 = 4$ cm, and $p_3 = 265$ kPa. Compute the rate of shaft work done for this device and its direction.

P3.127 A power plant on a river, as in Fig. P3.127, must eliminate 55 MW of waste heat to the river. The river

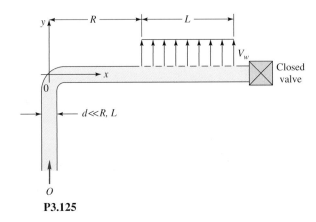

P3.125

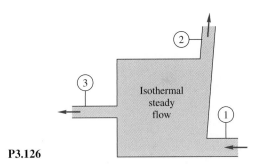

P3.126

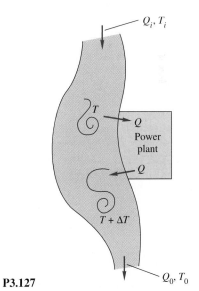

P3.127

conditions upstream are $Q_i = 2.5$ m³/s and $T_i = 18$°C. The river is 45 m wide and 2.7 m deep. If heat losses to the atmosphere and ground are negligible, estimate the downstream river conditions (Q_0, T_0).

P3.128   For the conditions of Prob. P3.127, if the power plant is to heat the nearby river water by no more than 12°C, what should be the minimum flow rate Q, in m^3/s, through the plant heat exchanger? How will the value of Q affect the downstream conditions (Q_0, T_0)?

P3.129   Multnomah Falls in the Columbia River Gorge has a sheer drop of 543 ft. Using the steady flow energy equation, estimate the water temperature change in °F caused by this drop.

P3.130   When the pump in Fig. P3.130 draws 220 m^3/h of water at 20°C from the reservoir, the total friction head loss is 5 m. The flow discharges through a nozzle to the atmosphere. Estimate the pump power in kW delivered to the water.

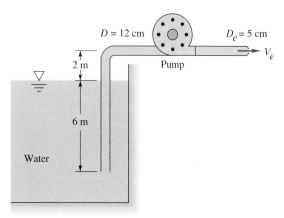

P3.130

P3.131   When the pump in Fig. P3.130 delivers 25 kW of power to the water, the friction head loss is 4 m. Estimate (a) the exit velocity V_e and (b) the flow rate Q.

P3.132   Consider a turbine extracting energy from a penstock in a dam, as in Fig. P3.132. For turbulent pipe flow (Chap. 6), the friction head loss is approximately $h_f = CQ^2$, where the constant C depends on penstock dimensions and the properties of water. Show that, for a given penstock geometry and variable river flow Q, the maximum turbine power possible in this case is $P_{max} = 2\rho g H Q/3$ and occurs when the flow rate is $Q = \sqrt{H/(3C)}$.

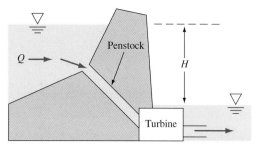

P3.132

P3.133   The long pipe in Fig. P3.133 is filled with water at 20°C. When valve A is closed, $p_1 - p_2 = 75$ kPa. When the valve is open and water flows at 500 m^3/h, $p_1 - p_2 = 160$ kPa. What is the friction head loss between 1 and 2, in m, for the flowing condition?

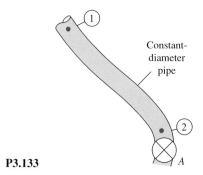

P3.133

P3.134   A 36-in-diameter pipeline carries oil (SG = 0.89) at 1 million barrels per day (bbl/day) (1 bbl = 42 U.S. gal). The friction head loss is 13 ft/1000 ft of pipe. It is planned to place pumping stations every 10 mi along the pipe. Estimate the horsepower that must be delivered to the oil by each pump.

P3.135   The *pump-turbine* system in Fig. P3.135 draws water from the upper reservoir in the daytime to produce power for a city. At night, it pumps water from lower to upper reservoirs to restore the situation. For a design flow rate of 15,000 gal/min in either direction, the friction head loss is 17 ft. Estimate the power in kW (a) extracted by the turbine and (b) delivered by the pump.

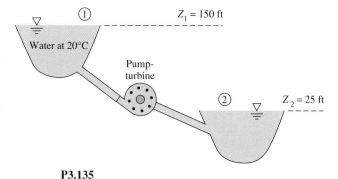

P3.135

P3.136   Water at 20°C is delivered from one reservoir to another through a long 8-cm-diameter pipe. The lower reservoir has a surface elevation $z_2 = 80$ m. The friction loss in the pipe is correlated by the formula $h_{loss} \approx 17.5(V^2/2g)$, where V is the average velocity in the pipe. If the steady flow rate through the pipe is 500 gallons per minute, estimate the surface elevation of the higher reservoir.

P3.137 A fireboat draws seawater (SG = 1.025) from a submerged pipe and discharges it through a nozzle, as in Fig. P3.137. The total head loss is 6.5 ft. If the pump efficiency is 75 percent, what horsepower motor is required to drive it?

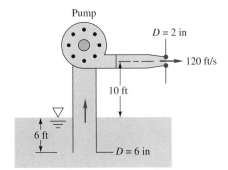

P3.137

*P3.138 Students in the fluid mechanics laboratory at Penn State use a very simple device to measure the viscosity of water as a function of temperature. The viscometer, shown in Fig. P3.138, consists of a tank, a long vertical capillary tube, a graduated cylinder, a thermometer, and a stopwatch. Because the tube has such a small diameter, the flow remains laminar. Because the tube is so long, entrance losses are negligible. It will be shown in Chap. 6 that the laminar head loss through a long pipe is given by $h_{f,\text{laminar}} = (32\mu LV)/(\rho g d^2)$, where V is the average speed through the pipe. (a) In a given experiment, diameter d, length L, and water level height H are known, and volume flow rate Q is measured with the stopwatch and graduated cylinder. The temperature of the water is also measured. The water density at this temperature is obtained by weighing a known volume of water. Write an expression for the viscosity of the water as a function of these variables. (b) Here are some actual data from an experiment: $T = 16.5°C$, $\rho = 998.7$ kg/m^3, $d = 0.041$ in, $Q = 0.310$ mL/s, $L = 36.1$ in, and $H = 0.153$ m. Calculate the viscosity of the water in kg/(m · s) based on these experimental data. (c) Compare the experimental result with the published value of μ at this temperature, and report a percentage error. (d) Compute the percentage error in the calculation of μ that would occur if a student forgot to include the kinetic energy flux correction factor in part (b) (compare results with and without inclusion of kinetic energy flux correction factor). Explain the importance (or lack of importance) of kinetic energy flux correction factor in a problem such as this.

P3.139 The horizontal pump in Fig. P3.139 discharges 20°C water at 57 m^3/h. Neglecting losses, what power in kW is delivered to the water by the pump?

P3.140 Steam enters a horizontal turbine at 350 lbf/in^2 absolute, 580°C, and 12 ft/s and is discharged at 110 ft/s and 25°C

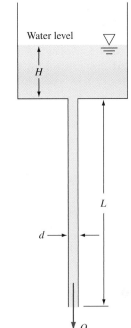

P3.138

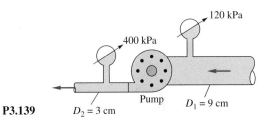

P3.139 $D_2 = 3$ cm

saturated conditions. The mass flow is 2.5 lbm/s, and the heat losses are 7 Btu/lb of steam. If head losses are negligible, how much horsepower does the turbine develop?

P3.141 Water at 20°C is pumped at 1500 gal/min from the lower to the upper reservoir, as in Fig. P3.141. Pipe friction losses are approximated by $h_f \approx 27V^2/(2g)$, where V is the average velocity in the pipe. If the pump is 75 percent efficient, what horsepower is needed to drive it?

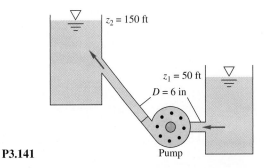

P3.141

P3.142 A typical pump has a head that, for a given shaft rotation rate, varies with the flow rate, resulting in a *pump performance curve* as in Fig. P3.142. Suppose that this pump is 75 percent efficient and is used for the system in Prob. 3.141. Estimate (*a*) the flow rate, in gal/min, and (*b*) the horsepower needed to drive the pump.

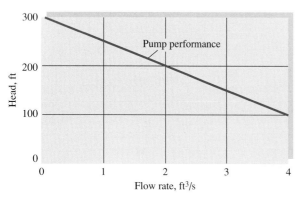

P3.142

P3.143 The insulated tank in Fig. P3.143 is to be filled from a high-pressure air supply. Initial conditions in the tank are $T = 20°C$ and $p = 200$ kPa. When the valve is opened, the initial mass flow rate into the tank is 0.013 kg/s. Assuming an ideal gas, estimate the initial rate of temperature rise of the air in the tank.

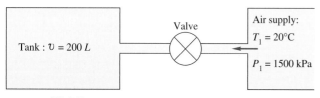

P3.143

P3.144 The pump in Fig. P3.144 creates a 20°C water jet oriented to travel a maximum horizontal distance. System friction head losses are 6.5 m. The jet may be approximated by the trajectory of frictionless particles. What power must be delivered by the pump?

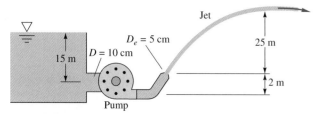

P3.144

P3.145 The large turbine in Fig. P3.145 diverts the river flow under a dam as shown. System friction losses are $h_f = 3.5V^2/(2g)$, where V is the average velocity in the supply pipe. For what river flow rate in m³/s will the power extracted be 25 MW? Which of the *two* possible solutions has a better "conversion efficiency"?

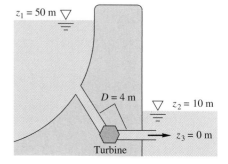

P3.145

P3.146 Kerosine at 20°C flows through the pump in Fig. P3.146 at 2.3 ft³/s. Head losses between 1 and 2 are 8 ft, and the pump delivers 8 hp to the flow. What should the mercury manometer reading h ft be?

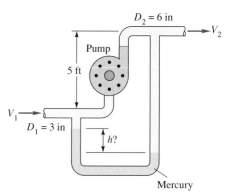

P3.146

P3.147 Repeat Prob. P3.49 by assuming that p_1 is unknown and using Bernoulli's equation with no losses. Compute the new bolt force for this assumption. What is the head loss between 1 and 2 for the data of Prob. P3.49?

P3.148 Extend the siphon analysis of Example 3.22 as follows. Let $p_1 = 1$ atm, and let the fluid be hot water at 60°C. Let $z_1, z_2,$ and z_4 be the same, with z_3 unknown. Find the value of z_3 for which the water might begin to vaporize.

P3.149 A jet of alcohol strikes the vertical plate in Fig. P3.149. A force $F \approx 425$ N is required to hold the plate stationary. Assuming there are no losses in the nozzle, estimate (*a*) the mass flow rate of alcohol and (*b*) the absolute pressure at section 1.

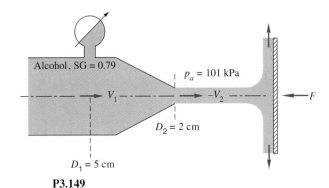

P3.149

P3.150 An airfoil at an angle of attack α, as in Fig. P3.150, provides lift by a Bernoulli effect, because the lower surface slows the flow (high pressure) and the upper surface speeds up the flow (low pressure). If the foil is 1.5 m long and 18 m wide into the paper, and the ambient air is 5000 m standard atmosphere, estimate the total lift if the average velocities on upper and lower surfaces are 215 m/s and 185 m/s, respectively. Neglect gravity. *Note:* For this case, the angle α is approximately 3°.

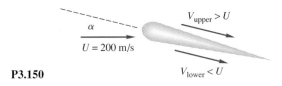

P3.150

P3.151 Water flows through a circular nozzle, exits into the air as a jet, and strikes a plate, as shown in Fig. P3.151. The force required to hold the plate steady is 70 N. Assuming steady, frictionless, one-dimensional flow, estimate (a) the velocities at sections (1) and (2) and (b) the mercury manometer reading h.

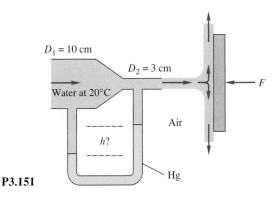

P3.151

P3.152 A free liquid jet, as in Fig. P3.152, has constant ambient pressure and small losses; hence from Bernoulli's equation $z + V^2/(2g)$ is constant along the jet. For the fire nozzle in the figure, what are (a) the minimum and (b) the maximum values of θ for which the water jet will clear the corner of the building? For which case will the jet velocity be higher when it strikes the roof of the building?

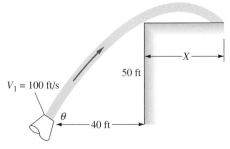

P3.152

P3.153 For the container of Fig. P3.153 use Bernoulli's equation to derive a formula for the distance X where the free jet leaving horizontally will strike the floor, as a function of h and H. For what ratio h/H will X be maximum? Sketch the three trajectories for $h/H = 0.4$, 0.5, and 0.6.

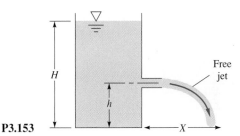

P3.153

P3.154 Water at 20°C, in the pressurized tank of Fig. P3.154, flows out and creates a vertical jet as shown. Assuming steady frictionless flow, determine the height H to which the jet rises.

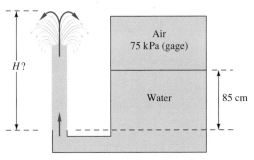

P3.154

P3.155   Bernoulli's 1738 treatise *Hydrodynamica* contains many excellent sketches of flow patterns related to his friction-less relation. One, however, redrawn here as Fig. P3.155, seems physically misleading. Can you explain what might be wrong with the figure?

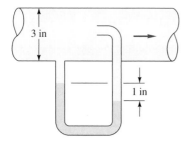

P3.155

P3.156   Extend Prob. P3.13 as follows: (*a*) Use Bernoulli's equation to estimate the elevation of the water surface above the exit of the bottom cone. (*b*) Then estimate the time required for the water surface to drop 20 cm in the cylindrical tank. If you fail to solve part (*a*), assume that the initial elevation above the exit is 52 cm. Neglect the possible contraction and nonuniformity of the exit jet mentioned in Example 3.21.

P3.157   The manometer fluid in Fig. P3.157 is mercury. Estimate the volume flow in the tube if the flowing fluid is (*a*) gasoline and (*b*) nitrogen, at 20°C and 1 atm.

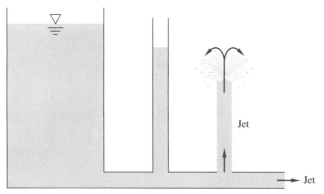

P3.157

P3.158   In Fig. P3.158 the flowing fluid is CO_2 at 20°C. Neglect losses. If $p_1 = 170$ kPa and the manometer fluid is Meriam red oil (SG = 0.827), estimate (*a*) p_2 and (*b*) the gas flow rate in m^3/h.

P3.159   The cylindrical water tank in Fig. P3.159 is being filled at a volume flow $Q_1 = 1.0$ gal/min, while the water also drains from a bottom hole of diameter $d = 6$ mm. At time $t = 0$, $h = 0$. Find and plot the variation $h(t)$ and the eventual maximum water depth

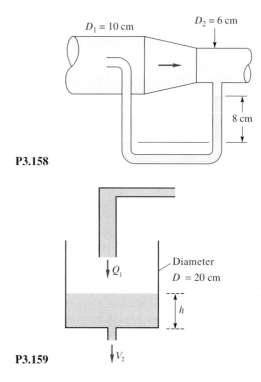

P3.158

P3.159

h_{max}. Assume that Bernoulli's steady-flow equation is valid.

P3.160   The air-cushion vehicle in Fig. P3.160 brings in sea-level standard air through a fan and discharges it at high velocity through an annular skirt of 3-cm clearance. If the vehicle weighs 50 kN, estimate (*a*) the required airflow rate and (*b*) the fan power in kW.

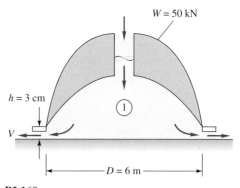

P3.160

P3.161   A necked-down section in a pipe flow, called a *venturi*, develops a low throat pressure that can aspirate fluid upward from a reservoir, as in Fig. P3.161. Using Bernoulli's equation with no losses, derive an expression for the velocity V_1 that is just sufficient to bring reservoir fluid into the throat.

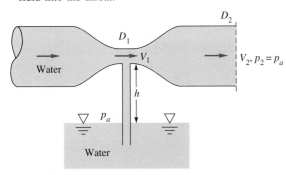

P3.161

P3.162 Suppose you are designing an air hockey table. The table is 3.0×6.0 ft in area, with $\frac{1}{16}$-in-diameter holes spaced every inch in a rectangular grid pattern (2592 holes total). The required jet speed from each hole is estimated to be 50 ft/s. Your job is to select an appropriate blower that will meet the requirements. Estimate the volumetric flow rate (in ft^3/min) and pressure rise (in lb/in^2) required of the blower. *Hint:* Assume that the air is stagnant in the large volume of the manifold under the table surface, and neglect any frictional losses.

P3.163 The liquid in Fig. P3.163 is kerosene at 20°C. Estimate the flow rate from the tank for (*a*) no losses and (*b*) pipe losses $h_f \approx 4.5V^2/(2g)$.

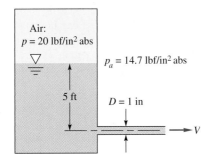

P3.163

P3.164 In Fig. P3.164 the open jet of water at 20°C exits a nozzle into sea-level air and strikes a stagnation tube as shown.

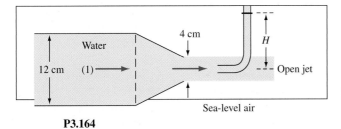

P3.164

If the pressure at the centerline at section 1 is 110 kPa, and losses are neglected, estimate (*a*) the mass flow in kg/s and (*b*) the height H of the fluid in the stagnation tube.

P3.165 A *venturi meter*, shown in Fig. P3.165, is a carefully designed constriction whose pressure difference is a measure of the flow rate in a pipe. Using Bernoulli's equation for steady incompressible flow with no losses, show that the flow rate Q is related to the manometer reading h by

$$Q = \frac{A_2}{\sqrt{1 - (D_2/D_1)^4}} \sqrt{\frac{2gh(\rho_M - \rho)}{\rho}}$$

where ρ_M is the density of the manometer fluid.

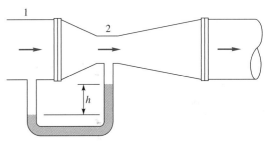

P3.165

P3.166 An open-circuit wind tunnel draws in sea-level standard air and accelerates it through a contraction into a 1-m by 1-m test section. A differential transducer mounted in the test section wall measures a pressure difference of 45 mm of water between the inside and outside. Estimate (*a*) the test section velocity in mi/h and (*b*) the absolute pressure on the front nose of a small model mounted in the test section.

P3.167 In Fig. P3.167 the fluid is gasoline at 20°C at a weight flux of 120 N/s. Assuming no losses, estimate the gage pressure at section 1.

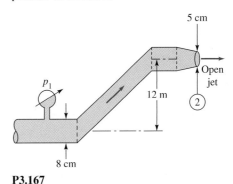

P3.167

P3.168 In Fig. P3.168 both fluids are at 20°C. If $V_1 = 1.7$ ft/s and losses are neglected, what should the manometer reading h ft be?

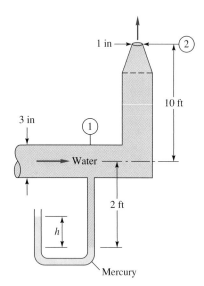

P3.168

P3.169 Extend the siphon analysis of Example 3.22 to account for friction in the tube, as follows. Let the friction head loss in the tube be correlated as $5.4(V_{tube})^2/(2g)$, which approximates turbulent flow in a 2-m-long tube. Calculate the exit velocity in m/s and the volume flow rate in cm^3/s, and compare to Example 3.22.

P3.170 If losses are neglected in Fig. P3.170, for what water level h will the flow begin to form vapor cavities at the throat of the nozzle?

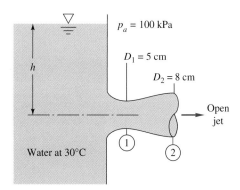

P3.170

*P3.171 For the 40°C water flow in Fig. P3.171, estimate the volume flow through the pipe, assuming no losses; then explain what is wrong with this seemingly innocent question. If the actual flow rate is $Q = 40$ m³/h, compute (a) the head loss in ft and (b) the constriction diameter D that causes cavitation, assuming that the throat divides the head loss equally and that changing the constriction causes no additional losses.

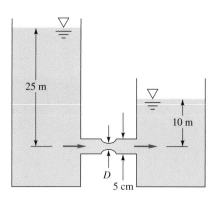

P3.171

P3.172 The 35°C water flow of Fig. P3.172 discharges to sea-level standard atmosphere. Neglecting losses, for what nozzle diameter D will cavitation begin to occur? To avoid cavitation, should you increase or decrease D from this critical value?

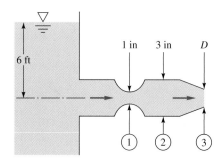

P3.172

P3.173 The horizontal wye fitting in Fig. P3.173 splits the 20°C water flow rate equally. If $Q_1 = 5$ ft³/s and $p_1 = 25$ lbf/in² (gage) and losses are neglected, estimate (a) p_2, (b) p_3, and (c) the vector force required to keep the wye in place.

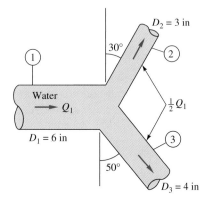

P3.173

P3.174 In Fig. P3.174 the piston drives water at 20°C. Neglecting losses, estimate the exit velocity V_2 ft/s. If D_2 is

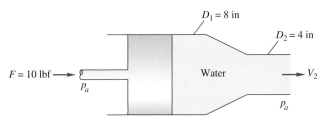

P3.174

further constricted, what is the maximum possible value of V_2?

P3.175 If the approach velocity is not too high, a hump in the bottom of a water channel causes a dip Δh in the water level, which can serve as a flow measurement. If, as shown in Fig. P3.175, $\Delta h = 10$ cm when the bump is 30 cm high, what is the volume flow Q_1 per unit width, assuming no losses? In general, is Δh proportional to Q_1?

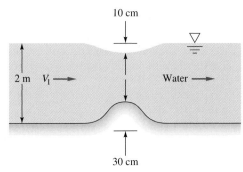

P3.175

P3.176 In the spillway flow of Fig. P3.176, the flow is assumed uniform and hydrostatic at sections 1 and 2. If losses are neglected, compute (a) V_2 and (b) the force per unit width of the water on the spillway.

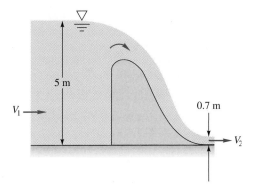

P3.176

P3.177 For the water channel flow of Fig. P3.177, $h_1 = 1.5$ m, $H = 4$ m, and $V_1 = 3$ m/s. Neglecting losses and assuming uniform flow at sections 1 and 2, find the downstream depth h_2, and show that *two* realistic solutions are possible.

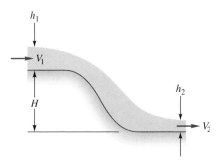

P3.177

P3.178 For the water channel flow of Fig. P3.178, $h_1 = 0.45$ ft, $H = 2.2$ ft, and $V_1 = 16$ ft/s. Neglecting losses and assuming uniform flow at sections 1 and 2, find the downstream depth h_2; show that *two* realistic solutions are possible.

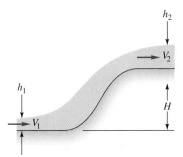

P3.178

*P3.179 A cylindrical tank of diameter D contains liquid to an initial height h_0. At time $t = 0$ a small stopper of diameter d is removed from the bottom. Using Bernoulli's equation with no losses, derive (a) a differential equation for the free-surface height $h(t)$ during draining and (b) an expression for the time t_0 to drain the entire tank.

*P3.180 The large tank of incompressible liquid in Fig. P3.180 is at rest when, at $t = 0$, the valve is opened to the atmosphere. Assuming $h \approx$ constant (negligible velocities and accelerations in the tank), use the unsteady frictionless Bernoulli equation to derive and solve a differential equation for $V(t)$ in the pipe.

*P3.181 Modify Prob. P3.180 as follows. Let the top of the tank be enclosed and under constant gage pressure p_0. Repeat the analysis to find $V(t)$ in the pipe.

P3.182 The incompressible flow form of Bernoulli's relation, Eq. (3.77), is accurate only for Mach numbers less than about 0.3. At higher speeds, variable density must be accounted for. The most common assumption for

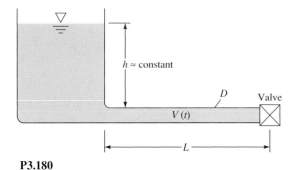

P3.180

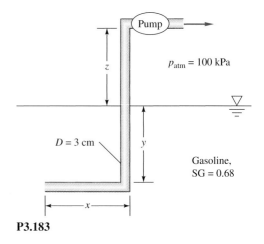

P3.183

compressible fluids is *isentropic flow of an ideal gas,* or $p = C\rho^k$, where $k = c_p/c_v$. Substitute this relation into Eq. (3.75), integrate, and eliminate the constant C. Compare your compressible result with Eq. (3.77) and comment.

P3.183 The pump in Fig. P3.183 draws gasoline at 20°C from a reservoir. Pumps are in big trouble if the liquid vaporizes (cavitates) before it enters the pump. (*a*) Neglecting losses and assuming a flow rate of 65 gal/min, find the limitations on (x, y, z) for avoiding cavitation. (*b*) If pipe friction losses are included, what additional limitations might be important?

P3.184 For the system of Prob P3.183, let the pump exhaust gasoline at 65 gal/min to the atmosphere through a 3-cm-

diameter opening, with no cavitation, when $x = 3$ m, $y = 2.5$ m, and $z = 2$ m. If the friction head loss is $h_{loss} \approx 3.7(V^2/2g)$, where V is the average velocity in the pipe, estimate the horsepower required to be delivered by the pump.

P3.185 Water at 20°C flows through a vertical tapered pipe at 163 m³/h. The entrance diameter is 12 cm, and the pipe diameter reduces by 3 mm for every 2-m rise in elevation. For frictionless flow, if the entrance pressure is 400 kPa, at what elevation will the fluid pressure be 100 kPa?

Word Problems

W3.1 Derive a control volume form of the *second* law of thermodynamics. Suggest some practical uses for your relation in analyzing real fluid flows.

W3.2 Suppose that it is desired to estimate volume flow Q in a pipe by measuring the axial velocity $u(r)$ at specific points. For cost reasons only *three* measuring points are to be used. What are the best radii selections for these three points?

W3.3 Consider water flowing by gravity through a short pipe connecting two reservoirs whose surface levels differ by an amount Δz. Why does the incompressible frictionless Bernoulli equation lead to an absurdity when the flow rate through the pipe is computed? Does the paradox have something to do with the length of the short pipe? Does the paradox disappear if we round the entrance and exit edges of the pipe?

W3.4 Use the steady flow energy equation to analyze flow through a water faucet whose supply pressure is p_0.

What physical mechanism causes the flow to vary continuously from zero to maximum as we open the faucet valve?

W3.5 Consider a long sewer pipe, half full of water, sloping downward at angle θ. Antoine Chézy in 1768 determined that the average velocity of such an open channel flow should be $V \approx C\sqrt{R \tan \theta}$, where R is the pipe radius and C is a constant. How does this famous formula relate to the steady flow energy equation applied to a length L of the channel?

W3.6 Put a table tennis ball in a funnel, and attach the small end of the funnel to an air supply. You probably won't be able to blow the ball either up or down out of the funnel. Explain why.

W3.7 How does a *siphon* work? Are there any limitations (such as how high or how low can you siphon water away from a tank)? Also, how far—could you use a flexible tube to siphon water from a tank to a point 100 ft away?

Fundamentals of Engineering Exam Problems

FE3.1   In Fig. FE3.1 water exits from a nozzle into atmospheric pressure of 101 kPa. If the flow rate is 160 gal/min, what is the average velocity at section 1?
(a) 2.6 m/s, (b) 0.81 m/s, (c) 93 m/s, (d) 23 m/s, (e) 1.62 m/s

FE3.2   In Fig. FE3.1 water exits from a nozzle into atmospheric pressure of 101 kPa. If the flow rate is 160 gal/min and friction is neglected, what is the gage pressure at section 1?
(a) 1.4 kPa, (b) 32 kPa, (c) 43 kPa, (d) 29 kPa, (e) 123 kPa

FE3.3   In Fig. FE3.1 water exits from a nozzle into atmospheric pressure of 101 kPa. If the exit velocity is $V_2 = 8$ m/s and friction is neglected, what is the axial flange force required to keep the nozzle attached to pipe 1?
(a) 11 N, (b) 56 N, (c) 83 N, (d) 123 N, (e) 110 N

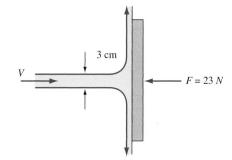

FE3.5

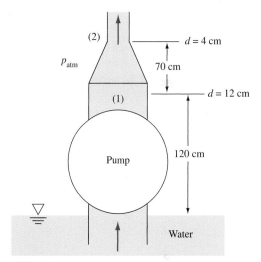

FE3.6

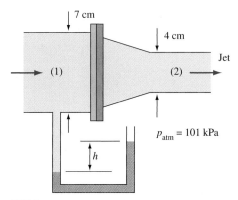

FE3.1

FE3.4   In Fig. FE3.1 water exits from a nozzle into atmospheric pressure of 101 kPa. If the manometer fluid has a specific gravity of 1.6 and $h = 66$ cm, with friction neglected, what is the average velocity at section 2?
(a) 4.55 m/s, (b) 2.4 m/s, (c) 2.95 m/s, (d) 5.55 m/s, (e) 3.4 m/s

FE3.5   A jet of water 3 cm in diameter strikes normal to a plate as in Fig. FE3.5. If the force required to hold the plate is 23 N, what is the jet velocity?
(a) 2.85 m/s, (b) 5.7 m/s, (c) 8.1 m/s, (d) 4.0 m/s, (e) 23 m/s

FE3.6   A fireboat pump delivers water to a vertical nozzle with a 3:1 diameter ratio, as in Fig. FE3.6. If friction is neglected and the flow rate is 500 gal/min, how high will the outlet water jet rise?
(a) 2.0 m, (b) 9.8 m, (c) 32 m, (d) 64 m, (e) 98 m

FE3.7   A fireboat pump delivers water to a vertical nozzle with a 3:1 diameter ratio, as in Fig. FE3.6. If friction is neglected and the pump increases the pressure at section 1 to 51 kPa (gage), what will be the resulting flow rate?
(a) 187 gal/min, (b) 199 gal/min, (c) 214 gal/min, (d) 359 gal/min, (e) 141 gal/min

FE3.8   A fireboat pump delivers water to a vertical nozzle with a 3:1 diameter ratio, as in Fig. FE3.6. If duct and nozzle friction are neglected and the pump provides 12.3 ft of head to the flow, what will be the outlet flow rate?
(a) 85 gal/min, (b) 120 gal/min, (c) 154 gal/min, (d) 217 gal/min, (e) 285 gal/min

FE3.9   Water flowing in a smooth 6-cm-diameter pipe enters a venturi contraction with a throat diameter of 3 cm. Upstream pressure is 120 kPa. If cavitation occurs in the throat at a flow rate of 155 gal/min, what is the estimated fluid vapor pressure, assuming ideal frictionless flow?
(a) 6 kPa, (b) 12 kPa, (c) 24 kPa, (d) 31 kPa, (e) 52 kPa

FE3.10   Water flowing in a smooth 6-cm-diameter pipe enters a venturi contraction with a throat diameter of 4 cm. Upstream pressure is 120 kPa. If the pressure in the throat is 50 kPa, what is the flow rate, assuming ideal frictionless flow?
(a) 7.5 gal/min, (b) 236 gal/min, (c) 263 gal/min, (d) 745 gal/min, (e) 1053 gal/min

Comprehensive Problems

C3.1   In a certain industrial process, oil of density ρ flows through the inclined pipe in Fig. C3.1. A U-tube manometer, with fluid density ρ_m, measures the pressure difference between points 1 and 2, as shown. The pipe flow is steady, so that the fluids in the manometer are stationary. (a) Find an analytic expression for $p_1 - p_2$ in terms of the system parameters. (b) Discuss the conditions on h necessary for there to be no flow in the pipe. (c) What about flow *up*, from 1 to 2? (d) What about flow *down*, from 2 to 1?

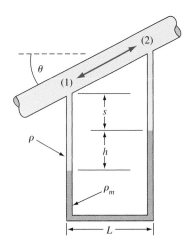

C3.1

C3.2   A rigid tank of volume $\mathcal{V} = 1.0$ m³ is initially filled with air at 20°C and $p_0 = 100$ kPa. At time $t = 0$, a vacuum pump is turned on and evacuates air at a constant volume flow rate $Q = 80$ L/min (regardless of the pressure). Assume an ideal gas and an isothermal process. (a) Set up a differential equation for this flow. (b) Solve this equation for t as a function of ($\mathcal{V}, Q, p, p_0$). (c) Compute the time in minutes to pump the tank down to $p = 20$ kPa. *Hint:* Your answer should lie between 15 and 25 min.

C3.3   Suppose the same steady water jet as in Prob. P3.40 (jet velocity 8 m/s and jet diameter 10 cm) impinges instead on a cup cavity as shown in Fig. C3.3. The water is turned 180° and exits, due to friction, at lower velocity, $V_e = 4$ m/s. (Looking from the left, the exit jet is a circular annulus of outer radius R and thickness h, flowing toward the viewer.) The cup has a radius of curvature of 25 cm. Find (a) the thickness h of the exit jet and (b) the force F required to hold the cupped object in place. (c) Compare part (b) to Prob. 3.40, where $F \approx 500$ N, and give a physical explanation as to why F has changed.

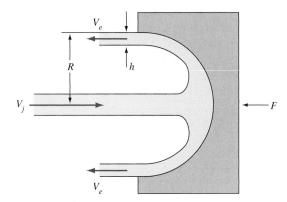

C3.3

C3.4   The air flow underneath an air hockey puck is very complex, especially since the air jets from the air hockey table impinge on the underside of the puck at various points nonsymmetrically. A reasonable approximation is that at any given time, the gage pressure on the bottom of the puck is halfway between zero (atmospheric pressure) and the stagnation pressure of the impinging jets. (Stagnation pressure is defined as $p_0 = \frac{1}{2}\rho V_{\text{jet}}^2$.) (a) Find the jet velocity V_{jet} required to support an air hockey puck of weight W and diameter d. Give your answer in terms of W, d, and the density ρ of the air. (b) For $W = 0.05$ lbf and $d = 2.5$ in, estimate the required jet velocity in ft/s.

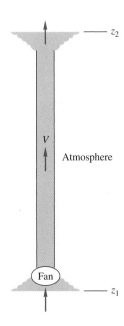

C3.5

C3.5 Neglecting friction sometimes leads to odd results. You are asked to analyze and discuss the following example in Fig. C3.5. A fan blows air through a duct from section 1 to section 2, as shown. Assume constant air density ρ. Neglecting frictional losses, find a relation between the required fan head h_p and the flow rate and the elevation change. Then explain what may be an unexpected result.

Design Project

D3.1 Let us generalize Probs. P3.141 and P3.142, in which a pump performance curve was used to determine the flow rate between reservoirs. The particular pump in Fig. P3.142 is one of a family of pumps of similar shape, whose dimensionless performance is as follows:

Head:

$$\phi \approx 6.04 - 161\zeta \qquad \phi = \frac{gh}{n^2 D_p^2} \quad \text{and} \quad \zeta = \frac{Q}{nD_p^3}$$

Efficiency:

$$\eta \approx 70\zeta - 91{,}500\zeta^3 \qquad \eta = \frac{\text{power to water}}{\text{power input}}$$

where h_p is the pump head (ft), n is the shaft rotation rate (r/s), and D_p is the impeller diameter (ft). The range of validity is $0 < \zeta < 0.027$. The pump of Fig. P3.142 had $D_p = 2$ ft in diameter and rotated at $n = 20$ r/s (1200 r/min). The solution to Prob. P3.142, namely, $Q \approx 2.57$ ft^3/s and $h_p \approx 172$ ft, corresponds to $\phi \approx 3.46$, $\zeta \approx 0.016$, $\eta \approx 0.75$ (or 75 percent), and power to the water = $\rho g Q h_p \approx 27{,}500$ ft · lbf/s (50 hp). Please check these numerical values before beginning this project.

Now revisit Prob. P3.142 an select a *low-cost* pump that rotates at a rate no slower than 600 r/min and delivers no less than 1.0 ft^3/s of water. Assume that the cost of the pump is linearly proportional to the power input required. Comment on any limitations to your results.

References

1. D. T. Greenwood and W. M. Greenfield, *Principles of Dynamics*, 2d ed., Prentice-Hall, Upper Saddle River, NJ, 1987.
2. T. von Kármán, *The Wind and Beyond*, Little, Brown, Boston, 1967.
3. J. P. Holman, *Heat Transfer*, 9th ed., McGraw-Hill, New York, 2001.
4. A. G. Hansen, *Fluid Mechanics*, Wiley, New York, 1967.
5. M. C. Potter, D. C. Wiggert, and M. Hondzo, *Mechanics of Fluids*, Brooks/Cole, Chicago, 2001.
6. R. E. Sonntag, C. Borgnakke, and G. J. Van Wylen, *Fundamentals of Thermodynamics*, 6th ed., John Wiley, New York, 2002.
7. Y. A. Cengel and M. A. Boles, *Thermodynamics: An Engineering Approach*, 5th ed., McGraw-Hill, New York, 2005.
8. J. D. Anderson, *Computational Fluid Dynamics: The Basics with Applications*, McGraw-Hill, New York, 1995.

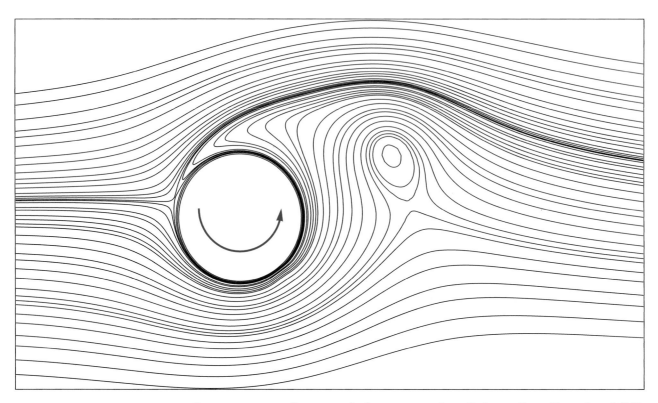

Instantaneous streamline pattern for flow past a rotating cylinder at a Reynolds number of 1000, after Nair et al. [19]. This pattern, which oscillates due to vortex shedding, was obtained by computational fluid dynamics (CFD) and agrees with flow visualization experiments [20]. The present chapter derives and discusses the equations of motion solved here by CFD. (*Figure courtesy of Professor Tapan K. Sengupta.*)

Chapter 4
Differential Relations
for Fluid Flow

Motivation.   In analyzing fluid motion, we might take one of two paths: (1) seeking an estimate of gross effects (mass flow, induced force, energy change) over a *finite* region or control volume or (2) seeking the point-by-point details of a flow pattern by analyzing an *infinitesimal* region of the flow. The former or gross-average viewpoint was the subject of Chap. 3.

This chapter treats the second in our trio of techniques for analyzing fluid motion: small-scale, or *differential,* analysis. That is, we apply our four basic conservation laws to an infinitesimally small control volume or, alternately, to an infinitesimal fluid system. In either case the results yield the basic *differential equations* of fluid motion. Appropriate *boundary conditions* are also developed.

In their most basic form, these differential equations of motion are quite difficult to solve, and very little is known about their general mathematical properties. However, certain things can be done that have great educational value. First, as shown in Chap. 5, the equations (even if unsolved) reveal the basic dimensionless parameters that govern fluid motion. Second, as shown in Chap. 6, a great number of useful solutions can be found if one makes two simplifying assumptions: (1) steady flow and (2) incompressible flow. A third and rather drastic simplification, frictionless flow, makes our old friend the Bernoulli equation valid and yields a wide variety of idealized, or *perfect-fluid,* possible solutions. These idealized flows are treated in Chap. 8, and we must be careful to ascertain whether such solutions are in fact realistic when compared with actual fluid motion. Finally, even the difficult general differential equations now yield to the approximating technique known as computational fluid dynamics (CFD) whereby the derivatives are simulated by algebraic relations between a finite number of grid points in the flow field, which are then solved on a computer. Reference 1 is an example of a textbook devoted entirely to numerical analysis of fluid motion.

4.1 The Acceleration Field of a Fluid

In Sec. 1.7 we established the cartesian vector form of a velocity field that varies in space and time:

$$\mathbf{V}(\mathbf{r},\ t) = \mathbf{i}u(x,\ y,\ z,\ t) + \mathbf{j}v(x,\ y,\ z,\ t) + \mathbf{k}w(x,\ y,\ z,\ t) \qquad (1.4)$$

This is the most important variable in fluid mechanics: Knowledge of the velocity vector field is nearly equivalent to *solving* a fluid flow problem. Our coordinates are fixed in space, and we observe the fluid as it passes by—as if we had scribed a set of coordinate lines on a glass window in a wind tunnel. This is the *eulerian* frame of reference, as opposed to the lagrangian frame, which follows the moving position of individual particles.

To write Newton's second law for an infinitesimal fluid system, we need to calculate the acceleration vector field **a** of the flow. Thus we compute the total time derivative of the velocity vector:

$$\mathbf{a} = \frac{d\mathbf{V}}{dt} = \mathbf{i}\frac{du}{dt} + \mathbf{j}\frac{dv}{dt} + \mathbf{k}\frac{dw}{dt}$$

Since each scalar component (u, v, w) is a function of the four variables (x, y, z, t), we use the chain rule to obtain each scalar time derivative. For example,

$$\frac{du(x, y, z, t)}{dt} = \frac{\partial u}{\partial t} + \frac{\partial u}{\partial x}\frac{dx}{dt} + \frac{\partial u}{\partial y}\frac{dy}{dt} + \frac{\partial u}{\partial z}\frac{dz}{dt}$$

But, by definition, dx/dt is the local velocity component u, and $dy/dt = v$, and $dz/dt = w$. The total derivative of u may thus be written in this compact form:

$$\frac{du}{dt} = \frac{\partial u}{\partial t} + u\frac{\partial u}{\partial x} + v\frac{\partial u}{\partial y} + w\frac{\partial u}{\partial z} = \frac{\partial u}{\partial t} + (\mathbf{V} \cdot \nabla)u \qquad (4.1)$$

Exactly similar expressions, with u replaced by v or w, hold for dv/dt or dw/dt. Summing these into a vector, we obtain the total acceleration:

$$\mathbf{a} = \frac{d\mathbf{V}}{dt} = \underbrace{\frac{\partial \mathbf{V}}{\partial t}}_{\text{Local}} + \underbrace{\left(u\frac{\partial \mathbf{V}}{\partial x} + v\frac{\partial \mathbf{V}}{\partial y} + w\frac{\partial \mathbf{V}}{\partial z} \right)}_{\text{Convective}} = \frac{\partial \mathbf{V}}{\partial t} + (\mathbf{V} \cdot \nabla)\mathbf{V} \qquad (4.2)$$

The term $\partial\mathbf{V}/\partial t$ is called the *local acceleration,* which vanishes if the flow is steady—that is, independent of time. The three terms in parentheses are called the *convective acceleration,* which arises when the particle moves through regions of spatially varying velocity, as in a nozzle or diffuser. Flows that are nominally "steady" may have large accelerations due to the convective terms.

Note our use of the compact dot product involving **V** and the gradient operator ∇:

$$u\frac{\partial}{\partial x} + v\frac{\partial}{\partial y} + w\frac{\partial}{\partial z} = \mathbf{V} \cdot \nabla \qquad \text{where} \qquad \nabla = \mathbf{i}\frac{\partial}{\partial x} + \mathbf{j}\frac{\partial}{\partial y} + \mathbf{k}\frac{\partial}{\partial z}$$

The total time derivative—sometimes called the *substantial* or *material* derivative—concept may be applied to any variable, such as the pressure:

$$\frac{dp}{dt} = \frac{\partial p}{\partial t} + u\frac{\partial p}{\partial x} + v\frac{\partial p}{\partial y} + w\frac{\partial p}{\partial z} = \frac{\partial p}{\partial t} + (\mathbf{V} \cdot \nabla)p \qquad (4.3)$$

Wherever convective effects occur in the basic laws involving mass, momentum, or energy, the basic differential equations become nonlinear and are usually more complicated than flows that do not involve convective changes.

We emphasize that this total time derivative follows a particle of fixed identity, making it convenient for expressing laws of particle mechanics in the eulerian fluid field description. The operator d/dt is sometimes assigned a special symbol such as D/Dt as a further reminder that it contains four terms and follows a fixed particle.

As another reminder of the special nature of d/dt, some writers give it the name *substantial derivative*.

EXAMPLE 4.1

Given the eulerian velocity vector field

$$\mathbf{V} = 3t\mathbf{i} + xz\mathbf{j} + ty^2\mathbf{k}$$

find the total acceleration of a particle.

Solution

- *Assumptions:* Given three known unsteady velocity components, $u = 3t$, $v = xz$, and $w = ty^2$.
- *Approach:* Carry out all the required derivatives with respect to (x, y, z, t), substitute into the total acceleration vector, Eq. (4.2), and collect terms.
- *Solution step 1:* First work out the local acceleration $\partial\mathbf{V}/\partial t$:

$$\frac{\partial\mathbf{V}}{\partial t} = \mathbf{i}\frac{\partial u}{\partial t} + \mathbf{j}\frac{\partial v}{\partial t} + \mathbf{k}\frac{\partial w}{\partial t} = \mathbf{i}\frac{\partial}{\partial t}(3t) + \mathbf{j}\frac{\partial}{\partial t}(xz) + \mathbf{k}\frac{\partial}{\partial t}(ty^2) = 3\mathbf{i} + 0\mathbf{j} + y^2\mathbf{k}$$

- *Solution step 2:* In a similar manner, the convective acceleration terms, from Eq. (4.2), are

$$u\frac{\partial\mathbf{V}}{\partial x} = (3t)\frac{\partial}{\partial x}(3t\mathbf{i} + xz\mathbf{j} + ty^2\mathbf{k}) = (3t)(0\mathbf{i} + z\mathbf{j} + 0\mathbf{k}) = 3tz\,\mathbf{j}$$

$$v\frac{\partial\mathbf{V}}{\partial y} = (xz)\frac{\partial}{\partial y}(3t\mathbf{i} + xz\mathbf{j} + ty^2\mathbf{k}) = (xz)(0\mathbf{i} + 0\mathbf{j} + 2ty\mathbf{k}) = 2txyz\,\mathbf{k}$$

$$w\frac{\partial\mathbf{V}}{\partial z} = (ty^2)\frac{\partial}{\partial z}(3t\mathbf{i} + xz\mathbf{j} + ty^2\mathbf{k}) = (ty^2)(0\mathbf{i} + x\mathbf{j} + 0\mathbf{k}) = txy^2\,\mathbf{j}$$

- *Solution step 3:* Combine all four terms above into the single "total" or "substantial" derivative:

$$\frac{d\mathbf{V}}{dt} = \frac{\partial\mathbf{V}}{\partial t} + u\frac{\partial\mathbf{V}}{\partial x} + v\frac{\partial\mathbf{V}}{\partial y} + w\frac{\partial\mathbf{V}}{\partial z} = (3\mathbf{i} + y^2\mathbf{k}) + 3tz\mathbf{j} + 2txyz\mathbf{k} + txy^2\mathbf{j}$$

$$= 3\mathbf{i} + (3tx + txy^2)\mathbf{j} + (y^2 + 2txyz)\mathbf{k} \qquad Ans.$$

- *Comments:* Assuming that $\mathbf{V}$ is valid everywhere as given, this total acceleration vector $d\mathbf{V}/dt$ applies to all positions and times within the flow field.

4.2 The Differential Equation of Mass Conservation

All the basic differential equations can be derived by considering either an elemental control volume or an elemental system. Here we choose an infinitesimal fixed control volume (dx, dy, dz), as in Fig. 4.1, and use our basic control volume relations from

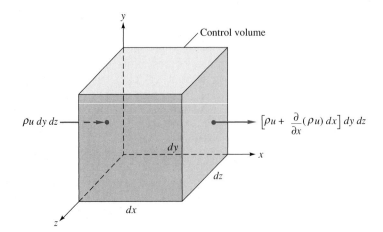

Fig. 4.1 Elemental cartesian fixed control volume showing the inlet and outlet mass flows on the x faces.

Chap. 3. The flow through each side of the element is approximately one-dimensional, and so the appropriate mass conservation relation to use here is

$$\int_{CV} \frac{\partial \rho}{\partial t}\, d\mathcal{V} + \sum_i (\rho_i A_i V_i)_{\text{out}} - \sum_i (\rho_i A_i V_i)_{\text{in}} = 0 \qquad (3.22)$$

The element is so small that the volume integral simply reduces to a differential term:

$$\int_{CV} \frac{\partial \rho}{\partial t}\, d\mathcal{V} \approx \frac{\partial \rho}{\partial t}\, dx\, dy\, dz$$

The mass flow terms occur on all six faces, three inlets and three outlets. We make use of the field or continuum concept from Chap. 1, where all fluid properties are considered to be uniformly varying functions of time and position, such as $\rho = \rho(x, y, z, t)$. Thus, if T is the temperature on the left face of the element in Fig. 4.1, the right face will have a slightly different temperature $T + (\partial T/\partial x)\, dx$. For mass conservation, if ρu is known on the left face, the value of this product on the right face is $\rho u + (\partial \rho u/\partial x)\, dx$.

Figure 4.1 shows only the mass flows on the x or left and right faces. The flows on the y (bottom and top) and the z (back and front) faces have been omitted to avoid cluttering up the drawing. We can list all these six flows as follows:

Face	Inlet mass flow	Outlet mass flow
x	$\rho u\, dy\, dz$	$\left[\rho u + \dfrac{\partial}{\partial x}(\rho u)\, dx\right] dy\, dz$
y	$\rho v\, dx\, dz$	$\left[\rho v + \dfrac{\partial}{\partial y}(\rho v)\, dy\right] dx\, dz$
z	$\rho w\, dx\, dy$	$\left[\rho w + \dfrac{\partial}{\partial z}(\rho w)\, dz\right] dx\, dy$

Introduce these terms into Eq. (3.22) and we have

$$\frac{\partial \rho}{\partial t}\, dx\, dy\, dz + \frac{\partial}{\partial x}(\rho u)\, dx\, dy\, dz + \frac{\partial}{\partial y}(\rho v)\, dx\, dy\, dz + \frac{\partial}{\partial z}(\rho w)\, dx\, dy\, dz = 0$$

The element volume cancels out of all terms, leaving a partial differential equation involving the derivatives of density and velocity:

$$\frac{\partial \rho}{\partial t} + \frac{\partial}{\partial x}(\rho u) + \frac{\partial}{\partial y}(\rho v) + \frac{\partial}{\partial z}(\rho w) = 0 \qquad (4.4)$$

This is the desired result: conservation of mass for an infinitesimal control volume. It is often called the *equation of continuity* because it requires no assumptions except that the density and velocity are continuum functions. That is, the flow may be either steady or unsteady, viscous or frictionless, compressible or incompressible.[1] However, the equation does not allow for any source or sink singularities within the element.

The vector gradient operator

$$\boldsymbol{\nabla} = \mathbf{i}\frac{\partial}{\partial x} + \mathbf{j}\frac{\partial}{\partial y} + \mathbf{k}\frac{\partial}{\partial z}$$

enables us to rewrite the equation of continuity in a compact form, not that it helps much in finding a solution. The last three terms of Eq. (4.4) are equivalent to the divergence of the vector $\rho \mathbf{V}$

$$\frac{\partial}{\partial x}(\rho u) + \frac{\partial}{\partial y}(\rho v) + \frac{\partial}{\partial z}(\rho w) \equiv \boldsymbol{\nabla} \cdot (\rho \mathbf{V}) \qquad (4.5)$$

so that the compact form of the continuity relation is

$$\frac{\partial \rho}{\partial t} + \boldsymbol{\nabla} \cdot (\rho \mathbf{V}) = 0 \qquad (4.6)$$

In this vector form the equation is still quite general and can readily be converted to other than cartesian coordinate systems.

Cylindrical Polar Coordinates

The most common alternative to the cartesian system is the *cylindrical polar* coordinate system, sketched in Fig. 4.2. An arbitrary point P is defined by a distance z along the axis, a radial distance r from the axis, and a rotation angle θ about the axis. The three independent orthogonal velocity components are an axial velocity v_z, a radial velocity v_r, and a circumferential velocity v_θ, which is positive counterclockwise—that is, in the direction of increasing θ. In general, all components, as well as pressure and density and other fluid properties, are continuous functions of r, θ, z, and t.

The divergence of any vector function $\mathbf{A}(r, \theta, z, t)$ is found by making the transformation of coordinates

$$r = (x^2 + y^2)^{1/2} \qquad \theta = \tan^{-1}\frac{y}{x} \qquad z = z \qquad (4.7)$$

---

[1] One case where Eq. (4.4) might need special care is *two-phase flow,* where the density is discontinuous between the phases. For further details on this case, see Ref. 2, for example.

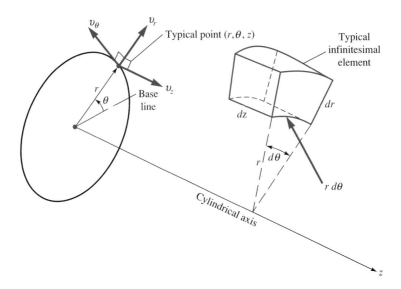

Fig. 4.2 Definition sketch for the cylindrical coordinate system.

and the result is given here without proof[2]

$$\nabla \cdot \mathbf{A} = \frac{1}{r}\frac{\partial}{\partial r}(rA_r) + \frac{1}{r}\frac{\partial}{\partial \theta}(A_\theta) + \frac{\partial}{\partial z}(A_z) \tag{4.8}$$

The general continuity equation (4.6) in cylindrical polar coordinates is thus

$$\frac{\partial \rho}{\partial t} + \frac{1}{r}\frac{\partial}{\partial r}(r\rho v_r) + \frac{1}{r}\frac{\partial}{\partial \theta}(\rho v_\theta) + \frac{\partial}{\partial z}(\rho v_z) = 0 \tag{4.9}$$

There are other orthogonal curvilinear coordinate systems, notably *spherical polar* coordinates, which occasionally merit use in a fluid mechanics problem. We shall not treat these systems here except in Prob. P4.12.

There are also other ways to derive the basic continuity equation (4.6) that are interesting and instructive. One example is the use of the divergence theorem. Ask your instructor about these alternative approaches.

Steady Compressible Flow

If the flow is steady, $\partial/\partial t \equiv 0$ and all properties are functions of position only. Equation (4.6) reduces to

Cartesian:
$$\frac{\partial}{\partial x}(\rho u) + \frac{\partial}{\partial y}(\rho v) + \frac{\partial}{\partial z}(\rho w) = 0$$

Cylindrical:
$$\frac{1}{r}\frac{\partial}{\partial r}(r\rho v_r) + \frac{1}{r}\frac{\partial}{\partial \theta}(\rho v_\theta) + \frac{\partial}{\partial z}(\rho v_z) = 0 \tag{4.10}$$

Since density and velocity are both variables, these are still nonlinear and rather formidable, but a number of special-case solutions have been found.

---

[2]See, for example, Ref. 3, p. 783.

Incompressible Flow

A special case that affords great simplification is incompressible flow, where the density changes are negligible. Then $\partial\rho/\partial t \approx 0$ regardless of whether the flow is steady or unsteady, and the density can be slipped out of the divergence in Eq. (4.6) and divided out. The result

$$\nabla \cdot \mathbf{V} = 0 \tag{4.11}$$

is valid for steady or unsteady incompressible flow. The two coordinate forms are

Cartesian:
$$\frac{\partial u}{\partial x} + \frac{\partial v}{\partial y} + \frac{\partial w}{\partial z} = 0 \tag{4.12a}$$

Cylindrical:
$$\frac{1}{r}\frac{\partial}{\partial r}(rv_r) + \frac{1}{r}\frac{\partial}{\partial \theta}(v_\theta) + \frac{\partial}{\partial z}(v_z) = 0 \tag{4.12b}$$

These are *linear* differential equations, and a wide variety of solutions are known, as discussed in Chaps. 6 to 8. Since no author or instructor can resist a wide variety of solutions, it follows that a great deal of time is spent studying incompressible flows. Fortunately, this is exactly what should be done, because most practical engineering flows are approximately incompressible, the chief exception being the high-speed gas flows treated in Chap. 9.

When is a given flow approximately incompressible? We can derive a nice criterion by playing a little fast and loose with density approximations. In essence, we wish to slip the density out of the divergence in Eq. (4.6) and approximate a typical term such as

$$\frac{\partial}{\partial x}(\rho u) \approx \rho\frac{\partial u}{\partial x} \tag{4.13}$$

This is equivalent to the strong inequality

$$\left| u\frac{\partial \rho}{\partial x}\right| \ll \left| \rho\frac{\partial u}{\partial x}\right|$$

or
$$\left| \frac{\delta\rho}{\rho}\right| \ll \left| \frac{\delta V}{V}\right| \tag{4.14}$$

As shown in Eq. (1.38), the pressure change is approximately proportional to the density change and the square of the speed of sound a of the fluid:

$$\delta p \approx a^2\,\delta\rho \tag{4.15}$$

Meanwhile, if elevation changes are negligible, the pressure is related to the velocity change by Bernoulli's equation (3.75):

$$\delta p \approx -\rho V\,\delta V \tag{4.16}$$

Combining Eqs. (4.14) to (4.16), we obtain an explicit criterion for incompressible flow:

$$\frac{V^2}{a^2} = \text{Ma}^2 \ll 1 \tag{4.17}$$

where $\text{Ma} = V/a$ is the dimensionless *Mach number* of the flow. How small is small? The commonly accepted limit is

$$\text{Ma} \le 0.3 \tag{4.18}$$

For air at standard conditions, a flow can thus be considered incompressible if the velocity is less than about 100 m/s (330 ft/s). This encompasses a wide variety of airflows: automobile and train motions, light aircraft, landing and takeoff of high-speed aircraft, most pipe flows, and turbomachinery at moderate rotational speeds. Further, it is clear that almost all liquid flows are incompressible, since flow velocities are small and the speed of sound is very large.[3]

Before attempting to analyze the continuity equation, we shall proceed with the derivation of the momentum and energy equations, so that we can analyze them as a group. A very clever device called the *stream function* can often make short work of the continuity equation, but we shall save it until Sec. 4.7.

One further remark is appropriate: The continuity equation is always important and must always be satisfied for a rational analysis of a flow pattern. Any newly discovered momentum or energy "solution" will ultimately crash in flames when subjected to critical analysis if it does not also satisfy the continuity equation.

EXAMPLE 4.2

Under what conditions does the velocity field

$$\mathbf{V} = (a_1x + b_1y + c_1z)\mathbf{i} + (a_2x + b_2y + c_2z)\mathbf{j} + (a_3x + b_3y + c_3z)\mathbf{k}$$

where a_1, b_1, etc. = const, represent an incompressible flow that conserves mass?

Solution

Recalling that $\mathbf{V} = u\mathbf{i} + v\mathbf{j} + w\mathbf{k}$, we see that $u = (a_1x + b_1y + c_1z)$, etc. Substituting into Eq. (4.12a) for incompressible continuity, we obtain

$$\frac{\partial}{\partial x}(a_1x + b_1y + c_1z) + \frac{\partial}{\partial y}(a_2x + b_2y + c_2z) + \frac{\partial}{\partial z}(a_3x + b_3y + c_3z) = 0$$

or
$$a_1 + b_2 + c_3 = 0 \qquad\qquad\qquad Ans.$$

At least two of constants a_1, b_2, and c_3 must have opposite signs. Continuity imposes no restrictions whatever on constants b_1, c_1, a_2, c_2, a_3, and b_3, which do not contribute to a volume increase or decrease of a differential element.

EXAMPLE 4.3

An incompressible velocity field is given by

$$u = a(x^2 - y^2) \qquad v \text{ unknown} \qquad w = b$$

where a and b are constants. What must the form of the velocity component v be?

---

[3] An exception occurs in geophysical flows, where a density change is imposed thermally or mechanically rather than by the flow conditions themselves. An example is fresh water layered upon saltwater or warm air layered upon cold air in the atmosphere. We say that the fluid is *stratified,* and we must account for vertical density changes in Eq. (4.6) even if the velocities are small.

Solution

Again Eq. (4.12a) applies:

$$\frac{\partial}{\partial x}(ax^2 - ay^2) + \frac{\partial v}{\partial y} + \frac{\partial b}{\partial z} = 0$$

or

$$\frac{\partial v}{\partial y} = -2ax \tag{1}$$

This is easily integrated partially with respect to y:

$$v(x, y, z, t) = -2axy + f(x, z, t) \qquad \textit{Ans.}$$

This is the only possible form for v that satisfies the incompressible continuity equation. The function of integration f is entirely arbitrary since it vanishes when v is differentiated with respect to y.[4]

EXAMPLE 4.4

A centrifugal impeller of 40-cm diameter is used to pump hydrogen at 15°C and 1-atm pressure. Estimate the maximum allowable impeller rotational speed to avoid compressibility effects at the blade tips.

Solution

- *Assumptions:*   The maximum fluid velocity is approximately equal to the impeller tip speed:

$$V_{max} \approx \Omega r_{max} \qquad \text{where } r_{max} = D/2 = 0.20 \text{ m}$$

- *Approach:*   Find the speed of sound of hydrogen and make sure that V_{max} is much less.
- *Property values:*   From Table A.4 for hydrogen, $R = 4124 \text{ m}^2/(\text{s}^2 - \text{K})$ and $k = 1.41$. From Eq. (1.39) at 15°C = 288K, compute the speed of sound:

$$a_{H_2} = \sqrt{kRT} = \sqrt{1.41[4124 \text{ m}^2/(\text{s}^2 - \text{K})](288 \text{ K})} \approx 1294 \text{ m/s}$$

- *Final solution step:*   Use our rule of thumb, Eq. (4.18), to estimate the maximum impeller speed:

$$V = \Omega r_{max} \leq 0.3a \qquad \text{or} \qquad \Omega(0.2 \text{ m}) \leq 0.3(1294 \text{ m/s})$$

$$\text{Solve for} \qquad \Omega \leq 1940 \frac{\text{rad}}{\text{s}} \approx 18{,}500 \frac{\text{rev}}{\text{min}} \qquad \textit{Ans.}$$

- *Comments:*   This is a high rate because the speed of sound of hydrogen, a light gas, is nearly four times greater than that of air. An impeller moving at this speed in air would create tip shock waves.

---

[4]This is a very realistic flow that simulates the turning of an inviscid fluid through a 60° angle; see Examples 4.7 and 4.9.

4.3 The Differential Equation of Linear Momentum

Having done it once in Sec. 4.2 for mass conservation, we can move along a little faster this time. We use the same elemental control volume as in Fig. 4.1, for which the appropriate form of the linear momentum relation is

$$\sum \mathbf{F} = \frac{\partial}{\partial t}\left(\int_{\mathrm{CV}} \mathbf{V}\rho\, d\mathcal{V} \right) + \sum (\dot{m}_i \mathbf{V}_i)_{\mathrm{out}} - \sum (\dot{m}_i \mathbf{V}_i)_{\mathrm{in}} \qquad (3.40)$$

Again the element is so small that the volume integral simply reduces to a derivative term:

$$\frac{\partial}{\partial t}(\mathbf{V}\rho\, d\mathcal{V}) \approx \frac{\partial}{\partial t}(\rho\mathbf{V})\, dx\, dy\, dz \qquad (4.19)$$

The momentum fluxes occur on all six faces, three inlets and three outlets. Referring again to Fig. 4.1, we can form a table of momentum fluxes by exact analogy with the discussion that led up to the equation for net mass flux:

Faces	Inlet momentum flux	Outlet momentum flux
x	$\rho u\mathbf{V}\ dy\ dz$	$\left[\rho u\mathbf{V} + \dfrac{\partial}{\partial x}(\rho u\mathbf{V})\, dx\right] dy\, dz$
y	$\rho v\mathbf{V}\ dx\ dz$	$\left[\rho v\mathbf{V} + \dfrac{\partial}{\partial y}(\rho v\mathbf{V})\, dy\right] dx\, dz$
z	$\rho w\mathbf{V}\ dx\ dy$	$\left[\rho w\mathbf{V} + \dfrac{\partial}{\partial z}(\rho w\mathbf{V})\, dz\right] dx\, dy$

Introduce these terms and Eq. (4.19) into Eq. (3.40), and get this intermediate result:

$$\sum \mathbf{F} = dx\, dy\, dz\left[\frac{\partial}{\partial t}(\rho\mathbf{V}) + \frac{\partial}{\partial x}(\rho u\mathbf{V}) + \frac{\partial}{\partial y}(\rho v\mathbf{V}) + \frac{\partial}{\partial z}(\rho w\mathbf{V})\right] \qquad (4.20)$$

Note that this is a vector relation. A simplification occurs if we split up the term in brackets as follows:

$$\frac{\partial}{\partial t}(\rho\mathbf{V}) + \frac{\partial}{\partial x}(\rho u\mathbf{V}) + \frac{\partial}{\partial y}(\rho v\mathbf{V}) + \frac{\partial}{\partial z}(\rho w\mathbf{V})$$

$$= \mathbf{V}\left[\frac{\partial\rho}{\partial t} + \boldsymbol{\nabla}\cdot(\rho\mathbf{V})\right] + \rho\left(\frac{\partial\mathbf{V}}{\partial t} + u\frac{\partial\mathbf{V}}{\partial x} + v\frac{\partial\mathbf{V}}{\partial y} + w\frac{\partial\mathbf{V}}{\partial z}\right) \qquad (4.21)$$

The term in brackets on the right-hand side is seen to be the equation of continuity, Eq. (4.6), which vanishes identically. The long term in parentheses on the right-hand side is seen from Eq. (4.2) to be the total acceleration of a particle that instantaneously occupies the control volume:

$$\frac{\partial\mathbf{V}}{\partial t} + u\frac{\partial\mathbf{V}}{\partial x} + v\frac{\partial\mathbf{V}}{\partial y} + w\frac{\partial\mathbf{V}}{\partial z} = \frac{d\mathbf{V}}{dt} \qquad (4.2)$$

Thus we have now reduced Eq. (4.20) to

$$\sum \mathbf{F} = \rho\frac{d\mathbf{V}}{dt}\, dx\, dy\, dz \qquad (4.22)$$

It might be good for you to stop and rest now and think about what we have just done. What is the relation between Eqs. (4.22) and (3.40) for an infinitesimal control volume? Could we have *begun* the analysis at Eq. (4.22)?

Equation (4.22) points out that the net force on the control volume must be of differential size and proportional to the element volume. These forces are of two types, *body* forces and *surface* forces. Body forces are due to external fields (gravity, magnetism, electric potential) that act on the entire mass within the element. The only body force we shall consider in this book is gravity. The gravity force on the differential mass $\rho\, dx\, dy\, dz$ within the control volume is

$$d\mathbf{F}_{\text{grav}} = \rho\mathbf{g}\, dx\, dy\, dz \tag{4.23}$$

where $\mathbf{g}$ may in general have an arbitrary orientation with respect to the coordinate system. In many applications, such as Bernoulli's equation, we take z "up," and $\mathbf{g} = -g\mathbf{k}$.

The surface forces are due to the stresses on the sides of the control surface. These stresses are the sum of hydrostatic pressure plus viscous stresses τ_{ij} that arise from motion with velocity gradients:

$$\sigma_{ij} = \begin{vmatrix} -p + \tau_{xx} & \tau_{yx} & \tau_{zx} \\ \tau_{xy} & -p + \tau_{yy} & \tau_{zy} \\ \tau_{xz} & \tau_{yz} & -p + \tau_{zz} \end{vmatrix} \tag{4.24}$$

The subscript notation for stresses is given in Fig. 4.3. Unlike velocity $\mathbf{V}$, which is a three-component *vector*, stresses σ_{ij} and τ_{ij} and strain rates ε_{ij} are nine-component *tensors* and require two subscripts to define each component. For further study of *tensor analysis,* see Refs. 6, 11, or 13.

It is not these stresses but their *gradients,* or differences, that cause a net force on the differential control surface. This is seen by referring to Fig. 4.4, which shows only

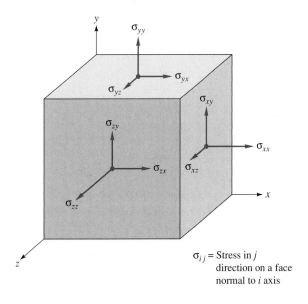

$\sigma_{ij} =$ Stress in j
direction on a face
normal to i axis

Fig. 4.3 Notation for stresses.

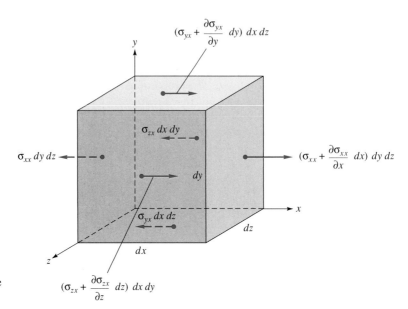

Fig. 4.4 Elemental cartesian fixed control volume showing the surface forces in the x direction only.

the x-directed stresses to avoid cluttering up the drawing. For example, the leftward force $\sigma_{xx}\,dy\,dz$ on the left face is balanced by the rightward force $\sigma_{xx}\,dy\,dz$ on the right face, leaving only the net rightward force $(\partial\sigma_{xx}/\partial x)\,dx\,dy\,dz$ on the right face. The same thing happens on the other four faces, so that the net surface force in the x direction is given by

$$dF_{x,\text{surf}} = \left[\frac{\partial}{\partial x}(\sigma_{xx}) + \frac{\partial}{\partial y}(\sigma_{yx}) + \frac{\partial}{\partial z}(\sigma_{zx})\right] dx\,dy\,dz \tag{4.25}$$

We see that this force is proportional to the element volume. Notice that the stress terms are taken from the *top row* of the array in Eq. (4.24). Splitting this row into pressure plus viscous stresses, we can rewrite Eq. (4.25) as

$$\frac{dF_x}{d\mathcal{V}} = -\frac{\partial p}{\partial x} + \frac{\partial}{\partial x}(\tau_{xx}) + \frac{\partial}{\partial y}(\tau_{yx}) + \frac{\partial}{\partial z}(\tau_{zx}) \tag{4.26}$$

where $d\mathcal{V} = dx\,dy\,dz$. In exactly similar manner, we can derive the y and z forces per unit volume on the control surface:

$$\frac{dF_y}{d\mathcal{V}} = -\frac{\partial p}{\partial y} + \frac{\partial}{\partial x}(\tau_{xy}) + \frac{\partial}{\partial y}(\tau_{yy}) + \frac{\partial}{\partial z}(\tau_{zy})$$

$$\frac{dF_z}{d\mathcal{V}} = -\frac{\partial p}{\partial z} + \frac{\partial}{\partial x}(\tau_{xz}) + \frac{\partial}{\partial y}(\tau_{yz}) + \frac{\partial}{\partial z}(\tau_{zz}) \tag{4.27}$$

Now we multiply Eqs. (4.26) and (4.27) by **i, j,** and **k,** respectively, and add to obtain an expression for the net vector surface force:

$$\left(\frac{d\mathbf{F}}{d\mathcal{V}}\right)_{\text{surf}} = -\boldsymbol{\nabla}p + \left(\frac{d\mathbf{F}}{d\mathcal{V}}\right)_{\text{viscous}} \tag{4.28}$$

where the viscous force has a total of nine terms:

$$\left(\frac{d\mathbf{F}}{d\mathcal{V}}\right)_{\text{viscous}} = \mathbf{i}\left(\frac{\partial \tau_{xx}}{\partial x} + \frac{\partial \tau_{yx}}{\partial y} + \frac{\partial \tau_{zx}}{\partial z}\right)$$
$$+ \mathbf{j}\left(\frac{\partial \tau_{xy}}{\partial x} + \frac{\partial \tau_{yy}}{\partial y} + \frac{\partial \tau_{zy}}{\partial z}\right)$$
$$+ \mathbf{k}\left(\frac{\partial \tau_{xz}}{\partial x} + \frac{\partial \tau_{yz}}{\partial y} + \frac{\partial \tau_{zz}}{\partial z}\right) \tag{4.29}$$

Since each term in parentheses in (4.29) represents the divergence of a stress component vector acting on the x, y, and z faces, respectively, Eq. (4.29) is sometimes expressed in divergence form:

$$\left(\frac{d\mathbf{F}}{d\mathcal{V}}\right)_{\text{viscous}} = \nabla \cdot \boldsymbol{\tau}_{ij} \tag{4.30}$$

where

$$\boldsymbol{\tau}_{ij} = \begin{bmatrix} \tau_{xx} & \tau_{yx} & \tau_{zx} \\ \tau_{xy} & \tau_{yy} & \tau_{zy} \\ \tau_{xz} & \tau_{yz} & \tau_{zz} \end{bmatrix} \tag{4.31}$$

is the viscous stress tensor acting on the element. The surface force is thus the sum of the *pressure gradient* vector and the divergence of the viscous stress tensor. Substituting into Eq. (4.22) and utilizing Eq. (4.23), we have the basic differential momentum equation for an infinitesimal element:

$$\rho \mathbf{g} - \nabla p + \nabla \cdot \boldsymbol{\tau}_{ij} = \rho \frac{d\mathbf{V}}{dt} \tag{4.32}$$

where

$$\frac{d\mathbf{V}}{dt} = \frac{\partial \mathbf{V}}{\partial t} + u\frac{\partial \mathbf{V}}{\partial x} + v\frac{\partial \mathbf{V}}{\partial y} + w\frac{\partial \mathbf{V}}{\partial z} \tag{4.33}$$

We can also express Eq. (4.32) in words:

Gravity force per unit volume + pressure force per unit volume
+ viscous force per unit volume = density × acceleration   (4.34)

Equation (4.32) is so brief and compact that its inherent complexity is almost invisible. It is a *vector* equation, each of whose component equations contains nine terms. Let us therefore write out the component equations in full to illustrate the mathematical difficulties inherent in the momentum equation:

$$\rho g_x - \frac{\partial p}{\partial x} + \frac{\partial \tau_{xx}}{\partial x} + \frac{\partial \tau_{yx}}{\partial y} + \frac{\partial \tau_{zx}}{\partial z} = \rho\left(\frac{\partial u}{\partial t} + u\frac{\partial u}{\partial x} + v\frac{\partial u}{\partial y} + w\frac{\partial u}{\partial z}\right)$$

$$\rho g_y - \frac{\partial p}{\partial y} + \frac{\partial \tau_{xy}}{\partial x} + \frac{\partial \tau_{yy}}{\partial y} + \frac{\partial \tau_{zy}}{\partial z} = \rho\left(\frac{\partial v}{\partial t} + u\frac{\partial v}{\partial x} + v\frac{\partial v}{\partial y} + w\frac{\partial v}{\partial z}\right) \tag{4.35}$$

$$\rho g_z - \frac{\partial p}{\partial z} + \frac{\partial \tau_{xz}}{\partial x} + \frac{\partial \tau_{yz}}{\partial y} + \frac{\partial \tau_{zz}}{\partial z} = \rho\left(\frac{\partial w}{\partial t} + u\frac{\partial w}{\partial x} + v\frac{\partial w}{\partial y} + w\frac{\partial w}{\partial z}\right)$$

This is the differential momentum equation in its full glory, and it is valid for any fluid in any general motion, particular fluids being characterized by particular viscous stress terms. Note that the last three "convective" terms on the right-hand side of each component equation in (4.35) are nonlinear, which complicates the general mathematical analysis.

Inviscid Flow: Euler's Equation

Equation (4.35) is not ready to use until we write the viscous stresses in terms of velocity components. The simplest assumption is frictionless flow $\tau_{ij} = 0$, for which Eq. (4.32) reduces to

$$\rho \mathbf{g} - \nabla p = \rho \frac{d\mathbf{V}}{dt} \tag{4.36}$$

This is *Euler's equation* for inviscid flow. We show in Sec. 4.9 that Euler's equation can be integrated along a streamline to yield the frictionless Bernoulli equation, (3.75) or (3.77). The complete analysis of inviscid flow fields, using continuity and the Bernoulli relation, is given in Chap. 8.

Newtonian Fluid: Navier-Stokes Equations

For a newtonian fluid, as discussed in Sec. 1.9, the viscous stresses are proportional to the element strain rates and the coefficient of viscosity. For incompressible flow, the generalization of Eq. (1.23) to three-dimensional viscous flow is[5]

$$\tau_{xx} = 2\mu \frac{\partial u}{\partial x} \qquad \tau_{yy} = 2\mu \frac{\partial v}{\partial y} \qquad \tau_{zz} = 2\mu \frac{\partial w}{\partial z}$$

$$\tau_{xy} = \tau_{yx} = \mu\left(\frac{\partial u}{\partial y} + \frac{\partial v}{\partial x}\right) \qquad \tau_{xz} = \tau_{zx} = \mu\left(\frac{\partial w}{\partial x} + \frac{\partial u}{\partial z}\right) \tag{4.37}$$

$$\tau_{yz} = \tau_{zy} = \mu\left(\frac{\partial v}{\partial z} + \frac{\partial w}{\partial y}\right)$$

where μ is the viscosity coefficient. Substitution into Eq. (4.35) gives the differential momentum equation for a newtonian fluid with constant density and viscosity:

$$\rho g_x - \frac{\partial p}{\partial x} + \mu\left(\frac{\partial^2 u}{\partial x^2} + \frac{\partial^2 u}{\partial y^2} + \frac{\partial^2 u}{\partial z^2}\right) = \rho \frac{du}{dt}$$

$$\rho g_y - \frac{\partial p}{\partial y} + \mu\left(\frac{\partial^2 v}{\partial x^2} + \frac{\partial^2 v}{\partial y^2} + \frac{\partial^2 v}{\partial z^2}\right) = \rho \frac{dv}{dt} \tag{4.38}$$

$$\rho g_z - \frac{\partial p}{\partial z} + \mu\left(\frac{\partial^2 w}{\partial x^2} + \frac{\partial^2 w}{\partial y^2} + \frac{\partial^2 w}{\partial z^2}\right) = \rho \frac{dw}{dt}$$

These are the incompressible flow *Navier-Stokes equations,* named after C. L. M. H. Navier (1785–1836) and Sir George G. Stokes (1819–1903), who are credited with their derivation. They are second-order nonlinear partial differential equations and are quite formidable, but solutions have been found to a variety of interesting viscous

[5]When compressibility is significant, additional small terms arise containing the element volume expansion rate and a *second* coefficient of viscosity; see Refs. 4 and 5 for details.

flow problems, some of which are discussed in Sec. 4.11 and in Chap. 6 (see also Refs. 4 and 5). For compressible flow, see Eq. (2.29) of Ref. 5.

Equations (4.38) have four unknowns: p, u, v, and w. They should be combined with the incompressible continuity relation [Eqs. (4.12)] to form four equations in these four unknowns. We shall discuss this again in Sec. 4.6, which presents the appropriate boundary conditions for these equations.

Even though the Navier-Stokes equations have only a limited number of known analytical solutions, they are amenable to fine-gridded computer modeling [1]. The field of CFD is maturing fast, with many commercial software tools available. It is possible now to achieve approximate, but realistic, CFD results for a wide variety of complex two- and three-dimensional viscous flows.

EXAMPLE 4.5

Take the velocity field of Example 4.3, with $b = 0$ for algebraic convenience

$$u = a(x^2 - y^2) \qquad v = -2axy \qquad w = 0$$

and determine under what conditions it is a solution to the Navier-Stokes momentum equations (4.38). Assuming that these conditions are met, determine the resulting pressure distribution when z is "up" ($g_x = 0$, $g_y = 0$, $g_z = -g$).

Solution

- *Assumptions:* Constant density and viscosity, steady flow (u and v independent of time).
- *Approach:* Substitute the known (u, v, w) into Eqs. (4.38) and solve for the pressure gradients. If a unique pressure function $p(x, y, z)$ can then be found, the given solution is exact.
- *Solution step 1:* Substitute (u, v, w) into Eqs. (4.38) in sequence:

$$\rho(0) - \frac{\partial p}{\partial x} + \mu(2a - 2a + 0) = \rho\left(u\frac{\partial u}{\partial x} + v\frac{\partial u}{\partial y}\right) = 2a^2\rho(x^3 + xy^2)$$

$$\rho(0) - \frac{\partial p}{\partial y} + \mu(0 + 0 + 0) = \rho\left(u\frac{\partial v}{\partial x} + v\frac{\partial v}{\partial y}\right) = 2a^2\rho(x^2y + y^3)$$

$$\rho(-g) - \frac{\partial p}{\partial z} + \mu(0 + 0 + 0) = \rho\left(u\frac{\partial w}{\partial x} + v\frac{\partial w}{\partial y}\right) = 0$$

Rearrange and solve for the three pressure gradients:

$$\frac{\partial p}{\partial x} = -2a^2\rho(x^3 + xy^2) \qquad \frac{\partial p}{\partial y} = -2a^2\rho(x^2y + y^3) \qquad \frac{\partial p}{\partial z} = -\rho g \qquad (1)$$

- *Comment 1:* The vertical pressure gradient is *hydrostatic*. (Could you have predicted this by noting in Eqs. (4.38) that $w = 0$?) However, the pressure is velocity-dependent in the xy plane.
- *Solution step 2:* To determine if the x and y gradients of pressure in Eq. (1) are compatible, evaluate the mixed derivative, $(\partial^2 p/\partial x\, \partial y)$; that is, cross-differentiate these two equations:

$$\frac{\partial}{\partial y}\left(\frac{\partial p}{\partial x}\right) = \frac{\partial}{\partial y}[-2a^2\rho(x^3 + xy^2)] = -4a^2\rho xy$$

$$\frac{\partial}{\partial x}\left(\frac{\partial p}{\partial y}\right) = \frac{\partial}{\partial x}[-2a^2\rho(x^2y + y^3)] = -4a^2\rho xy$$

- *Comment 2:* Since these are equal, the given velocity distribution is indeed an *exact* solution of the Navier-Stokes equations.
- *Solution step 3:* To find the pressure, integrate Eqs. (1), collect, and compare. Start with $\partial p/\partial x$. The procedure requires care! Integrate *partially* with respect to x, holding y and z constant:

$$p = \int \frac{\partial p}{\partial x}\bigg|_{y,z} \, dx = \int -2a^2\rho(x^3 + xy^2) \, dx\big|_{y,z} = -2a^2\rho\left(\frac{x^4}{4} + \frac{x^2y^2}{2}\right) + f_1(y, z) \qquad (2)$$

Note that the "constant" of integration f_1 is a *function* of the variables that were not integrated. Now differentiate Eq. (2) with respect to y and compare with $\partial p/\partial y$ from Eq. (1):

$$\frac{\partial p}{\partial y}\bigg|_{(2)} = -2a^2\rho\, x^2y + \frac{\partial f_1}{\partial y} = \frac{\partial p}{\partial y}\bigg|_{(1)} = -2a^2\rho(x^2y + y^3)$$

Compare: $\dfrac{\partial f_1}{\partial y} = -2a^2\rho\, y^3$ or $f_1 = \int \dfrac{\partial f_1}{\partial y} \, dy\big|_z = -2a^2\rho\,\dfrac{y^4}{4} + f_2(z)$

Collect terms: So far $p = -2a^2\rho\left(\dfrac{x^4}{4} + \dfrac{x^2y^2}{2} + \dfrac{y^4}{4}\right) + f_2(z)$ $\qquad (3)$

This time the "constant" of integration f_2 is a function of z only (the variable not integrated). Now differentiate Eq. (3) with respect to z and compare with $\partial p/\partial z$ from Eq. (1):

$$\frac{\partial p}{\partial z}\bigg|_{(3)} = \frac{df_2}{dz} = \frac{\partial p}{\partial z}\bigg|_{(1)} = -\rho g \qquad \text{or} \qquad f_2 = -\rho g z + C \qquad (4)$$

where C is a constant. This completes our three integrations. Combine Eqs. (3) and (4) to obtain the full expression for the pressure distribution in this flow:

$$p(x, y, z) = -\rho g z - \tfrac{1}{2}a^2\rho(x^4 + y^4 + 2x^2y^2) + C \qquad\qquad \textit{Ans.} \ (5)$$

This is the desired solution. Do you recognize it? Not unless you go back to the beginning and square the velocity components:

$$u^2 + v^2 + w^2 = V^2 = a^2(x^4 + y^4 + 2x^2y^2) \qquad (6)$$

Comparing with Eq. (5), we can rewrite the pressure distribution as

$$p + \tfrac{1}{2}\rho V^2 + \rho g z = C \qquad (7)$$

- *Comment:* This is Bernoulli's equation (3.77). That is no accident, because the velocity distribution given in this problem is one of a family of flows that are solutions to the Navier-Stokes equations and that satisfy Bernoulli's incompressible equation everywhere in the flow field. They are called *irrotational flows*, for which curl $\mathbf{V} = \nabla \times \mathbf{V} \equiv 0$. This subject is discussed again in Sec. 4.9.

4.4 The Differential Equation of Angular Momentum

Having now been through the same approach for both mass and linear momentum, we can go rapidly through a derivation of the differential angular momentum relation. The appropriate form of the integral angular momentum equation for a fixed control volume is

$$\sum \mathbf{M}_o = \frac{\partial}{\partial t}\left[\int_{\text{CV}} (\mathbf{r} \times \mathbf{V})\rho \, d\mathcal{V}\right] + \int_{\text{CS}} (\mathbf{r} \times \mathbf{V})\rho(\mathbf{V} \cdot \mathbf{n}) \, dA \qquad (3.55)$$

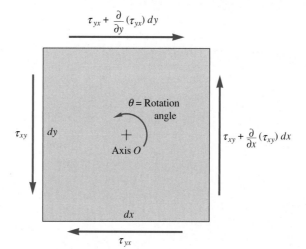

Fig. 4.5 Elemental cartesian fixed control volume showing shear stresses that may cause a net angular acceleration about axis O.

We shall confine ourselves to an axis through O that is parallel to the z axis and passes through the centroid of the elemental control volume. This is shown in Fig. 4.5. Let θ be the angle of rotation about O of the fluid within the control volume. The only stresses that have moments about O are the shear stresses τ_{xy} and τ_{yx}. We can evaluate the moments about O and the angular momentum terms about O. A lot of algebra is involved, and we give here only the result:

$$\left[\tau_{xy} - \tau_{yx} + \frac{1}{2}\frac{\partial}{\partial x}(\tau_{xy})\,dx - \frac{1}{2}\frac{\partial}{\partial y}(\tau_{yx})\,dy\right] dx\,dy\,dz$$

$$= \frac{1}{12}\rho(dx\,dy\,dz)(dx^2 + dy^2)\frac{d^2\theta}{dt^2} \quad (4.39)$$

Assuming that the angular acceleration $d^2\theta/dt^2$ is not infinite, we can neglect all higher-order differential terms, which leaves a finite and interesting result:

$$\tau_{xy} \approx \tau_{yx} \quad (4.40)$$

Had we summed moments about axes parallel to y or x, we would have obtained exactly analogous results:

$$\tau_{xz} \approx \tau_{zx} \qquad \tau_{yz} \approx \tau_{zy} \quad (4.41)$$

There is *no* differential angular momentum equation. Application of the integral theorem to a differential element gives the result, well known to students of stress analysis or strength of materials, that the shear stresses are symmetric: $\tau_{ij} = \tau_{ji}$. This is the only result of this section.[6] There is no differential equation to remember, which leaves room in your brain for the next topic, the differential energy equation.

---

[6]We are neglecting the possibility of a finite *couple* being applied to the element by some powerful external force field. See, for example, Ref. 6, p. 217.

4.5 The Differential Equation of Energy[7]

We are now so used to this type of derivation that we can race through the energy equation at a bewildering pace. The appropriate integral relation for the fixed control volume of Fig. 4.1 is

$$\dot{Q} - \dot{W}_s - \dot{W}_v = \frac{\partial}{\partial t}\left(\int_{CV} e\rho \, d\mathcal{V}\right) + \int_{CS}\left(e + \frac{p}{\rho}\right)\rho(\mathbf{V} \cdot \mathbf{n}) \, dA \qquad (3.63)$$

where $\dot{W}_s = 0$ because there can be no infinitesimal shaft protruding into the control volume. By analogy with Eq. (4.20), the right-hand side becomes, for this tiny element,

$$\dot{Q} - \dot{W}_v = \left[\frac{\partial}{\partial t}(\rho e) + \frac{\partial}{\partial x}(\rho u \zeta) + \frac{\partial}{\partial y}(\rho v \zeta) + \frac{\partial}{\partial z}(\rho w \zeta)\right]dx \, dy \, dz \qquad (4.42)$$

where $\zeta = e + p/\rho$. When we use the continuity equation by analogy with Eq. (4.21), this becomes

$$\dot{Q} - \dot{W}_v = \left(\rho\frac{de}{dt} + \mathbf{V} \cdot \nabla p + p\nabla \cdot \mathbf{V}\right)dx \, dy \, dz \qquad (4.43)$$

To evaluate $\dot{Q}$, we neglect radiation and consider only heat conduction through the sides of the element. The heat flow by conduction follows Fourier's law from Chap. 1.

$$\mathbf{q} = -k\nabla T \qquad (1.29a)$$

where k is the coefficient of thermal conductivity of the fluid. Figure 4.6 shows the heat flow passing through the x faces, the y and z heat flows being omitted for clarity. We can list these six heat flux terms:

Faces	Inlet heat flux	Outlet heat flux
x	$q_x \, dy \, dz$	$\left[q_x + \dfrac{\partial}{\partial x}(q_x) \, dx\right] dy \, dz$
y	$q_y \, dx \, dz$	$\left[q_y + \dfrac{\partial}{\partial y}(q_y) \, dy\right] dx \, dz$
z	$q_z \, dx \, dy$	$\left[q_z + \dfrac{\partial}{\partial z}(q_z) \, dz\right] dx \, dy$

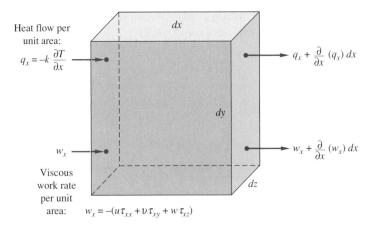

Heat flow per unit area:

$$q_x = -k\frac{\partial T}{\partial x}$$

Viscous work rate per unit area: $w_x = -(u\tau_{xx} + v\tau_{xy} + w\tau_{xz})$

Fig. 4.6 Elemental cartesian control volume showing heat flow and viscous work rate terms in the x direction.

[7]This section may be omitted without loss of continuity.

By adding the inlet terms and subtracting the outlet terms, we obtain the net heat added to the element:

$$\dot{Q} = -\left[\frac{\partial}{\partial x}(q_x) + \frac{\partial}{\partial y}(q_y) + \frac{\partial}{\partial z}(q_z)\right] dx \, dy \, dz = -\nabla \cdot \mathbf{q} \, dx \, dy \, dz \qquad (4.44)$$

As expected, the heat flux is proportional to the element volume. Introducing Fourier's law from Eq. (1.29), we have

$$\dot{Q} = \nabla \cdot (k \nabla T) \, dx \, dy \, dz \qquad (4.45)$$

The rate of work done by viscous stresses equals the product of the stress component, its corresponding velocity component, and the area of the element face. Figure 4.6 shows the work rate on the left x face is

$$\dot{W}_{v,\text{LF}} = w_x \, dy \, dz \qquad \text{where } w_x = -(u\tau_{xx} + v\tau_{xy} + w\tau_{xz}) \qquad (4.46)$$

(where the subscript LF stands for left face) and a slightly different work on the right face due to the gradient in w_x. These work fluxes could be tabulated in exactly the same manner as the heat fluxes in the previous table, with w_x replacing q_x, and so on. After outlet terms are subtracted from inlet terms, the net viscous work rate becomes

$$\begin{aligned}
\dot{W}_v &= -\left[\frac{\partial}{\partial x}(u\tau_{xx} + v\tau_{xy} + w\tau_{xz}) + \frac{\partial}{\partial y}(u\tau_{yx} + v\tau_{yy} + w\tau_{yz})\right.\\
&\quad \left.+ \frac{\partial}{\partial z}(u\tau_{zx} + v\tau_{zy} + w\tau_{zz})\right] dx \, dy \, dz \\
&= -\nabla \cdot (\mathbf{V} \cdot \boldsymbol{\tau}_{ij}) \, dx \, dy \, dz
\end{aligned} \qquad (4.47)$$

We now substitute Eqs. (4.45) and (4.47) into Eq. (4.43) to obtain one form of the differential energy equation:

$$\rho \frac{de}{dt} + \mathbf{V} \cdot \nabla p + p\nabla \cdot \mathbf{V} = \nabla \cdot (k \nabla T) + \nabla \cdot (\mathbf{V} \cdot \boldsymbol{\tau}_{ij})$$

$$\text{where } e = \hat{u} + \tfrac{1}{2}V^2 + gz \qquad (4.48)$$

A more useful form is obtained if we split up the viscous work term:

$$\nabla \cdot (\mathbf{V} \cdot \boldsymbol{\tau}_{ij}) \equiv \mathbf{V} \cdot (\nabla \cdot \boldsymbol{\tau}_{ij}) + \Phi \qquad (4.49)$$

where Φ is short for the *viscous-dissipation function*.[8] For a newtonian incompressible viscous fluid, this function has the form

$$\begin{aligned}
\Phi = \mu &\left[2\left(\frac{\partial u}{\partial x}\right)^2 + 2\left(\frac{\partial v}{\partial y}\right)^2 + 2\left(\frac{\partial w}{\partial z}\right)^2 + \left(\frac{\partial v}{\partial x} + \frac{\partial u}{\partial y}\right)^2\right.\\
&\left.+ \left(\frac{\partial w}{\partial y} + \frac{\partial v}{\partial z}\right)^2 + \left(\frac{\partial u}{\partial z} + \frac{\partial w}{\partial x}\right)^2\right]
\end{aligned} \qquad (4.50)$$

Since all terms are quadratic, viscous dissipation is always positive, so that a viscous flow always tends to lose its available energy due to dissipation, in accordance with the second law of thermodynamics.

---

[8]For further details, see, e.g., Ref. 5, p. 72.

Now substitute Eq. (4.49) into Eq. (4.48), using the linear momentum equation (4.32) to eliminate $\nabla \cdot \tau_{ij}$. This will cause the kinetic and potential energies to cancel, leaving a more customary form of the general differential energy equation:

$$\rho \frac{d\hat{u}}{dt} + p(\nabla \cdot \mathbf{V}) = \nabla \cdot (k\nabla T) + \Phi \qquad (4.51)$$

This equation is valid for a newtonian fluid under very general conditions of unsteady, compressible, viscous, heat-conducting flow, except that it neglects radiation heat transfer and internal *sources* of heat that might occur during a chemical or nuclear reaction.

Equation (4.51) is too difficult to analyze except on a digital computer [1]. It is customary to make the following approximations:

$$d\hat{u} \approx c_v\, dT \quad c_v, \mu, k, \rho \approx \text{const} \qquad (4.52)$$

Equation (4.51) then takes the simpler form, for $\nabla \cdot \mathbf{V} = 0$,

$$\rho c_v \frac{dT}{dt} = k\nabla^2 T + \Phi \qquad (4.53)$$

which involves temperature T as the sole primary variable plus velocity as a secondary variable through the total time-derivative operator:

$$\frac{dT}{dt} = \frac{\partial T}{\partial t} + u\frac{\partial T}{\partial x} + v\frac{\partial T}{\partial y} + w\frac{\partial T}{\partial z} \qquad (4.54)$$

A great many interesting solutions to Eq. (4.53) are known for various flow conditions, and extended treatments are given in advanced books on viscous flow [4, 5] and books on heat transfer [7, 8].

One well-known special case of Eq. (4.53) occurs when the fluid is at rest or has negligible velocity, where the dissipation Φ and convective terms become negligible:

$$\rho c_p \frac{\partial T}{\partial t} = k\nabla^2 T \qquad (4.55)$$

The change from c_v to c_p is correct and justified by the fact that, when pressure terms are neglected from a gas flow energy equation [4, 5], what remains is approximately an enthalpy change, not an internal energy change. This is called the *heat conduction equation* in applied mathematics and is valid for solids and fluids at rest. The solution to Eq. (4.55) for various conditions is a large part of courses and books on heat transfer.

This completes the derivation of the basic differential equations of fluid motion.

4.6 Boundary Conditions for the Basic Equations

There are three basic differential equations of fluid motion, just derived. Let us summarize them here:

Continuity:
$$\frac{\partial \rho}{\partial t} + \nabla \cdot (\rho \mathbf{V}) = 0 \qquad (4.56)$$

Momentum:
$$\rho \frac{d\mathbf{V}}{dt} = \rho \mathbf{g} - \nabla p + \nabla \cdot \tau_{ij} \qquad (4.57)$$

Energy:
$$\rho \frac{d\hat{u}}{dt} + p(\nabla \cdot \mathbf{V}) = \nabla \cdot (k\,\nabla T) + \Phi \qquad (4.58)$$

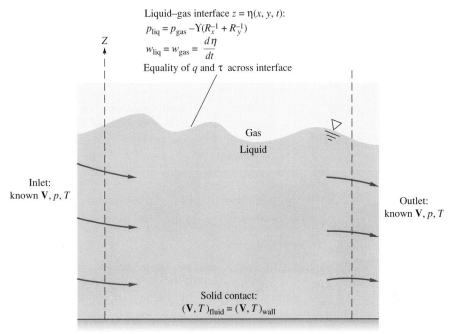

Liquid–gas interface $z = \eta(x, y, t)$:

$$p_{liq} = p_{gas} - Y(R_x^{-1} + R_y^{-1})$$

$$w_{liq} = w_{gas} = \frac{d\eta}{dt}$$

Equality of q and τ across interface

Gas

Liquid

Inlet:
known $\mathbf{V}, p, T$

Outlet:
known $\mathbf{V}, p, T$

Solid contact:
$(\mathbf{V}, T)_{fluid} = (\mathbf{V}, T)_{wall}$

Solid impermeable wall

Fig. 4.7 Typical boundary conditions in a viscous heat-conducting fluid flow analysis.

where Φ is given by Eq. (4.50). In general, the density is variable, so that these three equations contain five unknowns, ρ, V, p, $\hat{u}$, and T. Therefore we need two additional relations to complete the system of equations. These are provided by data or algebraic expressions for the state relations of the thermodynamic properties:

$$\rho = \rho(p, T) \qquad \hat{u} = \hat{u}(p, T) \tag{4.59}$$

For example, for a perfect gas with constant specific heats, we complete the system with

$$\rho = \frac{p}{RT} \qquad \hat{u} = \int c_v \, dT \approx c_v T + \text{const} \tag{4.60}$$

It is shown in advanced books [4, 5] that this system of equations (4.56) to (4.59) is well posed and can be solved analytically or numerically, subject to the proper boundary conditions.

What are the proper boundary conditions? First, if the flow is unsteady, there must be an *initial condition* or initial spatial distribution known for each variable:

At $t = 0$: $\qquad \rho, V, p, \hat{u}, T = \text{known } f(x, y, z) \tag{4.61}$

Thereafter, for all times t to be analyzed, we must know something about the variables at each *boundary* enclosing the flow.

Figure 4.7 illustrates the three most common types of boundaries encountered in fluid flow analysis: a solid wall, an inlet or outlet, and a liquid–gas interface.

First, for a solid, impermeable wall, there is no slip and no temperature jump in a viscous heat-conducting fluid:

Solid wall: $\qquad \mathbf{V}_{fllui} = \mathbf{V}_{wall} \qquad T_{fllui} = T_{wall} \tag{4.62}$

The only exception to Eq. (4.62) occurs in an extremely rarefied gas flow, where slippage can be present [5].

Second, at any inlet or outlet section of the flow, the complete distribution of velocity, pressure, and temperature must be known for all times:

$$\text{Inlet or outlet:} \qquad\qquad \text{Known } \mathbf{V},\, p,\, T \tag{4.63}$$

These inlet and outlet sections can be and often are at $\pm\infty$, simulating a body immersed in an infinite expanse of fluid.

Finally, the most complex conditions occur at a liquid–gas interface, or free surface, as sketched in Fig. 4.7. Let us denote the interface by

$$\text{Interface:} \qquad\qquad z = \eta(x, y, t) \tag{4.64}$$

Then there must be equality of vertical velocity across the interface, so that no holes appear between liquid and gas:

$$w_{\text{liq}} = w_{\text{gas}} = \frac{d\eta}{dt} = \frac{\partial\eta}{\partial t} + u\frac{\partial\eta}{\partial x} + v\frac{\partial\eta}{\partial y} \tag{4.65}$$

This is called the *kinematic boundary condition.*

There must be mechanical equilibrium across the interface. The viscous shear stresses must balance:

$$(\tau_{zy})_{\text{liq}} = (\tau_{zy})_{\text{gas}} \qquad (\tau_{zx})_{\text{liq}} = (\tau_{zx})_{\text{gas}} \tag{4.66}$$

Neglecting the viscous normal stresses, the pressures must balance at the interface except for surface tension effects:

$$p_{\text{liq}} = p_{\text{gas}} - \Upsilon(R_x^{-1} + R_y^{-1}) \tag{4.67}$$

which is equivalent to Eq. (1.34). The radii of curvature can be written in terms of the free surface position η:

$$\begin{aligned}
R_x^{-1} + R_y^{-1} = {} & \frac{\partial}{\partial x}\left[\frac{\partial\eta/\partial x}{\sqrt{1 + (\partial\eta/\partial x)^2 + (\partial\eta/\partial y)^2}}\right] \\
& + \frac{\partial}{\partial y}\left[\frac{\partial\eta/\partial y}{\sqrt{1 + (\partial\eta/\partial x)^2 + (\partial\eta/\partial y)^2}}\right]
\end{aligned} \tag{4.68}$$

Finally, the heat transfer must be the same on both sides of the interface, since no heat can be stored in the infinitesimally thin interface:

$$(q_z)_{\text{liq}} = (q_z)_{\text{gas}} \tag{4.69}$$

Neglecting radiation, this is equivalent to

$$\left(k\frac{\partial T}{\partial z}\right)_{\text{liq}} = \left(k\frac{\partial T}{\partial z}\right)_{\text{gas}} \tag{4.70}$$

This is as much detail as we wish to give at this level of exposition. Further and even more complicated details on fluid flow boundary conditions are given in Refs. 5 and 9.

Simplified Free Surface Conditions

In the introductory analyses given in this book, such as open-channel flows in Chap. 10, we shall back away from the exact conditions (4.65) to (4.69) and assume that the upper fluid is an "atmosphere" that merely exerts pressure on the lower fluid, with shear and heat conduction negligible. We also neglect nonlinear terms involving the slopes of the free surface. We then have a much simpler and linear set of conditions at the surface:

$$p_{liq} \approx p_{gas} - Y\left(\frac{\partial^2\eta}{\partial x^2} + \frac{\partial^2\eta}{\partial y^2}\right) \qquad w_{liq} \approx \frac{\partial\eta}{\partial t}$$

$$\left(\frac{\partial V}{\partial z}\right)_{liq} \approx 0 \qquad \left(\frac{\partial T}{\partial z}\right)_{liq} \approx 0 \tag{4.71}$$

In many cases, such as open-channel flow, we can also neglect surface tension, so that

$$p_{liq} \approx p_{atm} \tag{4.72}$$

These are the types of approximations that will be used in Chap. 10. The nondimensional forms of these conditions will also be useful in Chap. 5.

Incompressible Flow with Constant Properties

Flow with constant ρ, μ, and k is a basic simplification that will be used, for example, throughout Chap. 6. The basic equations of motion (4.56) to (4.58) reduce to

Continuity: $$\nabla \cdot V = 0 \tag{4.73}$$

Momentum: $$\rho\frac{dV}{dt} = \rho g - \nabla p + \mu\nabla^2 V \tag{4.74}$$

Energy: $$\rho c_p\frac{dT}{dt} = k\nabla^2 T + \Phi \tag{4.75}$$

Since ρ is constant, there are only three unknowns: p, V, and T. The system is closed.[9] Not only that, the system splits apart: Continuity and momentum are independent of T. Thus we can solve Eqs. (4.73) and (4.74) entirely separately for the pressure and velocity, using such boundary conditions as

Solid surface: $$V = V_{wall} \tag{4.76}$$

Inlet or outlet: $$\text{Known } V, p \tag{4.77}$$

Free surface: $$p \approx p_a \qquad w \approx \frac{\partial\eta}{\partial t} \tag{4.78}$$

Later, usually in another course,[10] we can solve for the temperature distribution from Eq. (4.75), which depends on velocity V through the dissipation Φ and the total time-derivative operator d/dt.

Inviscid Flow Approximations

Chapter 8 assumes inviscid flow throughout, for which the viscosity $\mu = 0$. The momentum equation (4.74) reduces to

$$\rho\frac{dV}{dt} = \rho g - \nabla p \tag{4.79}$$

---

[9]For this system, what are the thermodynamic equivalents to Eq. (4.59)?

[10]Since temperature is entirely *uncoupled* by this assumption, we may never get around to solving for it here and may ask you to wait until you take a course on heat transfer.

This is *Euler's equation;* it can be integrated along a streamline to obtain Bernoulli's equation (see Sec. 4.9). By neglecting viscosity we have lost the second-order derivative of **V** in Eq. (4.74); therefore we must relax one boundary condition on velocity. The only mathematically sensible condition to drop is the no-slip condition at the wall. We let the flow slip parallel to the wall but do not allow it to flow into the wall. The proper inviscid condition is that the normal velocities must match at any solid surface:

Inviscid flow:
$$(V_n)_{\text{fluid}} = (V_n)_{\text{wall}} \qquad (4.80)$$

In most cases the wall is fixed; therefore the proper inviscid flow condition is

$$V_n = 0 \qquad (4.81)$$

There is *no* condition whatever on the tangential velocity component at the wall in inviscid flow. The tangential velocity will be part of the solution to an inviscid flow analysis (see Chap. 8).

EXAMPLE 4.6

For steady incompressible laminar flow through a long tube, the velocity distribution is given by

$$v_z = U\left(1 - \frac{r^2}{R^2}\right) \qquad v_r = v_\theta = 0$$

where U is the maximum, or centerline, velocity and R is the tube radius. If the wall temperature is constant at T_w and the temperature $T = T(r)$ only, find $T(r)$ for this flow.

Solution

With $T = T(r)$, Eq. (4.75) reduces for steady flow to

$$\rho c_p v_r \frac{dT}{dr} = \frac{k}{r}\frac{d}{dr}\left(r\frac{dT}{dr}\right) + \mu\left(\frac{dv_z}{dr}\right)^2 \qquad (1)$$

But since $v_r = 0$ for this flow, the convective term on the left vanishes. Introduce v_z into Eq. (1) to obtain

$$\frac{k}{r}\frac{d}{dr}\left(r\frac{dT}{dr}\right)^2 = -\mu\left(\frac{dv_z}{dr}\right)^2 = -\frac{4U^2\mu r^2}{R^4} \qquad (2)$$

Multiply through by r/k and integrate once:

$$r\frac{dT}{dr} = -\frac{\mu U^2 r^4}{kR^4} + C_1 \qquad (3)$$

Divide through by r and integrate once again:

$$T = -\frac{\mu U^2 r^4}{4kR^4} + C_1 \ln r + C_2 \qquad (4)$$

Now we are in position to apply our boundary conditions to evaluate C_1 and C_2.

First, since the logarithm of zero is $-\infty$, the temperature at $r = 0$ will be infinite unless

$$C_1 = 0 \qquad (5)$$

Thus we eliminate the possibility of a logarithmic singularity. The same thing will happen if we apply the *symmetry* condition $dT/dr = 0$ at $r = 0$ to Eq. (3). The constant C_2 is then found by the wall-temperature condition at $r = R$:

$$T = T_w = -\frac{\mu U^2}{4k} + C_2$$

or

$$C_2 = T_w + \frac{\mu U^2}{4k} \qquad (6)$$

The correct solution is thus

$$T(r) = T_w + \frac{\mu U^2}{4k}\left(1 - \frac{r^4}{R^4}\right) \qquad Ans. \ (7)$$

which is a fourth-order parabolic distribution with a maximum value $T_0 = T_w + \mu U^2/(4k)$ at the centerline.

4.7 The Stream Function

We have seen in Sec. 4.6 that even if the temperature is uncoupled from our system of equations of motion, we must solve the continuity and momentum equations simultaneously for pressure and velocity. The *stream function* ψ is a clever device that allows us to satisfy the continuity equation and then solve the momentum equation directly for the single variable ψ.

The stream function idea works only if the continuity equation (4.56) can be reduced to *two* terms. In general, we have *four* terms:

Cartesian:
$$\frac{\partial \rho}{\partial t} + \frac{\partial}{\partial x}(\rho u) + \frac{\partial}{\partial y}(\rho v) + \frac{\partial}{\partial z}(\rho w) = 0 \qquad (4.82a)$$

Cylindrical:
$$\frac{\partial \rho}{\partial t} + \frac{1}{r}\frac{\partial}{\partial r}(r\rho v_r) + \frac{1}{r}\frac{\partial}{\partial \theta}(\rho v_\theta) + \frac{\partial}{\partial z}(\rho v_z) = 0 \qquad (4.82b)$$

First, let us eliminate unsteady flow, which is a peculiar and unrealistic application of the stream function idea. Reduce either of Eqs. (4.82) to any *two* terms. The most common application is incompressible flow in the *xy* plane:

$$\frac{\partial u}{\partial x} + \frac{\partial v}{\partial y} = 0 \qquad (4.83)$$

This equation is satisfied *identically* if a function $\psi(x, y)$ is defined such that Eq. (4.83) becomes

$$\frac{\partial}{\partial x}\left(\frac{\partial \psi}{\partial y}\right) + \frac{\partial}{\partial y}\left(-\frac{\partial \psi}{\partial x}\right) \equiv 0 \qquad (4.84)$$

Comparison of (4.83) and (4.84) shows that this new function ψ must be defined such that

$$u = \frac{\partial \psi}{\partial y} \qquad v = -\frac{\partial \psi}{\partial x} \qquad (4.85)$$

or

$$\mathbf{V} = \mathbf{i} \frac{\partial \psi}{\partial y} - \mathbf{j} \frac{\partial \psi}{\partial x}$$

Is this legitimate? Yes, it is just a mathematical trick of replacing two variables (u and v) by a single higher-order function ψ. The vorticity[11] or curl $\mathbf{V}$, is an interesting function:

$$\text{curl } \mathbf{V} = -\mathbf{k}\nabla^2 \psi \qquad \text{where} \qquad \nabla^2 \psi = \frac{\partial^2 \psi}{\partial x^2} + \frac{\partial^2 \psi}{\partial y^2} \qquad (4.86)$$

Thus, if we take the curl of the momentum equation (4.74) and utilize Eq. (4.86), we obtain a single equation for ψ for incompressible flow:

$$\frac{\partial \psi}{\partial y} \frac{\partial}{\partial x} (\nabla^2 \psi) - \frac{\partial \psi}{\partial x} \frac{\partial}{\partial y} (\nabla^2 \psi) = \nu \nabla^2 (\nabla^2 \psi) \qquad (4.87)$$

where $\nu = \mu/\rho$ is the kinematic viscosity. This is partly a victory and partly a defeat: Eq. (4.87) is scalar and has only one variable, ψ, but it now contains *fourth*-order derivatives and probably will require computer analysis. There will be four boundary conditions required on ψ. For example, for the flow of a uniform stream in the x direction past a solid body, the four conditions would be

At infinity:
$$\frac{\partial \psi}{\partial y} = U_\infty \qquad \frac{\partial \psi}{\partial x} = 0$$

$$(4.88)$$

At the body:
$$\frac{\partial \psi}{\partial y} = \frac{\partial \psi}{\partial x} = 0$$

Many examples of numerical solution of Eqs. (4.87) and (4.88) are given in Ref. 1.

One important application is inviscid, incompressible, *irrotational* flow[12] in the xy plane, where curl $\mathbf{V} \equiv 0$. Equations (4.86) and (4.87) reduce to

$$\nabla^2 \psi = \frac{\partial^2 \psi}{\partial x^2} + \frac{\partial^2 \psi}{\partial y^2} = 0 \qquad (4.89)$$

This is the second-order *Laplace equation* (Chap. 8), for which many solutions and analytical techniques are known. Also, boundary conditions like Eq. (4.88) reduce to

At infinity:
$$\psi = U_\infty y + \text{const} \qquad (4.90)$$

At the body:
$$\psi = \text{const}$$

[11]See Section 4.8.
[12]See Section 4.8.

It is well within our capability to find some useful solutions to Eqs. (4.89) and (4.90), which we shall do in Chap. 8.

Geometric Interpretation of ψ

The fancy mathematics above would serve alone to make the stream function immortal and always useful to engineers. Even better, though, ψ has a beautiful geometric interpretation: Lines of constant ψ are *streamlines* of the flow. This can be shown as follows. From Eq. (1.41) the definition of a streamline in two-dimensional flow is

$$\frac{dx}{u} = \frac{dy}{v}$$

or
$$u\,dy - v\,dx = 0 \qquad \text{streamline} \qquad (4.91)$$

Introducing the stream function from Eq. (4.85), we have

$$\frac{\partial \psi}{\partial x}\,dx + \frac{\partial \psi}{\partial y}\,dy = 0 = d\psi \qquad (4.92)$$

Thus the change in ψ is zero along a streamline, or

$$\psi = \text{const along a streamline} \qquad (4.93)$$

Having found a given solution $\psi(x, y)$, we can plot lines of constant ψ to give the streamlines of the flow.

There is also a physical interpretation that relates ψ to volume flow. From Fig. 4.8, we can compute the volume flow dQ through an element ds of control surface of unit depth:

$$dQ = (\mathbf{V} \cdot \mathbf{n})\,dA = \left(\mathbf{i}\frac{\partial \psi}{\partial y} - \mathbf{j}\frac{\partial \psi}{\partial x}\right) \cdot \left(\mathbf{i}\frac{dy}{ds} - \mathbf{j}\frac{dx}{ds}\right)ds(1)$$

$$= \frac{\partial \psi}{\partial x}\,dx + \frac{\partial \psi}{\partial y}\,dy = d\psi \qquad (4.94)$$

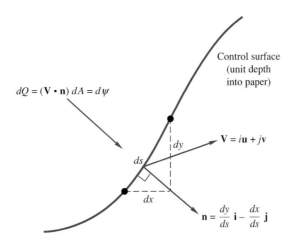

Fig. 4.8 Geometric interpretation of stream function: volume flow through a differential portion of a control surface.

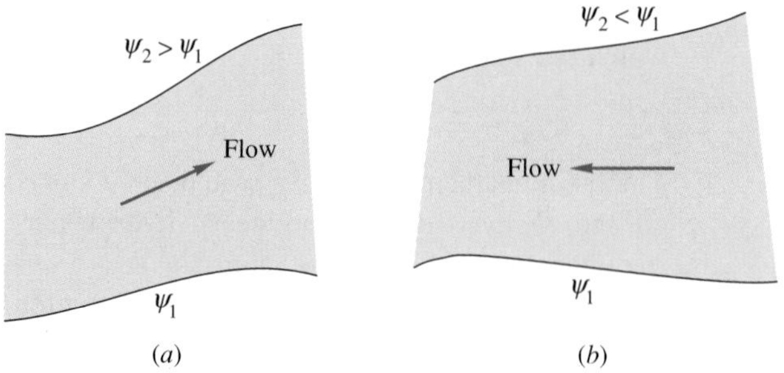

Fig. 4.9 Sign convention for flow in terms of change in stream function: (*a*) flow to the right if ψ_U is greater; (*b*) flow to the left if ψ_L is greater.

Thus the change in ψ across the element is numerically equal to the volume flow through the element. The volume flow between any two streamlines in the flow field is equal to the change in stream function between those streamlines:

$$Q_{1 \to 2} = \int_1^2 (\mathbf{V} \cdot \mathbf{n}) \, dA = \int_1^2 d\psi = \psi_2 - \psi_1 \qquad (4.95)$$

Further, the direction of the flow can be ascertained by noting whether ψ increases or decreases. As sketched in Fig. 4.9, the flow is to the right if ψ_U is greater than ψ_L, where the subscripts stand for upper and lower, as before; otherwise the flow is to the left.

Both the stream function and the velocity potential were invented by the French mathematician Joseph Louis Lagrange and published in his treatise on fluid mechanics in 1781.

EXAMPLE 4.7

If a stream function exists for the velocity field of Example 4.5

$$u = a(x^2 - y^2) \quad v = -2axy \quad w = 0$$

find it, plot it, and interpret it.

Solution

- *Assumptions:* Incompressible, two-dimensional flow.
- *Approach:* Use the definition of stream function derivatives, Eqs. (4.85), to find $\psi(x, y)$.
- *Solution step 1:* Note that this velocity distribution was also examined in Example 4.3. It satisfies continuity, Eq. (4.83), but let's check that; otherwise ψ will not exist:

$$\frac{\partial u}{\partial x} + \frac{\partial v}{\partial y} = \frac{\partial}{\partial x}[a(x^2 - y^2)] + \frac{\partial}{\partial y}(-2ay) = 2ax + (-2ax) \equiv 0 \qquad \text{checks}$$

Thus we are certain that a stream function exists.

• *Solution step 2:*   To find ψ, write out Eqs. (4.85) and integrate:

$$u = \frac{\partial \psi}{\partial y} = ax^2 - ay^2 \qquad (1)$$

$$v = -\frac{\partial \psi}{\partial x} = -2axy \qquad (2)$$

and work from either one toward the other. Integrate (1) partially

$$\psi = ax^2y - \frac{ay^3}{3} + f(x) \qquad (3)$$

Differentiate (3) with respect to x and compare with (2)

$$\frac{\partial \psi}{\partial x} = 2axy + f'(x) = 2axy \qquad (4)$$

Therefore $f'(x) = 0$, or $f =$ constant. The complete stream function is thus found:

$$\psi = a\left(x^2y - \frac{y^3}{3}\right) + C \qquad \text{Ans. (5)}$$

To plot this, set $C = 0$ for convenience and plot the function

$$3x^2y - y^3 = \frac{3\psi}{a} \qquad (6)$$

for constant values of ψ. The result is shown in Fig. E4.7a to be six 60° wedges of circulating motion, each with identical flow patterns except for the arrows. Once the streamlines are labeled, the flow directions follow from the sign convention of Fig. 4.9. How can the flow be interpreted? Since there is slip along all streamlines, no streamline can truly represent a solid surface in a viscous flow. However, the flow could represent the impingement of three incoming streams at 60, 180, and 300°. This would be a rather unrealistic yet exact solution to the Navier-Stokes equations, as we showed in Example 4.5.

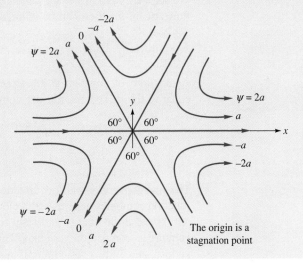

E4.7a

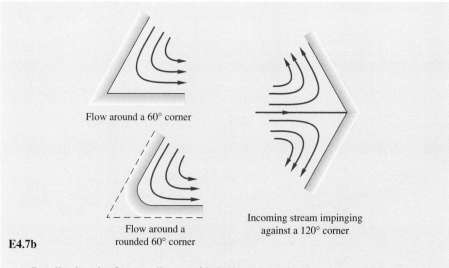

Flow around a 60° corner

Flow around a
rounded 60° corner

Incoming stream impinging
against a 120° corner

E4.7b

By allowing the flow to slip as a frictionless approximation, we could let any given streamline be a body shape. Some examples are shown in Fig. E4.7b.

A stream function also exists in a variety of other physical situations where only two coordinates are needed to define the flow. Three examples are illustrated here.

Steady Plane Compressible Flow

Suppose now that the density is variable but that $w = 0$, so that the flow is in the xy plane. Then the equation of continuity becomes

$$\frac{\partial}{\partial x}(\rho u) + \frac{\partial}{\partial y}(\rho v) = 0 \tag{4.96}$$

We see that this is in exactly the same form as Eq. (4.84). Therefore a compressible flow stream function can be defined such that

$$\rho u = \frac{\partial \psi}{\partial y} \qquad \rho v = -\frac{\partial \psi}{\partial x} \tag{4.97}$$

Again lines of constant ψ are streamlines of the flow, but the change in ψ is now equal to the *mass* flow, not the volume flow:

$$d\dot{m} = \rho(\mathbf{V} \cdot \mathbf{n})\, dA = d\psi$$

or $$\dot{m}_{1 \to 2} = \int_1^2 \rho(\mathbf{V} \cdot \mathbf{n})\, dA = \psi_2 - \psi_1 \tag{4.98}$$

The sign convention on flow direction is the same as in Fig. 4.9. This particular stream function combines density with velocity and must be substituted into not only momentum but also the energy and state relations (4.58) and (4.59) with pressure and temperature as companion variables. Thus the compressible stream function is not a great victory, and further assumptions must be made to effect an analytical solution to a typical problem (see, for instance, Ref. 5, chap. 7).

Incompressible Plane Flow in Polar Coordinates

Suppose that the important coordinates are r and θ, with $v_z = 0$, and that the density is constant. Then Eq. (4.82b) reduces to

$$\frac{1}{r}\frac{\partial}{\partial r}(rv_r) + \frac{1}{r}\frac{\partial}{\partial \theta}(v_\theta) = 0 \tag{4.99}$$

After multiplying through by r, we see that this is the analogous form of Eq. (4.84):

$$\frac{\partial}{\partial r}\left(\frac{\partial \psi}{\partial \theta}\right) + \frac{\partial}{\partial \theta}\left(-\frac{\partial \psi}{\partial r}\right) = 0 \tag{4.100}$$

By comparison of (4.99) and (4.100) we deduce the form of the incompressible polar coordinate stream function:

$$v_r = \frac{1}{r}\frac{\partial \psi}{\partial \theta} \qquad v_\theta = -\frac{\partial \psi}{\partial r} \tag{4.101}$$

Once again lines of constant ψ are streamlines, and the change in ψ is the *volume flow* $Q_{1\to2} = \psi_2 - \psi_1$. The sign convention is the same as in Fig. 4.9. This type of stream function is very useful in analyzing flows with cylinders, vortices, sources, and sinks (Chap. 8).

Incompressible Axisymmetric Flow

As a final example, suppose that the flow is three-dimensional (v_r, v_z) but with no circumferential variations, $v_\theta = \partial/\partial\theta = 0$ (see Fig. 4.2 for definition of coordinates). Such a flow is termed *axisymmetric,* and the flow pattern is the same when viewed on any meridional plane through the axis of revolution z. For incompressible flow, Eq. (4.82b) becomes

$$\frac{1}{r}\frac{\partial}{\partial r}(rv_r) + \frac{\partial}{\partial z}(v_z) = 0 \tag{4.102}$$

This doesn't seem to work: Can't we get rid of the one r outside? But when we realize that r and z are independent coordinates, Eq. (4.102) can be rewritten as

$$\frac{\partial}{\partial r}(rv_r) + \frac{\partial}{\partial z}(rv_z) = 0 \tag{4.103}$$

By analogy with Eq. (4.84), this has the form

$$\frac{\partial}{\partial r}\left(-\frac{\partial \psi}{\partial z}\right) + \frac{\partial}{\partial z}\left(\frac{\partial \psi}{\partial r}\right) = 0 \tag{4.104}$$

By comparing (4.103) and (4.104), we deduce the form of an incompressible axisymmetric stream function $\psi(r, z)$

$$v_r = -\frac{1}{r}\frac{\partial \psi}{\partial z} \qquad v_z = \frac{1}{r}\frac{\partial \psi}{\partial r} \tag{4.105}$$

Here again lines of constant ψ are streamlines, but there is a factor (2π) in the volume flow: $Q_{1\to2} = 2\pi(\psi_2 - \psi_1)$. The sign convention on flow is the same as in Fig. 4.9.

EXAMPLE 4.8

Investigate the stream function in polar coordinates

$$\psi = U \sin \theta \left(r - \frac{R^2}{r} \right) \tag{1}$$

where U and R are constants, a velocity and a length, respectively. Plot the streamlines. What does the flow represent? Is it a realistic solution to the basic equations?

Solution

The streamlines are lines of constant ψ, which has units of square meters per second. Note that $\psi/(UR)$ is dimensionless. Rewrite Eq. (1) in dimensionless form

$$\frac{\psi}{UR} = \sin \theta \left(\eta - \frac{1}{\eta} \right) \qquad \eta = \frac{r}{R} \tag{2}$$

Of particular interest is the special line $\psi = 0$. From Eq. (1) or (2) this occurs when (a) $\theta = 0$ or $180°$ and (b) $r = R$. Case (a) is the x axis, and case (b) is a circle of radius R, both of which are plotted in Fig. E4.8.

For any other nonzero value of ψ it is easiest to pick a value of r and solve for θ:

$$\sin \theta = \frac{\psi/(UR)}{r/R - R/r} \tag{3}$$

In general, there will be two solutions for θ because of the symmetry about the y axis. For example, take $\psi/(UR) = +1.0$:

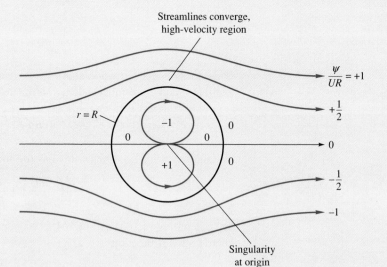

E4.8

Guess r/R	3.0	2.5	2.0	1.8	1.7	1.618
Compute θ	22°	28°	42°	53°	64°	90°
	158°	152°	138°	127°	116°	

This line is plotted in Fig. E4.8 and passes over the circle $r = R$. Be careful, though, because there is a second curve for $\psi/(UR) = +1.0$ for small $r < R$ below the x axis:

Guess r/R	0.618	0.6	0.5	0.4	0.3	0.2	0.1
Compute θ	$-90°$	$-70°$	$-42°$	$-28°$	$-19°$	$-12°$	$-6°$
		$-110°$	$-138°$	$-152°$	$-161°$	$-168°$	$-174°$

This second curve plots as a closed curve inside the circle $r = R$. There is a singularity of infinite velocity and indeterminate flow direction at the origin. Figure E4.8 shows the full pattern.

The given stream function, Eq. (1), is an exact and classic solution to the momentum equation (4.38) for frictionless flow. Outside the circle $r = R$ it represents two-dimensional inviscid flow of a uniform stream past a circular cylinder (Sec. 8.4). Inside the circle it represents a rather unrealistic trapped circulating motion of what is called a *line doublet*.

4.8 Vorticity and Irrotationality

The assumption of zero fluid angular velocity, or irrotationality, is a very useful simplification. Here we show that angular velocity is associated with the curl of the local velocity vector.

The differential relations for deformation of a fluid element can be derived by examining Fig. 4.10. Two fluid lines AB and BC, initially perpendicular at time t, move and deform so that at $t + dt$ they have slightly different lengths $A'B'$ and $B'C'$

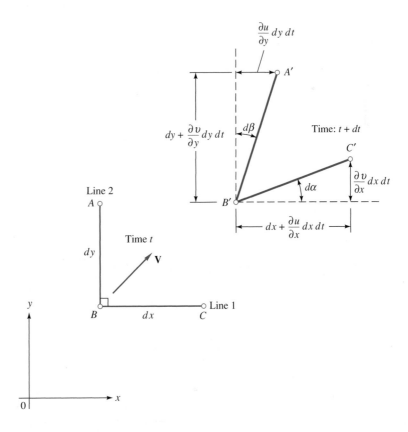

Fig. 4.10 Angular velocity and strain rate of two fluid lines deforming in the xy plane.

and are slightly off the perpendicular by angles $d\alpha$ and $d\beta$. Such deformation occurs kinematically because A, B, and C have slightly different velocities when the velocity field $\mathbf{V}$ has spatial gradients. All these differential changes in the motion of A, B, and C are noted in Fig. 4.10.

We define the angular velocity ω_z about the z axis as the average rate of counterclockwise turning of the two lines:

$$\omega_z = \frac{1}{2}\left(\frac{d\alpha}{dt} - \frac{d\beta}{dt}\right) \tag{4.106}$$

But from Fig. 4.10, $d\alpha$ and $d\beta$ are each directly related to velocity derivatives in the limit of small dt:

$$d\alpha = \lim_{dt \to 0}\left[\tan^{-1}\frac{(\partial v/\partial x)\,dx\,dt}{dx + (\partial u/\partial x)\,dx\,dt}\right] = \frac{\partial v}{\partial x}dt$$

$$d\beta = \lim_{dt \to 0}\left[\tan^{-1}\frac{(\partial u/\partial y)\,dy\,dt}{dy + (\partial v/\partial y)\,dy\,dt}\right] = \frac{\partial u}{\partial y}dt \tag{4.107}$$

Combining Eqs. (4.106) and (4.107) gives the desired result:

$$\omega_z = \frac{1}{2}\left(\frac{\partial v}{\partial x} - \frac{\partial u}{\partial y}\right) \tag{4.108}$$

In exactly similar manner we determine the other two rates:

$$\omega_x = \frac{1}{2}\left(\frac{\partial w}{\partial y} - \frac{\partial v}{\partial z}\right) \qquad \omega_y = \frac{1}{2}\left(\frac{\partial u}{\partial z} - \frac{\partial w}{\partial x}\right) \tag{4.109}$$

The vector $\omega = \mathbf{i}\omega_x + \mathbf{j}\omega_y + \mathbf{k}\omega_z$ is thus one-half the curl of the velocity vector

$$\omega = \frac{1}{2}(\text{curl } \mathbf{V}) = \frac{1}{2}\begin{vmatrix} \mathbf{i} & \mathbf{j} & \mathbf{k} \\ \dfrac{\partial}{\partial x} & \dfrac{\partial}{\partial y} & \dfrac{\partial}{\partial z} \\ u & v & w \end{vmatrix} \tag{4.110}$$

Since the factor of $\frac{1}{2}$ is annoying, many workers prefer to use a vector twice as large, called the *vorticity:*

$$\boldsymbol{\zeta} = 2\boldsymbol{\omega} = \text{curl } \mathbf{V} \tag{4.111}$$

Many flows have negligible or zero vorticity and are called *irrotational:*

$$\text{curl } \mathbf{V} \equiv 0 \tag{4.112}$$

The next section expands on this idea. Such flows can be incompressible or compressible, steady or unsteady.

We may also note that Fig. 4.10 demonstrates the *shear strain rate* of the element, which is defined as the rate of closure of the initially perpendicular lines:

$$\dot{\varepsilon}_{xy} = \frac{d\alpha}{dt} + \frac{d\beta}{dt} = \frac{\partial v}{\partial x} + \frac{\partial u}{\partial y} \tag{4.113}$$

When multiplied by viscosity μ, this equals the shear stress τ_{xy} in a newtonian fluid, as discussed earlier in Eqs. (4.37). Appendix D lists strain rate and vorticity components in cylindrical coordinates.

4.9 Frictionless Irrotational Flows

When a flow is both frictionless and irrotational, pleasant things happen. First, the momentum equation (4.38) reduces to Euler's equation:

$$\rho \frac{d\mathbf{V}}{dt} = \rho \mathbf{g} - \boldsymbol{\nabla}p \tag{4.114}$$

Second, there is a great simplification in the acceleration term. Recall from Sec. 4.1 that acceleration has two terms:

$$\frac{d\mathbf{V}}{dt} = \frac{\partial \mathbf{V}}{\partial t} + (\mathbf{V} \cdot \boldsymbol{\nabla})\mathbf{V} \tag{4.2}$$

A beautiful vector identity exists for the second term [11]:

$$(\mathbf{V} \cdot \boldsymbol{\nabla})\mathbf{V} \equiv \boldsymbol{\nabla}(\tfrac{1}{2}V^2) + \boldsymbol{\zeta} \times \mathbf{V} \tag{4.115}$$

where $\boldsymbol{\zeta} = \text{curl } \mathbf{V}$ from Eq. (4.111) is the fluid vorticity.

Now combine (4.114) and (4.115), divide by ρ, and rearrange on the left-hand side. Dot the entire equation into an arbitrary vector displacement $d\mathbf{r}$:

$$\left[\frac{\partial \mathbf{V}}{\partial t} + \boldsymbol{\nabla}\left(\frac{1}{2}V^2\right) + \boldsymbol{\zeta} \times \mathbf{V} + \frac{1}{\rho}\boldsymbol{\nabla}p - \mathbf{g} \right] \cdot d\mathbf{r} = 0 \tag{4.116}$$

Nothing works right unless we can get rid of the third term. We want

$$(\boldsymbol{\zeta} \times \mathbf{V}) \cdot (d\mathbf{r}) \equiv 0 \tag{4.117}$$

This will be true under various conditions:

1. $\mathbf{V}$ is zero; trivial, no flow (hydrostatics).
2. $\boldsymbol{\zeta}$ is zero; irrotational flow.
3. $d\mathbf{r}$ is perpendicular to $\boldsymbol{\zeta} \times \mathbf{V}$; this is rather specialized and rare.
4. $d\mathbf{r}$ is parallel to $\mathbf{V}$; we integrate *along a streamline* (see Sec. 3.7).

Condition 4 is the common assumption. If we integrate along a streamline in frictionless compressible flow and take, for convenience, $\mathbf{g} = -g\mathbf{k}$, Eq. (4.116) reduces to

$$\frac{\partial \mathbf{V}}{\partial t} \cdot d\mathbf{r} + d\left(\frac{1}{2}V^2\right) + \frac{dp}{\rho} + g\,dz = 0 \tag{4.118}$$

Except for the first term, these are exact differentials. Integrate between any two points 1 and 2 along the streamline:

$$\int_1^2 \frac{\partial V}{\partial t}\,ds + \int_1^2 \frac{dp}{\rho} + \frac{1}{2}(V_2^2 - V_1^2) + g(z_2 - z_1) = 0 \tag{4.119}$$

where ds is the arc length along the streamline. Equation (4.119) is Bernoulli's equation for frictionless unsteady flow along a streamline and is identical to Eq. (3.76). For incompressible steady flow, it reduces to

$$\frac{p}{\rho} + \frac{1}{2}V^2 + gz = \text{constant along streamline} \tag{4.120}$$

The constant may vary from streamline to streamline unless the flow is also irrotational (assumption 2). For irrotational flow $\zeta = 0$, the offending term Eq. (4.117) vanishes regardless of the direction of $d\mathbf{r}$, and Eq. (4.120) then holds all over the flow field with the same constant.

Velocity Potential

Irrotationality gives rise to a scalar function ϕ similar and complementary to the stream function ψ. From a theorem in vector analysis [11], a vector with zero curl must be the gradient of a scalar function

$$\text{If} \quad \nabla \times \mathbf{V} \equiv 0 \quad \text{then} \quad \mathbf{V} = \nabla\phi \tag{4.121}$$

where $\phi = \phi(x, y, z, t)$ is called the *velocity potential function*. Knowledge of ϕ thus immediately gives the velocity components

$$u = \frac{\partial\phi}{\partial x} \qquad v = \frac{\partial\phi}{\partial y} \qquad w = \frac{\partial\phi}{\partial z} \tag{4.122}$$

Lines of constant ϕ are called the *potential lines* of the flow.

Note that ϕ, unlike the stream function, is fully three-dimensional and not limited to two coordinates. It reduces a velocity problem with three unknowns u, v, and w to a single unknown potential ϕ; many examples are given in Chap. 8. The velocity potential also simplifies the unsteady Bernoulli equation (4.118) because if ϕ exists, we obtain

$$\frac{\partial\mathbf{V}}{\partial t} \cdot d\mathbf{r} = \frac{\partial}{\partial t}(\nabla\phi) \cdot d\mathbf{r} = d\left(\frac{\partial\phi}{\partial t}\right) \tag{4.123}$$

along any arbitrary direction. Equation (4.118) then becomes a relation between ϕ and p:

$$\frac{\partial\phi}{\partial t} + \int \frac{dp}{\rho} + \frac{1}{2}|\nabla\phi|^2 + gz = \text{const} \tag{4.124}$$

This is the unsteady irrotational Bernoulli equation. It is very important in the analysis of accelerating flow fields (see Refs. 10 and 15), but the only application in this text will be in Sec. 9.3 for steady flow.

Orthogonality of Streamlines and Potential Lines

If a flow is both irrotational and described by only two coordinates, ψ and ϕ both exist and the streamlines and potential lines are everywhere mutually perpendicular except at a stagnation point. For example, for incompressible flow in the xy plane, we would have

$$u = \frac{\partial\psi}{\partial y} = \frac{\partial\phi}{\partial x} \tag{4.125}$$

$$v = -\frac{\partial\psi}{\partial x} = \frac{\partial\phi}{\partial y} \tag{4.126}$$

Can you tell by inspection not only that these relations imply orthogonality but also that ϕ and ψ satisfy Laplace's equation?[13] A line of constant ϕ would be such that the change in ϕ is zero:

$$d\phi = \frac{\partial \phi}{\partial x} dx + \frac{\partial \phi}{\partial y} dy = 0 = u \, dx + v \, dy \tag{4.127}$$

Solving, we have

$$\left(\frac{dy}{dx}\right)_{\phi = \text{const}} = -\frac{u}{v} = -\frac{1}{(dy/dx)_{\psi = \text{const}}} \tag{4.128}$$

Equation (4.128) is the mathematical condition that lines of constant ϕ and ψ be mutually orthogonal. It may not be true at a stagnation point, where both u and v are zero, so that their ratio in Eq. (4.128) is indeterminate.

Generation of Rotationality[14]

This is the second time we have discussed Bernoulli's equation under different circumstances (the first was in Sec. 3.7). Such reinforcement is useful, since this is probably the most widely used equation in fluid mechanics. It requires frictionless flow with no shaft work or heat transfer between sections 1 and 2. The flow may or may not be irrotational, the latter being an easier condition, allowing a universal Bernoulli constant.

The only remaining question is this: *When* is a flow irrotational? In other words, when does a flow have negligible angular velocity? The exact analysis of fluid rotationality under arbitrary conditions is a topic for advanced study (for example, Ref. 10, sec. 8.5; Ref. 9, sec. 5.2; and Ref. 5, sec. 2.10). We shall simply state those results here without proof.

A fluid flow that is initially irrotational may become rotational if

1. There are significant viscous forces induced by jets, wakes, or solid boundaries. In this case Bernoulli's equation will not be valid in such viscous regions.
2. There are entropy gradients caused by curved shock waves (see Fig. 4.11*b*).
3. There are density gradients caused by *stratification* (uneven heating) rather than by pressure gradients.
4. There are significant *noninertial* effects such as the earth's rotation (the Coriolis acceleration).

In cases 2 to 4, Bernoulli's equation still holds along a streamline if friction is negligible. We shall not study cases 3 and 4 in this book. Case 2 will be treated briefly in Chap. 9 on gas dynamics. Primarily we are concerned with case 1, where rotation is induced by viscous stresses. This occurs near solid surfaces, where the no-slip condition creates a boundary layer through which the stream velocity drops to zero, and in jets and wakes, where streams of different velocities meet in a region of high shear.

Internal flows, such as pipes and ducts, are mostly viscous, and the wall layers grow to meet in the core of the duct. Bernoulli's equation does not hold in such flows unless it is modified for viscous losses.

---

[13]Equations (4.125) and (4.126) are called the *Cauchy-Riemann equations* and are studied in complex variable theory.

[14]This section may be omitted without loss of continuity.

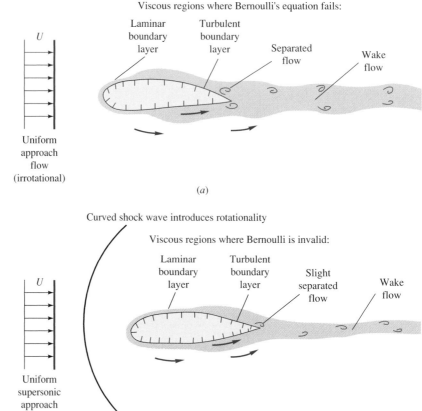

Fig. 4.11 Typical flow patterns illustrating viscous regions patched onto nearly frictionless regions: (*a*) low subsonic flow past a body ($U \ll a$); frictionless, irrotational potential flow outside the boundary layer (Bernoulli and Laplace equations valid); (*b*) supersonic flow past a body ($U > a$); frictionless, rotational flow outside the boundary layer (Bernoulli equation valid, potential flow invalid).

External flows, such as a body immersed in a stream, are partly viscous and partly inviscid, the two regions being patched together at the edge of the shear layer or boundary layer. Two examples are shown in Fig. 4.11. Figure 4.11*a* shows a low-speed subsonic flow past a body. The approach stream is irrotational; that is, the curl of a constant is zero, but viscous stresses create a rotational shear layer beside and downstream of the body. Generally speaking (see Chap. 7), the shear layer is laminar, or smooth, near the front of the body and turbulent, or disorderly, toward the rear. A separated, or deadwater, region usually occurs near the trailing edge, followed by an unsteady turbulent wake extending far downstream. Some sort of laminar or turbulent viscous theory must be applied to these viscous regions; they are then patched onto the outer flow, which is frictionless and irrotational. If the stream Mach number is less than about 0.3, we can combine Eq. (4.122) with the incompressible continuity equation (4.73):

$$\nabla \cdot \mathbf{V} = \nabla \cdot (\nabla \boldsymbol{\phi}) = 0$$

or

$$\nabla^2 \phi = 0 = \frac{\partial^2 \phi}{\partial x^2} + \frac{\partial^2 \phi}{\partial y^2} + \frac{\partial^2 \phi}{\partial z^2} \tag{4.129}$$

This is Laplace's equation in three dimensions, there being no restraint on the number of coordinates in potential flow. A great deal of Chap. 8 will be concerned with solving Eq. (4.129) for practical engineering problems; it holds in the entire region of Fig. 4.11a outside the shear layer.

Figure 4.11b shows a supersonic flow past a round-nosed body. A curved shock wave generally forms in front, and the flow downstream is *rotational* due to entropy gradients (case 2). We can use Euler's equation (4.114) in this frictionless region but not potential theory. The shear layers have the same general character as in Fig. 4.11a except that the separation zone is slight or often absent and the wake is usually thinner. Theory of separated flow is presently qualitative, but we can make quantitative estimates of laminar and turbulent boundary layers and wakes.

EXAMPLE 4.9

If a velocity potential exists for the velocity field of Example 4.5

$$u = a(x^2 - y^2) \qquad v = -2axy \quad w = 0$$

find it, plot it, and compare with Example 4.7.

Solution

Since $w = 0$, the curl of $\mathbf{V}$ has only one z component, and we must show that it is zero:

$$(\mathbf{\nabla} \times \mathbf{V})_z = 2\omega_z = \frac{\partial v}{\partial x} - \frac{\partial u}{\partial y} = \frac{\partial}{\partial x}(-2axy) - \frac{\partial}{\partial y}(ax^2 - ay^2)$$

$$= -2ay + 2ay = 0 \qquad \text{checks} \qquad\qquad Ans.$$

The flow is indeed irrotational. A velocity potential exists.

To find $\phi(x, y)$, set

$$u = \frac{\partial \phi}{\partial x} = ax^2 - ay^2 \qquad\qquad (1)$$

$$v = \frac{\partial \phi}{\partial y} = -2axy \qquad\qquad (2)$$

Integrate (1)

$$\phi = \frac{ax^3}{3} - axy^2 + f(y) \qquad\qquad (3)$$

Differentiate (3) and compare with (2)

$$\frac{\partial \phi}{\partial y} = -2axy + f'(y) = -2axy \qquad\qquad (4)$$

Therefore $f' = 0$, or $f =$ constant. The velocity potential is

$$\phi = \frac{ax^3}{3} - axy^2 + C \qquad\qquad Ans.$$

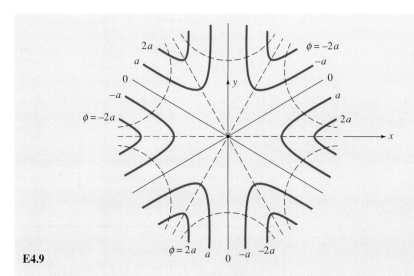

E4.9

Letting $C = 0$, we can plot the ϕ lines in the same fashion as in Example 4.7. The result is shown in Fig. E4.9 (no arrows on ϕ). For this particular problem, the ϕ lines form the same pattern as the ψ lines of Example 4.7 (which are shown here as dashed lines) but are displaced 30°. The ϕ and ψ lines are everywhere perpendicular except at the origin, a stagnation point, where they are 30° apart. We expected trouble at the stagnation point, and there is no general rule for determining the behavior of the lines at that point.

4.10 Some Illustrative Incompressible Viscous Flows

Inviscid flows do *not* satisfy the no-slip condition. They "slip" at the wall but do not flow through the wall. To look at fully viscous no-slip conditions, we must attack the complete Navier-Stokes equation (4.74), and the result is usually not at all irrotational, nor does a velocity potential exist. We look here at three cases: (1) flow between parallel plates due to a moving upper wall, (2) flow between parallel plates due to pressure gradient, and (3) flow between concentric cylinders when the inner one rotates. Other cases will be given as problem assignments or considered in Chap. 6. Extensive solutions for viscous flows are discussed in Refs. 4 and 5. All flows in this section are viscous and rotational.

Couette Flow between a Fixed and a Moving Plate

Consider two-dimensional incompressible plane ($\partial/\partial z = 0$) viscous flow between parallel plates a distance $2h$ apart, as shown in Fig. 4.12. We assume that the plates are very wide and very long, so that the flow is essentially axial, $u \neq 0$ but $v = w = 0$. The present case is Fig. 4.12a, where the upper plate moves at velocity V but there is no pressure gradient. Neglect gravity effects. We learn from the continuity equation (4.73) that

$$\frac{\partial u}{\partial x} + \frac{\partial v}{\partial y} + \frac{\partial w}{\partial z} = 0 = \frac{\partial u}{\partial x} + 0 + 0 \qquad \text{or} \qquad u = u(y) \text{ only}$$

Thus there is a single nonzero axial velocity component that varies only across the channel. The flow is said to be *fully developed* (far downstream of the entrance).

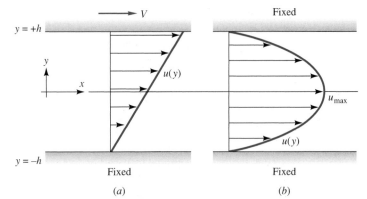

Fig. 4.12 Incompressible viscous flow between parallel plates: (*a*) no pressure gradient, upper plate moving; (*b*) pressure gradient $\partial p/\partial x$ with both plates fixed.

Substitute $u = u(y)$ into the x component of the Navier-Stokes momentum equation (4.74) for two-dimensional (x, y) flow:

$$\rho\left(u\frac{\partial u}{\partial x} + v\frac{\partial u}{\partial y}\right) = -\frac{\partial p}{\partial x} + \rho g_x + \mu\left(\frac{\partial^2 u}{\partial x^2} + \frac{\partial^2 u}{\partial y^2}\right)$$

or

$$\rho(0 + 0) = 0 + 0 + \mu\left(0 + \frac{d^2 u}{dy^2}\right) \tag{4.130}$$

Most of the terms drop out, and the momentum equation reduces to simply

$$\frac{d^2 u}{dy^2} = 0 \quad \text{or} \quad u = C_1 y + C_2$$

The two constants are found by applying the no-slip condition at the upper and lower plates:

At $y = +h$: $\qquad\qquad u = V = C_1 h + C_2$

At $y = -h$: $\qquad\qquad u = 0 = C_1(-h) + C_2$

or $\qquad\qquad C_1 = \dfrac{V}{2h} \quad \text{and} \quad C_2 = \dfrac{V}{2}$

Therefore the solution for this case (*a*), flow between plates with a moving upper wall, is

$$u = \frac{V}{2h}y + \frac{V}{2} \qquad -h \le y \le +h \tag{4.131}$$

This is *Couette flow* due to a moving wall: a linear velocity profile with no slip at each wall, as anticipated and sketched in Fig. 4.12a. Note that the origin has been placed in the center of the channel for convenience in case (*b*) which follows.

What we have just presented is a rigorous derivation of the more informally discussed flow of Fig. 1.6 (where y and h were defined differently).

Flow due to Pressure Gradient between Two Fixed Plates

Case (*b*) is sketched in Fig. 4.12b. Both plates are fixed ($V = 0$), but the pressure varies in the x direction. If $v = w = 0$, the continuity equation leads to the same conclusion as case (a)—namely, that $u = u(y)$ only. The x-momentum equation (4.130) changes only because the pressure is variable:

$$\mu \frac{d^2u}{dy^2} = \frac{\partial p}{\partial x} \tag{4.132}$$

Also, since $v = w = 0$ and gravity is neglected, the y- and z-momentum equations lead to

$$\frac{\partial p}{\partial y} = 0 \quad \text{and} \quad \frac{\partial p}{\partial z} = 0 \quad \text{or} \quad p = p(x) \text{ only}$$

Thus the pressure gradient in Eq. (4.132) is the total and only gradient:

$$\mu \frac{d^2u}{dy^2} = \frac{dp}{dx} = \text{const} < 0 \tag{4.133}$$

Why did we add the fact that dp/dx is *constant*? Recall a useful conclusion from the theory of separation of variables: If two quantities are equal and one varies only with y and the other varies only with x, then they must both equal the same constant. Otherwise they would not be independent of each other.

Why did we state that the constant is *negative*? Physically, the pressure must decrease in the flow direction in order to drive the flow against resisting wall shear stress. Thus the velocity profile $u(y)$ must have negative curvature everywhere, as anticipated and sketched in Fig. 4.12b.

The solution to Eq. (4.133) is accomplished by double integration:

$$u = \frac{1}{\mu} \frac{dp}{dx} \frac{y^2}{2} + C_1 y + C_2$$

The constants are found from the no-slip condition at each wall:

At $y = \pm h$: $\quad u = 0 \quad$ or $\quad C_1 = 0 \quad$ and $\quad C_2 = -\frac{dp}{dx} \frac{h^2}{2\mu}$

Thus the solution to case (b), flow in a channel due to pressure gradient, is

$$u = -\frac{dp}{dx} \frac{h^2}{2\mu} \left(1 - \frac{y^2}{h^2} \right) \tag{4.134}$$

The flow forms a *Poiseuille* parabola of constant negative curvature. The maximum velocity occurs at the centerline $y = 0$:

$$u_{max} = -\frac{dp}{dx} \frac{h^2}{2\mu} \tag{4.135}$$

Other (laminar) flow parameters are computed in the following example.

EXAMPLE 4.10

For case (b) in Fig. 4.12b, flow between parallel plates due to the pressure gradient, compute (a) the wall shear stress, (b) the stream function, (c) the vorticity, (d) the velocity potential, and (e) the average velocity.

Solution

All parameters can be computed from the basic solution, Eq. (4.134), by mathematical manipulation.

Part (a)

The wall shear follows from the definition of a newtonian fluid, Eq. (4.37):

$$\tau_w = \tau_{xy\,\text{wall}} = \mu\left(\frac{\partial u}{\partial y} + \frac{\partial v}{\partial x}\right)\bigg|_{y=\pm h} = \mu\frac{\partial}{\partial y}\left[\left(-\frac{dp}{dx}\right)\left(\frac{h^2}{2\mu}\right)\left(1 - \frac{y^2}{h^2}\right)\right]\bigg|_{y=\pm h}$$

$$= \pm\frac{dp}{dx}h = \mp\frac{2\mu u_{\text{max}}}{h} \qquad\qquad \textit{Ans. (a)}$$

The wall shear has the same magnitude at each wall, but by our sign convention of Fig. 4.3, the upper wall has negative shear stress.

Part (b)

Since the flow is plane, steady, and incompressible, a stream function exists:

$$u = \frac{\partial\psi}{\partial y} = u_{\text{max}}\left(1 - \frac{y^2}{h^2}\right) \qquad v = -\frac{\partial\psi}{\partial x} = 0$$

Integrating and setting $\psi = 0$ at the centerline for convenience, we obtain

$$\psi = u_{\text{max}}\left(y - \frac{y^3}{3h^2}\right) \qquad\qquad \textit{Ans. (b)}$$

At the walls, $y = \pm h$ and $\psi = \pm 2u_{\text{max}}h/3$, respectively.

Part (c)

In plane flow, there is only a single nonzero vorticity component:

$$\zeta_z = (\text{curl }\mathbf{V})_z = \frac{\partial v}{\partial x} - \frac{\partial u}{\partial y} = \frac{2u_{\text{max}}}{h^2}y \qquad\qquad \textit{Ans. (c)}$$

The vorticity is highest at the wall and is positive (counterclockwise) in the upper half and negative (clockwise) in the lower half of the fluid. Viscous flows are typically full of vorticity and are not at all irrotational.

Part (d)

From part (c), the vorticity is finite. Therefore the flow is not irrotational, and the velocity potential *does not exist*. \qquad\qquad \textit{Ans. (d)}

Part (e)

The average velocity is defined as $V_{\text{av}} = Q/A$, where $Q = \int u\,dA$ over the cross section. For our particular distribution $u(y)$ from Eq. (4.134), we obtain

$$V_{\text{av}} = \frac{1}{A}\int u\,dA = \frac{1}{b(2h)}\int_{-h}^{+h} u_{\text{max}}\left(1 - \frac{y^2}{h^2}\right)b\,dy = \frac{2}{3}u_{\text{max}} \qquad\qquad \textit{Ans. (e)}$$

In plane Poiseuille flow between parallel plates, the average velocity is two-thirds of the maximum (or centerline) value. This result could also have been obtained from the stream function derived in part (b). From Eq. (4.95),

$$Q_{\text{channel}} = \psi_{\text{upper}} - \psi_{\text{lower}} = \frac{2u_{\text{max}}h}{3} - \left(-\frac{2u_{\text{max}}h}{3}\right) = \frac{4}{3}u_{\text{max}}h \text{ per unit width}$$

whence $V_{\text{av}} = Q/A_{b=1} = (4u_{\text{max}}h/3)/(2h) = 2u_{\text{max}}/3$, the same result.

This example illustrates a statement made earlier: Knowledge of the velocity vector **V** [as in Eq. (4.134)] is essentially the *solution* to a fluid mechanics problem, since all other flow properties can then be calculated.

Fully Developed Laminar Pipe Flow

Perhaps the most useful exact solution of the Navier-Stokes equation is for incompressible flow in a straight circular pipe of radius R, first studied experimentally by G. Hagen in 1839 and J. L. Poiseuille in 1840. By *fully developed* we mean that the region studied is far enough from the entrance that the flow is purely axial, $v_z \neq 0$, while v_r and v_θ are zero. We neglect gravity and also assume axial symmetry—that is, $\partial/\partial\theta = 0$. The equation of continuity in cylindrical coordinates, Eq. (4.12b), reduces to

$$\frac{\partial}{\partial z}(v_z) = 0 \qquad \text{or} \qquad v_z = v_z(r) \qquad \text{only}$$

The flow proceeds straight down the pipe without radial motion. The r-momentum equation in cylindrical coordinates, Eq. (D.5), simplifies to $\partial p/\partial r = 0$, or $p = p(z)$ only. The z-momentum equation in cylindrical coordinates, Eq. (D.7), reduces to

$$\rho v_z \frac{\partial v_z}{\partial z} = -\frac{dp}{dz} + \mu \nabla^2 v_z = -\frac{dp}{dz} + \frac{\mu}{r} \frac{d}{dr}\left(r \frac{dv_z}{dr}\right)$$

The convective acceleration term on the left vanishes because of the previously given continuity equation. Thus the momentum equation may be rearranged as follows:

$$\frac{\mu}{r} \frac{d}{dr}\left(r \frac{dv_z}{dr}\right) = \frac{dp}{dz} = \text{const} < 0 \tag{4.136}$$

This is exactly the situation that occurred for flow between flat plates in Eq. (4.132). Again the "separation" constant is negative, and pipe flow will look much like the plate flow in Fig. 4.12b.

Equation (4.136) is linear and may be integrated twice, with the result

$$v_z = \frac{dp}{dz} \frac{r^2}{4\mu} + C_1 \ln(r) + C_2$$

where C_1 and C_2 are constants. The boundary conditions are no slip at the wall and finite velocity at the centerline:

$$\text{No slip at } r = R: \quad v_z = 0 = \frac{dp}{dz} \frac{R^2}{4\mu} + C_1 \ln(R) + C_2$$

$$\text{Finite velocity at } r = 0: \quad v_z = \text{finite} = 0 + C_1 \ln(0) + C_2$$

To avoid a logarithmic singularity, the centerline condition requires that $C_1 = 0$. Then, from no slip, $C_2 = (-dp/dz)(R^2/4\mu)$. The final, and famous, solution for fully developed *Hagen-Poiseuille flow* is

$$\boxed{v_z = \left(-\frac{dp}{dz}\right) \frac{1}{4\mu} (R^2 - r^2)} \tag{4.137}$$

The velocity profile is a paraboloid with a maximum at the centerline. Just as in Example 4.10, knowledge of the velocity distribution enables other parameters to be calculated:

$$V_{max} = v_z(r = 0) = \left(-\frac{dp}{dz}\right)\frac{R^2}{4\mu}$$

$$V_{avg} = \frac{1}{A}\int v_z \, dA = \frac{1}{\pi R^2}\int_0^R V_{max}\left(1-\frac{r^2}{R^2}\right)2\pi r \, dr = \frac{V_{max}}{2} = \left(-\frac{dp}{dz}\right)\frac{R^2}{8\mu}$$

$$Q = \int v_z \, dA = \int_0^R V_{max}\left(1-\frac{r^2}{R^2}\right)2\pi r \, dr = \pi R^2 V_{avg} = \frac{\pi R^4}{8\mu}\left(-\frac{dp}{dz}\right) = \frac{\pi R^4 \Delta p}{8\mu L}$$

$$\tau_{wall} = \mu\left|\frac{\partial v_z}{\partial r}\right|_{r=R} = \frac{4\mu V_{avg}}{R} = \frac{R}{2}\left(-\frac{dp}{dz}\right) = \frac{R}{2}\frac{\Delta p}{L} \qquad (4.138)$$

Note that we have substituted the equality $(-dp/dz) = \Delta p/L$, where Δp is the pressure drop along the entire length L of the pipe.

These formulas are valid as long as the flow is *laminar*—that is, when the dimensionless Reynolds number of the flow, $Re_D = \rho V_{avg}(2R)/\mu$, is less than about 2100. Note also that the formulas do not depend on density, the reason being that the convective acceleration of this flow is zero.

EXAMPLE 4.11

SAE 10W oil at 20°C flows at 1.1 m³/h through a horizontal pipe with $d = 2$ cm and $L = 12$ m. Find (*a*) the average velocity, (*b*) the Reynolds number, (*c*) the pressure drop, and (*d*) the power required.

Solution

- *Assumptions:* Laminar, steady, Hagen-Poiseuille pipe flow.
- *Approach:* The formulas of Eqs. (4.138) are appropriate for this problem. Note that $R = 0.01$ m.
- *Property values:* From Table A.3 for SAE 10W oil, $\rho = 870$ kg/m³ and $\mu = 0.104$ kg/(m − s).
- *Solution steps:* The average velocity follows easily from the flow rate and the pipe area:

$$V_{avg} = \frac{Q}{\pi R^2} = \frac{(1.1/3600) \text{ m}^3/\text{s}}{\pi(0.01\,\text{m})^2} = 0.973\frac{\text{m}}{\text{s}} \qquad \text{Ans. (a)}$$

We had to convert Q to m³/s. The (diameter) Reynolds number follows from the average velocity:

$$Re_d = \frac{\rho V_{avg} d}{\mu} = \frac{(870 \text{ kg/m}^3)(0.973 \text{ m/s})(0.02 \text{ m})}{0.104 \text{ kg/(m − s)}} = 163 \qquad \text{Ans. (b)}$$

This is less than the "transition" value of 2100; so the flow is indeed *laminar*, and the formulas are valid. The pressure drop is computed from the third of Eqs. (4.138):

$$Q = \frac{1.1}{3600}\frac{\text{m}^3}{\text{s}} = \frac{\pi R^4 \Delta p}{8\mu L} = \frac{\pi(0.01\,\text{m})^4 \Delta p}{8(0.104\,\text{kg/(m}-\text{s))}(12\text{m})} \quad \text{solve for} \ \ \Delta p = 97,100\,\text{Pa} \ Ans. \ (c)$$

When using SI units, the answer returns in pascals; no conversion factors are needed. Finally, the power required is the product of flow rate and pressure drop:

$$\text{Power} = Q\Delta p = \left(\frac{1.1}{3600}\,\text{m}^3/\text{s}\right)(97,100\,\text{N/m}^2) = 29.7\frac{\text{N}-\text{m}}{\text{s}} = 29.7\,\text{W} \quad Ans. \ (d)$$

- *Comments:* Pipe flow problems are straightforward algebraic exercises if the data are compatible. Note again that SI units can be used in the formulas without conversion factors.

Flow between Long Concentric Cylinders

Consider a fluid of constant (ρ, μ) between two concentric cylinders, as in Fig. 4.13. There is no axial motion or end effect $v_z = \partial/\partial z = 0$. Let the inner cylinder rotate at angular velocity Ω_i. Let the outer cylinder be fixed. There is circular symmetry, so the velocity does not vary with θ and varies only with r.

The continuity equation for this problem is Eq. (4.12b) with $v_z = 0$:

$$\frac{1}{r}\frac{\partial}{\partial r}(rv_r) + \frac{1}{r}\frac{\partial v_\theta}{\partial \theta} = 0 = \frac{1}{r}\frac{d}{dr}(rv_r) \quad \text{or} \quad rv_r = \text{const}$$

Note that v_θ does not vary with θ. Since $v_r = 0$ at both the inner and outer cylinders, it follows that $v_r = 0$ everywhere and the motion can only be purely circumferential, $v_\theta = v_\theta(r)$. The θ-momentum equation (D.6) becomes

$$\rho(\mathbf{V} \cdot \nabla)v_\theta + \frac{\rho v_r v_\theta}{r} = -\frac{1}{r}\frac{\partial p}{\partial \theta} + \rho g_\theta + \mu\left(\nabla^2 v_\theta - \frac{v_\theta}{r^2}\right)$$

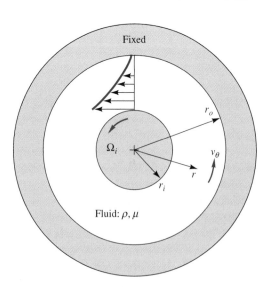

Fig. 4.13 Coordinate system for incompressible viscous flow between a fixed outer cylinder and a steadily rotating inner cylinder.

For the conditions of the present problem, all terms are zero except the last. Therefore the basic differential equation for flow between rotating cylinders is

$$\nabla^2 v_\theta = \frac{1}{r}\frac{d}{dr}\left(r\frac{dv_\theta}{dr}\right) = \frac{v_\theta}{r^2} \tag{4.139}$$

This is a linear second-order ordinary differential equation with the solution

$$v_\theta = C_1 r + \frac{C_2}{r}$$

The constants are found by the no-slip condition at the inner and outer cylinders:

Outer, at $r = r_o$: $\qquad\qquad v_\theta = 0 = C_1 r_o + \dfrac{C_2}{r_o}$

Inner, at $r = r_i$: $\qquad\qquad v_\theta = \Omega_i r_i = C_1 r_i + \dfrac{C_2}{r_i}$

The final solution for the velocity distribution is

Rotating inner cylinder: $\qquad v_\theta = \Omega_i r_i \dfrac{r_o/r - r/r_o}{r_o/r_i - r_i/r_o} \tag{4.140}$

The velocity profile closely resembles the sketch in Fig. 4.13. Variations of this case, such as a rotating outer cylinder, are given in the problem assignments.

Instability of Rotating Inner[15] Cylinder Flow

The classic *Couette flow* solution[16] of Eq. (4.140) describes a physically satisfying concave, two-dimensional, laminar flow velocity profile as in Fig. 4.13. The solution is mathematically exact for an incompressible fluid. However, it becomes unstable at a relatively low rate of rotation of the inner cylinder, as shown in 1923 in a classic paper by G. I. Taylor [17]. At a critical value of what is now called the dimensionless *Taylor number,* denoted Ta,

$$\text{Ta}_{\text{crit}} = \frac{r_i(r_o - r_i)^3 \Omega_i^2}{\nu^2} \approx 1700 \tag{4.141}$$

the plane flow of Fig. 4.13 vanishes and is replaced by a laminar *three-dimensional* flow pattern consisting of rows of nearly square alternating toroidal vortices. An experimental demonstration of toroidal "Taylor vortices" is shown in Fig. 4.14a, measured at Ta $\approx$ 1.16 Ta$_{\text{crit}}$ by Koschmieder [18]. At higher Taylor numbers, the vortices also develop a circumferential periodicity but are still laminar, as illustrated in Fig. 4.14b. At still higher Ta, turbulence ensues. This interesting instability reminds us that the Navier-Stokes equations, being nonlinear, do admit to multiple (nonunique) laminar solutions in addition to the usual instabilities associated with turbulence and chaotic dynamic systems.

---

[15]This section may be omitted without loss of continuity.
[16]Named after M. Couette, whose pioneering paper in 1890 established rotating cylinders as a method, still used today, for measuring the viscosity of fluids.

(a)

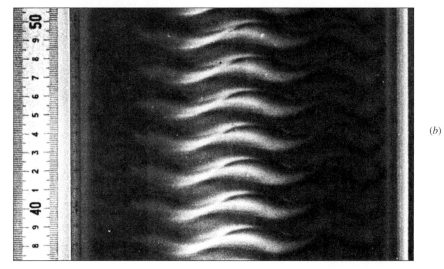

(b)

Fig. 4.14 Experimental verification of the instability of flow between a fixed outer and a rotating inner cylinder. (a) Toroidal Taylor vortices exist at 1.16 times the critical speed; (b) at 8.5 times the critical speed, the vortices are doubly periodic. (*Courtesy of Cambridge University Press—E.L. Koschmieder, "Turbulent Taylor Vortex Flow," Journal of Fluid Mechanics, vol. 93. pt. 3, 1979, pp. 515–527.*) This instability does not occur if only the outer cylinder rotates.

Summary

This chapter complements Chap. 3 by using an infinitesimal control volume to derive the basic partial differential equations of mass, momentum, and energy for a fluid. These equations, together with thermodynamic state relations for the fluid and appropriate boundary conditions, in principle can be solved for the complete flow field in any given fluid mechanics problem. Except for Chap. 9, in most of the problems to be studied here an incompressible fluid with constant viscosity is assumed.

In addition to deriving the basic equations of mass, momentum, and energy, this chapter introduced some supplementary ideas—the stream function, vorticity, irrotationality, and the velocity potential—which will be useful in coming chapters, especially Chap. 8. Temperature and density variations will be neglected except in Chap. 9, where compressibility is studied.

This chapter ended by discussing a few classical solutions for laminar viscous flows (Couette flow due to moving walls, Poiseuille duct flow due to pressure gradient, and flow between rotating cylinders). Whole books [4, 5, 9–11, 15] discuss classical approaches to fluid mechanics, and other texts [6, 12–14] extend these studies to the realm of continuum mechanics. This does not mean that all problems can be solved analytically. The new field of computational fluid dynamics [1] shows great promise of achieving approximate solutions to a wide variety of flow problems. In addition, when the geometry and boundary conditions are truly complex, experimentation (Chap. 5) is a preferred alternative.

Problems

Most of the problems herein are fairly straightforward. More difficult or open-ended assignments are labeled with an asterisk. Problems labeled with an EES icon will benefit from the use of the Engineering Equation Solver (EES), while problems labeled with a computer disk may require the use of a computer. The standard end-of-chapter problems P4.1 to P4.93 (categorized in the problem list here) are followed by word problems W4.1 to W4.10, fundamentals of engineering exam problems FE4.1 to FE4.3, and comprehensive problems C4.1 and C4.2.

Problem Distribution

Section	Topic	Problems
4.1	The acceleration of a fluid	P4.1–P4.8
4.2	The continuity equation	P4.9–P4.25
4.3	Linear momentum: Navier-Stokes	P4.26–P4.38
4.4	Angular momentum: couple stresses	P4.39
4.5	The differential energy equation	P4.40–P4.41
4.6	Boundary conditions	P4.42–P4.46
4.7	Stream function	P4.47–P4.55
4.8	Vorticity, irrotationality	P4.56–P4.60
4.9	Velocity potential	P4.61–P4.67
4.7 and 4.9	Stream function and velocity potential	P4.68–P4.78
4.10	Incompressible viscous flows	P4.79–P4.95

P4.1   An idealized velocity field is given by the formula

$$\mathbf{V} = 4tx\mathbf{i} - 2t^2y\mathbf{j} + 4xz\mathbf{k}$$

Is this flow field steady or unsteady? Is it two- or three-dimensional? At the point $(x, y, z) = (-1, 1, 0)$, compute (a) the acceleration vector and (b) any unit vector normal to the acceleration.

P4.2   Flow through the converging nozzle in Fig. P4.2 can be approximated by the one-dimensional velocity distribution

$$u \approx V_0\left(1 + \frac{2x}{L}\right) \quad v \approx 0 \quad w \approx 0$$

(a) Find a general expression for the fluid acceleration in the nozzle. (b) For the specific case $V_0 = 10$ ft/s and $L = 6$ in, compute the acceleration, in g's, at the entrance and at the exit.

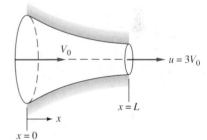

P4.2

P4.3   A two-dimensional velocity field is given by

$$\mathbf{V} = (x^2 - y^2 + x)\mathbf{i} - (2xy + y)\mathbf{j}$$

in arbitrary units. At $(x, y) = (1, 2)$, compute (a) the accelerations a_x and a_y, (b) the velocity component in the direction $\theta = 40°$, (c) the direction of maximum velocity, and (d) the direction of maximum acceleration.

P4.4   A simple flow model for a two-dimensional converging nozzle is the distribution

$$u = U_0\left(1 + \frac{x}{L}\right) \quad v = -U_0\frac{y}{L} \quad w = 0$$

(a) Sketch a few streamlines in the region $0 < x/L < 1$ and $0 < y/L < 1$, using the method of Section 1.11. (b) Find expressions for the horizontal and vertical accelerations. (c) Where is the largest resultant acceleration and its numerical value?

P4.5   The velocity field near a stagnation point (see Example 1.13) may be written in the form

$$u = \frac{U_0x}{L} \quad v = -\frac{U_0y}{L} \quad U_0 \text{ and } L \text{ are constants}$$

(a) Show that the acceleration vector is purely radial.
(b) For the particular case $L = 1.5$ m, if the acceleration at $(x, y) = (1$ m, 1 m) is 25 m/s², what is the value of U_0?

P4.6   Assume that flow in the converging nozzle of Fig. P4.2 has the form $\mathbf{V} = V_0[1 + (2x)/L]\mathbf{i}$. Compute (a) the fluid acceleration at $x = L$ and (b) the time required for a fluid particle to travel from $x = 0$ to $x = L$.

P4.7   Consider a sphere of radius R immersed in a uniform stream U_0, as shown in Fig. P4.7. According to the theory of Chap. 8, the fluid velocity along streamline AB is given by

$$\mathbf{V} = u\mathbf{i} = U_0\left(1 + \frac{R^3}{x^3}\right)\mathbf{i}$$

Find (a) the position of maximum fluid acceleration along AB and (b) the time required for a fluid particle to travel from A to B.

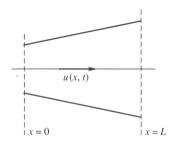

P4.7

P4.8   When a valve is opened, fluid flows in the expansion duct of Fig. P4.8 according to the approximation

$$\mathbf{V} = \mathbf{i}U\left(1 - \frac{x}{2L}\right)\tanh\frac{Ut}{L}$$

Find (a) the fluid acceleration at $(x, t) = (L, L/U)$ and (b) the time for which the fluid acceleration at $x = L$ is zero. Why does the fluid acceleration become negative after condition (b)?

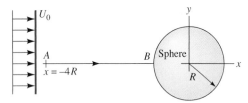

P4.8

P4.9   An idealized incompressible flow has the proposed three-dimensional velocity distribution

$$\mathbf{V} = 4xy^2\mathbf{i} + f(y)\mathbf{j} - zy^2\mathbf{k}$$

Find the appropriate form of the function $f(y)$ that satisfies the continuity relation.

P4.10   After discarding any constants of integration, determine the appropriate value of the unknown velocities u or v that satisfy the equation of two-dimensional incompressible continuity for

(a) $u = x^2y$ (b) $v = x^2y$
(c) $u = x^2 - xy$ (d) $v = y^2 - xy$

P4.11   Derive Eq. (4.12b) for cylindrical coordinates by considering the flux of an incompressible fluid in and out of the elemental control volume in Fig. 4.2.

P4.12   Spherical polar coordinates (r, θ, ϕ) are defined in Fig. P4.12. The cartesian transformations are

$$x = r\sin\theta\cos\phi$$
$$y = r\sin\theta\sin\phi$$
$$z = r\cos\theta$$

The cartesian incompressible continuity relation [Eq. (4.12a)] can be transformed to the spherical polar form

$$\frac{1}{r^2}\frac{\partial}{\partial r}(r^2v_r) + \frac{1}{r\sin\theta}\frac{\partial}{\partial\theta}(v_\theta\sin\theta) + \frac{1}{r\sin\theta}\frac{\partial}{\partial\phi}(v_\phi) = 0$$

What is the most general form of v_r when the flow is purely radial—that is, v_θ and v_ϕ are zero?

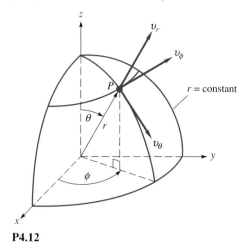

P4.12

P4.13   A two-dimensional velocity field is given by

$$u = -\frac{Ky}{x^2 + y^2} \qquad v = \frac{Kx}{x^2 + y^2}$$

where K is constant. Does this field satisfy incompressible continuity? Transform these velocities to polar components v_r and v_θ. What might the flow represent?

P4.25   An incompressible flow in polar coordinates is given by

$$v_r = K \cos \theta \left(1 - \frac{b}{r^2} \right)$$

$$v_\theta = -K \sin \theta \left(1 + \frac{b}{r^2} \right)$$

Does this field satisfy continuity? For consistency, what should the dimensions of constants K and b be? Sketch the surface where $v_r = 0$ and interpret.

*P4.26   Curvilinear, or streamline, coordinates are defined in Fig. P4.26, where n is normal to the streamline in the plane of the radius of curvature R. Euler's frictionless momentum equation (4.36) in streamline coordinates becomes

$$\frac{\partial V}{\partial t} + V \frac{\partial V}{\partial s} = -\frac{1}{\rho} \frac{\partial p}{\partial s} + g_s \qquad (1)$$

$$-V \frac{\partial \theta}{\partial t} - \frac{V^2}{R} = -\frac{1}{\rho} \frac{\partial p}{\partial n} + g_n \qquad (2)$$

Show that the integral of Eq. (1) with respect to s is none other than our old friend Bernoulli's equation (3.76).

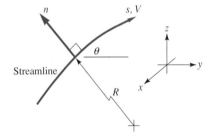

P4.26

P4.27   A frictionless, incompressible steady flow field is given by

$$\mathbf{V} = 2xy\mathbf{i} - y^2\mathbf{j}$$

in arbitrary units. Let the density be $\rho_0 = $ constant and neglect gravity. Find an expression for the pressure gradient in the x direction.

P4.28   If z is "up," what are the conditions on constants a and b for which the velocity field $u = ay$, $v = bx$, $w = 0$ is an exact solution to the continuity and Navier-Stokes equations for incompressible flow?

P4.29   Consider a steady, two-dimensional, incompressible flow of a newtonian fluid in which the velocity field is known: $u = -2xy$, $v = y^2 - x^2$, $w = 0$. (a) Does this flow satisfy conservation of mass? (b) Find the pressure field, $p(x, y)$ if the pressure at the point $(x = 0, y = 0)$ is equal to p_a.

P4.30   For the velocity distribution of Prob. P4.4, determine if (a) the equation of continuity and (b) the Navier-Stokes equation are satisfied. (c) If the latter is true, find the pressure distribution $p(x, y)$ when the pressure at the origin equals p_o.

P4.31   According to potential theory (Chap. 8) for the flow approaching a rounded two-dimensional body, as in Fig. P4.31, the velocity approaching the stagnation point is given by $u = U(1 - a^2/x^2)$, where a is the nose radius and U is the velocity far upstream. Compute the value and position of the maximum viscous normal stress along this streamline.

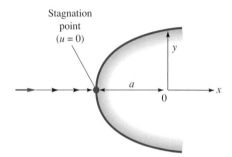

P4.31

Is this also the position of maximum fluid deceleration? Evaluate the maximum viscous normal stress if the fluid is SAE 30 oil at 20°C, with $U = 2$ m/s and $a = 6$ cm.

P4.32   The answer to Prob. P4.14 is $v_\theta = $ f(r) only. Do not reveal this to your friends if they are still working on Prob. P4.14. Show that this flow field is an exact solution to the Navier-Stokes equations (4.38) for only two special cases of the function f(r). Neglect gravity. Interpret these two cases physically.

P4.33   From Prob. P4.15 the purely radial polar coordinate flow that satisfies continuity is $v_r = f(\theta)/r$, where f is an arbitrary function. Determine what particular forms of $f(\theta)$ satisfy the full Navier-Stokes equations in polar coordinate form from Eqs. (D.5) and (D.6).

P4.34   A proposed three-dimensional incompressible flow field has the following vector form:

$$\mathbf{V} = Kx\mathbf{i} + Ky\mathbf{j} - 2Kz\mathbf{k}$$

(a) Determine if this field is a valid solution to continuity and Navier-Stokes. (b) If $\mathbf{g} = -g\mathbf{k}$, find the pressure field $p(x, y, z)$. (c) Is the flow irrotational?

P4.35   From the Navier-Stokes equations for incompressible flow in polar coordinates (App. D for cylindrical coordinates), find the most general case of purely circulating motion $v_\theta(r)$, $v_r = v_z = 0$, for flow with no slip between two fixed concentric cylinders, as in Fig. P4.35.

P4.14 For incompressible polar coordinate flow, what is the most general form of a purely circulatory motion, $v_\theta = v_\theta(r, \theta, t)$ and $v_r = 0$, that satisfies continuity?

P4.15 What is the most general form of a purely radial polar coordinate incompressible flow pattern, $v_r = v_r(r, \theta, t)$ and $v_\theta = 0$, that satisfies continuity?

P4.16 Consider the plane polar coordinate velocity distribution

$$v_r = \frac{C}{r} \qquad v_\theta = \frac{K}{r} \qquad v_z = 0$$

where C and K are constants. (a) Determine if the equation of continuity is satisfied. (b) By sketching some velocity vector directions, plot a single streamline for $C = K$. What might this flow field simulate?

P4.17 A reasonable approximation for the two-dimensional incompressible laminar boundary layer on the flat surface in Fig. P4.17 is

$$u = U\left(\frac{2y}{\delta} - \frac{y^2}{\delta^2}\right) \quad \text{for } y \leq \delta \quad \text{where } \delta = Cx^{1/2}, C = \text{const}$$

(a) Assuming a no-slip condition at the wall, find an expression for the velocity component $v(x, y)$ for $y \leq \delta$. (b) Then find the maximum value of v at the station $x = 1$ m, for the particular case of airflow, when $U = 3$ m/s and $\delta = 1.1$ cm.

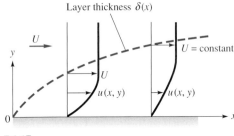

P4.17

P4.18 A piston compresses gas in a cylinder by moving at constant speed V, as in Fig. P4.18. Let the gas density and

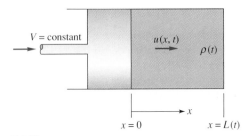

P4.18

length at $t = 0$ be ρ_0 and L_0, respectively. Let the gas velocity vary linearly from $u = V$ at the piston face to $u = 0$ at $x = L$. If the gas density varies only with time, find an expression for $\rho(t)$.

P4.19 An incompressible flow field has the cylindrical components $v_\theta = Cr$, $v_z = K(R^2 - r^2)$, $v_r = 0$, where C and K are constants and $r \leq R$, $z \leq L$. Does this flow satisfy continuity? What might it represent physically?

P4.20 A two-dimensional incompressible velocity field has $u = K(1 - e^{-ay})$, for $x \leq L$ and $0 \leq y \leq \infty$. What is the most general form of $v(x, y)$ for which continuity is satisfied and $v = v_0$ at $y = 0$? What are the proper dimensions for constants K and a?

P4.21 Air flows under steady, approximately one-dimensional conditions through the conical nozzle in Fig. P4.21. If the speed of sound is approximately 340 m/s, what is the minimum nozzle-diameter ratio D_e/D_0 for which we can safely neglect compressibility effects if $V_0 = (a)$ 10 m/s and (b) 30 m/s?

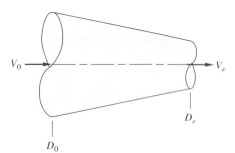

P4.21

P4.22 A flow field in the xy plane is described by $u = U_0 = $ constant, $v = V_0 = $ constant. Convert these velocities into plane polar coordinate velocities, v_r and v_θ.

P4.23 A tank volume $\mathcal{V}$ contains gas at conditions (ρ_0, p_0, T_0). At time $t = 0$ it is punctured by a small hole of area A. According to the theory of Chap. 9, the mass flow out of such a hole is approximately proportional to A and to the tank pressure. If the tank temperature is assumed constant and the gas is ideal, find an expression for the variation of density within the tank.

*P4.24 Reconsider Fig. P4.17 in the following general way. It is known that the boundary layer thickness $\delta(x)$ increases monotonically and that there is no slip at the wall ($y = 0$). Further, $u(x, y)$ merges smoothly with the outer stream flow, where $u \approx U = $ constant outside the layer. Use these facts to prove that (a) the component $v(x, y)$ is positive everywhere within the layer, (b) v increases parabolically with y very near the wall, and (c) v is a maximum at $y = \delta$.

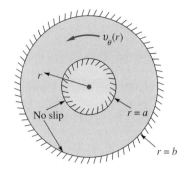

P4.35

P4.36 A constant-thickness film of viscous liquid flows in laminar motion down a plate inclined at angle θ, as in Fig. P4.36. The velocity profile is

$$u = Cy(2h - y) \quad v = w = 0$$

Find the constant C in terms of the specific weight and viscosity and the angle θ. Find the volume flux Q per unit width in terms of these parameters.

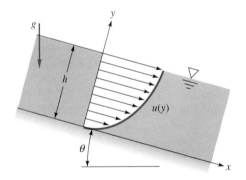

P4.36

***P4.37** A viscous liquid of constant ρ and μ falls due to gravity between two plates a distance $2h$ apart, as in Fig. P4.37. The flow is fully developed, with a single velocity component $w = w(x)$. There are no applied pressure gradients, only gravity. Solve the Navier-Stokes equation for the velocity profile between the plates.

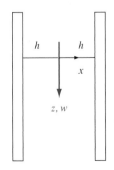

P4.37

P4.38 Show that the incompressible flow distribution, in cylindrical coordinates,

$$v_r = 0 \quad v_\theta = Cr^n \quad v_z = 0$$

where C is a constant, (*a*) satisfies the Navier-Stokes equation for only two values of n. Neglect gravity. (*b*) Knowing that $p = p(r)$ only, find the pressure distribution for each case, assuming that the pressure at $r = R$ is p_0. What might these two cases represent?

P4.39 Reconsider the angular momentum balance of Fig. 4.5 by adding a concentrated *body couple* C_z about the z axis [6]. Determine a relation between the body couple and shear stress for equilibrium. What are the proper dimensions for C_z? (Body couples are important in continuous media with microstructure, such as granular materials.)

P4.40 Problems involving viscous dissipation of energy are dependent on viscosity μ, thermal conductivity k, stream velocity U_0, and stream temperature T_0. Group these parameters into the dimensionless *Brinkman number,* which is proportional to μ.

P4.41 As mentioned in Sec. 4.10, the velocity profile for laminar flow between two plates, as in Fig. P4.41, is

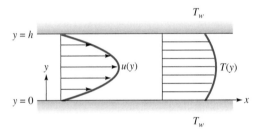

P4.41

$$u = \frac{4u_{max}y(h - y)}{h^2} \quad v = w = 0$$

If the wall temperature is T_w at both walls, use the incompressible flow energy equation (4.75) to solve for the temperature distribution $T(y)$ between the walls for steady flow.

P4.42 Suppose we wish to analyze the rotating, partly full cylinder of Fig. 2.23 as a *spin-up* problem, starting from rest and continuing until solid-body rotation is achieved. What are the appropriate boundary and initial conditions for this problem?

P4.43 For the draining liquid film of Fig. P4.36, what are the appropriate boundary conditions (*a*) at the bottom $y = 0$ and (*b*) at the surface $y = h$?

P4.44 Suppose that we wish to analyze the sudden pipe expansion flow of Fig. P3.59, using the full continuity and

Navier-Stokes equations. What are the proper boundary conditions to handle this problem?

P4.45   Suppose that we wish to analyze the U-tube oscillation flow of Fig. P3.96, using the full continuity and Navier-Stokes equations. What are the proper boundary conditions to handle this problem?

P4.46   Fluid from a large reservoir at temperature T_0 flows into a circular pipe of radius R. The pipe walls are wound with an electric resistance coil that delivers heat to the fluid at a rate q_w (energy per unit wall area). If we wish to analyze this problem by using the full continuity, Navier-Stokes, and energy equations, what are the proper boundary conditions for the analysis?

P4.47   A two-dimensional incompressible flow is given by the velocity field $\mathbf{V} = 3y\mathbf{i} + 2x\mathbf{j}$, in arbitrary units. Does this flow satisfy continuity? If so, find the stream function $\psi(x, y)$ and plot a few streamlines, with arrows.

P4.48   Consider the following two-dimensional incompressible flow, which clearly satisfies continuity:

$$u = U_0 = \text{constant}, \quad v = V_0 = \text{constant}$$

Find the stream function $\psi(r, \theta)$ of this flow using *polar coordinates*.

P4.49   Investigate the stream function $\psi = K(x^2 - y^2)$, $K = $ constant. Plot the streamlines in the full xy plane, find any stagnation points, and interpret what the flow could represent.

P4.50   Investigate the polar coordinate stream function $\psi = Kr^{1/2} \sin \frac{1}{2}\theta$, $K = $ constant. Plot the streamlines in the full xy plane, find any stagnation points, and interpret.

P4.51   For the velocity distribution of Prob. P4.4, determine if a stream function exists, and, if it does, find an expression for $\psi(x, y)$ and sketch the streamline which passes through the point $(x, y) = (L/2, L/2)$.

P4.52   A two-dimensional, incompressible, frictionless fluid is guided by wedge-shaped walls into a small slot at the origin, as in Fig. P4.52. The width into the paper is b,

and the volume flow rate is Q. At any given distance r from the slot, the flow is radial inward, with constant velocity. Find an expression for the polar coordinate stream function of this flow.

P4.53   For the fully developed laminar pipe flow solution of Eq. (4.137), find the axisymmetric stream function $\psi(r, z)$. Use this result to determine the average velocity $V = Q/A$ in the pipe as a ratio of u_{max}.

P4.54   An incompressible stream function is defined by

$$\psi(x, y) = \frac{U}{L^2}(3x^2y - y^3)$$

where U and L are (positive) constants. Where in this chapter are the streamlines of this flow plotted? Use this stream function to find the volume flow Q passing through the rectangular surface whose corners are defined by $(x, y, z) = (2L, 0, 0)$, $(2L, 0, b)$, $(0, L, b)$, and $(0, L, 0)$. Show the direction of Q.

P4.55   In Prob. P4.38 you were asked to find the pressure distribution $p(r)$ for solid-body rotation of a fluid, $v_\theta = Cr$. If z is "up," the result can be rewritten in the following manner:

$$\frac{p}{\gamma} + \frac{V^2}{2g} + z = \text{constant}$$

But this is Bernoulli's equation. How can that be? The flow is definitely not irrotational, and we are not following a streamline with this formula. Please explain this puzzle.

P4.56   Investigate the velocity potential $\phi = Kxy$, $K = $ constant. Sketch the potential lines in the full xy plane, find any stagnation points, and sketch in by eye the orthogonal streamlines. What could the flow represent?

P4.57   A two-dimensional incompressible flow field is defined by the velocity components

$$u = 2V\left(\frac{x}{L} - \frac{y}{L}\right) \qquad v = -2V\frac{y}{L}$$

where V and L are constants. If they exist, find the stream function and velocity potential.

P4.58   Show that the incompressible velocity potential in plane polar coordinates $\phi(r, \theta)$ is such that

$$v_r = \frac{\partial \phi}{\partial r} \qquad v_\theta = \frac{1}{r}\frac{\partial \phi}{\partial \theta}$$

Further show that the angular velocity about the z axis in such a flow would be given by

$$2\omega_z = \frac{1}{r}\frac{\partial}{\partial r}(rv_\theta) - \frac{1}{r}\frac{\partial}{\partial \theta}(v_r)$$

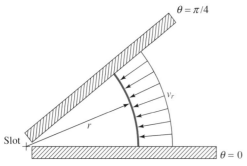

$\theta = \pi/4$

v_r

r

Slot

$\theta = 0$

P4.52

Finally show that ϕ as defined here satisfies Laplace's equation in polar coordinates for incompressible flow.

P4.59 Consider the two-dimensional incompressible velocity potential $\phi = xy + x^2 - y^2$. (a) Is it true that $\nabla^2\phi = 0$, and, if so, what does this mean? (b) If it exists, find the stream function $\psi(x, y)$ of this flow. (c) Find the equation of the streamline that passes through $(x, y) = (2, 1)$.

P4.60 Liquid drains from a small hole in a tank, as shown in Fig. P4.60, such that the velocity field set up is given by $v_r \approx 0$, $v_z \approx 0$, $v_\theta = KR^2/r$, where $z = H$ is the depth of the water far from the hole. Is this flow pattern rotational or irrotational? Find the depth z_C of the water at the radius $r = R$.

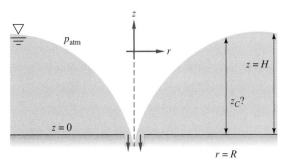

P4.60

P4.61 Investigate the polar coordinate velocity potential $\phi = Kr^{1/2} \cos\frac{1}{2}\theta$, $K = $ constant. Plot the potential lines in the full xy plane, sketch in by eye the orthogonal streamlines, and interpret.

P4.62 Show that the linear Couette flow between plates in Fig. 1.6 has a stream function but no velocity potential. Why is this so?

P4.63 Find the two-dimensional velocity potential $\phi(r, \theta)$ for the polar coordinate flow pattern $v_r = Q/r$, $v_\theta = K/r$, where Q and K are constants.

P4.64 Show that the velocity potential $\phi(r, z)$ in axisymmetric cylindrical coordinates (see Fig. 4.2) is defined such that

$$v_r = \frac{\partial\phi}{\partial r} \quad v_z = \frac{\partial\phi}{\partial z}$$

Further show that for incompressible flow this potential satisfies Laplace's equation in (r, z) coordinates.

P4.65 A two-dimensional incompressible flow is defined by

$$u = -\frac{Ky}{x^2 + y^2} \quad v = \frac{Kx}{x^2 + y^2}$$

where $K = $ constant. Is this flow irrotational? If so, find its velocity potential, sketch a few potential lines, and interpret the flow pattern.

P4.66 A plane polar coordinate velocity potential is defined by

$$\phi = \frac{K\cos\theta}{r} \quad K = \text{const}$$

Find the stream function for this flow, sketch some streamlines and potential lines, and interpret the flow pattern.

P4.67 A stream function for a plane, irrotational, polar coordinate flow is

$$\psi = C\theta - K\ln r \quad C \text{ and } K = \text{const}$$

Find the velocity potential for this flow. Sketch some streamlines and potential lines, and interpret the flow pattern.

P4.68 For the velocity distribution of Prob. P4.4, (a) determine if a velocity potential exists, and (b), if it does, find an expression for $\phi(x, y)$ and sketch the potential line which passes through the point $(x, y) = (L/2, L/2)$.

P4.69 A steady, two-dimensional flow has the following polar-coordinate velocity potential:

$$\phi = Cr\cos\theta + K\ln r$$

where C and K are constants. Determine the stream function $\psi(r, \theta)$ for this flow. For extra credit, let C be a velocity scale U and let $K = UL$, sketch what the flow might represent.

P4.70 A CFD model of steady two-dimensional incompressible flow has printed out the values of stream function $\psi(x, y)$, in m^2/s, at each of the four corners of a small 10-cm-by-10-cm cell, as shown in Fig. P4.70. Use these numbers to estimate the resultant velocity in the center of the cell and its angle α with respect to the x axis.

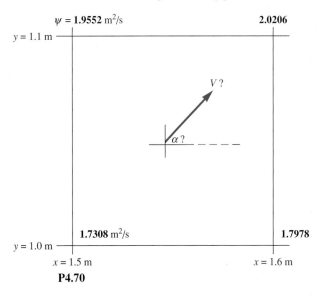

P4.70

P4.71 Consider the following two-dimensional function $f(x, y)$:

$$f = Ax^3 + Bxy^2 + Cx^2 + D \quad \text{where } A > 0$$

(a) Under what conditions, if any, on (A, B, C, D) can this function be a steady plane-flow velocity potential? (b) If you find a $\phi(x, y)$ to satisfy part (a), also find the associated stream function $\psi(x, y)$, if any, for this flow.

P4.72 Water flows through a two-dimensional narrowing wedge at 9.96 gal/min per meter of width into the paper (Fig. P4.72). If this inward flow is purely radial, find an expression, in SI units, for (a) the stream function and (b) the velocity potential of the flow. Assume one-dimensional flow. The included angle of the wedge is 45°.

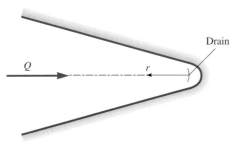

Drain

Q r

P4.72

P4.73 A CFD model of steady two-dimensional incompressible flow has printed out the values of velocity potential $\phi(x, y)$, in m²/s, at each of the four corners of a small 10-cm-by-10-cm cell, as shown in Fig. P4.73. Use these numbers to estimate the resultant velocity in the center of the cell and its angle α with respect to the x axis.

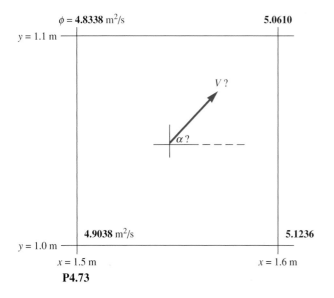

$\phi = 4.8338 \text{ m}^2/\text{s}$ 5.0610
$y = 1.1$ m

V ?

α ?

4.9038 m²/s   5.1236
$y = 1.0$ m

$x = 1.5$ m $x = 1.6$ m

P4.73

P4.74 Consider the two-dimensional incompressible polar-coordinate velocity potential

$$\phi = Br \cos\theta + B L \theta$$

where B is a constant and L is a constant length scale. (a) What are the dimensions of B? (b) Locate the only stagnation point in this flow field. (c) Prove that a stream function exists and then find the function $\psi(r, \theta)$.

P4.75 Given the following steady *axisymmetric* stream function:

$$\psi = \frac{B}{2}\left(r^2 - \frac{r^4}{2R^2}\right) \quad \text{where } B \text{ and } R \text{ are constants}$$

valid in the region $0 \leq r \leq R$ and $0 \leq z \leq L$. (a) What are the dimensions of the constant B? (b) Show whether this flow possesses a velocity potential, and, if so, find it. (c) What might this flow represent? [*Hint:* Examine the axial velocity v_z.]

*P4.76 A two-dimensional incompressible flow has the velocity potential

$$\phi = K(x^2 - y^2) + C \ln(x^2 + y^2)$$

where K and C are constants. In this discussion, avoid the origin, which is a singularity (infinite velocity). (a) Find the sole stagnation point of this flow, which is somewhere in the upper half plane. (b) Prove that a stream function exists, and then find $\psi(x, y)$, using the hint that $\int dx/(a^2 + x^2) = (1/a)\tan^{-1}(x/a)$.

P4.77 Investigate the polar coordinate stream function $\psi = Kr^{2/3} \sin(2\theta/3)$, $K = $ constant. Plot the streamlines in all except the bottom right quadrant, and interpret.

P4.78 In spherical polar coordinates, as in Fig. P4.12, the flow is called *axisymmetric* if $v_\phi \equiv 0$ and $\partial/\partial\phi \equiv 0$, so that $v_r = v_r(r, \theta)$ and $v_\theta = v_\theta(r, \theta)$. Show that a stream function $\psi(r, \theta)$ exists for this case and is given by

$$v_r = \frac{1}{r^2 \sin\theta}\frac{\partial\psi}{\partial\theta} \quad v_\theta = -\frac{1}{r \sin\theta}\frac{\partial\psi}{\partial r}$$

*P4.79 Study the combined effect of the two viscous flows in Fig. 4.12. That is, find $u(y)$ when the upper plate moves at speed V and there is also a constant pressure gradient (dp/dx). Is superposition possible? If so, explain why. Plot representative velocity profiles for (a) zero, (b) positive, and (c) negative pressure gradients for the same upper-wall speed V.

*P4.80 Oil, of density ρ and viscosity μ, drains steadily down the side of a vertical plate, as in Fig. P4.80. After a development region near the top of the plate, the oil film will become independent of z and of constant thickness δ. Assume that $w = w(x)$ only and that the atmosphere offers no shear resistance to the surface of the film. (a) Solve the Navier-Stokes equation for $w(x)$, and sketch its approximate shape. (b) Suppose that film thickness δ and the slope of the velocity profile at the wall $[\partial w/\partial x]_{wall}$ are measured with a laser-Doppler anemometer (Chap. 6). Find an expression for oil viscosity μ as a function of $(\rho, \delta, g, [\partial w/\partial x]_{wall})$.

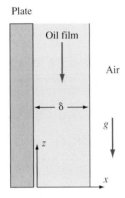

Plate

Oil film

Air

δ

g

z

x

P4.80

P4.81   Modify the analysis of Fig. 4.13 to find the velocity u_θ when the inner cylinder is fixed and the outer cylinder rotates at angular velocity Ω_0. May this solution be *added* to Eq. (4.140) to represent the flow caused when both inner and outer cylinders rotate? Explain your conclusion.

*P4.82   A solid circular cylinder of radius R rotates at angular velocity Ω in a viscous incompressible fluid that is at rest far from the cylinder, as in Fig. P4.82. Make simplifying assumptions and derive the governing differential equation and boundary conditions for the velocity field v_θ in the fluid. Do not solve unless you are obsessed with this problem. What is the steady-state flow field for this problem?

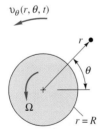

$v_\theta(r, \theta, t)$

r

θ

Ω

$r = R$

P4.82

P4.83   The flow pattern in bearing lubrication can be illustrated by Fig. P4.83, where a viscous oil (ρ, μ) is forced into the gap $h(x)$ between a fixed slipper block and a wall moving at velocity U. If the gap is thin, $h \ll L$, it can be shown that the pressure and velocity distributions are of the form $p = p(x)$, $u = u(y)$, $v = w = 0$. Neglecting gravity, reduce the Navier-Stokes equations (4.38) to a

single differential equation for $u(y)$. What are the proper boundary conditions? Integrate and show that

$$u = \frac{1}{2\mu}\frac{dp}{dx}(y^2 - yh) + U\left(1 - \frac{y}{h}\right)$$

where $h = h(x)$ may be an arbitrary, slowly varying gap width. (For further information on lubrication theory, see Ref. 16.)

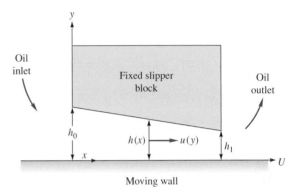

y

Oil inlet

Fixed slipper block

Oil outlet

h_0

$h(x)$

$u(y)$

h_1

x

U

Moving wall

P4.83

*P4.84   Consider a viscous film of liquid draining uniformly down the side of a vertical rod of radius a, as in Fig. P4.84. At some distance down the rod the film will approach a terminal or *fully developed* draining flow of constant outer radius b, with $v_z = v_z(r)$, $v_\theta = v_r = 0$. Assume that the atmosphere offers no shear resistance to the film motion. Derive a differential equation for v_z, state the proper boundary conditions, and solve for the film velocity distribution. How does the film radius b relate to the total film volume flow rate Q?

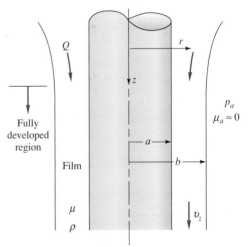

Q

r

z

p_a

$\mu_a \approx 0$

Fully developed region

Film

μ

ρ

a

b

v_z

P4.84

P4.85 A flat plate of essentially infinite width and breadth oscillates sinusoidally in its own plane beneath a viscous fluid, as in Fig. P4.85. The fluid is at rest far above the plate. Making as many simplifying assumptions as you can, set up the governing differential equation and boundary conditions for finding the velocity field u in the fluid. Do not solve (if you *can* solve it immediately, you might be able to get exempted from the balance of this course with credit).

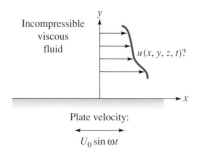

P4.85

P4.86 SAE 10 oil at 20°C flows between parallel plates 8 cm apart, as in Fig. P4.86. A mercury manometer, with wall pressure taps 1 m apart, registers a 6-cm height, as shown. Estimate the flow rate of oil for this condition.

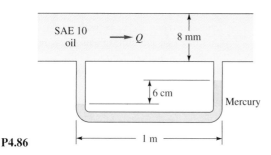

P4.86

P4.87 SAE 30W oil at 20°C flows through the 9-cm-diameter pipe in Fig. P4.87 at an average velocity of 4.3 m/s.

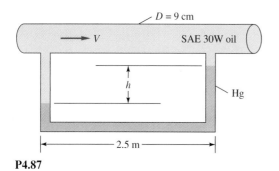

P4.87

(a) Verify that the flow is laminar. (b) Determine the volume flow rate in m³/h. (c) Calculate the expected reading h of the mercury manometer, in cm.

P4.88 The viscous oil in Fig. P4.88 is set into steady motion by a concentric inner cylinder moving axially at velocity U inside a fixed outer cylinder. Assuming constant pressure and density and a purely axial fluid motion, solve Eqs. (4.38) for the fluid velocity distribution $v_z(r)$. What are the proper boundary conditions?

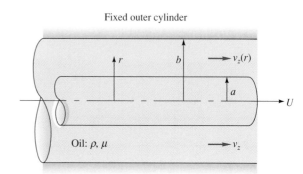

P4.88

*P4.89 Modify Prob. P4.88 so that the outer cylinder also moves to the *left* at constant speed V. Find the velocity distribution $v_z(r)$. For what ratio V/U will the wall shear stress be the same at both cylinder surfaces?

P4.90 SAE 10W oil at 20°C flows through a straight horizontal pipe. The pressure gradient is a constant 400 Pa/m. (a) What is the appropriate pipe diameter D in cm if the Reynolds number Re_D of the flow is to be exactly 1000? (b) For case a, what is the flow rate Q in m³/h?

*P4.91 Consider two-dimensional, incompressible, steady Couette flow (flow between two infinite parallel plates with the upper plate moving at constant speed and the lower plate stationary, as in Fig. 4.12a). Let the fluid be *nonnewtonian,* with its viscous stresses given by

$$\tau_{xx} = a\left(\frac{\partial u}{\partial x}\right)^c \qquad \tau_{yy} = a\left(\frac{\partial v}{\partial y}\right)^c \qquad \tau_{zz} = a\left(\frac{\partial w}{\partial z}\right)^c$$

$$\tau_{xy} = \tau_{yx} = \tfrac{1}{2}a\left(\frac{\partial u}{\partial y} + \frac{\partial v}{\partial x}\right)^c \qquad \tau_{xz} = \tau_{zx} = \tfrac{1}{2}a\left(\frac{\partial u}{\partial z} + \frac{\partial w}{\partial x}\right)^c$$

$$\tau_{yz} = \tau_{zy} = \tfrac{1}{2}a\left(\frac{\partial v}{\partial z} + \frac{\partial w}{\partial y}\right)^c$$

where a and c are constants of the fluid. Make all the same assumptions as in the derivation of Eq. (4.131). (a) Find the velocity profile $u(y)$. (b) How does the velocity profile for this case compare to that of a newtonian fluid?

P4.92 A tank of area A_0 is draining in laminar flow through a pipe of diameter D and length L, as shown in Fig. P4.92. Neglecting the exit jet kinetic energy and assuming the pipe flow is driven by the hydrostatic pressure at its entrance, derive a formula for the tank level $h(t)$ if its initial level is h_0.

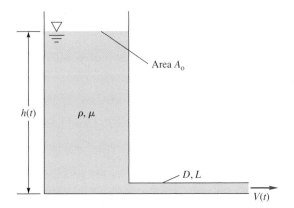

P4.92

P4.93 A number of straight 25-cm-long microtubes of diameter d are bundled together into a "honeycomb" whose total cross-sectional area is 0.0006 m^2. The pressure drop from entrance to exit is 1.5 kPa. It is desired that the total volume flow rate be 5 m^3/h of water at 20°C. (*a*) What is the appropriate microtube diameter? (*b*) How many microtubes are in the bundle? (*c*) What is the Reynolds number of each microtube?

P4.94 A long solid cylinder rotates steadily in a very viscous fluid, as in Fig. P4.94. Assuming laminar flow, solve the Navier-Stokes equation in polar coordinates to determine the resulting velocity distribution. The fluid is at rest far from the cylinder. [*Hint:* the cylinder does not induce any radial motion.]

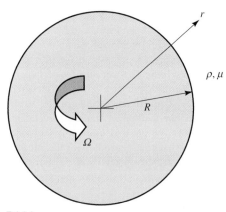

P4.94

*P4.95 Two immiscible liquids of equal thickness h are being sheared between a fixed and a moving plate, as in Fig. P4.95. Gravity is neglected, and there is no variation with x. Find an expression for (*a*) the velocity at the interface and (*b*) the shear stress in each fluid. Assume steady laminar flow.

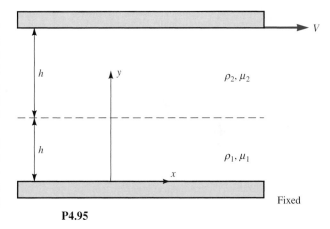

P4.95

Word Problems

W4.1 The total acceleration of a fluid particle is given by Eq. (4.2) in the eulerian system, where **V** is a known function of space and time. Explain how we might evaluate particle acceleration in the lagrangian frame, where particle position **r** is a known function of time and initial position, **r** = fcn(**r**$_0$, t). Can you give an illustrative example?

W4.2 Is it true that the continuity relation, Eq. (4.6), is valid for both viscous and inviscid, newtonian and nonnewtonian, compressible and incompressible flow? If so, are there *any* limitations on this equation?

W4.3 Consider a CD (compact disk) rotating at angular velocity Ω. Does it have *vorticity* in the sense of this chapter? If so, how much vorticity?

W4.4 How much acceleration can fluids endure? Are fluids like astronauts, who feel that $5g$ is severe? Perhaps use the flow pattern of Example 4.8, at $r = R$, to make some estimates of fluid acceleration magnitudes.

W4.5 State the conditions (there are more than one) under which the analysis of temperature distribution in a flow field can be completely uncoupled, so that a separate analysis for velocity and pressure is possible. Can we do this for both laminar and turbulent flow?

W4.6   Consider liquid flow over a dam or weir. How might the boundary conditions and the flow pattern change when we compare water flow over a large prototype to SAE 30 oil flow over a tiny scale model?

W4.7   What is the difference between the stream function ψ and our method of finding the streamlines from Sec. 1.11? Or are they essentially the same?

W4.8   Under what conditions do both the stream function ψ and the velocity potential ϕ exist for a flow field? When does one exist but not the other?

W4.9   How might the remarkable three-dimensional Taylor instability of Fig. 4.14 be predicted? Discuss a general procedure for examining the stability of a given flow pattern.

W4.10   Consider an irrotational, incompressible, axisymmetric ($\partial/\partial\theta = 0$) flow in ($r$, z) coordinates. Does a stream function exist? If so, does it satisfy Laplace's equation? Are lines of constant ψ equal to the flow streamlines? Does a velocity potential exist? If so, does it satisfy Laplace's equation? Are lines of constant ϕ everywhere perpendicular to the ψ lines?

Fundamentals of Engineering Exam Problems

This chapter is not a favorite of the people who prepare the FE Exam. Probably not a single problem from this chapter will appear on the exam, but if some did, they might be like these.

FE4.1   Given the steady, incompressible velocity distribution $\mathbf{V} = 3x\mathbf{i} + Cy\mathbf{j} + 0\mathbf{k}$, where C is a constant, if conservation of mass is satisfied, the value of C should be
(a) 3,   (b) 3/2,   (c) 0,   (d) −3/2,   (e) −3

FE4.2   Given the steady velocity distribution $\mathbf{V} = 3x\mathbf{i} + 0\mathbf{j} + $ $C y\mathbf{k}$, where C is a constant, if the flow is irrotational, the value of C should be
(a) 3,   (b) 3/2,   (c) 0,   (d) −3/2,   (e) −3

FE4.3   Given the steady, incompressible velocity distribution $\mathbf{V} = 3x\mathbf{i} + Cy\mathbf{j} + 0\mathbf{k}$, where C is a constant, the shear stress τ_{xy} at the point (x, y, z) is given by
(a) 3μ,   (b) $(3x + Cy)\mu$,   (c) 0,   (d) $C\mu$,
(e) $(3 + C)\mu$

Comprehensive Problem

C4.1   In a certain medical application, water at room temperature and pressure flows through a rectangular channel of length $L = 10$ cm, width $s = 1.0$ cm, and gap thickness $b = 0.30$ mm as in Fig. C4.1. The volume flow rate is sinusoidal with amplitude $\hat{Q} = 0.50$ mL/s and frequency $f = 20$ Hz, i.e., $Q = \hat{Q} \sin (2\pi ft)$.
(a) Calculate the maximum Reynolds number (Re = Vb/v) based on maximum average velocity and gap thickness. Channel flow like this remains laminar for Re less than about 2000. If Re is greater than about 2000, the flow will be turbulent. Is this flow laminar or turbulent? (b) In this problem, the frequency is low enough that at any given time, the flow can be solved as if it were steady at the given flow rate. (This is called a *quasi-steady assumption*.) At any arbitrary instant of time, find an expression for streamwise velocity u as a function of y, μ, dp/dx, and b, where dp/dx is the pressure gradient required to push the flow through the channel at volume flow rate Q. In addition, estimate the maximum magnitude of velocity component u. (c) At any instant of time, find a relationship between volume flow rate Q and pressure gradient dp/dx. Your answer should be given as an expression for Q as a function of dp/dx, s, b, and viscosity μ. (d) Estimate the wall shear stress, τ_w as a function of $\hat{Q}, f, \mu, b, s$, and time (t). (e) Finally, for the numbers given in the problem statement, estimate the amplitude of the wall shear stress, $\hat{\tau}_w$, in N/m^2.

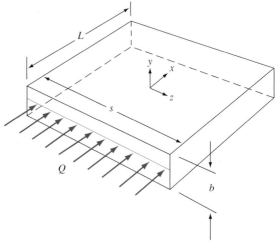

C4.1

C4.2   A belt moves upward at velocity V, dragging a film of viscous liquid of thickness h, as in Fig. C4.2. Near the belt, the film moves upward due to no slip. At its outer edge, the film moves downward due to gravity. Assuming that the only nonzero velocity is $v(x)$, with zero shear stress at the outer film edge, derive a formula for (*a*) $v(x)$, (*b*) the average velocity V_{avg} in the film, and (*c*) the velocity V_c for which there is no net flow either up or down. (*d*) Sketch $v(x)$ for case (*c*).

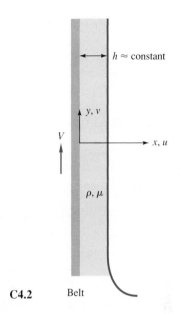

C4.2   Belt

References

1. J. D. Anderson, *Computational Fluid Dynamics: The Basics with Applications,* McGraw-Hill, New York, 1995.
2. C. E. Brennen, *Fundamentals of Multiphase Flow,* Cambridge University Press, New York, 2005. See also URL <http://caltechbook.library.caltech.edu/51/01/multiph.htm>
3. S. M. Selby, *CRC Handbook of Tables for Mathematics,* 4th ed., CRC Press Inc., Cleveland, OH, 1976.
4. H. Schlichting, *Boundary Layer Theory,* 7th ed., McGraw-Hill, New York, 1979.
5. F. M. White, *Viscous Fluid Flow,* 3d ed., McGraw-Hill, New York, 2005.
6. L. E. Malvern, *Introduction to Mechanics of a Continuous Medium,* Prentice-Hall, Upper Saddle River, NJ, 1997.
7. J. P. Holman, *Heat Transfer,* 9th ed., McGraw-Hill, New York, 2002.
8. W. M. Kays and M. E. Crawford, *Convective Heat and Mass Transfer,* 3d ed., McGraw-Hill, New York, 1993.
9. G. K. Batchelor, *An Introduction to Fluid Dynamics,* Cambridge University Press, Cambridge, England, 1967.
10. L. Prandtl and O. G. Tietjens, *Fundamentals of Hydro-and Aeromechanics,* Dover, New York, 1957.
11. R. Aris, *Vectors, Tensors, and the Basic Equations of Fluid Mechanics,* Dover, New York, 1989.
12. G. A. Holzapfel, *Nonlinear Solid Mechanics: A Continuum Approach for Engineering,* John Wiley, New York, 2000.
13. D. A. Danielson, *Vectors and Tensors in Engineering and Physics,* 2d ed., Westview (Perseus) Press, Boulder, CO, 1997.
14. R. I. Tanner, *Engineering Rheology,* 2d ed., Oxford University Press, New York, 2000.
15. H. Lamb, *Hydrodynamics,* 6th ed., Dover, New York, 1945.
16. G. W. Stakowiak and A. W. Batchelor, *Engineering Tribology,* 3d ed., Butterworth-Heinemann, Woburn, MA, 2005.
17. G. I. Taylor, "Stability of a Viscous Liquid Contained between Two Rotating Cylinders," *Philos. Trans. Roy. Soc. London Ser. A,* vol. 223, 1923, pp. 289–343.
18. E. L. Koschmieder, "Turbulent Taylor Vortex Flow," *J. Fluid Mech.,* vol. 93, pt. 3, 1979, pp. 515–527.
19. M. T. Nair, T. K. Sengupta, and U. S. Chauhan, "Flow Past Rotating Cylinders at High Reynolds Numbers Using Higher Order Upwind Scheme," *Computers and Fluids,* vol. 27, no. 1, 1998, pp. 47–70.
20. M. Constanceau and C. Menard, "Influence of Rotation on the Near-Wake Development behind an Impulsively Started Circular Cylinder," *J. Fluid Mechanics,* vol. 1258, 1985, pp. 399–446.

Experiments are at the heart of fluids engineering. Here a full-scale, 10-meter-diameter Grumman Corp. windmill is tested, by the National Renewable Energy Laboratory, in the NASA Ames 80 ft by 120 ft tunnel, which is the largest wind tunnel in the world. The turbine diameter is 10 m, and it rotates at 72 r/min. Smoke emitted from one blade shows the helical wake of the turbine. This experiment varied many dimensionless parameters: the Reynolds number based on blade chord length; the ratio of tip speed to wind speed; a Strouhal number based on blade pitch oscillations; and a parameter proportional to the rate of change of the blade pitch angle. *[From Ref. 37, Courtesy of American Society of Mechanical Engineers.]*

Chapter 5
Dimensional Analysis and Similarity

Motivation. In this chapter we discuss the planning, presentation, and interpretation of experimental data. We shall try to convince you that such data are best presented in *dimensionless* form. Experiments that might result in tables of output, or even multiple volumes of tables, might be reduced to a single set of curves—or even a single curve—when suitably nondimensionalized. The technique for doing this is *dimensional analysis*.

Chapter 3 presented large-scale control volume balances of mass, momentum, and energy, which led to global results: mass flow, force, torque, total work done, or heat transfer. Chapter 4 presented infinitesimal balances that led to the basic partial differential equations of fluid flow and some particular solutions for both inviscid and viscous (laminar) flow. These straight *analytical* techniques are limited to simple geometries and uniform boundary conditions. Only a fraction of engineering flow problems can be solved by direct analytical formulas.

Most practical fluid flow problems are too complex, both geometrically and physically, to be solved analytically. They must be tested by experiment or approximated by computational fluid dynamics (CFD) [2]. These results are typically reported as experimental or numerical data points and smoothed curves. Such data have much more generality if they are expressed in compact, economic form. This is the motivation for dimensional analysis. The technique is a mainstay of fluid mechanics and is also widely used in all engineering fields plus the physical, biological, medical, and social sciences. The present chapter shows how dimensional analysis improves the presentation of both data and theory.

5.1 Introduction

Basically, dimensional analysis is a method for reducing the number and complexity of experimental variables that affect a given physical phenomenon, by using a sort of compacting technique. If a phenomenon depends on n dimensional variables, dimensional analysis will reduce the problem to only k *dimensionless* variables, where the reduction $n - k = 1, 2, 3,$ or 4, depending on the problem complexity. Generally

$n - k$ equals the number of different dimensions (sometimes called basic or primary or fundamental dimensions) that govern the problem. In fluid mechanics, the four basic dimensions are usually taken to be mass M, length L, time T, and temperature Θ, or an $MLT\Theta$ system for short. Alternatively, one uses an $FLT\Theta$ system, with force F replacing mass.

Although its purpose is to reduce variables and group them in dimensionless form, dimensional analysis has several side benefits. The first is enormous savings in time and money. Suppose one knew that the force F on a particular body shape immersed in a stream of fluid depended only on the body length L, stream velocity V, fluid density ρ, and fluid viscosity μ; that is,

$$F = f(L, V, \rho, \mu) \tag{5.1}$$

Suppose further that the geometry and flow conditions are so complicated that our integral theories (Chap. 3) and differential equations (Chap. 4) fail to yield the solution for the force. Then we must find the function $f(L, V, \rho, \mu)$ experimentally or numerically.

Generally speaking, it takes about 10 points to define a curve. To find the effect of body length in Eq. (5.1), we have to run the experiment for 10 lengths L. For each L we need 10 values of V, 10 values of ρ, and 10 values of μ, making a grand total of 10^4, or 10,000, experiments. At \$100 per experiment—well, you see what we are getting into. However, with dimensional analysis, we can immediately reduce Eq. (5.1) to the equivalent form

$$\boxed{\frac{F}{\rho V^2 L^2} = g\left(\frac{\rho V L}{\mu}\right)}$$

or $$\boxed{C_F = g(\mathrm{Re})}$$

(5.2)

That is, the dimensionless *force coefficient* $F/(\rho V^2 L^2)$ is a function only of the dimensionless *Reynolds number* $\rho V L/\mu$. We shall learn exactly how to make this reduction in Secs. 5.2 and 5.3.

Note that Eq. (5.2) is just an *example,* not the full story, of forces caused by fluid flows. Some fluid forces have a very weak or negligible Reynolds number dependence in wide regions (Fig. 5.3*a*). There are also forces that depend upon Mach number (Fig. 7.20), Froude number (Fig. 7.19), or wall roughness (Fig. D5.2).

The function g is different mathematically from the original function f, but it contains all the same information. Nothing is lost in a dimensional analysis. And think of the savings: We can establish g by running the experiment for only 10 values of the single variable called the Reynolds number. We do not have to vary L, V, ρ, or μ separately but only the *grouping* $\rho V L/\mu$. This we do merely by varying velocity V in, say, a wind tunnel or drop test or water channel, and there is no need to build 10 different bodies or find 100 different fluids with 10 densities and 10 viscosities. The cost is now about \$1000, maybe less.

A second side benefit of dimensional analysis is that it helps our thinking and planning for an experiment or theory. It suggests dimensionless ways of writing equations before we spend money on computer analysis to find solutions. It suggests variables that can be discarded; sometimes dimensional analysis will immediately reject variables, and at other times it groups them off to the side, where a few simple tests will show them to be unimportant. Finally, dimensional analysis will often give a great deal of insight into the form of the physical relationship we are trying to study.

A third benefit is that dimensional analysis provides *scaling laws* that can convert data from a cheap, small *model* to design information for an expensive, large *prototype*. We do not build a million-dollar airplane and see whether it has enough lift force. We measure the lift on a small model and use a scaling law to predict the lift on the full-scale prototype airplane. There are rules we shall explain for finding scaling laws. When the scaling law is valid, we say that a condition of *similarity* exists between the model and the prototype. In the simple case of Eq. (5.1), similarity is achieved if the Reynolds number is the same for the model and prototype because the function *g* then requires the force coefficient to be the same also:

$$\text{If}\quad \text{Re}_m = \text{Re}_p \quad \text{then} \quad C_{Fm} = C_{Fp} \tag{5.3}$$

where subscripts *m* and *p* mean model and prototype, respectively. From the definition of force coefficient, this means that

$$\frac{F_p}{F_m} = \frac{\rho_p}{\rho_m}\left(\frac{V_p}{V_m}\right)^2\left(\frac{L_p}{L_m}\right)^2 \tag{5.4}$$

for data taken where $\rho_p V_p L_p / \mu_p = \rho_m V_m L_m / \mu_m$. Equation (5.4) is a scaling law: If you measure the model force at the model Reynolds number, the prototype force at the same Reynolds number equals the model force times the density ratio times the velocity ratio squared times the length ratio squared. We shall give more examples later.

Do you understand these introductory explanations? Be careful; learning dimensional analysis is like learning to play tennis: There are levels of the game. We can establish some ground rules and do some fairly good work in this brief chapter, but dimensional analysis in the broad view has many subtleties and nuances that only time, practice, and maturity enable you to master. Although dimensional analysis has a firm physical and mathematical foundation, considerable art and skill are needed to use it effectively.

EXAMPLE 5.1

A copepod is a water crustacean approximately 1 mm in diameter. We want to know the drag force on the copepod when it moves slowly in fresh water. A scale model 100 times larger is made and tested in glycerin at $V = 30$ cm/s. The measured drag on the model is 1.3 N. For similar conditions, what are the velocity and drag of the actual copepod in water? Assume that Eq. (5.2) applies and the temperature is 20°C.

Solution

- *Property values:*   From Table A.3, the densities and viscosities at 20°C are

Water (prototype):	$\mu_p = 0.001$ kg/(m−s)	$\rho_p = 998$ kg/m^3
Glycerin (model):	$\mu_m = 1.5$ kg/(m−s)	$\rho_m = 1263$ kg/m^3

- *Assumptions:*   Equation (5.2) is appropriate and *similarity* is achieved; that is, the model and prototype have the same Reynolds number and, therefore, the same force coefficient.

- *Approach:*   The length scales are $L_m = 100$ mm and $L_p = 1$ mm. Calculate the Reynolds number and force coefficient of the model and set them equal to prototype values:

$$\mathrm{Re}_m = \frac{\rho_m V_m L_m}{\mu_m} = \frac{(1263 \text{ kg/m}^3)(0.3 \text{ m/s})(0.1 \text{ m})}{1.5 \text{ kg/(m} - \text{s)}} = 25.3 = \mathrm{Re}_p = \frac{(998 \text{ kg/m}^3)V_p(0.001 \text{ m})}{0.001 \text{ kg/(m} - \text{s)}}$$

$$\text{Solve for } V_p = 0.0253 \text{ m/s} = 2.53 \text{ cm/s} \qquad\qquad Ans.$$

In like manner, using the prototype velocity just found, equate the force coefficients:

$$C_{Fm} = \frac{F_m}{\rho_m V_m^2 L_m^2} = \frac{1.3 \text{ N}}{(1263 \text{ kg/m}^3)(0.3 \text{ m/s})^2(0.1 \text{ m})^2} = 1.14$$

$$= C_{Fp} = \frac{F_p}{(998 \text{ kg/m}^3)(0.0253 \text{ m/s})^2(0.001 \text{ m})^2}$$

$$\text{Solve for }\quad F_p = 7.3\mathrm{E}\text{-}7\mathrm{N} \qquad\qquad Ans.$$

- *Comments:*   Assuming we modeled the Reynolds number correctly, the model test is a very good idea, as it would obviously be difficult to measure such a tiny copepod drag force.

Historically, the first person to write extensively about units and dimensional reasoning in physical relations was Euler in 1765. Euler's ideas were far ahead of his time, as were those of Joseph Fourier, whose 1822 book *Analytical Theory of Heat* outlined what is now called the *principle of dimensional homogeneity* and even developed some similarity rules for heat flow. There were no further significant advances until Lord Rayleigh's book in 1877, *Theory of Sound,* which proposed a "method of dimensions" and gave several examples of dimensional analysis. The final breakthrough that established the method as we know it today is generally credited to E. Buckingham in 1914 [1], whose paper outlined what is now called the *Buckingham Pi Theorem* for describing dimensionless parameters (see Sec. 5.3). However, it is now known that a Frenchman, A. Vaschy, in 1892 and a Russian, D. Riabouchinsky, in 1911 had independently published papers reporting results equivalent to the pi theorem. Following Buckingham's paper, P. W. Bridgman published a classic book in 1922 [3], outlining the general theory of dimensional analysis.

Dimensional analysis is so valuable and subtle, with both skill and art involved, that it has spawned a wide variety of textbooks and treatises. The writer is aware of more than 30 books on the subject, of which his engineering favorites are listed here [3–10]. Dimensional analysis is not confined to fluid mechanics, or even to engineering. Specialized books have been published on the application of dimensional analysis to metrology [11], astrophysics [12], economics [13], chemistry [14], hydrology [15], medications [16], clinical medicine [17], chemical processing pilot plants [18], social sciences [19], biomedical sciences [20], pharmacy [21], fractal geometry [22], and even the growth of plants [23]. Clearly this is a subject well worth learning for many career paths.

5.2 The Principle of Dimensional Homogeneity

In making the remarkable jump from the five-variable Eq. (5.1) to the two-variable Eq. (5.2), we were exploiting a rule that is almost a self-evident axiom in physics. This rule, the *principle of dimensional homogeneity* (PDH), can be stated as follows:

> If an equation truly expresses a proper relationship between variables in a physical process, it will be *dimensionally homogeneous;* that is, each of its additive terms will have the same dimensions.

All the equations that are derived from the theory of mechanics are of this form. For example, consider the relation that expresses the displacement of a falling body:

$$S = S_0 + V_0 t + \tfrac{1}{2}g t^2 \tag{5.5}$$

Each term in this equation is a displacement, or length, and has dimensions $\{L\}$. The equation is dimensionally homogeneous. Note also that any consistent set of units can be used to calculate a result.

Consider Bernoulli's equation for incompressible flow:

$$\frac{p}{\rho} + \frac{1}{2}V^2 + gz = \text{const} \tag{5.6}$$

Each term, including the constant, has dimensions of velocity squared, or $\{L^2 T^{-2}\}$. The equation is dimensionally homogeneous and gives proper results for any consistent set of units.

Students count on dimensional homogeneity and use it to check themselves when they cannot quite remember an equation during an exam. For example, which is it:

$$S = \tfrac{1}{2}g t^2? \qquad \text{or} \qquad S = \tfrac{1}{2}g^2 t? \tag{5.7}$$

By checking the dimensions, we reject the second form and back up our faulty memory. We are exploiting the principle of dimensional homogeneity, and this chapter simply exploits it further.

Variables and Constants

Equations (5.5) and (5.6) also illustrate some other factors that often enter into a dimensional analysis:

Dimensional variables are the quantities that actually vary during a given case and would be plotted against each other to show the data. In Eq. (5.5), they are S and t; in Eq. (5.6) they are p, V, and z. All have dimensions, and all can be nondimensionalized as a dimensional analysis technique.

Dimensional constants may vary from case to case but are held constant during a given run. In Eq. (5.5) they are S_0, V_0, and g, and in Eq. (5.6) they are ρ, g, and C. They all have dimensions and conceivably could be nondimensionalized, but they are normally used to help nondimensionalize the variables in the problem.

Pure constants have no dimensions and never did. They arise from mathematical manipulations. In both Eqs. (5.5) and (5.6) they are $\tfrac{1}{2}$ and the exponent 2, both of which came from an integration: $\int t\, dt = \tfrac{1}{2}t^2$, $\int V\, dV = \tfrac{1}{2}V^2$. Other common dimensionless constants are π and e. Also, the argument of any mathematical function, such as ln, exp, cos, or J_0, is dimensionless.

Angles and *revolutions* are dimensionless. The preferred unit for an angle is the radian, which makes it clear that an angle is a ratio. In like manner, a revolution is 2π radians.

Counting numbers are dimensionless. For example, if we triple the energy E to $3E$, the coefficient 3 is dimensionless.

Note that integration and differentiation of an equation may change the dimensions but not the homogeneity of the equation. For example, integrate or differentiate Eq. (5.5):

$$\int S \, dt = S_0 t + \tfrac{1}{2} V_0 t^2 + \tfrac{1}{6} g t^3 \tag{5.8a}$$

$$\frac{dS}{dt} = V_0 + gt \tag{5.8b}$$

In the integrated form (5.8a) every term has dimensions of $\{LT\}$, while in the derivative form (5.8b) every term is a velocity $\{LT^{-1}\}$.

Finally, some physical variables are naturally dimensionless by virtue of their definition as ratios of dimensional quantities. Some examples are strain (change in length per unit length), Poisson's ratio (ratio of transverse strain to longitudinal strain), and specific gravity (ratio of density to standard water density).

The motive behind dimensional analysis is that any dimensionally homogeneous equation can be written in an entirely equivalent nondimensional form that is more compact. Usually there are multiple methods of presenting one's dimensionless data or theory. Let us illustrate these concepts more thoroughly by using the falling-body relation (5.5) as an example.

Ambiguity: The Choice of Variables and Scaling Parameters[1]

Equation (5.5) is familiar and simple, yet illustrates most of the concepts of dimensional analysis. It contains five terms (S, S_0, V_0, t, g), which we may divide, in our thinking, into variables and parameters. The *variables* are the things we wish to plot, the basic output of the experiment or theory: in this case, S versus t. The *parameters* are those quantities whose effect on the variables we wish to know: in this case S_0, V_0, and g. Almost any engineering study can be subdivided in this manner.

To nondimensionalize our results, we need to know how many dimensions are contained among our variables and parameters: in this case, only two, length $\{L\}$ and time $\{T\}$. Check each term to verify this:

$$\{S\} = \{S_0\} = \{L\} \qquad \{t\} = \{T\} \qquad \{V_0\} = \{LT^{-1}\} \qquad \{g\} = \{LT^{-2}\}$$

Among our parameters, we therefore select two to be *scaling parameters* (also called *repeating variables*), used to define dimensionless variables. What remains will be the "basic" parameter(s) whose effect we wish to show in our plot. These choices will not affect the content of our data, only the form of their presentation. Clearly there is ambiguity in these choices, something that often vexes the beginning experimenter. But the ambiguity is deliberate. Its purpose is to show a particular effect, and the choice is yours to make.

For the falling-body problem, we select any two of the three parameters to be scaling parameters. Thus we have three options. Let us discuss and display them in turn.

[1]I am indebted to Prof. Jacques Lewalle of Syracuse University for suggesting, outlining, and clarifying this entire discussion.

Option 1: Scaling parameters S_0 and V_0: the effect of gravity g.

First use the scaling parameters (S_0, V_0) to define dimensionless (*) displacement and time. There is only one suitable definition for each:[2]

$$S^* = \frac{S}{S_0} \qquad t^* = \frac{V_0 t}{S_0} \tag{5.9}$$

Substitute these variables into Eq. (5.5) and clean everything up until each term is dimensionless. The result is our first option:

$$S^* = 1 + t^* + \frac{1}{2}\alpha t^{*2} \qquad \alpha = \frac{gS_0}{V_0^2} \tag{5.10}$$

This result is shown plotted in Fig. 5.1a. There is a single dimensionless parameter α, which shows here the effect of gravity. It cannot show the direct effects of S_0 and V_0, since these two are hidden in the ordinate and abscissa. We see that gravity increases the parabolic rate of fall for $t^* > 0$, but not the initial slope at $t^* = 0$. We would learn the same from falling-body data, and the plot, within experimental accuracy, would look like Fig. 5.1a.

Option 2: Scaling parameters V_0 and g: the effect of initial displacement S_0.

Now use the new scaling parameters (V_0, g) to define dimensionless (**) displacement and time. Again there is only one suitable definition:

$$S^{**} = \frac{Sg}{V_0^2} \qquad t^{**} = t\frac{g}{V_0} \tag{5.11}$$

Substitute these variables into Eq. (5.5) and clean everything up again. The result is our second option:

$$S^{**} = \alpha + t^{**} + \frac{1}{2}t^{**2} \qquad \alpha = \frac{gS_0}{V_0^2} \tag{5.12}$$

This result is plotted in Fig. 5.1b. The same single parameter α again appears and here shows the effect of initial *displacement*, which merely moves the curves upward without changing their shape.

Option 3: Scaling parameters S_0 and g: the effect of initial speed V_0.

Finally use the scaling parameters (S_0, g) to define dimensionless (***) displacement and time. Again there is only one suitable definition:

$$S^{***} = \frac{S}{S_0} \qquad t^{***} = t\left(\frac{g}{S_0}\right)^{1/2} \tag{5.13}$$

Substitute these variables into Eq. (5.5) and clean everything up as usual. The result is our third and final option:

$$S^{***} = 1 + \beta t^{***} + \frac{1}{2}t^{***2} \qquad \beta = \frac{1}{\sqrt{\alpha}} = \frac{V_0}{\sqrt{gS_0}} \tag{5.14}$$

---

[2]Make them *proportional* to S and t. Do not define dimensionless terms upside down: S_0/S or $S_0/(V_0 t)$. The plots will look funny, users of your data will be confused, and your supervisor will be angry. It is not a good idea.

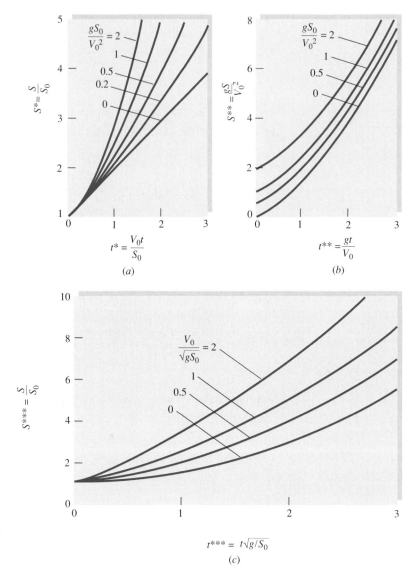

Fig. 5.1 Three entirely equivalent dimensionless presentations of the falling-body problem, Eq. (5.5): the effect of (a) gravity, (b) initial displacement, and (c) initial velocity. All plots contain the same information.

This final presentation is shown in Fig. 5.1c. Once again the parameter α appears, but we have redefined it upside down, $\beta = 1/\sqrt{\alpha}$, so that our display parameter V_0 is in the numerator and is linear. This is our free choice and simply improves the display. Figure 5.1c shows that initial *velocity* increases the falling displacement.

Note that, in all three options, the same parameter α appears but has a different meaning: dimensionless gravity, initial displacement, and initial velocity. The graphs, which contain exactly the same information, change their appearance to reflect these differences.

Whereas the original problem, Eq. (5.5), involved five quantities, the dimensionless presentations involve only three, having the form

$$S' = \text{fcn}(t', \alpha) \qquad \alpha = \frac{gS_0}{V_0^2} \qquad (5.15)$$

The reduction $5 - 3 = 2$ should equal the number of fundamental dimensions involved in the problem $\{L, T\}$. This idea led to the pi theorem (Sec. 5.3).

Selection of Scaling (Repeating) Variables

The selection of scaling variables is left to the user, but there are some guidelines. In Eq. (5.2), it is now clear that the scaling variables were ρ, V, and L, since they appear in both force coefficient and Reynolds number. We could then interpret data from Eq. (5.2) as the variation of dimensionless *force* versus dimensionless *viscosity,* since each appears in only one dimensionless group. Similarly, in Eq. (5.5) the scaling variables were selected from (S_0, V_0, g), not (S, t), because we wished to plot S versus t in the final result.

The following are some guidelines for selecting scaling variables:

1. They must *not* form a dimensionless group among themselves, but adding one more variable *will* form a dimensionless quantity. For example, test powers of ρ, V, and L:

   $$\rho^a V^b L^c = (ML^{-3})^a (L/T)^b (L)^c = M^0 L^0 T^0 \quad \text{only if} \quad a = 0, b = 0, c = 0$$

   In this case, we can see why this is so: Only ρ contains the dimension $\{M\}$, and only V contains the dimension $\{T\}$, so no cancellation is possible. If, now, we add μ to the scaling group, we will obtain the Reynolds number. If we add F to the group, we form the force coefficient.

2. Do not select output variables for your scaling parameters. In Eq. (5.1), certainly do not select F, which you wish to isolate for your plot. Nor was μ selected, for we wished to plot force versus viscosity.

3. If convenient, select *popular,* not obscure, scaling variables because they will appear in all of your dimensionless groups. Select density, not surface tension. Select body length, not surface roughness. Select stream velocity, not speed of sound.

The examples that follow will make this clear. Problem assignments might give hints.

Suppose we wish to study drag force versus *velocity.* Then we would not use V as a scaling parameter in Eq. (5.1). We would use (ρ, μ, L) instead, and the final dimensionless function would become

$$C_F' = \frac{\rho F}{\mu^2} = \text{fcn(Re)} \qquad \text{Re} = \frac{\rho V L}{\mu} \tag{5.16}$$

In plotting these data, we would not be able to discern the effect of ρ or μ, since they appear in both dimensionless groups. The grouping C_F' again would mean dimensionless force, and Re is now interpreted as either dimensionless velocity or size.[3] The plot would be quite different compared to Eq. (5.2), although it contains exactly the same information. The development of parameters such as C_F' and Re from the initial variables is the subject of the pi theorem (Sec. 5.3).

Some Peculiar Engineering Equations

The foundation of the dimensional analysis method rests on two assumptions: (1) The proposed physical relation is dimensionally homogeneous, and (2) all the relevant variables have been included in the proposed relation.

If a relevant variable is missing, dimensional analysis will fail, giving either algebraic difficulties or, worse, yielding a dimensionless formulation that does not

---

[3]We were lucky to achieve a size effect because in this case L, a scaling parameter, did not appear in the drag coefficient.

resolve the process. A typical case is Manning's open-channel formula, discussed in Example 1.4:

$$V = \frac{1.49}{n} R^{2/3} S^{1/2} \tag{1}$$

Since V is velocity, R is a radius, and n and S are dimensionless, the formula is not dimensionally homogeneous. This should be a warning that (1) the formula changes if the *units* of V and R change and (2) if valid, it represents a very special case. Equation (1) in Example 1.4 predates the dimensional analysis technique and is valid only for water in rough channels at moderate velocities and large radii in BG units.

Such dimensionally inhomogeneous formulas abound in the hydraulics literature. Another example is the Hazen-Williams formula [24] for volume flow of water through a straight smooth pipe:

$$Q = 61.9 D^{2.63} \left(\frac{dp}{dx} \right)^{0.54} \tag{5.17}$$

where D is diameter and dp/dx is the pressure gradient. Some of these formulas arise because numbers have been inserted for fluid properties and other physical data into perfectly legitimate homogeneous formulas. We shall not give the units of Eq. (5.17) to avoid encouraging its use.

On the other hand, some formulas are "constructs" that cannot be made dimensionally homogeneous. The "variables" they relate cannot be analyzed by the dimensional analysis technique. Most of these formulas are raw empiricisms convenient to a small group of specialists. Here are three examples:

$$B = \frac{25{,}000}{100 - R} \tag{5.18}$$

$$S = \frac{140}{130 + \text{API}} \tag{5.19}$$

$$0.0147 D_E - \frac{3.74}{D_E} = 0.26 t_R - \frac{172}{t_R} \tag{5.20}$$

Equation (5.18) relates the Brinell hardness B of a metal to its Rockwell hardness R. Equation (5.19) relates the specific gravity S of an oil to its density in degrees API. Equation (5.20) relates the viscosity of a liquid in D_E, or degrees Engler, to its viscosity t_R in Saybolt seconds. Such formulas have a certain usefulness when communicated between fellow specialists, but we cannot handle them here. Variables like Brinell hardness and Saybolt viscosity are not suited to an $MLT\Theta$ dimensional system.

5.3 The Pi Theorem

There are several methods of reducing a number of dimensional variables into a smaller number of dimensionless groups. The first scheme given here was proposed in 1914 by Buckingham [1] and is now called the *Buckingham Pi Theorem*. The name *pi* comes from the mathematical notation Π, meaning a product of variables. The dimensionless groups found from the theorem are power products denoted by Π_1, Π_2, Π_3, etc. The method allows the pi groups to be found in sequential order without resorting to free exponents.

The first part of the pi theorem explains what reduction in variables to expect:

> If a physical process satisfies the PDH and involves n dimensional variables, it can be reduced to a relation between only k dimensionless variables or Πs. The reduction $j = n - k$ equals the maximum number of variables that do not form a pi among themselves and is always less than or equal to the number of dimensions describing the variables.

Take the specific case of force on an immersed body: Eq. (5.1) contains five variables F, L, U, ρ, and μ described by three dimensions $\{MLT\}$. Thus $n = 5$ and $j \leq 3$. Therefore it is a good guess that we can reduce the problem to k pi groups, with $k = n - j \geq 5 - 3 = 2$. And this is exactly what we obtained: two dimensionless variables $\Pi_1 = C_F$ and $\Pi_2 = \text{Re}$. On rare occasions it may take more pi groups than this minimum (see Example 5.5).

The second part of the theorem shows how to find the pi groups one at a time:

> Find the reduction j, then select j scaling variables that do not form a pi among themselves.[4] Each desired pi group will be a power product of these j variables plus one additional variable, which is assigned any convenient nonzero exponent. Each pi group thus found is independent.

To be specific, suppose the process involves five variables:

$$v_1 = f(v_2, v_3, v_4, v_5)$$

Suppose there are three dimensions $\{MLT\}$ and we search around and find that indeed $j = 3$. Then $k = 5 - 3 = 2$ and we expect, from the theorem, two and only two pi groups. Pick out three convenient variables that do *not* form a pi, and suppose these turn out to be v_2, v_3, and v_4. Then the two pi groups are formed by power products of these three plus one additional variable, either v_1 or v_5:

$$\Pi_1 = (v_2)^a(v_3)^b(v_4)^c v_1 = M^0L^0T^0 \qquad \Pi_2 = (v_2)^a(v_3)^b(v_4)^c v_5 = M^0L^0T^0$$

Here we have arbitrarily chosen v_1 and v_5, the added variables, to have unit exponents. Equating exponents of the various dimensions is guaranteed by the theorem to give unique values of a, b, and c for each pi. And they are independent because only Π_1 contains v_1 and only Π_2 contains v_5. It is a very neat system once you get used to the procedure. We shall illustrate it with several examples.

Typically, six steps are involved:

1. List and count the n variables involved in the problem. If any important variables are missing, dimensional analysis will fail.

2. List the dimensions of each variable according to $\{MLT\Theta\}$ or $\{FLT\Theta\}$. A list is given in Table 5.1.

3. Find j. Initially guess j equal to the number of different dimensions present, and look for j variables that do not form a pi product. If no luck, reduce j by 1 and look again. With practice, you will find j rapidly.

4. Select j scaling parameters that do not form a pi product. Make sure they please you and have some generality if possible, because they will then appear

---

[4]Make a clever choice here because all pi groups will contain these j variables in various groupings.

Table 5.1 Dimensions of Fluid-Mechanics Properties

Quantity	Symbol	Dimensions	
		$MLT\Theta$	$FLT\Theta$
Length	L	L	L
Area	A	L^2	L^2
Volume	$\mathcal{V}$	L^3	L^3
Velocity	V	LT^{-1}	LT^{-1}
Acceleration	dV/dt	LT^{-2}	LT^{-2}
Speed of sound	a	LT^{-1}	LT^{-1}
Volume flow	Q	L^3T^{-1}	L^3T^{-1}
Mass flow	$\dot{m}$	MT^{-1}	FTL^{-1}
Pressure, stress	p, σ, τ	$ML^{-1}T^{-2}$	FL^{-2}
Strain rate	$\dot{\epsilon}$	T^{-1}	T^{-1}
Angle	θ	None	None
Angular velocity	ω, Ω	T^{-1}	T^{-1}
Viscosity	μ	$ML^{-1}T^{-1}$	FTL^{-2}
Kinematic viscosity	ν	L^2T^{-1}	L^2T^{-1}
Surface tension	Y	MT^{-2}	FL^{-1}
Force	F	MLT^{-2}	F
Moment, torque	M	ML^2T^{-2}	FL
Power	P	ML^2T^{-3}	FLT^{-1}
Work, energy	W, E	ML^2T^{-2}	FL
Density	ρ	ML^{-3}	FT^2L^{-4}
Temperature	T	Θ	Θ
Specific heat	c_p, c_v	$L^2T^{-2}\Theta^{-1}$	$L^2T^{-2}\Theta^{-1}$
Specific weight	γ	$ML^{-2}T^{-2}$	FL^{-3}
Thermal conductivity	k	$MLT^{-3}\Theta^{-1}$	$FT^{-1}\Theta^{-1}$
Expansion coefficient	β	Θ^{-1}	Θ^{-1}

in every one of your pi groups. Pick density or velocity or length. Do not pick surface tension, for example, or you will form six different independent Weber-number parameters and thoroughly annoy your colleagues.

5. Add one additional variable to your *j* repeating variables, and form a power product. Algebraically find the exponents that make the product dimensionless. Try to arrange for your output or *dependent* variables (force, pressure drop, torque, power) to appear in the numerator, and your plots will look better. Do this sequentially, adding one new variable each time, and you will find all $n - j = k$ desired pi products.

6. Write the final dimensionless function, and check the terms to make sure all pi groups are dimensionless.

EXAMPLE 5.2

Repeat the development of Eq. (5.2) from Eq. (5.1), using the pi theorem.

Solution

Step 1 Write the function and count variables:

$$F = f(L, U, \rho, \mu) \quad \text{there are five variables } (n = 5)$$

Step 2   List dimensions of each variable. From Table 5.1

F	L	U	ρ	μ
$\{MLT^{-2}\}$	$\{L\}$	$\{LT^{-1}\}$	$\{ML^{-3}\}$	$\{ML^{-1}T^{-1}\}$

Step 3   Find j. No variable contains the dimension Θ, and so j is less than or equal to 3 (MLT). We inspect the list and see that L, U, and ρ cannot form a pi group because only ρ contains mass and only U contains time. Therefore j does equal 3, and $n - j = 5 - 3 = 2 = k$. The pi theorem guarantees for this problem that there will be exactly two independent dimensionless groups.

Step 4   Select repeating j variables. The group L, U, ρ we found in step 3 will do fine.

Step 5   Combine L, U, ρ with one additional variable, in sequence, to find the two pi products.

First add force to find Π_1. You may select *any* exponent on this additional term as you please, to place it in the numerator or denominator to any power. Since F is the output, or dependent, variable, we select it to appear to the first power in the numerator:

$$\Pi_1 = L^a U^b \rho^c F = (L)^a (LT^{-1})^b (ML^{-3})^c (MLT^{-2}) = M^0 L^0 T^0$$

Equate exponents:

Length: $\qquad\qquad\qquad\qquad\qquad a + b - 3c + 1 = 0$

Mass: $\qquad\qquad\qquad\qquad\qquad\qquad c + 1 = 0$

Time: $\qquad\qquad\qquad\qquad\qquad -b \qquad -2 = 0$

We can solve explicitly for

$$a = -2 \qquad b = -2 \qquad c = -1$$

Therefore $\qquad\qquad \Pi_1 = L^{-2} U^{-2} \rho^{-1} F = \dfrac{F}{\rho U^2 L^2} = C_F$ $\qquad\qquad$ *Ans.*

This is exactly the right pi group as in Eq. (5.2). By varying the exponent on F, we could have found other equivalent groups such as $UL\rho^{1/2}/F^{1/2}$.

Finally, add viscosity to L, U, and ρ to find Π_2. Select any power you like for viscosity. By hindsight and custom, we select the power -1 to place it in the denominator:

$$\Pi_2 = L^a U^b \rho^c \mu^{-1} = L^a (LT^{-1})^b (ML^{-3})^c (ML^{-1}T^{-1})^{-1} = M^0 L^0 T^0$$

Equate exponents:

Length: $\qquad\qquad\qquad\qquad\qquad a + b - 3c + 1 = 0$

Mass: $\qquad\qquad\qquad\qquad\qquad\qquad c - 1 = 0$

Time: $\qquad\qquad\qquad\qquad\qquad -b \qquad + 1 = 0$

from which we find

$$a = b = c = 1$$

Therefore $\qquad\qquad \Pi_2 = L^1 U^1 \rho^1 \mu^{-1} = \dfrac{\rho U L}{\mu} = \text{Re}$ $\qquad\qquad$ *Ans.*

Step 6   We know we are finished; this is the second and last pi group. The theorem guarantees that the functional relationship must be of the equivalent form

$$\frac{F}{\rho U^2 L^2} = g\left(\frac{\rho U L}{\mu}\right) \qquad\qquad Ans.$$

which is exactly Eq. (5.2).

EXAMPLE 5.3

The power input P to a centrifugal pump is a function of the volume flow Q, impeller diameter D, rotational rate Ω, and the density ρ and viscosity μ of the fluid:

$$P = f(Q, D, \Omega, \rho, \mu)$$

Rewrite this as a dimensionless relationship. *Hint:* Use Ω, ρ, and D as repeating variables,

Solution

Step 1   Count the variables. There are six (don't forget the one on the left, P).

Step 2   List the dimensions of each variable from Table 5.1. Use the $\{FLT\Theta\}$ system:

P	Q	D	Ω	ρ	μ
$\{FLT^{-1}\}$	$\{L^3T^{-1}\}$	$\{L\}$	$\{T^{-1}\}$	$\{FT^2L^{-4}\}$	$\{FTL^{-2}\}$

Step 3   Find j. Lucky us, we were told to use (Ω, ρ, D) as repeating variables, so surely $j = 3$, the number of dimensions (FLT)? Check that these three do *not* form a pi group:

$$\Omega^a \rho^b D^c = (T^{-1})^a (FT^2L^{-4})^b (L)^c = F^0L^0T^0 \qquad \text{only if} \qquad a = 0, b = 0, c = 0$$

Yes, $j = 3$. This was not as obvious as the scaling group (L, U, ρ) in Example 5.2, but it is true. We now know, from the theorem, that adding one more variable will indeed form a pi group.

Step 4a   Combine (Ω, ρ, D) with power P to find the first pi group:

$$\Pi_1 = \Omega^a \rho^b D^c P = (T^{-1})^a (FT^2L^{-4})^b (L)^c (FLT^{-1}) = F^0L^0T^0$$

Equate exponents:

Force: $b \qquad\;\; + 1 = 0$

Length: $-4b + c + 1 = 0$

Time: $-a + 2b \qquad - 1 = 0$

Solve algebraically to obtain $a = -3$, $b = -1$, and $c = -5$. This first pi group, the output dimensionless variable, is called the *power coefficient* of a pump, C_P:

$$\Pi_1 = \Omega^{-3} \rho^{-1} D^{-5} P = \frac{P}{\rho \Omega^3 D^5} = C_P$$

Step 4b   Combine (Ω, ρ, D) with flow rate Q to find the second pi group:

$$\Pi_2 = \Omega^a \rho^b D^c Q = (T^{-1})^a (FT^2 L^{-4})^b (L)^c (L^3 T^{-1}) = F^0 L^0 T^0$$

After equating exponents, we now find $a = -1$, $b = 0$, and $c = -3$. This second pi group is called the *flow coefficient* of a pump, C_Q:

$$\Pi_2 = \Omega^{-1} \rho^0 D^{-3} Q = \frac{Q}{\Omega D^3} = C_Q$$

Step 4c   Combine (Ω, ρ, D) with viscosity μ to find the third and last pi group:

$$\Pi_3 = \Omega^a \rho^b D^c \mu = (T^{-1})^a (FT^2 L^{-4})^b (L)^c (FTL^{-2}) = F^0 L^0 T^0$$

This time, $a = -1$, $b = -1$, and $c = -2$; or $\Pi_3 = \mu/(\rho \Omega D^2)$, a sort of Reynolds number.

Step 5   The original relation between six variables is now reduced to three dimensionless groups:

$$\frac{P}{\rho \Omega^3 D^5} = f\left(\frac{Q}{\Omega D^3}, \frac{\mu}{\rho \Omega D^2} \right) \qquad\qquad Ans.$$

Comment: These three are the classical coefficients used to correlate pump power in Chap. 11.

EXAMPLE 5.4

At low velocities (laminar flow), the volume flow Q through a small-bore tube is a function only of the tube radius R, the fluid viscosity μ, and the pressure drop per unit tube length dp/dx. Using the pi theorem, find an appropriate dimensionless relationship.

Solution

Write the given relation and count variables:

$$Q = f\left(R, \mu, \frac{dp}{dx} \right) \quad \text{four variables } (n = 4)$$

Make a list of the dimensions of these variables from Table 5.1 using the $\{MLT\}$ system:

Q	R	μ	dp/dx
$\{L^3 T^{-1}\}$	$\{L\}$	$\{ML^{-1}T^{-1}\}$	$\{ML^{-2}T^{-2}\}$

There are three primary dimensions (M, L, T), hence $j \le 3$. By trial and error we determine that R, μ, and dp/dx cannot be combined into a pi group. Then $j = 3$, and $n - j = 4 - 3 = 1$. There is only *one* pi group, which we find by combining Q in a power product with the other three:

$$\Pi_1 = R^a \mu^b \left(\frac{dp}{dx} \right)^c Q^1 = (L)^a (ML^{-1}T^{-1})^b (ML^{-2}T^{-2})^c (L^3 T^{-1})$$

$$= M^0 L^0 T^0$$

Equate exponents:

Mass: $$b + c = 0$$

Length: $$a - b - 2c + 3 = 0$$

Time: $$-b - 2c - 1 = 0$$

Solving simultaneously, we obtain $a = -4$, $b = 1$, and $c = -1$. Then

$$\Pi_1 = R^{-4}\mu^1\left(\frac{dp}{dx}\right)^{-1}Q$$

or $$\Pi_1 = \frac{Q\mu}{R^4(dp/dx)} = \text{const} \qquad\qquad Ans.$$

Since there is only one pi group, it must equal a dimensionless constant. This is as far as dimensional analysis can take us. The laminar flow theory of Sec. 4.10 shows that the value of the constant is $-\frac{\pi}{8}$.

EXAMPLE 5.5

Assume that the tip deflection δ of a cantilever beam is a function of the tip load P, beam length L, area moment of inertia I, and material modulus of elasticity E; that is, $\delta = f(P, L, I, E)$. Rewrite this function in dimensionless form, and comment on its complexity and the peculiar value of j.

Solution

List the variables and their dimensions:

δ	P	L	I	E
$\{L\}$	$\{MLT^{-2}\}$	$\{L\}$	$\{L^4\}$	$\{ML^{-1}T^{-2}\}$

There are five variables ($n = 5$) and three primary dimensions (M, L, T), hence $j \leq 3$. But try as we may, we *cannot* find any combination of three variables that does not form a pi group. This is because $\{M\}$ and $\{T\}$ occur only in P and E and only in the same form, $\{MT^{-2}\}$. Thus we have encountered a special case of $j = 2$, which is less than the number of dimensions (M, L, T). To gain more insight into this peculiarity, you should rework the problem, using the (F, L, T) system of dimensions. You will find that only $\{F\}$ and $\{L\}$ occur in these variables, hence $j = 2$.

With $j = 2$, we select L and E as two variables that cannot form a pi group and then add other variables to form the three desired pis:

$$\Pi_1 = L^a E^b I^1 = (L)^a(ML^{-1}T^{-2})^b(L^4) = M^0L^0T^0$$

from which, after equating exponents, we find that $a = -4$, $b = 0$, or $\Pi_1 = I/L^4$. Then

$$\Pi_2 = L^a E^b P^1 = (L)^a(ML^{-1}T^{-2})^b(MLT^{-2}) = M^0L^0T^0$$

from which we find $a = -2$, $b = -1$, or $\Pi_2 = P/(EL^2)$, and

$$\Pi_3 = L^a E^b \delta^1 = (L)^a(ML^{-1}T^{-2})^b(L) = M^0L^0T^0$$

from which $a = -1$, $b = 0$, or $\Pi_3 = \delta/L$. The proper dimensionless function is $\Pi_3 = f(\Pi_2, \Pi_1)$, or

$$\frac{\delta}{L} = f\left(\frac{P}{EL^2}, \frac{I}{L^4}\right) \qquad \textit{Ans. (1)}$$

This is a complex three-variable function, but dimensional analysis alone can take us no further.

Comments: We can "improve" Eq. (1) by taking advantage of some physical reasoning, as Langhaar points out [4, p. 91]. For small elastic deflections, δ is proportional to load P and inversely proportional to moment of inertia I. Since P and I occur separately in Eq. (1), this means that Π_3 must be proportional to Π_2 and inversely proportional to Π_1. Thus, for these conditions,

$$\frac{\delta}{L} = (\text{const}) \frac{P}{EL^2} \frac{L^4}{I}$$

or
$$\delta = (\text{const}) \frac{PL^3}{EI} \qquad (2)$$

This could not be predicted by a pure dimensional analysis. Strength-of-materials theory predicts that the value of the constant is $\frac{1}{3}$.

An Alternate Step-by-Step Method by Ipsen (1960)[5]

The pi theorem method, just explained and illustrated, is often called the *repeating variable method* of dimensional analysis. Select the repeating variables, add one more, and you get a pi group. The writer likes it. This method is straightforward and systematically reveals all the desired pi groups. However, there are drawbacks: (1) All pi groups contain the same repeating variables and might lack variety or effectiveness, and (2) one must (sometimes laboriously) check that the selected repeating variables do *not* form a pi group among themselves (see Prob. P5.21).

Ipsen [5] suggests an entirely different procedure, a step-by-step method that obtains all of the pi groups at once, without any counting or checking. One simply successively eliminates each dimension in the desired function by division or multiplication. Let us illustrate with the same classical drag function proposed in Eq. (5.1). Underneath the variables, write out the dimensions of each quantity.

$$\begin{array}{cccccc} F & = \text{fcn}(L, & V, & \rho, & \mu) \\ \{MLT^{-2}\} & \{L\} & \{LT^{-1}\} & \{ML^{-3}\} & \{ML^{-1}T^{-1}\} \end{array} \qquad (5.1)$$

There are three dimensions, $\{MLT\}$. Eliminate them successively by division or multiplication by a variable. Start with mass $\{M\}$. Pick a variable that contains mass and divide it into all the other variables with mass dimensions. We select ρ, divide, and rewrite the function (5.1):

$$\begin{array}{cccccc} \dfrac{F}{\rho} & = \text{fcn}\left(L, & V, & \cancel{\rho}, & \dfrac{\mu}{\rho}\right) \\ \{L^4T^{-2}\} & \{L\} & \{LT^{-1}\} & \{ML^{-3}\} & \{L^2T^{-1}\} \end{array} \qquad (5.1a)$$

---

[5]These two methods (the pi theorem versus Ipsen) are quite different. Both are useful and interesting.

We did not divide into L or V, which do not contain $\{M\}$. Equation (5.1a) at first looks strange, but it contains five distinct variables and the same information as Eq. (5.1).

We see that ρ is no longer important because no other variable contains $\{M\}$. Thus *discard* ρ, and now there are only four variables. Next, eliminate time $\{T\}$ by dividing the time-containing variables by suitable powers of, say, V. The result is

$$\frac{F}{\rho V^2} = \text{fcn}\left(L, \quad \cancel{V}, \quad \frac{\mu}{\rho V}\right)$$

$$\{L^2\} \qquad \{L\} \quad \{LT^{-1}\} \quad \{L\}$$

(5.1b)

Now we see that V is no longer relevant since only V contains time $\{T\}$. Finally, eliminate $\{L\}$ through division by, say, appropriate powers of L itself:

$$\frac{F}{\rho V^2 L^2} = \text{fcn}\left(\cancel{L}, \quad \frac{\mu}{\rho V L}\right)$$

$$\{1\} \qquad \{L\} \quad \{1\}$$

(5.1c)

Now L by itself is no longer relevant and so discard it also. The result is equivalent to Eq. (5.2):

$$\frac{F}{\rho V^2 L^2} = \text{fcn}\left(\frac{\mu}{\rho V L}\right)$$

(5.2)

In Ipsen's step-by-step method, we find the force coefficient is a function solely of the Reynolds number. We did no counting and did not find j. We just successively eliminated each primary dimension by division with the appropriate variables.

Recall Example 5.5, where we discovered, awkwardly, that the number of repeating variables was *less* than the number of primary dimensions. Ipsen's method avoids this preliminary check. Recall the beam-deflection problem proposed in Example 5.5 and the various dimensions:

$$\delta = f(P, \quad L, \quad I, \quad E)$$

$$\{L\} \quad \{MLT^{-2}\} \quad \{L\} \quad \{L^4\} \quad \{ML^{-1}T^{-2}\}$$

For the first step, let us eliminate $\{M\}$ by dividing by E. We only have to divide into P:

$$\delta = f\left(\frac{P}{E}, \quad L, \quad I, \quad \cancel{E}\right)$$

$$\{L\} \quad \{L^2\} \quad \{L\} \quad \{L^4\} \quad \{ML^{-1}T^{-2}\}$$

We see that we may discard E as no longer relevant, and the dimension $\{T\}$ has vanished along with $\{M\}$. We need only eliminate $\{L\}$ by dividing by, say, powers of L itself:

$$\frac{\delta}{L} = \text{fcn}\left(\frac{P}{EL^2}, \quad \cancel{L}, \quad \frac{I}{L^4}\right)$$

$$\{1\} \qquad \{1\} \quad \{L\} \quad \{1\}$$

Discard L itself as now irrelevant, and we obtain *Answer* (1) to Example 5.5:

$$\frac{\delta}{L} = \text{fcn}\left(\frac{P}{EL^2}, \quad \frac{I}{L^4}\right)$$

Ipsen's approach is again successful. The fact that $\{M\}$ and $\{T\}$ vanished in the same division is proof that there are only *two* repeating variables this time, not the three that would be inferred by the presence of $\{M\}$, $\{L\}$, and $\{T\}$.

EXAMPLE 5.6

The leading-edge aerodynamic moment M_{LE} on a supersonic airfoil is a function of its chord length C, angle of attack a, and several air parameters: approach velocity V, density ρ, speed of sound a, and specific heat ratio k (Fig. E5.6). There is a very weak effect of air viscosity, which is neglected here.

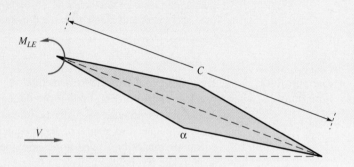

E5.6

Use Ipsen's method to rewrite this function in dimensionless form.

Solution

Write out the given function and list the variables' dimensions $\{MLT\}$ underneath:

$$M_{LE} = \text{fcn}(C, \quad \alpha, \quad V, \quad \rho, \quad a, \quad k)$$
$$\{ML^2/T^2\} \qquad \{L\} \quad \{1\} \quad \{L/T\} \quad \{M/L^3\} \quad \{L/T\} \quad \{1\}$$

Two of them, α and k, are already dimensionless. Leave them alone; they will be pi groups in the final function. You can eliminate any dimension. We choose mass $\{M\}$ and divide by ρ:

$$\frac{M_{LE}}{\rho} = \text{fcn}(C, \quad \alpha, \quad V, \quad \rlap{/}{\rho}, \quad a, \quad k)$$
$$\{L^5/T^2\} \qquad \{L\} \quad \{1\} \quad \{L/T\} \qquad \{L/T\} \quad \{1\}$$

Recall Ipsen's rules: Only divide into variables containing mass, in this case only M_{LE}, and then discard the divisor, ρ. Now eliminate time $\{T\}$ by dividing by appropriate powers of a:

$$\frac{M_{LE}}{\rho a^2} = \text{fcn}\left(C, \quad \alpha, \quad \frac{V}{a}, \quad \rlap{/}{a}, \quad k\right)$$
$$\{L^3\} \qquad \{L\} \quad \{1\} \quad \{1\} \qquad \{1\}$$

Finally, eliminate $\{L\}$ on the left side by dividing by C^3:

$$\frac{M_{LE}}{\rho a^2 C^3} = \text{fcn}\left(\mathcal{R}e, \; \alpha, \; \frac{V}{a}, \; k\right)$$
$$\;\{1\} \qquad\qquad\quad \{1\}\;\;\{1\}\;\;\{1\}$$

We end up with four pi groups and recognize V/a as the Mach number, Ma. In aerodynamics, the dimensionless moment is often called the *moment coefficient*, C_M. Thus our final result could be written in the compact form

$$C_M = \text{fcn}(\alpha, \text{Ma}, k) \qquad\qquad\qquad\qquad Ans.$$

Comments: Our analysis is fine, but experiment and theory and physical reasoning all indicate that M_{LE} varies more strongly with V than with a. Thus aerodynamicists commonly define the moment coefficient as $C_M = M_{LE}/(\rho V^2 C^3)$ or something similar. We will study the analysis of supersonic forces and moments in Chap. 9.

5.4 Nondimensionalization of the Basic Equations

We could use the pi theorem method of the previous section to analyze problem after problem after problem, finding the dimensionless parameters that govern in each case. Textbooks on dimensional analysis [for example, 5] do this. An alternative and very powerful technique is to attack the basic equations of flow from Chap. 4. Even though these equations cannot be solved in general, they will reveal basic dimensionless parameters, such as the Reynolds number, in their proper form and proper position, giving clues to when they are negligible. The boundary conditions must also be non-dimensionalized.

Let us briefly apply this technique to the incompressible flow continuity and momentum equations with constant viscosity:

Continuity: $$\nabla \cdot \mathbf{V} = 0 \qquad\qquad (5.21a)$$

Navier-Stokes: $$\rho \frac{d\mathbf{V}}{dt} = \rho \mathbf{g} - \nabla p + \mu \nabla^2 \mathbf{V} \qquad\qquad (5.21b)$$

Typical boundary conditions for these two equations are (Sect. 4.6)

Fixed solid surface: $$\mathbf{V} = 0$$

Inlet or outlet: $$\text{Known } \mathbf{V}, p \qquad\qquad (5.22)$$

Free surface, $z = \eta$: $$w = \frac{d\eta}{dt} \qquad p = p_a - Y(R_x^{-1} + R_y^{-1})$$

We omit the energy equation (4.75) and assign its dimensionless form in the problems (Prob. P5.43).

Equations (5.21) and (5.22) contain the three basic dimensions M, L, and T. All variables p, $\mathbf{V}$, x, y, z, and t can be nondimensionalized by using density and two reference constants that might be characteristic of the particular fluid flow:

$$\text{Reference velocity} = U \qquad \text{Reference length} = L$$

For example, U may be the inlet or upstream velocity and L the diameter of a body immersed in the stream.

Now define all relevant dimensionless variables, denoting them by an asterisk:

$$\mathbf{V}^* = \frac{\mathbf{V}}{U} \qquad \mathbf{\nabla}^* = L\mathbf{\nabla}$$

$$x^* = \frac{x}{L} \quad y^* = \frac{y}{L} \quad z^* = \frac{z}{L} \qquad R^* = \frac{R}{L} \tag{5.23}$$

$$t^* = \frac{tU}{L} \quad p^* = \frac{p + \rho gz}{\rho U^2}$$

All these are fairly obvious except for p^*, where we have introduced the piezometric pressure, assuming that z is up. This is a hindsight idea suggested by Bernoulli's equation (3.77).

Since ρ, U, and L are all constants, the derivatives in Eqs. (5.21) can all be handled in dimensionless form with dimensional coefficients. For example,

$$\frac{\partial u}{\partial x} = \frac{\partial(Uu^*)}{\partial(Lx^*)} = \frac{U}{L}\frac{\partial u^*}{\partial x^*}$$

Substitute the variables from Eqs. (5.23) into Eqs. (5.21) and (5.22) and divide through by the leading dimensional coefficient, in the same way as we handled Eq. (5.12). Here are the resulting dimensionless equations of motion:

Continuity:
$$\boxed{\mathbf{\nabla}^* \cdot \mathbf{V}^* = 0} \tag{5.24a}$$

Momentum:
$$\boxed{\frac{d\mathbf{V}^*}{dt^*} = -\mathbf{\nabla}^* p^* + \frac{\mu}{\rho UL}\mathbf{\nabla}^{*2}(\mathbf{V}^*)} \tag{5.24b}$$

The dimensionless boundary conditions are:

Fixed solid surface:
$$\boxed{\mathbf{V}^* = 0}$$

Inlet or outlet:
$$\text{Known } \mathbf{V}^*, p^*$$

Free surface, $z^* = \eta^*$:
$$\boxed{\begin{aligned} w^* &= \frac{d\eta^*}{dt^*} \\ p^* &= \frac{p_a}{\rho U^2} + \frac{gL}{U^2}z^* - \frac{Y}{\rho U^2 L}(R_x^{*-1} + R_y^{*-1}) \end{aligned}} \tag{5.25}$$

These equations reveal a total of four dimensionless parameters, one in the Navier-Stokes equation and three in the free-surface-pressure boundary condition.

Dimensionless Parameters

In the continuity equation there are no parameters. The Navier-Stokes equation contains one, generally accepted as the most important parameter in fluid mechanics:

$$\text{Reynolds number Re} = \frac{\rho UL}{\mu}$$

It is named after Osborne Reynolds (1842–1912), a British engineer who first proposed it in 1883 (Ref. 4 of Chap. 6). The Reynolds number is always important, with or without a free surface, and can be neglected only in flow regions away from high-velocity gradients—for example, away from solid surfaces, jets, or wakes.

The no-slip and inlet-exit boundary conditions contain no parameters. The free-surface-pressure condition contains three:

$$\text{Euler number (pressure coefficient) Eu} = \frac{p_a}{\rho U^2}$$

This is named after Leonhard Euler (1707–1783) and is rarely important unless the pressure drops low enough to cause vapor formation (cavitation) in a liquid. The Euler number is often written in terms of pressure differences: $\text{Eu} = \Delta p/(\rho U^2)$. If Δp involves vapor pressure p_v, it is called the *cavitation number* $\text{Ca} = (p_a - p_v)/(\rho U^2)$.

The second free-surface parameter is much more important:

$$\text{Froude number Fr} = \frac{U^2}{gL}$$

It is named after William Froude (1810–1879), a British naval architect who, with his son Robert, developed the ship-model towing-tank concept and proposed similarity rules for free-surface flows (ship resistance, surface waves, open channels). The Froude number is the dominant effect in free-surface flows and is totally unimportant if there is no free surface. Chapter 10 investigates Froude number effects in detail.

The final free-surface parameter is

$$\text{Weber number We} = \frac{\rho U^2 L}{Y}$$

It is named after Moritz Weber (1871–1951) of the Polytechnic Institute of Berlin, who developed the laws of similitude in their modern form. It was Weber who named Re and Fr after Reynolds and Froude. The Weber number is important only if it is of order unity or less, which typically occurs when the surface curvature is comparable in size to the liquid depth, such as in droplets, capillary flows, ripple waves, and very small hydraulic models. If We is large, its effect may be neglected.

If there is no free surface, Fr, Eu, and We drop out entirely, except for the possibility of cavitation of a liquid at very small Eu. Thus, in low-speed viscous flows with no free surface, the Reynolds number is the only important dimensionless parameter.

Compressibility Parameters

In high-speed flow of a gas there are significant changes in pressure, density, and temperature that must be related by an equation of state such as the perfect-gas law, Eq. (1.10). These thermodynamic changes introduce two additional dimensionless parameters mentioned briefly in earlier chapters:

$$\text{Mach number Ma} = \frac{U}{a} \qquad \text{Specific-heat ratio } k = \frac{c_p}{c_v}$$

The Mach number is named after Ernst Mach (1838–1916), an Austrian physicist. The effect of k is only slight to moderate, but Ma exerts a strong effect on compressible flow properties if it is greater than about 0.3. These effects are studied in Chap. 9.

Oscillating Flows

If the flow pattern is oscillating, a seventh parameter enters through the inlet boundary condition. For example, suppose that the inlet stream is of the form

$$u = U \cos \omega t$$

Nondimensionalization of this relation results in

$$\frac{u}{U} = u^* = \cos\left(\frac{\omega L}{U} t^*\right)$$

The argument of the cosine contains the new parameter

$$\text{Strouhal number St} = \frac{\omega L}{U}$$

The dimensionless forces and moments, friction, and heat transfer, and so on of such an oscillating flow would be a function of both Reynolds and Strouhal numbers. This parameter is named after V. Strouhal, a German physicist who experimented in 1878 with wires singing in the wind.

Some flows that you might guess to be perfectly steady actually have an oscillatory pattern that is dependent on the Reynolds number. An example is the periodic vortex shedding behind a blunt body immersed in a steady stream of velocity U. Figure 5.2a shows an array of alternating vortices shed from a circular cylinder immersed in a steady crossflow. This regular, periodic shedding is called a *Kármán vortex street,* after T. von Kármán, who explained it theoretically in 1912. The shedding occurs in the range $10^2 < \text{Re} < 10^7$, with an average Strouhal number $\omega d/(2\pi U) \approx 0.21$. Figure 5.2b shows measured shedding frequencies.

Resonance can occur if a vortex shedding frequency is near a body's structural vibration frequency. Electric transmission wires sing in the wind, undersea mooring lines gallop at certain current speeds, and slender structures flutter at critical wind or vehicle speeds. A striking example is the disastrous failure of the Tacoma Narrows suspension bridge in 1940, when wind-excited vortex shedding caused resonance with the natural torsional oscillations of the bridge. The problem was magnified by the bridge deck nonlinear stiffness, which occurred when the hangers went slack during the oscillation.

Other Dimensionless Parameters

We have discussed seven important parameters in fluid mechanics, and there are others. Four additional parameters arise from nondimensionalization of the energy equation (4.75) and its boundary conditions. These four (Prandtl number, Eckert number, Grashof number, and wall temperature ratio) are listed in Table 5.2 just in case you fail to solve Prob. P5.43. Another important and perhaps surprising parameter is the

(a)

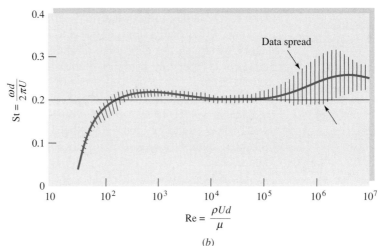

Fig. 5.2 Vortex shedding from a circular cylinder: (*a*) vortex street behind a circular cylinder *(Courtesy of U.S. Navy);* (*b*) experimental shedding frequencies *(data from Refs. 25 and 26).*

(b)

wall roughness ratio ϵ/L (in Table 5.2).[6] Slight changes in surface roughness have a striking effect in the turbulent flow or high-Reynolds-number range, as we shall see in Chap. 6 and in Fig. 5.3.

This book is primarily concerned with Reynolds-, Mach-, and Froude-number effects, which dominate most flows. Note that we discovered these parameters (except ϵ/L) simply by nondimensionalizing the basic equations without actually solving them.

---

[6]Roughness is easy to overlook because it is a slight geometric effect that does not appear in the equations of motion. It is a boundary condition that one might forget.

Table 5.2 Dimensionless Groups in Fluid Mechanics

Parameter	Definition	Qualitative ratio of effects	Importance
Reynolds number	$\mathrm{Re} = \dfrac{\rho U L}{\mu}$	$\dfrac{\text{Inertia}}{\text{Viscosity}}$	Almost always
Mach number	$\mathrm{Ma} = \dfrac{U}{a}$	$\dfrac{\text{Flow speed}}{\text{Sound speed}}$	Compressible flow
Froude number	$\mathrm{Fr} = \dfrac{U^2}{gL}$	$\dfrac{\text{Inertia}}{\text{Gravity}}$	Free-surface flow
Weber number	$\mathrm{We} = \dfrac{\rho U^2 L}{Y}$	$\dfrac{\text{Inertia}}{\text{Surface tension}}$	Free-surface flow
Rossby number	$\mathrm{Ro} = \dfrac{U}{\Omega_{\text{earth}} L}$	$\dfrac{\text{Flow velocity}}{\text{Coriolis effect}}$	Geophysical flows
Cavitation number (Euler number)	$\mathrm{Ca} = \dfrac{p - p_v}{\rho U^2}$	$\dfrac{\text{Pressure}}{\text{Inertia}}$	Cavitation
Prandtl number	$\mathrm{Pr} = \dfrac{\mu c_p}{k}$	$\dfrac{\text{Dissipation}}{\text{Conduction}}$	Heat convection
Eckert number	$\mathrm{Ec} = \dfrac{U^2}{c_p T_0}$	$\dfrac{\text{Kinetic energy}}{\text{Enthalpy}}$	Dissipation
Specific-heat ratio	$k = \dfrac{c_p}{c_v}$	$\dfrac{\text{Enthalpy}}{\text{Internal energy}}$	Compressible flow
Strouhal number	$\mathrm{St} = \dfrac{\omega L}{U}$	$\dfrac{\text{Oscillation}}{\text{Mean speed}}$	Oscillating flow
Roughness ratio	$\dfrac{\epsilon}{L}$	$\dfrac{\text{Wall roughness}}{\text{Body length}}$	Turbulent, rough walls
Grashof number	$\mathrm{Gr} = \dfrac{\beta \Delta T g L^3 \rho^2}{\mu^2}$	$\dfrac{\text{Buoyancy}}{\text{Viscosity}}$	Natural convection
Rayleigh number	$\mathrm{Ra} = \dfrac{\beta \Delta T g L^3 \rho c_p}{\mu k}$	$\dfrac{\text{Buoyancy}}{\text{Viscosity}}$	Natural convection
Temperature ratio	$\dfrac{T_w}{T_0}$	$\dfrac{\text{Wall temperature}}{\text{Stream temperature}}$	Heat transfer
Pressure coefficient	$C_p = \dfrac{p - p_\infty}{\frac{1}{2}\rho U^2}$	$\dfrac{\text{Static pressure}}{\text{Dynamic pressure}}$	Aerodynamics, hydrodynamics
Lift coefficient	$C_L = \dfrac{L}{\frac{1}{2}\rho U^2 A}$	$\dfrac{\text{Lift force}}{\text{Dynamic force}}$	Aerodynamics, hydrodynamics
Drag coefficient	$C_D = \dfrac{D}{\frac{1}{2}\rho U^2 A}$	$\dfrac{\text{Drag force}}{\text{Dynamic force}}$	Aerodynamics, hydrodynamics
Friction factor	$f = \dfrac{h_f}{(V^2/2g)(L/d)}$	$\dfrac{\text{Friction head loss}}{\text{Velocity head}}$	Pipe flow
Skin friction coefficient	$c_f = \dfrac{\tau_{\text{wall}}}{\rho V^2/2}$	$\dfrac{\text{Wall shear stress}}{\text{Dynamic pressure}}$	Boundary layer flow

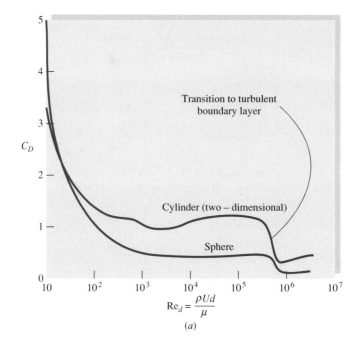

	Cylinder length effect $(10^4 < \text{Re} < 10^5)$	
	L/d	C_D
	∞	1.20
	40	0.98
	20	0.91
	10	0.82
	5	0.74
	3	0.72
	2	0.68
	1	0.64

Transition to turbulent boundary layer

Cylinder (two – dimensional)

Sphere

$$\text{Re}_d = \frac{\rho U d}{\mu}$$

(a)

Fig. 5.3 The proof of practical dimensional analysis: drag coefficients of a cylinder and sphere: (*a*) drag coefficient of a smooth cylinder and sphere (data from many sources); (*b*) increased roughness causes earlier transition to a turbulent boundary layer.

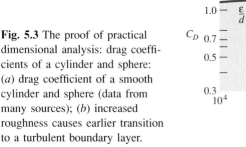
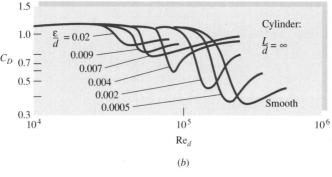

Cylinder:

$\frac{L}{d} = \infty$

$\frac{\varepsilon}{d} = 0.02$

0.009
0.007
0.004
0.002
0.0005

Smooth

Re_d

(b)

If the reader is not satiated with the 19 parameters given in Table 5.2, Ref. 29 contains a list of over 300 dimensionless parameters in use in engineering.

A Successful Application

Dimensional analysis is fun, but does it work? Yes, if all important variables are included in the proposed function, the dimensionless function found by dimensional analysis will collapse all the data onto a single curve or set of curves.

An example of the success of dimensional analysis is given in Fig. 5.3 for the measured drag on smooth cylinders and spheres. The flow is normal to the axis of the cylinder, which is extremely long, $L/d \to \infty$. The data are from many sources, for both liquids and gases, and include bodies from several meters in diameter down to fine wires and balls less than 1 mm in size. Both curves in Fig. 5.3*a* are entirely experimental; the analysis of immersed body drag is one of the weakest areas of modern fluid mechanics theory. Except for digital computer calculations, there is little theory for cylinder and sphere drag except *creeping flow,* Re < 1.

The Reynolds number of both bodies is based on diameter, hence the notation Re_d. But the drag coefficients are defined differently:

$$C_D = \begin{cases} \dfrac{\text{drag}}{\frac{1}{2}\rho U^2 L d} & \text{cylinder} \\[2ex] \dfrac{\text{drag}}{\frac{1}{2}\rho U^2 \frac{1}{4}\pi d^2} & \text{sphere} \end{cases} \tag{5.26}$$

They both have a factor $\frac{1}{2}$ because the term $\frac{1}{2}\rho U^2$ occurs in Bernoulli's equation, and both are based on the projected area—that is, the area one sees when looking toward the body from upstream. The usual definition of C_D is thus

$$C_D = \frac{\text{drag}}{\frac{1}{2}\rho U^2(\text{projected area})} \tag{5.27}$$

However, one should carefully check the definitions of C_D, Re, and the like before using data in the literature. Airfoils, for example, use the planform area.

Figure 5.3a is for long, smooth cylinders. If wall roughness and cylinder length are included as variables, we obtain from dimensional analysis a complex three-parameter function:

$$C_D = f\left(Re_d, \frac{\epsilon}{d}, \frac{L}{d}\right) \tag{5.28}$$

To describe this function completely would require 1000 or more experiments or CFD results. Therefore it is customary to explore the length and roughness effects separately to establish trends.

The table with Fig. 5.3a shows the length effect with zero wall roughness. As length decreases, the drag decreases by up to 50 percent. Physically, the pressure is "relieved" at the ends as the flow is allowed to skirt around the tips instead of deflecting over and under the body.

Figure 5.3b shows the effect of wall roughness for an infinitely long cylinder. The sharp drop in drag occurs at lower Re_d as roughness causes an earlier transition to a turbulent boundary layer on the surface of the body. Roughness has the same effect on sphere drag, a fact that is exploited in sports by deliberate dimpling of golf balls to give them less drag at their flight $Re_d \approx 10^5$. See Fig. D5.2.

Figure 5.3 is a typical experimental study of a fluid mechanics problem, aided by dimensional analysis. As time and money and demand allow, the complete three-parameter relation (5.28) could be filled out by further experiments.

EXAMPLE 5.7

A smooth cylinder, 1 cm in diameter and 20 cm long, is tested in a wind tunnel for a cross-flow of 45 m/s of air at 20°C and 1 atm. The measured drag is 2.2 ± 0.1 N. (*a*) Does this data point agree with the data in Fig. 5.3? (*b*) Can this data point be used to predict the drag of a chimney 1 m in diameter and 20 m high in winds at 20°C and 1 atm? If so, what is the recommended range of wind velocities and drag forces for this data point? (*c*) Why are the answers to part (*b*) always the same, regardless of the chimney height, as long as $L = 20d$?

Solution

(*a*) For air at 20°C and 1 atm, take $\rho = 1.2$ kg/m^3 and $\mu = 1.8$ E-5 kg/(m$-$s). Since the test cylinder is short, $L/d = 20$, it should be compared with the tabulated value $C_D \approx 0.91$ in the table to the right of Fig. 5.3*a*. First calculate the Reynolds number of the test cylinder:

$$\mathrm{Re}_d = \frac{\rho U d}{\mu} = \frac{(1.2 \text{ kg/m}^3)(45 \text{ m/s})(0.01 \text{ m})}{1.8\text{E}-5 \text{ kg/(m} - \text{s)}} = 30{,}000$$

Yes, this is in the range $10^4 < \mathrm{Re} < 10^5$ listed in the table. Now calculate the test drag coefficient:

$$C_{D,\text{test}} = \frac{F}{(1/2)\rho U^2 L d} = \frac{2.2 \text{ N}}{(1/2)(1.2 \text{ kg/m}^3)(45 \text{ m/s})^2(0.2 \text{ m})(0.01 \text{ m})} = 0.905$$

Yes, this is close, and certainly within the range of ±5 percent stated by the test results.

Ans. (*a*)

(*b*) Since the chimney has $L/d = 20$, we can use the data if the Reynolds number range is correct:

$$10^4 < \frac{(1.2 \text{ kg/m}^3)U_{\text{chimney}}(1 \text{ m})}{1.8 \text{ E}-5 \text{ kg/(m} \cdot \text{s)}} < 10^5 \quad \text{if} \quad 0.15\,\frac{\text{m}}{\text{s}} < U_{\text{chimney}} < 1.5\,\frac{\text{m}}{\text{s}}$$

These are negligible winds, so the test data point is not very useful.   *Ans.* (*b*)
The drag forces in this range are also negligibly small:

$$F_{\min} = C_D \frac{\rho}{2} U_{\min}^2 L d = (0.91)\left(\frac{1.2 \text{ kg/m}^3}{2}\right)(0.15 \text{ m/s})^2(20 \text{ m})(1 \text{ m}) = 0.25 \text{ N}$$

$$F_{\max} = C_D \frac{\rho}{2} U_{\max}^2 L d = (0.91)\left(\frac{1.2 \text{ kg/m}^3}{2}\right)(1.5 \text{ m/s})^2(20 \text{ m})(1 \text{ m}) = 25 \text{ N}$$

(*c*) Try this yourself. Choose any 20:1 size for the chimney, even something silly like 20 mm:1 mm. You will get the same results for U and F as in part (*b*) above. This is because the product Ud occurs in Re_d and, if $L = 20d$, the same product occurs in the drag force. For example, for $\mathrm{Re} = 10^4$,

$$Ud = 10^4\frac{\mu}{\rho} \quad \text{then } F = C_D \frac{\rho}{2} U^2 L d = C_D \frac{\rho}{2} U^2(20d)d = 20 C_D \frac{\rho}{2}(Ud)^2 = 20 C_D \frac{\rho}{2}\left(\frac{10^4 \mu}{\rho}\right)^2$$

The answer is always $F_{\min} = 0.25$ N. This is an algebraic quirk that seldom occurs.

EXAMPLE 5.8

Telephone wires are said to "sing" in the wind. Consider a wire of diameter 8 mm. At what sea-level wind velocity, if any, will the wire sing a middle C note?

Solution

For sea-level air take $\nu \approx 1.5\,\mathrm{E}-5\ \mathrm{m^2/s}$. For nonmusical readers, middle C is 262 Hz. Measured shedding rates are plotted in Fig. 5.2b. Over a wide range, the Strouhal number is approximately 0.2, which we can take as a first guess. Note that $(\omega/2\pi) = f$, the shedding frequency. Thus

$$\mathrm{St} = \frac{fd}{U} = \frac{(262\ \mathrm{s}^{-1})(0.008\ \mathrm{m})}{U} \approx 0.2$$

$$U \approx 10.5\ \frac{\mathrm{m}}{\mathrm{s}}$$

Now check the Reynolds number to see if we fall into the appropriate range:

$$\mathrm{Re}_d = \frac{Ud}{\nu} = \frac{(10.5\ \mathrm{m/s})(0.008\ \mathrm{m})}{1.5\,\mathrm{E}-5\ \mathrm{m^2/s}} \approx 5600$$

In Fig. 5.2b, at Re = 5600, maybe St is a little higher, at about 0.21. Thus a slightly improved estimate is

$$U_{\mathrm{wind}} = (262)(0.008)/(0.21) \approx 10.0\ \mathrm{m/s} \qquad\qquad Ans.$$

5.5 Modeling and Its Pitfalls

So far we have learned about dimensional homogeneity and the pi theorem method, using power products, for converting a homogeneous physical relation to dimensionless form. This is straightforward mathematically, but certain engineering difficulties need to be discussed.

First, we have more or less taken for granted that the variables that affect the process can be listed and analyzed. Actually, selection of the important variables requires considerable judgment and experience. The engineer must decide, for example, whether viscosity can be neglected. Are there significant temperature effects? Is surface tension important? What about wall roughness? Each pi group that is retained increases the expense and effort required. Judgment in selecting variables will come through practice and maturity; this book should provide some of the necessary experience.

Once the variables are selected and the dimensional analysis is performed, the experimenter seeks to achieve *similarity* between the model tested and the prototype to be designed. With sufficient testing, the model data will reveal the desired dimensionless function between variables:

$$\Pi_1 = f(\Pi_2, \Pi_3, \ldots \Pi_k) \qquad\qquad (5.29)$$

With Eq. (5.29) available in chart, graphical, or analytical form, we are in a position to ensure complete similarity between model and prototype. A formal statement would be as follows:

> Flow conditions for a model test are completely similar if all relevant dimensionless parameters have the same corresponding values for the model and the prototype.

This follows mathematically from Eq. (5.29). If $\Pi_{2m} = \Pi_{2p}$, $\Pi_{3m} = \Pi_{3p}$, and so forth, Eq. (5.29) guarantees that the desired output Π_{1m} will equal Π_{1p}. But this is easier said than done, as we now discuss. There are specialized texts on model testing [30–32].

Instead of complete similarity, the engineering literature speaks of particular types of similarity, the most common being geometric, kinematic, dynamic, and thermal. Let us consider each separately.

Geometric Similarity

Geometric similarity concerns the length dimension $\{L\}$ and must be ensured before any sensible model testing can proceed. A formal definition is as follows:

> A model and prototype are *geometrically similar* if and only if all body dimensions in all three coordinates have the same linear scale ratio.

Note that *all* length scales must be the same. It is as if you took a photograph of the prototype and reduced it or enlarged it until it fitted the size of the model. If the model is to be made one-tenth the prototype size, its length, width, and height must each be one-tenth as large. Not only that, but also its entire shape must be one-tenth as large, and technically we speak of *homologous* points, which are points that have the same relative location. For example, the nose of the prototype is homologous to the nose of the model. The left wingtip of the prototype is homologous to the left wingtip of the model. Then geometric similarity requires that all homologous points be related by the same linear scale ratio. This applies to the fluid geometry as well as the model geometry.

> All angles are preserved in geometric similarity. All flow directions are preserved. The orientations of model and prototype with respect to the surroundings must be identical.

Figure 5.4 illustrates a prototype wing and a one-tenth-scale model. The model lengths are all one-tenth as large, but its angle of attack with respect to the free stream is the same: 10° not 1°. All physical details on the model must be scaled, and some are rather subtle and sometimes overlooked:

1. The model nose radius must be one-tenth as large.
2. The model surface roughness must be one-tenth as large.

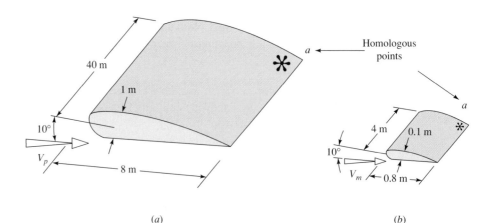

Fig. 5.4 Geometric similarity in model testing: (*a*) prototype; (*b*) one-tenth-scale model.

(*a*)

(*b*)

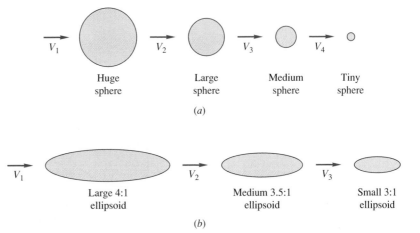

Fig. 5.5 Geometric similarity and dissimilarity of flows: (*a*) similar; (*b*) dissimilar.

3. If the prototype has a 5-mm boundary layer trip wire 1.5 m from the leading edge, the model should have a 0.5-mm trip wire 0.15 m from its leading edge.

4. If the prototype is constructed with protruding fasteners, the model should have homologous protruding fasteners one-tenth as large.

And so on. Any departure from these details is a violation of geometric similarity and must be justified by experimental comparison to show that the prototype behavior was not significantly affected by the discrepancy.

Models that appear similar in shape but that clearly violate geometric similarity should not be compared except at your own risk. Figure 5.5 illustrates this point. The spheres in Fig. 5.5*a* are all geometrically similar and can be tested with a high expectation of success if the Reynolds number, Froude number, or the like is matched. But the ellipsoids in Fig. 5.5*b* merely *look* similar. They actually have different linear scale ratios and therefore cannot be compared in a rational manner, even though they may have identical Reynolds and Froude numbers and so on. The data will not be the same for these ellipsoids, and any attempt to "compare" them is a matter of rough engineering judgment.

Kinematic Similarity

Kinematic similarity requires that the model and prototype have the same length scale ratio and the same time scale ratio. The result is that the velocity scale ratio will be the same for both. As Langhaar [4] states it:

> The motions of two systems are kinematically similar if homologous particles lie at homologous points at homologous times.

Length scale equivalence simply implies geometric similarity, but time scale equivalence may require additional dynamic considerations such as equivalence of the Reynolds and Mach numbers.

One special case is incompressible frictionless flow with no free surface, as sketched in Fig. 5.6*a*. These perfect-fluid flows are kinematically similar with independent length and time scales, and no additional parameters are necessary (see Chap. 8 for further details).

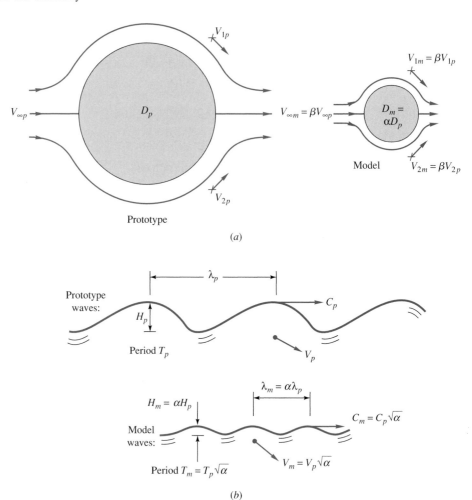

Fig. 5.6 Frictionless low-speed flows are kinematically similar: (a) Flows with no free surface are kinematically similar with independent length and time scale ratios; (b) free-surface flows are kinematically similar with length and time scales related by the Froude number.

Frictionless flows with a free surface, as in Fig. 5.6b, are kinematically similar if their Froude numbers are equal:

$$\text{Fr}_m = \frac{V_m^2}{gL_m} = \frac{V_p^2}{gL_p} = Fr_p \qquad (5.30)$$

Note that the Froude number contains only length and time dimensions and hence is a purely kinematic parameter that fixes the relation between length and time. From Eq. (5.30), if the length scale is

$$L_m = \alpha L_p \qquad (5.31)$$

where α is a dimensionless ratio, the velocity scale is

$$\frac{V_m}{V_p} = \left(\frac{L_m}{L_p}\right)^{1/2} = \sqrt{\alpha} \qquad (5.32)$$

and the time scale is

$$\frac{T_m}{T_p} = \frac{L_m/V_m}{L_p/V_p} = \sqrt{\alpha} \qquad (5.33)$$

These Froude-scaling kinematic relations are illustrated in Fig. 5.6*b* for wave motion modeling. If the waves are related by the length scale α, then the wave period, propagation speed, and particle velocities are related by $\sqrt{\alpha}$.

If viscosity, surface tension, or compressibility is important, kinematic similarity depends on the achievement of dynamic similarity.

Dynamic Similarity

Dynamic similarity exists when the model and the prototype have the same length scale ratio, time scale ratio, and force scale (or mass scale) ratio. Again geometric similarity is a first requirement; without it, proceed no further. Then dynamic similarity exists, simultaneous with kinematic similarity, if the model and prototype force and pressure coefficients are identical. This is ensured if

1. For compressible flow, the model and prototype Reynolds number and Mach number and specific-heat ratio are correspondingly equal.
2. For incompressible flow
    a. With no free surface: model and prototype Reynolds numbers are equal.
    b. With a free surface: model and prototype Reynolds number, Froude number, and (if necessary) Weber number and cavitation number are correspondingly equal.

Mathematically, Newton's law for any fluid particle requires that the sum of the pressure force, gravity force, and friction force equal the acceleration term, or inertia force,

$$\mathbf{F}_p + \mathbf{F}_g + \mathbf{F}_f = \mathbf{F}_i$$

The dynamic similarity laws listed above ensure that each of these forces will be in the same ratio and have equivalent directions between model and prototype. Figure 5.7

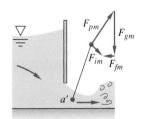

Fig. 5.7 Dynamic similarity in sluice gate flow. Model and prototype yield identical homologous force polygons if the Reynolds and Froude numbers are the same corresponding values: (*a*) prototype; (*b*) model.

(*a*)   (*b*)

shows an example for flow through a sluice gate. The force polygons at homologous points have exactly the same shape if the Reynolds and Froude numbers are equal (neglecting surface tension and cavitation, of course). Kinematic similarity is also ensured by these model laws.

Discrepancies in Water and Air Testing

The perfect dynamic similarity shown in Fig. 5.7 is more of a dream than a reality because true equivalence of Reynolds and Froude numbers can be achieved only by dramatic changes in fluid properties, whereas in fact most model testing is simply done with water or air, the cheapest fluids available.

First consider hydraulic model testing with a free surface. Dynamic similarity requires equivalent Froude numbers, Eq. (5.30), *and* equivalent Reynolds numbers:

$$\frac{V_m L_m}{\nu_m} = \frac{V_p L_p}{\nu_p} \tag{5.34}$$

But both velocity and length are constrained by the Froude number, Eqs. (5.31) and (5.32). Therefore, for a given length scale ratio α, Eq. (5.34) is true only if

$$\frac{\nu_m}{\nu_p} = \frac{L_m}{L_p} \frac{V_m}{V_p} = \alpha \sqrt{\alpha} = \alpha^{3/2} \tag{5.35}$$

For example, for a one-tenth-scale model, $\alpha = 0.1$ and $\alpha^{3/2} = 0.032$. Since ν_p is undoubtedly water, we need a fluid with only 0.032 times the kinematic viscosity of water to achieve dynamic similarity. Referring to Table 1.4, we see that this is impossible: Even mercury has only one-ninth the kinematic viscosity of water, and a mercury hydraulic model would be expensive and bad for your health. In practice, water is used for both the model and the prototype, and the Reynolds number similarity (5.34) is unavoidably violated. The Froude number is held constant since it is the dominant parameter in free-surface flows. Typically the Reynolds number of the model flow is too small by a factor of 10 to 1000. As shown in Fig. 5.8, the low-Reynolds-number model data are used to estimate by extrapolation the desired high-Reynolds-number prototype data. As the figure indicates, there is obviously considerable uncertainty in using such an extrapolation, but there is no other practical alternative in hydraulic model testing.

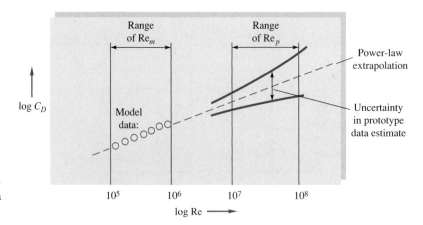

Fig. 5.8 Reynolds-number extrapolation, or scaling, of hydraulic data with equal Froude numbers.

Second, consider aerodynamic model testing in air with no free surface. The important parameters are the Reynolds number and the Mach number. Equation (5.34) should be satisfied, plus the compressibility criterion

$$\frac{V_m}{a_m} = \frac{V_p}{a_p} \tag{5.36}$$

Elimination of V_m/V_p between (5.34) and (5.36) gives

$$\frac{\nu_m}{\nu_p} = \frac{L_m}{L_p}\frac{a_m}{a_p} \tag{5.37}$$

Since the prototype is no doubt an air operation, we need a wind-tunnel fluid of low viscosity and high speed of sound. Hydrogen is the only practical example, but clearly it is too expensive and dangerous. Therefore wind tunnels normally operate with air as the working fluid. Cooling and pressurizing the air will bring Eq. (5.37) into better agreement but not enough to satisfy a length scale reduction of, say, one-tenth. Therefore Reynolds number scaling is also commonly violated in aerodynamic testing, and an extrapolation like that in Fig. 5.8 is required here also.

There are specialized monographs devoted entirely to wind tunnel testing: low speed [38], high speed [39], and a detailed general discussion [40]. The following example illustrates modeling discrepancies in aeronautical testing.

EXAMPLE 5.9

A prototype airplane, with a chord length of 1.6 m, is to fly at Ma = 2 at 10 km standard altitude. A one-eighth scale model is to be tested in a helium wind tunnel at 100°C and 1 atm. Find the helium test section velocity that will match (a) the Mach number or (b) the Reynolds number of the prototype. In each case criticize the lack of dynamic similarity. (c) What high pressure in the helium tunnel will match *both* the Mach and Reynolds numbers? (d) Why does part (c) *still* not achieve dynamic similarity?

Solution

For helium, from Table A.4, $R = 2077$ m^2/(s^2-K), $k = 1.66$, and estimate $\mu_{He} \approx 2.32$ E$-$5 kg/(m · s) from the power-law, $n = 0.67$, in the table. (a) Calculate the helium speed of sound and velocity:

$$a_{He} = \sqrt{(kRT)_{He}} = \sqrt{(1.66)(2077 \text{ m}^2/\text{s}^2\text{K}) \times (373 \text{ K})} = 1134 \text{ m/s}$$

$$\text{Ma}_{air} = \text{Ma}_{He} = 2.0 = \frac{V_{He}}{a_{He}} = \frac{V_{He}}{1134 \text{ m/s}}$$

$$V_{He} = 2268 \frac{\text{m}}{\text{s}} \qquad\qquad \textit{Ans. (a)}$$

For dynamic similarity, the Reynolds numbers should also be equal. From Table A.6 at an altitude of 10,000 m, read $\rho_{air} = 0.4125$ kg/m^3, $a_{air} = 299.5$ m/s, and estimate $\mu_{air} \approx 1.48$ E$-$5 kg/m · s from the power-law, $n = 0.7$, in Table A.4. The air velocity is $V_{air} = (\text{Ma})(a_{air}) = 2(299.5) = 599$ m/s. The model chord length is (1.6 m)/8 = 0.2 m. The helium density is $\rho_{He} = (p/RT)_{He} = (101{,}350 \text{ Pa})/[(2077 \text{ m}^2/\text{s}^2 \text{ K})(373 \text{ K})] = 0.131 \text{ kg/m}^3$. Now calculate the two Reynolds numbers:

$$\text{Re}_{C,\text{air}} = \frac{\rho V C}{\mu}\bigg|_{\text{air}} = \frac{(0.4125 \text{ kg/m}^3)(599 \text{ m/s})(1.6 \text{ m})}{1.48 \text{ E}-5 \text{ kg/(m} \cdot \text{s)}} = 26.6 \text{ E6}$$

$$\text{Re}_{C,\text{He}} = \frac{\rho V C}{\mu}\bigg|_{\text{He}} = \frac{(0.131 \text{ kg/m}^3)(2268 \text{ m/s})(0.2 \text{ m})}{2.32 \text{ E}-5 \text{ kg/(m} \cdot \text{s)}} = 2.56 \text{ E6}$$

The model Reynolds number is ten times less than the prototype. This is typical when using small-scale models. The test results must be extrapolated for Reynolds number effects. (b) Now ignore Mach number and let the model Reynolds number match the prototype:

$$\text{Re}_{He} = \text{Re}_{\text{air}} = 26.6 \text{ E6} = \frac{(0.131 \text{ kg/m}^3) V_{He}(0.2 \text{ m})}{2.32 \text{ E}-5 \text{ kg/(m} \cdot \text{s)}}$$

$$V_{He} = 23{,}600 \frac{\text{m}}{\text{s}} \qquad \qquad Ans. \ (b)$$

This is ridiculous: a hypersonic Mach number of 21, suitable for escaping from the earth's gravity. One should match the Mach numbers and correct for a lower Reynolds number. (c) Match both Reynolds and Mach numbers by increasing the helium density: Ma matches if

$$V_{He} = 2268 \frac{\text{m}}{\text{s}}$$

Then

$$\text{Re}_{He} = 26.6 \text{ E6} = \frac{\rho_{He}(2268 \text{ m/s})(0.2 \text{ m})}{2.32 \text{ E}-5 \text{ kg/(m} \cdot \text{s)}}$$

Solve for

$$\rho_{He} = 1.36 \frac{\text{kg}}{\text{m}^3} \quad p_{He} = \rho RT|_{He} = (1.36)(2077)(373) = 1.05 \text{ E6 Pa} \qquad Ans. \ (c)$$

A match is possible if we increase the tunnel pressure by a factor of ten, a daunting task. (d) Even with Ma and Re matched, we are *still* not dynamically similar because the two gases have different specific heat ratios: $k_{He} = 1.66$ and $k_{\text{air}} = 1.40$. This discrepancy will cause substantial differences in pressure, density, and temperature throughout supersonic flow.

Figure 5.9 shows a hydraulic model of the Bluestone Lake Dam in West Virginia. The model itself is located at the U.S. Army Waterways Experiment Station in Vicksburg, MS. The horizontal scale is 1:65, which is sufficient that the vertical scale can also be 1:65 without incurring significant surface tension (Weber number) effects. Velocities are scaled by the Froude number. However, the prototype Reynolds number, which is of order 1E7, cannot be matched here. The engineers set the Reynolds number at about 2E4, high enough for a reasonable approximation of prototype turbulent flow viscous effects. Note the intense turbulence below the dam. The downstream bed, or *apron,* of a dam must be strengthened structurally to avoid bed erosion.

Fig. 5.9 Hydraulic model of the Bluestone Lake Dam on the New River near Hinton, West Virginia. The model scale is 1:65 both vertically and horizontally, and the Reynolds number, though far below the prototype value, is set high enough for the flow to be turbulent. *(Courtesy of the U.S. Army Corps of Engineers Waterways Experiment Station.)*

For hydraulic models of larger scale, such as harbors, estuaries, and embayments, geometric similarity may be violated of necessity. The vertical scale will be distorted to avoid Weber number effects. For example, the horizontal scale may be 1:1000, while the vertical scale is only 1:100. Thus the model channel may be *deeper* relative to its horizontal dimensions. Since deeper passages flow more efficiently, the model channel bottom may be deliberately roughened to create the friction level expected in the prototype.

EXAMPLE 5.10

The pressure drop due to friction for flow in a long smooth pipe is a function of average flow velocity, density, viscosity, and pipe length and diameter: $\Delta p = \text{fcn}(V, \rho, \mu, L, D)$. We wish to know how Δp varies with V. (*a*) Use the pi theorem to rewrite this function in dimensionless form. (*b*) Then plot this function, using the following data for three pipes and three fluids:

D, cm	L, m	Q, m³/h	Δp, Pa	ρ, kg/m³	μ, kg/(m · s)	V, m/s*
1.0	5.0	0.3	4,680	680†	2.92 E-4†	1.06
1.0	7.0	0.6	22,300	680†	2.92 E-4†	2.12
1.0	9.0	1.0	70,800	680†	2.92 E-4†	3.54
2.0	4.0	1.0	2,080	998‡	0.0010‡	0.88
2.0	6.0	2.0	10,500	998‡	0.0010‡	1.77
2.0	8.0	3.1	30,400	998‡	0.0010‡	2.74
3.0	3.0	0.5	540	13,550§	1.56 E-3§	0.20
3.0	4.0	1.0	2,480	13,550§	1.56 E-3§	0.39
3.0	5.0	1.7	9,600	13,550§	1.56 E-3§	0.67

*$V = Q/A$, $A = \pi D^2/4$.
†Gasoline.
‡Water.
§Mercury.

(*c*) Suppose it is further known that Δp is proportional to L (which is quite true for long pipes with well-rounded entrances). Use this information to simplify and improve the pi theorem formulation. Plot the dimensionless data in this improved manner and comment on the results.

Solution

There are six variables with three primary dimensions involved $\{MLT\}$. Therefore we expect that $j = 6 - 3 = 3$ pi groups. We are correct, for we can find three variables that do not form a pi product, for example, (ρ, V, L). Carefully select three (j) repeating variables, but not including Δp or V, which we plan to plot versus each other. We select (ρ, μ, D), and the pi theorem guarantees that three independent power-product groups will occur:

$$\Pi_1 = \rho^a \mu^b D^c \, \Delta p \qquad \Pi_2 = \rho^d \mu^e D^f V \qquad \Pi_3 = \rho^g \mu^h D^i L$$

or
$$\Pi_1 = \frac{\rho D^2 \Delta p}{\mu^2} \qquad \Pi_2 = \frac{\rho V D}{\mu} \qquad \Pi_3 = \frac{L}{D}$$

We have omitted the algebra of finding $(a, b, c, d, e, f, g, h, i)$ by setting all exponents to zero M^0, L^0, T^0. Therefore we wish to plot the dimensionless relation

$$\frac{\rho D^2 \, \Delta p}{\mu^2} = \text{fcn}\left(\frac{\rho V D}{\mu}, \frac{L}{D}\right) \qquad\qquad Ans. (a)$$

We plot Π_1 versus Π_2 with Π_3 as a parameter. There will be nine data points. For example, the first row in the data here yields

$$\frac{\rho D^2 \, \Delta p}{\mu^2} = \frac{(680)(0.01)^2(4680)}{(2.92 \text{ E-4})^2} = 3.73 \text{ E9}$$

$$\frac{\rho V D}{\mu} = \frac{(680)(1.06)(0.01)}{2.92 \text{ E-4}} = 24,700 \qquad \frac{L}{D} = 500$$

The nine data points are plotted as the open circles in Fig. 5.10. The values of L/D are listed for each point, and we see a significant length effect. In fact, if we connect the only two points that have the same L/D (= 200), we could see (and cross-plot to verify) that Δp increases linearly with L, as stated in the last part of the problem. Since L occurs only in

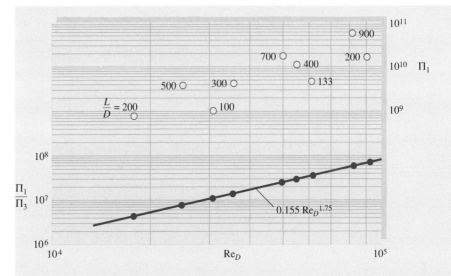

Fig. 5.10 Two different correlations of the data in Example 5.10: Open circles when plotting $\rho D^2 \,\Delta p/\mu^2$ versus Re_D, L/D is a parameter; once it is known that Δp is proportional to L, a replot (solid circles) of $\rho D^3 \,\Delta p/(L\mu^2)$ versus Re_D collapses into a single power-law curve.

$\Pi_3 = L/D$, the function $\Pi_1 = \mathrm{fcn}(\Pi_2, \Pi_3)$ must reduce to $\Pi_1 = (L/D)\,\mathrm{fcn}(\Pi_2)$, or simply a function involving only *two* parameters:

$$\frac{\rho D^3 \,\Delta p}{L\mu^2} = \mathrm{fcn}\!\left(\frac{\rho V D}{\mu}\right) \qquad \text{flow in a long pipe} \qquad \textit{Ans. (c)}$$

We now modify each data point in Fig. 5.10 by dividing it by its L/D value. For example, for the first row of data, $\rho D^3 \,\Delta p/(L\mu^2) = (3.73\ \mathrm{E}9)/500 = 7.46\ \mathrm{E}6$. We replot these new data points as solid circles in Fig. 5.10. They correlate almost perfectly into a straight-line power-law function:

$$\frac{\rho D^3 \,\Delta p}{L\mu^2} \approx 0.155\!\left(\frac{\rho V D}{\mu}\right)^{1.75} \qquad\qquad \textit{Ans. (c)}$$

All newtonian smooth pipe flows should correlate in this manner. This example is a variation of the first completely successful dimensional analysis, pipe-flow friction, performed by Prandtl's student Paul Blasius, who published a related plot in 1911. For this range of (turbulent flow) Reynolds numbers, the pressure drop increases approximately as $V^{1.75}$.

EXAMPLE 5.11

The smooth sphere data plotted in Fig. 5.3a represent dimensionless drag versus dimensionless *viscosity*, since (ρ, V, d) were selected as scaling or repeating variables. (a) Replot these data to display the effect of dimensionless *velocity* on the drag. (b) Use your new figure to predict the terminal (zero-acceleration) velocity of a 1-cm-diameter steel ball (SG = 7.86) falling through water at 20°C.

Solution

- *Assumptions:* Fig 5.3a is valid for any smooth sphere in that Reynolds number range.
- *Approach (a):* Form pi groups from the function $F = \mathrm{fcn}(d, V, \rho, \mu)$ in such a way that F is plotted versus V. The answer was already given as Eq. (5.16), but let us review the

steps. The proper scaling variables are (ρ, μ, d), which do *not* form a pi. Therefore $j = 3$, and we expect $n - j = 5 - 3 = 2$ pi groups. Skipping the algebra, they arise as follows:

$$\Pi_1 = \rho^a \mu^b d^c F = \frac{\rho F}{\mu^2} \qquad \Pi_2 = \rho^a \mu^b d^c V = \frac{\rho V d}{\mu} \qquad \textit{Ans. (a)}$$

We may replot the data of Fig. 5.3a in this new form, noting that $\Pi_1 \equiv (\pi/8)(C_D)(\text{Re})^2$. This replot is shown as Fig. 5.11. The drag increases rapidly with velocity up to transition, where there is a slight drop, after which it increases more than ever. If force is known, we may predict velocity from the figure, and vice versa.

- *Property values for part (b):* $\rho_{\text{water}} = 998 \ \text{kg/m}^3 \qquad \mu_{\text{water}} = 0.001 \ \text{kg/(m--s)}$

$$\rho_{\text{steel}} = 7.86 \rho_{\text{water}} = 7844 \ \text{kg/m}^3.$$

- *Solution to part (b):* For terminal velocity, the drag force equals the net weight of the sphere in water:

$$F = W_{\text{net}} = (\rho_s - \rho_w)g \frac{\pi}{6} d^3 = (7840 - 998)(9.81)\left(\frac{\pi}{6}\right)(0.01)^3 = 0.0351 \ \text{N}$$

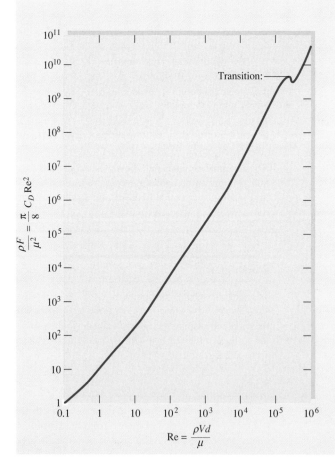

Fig. 5.11 Cross-plot of sphere drag data from Fig. 5.3a to show dimensionless force versus dimensionless velocity.

Therefore the ordinate of Fig. 5.11 is known:

Falling steel sphere: $\dfrac{\rho F}{\mu^2} = \dfrac{(998 \text{ kg/m}^3)(0.0351 \text{ N})}{[0.001 \text{ kg/(m} \cdot \text{s)}]^2} \approx 3.5 \text{ E7}$

From Fig. 5.11, at $\rho F / \mu^2 \approx 3.5$ E7, a magnifying glass reveals that $\text{Re}_d \approx 2$ E4. Then a crude estimate of the terminal fall velocity is

$$\frac{\rho V d}{\mu} \approx 20{,}000 \qquad \text{or} \qquad V \approx \frac{20{,}000[0.001 \text{ kg/(m} \cdot \text{s)}]}{(998 \text{ kg/m}^3)(0.01 \text{ m})} \approx 2.0 \frac{\text{m}}{\text{s}} \qquad Ans. \ (b)$$

- *Comments:*   Better accuracy could be obtained by expanding the scale of Fig. 5.11 in the region of the given force coefficient. However, there is considerable uncertainty in published drag data for spheres, so the predicted fall velocity is probably uncertain by at least ±10 percent.

    Note that we found the answer directly from Fig. 5.11. We could use Fig. 5.3a also but would have to iterate between the ordinate and abscissa to obtain the final result, since V is contained in both plotted variables.

Summary

Chapters 3 and 4 presented integral and differential methods of mathematical analysis of fluid flow. This chapter introduces the third and final method: experimentation, as supplemented by the technique of dimensional analysis. Tests and experiments are used both to strengthen existing theories and to provide useful engineering results when theory is inadequate.

The chapter begins with a discussion of some familiar physical relations and how they can be recast in dimensionless form because they satisfy the principle of dimensional homogeneity. A general technique, the pi theorem, is then presented for systematically finding a set of dimensionless parameters by grouping a list of variables that govern any particular physical process. A second technique, Ipsen's method, is also described. Alternately, direct application of dimensional analysis to the basic equations of fluid mechanics yields the fundamental parameters governing flow patterns: Reynolds number, Froude number, Prandtl number, Mach number, and others.

It is shown that model testing in air and water often leads to scaling difficulties for which compromises must be made. Many model tests do not achieve true dynamic similarity. The chapter ends by pointing out that classic dimensionless charts and data can be manipulated and recast to provide direct solutions to problems that would otherwise be quite cumbersome and laboriously iterative.

Problems

Most of the problems herein are fairly straightforward. More difficult or open-ended assignments are labeled with an asterisk. Problems labeled with an EES icon (for example, Prob. P5.61) will benefit from the use of the Engineering Equation Solver (EES), while problems labeled with a computer icon may require the use of a computer. The standard end-of-chapter problems P5.1 to P5.91 (categorized in the problem list here) are followed by word problems W5.1 to W5.10, fundamentals of engineering exam problems FE5.1 to FE5.10, comprehensive applied problems C5.1 to C5.5, and design projects D5.1 and D5.2.

P5.1 For axial flow through a circular tube, the Reynolds number for transition to turbulence is approximately 2300 [see Eq. (6.2)], based on the diameter and average velocity. If $d = 5$ cm and the fluid is kerosene at 20°C, find the volume flow rate in m^3/h that causes transition.

P5.2 A prototype automobile is designed for cold weather in Denver, CO ($-10°C$, 83 kPa). Its drag force is to be tested on a one-seventh-scale model in a wind tunnel at 20°C and 1 atm. If the model and prototype are to satisfy dynamic similarity, what prototype velocity, in mi/h, needs to be matched? Comment on your result.

P5.3 An airplane has a chord length $L = 1.2$ m and flies at a Mach number of 0.7 in the standard atmosphere. If its Reynolds number, based on chord length, is 7 E6, how high is it flying?

P5.4 When tested in water at 20°C flowing at 2 m/s, an 8-cm-diameter sphere has a measured drag of 5 N. What will be the velocity and drag force on a 1.5-m-diameter weather balloon moored in sea-level standard air under dynamically similar conditions?

P5.5 An automobile has a characteristic length and area of 8 ft and 60 ft^2, respectively. When tested in sea-level standard air, it has the following measured drag force versus speed:

V, mi/h	20	40	60
Drag, lbf	31	115	249

The same car travels in Colorado at 65 mi/h at an altitude of 3500 m. Using dimensional analysis, estimate (a) its drag force and (b) the horsepower required to overcome air drag.

*P5.6 SAE 10 oil at 20°C flows past an 8-cm-diameter sphere. At flow velocities of 1, 2, and 3 m/s, the measured sphere drag forces are 1.5, 5.3, and 11.2 N, respectively. Estimate the drag force if the same sphere is tested at a velocity of 15 m/s in glycerin at 20°C.

P5.7 A body is dropped on the moon ($g = 1.62$ m/s^2) with an initial velocity of 12 m/s. By using option 2 variables, Eq. (5.11), the ground impact occurs at $t^{**} = 0.34$ and $S^{**} = 0.84$. Estimate (a) the initial displacement, (b) the final displacement, and (c) the time of impact.

P5.8 The *Morton number* Mo, used to correlate bubble dynamics studies, is a dimensionless combination of acceleration of gravity g, viscosity μ, density ρ, and surface tension coefficient Y. If Mo is proportional to g, find its form.

P5.9 The *Richardson number,* Ri, which correlates the production of turbulence by buoyancy, is a dimensionless combination of the acceleration of gravity g, the fluid temperature T_0, the local temperature gradient $\partial T/\partial z$, and the local velocity gradient $\partial u/\partial z$. Determine the form of the Richardson number if it is proportional to g.

P5.10 Determine the dimension $\{MLT\Theta\}$ of the following quantities:

(a) $\rho u \dfrac{\partial u}{\partial x}$ (b) $\displaystyle\int_1^2 (p - p_0)\, dA$ (c) $\rho c_p \dfrac{\partial^2 T}{\partial x\, \partial y}$

(d) $\displaystyle\iiint \rho \dfrac{\partial u}{\partial t}\, dx\, dy\, dz$

All quantities have their standard meanings; for example, ρ is density.

P5.11 During World War II, Sir Geoffrey Taylor, a British fluid dynamicist, used dimensional analysis to estimate the wave speed of an atomic bomb explosion. He assumed that the blast wave radius R was a function of energy released E, air density ρ, and time t. Use dimensional reasoning to show how wave radius must vary with time.

P5.12 The *Stokes number,* St, used in particle dynamics studies, is a dimensionless combination of *five* variables: acceleration of gravity g, viscosity μ, density ρ, particle velocity U, and particle diameter D. (a) If St is proportional to μ and inversely proportional to g, find its form. (b) Show that St is actually the quotient of two more traditional dimensionless groups.

P5.13 The speed of propagation C of a capillary wave in deep water is known to be a function only of density ρ, wavelength λ, and surface tension Y. Find the proper functional relationship, completing it with a dimensionless constant. For a given density and wavelength, how does the propagation speed change if the surface tension is doubled?

P5.14 Consider flow in a pipe of diameter D through a pipe bend of radius R_b. The pressure loss Δp through the bend is a function of these two length scales, plus density ρ, viscosity μ, and average flow velocity V. (a) Use dimensional analysis to rewrite this function in terms of dimensionless pi groups. (b) In analyzing data for such pipe-bend losses (Chap. 6), the dimensionless loss is often correlated with the *Dean number*:

$$\text{De} = \text{Re}_D \sqrt{\frac{D}{2R_b}}$$

Can your dimensional analysis produce a similar group? If not, explain why not.

P5.15 The wall shear stress τ_w in a boundary layer is assumed to be a function of stream velocity U, boundary layer thickness δ, local turbulence velocity u', density ρ, and local pressure gradient dp/dx. Using (ρ, U, δ) as repeating variables, rewrite this relationship as a dimensionless function.

P5.16 Convection heat transfer data are often reported as a *heat transfer coefficient h*, defined by

$$\dot{Q} = hA\,\Delta T$$

where $\dot{Q}$ = heat flow, J/s
       A = surface area, m^2
       ΔT = temperature difference, K

The dimensionless form of h, called the *Stanton number,* is a combination of h, fluid density ρ, specific heat c_p, and flow velocity V. Derive the Stanton number if it is proportional to h. What are the units of h?

P5.17 The pressure drop per unit length $\Delta p/L$ in a porous, rotating duct (Really! See Ref. 35) depends on average velocity V, density ρ, viscosity μ, duct height h, wall injection velocity v_w, and rotation rate Ω. Using (ρ, V, h) as repeating variables, rewrite this relationship in dimensionless form.

P5.18 Under laminar conditions, the volume flow Q through a small triangular-section pore of side length b and length L is a function of viscosity μ, pressure drop per unit length $\Delta p/L$, and b. Using the pi theorem, rewrite this relation in dimensionless form. How does the volume flow change if the pore size b is doubled?

P5.19 The period of oscillation T of a water surface wave is assumed to be a function of density ρ, wavelength λ, depth h, gravity g, and surface tension Y. Rewrite this relationship in dimensionless form. What results if Y is negligible? *Hint:* Take λ, ρ, and g as repeating variables.

*P5.20 We can extend Prob. P5.18 to the case of laminar duct flow of a nonnewtonian fluid, for which the simplest relation for stress versus strain rate is the *power-law* approximation

$$\tau = C\left(\frac{d\theta}{dt}\right)^n$$

where θ is the angle of shear strain. This is the analog of Eq. (1.23). The constant C takes the place of viscos-

ity. If the exponent n is less than (greater than) unity, the material simulates a pseudoplastic (dilatant) fluid, as illustrated in Fig. 1.7. (*a*) Using the {*MLT*} system, determine the dimensions of C. (*b*) The analog of Prob. P5.18 for power-law laminar triangular-duct flow is $Q = \text{fcn}(C, \Delta p/L, b)$. Rewrite this function in the form of dimensionless Pi groups.

P5.21 In Example 5.1 we used the pi theorem to develop Eq. (5.2) from Eq. (5.1). Instead of merely listing the primary dimensions of each variable, some workers list the *powers* of each primary dimension for each variable in an array:

$$
\begin{array}{c}
 \\
M \\
L \\
T
\end{array}
\begin{array}{c}
\begin{array}{ccccc}
F & L & U & \rho & \mu
\end{array} \\
\left[
\begin{array}{ccccc}
1 & 0 & 0 & 1 & 1 \\
1 & 1 & 1 & -3 & -1 \\
-2 & 0 & -1 & 0 & -1
\end{array}
\right]
\end{array}
$$

This array of exponents is called the *dimensional matrix* for the given function. Show that the *rank* of this matrix (the size of the largest nonzero determinant) is equal to $j = n - k$, the desired reduction between original variables and the pi groups. This is a general property of dimensional matrices, as noted by Buckingham [1].

P5.22 When freewheeling, the angular velocity Ω of a windmill is found to be a function of the windmill diameter D, the wind velocity V, the air density ρ, the windmill height H as compared to the atmospheric boundary layer height L, and the number of blades N:

$$\Omega = \text{fcn}\left(D, V, \rho, \frac{H}{L}, N\right)$$

Viscosity effects are negligible. Find appropriate pi groups for this problem and rewrite the function in dimensionless form.

P5.23 The period T of vibration of a beam is a function of its length L, area moment of inertia I, modulus of elasticity E, density ρ, and Poisson's ratio σ. Rewrite this relation in dimensionless form. What further reduction can we make if E and I can occur only in the product form EI? *Hint:* Take L, ρ, and E as repeating variables.

P5.24 The lift force F on a missile is a function of its length L, velocity V, diameter D, angle of attack α, density ρ, viscosity μ, and speed of sound a of the air. Write out the dimensional matrix of this function and determine its rank. (See Prob. P5.21 for an explanation of this concept.) Rewrite the function in terms of pi groups.

P5.25 The thrust F of a propeller is generally thought to be a function of its diameter D and angular velocity Ω, the forward speed V, and the density ρ and viscosity μ of the fluid. Rewrite this relationship as a dimensionless function.

P5.26   A pendulum has an oscillation period T which is assumed to depend on its length L, bob mass m, angle of swing θ, and the acceleration of gravity. A pendulum 1 m long, with a bob mass of 200 g, is tested on earth and found to have a period of 2.04 s when swinging at 20°. (*a*) What is its period when it swings at 45°? A similarly constructed pendulum, with $L = 30$ cm and $m = 100$ g, is to swing on the moon ($g = 1.62$ m/s^2) at $\theta = 20°$. (*b*) What will be its period?

P5.27   In studying sand transport by ocean waves, A. Shields in 1936 postulated that the threshold wave-induced bottom shear stress τ required to move particles depends on gravity g, particle size d and density ρ_p, and water density ρ and viscosity μ. Find suitable dimensionless groups of this problem, which resulted in 1936 in the celebrated Shields sand transport diagram.

P5.28   A simply supported beam of diameter D, length L, and modulus of elasticity E is subjected to a fluid crossflow of velocity V, density ρ, and viscosity μ. Its center deflection δ is assumed to be a function of all these variables. (*a*) Rewrite this proposed function in dimensionless form. (*b*) Suppose it is known that δ is independent of μ, inversely proportional to E, and dependent only on ρV^2, not ρ and V separately. Simplify the dimensionless function accordingly. *Hint:* Take L, ρ, and V as repeating variables.

P5.29   When fluid in a pipe is accelerated linearly from rest, it begins as laminar flow and then undergoes transition to turbulence at a time t_{tr} that depends on the pipe diameter D, fluid acceleration a, density ρ, and viscosity μ. Arrange this into a dimensionless relation between t_{tr} and D.

P5.30   The wall shear stress τ_w for flow in a narrow annular gap between a fixed and a rotating cylinder is a function of density ρ, viscosity μ, angular velocity Ω, outer radius R, and gap width Δr. Using (ρ, Ω, R) as repeating variables, rewrite this relation in dimensionless form.

P5.31   The heat transfer rate per unit area q to a body from a fluid in natural or gravitational convection is a function of the temperature difference ΔT, gravity g, body length L, and three fluid properties: kinematic viscosity ν, conductivity k, and thermal expansion coefficient β. Rewrite in dimensionless form if it is known that g and β appear only as the product $g\beta$.

P5.32   A *weir* is an obstruction in a channel flow that can be calibrated to measure the flow rate, as in Fig. P5.32. The volume flow Q varies with gravity g, weir width b into the paper, and upstream water height H above the weir crest. If it is known that Q is proportional to b, use the pi theorem to find a unique functional relationship $Q(g, b, H)$.

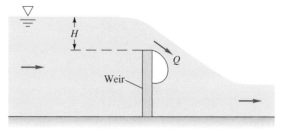

P5.32

P5.33   A spar buoy (see Prob. P2.113) has a period T of vertical (heave) oscillation that depends on the waterline cross-sectional area A, buoy mass m, and fluid specific weight γ. How does the period change due to doubling of (*a*) the mass and (*b*) the area? Instrument buoys should have long periods to avoid wave resonance. Sketch a possible long-period buoy design.

P5.34   To good approximation, the thermal conductivity k of a gas (see Ref. 30 of Chap. 1) depends only on the density ρ, mean free path l, gas constant R, and absolute temperature T. For air at 20°C and 1 atm, $k \approx 0.026$ W/(m · K) and $l \approx 6.5$ E-8 m. Use this information to determine k for hydrogen at 20°C and 1 atm if $l \approx 1.2$ E-7 m.

P5.35   The torque M required to turn the cone-plate viscometer in Fig. P5.35 depends on the radius R, rotation rate Ω, fluid viscosity μ, and cone angle θ. Rewrite this relation in dimensionless form. How does the relation simplify it if it is known that M is proportional to θ?

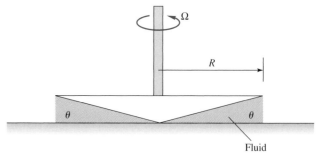

P5.35

P5.36   The rate of heat loss $\dot{Q}_{loss}$ through a window or wall is a function of the temperature difference between inside and outside ΔT, the window surface area A, and the R value of the window, which has units of (ft^2 · h · °F)/Btu. (*a*) Using the Buckingham Pi Theorem, find an expression for rate of heat loss as a function of the other three parameters in the problem. (*b*) If the temperature difference ΔT doubles, by what factor does the rate of heat loss increase?

P5.37 The volume flow Q through an orifice plate is a function of pipe diameter D, pressure drop Δp across the orifice, fluid density ρ and viscosity μ, and orifice diameter d. Using D, ρ, and Δp as repeating variables, express this relationship in dimensionless form.

P5.38 The size d of droplets produced by a liquid spray nozzle is thought to depend on the nozzle diameter D, jet velocity U, and the properties of the liquid ρ, μ, and Y. Rewrite this relation in dimensionless form. *Hint:* Take D, ρ, and U as repeating variables.

P5.39 In turbulent flow past a flat surface, the velocity u near the wall varies approximately logarithmically with distance y from the wall and also depends on viscosity μ, density ρ, and wall shear stress τ_w. For a certain airflow at 20°C and 1 atm, $\tau_w = 0.8$ Pa and $u = 15$ m/s at $y = 3.6$ mm. Use this information to estimate the velocity u at $y = 6$ mm.

P5.40 The time t_d to drain a liquid from a hole in the bottom of a tank is a function of the hole diameter d, the initial fluid volume v_0, the initial liquid depth h_0, and the density ρ and viscosity μ of the fluid. Rewrite this relation as a dimensionless function, using Ipsen's method.

P5.41 A certain axial flow turbine has an output torque M that is proportional to the volume flow rate Q and also depends on the density ρ, rotor diameter D, and rotation rate Ω. How does the torque change due to a doubling of (a) D and (b) Ω?

P5.42 When disturbed, a floating buoy will bob up and down at frequency f. Assume that this frequency varies with buoy mass m, waterline diameter d, and the specific weight γ of the liquid. (a) Express this as a dimensionless function. (b) If d and γ are constant and the buoy mass is halved, how will the frequency change?

P5.43 Nondimensionalize the energy equation (4.75) and its boundary conditions (4.62), (4.63), and (4.70) by defining $T^* = T/T_0$, where T_0 is the inlet temperature, assumed constant. Use other dimensionless variables as needed from Eqs. (5.23). Isolate all dimensionless parameters you find, and relate them to the list given in Table 5.2.

P5.44 The differential energy equation for incompressible two-dimensional flow through a "Darcy-type" porous medium is approximately

$$\rho c_p \frac{\sigma}{\mu} \frac{\partial p}{\partial x} \frac{\partial T}{\partial x} + \rho c_p \frac{\sigma}{\mu} \frac{\partial p}{\partial y} \frac{\partial T}{\partial y} + k \frac{\partial^2 T}{\partial y^2} = 0$$

where σ is the *permeability* of the porous medium. All other symbols have their usual meanings. (a) What are the appropriate dimensions for σ? (b) Nondimensionalize this equation, using (L, U, ρ, T_0) as scaling constants, and discuss any dimensionless parameters that arise.

P5.45 A model differential equation, for chemical reaction dynamics in a plug reactor, is as follows:

$$u \frac{\partial C}{\partial x} = D \frac{\partial^2 C}{\partial x^2} - kC - \frac{\partial C}{\partial t}$$

where u is the velocity, D is a diffusion coefficient, k is a reaction rate, x is distance along the reactor, and C is the (dimensionless) concentration of a given chemical in the reactor. (a) Determine the appropriate dimensions of D and k. (b) Using a characteristic length scale L and average velocity V as parameters, rewrite this equation in dimensionless form and comment on any pi groups appearing.

P5.46 The differential equation for compressible inviscid flow of a gas in the xy plane is

$$\frac{\partial^2 \phi}{\partial t^2} + \frac{\partial}{\partial t}(u^2 + v^2) + (u^2 - a^2)\frac{\partial^2 \phi}{\partial x^2}$$
$$+ (v^2 - a^2)\frac{\partial^2 \phi}{\partial y^2} + 2uv \frac{\partial^2 \phi}{\partial x\, \partial y} = 0$$

where ϕ is the velocity potential and a is the (variable) speed of sound of the gas. Nondimensionalize this relation, using a reference length L and the inlet speed of sound a_0 as parameters for defining dimensionless variables.

P5.47 The differential equation for small-amplitude vibrations $y(x, t)$ of a simple beam is given by

$$\rho A \frac{\partial^2 y}{\partial t^2} + EI \frac{\partial^4 y}{\partial x^4} = 0$$

where ρ = beam material density
A = cross-sectional area
I = area moment of inertia
E = Young's modulus

Use only the quantities ρ, E, and A to nondimensionalize y, x, and t, and rewrite the differential equation in dimensionless form. Do any parameters remain? Could they be removed by further manipulation of the variables?

P5.48 A smooth steel (SG = 7.86) sphere is immersed in a stream of ethanol at 20°C moving at 1.5 m/s. Estimate its drag in N from Fig. 5.3a. What stream velocity would quadruple its drag? Take $D = 2.5$ cm.

P5.49 The sphere in Prob. P5.48 is dropped in gasoline at 20°C. Ignoring its acceleration phase, what will its terminal (constant) fall velocity be, from Fig. 5.3a?

P5.50 When a microorganism moves in a viscous fluid, it turns out that fluid density has nearly negligible influence on the drag force felt by the microorganism. Such flows are called *creeping flows*. The only important parameters in the problem are the velocity of motion U, the viscosity of the fluid μ, and the length scale of the body. Here assume the microorganism's body diameter d as the appropriate length scale. (*a*) Using the Buckingham Pi Theorem, generate an expression for the drag force D as a function of the other parameters in the problem. (*b*) The drag coefficient discussed in this chapter $C_D = D/(\frac{1}{2}\rho U^2 A)$ is not appropriate for this kind of flow. Define instead a more appropriate drag coefficient, and call it C_c (for creeping flow). (*c*) For a spherically shaped microorganism, the drag force can be calculated exactly from the equations of motion for creeping flow. The result is $D = 3\pi\mu Ud$. Write expressions for both forms of the drag coefficient, C_c and C_D, for a sphere under conditions of creeping flow.

P5.51 A ship is towing a sonar array that approximates a submerged cylinder 1 ft in diameter and 30 ft long with its axis normal to the direction of tow. If the tow speed is 12 kn (1 kn = 1.69 ft/s), estimate the horsepower required to tow this cylinder. What will be the frequency of vortices shed from the cylinder? Use Figs. 5.2 and 5.3.

P5.52 A standard table tennis ball is smooth, weighs 2.6 g, and has a diameter of 1.5 in. If struck with an initial velocity of 85 mi/h, (*a*) what is the initial deceleration rate? (*b*) What is the estimated uncertainty of your result in part (*a*)?

P5.53 Vortex shedding can be used to design a *vortex flowmeter* (Fig. 6.33). A blunt rod stretched across the pipe sheds vortices whose frequency is read by the sensor downstream. Suppose the pipe diameter is 5 cm and the rod is a cylinder of diameter 8 mm. If the sensor reads 5400 counts per minute, estimate the volume flow rate of water in m^3/h. How might the meter react to other liquids?

P5.54 A fishnet is made of 1-mm-diameter strings knotted into 2×2 cm squares. Estimate the horsepower required to tow 300 ft^2 of this netting at 3 kn in seawater at 20°C. The net plane is normal to the flow direction.

P5.55 The radio antenna on a car begins to vibrate wildly at 8 Hz when the car is driven at 45 mi/h over a rutted road that approximates a sine wave of amplitude 2 cm and wavelength $\lambda = 2.5$ m. The antenna diameter is 4 mm. Is the vibration due to the road or to vortex shedding?

P5.56 Flow past a long cylinder of square cross-section results in more drag than the comparable round cylinder. Here are data taken in a water tunnel for a square cylinder of side length $b = 2$ cm:

V, m/s	1.0	2.0	3.0	4.0
Drag, N/(m of depth)	21	85	191	335

(*a*) Use these data to predict the drag force per unit depth of wind blowing at 6 m/s, in air at 20°C, over a tall square chimney of side length $b = 55$ cm. (*b*) Is there any uncertainty in your estimate?

P5.57 The simply supported 1040 carbon-steel rod of Fig. P5.57 is subjected to a crossflow stream of air at 20°C and 1 atm. For what stream velocity U will the rod center deflection be approximately 1 cm?

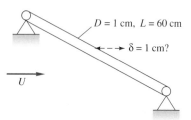

$D = 1$ cm, $L = 60$ cm
$\delta = 1$ cm?
U

P5.57

P5.58 For the steel rod of Prob. P5.57, at what airstream velocity U will the rod begin to vibrate laterally in resonance in its first mode (a half sine wave)? *Hint:* Consult a vibration text [34] under "lateral beam vibration."

P5.59 A long, slender, smooth flagpole bends alarmingly in 20 mi/h sea-level winds, causing patriotic citizens to gasp. An engineer claims that the pole will bend less if its surface is deliberately roughened. Is she correct, at least qualitatively?

*P5.60 The thrust F of a free propeller, either aircraft or marine, depends upon density ρ, the rotation rate n in r/s, the diameter D, and the forward velocity V. Viscous effects are slight and neglected here. Tests of a 25-cm-diameter model aircraft propeller, in a sea-level wind tunnel, yield the following thrust data at a velocity of 20 m/s:

Rotation rate, r/min	4800	6000	8000
Measured thrust, N	6.1	19	47

(*a*) Use this data to make a crude but effective dimensionless plot. (*b*) Use the dimensionless data to predict the thrust, in newtons, of a similar 1.6-m-diameter prototype propeller when rotating at 3800 r/min and flying at 225 mi/h at 4000-m standard altitude.

P5.61 If viscosity is neglected, typical pump flow results from Example 5.3 are shown in Fig. P5.61 for a model pump tested in water. The pressure rise decreases and the power required increases with the dimensionless flow coefficient. Curve-fit expressions are given for the data. Suppose a similar pump of 12-cm diameter is built to move gasoline at 20°C and a flow rate of 25 m^3/h. If the pump rotation speed is 30 r/s, find (*a*) the pressure rise and (*b*) the power required.

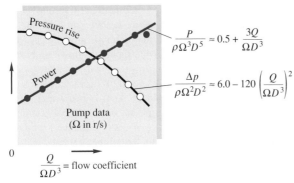

$$\frac{P}{\rho\Omega^3 D^5} \approx 0.5 + \frac{3Q}{\Omega D^3}$$

$$\frac{\Delta p}{\rho\Omega^2 D^2} \approx 6.0 - 120\left(\frac{Q}{\Omega D^3}\right)^2$$

$$\frac{Q}{\Omega D^3} = \text{flow coefficient}$$

P5.61

P5.62 A prototype water pump has an impeller diameter of 2 ft and is designed to pump 12 ft³/s at 750 r/min. A 1-ft-diameter model pump is tested in 20°C air at 1800 r/min, and Reynolds number effects are found to be negligible. For similar conditions, what will the volume flow of the model be in ft³/s? If the model pump requires 0.082 hp to drive it, what horsepower is required for the prototype?

*P5.63 The pressure drop per unit length $\Delta p/L$ in smooth pipe flow is known to be a function only of the average velocity V, diameter D, and fluid properties ρ and μ. The following data were obtained for flow of water at 20°C in an 8-cm-diameter pipe 50 m long:

Q, m³/s	0.005	0.01	0.015	0.020
Δp, Pa	5800	20,300	42,100	70,800

Verify that these data are slightly outside the range of Fig. 5.10. What is a suitable power-law curve fit for the present data? Use these data to estimate the pressure drop for flow of kerosene at 20°C in a smooth pipe of diameter 5 cm and length 200 m if the flow rate is 50 m³/h.

P5.64 The natural frequency ω of vibration of a mass M attached to a rod, as in Fig. P5.64, depends only on M

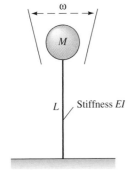

P5.64

and the stiffness EI and length L of the rod. Tests with a 2-kg mass attached to a 1040 carbon steel rod of diameter 12 mm and length 40 cm reveal a natural frequency of 0.9 Hz. Use these data to predict the natural frequency of a 1-kg mass attached to a 2024 aluminum alloy rod of the same size.

P5.65 In turbulent flow near a flat wall, the local velocity u varies only with distance y from the wall, wall shear stress τ_w, and fluid properties ρ and μ. The following data were taken in the University of Rhode Island wind tunnel for airflow, $\rho = 0.0023$ slug/ft³, $\mu = 3.81$ E-7 slug/(ft · s), and $\tau_w = 0.029$ lbf/ft²:

y, in	0.021	0.035	0.055	0.080	0.12	0.16
u, ft/s	50.6	54.2	57.6	59.7	63.5	65.9

(a) Plot these data in the form of dimensionless u versus dimensionless y, and suggest a suitable power-law curve fit. (b) Suppose that the tunnel speed is increased until $u = 90$ ft/s at $y = 0.11$ in. Estimate the new wall shear stress, in lbf/ft².

P5.66 A torpedo 8 m below the surface in 20°C seawater cavitates at a speed of 21 m/s when atmospheric pressure is 101 kPa. If Reynolds number and Froude number effects are negligible, at what speed will it cavitate when running at a depth of 20 m? At what depth should it be to avoid cavitation at 30 m/s?

P5.67 A student needs to measure the drag on a prototype of characteristic dimension d_p moving at velocity U_p in air at standard atmospheric conditions. He constructs a model of characteristic dimension d_m, such that the ratio d_p/d_m is some factor f. He then measures the drag on the model at dynamically similar conditions (also with air at standard atmospheric conditions). The student claims that the drag force on the prototype will be identical to that measured on the model. Is this claim correct? Explain.

P5.68 Consider flow over a very small object in a viscous fluid. Analysis of the equations of motion shows that the inertial terms are much smaller than the viscous and pressure terms. It turns out, then, that fluid density drops out of the equations of motion. Such flows are called *creeping flows*. The only important parameters in the problem are the velocity of motion U, the viscosity of the fluid μ, and the length scale of the body. For three-dimensional bodies, like spheres, creeping flow analysis yields very good results. It is uncertain, however, if such analysis can be applied to two-dimensional bodies such as a circular cylinder, since even though the diameter may be very small, the length of the cylinder is infinite for a two-dimensional flow. Let us see if dimensional analysis can help. (*a*) Using the Buckingham Pi Theorem, generate an expression for the two-dimensional drag $D_{2\text{-}D}$ as a function of the other parameters in the problem. Use cylinder diameter *d* as the appropriate length scale. Be careful—the two-dimensional drag has dimensions of force per unit length rather than simply force. (*b*) Is your result physically plausible? If not, explain why not. (*c*) It turns out that fluid density ρ *cannot* be neglected in analysis of creeping flow over two-dimensional bodies. Repeat the dimensional analysis, this time with ρ included as a parameter. Find the nondimensional relationship between the parameters in this problem.

P5.69 A simple flow measurement device for streams and channels is a notch, of angle α, cut into the side of a dam, as shown in Fig. P5.69. The volume flow Q depends only on α, the acceleration of gravity g, and the height δ of the upstream water surface above the notch vertex. Tests of a model notch, of angle $\alpha = 55°$, yield the following flow rate data:

δ, cm	10	20	30	40
Q, m³/h	8	47	126	263

(*a*) Find a dimensionless correlation for the data (*b*) Use the model data to predict the flow rate of a prototype notch, also of angle $\alpha = 55°$, when the upstream height δ is 3.2 m.

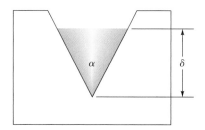

P5.69

P5.70 A diamond-shaped body, of characteristic length 9 in, has the following measured drag forces when placed in a wind tunnel at sea-level standard conditions:

V, ft/s	30	38	48	56	61
F, 1bf	1.25	1.95	3.02	4.05	4.81

Use these data to predict the drag force of a similar 15-in diamond placed at similar orientation in 20°C water flowing at 2.2 m/s.

P5.71 The pressure drop in a venturi meter (Fig. P3.165) varies only with the fluid density, pipe approach velocity, and diameter ratio of the meter. A model venturi meter tested in water at 20°C shows a 5-kPa drop when the approach velocity is 4 m/s. A geometrically similar prototype meter is used to measure gasoline at 20°C and a flow rate of 9 m³/min. If the prototype pressure gage is most accurate at 15 kPa, what should the upstream pipe diameter be?

P5.72 A one-fifteenth-scale model of a parachute has a drag of 450 lbf when tested at 20 ft/s in a water tunnel. If Reynolds number effects are negligible, estimate the terminal fall velocity at 5000-ft standard altitude of a parachutist using the prototype if chute and chutist together weigh 160 lbf. Neglect the drag coefficient of the parachutist.

P5.73 The power P generated by a certain windmill design depends on its diameter D, the air density ρ, the wind velocity V, the rotation rate Ω, and the number of blades n. (*a*) Write this relationship in dimensionless form. A model windmill, of diameter 50 cm, develops 2.7 kW at sea level when $V = 40$ m/s and when rotating at 4800 r/min. (*b*) What power will be developed by a geometrically and dynamically similar prototype, of diameter 5 m, in winds of 12 m/s at 2000 m standard altitude? (*c*) What is the appropriate rotation rate of the prototype?

P5.74 A one-tenth-scale model of a supersonic wing tested at 700 m/s in air at 20°C and 1 atm shows a pitching moment of 0.25 kN · m. If Reynolds number effects are negligible, what will the pitching moment of the prototype wing be if it is flying at the same Mach number at 8-km standard altitude?

P5.75 According to the web site *USGS Daily Water Data for the Nation*, the mean flow rate in the New River near Hinton, WV, is 10,100 ft³/s. If the hydraulic model in Fig. 5.9 is to match this condition with Froude number scaling, what is the proper model flow rate?

*P5.76 A 2-ft-long model of a ship is tested in a freshwater tow tank. The measured drag may be split into "friction" drag (Reynolds scaling) and "wave" drag (Froude scaling). The model data are as follows:

Tow speed, ft/s	0.8	1.6	2.4	3.2	4.0	4.8
Friction drag, lbf	0.016	0.057	0.122	0.208	0.315	0.441
Wave drag, lbf	0.002	0.021	0.083	0.253	0.509	0.697

The prototype ship is 150 ft long. Estimate its total drag when cruising at 15 kn in seawater at 20°C.

P5.77 A dam spillway is to be tested by using Froude scaling with a one-thirtieth-scale model. The model flow has an average velocity of 0.6 m/s and a volume flow of 0.05 m³/s. What will the velocity and flow of the prototype be? If the measured force on a certain part of the model is 1.5 N, what will the corresponding force on the prototype be?

P5.78   A prototype spillway has a characteristic velocity of 3 m/s and a characteristic length of 10 m. A small model is constructed by using Froude scaling. What is the minimum scale ratio of the model that will ensure that its minimum Weber number is 100? Both flows use water at 20°C.

P5.79   An East Coast estuary has a tidal period of 12.42 h (the semidiurnal lunar tide) and tidal currents of approximately 80 cm/s. If a one-five-hundredth-scale model is constructed with tides driven by a pump and storage apparatus, what should the period of the model tides be and what model current speeds are expected?

P5.80   A prototype ship is 35 m long and designed to cruise at 11 m/s (about 21 kn). Its drag is to be simulated by a 1-m-long model pulled in a tow tank. For Froude scaling find (a) the tow speed, (b) the ratio of prototype to model drag, and (c) the ratio of prototype to model power.

P5.81   An airplane, of overall length 55 ft, is designed to fly at 680 m/s at 8000-m standard altitude. A one-thirtieth-scale model is to be tested in a pressurized helium wind tunnel at 20°C. What is the appropriate tunnel pressure in atm? Even at this (high) pressure, exact dynamic similarity is not achieved. Why?

P5.82   A prototype ship is 400 ft long and has a wetted area of 30,000 ft^2. A one-eightieth-scale model is tested in a tow tank according to Froude scaling at speeds of 1.3, 2.0, and 2.7 kn (1 kn = 1.689 ft/s). The measured friction drag of the model at these speeds is 0.11, 0.24, and 0.41 lbf, respectively. What are the three prototype speeds? What is the estimated prototype friction drag at these speeds if we correct for Reynolds number discrepancy by extrapolation?

P5.83   A one-fortieth-scale model of a ship's propeller is tested in a tow tank at 1200 r/min and exhibits a power output of 1.4 ft · lbf/s. According to Froude scaling laws, what should the revolutions per minute and horsepower output of the prototype propeller be under dynamically similar conditions?

P5.84   A prototype ocean platform piling is expected to encounter currents of 150 cm/s and waves of 12-s period and 3-m height. If a one-fifteenth-scale model is tested in a wave channel, what current speed, wave period, and wave height should be encountered by the model?

*P5.85   As shown in Example 5.3, pump performance data can be nondimensionalized. Problem P5.62 gave typical dimensionless data for centrifugal pump "head," $H = \Delta p/\rho g$, as follows:

$$\frac{gH}{n^2 D^2} \approx 6.0 - 120\left(\frac{Q}{nD^3}\right)^2$$

where Q is the volume flow rate, n the rotation rate in r/s, and D the impeller diameter. This type of correlation allows one to compute H when (ρ, Q, D) are known. (a) Show how to rearrange these pi groups so that one can *size* the pump, that is, compute D directly when (Q, H, n) are known. (b) Make a crude but effective plot of your new function. (c) Apply part (b) to the following example: Find D when $H = 37$ m, $Q = 0.14$ m^3/s, and $n = 35$ r/s. Find the pump diameter for this condition.

P5.86   Solve Prob. P5.49 for glycerin at 20°C, using the modified sphere-drag plot of Fig. 5.11.

P5.87   In Prob. P5.61 it would be difficult to solve for Ω because it appears in all three of the dimensionless pump coefficients. Suppose that, in Prob. 5.61, Ω is unknown but $D = 12$ cm and $Q = 25$ m^3/h. The fluid is gasoline at 20°C. Rescale the coefficients, using the data of Prob. P5.61, to make a plot of dimensionless power versus dimensionless rotation speed. Enter this plot to find the maximum rotation speed Ω for which the power will not exceed 300 W.

P5.88   Modify Prob. P5.61 as follows: Let $\Omega = 32$ r/s and $Q = 24$ m^3/h for a geometrically similar pump. What is the maximum diameter if the power is not to exceed 340 W? Solve this problem by rescaling the data of Fig. P5.61 to make a plot of dimensionless power versus dimensionless diameter. Enter this plot directly to find the desired diameter.

P5.89   Knowing that Δp is proportional to L, rescale the data of Example 5.10 to plot dimensionless Δp versus dimensionless *diameter*. Use this plot to find the diameter required in the first row of data in Example 5.10 if the pressure drop is increased to 10 kPa for the same flow rate, length, and fluid.

P5.90   Knowing that Δp is proportional to L, rescale the data of Example 5.10 to plot dimensionless Δp versus dimensionless *viscosity*. Use this plot to find the viscosity required in the first row of data in Example 5.10 if the pressure drop is increased to 10 kPa for the same flow rate, length, and density.

*P5.91   The traditional "Moody-type" pipe friction correlation in Chap. 6 is of the form

$$f = \frac{2\Delta p D}{\rho V^2 L} = \text{fcn}\left(\frac{\rho V D}{\mu}, \frac{\varepsilon}{D}\right)$$

where D is the pipe diameter, L the pipe length, and ε the wall roughness. Note that pipe average velocity V is used on both sides. This form is meant to find Δp when V is known. (a) Suppose that Δp is known, and we wish to find V. Rearrange the above function so that V is isolated on the left-hand side. Use the following data, for $\varepsilon/D = 0.005$, to make a plot of your new function, with your velocity parameter as the ordinate of the plot.

f	0.0356	0.0316	0.0308	0.0305	0.0304
$\rho VD/\mu$	15,000	75,000	250,000	900,000	3,330,000

(*b*) Use your plot to determine V, in m/s, for the following pipe flow: $D = 5$ cm, $\varepsilon = 0.025$ cm, $L = 10$ m, for water flow at 20°C and 1 atm. The pressure drop Δp is 110 kPa.

Word Problems

W5.1 In 98 percent of data analysis cases, the "reducing factor" j, which lowers the number n of dimensional variables to $n - j$ dimensionless groups, exactly equals the number of relevant dimensions (M, L, T, Θ). In one case (Example 5.5) this was not so. Explain in words why this situation happens.

W5.2 Consider the following equation: 1 dollar bill ≈ 6 in. Is this relation dimensionally inconsistent? Does it satisfy the PDH? Why?

W5.3 In making a dimensional analysis, what rules do you follow for choosing your scaling variables?

W5.4 In an earlier edition, the writer asked the following question about Fig. 5.1: "Which of the three graphs is a more effective presentation?" Why was this a dumb question?

W5.5 This chapter discusses the difficulty of scaling Mach and Reynolds numbers together (an airplane) and Froude and Reynolds numbers together (a ship). Give an example of a flow that would combine Mach and Froude numbers. Would there be scaling problems for common fluids?

W5.6 What is different about a very *small* model of a weir or dam (Fig. P5.32) that would make the test results difficult to relate to the prototype?

W5.7 What else are you studying this term? Give an example of a popular equation or formula from another course (thermodynamics, strength of materials, or the like) that does not satisfy the principle of dimensional homogeneity. Explain what is wrong and whether it can be modified to be homogeneous.

W5.8 Some colleges (such as Colorado State University) have environmental wind tunnels that can be used to study phenomena like wind flow over city buildings. What details of scaling might be important in such studies?

W5.9 If the model scale ratio is $\alpha = L_m/L_p$, as in Eq. (5.31), and the Weber number is important, how must the model and prototype surface tension be related to α for dynamic similarity?

W5.10 For a typical incompressible velocity potential analysis in Chap. 8 we solve $\nabla^2 \phi = 0$, subject to known values of $\partial\phi/\partial n$ on the boundaries. What dimensionless parameters govern this type of motion?

Fundamentals of Engineering Exam Problems

FE5.1 Given the parameters (U, L, g, ρ, μ) that affect a certain liquid flow problem, the ratio $V^2/(Lg)$ is usually known as the
(*a*) velocity head, (*b*) Bernoulli head, (*c*) Froude number, (*d*) kinetic energy, (*e*) impact energy

FE5.2 A ship 150 m long, designed to cruise at 18 kn, is to be tested in a tow tank with a model 3 m long. The appropriate tow velocity is
(*a*) 0.19 m/s, (*b*) 0.35 m/s, (*c*) 1.31 m/s, (*d*) 2.55 m/s, (*e*) 8.35 m/s

FE5.3 A ship 150 m long, designed to cruise at 18 kn, is to be tested in a tow tank with a model 3 m long. If the model wave drag is 2.2 N, the estimated full-size ship wave drag is
(*a*) 5500 N, (*b*) 8700 N, (*c*) 38,900 N, (*d*) 61,800 N, (*e*) 275,000 N

FE5.4 A tidal estuary is dominated by the semidiurnal lunar tide, with a period of 12.42 h. If a 1:500 model of the estuary is tested, what should be the model tidal period?
(*a*) 4.0 s, (*b*) 1.5 min, (*c*) 17 min, (*d*) 33 min, (*e*) 64 min

FE5.5 A football, meant to be thrown at 60 mi/h in sea-level air ($\rho = 1.22$ kg/m³, $\mu = 1.78$ E-5 N · s/m²), is to be tested using a one-quarter scale model in a water tunnel ($\rho = 998$ kg/m³, $\mu = 0.0010$ N · s/m²). For dynamic similarity, what is the proper model water velocity?
(*a*) 7.5 mi/h, (*b*) 15.0 mi/h, (*c*) 15.6 mi/h, (*d*) 16.5 mi/h, (*e*) 30 mi/h

FE5.6 A football, meant to be thrown at 60 mi/h in sea-level air ($\rho = 1.22$ kg/m³, $\mu = 1.78$ E-5 N · m²), is to be tested using a one-quarter scale model in a water tunnel ($\rho = 998$ kg/m³, $\mu = 0.0010$ N · s/m²). For dynamic similarity, what is the ratio of prototype force to model force?
(*a*) 3.86:1, (*b*) 16:1, (*c*) 32:1, (*d*) 56:1, (*e*) 64:1

FE5.7 Consider liquid flow of density ρ, viscosity μ, and velocity U over a very small model spillway of length scale L, such that the liquid surface tension coefficient Y is important. The quantity $\rho U^2 L/Y$ in this case is important and is called the
(a) capillary rise, (b) Froude number, (c) Prandtl number, (d) Weber number, (e) Bond number

FE5.8 If a stream flowing at velocity U past a body of length L causes a force F on the body that depends only on U, L, and fluid viscosity μ, then F must be proportional to
(a) $\rho UL/\mu$, (b) $\rho U^2 L^2$, (c) $\mu U/L$, (d) μUL, (e) UL/μ

FE5.9 In supersonic wind tunnel testing, if different gases are used, dynamic similarity requires that the model and prototype have the same Mach number and the same
(a) Euler number, (b) speed of sound, (c) stagnation enthalpy, (d) Froude number, (e) specific-heat ratio

FE5.10 The Reynolds number for a 1-ft-diameter sphere moving at 2.3 mi/h through seawater (specific gravity 1.027, viscosity 1.07 E-3 N · s/m^2) is approximately
(a) 300, (b) 3000, (c) 30,000, (d) 300,000, (e) 3,000,000

Comprehensive Problems

C5.1 Estimating pipe wall friction is one of the most common tasks in fluids engineering. For long circular rough pipes in turbulent flow, wall shear τ_w is a function of density ρ, viscosity μ, average velocity V, pipe diameter d, and wall roughness height ϵ. Thus, functionally, we can write $\tau_w = \text{fcn}(\rho, \mu, V, d, \epsilon)$. (a) Using dimensional analysis, rewrite this function in dimensionless form. (b) A certain pipe has $d = 5$ cm and $\epsilon = 0.25$ mm. For flow of water at 20°C, measurements show the following values of wall shear stress:

Q, gal/min	1.5	3.0	6.0	9.0	12.0	14.0
τ_w, Pa	0.05	0.18	0.37	0.64	0.86	1.25

Plot these data using the dimensionless form obtained in part (a) and suggest a curve-fit formula. Does your plot reveal the entire functional relation obtained in part (a)?

C5.2 When the fluid exiting a nozzle, as in Fig. P3.49, is a gas, instead of water, compressibility may be important, especially if upstream pressure p_1 is large and exit diameter d_2 is small. In this case, the difference $p_1 - p_2$ is no longer controlling, and the gas mass flow $\dot{m}$ reaches a maximum value that depends on p_1 and d_2 and also on the absolute upstream temperature T_1 and the gas constant R. Thus, functionally, $\dot{m} = \text{fcn}(p_1, d_2, T_1, R)$. (a) Using dimensional analysis, rewrite this function in dimensionless form. (b) A certain pipe has $d_2 = 1$ mm. For flow of air, measurements show the following values of mass flow through the nozzle:

T_1, K	300	300	300	500	800
p_1, kPa	200	250	300	300	300
$\dot{m}$, kg/s	0.037	0.046	0.055	0.043	0.034

Plot these data in the dimensionless form obtained in part (a). Does your plot reveal the entire functional relation obtained in part (a)?

C5.3 Reconsider the fully developed draining vertical oil film problem (see Fig. P4.80) as an exercise in dimensional analysis. Let the vertical velocity be a function only of distance from the plate, fluid properties, gravity, and film thickness. That is, $w = \text{fcn}(x, \rho, \mu, g, \delta)$. (a) Use the pi theorem to rewrite this function in terms of dimensionless parameters. (b) Verify that the exact solution from Prob. P4.80 is consistent with your result in part (a).

C5.4 The Taco Inc. model 4013 centrifugal pump has an impeller of diameter $D = 12.95$ in. When pumping 20°C water at $\Omega = 1160$ r/min, the measured flow rate Q and pressure rise Δp are given by the manufacturer as follows:

Q, gal/min	200	300	400	500	600	700
Δp, lb/in^2	36	35	34	32	29	23

(a) Assuming that $\Delta p = \text{fcn}(\rho, Q, D, \Omega)$, use the pi theorem to rewrite this function in terms of dimensionless parameters and then plot the given data in dimensionless form. (b) It is desired to use the same pump, running at 900 r/min, to pump 20°C gasoline at 400 gal/min. According to your dimensionless correlation, what pressure rise Δp is expected, in lbf/in^2?

C5.5 Does an automobile radio antenna vibrate in resonance due to vortex shedding? Consider an antenna of length L and diameter D. According to beam vibration theory [see Kelly [34], p. 401], the first mode natural frequency of a solid circular cantilever beam is $\omega_n = 3.516[EI/(\rho AL^4)]^{1/2}$, where E is the modulus of elasticity, I is the area moment of inertia, ρ is the beam material density, and A is the beam cross-section area. (a) Show that ω_n is proportional to the antenna radius R. (b) If the antenna is steel, with $L = 60$ cm and $D = 4$ mm, estimate the natural vibration frequency, in Hz. (c) Compare with the shedding frequency if the car moves at 65 mi/h.

Design Projects

D5.1 We are given laboratory data, taken by Prof. Robert Kirchhoff and his students at the University of Massachusetts, for the spin rate of a 2-cup anemometer. The anemometer was made of ping-pong balls (d = 1.5 in) split in half, facing in opposite directions, and glued to thin ($\frac{1}{4}$-in) rods pegged to a center axle. (See Fig. P7.91 for a sketch.) There were four rods, of lengths l = 0.212, 0.322, 0.458, and 0.574 ft. The experimental data, for wind tunnel velocity U and rotation rate Ω, are as follows:

l = 0.212		l = 0.322		l = 0.458		l = 0.574	
U, ft/s	Ω, r/min	U, ft/s	Ω, r/min	U, ft/s	Ω, r/min	U, ft/s	Ω, r/min
18.95	435	18.95	225	20.10	140	23.21	115
22.20	545	23.19	290	26.77	215	27.60	145
25.90	650	29.15	370	31.37	260	32.07	175
29.94	760	32.79	425	36.05	295	36.05	195
38.45	970	38.45	495	39.03	327	39.60	215

Assume that the angular velocity Ω of the device is a function of wind speed U, air density ρ and viscosity μ, rod length l, and cup diameter d. For all data, assume air is at 1 atm and 20°C. Define appropriate pi groups for this problem, and plot the data in this dimensionless manner. Comment on the possible uncertainty of the results.

As a design application, suppose we are to use this anemometer geometry for a large-scale (d = 30 cm) airport wind anemometer. If wind speeds vary up to 25 m/s and we desire an average rotation rate Ω = 120 r/min, what should be the proper rod length? What are possible limitations of your design? Predict the expected Ω (in r/min) of your design as affected by wind speeds from 0 to 25 m/s.

D5.2 By analogy with the cylinder drag data in Fig. 5.3b, spheres also show a strong roughness effect on drag, at least in the Reynolds number range 4 E4 < Re$_D$ < 3 E5, which accounts for the dimpling of golf balls to increase their distance traveled. Some experimental data for roughened spheres [33] are given in Fig. D5.2. The figure

also shows typical golf ball data. We see that some roughened spheres are better than golf balls in some regions. For the present study, let us neglect the ball's *spin*, which causes the very important side-force or *Magnus effect* (see Fig. 8.15) and assume that the ball is hit without spin and follows the equations of motion for plane motion (x, z):

$$m\ddot{x} = -F \cos\theta \qquad m\ddot{z} = -F \sin\theta - W$$

where
$$F = C_D \frac{\rho}{2} \frac{\pi}{4} D^2 (\dot{x}^2 + \dot{z}^2) \qquad \theta = \tan^{-1} \frac{\dot{z}}{\dot{x}}$$

The ball has a particular C_D(Re$_D$) curve from Fig. D5.2 and is struck with an initial velocity V_0 and angle θ_0. Take the ball's average mass to be 46 g and its diameter to be 4.3 cm. Assuming sea-level air and a modest but finite range of initial conditions, integrate the equations of motion to compare the trajectory of "roughened spheres" to actual golf ball calculations. Can the rough sphere outdrive a normal golf ball for any conditions? What roughness-effect differences occur between a low-impact duffer and, say, Tiger Woods?

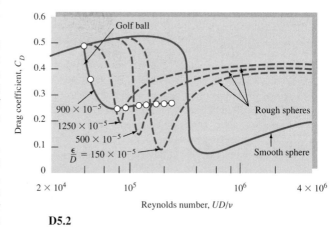

D5.2

References

1. E. Buckingham, "On Physically Similar Systems: Illustrations of the Use of Dimensional Equations," *Phys. Rev.,* vol. 4, no. 4, 1914, pp. 345–376.

2. J. D. Anderson, *Computational Fluid Dynamics: The Basics with Applications,* McGraw-Hill, New York, 1995.

3. P. W. Bridgman, *Dimensional Analysis,* Yale University Press, New Haven, CT, 1922, rev. ed., 1931.

4. H. L. Langhaar, *Dimensional Analysis and the Theory of Models,* Wiley, New York, 1951.

5. E. C. Ipsen, *Units, Dimensions, and Dimensionless Numbers,* McGraw-Hill, New York, 1960.

6. H. G. Hornung, *Dimensional Analysis: Examples of the Use of Symmetry,* Dover, New York, 2006.

7. E. S. Taylor, *Dimensional Analysis for Engineers,* Clarendon Press, Oxford, England, 1974.

8. G. I. Barenblatt, *Dimensional Analysis,* Gordon and Breach, New York, 1987.

9. L. I. Sedov and A. G. Volkovets, *Similarity and Dimensional Methods in Mechanics,* CRC Press, Boca Raton, FL, 1993.

10. T. Szirtes and P. Rozsa, *Applied Dimensional Analysis and Modeling,* McGraw-Hill, New York, 1997.

11. R. Esnault-Pelterie, *Dimensional Analysis and Metrology,* F. Rouge, Lausanne, Switzerland, 1950.

12. R. Kurth, *Dimensional Analysis and Group Theory in Astrophysics,* Pergamon, New York, 1972.

13. F. J. de-Jong, *Dimensional Analysis for Economists,* North Holland, Amsterdam, 1967.

14. R. Nakon, *Chemical Problem Solving Using Dimensional Analysis,* Prentice-Hall, Upper Saddle River, NJ, 1990.

15. D. R. Maidment (ed.), *Hydrologic and Hydraulic Modeling Support: With Geographic Information Systems,* Environmental Systems Research Institute, Redlands, CA, 2000.

16. A. M. Curren and L. D. Munday, *Dimensional Analysis for Meds,* 3d ed., Delmar Thomson Learning, Albany, NY, 2005.

17. G. P. Craig, *Clinical Calculations Made Easy: Solving Problems Using Dimensional Analysis,* 3d ed. Lippincott Williams and Wilkins, Baltimore, MD, 2004.

18. M. Zlokarnik, *Dimensional Analysis and Scale-Up in Chemical Engineering,* Springer-Verlag, New York, 1991.

19. W. G. Jacoby, *Data Theory and Dimensional Analysis,* Sage, Newbury Park, CA, 1991.

20. B. Schepartz, *Dimensional Analysis in the Biomedical Sciences,* Thomas, Springfield, IL, 1980.

21. A. J. Smith, *Dosage and Solution Calculations: The Dimensional Analysis Way,* Mosby, St. Louis, MO, 1989.

22. J. B. Bassingthwaighte et al., *Fractal Physiology,* Oxford Univ. Press, New York, 1994.

23. K. J. Niklas, *Plant Allometry: The Scaling of Form and Process,* Univ. of Chicago Press, Chicago, 1994.

24. *"Flow of Fluids through Valves, Fittings, and Pipes,"* Crane Valve Group, Long Beach, CA, 1957 (now updated as a CD-ROM; see <http://www.cranevalves.com>).

25. A. Roshko, "On the Development of Turbulent Wakes from Vortex Streets," *NACA Rep.* 1191, 1954.

26. G. W. Jones, Jr., "Unsteady Lift Forces Generated by Vortex Shedding about a Large, Stationary, Oscillating Cylinder at High Reynolds Numbers," *ASME Symp. Unsteady Flow,* 1968.

27. O. M. Griffin and S. E. Ramberg, "The Vortex Street Wakes of Vibrating Cylinders," *J. Fluid Mech.,* vol. 66, pt. 3, 1974, pp. 553–576.

28. *Encyclopedia of Science and Technology,* 9th ed., McGraw-Hill, New York, 2002.

29. H. A. Becker, *Dimensionless Parameters,* Wiley, New York, 1976.

30. V. P. Singh et al. (eds.), *Hydraulic Modeling,* Water Resources Publications LLC, Highlands Ranch, CO, 1999.

31. J. J. Sharp, *Hydraulic Modeling,* Butterworth, London, 1981.

32. R. Ettema, *Hydraulic Modeling: Concepts and Practice,* American Society of Civil Engineers, Reston, VA, 2000.

33. R. D. Blevins, *Applied Fluid Dynamics Handbook,* van Nostrand Reinhold, New York, 1984.

34. S. G. Kelly, *Fundamentals of Mechanical Vibration,* McGraw-Hill, New York, 1993.

35. B. Chaouat, "Simulations of Channel Flows with Effects of Spanwise Rotation or Wall Injection Using a Reynolds Stress Model," *Journal of Fluids Engineering,* vol. 123, March 2001, pp. 2–10.

36. G. I. Barenblatt, *Scaling,* Cambridge University Press, Cambridge, UK, 2003.

37. L. J. Fingersh, "Unsteady Aerodynamics Experiment," *Journal of Solar Energy Engineering,* vol. 123, Nov. 2001, p. 267.

38. J. B. Barlow, W. H. Rae, and A. Pope, *Low-Speed Wind Tunnel Testing,* Wiley, New York, 1999.

39. A. Pope and K. Goin, *High-Speed Wind Tunnel Testing,* Krieger Publishing, Melbourne, FL, 1978.

40. American Institute of Aeronautics and Astronautics, *Recommended Practice: Wind Tunnel Testing,* 2 vols., Reston, VA, 2003.

Valves and pipe angles on a gasoline tank form. Pipe flows are everywhere, often occurring in groups or networks. They are designed using the principles outlined in this chapter. *(Courtesy of Dr. E. R. Degginger/Color-Pic Inc.)*

Chapter 6
Viscous Flow in Ducts

Motivation. This chapter is completely devoted to an important practical fluids engineering problem: flow in ducts with various velocities, various fluids, and various duct shapes. Piping systems are encountered in almost every engineering design and thus have been studied extensively. There is a small amount of theory plus a large amount of experimentation.

The basic piping problem is this: Given the pipe geometry and its added components (such as fittings, valves, bends, and diffusers) plus the desired flow rate and fluid properties, what pressure drop is needed to drive the flow? Of course, it may be stated in alternative form: Given the pressure drop available from a pump, what flow rate will ensue? The correlations discussed in this chapter are adequate to solve most such piping problems.

6.1 Reynolds Number Regimes

Now that we have derived and studied the basic flow equations in Chap. 4, you would think that we could just whip off myriad beautiful solutions illustrating the full range of fluid behavior, of course expressing all these educational results in dimensionless form, using our new tool from Chap. 5, dimensional analysis.

The fact of the matter is that no general analysis of fluid motion yet exists. There are several dozen known particular solutions, there are many approximate digital computer solutions, and there are a great many experimental data. There is a lot of theory available if we neglect such important effects as viscosity and compressibility (Chap. 8), but there is no general theory and there may never be. The reason is that a profound and vexing change in fluid behavior occurs at moderate Reynolds numbers. The flow ceases being smooth and steady (*laminar*) and becomes fluctuating and agitated (*turbulent*). The changeover is called *transition* to turbulence. In Fig. 5.3*a* we saw that transition on the cylinder and sphere occurred at about $Re = 3 \times 10^5$, where the sharp drop in the drag coefficient appeared. Transition depends on many effects, such as wall roughness (Fig. 5.3*b*) or fluctuations in the inlet stream, but the primary parameter is the Reynolds number. There are a great many data on transition but only a small amount of theory [1 to 3].

Turbulence can be detected from a measurement by a small, sensitive instrument such as a hot-wire anemometer (Fig. 6.29*e*) or a piezoelectric pressure transducer. The

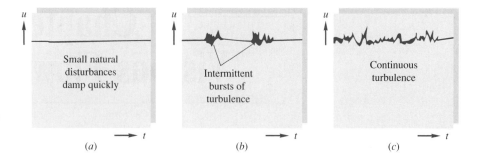

Fig. 6.1 The three regimes of viscous flow: (*a*) laminar flow at low Re; (*b*) transition at intermediate Re; (*c*) turbulent flow at high Re.

flow will appear steady on average but will reveal rapid, random fluctuations if turbulence is present, as sketched in Fig. 6.1. If the flow is laminar, there may be occasional natural disturbances that damp out quickly (Fig. 6.1*a*). If transition is occurring, there will be sharp bursts of intermittent turbulent fluctuation (Fig. 6.1*b*) as the increasing Reynolds number causes a breakdown or instability of laminar motion. At sufficiently large Re, the flow will fluctuate continually (Fig. 6.1*c*) and is termed *fully turbulent*. The fluctuations, typically ranging from 1 to 20 percent of the average velocity, are not strictly periodic but are random and encompass a continuous range, or spectrum, of frequencies. In a typical wind tunnel flow at high Re, the turbulent frequency ranges from 1 to 10,000 Hz, and the wavelength ranges from about 0.01 to 400 cm.

EXAMPLE 6.1

The accepted transition Reynolds number for flow in a circular pipe is $\mathrm{Re}_{d,\mathrm{crit}} \approx 2300$. For flow through a 5-cm-diameter pipe, at what velocity will this occur at 20°C for (*a*) airflow and (*b*) water flow?

Solution

Almost all pipe flow formulas are based on the *average* velocity $V = Q/A$, not centerline or any other point velocity. Thus transition is specified at $\rho V d/\mu \approx 2300$. With d known, we introduce the appropriate fluid properties at 20°C from Tables A.3 and A.4:

(*a*) Air:
$$\frac{\rho V d}{\mu} = \frac{(1.205 \text{ kg/m}^3)V(0.05 \text{ m})}{1.80 \text{ E-5 kg/(m} \cdot \text{s)}} = 2300 \quad \text{or} \quad V \approx 0.7 \frac{\text{m}}{\text{s}}$$

(*b*) Water:
$$\frac{\rho V d}{\mu} = \frac{(998 \text{ kg/m}^3)V(0.05 \text{ m})}{0.001 \text{ kg/(m} \cdot \text{s)}} = 2300 \quad \text{or} \quad V = 0.046 \frac{\text{m}}{\text{s}}$$

These are very low velocities, so most engineering air and water pipe flows are turbulent, not laminar. We might expect laminar duct flow with more viscous fluids such as lubricating oils or glycerin.

In free-surface flows, turbulence can be observed directly. Figure 6.2 shows liquid flow issuing from the open end of a tube. The low-Reynolds-number jet (Fig. 6.2*a*) is smooth and laminar, with the fast center motion and slower wall flow forming different

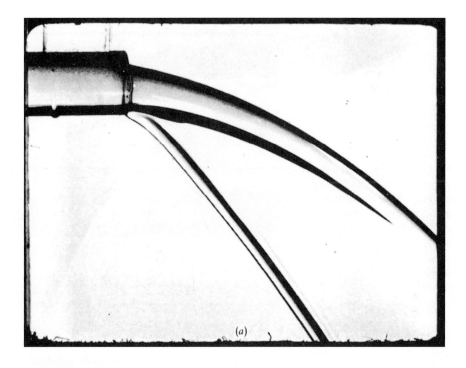

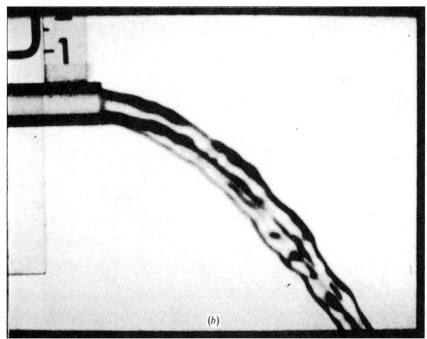

Fig. 6.2 Flow issuing at constant speed from a pipe: (*a*) high-viscosity, low-Reynolds-number, laminar flow; (*b*) low-viscosity, high-Reynolds-number, turbulent flow. *[National Committee for Fluid Mechanics Films, Education Development Center, Inc., © 1972.]*

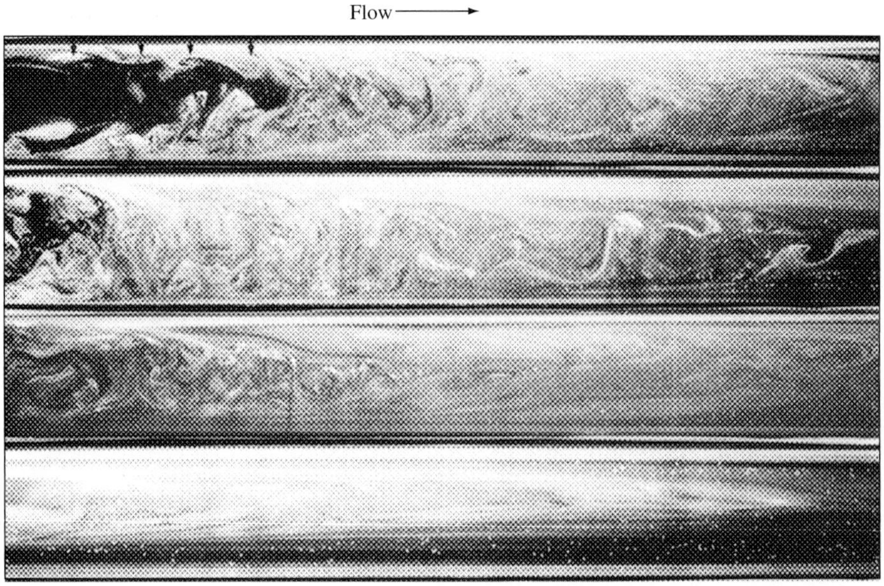

Flow ⟶

Fig. 6.3 Formation of a turbulent puff in pipe flow: (*a*) and (*b*) near the entrance; (*c*) somewhat downstream; (*d*) far downstream. (*Courtesy of Cambridge University Press–P. R. Bandyopadhyay, "Aspects of the Equilibrium Puff in Transitional Pipe Flow," Journal of Fluid Mechanics, vol. 163, 1986, pp. 439–458.*)

trajectories joined by a liquid sheet. The higher-Reynolds-number turbulent flow (Fig. 6.2*b*) is unsteady and irregular but, when averaged over time, is steady and predictable.

How did turbulence form inside the pipe? The laminar parabolic flow profile, which is similar to Eq. (4.146), became unstable and, at $\mathrm{Re}_d \approx 2300$, began to form "slugs" or "puffs" of intense turbulence. A puff has a fast-moving front and a slow-moving rear and may be visualized by experimenting with glass tube flow. Figure 6.3 shows a puff as photographed by Bandyopadhyay [45]. Near the entrance (Fig. 6.3*a* and *b*) there is an irregular laminar–turbulent interface, and vortex roll-up is visible. Further downstream (Fig. 6.3*c*) the puff becomes fully turbulent and very active, with helical motions visible. Far downstream (Fig. 6.3*d*) the puff is cone-shaped and less active, with a fuzzy ill-defined interface, sometimes called the "relaminarization" region.

A complete description of the statistical aspects of turbulence is given in Ref. 1, while theory and data on transition effects are given in Refs. 2 and 3. At this introductory level we merely point out that the primary parameter affecting transition is the Reynolds number. If $\mathrm{Re} = UL/\nu$, where U is the average stream velocity and L is the "width," or transverse thickness, of the shear layer, the following approximate ranges occur:

$0 < \mathrm{Re} < 1$:	highly viscous laminar "creeping" motion
$1 < \mathrm{Re} < 100$:	laminar, strong Reynolds number dependence
$100 < \mathrm{Re} < 10^3$:	laminar, boundary layer theory useful
$10^3 < \mathrm{Re} < 10^4$:	transition to turbulence
$10^4 < \mathrm{Re} < 10^6$:	turbulent, moderate Reynolds number dependence
$10^6 < \mathrm{Re} < \infty$:	turbulent, slight Reynolds number dependence

These representative ranges vary somewhat with flow geometry, surface roughness, and the level of fluctuations in the inlet stream. The great majority of our analyses are concerned with laminar flow or with turbulent flow, and one should not normally design a flow operation in the transition region.

Historical Outline

Since turbulent flow is more prevalent than laminar flow, experimenters have observed turbulence for centuries without being aware of the details. Before 1930 flow instruments were too insensitive to record rapid fluctuations, and workers simply reported mean values of velocity, pressure, force, and so on. But turbulence can change the mean values dramatically, as with the sharp drop in drag coefficient in Fig. 5.3. A German engineer named G. H. L. Hagen first reported in 1839 that there might be *two* regimes of viscous flow. He measured water flow in long brass pipes and deduced a pressure-drop law:

$$\Delta p = (\text{const}) \frac{LQ}{R^4} + \text{entrance effect} \qquad (6.1)$$

This is exactly our laminar flow scaling law from Example 5.4, but Hagen did not realize that the constant was proportional to the fluid viscosity.

The formula broke down as Hagen increased Q beyond a certain limit—that is, past the critical Reynolds number—and he stated in his paper that there must be a second mode of flow characterized by "strong movements of water for which Δp varies as the second power of the discharge. . . ." He admitted that he could not clarify the reasons for the change.

A typical example of Hagen's data is shown in Fig. 6.4. The pressure drop varies linearly with $V = Q/A$ up to about 1.1 ft/s, where there is a sharp change. Above about $V = 2.2$ ft/s the pressure drop is nearly quadratic with V. The actual power

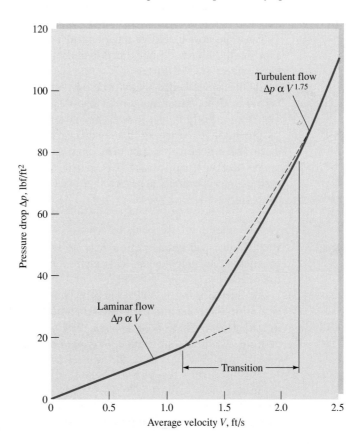

Fig. 6.4 Experimental evidence of transition for water flow in a $\frac{1}{4}$-in smooth pipe 10 ft long.

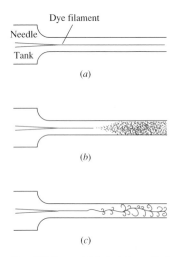

Dye filament

Needle

Tank

(a)

(b)

(c)

Fig. 6.5 Reynolds' sketches of pipe flow transition: (a) low-speed, laminar flow; (b) high-speed, turbulent flow; (c) spark photograph of condition (b). *(From Ref. 4.)*

$\Delta p \propto V^{1.75}$ seems impossible on dimensional grounds but is easily explained when the dimensionless pipe flow data (Fig. 5.10) are displayed.

In 1883 Osborne Reynolds, a British engineering professor, showed that the change depended on the parameter $\rho V d/\mu$, now named in his honor. By introducing a dye streak into a pipe flow, Reynolds could observe transition and turbulence. His sketches [4] of the flow behavior are shown in Fig. 6.5.

If we examine Hagen's data and compute the Reynolds number at $V = 1.1$ ft/s, we obtain $\mathrm{Re}_d = 2100$. The flow became fully turbulent, $V = 2.2$ ft/s, at $\mathrm{Re}_d = 4200$. The accepted design value for pipe flow transition is now taken to be

$$\mathrm{Re}_{d,\mathrm{crit}} \approx 2300 \qquad (6.2)$$

This is accurate for commercial pipes (Fig. 6.13), although with special care in providing a rounded entrance, smooth walls, and a steady inlet stream, $\mathrm{Re}_{d,\mathrm{crit}}$ can be delayed until much higher values. The study of transition in pipe flow, both experimentally and theoretically, continues to be a fascinating topic for researchers, as discussed in a recent review article [55]. *Note:* The value of 2300 is for transition in *pipes.* Other geometries, such as plates, airfoils, cylinders, and spheres, have completely different transition Reynolds numbers.

Transition also occurs in external flows around bodies such as the sphere and cylinder in Fig. 5.3. Ludwig Prandtl, a German engineering professor, showed in 1914 that the thin boundary layer surrounding the body was undergoing transition from laminar to turbulent flow. Thereafter the force coefficient of a body was acknowledged to be a function of the Reynolds number [Eq. (5.2)].

There are now extensive theories and experiments of laminar flow instability that explain why a flow changes to turbulence. Reference 5 is an advanced textbook on this subject.

Laminar flow theory is now well developed, and many solutions are known [2, 3], but no analyses can simulate the fine-scale random fluctuations of turbulent flow.[1] Therefore most turbulent flow theory is semiempirical, based on dimensional analysis and physical reasoning; it is concerned with the mean flow properties only and the mean of the fluctuations, not their rapid variations. The turbulent flow "theory" presented here in Chaps. 6 and 7 is unbelievably crude yet surprisingly effective. We shall attempt a rational approach that places turbulent flow analysis on a firm physical basis.

6.2 Internal versus External Viscous Flows

Both laminar and turbulent flow may be either internal (that is, "bounded" by walls) or external and unbounded. This chapter treats internal flows, and Chap. 7 studies external flows.

An internal flow is constrained by the bounding walls, and the viscous effects will grow and meet and permeate the entire flow. Figure 6.6 shows an internal flow in a long duct. There is an *entrance region* where a nearly inviscid upstream flow converges and enters the tube. Viscous boundary layers grow downstream, retarding the axial flow $u(r, x)$ at the wall and thereby accelerating the center core flow

---

[1]However, direct numerical simulation (DNS) of low-Reynolds-number turbulence is now quite common [32].

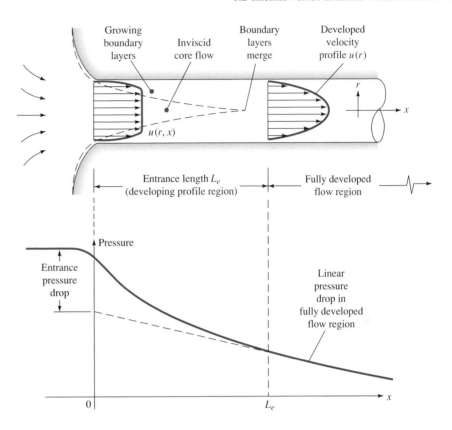

Fig. 6.6 Developing velocity profiles and pressure changes in the entrance of a duct flow.

to maintain the incompressible continuity requirement

$$Q = \int u\, dA = \text{const} \tag{6.3}$$

At a finite distance from the entrance, the boundary layers merge and the inviscid core disappears. The tube flow is then entirely viscous, and the axial velocity adjusts slightly further until at $x = L_e$ it no longer changes with x and is said to be *fully developed*, $u \approx u(r)$ only. Downstream of $x = L_e$ the velocity profile is constant, the wall shear is constant, and the pressure drops linearly with x, for either laminar or turbulent flow. All these details are shown in Fig. 6.6.

Dimensional analysis shows that the Reynolds number is the only parameter affecting entrance length. If

$$L_e = f(d, V, \rho, \mu) \qquad V = \frac{Q}{A}$$

then

$$\frac{L_e}{d} = g\!\left(\frac{\rho V d}{\mu}\right) = g(\text{Re}) \tag{6.4}$$

For laminar flow [2, 3], the accepted correlation is

$$\frac{L_e}{d} \approx 0.06\, \text{Re} \qquad \text{laminar} \tag{6.5}$$

The maximum laminar entrance length, at $\text{Re}_{d,\text{crit}} = 2300$, is $L_e = 138d$, which is the longest development length possible.

In turbulent flow the boundary layers grow faster, and L_e is relatively shorter, according to the approximation for smooth walls:

$$\frac{L_e}{d} \approx 4.4 \, \text{Re}_d^{1/6} \qquad \text{turbulent} \tag{6.6}$$

Some computed turbulent entrance lengths are thus

Re_d	4000	10^4	10^5	10^6	10^7	10^8
L_e/d	18	20	30	44	65	95

Now 44 diameters may seem "long," but typical pipe flow applications involve an L/d value of 1000 or more, in which case the entrance effect may be neglected and a simple analysis made for fully developed flow. This is possible for both laminar and turbulent flows, including rough walls and noncircular cross sections.

EXAMPLE 6.2

A $\frac{1}{2}$-in-diameter water pipe is 60 ft long and delivers water at 5 gal/min at 20°C. What fraction of this pipe is taken up by the entrance region?

Solution

Convert

$$Q = (5 \text{ gal/min}) \frac{0.00223 \text{ ft}^3/\text{s}}{1 \text{ gal/min}} = 0.0111 \text{ ft}^3/\text{s}$$

The average velocity is

$$V = \frac{Q}{A} = \frac{0.0111 \text{ ft}^3/\text{s}}{(\pi/4)[(\frac{1}{2}/12) \text{ ft}]^2} = 8.17 \text{ ft/s}$$

From Table 1.4 read for water $\nu = 1.01 \times 10^{-6} \text{ m}^2/\text{s} = 1.09 \times 10^{-5} \text{ ft}^2/\text{s}$. Then the pipe Reynolds number is

$$\text{Re}_d = \frac{Vd}{\nu} = \frac{(8.17 \text{ ft/s})[(\frac{1}{2}/12) \text{ ft}]}{1.09 \times 10^{-5} \text{ ft}^2/\text{s}} = 31,300$$

This is greater than 4000; hence the flow is fully turbulent, and Eq. (6.6) applies for entrance length:

$$\frac{L_e}{d} \approx 4.4 \, \text{Re}_d^{1/6} = (4.4)(31,300)^{1/6} = 25$$

The actual pipe has $L/d = (60 \text{ ft})/[(\frac{1}{2}/12)\text{ft}] = 1440$. Hence the entrance region takes up the fraction

$$\frac{L_e}{L} = \frac{25}{1440} = 0.017 = 1.7\% \qquad\qquad Ans.$$

This is a very small percentage, so that we can reasonably treat this pipe flow as essentially fully developed.

Shortness can be a virtue in duct flow if one wishes to maintain the inviscid core. For example, a "long" wind tunnel would be ridiculous, since the viscous core would invalidate the purpose of simulating free-flight conditions. A typical laboratory low-speed wind tunnel test section is 1 m in diameter and 5 m long, with $V = 30$ m/s. If we take $\nu_{air} = 1.51 \times 10^{-5}$ m²/s from Table 1.4, then $Re_d = 1.99 \times 10^6$ and, from Eq. (6.6), $L_e/d \approx 49$. The test section has $L/d = 5$, which is much shorter than the development length. At the end of the section the wall boundary layers are only 10 cm thick, leaving 80 cm of inviscid core suitable for model testing.

An external flow has no restraining walls and is free to expand no matter how thick the viscous layers on the immersed body may become. Thus, far from the body the flow is nearly inviscid, and our analytical technique, treated in Chap. 7, is to patch an inviscid-flow solution onto a viscous boundary-layer solution computed for the wall region. There is no external equivalent of fully developed internal flow.

6.3 Head Loss—The Friction Factor

When applying pipe flow formulas to practical problems, it is customary to use a control volume analysis. Consider incompressible steady flow between sections 1 and 2 of the inclined constant-area pipe in Fig. 6.7. The one-dimensional continuity relation, Eq. (3.30), reduces to

$$Q_1 = Q_2 = \text{const} \quad \text{or} \quad V_1 = V_2 = V$$

since the pipe is of constant area. The steady flow energy equation (3.71) becomes

$$\left(\frac{p}{\rho g} + \alpha \frac{V^2}{2g} + z\right)_1 = \left(\frac{p}{\rho g} + \alpha \frac{V^2}{2g} + z\right)_2 + h_f \tag{6.7}$$

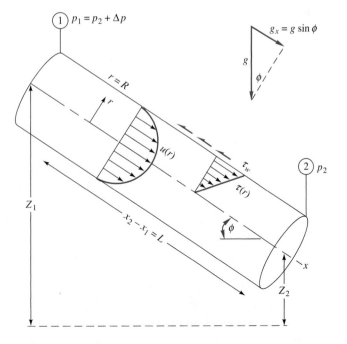

Fig. 6.7 Control volume of steady, fully developed flow between two sections in an inclined pipe.

since there is no pump or turbine between 1 and 2. For fully developed flow, the velocity profile shape is the same at sections 1 and 2. Thus $\alpha_1 = \alpha_2$ and, since $V_1 = V_2$, Eq. (6.7) reduces to head loss versus pressure drop and elevation change:

$$h_f = (z_1 - z_2) + \left(\frac{p_1}{\rho g} - \frac{p_2}{\rho g}\right) = \Delta z + \frac{\Delta p}{\rho g} \tag{6.8}$$

The pipe head loss equals the change in the sum of pressure and gravity head—that is, the change in height of the hydraulic grade line (HGL).

Finally, apply the momentum relation (3.40) to the control volume in Fig. 6.7, accounting for applied x-directed forces due to pressure, gravity, and shear:

$$\sum F_x = \Delta p(\pi R^2) + \rho g(\pi R^2)L \sin\phi - \tau_w(2\pi R)L = \dot{m}(V_2 - V_1) = 0 \quad (6.9a)$$

Rearrange this and we find that the head loss is also related to wall shear stress:

$$\Delta z + \frac{\Delta p}{\rho g} = h_f = \frac{2\tau_w}{\rho g}\frac{L}{R} = \frac{4\tau_w}{\rho g}\frac{L}{d} \tag{6.9b}$$

where we have substituted $\Delta z = L \sin\phi$ from the geometry of Fig. 6.7. Note that, regardless of whether the pipe is horizontal or tilted, the head loss is proportional to the wall shear stress.

How should we correlate the head loss for pipe flow problems? The answer was given a century and a half ago by Julius Weisbach, a German professor who in 1850 published the first modern textbook on hydrodynamics. Equation (6.9b) shows that h_f is proportional to (L/d), and data such as Hagen's in Fig. 6.6 show that, for turbulent flow, h_f is approximately proportional to V^2. The proposed correlation, still as effective today as in 1850, is

$$h_f = f\frac{L}{d}\frac{V^2}{2g} \quad \text{where} \quad f = \text{fcn}\left(\text{Re}_d, \frac{\varepsilon}{d}, \text{duct shape}\right) \tag{6.10}$$

The dimensionless parameter f is called the *Darcy friction factor,* after Henry Darcy (1803–1858), a French engineer whose pipe flow experiments in 1857 first established the effect of roughness on pipe resistance. The quantity ε is the wall roughness height, which is important in turbulent (but not laminar) pipe flow. We added the "duct shape" effect in Eq. (6.10) to remind us that square and triangular and other noncircular ducts have a somewhat different friction factor than a circular pipe. Actual data and theory for friction factors will be discussed in the sections that follow.

By equating Eqs. (6.9) and (6.10) we find an alternative form for friction factor:

$$f = \frac{8\tau_w}{\rho V^2} \tag{6.11}$$

For noncircular ducts, we must interpret τ_w to be an average value around the duct perimeter. For this reason Eq. (6.10) is preferred as a unified definition of the Darcy friction factor.

6.4 Laminar Fully Developed Pipe Flow

Analytical solutions can be readily derived for laminar flows, either circular or non-circular. Consider fully developed *Poiseuille* flow in a round pipe of diameter d, radius R. Complete analytical results were given in Sect. 4.10. Let us review those formulas here:

$$u = u_{max}\left(1 - \frac{r^2}{R^2}\right) \quad \text{where} \quad u_{max} = \left(-\frac{dp}{dx}\right)\frac{R^2}{4\mu} \quad \text{and} \quad \left(-\frac{dp}{dx}\right) = \left(\frac{\Delta p + \rho g \Delta z}{L}\right)$$

$$V = \frac{Q}{A} = \frac{u_{max}}{2} = \left(\frac{\Delta p + \rho g \Delta z}{L}\right)\frac{R^2}{8\mu}$$

$$Q = \int u\, dA = \pi R^2 V = \frac{\pi R^4}{8\mu}\left(\frac{\Delta p + \rho g \Delta z}{L}\right) \tag{6.12}$$

$$\tau_w = \left|\mu \frac{du}{dr}\right|_{r=R} = \frac{4\mu V}{R} = \frac{8\mu V}{d} = \frac{R}{2}\left(\frac{\Delta p + \rho g \Delta z}{L}\right)$$

$$h_f = \frac{32\mu L V}{\rho g d^2} = \frac{128\mu L Q}{\pi \rho g d^4}$$

The paraboloid velocity profile has an average velocity V which is one-half of the maximum velocity. The quantity Δp is the pressure *drop* in a pipe of length L; that is, (dp/dx) is negative. These formulas are valid whenever the pipe Reynolds number, $Re_d = \rho V d/\mu$, is less than about 2300. Note that τ_w is proportional to V (see Fig. 6.6) and is independent of density because the fluid acceleration is zero. Neither of these is true in turbulent flow.

With wall shear stress known, the Poiseuille flow friction factor is easily determined:

$$\boxed{f_{lam} = \frac{8\tau_{w,lam}}{\rho V^2} = \frac{8(8\mu V/d)}{\rho V^2} = \frac{64}{\rho V d/\mu} = \frac{64}{Re_d}} \tag{6.13}$$

In laminar flow, the pipe friction factor decreases inversely with Reynolds number. This famous formula is effective, but often the algebraic relations of Eqs. (6.12) are more direct for problems.

EXAMPLE 6.3

An oil with $\rho = 900$ kg/m^3 and $\nu = 0.0002$ m^2/s flows upward through an inclined pipe as shown in Fig. E6.3. The pressure and elevation are known at sections 1 and 2, 10 m apart. Assuming steady laminar flow, (*a*) verify that the flow is up, (*b*) compute h_f between 1 and 2, and compute (*c*) Q, (*d*) V, and (*e*) Re_d. Is the flow really laminar?

Solution

Part (a)

For later use, calculate

$$\mu = \rho\nu = (900 \text{ kg/m}^3)(0.0002 \text{ m}^2/\text{s}) = 0.18 \text{ kg/(m} \cdot \text{s)}$$

$$z_2 = \Delta L \sin 40° = (10 \text{ m})(0.643) = 6.43 \text{ m}$$

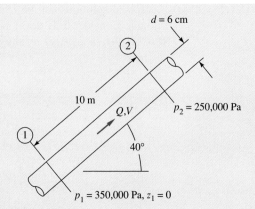

E6.3

The flow goes in the direction of falling HGL; therefore compute the hydraulic grade-line height at each section:

$$HGL_1 = z_1 + \frac{p_1}{\rho g} = 0 + \frac{350,000}{900(9.807)} = 39.65 \text{ m}$$

$$HGL_2 = z_2 + \frac{p_2}{\rho g} = 6.43 + \frac{250,000}{900(9.807)} = 34.75 \text{ m}$$

The HGL is lower at section 2; hence the flow is up from 1 to 2 as assumed.   *Ans. (a)*

Part (b)   The head loss is the change in HGL:

$$h_f = HGL_1 - HGL_2 = 39.65 \text{ m} - 34.75 \text{ m} = 4.9 \text{ m} \qquad \textit{Ans. (b)}$$

Half the length of the pipe is quite a large head loss.

Part (c)   We can compute Q from the various laminar flow formulas, notably Eq. (6.12):

$$Q = \frac{\pi \rho g d^4 h_f}{128 \mu L} = \frac{\pi (900)(9.807)(0.06)^4 (4.9)}{128(0.18)(10)} = 0.0076 \text{ m}^3/\text{s} \qquad \textit{Ans. (c)}$$

Part (d)   Divide Q by the pipe area to get the average velocity:

$$V = \frac{Q}{\pi R^2} = \frac{0.0076}{\pi (0.03)^2} = 2.7 \text{ m/s} \qquad \textit{Ans. (d)}$$

Part (e)   With V known, the Reynolds number is

$$Re_d = \frac{Vd}{\nu} = \frac{2.7(0.06)}{0.0002} = 810 \qquad \textit{Ans. (e)}$$

This is well below the transition value $Re_d = 2300$, so we are fairly certain the flow is laminar.

Notice that by sticking entirely to consistent SI units (meters, seconds, kilograms, newtons) for all variables we avoid the need for any conversion factors in the calculations.

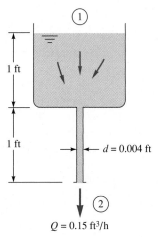

1 ft

1 ft   ← $d = 0.004$ ft

$Q = 0.15$ ft³/h

E6.4

EXAMPLE 6.4

A liquid of specific weight $\rho g = 58$ lb$_f$/ft^3 flows by gravity through a 1-ft tank and a 1-ft capillary tube at a rate of 0.15 ft^3/h, as shown in Fig. E6.4. Sections 1 and 2 are at atmospheric pressure. Neglecting entrance effects, compute the viscosity of the liquid.

Solution

- *System sketch:* Figure E6.4 shows $L = 1$ ft, $d = 0.004$ ft, and $Q = 0.15$ ft^3/h.
- *Assumptions:* Laminar, fully developed, incompressible (Poiseuille) pipe flow. Atmospheric pressure at sections 1 and 2. Negligible velocity at surface, $V_1 \approx 0$.
- *Approach:* Use continuity and energy to find the head loss and thence the viscosity.
- *Property values:* Given $\rho g = 58$ lbf/ft^3, figure out $\rho = 58/32.2 = 1.80$ slug/ft^3 if needed.
- *Solution step 1:* From continuity and the known flow rate, determine V_2:

$$V_2 = \frac{Q}{A_2} = \frac{Q}{(\pi/4)d^2} = \frac{(0.15/3600)\text{ft}^3/\text{s}}{(\pi/4)(0.004 \text{ ft})^2} = 3.32 \text{ ft/s}$$

Write the energy equation between 1 and 2, canceling terms, and find the head loss:

$$\frac{p_1}{\rho g} + \frac{\alpha_1 V_1^2}{2g} + z_1 = \frac{p_2}{\rho g} + \frac{\alpha_2 V_2^2}{2g} + z_2 + h_f$$

or $\qquad h_f = z_1 - z_2 - \dfrac{\alpha_2 V_2^2}{2g} = 2.0 \text{ ft} - 0 \text{ ft} - \dfrac{(2.0)(3.32 \text{ ft/s})^2}{2(32.2 \text{ ft/s}^2)} = 1.66 \text{ ft}$

- *Comment:* We introduced $\alpha_2 = 2.0$ for laminar pipe flow from Eq. (3.72). If we forgot α_2, we would have calculated $h_f = 1.83$ ft, a 10 percent error.
- *Solution step 2:* With head loss known, the viscosity follows from the laminar formula in Eqs. (6.12):

$$h_f = 1.66 \text{ ft} = \frac{32\,\mu L V}{(\rho g)d^2} = \frac{32\mu(1.0 \text{ ft})(3.32 \text{ ft/s})}{(58 \text{ lbf/ft}^3)(0.004 \text{ ft})^2} \quad \text{solve for } \mu = 1.45\text{E-5} \frac{\text{slug}}{\text{ft} - \text{s}} \quad Ans.$$

- *Comments:* We didn't need the value of ρ—the formula contains ρg, but who knew? Note also that L in this formula is the *pipe length* of 1 ft, not the total elevation change.
- *Final check:* Calculate the Reynolds number to see if it is less than 2300 for laminar flow:

$$\text{Re}_d = \frac{\rho V d}{\mu} = \frac{(1.80 \text{ slug/ft}^3)(3.32 \text{ ft/s})(0.004 \text{ ft})}{(1.45\text{E-5 slug/ft} - \text{s})} \approx 1650 \qquad \text{Yes, laminar.}$$

- *Comments:* So we did need ρ after all to calculate Re$_d$. *Unexpected comment:* For this head loss, there is a *second* (turbulent) solution, as we shall see in Example 6.8.

6.5 Turbulence Modeling

Throughout this chapter we assume constant density and viscosity and no thermal interaction, so that only the continuity and momentum equations are to be solved for velocity and pressure

Continuity: $\qquad\qquad\qquad \dfrac{\partial u}{\partial x} + \dfrac{\partial v}{\partial y} + \dfrac{\partial w}{\partial z} = 0$

$$(6.14)$$

Momentum: $\qquad\qquad\qquad \rho \dfrac{d\mathbf{V}}{dt} = -\nabla p + \rho \mathbf{g} + \mu \nabla^2 \mathbf{V}$

subject to no slip at the walls and known inlet and exit conditions. (We shall save our free-surface solutions for Chap. 10.)

We will not work with the differential energy relation, Eq. (4.53), in this chapter, but it is very important, both for heat transfer calculations and for general understanding of duct flow processes. There is work being done by pressure forces to drive the fluid through the duct. Where does this energy go? There is no work done by the wall shear stresses, because the velocity at the wall is zero. The answer is that pressure work is balanced by viscous dissipation in the interior of the flow. The integral of the dissipation function Φ, from Eq. (4.50), over the flow field will equal the pressure work. An example of this fundamental viscous flow energy balance is given in Problem C6.7.

Both laminar and turbulent flows satisfy Eqs. (6.14). For laminar flow, where there are no random fluctuations, we go right to the attack and solve them for a variety of geometries [2, 3], leaving many more, of course, for the problems.

Reynolds' Time-Averaging Concept

For turbulent flow, because of the fluctuations, every velocity and pressure term in Eqs. (6.14) is a rapidly varying random function of time and space. At present our mathematics cannot handle such instantaneous fluctuating variables. No single pair of random functions $\mathbf{V}(x, y, z, t)$ and $p(x, y, z, t)$ is known to be a solution to Eqs. (6.14). Moreover, our attention as engineers is toward the average or *mean* values of velocity, pressure, shear stress, and the like in a high-Reynolds-number (turbulent) flow. This approach led Osborne Reynolds in 1895 to rewrite Eqs. (6.14) in terms of mean or time-averaged turbulent variables.

The time mean $\bar{u}$ of a turbulent function $u(x, y, z, t)$ is defined by

$$\bar{u} = \frac{1}{T} \int_0^T u \, dt \tag{6.15}$$

where T is an averaging period taken to be longer than any significant period of the fluctuations themselves. The mean values of turbulent velocity and pressure are illustrated in Fig. 6.8. For turbulent gas and water flows, an averaging period $T \approx 5$ s is usually quite adequate.

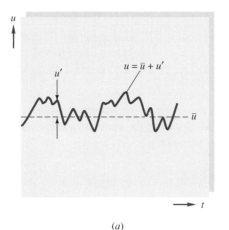

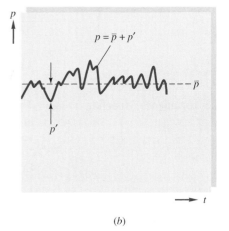

Fig. 6.8 Definition of mean and fluctuating turbulent variables: (*a*) velocity; (*b*) pressure.

(*a*)

(*b*)

The *fluctuation* u' is defined as the deviation of u from its average value

$$u' = u - \bar{u} \tag{6.16}$$

also shown in Fig. 6.8. It follows by definition that a fluctuation has zero mean value:

$$\bar{u'} = \frac{1}{T} \int_0^T (u - \bar{u}) \, dt = \bar{u} - \bar{u} = 0 \tag{6.17}$$

However, the mean square of a fluctuation is not zero and is a measure of the *intensity* of the turbulence:

$$\overline{u'^2} = \frac{1}{T} \int_0^T u'^2 \, dt \neq 0 \tag{6.18}$$

Nor in general are the mean fluctuation products such as $\overline{u'v'}$ and $\overline{u'p'}$ zero in a typical turbulent flow.

Reynolds's idea was to split each property into mean plus fluctuating variables:

$$u = \bar{u} + u' \quad v = \bar{v} + v' \quad w = \bar{w} + w' \quad p = \bar{p} + p' \tag{6.19}$$

Substitute these into Eqs. (6.14), and take the time mean of each equation. The continuity relation reduces to

$$\frac{\partial \bar{u}}{\partial x} + \frac{\partial \bar{v}}{\partial y} + \frac{\partial \bar{w}}{\partial z} = 0 \tag{6.20}$$

which is no different from a laminar continuity relation.

However, each component of the momentum equation (6.14*b*), after time averaging, will contain mean values plus three mean products, or *correlations*, of fluctuating velocities. The most important of these is the momentum relation in the mainstream, or x, direction, which takes the form

$$\rho \frac{d\bar{u}}{dt} = -\frac{\partial \bar{p}}{\partial x} + \rho g_x + \frac{\partial}{\partial x}\left(\mu \frac{\partial \bar{u}}{\partial x} - \rho \overline{u'^2} \right)$$
$$+ \frac{\partial}{\partial y}\left(\mu \frac{\partial \bar{u}}{\partial y} - \rho \overline{u'v'} \right) + \frac{\partial}{\partial z}\left(\mu \frac{\partial \bar{u}}{\partial z} - \rho \overline{u'w'} \right) \tag{6.21}$$

The three correlation terms $-\rho \overline{u'^2}$, $-\rho \overline{u'v'}$, and $-\rho \overline{u'w'}$, are called *turbulent stresses* because they have the same dimensions and occur right alongside the newtonian (laminar) stress terms $\mu(\partial \bar{u}/\partial x)$ and so on. Actually, they are convective acceleration terms (which is why the density appears), not stresses, but they have the mathematical effect of stress and are so termed almost universally in the literature.

The turbulent stresses are unknown a priori and must be related by experiment to geometry and flow conditions, as detailed in Refs. 1 to 3. Fortunately, in duct and boundary layer flow, the stress $-\rho \overline{u'v'}$, associated with direction y normal to the wall is dominant, and we can approximate with excellent accuracy a simpler streamwise momentum equation

$$\rho \frac{d\bar{u}}{dt} \approx -\frac{\partial \bar{p}}{\partial x} + \rho g_x + \frac{\partial \tau}{\partial y} \tag{6.22}$$

where

$$\tau = \mu \frac{\partial \bar{u}}{\partial y} - \rho \overline{u'v'} = \tau_{\text{lam}} + \tau_{\text{turb}} \tag{6.23}$$

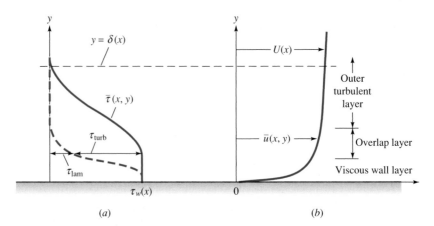

Fig. 6.9 Typical velocity and shear distributions in turbulent flow near a wall: (*a*) shear; (*b*) velocity.

Figure 6.9 shows the distribution of τ_{lam} and τ_{turb} from typical measurements across a turbulent shear layer near a wall. Laminar shear is dominant near the wall (the *wall layer*), and turbulent shear dominates in the *outer layer*. There is an intermediate region, called the *overlap layer,* where both laminar and turbulent shear are important. These three regions are labeled in Fig. 6.9.

In the outer layer τ_{turb} is two or three orders of magnitude greater than τ_{lam}, and vice versa in the wall layer. These experimental facts enable us to use a crude but very effective model for the velocity distribution $\bar{u}(y)$ across a turbulent wall layer.

The Logarithmic Overlap Law

We have seen in Fig. 6.9 that there are three regions in turbulent flow near a wall:

1.   Wall layer: Viscous shear dominates.
2.   Outer layer: Turbulent shear dominates.
3.   Overlap layer: Both types of shear are important.

From now on let us agree to drop the overbar from velocity $\bar{u}$. Let τ_w be the wall shear stress, and let δ and U represent the thickness and velocity at the edge of the outer layer, $y = \delta$.

For the wall layer, Prandtl deduced in 1930 that u must be independent of the shear layer thickness:

$$u = f(\mu, \tau_w, \rho, y) \tag{6.24}$$

By dimensional analysis, this is equivalent to

$$u^+ = \frac{u}{u^*} = F\left(\frac{yu^*}{\nu}\right) \qquad u^* = \left(\frac{\tau_w}{\rho}\right)^{1/2} \tag{6.25}$$

Equation (6.25) is called the *law of the wall,* and the quantity u^* is termed the *friction velocity* because it has dimensions $\{LT^{-1}\}$, although it is not actually a flow velocity.

Subsequently, Kármán in 1933 deduced that u in the outer layer is independent of molecular viscosity, but its deviation from the stream velocity U must depend on the layer thickness δ and the other properties:

$$(U - u)_{outer} = g(\delta, \tau_w, \rho, y) \tag{6.26}$$

Again, by dimensional analysis we rewrite this as

$$\frac{U - u}{u^*} = G\left(\frac{y}{\delta}\right) \tag{6.27}$$

where u^* has the same meaning as in Eq. (6.25). Equation (6.27) is called the *velocity-defect law* for the outer layer.

Both the wall law (6.25) and the defect law (6.27) are found to be accurate for a wide variety of experimental turbulent duct and boundary layer flows [1 to 3]. They are different in form, yet they must overlap smoothly in the intermediate layer. In 1937 C. B. Millikan showed that this can be true only if the overlap layer velocity varies logarithmically with y:

$$\boxed{\frac{u}{u^*} = \frac{1}{\kappa} \ln \frac{yu^*}{\nu} + B \qquad \text{overlap layer}} \tag{6.28}$$

Over the full range of turbulent smooth wall flows, the dimensionless constants κ and B are found to have the approximate values $\kappa \approx 0.41$ and $B \approx 5.0$. Equation (6.28) is called the *logarithmic overlap layer*.

Thus by dimensional reasoning and physical insight we infer that a plot of u versus $\ln y$ in a turbulent shear layer will show a curved wall region, a curved outer region, and a straight-line logarithmic overlap. Figure 6.10 shows that this

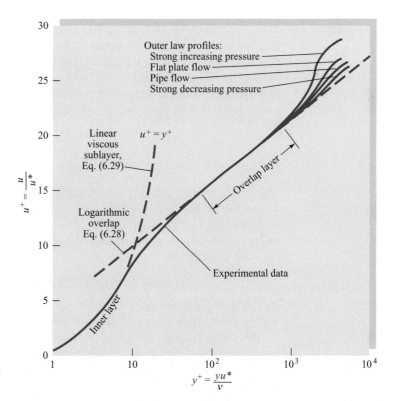

Fig. 6.10 Experimental verification of the inner, outer, and overlap layer laws relating velocity profiles in turbulent wall flow.

is exactly the case. The four outer-law profiles shown all merge smoothly with the logarithmic overlap law but have different magnitudes because they vary in external pressure gradient. The wall law is unique and follows the linear viscous relation

$$u^+ = \frac{u}{u^*} = \frac{yu^*}{\nu} = y^+ \tag{6.29}$$

from the wall to about $y^+ = 5$, thereafter curving over to merge with the logarithmic law at about $y^+ = 30$.

Believe it or not, Fig. 6.10, which is nothing more than a shrewd correlation of velocity profiles, is the basis for most existing "theory" of turbulent shear flows. Notice that we have not solved any equations at all but have merely expressed the streamwise velocity in a neat form.

There is serendipity in Fig. 6.10: The logarithmic law (6.28), instead of just being a short overlapping link, actually approximates nearly the entire velocity profile, except for the outer law when the pressure is increasing strongly downstream (as in a diffuser). The inner wall law typically extends over less than 2 percent of the profile and can be neglected. Thus we can use Eq. (6.28) as an excellent approximation to solve nearly every turbulent flow problem presented in this and the next chapter. Many additional applications are given in Refs. 2 and 3.

Advanced Modeling Concepts

Turbulence modeling is a very active field. Scores of papers have been published to more accurately simulate the turbulent stresses in Eq. (6.21) and their y and z components. This research, now available in advanced texts [1, 13, 19], goes well beyond the present book, which is confined to the use of the logarithmic law (6.28) for pipe and boundary layer problems. For example, L. Prandtl, who invented boundary layer theory in 1904, later proposed an *eddy viscosity* model of the Reynolds stress term in Eq. (6.23):

$$-\rho \overline{u'v'} = \tau_{\text{turb}} \approx \mu_t \frac{du}{dy} \qquad \text{where} \qquad \mu_t \approx \rho l^2 \left| \frac{du}{dy} \right| \tag{6.30}$$

The term μ_t, which is a property of the *flow*, not the fluid, is called the *eddy viscosity* and can be modeled in various ways. The most popular form is Eq. (6.30), where l is called the *mixing length* of the turbulent eddies (analogous to mean free path in molecular theory). Near a solid wall, l is approximately proportional to distance from the wall, and Kármán suggested

$$l \approx \kappa y \qquad \text{where} \qquad \kappa = \text{Kármán's constant} \approx 0.41 \tag{6.31}$$

As a homework assignment, Prob. P6.40, you may show that Eqs. (6.30) and (6.31) lead to the logarithmic law (6.28) near a wall.

Modern turbulence models approximate three-dimensional turbulent flows and employ additional partial differential equations for such quantities as the turbulence kinetic energy, the turbulent dissipation, and the six Reynolds stresses. For details, see Refs. 1, 13, and 19.

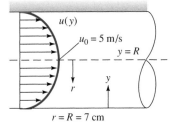

E6.5

$u(y)$

$u_0 = 5$ m/s

$y = R$

y

r

$r = R = 7$ cm

EXAMPLE 6.5

Air at 20°C flows through a 14-cm-diameter tube under fully developed conditions. The centerline velocity is $u_0 = 5$ m/s. Estimate from Fig. 6.10 (a) the friction velocity u^* and (b) the wall shear stress τ_w.

Solution

- *System sketch:* Figure E6.5 shows turbulent pipe flow with $u_0 = 5$ m/s and $R = 7$ cm.
- *Assumptions:* Figure 6.10 shows that the logarithmic law, Eq. (6.28), is accurate all the way to the center of the tube.
- *Approach:* Use Eq. (6.28) to estimate the unknown friction velocity u^*.
- *Property values:* For air at 20°C, $\rho = 1.205$ kg/m³ and $\nu = 1.51\text{E-}5$ m²/s.
- *Solution step:* Insert all the given data into Eq. (6.28) at $y = R$ (the centerline). The only unknown is u^*:

$$\frac{u_0}{u^*} = \frac{1}{\kappa}\ln\left(\frac{Ru^*}{\nu}\right) + B \quad \text{or} \quad \frac{5.0 \text{ m/s}}{u^*} = \frac{1}{0.41}\ln\left[\frac{(0.07 \text{ m})u^*}{1.51\text{E-}5 \text{ m}^2/\text{s}}\right] + 5$$

Although the logarithm makes it awkward, one can iterate this by hand to find u^*. Or one can open EES and type out a single statement of Eq. (6.28):

```
5.0/ustar = (1/0.41)*ln(0.07*ustar/1.51E-5)+5
```

Any nominal guess, e.g., $u^* = 1$, will do. EES immediately returns the correct solution:

$$u^* \approx 0.228 \text{ m/s} \qquad\qquad\qquad Ans. (a)$$

$$\tau_w = \rho u^{*2} = (1.205)(0.228)^2 \approx 0.062 \text{ Pa} \qquad Ans. (b)$$

- *Comments:* The logarithmic law solved everything! This is a powerful technique, using an experimental velocity correlation to approximate general turbulent flows. You may check that the Reynolds number Re_d is about 40,000, definitely turbulent flow.

6.6 Turbulent Pipe Flow

For turbulent pipe flow we need not solve a differential equation but instead proceed with the logarithmic law, as in Example 6.5. Assume that Eq. (6.28) correlates the local mean velocity $u(r)$ all the way across the pipe

$$\frac{u(r)}{u^*} \approx \frac{1}{\kappa}\ln\frac{(R-r)u^*}{\nu} + B \tag{6.32}$$

where we have replaced y by $R - r$. Compute the average velocity from this profile:

$$V = \frac{Q}{A} = \frac{1}{\pi R^2}\int_0^R u^*\left[\frac{1}{\kappa}\ln\frac{(R-r)u^*}{\nu} + B\right]2\pi r \, dr$$

$$= \frac{1}{2}u^*\left(\frac{2}{\kappa}\ln\frac{Ru^*}{\nu} + 2B - \frac{3}{\kappa}\right) \tag{6.33}$$

Introducing $\kappa = 0.41$ and $B = 5.0$, we obtain, numerically,

$$\frac{V}{u^*} \approx 2.44\ln\frac{Ru^*}{\nu} + 1.34 \tag{6.34}$$

This looks only marginally interesting until we realize that $V/u*$ is directly related to the Darcy friction factor:

$$\frac{V}{u*} = \left(\frac{\rho V^2}{\tau_w}\right)^{1/2} = \left(\frac{8}{f}\right)^{1/2} \tag{6.35}$$

Moreover, the argument of the logarithm in (6.34) is equivalent to

$$\frac{Ru*}{\nu} = \frac{\frac{1}{2}Vd}{\nu}\frac{u*}{V} = \frac{1}{2}\text{Re}_d\left(\frac{f}{8}\right)^{1/2} \tag{6.36}$$

Introducing (6.35) and (6.36) into Eq. (6.34), changing to a base-10 logarithm, and rearranging, we obtain

$$\frac{1}{f^{1/2}} \approx 1.99 \log(\text{Re}_d f^{1/2}) - 1.02 \tag{6.37}$$

In other words, by simply computing the mean velocity from the logarithmic law correlation, we obtain a relation between the friction factor and Reynolds number for turbulent pipe flow. Prandtl derived Eq. (6.37) in 1935 and then adjusted the constants slightly to fit friction data better:

$$\frac{1}{f^{1/2}} = 2.0 \log(\text{Re}_d f^{1/2}) - 0.8 \tag{6.38}$$

This is the accepted formula for a smooth-walled pipe. Some numerical values may be listed as follows:

Re_d	4000	10^4	10^5	10^6	10^7	10^8
f	0.0399	0.0309	0.0180	0.0116	0.0081	0.0059

Thus f drops by only a factor of 5 over a 10,000-fold increase in Reynolds number. Equation (6.38) is cumbersome to solve if Re_d is known and f is wanted. There are many alternative approximations in the literature from which f can be computed explicitly from Re_d:

$$f = \begin{cases} 0.316\,\text{Re}_d^{-1/4} & 4000 < \text{Re}_d < 10^5 \quad \text{H. Blasius (1911)} \\ \left(1.8 \log \dfrac{\text{Re}_d}{6.9}\right)^{-2} & \text{Ref. 9} \end{cases} \tag{6.39}$$

Blasius, a student of Prandtl, presented his formula in the first correlation ever made of pipe friction versus Reynolds number. Although his formula has a limited range, it illustrates what was happening in Fig. 6.4 to Hagen's 1839 pressure-drop data. For a horizontal pipe, from Eq. (6.39),

$$h_f = \frac{\Delta p}{\rho g} = f\frac{L}{d}\frac{V^2}{2g} \approx 0.316\left(\frac{\mu}{\rho V d}\right)^{1/4}\frac{L}{d}\frac{V^2}{2g}$$

or

$$\Delta p \approx 0.158\,L\rho^{3/4}\mu^{1/4}d^{-5/4}V^{7/4} \tag{6.40}$$

at low turbulent Reynolds numbers. This explains why Hagen's data for pressure drop begin to increase as the 1.75 power of the velocity, in Fig. 6.4. Note that Δp varies only slightly with viscosity, which is characteristic of turbulent flow. Introducing

$Q = \frac{1}{4}\pi d^2 V$ into Eq. (6.40), we obtain the alternative form

$$\Delta p \approx 0.241 L \rho^{3/4} \mu^{1/4} d^{-4.75} Q^{1.75} \tag{6.41}$$

For a given flow rate Q, the turbulent pressure drop decreases with diameter even more sharply than the laminar formula (6.12). Thus the quickest way to reduce required pumping pressure is to increase the pipe size, although, of course, the larger pipe is more expensive. Doubling the pipe size decreases Δp by a factor of about 27 for a given Q. Compare Eq. (6.40) with Example 5.7 and Fig. 5.10.

The maximum velocity in turbulent pipe flow is given by Eq. (6.32), evaluated at $r = 0$:

$$\frac{u_{max}}{u^*} \approx \frac{1}{\kappa} \ln \frac{R u^*}{\nu} + B \tag{6.42}$$

Combining this with Eq. (6.33), we obtain the formula relating mean velocity to maximum velocity:

$$\frac{V}{u_{max}} \approx (1 + 1.3\sqrt{f})^{-1} \tag{6.43}$$

Some numerical values are

Re_d	4000	10^4	10^5	10^6	10^7	10^8
V/u_{max}	0.794	0.814	0.852	0.877	0.895	0.909

The ratio varies with the Reynolds number and is much larger than the value of 0.5 predicted for all laminar pipe flow in Eq. (6.43). Thus a turbulent velocity profile, as shown in Fig. 6.11b, is very flat in the center and drops off sharply to zero at the wall.

Effect of Rough Walls

It was not known until experiments in 1800 by Coulomb [6] that surface roughness has an effect on friction resistance. It turns out that the effect is negligible for laminar pipe flow, and all the laminar formulas derived in this section are valid for rough

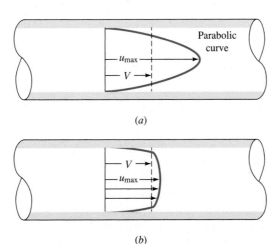

Fig. 6.11 Comparison of laminar and turbulent pipe flow velocity profiles for the same volume flow: (*a*) laminar flow; (*b*) turbulent flow.

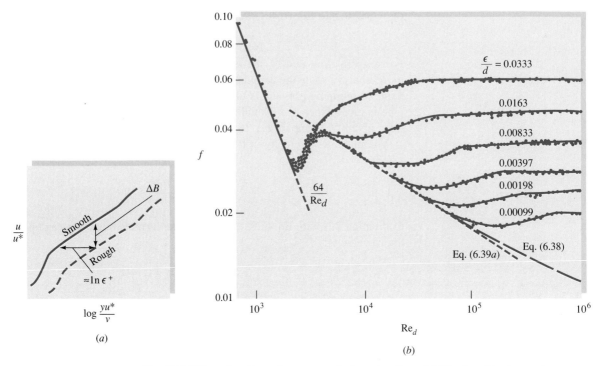

Fig. 6.12 Effect of wall roughness on turbulent pipe flow. (*a*) The logarithmic overlap velocity profile shifts down and to the right; (*b*) experiments with sand-grain roughness by Nikuradse [7] show a systematic increase of the turbulent friction factor with the roughness ratio.

walls also. But turbulent flow is strongly affected by roughness. In Fig. 6.10 the linear viscous sublayer extends out only to $y^+ = yu^*/\nu = 5$. Thus, compared with the diameter, the sublayer thickness y_s is only

$$\frac{y_s}{d} = \frac{5\nu/u^*}{d} = \frac{14.1}{\text{Re}_d f^{1/2}} \tag{6.44}$$

For example, at $\text{Re}_d = 10^5$, $f = 0.0180$, and $y_s/d = 0.001$, a wall roughness of about $0.001d$ will break up the sublayer and profoundly change the wall law in Fig. 6.10.

Measurements of $u(y)$ in turbulent rough-wall flow by Prandtl's student Nikuradse [7] show, as in Fig. 6.12*a*, that a roughness height ϵ will force the logarithm law profile outward on the abscissa by an amount approximately equal to $\ln \epsilon^+$, where $\epsilon^+ = \epsilon u^*/\nu$. The slope of the logarithm law remains the same, $1/\kappa$, but the shift outward causes the constant B to be less by an amount $\Delta B \approx (1/\kappa) \ln \epsilon^+$.

Nikuradse [7] simulated roughness by gluing uniform sand grains onto the inner walls of the pipes. He then measured the pressure drops and flow rates and correlated friction factor versus Reynolds number in Fig. 6.12*b*. We see that laminar friction is unaffected, but turbulent friction, after an *onset* point, increases monotonically with the roughness ratio ϵ/d. For any given ϵ/d, the friction factor becomes constant (*fully* *rough*) at high Reynolds numbers. These points of change are certain values of $\epsilon^+ = \epsilon u^*/\nu$:

$$\frac{\epsilon u^*}{\nu} < 5: \quad \textit{hydraulically smooth} \text{ walls, no effect of roughness on friction}$$

$$5 \leq \frac{\epsilon u^*}{\nu} \leq 70: \quad \textit{transitional} \text{ roughness, moderate Reynolds number effect}$$

$$\frac{\epsilon u^*}{\nu} > 70: \quad \textit{fully rough} \text{ flow, sublayer totally broken up and friction independent of Reynolds number}$$

For fully rough flow, $\epsilon^+ > 70$, the log law downshift ΔB in Fig. 6.12*a* is

$$\Delta B \approx \frac{1}{\kappa} \ln \epsilon^+ - 3.5 \tag{6.45}$$

and the logarithm law modified for roughness becomes

$$u^+ = \frac{1}{\kappa} \ln y^+ + B - \Delta B = \frac{1}{\kappa} \ln \frac{y}{\epsilon} + 8.5 \tag{6.46}$$

The viscosity vanishes, and hence fully rough flow is independent of the Reynolds number. If we integrate Eq. (6.46) to obtain the average velocity in the pipe, we obtain

$$\frac{V}{u^*} = 2.44 \ln \frac{d}{\epsilon} + 3.2$$

or
$$\frac{1}{f^{1/2}} = -2.0 \log \frac{\epsilon/d}{3.7} \quad \textit{fully rough flow} \tag{6.47}$$

There is no Reynolds number effect; hence the head loss varies exactly as the square of the velocity in this case. Some numerical values of friction factor may be listed:

ϵ/d	0.00001	0.0001	0.001	0.01	0.05
f	0.00806	0.0120	0.0196	0.0379	0.0716

The friction factor increases by 9 times as the roughness increases by a factor of 5000. In the transitional roughness region, sand grains behave somewhat differently from commercially rough pipes, so Fig. 6.12*b* has now been replaced by the Moody chart.

The Moody Chart

In 1939 to cover the transitionally rough range, Colebrook [9] combined the smooth wall [Eq. (6.38)] and fully rough [Eq. (6.47)] relations into a clever interpolation formula:

$$\frac{1}{f^{1/2}} = -2.0 \log\left(\frac{\epsilon/d}{3.7} + \frac{2.51}{\mathrm{Re}_d f^{1/2}}\right) \tag{6.48}$$

This is the accepted design formula for turbulent friction. It was plotted in 1944 by Moody [8] into what is now called the *Moody chart* for pipe friction (Fig. 6.13). The

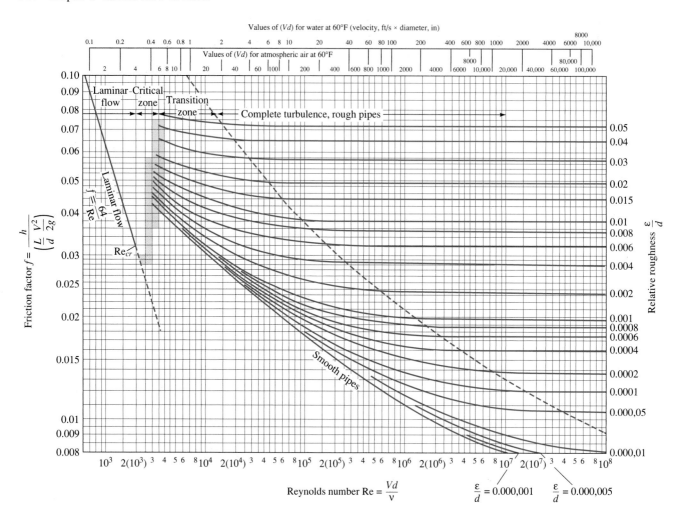

Fig. 6.13 The Moody chart for pipe friction with smooth and rough walls. This chart is identical to Eq. (6.48) for turbulent flow. *(From Ref. 8, by permission of the ASME.)*

Moody chart is probably the most famous and useful figure in fluid mechanics. It is accurate to ± 15 percent for design calculations over the full range shown in Fig. 6.13. It can be used for circular and noncircular (Sec. 6.6) pipe flows and for open-channel flows (Chap. 10). The data can even be adapted as an approximation to boundary layer flows (Chap. 7).

Equation (6.48) is cumbersome to evaluate for f if Re_d is known, although it easily yields to the EES Equation Solver. An alternative explicit formula given by Haaland [33] as

$$\frac{1}{f^{1/2}} \approx -1.8 \log \left[\frac{6.9}{Re_d} + \left(\frac{\epsilon/d}{3.7} \right)^{1.11} \right] \tag{6.49}$$

varies less than 2 percent from Eq. (6.48).

Table 6.1 Recommended
Roughness Values for Commercial
Ducts

Material	Condition	ϵ		Uncertainty, %
		ft	**mm**	
Steel	Sheet metal, new	0.00016	0.05	±60
	Stainless, new	0.000007	0.002	±50
	Commercial, new	0.00015	0.046	±30
	Riveted	0.01	3.0	±70
	Rusted	0.007	2.0	±50
Iron	Cast, new	0.00085	0.26	±50
	Wrought, new	0.00015	0.046	±20
	Galvanized, new	0.0005	0.15	±40
	Asphalted cast	0.0004	0.12	±50
Brass	Drawn, new	0.000007	0.002	±50
Plastic	Drawn tubing	0.000005	0.0015	±60
Glass	—	Smooth	Smooth	
Concrete	Smoothed	0.00013	0.04	±60
	Rough	0.007	2.0	±50
Rubber	Smoothed	0.000033	0.01	±60
Wood	Stave	0.0016	0.5	±40

The shaded area in the Moody chart indicates the range where transition from laminar to turbulent flow occurs. There are no reliable friction factors in this range, $2000 < \text{Re}_{d,} < 4000$. Notice that the roughness curves are nearly horizontal in the fully rough regime to the right of the dashed line.

From tests with commercial pipes, recommended values for average pipe roughness are listed in Table 6.1.

EXAMPLE 6.6[2]

Compute the loss of head and pressure drop in 200 ft of horizontal 6-in-diameter asphalted cast iron pipe carrying water with a mean velocity of 6 ft/s.

Solution

- *System sketch:*   See Fig. 6.7 for a horizontal pipe, with $\Delta z = 0$ and h_f proportional to Δp.
- *Assumptions:*   Turbulent flow, asphalted horizontal cast iron pipe, $d = 0.5$ ft, $L = 200$ ft.
- *Approach:*   Find Re_d and ϵ/d; enter the Moody chart, Fig. 6.13; find f, then h_f and Δp.
- *Property values:*   From Table A.3 for water, converting to BG units, $\rho = 998/515.38 = 1.94$ slug/ft^3, $\mu = 0.001/47.88 = 2.09\text{E-5}$ slug/(ft-s).
- *Solution step 1:*   Calculate Re_d and the roughness ratio. As a crutch, Moody provided water and air values of "Vd" at the top of Fig. 6.13 to find Re_d. No, let's calculate it ourselves:

$$\text{Re}_d = \frac{\rho V d}{\mu} = \frac{(1.94 \text{ slug/ft}^3)(6 \text{ ft/s})(0.5 \text{ ft})}{2.09\text{E-5 slug/(ft} \cdot \text{s)}} \approx 279{,}000 \qquad \text{(turbulent)}$$

---

[2]This example was given by Moody in his 1944 paper [8].

From Table 6.1, for asphalted cast iron, $\epsilon = 0.0004$ ft. Then calculate

$$\epsilon/d = (0.0004 \text{ ft})/(0.5 \text{ ft}) = 0.0008$$

- *Solution step 2:* Find the friction factor from the Moody chart or from Eq. (6.48). If you use the Moody chart, Fig. 6.13, you need practice. Find the line on the right side for $\epsilon/d = 0.0008$ and follow it back to the left until it hits the vertical line for $\text{Re}_d \approx 2.79\text{E}5$. Read, approximately, $f \approx 0.02$ [or compute $f = 0.0198$ from Eq. (6.48), perhaps using EES].
- *Solution step 3:* Calculate h_f from Eq. (6.10) and Δp from Eq. (6.8) for a horizontal pipe:

$$h_f = f\frac{L}{d}\frac{V^2}{2g} = (0.02)\left(\frac{200 \text{ ft}}{0.5 \text{ ft}}\right)\frac{(6 \text{ ft/s})^2}{2(32.2 \text{ ft/s}^2)} \approx 4.5 \text{ ft} \qquad\qquad Ans.$$

$$\Delta p = \rho g h_f = (1.94 \text{ slug/ft}^3)(32.2 \text{ ft/s}^2)(4.5 \text{ ft}) \approx 280 \text{ lbf/ft}^2 \qquad\qquad Ans.$$

- *Comments:* In giving this example, Moody [8] stated that this estimate, even for clean new pipe, can be considered accurate only to about ± 10 percent.

EXAMPLE 6.7

Oil, with $\rho = 900 \text{ kg/m}^3$ and $\nu = 0.00001 \text{ m}^2/\text{s}$, flows at 0.2 m^3/s through 500 m of 200-mm-diameter cast iron pipe. Determine (a) the head loss and (b) the pressure drop if the pipe slopes down at $10°$ in the flow direction.

Solution

First compute the velocity from the known flow rate:

$$V = \frac{Q}{\pi R^2} = \frac{0.2 \text{ m}^3/\text{s}}{\pi(0.1 \text{ m})^2} = 6.4 \text{ m/s}$$

Then the Reynolds number is

$$\text{Re}_d = \frac{Vd}{\nu} = \frac{(6.4 \text{ m/s})(0.2 \text{ m})}{0.00001 \text{ m}^2/\text{s}} = 128,000$$

From Table 6.1, $\epsilon = 0.26$ mm for cast iron pipe. Then

$$\frac{\epsilon}{d} = \frac{0.26 \text{ mm}}{200 \text{ mm}} = 0.0013$$

Enter the Moody chart on the right at $\epsilon/d = 0.0013$ (you will have to interpolate), and move to the left to intersect with $\text{Re} = 128,000$. Read $f \approx 0.0225$ [from Eq. (6.48) for these values we could compute $f = 0.0227$]. Then the head loss is

$$h_f = f\frac{L}{d}\frac{V^2}{2g} = (0.0225)\frac{500 \text{ m}}{0.2 \text{ m}}\frac{(6.4 \text{ m/s})^2}{2(9.81 \text{ m/s}^2)} = 117 \text{ m} \qquad\qquad Ans. (a)$$

From Eq. (6.9) for the inclined pipe,

$$h_f = \frac{\Delta p}{\rho g} + z_1 - z_2 = \frac{\Delta p}{\rho g} + L \sin 10°$$

or $\Delta p = \rho g[h_f - (500 \text{ m}) \sin 10°] = \rho g(117 \text{ m} - 87 \text{ m})$

$$= (900 \text{ kg/m}^3)(9.81 \text{ m/s}^2)(30 \text{ m}) = 265{,}000 \text{ kg/(m} \cdot \text{s}^2) = 265{,}000 \text{ Pa} \quad Ans. \ (b)$$

EXAMPLE 6.8

Repeat Example 6.4 to see whether there is any possible turbulent flow solution for a smooth-walled pipe.

Solution

In Example 6.4 we estimated a head loss $h_f \approx 1.66$ ft, assuming laminar exit flow ($\alpha \approx 2.0$). For this condition the friction factor is

$$f = h_f \frac{d}{L} \frac{2g}{V^2} = (1.66 \text{ ft}) \frac{(0.004 \text{ ft})(2)(32.2 \text{ ft/s}^2)}{(1.0 \text{ ft})(3.32 \text{ ft/s})^2} \approx 0.0388$$

For laminar flow, $\text{Re}_d = 64/f = 64/0.0388 \approx 1650$, as we showed in Example 6.4. However, from the Moody chart (Fig. 6.13), we see that $f = 0.0388$ also corresponds to a *turbulent* smooth-wall condition, at $\text{Re}_d \approx 4500$. If the flow actually were turbulent, we should change our kinetic energy factor to $\alpha \approx 1.06$ [Eq. (3.73)], whence the corrected $h_f \approx 1.82$ ft and $f \approx 0.0425$. With f known, we can estimate the Reynolds number from our formulas:

$$\text{Re}_d \approx 3250 \quad [\text{Eq. (6.38)}] \quad \text{or} \quad \text{Re}_d \approx 3400 \quad [\text{Eq. (6.39}b)]$$

So the flow *might* have been turbulent, in which case the viscosity of the fluid would have been

$$\mu = \frac{\rho V d}{\text{Re}_d} = \frac{1.80(3.32)(0.004)}{3300} = 7.2 \times 10^{-6} \text{ slug/(ft} \cdot \text{s)} \quad Ans.$$

This is about 55 percent less than our laminar estimate in Example 6.5. The moral is to keep the capillary-flow Reynolds number below about 1000 to avoid such duplicate solutions.

6.7 Four Types of Pipe Flow Problems

The Moody chart (Fig. 6.13) can be used to solve almost any problem involving friction losses in long pipe flows. However, many such problems involve considerable iteration and repeated calculations using the chart because the standard Moody chart is essentially a *head loss chart*. One is supposed to know all other variables, compute Re_d, enter the chart, find f, and hence compute h_f. This is one of four fundamental problems which are commonly encountered in pipe flow calculations:

1. Given d, L, and V or Q, ρ, μ, and g, compute the head loss h_f (head loss problem).
2. Given d, L, h_f, ρ, μ, and g, compute the velocity V or flow rate Q (flow rate problem).
3. Given Q, L, h_f, ρ, μ, and g, compute the diameter d of the pipe (sizing problem).
4. Given Q, d, h_f, ρ, μ, and g, compute the pipe length L.

Problems 1 and 4 are well suited to the Moody chart. We have to iterate to compute velocity or diameter because both d and V are contained in the ordinate *and* the abscissa of the chart.

There are two alternatives to iteration for problems of type 2 and 3: (*a*) preparation of a suitable new Moody-type chart (see Probs. P6.68 and P6.73); or (*b*) the use of *solver* software, especially the Engineering Equation Solver, known as EES [47], which gives the answer directly if the proper data are entered. Examples 6.9 and 6.11 include the EES approach to these problems.

Type 2 Problem: Find the Flow Rate

Even though velocity (or flow rate) appears in both the ordinate and the abscissa on the Moody chart, iteration for turbulent flow is nevertheless quite fast because f varies so slowly with Re_d. Alternately, in the spirit of Example 5.7, we could change the scaling variables to (ρ, μ, d) and thus arrive at dimensionless head loss versus dimensionless *velocity*. The result is[3]

$$\zeta = \text{fcn}(Re_d) \qquad \text{where} \qquad \zeta = \frac{gd^3 h_f}{L\nu^2} = \frac{f\,Re_d^2}{2} \tag{6.50}$$

Example 5.7 did this and offered the simple correlation $\zeta \approx 0.155\,Re_d^{1.75}$, which is valid for turbulent flow with smooth walls and $Re_d \leq 1$ E5.

A formula valid for all turbulent pipe flows is found by simply rewriting the Colebrook interpolation, Eq. (6.48), in the form of Eq. (6.50):

$$Re_d = -(8\zeta)^{1/2} \log\left(\frac{\epsilon/d}{3.7} + \frac{1.775}{\sqrt{\zeta}}\right) \qquad \zeta = \frac{gd^3 h_f}{L\nu^2} \tag{6.51}$$

Given ζ, we compute Re_d (and hence velocity) directly. Let us illustrate these two approaches with the following example.

EXAMPLE 6.9

Oil, with $\rho = 950$ kg/m^3 and $\nu = 2$ E-5 m^2/s, flows through a 30-cm-diameter pipe 100 m long with a head loss of 8 m. The roughness ratio is $\epsilon/d = 0.0002$. Find the average velocity and flow rate.

Direct Solution

First calculate the dimensionless head loss parameter:

$$\zeta = \frac{gd^3 h_f}{L\nu^2} = \frac{(9.81 \text{ m/s}^2)(0.3 \text{ m})^3(8.0 \text{ m})}{(100 \text{ m})(2 \text{ E-5 m}^2/\text{s})^2} = 5.30 \text{ E7}$$

Now enter Eq. (6.51) to find the Reynolds number:

$$Re_d = -[8(5.3 \text{ E7})]^{1/2} \log\left(\frac{0.0002}{3.7} + \frac{1.775}{\sqrt{5.3 \text{ E7}}}\right) = 72,600$$

---

[3]The parameter ζ was suggested by H. Rouse in 1942.

The velocity and flow rate follow from the Reynolds number:

$$V = \frac{\nu Re_d}{d} = \frac{(2\text{ E-5 m}^2/\text{s})(72{,}600)}{0.3\text{ m}} \approx 4.84\text{ m/s}$$

$$Q = V\frac{\pi}{4}d^2 = \left(4.84\frac{\text{m}}{\text{s}}\right)\frac{\pi}{4}(0.3\text{ m})^2 \approx 0.342\text{ m}^3/\text{s} \qquad\qquad Ans.$$

No iteration is required, but this idea falters if additional losses are present. Note that we never bothered to compute the friction factor.

Iterative Solution

By definition, the friction factor is known except for V:

$$f = h_f\frac{d}{L}\frac{2g}{V^2} = (8\text{ m})\left(\frac{0.3\text{ m}}{100\text{ m}}\right)\left[\frac{2(9.81\text{ m/s}^2)}{V^2}\right] \qquad\text{or}\qquad fV^2 \approx 0.471 \quad\text{(SI units)}$$

To get started, we only need to guess f, compute $V = \sqrt{0.471/f}$, then get Re_d, compute a better f from the Moody chart, and repeat. The process converges fairly rapidly. A good first guess is the "fully rough" value for $\epsilon/d = 0.0002$, or $f \approx 0.014$ from Fig. 6.13. The iteration would be as follows:

Guess $f \approx 0.014$, then $V = \sqrt{0.471/0.014} = 5.80$ m/s and $Re_d = Vd/\nu \approx 87{,}000$. At $Re_d = 87{,}000$ and $\epsilon/d = 0.0002$, compute $f_{\text{new}} \approx 0.0195$ [Eq. (6.48)].
New $f \approx 0.0195$, $V = \sqrt{0.471/0.0195} = 4.91$ m/s and $Re_d = Vd/\nu = 73{,}700$. At $Re_d = 73{,}700$ and $\epsilon/d = 0.0002$, compute $f_{\text{new}} \approx 0.0201$ [Eq. (6.48)].
Better $f \approx 0.0201$, $V = \sqrt{0.471/0.0201} = 4.84$ m/s and $Re_d \approx 72{,}600$. At $Re_d = 72{,}600$ and $\epsilon/d = 0.0002$, compute $f_{\text{new}} \approx 0.0201$ [Eq. (6.48)].

We have converged to three significant figures. Thus our iterative solution is

$$V = 4.84\text{ m/s}$$

$$Q = V\left(\frac{\pi}{4}\right)d^2 = (4.84)\left(\frac{\pi}{4}\right)(0.3)^2 \approx 0.342\text{ m}^3/\text{s} \qquad\qquad Ans.$$

The iterative approach is straightforward and not too onerous, so it is routinely used by engineers. Obviously this repetitive procedure is ideal for a personal computer.

Engineering Equation Solver (EES) Solution

In EES, one simply enters the data and the appropriate equations, letting the software do the rest. Correct units must of course be used. For the present example, the data could be entered as SI:

```
rho=950    nu=2E-5    d=0.3  L=100    epsod=0.0002    hf=8.0    g=9.81
```

The appropriate equations are the Moody formula (6.48) plus the definitions of Reynolds number, volume flow rate as determined from velocity, and the Darcy head loss formula (6.10):

```
Re=V*d/nu

Q=V*pi*d^2/4

f=(-2.0*log10(epsod/3.7+2.51/Re/f^0.5))^(-2)

hf=f*L/d*V^2/2/g
```

EES understands that "pi" represents 3.141593. Then hit "SOLVE" from the menu. If errors have been entered, EES will complain that the system cannot be solved and attempt to explain why. Otherwise, the software will iterate, and in this case EES prints the correct solution:

$$Q=0.342 \qquad V=4.84 \qquad f=0.0201 \qquad Re=72585$$

The units are spelled out in a separate list as [m, kg, s, N]. This elegant approach to engineering problem solving has one drawback—namely, that the user fails to check the solution for engineering viability. For example, are the data typed correctly? Is the Reynolds number turbulent?

Type 3 Problem: Find the Pipe Diameter

The Moody chart is especially awkward for finding the pipe size, since d occurs in all three parameters f, Re_d, and ϵ/d. Further, it depends on whether we know the velocity or the flow rate. We cannot know both, or else we could immediately compute $d = \sqrt{4Q/(\pi V)}$.

Let us assume that we know the flow rate Q. Note that this requires us to redefine the Reynolds number in terms of Q:

$$Re_d = \frac{Vd}{\nu} = \frac{4Q}{\pi d \nu} \tag{6.52}$$

Then, if we choose (Q, ρ, μ) as scaling parameters (to eliminate d), we obtain the functional relationship

$$Re_d = \frac{4Q}{\pi d \nu} = \text{fcn}\left(\frac{gh_f}{L\nu^5}, \frac{\epsilon \nu}{Q}\right)$$

and can thus solve d when the right-hand side is known. Unfortunately, the writer knows of no *formula* for this relation. Here it seems reasonable to forgo a plot or curve fitted formula and to simply set up the problem as an iteration in terms of the Moody chart variables. In this case we also have to set up the friction factor in terms of the flow rate:

$$f = h_f \frac{d}{L}\frac{2g}{V^2} = \frac{\pi^2}{8}\frac{gh_f d^5}{LQ^2} \tag{6.53}$$

The following two examples illustrate the iteration.

EXAMPLE 6.10

Work Example 6.9 backward, assuming that $Q = 0.342$ m³/s and $\epsilon = 0.06$ mm are known but that d (30 cm) is unknown. Recall $L = 100$ m, $\rho = 950$ kg/m³, $\nu = 2$ E-5 m²/s, and $h_f = 8$ m.

Iterative Solution

First write the diameter in terms of the friction factor:

$$f = \frac{\pi^2}{8}\frac{(9.81 \text{ m/s}^2)(8 \text{ m})d^5}{(100 \text{ m})(0.342 \text{ m}^3/\text{s})^2} = 8.28d^5 \qquad \text{or} \qquad d \approx 0.655f^{1/5} \tag{1}$$

in SI units. Also write the Reynolds number and roughness ratio in terms of the diameter:

$$\mathrm{Re}_d = \frac{4(0.342 \text{ m}^3/\text{s})}{\pi(2 \text{ E-5 m}^2/\text{s})d} = \frac{21{,}800}{d} \tag{2}$$

$$\frac{\epsilon}{d} = \frac{6 \text{ E-5 m}}{d} \tag{3}$$

Guess f, compute d from (1), then compute Re_d from (2) and ϵ/d from (3), and compute a better f from the Moody chart or Eq. (6.48). Repeat until (fairly rapid) convergence. Having no initial estimate for f, the writer guesses $f \approx 0.03$ (about in the middle of the turbulent portion of the Moody chart). The following calculations result:

$$f \approx 0.03 \qquad d \approx 0.655(0.03)^{1/5} \approx 0.325 \text{ m}$$

$$\mathrm{Re}_d \approx \frac{21{,}800}{0.325} \approx 67{,}000 \qquad \frac{\epsilon}{d} \approx 1.85 \text{ E-4}$$

Eq. (6.48): $\qquad f_{\text{new}} \approx 0.0203 \qquad$ then $\qquad d_{\text{new}} \approx 0.301 \text{ m}$

$$\mathrm{Re}_{d,\text{new}} \approx 72{,}500 \qquad \frac{\epsilon}{d} \approx 2.0 \text{ E-4}$$

Eq. (6.48): $\qquad f_{\text{better}} \approx 0.0201 \qquad$ and $\qquad d = 0.300 \text{ m}$ $\qquad\qquad$ *Ans.*

The procedure has converged to the correct diameter of 30 cm given in Example 6.9.

EES Solution

For an EES solution, enter the data and the appropriate equations. The diameter is unknown. Correct units must of course be used. For the present example, the data should use SI units:

```
rho=950    nu=2E-5    L=100    eps=6E-5    hf=8.0    g=9.81    Q=0.342
```

The appropriate equations are the Moody formula, the definition of Reynolds number, volume flow rate as determined from velocity, the Darcy head loss formula, and the roughness ratio:

```
Re=V*d/nu

Q=V*pi*d^2/4

f=(-2.0*log10(epsod/3.7+2.51/Re/f^0.5))^(-2)

hf=f*L/d*V^2/2/g

epsod=eps/d
```

Hit *Solve* from the menu. Unlike Example 6.9, this time EES complains that the system *cannot* be solved and reports "logarithm of a negative number." The reason is that we allowed EES to assume that f could be a negative number. Bring down *Variable Information* from the menu and change the limits of f so that it cannot be negative. EES agrees and iterates to this solution:

```
d=0.300    V=4.84    f=0.0201    Re=72,585
```

The unit system is spelled out as (m, kg, s, N). As always when using software, the user should check the solution for engineering viability. For example, is the Reynolds number turbulent? (Yes.)

EXAMPLE 6.11

Work Moody's problem, Example 6.6, backward to find the wall roughness ϵ if everything else is known: $V = 6$ ft/s, $d = 0.5$ ft, $L = 200$ ft, $\rho = 1.94$ slug/ft^3, $\mu = 2.09E-5$ slug/ft-s, $h_f = 4.5$ ft.

Solution

• *Analytic solution:*   This is not as bad as having the diameter unknown, because ϵ appears in only one parameter, ϵ/d. We can immediately calculate Q, Re_d, and the friction factor:

$$Q = V\,\pi R^2 = (6.0\text{ ft/s})\pi(0.25\text{ ft})^2 = 1.18\text{ ft}^3/\text{s}$$

$$\text{Re}_d = \frac{\rho V d}{\mu} = \frac{(1.94\text{ slug/ft}^3)(6\text{ ft/s})(0.5\text{ ft})}{2.09e-5\text{ slug/ft}-\text{s}} = 278{,}500$$

$$f = \frac{h_f}{(L/d)(V^2/2g)} = \frac{4.5\text{ ft}}{(200\text{ ft/0.5 ft})[(6\text{ ft/s})^2/2/(32.2\text{ ft/s}^2)]} = 0.0201$$

With f and Re_d known, we look on the Moody chart or solve Eq. (6.48) for the roughness ratio:

$$\frac{1}{\sqrt{f}} = -2.0\log_{10}\left(\frac{\epsilon/d}{3.7} + \frac{2.51}{\text{Re}_d\sqrt{f}}\right) \quad\text{or}\quad \frac{1}{\sqrt{0.0201}} = -2.0\log_{10}\left(\frac{\epsilon/d}{3.7} + \frac{2.51}{278500\sqrt{0.0201}}\right)$$

After a bit of ugly manipulation, we calculate $\epsilon/d = 0.000871$, or $\epsilon \approx 0.000435$ ft. *Ans.*

• *EES solution:*   Simply type in the data, in BG units (ft, s, lbf, slugs):

```
rho=1.94    mu=2.09E-5    d=0.5    V=6.0  L=200    hf=8.0    g=32.2
```

Then type in the same five defining formulas for pipe flow that we used in Example 6.11:

```
Re=rho*V*d/mu

Q=V*pi*d^2/4

f=(-2.0*log10 (epsod/3.7 + 2.51/Re/f^.5))^(-2)

hf=f*L/d*V^2/2/g

epsod=eps/d
```

With any reasonable guess for $\epsilon > 0$, EES promptly returns $\epsilon \approx 0.000435$ ft.   *Ans.*

• *Comments:*   Finding the roughness is not as hard as finding the diameter. The discrepancy from Moody's value of $\epsilon = 0.00040$ ft was caused by rounding off h_f to 4.5 ft.

Table 6.2 Nominal and Actual Sizes of Schedule 40 Wrought Steel Pipe*

Nominal size, in	Actual ID, in
$\frac{1}{8}$	0.269
$\frac{1}{4}$	0.364
$\frac{3}{8}$	0.493
$\frac{1}{2}$	0.622
$\frac{3}{4}$	0.824
1	1.049
$1\frac{1}{2}$	1.610
2	2.067
$2\frac{1}{2}$	2.469
3	3.068

*Nominal size within 1 percent for 4 in or larger.

In discussing pipe sizing problems, we should remark that commercial pipes are made only in certain sizes. Table 6.2 lists standard water pipe sizes in the United States. If the sizing calculation gives an intermediate diameter, the next largest pipe size should be selected.

Type 4 Problem: Find the Pipe Length

In designing piping systems, it is desirable to estimate the appropriate pipe length for a given pipe diameter, pump power, and flow rate. The pump head will match the piping head loss. If minor losses are neglected, the (horizontal) pipe length follows from Darcy's formula (6.10):

$$h_{\text{pump}} = \frac{\text{Power}}{\rho g Q} = h_f = f \frac{L}{d} \frac{V^2}{2g} \qquad (6.54)$$

With Q, and d, and ε known, we may compute Re_d and f, after which L is obtained from the formula. Note that pump efficiency varies strongly with flow rate (Chap. 11). Thus, it is important to match pipe length to the pump's region of maximum efficiency.

EXAMPLE 6.12

A pump delivers 0.6 hp to water at 68°F, flowing in a 6-in-diameter asphalted cast iron horizontal pipe at $V = 6$ ft/s. What is the proper pipe length to match these conditions?

Solution

- *Approach:* Find h_f from the known power and find f from Re_d and ε/d. Then find L.
- *Water properties:* For water at 68°F, Table A.3, converting to BG units, $\rho = 1.94$ slug/ft^3 and $\mu = 2.09\text{E}{-}5$ slug/(ft − s).
- *Pipe roughness:* From Table 6.1 for asphalted cast iron, $\varepsilon = 0.0004$ ft.
- *Solution step 1:* Find the pump head from the flow rate and the pump power:

$$Q = AV = \frac{\pi}{4}(0.5 \text{ ft})^2 \left(6\frac{\text{ft}}{\text{s}}\right) = 1.18 \frac{\text{ft}^3}{\text{s}}$$

$$h_{\text{pump}} = \frac{\text{Power}}{\rho g Q} = \frac{(0.6 \text{ hp})\left[550(\text{ft} \cdot \text{lbf})/(\text{s} \cdot \text{hp})\right]}{(1.94 \text{ slug/ft}^3)(32.2 \text{ ft/s}^2)(1.18 \text{ ft}^3/\text{s})} = 4.48 \text{ ft}$$

- *Solution step 2:* Compute the friction factor from the Colebrook formula, Eq. (6.48):

$$\text{Re}_d = \frac{\rho V d}{\mu} = \frac{(1.94)(6)(0.5)}{2.09 \text{ E}{-}5} = 278{,}500 \qquad \frac{\varepsilon}{d} = \frac{0.0004 \text{ ft}}{0.5 \text{ ft}} = 0.0008$$

$$\frac{1}{\sqrt{f}} \approx -2.0 \log_{10}\left(\frac{\varepsilon/d}{3.7} + \frac{2.51}{\text{Re}_d\sqrt{f}}\right) \quad \text{yields} \quad f = 0.0198$$

- *Solution step 3:* Find the pipe length from the Darcy formula (6.10):

$$h_p = h_f = 4.48 \text{ ft} = f\frac{L}{d}\frac{V^2}{2g} = (0.0198)\left(\frac{L}{0.5 \text{ ft}}\right)\frac{(6 \text{ ft/s})^2}{2(32.2 \text{ ft/s}^2)}$$

$$\text{Solve for} \quad L \approx 203 \text{ ft} \qquad\qquad Ans.$$

- *Comment:* This is Moody's problem (Example 6.6) turned around so that the length is unknown.

6.8 Flow in Noncircular Ducts[4]

If the duct is noncircular, the analysis of fully developed flow follows that of the circular pipe but is more complicated algebraically. For laminar flow, one can solve the exact equations of continuity and momentum. For turbulent flow, the logarithm law velocity profile can be used, or (better and simpler) the hydraulic diameter is an excellent approximation.

---

[4]This section may be omitted without loss of continuity.

The Hydraulic Diameter

For a noncircular duct, the control volume concept of Fig. 6.7 is still valid, but the cross-sectional area A does not equal πR^2 and the cross-sectional perimeter wetted by the shear stress $\mathcal{P}$ does not equal $2\pi R$. The momentum equation (6.9a) thus becomes

$$\Delta p\, A + \rho g A\, \Delta L \sin \phi - \bar{\tau}_w \mathcal{P}\, \Delta L = 0$$

or

$$h_f = \frac{\Delta p}{\rho g} + \Delta z = \frac{\bar{\tau}_w}{\rho g}\frac{\Delta L}{A/\mathcal{P}} \tag{6.55}$$

This is identical to Eq. (6.9b) except that (1) the shear stress is an average value integrated around the perimeter and (2) the length scale $A/\mathcal{P}$ takes the place of the pipe radius R. For this reason a noncircular duct is said to have a *hydraulic radius R_h*, defined by

$$R_h = \frac{A}{\mathcal{P}} = \frac{\text{cross-sectional area}}{\text{wetted perimeter}} \tag{6.56}$$

This concept receives constant use in open-channel flow (Chap. 10), where the channel cross section is almost never circular. If, by comparison to Eq. (6.11) for pipe flow, we define the friction factor in terms of average shear

$$f_{\text{NCD}} = \frac{8\bar{\tau}_w}{\rho V^2} \tag{6.57}$$

where NCD stands for noncircular duct and $V = Q/A$ as usual, Eq. (6.55) becomes

$$h_f = f\frac{L}{4R_h}\frac{V^2}{2g} = f\frac{L}{D_h}\frac{V^2}{2g} \tag{6.58}$$

This is equivalent to Eq. (6.10) for pipe flow except that d is replaced by $4R_h$. Therefore we customarily define the *hydraulic diameter* as

$$\boxed{D_h = \frac{4A}{\mathcal{P}} = \frac{4 \times \text{area}}{\text{wetted perimeter}} = 4R_h} \tag{6.59}$$

We should stress that the wetted perimeter includes all surfaces acted upon by the shear stress. For example, in a circular annulus, both the outer and the inner perimeter should be added. The fact that D_h equals $4R_h$ is just one of those things: Chalk it up to an engineer's sense of humor. Note that for the degenerate case of a circular pipe, $D_h = 4\pi R^2/(2\pi R) = 2R$, as expected.

We would therefore expect by dimensional analysis that this friction factor f, based on hydraulic diameter as in Eq. (6.58), would correlate with the Reynolds number and roughness ratio based on the hydraulic diameter

$$f = F\left(\frac{VD_h}{\nu}, \frac{\epsilon}{D_h}\right) \tag{6.60}$$

and this is the way the data are correlated. But we should not necessarily expect the Moody chart (Fig. 6.13) to hold exactly in terms of this new length scale. And it does

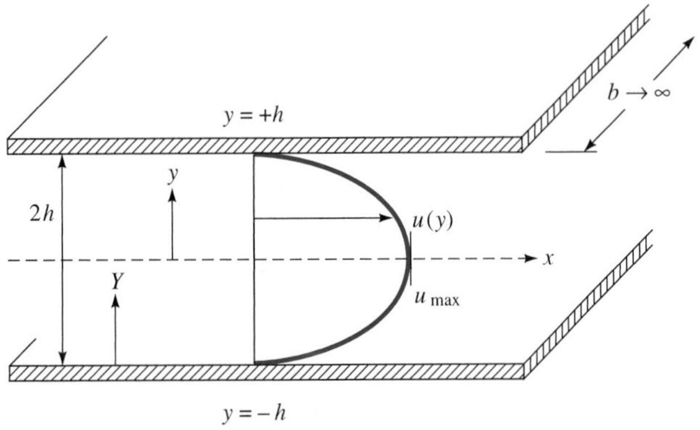

Fig. 6.14 Fully developed flow between parallel plates.

not, but it is surprisingly accurate:

$$f \approx \begin{cases} \dfrac{64}{\mathrm{Re}_{D_h}} & \pm 40\% \quad \text{laminar flow} \\[4mm] f_{\text{Moody}}\left(\mathrm{Re}_{D_{h.}} \dfrac{\epsilon}{D_h}\right) & \pm 15\% \quad \text{turbulent flow} \end{cases} \tag{6.61}$$

Now let us look at some particular cases.

Flow between Parallel Plates

Probably the simplest noncircular duct flow is fully developed flow between parallel plates a distance $2h$ apart, as in Fig. 6.14. As noted in the figure, the width $b \gg h$, so the flow is essentially two-dimensional; that is, $u = u(y)$ only. The hydraulic diameter is

$$D_h = \frac{4A}{\mathcal{P}} = \lim_{b \to \infty} \frac{4(2bh)}{2b + 4h} = 4h \tag{6.62}$$

that is, twice the distance between the plates. The pressure gradient is constant, $(-dp/dx) = \Delta p/L$, where L is the length of the channel along the x axis.

Laminar Flow Solution

The laminar solution was given in Sect. 4.10, in connection with Fig. 4.16b. Let us review those results here:

$$u = u_{\max}\left(1 - \frac{y^2}{h^2}\right) \quad \text{where} \quad u_{\max} = \frac{h^2}{2\mu}\frac{\Delta p}{L}$$

$$Q = \frac{2bh^3}{3\mu}\frac{\Delta p}{L}$$

$$V = \frac{Q}{A} = \frac{h^2}{3\mu}\frac{\Delta p}{L} = \frac{2}{3}u_{\max} \tag{6.63}$$

$$\tau_w = \mu\left|\frac{du}{dy}\right|_{y=h} = h\frac{\Delta p}{L} = \frac{3\mu V}{h}$$

$$h_f = \frac{\Delta p}{\rho g} = \frac{3\mu L V}{\rho g h^2}$$

Now use the head loss to establish the laminar friction factor:

$$f_{\text{lam}} = \frac{h_f}{(L/D_h)(V^2/2g)} = \frac{96\mu}{\rho V(4h)} = \frac{96}{\text{Re}_{D_h}} \tag{6.64}$$

Thus, if we could not work out the laminar theory and chose to use the approximation $f \approx 64/\text{Re}_{D_h}$, we would be 33 percent low. The hydraulic-diameter approximation is relatively crude in laminar flow, as Eq. (6.61) states.

Just as in circular-pipe flow, the laminar solution above becomes unstable at about $\text{Re}_{D_h} \approx 2000$; transition occurs and turbulent flow results.

Turbulent Flow Solution

For turbulent flow between parallel plates, we can again use the logarithm law, Eq. (6.28), as an approximation across the entire channel, using not y but a wall coordinate Y, as shown in Fig. 6.14:

$$\frac{u(Y)}{u^*} \approx \frac{1}{\kappa}\ln\frac{Yu^*}{\nu} + B \qquad 0 < Y < h \tag{6.65}$$

This distribution looks very much like the flat turbulent profile for pipe flow in Fig. 6.11b, and the mean velocity is

$$V = \frac{1}{h}\int_0^h u\, dY = u^*\left(\frac{1}{\kappa}\ln\frac{hu^*}{\nu} + B - \frac{1}{\kappa}\right) \tag{6.66}$$

Recalling that $V/u^* = (8/f)^{1/2}$, we see that Eq. (6.66) is equivalent to a parallel-plate friction law. Rearranging and cleaning up the constant terms, we obtain

$$\frac{1}{f^{1/2}} \approx 2.0 \log\left(\text{Re}_{D_h}f^{1/2}\right) - 1.19 \tag{6.67}$$

where we have introduced the hydraulic diameter $D_h = 4h$. This is remarkably close to the smooth-wall pipe friction law, Eq. (6.38). Therefore we conclude that the use of the hydraulic diameter in this turbulent case is quite successful. That turns out to be true for other noncircular turbulent flows also.

Equation (6.67) can be brought into exact agreement with the pipe law by rewriting it in the form

$$\frac{1}{f^{1/2}} = 2.0 \log\left(0.64\,\text{Re}_{D_h}f^{1/2}\right) - 0.8 \tag{6.68}$$

Thus the turbulent friction is predicted most accurately when we use an effective diameter D_{eff} equal to 0.64 times the hydraulic diameter. The effect on f itself is much less, about 10 percent at most. We can compare with Eq. (6.64) for laminar flow, which predicted

Parallel plates:
$$D_{\text{eff}} = \frac{64}{96}D_h = \frac{2}{3}D_h \tag{6.69}$$

This close resemblance ($0.64D_h$ versus $0.667D_h$) occurs so often in noncircular duct flow that we take it to be a general rule for computing turbulent friction in ducts:

$$D_{\text{eff}} = D_h = \frac{4A}{\mathscr{P}} \qquad \text{reasonable accuracy}$$

$$D_{\text{eff}} = D_h\frac{64}{(f\,\text{Re}_{D_h})\text{laminar theory}} \qquad \text{better accuracy} \tag{6.70}$$

Jones [10] shows that the effective-laminar-diameter idea collapses all data for rectangular ducts of arbitrary height-to-width ratio onto the Moody chart for pipe flow. We recommend this idea for all noncircular ducts.

EXAMPLE 6.13

Fluid flows at an average velocity of 6 ft/s between horizontal parallel plates a distance of 2.4 in apart. Find the head loss and pressure drop for each 100 ft of length for $\rho = 1.9$ slugs/ft^3 and (a) $\nu = 0.00002$ ft^2/s and (b) $\nu = 0.002$ ft^2/s. Assume smooth walls.

Solution

Part (a)

The viscosity $\mu = \rho\nu = 3.8 \times 10^{-5}$ slug/(ft · s). The spacing is $2h = 2.4$ in $= 0.2$ ft, and $D_h = 4h = 0.4$ ft. The Reynolds number is

$$\mathrm{Re}_{D_h} = \frac{VD_h}{\nu} = \frac{(6.0 \text{ ft/s})(0.4 \text{ ft})}{0.00002 \text{ ft}^2/\text{s}} = 120{,}000$$

The flow is therefore turbulent. For reasonable accuracy, simply look on the Moody chart (Fig. 6.13) for smooth walls:

$$f \approx 0.0173 \quad h_f \approx f\frac{L}{D_h}\frac{V^2}{2g} = 0.0173\frac{100}{0.4}\frac{(6.0)^2}{2(32.2)} \approx 2.42 \text{ ft} \qquad \textit{Ans. (a)}$$

Since there is no change in elevation,

$$\Delta p = \rho g h_f = 1.9(32.2)(2.42) = 148 \text{ lbf/ft}^2 \qquad \textit{Ans. (a)}$$

This is the head loss and pressure drop per 100 ft of channel. For more accuracy, take $D_{\mathrm{eff}} = \frac{64}{96}D_h$ from laminar theory; then

$$\mathrm{Re}_{\mathrm{eff}} = \frac{64}{96}(120{,}000) = 80{,}000$$

and from the Moody chart read $f \approx 0.0189$ for smooth walls. Thus a better estimate is

$$h_f = 0.0189\frac{100}{0.4}\frac{(6.0)^2}{2(32.2)} = 2.64 \text{ ft}$$

and

$$\Delta p = 1.9(32.2)(2.64) = 161 \text{ lbf/ft}^2 \qquad \textit{Better ans. (a)}$$

The more accurate formula predicts friction about 9 percent higher.

Part (b)

Compute $\mu = \rho\nu = 0.0038$ slug/(ft · s). The Reynolds number is $6.0(0.4)/0.002 = 1200$; therefore the flow is laminar, since Re is less than 2300.

You could use the laminar flow friction factor, Eq. (6.64)

$$f_{\mathrm{lam}} = \frac{96}{\mathrm{Re}_{D_h}} = \frac{96}{1200} = 0.08$$

from which

$$h_f = 0.08\frac{100}{0.4}\frac{(6.0)^2}{2(32.2)} = 11.2 \text{ ft}$$

and

$$\Delta p = 1.9(32.2)(11.2) = 684 \text{ lbf/ft}^2 \qquad \textit{Ans. (b)}$$

Alternately you can finesse the Reynolds number and go directly to the appropriate laminar flow formula, Eq. (6.63):

$$V = \frac{h^2}{3\mu}\frac{\Delta p}{L}$$

or
$$\Delta p = \frac{3(6.0 \text{ ft/s})[0.0038 \text{ slug/(ft} \cdot \text{s)}](100 \text{ ft})}{(0.1 \text{ ft})^2} = 684 \text{ slugs/(ft} \cdot \text{s}^2) = 684 \text{ lbf/ft}^2$$

and
$$h_f = \frac{\Delta p}{\rho g} = \frac{684}{1.9(32.2)} = 11.2 \text{ ft}$$

This is one of those (perhaps unexpected) problems where the laminar friction is greater than the turbulent friction.

Flow through a Concentric Annulus

Consider steady axial laminar flow in the annular space between two concentric cylinders, as in Fig. 6.15. There is no slip at the inner ($r = b$) and outer radius ($r = a$). For $u = u(r)$ only, the governing relation is Eq. (D.7):

$$\frac{d}{dr}\left(r\mu\frac{du}{dr}\right) = Kr \qquad K = \frac{d}{dx}(p + \rho gz) \tag{6.71}$$

Integrate this twice:

$$u = \frac{1}{4}r^2\frac{K}{\mu} + C_1 \ln r + C_2$$

The constants are found from the two no-slip conditions:

$$u(r = a) = 0 = \frac{1}{4}a^2\frac{K}{\mu} + C_1 \ln a + C_2$$

$$u(r = b) = 0 = \frac{1}{4}b^2\frac{K}{\mu} + C_1 \ln b + C_2$$

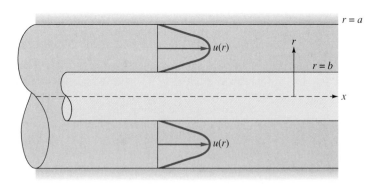

Fig. 6.15 Fully developed flow through a concentric annulus.

The final solution for the velocity profile is

$$u = \frac{1}{4\mu}\left[-\frac{d}{dx}(p + \rho g z)\right]\left[a^2 - r^2 + \frac{a^2 - b^2}{\ln(b/a)}\ln\frac{a}{r}\right] \qquad (6.72)$$

The volume flow is given by

$$Q = \int_b^a u\,2\pi r\,dr = \frac{\pi}{8\mu}\left[-\frac{d}{dx}(p + \rho g z)\right]\left[a^4 - b^4 - \frac{(a^2 - b^2)^2}{\ln(a/b)}\right] \qquad (6.73)$$

The velocity profile $u(r)$ resembles a parabola wrapped around in a circle to form a split doughnut, as in Fig. 6.15.

It is confusing to base the friction factor on the wall shear because there are two shear stresses, the inner stress being greater than the outer. It is better to define f with respect to the head loss, as in Eq. (6.58),

$$f = h_f\frac{D_h}{L}\frac{2g}{V^2} \qquad \text{where } V = \frac{Q}{\pi(a^2 - b^2)} \qquad (6.74)$$

The hydraulic diameter for an annulus is

$$D_h = \frac{4\pi(a^2 - b^2)}{2\pi(a + b)} = 2(a - b) \qquad (6.75)$$

It is twice the clearance, rather like the parallel-plate result of twice the distance between plates [Eq. (6.62)].

Substituting h_f, D_h, and V into Eq. (6.74), we find that the friction factor for laminar flow in a concentric annulus is of the form

$$f = \frac{64\zeta}{\mathrm{Re}_{D_h}} \qquad \zeta = \frac{(a - b)^2(a^2 - b^2)}{a^4 - b^4 - (a^2 - b^2)^2/\ln(a/b)} \qquad (6.76)$$

The dimensionless term ζ is a sort of correction factor for the hydraulic diameter. We could rewrite Eq. (6.76) as

Table 6.3 Laminar Friction Factors for a Concentric Annulus

b/a	$f\,\mathrm{Re}_{D_h}$	$D_{\mathrm{eff}}/D_h = 1/\zeta$
0.0	64.0	1.000
0.00001	70.09	0.913
0.0001	71.78	0.892
0.001	74.68	0.857
0.01	80.11	0.799
0.05	86.27	0.742
0.1	89.37	0.716
0.2	92.35	0.693
0.4	94.71	0.676
0.6	95.59	0.670
0.8	95.92	0.667
1.0	96.0	0.667

Concentric annulus: $\qquad f = \dfrac{64}{\mathrm{Re}_{\mathrm{eff}}} \qquad \mathrm{Re}_{\mathrm{eff}} = \dfrac{1}{\zeta}\mathrm{Re}_{D_h} \qquad (6.77)$

Some numerical values of $f\,\mathrm{Re}_{D_h}$ and $D_{\mathrm{eff}}/D_h = 1/\zeta$ are given in Table 6.3. Again, laminar annular flow becomes unstable at $\mathrm{Re}_{D_h} \approx 2000$.

For turbulent flow through a concentric annulus, the analysis might proceed by patching together two logarithmic law profiles, one going out from the inner wall to meet the other coming in from the outer wall. We omit such a scheme here and proceed directly to the friction factor. According to the general rule proposed in Eq. (6.61), turbulent friction is predicted with excellent accuracy by replacing d in the Moody chart by $D_{\mathrm{eff}} = 2(a - b)/\zeta$, with values listed in Table 6.3.[5] This idea includes roughness also (replace ϵ/d in the chart by $\epsilon/D_{\mathrm{eff}}$). For a quick design number with about 10 percent accuracy, one can simply use the hydraulic diameter $D_h = 2(a - b)$.

---

[5]Jones and Leung [44] show that data for annular flow also satisfy the effective-laminar-diameter idea.

EXAMPLE 6.14

What should the reservoir level h be to maintain a flow of 0.01 m³/s through the commercial steel annulus 30 m long shown in Fig. E6.14? Neglect entrance effects and take $\rho = 1000$ kg/m³ and $\nu = 1.02 \times 10^{-6}$ m²/s for water.

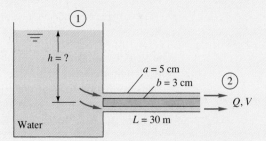

E6.14

Solution

- *Assumptions:*   Fully developed annulus flow, minor losses neglected.
- *Approach:*   Determine the Reynolds number, then find f and h_f and thence h.
- *Property values:*   Given $\rho = 1000$ kg/m³ and $\nu = 1.02$E-6 m²/s.
- *Solution step 1:*   Calculate the velocity, hydraulic diameter, and Reynolds number:

$$V = \frac{Q}{A} = \frac{0.01 \text{ m}^3/\text{s}}{\pi[(0.05 \text{ m})^2 - (0.03 \text{ m})^2]} = 1.99 \frac{\text{m}}{\text{s}}$$

$$D_h = 2(a - b) = 2(0.05 \text{ m} - 0.03 \text{ m}) = 0.04 \text{ m}$$

$$\text{Re}_{D_h} = \frac{VD_h}{\nu} = \frac{(1.99 \text{ m/s})(0.04 \text{ m})}{1.02\text{E-6 m}^2/\text{s}} = 78,000 \qquad \text{(turbulent flow)}$$

- *Solution step 2:*   Apply the steady flow energy equation between sections 1 and 2:

$$\frac{p_1}{\rho g} + \frac{\alpha_1 V_1^2}{2g} + z_1 = \frac{p_2}{\rho g} + \frac{\alpha_2 V_2^2}{2g} + z_2 + h_f$$

or
$$h = \frac{\alpha_2 V_2^2}{2g} + h_f = \frac{V_2^2}{2g}\left(\alpha_2 + f\frac{L}{D_h}\right) \qquad (1)$$

Note that $z_1 = h$. For turbulent flow, from Eq. (3.43c), we estimate $\alpha_2 \approx 1.03$

- *Solution step 3:*   Determine the roughness ratio and the friction factor. From Table 6.1, for (new) commercial steel pipe, $\epsilon = 0.046$ mm. Then

$$\frac{\epsilon}{D_h} = \frac{0.046 \text{ mm}}{40 \text{ mm}} = 0.00115$$

For a reasonable estimate, use Re_{D_h} to estimate the friction factor from Eq. (6.48):

$$\frac{1}{\sqrt{f}} \approx -2.0 \log_{10}\left(\frac{0.00115}{3.7} + \frac{2.51}{78,000\sqrt{f}}\right) \qquad \text{solve for } f \approx 0.0232$$

For slightly better accuracy, we could use $D_{\text{eff}} = D_h/\zeta$. From Table 6.3, for $b/a = 3/5$, $1/\zeta = 0.67$. Then $D_{\text{eff}} = 0.67(40 \text{ mm}) = 26.8$ mm, whence $\text{Re}_{D_{\text{eff}}} = 52,300$, $\epsilon/D_{\text{eff}} = 0.00172$, and

$f_{\text{eff}} \approx 0.0257$. Using the latter estimate, we find the required reservoir level from Eq. (1):

$$h = \frac{V_2^2}{2g}\left(\alpha_2 + f_{\text{eff}}\frac{L}{D_h}\right) = \frac{(1.99 \text{ m/s})^2}{2(9.81 \text{ m/s})^2}\left[1.03 + 0.0257\frac{30 \text{ m}}{0.04 \text{ m}}\right] \approx 4.1 \text{ m} \qquad Ans.$$

- *Comments:* Note that we do *not* replace D_h by D_{eff} in the head loss term fL/D_h, which comes from a momentum balance and *requires* hydraulic diameter. If we used the simpler friction estimate, $f \approx 0.0232$, we would obtain $h \approx 3.72$ m, or about 9 percent lower.

Other Noncircular Cross Sections

In principle, any duct cross section can be solved analytically for the laminar flow velocity distribution, volume flow, and friction factor. This is because any cross section can be mapped onto a circle by the methods of complex variables, and other powerful analytical techniques are also available. Many examples are given by White [3, pp. 112–115], Berker [11], and Olson and Wright [12, pp. 315–317]. Reference 34 is devoted entirely to laminar duct flow.

In general, however, most unusual duct sections have strictly academic and not commercial value. We list here only the rectangular and isosceles-triangular sections, in Table 6.4, leaving other cross sections for you to find in the references.

For turbulent flow in a duct of unusual cross section, one should replace d by D_h on the Moody chart if no laminar theory is available. If laminar results are known, such as Table 6.4, replace d by $D_{\text{eff}} = [64/(f\,\text{Re})]D_h$ for the particular geometry of the duct.

For laminar flow in rectangles and triangles, the wall friction varies greatly, being largest near the midpoints of the sides and zero in the corners. In turbulent flow through the same sections, the shear is nearly constant along the sides, dropping off sharply to zero in the corners. This is because of the phenomenon of turbulent *secondary flow,* in which there are nonzero mean velocities v and w in the plane of the cross section. Some measurements of axial velocity and secondary flow patterns are shown in Fig. 6.16, as sketched by Nikuradse in his 1926 dissertation. The

Table 6.4 Laminar Friction Constants $f\,$Re for Rectangular and Triangular Ducts

Rectangular		Isosceles triangle	
b/a	$f\text{Re}_{D_h}$	θ, deg	$f\text{Re}_{D_h}$
0.0	96.00	0	48.0
0.05	89.91	10	51.6
0.1	84.68	20	52.9
0.125	82.34	30	53.3
0.167	78.81	40	52.9
0.25	72.93	50	52.0
0.4	65.47	60	51.1
0.5	62.19	70	49.5
0.75	57.89	80	48.3
1.0	56.91	90	48.0

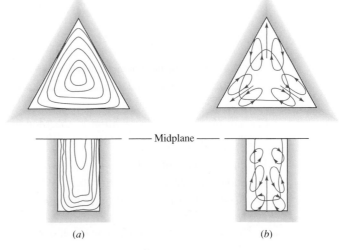

Midplane

(a)   (b)

Fig. 6.16 Illustration of secondary turbulent flow in noncircular ducts: (*a*) axial mean velocity contours; (*b*) secondary flow in-plane cellular motions. *(After J. Nikuradse, dissertation, Göttingen, 1926.)*

secondary flow "cells" drive the mean flow toward the corners, so that the axial velocity contours are similar to the cross section and the wall shear is nearly constant. This is why the hydraulic-diameter concept is so successful for turbulent flow. Laminar flow in a straight noncircular duct has no secondary flow. An accurate theoretical prediction of turbulent secondary flow has yet to be achieved, although numerical models are often successful [36].

EXAMPLE 6.15

Air, with $\rho = 0.00237$ slug/ft^3 and $\nu = 0.000157$ ft^2/s, is forced through a horizontal square 9-by 9-in duct 100 ft long at 25 ft^3/s. Find the pressure drop if $\epsilon = 0.0003$ ft.

Solution

Compute the mean velocity and hydraulic diameter:

$$V = \frac{25 \text{ ft}^3/\text{s}}{(0.75 \text{ ft})^2} = 44.4 \text{ ft/s}$$

$$D_h = \frac{4A}{\mathcal{P}} = \frac{4(81 \text{ in}^2)}{36 \text{ in}} = 9 \text{ in} = 0.75 \text{ ft}$$

From Table 6.4, for $b/a = 1.0$, the effective diameter is

$$D_{\text{eff}} = \frac{64}{56.91} D_h = 0.843 \text{ ft}$$

whence
$$\text{Re}_{\text{eff}} = \frac{V D_{\text{eff}}}{\nu} = \frac{44.4(0.843)}{0.000157} = 239,000$$

$$\frac{\epsilon}{D_{\text{eff}}} = \frac{0.0003}{0.843} = 0.000356$$

From the Moody chart, read $f = 0.0177$. Then the pressure drop is

$$\Delta p = \rho g h_f = \rho g \left(f \frac{L}{D_h} \frac{V^2}{2g} \right) = 0.00237(32.2) \left[0.0177 \frac{100}{0.75} \frac{44.4^2}{2(32.2)} \right]$$

or
$$\Delta p = 5.5 \text{ lbf/ft}^2 \qquad\qquad Ans.$$

Pressure drop in air ducts is usually small because of the low density.

6.9 Minor Losses in Pipe Systems[6]

For any pipe system, in addition to the Moody-type friction loss computed for the length of pipe, there are additional so-called *minor losses* due to

1. Pipe entrance or exit.
2. Sudden expansion or contraction.
3. Bends, elbows, tees, and other fittings.

---

[6]This section may be omitted without loss of continuity.

4.   Valves, open or partially closed.

5.   Gradual expansions or contractions.

The losses may not be so minor; for example, a partially closed valve can cause a greater pressure drop than a long pipe.

Since the flow pattern in fittings and valves is quite complex, the theory is very weak. The losses are commonly measured experimentally and correlated with the pipe flow parameters. The data, especially for valves, are somewhat dependent on the particular manufacturer's design, so that the values listed here must be taken as average design estimates [15, 16, 35, 43, 46].

The measured minor loss is usually given as a ratio of the head loss $h_m = \Delta p/(\rho g)$ through the device to the velocity head $V^2/(2g)$ of the associated piping system:

$$\text{Loss coefficient } K = \frac{h_m}{V^2/(2g)} = \frac{\Delta p}{\frac{1}{2}\rho V^2} \tag{6.78}$$

Although K is dimensionless, it often is not correlated in the literature with the Reynolds number and roughness ratio but rather simply with the raw size of the pipe in, say, inches. Almost all data are reported for turbulent flow conditions.

A single pipe system may have many minor losses. Since all are correlated with $V^2/(2g)$, they can be summed into a single total system loss if the pipe has constant diameter:

$$\boxed{\Delta h_{\text{tot}} = h_f + \Sigma h_m = \frac{V^2}{2g}\left(\frac{fL}{d} + \Sigma K\right)} \tag{6.79}$$

Note, however, that we must sum the losses separately if the pipe size changes so that V^2 changes. The length L in Eq. (6.79) is the total length of the pipe axis.

There are many different valve designs in commercial use. Figure 6.17 shows five typical designs: (a) the *gate,* which slides down across the section; (b) the *globe,* which closes a hole in a special insert; (c) the *angle,* similar to a globe but with a 90° turn; (d) the *swing-check* valve, which allows only one-way flow; and (e) the *disk,* which closes the section with a circular gate. The globe, with its tortuous flow path, has the highest losses when fully open. Many excellent details about these and other valves are given in the handbooks by Skousen [35] and Zappe [52].

Table 6.5 lists loss coefficients K for four types of valve, three angles of elbow fitting, and two tee connections. Fittings may be connected by either internal screws or flanges, hence the two listings. We see that K generally decreases with pipe size, which is consistent with the higher Reynolds number and decreased roughness ratio of large pipes. We stress that Table 6.5 represents losses *averaged among various manufacturers,* so there is an uncertainty as high as ±50 percent.

In addition, most of the data in Table 6.5 are relatively old [15, 16] and therefore based on fittings manufactured in the 1950s. Modern forged and molded fittings may yield somewhat different loss factors, often less than listed in Table 6.5. An example, shown in Fig. 6.18a, gives recent data [48] for fairly short (bend-radius/elbow-diameter = 1.2) flanged 90° elbows. The elbow diameter was 1.69 in. Notice first that K is plotted versus Reynolds number, rather than versus the raw (dimensional) pipe diameters in Table 6.5, and therefore Fig. 6.18a has more generality. Then notice

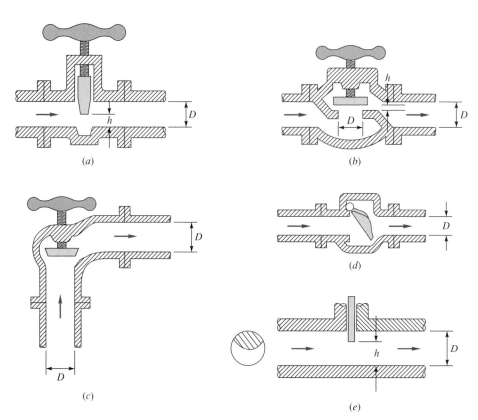

Fig. 6.17 Typical commercial valve geometries: (*a*) gate valve; (*b*) globe valve; (*c*) angle valve; (*d*) swing-check valve; (*e*) disk-type gate valve.

that the K values of 0.23 ± 0.05 are significantly less than the values for 90° elbows in Table 6.5, indicating smoother walls and/or better design. One may conclude that (1) Table 6.5 data are probably conservative and (2) loss factors are highly dependent on actual design and manufacturing factors, with Table 6.5 serving only as a rough guide.

The valve losses in Table 6.5 are for the fully open condition. Losses can be much higher for a partially open valve. Figure 6.18*b* gives average losses for three valves as a function of "percentage open," as defined by the opening-distance ratio h/D (see Fig. 6.17 for the geometries). Again we should warn of a possible uncertainty of ± 50 percent. Of all minor losses, valves, because of their complex geometry, are most sensitive to manufacturers' design details. For more accuracy, the particular design and manufacturer should be consulted [35].

The *butterfly* valve of Fig. 6.19*a* is a stem-mounted disk that, when closed, seats against an O-ring or compliant seal near the pipe surface. A single 90° turn opens the valve completely, hence the design is ideal for controllable quick-opening and quick-closing situations such as occur in fire protection and the electric power industry. However, considerable dynamic torque is needed to close these valves, and losses are high when the valves are nearly closed.

Figure 6.19*b* shows butterfly-valve loss coefficients as a function of the opening angle θ for turbulent flow conditions ($\theta = 0$ is closed). The losses are huge when the opening is small, and K drops off nearly exponentially with the opening angle. There

Table 6.5 Resistance Coefficients $K = h_m/[V^2/(2g)]$ for Open Valves, Elbows, and Tees

	Nominal diameter, in								
	Screwed				Flanged				
	$\frac{1}{2}$	1	2	4	1	2	4	8	20
Valves (fully open):									
Globe	14	8.2	6.9	5.7	13	8.5	6.0	5.8	5.5
Gate	0.30	0.24	0.16	0.11	0.80	0.35	0.16	0.07	0.03
Swing check	5.1	2.9	2.1	2.0	2.0	2.0	2.0	2.0	2.0
Angle	9.0	4.7	2.0	1.0	4.5	2.4	2.0	2.0	2.0
Elbows:									
45° regular	0.39	0.32	0.30	0.29					
45° long radius					0.21	0.20	0.19	0.16	0.14
90° regular	2.0	1.5	0.95	0.64	0.50	0.39	0.30	0.26	0.21
90° long radius	1.0	0.72	0.41	0.23	0.40	0.30	0.19	0.15	0.10
180° regular	2.0	1.5	0.95	0.64	0.41	0.35	0.30	0.25	0.20
180° long radius					0.40	0.30	0.21	0.15	0.10
Tees:									
Line flow	0.90	0.90	0.90	0.90	0.24	0.19	0.14	0.10	0.07
Branch flow	2.4	1.8	1.4	1.1	1.0	0.80	0.64	0.58	0.41

is a factor of 2 spread among the various manufacturers. Note that K in Fig. 6.19b is, as usual, based on the average *pipe* velocity $V = Q/A$, not on the increased velocity of the flow as it passes through the narrow valve passage.

A bend or curve in a pipe, as in Fig. 6.20, always induces a loss larger than the simple straight-pipe Moody friction loss, due to flow separation on the curved walls and a swirling secondary flow arising from the centripetal accelaration. The smooth-wall loss

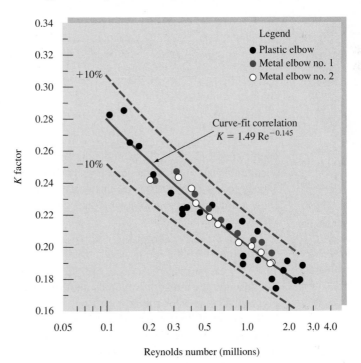

Fig. 6.18a Recent measured loss coefficients for 90° elbows. These values are less than those reported in Table 6.5. *(From Ref. 48, courtesy of R. D. Coffield.)*

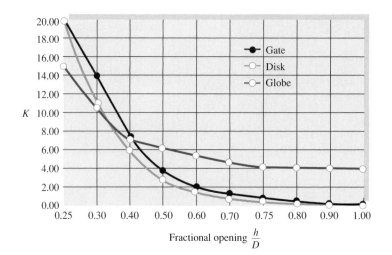

Fig. 6.18b Average loss coefficients for partially open valves (see sketches in Fig. 6.17).

(a)

Fig. 6.19 Performance of butterfly valves: (a) typical geometry (*Courtesy of Tyco Engineered Products and Services*) (b) loss coefficients for three different manufacturers.

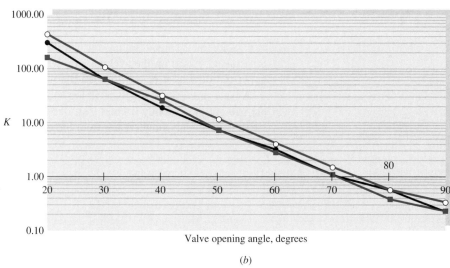

(b)

coefficients K in Fig. 6.20, from the data of Ito [49], are for *total* loss, including Moody friction effects. The separation and secondary flow losses decrease with R/d, while the Moody losses increase because the bend length increases. The curves in Fig. 6.20 thus show a minimum where the two effects cross. Ito [49] gives a curve-fit formula for the 90° bend in turbulent flow:

$$90° \text{ bend: } K \approx 0.388\alpha\left(\frac{R}{d}\right)^{0.84}\text{Re}_D{}^{-0.17} \text{ where } \alpha = 0.95 + 4.42\left(\frac{R}{d}\right)^{-1.96} \geq 1 \quad (6.80a)$$

The formula accounts for Reynolds number, which equals 200,000 in Fig. 6.20. Comprehensive reviews of curved-pipe flow, for both laminar and turbulent flow, are given by Berger et al. [53] and for 90° bends by Spedding et al. [54].

As shown in Fig. 6.21, entrance losses are highly dependent on entrance geometry, but exit losses are not. Sharp edges or protrusions in the entrance cause large zones of flow separation and large losses. A little rounding goes a long way, and a

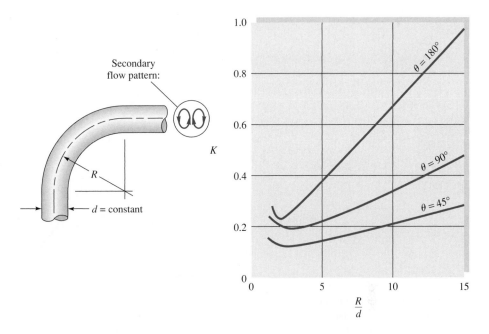

Fig. 6.20 Resistance coefficients for smooth-walled 45°, 90°, and 180° bends, at $Re_d = 200{,}000$, after Ito [49].

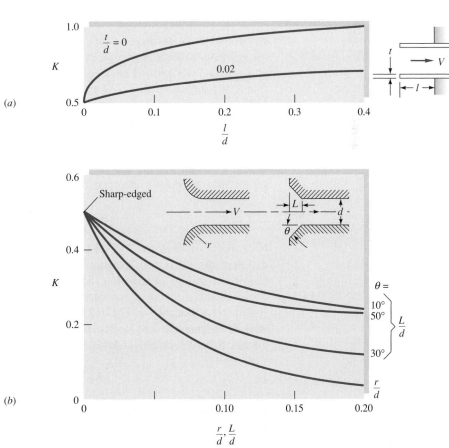

Fig. 6.21 Entrance and exit loss coefficients: (a) reentrant inlets; (b) rounded and beveled inlets. Exit losses are $K \approx 1.0$ for all shapes of exit (reentrant, sharp, beveled, or rounded). (*From Ref. 37.*)

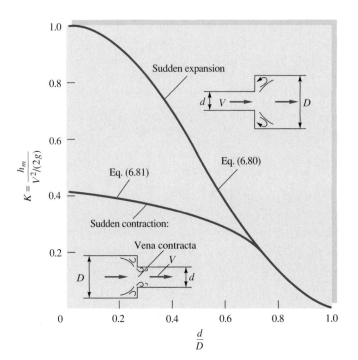

Fig. 6.22 Sudden expansion and contraction losses. Note that the loss is based on velocity head in the small pipe.

well-rounded entrance $(r = 0.2d)$ has a nearly negligible loss $K = 0.05$. At a submerged exit, on the other hand, the flow simply passes out of the pipe into the large downstream reservoir and loses all its velocity head due to viscous dissipation. Therefore $K = 1.0$ for all *submerged exits,* no matter how well rounded.

If the entrance is from a finite reservoir, it is termed a *sudden contraction* (SC) between two sizes of pipe. If the exit is to finite-sized pipe, it is termed a *sudden expansion* (SE). The losses for both are graphed in Fig. 6.22. For the sudden expansion, the shear stress in the corner separated flow, or deadwater region, is negligible, so that a control volume analysis between the expansion section and the end of the separation zone gives a theoretical loss:

$$K_{SE} = \left(1 - \frac{d^2}{D^2}\right)^2 = \frac{h_m}{V^2/(2g)} \tag{6.80}$$

Note that K is based on the velocity head in the small pipe. Equation (6.80) is in excellent agreement with experiment.

For the sudden contraction, however, flow separation in the downstream pipe causes the main stream to contract through a minimum diameter d_{min}, called the *vena contracta,* as sketched in Fig. 6.22. Because the theory of the vena contracta is not well developed, the loss coefficient in the figure for sudden contraction is experimental. It fits the empirical formula

$$K_{SC} \approx 0.42\left(1 - \frac{d^2}{D^2}\right) \tag{6.81}$$

up to the value $d/D = 0.76$, above which it merges into the sudden-expansion prediction, Eq. (6.80).

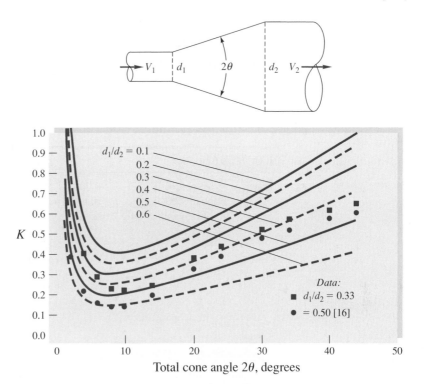

Fig. 6.23 Flow losses in a gradual conical expansion region, as calculated from Gibson's suggestion [15, 50], Eq. (6.82), for a smooth wall.

Gradual Expansion—The Diffuser

As flow enters a gradual expansion or *diffuser*, such as the conical geometry of Fig. 6.23, the velocity drops and the pressure rises. Head loss can be large, due to flow separation on the walls, if the cone angle is too great. A thinner entrance boundary layer, as in Fig. 6.6, causes a slightly smaller loss than a fully developed inlet flow. The flow loss is a combination of nonideal pressure recovery plus wall friction. Some correlating curves are shown in Fig. 6.23. The loss coefficient K is based on the velocity head in the inlet (small) pipe and depends upon cone angle 2θ and the diffuser diameter ratio d_1/d_2. There is scatter in the reported data [15, 16]. The curves in Fig. 6.23 are based on a correlation by A. H. Gibson [50], cited in Ref. 15:

$$K_{\text{diffuser}} = \frac{h_m}{V_1^2/(2\ g)} \approx 2.61 \sin \theta \left(1 - \frac{d^2}{D^2} \right)^2 + f_{\text{avg}} \frac{L}{d_{\text{avg}}} \qquad \text{for} \quad 2\theta \leq 45° \qquad (6.82)$$

For large angles, $2\theta > 45°$, drop the coefficient $(2.61 \sin\theta)$, which leaves us with a loss equivalent to the sudden expansion of Eq. (6.80). As seen, the formula is in reasonable agreement with the data from Ref. 16. The minimum loss lies in the region $5° < 2\theta < 15°$, which is the best geometry for an efficient diffuser. For angles less than $5°$, the diffuser is too long and has too much friction. Angles greater than $15°$ cause flow separation, resulting in poor pressure recovery. Professor Gordon Holloway provided the writer a recent example, where an improved diffuser design reduced the power requirement of a wind tunnel by 40 percent (100 hp decrease!). We shall look again at diffusers in Sec. 6.11, using the data of Ref. 14.

For a gradual *contraction,* the loss is very small, as seen from the following experimental values [15]:

Contraction cone angle 2θ, deg	30	45	60
K for gradual contraction	0.02	0.04	0.07

References 15, 16, 43, and 46 contain additional data on minor losses.

EXAMPLE 6.16

Water, $\rho = 1.94$ slugs/ft^3 and $\nu = 0.000011$ ft^2/s, is pumped between two reservoirs at 0.2 ft^3/s through 400 ft of 2-in-diameter pipe and several minor losses, as shown in Fig. E6.16. The roughness ratio is $\epsilon/d = 0.001$. Compute the pump horsepower required.

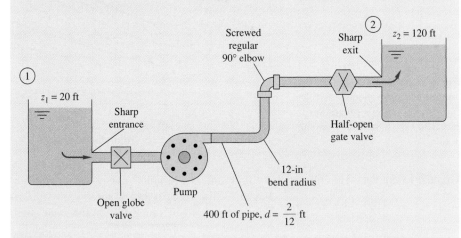

E6.16

Solution

Write the steady flow energy equation between sections 1 and 2, the two reservoir surfaces:

$$\frac{p_1}{\rho g} + \frac{V_1^2}{2g} + z_1 = \left(\frac{p_2}{\rho g} + \frac{V_2^2}{2g} + z_2\right) + h_f + \sum h_m - h_p$$

where h_p is the head increase across the pump. But since $p_1 = p_2$ and $V_1 = V_2 \approx 0$, solve for the pump head:

$$h_p = z_2 - z_1 + h_f + \sum h_m = 120 \text{ ft} - 20 \text{ ft} + \frac{V^2}{2g}\left(\frac{fL}{d} + \sum K\right) \qquad (1)$$

Now with the flow rate known, calculate

$$V = \frac{Q}{A} = \frac{0.2 \text{ ft}^3/\text{s}}{\frac{1}{4}\pi\left(\frac{2}{12} \text{ ft}\right)^2} = 9.17 \text{ ft/s}$$

Now list and sum the minor loss coefficients:

Loss	K
Sharp entrance (Fig. 6.21)	0.5
Open globe valve (2 in, Table 6.5)	6.9
12-in bend (Fig. 6.20)	0.25
Regular 90° elbow (Table 6.5)	0.95
Half-closed gate valve (from Fig. 6.18b)	2.7
Sharp exit (Fig. 6.21)	1.0
	$\Sigma\, K = 12.3$

Calculate the Reynolds number and pipe friction factor:

$$\text{Re}_d = \frac{Vd}{\nu} = \frac{9.17(\frac{2}{12})}{0.000011} = 139{,}000$$

For $\epsilon/d = 0.001$, from the Moody chart read $f = 0.0216$. Substitute into Eq. (1):

$$h_p = 100\ \text{ft} + \frac{(9.17\ \text{ft/s})^2}{2(32.2\ \text{ft/s}^2)}\left[\frac{0.0216(400)}{\frac{2}{12}} + 12.3\right]$$

$$= 100\ \text{ft} + 84\ \text{ft} = 184\ \text{ft} \quad \text{pump head}$$

The pump must provide a power to the water of

$$P = \rho g Q h_p = [1.94(32.2)\ \text{lbf/ft}^3](0.2\ \text{ft}^3/\text{s})(184\ \text{ft}) = 2300\ \text{ft} \cdot \text{lbf/s}$$

The conversion factor is 1 hp = 550 ft · lbf/s. Therefore

$$P = \frac{2300}{550} = 4.2\ \text{hp} \qquad\qquad\qquad Ans.$$

Allowing for an efficiency of 70 to 80 percent, a pump is needed with an input of about 6 hp.

6.10 Multiple-Pipe Systems[7]

If you can solve the equations for one-pipe systems, you can solve them all; but when systems contain two or more pipes, certain basic rules make the calculations very smooth. Any resemblance between these rules and the rules for handling electric circuits is not coincidental.

Figure 6.24 shows three examples of multiple-pipe systems.

Pipes in Series

The first is a set of three (or more) pipes in series. Rule 1 is that the flow rate is the same in all pipes:

$$Q_1 = Q_2 = Q_3 = \text{const} \tag{6.83}$$

or

$$V_1 d_1^2 = V_2 d_2^2 = V_3 d_3^2 \tag{6.84}$$

Rule 2 is that the total head loss through the system equals the sum of the head loss in each pipe:

$$\Delta h_{A\rightarrow B} = \Delta h_1 + \Delta h_2 + \Delta h_3 \tag{6.85}$$

[7]This section may be omitted without loss of continuity.

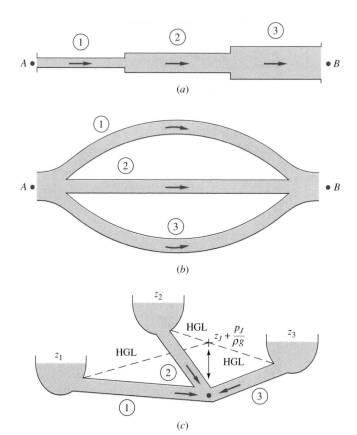

Fig. 6.24 Examples of multiple-pipe systems: (*a*) pipes in series; (*b*) pipes in parallel; (*c*) the three-reservoir junction problem.

In terms of the friction and minor losses in each pipe, we could rewrite this as

$$\Delta h_{A\rightarrow B} = \frac{V_1^2}{2g}\left(\frac{f_1 L_1}{d_1} + \sum K_1\right) + \frac{V_2^2}{2g}\left(\frac{f_2 L_2}{d_2} + \sum K_2\right)$$
$$+ \frac{V_3^2}{2g}\left(\frac{f_3 L_3}{d_3} + \sum K_3\right) \tag{6.86}$$

and so on for any number of pipes in the series. Since V_2 and V_3 are proportional to V_1 from Eq. (6.84), Eq. (6.86) is of the form

$$\Delta h_{A\rightarrow B} = \frac{V_1^2}{2g}(\alpha_0 + \alpha_1 f_1 + \alpha_2 f_2 + \alpha_3 f_3) \tag{6.87}$$

where the α_i are dimensionless constants. If the flow rate is given, we can evaluate the right-hand side and hence the total head loss. If the head loss is given, a little iteration is needed, since f_1, f_2, and f_3 all depend on V_1 through the Reynolds number. Begin by calculating f_1, f_2, and f_3, assuming fully rough flow, and the solution for V_1 will converge with one or two iterations. EES is ideal for this purpose.

EXAMPLE 6.17

Given is a three-pipe series system, as in Fig. 6.24a. The total pressure drop is $p_A - p_B = 150{,}000$ Pa, and the elevation drop is $z_A - z_B = 5$ m. The pipe data are

Pipe	L, m	d, cm	ϵ, mm	ϵ/d
1	100	8	0.24	0.003
2	150	6	0.12	0.002
3	80	4	0.20	0.005

The fluid is water, $\rho = 1000$ kg/m^3 and $\nu = 1.02 \times 10^{-6}$ m^2/s. Calculate the flow rate Q in m^3/h through the system.

Solution

The total head loss across the system is

$$\Delta h_{A \to B} = \frac{p_A - p_B}{\rho g} + z_A - z_B = \frac{150{,}000}{1000(9.81)} + 5 \text{ m} = 20.3 \text{ m}$$

From the continuity relation (6.84) the velocities are

$$V_2 = \frac{d_1^2}{d_2^2} V_1 = \frac{16}{9} V_1 \qquad V_3 = \frac{d_1^2}{d_3^2} V_1 = 4V_1$$

and

$$\text{Re}_2 = \frac{V_2 d_2}{V_1 d_1} \text{Re}_1 = \frac{4}{3} \text{Re}_1 \qquad \text{Re}_3 = 2\text{Re}_1$$

Neglecting minor losses and substituting into Eq. (6.86), we obtain

$$\Delta h_{A \to B} = \frac{V_1^2}{2g} \left[1250 f_1 + 2500 \left(\frac{16}{9} \right)^2 f_2 + 2000(4)^2 f_3 \right]$$

or

$$20.3 \text{ m} = \frac{V_1^2}{2g} (1250 f_1 + 7900 f_2 + 32{,}000 f_3) \tag{1}$$

This is the form that was hinted at in Eq. (6.87). It seems to be dominated by the third pipe loss $32{,}000 f_3$. Begin by estimating f_1, f_2, and f_3 from the Moody-chart fully rough regime:

$$f_1 = 0.0262 \qquad f_2 = 0.0234 \qquad f_3 = 0.0304$$

Substitute in Eq. (1) to find $V_1^2 \approx 2g(20.3)/(33 + 185 + 973)$. The first estimate thus is $V_1 = 0.58$ m/s, from which

$$\text{Re}_1 \approx 45{,}400 \qquad \text{Re}_2 = 60{,}500 \qquad \text{Re}_3 = 90{,}800$$

Hence, from the Moody chart,

$$f_1 = 0.0288 \qquad f_2 = 0.0260 \qquad f_3 = 0.0314$$

Substitution into Eq. (1) gives the better estimate

$$V_1 = 0.565 \text{ m/s} \qquad Q = \tfrac{1}{4}\pi d_1^2 V_1 = 2.84 \times 10^{-3}\,\text{m}^3/\text{s}$$

or
$$Q = 10.2 \text{ m}^3/\text{h} \qquad\qquad\qquad\qquad\qquad Ans.$$

A second iteration gives $Q = 10.22$ m^3/h, a negligible change.

Pipes in Parallel

The second multiple-pipe system is the *parallel* flow case shown in Fig. 6.24b. Here the pressure drop is the same in each pipe, and the total flow is the sum of the individual flows:

$$\Delta h_{A\to B} = \Delta h_1 = \Delta h_2 = \Delta h_3 \tag{6.88a}$$

$$Q = Q_1 + Q_2 + Q_3 \tag{6.88b}$$

If the total head loss is known, it is straightforward to solve for Q_i in each pipe and sum them, as will be seen in Example 6.18. The reverse problem, of determining ΣQ_i when h_f is known, requires iteration. Each pipe is related to h_f by the Moody relation $h_f = f(L/d)(V^2/2g) = fQ^2/C$, where $C = \pi^2 g d^5/8L$. Thus each pipe has nearly quadratic nonlinear parallel resistance, and head loss is related to total flow rate by

$$h_f = \frac{Q^2}{(\Sigma\sqrt{C_i/f_i})^2} \qquad \text{where } C_i = \frac{\pi^2 g d_i^5}{8L_i} \tag{6.89}$$

Since the f_i vary with Reynolds number and roughness ratio, one begins Eq. (6.89) by guessing values of f_i (fully rough values are recommended) and calculating a first estimate of h_f. Then each pipe yields a flow-rate estimate $Q_i \approx (C_i h_f/f_i)^{1/2}$ and hence a new Reynolds number and a better estimate of f_i. Then repeat Eq. (6.89) to convergence.

It should be noted that both of these parallel-pipe cases—finding either ΣQ or h_f— are easily solved by EES if reasonable initial guesses are given.

EXAMPLE 6.18

Assume that the same three pipes in Example 6.17 are now in parallel with the same total head loss of 20.3 m. Compute the total flow rate Q, neglecting minor losses.

Solution

From Eq. (6.88a) we can solve for each V separately:

$$20.3 \text{ m} = \frac{V_1^2}{2g} 1250 f_1 = \frac{V_2^2}{2g} 2500 f_2 = \frac{V_3^2}{2g} 2000 f_3 \tag{1}$$

Guess fully rough flow in pipe 1: $f_1 = 0.0262$, $V_1 = 3.49$ m/s; hence $\text{Re}_1 = V_1 d_1/\nu = 273{,}000$. From the Moody chart read $f_1 = 0.0267$; recompute $V_1 = 3.46$ m/s, $Q_1 = 62.5$ m^3/h. [This problem can also be solved from Eq. (6.51).]

Next guess for pipe 2: $f_2 \approx 0.0234$, $V_2 \approx 2.61$ m/s; then $Re_2 = 153,000$, and hence $f_2 = 0.0246$, $V_2 = 2.55$ m/s, $Q_2 = 25.9$ m^3/h.

Finally guess for pipe 3: $f_3 \approx 0.0304$, $V_3 \approx 2.56$ m/s; then $Re_3 = 100,000$, and hence $f_3 = 0.0313$, $V_3 = 2.52$ m/s, $Q_3 = 11.4$ m^3/h.

This is satisfactory convergence. The total flow rate is

$$Q = Q_1 + Q_2 + Q_3 = 62.5 + 25.9 + 11.4 = 99.8 \text{ m}^3/\text{h} \qquad Ans.$$

These three pipes carry 10 times more flow in parallel than they do in series.

This example is ideal for EES. One enters the pipe data (L_i, d_i, ϵ_i); the fluid properties (ρ, μ); the definitions $Q_i = (\pi/4)d_i^2 V_i$, $Re_i = \rho V_i d_i/\mu$, and $h_f = f_i (L_i/d_i) (V_i^2/2g)$; plus the Colebrook formula (6.48) for each friction factor f_i. There is no need to use resistance ideas such as Eq. (6.89). Specify that $f_i > 0$ and $Re_i > 4000$. Then, if one enters $Q = \Sigma Q_i = (99.8/3600)$ m^3/s, EES quickly solves for $h_f = 20.3$ m. Conversely, if one enters $h_f = 20.3$ m, EES solves for $Q = 99.8$ m^3/h.

Three-Reservoir Junction

Consider the third example of a *three-reservoir pipe junction*, as in Fig. 6.24c. If all flows are considered positive toward the junction, then

$$Q_1 + Q_2 + Q_3 = 0 \qquad (6.90)$$

which obviously implies that one or two of the flows must be away from the junction. The pressure must change through each pipe so as to give the same static pressure p_J at the junction. In other words, let the HGL at the junction have the elevation

$$h_J = z_J + \frac{p_J}{\rho g}$$

where p_J is in gage pressure for simplicity. Then the head loss through each, assuming $p_1 = p_2 = p_3 = 0$ (gage) at each reservoir surface, must be such that

$$\Delta h_1 = \frac{V_1^2}{2g} \frac{f_1 L_1}{d_1} = z_1 - h_J$$

$$\Delta h_2 = \frac{V_2^2}{2g} \frac{f_2 L_2}{d_2} = z_2 - h_J \qquad (6.91)$$

$$\Delta h_3 = \frac{V_3^2}{2g} \frac{f_3 L_3}{d_3} = z_3 - h_J$$

We guess the position h_J and solve Eqs. (6.91) for V_1, V_2, and V_3 and hence Q_1, Q_2, and Q_3, iterating until the flow rates balance at the junction according to Eq. (6.90). If we guess h_J too *high*, the sum $Q_1 + Q_2 + Q_3$ will be *negative* and the remedy is to reduce h_J, and vice versa.

EXAMPLE 6.19

Take the same three pipes as in Example 6.17, and assume that they connect three reservoirs at these surface elevations

$$z_1 = 20 \text{ m} \qquad z_2 = 100 \text{ m} \qquad z_3 = 40 \text{ m}$$

Find the resulting flow rates in each pipe, neglecting minor losses.

Solution

As a first guess, take h_J equal to the middle reservoir height, $z_3 = h_J = 40$ m. This saves one calculation ($Q_3 = 0$) and enables us to get the lay of the land:

Reservoir	h_J, m	$z_i - h_J$, m	f_i	V_i, m/s	Q_i, m³/h	L_i/d_i
1	40	−20	0.0267	−3.43	−62.1	1250
2	40	60	0.0241	4.42	45.0	2500
3	40	0		0	0	2000
					$\Sigma Q = -17.1$	

Since the sum of the flow rates toward the junction is negative, we guessed h_J too high. Reduce h_J to 30 m and repeat:

Reservoir	h_J, m	$z_i - h_J$, m	f_i	V_i, m/s	Q_i, m³/h
1	30	−10	0.0269	−2.42	−43.7
2	30	70	0.0241	4.78	48.6
3	30	10	0.0317	1.76	8.0
					$\Sigma Q = 12.9$

This is positive ΣQ, and so we can linearly interpolate to get an accurate guess: $h_J \approx 34.3$ m. Make one final list:

Reservoir	h_J, m	$z_i - h_J$, m	f_i	V_i, m/s	Q_i, m³/h
1	34.3	−14.3	0.0268	−2.90	−52.4
2	34.3	65.7	0.0241	4.63	47.1
3	34.3	5.7	0.0321	1.32	6.0
					$\Sigma Q = 0.7$

This is close enough; hence we calculate that the flow rate is 52.4 m³/h toward reservoir 3, balanced by 47.1 m³/h away from reservoir 1 and 6.0 m³/h away from reservoir 3.

One further iteration with this problem would give $h_J = 34.53$ m, resulting in $Q_1 = -52.8$, $Q_2 = 47.0$, and $Q_3 = 5.8$ m³/h, so that $\Sigma Q = 0$ to three-place accuracy. Pedagogically speaking, we would then be exhausted.

Pipe Networks

The ultimate case of a multipipe system is the *piping network* illustrated in Fig. 6.25. This might represent a water supply system for an apartment or subdivision or even a city. This network is quite complex algebraically but follows the same basic rules:

1. The net flow into any junction must be zero.
2. The net pressure change around any closed loop must be zero. In other words, the HGL at each junction must have one and only one elevation.
3. All pressure changes must satisfy the Moody and minor-loss friction correlations.

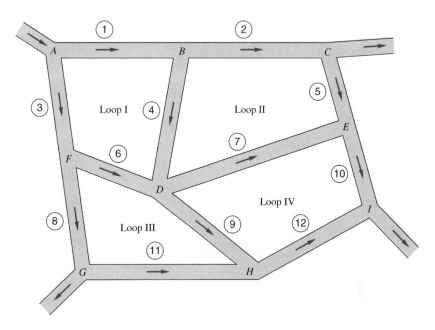

Fig. 6.25 Schematic of a piping network.

By supplying these rules to each junction and independent loop in the network, one obtains a set of simultaneous equations for the flow rates in each pipe leg and the HGL (or pressure) at each junction. Solution may then be obtained by numerical iteration, as first developed in a hand calculation technique by Prof. Hardy Cross in 1936 [17]. Computer solution of pipe network problems is now quite common and covered in at least one specialized text [18]. Network analysis is quite useful for real water distribution systems if well calibrated with the actual system head loss data.

6.11 Experimental Duct Flows: Diffuser Performance

The Moody chart is such a great correlation for tubes of any cross section with any roughness or flow rate that we may be deluded into thinking that the world of internal flow prediction is at our feet. Not so. The theory is reliable only for ducts of constant cross section. As soon as the section varies, we must rely principally on experiment to determine the flow properties. As mentioned many times before, experimentation is a vital part of fluid mechanics.

Literally thousands of papers in the literature report experimental data for specific internal and external viscous flows. We have already seen several examples:

1. Vortex shedding from a cylinder (Fig. 5.2).
2. Drag of a sphere and a cylinder (Fig. 5.3).
3. Hydraulic model of a dam spillway (Fig. 5.9).
4. Rough-wall pipe flows (Fig. 6.12).
5. Secondary flow in ducts (Fig. 6.16).
6. Minor duct loss coefficients (Sec. 6.9).

Chapter 7 will treat a great many more external flow experiments, especially in Sec. 7.6. Here we shall show data for one type of internal flow, the diffuser.

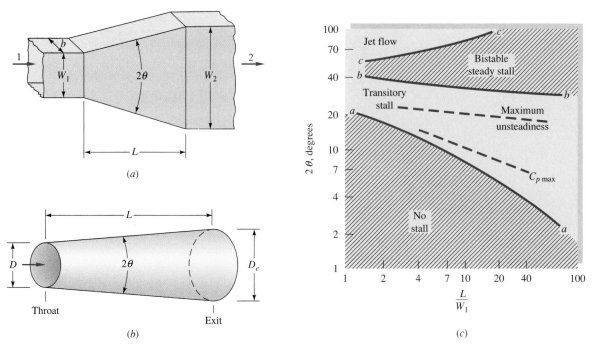

Fig. 6.26 Diffuser geometry and typical flow regimes: (*a*) geometry of a flat-walled diffuser; (*b*) geometry of a conical diffuser; (*c*) flat diffuser stability map. *(From Ref. 14, by permission of Creare, Inc.)*

Diffuser Performance

A diffuser, shown in Fig. 6.26*a* and *b*, is an expansion or area increase intended to reduce velocity in order to recover the pressure head of the flow. Rouse and Ince [6] relate that it may have been invented by customers of the early Roman (about 100 A.D.) water supply system, where water flowed continuously and was billed according to pipe size. The ingenious customers discovered that they could increase the flow rate at no extra cost by flaring the outlet section of the pipe.

Engineers have always designed diffusers to increase pressure and reduce kinetic energy of ducted flows, but until about 1950, diffuser design was a combination of art, luck, and vast amounts of empiricism. Small changes in design parameters caused large changes in performance. The Bernoulli equation seemed highly suspect as a useful tool.

Neglecting losses and gravity effects, the incompressible Bernoulli equation predicts that

$$p + \tfrac{1}{2}\rho V^2 = p_0 = \text{const} \tag{6.92}$$

where p_0 is the stagnation pressure the fluid would achieve if the fluid were slowed to rest ($V = 0$) without losses.

The basic output of a diffuser is the *pressure-recovery coefficient* C_p, defined as

$$C_p = \frac{p_e - p_t}{p_{0t} - p_t} \tag{6.93}$$

where subscripts e and t mean the exit and the throat (or inlet), respectively. Higher C_p means better performance.

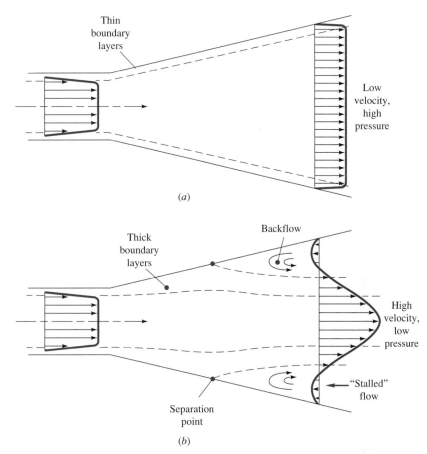

Fig. 6.27 Diffuser performance: (*a*) ideal pattern with good performance; (*b*) actual measured pattern with boundary layer separation and resultant poor performance.

Consider the flat-walled diffuser in Fig. 6.26*a*, where section 1 is the inlet and section 2 the exit. Application of Bernoulli's equation (6.92) to this diffuser predicts that

$$p_{01} = p_1 + \tfrac{1}{2}\rho V_1^2 = p_2 + \tfrac{1}{2}\rho V_2^2 = p_{02}$$

or

$$C_{p,\text{frictionless}} = 1 - \left(\frac{V_2}{V_1}\right)^2 \tag{6.94}$$

Meanwhile, steady one-dimensional continuity would require that

$$Q = V_1 A_1 = V_2 A_2 \tag{6.95}$$

Combining (6.94) and (6.95), we can write the performance in terms of the *area ratio* $AR = A_2/A_1$, which is a basic parameter in diffuser design:

$$C_{p,\text{frictionless}} = 1 - (AR)^{-2} \tag{6.96}$$

A typical design would have AR = 5:1, for which Eq. (6.96) predicts $C_p = 0.96$, or nearly full recovery. But, in fact, measured values of C_p for this area ratio [14] are only as high as 0.86 and can be as low as 0.24.

The basic reason for the discrepancy is flow separation, as sketched in Fig. 6.27. The increasing pressure in the diffuser is an unfavorable gradient (Sec. 7.5), which causes the viscous boundary layers to break away from the walls and greatly reduces the performance. Computational fluid dynamics (CFD) can now predict this behavior.

As an added complication to boundary layer separation, the flow patterns in a diffuser are highly variable and were considered mysterious and erratic until 1955, when Kline revealed the structure of these patterns with flow visualization techniques in a simple water channel.

A complete *stability map* of diffuser flow patterns was published in 1962 by Fox and Kline [21], as shown in Fig. 6.26c. There are four basic regions. Below line *aa* there is steady viscous flow, no separation, and moderately good performance. Note that even a very short diffuser will separate, or stall, if its half-angle is greater than 10°.

Between lines *aa* and *bb* is a transitory stall pattern with strongly unsteady flow. Best performance (highest C_p) occurs in this region. The third pattern, between *bb* and *cc*, is steady bistable stall from one wall only. The stall pattern may flip-flop from one wall to the other, and performance is poor.

The fourth pattern, above line *cc*, is *jet flow*, where the wall separation is so gross and pervasive that the mainstream ignores the walls and simply passes on through at nearly constant area. Performance is extremely poor in this region.

Dimensional analysis of a flat-walled or conical diffuser shows that C_p should depend on the following parameters:

1. Any two of the following geometric parameters:
   a. Area ratio AR = A_2/A_1 or $(D_e/D)^2$
   b. Divergence angle 2θ
   c. Slenderness L/W_1 or L/D
2. Inlet Reynolds number $\text{Re}_t = V_1 W_1/\nu$ or $\text{Re}_t = V_1 D/\nu$
3. Inlet Mach number $\text{Ma}_t = V_1/a_1$
4. Inlet boundary layer *blockage factor* $B_t = A_{\text{BL}}/A_1$, where A_{BL} is the wall area blocked, or displaced, by the retarded boundary layer flow in the inlet (typically B_t varies from 0.03 to 0.12)

A flat-walled diffuser would require an additional shape parameter to describe its cross section:

5. Aspect ratio AS = b/W_1

Even with this formidable list, we have omitted five possible important effects: inlet turbulence, inlet swirl, inlet profile vorticity, superimposed pulsations, and downstream obstruction, all of which occur in practical machinery applications.

The three most important parameters are AR, θ, and B. Typical performance maps for diffusers are shown in Fig. 6.28. For this case of 8 to 9 percent blockage, both the flat-walled and conical types give about the same maximum performance, $C_p = 0.70$, but at different divergence angles (9° flat versus 4.5° conical). Both types fall far short of the Bernoulli estimates of $C_p = 0.93$ (flat) and 0.99 (conical), primarily because of the blockage effect.

From the data of Ref. 14 we can determine that, in general, performance decreases with blockage and is approximately the same for both flat-walled and conical diffusers,

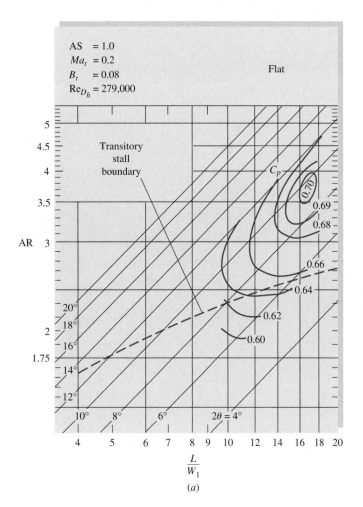

AS = 1.0
Ma_t = 0.2
B_t = 0.08
Re_{D_h} = 279,000

Flat

Transitory stall boundary

C_p

Fig. 6.28a Typical performance maps for flat-wall and conical diffusers at similar operating conditions: flat wall. *(From Ref. 14, by permission of Creare, Inc.)*

as shown in Table 6.6. In all cases, the best conical diffuser is 10 to 80 percent longer than the best flat-walled design. Therefore, if length is limited in the design, the flat-walled design will give the better performance depending on duct cross section.

The experimental design of a diffuser is an excellent example of a successful attempt to minimize the undesirable effects of adverse pressure gradient and flow separation.

Table 6.6 Maximum Diffuser Performance Data [14]

Inlet blockage B_t	Flat-walled		Conical	
	C_p,max	L/W_1	C_p,max	L/d
0.02	0.86	18	0.83	20
0.04	0.80	18	0.78	22
0.06	0.75	19	0.74	24
0.08	0.70	20	0.71	26
0.10	0.66	18	0.68	28
0.12	0.63	16	0.65	30

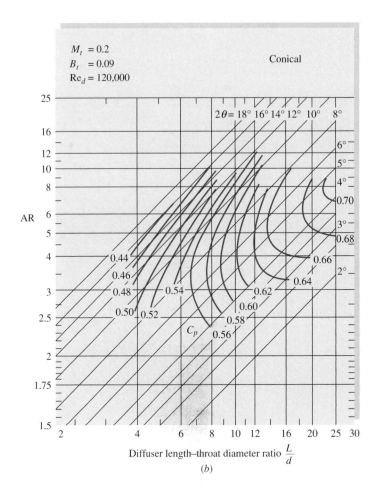

Fig. 6.28b Typical performance maps for flat-wall and conical diffusers at similar operating conditions: conical wall. *(From Ref. 14, by permission of Creare, Inc.)*

6.12 Fluid Meters

Almost all practical fluid engineering problems are associated with the need for an accurate flow measurement. There is a need to measure *local* properties (velocity, pressure, temperature, density, viscosity, turbulent intensity), *integrated* properties (mass flow and volume flow), and *global* properties (visualization of the entire flow field). We shall concentrate in this section on velocity and volume flow measurements.

We have discussed pressure measurement in Sec. 2.10. Measurement of other thermodynamic properties, such as density, temperature, and viscosity, is beyond the scope of this text and is treated in specialized books such as Refs. 22 and 23. Global visualization techniques were discussed in Sec. 1.11 for low-speed flows, and the special optical techniques used in high-speed flows are treated in Ref. 34 of Chap. 1. Flow measurement schemes suitable for open-channel and other free-surface flows are treated in Chap. 10.

Local Velocity Measurements

Velocity averaged over a small region, or point, can be measured by several different physical principles, listed in order of increasing complexity and sophistication:

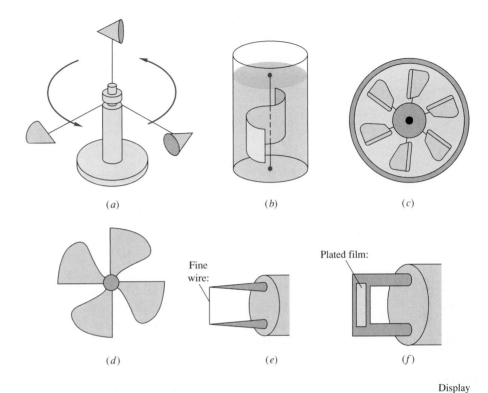

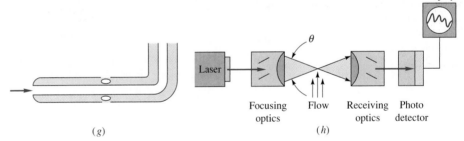

Fig. 6.29 Eight common velocity meters: (*a*) three-cup anemometer; (*b*) Savonius rotor; (*c*) turbine mounted in a duct; (*d*) free-propeller meter; (*e*) hot-wire anemometer; (*f*) hot-film anemometer; (*g*) pitot-static tube; (*h*) laser-doppler anemometer.

1. Trajectory of floats or neutrally buoyant particles.
2. Rotating mechanical devices:
   a. Cup anemometer.
   b. Savonius rotor.
   c. Propeller meter.
   d. Turbine meter.
3. Pitot-static tube (Fig. 6.30).
4. Electromagnetic current meter.
5. Hot wires and hot films.
6. Laser-doppler anemometer (LDA).

Some of these meters are sketched in Fig. 6.29.

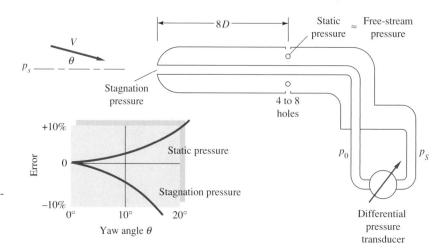

Fig. 6.30 Pitot-static tube for combined measurement of static and stagnation pressure in a moving stream.

Floats or Buoyant Particles. A simple but effective estimate of flow velocity can be found from visible particles entrained in the flow. Examples include flakes on the surface of a channel flow, small neutrally buoyant spheres mixed with a liquid, or hydrogen bubbles. Sometimes gas flows can be estimated from the motion of entrained dust particles. One must establish whether the particle motion truly simulates the fluid motion. Floats are commonly used to track the movement of ocean waters and can be designed to move at the surface, along the bottom, or at any given depth [24]. Many official tidal current charts [25] were obtained by releasing and timing a floating spar attached to a length of string. One can release whole groups of spars to determine a flow pattern.

Rotating Sensors. The rotating devices of Fig. 6.29a to d can be used in either gases or liquids, and their rotation rate is approximately proportional to the flow velocity. The cup anemometer (Fig. 6.29a) and Savonius rotor (Fig. 6.29b) always rotate the same way, regardless of flow direction. They are popular in atmospheric and oceanographic applications and can be fitted with a direction vane to align themselves with the flow. The ducted-propeller (Fig. 6.29c) and free-propeller (Fig. 6.29d) meters must be aligned with the flow parallel to their axis of rotation. They can sense reverse flow because they will then rotate in the opposite direction. All these rotating sensors can be attached to counters or sensed by electromagnetic or slip-ring devices for either a continuous or a digital reading of flow velocity. All have the disadvantage of being relatively large and thus not representing a "point."

Pitot-Static Tube. A slender tube aligned with the flow (Figs. 6.29g and 6.30) can measure local velocity by means of a pressure difference. It has sidewall holes to measure the static pressure p_s in the moving stream and a hole in the front to measure the *stagnation* pressure p_0, where the stream is decelerated to zero velocity. Instead of measuring p_0 or p_s separately, it is customary to measure their difference with, say, a transducer, as in Fig. 6.30.

If $\mathrm{Re}_D > 1000$, where D is the probe diameter, the flow around the probe is nearly frictionless and Bernoulli's relation, Eq. (3.77), applies with good accuracy. For incompressible flow

$$p_s + \tfrac{1}{2}\rho V^2 + \rho g z_s \approx p_0 + \tfrac{1}{2}\rho(0)^2 + \rho g z_0$$

Assuming that the elevation pressure difference $\rho g(z_s - z_0)$ is negligible, this reduces to

$$V \approx \left[2\frac{(p_0 - p_s)}{\rho} \right]^{1/2} \tag{6.97}$$

This is the *Pitot formula,* named after the French engineer, Henri de Pitot, who designed the device in 1732.

The primary disadvantage of the pitot tube is that it must be aligned with the flow direction, which may be unknown. For yaw angles greater than $5°$, there are substantial errors in both the p_0 and p_s measurements, as shown in Fig. 6.30. The pitot-static tube is useful in liquids and gases; for gases a compressibility correction is necessary if the stream Mach number is high (Chap. 9). Because of the slow response of the fluid-filled tubes leading to the pressure sensors, it is not useful for unsteady flow measurements. It does resemble a point and can be made small enough to measure, for example, blood flow in arteries and veins. It is not suitable for low-velocity measurement in gases because of the small pressure differences developed. For example, if $V = 1$ ft/s in standard air, from Eq. (6.97) we compute $p_0 - p$ equal to only 0.001 lbf/ft^2 (0.048 Pa). This is beyond the resolution of most pressure gages.

Electromagnetic Meter.   If a magnetic field is applied across a conducting fluid, the fluid motion will induce a voltage across two electrodes placed in or near the flow. The electrodes can be streamlined or built into the wall, and they cause little or no flow resistance. The output is very strong for highly conducting fluids such as liquid metals. Seawater also gives good output, and electromagnetic current meters are in common use in oceanography. Even low-conductivity freshwater can be measured by amplifying the output and insulating the electrodes. Commercial instruments are available for most liquid flows but are relatively costly. Electromagnetic flowmeters are treated in Ref. 26.

Hot-Wire Anemometer.   A very fine wire ($d = 0.01$ mm or less) heated between two small probes, as in Fig. 6.29e, is ideally suited to measure rapidly fluctuating flows such as the turbulent boundary layer. The idea dates back to work by L. V. King in 1914 on heat loss from long thin cylinders. If electric power is supplied to heat the cylinder, the loss varies with flow velocity across the cylinder according to *King's law*

$$q = I^2 R \approx a + b(\rho V)^n \tag{6.98}$$

where $n \approx \frac{1}{3}$ at very low Reynolds numbers and equals $\frac{1}{2}$ at high Reynolds numbers. The hot wire normally operates in the high-Reynolds-number range but should be calibrated in each situation to find the best-fit a, b, and n. The wire can be operated either at constant current I, so that resistance R is a measure of V, or at constant resistance R (constant temperature), with I a measure of velocity. In either case, the output is a nonlinear function of V, and the equipment should contain a *linearizer* to produce convenient velocity data. Many varieties of commercial hot-wire equipment are available, as are do-it-yourself designs [27]. Excellent detailed discussions of the hot wire are given in Ref. 28.

Because of its frailty, the hot wire is not suited to liquid flows, whose high density and entrained sediment will knock the wire right off. A more stable yet quite sensitive alternative for liquid flow measurement is the hot-film anemometer (Fig. 6.29f). A thin metallic film, usually platinum, is plated onto a relatively thick support, which can be a wedge, a cone, or a cylinder. The operation is similar to the hot wire. The cone gives best response but is liable to error when the flow is yawed to its axis.

Hot wires can easily be arranged in groups to measure two- and three-dimensional velocity components.

Laser-Doppler Anemometer.   In the LDA a laser beam provides highly focused, coherent monochromatic light that is passed through the flow. When this light is scattered from a moving particle in the flow, a stationary observer can detect a change, or *doppler shift*, in the frequency of the scattered light. The shift Δf is proportional to the velocity of the particle. There is essentially zero disturbance of the flow by the laser.

Figure 6.29*h* shows the popular dual-beam mode of the LDA. A focusing device splits the laser into two beams, which cross the flow at an angle θ. Their intersection, which is the measuring volume or resolution of the measurement, resembles an ellipsoid about 0.5 mm wide and 0.1 mm in diameter. Particles passing through this measuring volume scatter the beams; they then pass through receiving optics to a photodetector, which converts the light to an electric signal. A signal processor then converts electric frequency to a voltage that can be either displayed or stored. If λ is the wavelength of the laser light, the measured velocity is given by

$$V = \frac{\lambda \, \Delta f}{2 \sin{(\theta/2)}} \tag{6.99}$$

Multiple components of velocity can be detected by using more than one photodetector and other operating modes. Either liquids or gases can be measured as long as scattering particles are present. In liquids, normal impurities serve as scatterers, but gases may have to be seeded. The particles may be as small as the wavelength of the light. Although the measuring volume is not as small as with a hot wire, the LDA is capable of measuring turbulent fluctuations.

The advantages of the LDA are as follows:

1.   No disturbance of the flow.
2.   High spatial resolution of the flow field.
3.   Velocity data that are independent of the fluid thermodynamic properties.
4.   An output voltage that is linear with velocity.
5.   No need for calibration.

The disadvantages are that both the apparatus and the fluid must be transparent to light and that the cost is high (a basic system shown in Fig. 6.29*h* begins at about $50,000).

Once installed, an LDA can map the entire flow field in minutest detail. To truly appreciate the power of the LDA, one should examine, for instance, the amazingly detailed three-dimensional flow profiles measured by Eckardt [29] in a high-speed centrifugal compressor impeller. Extensive discussions of laser velocimetry are given in Refs. 38 and 39.

Particle Image Velocimetry.   This popular new idea, called PIV for short, measures not just a single point but instead maps the entire field of flow. An illustration was shown in Fig. 1.17*b*. The flow is seeded with neutrally buoyant particles. A planar laser light sheet across the flow is pulsed twice and photographed twice. If $\Delta \mathbf{r}$ is the particle displacement vector over a short time Δt, an estimate of its velocity is $\mathbf{V} \approx \Delta \mathbf{r}/\Delta t$.

A dedicated computer applies this formula to a whole cloud of particles and thus maps the flow field. One can also use the data to calculate velocity gradient and vorticity fields. Since the particles all look alike, other cameras may be needed to identify them. Three-dimensional velocities can be measured by two cameras in a stereoscopic arrangement. The PIV method is not limited to stop-action. New high-speed cameras (up to 10,000 frames per second) can record movies of unsteady flow fields. For further details, see the monograph by M. Raffel [51].

EXAMPLE 6.20

The pitot-static tube of Fig. 6.30 uses mercury as a manometer fluid. When it is placed in a water flow, the manometer height reading is $h = 8.4$ in. Neglecting yaw and other errors, what is the flow velocity V in ft/s?

Solution

From the two-fluid manometer relation (2.33), with $z_A = z_2$, the pressure difference is related to h by

$$p_0 - p_s = (\gamma_M - \gamma_w)h$$

Taking the specific weights of mercury and water from Table 2.1, we have

$$p_0 - p_s = (846 - 62.4 \text{ lbf/ft}^3)\frac{8.4}{12}\text{ft} = 549 \text{ lbf/ft}^2$$

The density of water is $62.4/32.2 = 1.94$ slugs/ft^3. Introducing these values into the pitot-static formula (6.97), we obtain

$$V = \left[\frac{2(549 \text{ lbf/ft}^2)}{1.94 \text{ slugs/ft}^3}\right]^{1/2} = 23.8 \text{ ft/s} \qquad\qquad Ans.$$

Since this is a low-speed flow, no compressibility correction is needed.

Volume Flow Measurements

It is often desirable to measure the integrated mass, or volume flow, passing through a duct. Accurate measurement of flow is vital in billing customers for a given amount of liquid or gas passing through a duct. The different devices available to make these measurements are discussed in great detail in the ASME text on fluid meters [30]. These devices split into two classes: mechanical instruments and head loss instruments.

The mechanical instruments measure actual mass or volume of fluid by trapping it and counting it. The various types of measurement are

1. Mass measurement
   a. Weighing tanks
   b. Tilting traps
2. Volume measurement
   a. Volume tanks
   b. Reciprocating pistons

   c. Rotating slotted rings

   d. Nutating disc

   e. Sliding vanes

   f. Gear or lobed impellers

   g. Reciprocating bellows

   h. Sealed-drum compartments

The last three of these are suitable for gas flow measurement.

    The head loss devices obstruct the flow and cause a pressure drop, which is a measure of flux:

1. Bernoulli-type devices
   a. Thin-plate orifice
   b. Flow nozzle
   c. Venturi tube
2. Friction loss devices
   a. Capillary tube
   b. Porous plug

The friction loss meters cause a large nonrecoverable head loss and obstruct the flow too much to be generally useful.

    Six other widely used meters operate on different physical principles:

1. Turbine meter
2. Vortex meter
3. Ultrasonic flowmeter
4. Rotameter
5. Coriolis mass flowmeter
6. Laminar flow element

Nutating Disc Meter.   For measuring liquid *volumes,* as opposed to volume rates, the most common devices are the nutating disc and the turbine meter. Figure 6.31 shows

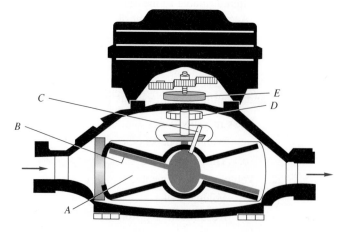

Fig. 6.31 Cutaway sketch of a nutating disc fluid meter. *A:* metered-volume chamber; *B:* nutating disc; *C:* rotating spindle; *D:* drive magnet; *E:* magnetic counter sensor. *(Courtesy of Badger Meter, Inc., Milwaukee, Wisconsin.)*

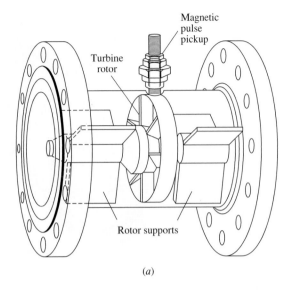

(a)

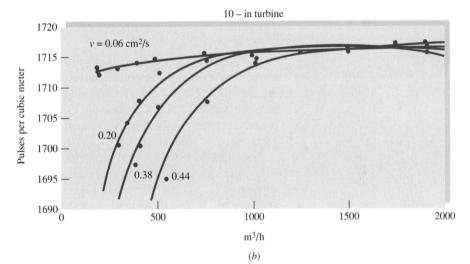

10 – in turbine

(b)

Fig. 6.32 The turbine meter widely used in the oil, gas, and water supply industries: (a) basic design; (b) typical calibration curve for a range of crude oils. *(Daniel Industries, Houston, TX.)*

a cutaway sketch of *a nutating disc meter,* widely used in both water and gasoline delivery systems. The mechanism is clever and perhaps beyond the writer's capability to explain. The metering chamber is a slice of a sphere and contains a rotating disc set at an angle to the incoming flow. The fluid causes the disc to *nutate* (spin eccentrically), and one revolution corresponds to a certain fluid volume passing through. Total volume is obtained by counting the number of revolutions.

Turbine Meter.   The turbine meter, sometimes called a *propeller meter,* is a freely rotating propeller that can be installed in a pipeline. A typical design is shown in Fig. 6.32a. There are flow straighteners upstream of the rotor, and the rotation

Fig. 6.33 A Commercial handheld wind velocity turbine meter. *(Courtesy of Nielsen-Kellerman Company.)*

is measured by electric or magnetic pickup of pulses caused by passage of a point on the rotor. The rotor rotation is approximately proportional to the volume flow in the pipe.

Like the nutating disc, a major advantage of the turbine meter is that each pulse corresponds to a finite incremental volume of fluid, and the pulses are digital and can be summed easily. Liquid flow turbine meters have as few as two blades and produce a constant number of pulses per unit fluid volume over a 5:1 flow rate range with ± 0.25 percent accuracy. Gas meters need many blades to produce sufficient torque and are accurate to ± 1 percent.

Since turbine meters are very individualistic, flow calibration is an absolute necessity. A typical liquid meter calibration curve is shown in Fig. 6.32b. Researchers attempting to establish universal calibration curves have met with little practical success as a result of manufacturing variabilities.

Turbine meters can also be used in unconfined flow situations, such as winds or ocean currents. They can be compact, even microsize with two or three component directions. Figure 6.33 illustrates a handheld wind velocity meter that uses a seven-bladed turbine with a calibrated digital output. The accuracy of this device is quoted at ± 2 percent.

Vortex Flowmeters.   Recall from Fig. 5.2 that a bluff body placed in a uniform crossflow sheds alternating vortices at a nearly uniform Strouhal number St = fL/U, where U is the approach velocity and L is a characteristic body width. Since L and St are constant, this means that the shedding frequency is proportional to velocity:

$$f = (\text{const})(U) \tag{6.100}$$

The vortex meter introduces a shedding element across a pipe flow and picks up the shedding frequency downstream with a pressure, ultrasonic, or heat transfer type of sensor. A typical design is shown in Fig. 6.34.

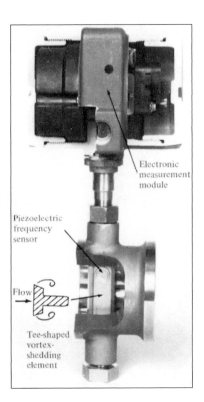

Electronic
measurement
module

Piezoelectric
frequency
sensor

Flow

Tee-shaped
vortex-
shedding
element

Fig. 6.34 A vortex flowmeter.
(Courtesy of Invensys p/c.)

The advantages of a vortex meter are as follows:

1. Absence of moving parts.
2. Accuracy to ± 1 percent over a wide flow rate range (up to 100:1).
3. Ability to handle very hot or very cold fluids.
4. Requirement of only a short pipe length.
5. Calibration insensitive to fluid density or viscosity.

For further details see Ref. 40.

Ultrasonic Flowmeters.   The sound-wave analog of the laser velocimeter of Fig. 6.29h is the ultrasonic flowmeter. Two examples are shown in Fig. 6.35. The pulse-type flowmeter is shown in Fig. 6.35a. Upstream piezoelectric transducer A is excited with a short sonic pulse that propagates across the flow to downstream transducer B. The arrival at B triggers another pulse to be created at A, resulting in a regular pulse frequency f_A. The same process is duplicated in the reverse direction from B to A, creating frequency f_B. The difference $f_A - f_B$ is proportional to the flow rate. Figure 6.35b shows a doppler-type arrangement, where sound waves from transmitter T are scattered by particles or contaminants in the flow to receiver R. Comparison of the two signals reveals a doppler frequency shift that is proportional to the flow rate. Ultrasonic meters are nonintrusive and can be directly attached to pipe flows in the field

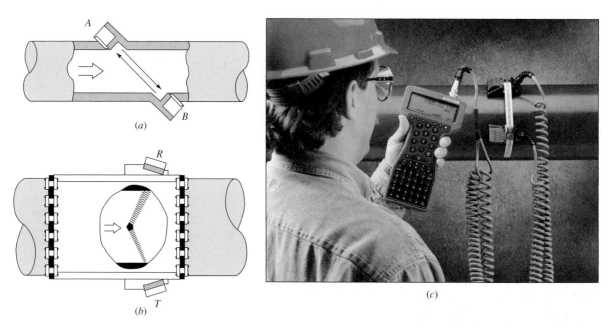

(a)

(b)

(c)

Fig. 6.35 Ultrasonic flowmeters: (*a*) pulse type; (*b*) doppler-shift type *(from Ref. 41)*; (*c*) a portable noninvasive installation *(Courtesy of Thermo Polysonics, Houston, TX.)*

(Fig. 6.35*c*). Their quoted uncertainty of ± 1 to 2 percent can rise to ± 5 percent or more due to irregularities in velocity profile, fluid temperature, or Reynolds number. For further details see Ref. 41.

Rotameter.   The variable-area transparent *rotameter* of Fig. 6.36 has a float that, under the action of flow, rises in the vertical tapered tube and takes a certain equilibrium position for any given flow rate. A student exercise for the forces on the float would yield the approximate relation

$$Q = C_d A_a \left(\frac{2W_{\text{net}}}{A_{\text{float}} \rho_{\text{fluid}}} \right)^{1/2} \tag{6.101}$$

where W_{net} is the float's net weight in the fluid, $A_a = A_{\text{tube}} - A_{\text{float}}$ is the annular area between the float and the tube, and C_d is a dimensionless discharge coefficient of order unity, for the annular constricted flow. For slightly tapered tubes, A_a varies nearly linearly with the float position, and the tube may be calibrated and marked with a flow rate scale, as in Fig. 6.36. The rotameter thus provides a readily visible measure of the flow rate. Capacity may be changed by using different-sized floats. Obviously the tube must be vertical, and the device does not give accurate readings for fluids containing high concentrations of bubbles or particles.

Coriolis Mass Flowmeter.   Most commercial meters measure *volume* flow, with mass flow then computed by multiplying by the nominal fluid density. An attractive modern alternative is a *mass* flowmeter, which operates on the principle of the Coriolis acceleration associated with noninertial coordinates [recall Fig. 3.12 and the Coriolis term $2\Omega \times V$ in Eq. (3.48)]. The output of the meter is directly proportional to mass flow.

Figure 6.37 is a schematic of a Coriolis device, to be inserted into a piping system. The flow enters a double-loop, double-tube arrangement, which is electromagnetically vibrated at a high natural frequency (amplitude < 1 mm and frequency > 100 Hz). The up flow induces inward loop motion, while the down flow creates outward loop motion, both due to the Coriolis effect. Sensors at both ends register a phase difference that is proportional to mass flow. Quoted accuracy is approximately ± 0.2 percent of full scale.

Laminar Flow Element.   In many, perhaps most, commercial flowmeters, the flow through the meter is turbulent and the variation of flow rate with pressure drop is nonlinear. In laminar duct flow, however, Q is linearly proportional to Δp, as in Eq. (6.12): $Q = [\pi R^4/(8\mu L)]\ \Delta p$. Thus a *laminar* flow sensing element is attractive, since its calibration will be linear. To ensure laminar flow for what otherwise would be a turbulent condition, all or part of the fluid is directed into small passages, each of which has a low (laminar) Reynolds number. A honeycomb is a popular design.

Figure 6.38 uses axial flow through a narrow annulus to effect laminar flow. The theory again predicts $Q \propto \Delta p$, as in Eq. (6.73). However, the flow is very sensitive to passage size; for example, halving the annulus clearance increases Δp more than eight times. Careful calibration is thus necessary. In Fig. 6.38 the laminar flow concept has been synthesized into a complete mass flow system, with temperature control, differential pressure measurement, and a microprocessor all self-contained. The accuracy of this device is rated at ± 0.2 percent.

Bernoulli Obstruction Theory.   Consider the generalized flow obstruction shown in Fig. 6.39. The flow in the basic duct of diameter D is forced through an obstruction

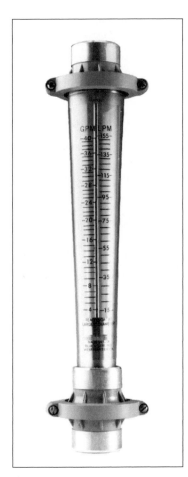

Fig. 6.36 A commercial rotameter. The float rises in the tapered tube to an equilibrium position, which is a measure of the fluid flow rate. *(Courtesy of Blue White Industries, Huntington Beach, CA.)*

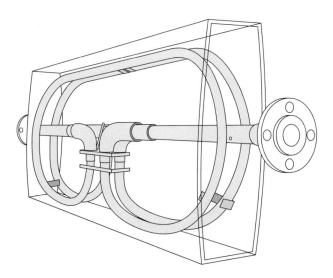

Fig. 6.37 A Coriolis mass flowmeter. *(Courtesy of ABB Instrumentation, Inc.)*

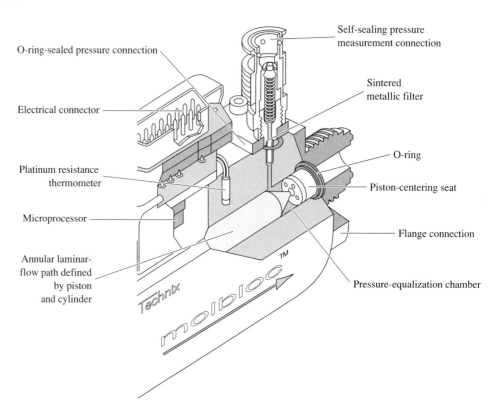

Self-sealing pressure measurement connection

O-ring-sealed pressure connection

Electrical connector

Sintered metallic filter

Platinum resistance thermometer

O-ring

Piston-centering seat

Microprocessor

Flange connection

Annular laminar-flow path defined by piston and cylinder

Pressure-equalization chamber

Fig. 6.38 A complete flowmeter system using a laminar flow element (in this case a narrow annulus). The flow rate is linearly proportional to the pressure drop. *(Courtesy of Martin Girard, DH Instruments, Inc.)*

of diameter d; the β ratio of the device is a key parameter:

$$\beta = \frac{d}{D} \tag{6.102}$$

After leaving the obstruction, the flow may neck down even more through a vena contracta of diameter $D_2 < d$, as shown. Apply the Bernoulli and continuity equations for incompressible steady frictionless flow to estimate the pressure change:

Continuity:
$$Q = \frac{\pi}{4}D^2V_1 = \frac{\pi}{4}D_2^2V_2$$

Bernoulli:
$$p_0 = p_1 + \tfrac{1}{2}\rho V_1^2 = p_2 + \tfrac{1}{2}\rho V_2^2$$

Eliminating V_1, we solve these for V_2 or Q in terms of the pressure change $p_1 - p_2$:

$$\frac{Q}{A_2} = V_2 \approx \left[\frac{2(p_1 - p_2)}{\rho(1 - D_2^4/D^4)}\right]^{1/2} \tag{6.103}$$

But this is surely inaccurate because we have neglected friction in a duct flow, where we know friction will be very important. Nor do we want to get into the business of measuring vena contracta ratios D_2/d for use in (6.103). Therefore we assume that

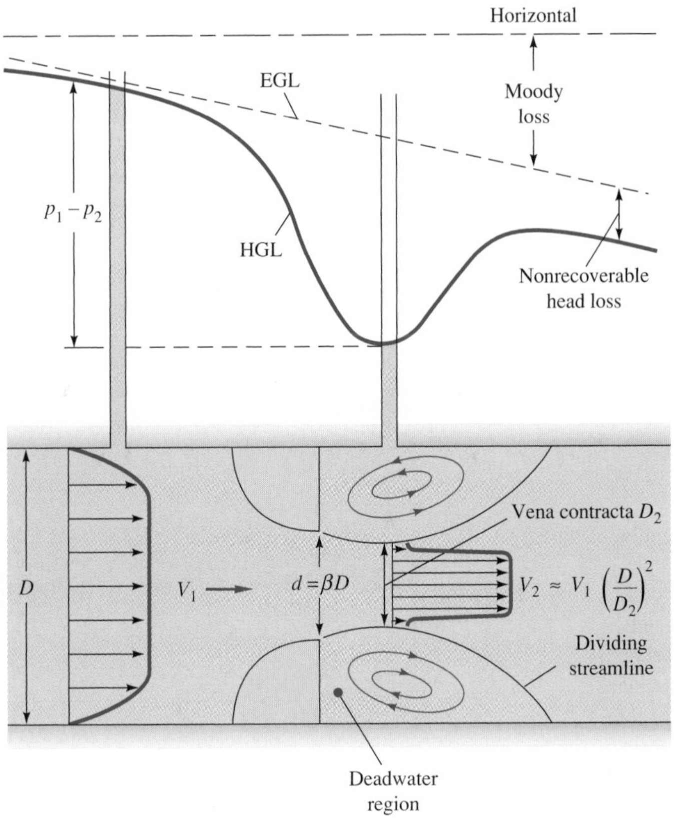

Fig. 6.39 Velocity and pressure change through a generalized Bernoulli obstruction meter.

$D_2/D \approx \beta$ and then calibrate the device to fit the relation

$$Q = A_t V_t = C_d A_t \left[\frac{2(p_1 - p_2)/\rho}{1 - \beta^4} \right]^{1/2} \tag{6.104}$$

where subscript t denotes the throat of the obstruction. The dimensionless *discharge coefficient* C_d accounts for the discrepancies in the approximate analysis. By dimensional analysis for a given design we expect

$$C_d = f(\beta, \text{Re}_D) \quad \text{where} \quad \text{Re}_D = \frac{V_1 D}{\nu} \tag{6.105}$$

The geometric factor involving β in (6.104) is called the *velocity-of-approach factor*:

$$E = (1 - \beta^4)^{-1/2} \tag{6.106}$$

One can also group C_d and E in Eq. (6.104) to form the dimensionless *flow coefficient* α:

$$\alpha = C_d E = \frac{C_d}{(1 - \beta^4)^{1/2}} \tag{6.107}$$

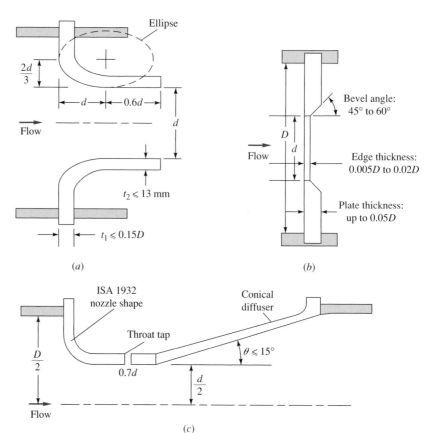

Fig. 6.40 International standard shapes for the three primary Bernoulli obstruction-type meters: (*a*) long-radius nozzle; (*b*) thin-plate orifice; (*c*) venturi nozzle. *(From Ref. 31 by permission of the International Organization for Standardization.)*

Thus Eq. (6.104) can be written in the equivalent form

$$Q = \alpha A_t \left[\frac{2(p_1 - p_2)}{\rho} \right]^{1/2} \tag{6.108}$$

Obviously the flow coefficient is correlated in the same manner:

$$\alpha = f(\beta, \text{Re}_D) \tag{6.109}$$

Occasionally one uses the throat Reynolds number instead of the approach Reynolds number:

$$\text{Re}_d = \frac{V_t d}{\nu} = \frac{\text{Re}_D}{\beta} \tag{6.110}$$

Since the design parameters are assumed known, the correlation of α from Eq. (6.109) or of C_d from Eq. (6.105) is the desired solution to the fluid metering problem.

The mass flow is related to Q by

$$\dot{m} = \rho Q \tag{6.111}$$

and is thus correlated by exactly the same formulas.

Figure 6.40 shows the three basic devices recommended for use by the International Organization for Standardization (ISO) [31]: the orifice, nozzle, and venturi tube.

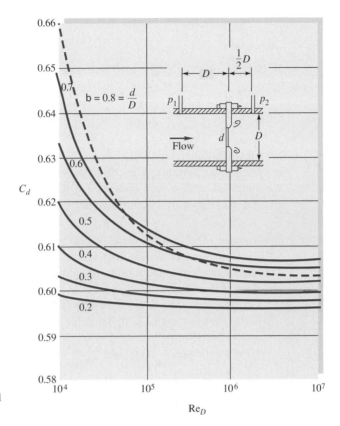

Fig. 6.41 Discharge coefficient for a thin-plate orifice with $D: \frac{1}{2}D$ taps, plotted from Eqs. (6.112) and (6.113b).

Thin-Plate Orifice. The thin-plate orifice, Fig. 6.40b, can be made with β in the range of 0.2 to 0.8, except that the hole diameter d should not be less than 12.5 mm. To measure p_1 and p_2, three types of tappings are commonly used:

1. Corner taps where the plate meets the pipe wall.
2. $D: \frac{1}{2}D$ taps: pipe-wall taps at D upstream and $\frac{1}{2}D$ downstream.
3. Flange taps: 1 in (25 mm) upstream and 1 in (25 mm) downstream of the plate, regardless of the size D.

Types 1 and 2 approximate geometric similarity, but since the flange taps 3 do not, they must be correlated separately for every single size of pipe in which a flange-tap plate is used [30, 31].

Figure 6.41 shows the discharge coefficient of an orifice with $D: \frac{1}{2}D$ or type 2 taps in the Reynolds number range $\mathrm{Re}_D = 10^4$ to 10^7 of normal use. Although detailed charts such as Fig. 6.41 are available for designers [30], the ASME recommends use of the curve-fit formulas developed by the ISO [31]. The basic form of the curve fit is [42]

$$C_d = f(\beta) + 91.71\beta^{2.5}\mathrm{Re}_D^{-0.75} + \frac{0.09\beta^4}{1 - \beta^4}F_1 - 0.0337\beta^3 F_2 \qquad (6.112)$$

where
$$f(\beta) = 0.5959 + 0.0312\beta^{2.1} - 0.184\beta^8$$

The correlation factors F_1 and F_2 vary with tap position:

Corner taps: $$F_1 = 0 \quad F_2 = 0 \tag{6.113a}$$

$D : \frac{1}{2}D$ taps: $$F_1 = 0.4333 \quad F_2 = 0.47 \tag{6.113b}$$

Flange taps: $$F_2 = \frac{1}{D\ (\text{in})} \quad F_1 = \begin{cases} \dfrac{1}{D\ (\text{in})} & D > 2.3 \text{ in} \\[2mm] 0.4333 & 2.0 \le D \le 2.3 \text{ in} \end{cases} \tag{6.113c}$$

Note that the flange taps (6.113c), not being geometrically similar, use raw diameter in inches in the formula. The constants will change if other diameter units are used. We cautioned against such dimensional formulas in Example 1.4 and Eq. (5.17) and give Eq. (6.113c) only because flange taps are widely used in the United States.

Flow Nozzle.   The flow nozzle comes in two types, a long-radius type shown in Fig. 6.40a and a short-radius type (not shown) called the ISA 1932 nozzle [30, 31]. The flow nozzle, with its smooth rounded entrance convergence, practically eliminates the vena contracta and gives discharge coefficients near unity. The nonrecoverable loss is still large because there is no diffuser provided for gradual expansion.

The ISO recommended correlation for long-radius-nozzle discharge coefficient is

$$C_d \approx 0.9965 - 0.00653\beta^{1/2}\left(\frac{10^6}{\text{Re}_D}\right)^{1/2} = 0.9965 - 0.00653\left(\frac{10^6}{\text{Re}_d}\right)^{1/2} \tag{6.114}$$

The second form is independent of the β ratio and is plotted in Fig. 6.42. A similar ISO correlation is recommended for the short-radius ISA 1932 flow nozzle:

$$C_d \approx 0.9900 - 0.2262\beta^{4.1}$$
$$+ (0.000215 - 0.001125\beta + 0.00249\beta^{4.7})\left(\frac{10^6}{\text{Re}_D}\right)^{1.15} \tag{6.115}$$

Flow nozzles may have β values between 0.2 and 0.8.

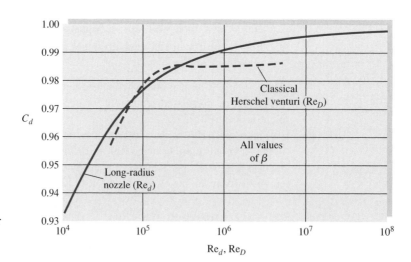

Fig. 6.42 Discharge coefficient for long-radius nozzle and classical Herschel-type venturi.

Venturi Meter.   The third and final type of obstruction meter is the venturi, named in honor of Giovanni Venturi (1746–1822), an Italian physicist who first tested conical expansions and contractions. The original, or *classical,* venturi was invented by a U.S. engineer, Clemens Herschel, in 1898. It consisted of a 21° conical contraction, a straight throat of diameter d and length d, then a 7 to 15° conical expansion. The discharge coefficient is near unity, and the nonrecoverable loss is very small. Herschel venturis are seldom used now.

The modern venturi nozzle, Fig. 6.40c, consists of an ISA 1932 nozzle entrance and a conical expansion of half-angle no greater than 15°. It is intended to be operated in a narrow Reynolds number range of 1.5×10^5 to 2×10^6. Its discharge coefficient, shown in Fig. 6.43, is given by the ISO correlation formula

$$C_d \approx 0.9858 - 0.196\beta^{4.5} \tag{6.116}$$

It is independent of Re_D within the given range. The Herschel venturi discharge varies with Re_D but not with β, as shown in Fig. 6.42. Both have very low net losses.

The choice of meter depends on the loss and the cost and can be illustrated by the following table:

Type of meter	Net head loss	Cost
Orifice	Large	Small
Nozzle	Medium	Medium
Venturi	Small	Large

As so often happens, the product of inefficiency and initial cost is approximately constant.

The average nonrecoverable head losses for the three types of meters, expressed as a fraction of the throat velocity head $V_t^2/(2g)$, are shown in Fig. 6.44. The orifice

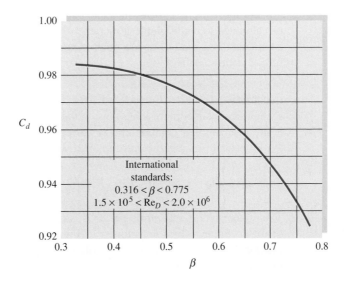

Fig. 6.43 Discharge coefficient for a venturi nozzle.

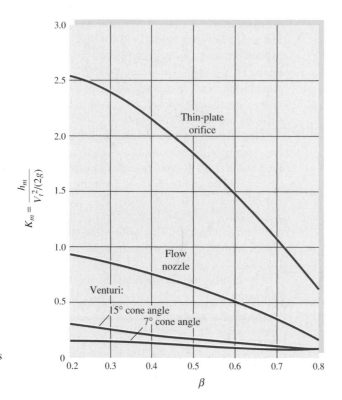

Fig. 6.44 Nonrecoverable head loss in Bernoulli obstruction meters. *(Adapted from Ref. 30.)*

has the greatest loss and the venturi the least, as discussed. The orifice and nozzle simulate partially closed valves as in Fig. 6.18b, while the venturi is a very minor loss. When the loss is given as a fraction of the measured *pressure drop,* the orifice and nozzle have nearly equal losses, as Example 6.21 will illustrate.

The other types of instruments discussed earlier in this section can also serve as flowmeters if properly constructed. For example, a hot wire mounted in a tube can be calibrated to read volume flow rather than point velocity. Such hot-wire meters are commercially available, as are other meters modified to use velocity instruments. For further details see Ref. 30.

Compressible Gas Flow Correction Factor. The orifice/nozzle/venturi formulas in this section assume incompressible flow. If the fluid is a gas, and the pressure ratio (p_2/p_1) is not near unity, a compressibility correction is needed. Equation (6.104) is rewritten in terms of mass flow and the upstream density ρ_1:

$$\dot{m} = C_d Y A_t \sqrt{\frac{2\rho_1(p_1 - p_2)}{1 - \beta^4}} \quad \text{where} \quad \beta = \frac{d}{D} \tag{6.117}$$

The dimensionless *expansion factor Y* is a function of pressure ratio, β, and the type of meter. Some values are plotted in Fig. 6.45. The orifice, with its strong jet contraction, has a different factor from the venturi or the flow nozzle, which are designed to eliminate contraction.

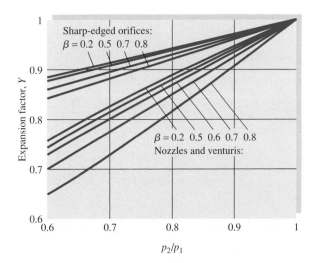

Fig. 6.45 Compressible flow expansion factor Y for flowmeters.

EXAMPLE 6.21

We want to meter the volume flow of water ($\rho = 1000$ kg/m^3, $\nu = 1.02 \times 10^{-6}$ m^2/s) moving through a 200-mm-diameter pipe at an average velocity of 2.0 m/s. If the differential pressure gage selected reads accurately at $p_1 - p_2 = 50{,}000$ Pa, what size meter should be selected for installing (*a*) an orifice with D: $\frac{1}{2}D$ taps, (*b*) a long-radius flow nozzle, or (*c*) a venturi nozzle? What would be the nonrecoverable head loss for each design?

Solution

Here the unknown is the β ratio of the meter. Since the discharge coefficient is a complicated function of β, iteration or EES will be necessary. We are given $D = 0.2$ m and $V_1 = 2.0$ m/s. The pipe-approach Reynolds number is thus

$$\mathrm{Re}_D = \frac{V_1 D}{\nu} = \frac{(2.0)(0.2)}{1.02 \times 10^{-6}} = 392{,}000$$

For all three cases [(*a*) to (*c*)] the generalized formula (6.108) holds:

$$V_t = \frac{V_1}{\beta^2} = \alpha \left[\frac{2(p_1 - p_2)}{\rho} \right]^{1/2} \qquad \alpha = \frac{C_d}{(1 - \beta^4)^{1/2}} \qquad (1)$$

where the given data are $V_1 = 2.0$ m/s, $\rho = 1000$ kg/m^3, and $\Delta p = 50{,}000$ Pa. Inserting these known values into Eq. (1) gives a relation between β and α:

$$\frac{2.0}{\beta^2} = \alpha \left[\frac{2(50{,}000)}{1000} \right]^{1/2} \qquad \text{or} \qquad \beta^2 = \frac{0.2}{\alpha} \qquad (2)$$

The unknowns are β (or α) and C_d. Parts (*a*) to (*c*) depend on the particular chart or formula needed for $C_d = \mathrm{fcn}(\mathrm{Re}_D, \beta)$. We can make an initial guess $\beta \approx 0.5$ and iterate to convergence.

Part (a)   For the orifice with $D: \frac{1}{2}D$ taps, use Eq. (6.112) or Fig. 6.41. The iterative sequence is

$$\beta_1 \approx 0.5, C_{d1} \approx 0.604, \alpha_1 \approx 0.624, \beta_2 \approx 0.566, C_{d2} \approx 0.606, \alpha_2 \approx 0.640, \beta_3 = 0.559$$

We have converged to three figures. The proper orifice diameter is

$$d = \beta D = 112 \text{ mm} \qquad\qquad Ans. (a)$$

Part (b)   For the long-radius flow nozzle, use Eq. (6.114) or Fig. 6.42. The iterative sequence is

$$\beta_1 \approx 0.5, C_{d1} \approx 0.9891, \alpha_1 \approx 1.022, \beta_2 \approx 0.442, C_{d2} \approx 0.9896, \alpha_2 \approx 1.009, \beta_3 = 0.445$$

We have converged to three figures. The proper nozzle diameter is

$$d = \beta D = 89 \text{ mm} \qquad\qquad Ans. (b)$$

Part (c)   For the venturi nozzle, use Eq. (6.116) or Fig. 6.43. The iterative sequence is

$$\beta_1 \approx 0.5, C_{d1} \approx 0.977, \alpha_1 \approx 1.009, \beta_2 \approx 0.445, C_{d2} \approx 0.9807, \alpha_2 \approx 1.0004, \beta_3 = 0.447$$

We have converged to three figures. The proper venturi diameter is

$$d = \beta D = 89 \text{ mm} \qquad\qquad Ans. (c)$$

Comments: These meters are of similar size, but their head losses are not the same. From Fig. 6.44 for the three different shapes we may read the three K factors and compute

$$h_{m,\text{orifice}} \approx 3.5 \text{ m} \quad h_{m,\text{nozzle}} \approx 3.6 \text{ m} \quad h_{m,\text{venturi}} \approx 0.8 \text{ m}$$

The venturi loss is only about 22 percent of the orifice and nozzle losses.

Solution

The iteration encountered in this example is ideal for the EES. Input the data in SI units:

```
Rho=1000    Nu=1.02E-6    D=0.2    V=2.0    DeltaP=50000
```

Then write out the basic formulas for Reynolds number, throat velocity and flow coefficient:

```
Re=V*D/Nu

Vt=V/Beta^2

Alpha=Cd/(1-Beta^4)^0.5

Vt=Alpha*SQRT(2*DeltaP/Rho)
```

Finally, input the proper formula for the discharge coefficient. For example, for the flow nozzle,

```
Cd=0.9965-0.00653*Beta^0.5*(1E6/Re)^0.5
```

When asked to Solve the equation, EES at first complains of dividing by zero. One must then tighten up the Variable Information by not allowing β, α, or C_d to be negative and, in particular, by confining β to its practical range $0.2 < \beta < 0.9$. EES then readily announces correct answers for the flow nozzle:

```
Alpha=1.0096    Cd=0.9895    Beta=0.4451
```

EXAMPLE 6.22

A long-radius nozzle of diameter 6 cm is used to meter air flow in a 10-cm-diameter pipe. Upstream conditions are p_1 = 200 kPa and T_1 = 100°C. If the pressure drop through the nozzle is 60 kPa, estimate the flow rate in m³/s.

Solution

- *Assumptions:* The pressure drops 30 percent, so we need the compressibility factor Y, and Eq. (6.117) is applicable to this problem.
- *Approach:* Find ρ_1 and C_d and apply Eq. (6.117) with β = 6/10 = 0.6.
- *Property values:* Given p_1 and T_1, $\rho_1 = p_1/RT_1$ = (200,000)/[287(100 + 273)] = 1.87 kg/m³. The downstream pressure is p_2 = 200−60 = 140 kPa, hence p_2/p_1 = 0.7. At 100°C, from Table A.2, the viscosity of air is 2.17E-5 kg/m-s
- *Solution steps:* Initially apply Eq. (6.117) by guessing, from Fig. 6.42, that $C_d \approx$ 0.98. From Fig. 6.45, for a nozzle with p_2/p_1 = 0.7 and β = 0.6, read $Y \approx$ 0.80. Then

$$\dot{m} = C_d Y A_t \sqrt{\frac{2\rho_1(p_1 - p_2)}{1 - \beta^4}} \approx (0.98)(0.80)\frac{\pi}{4}(0.06 \text{ m})^2 \sqrt{\frac{2(1.87 \text{ kg/m}^3)(60,000 \text{ Pa})}{1 - (0.6)}}$$

$$\approx 1.13 \frac{\text{kg}}{\text{s}}$$

Now estimate Re_d, putting it in the convenient mass flow form:

$$\text{Re}_d = \frac{\rho V d}{\mu} = \frac{4\dot{m}}{\pi\mu d} = \frac{4(1.13 \text{ kg/s})}{\pi(2.17 \text{ E-5 kg/m} - \text{s})(0.06 \text{ m})} \approx 1.11\text{E6}$$

Returning to Fig. 6.42, we could read a slightly better $C_d \approx$ 0.99. Thus our final estimate is

$$\dot{m} \approx 1.14 \text{ kg/s} \qquad\qquad Ans.$$

- *Comments:* Figure 6.45 is not just a "chart" for engineers to use casually. It is based on the compressible flow theory of Chap. 9. There, we may reassign this example as a *theory*.

Summary

This chapter has been concerned with internal pipe and duct flows, which are probably the most common problems encountered in engineering fluid mechanics. Such flows are very sensitive to the Reynolds number and change from laminar to transitional to turbulent flow as the Reynolds number increases.

The various Reynolds number regimes were outlined, and a semiempirical approach to turbulent flow modeling was presented. The chapter then made a detailed analysis of flow through a straight circular pipe, leading to the famous Moody chart (Fig. 6.13) for the friction factor. Possible uses of the Moody chart were discussed for flow rate and sizing problems, as well as the application of the Moody chart to noncircular ducts using an equivalent duct "diameter." The addition of minor losses due to valves, elbows, fittings, and other devices was presented in the form of loss coefficients to be incorporated along with Moody-type friction losses. Multiple-pipe systems were discussed briefly and were seen to be quite complex algebraically and appropriate for computer solution.

Diffusers are added to ducts to increase pressure recovery at the exit of a system. Their behavior was presented as experimental data, since the theory of real diffusers is still not well developed. The chapter ended with a discussion of flowmeters, especially the pitot-static tube and the Bernoulli obstruction type of meter. Flowmeters also require careful experimental calibration.

Problems

Most of the problems herein are fairly straightforward. More difficult or open-ended assignments are labeled with an asterisk. Problems labeled with an EES icon will benefit from the use of the Engineering Equation Solver (EES), while problems labeled with a computer disk may require the use of a computer. The standard end-of-chapter problems P6.1 to P6.162 (categorized in the problem list here) are followed by word problems W6.1 to W6.4, fundamentals of engineering exam problems FE6.1 to FE6.15, comprehensive problems C6.1 to C6.8, and design projects D6.1 and D6.2.

Problem Distribution

Section	Topic	Problems
6.1	Reynolds number regimes	P6.1–P6.5
6.2	Internal and external flow	P6.6–P6.8
6.3	Head loss—friction factor	P6.9–P6.11
6.4	Laminar pipe flow	P6.12–P6.33
6.5	Turbulence modeling	P6.34–P6.40
6.6	Turbulent pipe flow	P6.41–P6.62
6.7	Flow rate and sizing problems	P6.63–P6.85
6.8	Noncircular ducts	P6.86–P6.98
6.9	Minor losses	P6.99–P6.110
6.10	Series and parallel pipe systems	P6.111–P6.120
6.10	Three-reservoir and pipe network systems	P6.121–P6.130
6.11	Diffuser performance	P6.131–P6.134
6.12	The pitot-static tube	P6.135–P6.139
6.12	Flowmeters: the orifice plate	P6.140–P6.148
6.12	Flowmeters: the flow nozzle	P6.149–P6.153
6.12	Flowmeters: the venturi meter	P6.154–P6.159
6.12	Flowmeters: butterfly valves	P6.160
6.12	Flowmeters: compressibility correction	P6.161–P6.162

P6.1 An engineer claims that the flow of SAE 30W oil, at 20°C, through a 5-cm-diameter smooth pipe at 1 million N/h, is laminar. Do you agree? A million newtons is a lot, so this sounds like an awfully high flow rate.

P6.2 Air at approximately 1 atm flows through a horizontal 4-cm-diameter pipe. (a) Find a formula for Q_{max}, the maximum volume flow for which the flow remains laminar, and plot Q_{max} versus temperature in the range $0°C \leq T \leq 500°C$. (b) Is your plot linear? If not, explain.

P6.3 For a thin wing moving parallel to its chord line, transition to a turbulent boundary layer occurs at a "local" Reynolds number Re_x, where x is the distance from the leading edge of the wing. The critical Reynolds number depends on the intensity of turbulent fluctuations in the stream and equals 2.8E6 if the stream is very quiet. A semiempirical correlation for this case [3, p. 385] is

$$Re_{x_{crit}}^{1/2} \approx \frac{-1 + (1 + 13.25\zeta^2)^{1/2}}{0.00392\zeta^2}$$

where ζ is the tunnel turbulence intensity in percent. If $V = 20$ m/s in air at 20°C, use this formula to plot the transition position on the wing versus stream turbulence for ζ between 0 and 2 percent. At what value of ζ is x_{crit} decreased 50 percent from its value at $\zeta = 0$?

P6.4 For flow of SAE 30 oil through a 5-cm-diameter pipe, from Fig. A.1, for what flow rate in m³/h would we expect transition to turbulence at (a) 20°C and (b) 100°C?

P6.5 In flow past a body or wall, early transition to turbulence can be induced by placing a trip wire on the wall across the flow, as in Fig. P6.5. If the trip wire in Fig. P6.5 is placed where the local velocity is U, it will trigger turbulence if $Ud/\nu = 850$, where d is the wire diameter [3, p. 388]. If the sphere diameter is 20 cm and transition is observed at $Re_D = 90,000$, what is the diameter of the trip wire in mm?

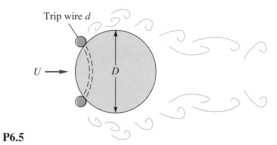

P6.5

P6.6 For flow of a uniform stream parallel to a sharp flat plate, transition to a turbulent boundary layer on the plate may occur at $Re_x = \rho Ux/\mu \approx 1E6$, where U is the approach velocity and x is distance along the plate. If $U = 2.5$ m/s, determine the distance x for the following fluids at 20°C and 1 atm: (a) hydrogen, (b) air, (c) gasoline, (d) water, (e) mercury, and (f) glycerin.

P6.7   Cola, approximated as pure water at 20°C, is to fill an 8-oz container (1 U.S. gal = 128 fl oz) through a 5-mm-diameter tube. Estimate the minimum filling time if the tube flow is to remain laminar. For what cola (water) temperature would this minimum time be 1 min?

P6.8   When water at 20°C is in steady turbulent flow through an 8-cm-diameter pipe, the wall shear stress is 72 Pa. What is the axial pressure gradient $(\partial p/\partial x)$ if the pipe is (a) horizontal and (b) vertical with the flow up?

P6.9   A light liquid $(\rho \approx 950 \text{ kg/m}^3)$ flows at an average velocity of 10 m/s through a horizontal smooth tube of diameter 5 cm. The fluid pressure is measured at 1-m intervals along the pipe, as follows:

x, m	0	1	2	3	4	5	6
p, kPa	304	273	255	240	226	213	200

Estimate (a) the total head loss, in meters; (b) the wall shear stress in the fully developed section of the pipe; and (c) the overall friction factor.

P6.10   Water at 20°C flows through an inclined 8-cm-diameter pipe. At sections A and B the following data are taken: $p_A = 186$ kPa, $V_A = 3.2$ m/s, $z_A = 24.5$ m, and $p_B = 260$ kPa, $V_B = 3.2$ m/s, $z_B = 9.1$ m. Which way is the flow going? What is the head loss in meters?

P6.11   Water at 20°C flows upward at 4 m/s in a 6-cm-diameter pipe. The pipe length between points 1 and 2 is 5 m, and point 2 is 3 m higher. A mercury manometer, connected between 1 and 2, has a reading $h = 135$ mm, with p_1 higher. (a) What is the pressure change $(p_1 - p_2)$? (b) What is the head loss, in meters? (c) Is the manometer reading proportional to head loss? Explain. (d) What is the friction factor of the flow?

In Probs. 6.12 to 6.99, neglect minor losses.

P6.12   A 5-mm-diameter capillary tube is used as a viscometer for oils. When the flow rate is 0.071 m^3/h, the measured pressure drop per unit length is 375 kPa/m. Estimate the viscosity of the fluid. Is the flow laminar? Can you also estimate the density of the fluid?

P6.13   A soda straw is 20 cm long and 2 mm in diameter. It delivers cold cola, approximated as water at 10°C, at a rate of 3 cm^3/s. (a) What is the head loss through the straw? What is the axial pressure gradient $\partial p/\partial x$ if the flow is (b) vertically up or (c) horizontal? Can the human lung deliver this much flow?

P6.14   Water at 20°C is to be siphoned through a tube 1 m long and 2 mm in diameter, as in Fig. P6.14. Is there any height H for which the flow might not be laminar? What is the flow rate if $H = 50$ cm? Neglect the tube curvature.

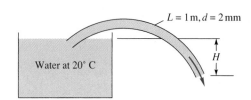

P6.14

P6.15   Professor Gordon Holloway and his students at the University of New Brunswick went to a fast-food emporium and tried to drink chocolate shakes $(\rho \approx 1200 \text{ kg/m}^3, \mu \approx 6 \text{ kg/m-s})$ through fat straws 8 mm in diameter and 30 cm long. (a) Verify that their human lungs, which can develop approximately 3000 Pa of vacuum pressure, would be unable to drink the milkshake through the vertical straw. (b) A student cut 15 cm from his straw and proceeded to drink happily. What rate of milkshake flow was produced by this strategy?

P6.16   Glycerin at 20°C is to be pumped through a horizontal smooth pipe at 3.1 m^3/s. It is desired that (1) the flow be laminar and (2) the pressure drop be no more than 100 Pa/m. What is the minimum pipe diameter allowable?

P6.17   A *capillary viscometer* measures the time required for a specified volume v of liquid to flow through a small-bore glass tube, as in Fig. P6.17. This transit time is then correlated with fluid viscosity. For the system shown, (a) derive an approximate formula for the time required, assuming laminar flow with no entrance and exit losses. (b) If $L = 12$ cm, $l = 2$ cm, $v = 8$ cm^3, and the fluid is water at 20°C, what capillary diameter D will result in a transit time t of 6 seconds?

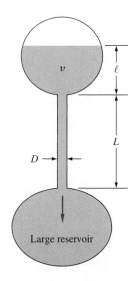

P6.17

P6.18 To determine the viscosity of a liquid of specific gravity 0.95, you fill, to a depth of 12 cm, a large container that drains through a 30-cm-long vertical tube attached to the bottom. The tube diameter is 2 mm, and the rate of draining is found to be 1.9 cm³/s. What is your estimate of the fluid viscosity? Is the tube flow laminar?

P6.19 An oil (SG = 0.9) issues from the pipe in Fig. P6.19 at Q = 35 ft³/h. What is the kinematic viscosity of the oil in ft³/s? Is the flow laminar?

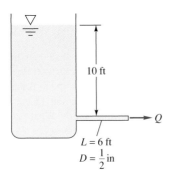

$L = 6$ ft

$D = \dfrac{1}{2}$ in

P6.19

P6.20 The oil tanks in Tinyland are only 160 cm high, and they discharge to the Tinyland oil truck through a smooth tube 4 mm in diameter and 55 cm long. The tube exit is open to the atmosphere and 145 cm below the tank surface. The fluid is medium fuel oil, ρ = 850 kg/m³ and μ = 0.11 kg/(m-s). Estimate the oil flow rate in cm³/h.

P6.21 In Tinyland, houses are less than a foot high! The rainfall is laminar! The drainpipe in Fig. P6.21 is only 2 mm in diameter. (*a*) When the gutter is full, what is the rate of draining? (*b*) The gutter is designed for a sudden rainstorm of up to 5 mm per hour. For this condition, what is the maximum roof area that can be drained successfully? (*c*) What is Re$_d$?

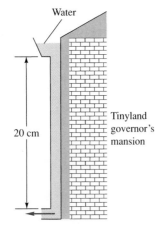

Water

20 cm

Tinyland governor's mansion

P6.21

P6.22 A steady push on the piston in Fig. P6.22 causes a flow rate Q = 0.15 cm³/s through the needle. The fluid has ρ = 900 kg/m³ and μ = 0.002 kg/(m · s). What force F is required to maintain the flow?

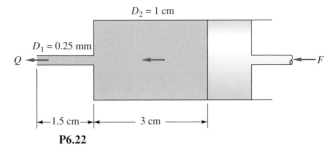

$D_2 = 1$ cm

$D_1 = 0.25$ mm

Q

F

←1.5 cm→ ← 3 cm →

P6.22

P6.23 SAE 10 oil at 20°C flows in a vertical pipe of diameter 2.5 cm. It is found that the pressure is constant throughout the fluid. What is the oil flow rate in m³/h? Is the flow up or down?

P6.24 Two tanks of water at 20°C are connected by a capillary tube 4 mm in diameter and 3.5 m long. The surface of tank 1 is 30 cm higher than the surface of tank 2. (*a*) Estimate the flow rate in m³/h. Is the flow laminar? (*b*) For what tube diameter will Re$_d$ be 500?

P6.25 For the configuration shown in Fig. P6.25, the fluid is ethyl alcohol at 20°C, and the tanks are very wide. Find the flow rate which occurs in m³/h. Is the flow laminar?

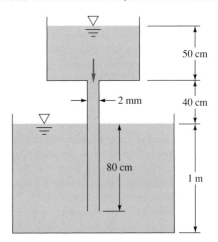

50 cm

2 mm

40 cm

80 cm

1 m

P6.25

P6.26 Two oil tanks are connected by two 9-m-long pipes, as in Fig. P6.26. Pipe 1 is 5 cm in diameter and is 6 m higher than pipe 2. It is found that the flow rate in pipe 2 is twice as large as the flow in pipe 1. (*a*) What is the diameter of pipe 2? (*b*) Are both pipe flows laminar? (*c*) What is the flow rate in pipe 2 (m³/s)? Neglect minor losses.

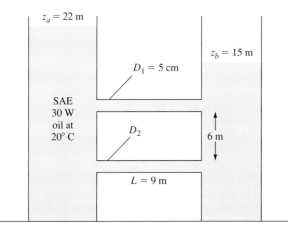

P6.26

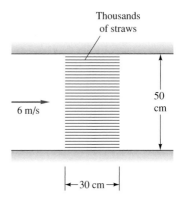

P6.28

*P6.27 Let us attack Prob. P6.25 in symbolic fashion, using Fig. P6.27. All parameters are constant except the upper tank depth $Z(t)$. Find an expression for the flow rate $Q(t)$ as a function of $Z(t)$. Set up a differential equation, and solve for the time t_0 to drain the upper tank completely. Assume quasi-steady laminar flow.

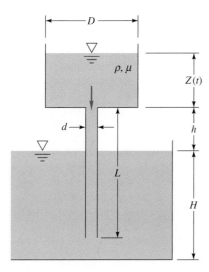

P6.27

P6.28 For straightening and smoothing an airflow in a 50-cm-diameter duct, the duct is packed with a "honeycomb" of thin straws of length 30 cm and diameter 4 mm, as in Fig. P6.28. The inlet flow is air at 110 kPa and 20°C, moving at an average velocity of 6 m/s. Estimate the pressure drop across the honeycomb.

P6.29 Oil, with $\rho = 890$ kg/m^3 and $\mu = 0.07$ kg/m-s, flows through a horizontal pipe 15 m long. The power delivered to the flow is 1 hp. (a) What is the appropriate

pipe diameter if the flow is at the laminar transition point? For this condition, what are (b) Q in m^3/h; and (c) τ_w in kPa?

P6.30 SAE 10 oil at 20°C flows through the 4-cm-diameter vertical pipe of Fig. P6.30. For the mercury manometer reading $h = 42$ cm shown, (a) calculate the volume flow rate in m^3/h and (b) state the direction of flow.

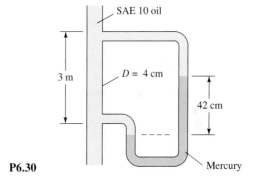

P6.30

P6.31 A *laminar flow element* (LFE) (Meriam Instrument Co.) measures low gas-flow rates with a bundle of capillary tubes or ducts packed inside a large outer tube. Consider oxygen at 20°C and 1 atm flowing at 84 ft^3/min in a 4-in-diameter pipe. (a) Is the flow turbulent when approaching the element? (b) If there are 1000 capillary tubes, $L = 4$ in, select a tube diameter to keep Re$_d$ below 1500 and also to keep the tube pressure drop no greater than 0.5 lbf/in^2. (c) Do the tubes selected in part (b) fit nicely within the approach pipe?

P6.32 SAE 30 oil at 20°C flows in the 3-cm-diameter pipe in Fig. P6.32, which slopes at 37°. For the pressure measurements shown, determine (a) whether the flow is up or down and (b) the flow rate in m^3/h.

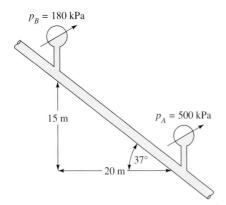

P6.32

P6.33   In Prob. P6.32 suppose it is desired to add a pump between A and B to drive the oil upward from A to B at a rate of 3 kg/s. At 100 percent efficiency, what pump power is required?

P6.34   Derive the time-averaged x-momentum equation (6.21) by direct substitution of Eqs. (6.19) into the momentum equation (6.14). It is convenient to write the convective acceleration as

$$\frac{du}{dt} = \frac{\partial}{\partial x}(u^2) + \frac{\partial}{\partial y}(uv) + \frac{\partial}{\partial z}(uw)$$

which is valid because of the continuity relation, Eq. (6.14).

P6.35   By analogy with Eq. (6.21), write the turbulent mean-momentum differential equation for (a) the y direction and (b) the z direction. How many turbulent stress terms appear in each equation? How many unique turbulent stresses are there for the total of three directions?

P6.36   The following turbulent flow velocity data $u(y)$, for air at 75°F and 1 atm near a smooth flat wall, were taken in the University of Rhode Island wind tunnel:

y, in	0.025	0.035	0.047	0.055	0.065
u, ft/s	51.2	54.2	56.8	57.6	59.1

Estimate (a) the wall shear stress and (b) the velocity u at $y = 0.22$ in.

P6.37   Two infinite plates a distance h apart are parallel to the xz plane with the upper plate moving at speed V, as in Fig. P6.37. There is a fluid of viscosity μ and constant pressure between the plates. Neglecting gravity and assuming incompressible turbulent flow $u(y)$ between the plates, use the logarithmic law and appropriate boundary conditions to derive a formula for dimensionless wall shear stress versus dimensionless plate velocity. Sketch a typical shape of the profile $u(y)$.

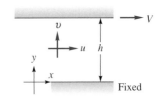

P6.37

P6.38   Suppose in Fig. P6.37 that $h = 3$ cm, the fluid in water at 20°C, and the flow is turbulent, so that the logarithmic law is valid. If the shear stress in the fluid is 15 Pa, what is V in m/s?

P6.39   By analogy with laminar shear, $\tau = \mu\,du/dy$, T. V. Boussinesq in 1877 postulated that turbulent shear could also be related to the mean velocity gradient $\tau_{\mathrm{turb}} = \epsilon\,du/dy$, where ϵ is called the *eddy viscosity* and is much larger than μ. If the logarithmic overlap law, Eq. (6.28), is valid with $\tau_{\mathrm{turb}} \approx \tau_w$, show that $\epsilon \approx \kappa\rho u^* y$.

P6.40   Theodore von Kármán in 1930 theorized that turbulent shear could be represented by $\tau_{\mathrm{turb}} = \epsilon\,du/dy$, where $\epsilon = \rho\kappa^2 y^2 |du/dy|$ is called the *mixing-length eddy viscosity* and $\kappa \approx 0.41$ is Kármán's dimensionless *mixing-length constant* [2, 3]. Assuming that $\tau_{\mathrm{turb}} \approx \tau_w$ near the wall, show that this expression can be integrated to yield the logarithmic overlap law, Eq. (6.28).

P6.41   Two reservoirs, which differ in surface elevation by 40 m, are connected by 350 m of new pipe of diameter 8 cm. If the desired flow rate is at least 500 N/s of water at 20°C, can the pipe material be made of (a) galvanized iron, (b) commercial steel, or (c) cast iron? Neglect minor losses.

*P6.42   It is clear by comparing Figs. 6.12b and 6.13 that the effects of sand roughness and commercial (manufactured) roughness are not quite the same. Take the special case of commercial roughness ratio $\epsilon/d = 0.001$ in Fig. 6.13, and replot in the form of the wall-law shift ΔB (Fig. 6.12a) versus the logarithm of $\epsilon^+ = \epsilon u^*/\nu$. Compare your plot with Eq. (6.45).

P6.43   A reservoir supplies water through 100 m of 30-cm-diameter cast iron pipe to a turbine that extracts 80 hp from the flow. The water then exhausts to the atmosphere.

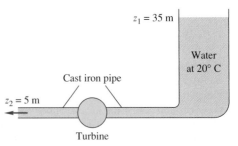

P6.43

Neglect minor losses. (a) Assuming that $f \approx 0.019$, find the flow rate (which results in a cubic polynomial). Explain why there are *two* legitimate solutions. (b) For extra credit, solve for the flow rates using the actual friction factors.

P6.44 Mercury at 20°C flows through 4 m of 7-mm-diameter glass tubing at an average velocity of 5 m/s. Estimate the head loss in m and the pressure drop in kPa.

P6.45 Oil, SG = 0.88 and ν = 4 E-5 m^2/s, flows at 400 gal/min through a 6-in asphalted cast iron pipe. The pipe is 0.5 mi long and slopes upward at 8° in the flow direction. Compute the head loss in ft and the pressure change.

P6.46 Kerosene at 20°C is pumped at 0.15 m^3/s through 20 km of 16-cm diameter cast iron horizontal pipe. Compute the input power in kW required if the pumps are 85 percent efficient.

P6.47 The gutter and smooth drainpipe in Fig. P6.47 remove rainwater from the roof of a building. The smooth drainpipe is 7 cm in diameter. (a) When the gutter is full, estimate the rate of draining. (b) The gutter is designed for a sudden rainstorm of up to 5 inches per hour. For this condition, what is the maximum roof area that can be drained successfully?

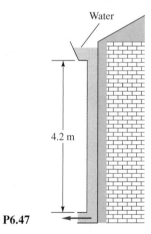

Water

4.2 m

P6.47

P6.48 Show that if Eq. (6.33) is accurate, the position in a turbulent pipe flow where local velocity u equals average velocity V occurs exactly at $r = 0.777R$, independent of the Reynolds number.

P6.49 The tank–pipe system of Fig. P6.49 is to deliver at least 11 m^3/h of water at 20°C to the reservoir. What is the maximum roughness height ϵ allowable for the pipe?

P6.50 Ethanol at 20°C flows at 125 U.S. gal/min through a horizontal cast iron pipe with L = 12 m and d = 5 cm. Neglecting entrance effects, estimate (a) the pressure gradient dp/dx, (b) the wall shear stress τ_w, and (c) the

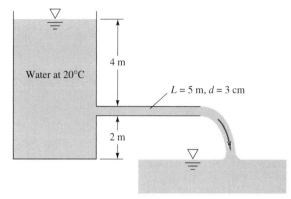

Water at 20°C

4 m

$L = 5$ m, $d = 3$ cm

2 m

P6.49

percentage reduction in friction factor if the pipe walls are polished to a smooth surface.

P6.51 The viscous sublayer (Fig. 6.9) is normally less than 1 percent of the pipe diameter and therefore very difficult to probe with a finite-sized instrument. In an effort to generate a thick sublayer for probing, Pennsylvania State University in 1964 built a pipe with a flow of glycerin. Assume a smooth 12-in-diameter pipe with V = 60 ft/s and glycerin at 20°C. Compute the sublayer thickness in inches and the pumping horsepower required at 75 percent efficiency if L = 40 ft.

P6.52 The pipe flow in Fig. P6.52 is driven by pressurized air in the tank. What gage pressure p_1 is needed to provide a 20°C water flow rate Q = 60 m^3/h?

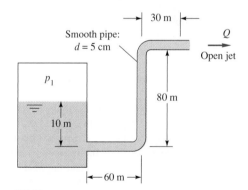

30 m

Smooth pipe: $d = 5$ cm

Q

Open jet

p_1

80 m

10 m

60 m

P6.52

*P6.53 In Fig. P6.52 suppose p_1 = 700 kPa and the fluid specific gravity is 0.68. If the flow rate is 27 m^3/h, estimate the viscosity of the fluid. What fluid in Table A.3 is the likely suspect?

*P6.54 A swimming pool W by Y by h deep is to be emptied by gravity through the long pipe shown in Fig. P6.54. Assuming an average pipe friction factor f_{av} and neglecting minor losses, derive a formula for the time to empty the tank from an initial level h_o.

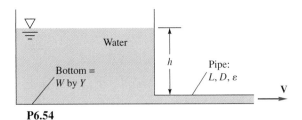

P6.54

P6.55 The reservoirs in Fig. P6.55 contain water at 20°C. If the pipe is smooth with $L = 4500$ m and $d = 4$ cm, what will the flow rate in m^3/h be for $\Delta z = 100$ m?

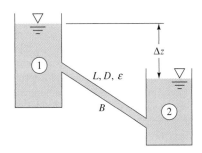

P6.55

P6.56 Consider a horizontal 4-ft-diameter galvanized iron pipe simulating the Alaskan pipeline. The oil flow is 70 million U.S. gallons per day, at a density of 910 kg/m^3 and viscosity of 0.01 kg/(m · s) (see Fig. A.1 for SAE 30 oil at 100°C). Each pump along the line raises the oil pressure to 8 MPa, which then drops, due to head loss, to 400 kPa at the entrance to the next pump. Estimate (a) the appropriate distance between pumping stations and (b) the power required if the pumps are 88 percent efficient.

P6.57 Apply the analysis of Prob. P6.54 to the following data. Let $W = 5$ m, $Y = 8$ m, $h_o = 2$ m, $L = 15$ m, $D = 5$ cm, and $\epsilon = 0$. (a) By letting $h = 1.5$ m and 0.5 m as representative depths, estimate the average friction factor. Then (b) estimate the time to drain the pool.

P6.58 In Fig. P6.55 assume that the pipe is cast iron with $L = 550$ m, $d = 7$ cm, and $\Delta z = 100$ m. If an 80 percent efficient pump is placed at point B, what input power is required to deliver 160 m^3/h of water upward from reservoir 2 to 1?

P6.59 The following data were obtained for flow of 20°C water at 20 m^3/h through a badly corroded 5-cm-diameter pipe that slopes downward at an angle of 8°: $p_1 = 420$ kPa, $z_1 = 12$ m, $p_2 = 250$ kPa, $z_2 = 3$ m. Estimate (a) the roughness ratio of the pipe and (b) the

percentage change in head loss if the pipe were smooth and the flow rate the same.

P6.60 In the spirit of Haaland's explicit pipe friction factor approximation, Eq. (6.49), Jeppson [20] proposed the following explicit formula:

$$\frac{1}{\sqrt{f}} \approx -2.0 \log_{10}\left(\frac{\epsilon/d}{3.7} + \frac{5.74}{Re_d^{0.9}}\right)$$

(a) Is this identical to Haaland's formula with just a simple rearrangement? Explain. (b) Compare Jeppson's formula to Haaland's for a few representative values of (turbulent) Re_d and ϵ/d and their errors compared to the Colebrook formula (6.48). Discuss briefly.

P6.61 What level h must be maintained in Fig. P6.61 to deliver a flow rate of 0.015 ft^3/s through the $\frac{1}{2}$-in commercial steel pipe?

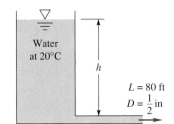

P6.61

P6.62 Water at 20°C is to be pumped through 2000 ft of pipe from reservoir 1 to 2 at a rate of 3 ft^3/s, as shown in Fig. P6.62. If the pipe is cast iron of diameter 6 in and the pump is 75 percent efficient, what horsepower pump is needed?

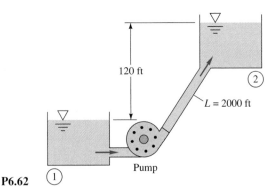

P6.62

P6.63 A tank contains 1 m^3 of water at 20°C and has a drawn-capillary outlet tube at the bottom, as in Fig. P6.63. Find the outlet volume flux Q in m^3/h at this instant.

P6.64 For the system in Fig. P6.63, solve for the flow rate in m^3/h if the fluid is SAE 10 oil at 20°C. Is the flow laminar or turbulent?

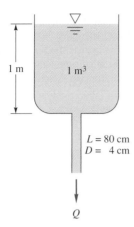

P6.63

P6.65   In Prob. P6.63 the initial flow is turbulent. As the water drains out of the tank, will the flow revert to laminar motion as the tank becomes nearly empty? If so, at what tank depth? Estimate the time, in h, to drain the tank completely.

P6.66   Ethyl alcohol at 20°C flows through a 10-cm horizontal drawn tube 100 m long. The fully developed wall shear stress is 14 Pa. Estimate (*a*) the pressure drop, (*b*) the volume flow rate, and (*c*) the velocity u at $r = 1$ cm.

P6.67   A straight 10-cm commercial-steel pipe is 1 km long and is laid on a constant slope of 5°. Water at 20°C flows downward, due to gravity only. Estimate the flow rate in m^3/h. What happens if the pipe length is 2 km?

P6.68   The Moody chart, Fig. 6.13, is best for finding head loss (or Δp) when Q, V, d, and L are known. It is awkward for the second type of problem, finding Q when h_f or Δp is known (see Example 6.9). Prepare a modified Moody chart whose abscissa is independent of Q and V, using ϵ/d as a parameter, from which one can immediately read the ordinate to find (dimensionless) Q or V. Use your chart to solve Example 6.9.

P6.69   For Prob. P6.62 suppose the only pump available can deliver 80 hp to the fluid. What is the proper pipe size in inches to maintain the 3 ft^3/s flow rate?

P6.70   Water at 68°F flows through 200 ft of a horizontal 6-in-diameter asphalted cast iron pipe. (*a*) If the head loss is 4.5 ft, find the average velocity and the flow rate. (*b*) Does this input data seem familiar to you?

*P6.71   It is desired to solve Prob. 6.62 for the most economical pump and cast iron pipe system. If the pump costs $125 per horsepower delivered to the fluid and the pipe costs $7000 per inch of diameter, what are the minimum cost and the pipe and pump size to maintain the 3 ft^3/s flow rate? Make some simplifying assumptions.

P6.72   Modify Prob. P6.57 by letting the diameter be unknown. Find the proper pipe diameter for which the pool will drain in about two hours flat.

P6.73   The Moody chart, Fig. 6.13, is best for finding head loss (or Δp) when Q, V, d, and L are known. It is awkward for the third type of problem, finding d when h_f (or Δp) and Q are known (see Example 6.10). Prepare a modified Moody chart whose abscissa is independent of d, using as a parameter ϵ nondimensionalized without d, from which one can immediately read the (dimensionless) ordinate to find d. Use your chart to solve Example 6.10.

P6.74   Two reservoirs, which differ in surface elevation by 40 m, are connected by a new commercial steel pipe of diameter 8 cm. If the desired flow rate is 200 N/s of water at 20°C, what is the proper length of the pipe?

P6.75   You wish to water your garden with 100 ft of $\frac{5}{8}$-in-diameter hose whose roughness is 0.011 in. What will be the delivery, in ft^3/s, if the gage pressure at the faucet is 60 lbf/in^2? If there is no nozzle (just an open hose exit), what is the maximum horizontal distance the exit jet will carry?

*P6.76   The small turbine in Fig. P6.76 extracts 400 W of power from the water flow. Both pipes are wrought iron. Compute the flow rate Q m^3/h. Sketch the EGL and HGL accurately.

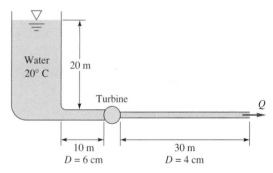

P6.76

*P6.77   Modify Prob. P6.76 into an economic analysis, as follows. Let the 40 m of wrought iron pipe have a uniform diameter d. Let the steady water flow available be $Q = 30$ m^3/h. The cost of the turbine is $4 per watt developed, and the cost of the piping is $75 per centimeter of diameter. The power generated may be sold for $0.08 per kilowatt-hour. Find the proper pipe diameter for minimum *payback time*—that is, the minimum time for which the power sales will equal the initial cost of the system.

P6.78   In Fig. P6.78 the connecting pipe is commercial steel 6 cm in diameter. Estimate the flow rate, in m^3/h, if the fluid is water at 20°C. Which way is the flow?

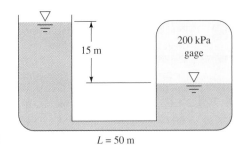

P6.78 $L = 50$ m

P6.79   A garden hose is to be used as the return line in a water-fall display at a mall. In order to select the proper pump, you need to know the roughness height inside the garden hose. Unfortunately, roughness information is not supplied by the hose manufacturer. So you devise a simple experiment to measure the roughness. The hose is attached to the drain of an above-ground swimming pool, the surface of which is 3.0 m above the hose outlet. You estimate the minor loss coefficient of the entrance region as 0.5, and the drain valve has a minor loss equivalent length of 200 diameters when fully open. Using a bucket and stopwatch, you open the valve and measure the flow rate to be 2.0×10^{-4} m^3/s for a hose that is 10.0 m long and has an inside diameter of 1.50 cm. Estimate the roughness height in mm inside the hose.

P6.80   The head-versus-flow-rate characteristics of a centrifugal pump are shown in Fig. P6.80. If this pump drives water at 20°C through 120 m of 30-cm-diameter cast iron pipe, what will be the resulting flow rate, in m^3/s?

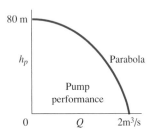

P6.80   0 Q 2m^3/s

P6.81   The pump in Fig. P6.80 is used to deliver gasoline at 20°C through 350 m of 30-cm-diameter galvanized iron pipe. Estimate the resulting flow rate, in m^3/s. (Note that the pump head is now in meters of gasoline.)

P6.82   The pump in Fig. P6.80 has its maximum efficiency at a head of 45 m. If it is used to pump ethanol at 20°C through 200 m of commercial steel pipe, what is the proper pipe diameter for maximum pump efficiency?

P6.83   For the system of Fig. P6.55, let $\Delta z = 80$ m and $L = 185$ m of cast iron pipe. What is the pipe diameter for which the flow rate will be 7 m^3/h?

P6.84   It is desired to deliver 60 m^3/h of water at 20°C through a horizontal asphalted cast iron pipe. Estimate the pipe diameter that will cause the pressure drop to be exactly 40 kPa per 100 m of pipe length.

P6.85   Repeat Prob. 6.26 using *water* at 20°C as the fluid. This is slightly more laborious than the earlier problem, but the basic concepts are just the same. Again, neglect minor losses.

P6.86   SAE 10 oil at 20°C flows at an average velocity of 2 m/s between two smooth parallel horizontal plates 3 cm apart. Estimate (*a*) the centerline velocity, (*b*) the head loss per meter, and (*c*) the pressure drop per meter.

P6.87   A commercial steel annulus 40 ft long, with $a = 1$ in and $b = \frac{1}{2}$ in, connects two reservoirs that differ in surface height by 20 ft. Compute the flow rate in ft^3/s through the annulus if the fluid is water at 20°C.

P6.88   An oil cooler consists of multiple parallel-plate passages, as shown in Fig. P6.88. The available pressure drop is 6 kPa, and the fluid is SAE 10W oil at 20°C. If the desired total flow rate is 900 m^3/h, estimate the appropriate number of passages. The plate walls are hydraulically smooth.

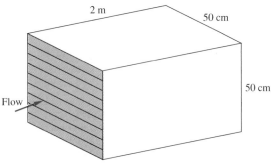

P6.88

P6.89   An annulus of narrow clearance causes a very large pressure drop and is useful as an accurate measurement of viscosity. If a smooth annulus 1 m long with $a = 50$ mm and $b = 49$ mm carries an oil flow at 0.001 m^3/s, what is the oil viscosity if the pressure drop is 250 kPa?

P6.90   A 90-ft-long sheet-steel duct carries air at approximately 20°C and 1 atm. The duct cross section is an equilateral triangle whose side measures 9 in. If a blower can supply 1 hp to the flow, what flow rate, in ft^3/s, will result?

P6.91   Heat exchangers often consist of many triangular passages. Typical is Fig. P6.91, with $L = 60$ cm and an isosceles-triangle cross section of side length $a = 2$ cm and included angle $\beta = 80°$. If the average velocity is $V = 2$ m/s and the fluid is SAE 10 oil at 20°C, estimate the pressure drop.

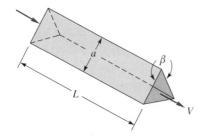

P6.91

P6.92 A large room uses a fan to draw in atmospheric air at 20°C through a 30-cm by 30-cm commercial-steel duct 12 m long, as in Fig. P6.92. Estimate (*a*) the air flow rate in m^3/h if the room pressure is 10 Pa vacuum and (*b*) the room pressure if the flow rate is 1200 m^3/h. Neglect minor losses.

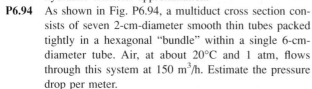

P6.92

P6.93 In Moody's Example 6.6, the 6-inch diameter, 200-ft-long asphalted cast iron pipe has a pressure drop of about 280 lbf/ft^2 when the average water velocity is 6 ft/s. Compare this to an *annular* cast iron pipe with an inner diameter of 6 in and the same annular average velocity of 6 ft/s. (*a*) What outer diameter would cause the flow to have the same pressure drop of 280 lbf/ft^2? (*b*) How do the cross-section areas compare, and why? Use the hydraulic diameter approximation.

P6.94 As shown in Fig. P6.94, a multiduct cross section consists of seven 2-cm-diameter smooth thin tubes packed tightly in a hexagonal "bundle" within a single 6-cm-diameter tube. Air, at about 20°C and 1 atm, flows through this system at 150 m^3/h. Estimate the pressure drop per meter.

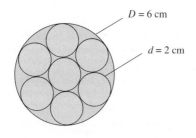

P6.94

P6.95 A wind tunnel is made of wood and is 28 m long, with a rectangular section 50 cm by 80 cm. It draws in sea-level standard air with a fan. If the fan delivers 7 kW of power to the air, estimate (*a*) the average velocity and (*b*) the pressure drop in the wind tunnel.

P6.96 Water at 20°C is flowing through a 20-cm-square smooth duct at a (turbulent) Reynolds number of 100,000. For a "laminar flow element" measurement, it is desired to pack the pipe with a honeycomb array of small square passages (see Fig. P6.28 for an example). What passage width *h* will ensure that the flow in each tube will be laminar (Reynolds number less than 2000)?

P6.97 A heat exchanger consists of multiple parallel-plate passages, as shown in Fig. P6.97. The available pressure drop is 2 kPa, and the fluid is water at 20°C. If the desired total flow rate is 900 m^3/h, estimate the appropriate number of passages. The plate walls are hydraulically smooth.

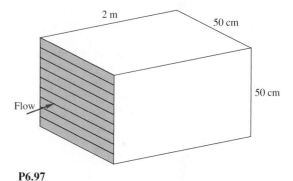

P6.97

P6.98 A rectangular heat exchanger is to be divided into smaller sections using sheets of commercial steel 0.4 mm thick, as sketched in Fig. P6.98. The flow rate is 20 kg/s of water at 20°C. Basic dimensions are $L = 1$ m, $W = 20$ cm, and $H = 10$ cm. What is the proper number of *square* sections if the overall pressure drop is to be no more than 1600 Pa?

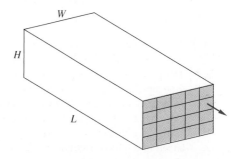

P6.98

P6.99 In Sec. 6.11 it was mentioned that Roman aqueduct customers obtained extra water by attaching a diffuser to their pipe exits. Fig. P6.99 shows a simulation: a smooth inlet pipe, with or without a 15° conical diffuser expanding to a 5-cm-diameter exit. The pipe entrance is sharp-edged. Calculate the flow rate (a) without and (b) with the diffuser.

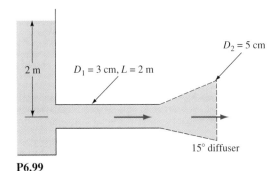

P6.99

*P6.100 Repeat Prob. P6.92 by including minor losses due to a sharp-edged entrance, the exit into the room, and an open gate valve. If the room pressure is 10 Pa vacuum, by what percentage is the air flow rate decreased from part (a) of Prob. P6.92?

P6.101 In Fig. P6.101 a thick filter is being tested for losses. The flow rate in the pipe is 7 m³/min, and the upstream pressure is 120 kPa. The fluid is air at 20°C. Using the water manometer reading, estimate the loss coefficient K of the filter.

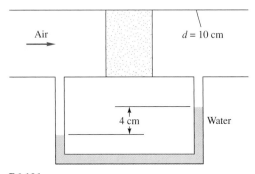

P6.101

*P6.102 A 70 percent efficient pump delivers water at 20°C from one reservoir to another 20 ft higher, as in Fig. P6.102. The piping system consists of 60 ft of galvanized iron 2-in pipe, a reentrant entrance, two screwed 90° long-radius elbows, a screwed-open gate valve, and a sharp exit. What is the input power required in horsepower with and without a 6° well-designed conical expansion added to the exit? The flow rate is 0.4 ft³/s.

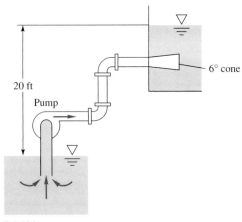

P6.102

P6.103 The reservoirs in Fig. P6.103 are connected by cast iron pipes joined abruptly, with sharp-edged entrance and exit. Including minor losses, estimate the flow of water at 20°C if the surface of reservoir 1 is 45 ft higher than that of reservoir 2.

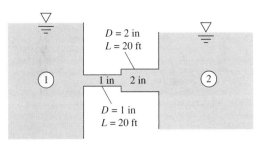

P6.103

*P6.104 Reconsider the air hockey table of Prob. P3.162 but with inclusion of minor losses. The table is 3.0 × 6.0 ft in area, with $\frac{1}{16}$-in-diameter holes spaced every inch in a rectangular grid pattern (2592 holes total). The required jet speed from each hole is estimated to be $V_{\text{jet}} = 50$ ft/s. Your job is to select an appropriate blower that will meet the requirements. *Hint:* Assume that the air is stagnant in the large volume of the manifold under the table surface, and assume sharp edge inlets at each hole. (a) Estimate the pressure rise (in lb/in²) required of the blower. (b) Compare your answer to the previous calculation in which minor losses were ignored. Are minor losses significant in this application?

P6.105 The system in Fig. P6.105 consists of 1200 m of 5 cm cast iron pipe, two 45° and four 90° flanged long-radius elbows, a fully open flanged globe valve, and a sharp exit into a reservoir. If the elevation at point 1 is 400 m, what gage pressure is required at point 1 to deliver 0.005 m³/s of water at 20°C into the reservoir?

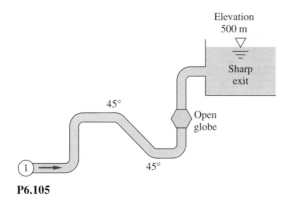

P6.105

P6.106 The water pipe in Fig. P6.106 slopes upward at 30°. The pipe has a 1-in diameter and is smooth. The flanged globe valve is fully open. If the mercury manometer shows a 7-in deflection, what is the flow rate in ft³/s?

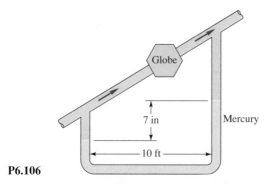

P6.106

P6.107 In Fig. P6.107 the pipe is galvanized iron. Estimate the percentage increase in the flow rate (a) if the pipe entrance

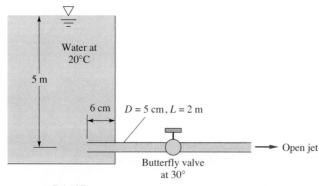

P6.107

is cut off flush with the wall and (b) if the butterfly valve is opened wide.

P6.108 The water pump in Fig. P6.108 maintains a pressure of 6.5 psig at point 1. There is a filter, a half-open disk valve, and two regular screwed elbows. There are 80 ft of 4-in diameter commercial steel pipe. (a) If the flow rate is 0.4 ft³/s, what is the loss coefficient of the filter? (b) If the disk valve is wide open and $K_{\text{fiilte}} = 7$, what is the resulting flow rate?

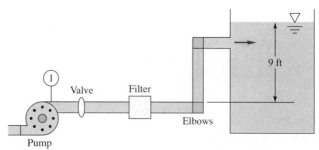

P6.108

P6.109 In Fig. P6.109 there are 125 ft of 2-in pipe, 75 ft of 6-in pipe, and 150 ft of 3-in pipe, all cast iron. There are three 90° elbows and an open globe valve, all flanged. If the exit elevation is zero, what horsepower is extracted by the turbine when the flow rate is 0.16 ft³/s of water at 20°C?

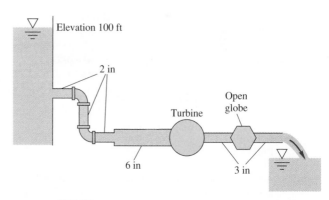

P6.109

P6.110 In Fig. P6.110 the pipe entrance is sharp-edged. If the flow rate is 0.004 m³/s, what power, in W, is extracted by the turbine?

P6.111 For the parallel-pipe system of Fig. P6.111, each pipe is cast iron, and the pressure drop $p_1 - p_2 = 3$ lbf/in². Compute the total flow rate between 1 and 2 if the fluid is SAE 10 oil at 20°C.

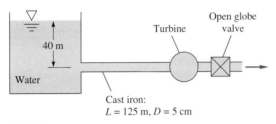

P6.110

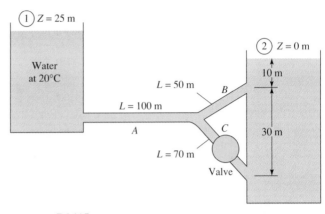

P6.115

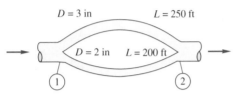

P6.111

P6.112   If the two pipes in Fig. P6.111 are instead laid in series with the same total pressure drop of 3 lbf/in², what will the flow rate be? The fluid is SAE 10 oil at 20°C.

P6.113   The parallel galvanized iron pipe system of Fig. P6.113 delivers water at 20°C with a total flow rate of 0.036 m³/s. If the pump is wide open and not running, with a loss coefficient $K = 1.5$, determine (*a*) the flow rate in each pipe and (*b*) the overall pressure drop.

P6.116   For the series-parallel system of Fig. P6.116, all pipes are 8-cm-diameter asphalted cast iron. If the total pressure drop $p_1 - p_2 = 750$ kPa, find the resulting flow rate Q m³/h for water at 20°C. Neglect minor losses.

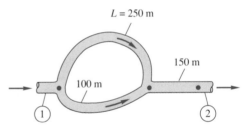

P6.116

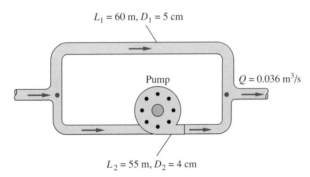

P6.113

P6.117   A blower delivers air at 3000 m³/h to the duct circuit in Fig. P6.117. Each duct is commercial steel and of square cross section, with side lengths $a_1 = a_3 = 20$ cm and $a_2 = a_4 = 12$ cm Assuming sea-level air conditions, estimate the power required if the blower has an efficiency of 75 percent. Neglect minor losses.

*P6.114   A blower supplies standard air to a plenum that feeds two horizontal square sheet-metal ducts with sharp-edged entrances. One duct is 100 ft long, with a cross-section 6 in by 6 in. The second duct is 200 ft long. Each duct exhausts to the atmosphere. When the plenum pressure is 5.0 lbf/ft²(gage) the volume flow in the longer duct is three times the flow in the shorter duct. Estimate both volume flows and the cross-section size of the longer duct.

P6.115   In Fig. P6.115 all pipes are 8-cm-diameter cast iron. Determine the flow rate from reservoir 1 if valve C is (*a*) closed and (*b*) open, $K = 0.5$.

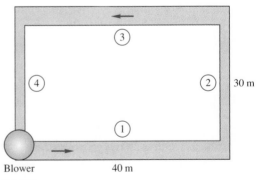

P6.117

P6.118 For the piping system of Fig. P6.118, all pipes are concrete with a roughness of 0.04 in. Neglecting minor losses, compute the overall pressure drop $p_1 - p_2$ in lbf/in^2 if $Q = 20$ ft^3/s. The fluid is water at 20°C.

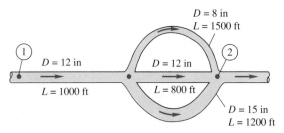

P6.118

P6.119 Modify Prob. P6.118 as follows. Let the pressure drop $p_1 - p_2$ be 98 lbf/in^2. Neglecting minor losses, determine the flow rate in m^3/h.

P6.120 Three cast iron pipes are laid in parallel with these dimensions:

Pipe	Length, m	Diameter, cm
1	800	12
2	600	8
3	900	10

The total flow rate is 200 m^3/h of water at 20°C. Determine (a) the flow rate in each pipe and (b) the pressure drop across the system.

P6.121 Consider the three-reservoir system of Fig. P6.121 with the following data:

$$L_1 = 95 \text{ m} \quad L_2 = 125 \text{ m} \quad L_3 = 160 \text{ m}$$
$$z_1 = 25 \text{ m} \quad z_2 = 115 \text{ m} \quad z_3 = 85 \text{ m}$$

All pipes are 28-cm-diameter unfinished concrete ($\epsilon = 1$ mm). Compute the steady flow rate in all pipes for water at 20°C.

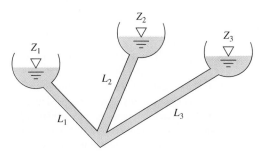

P6.121

P6.122 Modify Prob. P6.121 as follows. Reduce the diameter to 15 cm (with $\epsilon = 1$ mm), and compute the flow rates for water at 20°C. These flow rates distribute in nearly the same manner as in Prob. P6.121 but are about 5.2 times lower. Can you explain this difference?

P6.123 Modify Prob. P6.121 as follows. All data are the same except that z_3 is unknown. Find the value of z_3 for which the flow rate in pipe 3 is 0.2 m^3/s toward the junction. (This problem requires iteration and is best suited to a computer.)

P6.124 The three-reservoir system in Fig. P6.124 delivers water at 20°C. The system data are as follows:

$$D_1 = 8 \text{ in} \quad D_2 = 6 \text{ in} \quad D_3 = 9 \text{ in}$$
$$L_1 = 1800 \text{ ft} \quad L_2 = 1200 \text{ ft} \quad L_3 = 1600 \text{ ft}$$

All pipes are galvanized iron. Compute the flow rate in all pipes.

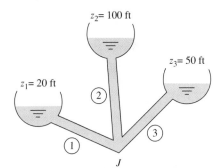

P6.124

P6.125 Suppose that the three cast iron pipes in Prob. P6.120 are instead connected to meet smoothly at a point B, as shown in Fig. P6.125. The inlet pressures in each pipe are

$$p_1 = 200 \text{ kPa} \quad p_2 = 160 \text{ kPa} \quad p_3 = 100 \text{ kPa.}$$

The fluid is water at 20°C. Neglect minor losses. Estimate the flow rate in each pipe and whether it is toward or away from point B.

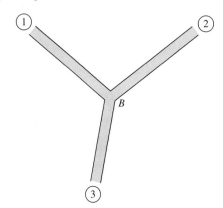

P6.125

P6.126 Modify Prob. P6.124 as follows. Let all data be the same except that pipe 1 is fitted with a butterfly valve (Fig. 6.19b). Estimate the proper valve opening angle (in degrees) for the flow rate through pipe 1 to be reduced to 1.5 ft³/s toward reservoir 1. (This problem requires iteration and is best suited to a computer.)

P6.127 In the five-pipe horizontal network of Fig. P6.127, assume that all pipes have a friction factor $f = 0.025$. For the given inlet and exit flow rate of 2 ft³/s of water at 20°C, determine the flow rate and direction in all pipes. If $p_A = 120$ lbf/in² gage, determine the pressures at points B, C, and D.

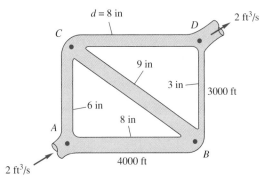

P6.127

P6.128 Modify Prob. P6.127 as follows. Let the inlet flow rate at A and the exit flow at D be unknown. Let $p_A - p_B = 100$ lbf/in². Compute the flow rate in all five pipes.

P6.129 In Fig. P6.129 all four horizontal cast iron pipes are 45 m long and 8 cm in diameter and meet at junction a, delivering water at 20°C. The pressures are known at four points as shown:

$$p_1 = 950 \text{ kPa} \quad p_2 = 350 \text{ kPa}$$
$$p_3 = 675 \text{ kPa} \quad p_4 = 100 \text{ kPa}$$

Neglecting minor losses, determine the flow rate in each pipe.

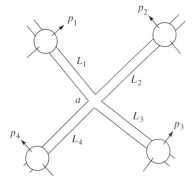

P6.129

P6.130 In Fig. P6.130 lengths AB and BD are 2000 and 1500 ft, respectively. The friction factor is 0.022 everywhere, and $p_A = 90$ lbf/in² gage. All pipes have a diameter of 6 in. For water at 20°C, determine the flow rate in all pipes and the pressures at points B, C, and D.

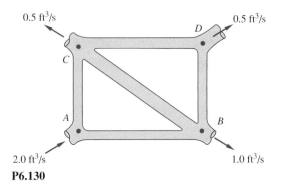

P6.130

P6.131 A water tunnel test section has a 1-m diameter and flow properties $V = 20$ m/s, $p = 100$ kPa, and $T = 20°C$. The boundary layer blockage at the end of the section is 9 percent. If a conical diffuser is to be added at the end of the section to achieve maximum pressure recovery, what should its angle, length, exit diameter, and exit pressure be?

P6.132 For Prob. P6.131 suppose we are limited by space to a total diffuser length of 10 m. What should the diffuser angle, exit diameter, and exit pressure be for maximum recovery?

P6.133 A wind tunnel test section is 3 ft square with flow properties $V = 150$ ft/s, $p = 15$ lbf/in² absolute, and $T = 68°F$. Boundary layer blockage at the end of the test section is 8 percent. Find the angle, length, exit height, and exit pressure of a flat-walled diffuser added onto the section to achieve maximum pressure recovery.

P6.134 For Prob. P6.133 suppose we are limited by space to a total diffuser length of 30 ft. What should the diffuser angle, exit height, and exit pressure be for maximum recovery?

P6.135 An airplane uses a pitot-static tube as a velocimeter. The measurements, with their uncertainties, are a static temperature of $(-11 \pm 3)°C$, a static pressure of 60 ± 2 kPa, and a pressure difference $(p_o - p_s) = 3200 \pm 60$ Pa. (a) Estimate the airplane's velocity and its uncertainty. (b) Is a compressibility correction needed?

P6.136 For the pitot-static pressure arrangement of Fig. P6.136, the manometer fluid is (colored) water at 20°C. Estimate (a) the centerline velocity, (b) the pipe volume flow, and (c) the (smooth) wall shear stress.

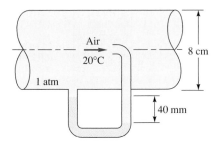

P6.136

P6.137 For the 20°C water flow of Fig. P6.137, use the pitot-static arrangement to estimate (a) the centerline velocity and (b) the volume flow in the 5-in-diameter smooth pipe. (c) What error in flow rate is caused by neglecting the 1-ft elevation difference?

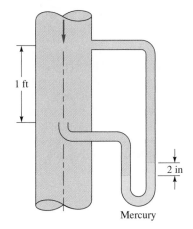

P6.137 Mercury

P6.138 An engineer who took college fluid mechanics on a pass–fail basis has placed the static pressure hole far upstream of the stagnation probe, as in Fig. P6.138, thus contaminating the pitot measurement ridiculously with pipe friction losses. If the pipe flow is air at 20°C and 1 atm and the manometer fluid is Meriam red oil (SG = 0.827), estimate the air centerline velocity for the given manometer reading of 16 cm. Assume a smooth-walled tube.

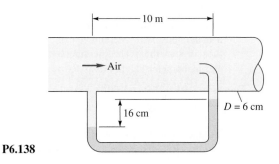

P6.138

P6.139 Professor Walter Tunnel needs to measure the flow velocity in a water tunnel. Due to budgetary restrictions, he cannot afford a pitot-static probe, but instead inserts a total head probe and a static pressure probe, as shown in Fig. P6.139, a distance h_1 apart from each other. Both probes are in the main free stream of the water tunnel, unaffected by the thin boundary layers on the sidewalls. The two probes are connected as shown to a U-tube manometer. The densities and vertical distances are shown in Fig. P6.139. (a) Write an expression for velocity V in terms of the parameters in the problem. (b) Is it critical that distance h_1 be measured accurately? (c) How does the expression for velocity V differ from that which would be obtained if a pitot-static probe had been available and used with the same U-tube manometer?

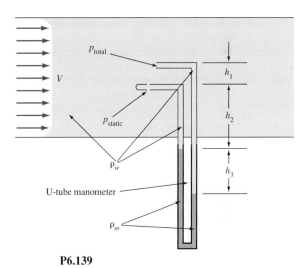

P6.139

P6.140 Kerosene at 20°C flows at 18 m³/h in a 5-cm-diameter pipe. If a 2-cm-diameter thin-plate orifice with corner taps is installed, what will the measured pressure drop be, in Pa?

P6.141 Gasoline at 20°C flows at 105 m³/h in a 10-cm-diameter pipe. We wish to meter the flow with a thin-plate orifice and a differential pressure transducer that reads best at about 55 kPa. What is the proper β ratio for the orifice?

P6.142 The shower head in Fig. P6.142 delivers water at 50°C. An orifice-type flow reducer is to be installed. The upstream pressure is constant at 400 kPa. What flow rate, in gal/min, results without the reducer? What reducer orifice diameter would decrease the flow by 40 percent?

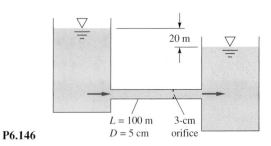

$D = 1.5$ cm

$p = 400$ kPa

Flow reducer

45 holes, 1.5-mm diameter

P6.142

P6.143 A 10-cm-diameter smooth pipe contains an orifice plate with $D: \frac{1}{2}D$ taps and $\beta = 0.5$. The measured orifice pressure drop is 75 kPa for water flow at 20°C. Estimate the flow rate, in m^3/h. What is the nonrecoverable head loss?

P6.144 Accurate solution of Prob. P6.143, using Fig. 6.41, requires iteration because both the ordinate and the abscissa of this figure contain the unknown flow rate Q. In the spirit of Example 5.11, rescale the variables and construct a new plot in which Q may be read directly from the ordinate. Solve Prob. P6.143 with your new chart.

P6.145 The 1-m-diameter tank in Fig. P6.145 is initially filled with gasoline at 20°C. There is a 2-cm-diameter orifice in the bottom. If the orifice is suddenly opened, estimate the time for the fluid level $h(t)$ to drop from 2.0 to 1.6 m.

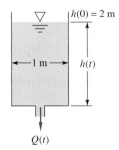

$h(0) = 2$ m

1 m

$h(t)$

P6.145 $Q(t)$

P6.146 A pipe connecting two reservoirs, as in Fig. P6.146, contains a thin-plate orifice. For water flow at 20°C, estimate (a) the volume flow through the pipe and (b) the pressure drop across the orifice plate.

P6.147 Air flows through a 6-cm-diameter smooth pipe that has a 2-m-long perforated section containing 500 holes (diameter 1 mm), as in Fig. P6.147. Pressure outside the pipe is sea-level standard air. If $p_1 = 105$ kPa and $Q_1 = 110$ m^3/h, estimate p_2 and Q_2, assuming that the holes

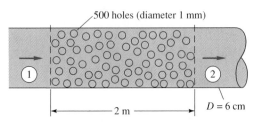

$L = 100$ m
$D = 5$ cm

20 m

3-cm orifice

P6.146

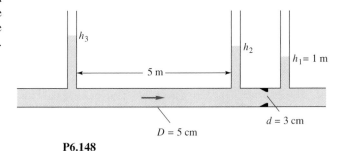

500 holes (diameter 1 mm)

1

2 m

2

$D = 6$ cm

P6.147

are approximated by thin-plate orifices. *Hint:* A momentum control volume may be very useful.

P6.148 A smooth pipe containing ethanol at 20°C flows at 7 m^3/h through a Bernoulli obstruction, as in Fig. P6.148. Three piezometer tubes are installed, as shown. If the obstruction is a thin-plate orifice, estimate the piezometer levels (a) h_2 and (b) h_3.

h_3

5 m

h_2

$h_1 = 1$ m

$D = 5$ cm

$d = 3$ cm

P6.148

P6.149 In a laboratory experiment, air at 20°C flows from a large tank through a 2-cm-diameter smooth pipe into a sea-level atmosphere, as in Fig. P6.149. The flow is metered by a long-radius nozzle of 1-cm diameter, using a manometer with Meriam red oil (SG = 0.827). The pipe is 8 m long. The measurements of tank pressure and manometer height are as follows:

p_{tank}, Pa(gage):	60	320	1200	2050	2470	3500	4900
h_{mano}, mm:	6	38	160	295	380	575	820

Use these data to calculate the flow rates Q and Reynolds numbers Re_D and make a plot of measured flow rate versus tank pressure. Is the flow laminar or turbulent? Compare the data with theoretical results obtained from the Moody chart, including minor losses. Discuss.

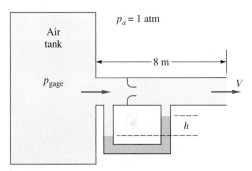

P6.149

P6.150 Gasoline at 20°C flows at 0.06 m³/s through a 15-cm pipe and is metered by a 9-cm long-radius flow nozzle (Fig. 6.40a). What is the expected pressure drop across the nozzle?

P6.151 An engineer needs to monitor a flow of 20°C gasoline at about 250 ± 25 gal/min through a 4-in-diameter smooth pipe. She can use an orifice plate, a long-radius flow nozzle, or a venturi nozzle, all with 2-in-diameter throats. The only differential pressure gage available is accurate in the range 6 to 10 lbf/in². Disregarding flow losses, which device is best?

P6.152 Kerosene at 20°C flows at 20 m³/h in an 8-cm-diameter pipe. The flow is to be metered by an ISA 1932 flow nozzle so that the pressure drop is 7000 Pa. What is the proper nozzle diameter?

P6.153 Two water tanks, each with base area of 1 ft², are connected by a 0.5-in-diameter long-radius nozzle as in Fig. P6.153. If $h = 1$ ft as shown for $t = 0$, estimate the time for $h(t)$ to drop to 0.25 ft.

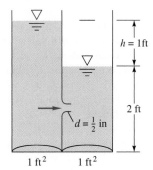

P6.153

*P6.154 Water at 20°C flows through the orifice in Fig. P6.154, which is monitored by a mercury manometer. If $d = 3$

cm, (a) what is h when the flow rate is 20 m³/h and (b) what is Q in m³/h when $h = 58$ cm?

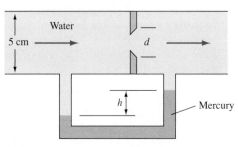

P6.154

P6.155 It is desired to meter a flow of 20°C gasoline in a 12-cm-diameter pipe, using a modern venturi nozzle. In order for international standards to be valid (Fig. 6.43), what is the permissible range of (a) flow rates, (b) nozzle diameters, and (c) pressure drops? (d) For the highest pressure-drop condition, would compressibility be a problem?

P6.156 Ethanol at 20°C flows down through a modern venturi nozzle as in Fig. P6.156. If the mercury manometer reading is 4 in, as shown, estimate the flow rate, in gal/min.

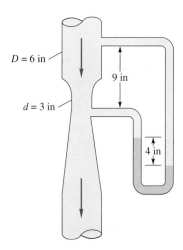

P6.156

P6.157 Modify Prob. P6.156 if the fluid is air at 20°C, entering the venturi at a pressure of 18 lbf/in². Should a compressibility correction be used?

P6.158 Water at 20°C flows in a long horizontal commercial steel 6-cm-diameter pipe that contains a classical Herschel venturi with a 4-cm throat. The venturi is connected to a mercury manometer whose reading is $h = 40$ cm. Estimate (a) the flow rate, in m³/h, and (b) the total pressure difference between points 50 cm upstream and 50 cm downstream of the venturi.

P6.159 A modern venturi nozzle is tested in a laboratory flow with water at 20°C. The pipe diameter is 5.5 cm, and the venturi throat diameter is 3.5 cm. The flow rate is measured by a weigh tank and the pressure drop by a water–mercury manometer. The mass flow rate and manometer readings are as follows:

$\dot{m}$, kg/s	0.95	1.98	2.99	5.06	8.15
h, mm	3.7	15.9	36.2	102.4	264.4

Use these data to plot a calibration curve of venturi discharge coefficient versus Reynolds number. Compare with the accepted correlation, Eq. (6.114).

*P6.160 The butterfly-valve losses in Fig. 6.19b may be viewed as a type of Bernoulli obstruction device, as in Fig. 6.39. The "throat area" A_t in Eq. (6.104) can be interpreted as the two slivers of opening around the butterfly disk when viewed from upstream. First fit the average loss K_{mean} versus the opening angle in Fig. 6.19b to an exponential curve. Then use your curve fit to compute the "discharge coefficient" of a butterfly valve as a function of the opening angle. Plot the results and compare them to those for a typical flowmeter.

P6.161 Air flows at high speed through a Herschel venturi monitored by a mercury manometer, as shown in Fig. P6.161. The upstream conditions are 150 kPa and 80°C. If $h = 37$ cm, estimate the mass flow in kg/s. (*Hint:* The flow is compressible.)

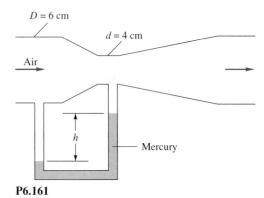

P6.161

P6.162 Modify Prob. P6.161 as follows. Find the manometer reading h for which the mass flow through the venturi is approximately 0.4 kg/s. (*Hint:* The flow is compressible.)

Word Problems

W6.1 In fully developed straight-duct flow, the velocity profiles do not change (why?), but the pressure drops along the pipe axis. Thus there is pressure work done on the fluid. If, say, the pipe is insulated from heat loss, where does this energy go? Make a thermodynamic analysis of the pipe flow.

W6.2 From the Moody chart (Fig. 6.13), rough surfaces, such as sand grains or ragged machining, do not affect laminar flow. Can you explain why? They *do* affect turbulent flow. Can you develop, or suggest, an analytical–physical model of turbulent flow near a rough surface that might be used to predict the known increase in pressure drop?

W6.3 Differentiation of the laminar pipe flow solution, Eq. (6.40), shows that the fluid shear stress $\tau(r)$ varies linearly from zero at the axis to τ_w at the wall. It is claimed that this is also true, at least in the time mean, for fully developed *turbulent* flow. Can you verify this claim analytically?

W6.4 A porous medium consists of many tiny tortuous passages, and Reynolds numbers based on pore size are usually very low, of order unity. In 1856 H. Darcy proposed that the pressure gradient in a porous medium was directly proportional to the volume-averaged velocity **V** of the fluid:

$$\nabla p = -\frac{\mu}{K}\mathbf{V}$$

where K is termed the *permeability* of the medium. This is now called *Darcy's law* of porous flow. Can you make a Poiseuille flow model of porous-media flow that verifies Darcy's law? Meanwhile, as the Reynolds number increases, so that $VK^{1/2}/\nu > 1$, the pressure drop becomes nonlinear, as was shown experimentally by P. H. Forscheimer as early as 1782. The flow is still decidedly laminar, yet the pressure gradient is quadratic:

$$\nabla p = -\frac{\mu}{K}\mathbf{V} - C|V|\mathbf{V} \quad \text{Darcy-Forscheimer law}$$

where C is an empirical constant. Can you explain the reason for this nonlinear behavior?

Fundamentals of Engineering Exam Problems

FE6.1 In flow through a straight, smooth pipe, the diameter Reynolds number for transition to turbulence is generally taken to be
(a) 1500, (b) 2300, (c) 4000, (d) 250,000, (e) 500,000

FE6.2 For flow of water at 20°C through a straight, smooth pipe at 0.06 m^3/h, the pipe diameter for which transition to turbulence occurs is approximately
(a) 1.0 cm, (b) 1.5 cm, (c) 2.0 cm, (d) 2.5 cm, (e) 3.0 cm

FE6.3 For flow of oil [$\mu = 0.1$ kg/(m · s), SG = 0.9] through a long, straight, smooth 5-cm-diameter pipe at 14 m^3/h, the pressure drop per meter is approximately
(a) 2200 Pa, (b) 2500 Pa, (c) 10,000 Pa, (d) 160 Pa, (e) 2800 Pa

FE6.4 For flow of water at a Reynolds number of 1.03 E6 through a 5-cm-diameter pipe of roughness height 0.5 mm, the approximate Moody friction factor is
(a) 0.012, (b) 0.018, (c) 0.038, (d) 0.049, (e) 0.102

FE6.5 Minor losses through valves, fittings, bends, contractions, and the like are commonly modeled as proportional to (a) total head, (b) static head, (c) velocity head, (d) pressure drop, (e) velocity

FE6.6 A smooth 8-cm-diameter pipe, 200 m long, connects two reservoirs, containing water at 20°C, one of which has a surface elevation of 700 m and the other a surface elevation of 560 m. If minor losses are neglected, the expected flow rate through the pipe is
(a) 0.048 m^3/h, (b) 2.87 m^3/h, (c) 134 m^3/h, (d) 172 m^3/h, (e) 385 m^3/h

FE6.7 If, in Prob. FE6.6 the pipe is rough and the actual flow rate is 90 m^3/h, then the expected average roughness height of the pipe is approximately
(a) 1.0 mm, (b) 1.25 mm, (c) 1.5 mm, (d) 1.75 mm, (e) 2.0 mm

FE6.8 Suppose in Prob. FE6.6 the two reservoirs are connected, not by a pipe, but by a sharp-edged orifice of diameter 8 cm. Then the expected flow rate is approximately

(a) 90 m^3/h, (b) 579 m^3/h, (c) 748 m^3/h, (d) 949 m^3/h, (e) 1048 m^3/h

FE6.9 Oil [$\mu = 0.1$ kg/(m · s), SG = 0.9] flows through a 50-m-long smooth 8-cm-diameter pipe. The maximum pressure drop for which laminar flow is expected is approximately
(a) 30 kPa, (b) 40 kPa, (c) 50 kPa, (d) 60 kPa, (e) 70 kPa

FE6.10 Air at 20°C and approximately 1 atm flows through a smooth 30-cm-square duct at 1500 ft^3/min. The expected pressure drop per meter of duct length is
(a) 1.0 Pa, (b) 2.0 Pa, (c) 3.0 Pa, (d) 4.0 Pa, (e) 5.0 Pa

FE6.11 Water at 20°C flows at 3 m^3/h through a sharp-edged 3-cm-diameter orifice in a 6-cm-diameter pipe. Estimate the expected pressure drop across the orifice.
(a) 440 Pa, (b) 680 Pa, (c) 875 Pa, (d) 1750 Pa, (e) 1870 Pa

FE6.12 Water flows through a straight 10-cm-diameter pipe at a diameter Reynolds number of 250,000. If the pipe roughness is 0.06 mm, what is the approximate Moody friction factor?
(a) 0.015, (b) 0.017, (c) 0.019, (d) 0.026, (e) 0.032

FE6.13 What is the hydraulic diameter of a rectangular air-ventilation duct whose cross section is 1 m by 25 cm?
(a) 25 cm, (b) 40 cm, (c) 50 cm, (d) 75 cm, (e) 100 cm

FE6.14 Water at 20°C flows through a pipe at 300 gal/min with a friction head loss of 45 ft. What is the power required to drive this flow?
(a) 0.16 kW, (b) 1.88 kW, (c) 2.54 kW, (d) 3.41 kW, (e) 4.24 kW

FE6.15 Water at 20°C flows at 200 gal/min through a pipe 150 m long and 8 cm in diameter. If the friction head loss is 12 m, what is the Moody friction factor?
(a) 0.010, (b) 0.015, (c) 0.020, (d) 0.025, (e) 0.030

Comprehensive Problems

C6.1 A pitot-static probe will be used to measure the velocity distribution in a water tunnel at 20°C. The two pressure lines from the probe will be connected to a U-tube manometer that uses a liquid of specific gravity 1.7. The maximum velocity expected in the water tunnel is 2.3 m/s. Your job is to select an appropriate U-tube from a manufacturer that supplies manometers of heights 8, 12, 16, 24, and 36 in. The cost increases significantly with manometer height. Which of these should you purchase?

*C6.2 A pump delivers a steady flow of water (ρ, μ) from a large tank to two other higher-elevation tanks, as shown in Fig. C6.2. The same pipe of diameter d and roughness ϵ is used throughout. All minor losses *except through the valve* are neglected, and the partially closed valve has a loss coefficient K_{valve}. Turbulent flow may be assumed with all kinetic energy flux correction coefficients equal to 1.06. The pump net head H is a known function of Q_A and hence also of $V_A = Q_A/A_{pipe}$; for example, $H = a - bV_A^2$, where a

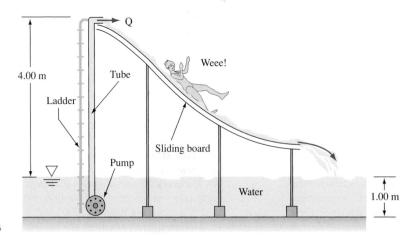

C6.2

and b are constants. Subscript J refers to the junction point at the tee where branch A splits into B and C. Pipe length L_C is much longer than L_B. It is desired to predict the pressure at J, the three pipe velocities and friction factors, and the pump head. Thus there are eight variables: H, V_A, V_B, V_C, f_A, f_B, f_C, p_J. Write down the eight equations needed to resolve this problem, but *do not solve*, since an elaborate iteration procedure, or an equation solver such as EES, would be required.

C6.3 A small water slide is to be installed inside a swimming pool. See Fig. C6.3. The slide manufacturer recommends a continuous water flow rate Q of 1.39×10^{-3} m³/s (about 22 gal/min) down the slide, to ensure that the customers do not burn their bottoms. A pump is to be installed under the slide, with a 5.00-m-long, 4.00-cm-diameter hose supplying swimming pool water for the slide. The pump is 80 percent efficient and will rest fully submerged 1.00 m below the water surface. The roughness inside the hose is about 0.0080 cm. The

hose discharges the water at the top of the slide as a free jet open to the atmosphere. The hose outlet is 4.00 m above the water surface. For fully developed turbulent pipe flow, the kinetic energy flux correction factor is about 1.06. Ignore any minor losses here. Assume that $\rho = 998$ kg/m³ and $v = 1.00 \times 10^{-6}$ m²/s for this water. Find the brake horsepower (that is, the actual shaft power in watts) required to drive the pump.

***C6.4** Suppose you build a rural house where you need to run a pipe to the nearest water supply, which is fortunately at an elevation of about 1000 m above that of your house. The pipe will be 6.0 km long (the distance to the water supply), and the gage pressure at the water supply is 1000 kPa. You require a minimum of 3.0 gal/min of water when the end of your pipe is open to the atmosphere. To minimize cost, you want to buy the smallest-diameter pipe possible. The pipe you will use is extremely smooth. (*a*) Find the total head loss from the pipe inlet to its exit. Neglect any minor losses due to valves, elbows, entrance lengths, and so on, since

C6.3

the length is so long here and major losses dominate. Assume the outlet of the pipe is open to the atmosphere. (*b*) Which is more important in this problem, the head loss due to elevation difference or the head loss due to pressure drop in the pipe? (*c*) Find the minimum required pipe diameter.

C6.5 Water at room temperature flows at the *same* volume flow rate, $Q = 9.4 \times 10^{-4}$ m³/s, through two ducts, one a round pipe and one an annulus. The cross-sectional area A of the two ducts is identical, and all walls are made of commercial steel. Both ducts are the same length. In the cross sections shown in Fig. C6.5 $R = 15.0$ mm and $a = 25.0$ mm.

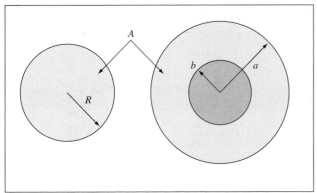

C6.5

(*a*) What is the radius b such that the cross-sectional areas of the two ducts are identical? (*b*) Compare the frictional head loss h_f per unit length of pipe for the two cases, assuming fully developed flow. For the annulus, do both a quick estimate (using the hydraulic diameter) and a more accurate estimate (using the effective diameter correction), and compare. (*c*) If the losses are different for the two cases, explain why. Which duct, if any, is more "efficient"?

Design Projects

D6.1 A hydroponic garden uses the 10-m-long perforated-pipe system in Fig. D6.1 to deliver water at 20°C. The pipe is 5 cm in diameter and contains a circular hole every 20 cm. A pump delivers water at 75 kPa (gage) at the entrance, while the other end of the pipe is closed. If you attempted, for example, Prob. P3.125, you know that the pressure near the closed end of a perforated "manifold" is surprisingly high, and there will be too much flow through the holes near that end. One remedy is to vary the hole size along the pipe

C6.6 John Laufer (*NACA Tech Rep.* 1174, 1954) gave velocity data 20°C airflow in a smooth 24.7-cm-diameter pipe at Re ≈ 5 E5:

u/u_{CL}	1.0	0.997	0.988	0.959	0.908	0.847	0.818	0.771	0.690
r/R	0.0	0.102	0.206	0.412	0.617	0.784	0.846	0.907	0.963

The centerline velocity u_{CL} was 30.5 m/s. Determine (*a*) the average velocity by numerical integration and (*b*) the wall shear stress from the log law approximation. Compare with the Moody chart and with Eq. (6.43).

C6.7 Consider energy exchange in fully developed laminar flow between parallel plates, as in Eqs. (6.63). Let the pressure drop over a length L be Δp. Calculate the rate of work done by this pressure drop on the fluid in the region $(0 < x < L, -h < y < +h)$ and compare with the integrated energy dissipated due to the viscous function Φ from Eq. (4.50) over this same region. The two should be equal. Explain why this is so. Can you relate the viscous drag force and the wall shear stress to this energy result?

C6.8 This text has presented the traditional correlations for the turbulent smooth-wall friction factor, Eq. (6.38), and the law-of-the-wall, Eq. (6.28). Recently, groups at Princeton and Oregon [56] have made new friction measurements and suggest the following smooth-wall friction law:

$$\frac{1}{\sqrt{f}} = 1.930 \log_{10}(\mathrm{Re}_D \sqrt{f}) - 0.537$$

In earlier work, they also report that better values for the constants κ and B in the log-law, Eq. (6.28), are $\kappa \approx 0.421 \pm 0.002$ and $B \approx 5.62 \pm 0.08$. (*a*) Calculate a few values of f in the range 1E4 ≤ Re$_D$ ≤ 1E8 and see how the two formulas differ. (*b*) Read Ref. 56 and briefly check the five papers in its bibliography. Report to the class on the general results of this work.

axis. Make a design analysis, perhaps using a personal computer, to pick the optimum hole size distribution that will make the discharge flow rate as uniform as possible along the pipe axis. You are constrained to pick hole sizes that correspond only to commercial (numbered) metric drill-bit sizes available to the typical machine shop.

D6.2 It is desired to design a pump-piping system to keep a 1-million-gallon capacity water tank filled. The plan is to use a modified (in size and speed) version of the

D6.1

model 1206 centrifugal pump manufactured by Taco Inc., Cranston, Rhode Island. Test data have been provided to us by Taco Inc. for a small model of this pump: $D = 5.45$ in, $\Omega = 1760$ r/min, tested with water at 20°C:

Q, gal/min	0	5	10	15	20	25	30	35	40	45	50	55	60
H, ft	28	28	29	29	28	28	27	26	25	23	21	18	15
Efficiency, %	0	13	25	35	44	48	51	53	54	55	53	50	45

The tank is to be filled daily with rather chilly (10°C) groundwater from an aquifer, which is 0.8 mi from the tank and 150 ft lower than the tank. Estimated daily water use is 1.5 million gal/day. Filling time should not exceed 8 h per day. The piping system should have four "butterfly" valves with variable openings (see Fig. 6.19), 10 elbows of various angles, and galvanized iron pipe of a size to be selected in the design. The design

should be economical—both in capital costs and operating expense. Taco Inc. has provided the following cost estimates for system components:

Pump and motor	$3500 plus $1500 per inch of impeller size
Pump speed	Between 900 and 1800 r/min
Valves	$300 + $200 per inch of pipe size
Elbows	$50 plus $50 per inch of pipe size
Pipes	$1 per inch of diameter per foot of length
Electricity cost	10¢ per kilowatt-hour

Your design task is to select an economical pipe size and pump impeller size and speed for this task, using the pump test data in nondimensional form (see Prob. P5.61) as design data. Write a brief report (five to six pages) showing your calculations and graphs.

References

1. P. S. Bernard and J. M. Wallace, *Turbulent Flow: Analysis, Measurement, and Prediction,* Wiley, New York, 2002.

2. H. Schlichting, *Boundary Layer Theory,* 7th ed., McGraw-Hill, New York, 1979.

3. F. M. White, *Viscous Fluid Flow,* 3d ed., McGraw-Hill, New York, 2005.

4. O. Reynolds, "An Experimental Investigation of the Circumstances which Determine Whether the Motion of Water Shall Be Direct or Sinuous and of the Law of Resistance in Parallel Channels," *Phil. Trans. R. Soc.,* vol. 174, 1883, pp. 935–982.

5. P. G. G. Drazin, *Introduction to Hydrodynamic Stability,* Cambridge University Press, New York, 2002.

6. H. Rouse and S. Ince, *History of Hydraulics,* Iowa Institute of Hydraulic Research, State University of Iowa, Iowa City, 1957.

7. J. Nikuradse, "Strömungsgesetze in Rauhen Rohren," *VDI Forschungsh.* 361, 1933; English trans., *NACA Tech. Mem.* 1292.

8. L. F. Moody, "Friction Factors for Pipe Flow," *ASME Trans.,* vol. 66, pp. 671–684, 1944.

9. C. F. Colebrook, "Turbulent Flow in Pipes, with Particular Reference to the Transition between the Smooth and Rough Pipe Laws," *J. Inst. Civ. Eng. Lond.,* vol. 11, 1938–1939, pp. 133–156.

10. O. C. Jones, Jr., "An Improvement in the Calculations of Turbulent Friction in Rectangular Ducts," *J. Fluids Eng.,* June 1976, pp. 173–181.

11. R. Berker, *Handbuch der Physik,* vol. 7, no. 2, pp. 1–384, Springer-Verlag, Berlin, 1963.

12. R. M. Olson and S. J. Wright, *Essentials of Engineering Fluid Mechanics,* 5th ed., Harper & Row, New York, 1990.

13. P. A. Durbin and R. B. A. Pettersson, *Statistical Theory and Modeling for Turbulent Flows*, Wiley, New York, 2001.

14. P. W. Runstadler, Jr., et al., "Diffuser Data Book," *Creare Inc. Tech. Note* 186, Hanover, NH, 1975.

15. "*Flow of Fluids through Valves, Fittings, and Pipes,*" Tech. Paper 410, Crane Valve Group, Long Beach, CA, 1957 (now updated as a CD-ROM; see < http://www.cranevalves. com >).

16. E. F. Brater, H. W. King, J. E. Lindell, and C. Y. Wei, *Handbook of Hydraulics*, 7th ed., McGraw-Hill, New York, 1996.

17. H. Cross, "Analysis of Flow in Networks of Conduits or Conductors," *Univ. Ill. Bull.* 286, November 1936.

18. B. E. Larock, R. W. Jeppson, and G. Z. Watters, *Hydraulics of Pipeline Systems*, CRC Press, Boca Raton, FL, 1999.

19. G. Hewitt and C. Vasillicos, *Prediction of Turbulent Flows*, Cambridge University Press, New York, 2005.

20. R. W. Jeppson, *Analysis of Flow in Pipe Networks*, Butterworth-Heinemann, Woburn, MA, 1976.

21. R. W. Fox and S. J. Kline, "Flow Regime Data and Design Methods for Curved Subsonic Diffusers," *J. Basic Eng.*, vol. 84, 1962, pp. 303–312.

22. R. C. Baker, *Introductory Guide to Flow Measurement*, 2d ed., Wiley, New York, 2002.

23. R. W. Miller, *Flow Measurement Engineering Handbook*, 3d edition, McGraw-Hill, New York, 1997.

24. B. Warren and C. Wunsch (eds.), *Evolution of Physical Oceanography*, M.I.T. Press, Cambridge, MA, 1981.

25. U.S. Department of Commerce, *Tidal Current Tables*, National Oceanographic and Atmospheric Administration, Washington, DC, 1971.

26. J. A. Shercliff, *Electromagnetic Flow Measurement*, Cambridge University Press, New York, 1962.

27. J. A. Miller, "A Simple Linearized Hot-Wire Anemometer," *J. Fluids Eng.*, December 1976, pp. 749–752.

28. R. J. Goldstein (ed.), *Fluid Mechanics Measurements*, 2d ed., Hemisphere, New York, 1996.

29. D. Eckardt, "Detailed Flow Investigations within a High Speed Centrifugal Compressor Impeller," *J. Fluids Eng.*, September 1976, pp. 390–402.

30. H. S. Bean (ed.), *Fluid Meters: Their Theory and Application*, 6th ed., American Society of Mechanical Engineers, New York, 1971.

31. "Measurement of Fluid Flow by Means of Orifice Plates, Nozzles, and Venturi Tubes Inserted in Circular Cross Section Conduits Running Full," *Int. Organ. Stand. Rep.* DIS-5167, Geneva, April 1976.

32. B. Geurts, *Elements of Direct and Large-Eddy Simulation*, R. T. Edwards, Flourtown, PA, 2003.

33. S. E. Haaland, "Simple and Explicit Formulas for the Friction Factor in Turbulent Pipe Flow," *J. Fluids Eng.*, March 1983, pp. 89–90.

34. R. K. Shah and A. L. London, *Laminar Flow Forced Convection in Ducts*, Academic, New York, 1979.

35. P. L. Skousen, *Valve Handbook*, McGraw-Hill, New York, 1998.

36. W. Li, W.-X. Chen, and S.-Z. Xie, "Numerical Simulation of Stress-Induced Secondary Flows with Hybrid Finite Analytic Method," *Journal of Hydrodynamics*, vol. 14, no. 4, December 2002, pp. 24–30.

37. *ASHRAE Handbook—2001 Fundamentals*, ASHRAE, Atlanta, GA, 2001.

38. F. Durst, A. Melling, and J. H. Whitelaw, *Principles and Practice of Laser-Doppler Anemometry*, 2d ed., Academic, New York, 1981.

39. A. P. Lisitsyn et al., *Laser Doppler and Phase Doppler Measurement Techniques*, Springer-Verlag, New York, 2003.

40. J. E. Amadi-Echendu, H. Zhu, and E. H. Higham, "Analysis of Signals from Vortex Flowmeters," *Flow Measurement and Instrumentation*, vol. 4, no. 4, Oct. 1993, pp. 225–231.

41. G. Vass, "Ultrasonic Flowmeter Basics," *Sensors*, vol. 14, no. 10, Oct. 1997, pp. 73–78.

42. ASME Fluid Meters Research Committee, "The ISO-ASME Orifice Coefficient Equation," *Mech. Eng.* July 1981, pp. 44–45.

43. R. D. Blevins, *Applied Fluid Dynamics Handbook*, Van Nostrand Reinhold, New York, 1984.

44. O. C. Jones, Jr., and J. C. M. Leung, "An Improvement in the Calculation of Turbulent Friction in Smooth Concentric Annuli," *J. Fluids Eng.*, December 1981, pp. 615–623.

45. P. R. Bandyopadhyay, "Aspects of the Equilibrium Puff in Transitional Pipe Flow," *J. Fluid Mech.*, vol. 163, 1986, pp. 439–458.

46. I. E. Idelchik, *Handbook of Hydraulic Resistance*, 3d ed., CRC Press, Boca Raton, FL, 1993.

47. S. Klein and W. Beckman, *Engineering Equation Solver (EES)*, University of Wisconsin, Madison, WI, 2002.

48. R. D. Coffield, P. T. McKeown, and R. B. Hammond, "Irrecoverable Pressure Loss Coefficients for Two Elbows in Series with Various Orientation Angles and Separation Distances," *Report WAPD-T-3117*, Bettis Atomic Power Laboratory, West Mifflin, PA, 1997.

49. H. Ito, "Pressure Losses in Smooth Pipe Bends," *Journal of Basic Engineering*, March 1960, pp. 131–143.

50. A. H. Gibson, "On the Flow of Water through Pipes and Passages," *Proc. Roy. Soc. London*, Ser. A, vol. 83, 1910, pp. 366–378.

51. M. Raffel, *Particle Image Velocimetry: A Practical Guide*, Wiley, New York, 2000.

52. R. W. Zappe, *Valve Selection Handbook*, 4th ed., Elsevier, New York, 1998.

53. S. A. Berger, L. Talbot, and L.-S. Yao, "Flow in Curved Pipes," *Annual Review of Fluid Mechanics*, vol. 15, 1983, pp. 461–512.

54. P. L. Spedding, E. Benard, and G. M. McNally, "Fluid Flow through 90° Bends," *Developments in Chemical Engineering and Mineral Processing*, vol. 12, nos. 1–2, 2004, pp. 107–128.

55. R. R. Kerswell, "Recent Progress in Understanding the Transition to Turbulence in a Pipe," *Nonlinearity*, vol. 18, 2005, pp. R17–R44.

56. B. J. McKeon et al., "Friction Factors for Smooth Pipe Flow," *J. Fluid Mech.*, vol. 511, 2004, pp. 41–44.

The NASA *Helios* prototype solar-powered wing on its first test flight over the Pacific Ocean on July 14, 2001. Each of the 14 solar panels powers a 2-horsepower electric motor, which drives one of the fourteen propellers. This remarkable vehicle has reached an altitude of 81,100 ft (24,700 m) and, under ideal weather conditions, can go even higher. It is also ideal for checking the "wing theory" of the present chapter, since we don't have to worry about tail surfaces or fuselages or large engine nacelles. *(NASA)*

Chapter 7
Flow Past
Immersed Bodies

Motivation. This chapter is devoted to "external" flows around bodies immersed in a fluid stream. Such a flow will have viscous (shear and no-slip) effects near the body surfaces and in its wake, but will typically be nearly inviscid far from the body. These are unconfined *boundary layer* flows.

Chapter 6 considered "internal" flows confined by the walls of a duct. In that case the viscous boundary layers grow from the sidewalls, meet downstream, and fill the entire duct. Viscous shear is the dominant effect. For example, the Moody chart of Fig. 6.13 is essentially a correlation of wall shear stress for long ducts of constant cross section.

External flows are unconfined, free to expand no matter how thick the viscous layers grow. Although boundary layer theory (Sec. 7.3) and computational fluid dynamics (CFD) [4] are helpful in understanding external flows, complex body geometries usually require experimental data on the forces and moments caused by the flow. Such immersed-body flows are commonly encountered in engineering studies: *aerodynamics* (airplanes, rockets, projectiles), *hydrodynamics* (ships, submarines, torpedos), *transportation* (automobiles, trucks, cycles), *wind engineering* (buildings, bridges, water towers, wind turbines), and *ocean engineering* (buoys, breakwaters, pilings, cables, moored instruments). This chapter provides data and analysis to assist in such studies.

7.1 Reynolds Number and Geometry Effects

The technique of boundary layer (BL) analysis can be used to compute viscous effects near solid walls and to "patch" these onto the outer inviscid motion. This patching is more successful as the body Reynolds number becomes larger, as shown in Fig. 7.1.

In Fig. 7.1 a uniform stream U moves parallel to a sharp flat plate of length L. If the Reynolds number UL/ν is low (Fig. 7.1a), the viscous region is very broad and extends far ahead and to the sides of the plate. The plate retards the oncoming stream greatly, and small changes in flow parameters cause large changes in the pressure

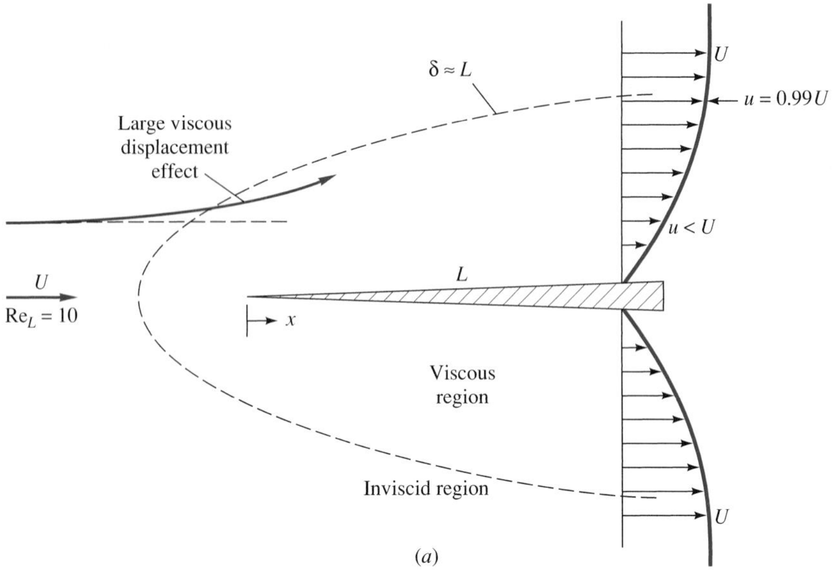

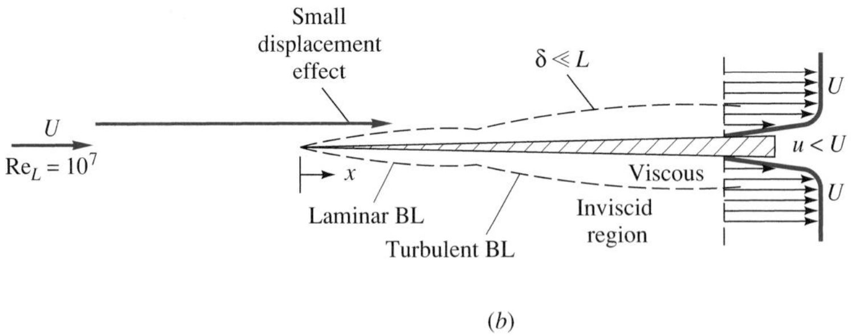

Fig. 7.1 Comparison of flow past a sharp flat plate at low and high Reynolds numbers: (a) laminar, low-Re flow; (b) high-Re flow.

distribution along the plate. Thus, although in principle it should be possible to patch the viscous and inviscid layers in a mathematical analysis, their interaction is strong and nonlinear [1 to 3]. There is no existing simple theory for external flow analysis at Reynolds numbers from 1 to about 1000. Such thick-shear-layer flows are typically studied by experiment or by numerical modeling of the flow field on a computer [4].

A high-Reynolds-number flow (Fig. 7.1b) is much more amenable to boundary layer patching, as first pointed out by Prandtl in 1904. The viscous layers, either laminar or turbulent, are very thin, thinner even than the drawing shows. We define the boundary layer thickness δ as the locus of points where the velocity u parallel to the plate reaches 99 percent of the external velocity U. As we shall see in Sec. 7.4, the accepted formulas for flat-plate flow, and their approximate ranges, are

$$\frac{\delta}{x} \approx \begin{cases} \dfrac{5.0}{\mathrm{Re}_x^{1/2}} & \text{laminar} \quad 10^3 < \mathrm{Re}_x < 10^6 & (7.1a) \\[2mm] \dfrac{0.16}{\mathrm{Re}_x^{1/7}} & \text{turbulent} \quad 10^6 < \mathrm{Re}_x & (7.1b) \end{cases}$$

where $Re_x = Ux/\nu$ is called the *local Reynolds number* of the flow along the plate surface. The turbulent flow formula applies for Re_x greater than approximately 10^6.

Some computed values from Eq. (7.1) are

Re_x	10^4	10^5	10^6	10^7	10^8
$(\delta/x)_{lam}$	0.050	0.016	0.005		
$(\delta/x)_{turb}$			0.022	0.016	0.011

The blanks indicate that the formula is not applicable. In all cases these boundary layers are so thin that their displacement effect on the outer inviscid layer is negligible. Thus the pressure distribution along the plate can be computed from inviscid theory as if the boundary layer were not even there. This external pressure field then "drives" the boundary layer flow, acting as a forcing function in the momentum equation along the surface. We shall explain this boundary layer theory in Secs. 7.4 and 7.5.

For slender bodies, such as plates and airfoils parallel to the oncoming stream, we conclude that this assumption of negligible interaction between the boundary layer and the outer pressure distribution is an excellent approximation.

For a blunt-body flow, however, even at very high Reynolds numbers, there is a discrepancy in the viscous–inviscid patching concept. Figure 7.2 shows two sketches of flow past a two- or three-dimensional blunt body. In the idealized

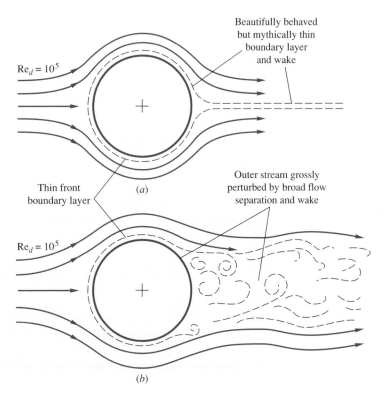

Fig. 7.2 Illustration of the strong interaction between viscous and inviscid regions in the rear of blunt-body flow: (*a*) idealized and definitely false picture of blunt-body flow; (*b*) actual picture of blunt-body flow.

sketch (7.2*a*), there is a thin film of boundary layer about the body and a narrow sheet of viscous wake in the rear. The patching theory would be glorious for this picture, but it is false. In the actual flow (Fig. 7.2*b*), the boundary layer is thin on the front, or windward, side of the body, where the pressure decreases along the surface (*favorable* pressure gradient). But in the rear the boundary layer encounters increasing pressure (*adverse* pressure gradient) and breaks off, or separates, into a broad, pulsating wake. (See Fig. 5.2*a* for a photograph of a specific example.) The mainstream is deflected by this wake, so that the external flow is quite different from the prediction from inviscid theory with the addition of a thin boundary layer.

The theory of strong interaction between blunt-body viscous and inviscid layers is not well developed. Flows like that of Fig. 7.2*b* are normally studied experimentally or with CFD [4]. Reference 5 is an example of efforts to improve the theory of separated flows. Reference 6 is another textbook devoted to separated flow.

EXAMPLE 7.1

A long, thin flat plate is placed parallel to a 20-ft/s stream of water at 68°F. At what distance x from the leading edge will the boundary layer thickness be 1 in?

Solution

- *Assumptions:* Flat-plate flow, with Eqs. (7.1) applying in their appropriate ranges.
- *Approach:* Guess laminar flow first. If contradictory, try turbulent flow.
- *Property values:* From Table A.1 for water at 68°F, $\nu \approx 1.082\text{E-}5 \text{ ft}^2/\text{s}$.
- *Solution step 1:* With $\delta = 1$ in $= 1/12$ ft, try laminar flow, Eq. (7.1*a*):

$$\frac{\delta}{x}\bigg|_{\text{lam}} = \frac{5}{(Ux/\nu)^{1/2}} \quad \text{or} \quad \frac{1/12 \text{ ft}}{x} = \frac{5}{\left[(20 \text{ ft/s})x/(1.082\text{E-}5 \text{ ft}^2/\text{s})\right]^{1/2}}$$

$$\text{Solve for} \quad x \approx 513 \text{ ft}$$

Pretty long plate! This does not sound right. Check the local Reynolds number:

$$\text{Re}_x = \frac{Ux}{\nu} = \frac{(20 \text{ ft/s})(513 \text{ ft})}{1.082\text{E-}5 \text{ ft}^2/\text{s}} = 9.5\text{E}8 \qquad (!)$$

This is impossible, since laminar boundary layer flow only persists up to about 10^6 (or, with special care to avoid disturbances, up to 3×10^6).

- *Solution step 2:* Try turbulent flow, Eq. (7.1*b*):

$$\frac{\delta}{x} = \frac{0.16}{(Ux/\nu)^{1/7}} \quad \text{or} \quad \frac{1/12 \text{ ft}}{x} = \frac{0.16}{\left[(20 \text{ ft/s})x/(1.082\text{E-}5 \text{ ft}^2/\text{s})\right]^{1/7}}$$

$$\text{Solve for} \quad x \approx 5.17 \text{ ft} \qquad\qquad \textit{Ans.}$$

Check $\text{Re}_x = (20 \text{ ft/s})(5.17 \text{ ft})/(1.082\text{E-}5 \text{ ft}^2/\text{s}) = 9.6\text{E}6 > 10^6$. OK, turbulent flow.

- *Comments:* The flow is turbulent, and the inherent ambiguity of the theory is resolved.

7.2 Momentum Integral Estimates

When we derived the momentum integral relation, Eq. (3.37), and applied it to a flat-plate boundary layer in Example 3.11, we promised to consider it further in Chap. 7. Well, here we are! Let us review the problem, using Fig. 7.3.

A shear layer of unknown thickness grows along the sharp flat plate in Fig. 7.3. The no-slip wall condition retards the flow, making it into a rounded profile $u(y)$, which merges into the external velocity $U = $ constant at a "thickness" $y = \delta(x)$. By utilizing the control volume of Fig. 3.11, we found (without making any assumptions about laminar versus turbulent flow) in Example 3.11 that the drag force on the plate is given by the following momentum integral across the exit plane:

$$D(x) = \rho b \int_0^{\delta(x)} u(U - u)\, dy \tag{7.2}$$

where b is the plate width into the paper and the integration is carried out along a vertical plane $x = $ constant. You should review the momentum integral relation (3.37) and its use in Example 3.11.

Kármán's Analysis of the Flat Plate

Equation (7.2) was derived in 1921 by Kármán [7], who wrote it in the convenient form of the *momentum thickness* θ:

$$D(x) = \rho b U^2 \theta \qquad \theta = \int_0^{\delta} \frac{u}{U}\left(1 - \frac{u}{U}\right) dy \tag{7.3}$$

Momentum thickness is thus a measure of total plate drag. Kármán then noted that the drag also equals the integrated wall shear stress along the plate:

$$D(x) = b \int_0^x \tau_w(x)\, dx$$

or

$$\frac{dD}{dx} = b\tau_w \tag{7.4}$$

Meanwhile, the derivative of Eq. (7.3), with $U = $ constant, is

$$\frac{dD}{dx} = \rho b U^2 \frac{d\theta}{dx}$$

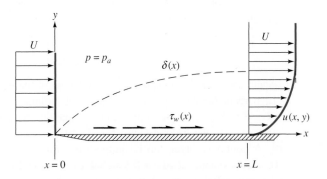

Fig. 7.3 Growth of a boundary layer on a flat plate.

By comparing this with Eq. (7.4) Kármán arrived at what is now called the *momentum integral relation* for flat-plate boundary layer flow:

$$\tau_w = \rho U^2 \frac{d\theta}{dx} \tag{7.5}$$

It is valid for either laminar or turbulent flat-plate flow.

To get a numerical result for laminar flow, Kármán assumed that the velocity profiles had an approximately parabolic shape

$$u(x, y) \approx U\left(\frac{2y}{\delta} - \frac{y^2}{\delta^2}\right) \qquad 0 \le y \le \delta(x) \tag{7.6}$$

which makes it possible to estimate both momentum thickness and wall shear:

$$\theta = \int_0^\delta \left(\frac{2y}{\delta} - \frac{y^2}{\delta^2}\right)\left(1 - \frac{2y}{\delta} + \frac{y^2}{\delta^2}\right) dy \approx \frac{2}{15}\delta$$

$$\tau_w = \mu \frac{\partial u}{\partial y}\bigg|_{y=0} \approx \frac{2\mu U}{\delta} \tag{7.7}$$

By substituting (7.7) into (7.5) and rearranging we obtain

$$\delta \, d\delta \approx 15\frac{\nu}{U}\, dx \tag{7.8}$$

where $\nu = \mu/\rho$. We can integrate from 0 to x, assuming that $\delta = 0$ at $x = 0$, the leading edge:

$$\frac{1}{2}\delta^2 = \frac{15\nu x}{U}$$

or
$$\frac{\delta}{x} \approx 5.5\left(\frac{\nu}{Ux}\right)^{1/2} = \frac{5.5}{\mathrm{Re}_x^{1/2}} \tag{7.9}$$

This is the desired thickness estimate. It is all approximate, of course, part of Kármán's *momentum integral theory* [7], but it is startlingly accurate, being only 10 percent higher than the known accepted solution for laminar flat-plate flow, which we gave as Eq. (7.1a).

By combining Eqs. (7.9) and (7.7) we also obtain a shear stress estimate along the plate:

$$c_f = \frac{2\tau_w}{\rho U^2} \approx \left(\frac{\frac{8}{15}}{\mathrm{Re}_x}\right)^{1/2} = \frac{0.73}{\mathrm{Re}_x^{1/2}} \tag{7.10}$$

Again this estimate, in spite of the crudeness of the profile assumption [Eq. (7.6)] is only 10 percent higher than the known exact laminar-plate-flow solution $c_f = 0.664/\mathrm{Re}_x^{1/2}$, treated in Sec. 7.4. The dimensionless quantity c_f, called the *skin friction coefficient*, is analogous to the friction factor f in ducts.

A boundary layer can be judged as "thin" if, say, the ratio δ/x is less than about 0.1. This occurs at $\delta/x = 0.1 = 5.0/\mathrm{Re}_x^{1/2}$ or at $\mathrm{Re}_x = 2500$. For Re_x less than 2500 we can estimate that boundary layer theory fails because the thick layer has a significant effect on the outer inviscid flow. The upper limit on Re_x for laminar flow is about 3×10^6, where measurements on a smooth flat plate [8] show that the flow undergoes transition to a turbulent boundary layer. From 3×10^6 upward the turbulent Reynolds number may be arbitrarily large, and a practical limit at present is 5×10^{10} for oil supertankers.

Displacement Thickness

Another interesting effect of a boundary layer is its small but finite displacement of the outer streamlines. As shown in Fig. 7.4, outer streamlines must deflect outward a distance $\delta^*(x)$ to satisfy conservation of mass between the inlet and outlet:

$$\int_0^h \rho U b \, dy = \int_0^\delta \rho u b \, dy \qquad \delta = h + \delta^* \qquad (7.11)$$

The quantity δ^* is called the *displacement thickness* of the boundary layer. To relate it to $u(y)$, cancel ρ and b from Eq. (7.11), evaluate the left integral, and slyly add and subtract U from the right integrand:

$$U h = \int_0^\delta (U + u - U) \, dy = U(h + \delta^*) + \int_0^\delta (u - U) \, dy$$

or

$$\delta^* = \int_0^\delta \left(1 - \frac{u}{U}\right) dy \qquad (7.12)$$

Thus the ratio of δ^*/δ varies only with the dimensionless velocity profile shape u/U.

Introducing our profile approximation (7.6) into (7.12), we obtain by integration this approximate result:

$$\delta^* \approx \frac{1}{3}\delta \qquad \frac{\delta^*}{x} \approx \frac{1.83}{Re_x^{1/2}} \qquad (7.13)$$

These estimates are only 6 percent away from the exact solutions for laminar flat-plate flow given in Sec. 7.4: $\delta^* = 0.344\delta = 1.721x/Re_x^{1/2}$. Since δ^* is much smaller than x for large Re_x and the outer streamline slope V/U is proportional to δ^*, we conclude that the velocity normal to the wall is much smaller than the velocity parallel to the wall. This is a key assumption in boundary layer theory (Sec. 7.3).

We also conclude from the success of these simple parabolic estimates that Kármán's momentum integral theory is effective and useful. Many details of this theory are given in Refs. 1 to 3.

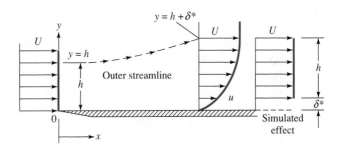

Fig. 7.4 Displacement effect of a boundary layer.

EXAMPLE 7.2

Are low-speed, small-scale air and water boundary layers really thin? Consider flow at $U = 1$ ft/s past a flat plate 1 ft long. Compute the boundary layer thickness at the trailing edge for (a) air and (b) water at 68°F.

Solution

Part (a)

From Table A.2, $\nu_{air} \approx 1.61$ E-4 ft^2/s. The trailing-edge Reynolds number thus is

$$\text{Re}_L = \frac{UL}{\nu} = \frac{(1 \text{ ft/s})(1 \text{ ft})}{1.61 \text{ E-4 ft}^2/\text{s}} = 6200$$

Since this is less than 10^6, the flow is presumed laminar, and since it is greater than 2500, the boundary layer is reasonably thin. From Eq. (7.1a), the predicted laminar thickness is

$$\frac{\delta}{x} = \frac{5.0}{\sqrt{6200}} = 0.0634$$

or, at $x = 1$ ft, $\delta = 0.0634 \text{ ft} \approx 0.76 \text{ in}$ *Ans. (a)*

Part (b)

From Table A.1 $\nu_{water} \approx 1.08$ E-5 ft^2/s. The trailing-edge Reynolds number is

$$\text{Re}_L = \frac{(1 \text{ ft/s})(1 \text{ ft})}{1.08 \text{ E-5 ft}^2/\text{s}} \approx 92,600$$

This again satisfies the laminar and thinness conditions. The boundary layer thickness is

$$\frac{\delta}{x} \approx \frac{5.0}{\sqrt{92,600}} = 0.0164$$

or, at $x = 1$ ft, $\delta = 0.0164 \text{ ft} \approx 0.20 \text{ in}$ *Ans. (b)*

Thus, even at such low velocities and short lengths, both airflows and water flows satisfy the boundary layer approximations.

7.3 The Boundary Layer Equations

In Chaps. 4 and 6 we learned that there are several dozen known analytical laminar flow solutions [1 to 3]. None are for external flow around immersed bodies, although this is one of the primary applications of fluid mechanics. No exact solutions are known for turbulent flow, whose analysis typically uses empirical modeling laws to relate time-mean variables.

There are presently three techniques used to study external flows: (1) numerical (computer) solutions, (2) experimentation, and (3) boundary layer theory.

Computational fluid dynamics is now well developed and described in advanced texts such as that by Blazek [4]. Thousands of computer solutions and models have been published; execution times, mesh sizes, and graphical presentations are improving each year. Both laminar and turbulent flow solutions have been published, and turbulence modeling is a current research topic [9]. Except for a brief discussion of computer analysis in Chap. 8, the topic of CFD is beyond our scope here.

Experimentation is the most common method of studying external flows. Chapter 5 outlined the technique of dimensional analysis, and we shall give many nondimensional experimental data for external flows in Sec. 7.6.

The third tool is boundary layer theory, first formulated by Ludwig Prandtl in 1904. We shall follow Prandtl's ideas here and make certain order-of-magnitude assumptions to greatly simplify the Navier-Stokes equations (4.38) into boundary layer equations that are solved relatively easily and patched onto the outer inviscid flow field.

One of the great achievements of boundary layer theory is its ability to predict the flow separation that occurs in adverse (positive) pressure gradients, as illustrated in Fig. 7.2b. Before 1904, when Prandtl published his pioneering paper, no one realized that such thin shear layers could cause such a gross effect as flow separation. Even today, however, boundary layer theory cannot accurately predict the behavior of the separated-flow region and its interaction with the outer flow. Modern research [4, 9] has focused on detailed CFD simulations of separated flow, and the resultant wakes, to gain further insight.

Derivation for Two-Dimensional Flow

We consider only steady two-dimensional incompressible viscous flow with the x direction along the wall and y normal to the wall, as in Fig. 7.3.[1] We neglect gravity, which is important only in boundary layers where fluid buoyancy is dominant [2, sec. 4.14]. From Chap. 4, the complete equations of motion consist of continuity and the x- and y-momentum relations:

$$\frac{\partial u}{\partial x} + \frac{\partial v}{\partial y} = 0 \tag{7.14a}$$

$$\rho\left(u\frac{\partial u}{\partial x} + v\frac{\partial u}{\partial y}\right) = -\frac{\partial p}{\partial x} + \mu\left(\frac{\partial^2 u}{\partial x^2} + \frac{\partial^2 u}{\partial y^2}\right) \tag{7.14b}$$

$$\rho\left(u\frac{\partial v}{\partial x} + v\frac{\partial v}{\partial y}\right) = -\frac{\partial p}{\partial y} + \mu\left(\frac{\partial^2 v}{\partial x^2} + \frac{\partial^2 v}{\partial y^2}\right) \tag{7.14c}$$

These should be solved for u, v, and p subject to typical no-slip, inlet, and exit boundary conditions, but in fact they are too difficult to handle for most external flows except with CFD.

In 1904 Prandtl correctly deduced that a shear layer must be very thin if the Reynolds number is large, so that the following approximations apply:

Velocities:
$$v \ll u \tag{7.15a}$$

Rates of change:
$$\frac{\partial u}{\partial x} \ll \frac{\partial u}{\partial y} \qquad \frac{\partial v}{\partial x} \ll \frac{\partial v}{\partial y} \tag{7.15b}$$

Reynolds number:
$$\mathrm{Re}_x = \frac{Ux}{v} \gg 1 \tag{7.15c}$$

[1]For a curved wall, x can represent the arc length along the wall and y can be everywhere normal to x with negligible change in the boundary layer equations as long as the radius of curvature of the wall is large compared with the boundary layer thickness [1 to 3].

Our discussion of displacement thickness in the previous section was intended to justify these assumptions.

Applying these approximations to Eq. (7.14c) results in a powerful simplification:

$$\rho\left(u\frac{\partial v}{\partial x}\right) + \rho\left(v\frac{\partial v}{\partial y}\right) = -\frac{\partial p}{\partial y} + \mu\left(\frac{\partial^2 v}{\partial x^2}\right) + \mu\left(\frac{\partial^2 v}{\partial y^2}\right)$$

$$\underset{\text{small}}{} \qquad \underset{\text{small}}{} \qquad \underset{\text{very small}}{} \qquad \underset{\text{small}}{}$$

$$\frac{\partial p}{\partial y} \approx 0 \qquad \text{or} \qquad p \approx p(x) \quad \text{only} \tag{7.16}$$

In other words, the y-momentum equation can be neglected entirely, and the pressure varies only *along* the boundary layer, not through it. The pressure gradient term in Eq. (7.14b) is assumed to be known in advance from Bernoulli's equation applied to the outer inviscid flow:

$$\frac{\partial p}{\partial x} = \frac{dp}{dx} = -\rho U \frac{dU}{dx} \tag{7.17}$$

Presumably we have already made the inviscid analysis and know the distribution of $U(x)$ along the wall (Chap. 8).

Meanwhile, one term in Eq. (7.14b) is negligible due to Eqs. (7.15):

$$\frac{\partial^2 u}{\partial x^2} \ll \frac{\partial^2 u}{\partial y^2} \tag{7.18}$$

However, neither term in the continuity relation (7.14a) can be neglected—another warning that continuity is always a vital part of any fluid flow analysis.

The net result is that the three full equations of motion (7.14) are reduced to Prandtl's two boundary layer equations for two-dimensional incompressible flow:

Continuity:
$$\frac{\partial u}{\partial x} + \frac{\partial v}{\partial y} = 0 \tag{7.19a}$$

Momentum along wall:
$$u\frac{\partial u}{\partial x} + v\frac{\partial u}{\partial y} \approx U\frac{dU}{dx} + \frac{1}{\rho}\frac{\partial \tau}{\partial y} \tag{7.19b}$$

where
$$\tau = \begin{cases} \mu\dfrac{\partial u}{\partial y} & \text{laminar flow} \\[2ex] \mu\dfrac{\partial u}{\partial y} - \overline{\rho u'v'} & \text{turbulent flow} \end{cases}$$

These are to be solved for $u(x, y)$ and $v(x, y)$, with $U(x)$ assumed to be a known function from the outer inviscid flow analysis. There are two boundary conditions on u and one on v:

At $y = 0$ (wall): $\qquad\qquad\qquad u = v = 0 \qquad$ (no slip) $\qquad\qquad$ (7.20a)

As $y = \delta(x)$ (outer stream): $\qquad u = U(x) \qquad$ (patching) $\qquad\qquad$ (7.20b)

Unlike the Navier-Stokes equations (7.14), which are mathematically elliptic and must be solved simultaneously over the entire flow field, the boundary layer equations (7.19) are mathematically parabolic and are solved by beginning at the leading edge and marching downstream as far as you like, stopping at the separation point or earlier if you prefer.[2]

The boundary layer equations have been solved for scores of interesting cases of internal and external flow for both laminar and turbulent flow, utilizing the inviscid distribution $U(x)$ appropriate to each flow. Full details of boundary layer theory and results and comparison with experiment are given in Refs. 1 to 3. Here we shall confine ourselves primarily to flat-plate solutions (Sec. 7.4).

7.4 The Flat-Plate Boundary Layer

The classic and most often used solution of boundary layer theory is for flat-plate flow, as in Fig. 7.3, which can represent either laminar or turbulent flow.

Laminar Flow

For laminar flow past the plate, the boundary layer equations (7.19) can be solved exactly for u and v, assuming that the free-stream velocity U is constant ($dU/dx = 0$). The solution was given by Prandtl's student Blasius, in his 1908 dissertation from Göttingen. With an ingenious coordinate transformation, Blasius showed that the dimensionless velocity profile u/U is a function only of the single composite dimensionless variable $(y)[U/(\nu x)]^{1/2}$:

$$\frac{u}{U} = f'(\eta) \quad \eta = y\left(\frac{U}{\nu x}\right)^{1/2} \tag{7.21}$$

where the prime denotes differentiation with respect to η. Substitution of (7.21) into the boundary layer equations (7.19) reduces the problem, after much algebra, to a single third-order nonlinear ordinary differential equation for f [1–3]:

$$f''' + \tfrac{1}{2}ff'' = 0 \tag{7.22}$$

The boundary conditions (7.20) become

At $y = 0$: $\qquad\qquad\qquad\qquad\qquad f(0) = f'(0) = 0 \tag{7.23a}$

As $y \to \infty$: $\qquad\qquad\qquad\qquad\qquad f'(\infty) \to 1.0 \tag{7.23b}$

This is the *Blasius equation,* for which accurate solutions have been obtained only by numerical integration. Some tabulated values of the velocity profile shape $f'(\eta) = u/U$ are given in Table 7.1.

Since u/U approaches 1.0 only as $y \to \infty$, it is customary to select the boundary layer thickness δ as that point where $u/U = 0.99$. From the table, this occurs at $\eta \approx 5.0$:

$$\delta_{99\%}\left(\frac{U}{\nu x}\right)^{1/2} \approx 5.0$$

or

$$\boxed{\dfrac{\delta}{x} \approx \dfrac{5.0}{\mathrm{Re}_x^{1/2}} \qquad \text{Blasius (1908)}} \tag{7.24}$$

---

[2]For further mathematical details, see Ref. 2, sec. 2.8.

Table 7.1 The Blasius Velocity Profile [1 to 3]

$y[U/(\nu x)]^{1/2}$	u/U	$y[U/(\nu x)]^{1/2}$	u/U
0.0	0.0	2.8	0.81152
0.2	0.06641	3.0	0.84605
0.4	0.13277	3.2	0.87609
0.6	0.19894	3.4	0.90177
0.8	0.26471	3.6	0.92333
1.0	0.32979	3.8	0.94112
1.2	0.39378	4.0	0.95552
1.4	0.45627	4.2	0.96696
1.6	0.51676	4.4	0.97587
1.8	0.57477	4.6	0.98269
2.0	0.62977	4.8	0.98779
2.2	0.68132	5.0	0.99155
2.4	0.72899	∞	1.00000
2.6	0.77246		

With the profile known, Blasius, of course, could also compute the wall shear and displacement thickness:

$$c_f = \frac{0.664}{\mathrm{Re}_x^{1/2}} \qquad \frac{\delta^*}{x} = \frac{1.721}{\mathrm{Re}_x^{1/2}} \tag{7.25}$$

Notice how close these are to our integral estimates, Eqs. (7.9), (7.10), and (7.13). When c_f is converted to dimensional form, we have

$$\tau_w(x) = \frac{0.332\rho^{1/2}\mu^{1/2}U^{1.5}}{x^{1/2}}$$

The wall shear drops off with $x^{1/2}$ because of boundary layer growth and varies as velocity to the 1.5 power. This is in contrast to laminar pipe flow, where $\tau_w \propto U$ and is independent of x.

If $\tau_w(x)$ is substituted into Eq. (7.4), we compute the total drag force:

$$D(x) = b\int_0^x \tau_w(x)\,dx = 0.664b\rho^{1/2}\mu^{1/2}U^{1.5}x^{1/2} \tag{7.26}$$

The drag increases only as the square root of the plate length. The nondimensional *drag coefficient* is defined as

$$C_D = \frac{2D(L)}{\rho U^2 bL} = \frac{1.328}{\mathrm{Re}_L^{1/2}} = 2c_f(L) \tag{7.27}$$

Thus, for laminar plate flow, C_D equals twice the value of the skin friction coefficient at the trailing edge. This is the drag on one side of the plate.

Kármán pointed out that the drag could also be computed from the momentum relation (7.2). In dimensionless form, Eq. (7.2) becomes

$$C_D = \frac{2}{L}\int_0^\delta \frac{u}{U}\left(1 - \frac{u}{U}\right)dy \tag{7.28}$$

This can be rewritten in terms of the momentum thickness at the trailing edge:

$$C_D = \frac{2\theta(L)}{L} \tag{7.29}$$

Computation of θ from the profile u/U or from C_D gives

$$\frac{\theta}{x} = \frac{0.664}{Re_x^{1/2}} \qquad \text{laminar flat plate} \tag{7.30}$$

Since δ is so ill defined, the momentum thickness, being definite, is often used to correlate data taken for a variety of boundary layers under differing conditions. The ratio of displacement to momentum thickness, called the dimensionless-profile *shape factor,* is also useful in integral theories. For laminar flat-plate flow

$$H = \frac{\delta^*}{\theta} = \frac{1.721}{0.664} = 2.59 \tag{7.31}$$

A large shape factor then implies that boundary layer separation is about to occur.

If we plot the Blasius velocity profile from Table 7.1 in the form of u/U versus y/δ, we can see why the simple integral theory guess, Eq. (7.6), was such a great success. This is done in Fig. 7.5. The simple parabolic approximation is not far from the true Blasius profile; hence its momentum thickness is within 10 percent of the true value. Also shown in Fig. 7.5 are three typical turbulent flat-plate velocity profiles. Notice how strikingly different in shape they are from the laminar profiles.

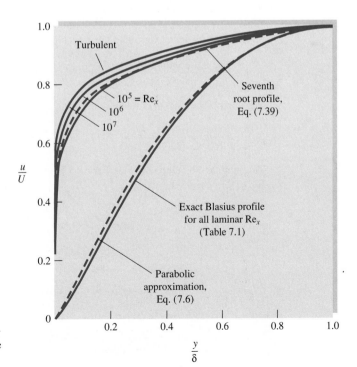

Fig. 7.5 Comparison of dimensionless laminar and turbulent flat-plate velocity profiles.

Instead of decreasing parabolically to zero, the turbulent profiles are very flat and then drop off sharply at the wall. As you might guess, they follow the logarithmic law shape and thus can be analyzed by momentum integral theory if this shape is properly represented.

Transition to Turbulence

The laminar flat-plate boundary layer eventually becomes turbulent, but there is no unique value for this change to occur. With care in polishing the wall and keeping the free stream quiet, one can delay the transition Reynolds number to $Re_{x,tr} \approx 3$ E6 [8]. However, for typical commercial surfaces and gusty free streams, a more realistic value is

$$Re_{x,tr} \approx 5 \text{ E5.}$$

EXAMPLE 7.3

A sharp flat plate with $L = 50$ cm and $b = 3$ m is parallel to a stream of velocity 2.5 m/s. Find the drag on *one side* of the plate, and the boundary thickness δ at the trailing edge, for (*a*) air and (*b*) water at 20°C and 1 atm.

Solution

- *Assumptions:*   Laminar flat-plate flow, but we should check the Reynolds numbers.
- *Approach:*   Find the Reynolds number and use the appropriate boundary layer formulas.
- *Property values:*   From Table A.2 for air at 20°C, $\rho = 1.2$ kg/m^3, $\nu = 1.5$E-5 m^2/s. From Table A.1 for water at 20°C, $\rho = 998$ kg/m^3, $\nu = 1.005$E-6 m^2/s.
- (*a*) *Solution for air:*   Calculate the Reynolds number at the trailing edge:

$$Re_L = \frac{UL}{\nu_{air}} = \frac{(2.5 \text{ m/s})(0.5 \text{ m})}{1.5\text{E-5 m}^2/\text{s}} = 83{,}300 < 5\text{E5 therefore assuredly laminar}$$

The appropriate thickness relation is Eq. (7.24):

$$\frac{\delta}{L} = \frac{5}{Re_L^{1/2}} = \frac{5}{(83{,}300)^{1/2}} = 0.0173, \text{ or } \delta_{x=L} = 0.0173(0.5 \text{ m}) \cong 0.0087 \text{ m} \qquad Ans. (a)$$

The laminar boundary layer is only 8.7 mm thick. The drag coefficient follows from Eq. (7.27):

$$C_D = \frac{1.328}{Re_L^{1/2}} = \frac{1.328}{(83{,}300)^{1/2}} = 0.0046$$

$$\text{or } D_{\text{one side}} = C_D \frac{\rho}{2} U^2 bL = (0.0046) \frac{1.2 \text{ kg/m}^3}{2} (2.5 \text{ m/s})^2 (3 \text{ m})(0.5 \text{ m}) \approx 0.026 \text{ N} \quad Ans. (a)$$

- *Comment (a):*   This is purely *friction* drag and is very small for gases at low velocities.
- (*b*) *Solution for water:*   Again calculate the Reynolds number at the trailing edge:

$$Re_L = \frac{UL}{\nu_{water}} = \frac{(2.5 \text{ m/s})(0.5 \text{ m})}{1.005\text{E-6 m}^2/\text{s}} = 1.24\text{E6} > 5\text{E5 therefore it might be turbulent}$$

This is a quandary. If the plate is rough or encounters disturbances, the flow at the trailing edge will be turbulent. Let us assume a smooth, undisturbed plate, which will remain laminar. Then again the appropriate thickness relation is Eq. (7.24):

$$\frac{\delta}{L} = \frac{5}{\text{Re}_L^{1/2}} = \frac{5}{(1.24\text{E}6)^{1/2}} = 0.00448 \quad \text{or} \quad \delta_{x=L} = 0.00448(0.5 \text{ m}) \cong 0.0022 \text{ m} \quad \textit{Ans. (b)}$$

This is four times thinner than the air result in part (a), due to the high laminar Reynolds number. Again the drag coefficient follows from Eq. (7.27):

$$C_D = \frac{1.328}{\text{Re}_L^{1/2}} = \frac{1.328}{(1.24\text{E}6)^{1/2}} = 0.0012$$

$$\text{or} \quad D_{\text{one side}} = C_D \frac{\rho}{2} U^2 bL = (0.0012)\frac{998 \text{ kg/m}^3}{2}(2.5 \text{ m/s})^2(3 \text{ m})(0.5 \text{ m}) \approx 5.6 \text{ N}$$

$$\textit{Ans. (b)}$$

- *Comment (b):* The drag is 215 times larger for water, although C_D is lower, reflecting that water is 56 times more viscous and 830 times denser than air. From Eq. (7.26), for the same U and x, the water drag should be $(56)^{1/2}(830)^{1/2} \approx 215$ times higher. *Note:* If transition to turbulence had occurred at $\text{Re}_x = 5\text{E}5$ (at about $x = 20$ cm), the drag would be about 2.5 times higher, and the trailing edge thickness about four times higher than for fully laminar flow.

Turbulent Flow

There is no exact theory for turbulent flat-plate flow, although there are many elegant computer solutions of the boundary layer equations using various empirical models for the turbulent eddy viscosity [9]. The most widely accepted result is simply an integal analysis similar to our study of the laminar profile approximation (7.6).

We begin with Eq. (7.5), which is valid for laminar or turbulent flow. We write it here for convenient reference:

$$\tau_w(x) = \rho U^2 \frac{d\theta}{dx} \tag{7.32}$$

From the definition of c_f, Eq. (7.10), this can be rewritten as

$$c_f = 2\frac{d\theta}{dx} \tag{7.33}$$

Now recall from Fig. 7.5 that the turbulent profiles are nowhere near parabolic. Going back to Fig. 6.10, we see that flat-plate flow is very nearly logarithmic, with a slight outer wake and a thin viscous sublayer. Therefore, just as in turbulent pipe flow, we assume that the logarithmic law (6.28) holds all the way across the boundary layer

$$\frac{u}{u^*} \approx \frac{1}{\kappa}\ln\frac{yu^*}{\nu} + B \qquad u^* = \left(\frac{\tau_w}{\rho}\right)^{1/2} \tag{7.34}$$

with, as usual, $\kappa = 0.41$ and $B = 5.0$. At the outer edge of the boundary layer, $y = \delta$ and $u = U$, and Eq. (7.34) becomes

$$\frac{U}{u^*} = \frac{1}{\kappa}\ln\frac{\delta u^*}{\nu} + B \tag{7.35}$$

But the definition of the skin friction coefficient, Eq. (7.10), is such that the following identities hold:

$$\frac{U}{u^*} \equiv \left(\frac{2}{c_f}\right)^{1/2} \qquad \frac{\delta u^*}{\nu} \equiv \text{Re}_\delta \left(\frac{c_f}{2}\right)^{1/2} \tag{7.36}$$

Therefore Eq. (7.35) is a *skin friction law* for turbulent flat-plate flow:

$$\left(\frac{2}{c_f}\right)^{1/2} \approx 2.44 \ln \left[\text{Re}_\delta \left(\frac{c_f}{2}\right)^{1/2} \right] + 5.0 \tag{7.37}$$

It is a complicated law, but we can at least solve for a few values and list them:

Re_δ	10^4	10^5	10^6	10^7
c_f	0.00493	0.00315	0.00217	0.00158

Following a suggestion of Prandtl, we can forget the complex log friction law (7.37) and simply fit the numbers in the table to a power-law approximation:

$$c_f \approx 0.02 \, \text{Re}_\delta^{-1/6} \tag{7.38}$$

This we shall use as the left-hand side of Eq. (7.33). For the right-hand side, we need an estimate for $\theta(x)$ in terms of $\delta(x)$. If we use the logarithmic law profile (7.34), we shall be up to our hips in logarithmic integrations for the momentum thickness. Instead we follow another suggestion of Prandtl, who pointed out that the turbulent profiles in Fig. 7.5 can be approximated by a one-seventh-power law:

$$\left(\frac{u}{U}\right)_{\text{turb}} \approx \left(\frac{y}{\delta}\right)^{1/7} \tag{7.39}$$

This is shown as a dashed line in Fig. 7.5. It is an excellent fit to the low-Reynolds-number turbulent data, which were all that were available to Prandtl at the time. With this simple approximation, the momentum thickness (7.28) can easily be evaluated:

$$\theta \approx \int_0^\delta \left(\frac{y}{\delta}\right)^{1/7} \left[1 - \left(\frac{y}{\delta}\right)^{1/7} \right] dy = \frac{7}{72} \delta \tag{7.40}$$

We accept this result and substitute Eqs. (7.38) and (7.40) into Kármán's momentum law (7.33):

$$c_f = 0.02 \, \text{Re}_\delta^{-1/6} = 2 \frac{d}{dx}\left(\frac{7}{72} \delta\right)$$

or $\qquad\qquad \text{Re}_\delta^{-1/6} = 9.72 \frac{d\delta}{dx} = 9.72 \frac{d(\text{Re}_\delta)}{d(\text{Re}_x)} \tag{7.41}$

Separate the variables and integrate, assuming $\delta = 0$ at $x = 0$:

$$\boxed{\text{Re}_\delta \approx 0.16 \, \text{Re}_x^{6/7} \qquad \text{or} \qquad \frac{\delta}{x} \approx \frac{0.16}{\text{Re}_x^{1/7}}} \tag{7.42}$$

Thus the thickness of a turbulent boundary layer increases as $x^{6/7}$, far more rapidly than the laminar increase $x^{1/2}$. Equation (7.42) is the solution to the problem, because all other parameters are now available. For example, combining Eqs. (7.42) and (7.38), we obtain the friction variation

$$c_f \approx \frac{0.027}{\mathrm{Re}_x^{1/7}} \tag{7.43}$$

Writing this out in dimensional form, we have

$$\tau_{w,\text{turb}} \approx \frac{0.0135 \mu^{1/7} \rho^{6/7} U^{13/7}}{x^{1/7}} \tag{7.44}$$

Turbulent plate friction drops slowly with x, increases nearly as ρ and U^2, and is rather insensitive to viscosity.

We can evaluate the drag coefficient from Eq. (7.29):

$$C_D = \frac{0.031}{\mathrm{Re}_L^{1/7}} = \frac{7}{6} c_f(L) \tag{7.45}$$

Then C_D is only 16 percent greater than the trailing-edge skin friction coefficient [compare with Eq. (7.27) for laminar flow].

The displacement thickness can be estimated from the one-seventh-power law and Eq. (7.12):

$$\delta^* \approx \int_0^\delta \left[1 - \left(\frac{y}{\delta} \right)^{1/7} \right] dy = \frac{1}{8} \delta \tag{7.46}$$

The turbulent flat-plate shape factor is approximately

$$H = \frac{\delta^*}{\theta} = \frac{\frac{1}{8}}{\frac{7}{72}} = 1.3 \tag{7.47}$$

These are the basic results of turbulent flat-plate theory.

Figure 7.6 shows flat-plate drag coefficients for both laminar and turbulent flow conditions. The smooth-wall relations (7.27) and (7.45) are shown, along with the effect of wall roughness, which is quite strong. The proper roughness parameter here is x/ϵ or L/ϵ, by analogy with the pipe parameter ϵ/d. In the fully rough regime, C_D is independent of the Reynolds number, so that the drag varies exactly as U^2 and is independent of μ. Reference 2 presents a theory of rough flat-plate flow, and Ref. 1 gives a curve fit for skin friction and drag in the fully rough regime:

$$c_f \approx \left(2.87 + 1.58 \log \frac{x}{\epsilon} \right)^{-2.5} \tag{7.48a}$$

$$C_D \approx \left(1.89 + 1.62 \log \frac{L}{\epsilon} \right)^{-2.5} \tag{7.48b}$$

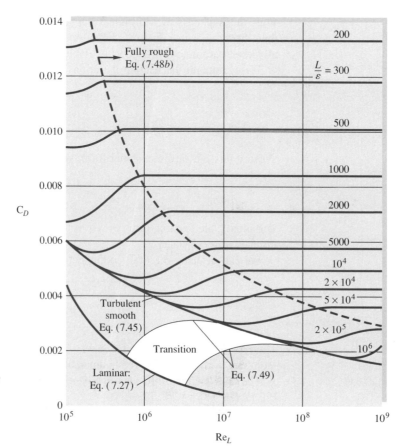

Fig. 7.6 Drag coefficient of laminar and turbulent boundary layers on smooth and rough flat plates. This chart is the flat-plate analog of the Moody diagram of Fig. 6.13.

Equation (7.48b) is plotted to the right of the dashed line in Fig. 7.6. The figure also shows the behavior of the drag coefficient in the transition region $5 \times 10^5 < \text{Re}_L < 8 \times 10^7$, where the laminar drag at the leading edge is an appreciable fraction of the total drag. Schlichting [1] suggests the following curve fits for these transition drag curves, depending on the Reynolds number Re_{trans} where transition begins:

$$C_D \approx \begin{cases} \dfrac{0.031}{\text{Re}_L^{1/7}} - \dfrac{1440}{\text{Re}_L} & \text{Re}_{\text{trans}} = 5 \times 10^5 \qquad (7.49a) \\[3mm] \dfrac{0.031}{\text{Re}_L^{1/7}} - \dfrac{8700}{\text{Re}_L} & \text{Re}_{\text{trans}} = 3 \times 10^6 \qquad (7.49b) \end{cases}$$

EXAMPLE 7.4

A hydrofoil 1.2 ft long and 6 ft wide is placed in a seawater flow of 40 ft/s, with $\rho = 1.99$ slugs/ft^3 and $\nu = 0.000011$ ft^2/s. (a) Estimate the boundary layer thickness at the end of the plate. Estimate the friction drag for (b) turbulent smooth-wall flow from the leading edge, (c) laminar turbulent flow with $Re_{trans} = 5 \times 10^5$, and (d) turbulent rough-wall flow with $\epsilon = 0.0004$ ft.

Solution

Part (a)

The Reynolds number is

$$Re_L = \frac{UL}{\nu} = \frac{(40 \text{ ft/s})(1.2 \text{ ft})}{0.000011 \text{ ft}^2/\text{s}} = 4.36 \times 10^6$$

Thus the trailing-edge flow is certainly turbulent. The maximum boundary layer thickness would occur for turbulent flow starting at the leading edge. From Eq. (7.42),

$$\frac{\delta(L)}{L} = \frac{0.16}{(4.36 \times 10^6)^{1/7}} = 0.018$$

or
$$\delta = 0.018(1.2 \text{ ft}) = 0.0216 \text{ ft} \qquad \textit{Ans. (a)}$$

This is 7.5 times thicker than a fully laminar boundary layer at the same Reynolds number.

Part (b)

For fully turbulent smooth-wall flow, the drag coefficient on one side of the plate is, from Eq. (7.45),

$$C_D = \frac{0.031}{(4.36 \times 10^6)^{1/7}} = 0.00349$$

Then the drag on both sides of the foil is approximately

$$D = 2C_D(\tfrac{1}{2}\rho U^2)bL = 2(0.00349)(\tfrac{1}{2})(1.99)(40)^2(6.0)(1.2) = 80 \text{ lbf} \qquad \textit{Ans. (b)}$$

Part (c)

With a laminar leading edge and $Re_{trans} = 5 \times 10^5$, Eq. (7.49a) applies:

$$C_D = 0.00349 - \frac{1440}{4.36 \times 10^6} = 0.00316$$

The drag can be recomputed for this lower drag coefficient:

$$D = 2C_D(\tfrac{1}{2}\rho U^2)bL = 72 \text{ lbf} \qquad \textit{Ans. (c)}$$

Part (d)

Finally, for the rough wall, we calculate

$$\frac{L}{\epsilon} = \frac{1.2 \text{ ft}}{0.0004 \text{ ft}} = 3000$$

From Fig. 7.6 at $Re_L = 4.36 \times 10^6$, this condition is just inside the fully rough regime. Equation (7.48b) applies:

$$C_D = (1.89 + 1.62 \log 3000)^{-2.5} = 0.00644$$

and the drag estimate is

$$D = 2C_D(\tfrac{1}{2}\rho U^2)bL = 148 \text{ lbf} \qquad \textit{Ans. (d)}$$

This small roughness nearly doubles the drag. It is probable that the total hydrofoil drag is still another factor of 2 larger because of trailing-edge flow separation effects.

7.5 Boundary Layers with Pressure Gradient[3]

The flat-plate analysis of the previous section should give us a good feeling for the behavior of both laminar and turbulent boundary layers, except for one important effect: flow separation. Prandtl showed that separation like that in Fig. 7.2b is caused by excessive momentum loss near the wall in a boundary layer trying to move downstream against increasing pressure, $dp/dx > 0$, which is called an *adverse pressure gradient*. The opposite case of decreasing pressure, $dp/dx < 0$, is called a *favorable gradient,* where flow separation can never occur. In a typical immersed-body flow, such as in Fig. 7.2b, the favorable gradient is on the front of the body and the adverse gradient is in the rear, as discussed in detail in Chap. 8.

We can explain flow separation with a geometric argument about the second derivative of velocity u at the wall. From the momentum equation (7.19b) at the wall, where $u = v = 0$, we obtain

$$\left.\frac{\partial \tau}{\partial y}\right|_{\text{wall}} = \left.\mu \frac{\partial^2 u}{\partial y^2}\right|_{\text{wall}} = -\rho U \frac{dU}{dx} = \frac{dp}{dx}$$

or

$$\left.\frac{\partial^2 u}{\partial y^2}\right|_{\text{wall}} = \frac{1}{\mu}\frac{dp}{dx} \tag{7.50}$$

for either laminar or turbulent flow. Thus in an adverse gradient the second derivative of velocity is positive at the wall; yet it must be negative at the outer layer $(y = \delta)$ to merge smoothly with the mainstream flow $U(x)$. It follows that the second derivative must pass through zero somewhere in between, at a point of inflection, and any boundary layer profile in an adverse gradient must exhibit a characteristic S shape.

Figure 7.7 illustrates the general case. In a favorable gradient (Fig. 7.7a) the profile is very rounded, there is no point of inflection, there can be no separation, and laminar profiles of this type are very resistant to a transition to turbulence [1 to 3].

In a zero pressure gradient (Fig. 7.7b), such as a flat-plate flow, the point of inflection is at the wall itself. There can be no separation, and the flow will undergo transition at Re_x no greater than about 3×10^6, as discussed earlier.

In an adverse gradient (Fig. 7.7c to e), a point of inflection (PI) occurs in the boundary layer, its distance from the wall increasing with the strength of the adverse gradient. For a weak gradient (Fig. 7.7c) the flow does not actually separate, but it is vulnerable to transition to turbulence at Re_x as low as 10^5 [1, 2]. At a moderate gradient, a critical condition (Fig. 7.7d) is reached where the wall shear is exactly zero $(\partial u/\partial y = 0)$. This is defined as the *separation point* $(\tau_w = 0)$, because any stronger gradient will actually cause backflow at the wall (Fig. 7.7e): the boundary layer thickens greatly, and the main flow breaks away, or separates, from the wall (Fig. 7.2b).

The flow profiles of Fig. 7.7 usually occur in sequence as the boundary layer progresses along the wall of a body. For example, in Fig. 7.2a, a favorable gradient occurs on the front of the body, zero pressure gradient occurs just upstream of the shoulder, and an adverse gradient occurs successively as we move around the rear of the body.

---

[3]This section may be omitted without loss of continuity.

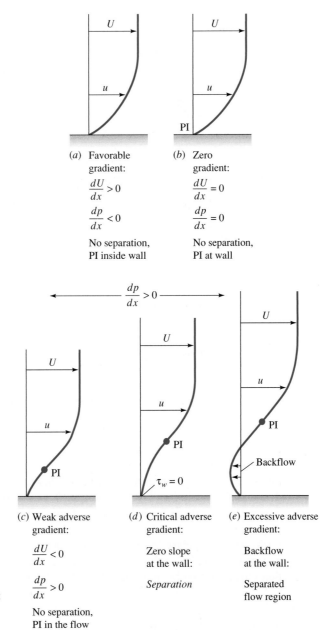

Fig. 7.7 Effect of pressure gradient on boundary layer profiles; PI = point of inflection.

(a) Favorable gradient:

$$\frac{dU}{dx} > 0$$

$$\frac{dp}{dx} < 0$$

No separation, PI inside wall

(b) Zero gradient:

$$\frac{dU}{dx} = 0$$

$$\frac{dp}{dx} = 0$$

No separation, PI at wall

$$\frac{dp}{dx} > 0$$

(c) Weak adverse gradient:

$$\frac{dU}{dx} < 0$$

$$\frac{dp}{dx} > 0$$

No separation, PI in the flow

(d) Critical adverse gradient:

Zero slope at the wall:

Separation

(e) Excessive adverse gradient:

Backflow at the wall:

Separated flow region

A second practical example is the flow in a duct consisting of a nozzle, throat, and diffuser, as in Fig. 7.8. The nozzle flow is a favorable gradient and never separates, nor does the throat flow where the pressure gradient is approximately zero. But the expanding-area diffuser produces low velocity and increasing pressure, an adverse gradient. If the diffuser angle is too large, the adverse gradient is excessive, and the boundary layer will separate at one or both walls, with backflow,

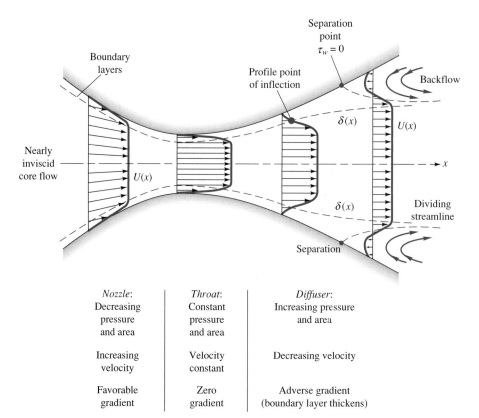

Fig. 7.8 Boundary layer growth and separation in a nozzle–diffuser configuration.

Nozzle: Decreasing pressure and area	*Throat:* Constant pressure and area	*Diffuser:* Increasing pressure and area
Increasing velocity	Velocity constant	Decreasing velocity
Favorable gradient	Zero gradient	Adverse gradient (boundary layer thickens)

increased losses, and poor pressure recovery. In the diffuser literature [10] this condition is called *diffuser stall,* a term used also in airfoil aerodynamics (Sec. 7.6) to denote airfoil boundary layer separation. Thus the boundary layer behavior explains why a large-angle diffuser has heavy flow losses (Fig. 6.23) and poor performance (Fig. 6.28).

Presently boundary layer theory can compute only up to the separation point, after which it is invalid. Techniques are now developed for analyzing the strong interaction effects caused by separated flows [5, 6].

Laminar Integral Theory[4]

Both laminar and turbulent theories can be developed from Kármán's general two-dimensional boundary layer integral relation [2, 7], which extends Eq. (7.33) to variable $U(x)$ by integration across the boundary layer:

$$\frac{\tau_w}{\rho U^2} = \frac{1}{2}c_f = \frac{d\theta}{dx} + (2 + H)\frac{\theta}{U}\frac{dU}{dx} \qquad (7.51)$$

[4]This section may be omitted without loss of continuity.

where $\theta(x)$ is the momentum thickness and $H(x) = \delta^*(x)/\theta(x)$ is the shape factor. From Eq. (7.17) negative dU/dx is equivalent to positive dp/dx—that is, an adverse gradient.

We can integrate Eq. (7.51) to determine $\theta(x)$ for a given $U(x)$ if we correlate c_f and H with the momentum thickness. This has been done by examining typical velocity profiles of laminar and turbulent boundary layer flows for various pressure gradients. Some examples are given in Fig. 7.9, showing that the shape factor H is a good indicator of the pressure gradient. The higher the H, the stronger the adverse gradient, and separation occurs approximately at

$$H \approx \begin{cases} 3.5 & \text{laminar flow} \\ 2.4 & \text{turblent flow} \end{cases} \tag{7.52}$$

The laminar profiles (Fig. 7.9a) clearly exhibit the S shape and a point of inflection with an adverse gradient. But in the turbulent profiles (Fig. 7.9b) the points of inflection are typically buried deep within the thin viscous sublayer, which can hardly be seen on the scale of the figure.

There are scores of turbulent theories in the literature, but they are all complicated algebraically and will be omitted here. The reader is referred to advanced texts [1–3, 9].

For laminar flow, a simple and effective method was developed by Thwaites [11], who found that Eq. (7.51) can be correlated by a single dimensionless momentum

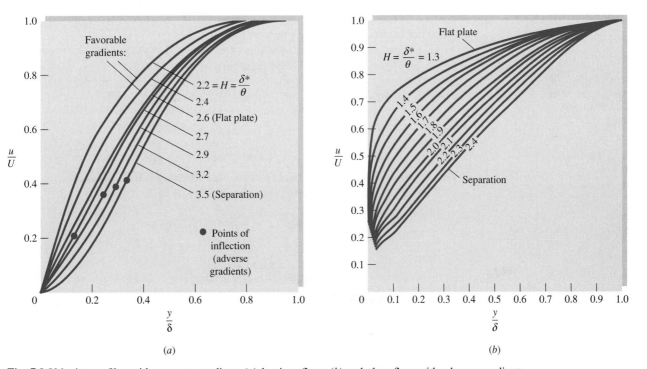

Fig. 7.9 Velocity profiles with pressure gradient: (a) laminar flow; (b) turbulent flow with adverse gradients.

thickness variable λ, defined as

$$\lambda = \frac{\theta^2}{\nu}\frac{dU}{dx} \tag{7.53}$$

Using a straight-line fit to his correlation, Thwaites was able to integrate Eq. (7.51) in closed form, with the result

$$\theta^2 = \theta_0^2\left(\frac{U_0}{U}\right)^6 + \frac{0.45\nu}{U^6}\int_0^x U^5\,dx \tag{7.54}$$

where θ_0 is the momentum thickness at $x = 0$ (usually taken to be zero). Separation ($c_f = 0$) was found to occur at a particular value of λ:

Separation: $\lambda = -0.09$ \qquad (7.55)

Finally, Thwaites correlated values of the dimensionless shear stress $S = \tau_w\theta/(\mu U)$ with λ, and his graphed result can be curve-fitted as follows:

$$S(\lambda) = \frac{\tau_w\theta}{\mu U} \approx (\lambda + 0.09)^{0.62} \tag{7.56}$$

This parameter is related to the skin friction by the identity

$$S \equiv \tfrac{1}{2}c_f\,\mathrm{Re}_\theta \tag{7.57}$$

Equations (7.54) to (7.56) constitute a complete theory for the laminar boundary layer with variable $U(x)$, with an accuracy of ±10 percent compared with computer solutions of the laminar-boundary-layer equations (7.19). Complete details of Thwaites's and other laminar theories are given in Ref. 2.

As a demonstration of Thwaites's method, take a flat plate, where $U = $ constant, $\lambda = 0$, and $\theta_0 = 0$. Equation (7.54) integrates to

$$\theta^2 = \frac{0.45\nu x}{U}$$

or \qquad\qquad $$\frac{\theta}{x} = \frac{0.671}{\mathrm{Re}_x^{1/2}} \tag{7.58}$$

This is within 1 percent of Blasius's numerical solution, Eq. (7.30).

With $\lambda = 0$, Eq. (7.56) predicts the flat-plate shear to be

$$\frac{\tau_w\theta}{\mu U} = (0.09)^{0.62} = 0.225$$

or \qquad\qquad $$c_f = \frac{2\tau_w}{\rho U^2} = \frac{0.671}{\mathrm{Re}_x^{1/2}} \tag{7.59}$$

This is also within 1 percent of the Blasius result, Eq. (7.25). However, the general accuracy of this method is poorer than 1 percent because Thwaites actually "tuned" his correlation constants to make them agree with exact flat-plate theory.

We shall not compute any more boundary layer details here; but as we go along, investigating various immersed-body flows, especially in Chap. 8, we shall use Thwaites' method to make qualitative assessments of the boundary layer behavior.

EXAMPLE 7.5

In 1938 Howarth proposed a linearly decelerating external velocity distribution

$$U(x) = U_0\left(1 - \frac{x}{L}\right) \tag{1}$$

as a theoretical model for laminar-boundary-layer study. (*a*) Use Thwaites's method to compute the separation point x_{sep} for $\theta_0 = 0$, and compare with the exact computer solution $x_{sep}/L = 0.119863$ given by H. Wipperman in 1966. (*b*) Also compute the value of $c_f = 2\tau_w/(\rho U^2)$ at $x/L = 0.1$.

Solution

Part (a)

First note that $dU/dx = -U_0/L = $ constant: Velocity decreases, pressure increases, and the pressure gradient is adverse throughout. Now integrate Eq. (7.54):

$$\theta^2 = \frac{0.45\nu}{U_0^6(1 - x/L)^6}\int_0^x U_0^5\left(1 - \frac{x}{L}\right)^5 dx = 0.075\frac{\nu L}{U_0}\left[\left(1 - \frac{x}{L}\right)^{-6} - 1\right] \tag{2}$$

Then the dimensionless factor λ is given by

$$\lambda = \frac{\theta^2}{\nu}\frac{dU}{dx} = -\frac{\theta^2 U_0}{\nu L} = -0.075\left[\left(1 - \frac{x}{L}\right)^{-6} - 1\right] \tag{3}$$

From Eq. (7.55) we set this equal to -0.09 for separation:

$$\lambda_{sep} = -0.09 = -0.075\left[\left(1 - \frac{x_{sep}}{L}\right)^{-6} - 1\right]$$

or

$$\frac{x_{sep}}{L} = 1 - (2.2)^{-1/6} = 0.123 \qquad\qquad Ans. (a)$$

This is less than 3 percent higher than Wipperman's exact solution, and the computational effort is very modest.

Part (b)

To compute c_f at $x/L = 0.1$ (just before separation), we first compute λ at this point, using Eq. (3):

$$\lambda(x = 0.1L) = -0.075[(1 - 0.1)^{-6} - 1] = -0.0661$$

Then from Eq. (7.56) the shear parameter is

$$S(x = 0.1L) = (-0.0661 + 0.09)^{0.62} = 0.099 = \tfrac{1}{2}c_f\text{Re}_\theta \tag{4}$$

We can compute Re_θ in terms of Re_L from Eq. (2) or (3):

$$\frac{\theta^2}{L^2} = \frac{0.0661}{UL/\nu} = \frac{0.0661}{\text{Re}_L}$$

or

$$\text{Re}_\theta = 0.257\,\text{Re}_L^{1/2} \qquad \text{at } \frac{x}{L} = 0.1$$

Substitute into Eq. (4):

$$0.099 = \tfrac{1}{2} c_f (0.257 \, \mathrm{Re}_L^{1/2})$$

or
$$c_f = \frac{0.77}{\mathrm{Re}_L^{1/2}} \qquad \mathrm{Re}_L = \frac{UL}{\nu}$$
  Ans. (b)

We cannot actually compute c_f without the value of, say, $U_0 L / \nu$.

7.6 Experimental External Flows

Boundary layer theory is very interesting and illuminating and gives us a great qualitative grasp of viscous flow behavior; but, because of flow separation, the theory does not generally allow a quantitative computation of the complete flow field. In particular, there is at present no satisfactory theory, except CFD results, for the forces on an arbitrary body immersed in a stream flowing at an arbitrary Reynolds number. Therefore experimentation is the key to treating external flows.

Literally thousands of papers in the literature report experimental data on specific external viscous flows. This section gives a brief description of the following external flow problems:

1. Drag of two- and three-dimensional bodies:
   a. Blunt bodies.
   b. Streamlined shapes.
2. Performance of lifting bodies:
   a. Airfoils and aircraft.
   b. Projectiles and finned bodies.
   c. Birds and insects.

For further reading see the goldmine of data compiled in Hoerner [12]. In later chapters we shall study data on supersonic airfoils (Chap. 9), open-channel friction (Chap. 10), and turbomachinery performance (Chap. 11).

Drag of Immersed Bodies

Any body of any shape when immersed in a fluid stream will experience forces and moments from the flow. If the body has arbitrary shape and orientation, the flow will exert forces and moments about all three coordinate axes, as shown in Fig. 7.10. It is customary to choose one axis parallel to the free stream and positive downstream. The force on the body along this axis is called *drag,* and the moment about that axis the *rolling moment.* The drag is essentially a flow loss and must be overcome if the body is to move against the stream.

A second and very important force is perpendicular to the drag and usually performs a useful job, such as bearing the weight of the body. It is called the *lift.* The moment about the lift axis is called *yaw.*

The third component, neither a loss nor a gain, is the *side force,* and about this axis is the *pitching moment.* To deal with this three-dimensional force-moment situation is more properly the role of a textbook on aerodynamics [for example, 13]. We shall limit the discussion here to lift and drag.

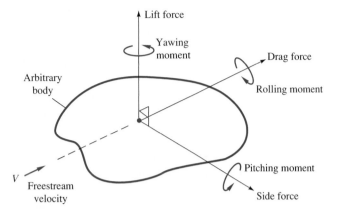

Fig. 7.10 Definition of forces and moments on a body immersed in a uniform flow.

When the body has symmetry about the lift–drag axis, as with airplanes, ships, and cars moving directly into a stream, the side force, yaw, and roll vanish, and the problem reduces to a two-dimensional case: two forces, lift and drag, and one moment, pitch.

A final simplification often occurs when the body has two planes of symmetry, as in Fig. 7.11. A wide variety of shapes such as cylinders, wings, and all bodies of revolution satisfy this requirement. If the free stream is parallel to the intersection of these two planes, called the *principal chord line of the body,* the body experiences drag only, with no lift, side force, or moments.[5] This type of degenerate one-force drag data is what is most commonly reported in the literature, but if the free stream is not parallel to the chord line, the body will have an unsymmetric orientation and all three forces and three moments can arise in principle.

In low-speed flow past geometrically similar bodies with identical orientation and relative roughness, the drag coefficient should be a function of the body Reynolds number:

$$C_D = f(\text{Re}) \tag{7.60}$$

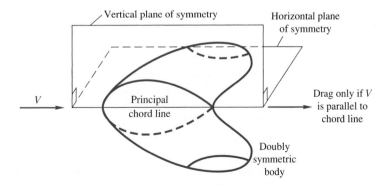

Fig. 7.11 Only the drag force occurs if the flow is parallel to both planes of symmetry.

---

[5]In bodies with shed vortices, such as the cylinder in Fig. 5.2, there may be *oscillating* lift, side force, and moments, but their mean value is zero.

The Reynolds number is based upon the free-stream velocity V and a characteristic length L of the body, usually the chord or body length parallel to the stream:

$$\text{Re} = \frac{VL}{\nu} \qquad (7.61)$$

For cylinders, spheres, and disks, the characteristic length is the diameter D.

Characteristic Area

Drag coefficients are defined by using a characteristic area A, which may differ depending on the body shape:

$$\boxed{C_D = \frac{\text{drag}}{\frac{1}{2}\rho V^2 A}} \qquad (7.62)$$

The factor $\frac{1}{2}$ is our traditional tribute to Euler and Bernoulli. The area A is usually one of three types:

1. *Frontal area,* the body as seen from the stream; suitable for thick, stubby bodies, such as spheres, cylinders, cars, trucks, missiles, projectiles, and torpedoes.
2. *Planform area,* the body area as seen from above; suitable for wide, flat bodies such as wings and hydrofoils.
3. *Wetted area,* customary for surface ships and barges.

In using drag or other fluid force data, it is important to note what length and area are being used to scale the measured coefficients.

Friction Drag and Pressure Drag

As we have mentioned, the theory of drag is weak and inadequate, except for the flat plate. This is because of flow separation. Boundary layer theory can predict the separation point but cannot accurately estimate the (usually low) pressure distribution in the separated region. The difference between the high pressure in the front stagnation region and the low pressure in the rear separated region causes a large drag contribution called *pressure drag.* This is added to the integrated shear stress or *friction drag* of the body, which it often exceeds:

$$C_D = C_{D,\text{press}} + C_{D,\text{fric}} \qquad (7.63)$$

The relative contribution of friction and pressure drag depends upon the body's shape, especially its thickness. Figure 7.12 shows drag data for a streamlined cylinder of very large depth into the paper. At zero thickness the body is a flat plate and exhibits 100 percent friction drag. At thickness equal to the chord length, simulating a circular cylinder, the friction drag is only about 3 percent. Friction and pressure drag are about equal at thickness $t/c = 0.25$. Note that C_D in Fig. 7.12b looks quite different when based on frontal area instead of planform area, planform being the usual choice for this body shape. The two curves in Fig. 7.12b represent exactly the same drag data.

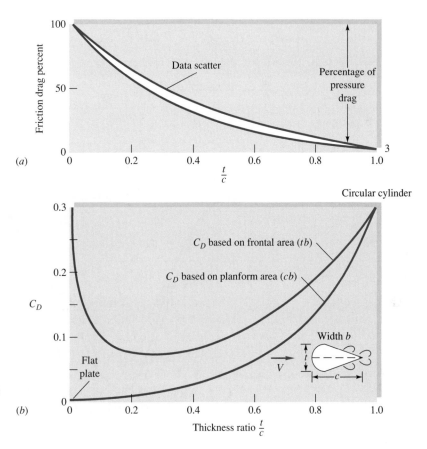

Fig. 7.12 Drag of a streamlined two-dimensional cylinder at $Re_c = 10^6$: (*a*) effect of thickness ratio on percentage of friction drag; (*b*) total drag versus thickness when based on two different areas.

Figure 7.13 illustrates the dramatic effect of separated flow and the subsequent failure of boundary layer theory. The theoretical inviscid pressure distribution on a circular cylinder (Chap. 8) is shown as the dashed line in Fig. 7.13*c*:

$$C_p = \frac{p - p_\infty}{\frac{1}{2}\rho V^2} = 1 - 4\sin^2\theta$$

where p_∞ and V are the pressure and velocity, respectively, in the free stream. The actual laminar and turbulent boundary layer pressure distributions in Fig. 7.13*c* are startlingly different from those predicted by theory. Laminar flow is very vulnerable to the adverse gradient on the rear of the cylinder, and separation occurs at $\theta = 82°$, which certainly could not have been predicted from inviscid theory. The broad wake and very low pressure in the separated laminar region cause the large drag $C_D = 1.2$.

The turbulent boundary layer in Fig. 7.13*b* is more resistant, and separation is delayed until $\theta = 120°$, with a resulting smaller wake, higher pressure on the rear, and 75 percent less drag, $C_D = 0.3$. This explains the sharp drop in drag at transition in Fig. 5.3.

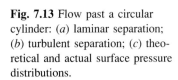

Separation

$$\theta$$

V
p_∞

82°

$C_D = 1.2$

Broad
wake

(a)

Separation

$$\theta$$

V
p_∞

120°

$C_D = 0.3$

Narrow
wake

(b)

$C_p = \dfrac{p - p_\infty}{\rho V^2 / 2}$

1.0

0.0

Turbulent

Laminar

−1.0

−2.0

Inviscid
theory

$C_p = 1 - 4 \sin^2 \theta$

−3.0

0°   45°   90°   135°   180°

θ

(c)

Fig. 7.13 Flow past a circular cylinder: (*a*) laminar separation; (*b*) turbulent separation; (*c*) theoretical and actual surface pressure distributions.

The same sharp difference between vulnerable laminar separation and resistant turbulent separation can be seen for a sphere in Fig. 7.14. The laminar flow (Fig. 7.14*a*) separates at about 80°, $C_D = 0.5$, while the turbulent flow (Fig. 7.14*b*) separates at 120°, $C_D = 0.2$. Here the Reynolds numbers are exactly the same, and the turbulent boundary layer is induced by a patch of sand roughness at the nose of the ball. Golf balls fly in this range of Reynolds numbers, which is why they are deliberately dimpled—to induce a turbulent boundary layer and lower drag. Again we would find the actual pressure distribution on the sphere to be quite different from that predicted by inviscid theory.

In general, we cannot overstress the importance of body streamlining to reduce drag at Reynolds numbers above about 100. This is illustrated in Fig. 7.15. The rectangular cylinder (Fig. 7.15*a*) has separation at all sharp corners and very high drag. Rounding its nose (Fig. 7.15*b*) reduces drag by about 45 percent, but C_D is still high. Streamlining its rear to a sharp trailing edge (Fig. 7.15*c*) reduces its drag another 85 percent to a practical minimum for the given thickness. As a dramatic contrast, the circular cylinder (Fig. 7.15*d*) has one-eighth the thickness and one-three-hundredth the cross

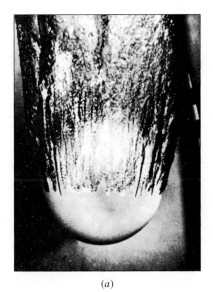 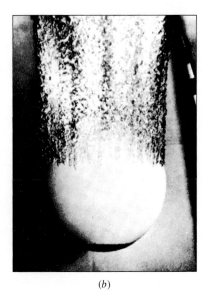

Fig. 7.14 Strong differences in laminar and turbulent separation on an 8.5-in bowling ball entering water at 25 ft/s: (*a*) smooth ball, laminar boundary layer; (*b*) same entry, turbulent flow induced by patch of nose-sand roughness. (*NAVAIR Weapons Division Historical Archives.*)

(*a*)   (*b*)

section (*c*) (Fig. 7.15*c*), yet it has the same drag. For high-performance vehicles and other moving bodies, the name of the game is drag reduction, for which intense research continues for both aerodynamic and hydrodynamic applications [20, 39].

Two-Dimensional Bodies

The drag of some representative wide-span (nearly two-dimensional) bodies is shown versus the Reynolds number in Fig. 7.16*a*. All bodies have high C_D at very low (*creeping flow*) Re $\leq$ 1.0, while they spread apart at high Reynolds numbers according to their degree of streamlining. All values of C_D are based on the planform area except the plate normal to the flow. The birds and the sailplane are, of course, not very two-dimensional, having only modest span length. Note that birds are not nearly as efficient as modern sailplanes or airfoils [14, 15].

Creeping Flow

In 1851 G. G. Stokes showed that, if the Reynolds number is very small, Re $\ll$ 1, the acceleration terms in the Navier-Stokes equations (7.14*b*, *c*) are negligible. The flow is termed *creeping flow*, or Stokes flow, and is a balance between pressure

Fig. 7.15 The importance of streamlining in reducing drag of a body (C_D based on frontal area): (*a*) rectangular cylinder; (*b*) rounded nose; (*c*) rounded nose and streamlined sharp trailing edge; (*d*) circular cylinder with the same drag as case (*c*).

$V \longrightarrow$ $C_D = 2.0$ $V \longrightarrow$ $C_D = 1.1$

(*a*)   (*b*)

$V \longrightarrow$ $C_D = 0.15$ $V \longrightarrow$

(*c*)   (*d*)

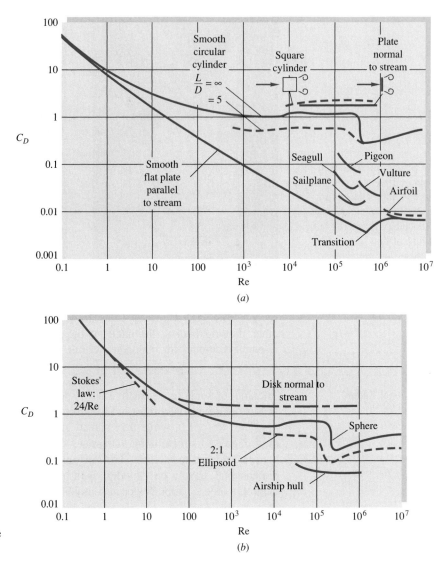

Fig. 7.16 Drag coefficients of smooth bodies at low Mach numbers: (*a*) two-dimensional bodies; (*b*) three-dimensional bodies. Note the Reynolds number independence of blunt bodies at high Re.

gradient and viscous stresses. Continuity and momentum reduce to two linear equations for velocity and pressure:

$$\text{Re} \ll 1: \qquad \nabla \cdot \mathbf{V} = 0 \qquad \text{and} \qquad \nabla p \approx \mu \nabla^2 \mathbf{V}$$

If the geometry is simple (for example, a sphere or disk) closed-form solutions can be found and the body drag can be computed [2]. Stokes himself provided the sphere drag formula:

$$F_{\text{sphere}} = 3\pi \mu U d$$

or

$$C_D = \frac{F}{\frac{1}{2}\rho U^2 \, \frac{\pi}{4} d^2} = \frac{24}{\rho U d / \mu} = \frac{24}{\text{Re}_d} \qquad (7.64)$$

This relation is plotted in Fig. 7.16*b* and is seen to be accurate for about $\text{Re}_d \lesssim 1$.

Table 7.2 gives a few data on drag, based on frontal area, of two-dimensional bodies of various cross section, at Re $\geq 10^4$. The sharp-edged bodies, which tend to cause flow separation regardless of the character of the boundary layer, are insensitive to the Reynolds number. The elliptic cylinders, being smoothly rounded, have the

Table 7.2 Drag of Two-Dimensional Bodies at Re $\geq 10^4$

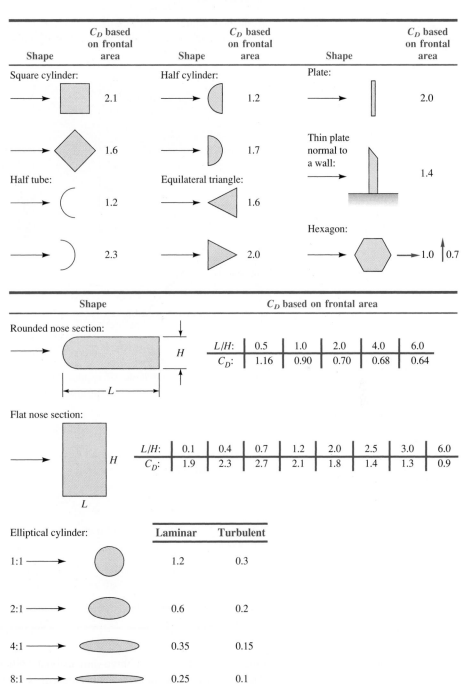

Shape	C_D based on frontal area	Shape	C_D based on frontal area	Shape	C_D based on frontal area
Square cylinder:	2.1	Half cylinder:	1.2	Plate:	2.0
	1.6		1.7	Thin plate normal to a wall:	1.4
Half tube:	1.2	Equilateral triangle:	1.6		
	2.3		2.0	Hexagon:	1.0 / 0.7

Shape	C_D based on frontal area						
Rounded nose section:	L/H:	0.5	1.0	2.0	4.0	6.0	
	C_D:	1.16	0.90	0.70	0.68	0.64	

Flat nose section:	L/H:	0.1	0.4	0.7	1.2	2.0	2.5	3.0	6.0
	C_D:	1.9	2.3	2.7	2.1	1.8	1.4	1.3	0.9

Elliptical cylinder:	Laminar	Turbulent
1:1	1.2	0.3
2:1	0.6	0.2
4:1	0.35	0.15
8:1	0.25	0.1

laminar-to-turbulent transition effect of Figs. 7.13 and 7.14 and are therefore quite sensitive to whether the boundary layer is laminar or turbulent.

EXAMPLE 7.6

A square 6-in piling is acted on by a water flow of 5 ft/s that is 20 ft deep, as shown in Fig. E7.6. Estimate the maximum bending exerted by the flow on the bottom of the piling.

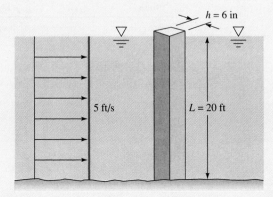

E7.6

Solution

Assume seawater with $\rho = 1.99$ slugs/ft^3 and kinematic viscosity $\nu = 0.000011$ ft^2/s. With a piling width of 0.5 ft, we have

$$\mathrm{Re}_h = \frac{(5 \text{ ft/s})(0.5 \text{ ft})}{0.000011 \text{ ft}^2/\text{s}} = 2.3 \times 10^5$$

This is the range where Table 7.2 applies. The worst case occurs when the flow strikes the flat side of the piling, $C_D \approx 2.1$. The frontal area is $A = Lh = (20 \text{ ft})(0.5 \text{ ft}) = 10$ ft^2. The drag is estimated by

$$F = C_D(\tfrac{1}{2}\rho V^2 A) \approx 2.1(\tfrac{1}{2})(1.99 \text{ slugs/ft}^3)(5 \text{ ft/s})^2(10 \text{ ft}^2) = 522 \text{ lbf}$$

If the flow is uniform, the center of this force should be at approximately middepth. Therefore the bottom bending moment is

$$M_0 \approx \frac{FL}{2} = 522(10) = 5220 \text{ ft} \cdot \text{lbf} \qquad\qquad Ans.$$

According to the flexure formula from strength of materials, the bending stress at the bottom would be

$$S = \frac{M_0 y}{I} = \frac{(5220 \text{ ft} \cdot \text{lb})(0.25 \text{ ft})}{\tfrac{1}{12}(0.5 \text{ ft})^4} = 251{,}000 \text{ lbf/ft}^2 = 1740 \text{ lbf/in}^2$$

to be multiplied, of course, by the stress concentration factor due to the built-in end conditions.

Three-Dimensional Bodies

Some drag coefficients of three-dimensional bodies are listed in Table 7.3 and Fig. 7.16b. Again we can conclude that sharp edges always cause flow separation and high

Table 7.3 Drag of Three-Dimensional Bodies at Re $\geq 10^4$

Body	C_D based on frontal area	Body	C_D based on frontal area							

Cube:

1.07

0.81

Cup:

1.4

0.4

Disk:

1.17

Parachute (Low porosity):

1.2

Streamlined train (approximately 5 cars):

$C_D A \approx 8.5 \text{ m}^2$

Bicycle:

Upright: $C_D A \approx 0.51 \text{ m}^2$; Racing: $C_D A \approx 0.30 \text{ m}^2$

Cone:

θ:	10°	20°	30°	40°	60°	75°	90°
C_D:	0.30	0.40	0.55	0.65	0.80	1.05	1.15

Short cylinder, laminar flow:

L/D:	1	2	3	5	10	20	40	∞
C_D:	0.64	0.68	0.72	0.74	0.82	0.91	0.98	1.20

Porous parabolic dish [23]:

Porosity:	0	0.1	0.2	0.3	0.4	0.5
← C_D:	1.42	1.33	1.20	1.05	0.95	0.82
→ C_D:	0.95	0.92	0.90	0.86	0.83	0.80

Average person:

$C_D A \approx 9 \text{ ft}^2$ $\uparrow C_D A \approx 1.2 \text{ ft}^2$

Pine and spruce trees [24]:

U, m/s:	10	20	30	40
C_D:	1.2 ± 0.2	1.0 ± 0.2	0.7 ± 0.2	0.5 ± 0.2

Tractor-trailer truck:

Without deflector: 0.96; with deflector: 0.76

Body	Ratio	C_D based on frontal area		Body	Ratio	C_D based on frontal area

Rectangular plate:

b/h	1	1.18
	5	1.2
	10	1.3
	20	1.5
	∞	2.0

Ellipsoid:

		Laminar	Turbulent
L/d	0.75	0.5	0.2
	1	0.47	0.2
	2	0.27	0.13
	4	0.25	0.1
	8	0.2	0.08

Flat-faced cylinder:

L/d	0.5	1.15
	1	0.90
	2	0.85
	4	0.87
	8	0.99

Buoyant rising sphere [50], $C_D \approx 0.95$

$135 < \text{Re}_d < 1\text{E}5$

drag that is insensitive to the Reynolds number. Rounded bodies like the ellipsoid have drag that depends on the point of separation, so that both the Reynolds number and the character of the boundary layer are important. Body length will generally decrease pressure drag by making the body relatively more slender, but sooner or later the friction drag will catch up. For the flat-faced cylinder in Table 7.3, pressure drag decreases with L/d but friction increases, so that minimum drag occurs at about $L/d = 2$.

Buoyant Rising Light Spheres

The sphere data in Fig. 7.16*b* are for fixed models in wind tunnels and from falling sphere tests and indicate a drag coefficient of about 0.5 in the range $1E3 < Re_d < 1E5$. It was recently pointed out [50] that this is *not* the case for a freely rising buoyant sphere or bubble. If the sphere is light, $\rho_{sphere} < 0.8 \, \rho_{fluid}$, a wake instability arises in the range $135 < Re_d < 1E5$. The sphere then spirals upward at an angle of about $60°$ from the horizontal. The drag coefficient is approximately doubled, to an average value $C_D \approx 0.95$, as listed in Table 7.3 [50]. For a heavier body, $\rho_{sphere} \approx \rho_{fluid}$, the buoyant sphere rises vertically and the drag coefficient follows the standard curve in Fig. 7.16*b*.

EXAMPLE 7.7

According to Ref. 12, the drag coefficient of a blimp, based on surface area, is approximately 0.006 if $Re_L > 10^6$. A certain blimp is 75 m long and has a surface area of 3400 m². Estimate the power required to propel this blimp at 18 m/s at a standard altitude of 1000m.

Solution

- *Assumptions:* We hope the Reynolds number will be high enough that the given data are valid.
- *Approach:* Determine if $Re_L > 10^6$ and, if so, compute the drag and the power required.
- *Property values:* Table A.6 at $z = 1000$ m: $\rho = 1.112$ kg/m³, $T = 282$ K, thus $\mu \approx$ 1.75E-5 kg/m-s.
- *Solution steps:* Determine the Reynolds number of the blimp:

$$Re_L = \frac{\rho UL}{\mu} = \frac{(1.112 \text{ kg/m}^3)(18 \text{ m/s})(75 \text{ m})}{1.75\text{E-5 kg/m-s}} = 8.6\text{E7} > 10^6 \qquad \text{OK}$$

The given drag coefficient is valid. Compute the blimp drag and the power = (drag) × (velocity):

$$F = C_D \frac{\rho}{2} U^2 A_{wet} = (0.006) \frac{1.112 \text{ kg/m}^3}{2} (18 \text{ m/s})^2 (3400 \text{ m}^2) = 3675 \text{ N}$$

$$\text{Power} = FV = (3675 \text{ N})(18 \text{ m/s}) = 66{,}000 \text{ W} \quad (89 \text{ hp}) \qquad\qquad Ans.$$

- *Comments:* These are nominal estimates. Drag is highly dependent on both shape and Reynolds number, and the coefficient $C_D = 0.006$ has considerable uncertainty.

Aerodynamic Forces on Road Vehicles

Automobiles and trucks are now the subject of much research on aerodynamic forces, both lift and drag [21]. At least one textbook is devoted to the subject [22].

A very readable description of race car drag is given by Katz [51]. Consumer, manufacturer, and government interest has cycled between high speed/high horsepower and lower speed/lower drag. Better streamlining of car shapes has resulted over the years in a large decrease in the automobile drag coefficient, as shown in Fig. 7.17a. Modern cars have an average drag coefficient of about 0.3, based on the frontal area. Since the frontal area has also decreased sharply, the actual raw drag *force* on cars has dropped even more than indicated in Fig. 7.17a. The theoretical minimum shown in the figure, $C_D \approx 0.15$, is about right for a commercial automobile, but lower values are possible for experimental vehicles, see Prob. P7.109. Note that basing C_D on the frontal area is awkward, since one would need an accurate drawing of the automobile to estimate its frontal area. For this reason, some

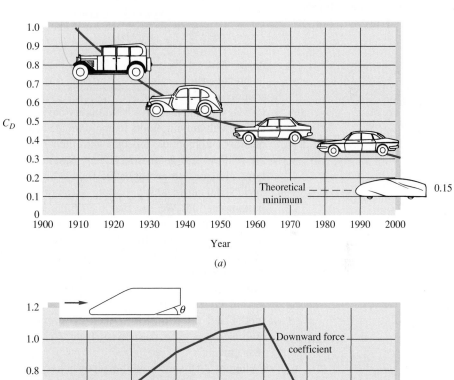

Fig. 7.17 Aerodynamics of automobiles: (*a*) the historical trend for drag coefficients (from Ref. 21); (*b*) effect of bottom rear upsweep angle on drag and downward lift force (from Ref. 25).

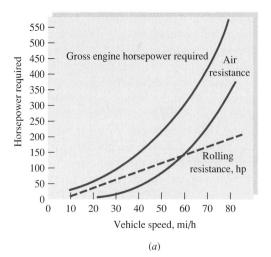

Fig. 7.18 Drag reduction of a tractor-trailer truck: (*a*) horsepower required to overcome resistance; (*b*) deflector added to cab reduces air drag by 20 percent. (*Uniroyal Inc.*)

(*a*)

(*b*)

technical articles simply report the raw drag in newtons or pound-force, or the product $C_D A$.

Many companies and laboratories have automotive wind tunnels, some full-scale and/or with moving floors to approximate actual kinematic similarity. The blunt shapes of most automobiles, together with their proximity to the ground, cause a wide variety of flow and geometric effects. Simple changes in part of the shape can have a large influence on aerodynamic forces. Figure 7.17*b* shows force data by Bearman et al. [25] for an idealized smooth automobile shape with upsweep in the rear of the bottom section. We see that by simply adding an upsweep angle of 25°, we can quadruple the downward force, gaining tire traction at the expense of doubling the drag. For this study, the effect of a moving floor was small—about a 10 percent increase in both drag and lift compared to a fixed floor.

It is difficult to quantify the exact effect of geometric changes on automotive forces, since, for example, changes in a windshield shape might interact with downstream flow over the roof and trunk. Nevertheless, based on correlation of many model and full-scale tests, Ref. 26 proposes a formula for automobile drag that adds separate effects such as front ends, cowls, fenders, windshield, roofs, and rear ends.

Figure 7.18 shows the horsepower required to drive a typical tractor-trailer truck at speeds up to 80 mi/h (117 ft/s or 36 m/s). The rolling resistance increases linearly and the air drag quadratically with speed ($C_D \approx 1.0$). The two are about equally important at 55 mi/h. As shown in Fig. 7.18*b*, air drag can be reduced by attaching a deflector to the top of the tractor. If the angle of the deflector is adjusted to carry the flow smoothly over the top and around the sides of the trailer, the reduction in C_D is about 20 percent. Thus, at 55 mi/h the total resistance is reduced 10 percent, with a corresponding reduction in fuel costs and/or trip time for the trucker. Further reduction occurs when the deflector is lengthened to cover the gap between cab and trailer. This type of applied fluids engineering can be a large factor in many of the conservation-oriented transportation problems of the future.

EXAMPLE 7.8

A high-speed car with $m = 2000$ kg, $C_D = 0.3$, and $A = 1$ m^2 deploys a 2-m parachute to slow down from an initial velocity of 100 m/s (Fig. E7.8). Assuming constant C_D, brakes free, and no rolling resistance, calculate the distance and velocity of the car after 1, 10, 100, and 1000 s. For air assume $\rho = 1.2$ kg/m^3, and neglect interference between the wake of the car and the parachute.

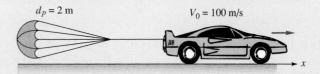

$d_p = 2$ m $V_0 = 100$ m/s

E7.8

Solution

Newton's law applied in the direction of motion gives

$$F_x = m\frac{dV}{dt} = -F_c - F_p = -\frac{1}{2}\rho V^2 (C_{Dc}A_c + C_{Dp}A_p)$$

where subscript c denotes the car and subscript p the parachute. This is of the form

$$\frac{dV}{dt} = -\frac{K}{m}V^2 \qquad K = \sum C_D A \frac{\rho}{2}$$

Separate the variables and integrate:

$$\int_{v_0}^{v}\frac{dV}{V^2} = -\frac{K}{m}\int_0^t dt$$

or

$$V_0^{-1} - V^{-1} = -\frac{K}{m}t$$

Rearrange and solve for the velocity V:

$$V = \frac{V_0}{1 + (K/m)V_0 t} \qquad K = \frac{(C_{Dc}A_c + C_{Dp}A_p)\rho}{2} \tag{1}$$

We can integrate this to find the distance traveled:

$$S = \frac{V_0}{\alpha}\ln(1 + \alpha t) \qquad \alpha = \frac{K}{m}V_0 \tag{2}$$

Now work out some numbers. From Table 7.3, $C_{Dp} \approx 1.2$; hence

$$C_{Dc}A_c + C_{Dp}A_p = 0.3(1\ \text{m}^2) + 1.2\frac{\pi}{4}(2\ \text{m})^2 = 4.07\ \text{m}^2$$

Then

$$\frac{K}{m}V_0 = \frac{\frac{1}{2}(4.07\ \text{m}^2)(1.2\ \text{kg/m}^3)(100\ \text{m/s})}{2000\ \text{kg}} = 0.122\ \text{s}^{-1} = \alpha$$

Now make a table of the results for V and S from Eqs. (1) and (2):

t, s	1	10	100	1000
V, m/s	89	45	7.6	0.8
S, m	94	654	2110	3940

Air resistance alone will not stop a body completely. If you don't apply the brakes, you'll be halfway to the Yukon Territory and still going.

Other Methods of Drag Reduction

Sometimes drag is good, for example, when using a parachute. Do not jump out of an airplane holding a flat plate parallel to your motion (see Prob. P7.81). Mostly, though, drag is bad and should be reduced. The classical method of drag reduction is *streamlining* (Figs. 7.15 and 7.18). For example, nose fairings and body panels have produced motorcycles that can travel over 200 mi/h. More recent research has uncovered other methods that hold great promise, especially for turbulent flows.

1. Oil pipelines introduce an *annular strip* of water to reduce the pumping power [36]. The low-viscosity water rides the wall and reduces friction up to 60 percent.
2. Turbulent friction in liquid flows is reduced up to 60 percent by dissolving small amounts of a *high-molecular-weight polymer additive* [37]. Without changing pumps, the Trans-Alaska Pipeline System (TAPS) increased oil flow 50 percent by injecting small amounts of polymer dissolved in kerosene.
3. Stream-oriented surface *vee-groove microriblets* can reduce turbulent friction up to 8 percent [38]. Riblet heights are of order 1 mm and were used on the Stars and Stripes yacht hull in the Americas Cup races. Riblets are also effective on aircraft skins.
4. Small, near-wall *large-eddy breakup devices* (LEBUs) reduce local turbulent friction up to 10 percent [39]. However, one must add these small structures to the surface and LEBU drag may be significant.
5. Air *microbubbles* injected at the wall of a water flow create a low-shear bubble blanket [40]. At high void fractions, drag reduction can be 80 percent.
6. Spanwise (transverse) *wall oscillation* may reduce turbulent friction up to 30 percent [41].
7. *Active flow control*, especially of turbulent flows, is the wave of the future, as reviewed in Ref. 47. These methods generally require expenditure of energy but can be worth it. For example, tangential blowing at the rear of an auto [48] evokes the *Coanda effect*, in which the separated near-wake flow attaches itself to the body surface and reduces auto drag up to 10 percent.

Drag reduction is presently an area of intense and fruitful research and applies to many types of airflows [53] and water flows for both vehicles and conduits.

Drag of Surface Ships

The drag data given so far, such as Tables 7.2 and 7.3, are for bodies "fully immersed" in a free stream—that is, with no free surface. If, however, the body moves at or near a free liquid surface, *wave-making drag* becomes important and is dependent on both the Reynolds number and the Froude number. To move through a water surface, a ship must create waves on both sides. This implies putting energy into the water surface and requires a finite drag force to keep the ship moving, even in a frictionless fluid. The total drag of a ship can then be approximated as the sum of friction drag and wave-making drag:

$$F \approx F_{\text{fric}} + F_{\text{wave}} \qquad \text{or} \qquad C_D \approx C_{D,\text{fric}} + C_{D,\text{wave}}$$

The friction drag can be estimated by the (turbulent) flat-plate formula, Eq. (7.45), based on the below-water or *wetted area* of the ship.

Reference 27 is an interesting review of both theory and experiment for wake-making surface ship drag. Generally speaking, the bow of the ship creates a wave system whose wavelength is related to the ship speed but not necessarily to the ship length. If the stern of the ship is a wave *trough,* the ship is essentially climbing uphill and has high wave drag. If the stern is a wave crest, the ship is nearly level and has lower drag. The criterion for these two conditions results in certain approximate Froude numbers [27]:

$$\text{Fr} = \frac{V}{\sqrt{gL}} \approx \frac{0.53}{\sqrt{N}} \qquad \begin{array}{l} \text{high drag if } N = 1, 3, 5, 7, \ldots; \\ \text{low drag if } N = 2, 4, 6, 8, \ldots \end{array} \tag{7.65}$$

where V is the ship's speed, L is the ship's length along the centerline, and N is the number of half-lengths, from bow to stern, of the drag-making wave system. The wave drag will increase with the Froude number and oscillate between lower drag (Fr $\approx$ 0.38, 0.27, 0.22, . . .) and higher drag (Fr $\approx$ 0.53, 0.31, 0.24, . . .) with negligible variation for Fr $<$ 0.2. Thus it is best to design a ship to cruise at $N = 2, 4, 6, 8$. Shaping the bow and stern can further reduce wave-making drag.

Figure 7.19 shows the data of Inui [27] for a model ship. The main hull, curve A, shows peaks and valleys in wave drag at the appropriate Froude numbers > 0.2. Introduction of a *bulb* protrusion on the bow, curve B, greatly reduces the drag. Adding a second bulb to the stern, curve C, is still better, and Inui recommends that the design speed of this two-bulb ship be at $N = 4$, Fr $\approx$ 0.27, which is a nearly "waveless" condition. In this figure $C_{D,\text{wave}}$ is defined as $2F_{\text{wave}}/(\rho V^2 L^2)$ instead of using the wetted area.

The solid curves in Fig. 7.19 are based on potential flow theory for the below-water hull shape. Chapter 8 is an introduction to potential flow theory. Modern computers can be programmed for numerical CFD solutions of potential flow over the hulls of ships, submarines, yachts, and sailboats, including boundary layer effects driven by the potential flow [28]. Thus theoretical prediction of flow past surface ships is now at a fairly high level. See also Ref. 15.

Body Drag at High Mach Numbers

All the data presented to this point are for nearly incompressible flows, with Mach numbers assumed less than about 0.3. Beyond this value compressibility can be very important, with $C_D = \text{fcn}(\text{Re}, \text{Ma})$. As the stream Mach number increases, at some subsonic value $M_{\text{crit}} < 1$ that depends on the body's bluntness and thickness, the local velocity at some point near the body surface will become sonic. If Ma increases beyond Ma_{crit}, shock waves form, intensify, and spread, raising surface

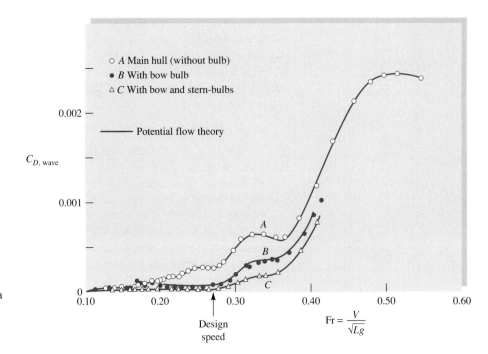

$\circ$ *A* Main hull (without bulb)
$\bullet$ *B* With bow bulb
$\triangle$ *C* With bow and stern-bulbs

Potential flow theory

$C_{D,\,\text{wave}}$

0.002

0.001

0

A

B

C

Design
speed

$\text{Fr} = \dfrac{V}{\sqrt{Lg}}$

0.10   0.20   0.30   0.40   0.50   0.60

Fig. 7.19 Wave-making drag on a ship model. (*After Inui [27].*) *Note:* The drag coefficient is defined as $C_{DW} = 2F/(\rho V^2 L^2)$.

pressures near the front of the body and therefore increasing the pressure drag. The effect can be dramatic with C_D increasing tenfold, and 70 years ago this sharp increase was called the *sonic barrier,* implying that it could not be surmounted. Of course, it can be—the rise in C_D is finite, as supersonic bullets have proved for centuries.

Figure 7.20 shows the effect of the Mach number on the drag coefficient of various body shapes tested in air.[6] We see that compressibility affects blunt bodies earlier, with Ma_{crit} equal to 0.4 for cylinders, 0.6 for spheres, and 0.7 for airfoils and pointed projectiles. Also the Reynolds number (laminar versus turbulent boundary layer flow) has a large effect below Ma_{crit} for spheres and cylinders but becomes unimportant above $\text{Ma} \approx 1$. In contrast, the effect of the Reynolds number is small for airfoils and projectiles and is not shown in Fig. 7.20. A general statement might divide Reynolds and Mach number effects as follows:

$$\text{Ma} \leq 0.3: \quad \text{Reynolds number important, Mach number unimportant}$$
$$0.3 < \text{Ma} < 1: \quad \text{both Reynolds and Mach numbers important}$$
$$\text{Ma} > 1.0: \quad \text{Reynolds number unimportant, Mach number important}$$

At supersonic speeds, a broad *bow shock wave* forms in front of the body (see Figs. 9.10*b* and 9.19), and the drag is mainly due to high shock-induced pressures on the front. Making the bow a sharp point can sharply reduce the drag (Fig. 9.28) but does

---

[6]There is a slight effect of the specific-heat ratio *k*, which would appear if other gases were tested.

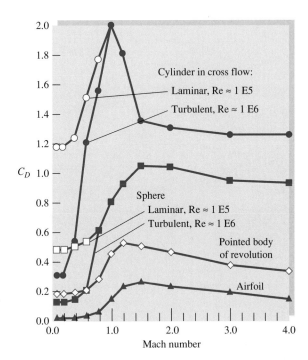

Fig. 7.20 Effect of the Mach number on the drag of various body shapes. (*Data from Refs. 23 and 29.*)

not eliminate the bow shock. Chapter 9 gives a brief treatment of compressible flow. References 30 and 31 are more advanced textbooks devoted entirely to compressible flow.

Biological Drag Reduction

A great deal of engineering effort goes into designing immersed bodies to reduce their drag. Most such effort concentrates on rigid-body shapes. A different process occurs in nature, as organisms adapt to survive high winds or currents, as reported in a series of papers by S. Vogel [33, 34]. A good example is a tree, whose flexible structure allows it to reconfigure in high winds and thus reduce drag and damage. Tree root systems have evolved in several ways to resist wind-induced bending moments, and trunk cross sections have become resistant to bending but relatively easy to twist and reconfigure. We saw this in Table 7.3, where tree drag coefficients [24] reduced by 60 percent as wind velocity increased. The shape of the tree changes to offer less resistance.

The individual branches and leaves of a tree also curl and cluster to reduce drag. Figure 7.21 shows the results of wind tunnel experiments by Vogel [33]. A tulip tree leaf, Fig. 7.21*a*, broad and open in low wind, curls into a conical low-drag shape as wind increases. A compound black walnut leaf group, Fig. 7.21*b*, clusters into a low-drag shape at high wind speed. Although drag coefficients were reduced up to 50 percent by flexibility, Vogel points out that rigid structures are sometimes just as effective. An interesting recent symposium [35] was devoted entirely to the solid mechanics and fluid mechanics of biological organisms.

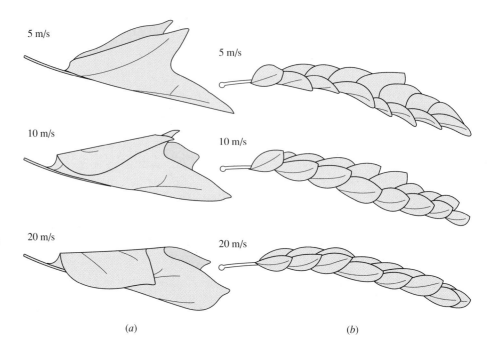

5 m/s

5 m/s

10 m/s

10 m/s

20 m/s

20 m/s

Fig. 7.21 Biological adaptation to wind forces: (*a*) a tulip tree leaf curls into a conical shape at high velocity; (*b*) black walnut leaves cluster into a low-drag shape as wind increases. (*From Vogel, Ref. 33.*)

(*a*)

(*b*)

Forces on Lifting Bodies

Lifting bodies (airfoils, hydrofoils, or vanes) are intended to provide a large force normal to the free stream and as little drag as possible. Conventional design practice has evolved a shape not unlike a bird's wing—that is, relatively thin ($t/c \le 0.24$) with a rounded leading edge and a sharp trailing edge. A typical shape is sketched in Fig. 7.22.

For our purposes we consider the body to be symmetric, as in Fig. 7.11, with the free-stream velocity in the vertical plane. If the chord line between the leading and trailing edge is not a line of symmetry, the airfoil is said to be *cambered*. The camber line is the line midway between the upper and lower surfaces of the vane.

The angle between the free stream and the chord line is called the *angle of attack* α. The lift L and the drag D vary with this angle. The dimensionless forces are defined

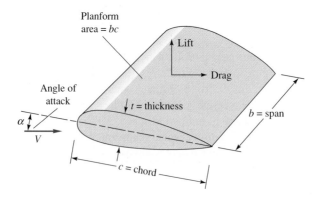

Fig. 7.22 Definition sketch for a lifting vane.

with respect to the planform area $A_p = bc$:

Lift coefficient:
$$C_L = \frac{L}{\frac{1}{2}\rho V^2 A_p}$$
(7.66a)

Drag coefficient:
$$C_D = \frac{D}{\frac{1}{2}\rho V^2 A_p}$$
(7.66b)

If the chord length is not constant, as in the tapered wings of modern aircraft, $A_p = \int c\, db$.

For low-speed flow with a given roughness ratio, C_L and C_D should vary with α and the chord Reynolds number:

$$C_L = f(\alpha, \mathrm{Re}_c) \qquad \text{or} \qquad C_D = f(\alpha, \mathrm{Re}_c)$$

where $\mathrm{Re}_c = Vc/\nu$. The Reynolds numbers are commonly in the turbulent boundary layer range and have a modest effect.

The rounded leading edge prevents flow separation there, but the sharp trailing edge causes a tangential wake motion that generates the lift. Figure 7.23 shows what happens when a flow starts up past a lifting vane or an airfoil.

Just after start-up in Fig. 7.23a the streamline motion is irrotational and inviscid. The rear stagnation point, assuming a positive angle of attack, is on the upper surface, and there is no lift; but the flow cannot long negotiate the sharp turn at the trailing edge: it separates, and a *starting vortex* forms in Fig. 7.23b. This starting vortex is shed downstream in Fig. 7.23c and d, and a smooth streamline flow develops over the wing, leaving the foil in a direction approximately parallel to the chord line. Lift at this time is fully developed, and the starting vortex is gone. Should the flow now cease, a *stopping vortex* of opposite (clockwise) sense will form and be shed. During flight, increases or decreases in lift will cause incremental starting or stopping

Fig. 7.23 Transient stages in the development of lift: (*a*) start-up: rear stagnation point on the upper surface: no lift; (*b*) sharp trailing edge induces separation, and a starting vortex forms: slight lift; (*c*) starting vortex is shed, and streamlines flow smoothly from trailing edge: lift is now 80 percent developed; (*d*) starting vortex now shed far behind, trailing edge now very smooth: lift fully developed.

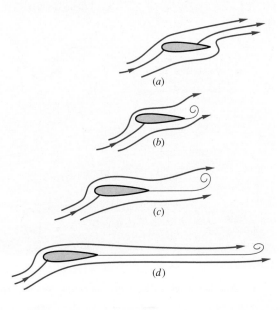

vortices, always with the effect of maintaining a smooth parallel flow at the trailing edge. We pursue this idea mathematically in Chap. 8.

At a low angle of attack, the rear surfaces have an adverse pressure gradient but not enough to cause significant boundary layer separation. The flow pattern is smooth, as in Fig. 7.23d, and drag is small and lift excellent. As the angle of attack is increased, the upper-surface adverse gradient becomes stronger, and generally a *separation bubble* begins to creep forward on the upper surface.[7] At a certain angle α = 15 to 20°, the flow is separated completely from the upper surface, as in Fig. 7.24. The airfoil is said to be *stalled*: Lift drops off markedly, drag increases markedly, and the foil is no longer flyable.

Early airfoils were thin, modeled after birds' wings. The German engineer Otto Lilienthal (1848–1896) experimented with flat and cambered plates on a rotating arm. He and his brother Gustav flew the world's first glider in 1891. Horatio Frederick Phillips (1845–1912) built the first wind tunnel in 1884 and measured the lift and drag of cambered vanes. The first theory of lift was proposed by Frederick W. Lanchester shortly afterward. Modern airfoil theory dates from 1905, when the Russian hydrodynamicist N. E. Joukowsky (1847–1921) developed a circulation theorem (Chap. 8) for computing airfoil lift for arbitrary camber and thickness. With this basic theory, as extended and developed by Prandtl and Kármán and their students, it is now possible to design a low-speed airfoil to satisfy particular surface pressure distributions

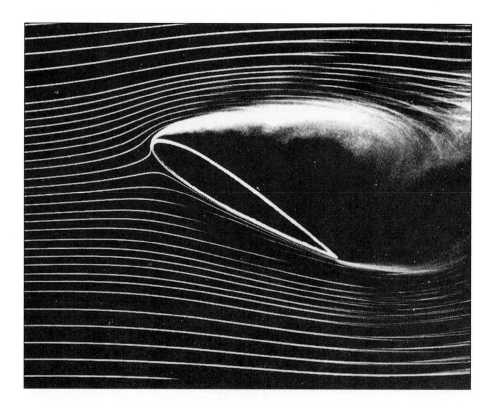

Fig. 7.24 At high angle of attack, smoke flow visualization shows stalled flow on the upper surface of a lifting vane. (*National Committee for Fluid Mechanics Films, Education Development Center, Inc., © 1972.*)

[7]For some airfoils the bubble leaps, not creeps, forward and stall occurs rapidly and dangerously.

and boundary layer characteristics. There are whole families of airfoil designs, notably those developed in the United States under the sponsorship of the NACA (now NASA). Extensive theory and data on these airfoils are contained in Ref. 16. We shall discuss this further in Chap. 8. The history of aeronautics is a rich and engaging topic and highly recommended to the reader [43, 44].

Figure 7.25 shows the lift and drag on a symmetric airfoil denoted as the NACA 0009 foil, the last digit indicating the thickness of 9 percent. With no flap extended, this airfoil, as expected, has zero lift at zero angle of attack. Up to about 12° the lift coefficient increases linearly with a slope of 0.1 per degree, or 6.0 per radian. This is in agreement with the theory outlined in Chap. 8:

$$C_{L,\text{theory}} \approx 2\pi \sin\left(\alpha + \frac{2h}{c}\right) \tag{7.67}$$

where h/c is the maximum camber expressed as a fraction of the chord. The NACA 0009 has zero camber; hence $C_L = 2\pi \sin \alpha \approx 0.11\alpha$, where α is in degrees. This is excellent agreement.

The drag coefficient of the smooth-model airfoils in Fig. 7.25 is as low as 0.005, which is actually lower than both sides of a flat plate in turbulent flow. This is misleading inasmuch as a commercial foil will have roughness effects; for example, a paint job will double the drag coefficient.

The effect of increasing Reynolds number in Fig. 7.25 is to increase the maximum lift and stall angle (without changing the slope appreciably) and to reduce the drag coefficient. This is a salutary effect since the prototype will probably be at a higher Reynolds number than the model (10^7 or more).

For takeoff and landing, the lift is greatly increased by deflecting a split flap, as shown in Fig. 7.25. This makes the airfoil unsymmetric (or effectively cambered)

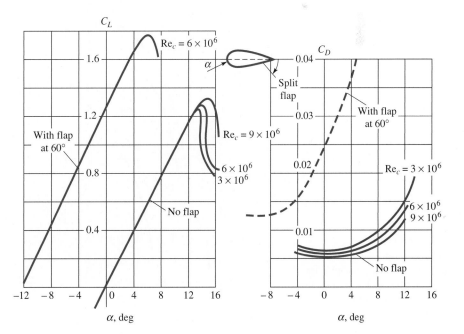

Fig. 7.25 Lift and drag of a symmetric NACA 0009 airfoil of infinite span, including effect of a split-flap deflection. Note that roughness can increase C_D from 100 to 300 percent.

and changes the zero-lift point to $\alpha = -12°$. The drag is also greatly increased by the flap, but the reduction in takeoff and landing distance is worth the extra power needed.

A lifting craft cruises at low angle of attack, where the lift is much larger than the drag. Maximum lift-to-drag ratios for the common airfoils lie between 20 and 50.

Some airfoils, such as the NACA 6 series, are shaped to provide favorable gradients over much of the upper surface at low angles. Thus separation is small, and transition to turbulence is delayed; the airfoil retains a good length of laminar flow even at high Reynolds numbers. The lift-drag *polar plot* in Fig. 7.26 shows the NACA 0009 data from Fig. 7.25 and a laminar flow airfoil, NACA 63-009, of the same thickness. The laminar flow airfoil has a low-drag bucket at small angles but also suffers lower stall angle and lower maximum lift coefficient. The drag is 30 percent less in the bucket, but the bucket disappears if there is significant surface roughness.

All the data in Figs. 7.25 and 7.26 are for infinite span—that is, a two-dimensional flow pattern about wings without tips. The effect of finite span can be correlated with the dimensionless slenderness, or *aspect ratio,* denoted (AR):

$$\text{AR} = \frac{b^2}{A_p} = \frac{b}{\bar{c}} \tag{7.68}$$

where $\bar{c}$ is the average chord length. Finite-span effects are shown in Fig. 7.27. The lift slope decreases, but the zero-lift angle is the same; and the drag increases, but the zero-lift drag is the same. The theory of finite-span airfoils [16] predicts that the effective angle of attack increases, as in Fig. 7.27, by the amount

$$\Delta\alpha \approx \frac{C_L}{\pi\text{AR}} \tag{7.69}$$

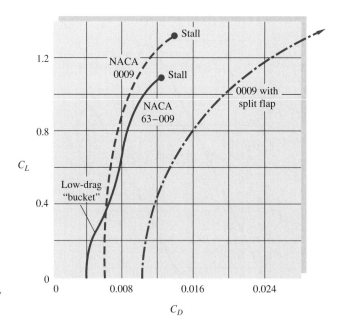

Fig. 7.26 Lift-drag polar plot for standard (0009) and a laminar flow (63-009) NACA airfoil.

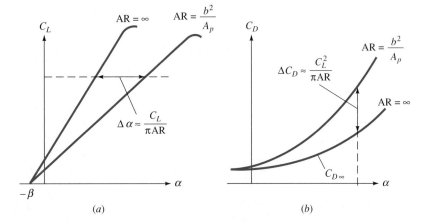

Fig. 7.27 Effect of finite aspect ratio on lift and drag of an airfoil: (*a*) effective angle increase; (*b*) induced drag increase.

When applied to Eq. (7.67), the finite-span lift becomes

$$C_L \approx \frac{2\pi \sin (\alpha + 2h/c)}{1 + 2/AR} \tag{7.70}$$

The associated drag increase is $\Delta C_D \approx C_L \sin \Delta\alpha \approx C_L \Delta\alpha$, or

$$C_D \approx C_{D\infty} + \frac{C_L^2}{\pi AR} \tag{7.71}$$

where $C_{D\infty}$ is the drag of the infinite-span airfoil, as sketched in Fig. 7.25. These correlations are in good agreement with experiments on finite-span wings [16].

The existence of a maximum lift coefficient implies the existence of a minimum speed, or *stall speed,* for a craft whose lift supports its weight:

$$L = W = C_{L,max}(\tfrac{1}{2}\rho V_s^2 A_p)$$

or

$$V_s = \left(\frac{2W}{C_{L,max}\,\rho A_p}\right)^{1/2} \tag{7.72}$$

The stall speed of typical aircraft varies between 60 and 200 ft/s, depending on the weight and value of $C_{L,max}$. The pilot must hold the speed greater than about $1.2V_s$ to avoid the instability associated with complete stall.

The split flap in Fig. 7.25 is only one of many devices used to secure high lift at low speeds. Figure 7.28*a* shows six such devices whose lift performance is given in Fig. 7.28*b* along with a standard (*A*) and laminar flow (*B*) airfoil. The double-slotted flap achieves $C_{L,max} \approx 3.4$, and a combination of this plus a leading-edge slat can achieve $C_{L,max} \approx 4.0$. These are not scientific curiosities; for instance, the Boeing 727 commercial jet aircraft uses a triple-slotted flap plus a leading-edge slat during landing.

Also shown as *C* in Fig. 7.28*b* is the Kline-Fogleman airfoil [17], not yet a reality. The designers are amateur model-plane enthusiasts who did not know that

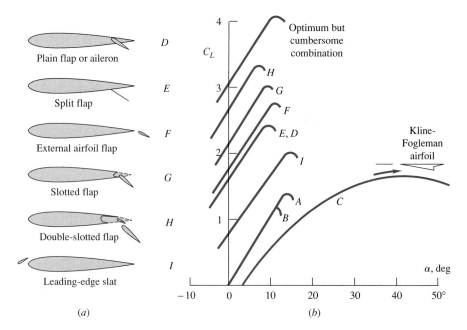

Fig. 7.28 Performance of airfoils with and without high-lift devices: A = NACA 0009; B = NACA 63-009; C = Kline-Fogleman airfoil (*from Ref. 17*); D to I shown in (*a*): (*a*) types of high-lift devices; (*b*) lift coefficients for various devices.

conventional aerodynamic wisdom forbids a sharp leading edge and a step cutout from the trailing edge. The Kline-Fogleman airfoil has relatively high drag but shows an amazing continual increase in lift out to $\alpha = 45°$. In fact, we may fairly say that this airfoil does not stall and provides smooth performance over a tremendous range of flight conditions. No explanation for this behavior is known to the writer. This airfoil is under study and may or may not have any commercial value.

Another violation of conventional aerodynamic wisdom is that military aircraft are beginning to fly, briefly, *above the stall point*. Fighter pilots are learning to make quick maneuvers in the stalled region as detailed in Ref. 32. Some planes can even *fly continuously* while stalled—the Grumman X-29 experimental aircraft recently set a record by flying at $\alpha = 67°$.

New Aircraft Designs

The Kline-Fogleman airfoil in Fig. 7.28 is a departure from conventional aerodynamics, but there have been other striking departures, as detailed in a recent article [42]. These new aircraft, conceived presently as small models, have a variety of configurations, as shown in Fig. 7.29: ring-wing, cruciform, flying saucer, and flap-wing. A saucer configuration (Fig. 7.29*c*), with a diameter of 40 in, has been successfully flown by radio control, and its inventor, Jack M. Jones, plans for a 20-ft two-passenger version. Another 18-in-span microplane called the Bat (not shown), made by MLB Co., flies for 20 min at 40 mi/h and contains a video camera for surveillance. New engines have been reduced to a 10- by 3-mm size, producing 20 W of power. At the other end of the size spectrum, Airbus has just put into service the new A380 jumbo jet. It is 73 m long, 80 m in wingspan, and carries up to 840 passengers for a range of up to 9300 miles.

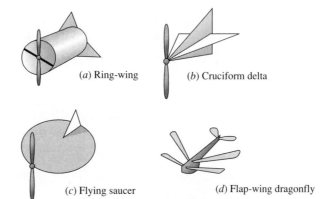

Fig. 7.29 New aircraft designs do not necessarily look like your typical jetliner. *(From Ref. 42.)*

(a) Ring-wing

(b) Cruciform delta

(c) Flying saucer

(d) Flap-wing dragonfly

Further information on the performance of lifting craft can be found in Refs. 12, 13, and 16. We discuss this matter again briefly in Chap. 8.

EXAMPLE 7.9

An aircraft weighs 75,000 lbf, has a planform area of 2500 ft^2, and can deliver a constant thrust of 12,000 lb. It has an aspect ratio of 7, and $C_{D\infty} \approx 0.02$. Neglecting rolling resistance, estimate the takeoff distance at sea level if takeoff speed equals 1.2 times stall speed. Take $C_{L,max} = 2.0$.

Solution

The stall speed from Eq. (7.72), with sea-level density $\rho = 0.00237$ slug/ft^3, is

$$V_s = \left(\frac{2W}{C_{L,max}\rho A_p}\right)^{1/2} = \left[\frac{2(75,000)}{2.0(0.00237)(2500)}\right]^{1/2} = 112.5 \text{ ft/s}$$

Hence takeoff speed $V_0 = 1.2V_s = 135$ ft/s. The drag is estimated from Eq. (7.71) for AR = 7 as

$$C_D \approx 0.02 + \frac{C_L^2}{7\pi} = 0.02 + 0.0455C_L^2$$

A force balance in the direction of takeoff gives

$$F_s = m\frac{dV}{dt} = \text{thrust} - \text{drag} = T - kV^2 \qquad k = \tfrac{1}{2}C_D\rho A_p \qquad (1)$$

Since we are looking for distance, not time, we introduce $dV/dt = V\,dV/ds$ into Eq. (1), separate variables, and integrate:

$$\int_0^{s_o} dS = \frac{m}{2}\int_0^{V_0} \frac{d(V^2)}{T - kV^2} \qquad k \approx \text{const}$$

or

$$S_0 = \frac{m}{2k}\ln\frac{T}{T - kV_0^2} = \frac{m}{2k}\ln\frac{T}{T - D_0} \qquad (2)$$

where $D_0 = kV_0^2$ is the takeoff drag. Equation (2) is the desired theoretical relation for take-off distance. For the particular numerical values, take

$$m = \frac{75,000}{32.2} = 2329 \text{ slugs}$$

$$C_{L_0} = \frac{W}{\frac{1}{2}\rho V_0^2 A_p} = \frac{75,000}{\frac{1}{2}(0.00237)(135)^2(2500)} = 1.39$$

$$C_{D_0} = 0.02 + 0.0455(C_{L_0})^2 = 0.108$$

$$k \approx \tfrac{1}{2}C_{D_0}\rho A_p = (\tfrac{1}{2})(0.108)(0.00237)(2500) = 0.319 \text{ slug/ft}$$

$$D_0 = kV_0^2 = 5820 \text{ lbf}$$

Then Eq. (2) predicts that

$$S_0 = \frac{2329 \text{ slugs}}{2(0.319 \text{ slug/ft})} \ln \frac{12,000}{12,000 - 5820} = 3650 \ln 1.94 = 2420 \text{ ft} \qquad \textit{Ans.}$$

A more exact analysis accounting for variable k [13] gives the same result to within 1 percent.

EXAMPLE 7.10

For the aircraft of Example 7.9, if maximum thrust is applied during flight at 6000 m standard altitude, estimate the resulting velocity of the plane, in mi/h.

Solution

- *Assumptions:* Given $W = 75,000$ lbf, $A_p = 2500$ ft^2, $T = 12,000$ lbf, $AR = 7$, $C_{D\infty} = 0.02$.
- *Approach:* Set lift equal to weight and drag equal to thrust and solve for the velocity.
- *Property values:* From Table A.6, at $z = 6000$ m, $\rho = 0.6596$ kg/m^3 = 0.00128 slug/ft^3.
- *Solution steps:* Write out the formulas for lift and drag. The unknowns will be C_L and V.

$$W = 75000 \text{ lbf} = \text{lift} = C_L \frac{\rho}{2}V^2 A_p = C_L \frac{0.00128 \text{ slug/ft}^3}{2}V^2(2500 \text{ ft}^2)$$

$$T = 12000 \text{ lbf} = \text{drag} = \left(C_{D\infty} + \frac{C_L^2}{\pi AR}\right)\frac{\rho}{2}V^2 A_p$$

$$= \left[0.02 + \frac{C_L^2}{\pi(7)}\right]\frac{0.00128 \text{ slug/ft}^3}{2} V^2(2500 \text{ ft}^2)$$

This looks like a job for EES, but in fact some clever manipulation (dividing W by T) would reveal a quadratic equation for C_L. In either case, the final solution is

$$C_L = 0.13 \qquad V \approx 600 \text{ ft/s} = 410 \text{ mi/h} \qquad \textit{Ans.}$$

- *Comments:* These are *preliminary design* estimates that do not depend on airfoil shape.

Summary

This chapter has dealt with viscous effects in external flow past bodies immersed in a stream. When the Reynolds number is large, viscous forces are confined to a thin boundary layer and wake in the vicinity of the body. Flow outside these "shear layers" is essentially inviscid and can be predicted by potential theory and Bernoulli's equation.

The chapter began with a discussion of the flat-plate boundary layer and the use of momentum integral estimates to predict the wall shear, friction drag, and thickness of such layers. These approximations suggest how to eliminate certain small terms in the Navier-Stokes equations, resulting in Prandtl's boundary layer equations for laminar and turbulent flow. Section 7.4 then solved the boundary layer equations to give very accurate formulas for flat-plate flow at high Reynolds numbers. Rough-wall effects were included, and Sec. 7.5 gave a brief introduction to pressure gradient effects. An adverse (decelerating) gradient was seen to cause flow separation, where the boundary layer breaks away from the surface and forms a broad, low-pressure wake.

Boundary layer theory fails in separated flows, which are commonly studied by experiment or CFD. Section 7.6 gave data on drag coefficients of various two- and three-dimensional body shapes. The chapter ended with a brief discussion of lift forces generated by lifting bodies such as airfoils and hydrofoils. Airfoils also suffer flow separation or *stall* at high angles of incidence.

Problems

Most of the problems herein are fairly straightforward. More difficult or open-ended assignments are labeled with an asterisk. Problems labeled with an EES icon will benefit from the use of the Engineering Equation Solver (EES), while problems labeled with a computer disk may require the use of a computer. The standard end-of-chapter problems P7.1 to P7.125 (categorized in the problem list here) are followed by word problems W7.1 to W7.12, fundamentals of engineering exam problems FE7.1 to FE7.10, comprehensive problems C7.1 to C7.5, and design project D7.1.

Problem Distribution

Section	Topic	Problems
7.1	Reynolds number and geometry	P7.1–P7.5
7.2	Momentum integral estimates	P7.6–P7.12
7.3	The boundary layer equations	P7.13–P7.15
7.4	Laminar flat-plate flow	P7.16–P7.29
7.4	Turbulent flat-plate flow	P7.30–P7.46
7.5	Boundary layers with pressure gradient	P7.47–P7.50
7.6	Drag of bodies	P7.51–P7.114
7.6	Lifting bodies—airfoils	P7.115–P7.125

P7.1 An ideal gas, at 20°C and 1 atm, flows at 12 m/s past a thin flat plate. At a position 60 cm downstream of the leading edge, the boundary layer thickness is 5 mm. Which of the 13 gases in Table A.4 is this likely to be?

P7.2 Air, equivalent to that at a standard altitude of 4000 m, flows at 450 mi/h past a wing that has a thickness of 18 cm, a chord length of 1.5 m, and a wingspan of 12 m. What is the appropriate value of the Reynolds number for correlating the lift and drag of this wing? Explain your selection.

P7.3 Equation (7.1b) assumes that the boundary layer on the plate is turbulent from the leading edge onward. Devise a scheme for determining the boundary layer thickness more accurately when the flow is laminar up to a point $Re_{x,crit}$ and turbulent thereafter. Apply this scheme to computation of the boundary layer thickness at $x = 1.5$ m in 40 m/s flow of air at 20°C and 1 atm past a flat plate. Compare your result with Eq. (7.1b). Assume $Re_{x,crit} \approx 1.2$ E6.

P7.4 A smooth ceramic sphere (SG = 2.6) is immersed in a flow of water at 20°C and 25 cm/s. What is the sphere diameter if it is encountering (a) creeping motion, $Re_d = 1$ or (b) transition to turbulence, $Re_d = 250,000$?

P7.5 SAE 30 oil at 20°C flows at 1.8 ft³/s from a reservoir into a 6-in-diameter pipe. Use flat-plate theory to estimate the position x where the pipe wall boundary layers meet in the center. Compare with Eq. (6.5), and give some explanations for the discrepancy.

P7.6   For the laminar parabolic boundary layer profile of Eq. (7.6), compute the shape factor H and compare with the exact Blasius result, Eq. (7.31).

P7.7   Air at 20°C and 1 atm enters a 40-cm-square duct as in Fig. P7.7. Using the "displacement thickness" concept of Fig. 7.4, estimate (a) the mean velocity and (b) the mean pressure in the core of the flow at the position $x = 3$ m. (c) What is the average gradient, in Pa/m, in this section?

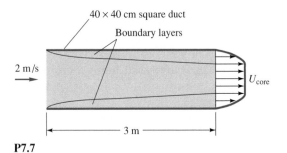

40 × 40 cm square duct

Boundary layers

2 m/s

U_{core}

3 m

P7.7

P7.8   Air, $\rho = 1.2$ kg/m^3 and $\mu = 1.8$ E-5 kg/(m · s), flows at 10 m/s past a flat plate. At the trailing edge of the plate, the following velocity profile data are measured:

y, mm	0	0.5	1.0	2.0	3.0	4.0	5.0	6.0
u, m/s	0	1.75	3.47	6.58	8.70	9.68	10.0	10.0

If the upper surface has an area of 0.6 m^2, estimate, using momentum concepts, the friction drag, in N, on the upper surface.

P7.9   Repeat the flat-plate momentum analysis of Sec. 7.2 by replacing the parabolic profile, Eq. (7.6), with a more accurate sinusoidal profile:

$$\frac{u}{U} = \sin \frac{\pi y}{2\delta}$$

Compute momentum-integral estimates of c_f, θ/x, $\delta*/x$, and H.

P7.10   Repeat Prob. P7.9, using the polynomial profile suggested by K. Pohlhausen in 1921:

$$\frac{u}{U} \approx 2\frac{y}{\delta} - 2\frac{y^3}{\delta^3} + \frac{y^4}{\delta^4}$$

Does this profile satisfy the boundary conditions of laminar flat-plate flow?

P7.11   Air at 20°C and 1 atm flows at 2 m/s past a sharp flat plate. Assuming that Kármán's parabolic-profile analysis,

Eqs. (7.6–7.10), is accurate, estimate (a) the local velocity u and (b) the local shear stress τ at the position $(x, y) = (50$ cm, 5 mm).

P7.12   The velocity profile shape $u/U \approx 1 - \exp(-4.605y/\delta)$ is a smooth curve with $u = 0$ at $y = 0$ and $u = 0.99U$ at $y = \delta$ and thus would seem to be a reasonable substitute for the parabolic flat-plate profile of Eq. (7.3). Yet when this new profile is used in the integral analysis of Sec. 7.3, we get the lousy result $\delta/x \approx 9.2/\text{Re}_x^{1/2}$, which is 80 percent high. What is the reason for the inaccuracy? (*Hint:* The answer lies in evaluating the laminar boundary layer momentum equation (7.19b) at the wall, $y = 0$.)

P7.13   Derive modified forms of the laminar boundary layer equations (7.19) for the case of axisymmetric flow along the outside of a circular cylinder of constant radius R, as in Fig. P7.13. Consider the two special cases (a) $\delta \ll R$ and (b) $\delta \approx R$. What are the proper boundary conditions?

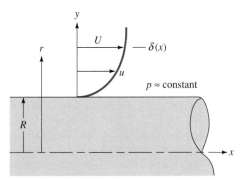

y

U

$\delta(x)$

r

u

$p \approx$ constant

R

x

P7.13

P7.14   Show that the two-dimensional laminar flow pattern with $dp/dx = 0$

$$u = U_0(1 - e^{Cy}) \qquad v = v_0 < 0$$

is an exact solution to the boundary layer equations (7.19). Find the value of the constant C in terms of the flow parameters. Are the boundary conditions satisfied? What might this flow represent?

P7.15   Discuss whether fully developed laminar incompressible flow between parallel plates, Eq. (4.143) and Fig. 4.16b, represents an exact solution to the boundary layer equations (7.19) and the boundary conditions (7.20). In what sense, if any, are duct flows also boundary layer flows?

P7.16 A thin flat plate 55 by 110 cm is immersed in a 6-m/s stream of SAE 10 oil at 20°C. Compute the total friction drag if the stream is parallel to (a) the long side and (b) the short side.

P7.17 Helium at 20°C and low pressure flows past a thin flat plate 1 m long and 2 m wide. It is desired that the total friction drag of the plate be 0.5 N. What is the appropriate absolute pressure of the helium if $U = 35$ m/s?

P7.18 The approximate answers to Prob. P7.11 are $u \approx 1.44$ m/s and $\tau \approx 0.0036$ Pa at $x = 50$ cm and $y = 5$ mm. (Do not reveal this to your friends who are working on Prob. P7.11.) Repeat that problem by using the exact Blasius flat-plate boundary layer solution.

P7.19 Air at 20°C and 1 atm flows at 50 ft/s past a thin flat plate whose area (bL) is 24 ft². If the total friction drag is 0.3 lbf, what are the length and width of the plate?

P7.20 Air at 20°C and 1 atm flows at 20 m/s past the flat plate in Fig. P7.20. A pitot stagnation tube, placed 2 mm from the wall, develops a manometer head $h = 16$ mm of Meriam red oil, SG = 0.827. Use this information to estimate the downstream position x of the pitot tube. Assume laminar flow.

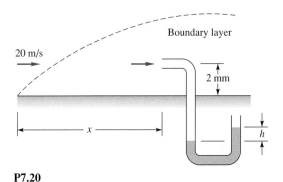

P7.20

P7.21 For the experimental setup of Fig. P7.20, suppose the stream velocity is unknown and the pitot stagnation tube is traversed across the boundary layer of air at 1 atm and 20°C. The manometer fluid is Meriam red oil, and the following readings are made:

y, mm	0.5	1.0	1.5	2.0	2.5	3.0	3.5	4.0	4.5	5.0
h, mm	1.2	4.6	9.8	15.8	21.2	25.3	27.8	29.0	29.7	29.7

Using these data only (not the Blasius theory) estimate (a) the stream velocity, (b) the boundary layer thickness, (c) the wall shear stress, and (d) the total friction drag between the leading edge and the position of the pitot tube.

P7.22 For the Blasius flat-plate problem, Eqs. (7.21) to (7.23), does a two-dimensional stream function $\psi (x, y)$ exist? If so, determine the correct *dimensionless* form for ψ, assuming that $\psi = 0$ at the wall, $y = 0$.

P7.23 Suppose you buy a 4- by 8-ft sheet of plywood and put it on your roof rack. (See Fig. P7.23.) You drive home at 35 mi/h. (a) Assuming the board is perfectly aligned with the airflow, how thick is the boundary layer at the end of the board? (b) Estimate the drag on the sheet of plywood if the boundary layer remains laminar. (c) Estimate the drag on the sheet of plywood if the boundary layer is turbulent (assume the wood is smooth), and compare the result to that of the laminar boundary layer case.

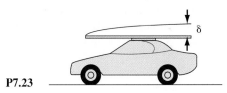

P7.23

*P7.24 Air at 20°C and 1 atm flows past the flat plate in Fig. P7.24 under laminar conditions. There are two equally spaced pitot stagnation tubes, each placed 2 mm from the wall. The manometer fluid is water at 20°C. If $U = 15$ m/s and $L = 50$ cm, determine the values of the manometer readings h_1 and h_2, in mm.

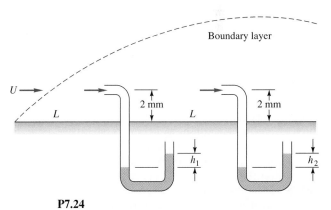

P7.24

P7.25 Consider the smooth square 10-cm-by-10-cm duct in Fig. P7.25. The fluid is air at 20°C and 1 atm, flowing at $V_{avg} = 24$ m/s. It is desired to increase the pressure drop over the 1-m length by adding sharp 8-mm-long flat plates across the duct, as shown. (a) Estimate the pressure drop if there are no plates. (b) Estimate how many plates are needed to generate an additional 100 Pa of pressure drop.

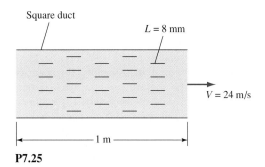

P7.25

P7.26 Consider laminar boundary layer flow past the square-plate arrangements in Fig. P7.26. Compared to the friction drag of a single plate 1, how much larger is the drag of four plates together as in configurations (a) and (b)? Explain your results.

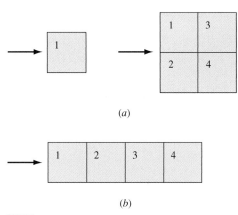

(a)

(b)

P7.26

P7.27 Consider flow at 2 m/s past a thin flat plate. At a position 40 cm downstream from the leading edge, estimate the wall shear stress for (a) air and (b) water at 20°C and 1 atm. (c) How can you quickly show why the result for (b) is so much (215 times) larger?

P7.28 Flow straighteners are arrays of narrow ducts placed in wind tunnels to remove swirl and other in-plane secondary velocities. They can be idealized as square boxes constructed by vertical and horizontal plates, as in Fig. P7.28. The cross section is a by a, and the box length is L. Assuming laminar flat-plate flow and an array of $N \times N$ boxes, derive a formula for (a) the total drag on the bundle of boxes and (b) the effective pressure drop across the bundle.

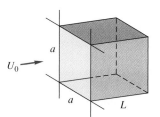

P7.28

P7.29 Let the flow straighteners in Fig. P7.28 form an array of 20×20 boxes of size $a = 4$ cm and $L = 25$ cm. If the approach velocity is $U_0 = 12$ m/s and the fluid is sea-level standard air, estimate (a) the total array drag and (b) the pressure drop across the array. Compare with Sec. 6.8.

P7.30 In Ref. 56 of Chap. 6, McKeon et al. propose new, more accurate values for the turbulent log-law constants, $\kappa = 0.421$ and $B = 5.62$. Use these constants, and the one-seventh power-law, to repeat the analysis that led to the formula for turbulent boundary layer thickness, Eq. (7.42). By what percent is δ/x in your new formula different from that in Eq. (7.42)? Comment.

P7.31 The centerboard on a sailboat is 3 ft long parallel to the flow and protrudes 7 ft down below the hull into seawater at 20°C. Using flat-plate theory for a smooth surface, estimate its drag if the boat moves at 10 knots. Assume $Re_{x,tr} = 5E5$.

P7.32 A flat plate of length L and height δ is placed at a wall and is parallel to an approaching boundary layer, as in Fig. P7.32. Assume that the flow over the plate is fully turbulent and that the approaching flow is a one-seventh-power law:

$$u(y) = U_0 \left(\frac{y}{\delta}\right)^{1/7}$$

Using strip theory, derive a formula for the drag coefficient of this plate. Compare this result with the drag of the same plate immersed in a uniform stream U_0.

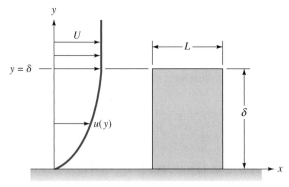

P7.32

P7.33 An alternate analysis of turbulent flat-plate flow was given by Prandtl in 1927, using a wall shear stress formula from pipe flow:

$$\tau_w = 0.0225\rho U^2 \left(\frac{\nu}{U\delta}\right)^{1/4}$$

Show that this formula can be combined with Eqs. (7.33) and (7.40) to derive the following relations for turbulent flat-plate flow:

$$\frac{\delta}{x} = \frac{0.37}{\mathrm{Re}_x^{1/5}} \quad c_f = \frac{0.0577}{\mathrm{Re}_x^{1/5}} \quad C_D = \frac{0.072}{\mathrm{Re}_L^{1/5}}$$

These formulas are limited to Re_x between 5×10^5 and 10^7.

*P7.34 A thin equilateral-triangle plate is immersed parallel to a 12 m/s stream of water at 20°C, as in Fig. P7.34. Assuming $\mathrm{Re}_{tr} = 5 \times 10^5$, estimate the drag of this plate.

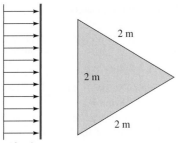

2 m

2 m

2 m

P7.34 12 m/s

P7.35 The solutions to Prob. P7.26 are (a) $F = 2.83F_{\text{1-plate}}$ and (b) $F = 2.0F_{\text{1-plate}}$. (Do not reveal these results to your friends.) Repeat Prob. P7.26 assuming the boundary layer flow is *turbulent,* and comment on the striking increase in numerical values.

P7.36 A ship is 125 m long and has a wetted area of 3500 m². Its propellers can deliver a maximum power of 1.1 MW to seawater at 20°C. If all drag is due to friction, estimate the maximum ship speed, in kn.

P7.37 Air at 20°C and 1 atm flows past a long flat plate, at the end of which is placed a narrow scoop, as shown in Fig. P7.37. (a) Estimate the height h of the scoop if it is to extract 4 kg/s per meter of width into the paper. (b) Find the drag on the plate up to the inlet of the scoop, per meter of width.

30 m/s

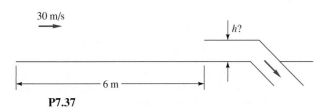

h?

6 m

P7.37

P7.38 Atmospheric boundary layers are very thick but follow formulas very similar to those of flat-plate theory. Consider wind blowing at 10 m/s at a height of 80 m above a smooth beach. Estimate the wind shear stress, in Pa, on the beach if the air is standard sea-level conditions. What will the wind velocity striking your nose be if (a) you are standing up and your nose is 170 cm off the ground and (b) you are lying on the beach and your nose is 17 cm off the ground?

P7.39 A hydrofoil 50 cm long and 4 m wide moves at 28 kn in seawater at 20°C. Using flat-plate theory with $\mathrm{Re}_{tr} = 5\,E5$, estimate its drag, in N, for (a) a smooth wall and (b) a rough wall, $\epsilon = 0.3$ mm.

P7.40 Hoerner [12, p. 3.25] states that the drag coefficient of a flag in winds, based on total wetted area $2bL$, is approximated by $C_D \approx 0.01 + 0.05L/b$, where L is the flag length in the flow direction. Test Reynolds numbers Re_L were 1 E6 or greater. (a) Explain why, for $L/b \geq 1$, these drag values are much higher than for a flat plate. Assuming sea-level standard air at 50 mi/h, with area $bL = 4$ m², find (b) the proper flag dimensions for which the total drag is approximately 400 N.

P7.41 Repeat Prob. P7.20 with the sole change that the pitot probe is now 10 mm from the wall (5 times higher). Show that the flow there cannot possibly be laminar, and use smooth-wall turbulent flow theory to estimate the position x of the probe, in m.

P7.42 A wind tunnel with a square test section, 75 cm by 75 cm, is 5 m long. It delivers air, at 20°C and 1 atm, to the entrance, where the velocity is a uniform 60 mi/h. Assuming turbulent flow and the displacement-thickness concept, estimate (a) the velocity at the exit, in mi/h, and (b) the average pressure gradient in Pa/m. (c) Is the pressure gradient favorable or adverse?

P7.43 In the flow of air at 20°C and 1 atm past a flat plate in Fig. P7.43, the wall shear is to be determined at position x by a *floating element* (a small area connected to a strain-gage force measurement). At $x = 2$ m, the element indicates a shear stress of 2.1 Pa. Assuming turbulent flow from the leading edge, estimate (a) the stream velocity U, (b) the boundary layer thickness δ at

$U \longrightarrow$

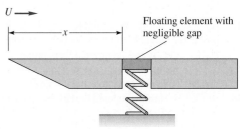

Floating element with negligible gap

x

P7.43

the element, and (c) the boundary layer velocity u, in m/s, at 5 mm above the element.

P7.44   Extensive measurements of wall shear stress and local velocity for turbulent airflow on the flat surface of the University of Rhode Island wind tunnel have led to the following proposed correlation:

$$\frac{\rho y^2 \tau_w}{\mu^2} \approx 0.0207 \left(\frac{uy}{\nu}\right)^{1.77}$$

Thus, if y and u(y) are known at a point in a flat-plate boundary layer, the wall shear may be computed directly. If the answer to part (c) of Prob. P7.43 is $u \approx 26.3$ m/s, determine the shear stress and compare with Prob. P7.43. Discuss.

P7.45   A thin sheet of fiberboard weighs 90 N and lies on a rooftop, as shown in Fig. P7.45. Assume ambient air at 20°C and 1 atm. If the coefficient of solid friction between board and roof is $\sigma \approx 0.12$, what wind velocity will generate enough fluid friction to dislodge the board?

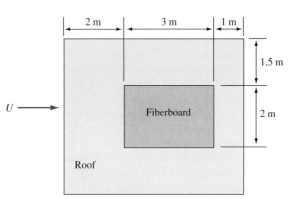

P7.45

P7.46   A ship is 150 m long and has a wetted area of 5000 m². If it is encrusted with barnacles, the ship requires 7000 hp to overcome friction drag when moving in seawater at 15 kn and 20°C. What is the average roughness of the barnacles? How fast would the ship move with the same power if the surface were smooth? Neglect wave drag.

P7.47   As a case similar to Example 7.5, Howarth also proposed the adverse gradient velocity distribution $U = U_0(1 - x^2/L^2)$ and computed separation at $x_{sep}/L = 0.271$ by a series expansion method. Compute separation by Thwaites's method and compare.

P7.48   In 1957 H. Görtler proposed the adverse gradient test cases

$$U = \frac{U_0}{(1 + x/L)^n}$$

and computed separation for laminar flow at $n = 1$ to be $x_{sep}/L = 0.159$. Compare with Thwaites's method, assuming $\theta_0 = 0$.

P7.49   Based strictly on your understanding of flat-plate theory plus adverse and favorable pressure gradients, explain the direction (left or right) for which airflow past the slender airfoil shape in Fig. P7.49 will have lower total (friction + pressure) drag.

P7.49

P7.50   Consider the flat-walled diffuser in Fig. P7.50, which is similar to that of Fig. 6.26a with constant width b. If x is measured from the inlet and the wall boundary layers are thin, show that the core velocity U(x) in the diffuser is given approximately by

$$U = \frac{U_0}{1 + (2x \tan \theta)/W}$$

where W is the inlet height. Use this velocity distribution with Thwaites's method to compute the wall angle θ for which laminar separation will occur in the exit plane when diffuser length $L = 2W$. Note that the result is independent of the Reynolds number.

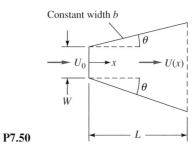

P7.50

7.51   The derivation of Eq. (7.42) for δ/x in turbulent flow used very simple velocity correlations, Eqs. (7.38) and (7.39). A unified law-of-the-wall was given by Spalding [54]:

$$y^+ = u^+ + e^{-\kappa B}\left[e^{\kappa u^+} - 1 - \kappa u^+ - \frac{(\kappa u^+)^2}{2} - \frac{(\kappa u^+)^3}{6}\right]$$

where $u^+ = u/u^*$ and $y^+ = yu^*/\nu$, while κ and B are the log-law constants in Eq. (7.34). This clever inverse formula fits all the data in Fig. 6.10 out to the edge of the logarithmic region. How can Spalding's formula be used to improve Eqs. (7.38) and (7.40) and thus lead to improved turbulent flat-plate-flow relations?

P7.52 Clift et al. [46] give the formula $F \approx (6\pi/5)(4 + a/b)\mu Ub$ for the drag of a prolate spheroid in *creeping motion,* as shown in Fig. P7.52. The half-thickness b is 4 mm. If the fluid is SAE 50W oil at 20°C, (*a*) check that $Re_b < 1$ and (*b*) estimate the spheroid length if the drag is 0.02 N.

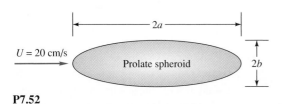

P7.52

P7.53 From Table 7.2, the drag coefficient of a wide plate normal to a stream is approximately 2.0. Let the stream conditions be U_∞ and p_∞. If the average pressure on the front of the plate is approximately equal to the free-stream stagnation pressure, what is the average pressure on the rear?

P7.54 A chimney at sea level is 2 m in diameter and 40 m high. When it is subjected to 50 mi/h storm winds, what is the estimated wind-induced bending moment about the bottom of the chimney?

P7.55 A ship tows a submerged cylinder, which is 1.5 m in diameter and 22 m long, at 5 m/s in fresh water at 20°C. Estimate the towing power, in kW, required if the cylinder is (*a*) parallel and (*b*) normal to the tow direction.

P7.56 A delivery vehicle carries a long sign on top, as in Fig. P7.56. If the sign is very thin and the vehicle moves at 65 mi/h, (*a*) estimate the force on the sign with no crosswind and (*b*) discuss the effect of a crosswind.

P7.56

P7.57 The main cross-cable between towers of a coastal suspension bridge is 60 cm in diameter and 90 m long. Estimate the total drag force on this cable in crosswinds of 50 mi/h. Are these laminar flow conditions?

P7.58 A long cylinder of rectangular cross section, 5 cm high and 30 cm long, is immersed in water at 20°C flowing at 12 m/s parallel to the long side of the rectangle. Estimate the drag force on the cylinder, per unit length, if the rectangle (*a*) has a flat face or (*b*) has a rounded nose.

*P7.59 Joe can pedal his bike at 10 m/s on a straight level road with no wind. The rolling resistance of his bike is 0.80 N·s/m— that is, 0.80 N of force per m/s of speed. The drag area ($C_D A$) of Joe and his bike is 0.422 m². Joe's mass is 80 kg and that of the bike is 15 kg. He now encounters a head wind of 5.0 m/s. (*a*) Develop an equation for the speed at which Joe can pedal into the wind. [*Hint:* A cubic equation for V will result.] (*b*) Solve for V; that is, how fast can Joe ride into the head wind? (*c*) Why is the result not simply $10 - 5.0 = 5.0$ m/s, as one might first suspect?

P7.60 A fishnet consists of 1-mm-diameter strings overlapped and knotted to form 1 by 1-cm squares. Estimate the drag of 1 m² of such a net when towed normal to its plane at 3 m/s in 20°C seawater. What horsepower is required to tow 400 ft² of this net?

P7.61 A filter may be idealized as an array of cylindrical fibers normal to the flow, as in Fig. P7.61. Assuming that the fibers are uniformly distributed and have drag coefficients given by Fig. 7.16a, derive an approximate expression for the pressure drop Δp through a filter of thickness L.

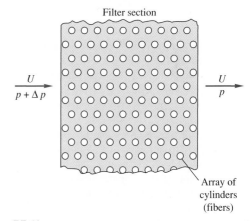

P7.61

P7.62 A sea-level smokestack is 52 m high and has a square cross section. Its supports can withstand a maximum side force of 90 kN. If the stack is to survive 90-mi/h hurricane winds, what is its maximum possible width?

P7.63 For those who think electric cars are sissy, Keio University in Japan has tested a 22-ft-long prototype whose eight electric motors generate a total of 590 horsepower. The "Kaz" cruises at 180 mi/hr (see *Popular Science,*

August 2001, p. 15). If the drag coefficient is 0.35 and the frontal area is 26 ft^2, what percentage of this power is expended against sea-level air drag?

P7.64 A parachutist jumps from a plane, using an 8.5-m-diameter chute in the standard atmosphere. The total mass of the chutist and the chute is 90 kg. Assuming an open chute and quasi-steady motion, estimate the time to fall from 2000- to 1000-m altitude.

P7.65 As soldiers get bigger and packs get heavier, a parachutist and load can weigh as much as 400 lbf. The standard 28-ft parachute may descend too fast for safety. For heavier loads, the U. S. Army Natick Center has developed a 28-ft, higher-drag, less porous XT-11 parachute (see <http://www.natick.army.mil>). This parachute has a sea-level descent speed of 16 ft/s with a 400-lbf load. (*a*) What is the drag coefficient of the XT-11? (*b*) How fast would the standard chute descend at sea level with such a load?

P7.66 A sphere of density ρ_s and diameter D is dropped from rest in a fluid of density ρ and viscosity μ. Assuming a constant drag coefficient C_{d_0}, derive a differential equation for the fall velocity $V(t)$ and show that the solution is

$$V = \left[\frac{4gD(S-1)}{3C_{d_0}}\right]^{1/2} \tanh Ct$$

$$C = \left[\frac{3gC_{d_0}(S-1)}{4S^2D}\right]^{1/2}$$

where $S = \rho_s/\rho$ is the specific gravity of the sphere material.

P7.67 A world-class bicycle rider can generate one-half horsepower for long periods. If racing at sea level, estimate the velocity this cyclist can maintain. Neglect rolling friction.

P7.68 A baseball weighs 145 g and is 7.35 cm in diameter. It is dropped from rest from a 35-m-high tower at approximately sea level. Assuming a laminar flow drag coefficient, estimate (*a*) its terminal velocity and (*b*) whether it will reach 99 percent of its terminal velocity before it hits the ground.

P7.69 Two baseballs from Prob. P7.68 are connected to a rod 7 mm in diameter and 56 cm long, as in Fig. P7.69. What power, in W, is required to keep the system spinning at 400 r/min? Include the drag of the rod, and assume sea-level standard air.

P7.70 The Army's new ATPS personnel parachute is said to be able to bring a 400-lbf load, trooper plus pack, to ground at 16 ft/s in "mile-high" Denver, Colorado. If we assume that Table 7.3 is valid, what is the approximate diameter of this new parachute?

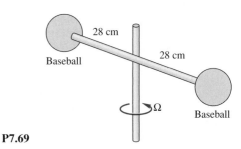

P7.69

P7.71 A football weighs 0.91 lbf and approximates an ellipsoid 6 in in diameter and 12 in long (Table 7.3). It is thrown upward at a 45° angle with an initial velocity of 80 ft/s. Neglect spin and lift. Assuming turbulent flow, estimate the horizontal distance traveled, (*a*) neglecting drag and (*b*) accounting for drag with a numerical (computer) model.

P7.72 A settling tank for a municipal water supply is 2.5 m deep, and 20°C water flows through continuously at 35 cm/s. Estimate the minimum length of the tank that will ensure that all sediment (SG = 2.55) will fall to the bottom for particle diameters greater than (*a*) 1 mm and (*b*) 100 μm.

P7.73 A balloon is 4 m in diameter and contains helium at 125 kPa and 15°C. Balloon material and payload weigh 200 N, not including the helium. Estimate (*a*) the terminal ascent velocity in sea-level standard air, (*b*) the final standard altitude (neglecting winds) at which the balloon will come to rest, and (*c*) the minimum diameter (< 4 m) for which the balloon will just barely begin to rise in sea-level standard air.

P7.74 It is difficult to define the "frontal area" of a motorcycle due to its complex shape. One then measures the *drag area* (that is, C_DA) in area units. Hoerner [12] reports the drag area of a typical motorcycle, including the (upright) driver, as about 5.5 ft^2. Rolling friction is typically about 0.7 lbf per mi/h of speed. If that is the case, estimate the maximum sea-level speed (in mi/h) of the new Harley-Davidson V-Rod™ cycle, whose liquid-cooled engine produces 115 hp.

P7.75 The helium-filled balloon in Fig. P7.75 is tethered at 20°C and 1 atm with a string of negligible weight and drag. The diameter is 50 cm, and the balloon material

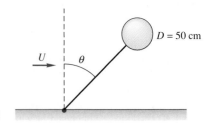

P7.75

weighs 0.2 N, not including the helium. The helium pressure is 120 kPa. Estimate the tilt angle θ if the airstream velocity U is (a) 5 m/s or (b) 20 m/s.

P7.76 The recent movie *The World's Fastest Indian* tells the story of Burt Munro, a New Zealander who, in 1937, set a motorcycle record of 201 mi/h on the Bonneville Salt Flats. Using the data of Prob. P7.74, (a) estimate the horsepower needed to drive this fast. (b) What horsepower would have gotten Burt up to 250 mi/h?

P7.77 To measure the drag of an upright person, without violating human subject protocols, a life-sized mannequin is attached to the end of a 6-m rod and rotated at $\Omega = 80$ rev/min, as in Fig. P7.77. The power required to maintain the rotation is 60 kW. By including rod drag power, which is significant, estimate the drag area $C_D A$ of the mannequin, in m².

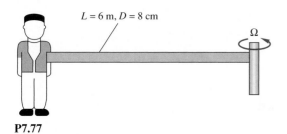

$L = 6$ m, $D = 8$ cm

Ω

P7.77

P7.78 Apply Prob. P7.61 to a filter consisting of 300-μm-diameter fibers packed 250 per square centimeter in the plane of Fig. P7.61. For air at 20°C and 1 atm flowing at 1.5 m/s, estimate the pressure drop if the filter is 5 cm thick.

P7.79 Assume that a radioactive dust particle approximates a sphere of density 2400 kg/m³. How long, in days, will it take such a particle to settle to sea level from an altitude of 12 km if the particle diameter is (a) 1 μm or (b) 20 μm?

P7.80 A heavy sphere attached to a string should hang at an angle θ when immersed in a stream of velocity U, as in Fig. P7.80. Derive an expression for θ as a function of the sphere and flow properties. What is θ if the sphere is steel (SG = 7.86) of diameter 3 cm and the flow is sea-level standard air at $U = 40$ m/s? Neglect the string drag.

P7.81 A typical U.S. Army parachute has a projected diameter of 28 ft. For a payload mass of 80 kg, (a) what terminal velocity will result at 1000-m standard altitude? For the same velocity and net payload, what size drag-producing "chute" is required if one uses a square flat plate held (b) vertically and (c) horizontally? (Neglect the fact that flat shapes are not dynamically stable in free fall.)

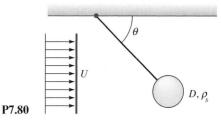

θ

U

D, ρ_s

P7.80

P7.82 Skydivers, flying over sea-level ground, typically jump at about 8000 ft altitude and free-fall spread-eagled until they open their chutes at about 2000 ft. They take about 10 s to reach terminal velocity. Estimate how many seconds of free-fall they enjoy if (a) they fall spread-eagled or (b) they fall feet first? Assume a total skydiver weight of 220 lbf.

P7.83 A high-speed car has a drag coefficient of 0.3 and a frontal area of 1 m². A parachute is to be used to slow this car from 80 to 40 m/s in 8 s. What should the chute diameter be? What distance will be traveled during this deceleration? Take $m = 2000$ kg.

P7.84 A Ping-Pong ball weighs 2.6 g and has a diameter of 3.8 cm. It can be supported by an air jet from a vacuum cleaner outlet, as in Fig. P7.84. For sea-level standard air, what jet velocity is required?

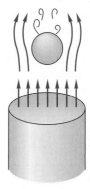

P7.84

***P7.85** An aluminum cylinder (SG = 2.7) slides concentrically down a taut 1-mm-diameter wire as shown in Fig. P7.85. Its length $L = 8$ cm, and its radius $R = 1$ cm. A 2-mm-diameter hole down the cylinder center is lubricated by SAE 30 oil at 20°C. Estimate the terminal fall velocity V of the cylinder if ambient air drag is (a) neglected and (b) included. Assume air at 1 atm and 20°C.

P7.86 Hoerner [Ref. 12, pp. 3–25] states that the drag coefficient of a flag of 2:1 aspect ratio is 0.11 based on planform area. The University of Rhode Island has an aluminum flagpole 25 m high and 14 cm in diameter. It flies equal-sized national and state flags together. If the

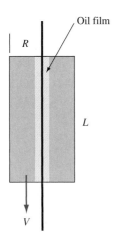

P7.85

P7.88

fracture stress of aluminum is 210 MPa, what is the maximum flag size that can be used yet avoids breaking the flagpole in hurricane (75 mi/h) winds? (Neglect the drag of the flagpole.)

P7.87 A tractor-trailer truck has a drag area $C_D A = 8$ m^2 bare and 6.7 m^2 with an aerodynamic deflector (Fig. 7.18b). Its rolling resistance is 50 N for each mile per hour of speed. Calculate the total horsepower required at sea level with and without the deflector if the truck moves at (a) 55 mi/h and (b) 75 mi/h.

P7.88 A pickup truck has a clean drag area $C_D A$ of 35 ft^2. Estimate the horsepower required to drive the truck at 55 mi/h (a) clean and (b) with the 3- by 6-ft sign in Fig. P7.88 installed if the rolling resistance is 150 lbf at sea level.

P7.89 The new AMTRAK high-speed Acela train can reach 150 mi/h, which presently it seldom does because of the curvy coastline tracks in New England. If 75 percent of the power expended at this speed is due to air drag, estimate the total horsepower required by the Acela.

P7.90 In the great hurricane of 1938, winds of 85 mi/h blew over a boxcar in Providence, Rhode Island. The boxcar was 10 ft high, 40 ft long, and 6 ft wide, with a 3-ft clearance above tracks 4.8 ft apart. What wind speed would topple a boxcar weighing 40,000 lbf?

*P7.91 A cup anemometer uses two 5-cm-diameter hollow hemispheres connected to 15-cm rods, as in Fig. P7.91. Rod drag is negligible, and the central bearing has a retarding torque of 0.004 N·m. Making simplifying assumptions to average out the time-varying geometry, estimate and plot the variation of anemometer rotation rate Ω with wind velocity U in the range $0 < U < 25$ m/s for sea-level standard air.

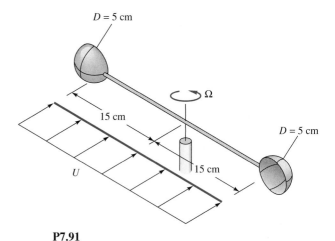

P7.91

P7.92 A 1500-kg automobile uses its drag area $C_D A = 0.4$ m^2, plus brakes and a parachute, to slow down from 50 m/s. Its brakes apply 5000 N of resistance. Assume sea-level standard air. If the automobile must stop in 8 s, what diameter parachute is appropriate?

P7.93 A hot-film probe is mounted on a cone-and-rod system in a sea-level airstream of 45 m/s, as in Fig. P7.93. Estimate the maximum cone vertex angle allowable if the flow-induced bending moment at the root of the rod is not to exceed 30 N · cm.

P7.94 A rotary mixer consists of two 1-m-long half-tubes rotating around a central arm, as in Fig. P7.94. Using the drag from Table 7.2, derive an expression for the torque T required to drive the mixer at angular velocity Ω in a fluid of density ρ. Suppose that the fluid is water at 20°C and the maximum driving power available is 20 kW. What is the maximum rotation speed Ω r/min?

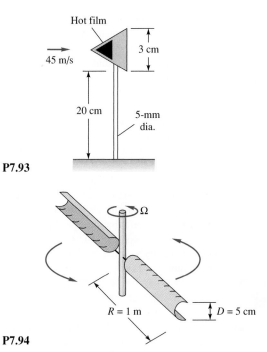

P7.93

P7.94

P7.95 An airplane weighing 28 kN, with a drag area $C_D A \approx 5$ m², lands at sea level at 55 m/s and deploys a drag parachute 3 m in diameter. No other brakes are applied. (a) How long will it take the plane to slow down to 20 m/s? (b) How far will it have traveled in that time?

*P7.96 A Savonius rotor (Fig. 6.29b) can be approximated by the two open half-tubes in Fig. P7.96 mounted on a central axis. If the drag of each tube is similar to that in Table 7.2, derive an approximate formula for the rotation rate Ω as a function of U, D, L, and the fluid properties (ρ, μ).

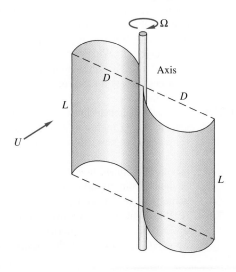

P7.96

P7.97 A simple measurement of automobile drag can be found by an unpowered *coastdown* on a level road with no wind. Assume constant rolling resistance. For an automobile of mass 1500 kg and frontal area 2 m², the following velocity-versus-time data are obtained during a coastdown:

t, s	0	10	20	30	40
V, m/s	27.0	24.2	21.8	19.7	17.9

Estimate (a) the rolling resistance and (b) the drag coefficient. This problem is well suited for computer analysis but can be done by hand also.

*P7.98 A buoyant ball of specific gravity SG < 1 dropped into water at inlet velocity V_0 will penetrate a distance h and then pop out again, as in Fig. P7.98. Make a dynamic analysis of this problem, assuming a constant drag coefficient, and derive an expression for h as a function of the system properties. How far will a 5-cm-diameter ball with SG = 0.5 and $C_D \approx 0.47$ penetrate if it enters at 10 m/s?

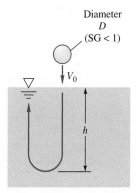

P7.98

P7.99 Two steel balls (SG = 7.86) are connected by a thin hinged rod of negligible weight and drag, as in Fig. P7.99. A stop prevents the rod from rotating counterclockwise. Estimate the sea-level air velocity U for which the rod will first begin to rotate clockwise.

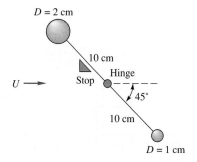

P7.99

P7.100 A tractor-trailer truck is coasting freely, with no brakes, down an 8° slope at 1000-m standard altitude. Rolling resistance is 120 N for every m/s of speed. Its frontal area is 9 m^2, and the weight is 65 kN. Estimate the terminal coasting velocity, in mi/h, for (a) no deflector and (b) a deflector installed.

P7.101 Icebergs can be driven at substantial speeds by the wind. Let the iceberg be idealized as a large, flat cylinder, $D \gg L$, with one-eighth of its bulk exposed, as in Fig. P7.101. Let the seawater be at rest. If the upper and lower drag forces depend on relative velocities between the iceberg and the fluid, derive an approximate expression for the steady iceberg speed V when driven by wind velocity U.

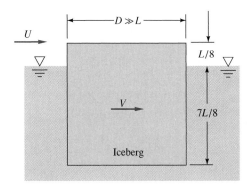

P7.101

P7.102 Sand particles (SG = 2.7), approximately spherical with diameters from 100 to 250 μm, are introduced into an upward-flowing stream of water at 20°C. What is the minimum water velocity that will carry all the sand particles *upward*?

P7.103 When immersed in a uniform stream V, a heavy rod hinged at A will hang at *Pode's angle* θ, after an analysis by L. Pode in 1951 (Fig. P7.103). Assume that the cylinder has normal drag coefficient C_{DN} and tangential coefficient C_{DT} that relate the drag forces to V_N and V_T, respectively. Derive an expression for Pode's angle as a function of the flow and rod parameters. Find θ for a steel rod, L = 40 cm, D = 1 cm, hanging in sea-level air at V = 35 m/s.

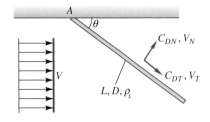

P7.103

P7.104 The Russian Typhoon-class submarine is 170 m long, with a maximum diameter of 23 m. Its propulsor can deliver up to 80,000 hp to the seawater. Model the submarine as an 8:1 ellipsoid and estimate the maximum speed, in knots, of this ship.

P7.105 A ship 50 m long, with a wetted area of 800 m^2, has the hull shape tested in Fig. 7.19. There are no bow or stern bulbs. The total propulsive power available is 1 MW. For seawater at 20°C, plot the ship's velocity V kn versus power P for $0 < P < 1$ MW. What is the most efficient setting?

P7.106 A smooth steel 1-cm-diameter sphere ($W \approx 0.04$ N) is fired vertically at sea level at the initial supersonic velocity V_0 = 1000 m/s. Its drag coefficient is given by Fig. 7.20. Assuming that the speed of sound is constant at $a \approx 343$ m/s, compute the maximum altitude of the projectile (a) by a simple analytical estimate and (b) by a computer program.

P7.107 Repeat Prob. P7.106 if the body is a 9-mm steel bullet ($W \approx 0.07$ N) that approximates the "pointed body of revolution" in Fig. 7.20.

P7.108 The data in Fig. P7.108 are for the lift and drag of a spinning sphere from Ref. 45. Suppose that a tennis ball ($W \approx 0.56$ N, $D \approx 6.35$ cm) is struck at sea level with initial velocity V_0 = 30 m/s and "topspin" (front of the ball rotating downward) of 120 r/s. If the initial height of the ball is 1.5 m, estimate the horizontal distance traveled before it strikes the ground.

P7.109 The world record for automobile mileage, 12,665 miles per gallon, was set in 2005 by the PAC-CAR II in Fig. P7.109, built by students at the Swiss Federal Institute of Technology in Zurich [52]. This little car, with an empty weight of 64 lbf and a height of only 2.5 ft, traveled a 21-km course at 30 km/hr to set the record. It has a reported drag coefficient of 0.075 (comparable to an airfoil), based upon a frontal area of 3 ft^2. (a) What is the drag of this little car when on the course? (b) What horsepower is required to propel it? (c) Do a bit of research and explain why a value of miles per gallon is completely misleading in this particular case.

P7.110 A baseball pitcher throws a curveball with an initial velocity of 65 mi/h and a spin of 6500 r/min about a vertical axis. A baseball weighs 0.32 lbf and has a diameter of 2.9 in. Using the data of Fig. P7.108 for turbulent flow, estimate how far such a curveball will have deviated from its straight-line path when it reaches home plate 60.5 ft away.

*P7.111 A table tennis ball has a mass of 2.6 g and a diameter of 3.81 cm. It is struck horizontally at an initial velocity of 20 m/s while it is 50 cm above the table, as in

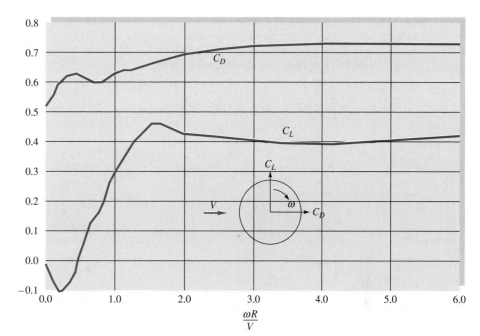

P7.108 Drag and lift coefficients for a rotating sphere at $Re_D \approx 10^5$, from Ref. 45. *(Reproduced by permission of the American Society of Mechanical Engineers.)*

P7.109 The world's best mileage set by PAC-Car II of ETH Zurich.

Fig. P7.111. For sea-level air, what spin, in r/min, will cause the ball to strike the opposite edge of the table, 4 m away? Make an analytical estimate, using Fig. P7.108, and account for the fact that the ball decelerates during flight.

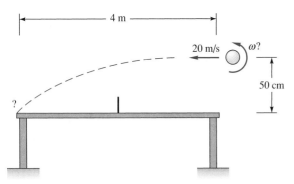

P7.111

P7.112 A smooth wooden sphere (SG = 0.65) is connected by a thin rigid rod to a hinge in a wind tunnel, as in Fig. P7.112. Air at 20°C and 1 atm flows and levitates the sphere. (a) Plot the angle θ versus sphere diameter d in the range 1 cm $\le d \le$ 15 cm. (b) Comment on the feasibility of this configuration. Neglect rod drag.

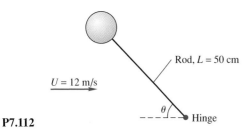

P7.112

P7.113 An automobile has a mass of 1000 kg and a drag area $C_D A = 0.7$ m². The rolling resistance of 70 N is approximately constant. The car is coasting without brakes at 90 km/h as it begins to climb a hill of 10 percent grade (slope = $\tan^{-1} 0.1 = 5.71°$). How far up the hill will the car come to a stop?

*P7.114 Suppose that the car in Prob. P7.113 is placed at the top of the 10 percent grade hill and released from rest to coast down without brakes. What will be its speed, in km/h, after dropping a vertical distance of 20 m?

P7.115 The Cessna Citation executive jet weighs 67 kN and has a wing area of 32 m². It cruises at 10 km standard altitude with a lift coefficient of 0.21 and a drag coefficient of 0.015. Estimate (a) the cruise speed in mi/h and (b) the horsepower required to maintain cruise velocity.

P7.116 An airplane weighs 180 kN and has a wing area of 160 m² and a mean chord of 4 m. The airfoil properties are given by Fig. 7.25. If the plane is designed to land at $V_0 = 1.2V_{stall}$, using a split flap set at 60°, (a) what is the proper landing speed in mi/h? (b) What power is required for takeoff at the same speed?

P7.117 Suppose that the airplane of Prob. P7.116 takes off at sea level without benefit of flaps, with C_L constant so that the takeoff speed is 100 mi/h. Estimate the takeoff distance if the thrust is 10 kN. How much thrust is needed to make the takeoff distance 1250 m?

*P7.118 Suppose that the airplane of Prob. P7.116 is fitted with all the best high-lift devices of Fig. 7.28. What is its minimum stall speed in mi/h? Estimate the stopping distance if the plane lands at $V_0 = 1.25V_{stall}$ with constant $C_L = 3.0$ and $C_D = 0.2$ and the braking force is 20 percent of the weight on the wheels.

P7.119 A transport plane has a mass of 45,000 kg, a wing area of 160 m², and an aspect ratio of 7. Assume all lift and drag due to the wing alone, with $C_{D\infty} = 0.020$ and $C_{L,max} = 1.5$. If the aircraft flies at 9000 m standard altitude, make a plot of drag (in N) versus speed (from stall to 240 m/s) and determine the optimum cruise velocity (minimum drag per unit speed).

P7.120 Show that if Eqs. (7.70) and (7.71) are valid, the maximum lift-to-drag ratio occurs when $C_D = 2C_{D\infty}$. What are $(L/D)_{max}$ and α for a symmetric wing when AR = 5 and $C_{D\infty} = 0.009$?

P7.121 In gliding (unpowered) flight, the lift and drag are in equilibrium with the weight. Show that if there is no wind, the aircraft sinks at an angle

$$\tan \theta \approx \frac{\text{drag}}{\text{lift}}$$

For a sailplane of mass 200 kg, wing area 12 m², and aspect ratio 11, with an NACA 0009 airfoil, estimate (a) the stall speed, (b) the minimum gliding angle, and (c) the maximum distance it can glide in still air when it is 1200 m above level ground.

P7.122 A boat of mass 2500 kg has two hydrofoils, each of chord 30 cm and span 1.5 m, with $C_{L,max} = 1.2$ and $C_{D\infty} = 0.08$. Its engine can deliver 130 kW to the water. For seawater at 20°C, estimate (a) the minimum speed for which the foils support the boat and (b) the maximum speed attainable.

P7.123 In prewar days there was a controversy, perhaps apocryphal, about whether the bumblebee has a legitimate aerodynamic right to fly. The average bumblebee (*Bombus terrestris*) weighs 0.88 g, with a wing span of 1.73 cm and a wing area of 1.26 cm². It can indeed fly at 10 m/s. Using fixed-wing theory, what is the lift coefficient of the bee at this speed? Is this reasonable for typical airfoils?

*P7.124 The bumblebee can hover at zero speed by flapping its wings. Using the data of Prob. P7.123, devise a theory for flapping wings where the downstroke approximates a short flat plate normal to the flow (Table 7.3) and the upstroke is feathered at nearly zero drag. How many flaps per second of such a model wing are needed to support the bee's weight? (Actual measurements of bees show a flapping rate of 194 Hz.)

P7.125 In 2001 a commercial aircraft lost all power while flying at 33,000 ft over the open Atlantic Ocean, about 60 miles from the Azores Islands. The pilots, with admirable skill, put the plane into a shallow glide and successfully landed in the Azores. Assume that the airplane satisfies Eqs. (7.70) and (7.71), with AR = 7, $C_{d\infty} = 0.02$, and a symmetric airfoil. Estimate its optimum glide distance with a mathematically perfect pilot.

Word Problems

W7.1 How do you *recognize* a boundary layer? Cite some physical properties and some measurements that reveal appropriate characteristics.

W7.2 In Chap. 6 the Reynolds number for transition to turbulence in pipe flow was about $Re_{tr} \approx 2300$, whereas in flat-plate flow $Re_{tr} \approx 1$ E6, nearly three orders of magnitude higher. What accounts for the difference?

W7.3 Without writing any equations, give a verbal description of boundary layer displacement thickness.

W7.4 Describe, in words only, the basic ideas behind the "boundary layer approximations."

W7.5 What is an *adverse* pressure gradient? Give three examples of flow regimes where such gradients occur.

W7.6 What is a *favorable* pressure gradient? Give three examples of flow regimes where such gradients occur.

W7.7 The drag of an airfoil (Fig. 7.12) increases considerably if you turn the sharp edge around 180° to face the stream. Can you explain this?

W7.8 In Table 7.3, the drag coefficient of a spruce tree decreases sharply with wind velocity. Can you explain this?

W7.9 Thrust is required to propel an airplane at a finite forward velocity. Does this imply an energy loss to the system? Explain the concepts of thrust and drag in terms of the first law of thermodynamics.

W7.10 How does the concept of *drafting,* in automobile and bicycle racing, apply to the material studied in this chapter?

W7.11 The circular cylinder of Fig. 7.13 is doubly symmetric and therefore should have no lift. Yet a lift sensor would definitely reveal a finite root-mean-square value of lift. Can you explain this behavior?

W7.12 Explain in words why a thrown spinning ball moves in a curved trajectory. Give some physical reasons why a side force is developed in addition to the drag.

Fundamentals of Engineering Exam Problems

FE7.1 A smooth 12-cm-diameter sphere is immersed in a stream of 20°C water moving at 6 m/s. The appropriate Reynolds number of this sphere is approximately
(*a*) 2.3 E5, (*b*) 7.2 E5, (*c*) 2.3 E6, (*d*) 7.2 E6, (*e*) 7.2 E7

FE7.2 If, in Prob. FE7.1, the drag coefficient based on frontal area is 0.5, what is the drag force on the sphere?
(*a*) 17 N, (*b*) 51 N, (*c*) 102 N, (*d*) 130 N, (*e*) 203 N

FE7.3 If, in Prob. FE7.1, the drag coefficient based on frontal area is 0.5, at what terminal velocity will an aluminum sphere (SG = 2.7) fall in still water?
(*a*) 2.3 m/s, (*b*) 2.9 m/s, (*c*) 4.6 m/s, (*d*) 6.5 m/s, (*e*) 8.2 m/s

FE7.4 For flow of sea-level standard air at 4 m/s parallel to a thin flat plate, estimate the boundary layer thickness at $x = 60$ cm from the leading edge:
(*a*) 1.0 mm, (*b*) 2.6 mm, (*c*) 5.3 mm, (*d*) 7.5 mm, (*e*) 20.2 mm

FE7.5 In Prob. FE7.4, for the same flow conditions, what is the wall shear stress at $x = 60$ cm from the leading edge?
(*a*) 0.053 Pa, (*b*) 0.11 Pa, (*c*) 0.16 Pa, (*d*) 0.32 Pa, (*e*) 0.64 Pa

FE7.6 Wind at 20°C and 1 atm blows at 75 km/h past a flagpole 18 m high and 20 cm in diameter. The drag coefficient, based on frontal area, is 1.15. Estimate the wind-induced bending moment at the base of the pole.
(*a*) 9.7 kN · m, (*b*) 15.2 kN · m, (*c*) 19.4 kN · m, (*d*) 30.5 kN · m, (*e*) 61.0 kN · m

FE7.7 Consider wind at 20°C and 1 atm blowing past a chimney 30 m high and 80 cm in diameter. If the chimney may fracture at a base bending moment of 486 kN·m, and its drag coefficient based on frontal area is 0.5, what is the approximate maximum allowable wind velocity to avoid fracture?
(*a*) 50 mi/h, (*b*) 75 mi/h, (*c*) 100 mi/h, (*d*) 125 mi/h, (*e*) 150 mi/h

FE7.8 A dust particle of density 2600 kg/m³, small enough to satisfy Stokes's drag law, settles at 1.5 mm/s in air at 20°C and 1 atm. What is its approximate diameter?
(a) 1.8 μm, (b) 2.9 μm, (c) 4.4 μm, (d) 16.8 μm, (e) 234 μm

FE7.9 An airplane has a mass of 19,550 kg, a wing span of 20 m, and an average wing chord of 3 m. When fly-ing in air of density 0.5 kg/m³, its engines provide a thrust of 12 kN against an overall drag coefficient of 0.025. What is its approximate velocity?
(a) 250 mi/h, (b) 300 mi/h, (c) 350 mi/h, (d) 400 mi/h, (e) 450 mi/h

FE7.10 For the flight conditions of the airplane in Prob. FE7.9 above, what is its approximate lift coefficient?
(a) 0.1, (b) 0.2, (c) 0.3, (d) 0.4, (e) 0.5

Comprehensive Problems

C7.1 Jane wants to estimate the drag coefficient of herself on her bicycle. She measures the projected frontal area to be 0.40 m² and the rolling resistance to be 0.80 N·s/m. The mass of the bike is 15 kg, while the mass of Jane is 80 kg. Jane coasts down a long hill that has a constant 4° slope. (See Fig. C7.1.) She reaches a terminal (steady state) speed of 14 m/s down the hill. Estimate the aerodynamic drag coefficient C_D of the rider and bicycle combination.

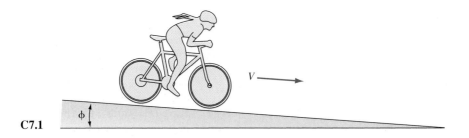

C7.1

C7.2 Air at 20°C and 1 atm flows at $V_{avg} = 5$ m/s between long, smooth parallel heat exchanger plates 10 cm apart, as in Fig. C7.2. It is proposed to add a number of widely spaced 1-cm-long interrupter plates to increase the heat transfer, as shown. Although the flow in the channel is turbulent, the boundary layers over the interrupter plates are essentially laminar. Assume all plates are 1 m wide into the paper. Find (a) the pressure drop in Pa/m without the small plates present. Then find (b) the number of small plates per meter of channel length that will cause the pressure drop to rise to 10.0 Pa/m.

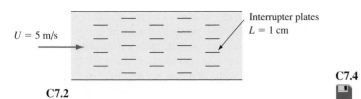

Interrupter plates
$L = 1$ cm

$U = 5$ m/s

C7.2

C7.3 A new pizza store is planning to open. It will, of course, offer free delivery, and therefore need a small delivery car with a large sign attached. The sign (a flat plate) is 1.5 ft high and 5 ft long. The boss (having no feel for fluid mechanics) mounts the sign bluntly facing the wind. One of his drivers is taking fluid mechanics and tells his boss he can save lots of money by mounting the sign parallel to the wind. (See Fig. C7.3.) (a) Calculate the drag (in lbf) on the *sign alone* at 40 mi/h (58.7 ft/s) in *both orienta-tions*. (b) Suppose the car without any sign has a drag coef-ficient of 0.4 and a frontal area of 40 ft². For $V = 40$ mi/h, calculate the *total* drag of the car–sign combination for both orientations. (c) If the car has a rolling resistance of 40 lbf at 40 mi/h, calculate the horsepower required by the engine to drive the car at 40 mi/h in both orientations. (d) Finally, if the engine can deliver 10 hp for 1 h on a gallon of gasoline, calculate the fuel efficiency in mi/gal for both orientations at 40 mi/h.

C7.4 Consider a pendulum with an unusual bob shape: a hemi-spherical cup of diameter D whose axis is in the plane of oscillation, as in Fig. C7.4. Neglect the mass and drag of the rod L. (a) Set up the differential equation for the oscillation $\theta(t)$, including different cup drag (air density ρ) in each direc-

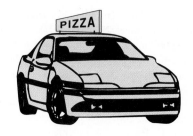

C7.3

tion, and (*b*) nondimensionalize this equation. (*c*) Determine the natural frequency of oscillation for small $\theta \ll 1$ rad. (*d*) For the special case $L = 1$ m, $D = 10$ cm, $m = 50$ g, and air at 20°C and 1 atm, with $\theta(0) = 30°$, find (numerically) the time required for the oscillation amplitude to drop to 1°.

C7.5 Program a method of numerical solution of the Blasius flat-plate relation, Eq. (7.22), subject to the conditions in Eqs. (7.23). You will find that you cannot get started without knowing the initial second derivative $f''(0)$, which lies between 0.2 and 0.5. Devise an iteration scheme that starts at $f''(0) \approx 0.2$ and converges to the correct value. Print out $u/U = f'(\eta)$ and compare with Table 7.1.

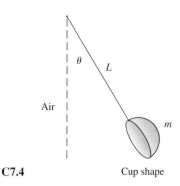

C7.4   Cup shape

Design Project

D7.1 It is desired to design a cup anemometer for wind speed, similar to Fig. P7.91, with a more sophisticated approach than the "average-torque" method of Prob. P7.91. The design should achieve an approximately linear relation between wind velocity and rotation rate in the range $20 < U < 40$ mi/h, and the anemometer should rotate at about 6 r/s at $U = 30$ mi/h. All specifications—cup diameter D, rod length L, rod diameter d, the bearing type, and all materials—are to be selected through your analysis. Make suitable assumptions about the instantaneous drag of the cups and rods at any given angle $\theta(t)$ of the system. Compute the instantaneous torque $T(t)$, and find and integrate the instantaneous angular acceleration of the device. Develop a complete theory for rotation rate versus wind speed in the range $0 < U < 50$ mi/h. Try to include actual commercial bearing friction properties.

References

1. H. Schlichting, *Boundary Layer Theory,* 7th ed., McGraw-Hill, New York, 1979.
2. F. M. White, *Viscous Fluid Flow,* 3d ed., McGraw-Hill, New York, 2005.
3. J. Cousteix, *Modeling and Computation of Boundary-Layer Flows,* 2d ed., Springer-Verlag, New York, 2005.
4. J. Blazek, *Computational Fluid Dynamics Principles*, 2d ed., Elsevier Science and Technology Books, New York, 2005.
5. V. V. Sychev et al., *Asymptotic Theory of Separated Flows,* Cambridge University Press, New York, 1998.
6. I. J. Sobey, *Introduction to Interactive Boundary Layer Theory,* Oxford University Press, New York, 2001.
7. T. von Kármán, "On Laminar and Turbulent Friction," *Z. Angew. Math. Mech.,* vol. 1, 1921, pp. 235–236.
8. G. B. Schubauer and H. K. Skramstad, "Laminar Boundary Layer Oscillations and Stability of Laminar Flow," *Natl. Bur. Stand. Res. Pap.* 1772, April 1943 (see also *J. Aero. Sci.,* vol. 14, 1947, pp. 69–78, and *NACA Rep.* 909, 1947).
9. P. S. Bernard and J. M. Wallace, *Turbulent Flow: Analysis, Measurement, and Prediction,* Wiley, New York, 2002.

10. P. W. Runstadler, Jr., et al., "Diffuser Data Book," Creare Inc., *Tech. Note* 186, Hanover, NH, May 1975.

11. B. Thwaites, "Approximate Calculation of the Laminar Boundary Layer," *Aeronaut. Q.,* vol. 1, 1949, pp. 245–280.

12. S. F. Hoerner, *Fluid Dynamic Drag,* published by the author, Midland Park, NJ, 1965.

13. J. D. Anderson, *Fundamentals of Aerodynamics,* 4th ed., McGraw-Hill, New York, 2007.

14. V. Tucker and G. C. Parrott, "Aerodynamics of Gliding Flight of Falcons and Other Birds," *J. Exp. Biol.,* vol. 52, 1970, pp. 345–368.

15. E. C. C. Tupper, *Introduction to Naval Architecture,* 4th ed., Elsevier, New York, 2004.

16. I. H. Abbott and A. E. von Doenhoff, *Theory of Wing Sections,* Dover, New York, 1981.

17. R. L. Kline and F. F. Fogleman, Airfoil for Aircraft, U.S. Patent 3,706,430 Dec. 19, 1972.

18. S. Childress, *Mechanics of Swimming and Flying,* Cambridge University Press, New York, 1981.

19. National Committee for Fluid Mechanics Films, *Illustrated Experiments in Fluid Mechanics,* M.I.T. Press, Cambridge, MA, 1972.

20. D. M. Bushnell and J. Hefner (Eds.), *Viscous Drag Reduction in Boundary Layers,* American Institute of Aeronautics & Astronautics, Reston, VA, 1990.

21. *Vehicle Aerodynamics: Wake Flows, Computational Fluid Dynamics, and Aerodynamic Testing,* SAE Special Publications, Int. Congress and Exposition, Detroit, MI, 1992.

22. W. H. Hucho, *Aerodynamics of Road Vehicles,* 4th ed., Soc. of Automotive Engineers, Warrendale, PA, 1998.

23. R. D. Blevins, *Applied Fluid Dynamics Handbook,* van Nostrand Reinhold, New York, 1984.

24. R. C. Johnson, Jr., G. E. Ramey, and D. S. O'Hagen, "Wind Induced Forces on Trees," *J. Fluids Eng.,* vol. 104, March 1983, pp. 25–30.

25. P. W. Bearman et al., "The Effect of a Moving Floor on Wind-Tunnel Simulation of Road Vehicles," Paper No. 880245, SAE Transactions, *J. Passenger Cars,* vol. 97, sec. 4, 1988, pp. 4.200–4.214.

26. *CRC Handbook of Tables for Applied Engineering Science,* 2d ed., CRC Press, Boca Raton, FL, 1973.

27. T. Inui, "Wavemaking Resistance of Ships," *Trans. Soc. Nav. Arch. Marine Engrs.,* vol. 70, 1962, pp. 283–326.

28. L. Larsson, "CFD in Ship Design—Prospects and Limitations," *Ship Technology Research,* vol. 44, no. 3, July 1997, pp. 133–154.

29. R. L. Street, G. Z. Watters, and J. K. Vennard, *Elementary Fluid Mechanics,* 7th ed., Wiley, New York, 1995.

30. P. H. Oosthuizen and W. Carscallen, *Compressible Fluid Flow,* McGraw-Hill, New York, 2003.

31. J. D. Anderson, Jr., *Hypersonic and High Temperature Gas Dynamics,* AIAA, Reston, VA, 2000.

32. J. Rom, *High Angle of Attack Aerodynamics: Subsonic, Transonic, and Supersonic Flows,* Springer-Verlag, New York, 1992.

33. S. Vogel, "Drag and Reconfiguration of Broad Leaves in High Winds," *J. Exp. Bot.,* vol. 40, no. 217, August 1989, pp. 941–948.

34. S. Vogel, *Life in Moving Fluids,* Princeton University Press 2d ed., Princeton, NJ, 1996.

35. J. A. C. Humphrey (ed.), *Proceedings 2d International Symposium on Mechanics of Plants, Animals, and Their Environment,* Engineering Foundation, New York, January 2000.

36. D. D. Joseph, R. Bai, K. P. Chen, and Y. Y. Renardy, "Core-Annular Flows," *Annu. Rev. Fluid Mech.,* vol. 29, 1997, pp. 65–90.

37. J. W. Hoyt and R. H. J. Sellin, "Scale Effects in Polymer Solution Pipe Flow," *Experiments in Fluids,* vol. 15, no. 1, June 1993, pp. 70–74.

38. S. Nakao, "Application of V-Shape Riblets to Pipe Flows," *J. Fluids Eng.,* vol. 113, December 1991, pp. 587–590.

39. P. R. Bandyopadhyay, "Review: Mean Flow in Turbulent Boundary Layers Disturbed to Alter Skin Friction," *J. Fluids Eng.,* vol. 108, 1986, pp. 127–140.

40. C. L. Merkle and S. Deutsch, "Microbubble Drag Reduction in Liquid Turbulent Boundary Layers," *Applied Mechanics Reviews,* vol. 45, no. 3 part 1, March 1992, pp. 103–127.

41. K. S. Choi and G. E. Karniadakis, "Mechanisms on Transverse Motions in Turbulent Wall Flows," *Annual Review of Fluid Mechanics,* vol. 35, 2003, pp. 45–62.

42. J. G. Chandler, "Microplanes," *Popular Science,* January 1998, pp. 54–59.

43. *Evolution of Flight,* Internet URL <http://www.flight100.org>.

44. J. D. Anderson Jr., *A History of Aerodynamics,* Cambridge University Press, New York, 1999.

45. Y. Tsuji, Y. Morikawa, and O. Mizuno, "Experimental Measurement of the Magnus Force on a Rotating Sphere at Low Reynolds Numbers," *Journal of Fluids Engineering,* vol. 107, 1985, pp. 484–488.

46. R. Clift, J. R. Grace, and M. E. Weber, *Bubbles, Drops and Particles,* Dover, NY, 2005.

47. M. Gad-el-Hak, "Flow Control: The Future," *Journal of Aircraft,* vol. 38, no. 3, 2001, pp. 402–418.

48. D. Geropp and H. J. Odenthal, "Drag Reduction of Motor Vehicles by Active Flow Control Using the Coanda Effect," *Experiments in Fluids,* vol. 28, no. 1, 2000, pp. 74–85.

49. Z. Zapryanov and S. Tabakova, *Dynamics of Bubbles, Drops, and Rigid Particles,* Kluwer Academic Pub., New York, 1998.

50. D. G. Karamanev, and L. N. Nikolov, "Freely Rising Spheres Do Not Obey Newton's Law for Free Settling," *AIChE Journal,* vol. 38, no. 1, Nov. 1992, pp. 1843–1846.

51. Katz J., *Race-Car Aerodynamics,* Robert Bentley Inc., Cambridge, MA, 1995.

52. A. S. Brown, "More than 12,000 Miles to the Gallon," *Mechanical Engineering,* January 2006, p. 64.

53. D. M. Bushnell, "Aircraft Drag Reduction: A Review," *Proceedings of the Institution of Mechanical Engineers, Part G: Journal of Aerospace Engineering,* vol. 217, no. 1, 2003, pp. 1–18.

54. D. B. Spalding, "A Single Formula for the Law of the Wall," *J. Appl. Mechanics,* vol. 28, no. 3, 1961, pp. 444–458.

Exact analytical streamlines for potential flow past a symmetric airfoil at an angle of attack. Perhaps the most glorious achievement of potential theory is the lift on an airfoil. Note the closely packed streamlines on the upper surface (high velocity, low pressure), and vice versa on the lower surface. The proper theory uses the *Kutta condition* (see Fig. 8.22), which requires the flow to leave the sharp trailing edge smoothly and parallel to the chord line. Potential theory does not account for boundary layer separation (stall) at high angles of incidence.

Chapter 8
Potential Flow
and Computational
Fluid Dynamics

Motivation. The basic partial differential equations of mass, momentum, and energy were discussed in Chap. 4. A few solutions were then given for incompressible *viscous* flow in Sec. 4.10. The viscous solutions were limited to simple geometries and unidirectional flows, where the difficult nonlinear convective terms were neglected. Potential flows are not limited by such nonlinear terms. Then, in Chap. 7, we found an approximation: patching *boundary layer flows* onto an outer inviscid flow pattern. For more complicated viscous flows, we found no theory or solutions, just experimental data.

The purposes of the present chapter are (1) to explore examples of potential theory and (2) to indicate some flows that can be approximated by computational fluid dynamics (CFD). The combination of these two gives us a good picture of incompressible-flow theory and its relation to experiment. One of the most important applications of potential-flow theory is to aerodynamics and marine hydrodynamics. First, however, we will review and extend the concepts of Ch. 4.

8.1 Introduction and Review

Figure 8.1 reminds us of the problems to be faced. A free stream approaches two closely spaced bodies, creating an "internal" flow between them and "external" flows above and below them. The fronts of the bodies are regions of favorable gradient (decreasing pressure along the surface), and the boundary layers will be attached and thin: Inviscid theory will give excellent results for the outer flow if $Re > 10^4$. For the internal flow between bodies, the boundary layers will grow and eventually meet, and the inviscid core vanishes. Inviscid theory works well in a "short" duct $L/D < 10$, such as the nozzle of a wind tunnel. For longer ducts we must estimate boundary layer growth and be cautious about using inviscid theory.

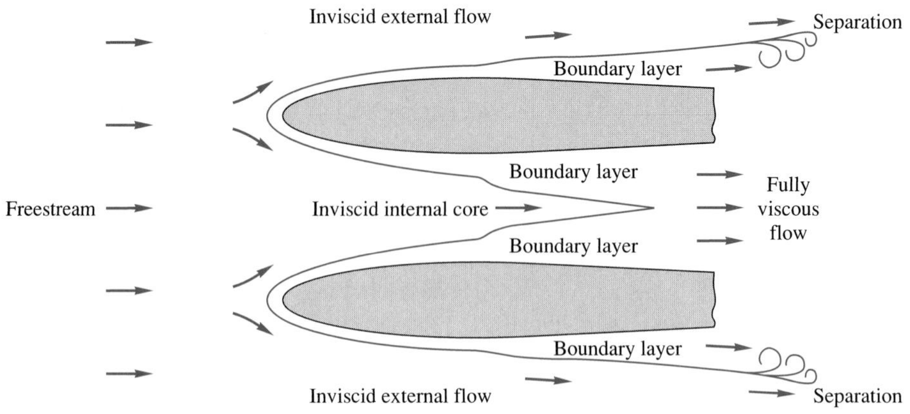

Fig. 8.1 Patching viscous and inviscid flow regions. Potential theory in this chapter does not apply to the boundary layer regions

For the external flows above and below the bodies in Fig. 8.1, inviscid theory should work well for the outer flows, until the surface pressure gradient becomes adverse (increasing pressure) and the boundary layer separates or stalls. After the separation point, boundary layer theory becomes inaccurate, and the outer flow streamlines are deflected and have a strong interaction with the viscous near-wall regions. The theoretical analysis of separated-flow regions is an active research area at present.

Review of Velocity Potential Concepts

Recall from Sec. 4.9 that if viscous effects are neglected, low-speed flows are irrotational, $\nabla \times \mathbf{V} = 0$, and the velocity potential ϕ exists, such that

$$\mathbf{V} = \nabla\phi \qquad \text{or} \qquad u = \frac{\partial\phi}{\partial x} \qquad v = \frac{\partial\phi}{\partial y} \qquad w = \frac{\partial\phi}{\partial z} \tag{8.1}$$

The continuity equation (4.73), $\nabla \cdot \mathbf{V} = 0$, reduces to Laplace's equation for ϕ:

$$\nabla^2\phi = \frac{\partial^2\phi}{\partial x^2} + \frac{\partial^2\phi}{\partial y^2} + \frac{\partial^2\phi}{\partial z^2} = 0 \tag{8.2}$$

and the momentum equation (4.74) reduces to Bernoulli's equation:

$$\frac{\partial\phi}{\partial t} + \frac{p}{\rho} + \frac{1}{2}V^2 + gz = \text{const} \quad \text{where } V = |\nabla\phi| \tag{8.3}$$

Typical boundary conditions are known free-stream conditions

Outer boundaries: $\qquad\qquad$ Known $\dfrac{\partial\phi}{\partial x}, \dfrac{\partial\phi}{\partial y}, \dfrac{\partial\phi}{\partial z}$ $\qquad\qquad$ (8.4)

and no velocity normal to the boundary at the body surface:

Solid surfaces: $\qquad \dfrac{\partial\phi}{\partial n} = 0 \qquad$ where n is perpendicular to body $\qquad$ (8.5)

Unlike the no-slip condition in viscous flow, here there is *no* condition on the tangential surface velocity $V_s = \partial\phi/\partial s$, where s is the coordinate along the surface. This velocity is determined as part of the solution to the problem.

Occasionally the problem involves a free surface, for which the boundary pressure is known and equal to p_a, usually a constant. The Bernoulli equation (8.3) then supplies a relation at the surface between V and the elevation z of the surface. For steady flow,

Free surface:
$$V^2 = |\boldsymbol{\nabla}\phi|^2 = \text{const} - 2gz_{\text{surf}} \tag{8.6}$$

It should be clear to the reader that this use of Laplace's equation, with known values of the derivative of ϕ along the boundaries, is much easier than a direct attack using the fully viscous Navier-Stokes equations. The analysis of Laplace's equation is very well developed and is termed *potential theory,* with whole books written about its application to fluid mechanics [1 to 4]. There are many analytical techniques, including superposition of elementary functions, conformal mapping [4], numerical finite differences [5], numerical finite elements [6], numerical boundary elements [7], and electric or mechanical analogs [8] that are now outdated. Having found $\phi(x, y, z, t)$ from such an analysis, we then compute $\mathbf{V}$ by direct differentiation in Eq. (8.1), after which we compute p from Eq. (8.3). The procedure is quite straightforward, and many interesting albeit idealized results can be obtained. A beautiful collection of computer-generated potential flow sketches is given by Kirchhoff [43].

Review of Stream Function Concepts

Recall from Sec. 4.7 that if a flow is described by only two coordinates, the stream function ψ also exists as an alternate approach. For plane incompressible flow in xy coordinates, the correct form is

$$u = \frac{\partial\psi}{\partial y} \qquad v = -\frac{\partial\psi}{\partial x} \tag{8.7}$$

The condition of irrotationality reduces to Laplace's equation for ψ also:

$$2\omega_z = 0 = \frac{\partial v}{\partial x} - \frac{\partial u}{\partial y} = \frac{\partial}{\partial x}\left(-\frac{\partial\psi}{\partial x}\right) - \frac{\partial}{\partial y}\left(\frac{\partial\psi}{\partial y}\right)$$

or
$$\boxed{\frac{\partial^2\psi}{\partial x^2} + \frac{\partial^2\psi}{\partial y^2} = 0} \tag{8.8}$$

The boundary conditions again are known velocity in the stream and no flow through any solid surface:

Free stream:
$$\text{Known } \frac{\partial\psi}{\partial x}, \frac{\partial\psi}{\partial y} \tag{8.9a}$$

Solid surface:
$$\psi_{\text{body}} = \text{const} \tag{8.9b}$$

Equation (8.9b) is particularly interesting because *any* line of constant ψ in a flow can therefore be interpreted as a body shape and may lead to interesting applications.

For the applications in this chapter, we may compute either ϕ or ψ or both, and the solution will be an *orthogonal flow net* as in Fig. 8.2. Once found, either set of lines may be considered the ϕ lines, and the other set will be the ψ lines. Both sets of lines are laplacian and could be useful.

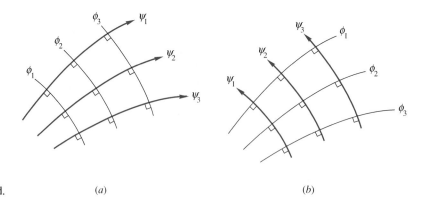

Fig. 8.2 Streamlines and potential lines are orthogonal and may reverse roles if results are useful: (*a*) typical inviscid flow pattern; (*b*) same as (*a*) with roles reversed.

(*a*)   (*b*)

Plane Polar Coordinates

Many solutions in this chapter are conveniently expressed in polar coordinates (r, θ). Both the velocity components and the differential relations for ϕ and ψ are then changed, as follows:

$$v_r = \frac{\partial \phi}{\partial r} = \frac{1}{r}\frac{\partial \psi}{\partial \theta} \qquad v_\theta = \frac{1}{r}\frac{\partial \phi}{\partial \theta} = -\frac{\partial \psi}{\partial r} \tag{8.10}$$

Laplace's equation takes the form

$$\frac{1}{r}\frac{\partial}{\partial r}\left(r\frac{\partial \phi}{\partial r}\right) + \frac{1}{r^2}\frac{\partial^2 \phi}{\partial \theta^2} = 0 \tag{8.11}$$

Exactly the same equation holds for the polar-coordinate form of $\psi(r, \theta)$.

An intriguing facet of potential flow with no free surface is that the governing equations (8.2) and (8.8) contain no parameters, nor do the boundary conditions. Therefore the solutions are purely geometric, depending only on the body shape, the free-stream orientation, and—surprisingly—the position of the rear stagnation point.[1] There is no Reynolds, Froude, or Mach number to complicate the dynamic similarity. Inviscid flows are kinematically similar without additional parameters—recall Fig. 5.6*a*.

8.2 Elementary Plane Flow Solutions

The present chapter is a detailed introductory study of inviscid incompressible flows, especially those that possess both a stream function and a velocity potential. Many solutions make use of the superposition principle, so we begin with the three elementary building blocks illustrated in Fig. 8.3: (*a*) a uniform stream in the *x* direction, (*b*) a line source or sink at the origin, and (*c*) a line vortex at the origin.

Uniform Stream in the *x* Direction

A uniform stream $\mathbf{V} = \mathbf{i}U$, as in Fig. 8.3*a*, possesses both a stream function and a velocity potential, which may be found as follows:

$$u = U = \frac{\partial \phi}{\partial x} = \frac{\partial \psi}{\partial y} \qquad v = 0 = \frac{\partial \phi}{\partial y} = -\frac{\partial \psi}{\partial x}$$

---

[1]The rear stagnation condition establishes the net amount of "circulation" about the body, giving rise to a lift force. Otherwise the solution could not be unique. See Sec. 8.4.

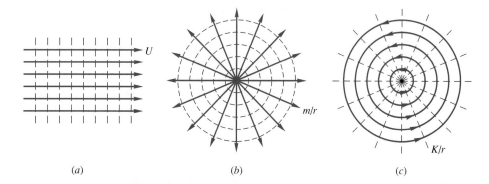

Fig. 8.3 Three elementary plane potential flows. Solid lines are streamlines; dashed lines are potential lines.

(a)   (b)   (c)

We may integrate each expression and discard the constants of integration, which do not affect the velocities in the flow. The results are

Uniform stream $\mathbf{i}U$: $\qquad\qquad \psi = Uy \qquad \phi = Ux$ $\qquad\qquad\qquad$ (8.12)

The streamlines are horizontal straight lines (y = const), and the potential lines are vertical (x = const)—that is, orthogonal to the streamlines, as expected.

Line Source or Sink at the Origin

Suppose that the z axis were a sort of thin pipe manifold through which fluid issued at total rate Q uniformly along its length b. Looking at the xy plane, we would see a cylindrical radial outflow or *line source,* as sketched in Fig. 8.3b. Plane polar coordinates are appropriate (see Fig. 4.2), and there is no circumferential velocity. At any radius r, the velocity is

$$v_r = \frac{Q}{2\pi rb} = \frac{m}{r} = \frac{1}{r}\frac{\partial \psi}{\partial \theta} = \frac{\partial \phi}{\partial r} \qquad v_\theta = 0 = -\frac{\partial \psi}{\partial r} = \frac{1}{r}\frac{\partial \phi}{\partial \theta}$$

where we have used the polar coordinate forms of the stream function and the velocity potential. Integrating and again discarding the constants of integration, we obtain the proper functions for this simple radial flow:

Line source or sink: $\qquad\qquad \psi = m\theta \qquad \phi = m \ln r$ $\qquad\qquad\qquad$ (8.13)

where $m = Q/(2\pi b)$ is a constant, positive for a source, negative for a sink. As shown in Fig. 8.3b, the streamlines are radial spokes (constant θ), and the potential lines are circles (constant r).

Line Irrotational Vortex

A (two-dimensional) line vortex is a purely circulating steady motion, $v_\theta = f(r)$ only, $v_r = 0$. This satisfies the continuity equation identically, as may be checked from Eq. (4.12b). We may also note that a variety of velocity distributions $v_\theta(r)$ satisfy the θ momentum equation of a viscous fluid, Eq. (D.6). We may show, as a problem exercise, that only one function $v_\theta(r)$ is *irrotational;* that is, curl $\mathbf{V} = 0$, and $v_\theta = K/r$, where K is a constant. This is sometimes called a *free vortex,* for which the stream function and velocity may be found:

$$v_r = 0 = \frac{1}{r}\frac{\partial \psi}{\partial \theta} = \frac{\partial \phi}{\partial r} \qquad v_\theta = \frac{K}{r} = -\frac{\partial \psi}{\partial r} = \frac{1}{r}\frac{\partial \phi}{\partial \theta}$$

We may again integrate to determine the appropriate functions:

$$\psi = -K \ln r \qquad \phi = K\theta \tag{8.14}$$

where K is a constant called the *strength* of the vortex. As shown in Fig. 8.3c, the stream-lines are circles (constant r), and the potential lines are radial spokes (constant θ). Note the similarity between Eqs. (8.13) and (8.14). A free vortex is a sort of reversed image of a source. The "bathtub vortex," formed when water drains through a bottom hole in a tank, is a good approximation to the free-vortex pattern.

Superposition: Source Plus an Equal Sink

Each of the three elementary flow patterns in Fig. 8.3 is an incompressible irrotational flow and therefore satisfies both plane "potential flow" equations $\nabla^2 \psi = 0$ and $\nabla^2 \phi = 0$. Since these are linear partial differential equations, any *sum* of such basic solutions is also a solution. Some of these composite solutions are quite interesting and useful.

For example, consider a source $+m$ at $(x, y) = (-a, 0)$, combined with a sink of equal strength $-m$, placed at $(+a, 0)$, as in Fig. 8.4. The resulting stream function is simply the sum of the two. In cartesian coordinates,

$$\psi = \psi_{\text{source}} + \psi_{\text{sink}} = m \tan^{-1} \frac{y}{x + a} - m \tan^{-1} \frac{y}{x - a}$$

Similarly, the composite velocity potential is

$$\phi = \phi_{\text{source}} + \phi_{\text{sink}} = \frac{1}{2} m \ln \left[(x + a)^2 + y^2 \right] - \frac{1}{2} m \ln \left[(x - a)^2 + y^2 \right]$$

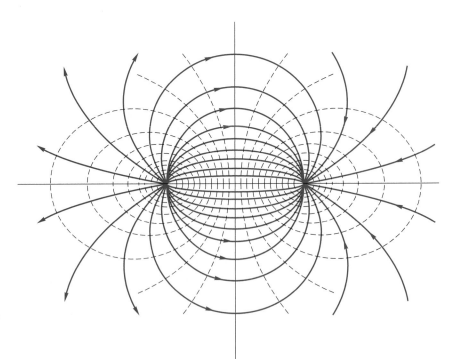

Fig. 8.4 Potential flow due to a line source plus an equal line sink, from Eq. (8.15). Solid lines are streamlines; dashed lines are potential lines.

By using trigonometric and logarithmic identities, these may be simplified to

Source plus sink:
$$\psi = -m \tan^{-1} \frac{2ay}{x^2 + y^2 - a^2}$$

$$\phi = \frac{1}{2} m \ln \frac{(x + a)^2 + y^2}{(x - a)^2 + y^2}$$

(8.15)

These lines are plotted in Fig. 8.4 and are seen to be two families of orthogonal circles, with the streamlines passing through the source and sink and the potential lines encircling them. They are harmonic (laplacian) functions that are exactly analogous in electromagnetic theory to the electric current and electric potential patterns of a magnet with poles at $(\pm a, 0)$.

Sink Plus a Vortex at the Origin

An interesting flow pattern, approximated in nature, occurs by superposition of a sink and a vortex, both centered at the origin. The composite stream function and velocity potential are

Sink plus vortex:
$$\psi = m\theta - K \ln r \qquad \phi = m \ln r + K\theta$$

(8.16)

When plotted, these form two orthogonal families of logarithmic spirals, as shown in Fig. 8.5. This is a fairly realistic simulation of a tornado (where the sink flow moves up the z axis into the atmosphere) or a rapidly draining bathtub vortex. At the center of a real (viscous) vortex, where Eq. (8.16) predicts infinite velocity, the actual circulating flow is highly *rotational* and approximates solid-body rotation $v_\theta \approx Cr$.

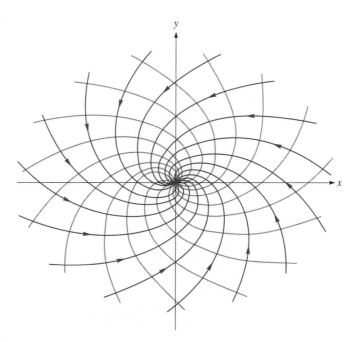

Fig. 8.5 Superposition of a sink plus a vortex, Eq. (8.16), simulates a tornado.

Uniform Stream Plus a Sink at the Origin: The Rankine Half-Body

If we superimpose a uniform x-directed stream against an isolated source, a half-body shape appears. If the source is at the origin, the combined stream function is, in polar coordinates,

Uniform stream plus source: $\psi = Ur \sin \theta + m\theta$ (8.17)

We can set this equal to various constants and plot the streamlines, as shown in Fig. 8.6. A curved, roughly elliptical, *half-body* shape appears, which separates the source flow from the stream flow. The body shape, which is named after the Scottish engineer W. J. M. Rankine (1820–1872), is formed by the particular streamlines $\psi = \pm\pi m$. The half-width of the body far downstream is $\pi m/U$. The upper surface may be plotted from the relation

$$r = \frac{m(\pi - \theta)}{U \sin \theta}$$ (8.18)

It is not a true ellipse. The nose of the body, which is a "stagnation" point where $V = 0$, stands at $(x, y) = (-a, 0)$, where $a = m/U$. The streamline $\psi = 0$ also crosses this point—recall that streamlines can cross only at a stagnation point.

The cartesian velocity components are found by differentiation:

$$u = \frac{\partial \psi}{\partial y} = U + \frac{m}{r}\cos\theta \qquad v = -\frac{\partial \psi}{\partial x} = \frac{m}{r}\sin\theta$$ (8.19)

Setting $u = v = 0$, we find a single stagnation point at $\theta = 180°$ and $r = m/U$, or $(x, y) = (-m/U, 0)$, as stated. The resultant velocity at any point is

$$V^2 = u^2 + v^2 = U^2\left(1 + \frac{a^2}{r^2} + \frac{2a}{r}\cos\theta\right)$$ (8.20)

where we have substituted $m = Ua$. If we evaluate the velocities along the upper surface $\psi = \pi m$, we find a maximum value $U_{s,\max} \approx 1.26U$ at $\theta = 63°$. This point is labeled in Fig. 8.6 and, by Bernoulli's equation, is the point of minimum pressure on the body surface. After this point, the surface flow decelerates, the pressure rises, and the viscous layer grows thicker and more susceptible to "flow separation," as we saw in Chap. 7.

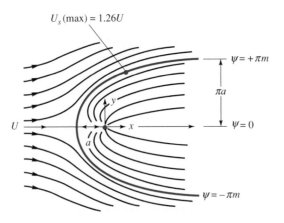

Fig. 8.6 Superposition of a source plus a uniform stream forms a Rankine half-body.

EXAMPLE 8.1

The bottom of a river has a 4-m-high bump that approximates a Rankine half-body, as in Fig. E8.1. The pressure at point B on the bottom is 130 kPa, and the river velocity is 2.5 m/s. Use inviscid theory to estimate the water pressure at point A on the bump, which is 2 m above point B.

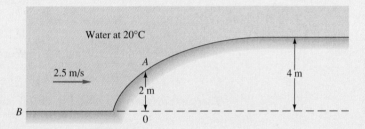

Water at 20°C

2.5 m/s

A

2 m

4 m

E8.1 B 0

Solution

As in all inviscid theories, we ignore the low-velocity boundary layers that form on solid surfaces due to the no-slip condition. From Eq. (8.18) and Fig. 8.6, the downstream bump half-height equals πa. Therefore, for our case, $a = (4 \text{ m})/\pi = 1.27$ m. We have to find the spot where the bump height is half that much, $h = 2 \text{ m} = \pi a/2$. From Eq. (8.18) we may compute

$$r = h_A = \frac{a(\pi - \theta)}{\sin \theta} = \frac{\pi}{2}a \qquad \text{or} \qquad \theta = \frac{\pi}{2} = 90°$$

Thus point A in Fig. E8.1 is directly above the (initially unknown) origin of coordinates (labeled O in Fig. E8.1) and is 1.27 m to the right of the nose of the bump. With $r = \pi a/2$ and $\theta = \pi/2$ known, we compute the velocity at point A from Eq. (8.20):

$$V_A^2 = U^2\left[1 + \frac{a^2}{(\pi a/2)^2} + \frac{2a}{\pi a/2}\cos\frac{\pi}{2}\right] = 1.405U^2$$

or $V_A \approx 1.185U = 1.185(2.5 \text{ m/s}) = 2.96$ m/s

For water at 20°C, take $\rho = 998$ kg/m² and $\gamma = 9790$ N/m³. Now, since the velocity and elevation are known at point A, we are in a position to use Bernoulli's inviscid, incompressible flow equation (4.120) to estimate p_A from the known properties at point B (on the same streamline):

$$\frac{p_A}{\gamma} + \frac{V_A^2}{2g} + z_A \approx \frac{p_B}{\gamma} + \frac{V_B^2}{2g} + z_B$$

or $$\frac{p_A}{9790 \text{ N/m}^3} + \frac{(2.96 \text{ m/s})^2}{2(9.81 \text{ m/s}^2)} + 2 \text{ m} \approx \frac{130,000}{9790} + \frac{(2.5)^2}{2(9.81)} + 0$$

Solving, we find

$$p_A = (13.60 - 2.45)(9790) \approx 109,200 \text{ Pa} \qquad\qquad Ans.$$

If the approach velocity is uniform, this should be a pretty good approximation, since water is relatively inviscid and its boundary layers are thin.

Uniform Stream at an Angle α

If the uniform stream is written in plane polar coordinates, it becomes

Uniform stream $\mathbf{i}U$:
$$\psi = Ur \sin \theta \qquad \phi = Ur \cos \theta \qquad (8.21)$$

This makes it easier to superimpose, say, a stream and a source or vortex by using the same coordinates. If the uniform stream is moving at angle α with respect to the x axis—that is,

$$u = U \cos \alpha = \frac{\partial \psi}{\partial y} = \frac{\partial \phi}{\partial x} \qquad v = U \sin \alpha = -\frac{\partial \psi}{\partial x} = \frac{\partial \phi}{\partial y}$$

then by integration we obtain the correct functions for flow at an angle:

$$\psi = U(y \cos \alpha - x \sin \alpha) \qquad \phi = U(x \cos \alpha + y \sin \alpha) \qquad (8.22)$$

These expressions are useful in airfoil angle-of-attack problems (Sec. 8.7).

Circulation

The line vortex flow is irrotational everywhere except at the origin, where the vorticity $\nabla \times \mathbf{V}$ is infinite. This means that a certain line integral called the *fluid circulation* Γ does not vanish when taken around a vortex center.

With reference to Fig. 8.7, the circulation is defined as the counterclockwise line integral, around a closed curve C, of arc length ds times the velocity component tangent to the curve:

$$\Gamma = \oint_C V \cos \alpha \, ds = \int_C \mathbf{V} \cdot d\mathbf{s} = \int_C (u \, dx + v \, dy + w \, dz) \qquad (8.23)$$

From the definition of ϕ, $\mathbf{V} \cdot d\mathbf{s} = \nabla \phi \cdot d\mathbf{s} = d\phi$ for an irrotational flow; hence normally Γ in an irrotational flow would equal the final value of ϕ minus the initial value of ϕ. Since we start and end at the same point, we compute $\Gamma = 0$, but not for vortex flow: With $\phi = K\theta$ from Eq. (8.14) there is a change in ϕ of amount $2\pi K$ as we make one complete circle:

Path enclosing a vortex:
$$\Gamma = 2\pi K \qquad (8.24)$$

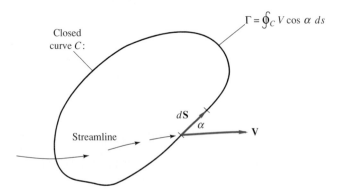

Fig. 8.7 Definition of the fluid circulation Γ.

Alternatively the calculation can be made by defining a circular path of radius r around the vortex center, from Eq. (8.23):

$$\Gamma = \int_C v_\theta \, ds = \int_0^{2\pi} \frac{K}{r} r \, d\theta = 2\pi K$$

In general, Γ denotes the net algebraic strength of all the vortex filaments contained within the closed curve. In the next section we shall see that a region of finite circulation within a flowing stream will be subjected to a lift force proportional to both U_∞ and Γ.

One can show, by using Eq. (8.23), that a source or sink creates no circulation. If there are no vortices present, the circulation will be zero for any path enclosing any number of sources and sinks.

8.3 Superposition of Plane Flow Solutions

We can now form a variety of interesting potential flows by summing the velocity potential and stream functions of a uniform stream, source or sink, and vortex. Most of the results are classic, of course, needing only a brief treatment here. Superposition is valid because the basic equations, (8.2) and (8.8), are linear.

Graphical Method of Superposition

A simple means of accomplishing $\psi_{\text{tot}} = \Sigma \, \psi_i$ graphically is to plot the individual stream functions separately and then look at their intersections. The value of ψ_{tot} at each intersection is the sum of the individual values ψ_i that cross there. Connecting intersections with the same value of ψ_{tot} creates the desired superimposed flow streamlines.

A simple example is shown in Fig. 8.8, summing two families of streamlines ψ_a and ψ_b. The individual components are plotted separately, and four typical intersections are shown. Dashed lines are then drawn through intersections representing the same sum of $\psi_a + \psi_b$. These dashed lines are the desired solution. Often this graphical method is a quick means of evaluating the proposed superposition before a full-blown numerical plot routine is executed.

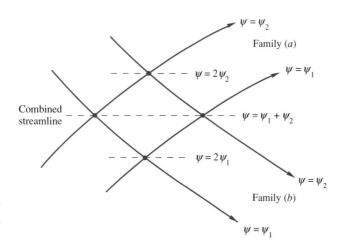

Fig. 8.8 Intersections of elementary streamlines can be joined to form a combined streamline.

Boundary Layer Separation on a Half-Body

Although the inviscid flow patterns seen in Fig. 8.9a and c are mirror images, their viscous (boundary layer) behavior is different. The body shape and the velocity along the surface are

$$V^2 = U_\infty^2 \left(1 + \frac{a^2}{r^2} + \frac{2a}{r} \cos \theta \right) \quad \text{along} \quad r = \frac{m(\pi - \theta)}{U_\infty \sin \theta} \quad (8.25)$$

The computed surface velocities are plotted along the half-body contours in Fig. 8.9b and d as a function of arc length s/a measured from the stagnation point. These plots are also mirror images. However, if the nose is in front, Fig. 8.9b, the pressure gradient there is *favorable* (decreasing pressure along the surface). In contrast, the pressure gradient is *adverse* (increasing pressure along the surface) when the nose is in the rear, Fig. 8.9d, and boundary layer separation may occur.

Application to Fig. 8.9b of Thwaites's laminar boundary method from Eqs. (7.54) and (7.56) reveals that separation does not occur on the front nose of the half-body. Therefore Fig. 8.9a is a very realistic picture of streamlines past a half-body nose. In contrast, when applied to the tail, Fig. 8.9c, Thwaites's method predicts separation at about $s/a \approx -2.2$, or $\theta \approx 110°$. Thus, if a half-body is a solid surface, Fig. 8.9c is *not* realistic and a broad separated wake will form. However, if the half-body tail is a *fluid line* separating the sink-directed flow from the outer stream, as in Example 8.2, then Fig. 8.9c is quite realistic and useful. Computations for turbulent boundary layer theory would be similar: separation on the tail, no separation on the nose.

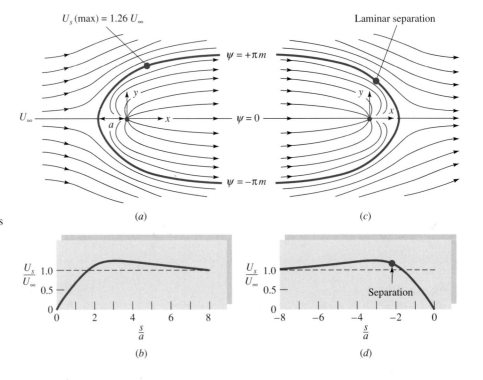

Fig. 8.9 The Rankine half-body; pattern (c) is not found in a real fluid because of boundary layer separation. (a) Uniform stream plus a source equals a half-body; stagnation point at $x = -a = -m/U_\infty$. (b) Slight adverse gradient for s/a greater than 3.0: no separation. (c) Uniform stream plus a sink equals the rear of a half-body; stagnation point at $x = a = m/U_\infty$. (d) Strong adverse gradient for $s/a > -3.0$: separation.

EXAMPLE 8.2

An offshore power plant cooling-water intake sucks in 1500 ft³/s in water 30 ft deep, as in Fig. E8.2. If the tidal velocity approaching the intake is 0.7 ft/s, (a) how far downstream does the intake effect extend and (b) how much width L of tidal flow is entrained into the intake?

Solution

Recall from Eq. (8.13) that the sink strength m is related to the volume flow Q and the depth b into the paper:

$$m = \frac{Q}{2\pi b} = \frac{1500 \text{ ft}^3/\text{s}}{2\pi(30 \text{ ft})} = 7.96 \text{ ft}^2/\text{s}$$

Therefore from Fig. 8.9 the desired lengths a and L are

$$a = \frac{m}{U_\infty} = \frac{7.96 \text{ ft}^2/\text{s}}{0.7 \text{ ft/s}} = 11.4 \text{ ft} \qquad \qquad \textit{Ans. (a)}$$

$$L = 2\pi a = 2\pi(11.4 \text{ ft}) = 71 \text{ ft} \qquad \qquad \textit{Ans. (b)}$$

E8.2 *(margin figure)*
Half-body shape
Intake   a?
L?
1500 ft³/s
0.7 ft/s   Top view

Flow Past a Vortex

Consider a uniform stream U_∞ in the x direction flowing past a vortex of strength K with center at the origin. By superposition the combined stream function is

$$\psi = \psi_{\text{stream}} + \psi_{\text{vortex}} = U_\infty r \sin \theta - K \ln r \qquad (8.26)$$

The velocity components are given by

$$v_r = \frac{1}{r}\frac{\partial \psi}{\partial \theta} = U_\infty \cos \theta \qquad v_\theta = -\frac{\partial \psi}{\partial r} = -U_\infty \sin \theta + \frac{K}{r} \qquad (8.27)$$

The streamlines are plotted in Fig. 8.10 by the graphical method, intersecting the circular streamlines of the vortex with the horizontal lines of the uniform stream.

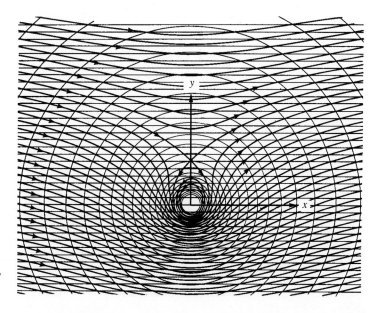

Fig. 8.10 Flow of a uniform stream past a vortex constructed by the graphical method.

By setting $v_r = v_\theta = 0$ from (8.27) we find a stagnation point at $\theta = 90°$, $r = a = K/U_\infty$, or $(x, y) = (0, a)$. This is where the counterclockwise vortex velocity K/r exactly cancels the stream velocity U_∞.

Probably the most interesting thing about this example is that there is a nonzero lift force normal to the stream on the surface of any region enclosing the vortex, but we postpone this discussion until the next section.

An Infinite Row of Vortices

Consider an infinite row of vortices of equal strength K and equal spacing a, as in Fig. 8.11a. This case is included here to illustrate the interesting concept of a *vortex sheet*.

From Eq. (8.14), the ith vortex in Fig. 8.11a has a stream function $\psi_i = -K \ln r_i$, so that the total infinite row has a combined stream function

$$\psi = -K \sum_{i=1}^{\infty} \ln r_i$$

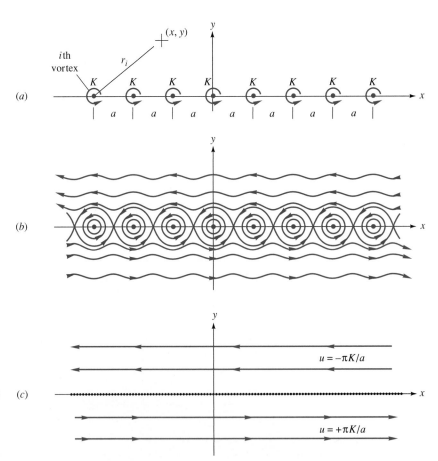

Fig. 8.11 Superposition of vortices: (a) an infinite row of equal strength; (b) streamline pattern for part (a); (c) vortex sheet: part (b) viewed from afar.

It can be shown [2, Sec. 4.51] that this infinite sum of logarithms is equivalent to a closed-form function:

$$\psi = -\tfrac{1}{2}K \ln\left[\frac{1}{2}\left(\cosh\frac{2\pi y}{a} - \cos\frac{2\pi x}{a}\right)\right] \tag{8.28}$$

Since the proof uses the complex variable $z = x + iy$, $i = (-1)^{1/2}$, we are not going to show the details here.

The streamlines from Eq. (8.28) are plotted in Fig. 8.11b, showing what is called a *cat's-eye* pattern of enclosed flow cells surrounding the individual vortices. Above the cat's eyes the flow is entirely to the left, and below the cat's eyes the flow is to the right. Moreover, these left and right flows are uniform if $|y| \gg a$, which follows by differentiating Eq. (8.28):

$$u = \frac{\partial\psi}{\partial y}\bigg|_{|y|\gg a} = \pm\frac{\pi K}{a}$$

where the plus sign applies below the row and the minus sign above the row. This uniform left and right streaming is sketched in Fig. 8.11c. We stress that this effect is induced by the row of vortices: There is no uniform stream approaching the row in this example.

The Vortex Sheet

When Fig. 8.11b is viewed from afar, the streaming motion is uniform left above and uniform right below, as in Fig. 8.11c, and the vortices are packed so closely together that they are smudged into a continuous *vortex sheet*. The strength of the sheet is defined as

$$\gamma = \frac{2\pi K}{a} \tag{8.29}$$

and in the general case γ can vary with x. The circulation about any closed curve that encloses a short length dx of the sheet would be, from Eqs. (8.23) and (8.29),

$$d\Gamma = u_l\, dx - u_u\, dx = (u_l - u_u)\, dx = \frac{2\pi K}{a}\, dx = \gamma\, dx \tag{8.30}$$

where the subscripts l and u stand for lower and upper, respectively. Thus the sheet strength $\gamma = d\Gamma/dx$ is the circulation per unit length of the sheet. Thus when a vortex sheet is immersed in a uniform stream, γ is proportional to the lift per unit length of any surface enclosing the sheet.

Note that there is no velocity normal to the sheet at the sheet surface. Therefore a vortex sheet can simulate a thin-body shape, like a plate or thin airfoil. This is the basis of the thin airfoil theory mentioned in Sec. 8.7.

The Doublet

As we move far away from the source–sink pair of Fig. 8.4, the flow pattern begins to resemble a family of circles tangent to the origin, as in Fig. 8.12. This limit of vanishingly small distance a is called a *doublet*. To keep the flow strength large enough to exhibit decent velocities as a becomes small, we specify that the product

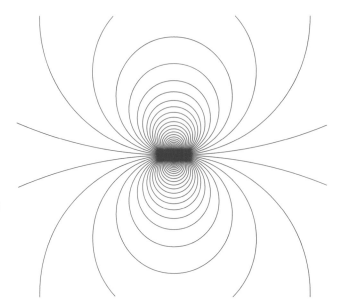

Fig. 8.12 A doublet, or source–sink pair, is the limiting case of Fig. 4.13 viewed from afar. Streamlines are circles tangent to the *x* axis at the origin. This figure was prepared using the *contour* feature of MATLAB [34, 35].

2*am* remain constant. Let us call this constant λ. Then the stream function of a doublet is

$$\psi = \lim_{\substack{a \to 0 \\ 2am = \lambda}} \left(-m \tan^{-1} \frac{2ay}{x^2 + y^2 - a^2} \right) = -\frac{2amy}{x^2 + y^2} = -\frac{\lambda y}{x^2 + y^2} \qquad (8.31)$$

We have used the fact that $\tan^{-1} \alpha \approx \alpha$ as α becomes small. The quantity λ is called the *strength* of the doublet.

Equation (8.31) can be rearranged to yield

$$x^2 + \left(y + \frac{\lambda}{2\psi} \right)^2 = \left(\frac{\lambda}{2\psi} \right)^2$$

so that, as advertised, the streamlines are circles tangent to the origin with centers on the *y* axis. This pattern is sketched in Fig. 8.12.

Although the author has in the past laboriously sketched streamlines by hand, this is no longer necessary. Figure 8.12 was computer-drawn, using the *contour* feature of the student version of MATLAB [34]. Simply set up a grid of points, spell out the stream function, and call for a contour. For Fig. 8.12, the actual statements were

```
[X,Y] = meshgrid(-1:.02:1);

PSI = -Y./(X.^2 + Y.^2);

contour(X,Y,PSI,100)
```

This would produce 100 contour lines of ψ from Eq. (8.31), with $\lambda = 1$ for convenience. The plot would include grid lines, scale markings, and a surrounding box, and the circles might look a bit elliptical. These blemishes can be eliminated with three statements of cosmetic improvement:

```
axis square

grid off

axis off
```

The final plot, Fig. 8.12, has no markings but the streamlines themselves. MATLAB is thus a recommended tool and, in addition, has scores of other uses. All this chapter's problem assignments that call for "sketch the streamlines/potential lines" can be completed using this contour feature. For further details, consult Ref. 34.

In a similar manner the velocity potential of a doublet is found by taking the limit of Eq. (8.15) as $a \to 0$ and $2am = \lambda$:

$$\phi_{\text{doublet}} = \frac{\lambda x}{x^2 + y^2}$$

or

$$\left(x - \frac{\lambda}{2\phi} \right)^2 + y^2 = \left(\frac{\lambda}{2\phi} \right)^2 \tag{8.32}$$

The potential lines are circles tangent to the origin with centers on the x axis. Simply turn Fig. 8.12 clockwise $90°$ to visualize the ϕ lines, which are everywhere normal to the streamlines.

The doublet functions can also be written in polar coordinates:

$$\psi = -\frac{\lambda \sin \theta}{r} \qquad \phi = \frac{\lambda \cos \theta}{r} \tag{8.33}$$

These forms are convenient for the cylinder flows of the next section.

8.4 Plane Flow Past Closed-Body Shapes

A variety of closed-body external flows can be constructed by superimposing a uniform stream with sources, sinks, and vortices. The body shape will be closed only if the net source outflow equals the net sink inflow.

The Rankine Oval

A cylindrical shape called a *Rankine oval,* which is long compared with its height, is formed by a source–sink pair aligned parallel to a uniform stream, as in Fig. 8.13a.

From Eqs. (8.12) and (8.15) the combined stream function is

$$\psi = U_\infty y - m \tan^{-1} \frac{2ay}{x^2 + y^2 - a^2} = U_\infty r \sin \theta + m(\theta_1 - \theta_2) \tag{8.34}$$

When streamlines of constant ψ are plotted from Eq. (8.34), an oval body shape appears, as in Fig. 8.13b. The half-length L and half-height h of the oval depend on the relative strength of source and stream—that is, the ratio $m/(U_\infty a)$, which equals 1.0 in Fig. 8.13b. The circulating streamlines inside the oval are uninteresting and not usually shown. The oval is the line $\psi = 0$.

There are stagnation points at the front and rear, $x = \pm L$, and points of maximum velocity and minimum pressure at the shoulders, $y = \pm h$, of the oval. All these

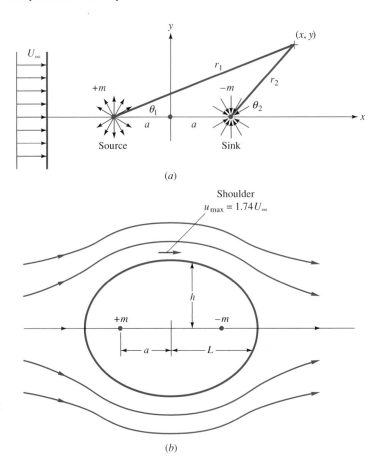

Fig. 8.13 Flow past a Rankine oval:
(*a*) uniform stream plus a source–
sink pair; (*b*) oval shape and
streamlines for $m/(U_\infty a) = 1.0$.

parameters are a function of the basic dimensionless parameter $m/(U_\infty a)$, which we can determine from Eq. (8.34):

$$\frac{h}{a} = \cot \frac{h/a}{2m/(U_\infty a)} \qquad \frac{L}{a} = \left(1 + \frac{2m}{U_\infty a}\right)^{1/2}$$

$$\frac{u_{max}}{U_\infty} = 1 + \frac{2m/(U_\infty a)}{1 + h^2/a^2}$$

(8.35)

As we increase $m/(U_\infty a)$ from zero to large values, the oval shape increases in size and thickness from a flat plate of length $2a$ to a huge, nearly circular cylinder. This is shown in Table 8.1. In the limit as $m/(U_\infty a) \to \infty$, $L/h \to 1.0$ and $u_{max}/U_\infty \to 2.0$, which is equivalent to flow past a circular cylinder.

All the Rankine ovals except very thin ones have a large adverse pressure gradient on their leeward surface. Thus boundary layer separation will occur in the rear with a broad wake flow, and the inviscid pattern is unrealistic in that region.

**Flow Past a Circular Cylinder
with Circulation**

From Table 8.1 at large source strength the Rankine oval becomes a large circle, much greater in diameter than the source–sink spacing $2a$. Viewed on the scale of

Table 8.1 Rankine Oval Parameters from Eq. (8.30)

$m/(U_\infty a)$	h/a	L/a	L/h	u_{max}/U_∞
0.0	0.0	1.0	∞	1.0
0.01	0.031	1.010	32.79	1.020
0.1	0.263	1.095	4.169	1.187
1.0	1.307	1.732	1.326	1.739
10.0	4.435	4.583	1.033	1.968
100.0	14.130	14.177	1.003	1.997
∞	∞	∞	1.000	2.000

the cylinder, this is equivalent to a uniform stream plus a doublet. We also throw in a vortex at the doublet center, which does not change the shape of the cylinder.

Thus the stream function for flow past a circular cylinder with circulation, centered at the origin, is a uniform stream plus a doublet plus a vortex:

$$\psi = U_\infty r \sin\theta - \frac{\lambda \sin\theta}{r} - K \ln r + \text{const} \tag{8.36}$$

The doublet strength λ has units of velocity times length squared. For convenience, let $\lambda = U_\infty a^2$, where a is a length, and let the arbitrary constant in Eq. (8.36) equal $K \ln a$. Then the stream function becomes

$$\psi = U_\infty \sin\theta\left(r - \frac{a^2}{r}\right) - K \ln\frac{r}{a} \tag{8.37}$$

The streamlines are plotted in Fig. 8.14 for four different values of the dimensionless vortex strength $K/(U_\infty a)$. For all cases the line $\psi = 0$ corresponds to the circle $r = a$—that is, the shape of the cylindrical body. As circulation $\Gamma = 2\pi K$ increases, the velocity becomes faster and faster below the cylinder and slower and slower above it. The velocity components in the flow are given by

$$v_r = \frac{1}{r}\frac{\partial\psi}{\partial\theta} = U_\infty \cos\theta\left(1 - \frac{a^2}{r^2}\right)$$

$$v_\theta = -\frac{\partial\psi}{\partial r} = -U_\infty \sin\theta\left(1 + \frac{a^2}{r^2}\right) + \frac{K}{r} \tag{8.38}$$

The velocity at the cylinder surface $r = a$ is purely tangential, as expected:

$$v_r(r = a) = 0 \quad v_\theta(r = a) = -2U_\infty \sin\theta + \frac{K}{a} \tag{8.39}$$

For small K, two stagnation points appear on the surface at angles θ_s where $v_\theta = 0$; or, from Eq. (8.39),

$$\sin\theta_s = \frac{K}{2U_\infty a} \tag{8.40}$$

Figure 8.14a is for $K = 0$, $\theta_s = 0$ and 180°, or doubly symmetric inviscid flow past a cylinder with no circulation. Figure 8.14b is for $K/(U_\infty a) = 1$, $\theta_s = 30$ and 150°; and

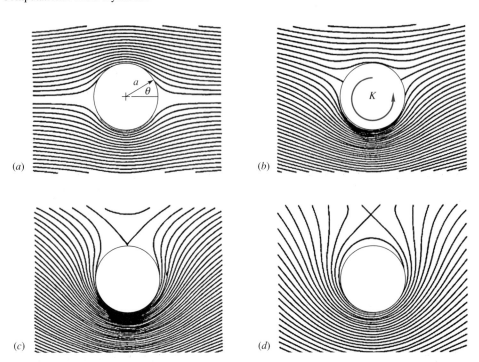

Fig. 8.14 Flow past a circular cylinder with circulation for values of $K/(U_\infty a)$ of (a) 0, (b) 1.0, (c) 2.0, and (d) 3.0.

Fig. 8.14c is the limiting case where the two stagnation points meet at the top, $K/(U_\infty a) = 2$, $\theta_s = 90°$.

For $K > 2U_\infty a$, Eq. (8.40) is invalid, and the single stagnation point is above the cylinder, as in Fig. 8.14d, at a point $y = h$ given by

$$\frac{h}{a} = \frac{1}{2}[\beta + (\beta^2 - 4)^{1/2}] \qquad \beta = \frac{K}{U_\infty a} > 2$$

In Fig. 8.14d, $K/(U_\infty a) = 3.0$, and $h/a = 2.6$.

The Kutta-Joukowski Lift Theorem

The cylinder flows with circulation, Figs. 8.14b to d, develop an inviscid downward *lift* normal to the free stream, called the *Magnus-Robins force*. This lift is proportional to stream velocity and vortex strength. Its discovery, by experiment, has long been attributed to the German physicist Gustav Magnus, who observed it in 1853. It is now known [40, 45] that the brilliant British engineer Benjamin Robins first reported a lift force on a spinning body in 1761. We see from the streamline patterns that the velocity on top of the cylinder is less, and, thus, from Bernoulli's equation, the pressure is higher. On the bottom, we see tightly packed streamlines, high velocity, and low pressure; viscosity is neglected. Inviscid theory predicts this force.

The surface velocity is given by Eq. (8.39). From Bernoulli's equation (8.3), neglecting gravity, the surface pressure p_s is given by

$$p_\infty + \frac{1}{2}\rho U_\infty^2 = p_s + \frac{1}{2}\rho\left(-2U_\infty \sin\theta + \frac{K}{a}\right)^2$$

or $\qquad p_s = p_\infty + \tfrac{1}{2}\rho U_\infty^2(1 - 4\sin^2\theta + 4\beta\sin\theta - \beta^2)$ $\qquad\qquad$ (8.41) where $\beta = K/(U_\infty a)$ and p_∞ is the free-stream pressure. If b is the cylinder depth into the paper, the drag D is the integral over the surface of the horizontal component of pressure force:

$$D = -\int_0^{2\pi} (p_s - p_\infty) \cos\theta \, ba \, d\theta$$

where $p_s - p_\infty$ is substituted from Eq. (8.41). But the integral of $\cos\theta$ times any power of $\sin\theta$ over a full cycle 2π is identically zero. Thus we obtain the (perhaps surprising) result

$$D(\text{cylinder with circulation}) = 0 \qquad (8.42)$$

This is a special case of d'Alembert's paradox, mentioned in Sec. 1.2:

> According to inviscid theory, the drag of any body of any shape immersed in a uniform stream is identically zero.

D'Alembert published this result in 1752 and pointed out himself that it did not square with the facts for real fluid flows. This unfortunate paradox caused everyone to over-react and reject all inviscid theory until 1904, when Prandtl first pointed out the profound effect of the thin viscous boundary layer on the flow pattern in the rear, as in Fig. 7.2b, for example.

The lift force L normal to the stream, taken positive upward, is given by summation of vertical pressure forces:

$$L = -\int_0^{2\pi} (p_s - p_\infty) \sin\theta \, ba \, d\theta$$

Since the integral over 2π of any odd power of $\sin\theta$ is zero, only the third term in the parentheses in Eq. (8.41) contributes to the lift:

$$L = -\frac{1}{2}\rho U_\infty^2 \frac{4K}{aU_\infty} ba \int_0^{2\pi} \sin^2\theta \, d\theta = -\rho U_\infty(2\pi K)b$$

or

$$\boxed{\frac{L}{b} = -\rho U_\infty \Gamma} \qquad (8.43)$$

Notice that the lift seems independent of the radius a of the cylinder. Actually, though, as we shall see in Sec. 8.7, the circulation Γ depends on the body size and orientation through a physical requirement.

Equation (8.43) was generalized by W. M. Kutta in 1902 and independently by N. Joukowski in 1906 as follows:

> According to inviscid theory, the lift per unit depth of any cylinder of any shape immersed in a uniform stream equals $\rho u_\infty \Gamma$, where Γ is the total net circulation contained within the body shape. The direction of the lift is 90° from the stream direction, rotating opposite to the circulation.

The problem in airfoil analysis, Sec. 8.7, is thus to determine the circulation Γ as a function of airfoil shape and orientation.

Lift and Drag of Rotating Cylinders[2]

The flows in Fig. 8.14 are mathematical: a doublet plus a vortex plus a uniform stream. The physical realization could be a rotating cylinder in a free stream. The no-slip condition would cause the fluid in contact with the cylinder to move tangentially at velocity $v_\theta = a\omega$, setting up a net circulation Γ. Measurement of forces on a spinning cylinder is very difficult, and no reliable drag data are known to the author. However, Tokumaru and Dimotakis [22] used a clever auxiliary scheme to measure lift forces at $Re_D = 3800$.

Figure 8.15 shows lift and drag coefficients, based on frontal area ($2ab$), for a rotating cylinder at $Re_D = 3800$. The drag curve is from CFD calculations [41]. Reported CFD drag results, from several different authors, are quite controversial because they do not agree, even qualitatively. The writer feels that Ref. 41 gives the most reliable results. Note that the experimental C_L increases to a value of 15.3 at $a\omega/U_\infty = 10$. This contradicts an early theory of Prandtl, in 1926, that the maximum possible value of C_L would be $4\pi \approx 12.6$, corresponding to the flow conditions in Fig. 8.14c. The inviscid theory for lift would be:

$$C_L = \frac{L}{\frac{1}{2}\rho U_\infty^2 (2ba)} = \frac{2\pi\rho U_\infty K b}{\rho U_\infty^2 ba} = \frac{2\pi v_{\theta s}}{U_\infty} \tag{8.44}$$

where $v_{\theta s} = K/a$ is the peripheral speed of the cylinder.

Figure 8.15 shows that the theoretical lift from Eq. (8.44) is much too high, but the measured lift is quite respectable, much larger in fact than a typical airfoil of the same chord length, as in Fig. 7.25. Thus rotating cylinders have practical possibilities. The Flettner rotor ship built in Germany in 1924 employed rotating vertical cylinders that developed a thrust due to any winds blowing past the ship. The Flettner design did not gain any popularity, but such inventions may be more attractive in this era of high energy costs.

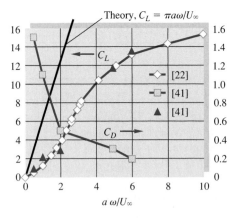

Fig. 8.15 Drag and lift of a rotating cylinder of large aspect ratio at $Re_D = 3800$, after Tokumaru and Dimotakis [22] and Sengupta et al. [41].

[2]The writer is indebted to Prof. T. K. Sengupta of I.I.T. Kanpur for data and discussion for this subsection.

EXAMPLE 8.3

The experimental Flettner rotor sailboat at the University of Rhode Island is shown in Fig. E8.3. The rotor is 2.5 ft in diameter and 10 ft long and rotates at 220 r/min. It is driven by a small lawnmower engine. If the wind is a steady 10 kn and boat relative motion is neglected, what is the maximum thrust expected for the rotor? Assume standard air and water density.

Solution

Convert the rotation rate to $\omega = 2\pi(220)/60 = 23.04$ rad/s. The wind velocity is 10 kn = 16.88 ft/s, so the velocity ratio is

$$\frac{a\omega}{U_\infty} = \frac{(1.25\ \text{ft})(23.04\ \text{rad/s})}{16.88\ \text{ft/s}} = 1.71$$

Using Fig. 8.15, we read $C_D \approx 0.7$ and $C_L \approx 2.5$. From Table A.6, standard air density in BG units is 0.00238 slug/ft³. Then the estimated rotor lift and drag are

$$L = C_L \frac{1}{2}\rho U_\infty^2\, 2ba = (2.5)\frac{1}{2}\left(0.00238\frac{\text{slug}}{\text{ft}^3}\right)\left(16.88\frac{\text{ft}}{\text{s}}\right)^2 2(10\ \text{ft})(1.25\ \text{ft}) = 21.2\ \text{lbf}$$

$$D = C_D \frac{1}{2}\rho U_\infty^2\, 2ba = (0.7)\frac{1}{2}\left(0.00238\frac{\text{slug}}{\text{ft}^3}\right)\left(16.88\frac{\text{ft}}{\text{s}}\right)^2 2(10\ \text{ft})(1.25\ \text{ft}) = 5.9\ \text{lbf}$$

The maximum thrust available to the sailboat is the resultant of these two:

$$F = [(21.2)^2 + (5.9)^2] = 22.0\ \text{lbf} \qquad\qquad Ans.$$

Note that water density did not enter into this calculation, which is a force due to *air*. If aligned along the boat's keel, this thrust will drive the boat through the water at a speed of about 4 kn.

E8.3   *(Courtesy of R. C. Lessmann, University of Rhode Island)*

Comment: For the sake of a numerical example, we have done something improper here. We have used data for $Re_D = 3800$ to estimate forces when the rotor $Re_D \approx 260{,}000$. Do not do this in your real job after you graduate!

The Kelvin Oval

A family of body shapes taller than they are wide can be formed by letting a uniform stream flow normal to a vortex pair. If U_∞ is to the right, the negative vortex $-K$ is placed at $y = +a$ and the counterclockwise vortex $+K$ placed at $y = -a$, as in Fig. 8.16. The combined stream function is

$$\psi = U_\infty y - \frac{1}{2} K \ln \frac{x^2 + (y + a)^2}{x^2 + (y - a)^2} \qquad (8.45)$$

The body shape is the line $\psi = 0$, and some of these shapes are shown in Fig. 8.16. For $K/(U_\infty a) > 10$ the shape is within 1 percent of a Rankine oval (Fig. 8.13) turned 90°, but for small $K/(U_\infty a)$ the waist becomes pinched in, and a figure-eight shape occurs at 0.5. For $K/(U_\infty a) < 0.5$ the stream blasts right between the vortices and isolates two more or less circular body shapes, one surrounding each vortex.

A closed body of practically any shape can be constructed by proper superposition of sources, sinks, and vortices. See the advanced work in Refs. 1 to 3 for further details. A summary of elementary potential flows is given in Table 8.2.

Potential Flow Analogs

For complicated potential flow geometries, one can resort to other methods than superposition of sources, sinks, and vortices. A variety of devices simulate solutions to Laplace's equation.

From 1897 to 1900 Hele-Shaw [9] developed a technique whereby laminar flow between very closely spaced parallel plates simulated potential flow when viewed

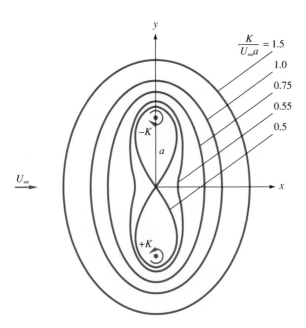

Fig. 8.16 Kelvin oval body shapes as a function of the vortex strength parameter $K/(U_\infty a)$; outer streamlines not shown.

Table 8.2 Summary of Plane Incompressible Potential Flows

Type of flow	Potential functions		Remarks
Stream $\mathbf{i}U$	$\psi = Uy$	$\phi = Ux$	See Fig. 8.3a
Line source ($m > 0$) or sink ($m < 0$)	$\psi = m\theta$	$\phi = m \ln r$	See Fig. 8.3b
Line vortex	$\psi = -K \ln r$	$\phi = K\theta$	See Fig. 8.3c
Half-body	$\psi = Ur \sin \theta + m\theta$		
	$\phi = Ur \cos \theta + m \ln r$		See Fig. 8.9
Doublet	$\psi = \dfrac{-\lambda \sin \theta}{r}$	$\phi = \dfrac{\lambda \cos \theta}{r}$	See Fig. 8.12
Rankine oval	$\psi = Ur \sin \theta + m(\theta_1 - \theta_2)$		See Fig. 8.13
Cylinder with circulation	$\psi = U \sin \theta \left(r - \dfrac{a^2}{r} \right) - K \ln \dfrac{r}{a}$		See Fig. 8.14

from above the plates. Obstructions simulate body shapes, and dye streaks represent the streamlines. The Hele-Shaw apparatus makes an excellent laboratory demonstration of potential flow [10, pp. 197–198, 219–220]. Figure 8.17a illustrates Hele-Shaw (potential) flow through an array of cylinders, a flow pattern that would be difficult to analyze just using Laplace's equation. However beautiful this array pattern may be, it is not a good approximation to real (laminar viscous) array flow. Figure 8.17b shows experimental streakline patterns for a similar staggered-array flow at Re $\approx$ 6400. We see that the interacting wakes of the real flow (Fig. 8.17b) cause intensive mixing and transverse motion, not the smooth streaming passage of the potential flow model (Fig. 8.17a). The moral is that this is an internal flow with multiple bodies and, therefore, not a good candidate for a realistic potential flow model.

Other flow-mapping techniques are discussed in Ref. 8. Electromagnetic fields also satisfy Laplace's equation, with voltage analogous to velocity potential and current lines analogous to streamlines. At one time commercial analog field plotters were available, using thin conducting paper cut to the shape of the flow geometry. Potential lines (voltage contours) were plotted by probing the paper with a potentiometer pointer. Hand-sketching "curvilinear square" techniques were also popular. The availability and the simplicity of computer potential flow methods [5 to 7] have made analog models obsolete.

EXAMPLE 8.4

A Kelvin oval from Fig. 8.16 has $K/(U_\infty a) = 1.0$. Compute the velocity at the top shoulder of the oval in terms of U_∞.

Solution

We must locate the shoulder $y = h$ from Eq. (8.45) for $\psi = 0$ and then compute the velocity by differentiation. At $\psi = 0$ and $y = h$ and $x = 0$, Eq. (8.45) becomes

$$\frac{h}{a} = \frac{K}{U_\infty a} \ln \frac{h/a + 1}{h/a - 1}$$

With $K/(U_\infty a) = 1.0$ and the initial guess $h/a \approx 1.5$ from Fig. 8.16, we iterate and find the location $h/a = 1.5434$.

By inspection $v = 0$ at the shoulder because the streamline is horizontal. Therefore the shoulder velocity is, from Eq. (8.45),

$$u\bigg|_{y=h} = \frac{\partial \psi}{\partial y}\bigg|_{y=h} = U_\infty + \frac{K}{h-a} - \frac{K}{h+a}$$

Introducing $K = U_\infty a$ and $h = 1.5434a$, we obtain

$$u_{\text{shoulder}} = U_\infty(1.0 + 1.84 - 0.39) = 2.45U_\infty \qquad\qquad Ans.$$

Because they are short-waisted compared with a circular cylinder, all the Kelvin ovals have shoulder velocity greater than the cylinder result $2.0U_\infty$ from Eq. (8.39).

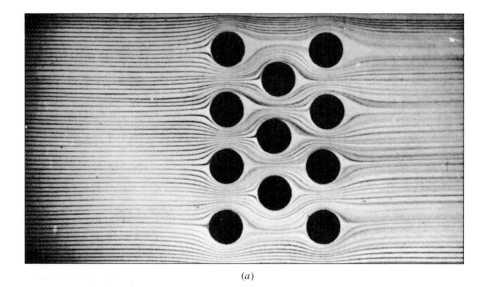

(a)

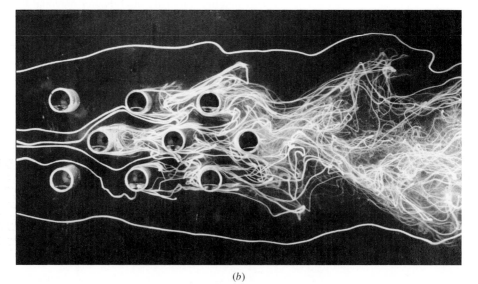

Fig. 8.17 Flow past a staggered array of cylinders: (a) potential flow model using the Hele-Shaw apparatus *(TQ Education and Training Ltd.)*; (b) experimental streaklines for actual staggered-array flow at $\text{Re}_D \approx 6400$. *(From Ref. 36, Courtesy of Jack Hoyt, with the permission of the American Society of Mechanical Engineers.)*

(b)

8.5 Other Plane Potential Flows[3]

References 2 to 4 treat many other potential flows of interest in addition to the cases presented in Secs. 8.3 and 8.4. In principle, any plane potential flow can be solved by the method of *conformal mapping,* by using the complex variable

$$z = x + iy \qquad i = (-1)^{1/2}$$

It turns out that any arbitrary analytic function of this complex variable z has the remarkable property that both its real and its imaginary parts are solutions of Laplace's equation. If

$$f(z) = f(x + iy) = f_1(x, y) + i f_2(x, y)$$

then

$$\frac{\partial^2 f_1}{\partial x^2} + \frac{\partial^2 f_1}{\partial y^2} = 0 = \frac{\partial^2 f_2}{\partial x^2} + \frac{\partial^2 f_2}{\partial y^2} \tag{8.46}$$

We shall assign the proof of this as Prob. W8.4. Even more remarkable if you have never seen it before is that lines of constant f_1 will be everywhere perpendicular to lines of constant f_2:

$$\left(\frac{dy}{dx}\right)_{f_1 = C} = -\frac{1}{(dy/dx)_{f_2 = C}} \tag{8.47}$$

This is true for totally arbitrary $f(z)$ as long as this function is analytic; that is, it must have a unique derivative df/dz at every point in the region.

The net result of Eqs. (8.46) and (8.47) is that the functions f_1 and f_2 can be interpreted to be the potential lines and streamlines of an inviscid flow. By long custom we let the real part of $f(z)$ be the velocity potential and the imaginary part be the stream function:

$$f(z) = \phi(x, y) + i\psi(x, y) \tag{8.48}$$

We try various functions $f(z)$ and see whether any interesting flow pattern results. Of course, most of them have already been found, and we simply report on them here.

We shall not go into the details, but there are excellent treatments of this complex-variable technique on both an introductory [4] and a more advanced [2, 3] level. The method is less important now because of the popularity of computer techniques.

As a simple example, consider the linear function

$$f(z) = U_\infty z = U_\infty x + i U_\infty y$$

It follows from Eq. (8.48) that $\phi = U_\infty x$ and $\psi = U_\infty y$, which, we recall from Eq. (8.12), represents a uniform stream in the x direction. Once you get used to the complex variable, the solution practically falls in your lap.

To find the velocities, you may either separate ϕ and ψ from $f(z)$ and differentiate, or differentiate f directly:

$$\frac{df}{dz} = \frac{\partial \phi}{\partial x} + i \frac{\partial \psi}{\partial x} = -i \frac{\partial \phi}{\partial y} + \frac{\partial \psi}{\partial y} = u - iv \tag{8.49}$$

Thus the real part of df/dz equals $u(x, y)$, and the imaginary part equals $-v(x, y)$. To get a practical result, the derivative df/dz must exist and be unique, hence the requirement

---

[3]This section may be omitted without loss of continuity.

that f be an analytic function. For $f(z) = U_\infty z$, $df/dz = U_\infty = u$, since it is real, and $v = 0$, as expected.

Sometimes it is convenient to use the polar coordinate form of the complex variable

$$z = x + iy = re^{i\theta} = r\cos\theta + ir\sin\theta$$

where

$$r = (x^2 + y^2)^{1/2} \qquad \theta = \tan^{-1}\frac{y}{x}$$

This form is especially convenient when powers of z occur.

Uniform Stream at an Angle of Attack

All the elementary plane flows of Sec. 8.2 have a complex-variable formulation. The uniform stream U_∞ at an angle of attack α has the complex potential

$$f(z) = U_\infty z e^{-i\alpha} \tag{8.50}$$

Compare this form with Eq. (8.22).

Line Source at a Point z_0

Consider a line source of strength m placed off the origin at a point $z_0 = x_0 + iy_0$. Its complex potential is

$$f(z) = m \ln (z - z_0) \tag{8.51}$$

This can be compared with Eq. (8.13), which is valid only for the source at the origin. For a line sink, the strength m is negative.

Line Vortex at a Point z_0

If a line vortex of strength K is placed at point z_0, its complex potential is

$$f(z) = -iK \ln (z - z_0) \tag{8.52}$$

to be compared with Eq. (8.14). Also compare to Eq. (8.51) to see that we reverse the meaning of ϕ and ψ simply by multiplying the complex potential by $-i$.

Flow around a Corner of Arbitrary Angle

Corner flow is an example of a pattern that cannot be conveniently produced by superimposing sources, sinks, and vortices. It has a strikingly simple complex representation:

$$f(z) = Az^n = Ar^n e^{in\theta} = Ar^n \cos n\theta + iAr^n \sin n\theta$$

where A and n are constants.

It follows from Eq. (8.48) that for this pattern

$$\phi = Ar^n \cos n\theta \qquad \psi = Ar^n \sin n\theta \tag{8.53}$$

Streamlines from Eq. (8.53) are plotted in Fig. 8.18 for five different values of n. The flow is seen to represent a stream turning through an angle $\beta = \pi/n$. Patterns in Fig. 8.18d and e are not realistic on the downstream side of the corner, where separation will occur due to the adverse pressure gradient and sudden change of direction. In general, separation always occurs downstream of salient, or protruding corners, except in creeping flows at low Reynolds number $\text{Re} < 1$.

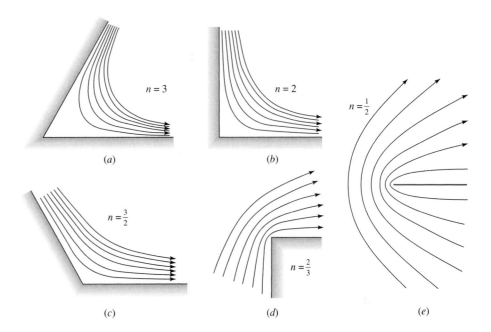

Fig. 8.18 Streamlines for corner flow, Eq. (8.53), for corner angle β of (*a*) 60°, (*b*) 90°, (*c*) 120°, (*d*) 270°, and (*e*) 360°.

Since $360° = 2\pi$ is the largest possible corner, the patterns for $n < \frac{1}{2}$ do not represent corner flow.

If we expand the plot of Fig. 8.18*a* to *c* to double size, we can represent stagnation flow toward a corner of angle $2\beta = 2\pi/n$. This is done in Fig. 8.19 for $n = 3$, 2, and 1.5. These are very realistic flows; although they slip at the wall, they can be patched to boundary layer theories very successfully. We took a brief look at corner flows before, in Examples 4.5 and 4.9 and in Probs. P4.49 to P4.51.

Flow Normal to a Flat Plate

We treat this case separately because the Kelvin ovals of Fig. 8.16 failed to degenerate into a flat plate as K became small. The flat plate normal to a uniform stream is an extreme case worthy of our attention.

Although the result is quite simple, the derivation is very complicated and is given, for example, in Ref. 2, Sec. 9.3. There are three changes of complex variable, or *mappings*, beginning with the basic cylinder flow solution of Fig. 8.14*a*. First the uniform stream is rotated to be vertical upward, then the cylinder is squeezed down into a plate shape, and finally the free stream is rotated back to the horizontal. The final result for complex potential is

$$f(z) = \phi + i\psi = U_\infty(z^2 + a^2)^{1/2} \tag{8.54}$$

where $2a$ is the height of the plate. To isolate ϕ or ψ, square both sides and separate real and imaginary parts:

$$\phi^2 - \psi^2 = U_\infty^2(x^2 - y^2 + a^2) \qquad \phi\psi = U_\infty^2 xy$$

We can solve for ψ to determine the streamlines

$$\psi^4 + \psi^2 U_\infty^2(x^2 - y^2 + a^2) = U_\infty^4 x^2 y^2 \tag{8.55}$$

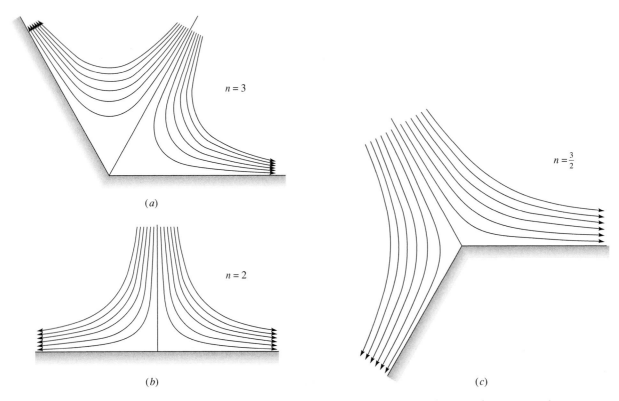

Fig. 8.19 Streamlines for stagnation flow from Eq. (8.53) for corner angle 2β of (a) 120°, (b) 180°, and (c) 240°.

Equation (8.55) is plotted in Fig. 8.20a, revealing a doubly symmetric pattern of streamlines that approach very closely to the plate and then deflect up and over, with very high velocities and low pressures near the plate tips.

The velocity v_s along the plate surface is found by computing df/dz from Eq. (8.54) and isolating the imaginary part:

$$\left.\frac{v_s}{U_\infty}\right|_{\text{plate surface}} = \frac{y/a}{(1 - y^2/a^2)^{1/2}} \tag{8.56}$$

Some values of surface velocity can be tabulated as follows:

y/a	0.0	0.2	0.4	0.6	0.707	0.8	0.9	1.0
v_s/U_∞	0.0	0.204	0.436	0.750	1.00	1.33	2.07	∞

The origin is a stagnation point; then the velocity grows linearly at first and very rapidly near the tip, with both velocity and acceleration being infinite at the tip.

As you might guess, Fig. 8.20a is not realistic. In a real flow the sharp salient edge causes separation, and a broad, low-pressure wake forms in the lee, as in Fig. 8.20b. Instead of being zero, the drag coefficient is very large, $C_D \approx 2.0$ from Table 7.2.

A discontinuous potential flow theory that accounts for flow separation was devised by Helmholtz in 1868 and Kirchhoff in 1869. This free-streamline solution is shown in Fig. 8.20c, with the streamline that breaks away from the tip having a constant velocity

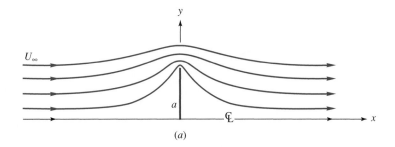

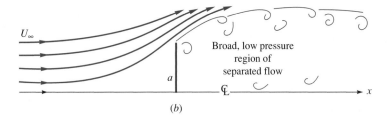

Fig. 8.20 Streamlines in upper half-plane for flow normal to a flat plate of height $2a$: (*a*) continuous potential flow theory, Eq. (8.55); (*b*) actual measured flow pattern; (*c*) discontinuous potential theory with $k \approx 1.5$.

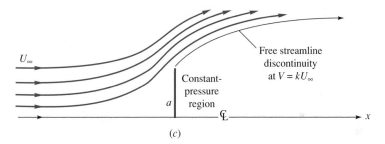

$V = kU_\infty$. From Bernoulli's equation the pressure in the dead-water region behind the plate will equal $p_r = p_\infty + \frac{1}{2}\rho U_\infty^2(1 - k^2)$ to match the pressure along the free stream-line. For $k = 1.5$ this Helmholtz-Kirchoff theory predicts $p_r = p_\infty - 0.625\rho U_\infty^2$ and an average pressure on the front $p_f = p_\infty + 0.375\rho U_\infty^2$, giving an overall drag coefficient of 2.0, in agreement with experiment. However, the coefficient k is a priori unknown and must be tuned to experimental data, so free-streamline theory can be considered only a qualified success. For further details see Ref. 2, Sec. 11.2.

8.6 Images[4]

The previous solutions have all been for unbounded flows, such as a circular cylinder immersed in a broad expanse of uniformly streaming fluid, Fig. 8.14*a*. However, many practical problems involve a nearby rigid boundary constraining the flow, such as (1) groundwater flow near the bottom of a dam, (2) an airfoil near the ground, simulating landing or takeoff, or (3) a cylinder mounted in a wind tunnel with narrow

---

[4]This section may be omitted without loss of continuity.

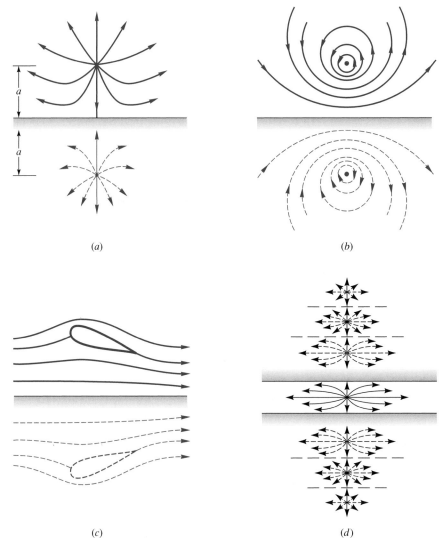

Fig. 8.21 Constraining walls can be created by image flows: (*a*) source near a wall with identical image source; (*b*) vortex near a wall with image vortex of opposite sense; (*c*) airfoil in ground effect with image airfoil of opposite circulation; (*d*) source between two walls requiring an infinite row of images.

walls. In such cases the basic unbounded potential flow solutions can be modified for wall effects by the method of *images*.

Consider a line source placed a distance *a* from a wall, as in Fig. 8.21*a*. To create the desired wall, an image source of identical strength is placed the same distance below the wall. By symmetry the two sources create a plane surface streamline between them, which is taken to be the wall.

In Fig. 8.21*b* a vortex near a wall requires an image vortex the same distance below but of *opposite* rotation. We have shaded in the wall, but of course the pattern could also be interpreted as the flow near a vortex pair in an unbounded fluid.

In Fig. 8.21*c* an airfoil in a uniform stream near the ground is created by an image airfoil below the ground of opposite circulation and lift. This looks easy, but actually it is not because the airfoils are so close together that they interact and distort each other's shapes. A rule of thumb is that nonnegligible shape distortion occurs if the body shape is within two chord lengths of the wall. To eliminate distortion, a whole series of "corrective" images must be added to the flow to recapture the shape of the original isolated airfoil. Reference 2, Sec. 7.75, has a good discussion of this procedure, which usually requires computer summation of the multiple images needed.

Figure 8.21d shows a source constrained between two walls. One wall required only one image in Fig. 8.21a, but *two* walls require an infinite array of image sources above and below the desired pattern, as shown. Usually computer summation is necessary, but sometimes a closed-form summation can be achieved, as in the infinite vortex row of Eq. (8.28).

EXAMPLE 8.5

For the source near a wall as in Fig. 8.21a, the wall velocity is zero between the sources, rises to a maximum moving out along the wall, and then drops to zero far from the sources. If the source strength is 8 m²/s, how far from the wall should the source be to ensure that the maximum velocity along the wall will be 5 m/s?

Solution

At any point x along the wall, as in Fig. E8.5, each source induces a radial outward velocity $v_r = m/r$, which has a component $v_r \cos \theta$ along the wall. The total wall velocity is thus

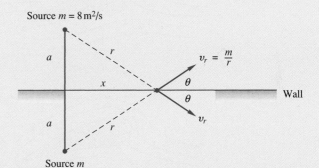

E8.5

$$u_{wall} = 2v_r \cos \theta$$

From the geometry of Fig. E8.5, $r = (x^2 + a^2)^{1/2}$ and $\cos \theta = x/r$. Then the total wall velocity can be expressed as

$$u = \frac{2mx}{x^2 + a^2}$$

This is zero at $x = 0$ and at $x \to \infty$. To find the maximum velocity, differentiate and set equal to zero:

$$\frac{du}{dx} = 0 \quad \text{at} \quad x = a \quad \text{and} \quad u_{max} = \frac{m}{a}$$

We have omitted a bit of algebra in giving these results. For the given source strength and maximum velocity, the proper distance a is

$$a = \frac{m}{u_{max}} = \frac{8 \text{ m}^2/\text{s}}{5 \text{ m/s}} = 1.6 \text{ m} \qquad Ans.$$

For $x > a$, there is an adverse pressure gradient along the wall, and boundary layer theory should be used to predict separation.

8.7 Airfoil Theory[5]

As mentioned in conjunction with the Kutta-Joukowski lift theorem, Eq. (8.43), the problem in airfoil theory is to determine the net circulation Γ as a function of airfoil shape and free-stream angle of attack α.

The Kutta Condition

Even if the airfoil shape and free-stream angle of attack are specified, the potential flow theory solution is nonunique: An infinite family of solutions can be found corresponding to different values of circulation Γ. Four examples of this nonuniqueness were shown for the cylinder flows in Fig. 8.14. The same is true of the airfoil, and Fig. 8.22 shows three mathematically acceptable "solutions" to a given airfoil flow for small (Fig. 8.22a), large (Fig. 8.22b), and medium (Fig. 8.22c) net circulation.

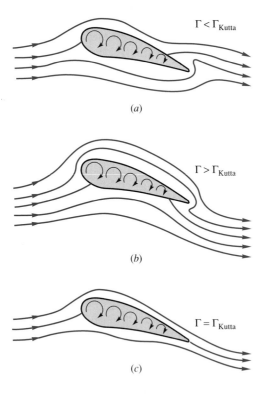

Fig. 8.22 The Kutta condition properly simulates the flow about an airfoil; (a) too little circulation, stagnation point on rear upper surface; (b) too much, stagnation point on rear lower surface; (c) just right, Kutta condition requires smooth flow at trailing edge.

[5]This section may be omitted without loss of continuity.

You can guess which case best simulates a real airfoil from the earlier discussion of transient lift development in Fig. 7.23. It is the case (Fig. 8.22c), where the upper and lower flows meet and leave the trailing edge smoothly. If the trailing edge is rounded slightly, there will be a stagnation point there. If the trailing edge is sharp, approximating most airfoil designs, the upper- and lower-surface flow velocities will be equal as they meet and leave the airfoil.

This statement of the physically proper value of Γ is generally attributed to W. M. Kutta, hence the name *Kutta condition,* although some texts give credit to Joukowski and/or Chaplygin. All airfoil theories use the Kutta condition, which is in good agreement with experiment. It turns out that the correct circulation Γ_{Kutta} depends on flow velocity, angle of attack, and airfoil shape.

Flat-Plate Airfoil Vortex Sheet Theory

The flat plate is the simplest airfoil, having no thickness or "shape," but even its theory is not so simple. The problem can be solved by a complex-variable mapping [2, p. 480], but here we shall use a vortex sheet approach. Figure 8.23a shows a flat plate of length C simulated by a vortex sheet of variable strength $\gamma(x)$. The free stream U_∞ is at an angle of attack α with respect to the plate chord line.

To make the lift "up" with flow from left to right as shown, we specify here that the circulation is positive clockwise. Recall from Fig. 8.11c that there is a jump in tangential velocity across a sheet equal to the local strength:

$$u_u - u_l = \gamma(x)$$

If we omit the free stream, the sheet should cause a rightward flow $\delta u = +\frac{1}{2}\gamma$ on the upper surface and an equal and opposite leftward flow on the lower surface, as shown in Fig. 8.23a. The Kutta condition for this sharp trailing edge requires that this velocity difference vanish at the trailing edge to keep the exit flow smooth and parallel:

$$\gamma(C) = 0 \tag{8.57}$$

The proper solution must satisfy this condition, after which the total lift can be computed by summing the sheet strength over the whole airfoil. From Eq. (8.43) for a foil of depth b:

$$L = \rho U_\infty b \Gamma \qquad \Gamma = \int_0^C \gamma(x)\,dx \tag{8.58}$$

An alternative way to compute lift is from the dimensionless pressure coefficient C_p on the upper and lower surfaces:

$$C_{p_{u,l}} = \frac{p_{u,l} - p_\infty}{\frac{1}{2}\rho U_\infty^2} = 1 - \frac{U_{u,l}^2}{U_\infty^2} \tag{8.59}$$

where the last expression follows from Bernoulli's equation. The surface velocity squared is given by combining the uniform stream and the vortex sheet velocity components from Fig. 8.23a:

$$U_{u,l}^2 = (U_\infty \cos\alpha \pm \delta u)^2 + (U_\infty \sin\alpha)^2$$
$$= U_\infty^2 \pm 2U_\infty \delta u \cos\alpha + \delta u^2 \approx U_\infty^2\left(1 \pm \frac{2\,\delta u}{U_\infty}\right) \tag{8.60}$$

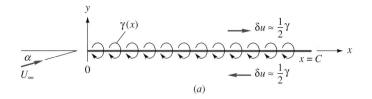

(a)

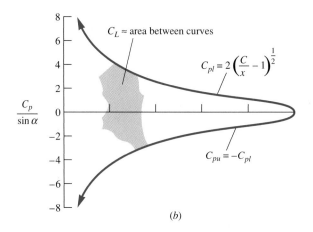

(b)

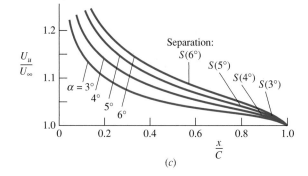

(c)

Fig. 8.23 Vortex sheet solution for the flat-plate airfoil; (a) sheet geometry; (b) theoretical pressure coefficient on upper and lower surfaces; (c) upper-surface velocity with laminar separation points S.

where we have made the approximations $\delta u \ll U_\infty$ and $\cos \alpha \approx 1$ in the last expression, assuming a small angle of attack. Equations (8.59) and (8.60) combine to the first-order approximation

$$C_{p_{u,l}} = \mp \frac{2\,\delta u}{U_\infty} = \mp \frac{\gamma}{U_\infty} \qquad (8.61)$$

The lift force is the integral of the pressure difference over the length of the airfoil, assuming depth b:

$$L = \int_0^C (p_l - p_u)b\,dx$$

or

$$C_L = \frac{L}{\frac{1}{2}\rho U_\infty^2 bC} = \int_0^1 (C_{p_l} - C_{p_u})\frac{dx}{C} = 2\int_0^1 \frac{\gamma}{U_\infty}\,d\left(\frac{x}{C}\right) \qquad (8.62)$$

Equations (8.58) and (8.62) are entirely equivalent within the small-angle approximations.

The sheet strength $\gamma(x)$ is computed from the requirement that the net normal velocity $v(x)$ be zero at the sheet ($y = 0$), since the sheet represents a solid plate or stream surface. Consider a small piece of sheet $\gamma \, dx$ located at position x_0. The velocity v at point x on the sheet is that of an infinitesimal line vortex of strength $d\Gamma = -\gamma \, dx$:

$$dv \bigg|_x = \frac{d\Gamma}{2\pi r}\bigg|_{x_0 \to x} = \frac{-\gamma \, dx}{2\pi (x_0 - x)}$$

The total normal velocity induced by the entire sheet at point x is thus

$$v_{\text{sheet}} = -\int_0^C \frac{-\gamma \, dx}{2\pi(x_0 - x)} \tag{8.63}$$

Meanwhile, from Fig. 8.23a, the uniform stream induces a constant normal velocity at every point on the sheet given by

$$v_{\text{stream}} = U_\infty \sin \alpha$$

Setting the sum of v_{sheet} and v_{stream} equal to zero gives the integral equation

$$\int_0^C \frac{\gamma \, dx}{x_0 - x} = 2\pi U_\infty \sin \alpha \tag{8.64}$$

to be solved for $\gamma(x)$ subject to the Kutta condition $\gamma(C) = 0$ from Eq. (8.57).

Although Eq. (8.64) is quite formidable (and not only for beginners), in fact it was solved long ago by using integral formulas developed by Poisson in the nineteenth century. The sheet strength that satisfies Eq. (8.64) is

$$\gamma(x) = 2U_\infty \sin \alpha \left(\frac{C}{x} - 1\right)^{1/2} \tag{8.65}$$

From Eq. (8.61) the surface pressure coefficients are thus

$$C_{p_{u,l}} = \mp 2 \sin \alpha \left(\frac{C}{x} - 1\right)^{1/2} \tag{8.66}$$

Details of the calculations are given in advanced texts [for example, 11, Chap. 4].

The pressure coefficients from Eq. (8.66) are plotted in Fig. 8.23b, showing that the upper surface has pressure continually increasing with x—that is, an adverse gradient. The upper-surface velocity $U_u \approx U_\infty + \delta u = U_\infty + \frac{1}{2}\gamma$ is plotted in Fig. 8.23c for various angles of attack. Above $\alpha = 5°$ the sheet contribution δu is about 20 percent of U_∞ so that the small-disturbance assumption is violated. Figure 8.23c also shows separation points computed by Thwaites's laminar boundary layer method, Eqs. (7.54) and (7.55). The prediction is that a flat plate would be extensively stalled on the upper surface for $\alpha > 6°$, which is approximately correct.

The lift coefficient of the airfoil is proportional to the area between c_{p_l} and c_{p_u} in Fig. 8.23b, from Eq. (8.62):

$$C_L = 2\int_0^1 \frac{\gamma}{U} \, d\left(\frac{x}{C}\right) = 4 \sin \alpha \int_0^1 \left(\frac{C}{x} - 1\right)^{1/2} d\left(\frac{x}{C}\right) = 2\pi \sin \alpha \approx 2\pi\alpha \tag{8.67}$$

This is a classic result that was alluded to earlier in Eq. (7.70) without proof.

Also of interest is the moment coefficient about the leading edge (LE) of the airfoil, taken as positive counterclockwise:

$$C_{M_{\text{LE}}} = \frac{M_{\text{LE}}}{\frac{1}{2}\rho U_\infty^2 bC^2} = \int_0^1 (C_{p_l} - C_{p_u})\frac{x}{C}\,d\left(\frac{x}{C}\right) = \frac{\pi}{2}\sin\alpha = \frac{1}{4}C_L \quad (8.68)$$

Thus the *center of pressure* (CP), or position of the resultant lift force, is at the one-quarter-chord point:

$$\left(\frac{x}{C}\right)_{\text{CP}} = \frac{1}{4} \quad (8.69)$$

This theoretical result is independent of the angle of attack.

These results can be compared with experimental results for NACA airfoils in Fig. 8.24. The thinnest NACA airfoil is $t/C = 0.06$, and the thickest is 24 percent, or $t/C = 0.24$. The lift curve slope $dC_L/d\alpha$ is within 9 percent of the theoretical value of 2π for all the various airfoil families at all thicknesses. Increasing thickness tends to increase both $C_{L,\text{max}}$ and the stall angle. The stall angle at $t/C = 0.06$ is about 8° and would be even less for a flat plate, verifying the boundary layer separation estimates in Fig. 8.23c. Best performance is usually at about the 12 percent thickness point for any airfoil.

Potential Theory for Thick Cambered Airfoils

The theory of thick cambered airfoils is covered in advanced texts [for example, 2 to 4]; Ref. 13 has a thorough and comprehensive review of both inviscid and viscous aspects of airfoil behavior.

Basically the theory uses a complex-variable mapping that transforms the flow about a cylinder with circulation in Fig. 8.14 into flow about a foil shape with circulation. The circulation is then adjusted to match the Kutta condition of smooth exit flow from the trailing edge.

Regardless of the exact airfoil shape, the inviscid mapping theory predicts that the correct circulation for any thick cambered airfoil is

$$\Gamma_{\text{Kutta}} = \pi bCU_\infty\left(1 + 0.77\frac{t}{C}\right)\sin(\alpha + \beta) \quad (8.70)$$

where $\beta = \tan^{-1}(2h/C)$ and h is the maximum camber, or maximum deviation of the airfoil midline from its chord line, as in Fig. 8.25a.

The lift coefficient of the infinite-span airfoil is thus

$$C_L = \frac{\rho U_\infty \Gamma}{\frac{1}{2}\rho U_\infty^2 bC} = 2\pi\left(1 + 0.77\frac{t}{C}\right)\sin(\alpha + \beta) \quad (8.71)$$

This reduces to Eq. (8.67) when the thickness and camber are zero. Figure 8.24 shows that the thickness effect $1 + 0.77t/C$ is not verified by experiment. Some airfoils increase lift with thickness, others decrease, and none approach the theory very closely, the primary reason being the boundary layer growth on the upper surface affecting the airfoil "shape." Thus it is customary to drop the thickness effect from the theory:

$$C_L \approx 2\pi\sin(\alpha + \beta) \quad (8.72)$$

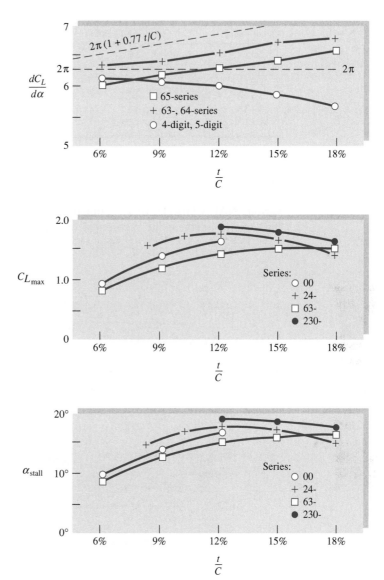

Fig. 8.24 Lift characteristics of smooth NACA airfoils as a function of thickness ratio, for infinite aspect ratio. *(From Ref. 12.)*

The theory correctly predicts that a cambered airfoil will have finite lift at zero angle of attack and zero lift (ZL) at an angle

$$\alpha_{ZL} = -\beta = -\tan^{-1}\frac{2h}{C} \tag{8.73}$$

Equation (8.73) overpredicts the measured zero-lift angle by 1° or so, as shown in Table 8.3. The measured values are essentially independent of thickness. The designation XX in the NACA series indicates the thickness in percent, and the other digits refer to camber and other details. For example, the 2415 airfoil has 2 percent maximum camber (the first digit) occurring at 40 percent chord (the second digit) with

Table 8.3 Zero-Lift Angle of
NACA Airfoils

Airfoil series	Camber h/C, %	Measured α_{ZL}, deg	Theory $-\beta$, deg
24XX	2.0	−2.1	−2.3
44XX	4.0	−4.0	−4.6
230XX	1.8	−1.3	−2.1
63-2XX	2.2	−1.8	−2.5
63-4XX	4.4	−3.1	−5.0
64-1XX	1.1	−0.8	−1.2

15 percent maximum thickness (the last two digits). The maximum thickness need not occur at the same position as the maximum camber.

Figure 8.25*b* shows the measured position of the center of pressure of the various NACA airfoils, both symmetric and cambered. In all cases x_{CP} is within 0.02 chord length of the theoretical quarter-chord point predicted by Eq. (8.69).

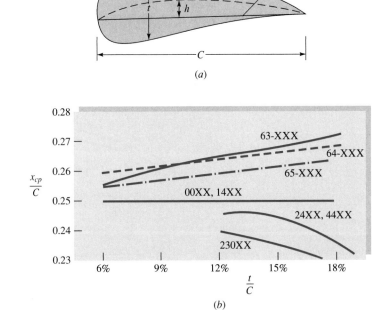

(a)

(b)

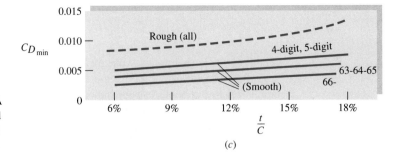

Fig. 8.25 Characteristics of NACA airfoils: (*a*) typical thick cambered airfoil; (*b*) center-of-pressure data; and (*c*) minimum drag coefficient.

(c)

The standard cambered airfoils (24, 44, and 230 series) lie slightly forward of $x/C = 0.25$ and the low-drag (60 series) foils slightly aft. The symmetric airfoils are at 0.25.

Figure 8.25c shows the minimum drag coefficient of NACA airfoils as a function of thickness. As mentioned earlier in conjunction with Fig. 7.25, these foils when smooth actually have less drag than turbulent flow parallel to a flat plate, especially the low-drag 60 series. However, for standard surface roughness all foils have about the same minimum drag, roughly 30 percent greater than that for a smooth flat plate.

Wings of Finite Span

The results of airfoil theory and experiment in the previous subsection were for two-dimensional, or infinite-span, wings. But all real wings have tips and are therefore of finite span or finite aspect ratio AR, defined by

$$\text{AR} = \frac{b^2}{A_p} = \frac{b}{C} \tag{8.74}$$

where b is the span length from tip to tip and A_p is the planform area of the wing as seen from above. The lift and drag coefficients of a finite-aspect-ratio wing depend strongly on the aspect ratio and slightly on the planform shape of the wing.

Vortices cannot end in a fluid; they must either extend to the boundary or form a closed loop. Figure 8.26a shows how the vortices that provide the wing circulation bend downstream at finite wing tips and extend far behind the wing to join the starting vortex (Fig. 7.23) downstream. The strongest vortices are shed from the tips, but some are shed from the body of the wing, as sketched schematically in Fig. 8.26b. The effective circulation $\Gamma(y)$ of these trailing shed vortices is zero at the tips and usually a maximum at the center plane, or root, of the wing. In 1918 Prandtl successfully modeled this flow by replacing the wing by a single lifting line and a continuous sheet of semi-infinite trailing vortices of strength $\gamma(y) = d\Gamma/dy$, as in Fig. 8.26c. Each elemental piece of trailing sheet $\gamma(\eta)\,d\eta$ induces a downwash, or downward velocity, $dw(y)$, given by

$$dw(y) = \frac{\gamma(\eta)\,d\eta}{4\pi(y - \eta)}$$

at position y on the lifting line. Note the denominator term 4π rather than 2π because the trailing vortex extends only from 0 to ∞ rather than from $-\infty$ to $+\infty$.

The total downwash $w(y)$ induced by the entire trailing vortex system is thus

$$w(y) = \frac{1}{4\pi} \int_{-(1/2)b}^{(1/2)b} \frac{\gamma(\eta)\,d\eta}{y - \eta} \tag{8.75}$$

When the downwash is vectorially added to the approaching free stream U_∞, the effective angle of attack at this section of the wing is reduced to

$$\alpha_{\text{eff}} = \alpha - \alpha_i \qquad \alpha_i = \tan^{-1} \frac{w}{U_\infty} \approx \frac{w}{U_\infty} \tag{8.76}$$

where we have used a small-amplitude approximation $w \ll U_\infty$.

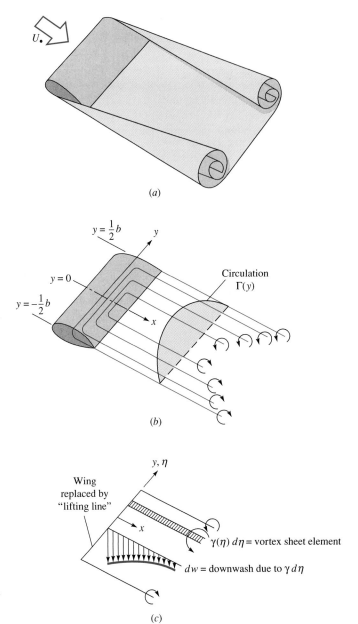

Fig. 8.26 Lifting-line theory for a finite wing: (*a*) actual trailing-vortex system behind a wing; (*b*) simulation by vortex system "bound" to the wing; (*c*) downwash on the wing due to an element of the trailing-vortex system.

The final step is to assume that the local circulation $\Gamma(y)$ is equal to that of a two-dimensional wing of the same shape and same effective angle of attack. From thin-airfoil theory, Eqs. (8.58) and (8.67), we have the estimate

$$C_L = \frac{\rho U_\infty \Gamma b}{\frac{1}{2}\rho U_\infty^2 bC} \approx 2\pi\alpha_{\text{eff}}$$

or

$$\Gamma \approx \pi C U_\infty \alpha_{\text{eff}} \tag{8.77}$$

Combining Eqs. (8.75) and (8.77), we obtain Prandtl's lifting-line theory for a finite-span wing:

$$\Gamma(y) = \pi C(y)U_\infty \left[\alpha(y) - \frac{1}{4\pi U_\infty} \int_{-(1/2)b}^{(1/2)b} \frac{(d\Gamma/d\eta)\, d\eta}{y - \eta} \right] \qquad (8.78)$$

This is an integrodifferential equation to be solved for $\Gamma(y)$ subject to the conditions $\Gamma(\frac{1}{2}b) = \Gamma(-\frac{1}{2}b) = 0$. It is similar to the thin-airfoil integral equation (8.64) and even more formidable. Once it is solved, the total wing lift and induced drag are given by

$$L = \rho U_\infty \int_{-(1/2)b}^{(1/2)b} \Gamma(y)\, dy \qquad D_i = \rho U_\infty \int_{-(1/2)b}^{(1/2)b} \Gamma(y)\alpha_i(y)\, dy \qquad (8.79)$$

Here is a case where the drag is not zero in a frictionless theory because the downwash causes the lift to slant backward by angle α_i so that it has a drag component parallel to the free-stream direction, $dDi = dL \sin \alpha_i \approx dL\alpha_i$.

The complete solution to Eq. (8.78) for arbitrary wing planform $C(y)$ and arbitrary twist $\alpha(y)$ is treated in advanced texts [for example, 11]. It turns out that there is a simple representative solution for an untwisted wing of elliptical planform:

$$C(y) = C_0 \left[1 - \left(\frac{2y}{b} \right)^2 \right]^{1/2}$$

The area and aspect ratio of this wing are

$$A_p = \int_{-(1/2)b}^{(1/2)b} C\, dy = \frac{1}{4}\pi b C_0 \qquad \text{AR} = \frac{4b}{\pi C_0} \qquad (8.80)$$

The solution to Eq. (8.78) for this $C(y)$ is an elliptical circulation distribution of exactly similar shape:

$$\Gamma(y) = \Gamma_0 \left[1 - \left(\frac{2y}{b} \right)^2 \right]^{1/2}$$

Substituting into Eq. (8.78) and integrating give a relation between Γ_0 and C_0:

$$\Gamma_0 = \frac{\pi C_0 U_\infty \alpha}{1 + 2/\text{AR}}$$

where α is assumed constant across the untwisted wing.

Substitution into Eq. (8.79) gives the elliptical wing lift:

$$L = \tfrac{1}{4}\pi^2 b C_0 \rho U_\infty^2 \alpha/(1 + 2/\text{AR})$$

or

$$C_L = \frac{2\pi\alpha}{1 + 2/\text{AR}} \qquad (8.81)$$

If we generalize this to a thick cambered finite wing of approximately elliptical planform, we obtain

$$\boxed{C_L = \frac{2\pi \sin(\alpha + \beta)}{1 + 2/\text{AR}}} \qquad (8.82)$$

This result was given without proof as Eq. (7.70). From Eq. (8.75) the computed downwash for the elliptical wing is constant:

$$w(y) = \frac{2U_\infty \alpha}{2 + \text{AR}} = \text{const} \tag{8.83}$$

Finally, the induced drag coefficient from Eq. (8.76) is

$$C_{Di} = C_L \frac{w}{U_\infty} = \frac{C_L^2}{\pi \text{AR}} \tag{8.84}$$

This was given without proof as Eq. (7.71).

Figure 8.27 shows the effectiveness of this theory when tested against a nonelliptical cambered wing by Prandtl in 1921 [14]. Figures 8.27a and b show the measured lift curves and drag polars for five different aspect ratios. Note the increase in stall angle and drag and the decrease in lift slope as the aspect ratio decreases.

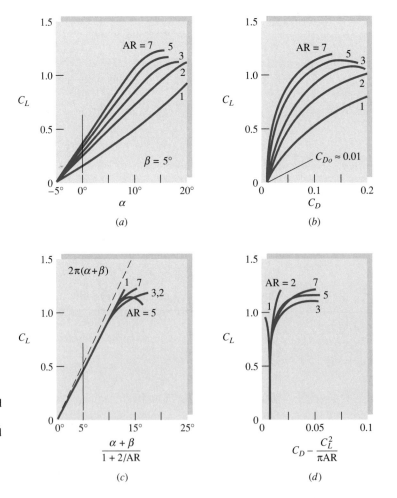

Fig. 8.27 Comparison of theory and experiment for a finite wing: (a) measured lift [14]; (b) measured drag polar [14]; (c) lift reduced to infinite aspect ratio; (d) drag polar reduced to infinite aspect ratio.

Figure 8.27c shows the lift data replotted against effective angle of attack $\alpha_{\text{eff}} = (\alpha + \beta)/(1 + 2/\text{AR})$, as predicted by Eq. (8.82). These curves should be equivalent to an infinite-aspect-ratio wing, and they do collapse together except near stall. Their common slope $dC_L/d\alpha$ is about 10 percent less than the theoretical value 2π, but this is consistent with the thickness and shape effects noted in Fig. 8.24.

Figure 8.27d shows the drag data replotted with the theoretical induced drag $C_{Di} = C_L^2/(\pi\text{AR})$ subtracted out. Again, except near stall, the data collapse onto a single line of nearly constant infinite-aspect-ratio drag $C_{D0} \approx 0.01$. We conclude that the finite-wing theory is very effective and may be used for design calculations.

Aircraft Trailing Vortices

The trailing vortices in Fig. 8.26a are real, not just mathematical abstractions. On commercial aircraft, such vortices are long, strong, and lingering. They can stretch for miles behind a large aircraft and endanger the following planes by inducing drastic rolling moments. The vortex persistence governs the separation distance between planes at an airport and thus determines airport capacity. An example of strong trailing vortices is shown in Fig. 8.28. There is a continuing research effort to alleviate trailing vortices by breaking them up or otherwise causing them to decay. See the review article by Spalart [46].

Fig. 8.28 Wingtip vortices from a smoke-visualization test of a Boeing 737. Vortices from large airplanes can be extremely dangerous to any following aircraft, especially small planes. This test was part of a research effort to alleviate these swirling wakes. [*NASA photo*]

8.8 Axisymmetric Potential Flow[6]

The same superposition technique that worked so well for plane flow in Sec. 8.3 is also successful for axisymmetric potential flow. We give some brief examples here.

Most of the basic results carry over from plane to axisymmetric flow with only slight changes owing to the geometric differences. Consider the following related flows:

Basic plane flow	Counterpart axisymmetric flow
Uniform stream	Uniform stream
Line source or sink	Point source or sink
Line doublet	Point doublet
Line vortex	No counterpart
Rankine half-body cylinder	Rankine half-body of revolution
Rankine oval cylinder	Rankine oval of revolution
Circular cylinder	Sphere
Symmetric airfoil	Tear-shaped body

Since there is no such thing as a point vortex, we must forgo the pleasure of studying circulation effects in axisymmetric flow. However, as any smoker knows, there is an axisymmetric ring vortex, and there are also ring sources and ring sinks, which we leave to advanced texts [for example, 3].

Spherical Polar Coordinates

Axisymmetric potential flows are conveniently treated in the spherical polar coordinates of Fig. 8.29. There are only two coordinates (r, θ), and flow properties are constant on a circle of radius $r \sin \theta$ about the x axis.

The equation of continuity for incompressible flow in these coordinates is

$$\frac{\partial}{\partial r}(r^2 v_r \sin \theta) + \frac{\partial}{\partial \theta}(r v_\theta \sin \theta) = 0 \tag{8.85}$$

where v_r and v_θ are radial and tangential velocity as shown. Thus a spherical polar stream function[7] exists such that

$$v_r = -\frac{1}{r^2 \sin \theta}\frac{\partial \psi}{\partial \theta} \qquad v_\theta = \frac{1}{r \sin \theta}\frac{\partial \psi}{\partial r} \tag{8.86}$$

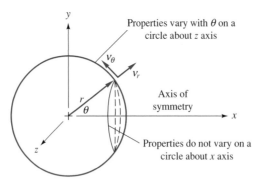

Fig. 8.29 Spherical polar coordinates for axisymmetric flow.

[6]This section may be omitted without loss of continuity.
[7]It is often called *Stokes's stream function*, having been used in a paper Stokes wrote in 1851 on viscous sphere flow.

In like manner a velocity potential $\phi(r, \theta)$ exists such that

$$v_r = \frac{\partial \phi}{\partial r} \qquad v_\theta = \frac{1}{r} \frac{\partial \phi}{\partial \theta} \tag{8.87}$$

These formulas serve to deduce the ψ and ϕ functions for various elementary axisymmetric potential flows.

Uniform Stream in the x Direction

A stream U_∞ in the x direction has components

$$v_r = U_\infty \cos \theta \qquad v_\theta = -U_\infty \sin \theta$$

Substitution into Eqs. (8.86) and (8.87) and integrating give

Uniform stream: $\qquad \psi = -\tfrac{1}{2} U_\infty r^2 \sin^2 \theta \qquad \phi = U_\infty r \cos \theta \tag{8.88}$

As usual, arbitrary constants of integration have been neglected.

Point Source or Sink

Consider a volume flux Q issuing from a point source. The flow will spread out radially and at radius r will equal Q divided by the area $4\pi r^2$ of a sphere. Thus

$$v_r = \frac{Q}{4\pi r^2} = \frac{m}{r^2} \qquad v_\theta = 0 \tag{8.89}$$

with $m = Q/(4\pi)$ for convenience. Integrating (8.86) and (8.87) gives

Point source $\qquad \psi = m \cos \theta \qquad \phi = -\frac{m}{r} \tag{8.90}$

For a point sink, change m to $-m$ in Eq. (8.90).

Point Doublet

Exactly as in Fig. 8.12, place a source at $(x, y) = (-a, 0)$ and an equal sink at $(+a, 0)$, taking the limit as a becomes small with the product $2am = \lambda$ held constant:

$$\psi_{\text{doublet}} = \lim_{\substack{a \to 0 \\ 2am = \lambda}} (m \cos \theta_{\text{source}} - m \cos \theta_{\text{sink}}) = \frac{\lambda \sin^2 \theta}{r} \tag{8.91}$$

We leave the proof of this limit as a problem. The point-doublet velocity potential is

$$\phi_{\text{doublet}} = \lim_{\substack{a \to 0 \\ 2am = \lambda}} \left(-\frac{m}{r_{\text{source}}} + \frac{m}{r_{\text{sink}}} \right) = \frac{\lambda \cos \theta}{r^2} \tag{8.92}$$

The streamlines and potential lines are shown in Fig. 8.30. Unlike the plane doublet flow of Fig. 8.12, neither set of lines represents perfect circles.

Uniform Stream plus a Point Source

By combining Eqs. (8.88) and (8.90), we obtain the stream function for a uniform stream plus a point source at the origin:

$$\psi = -\tfrac{1}{2} U_\infty r^2 \sin^2 \theta + m \cos \theta \tag{8.93}$$

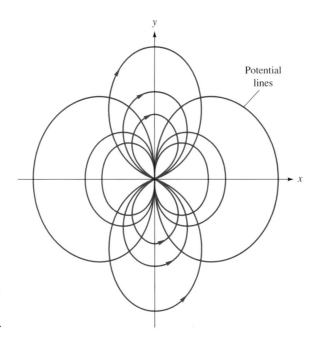

Fig. 8.30 Streamlines and potential lines due to a point doublet at the origin, from Eqs. (8.91) and (8.92).

From Eq. (8.86) the velocity components are, by differentiation,

$$v_r = U_\infty \cos \theta + \frac{m}{r^2} \qquad v_\theta = -U_\infty \sin \theta \qquad (8.94)$$

Setting these equal to zero reveals a stagnation point at $\theta = 180°$ and $r = a = (m/U_\infty)^{1/2}$, as shown in Fig. 8.31. If we let $m = U_\infty a^2$, the stream function can be rewritten as

$$\frac{\psi}{U_\infty a^2} = \cos \theta - \frac{1}{2}\left(\frac{r}{a}\right)^2 \sin^2 \theta \qquad (8.95)$$

The stream surface that passes through the stagnation point $(r, \theta) = (a, \pi)$ has the value $\psi = -U_\infty a^2$ and forms a half-body of revolution enclosing the point source, as

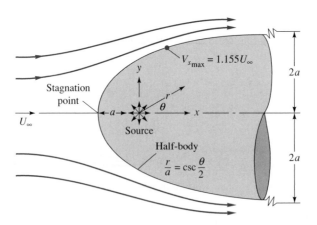

Fig. 8.31 Streamlines for a Rankine half-body of revolution.

shown in Fig. 8.31. This half-body can be used to simulate a pitot tube. Far downstream the half-body approaches the constant radius $R = 2a$ about the x axis. The maximum velocity and minimum pressure along the half-body surface occur at $\theta = 70.5°$, $r = a\sqrt{3}$, $V_s = 1.155U_\infty$. Downstream of this point there is an adverse gradient as V_s slowly decelerates to U_∞, but boundary layer theory indicates no flow separation. Thus Eq. (8.95) is a very realistic simulation of a real half-body flow. But when the uniform stream is added to a sink to form a half-body rear surface, similar to Fig. 8.9c, separation is predicted and the rear inviscid pattern is not realistic.

Uniform Stream plus a Point Doublet

From Eqs. (8.88) and (8.91), combination of a uniform stream and a point doublet at the origin gives

$$\psi = -\frac{1}{2}U_\infty r^2 \sin^2\theta + \frac{\lambda}{r}\sin^2\theta \tag{8.96}$$

Examination of this relation reveals that the stream surface $\psi = 0$ corresponds to the sphere of radius

$$r = a = \left(\frac{2\lambda}{U_\infty}\right)^{1/3} \tag{8.97}$$

This is exactly analogous to the cylinder flow of Fig. 8.14a formed by combining a uniform stream and a line doublet.

Letting $\lambda = \frac{1}{2}U_\infty a^3$ for convenience, we rewrite Eq. (8.96) as

$$\frac{\psi}{\frac{1}{2}U_\infty a^2} = -\sin^2\theta\left(\frac{r^2}{a^2} - \frac{a}{r}\right) \tag{8.98}$$

The streamlines for this sphere flow are plotted in Fig. 8.32. By differentiation from Eq. (8.86) the velocity components are

$$v_r = U_\infty \cos\theta\left(1 - \frac{a^3}{r^3}\right) \qquad v_\theta = -\frac{1}{2}U_\infty \sin\theta\left(2 + \frac{a^3}{r^3}\right) \tag{8.99}$$

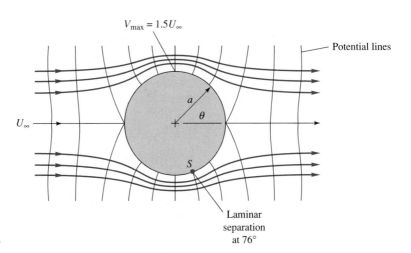

Fig. 8.32 Streamlines and potential lines for inviscid flow past a sphere.

We see that the radial velocity vanishes at the sphere surface $r = a$, as expected. There is a stagnation point at the front (a, π) and the rear $(a, 0)$ of the sphere. The maximum velocity occurs at the shoulder $(a, \pm\frac{1}{2}\pi)$, where $v_r = 0$ and $v_\theta = \mp 1.5 U_\infty$. The surface velocity distribution is

$$V_s = -v_\theta|_{r=a} = \tfrac{3}{2} U_\infty \sin \theta \tag{8.100}$$

Note the similarity to the cylinder surface velocity equal to $2U_\infty \sin \theta$ from Eq. (8.39) with zero circulation.

Equation (8.100) predicts, as expected, an adverse pressure gradient on the rear $(\theta < 90°)$ of the sphere. If we use this distribution with laminar boundary layer theory [for example, 15, p. 294], separation is computed to occur at about $\theta = 76°$, so that in the actual flow pattern of Fig. 7.14 a broad wake forms in the rear. This wake interacts with the free stream and causes Eq. (8.100) to be inaccurate even in the front of the sphere. The measured maximum surface velocity is equal only to about $1.3U_\infty$ and occurs at about $\theta = 107°$ (see Ref. 15, Sec. 4.10.4, for further details).

The Concept of Hydrodynamic Mass

When a body moves through a fluid, it must push a finite mass of fluid out of the way. If the body is accelerated, the surrounding fluid must also be accelerated. The body behaves as if it were heavier by an amount called the *hydrodynamic mass* (also called the *added* or *virtual mass*) of the fluid. If the instantaneous body velocity is $\mathbf{U}(t)$, the summation of forces must include this effect:

$$\Sigma \mathbf{F} = (m + m_h)\frac{d\mathbf{U}}{dt} \tag{8.101}$$

where m_h, the hydrodynamic mass, is a function of body shape, the direction of motion, and (to a lesser extent) flow parameters such as the Reynolds number.

According to potential theory [2, Sec. 6.4; 3, Sec. 9.22], m_h depends only on the shape and direction of motion and can be computed by summing the total kinetic energy of the fluid relative to the body and setting this equal to an equivalent body energy:

$$\text{KE}_{\text{fluid}} = \int \tfrac{1}{2} dm\, V_{\text{rel}}^2 = \tfrac{1}{2} m_h U^2 \tag{8.102}$$

The integration of fluid kinetic energy can also be accomplished by a body-surface integral involving the velocity potential [16, Sec. 11].

Consider the previous example of a sphere immersed in a uniform stream. By subtracting out the stream velocity we can replot the flow as in Fig. 8.33, showing the streamlines relative to the moving sphere. Note the similarity to the doublet flow in Fig. 8.30. The relative velocity components are found by subtracting U from Eqs. (8.99):

$$v_r = -\frac{Ua^3 \cos \theta}{r^3} \qquad v_\theta = -\frac{Ua^3 \sin \theta}{2r^3}$$

The element of fluid mass, in spherical polar coordinates, is

$$dm = \rho(2\pi r \sin \theta) r\, dr\, d\theta$$

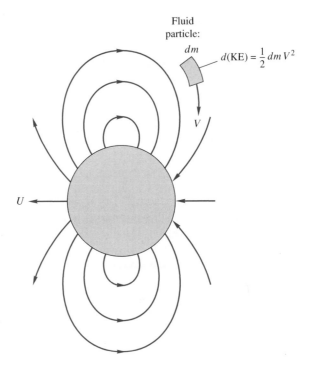

Fig. 8.33 Potential flow streamlines relative to a moving sphere. Compare with Figs. 8.30 and 8.32.

When dm and $V_{\text{rel}}^2 = v_r^2 + v_\theta^2$ are substituted into Eq. (8.102), the integral can be evaluated:

$$\text{KE}_{\text{fluid}} = \tfrac{1}{3}\rho\pi a^3 U^2$$

or

$$m_h(\text{sphere}) = \tfrac{2}{3}\rho\pi a^3 \tag{8.103}$$

Thus, according to potential theory, the hydrodynamic mass of a sphere equals one-half of its displaced mass, independent of the direction of motion.

A similar result for a cylinder moving normal to its axis can be computed from Eqs. (8.38) after subtracting out the stream velocity. The result is

$$m_h(\text{cylinder}) = \rho\pi a^2 L \tag{8.104}$$

for a cylinder of length L, assuming two-dimensional motion. The cylinder's hydrodynamic mass equals its displaced mass.

Tables of hydrodynamic mass for various body shapes and directions of motion are given by Patton [17]. See also Ref. 21.

8.9 Numerical Analysis

When potential flow involves complicated geometries or unusual stream conditions, the classical superposition scheme of Secs. 8.3 and 8.4 becomes less attractive. Conformal mapping of body shapes, by using the complex-variable technique of Sec. 8.5, is no longer popular. Numerical analysis is the appropriate modern approach, and at least three different approaches are in use:

1. The finite element method (FEM) [6, 19]
2. The finite difference method (FDM) [5, 20, 23–27]

3.   *a.*   Integral methods with distributed singularities [18]

   b.   The boundary element method [7, 38]

Methods 3*a* and 3*b* are closely related, having first been developed on an ad hoc basis by aerodynamicists in the 1960s [18] and then generalized into a multipurpose applied mechanics technique in the 1970s [7].

   Methods 1 (or FEM) and 2 (or FDM), though strikingly different in concept, are comparable in scope, mesh size, and general accuracy. We concentrate here on the latter method for illustration purposes.

The Finite Element Method

The finite element method [19] is applicable to all types of linear and nonlinear partial differential equations in physics and engineering. The computational domain is divided into small regions, usually triangular or quadrilateral. These regions are delineated with a finite number of *nodes* where the field variables—temperature, velocity, pressure, stream function, and so on—are to be calculated. The solution in each region is approximated by an algebraic combination of local nodal values. Then the approximate functions are integrated over the region, and their error is minimized, often by using a weighting function. This process yields a set of N algebraic equations for the N unknown nodal values. The nodal equations are solved simultaneously, by matrix inversion or iteration. For further details see Ref. 6 or 19.

The Finite Difference Method

Although textbooks on numerical analysis [5, 20] apply finite difference techniques to many different problems, here we concentrate on potential flow. The idea of FDM is to approximate the partial derivatives in a physical equation by "differences" between nodal values spaced a finite distance apart—a sort of numerical calculus. The basic partial differential equation is thus replaced by a set of algebraic equations for the nodal values. For potential (inviscid) flow, these algebraic equations are linear, but they are generally nonlinear for viscous flows. The solution for nodal values is obtained by iteration or matrix inversion. Nodal spacings need not be equal.

   Here we illustrate the two-dimensional Laplace equation, choosing for convenience the stream-function form

$$\frac{\partial^2 \psi}{\partial x^2} + \frac{\partial^2 \psi}{\partial y^2} = 0 \tag{8.105}$$

subject to known values of ψ along any body surface and known values of $\partial\psi/\partial x$ and $\partial\psi/\partial y$ in the free stream.

   Our finite difference technique divides the flow field into equally spaced nodes, as shown in Fig. 8.34. To economize on the use of parentheses or functional notation, subscripts i and j denote the position of an arbitrary, equally spaced node, and $\psi_{i,j}$ denotes the value of the stream function at that node:

$$\psi_{i,j} = \psi(x_0 + i\,\Delta x, \ y_0 + j\,\Delta y)$$

Thus $\psi_{i+1,j}$ is just to the right of $\psi_{i,j}$, and $\psi_{i,j+1}$ is just above.

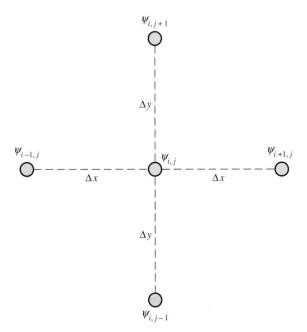

Fig. 8.34 Definition sketch for a two-dimensional rectangular finite difference grid.

An algebraic approximation for the derivative $\partial\psi/\partial x$ is

$$\frac{\partial\psi}{\partial x} \approx \frac{\psi(x+\Delta x, y) - \psi(x, y)}{\Delta x}$$

A similar approximation for the second derivative is

$$\frac{\partial^2\psi}{\partial x^2} \approx \frac{1}{\Delta x}\left[\frac{\psi(x+\Delta x, y) - \psi(x, y)}{\Delta x} - \frac{\psi(x, y) - \psi(x-\Delta x, y)}{\Delta x}\right]$$

The subscript notation makes these expressions more compact:

$$\frac{\partial\psi}{\partial x} \approx \frac{1}{\Delta x}(\psi_{i+1,j} - \psi_{i,j})$$

$$\frac{\partial^2\psi}{\partial x^2} \approx \frac{1}{\Delta x^2}(\psi_{i+1,j} - 2\psi_{i,j} + \psi_{i-1,j})$$

(8.106)

These formulas are exact in the calculus limit as $\Delta x \to 0$, but in numerical analysis we keep Δx and Δy finite, hence the term *finite differences*.

In an exactly similar manner we can derive the equivalent difference expressions for the *y* direction:

$$\frac{\partial\psi}{\partial y} \approx \frac{1}{\Delta y}(\psi_{i,j+1} - \psi_{i,j})$$

$$\frac{\partial^2\psi}{\partial y^2} \approx \frac{1}{\Delta y^2}(\psi_{i,j+1} - 2\psi_{i,j} + \psi_{i,j-1})$$

(8.107)

The use of subscript notation allows these expressions to be programmed directly into a scientific computer language such as BASIC or FORTRAN.

When (8.106) and (8.107) are substituted into Laplace's equation (8.105), the result is the algebraic formula

$$2(1 + \beta)\psi_{i,j} \approx \psi_{i-1,j} + \psi_{i+1,j} + \beta(\psi_{i,j-1} + \psi_{i,j+1}) \tag{8.108}$$

where $\beta = (\Delta x/\Delta y)^2$ depends on the mesh size selected. This finite difference model of Laplace's equation states that every nodal stream-function value $\psi_{i,j}$ is a linear combination of its four nearest neighbors.

The most commonly programmed case is a square mesh ($\beta = 1$), for which Eq. (8.108) reduces to

$$\boxed{\psi_{i,j} \approx \tfrac{1}{4}(\psi_{i,j+1} + \psi_{i,j-1} + \psi_{i+1,j} + \psi_{i-1,j})} \tag{8.109}$$

Thus, for a square mesh, each nodal value equals the arithmetic average of the four neighbors shown in Fig. 8.29. The formula is easily remembered and easily programmed. If P(I, J) is a subscripted variable stream function, the BASIC or FORTRAN statement of (8.109) is

```
P(I, J) = 0.25 * (P(I, J + 1) + P(I, J − 1) + P(I + 1, J) + P(I − 1, J))
```
$$\tag{8.110}$$

This is applied in iterative fashion sweeping over each of the internal nodes (I, J), with known values of P specified at each of the surrounding boundary nodes. Any initial guesses can be specified for the internal nodes P(I, J), and the iteration process will converge to the final algebraic solution in a finite number of sweeps. The numerical error, compared with the exact solution of Laplace's equation, is proportional to the square of the mesh size.

Convergence can be speeded up by the *successive overrelaxation* (SOR) method, discussed by Cebeci [5]. The modified SOR form of the iteration is

```
P(I, J) = P(I, J) + 0.25 * A * (P(I, J + 1) + P(I, J − 1)
          +  P(I + 1, J) + P(I − 1, J) − 4 * P(I, J))
```
$$\tag{8.111}$$

The recommended value of the SOR convergence factor A is about 1.7. Note that the value $A = 1.0$ reduces Eq. (8.111) to (8.110).

Let us illustrate the finite difference method with an example.

EXAMPLE 8.6

Make a numerical analysis, using $\Delta x = \Delta y = 0.2$ m, of potential flow in the duct expansion shown in Fig. 8.35. The flow enters at a uniform 10 m/s, where the duct width is 1 m, and is assumed to leave at a uniform velocity of 5 m/s, where the duct width is 2 m. There is a straight section 1 m long, a 45° expansion section, and a final straight section 1 m long.

Solution

Using the mesh shown in Fig. 8.35 results in 45 boundary nodes and 91 internal nodes, with i varying from 1 to 16 and j varying from 1 to 11. The internal points are modeled by

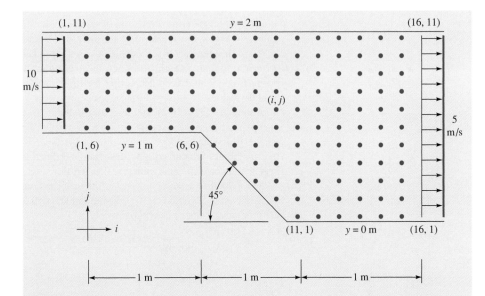

Fig. 8.35 Numerical model of potential flow through a two-dimensional 45° expansion. The nodal points shown are 20 cm apart. There are 45 boundary nodes and 91 internal nodes.

Eq. (8.110). For convenience, let the stream function be zero along the lower wall. Then since the volume flow is $(10 \text{ m/s})(1 \text{ m}) = 10 \text{ m}^2/\text{s}$ per unit depth, the stream function must equal $10 \text{ m}^2/\text{s}$ along the upper wall. Over the entrance and exit planes, the stream function must vary linearly to give uniform velocities:

Inlet: $\qquad\qquad\qquad\qquad \psi(1, J) = 2 * (J - 6) \qquad$ for J = 7 to 10

Exit: $\qquad\qquad\qquad\qquad \psi(16, J) = J - 1 \qquad$ for J = 2 to 10

All these boundary values must be input to the program and are shown printed in Fig. 8.36.

Initial guesses are stored for the internal points, say, zero or an average value of 5.0 m^2/s. The program then starts at any convenient point, such as the upper left (2, 10), and evaluates Eq. (8.110) at every internal point, repeating this sweep iteratively until there are no further changes (within some selected maximum change) in the nodal values. The results

ψ = 10.00	10.00	10.00	10.00	10.00	10.00	10.00	10.00	10.00	10.00	10.00	10.00	10.00	10.00	10.00	10.00
8.00	8.02	8.04	8.07	8.12	8.20	8.30	8.41	8.52	8.62	8.71	8.79	8.85	8.91	8.95	9.00
6.00	6.03	6.06	6.12	6.22	6.37	6.58	6.82	7.05	7.26	7.44	7.59	7.71	7.82	7.91	8.00
4.00	4.03	4.07	4.13	4.26	4.48	4.84	5.24	5.61	5.93	6.19	6.41	6.59	6.74	6.88	7.00
2.00	2.02	2.05	2.09	2.20	2.44	3.08	3.69	4.22	4.65	5.00	5.28	5.50	5.69	5.85	6.00
ψ = 0.00	0.00	0.00	0.00	0.00	0.00	1.33	2.22	2.92	3.45	3.87	4.19	4.45	4.66	4.84	5.00
						0.00	1.00	1.77	2.37	2.83	3.18	3.45	3.66	3.84	4.00
							0.00	0.80	1.42	1.90	2.24	2.50	2.70	2.86	3.00
								0.00	0.63	1.09	1.40	1.61	1.77	1.89	2.00
									0.00	0.44	0.66	0.79	0.87	0.94	1.00
										0.00	0.00	0.00	0.00	0.00	0.00

Fig. 8.36 Stream-function nodal values for the potential flow of Fig. 8.35. Boundary values are known inputs. Internal nodes are solutions to Eq. (8.110).

are the finite difference simulation of this potential flow for this mesh size; they are shown printed in Fig. 8.36 to three-digit accuracy. The reader should test a few nodes in Fig. 8.36 to verify that Eq. (8.110) is satisfied everywhere. The numerical accuracy of these printed values is difficult to estimate, since there is no known exact solution to this problem. In practice, one would keep decreasing the mesh size until there were no significant changes in nodal values.

Although Fig. 8.36 is the computer solution to the problem, these numbers must be manipulated to yield practical engineering results. For example, one can interpolate these numbers to sketch various streamlines of the flow. This is done in Fig. 8.37*a*. We see that the streamlines are curved both upstream and downstream of the corner regions, especially near the lower wall. This indicates that the flow is not one-dimensional.

The velocities at any point in the flow can be computed from finite difference formulas such as Eqs. (8.106) and (8.107). For example, at the point (I, J) = (3, 6), from Eq. (8.107), the horizontal velocity is approximately

$$u(3, 6) \approx \frac{\psi(3, 7) - \psi(3, 6)}{\Delta y} = \frac{2.09 - 0.00}{0.2} = 10.45 \text{ m/s}$$

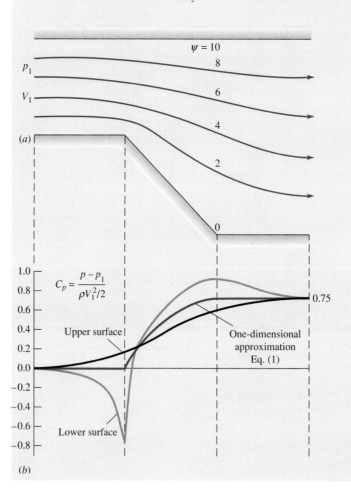

Fig. 8.37 Useful results computed from Fig. 8.36: (*a*) streamlines of the flow; (*b*) pressure coefficient distribution along each wall.

and the vertical velocity is zero from Eq. (8.106). Directly above this on the upper wall, we estimate

$$u(3, 11) \approx \frac{\psi(3, 11) - \psi(3, 10)}{\Delta y} = \frac{10.00 - 8.07}{0.2} = 9.65 \text{ m/s}$$

The flow is not truly one-dimensional in the entrance duct. The lower wall, which contains the diverging section, accelerates the fluid, while the flat upper wall is actually decelerating the fluid.

Another output function, useful in making boundary layer analyses of the wall regions, is the pressure distribution along the walls. If p_1 and V_1 are the pressure and velocity at the entrance (I = 1), conditions at any other point are computed from Bernoulli's equation (8.3), neglecting gravity:

$$p + \tfrac{1}{2}\rho V^2 = p_1 + \tfrac{1}{2}\rho V_1^2$$

which can be rewritten as a dimensionless pressure coefficient:

$$C_p = \frac{p - p_1}{\tfrac{1}{2}\rho V_1^2} = 1 - \left(\frac{V}{V_1}\right)^2$$

This determines p after V is computed from the stream-function differences in Fig. 8.36.

Figure 8.37b shows the computed wall pressure distributions as compared with the one-dimensional continuity approximation $V_1 A_1 \approx V(x)A(x)$, or

$$C_p(\text{one-dim}) \approx 1 - \left(\frac{A_1}{A}\right)^2 \tag{1}$$

The one-dimensional approximation, which is rather crude for this large (45°) expansion, lies between the upper and lower wall pressures. One-dimensional theory would be much more accurate for a 10° expansion.

Analyzing Fig. 8.37b, we predict that boundary layer separation will probably occur on the lower wall between the corners, where pressure is strongly rising (highly adverse gradient). Therefore potential theory is probably not too realistic for this flow, where viscous effects are strong. (Recall Figs. 6.27 and 7.8.)

Potential theory is *reversible;* that is, when we reverse the flow arrows in Fig. 8.37a, then Fig. 8.37b is still valid and would represent a 45° *contraction* flow. The pressure would fall on both walls (no separation) from $x = 3$ m to $x = 1$ m. Between $x = 1$ m and $x = 0$, the pressure rises on the lower surface, indicating possible separation, probably just downstream of the corner.

This example should give the reader an idea of the usefulness and generality of numerical analysis of fluid flows.

The Boundary Element Method

A relatively new technique for numerical solution of partial differential equations is the *boundary element method* (BEM). Reference 7 is an introductory textbook outlining the concepts of BEM. There are no interior elements. Rather, all nodes are placed on the boundary of the domain, as in Fig. 8.38. The "element" is a small piece of the boundary surrounding the node. The "strength" of the element can be either constant or variable.

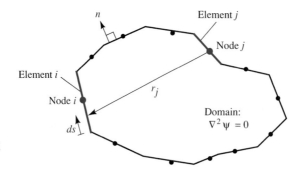

Fig. 8.38 Boundary elements of constant strength in plane potential flow.

For plane potential flow, the method takes advantage of the particular solution

$$\psi^* = \frac{1}{2\pi} \ln \frac{1}{r} \tag{8.112}$$

which satisfies Laplace's equation, $\nabla^2 \psi = 0$. Each element i is assumed to have a different strength ψ_i. Then r represents the distance from that element to any other point in the flow field. Summing all these elemental effects, with proper boundary conditions, will give the total solution to the potential flow problem.

At each element of the boundary, we typically know either the value of ψ or the value of $\partial\psi/\partial n$, where n is normal to the boundary. (Mixed combinations of ψ and $\partial\psi/\partial n$ are also possible but are not discussed here.) The correct strengths ψ_i are such that these boundary conditions are satisfied at every element. Summing these effects over N elements requires integration by parts plus a careful evaluation of the (singular) effect of element i upon itself. The mathematical details are given in Ref. 7. The result is a set of N algebraic equations for the unknown boundary values. In the case of elements of constant strength, the final expression is

$$\frac{1}{2}\psi_i + \sum_{j=1}^{N} \psi_j \left(\int_j \frac{\partial\psi^*}{\partial n} ds \right) = \sum_{j=1}^{N} \left(\frac{\partial\psi}{\partial n} \right)_j \left(\int_j \psi^* ds \right) \qquad i = 1 \text{ to } N \tag{8.113}$$

The integrals, which involve the logarithmic particular solution ψ^* from Eq. (8.112), are evaluated numerically for each element.

Equations (8.113) contain $2N$ element values, ψ_i and $(\partial\psi/\partial n)_i$, of which N are known from the given boundary conditions. The remaining N are solved simultaneously from Eqs. (8.113). Generally this completes the analysis—only the boundary solution is computed, and interior points are not studied. In most cases, the boundary velocity and pressure are all that is needed.

We illustrated the method with stream function ψ. Naturally the entire technique also applies to velocity potential ϕ, if we are given proper conditions on ϕ or $\partial\phi/\partial n$ at each boundary element. The method is readily extended to three dimensions [7, 38].

Reference 7 is a general introduction to boundary elements, while Ref. 38 emphasizes programming methods. Meanwhile, research continues. Dargush and Grigoriev [42] have developed a multilevel boundary element method for steady Stokes or *creeping* flows (see Sec. 7.6) in irregular geometries. Their scheme avoids the heavy memory and CPU-time requirements of most boundary element methods. They estimate that CPU time is reduced by a factor of 700,000 and required memory is reduced by a factor of 16,000.

Viscous Flow Computer Models

Our previous finite difference model of Laplace's equation, as in Eq. (8.109), was very well behaved and converged nicely with or without overrelaxation. Much more care is needed to model the full Navier-Stokes equations. The challenges are quite different, and they have been met to a large extent, so there are now many textbooks [5, 20, 23 to 27] on (fully viscous) *computational fluid dynamics* (CFD). This is not a textbook on CFD, but we will address some of the issues in this section.

One-Dimensional Unsteady Flow

We begin with a simplified problem, showing that even a single viscous term introduces new effects and possible instabilities. Recall (or review) Prob. P4.85, where a wall moves and drives a viscous fluid parallel to itself. Gravity is neglected. Let the wall be the plane $y = 0$, moving at a speed $U_0(t)$, as in Fig. 8.39. A uniform vertical grid, of spacing Δy, has nodes n at which the local velocity u_n^j is to be calculated, where superscript j denotes the time-step $j\Delta t$. The wall is $n = 1$. If $u = u(y, t)$ only and $v = w = 0$, continuity, $\nabla \cdot \mathbf{V} = 0$, is satisfied, and we need only solve the x-momentum Navier-Stokes equation:

$$\frac{\partial u}{\partial t} = \nu \frac{\partial^2 u}{\partial y^2} \tag{8.114}$$

where $\nu = \mu/\rho$. Utilizing the same finite difference approximations as in Eq. (8.106), we may model Eq. (8.114) algebraically as a forward time difference and a central spatial difference:

$$\frac{u_n^{j+1} - u_n^j}{\Delta t} \approx \nu \frac{u_{n+1}^j - 2u_n^j + u_{n-1}^j}{\Delta y^2}$$

Rearrange and find that we can solve explicitly for u_n at the next time-step $j + 1$:

$$u_n^{j+1} \approx (1 - 2\sigma)\, u_n^j + \sigma(u_{n-1}^j + u_{n+1}^j) \qquad \sigma = \frac{\nu \Delta t}{\Delta y^2} \tag{8.115}$$

Thus u at node n at the next time-step $j + 1$ is a weighted average of three previous values, similar to the "four-nearest-neighbors" average in the laplacian model of

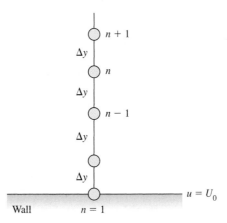

Fig. 8.39 An equally spaced finite difference mesh for one-dimensional viscous flow [Eq. (8.114)].

Eq. (8.109). Since the new velocity is calculated immediately, Eq. (8.115) is called an *explicit* model. It differs from the well-behaved laplacian model, however, because it may be *unstable*. The weighting coefficients in Eq. (8.115) must all be positive to avoid divergence. Now σ is positive, but $(1 - 2\sigma)$ may not be. Therefore, our explicit viscous flow model has a stability requirement:

$$\sigma = \frac{\nu \Delta t}{\Delta y^2} \leq \frac{1}{2} \qquad (8.116)$$

Normally one would first set up the mesh size Δy in Fig. 8.39, after which Eq. (8.116) would limit the time-step Δt. The solutions for nodal values would then be stable, but not necessarily that accurate. The mesh sizes Δy and Δt could be reduced to increase accuracy, similar to the case of the potential flow laplacian model (8.109).

For example, to solve Prob. P4.85 numerically, one sets up a mesh with plenty of nodes (30 or more Δy within the expected viscous layer); selects Δt according to Eq. (8.116); and sets two boundary conditions for all j: $u_1 = U_0 \sin \omega t$[8] and $u_N = 0$, where N is the outermost node. For initial conditions, perhaps assume the fluid initially at rest: $u_n^1 = 0$ for $2 \leq n \leq N - 1$. Sweeping the nodes $2 \leq n \leq N - 1$ using Eq. (8.115) (an Excel spreadsheet is excellent for this), one generates numerical values of u_n^j for as long as one desires. After an initial transient, the final "steady" fluid oscillation will approach the classical solution in viscous flow textbooks [15]. Try Prob. P8.115 to demonstrate this.

An Alternative Implicit Approach

In many finite difference problems, a stability limitation such as Eq. (8.116) requires an extremely small time-step. To allow larger steps, one can recast the model in an implicit fashion by evaluating the second-derivative model in Eq. (8.114) at the *next* time-step:

$$\frac{u_n^{j+1} - u_n^j}{\Delta t} \approx \nu \frac{u_{n+1}^{j+1} - 2u_n^{j+1} + u_{n-1}^{j+1}}{\Delta y^2}$$

This rearrangement is unconditionally stable for any σ, but now we have *three* unknowns:

$$-\sigma u_{n-1}^{j+1} + (1 + 2\sigma)u_n^{j+1} - \sigma u_{n+1}^{j+1} \approx u_n^j \qquad (8.117)$$

This is an *implicit* model, meaning that one must solve a large system of algebraic equations for the new nodal values at time $j + 1$. Fortunately, the system is narrowly banded, with the unknowns confined to the principal diagonal and its two nearest diagonals. In other words, the coefficient matrix of Eq. (8.117) is *tridiagonal*, a happy event. A direct method, called the *tridiagonal matrix algorithm* (TDMA), is available and explained in most CFD texts [20, 23 to 27]. If you have not learned TDMA yet, Eq. (8.117) converges satisfactorily by rearrangement and iteration:

$$u_n^{j+1} \approx \frac{u_n^j + \sigma(u_{n-1}^{j+1} + u_{n+1}^{j+1})}{1 + 2\sigma} \qquad (8.118)$$

[8]Finite differences are not analytical; one must set U_0 and ω equal to numerical values.

At each time-step $j + 1$, sweep the nodes $2 \leq n \leq N - 1$ over and over, using Eq. (8.118), until the nodal values have converged. This implicit method is stable for any σ, however large. To ensure accuracy, though, one should keep Δt and Δy small compared to the basic time and length scales of the problem. This author's habit is to keep Δt and Δy small enough that nodal values change no more than 10 percent from one (n, j) to the next.

EXAMPLE 8.7

SAE 30 oil at 20°C is at rest near a wall when the wall suddenly begins moving at a constant 1 m/s. Using the explicit model of Eq. (8.114), estimate the oil velocity at $y = 3$ cm after 1 second of wall motion.

Solution

For SAE 30 oil, from Table A.3, $\nu = 0.29/891 = 3.25$ E-4 m^2/s. For convenience in putting a node exactly at $y = 3$ cm, choose $\Delta y = 0.01$ m. The stability limit (8.116) is $\nu\Delta t/\Delta y^2 < 0.5$, or $\Delta t < 0.154$ s. Again for convenience, to hit $t = 1$ s on the nose, choose $\Delta t = 0.1$ s, or $\sigma = 0.3255$ and $(1 - 2\sigma) = 0.3491$. Then our explicit algebraic model (8.115) for this problem is

$$u_n^{j+1} \approx 0.3491\, u_n^j + 0.3255(u_{n-1}^j + u_{n-1}^j) \tag{1}$$

We apply this relation from $n = 2$ out to at least $n = N = 15$, to make sure that the desired value of u at $n = 3$ is accurate. The wall no-slip boundary requires $u_1^j = 1.0$ m/s = constant for all j. The outer boundary condition is $u_N = 0$. The initial conditions are $u_n^1 = 0$ for $n \geq 2$. We then apply Eq. (1) repeatedly for $n \geq 2$ until we reach $j = 11$, which corresponds to $t = 1$ s. This is easily programmed on a spreadsheet such as Excel. Here we print out only $j = 1$, 6, and 11 as follows:

j	t	u_1	u_2	u_3	u_4	u_5	u_6	u_7	u_8	u_9	u_{10}	u_{11}
1	0.000	1.000	0.000	0.000	0.000	0.000	0.000	0.000	0.000	0.000	0.000	0.000
6	0.500	1.000	0.601	0.290	0.107	0.027	0.004	0.000	0.000	0.000	0.000	0.000
11	1.000	1.000	0.704	0.446	**0.250**	0.123	0.052	0.018	0.005	0.001	0.000	0.000

Note: Units for t and u's are s and m/s, respectively.

Our numerical estimate is $u_4^{11} = u(3$ cm, 1 s$) \approx 0.250$ m/s, which is about 4 percent high—this problem has a known exact solution, $u = 0.241$ m/s [15]. We could improve the accuracy indefinitely by decreasing Δy and Δt.

Steady Two-Dimensional Laminar Flow

The previous example, unsteady one-dimensional flow, had only one viscous term and no convective accelerations. Let us look briefly at incompressible two-dimensional steady flow, which has four of each type of term, plus a nontrivial continuity equation:

Continuity:
$$\frac{\partial u}{\partial x} + \frac{\partial v}{\partial y} = 0 \tag{8.119a}$$

x momentum:
$$u\frac{\partial u}{\partial x} + v\frac{\partial u}{\partial y} = -\frac{1}{\rho}\frac{\partial p}{\partial x} + \nu\left(\frac{\partial^2 u}{\partial x^2} + \frac{\partial^2 u}{\partial y^2}\right) \qquad (8.119b)$$

y momentum
$$u\frac{\partial v}{\partial x} + v\frac{\partial v}{\partial y} = -\frac{1}{\rho}\frac{\partial p}{\partial y} + \nu\left(\frac{\partial^2 v}{\partial x^2} + \frac{\partial^2 v}{\partial y^2}\right) \qquad (8.119c)$$

These equations, to be solved for (u, v, p) as functions of (x, y), are familiar to us from analytical solutions in Chaps. 4 and 6. However, to a numerical analyst, they are odd, because there is no *pressure equation*—that is, a differential equation for which the dominant derivatives involve p. This situation has led to several different "pressure adjustment" schemes in the literature [20, 23 to 27], most of which manipulate the continuity equation to insert a pressure correction.

A second difficulty in Eqs. (8.119b and c) is the presence of nonlinear convective accelerations such as $u(\partial u/\partial x)$, which creates asymmetry in viscous flows. Early attempts, which modeled such terms with a central difference, led to numerical instability. The remedy is to relate convection finite differences solely to the *upwind* flow entering the cell, ignoring the downwind cell. For example, the derivative $\partial u/\partial x$ could be modeled, for a given cell, as $(u_{\text{upwind}} - u_{\text{cell}})/\Delta x$. Such improvements have made fully viscous CFD an effective tool, with various commercial user-friendly codes available. For details beyond our scope, see Refs. 20 and 23 to 27.

Mesh generation and gridding have also become quite refined in modern CFD. Figure 8.40 illustrates a CFD solution of two-dimensional flow past an NACA 66(MOD) hydrofoil [28]. The gridding in Fig. 8.40a is of the C type, which wraps around the leading edge and trails off behind the foil, thus capturing the important near-wall and wake details without wasting nodes in front or to the sides. The grid size is 262 by 91.

The CFD model for this hydrofoil flow is also quite sophisticated: a full Navier-Stokes solver with turbulence modeling [29] and allowance for cavitation bubble formation when surface pressures drop below the local vapor pressure. Figure 8.40b compares computed and experimental surface pressure coefficients for an angle of attack of 1°. The dimensionless pressure coefficient is defined as $C_p = (p_{\text{surface}} - p_\infty)/(\rho V_\infty^2/2)$. The agreement is excellent, as indeed it is also for cases where the hydrofoil cavitates [28]. Clearly, when properly implemented for the proper flow cases, CFD can be an extremely effective tool for engineers.

Commercial CFD Codes

The arrival of the third millennium has seen an enormous emphasis on computer applications in nearly every field, fluid mechanics being a prime example. It is now possible, at least for moderately complex geometries and flow patterns, to model on a computer, approximately, the equations of motion of fluid flow, with dedicated CFD textbooks available [5, 20, 23 to 27]. The flow region is broken into a fine grid of elements and nodes, which algebraically simulate the basic partial differential equations of flow. While simple two-dimensional flow simulations have long been reported and can be programmed as student exercises, three-dimensional flows, involving thousands or even millions of grid points, are now solvable with the modern supercomputer.

Although elementary computer modeling was treated briefly here, the general topic of CFD is essentially for advanced study or professional practice. The big

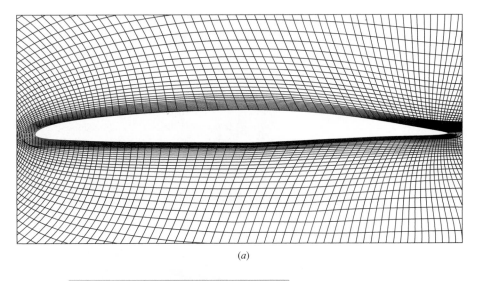

(a)

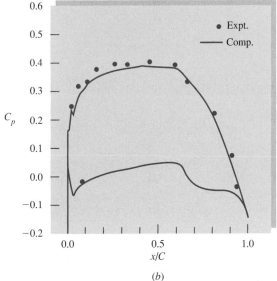

(b)

Fig. 8.40 CFD results for water flow past an NASA 66(MOD) hydrofoil *(from Ref. 28, with permission of the American Society of Mechanical Engineers):* (a) C gridding, 262 by 91 nodes; (b) surface pressures at $\alpha = 1°$.

change over the past decade is that engineers, rather than laboriously programming CFD problems themselves, can now take advantage of any of several commercial CFD codes. These extensive software packages allow engineers to construct a geometry and boundary conditions to simulate a given viscous flow problem. The software then grids the flow region and attempts to compute flow properties at each grid element. The convenience is great; the danger is also great. That is, computations are not merely automatic, like when using a hand calculator, but rather require care and concern from the user. Convergence and accuracy are real problems for the modeler. Use of the codes requires some art and experience. In particular, when the flow Reynolds number, $Re = \rho V L / \mu$, goes from moderate

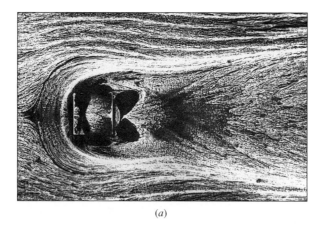

(a)

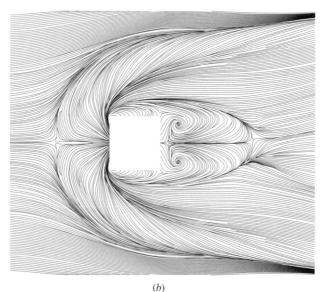

(b)

Fig. 8.41 Flow over a surface-mounted cube creates a complex and perhaps unexpected pattern: (a) experimental oil-streak visualization of surface flow at Re = 40,000 (based on cube height) *(Courtesy of Robert Martinuzzi with the permission of the American Society of Mechanical Engineers); (b) computational large-eddy simulation of the surface flow in (a) (from Ref. 32, courtesy of Kishan Shah, Stanford University);* and *(c)* a side view of the flow in (a) visualized by smoke generation and a laser light sheet *(Courtesy of Robert Martinuzzi with the permission of the American Society of Mechanical Engineers).*

(c)

(laminar flow) to high (turbulent flow), the accuracy of the simulation is no longer assured in any real sense. The reason is that turbulent flows are not completely resolved by the full equations of motion, and one resorts to using approximate turbulence models.

Turbulence models [29] are developed for particular geometries and flow conditions and may be inaccurate or unrealistic for others. This is discussed by Freitas [30], who compared eight different commercial code calculations (FLOW-3D, FLOTRAN, STAR-CD, N3S, CFD-ACE, FLUENT, CFDS-FLOW3D, and NISA/3D-FLUID) with experimental results for five benchmark flow experiments. Calculations were made by the vendors themselves. Freitas concludes that commercial codes, though promising in general, can be inaccurate for certain laminar and turbulent flow situations. Further research is recommended before engineers can truly rely upon such software to give generally accurate fluid flow predictions.

An example of erratic CFD results has already been mentioned here, namely, the drag and lift of a rotating cylinder, Fig. 8.15. Perhaps because the flow itself is physically unstable [41, 44], results computed by different workers are strikingly different: Some predicted forces are high, some low, some increase, some decrease. The text by Sengupta [27] discusses why several popular CFD schemes can be unreliable.

In spite of this warning to treat CFD codes with care, one should also realize that the results of a given CFD simulation can be spectacular. Figure 8.41 illustrates turbulent flow past a cube mounted on the floor of a channel whose clearance is twice the cube height. Compare Fig. 8.41*a*, a top view of the experimental surface flow [31] as visualized by oil streaks, with Fig. 8.41*b*, a CFD supercomputer result using the method of large-eddy simulation [32, 33]. The agreement is remarkable. The C-shaped flow pattern in front of the cube is caused by formation of a horseshoe vortex, as seen in a side view of the experiment [31] in Fig. 8.41*c*. Horseshoe vortices commonly result when surface shear flows meet an obstacle. We conclude that CFD has a tremendous potential for flow prediction.

Summary

This chapter has analyzed a highly idealized but very useful type of flow: inviscid, incompressible, irrotational flow, for which Laplace's equation holds for the velocity potential (8.1) and for the plane stream function (8.7). The mathematics is well developed, and solutions of potential flows can be obtained for practically any body shape.

Some solution techniques outlined here are (1) superposition of elementary line or point solutions in both plane and axisymmetric flow, (2) the analytic functions of a complex variable, (3) use of variable-strength vortex sheets, and (4) numerical analysis on a computer. Potential theory is especially useful and accurate for thin bodies such as airfoils. The only requirement is that the boundary layer be thin—in other words, that the Reynolds number be large.

For blunt bodies or highly divergent flows, potential theory serves as a first approximation, to be used as input to a boundary layer analysis. The reader should consult the advanced texts [for example, 2 to 4, 11 to 13] for further applications of potential theory. Section 8.9 discussed computational methods for viscous (nonpotential) flows.

Problems

Most of the problems herein are fairly straightforward. More difficult or open-ended assignments are labeled with an asterisk. Problems labeled with an EES icon will benefit from the use of the Engineering Equation Solver (EES), while problems labeled with a computer disk may require the use of a computer. The standard end-of-chapter problems P8.1 to P8.115 (categorized in the problem list here) are followed by word problems W8.1 to W8.7, comprehensive problems C8.1 to C8.7, and design projects D8.1 to D8.3.

Problem Distribution

Section	Topic	Problems
8.1	Introduction and review	P8.1–P8.7
8.2	Elementary plane flow solutions	P8.8–P8.17
8.3	Superposition of plane flows	P8.18–P8.34
8.4	Plane flow past closed-body shapes	P8.35–P8.59
8.5	The complex potential	P8.60–P8.71
8.6	Images	P8.72–P8.79
8.7	Airfoil theory: Two-dimensional	P8.80–P8.84
8.7	Airfoil theory: Finite-span wings	P8.85–P8.90
8.8	Axisymmetric potential flow	P8.91–P8.103
8.8	Hydrodynamic mass	P8.104–P8.105
8.9	Numerical methods	P8.106–P8.115

P8.1 Prove that the streamlines $\psi(r, \theta)$ in polar coordinates from Eqs. (8.10) are orthogonal to the potential lines $\phi(r, \theta)$.

P8.2 The steady plane flow in Fig. P8.2 has the polar velocity components $v_\theta = \Omega r$ and $v_r = 0$. Determine the circulation Γ around the path shown.

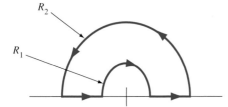

P8.2

P8.3 Using cartesian coordinates, show that each velocity component (u, v, w) of a potential flow satisfies Laplace's equation separately.

P8.4 Is the function $1/r$ a legitimate velocity potential in plane polar coordinates? If so, what is the associated stream function $\psi(r, \theta)$?

P8.5 Consider the velocity distribution $\mu = Ax$, $v = By$, $w = 0$. Find the conditions on A and B, if any, for which the flow has (a) a velocity potential and (b) a stream function.

P8.6 Given the plane polar coordinate velocity potential $\phi = Br^2 \cos(2\theta)$, where B is a constant. (a) Show that a stream function also exists. (b) Find the algebraic form of $\psi(r, \theta)$. (c) Find any stagnation points in this flow field.

P8.7 Consider a flow with constant density and viscosity. If the flow possesses a velocity potential as defined by Eq. (8.1), show that it exactly satisfies the full Navier-Stokes equations (4.38). If this is so, why for inviscid theory do we back away from the full Navier-Stokes equations?

P8.8 For the velocity distribution of Prob. P8.5, evaluate the circulation Γ around the rectangular closed curve defined by $(x, y) = (1, 1)$, $(3, 1)$, $(3, 2)$, and $(1, 2)$. Interpret your result, especially vis-à-vis the velocity potential.

P8.9 Consider the two-dimensional flow $u = -Ax$, $v = Ay$, where A is a constant. Evaluate the circulation Γ around the rectangular closed curve defined by $(x, y) = (1, 1)$, $(4, 1)$, $(4, 3)$, and $(1, 3)$. Interpret your result, especially vis-à-vis the velocity potential.

P8.10 A two-dimensional Rankine half-body, 8 cm thick, is placed in a water tunnel at 20°C. The water pressure far upstream along the body centerline is 105 kPa. What is the nose radius of the half-body? At what tunnel flow velocity will cavitation bubbles begin to form on the surface of the body?

P8.11 A power plant discharges cooling water through the manifold in Fig. P8.11, which is 55 cm in diameter and 8 m high and is perforated with 25,000 holes 1 cm in diameter. Does this manifold simulate a line source? If so, what is the equivalent source strength m?

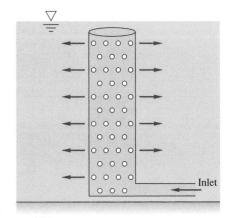

P8.11

P8.12 Consider the flow due to a vortex of strength K at the origin. Evaluate the circulation from Eq. (8.23) about the clockwise path from $(r, \theta) = (a, 0)$ to $(2a, 0)$ to $(2a, 3\pi/2)$ to $(a, 3\pi/2)$ and back to $(a, 0)$. Interpret the result.

P8.13 A small fish pond is approximated by a half-body shape, as shown in Fig. P8.13. Point O, which is 0.5 m from the left edge of the pond, is a water source delivering about 0.35 m³/s per meter of depth into the paper. Find the point B along the axis where the water velocity is approximately 25 cm/s.

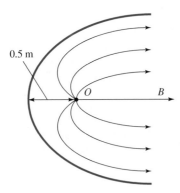

P8.13

P8.14 A tornado may be modeled as the circulating flow shown in Fig. P8.14, with $v_r = v_z = 0$ and $v_\theta(r)$ such that

$$v_\theta = \begin{cases} \omega r & r \le R \\ \dfrac{\omega R^2}{r} & r > R \end{cases}$$

Determine whether this flow pattern is irrotational in either the inner or outer region. Using the r-momentum equation (D.5) of App. D, determine the pressure distribution $p(r)$ in the tornado, assuming $p = p_\infty$ as $r \to \infty$. Find the location and magnitude of the lowest pressure.

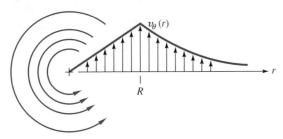

P8.14

P8.15 A category-3 hurricane on the Saffir-Simpson scale (<www.encyclopedia.com>) has a maximum velocity of 130 mi/h. Let the match-point radius be $R = 18$ km (see Fig. P8.14). Assuming sea-level standard conditions at

large r, (a) find the minimum pressure; (b) find the pressure at the match-point.

P8.16 Air flows at 1.2 m/s along a flat surface when it encounters a jet of air issuing from the horizontal wall at point A, as in Fig. P8.16. The jet volume flow is 0.4 m³/s per unit depth into the paper. If the jet is approximated as an inviscid line source, (a) locate the stagnation point S on the wall. (b) How far vertically will the jet flow extend into the stream?

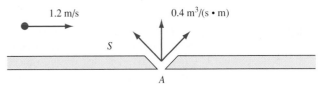

P8.16

P8.17 Find the position (x, y) on the upper surface of the half-body in Fig. 8.9a for which the local velocity equals the uniform stream velocity. What should be the pressure at this point?

P8.18 Plot the streamlines and potential lines of the flow due to a line source of strength m at $(a, 0)$ plus a source $3m$ at $(-a, 0)$. What is the flow pattern viewed from afar?

P8.19 Plot the streamlines and potential lines of the flow due to a line source of strength $3m$ at $(a, 0)$ plus a sink $-m$ at $(-a, 0)$. What is the pattern viewed from afar?

P8.20 Plot the streamlines of the flow due to a line vortex $+K$ at $(0, +a)$ and a vortex $-K$ at $(0, -a)$. What is the pattern viewed from afar?

*P8.21 Find the stream function and plot some streamlines for the combination of a line source $2m$ at $(x, y) = (+a, 0)$ and a line source m at $(-a, 0)$. Are there any stagnation points in the flow field?

P8.22 Consider inviscid stagnation flow, $\psi = Kxy$ (see Fig. 8.19b), superimposed with a source at the origin of strength m. Plot the resulting streamlines in the upper half-plane, using the length scale $a = (m/K)^{1/2}$. Give a physical interpretation of the flow pattern.

P8.23 Find the resultant velocity vector induced at point A in Fig. P8.23 by the uniform stream, vortex, and line source.

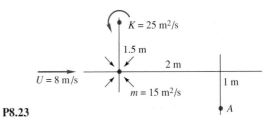

P8.23

P8.24 Line sources of equal strength $m = Ua$, where U is a reference velocity, are placed at $(x, y) = (0, a)$ and $(0, -a)$. Sketch the stream and potential lines in the upper half plane. Is $y = 0$ a "wall"? If so, sketch the pressure coefficient

$$C_p = \frac{p - p_0}{\frac{1}{2}\rho U^2}$$

along the wall, where p_0 is the pressure at $(0, 0)$. Find the minimum pressure point and indicate where flow separation might occur in the boundary layer.

P8.25 Let the vortex/sink flow of Eq. (4.134) simulate a tornado as in Fig. P8.26. Suppose that the circulation about the tornado is $\Gamma = 8500$ m²/s and that the pressure at $r = 40$ m is 2200 Pa less than the far-field pressure. Assuming inviscid flow at sea-level density, estimate (a) the appropriate sink strength $-m$, (b) the pressure at $r = 15$ m, and (c) the angle β at which the streamlines cross the circle at $r = 40$ m (see Fig. P8.25).

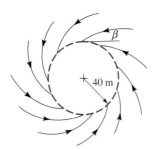

P8.25

P8.26 A coastal power plant takes in cooling water through a vertical perforated manifold, as in Fig. P8.26. The total volume flow intake is 110 m³/s. Currents of 25 cm/s flow past the manifold, as shown. Estimate (a) how far downstream and (b) how far normal to the paper the effects of the intake are felt in the ambient 8-m-deep waters.

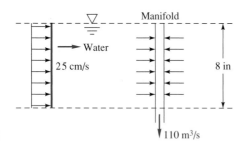

P8.26

P8.27 Water at 20C flows past a half-body as shown in Fig. P8.27. Measured pressures at points A and B are 160 kPa and 90 kPa, respectively, with uncertainties

of 3 kPa each. Estimate the stream velocity and its uncertainty.

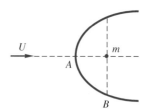

P8.27

P8.28 Sources of equal strength m are placed at the four symmetric positions $(x, y) = (a, a), (-a, a), (-a, -a)$, and $(a, -a)$. Sketch the streamline and potential line patterns. Do any plane "walls" appear?

P8.29 A uniform water stream, $U_\infty = 20$ m/s and $\rho = 998$ kg/m³, combines with a source at the origin to form a half-body. At $(x, y) = (0, 1.2$ m$)$, the pressure is 12.5 kPa less than p_∞. (a) Is this point outside the body? Estimate (b) the appropriate source strength m and (c) the pressure at the nose of the body.

P8.30 A tornado is simulated by a line sink $m = -1000$ m²/s plus a line vortex $K = 1600$ m²/s. Find the angle between any streamline and a radial line, and show that it is independent of both r and θ. If this tornado forms in sea-level standard air, at what radius will the local pressure be equivalent to 29 inHg?

P8.31 A Rankine half-body is formed as shown in Fig. P8.31. For the stream velocity and body dimension shown, compute (a) the source strength m in m²/s, (b) the distance a, (c) the distance h, and (d) the total velocity at point A.

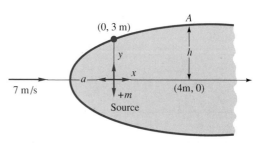

P8.31

P8.32 Sketch the streamlines, especially the body shape, due to equal line sources $+m$ at $(-a, 0)$ and $(+a, 0)$ plus a uniform stream $U_\infty = ma$.

P8.33 Sketch the streamlines, especially the body shape, due to equal line sources $+m$ at $(0, +a)$ and $(0, -a)$ plus a uniform stream $U_\infty = ma$.

P8.34 Consider three equally spaced sources of strength m placed at $(x, y) = (0, +a), (0, 0)$, and $(0, -a)$. Sketch the resulting streamlines, noting the position of any stagnation points. What would the pattern look like from afar?

P8.35 Consider three equal sources m in a triangular configuration: one at $(a/2, 0)$, one at $(-a/2, 0)$, and one at $(0, a)$. Plot the streamlines for this flow. Are there any stagnation points? *Hint:* Try the MATLAB contour command [34].

P8.36 When a line source–sink pair with $m = 2$ m^2/s is combined with a uniform stream, it forms a Rankine oval whose minimum dimension is 40 cm. If $a = 15$ cm, what are the stream velocity and the velocity at the shoulder? What is the maximum dimension?

P8.37 A Rankine oval 2 m long and 1 m high is immersed in a stream $U_\infty = 10$ m/s, as in Fig. P8.37. Estimate (*a*) the velocity at point A and (*b*) the location of point B where a particle approaching the stagnation point achieves its maximum deceleration.

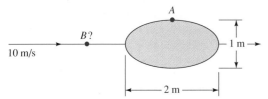

P8.37

P8.38 A uniform stream U in the x direction combines with a source m at $(a, 0)$ and a sink $-m$ at $(-a, 0)$. Plot the resulting streamlines and note any stagnation points.

P8.39 For the Rankine oval of Fig. P8.37, if the fluid is water at 20°C and the pressure far upstream along the body centerline is 115 kPa, determine the freestream velocity U_∞ for which the fluid will cavitate at point A.

P8.40 Modify the Rankine oval in Fig. P8.37 so that the stream velocity and body length are the same but the thickness is unknown (not 1 m). The fluid is water at 30°C and the pressure far upstream along the body centerline is 108 kPa. Find the body thickness for which cavitation will occur at point A.

P8.41 A Kelvin oval is formed by a line–vortex pair with $K = 9$ m^2/s, $a = 1$ m, and $U = 10$ m/s. What are the height, width, and shoulder velocity of this oval?

P8.42 For what value of $K/(U_\infty a)$ does the velocity at the shoulder of a Kelvin oval equal $4U_\infty$? What is the height h/a of this oval?

P8.43 Consider water at 20°C flowing at 6 m/s past a 1-m-diameter circular cylinder. What doublet strength λ in m^3/s is required to simulate this flow? If the stream pressure is 200 kPa, use inviscid theory to estimate the surface pressure at θ equal to (*a*) 180°, (*b*) 135°, and (*c*) 90°.

P8.44 Suppose that circulation is added to the cylinder flow of Prob. P8.43 sufficient to place the stagnation points at θ equal to 35° and 145°. What is the required vortex strength K in m^2/s? Compute the resulting pressure and surface velocity at (*a*) the stagnation points and (*b*) the upper and lower shoulders. What will the lift per meter of cylinder width be?

P8.45 If circulation K is added to the cylinder flow in Prob. P8.43, (*a*) for what value of K will the flow begin to cavitate at the surface? (*b*) Where on the surface will cavitation begin? (*c*) For this condition, where will the stagnation points lie?

P8.46 A cylinder is formed by bolting two semicylindrical channels together on the inside, as shown in Fig. P8.46. There are 10 bolts per meter of width on each side, and the inside pressure is 50 kPa (gage). Using potential theory for the outside pressure, compute the tension force in each bolt if the fluid outside is sea-level air.

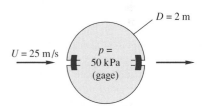

P8.46

P8.47 A circular cylinder is fitted with two surface-mounted pressure sensors, to measure p_a at $\theta = 180°$ and p_b at $\theta = 105°$. The intention is to use the cylinder as a stream velocimeter. Using inviscid theory, derive a formula for estimating U_∞ in terms of p_a, p_b, ρ, and the cylinder radius a.

*P8.48 Wind at U_∞ and p_∞ flows past a Quonset hut which is a half-cylinder of radius a and length L (Fig. P8.48). The internal pressure is p_i. Using inviscid theory, derive an expression for the upward force on the hut due to the difference between p_i and p_s.

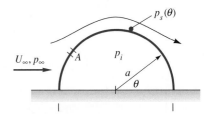

P8.48

P8.49 In strong winds the force in Prob. P8.48 can be quite large. Suppose that a hole is introduced in the hut roof at point A to make p_i equal to the surface pressure there.

At what angle θ should hole A be placed to make the net wind force zero?

P8.50 It is desired to simulate flow past a two-dimensional ridge or bump by using a streamline that passes above the flow over a cylinder, as in Fig. P8.50. The bump is to be $a/2$ high, where a is the cylinder radius. What is the elevation h of this streamline? What is U_{max} on the bump compared with stream velocity U?

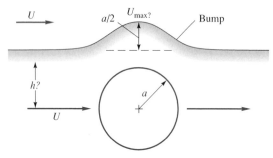

P8.50

P8.51 Consider inviscid flow along the streamline approaching the front stagnation point of a cylinder, as in Fig. 8.14a. The circulation is zero. Find (a) the maximum fluid deceleration along this streamline and (b) its position.

P8.52 The Flettner rotor sailboat in Fig. E8.3 has a water drag coefficient of 0.006 based on a wetted area of 45 ft². If the rotor spins at 220 r/min, find the maximum boat velocity that can be achieved in a 15-mi/h wind. What is the optimum angle between the boat and the wind?

P8.53 Modify Prob. P8.52 as follows. For the same sailboat data, find the wind velocity, in mi/h, that will drive the boat at an optimum speed of 8 kn parallel to its keel.

P8.54 The original Flettner rotor ship was approximately 100 ft long, displaced 800 tons, and had a wetted area of 3500 ft². As sketched in Fig. P8.54, it had two rotors

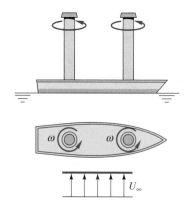

P8.54

50 ft high and 9 ft in diameter rotating at 750 r/min, which is far outside the range of Fig. 8.15. The measured lift and drag coefficients for each rotor were about 10 and 4, respectively. If the ship is moored and subjected to a crosswind of 25 ft/s, as in Fig. P8.54, what will the wind force parallel and normal to the ship centerline be? Estimate the power required to drive the rotors.

P8.55 Assume that the Flettner rotorship of Fig. P8.54 has a water resistance coefficient of 0.005. How fast will the ship sail in seawater at 20°C in a 20-ft/s wind if the keel aligns itself with the resultant force on the rotors? *Hint:* This is a problem in relative velocities.

P8.56 A proposed free-stream velocimeter would use a cylinder with pressure taps at $\theta = 180°$ and at $150°$. The pressure difference would be a measure of stream velocity U_∞. However, the cylinder must be aligned so that one tap exactly faces the free stream. Let the misalignment angle be δ; that is, the two taps are at $(180° + \delta)$ and $(150° + \delta)$. Make a plot of the percentage error in velocity measurement in the range $-20° < \delta < +20°$ and comment on the idea.

P8.57 In principle, it is possible to use rotating cylinders as aircraft wings. Consider a cylinder 30 cm in diameter, rotating at 2400 r/min. It is to lift a 55-kN airplane cruising at 100 m/s. What should the cylinder length be? How much power is required to maintain this speed? Neglect end effects on the rotating wing.

P8.58 Plot the streamlines due to the combined flow of a line sink $-m$ at the origin plus line sources $+m$ at $(a, 0)$ and $(4a, 0)$. *Hint:* A cylinder of radius $2a$ will appear.

P8.59 Consider inviscid flow past a cylinder with zero circulation, as in Fig. 8.14a. Find (a) the point on the front surface where the fluid acceleration a_{max} is maximum and (b) the magnitude of a_{max}. (c) If the stream velocity is 1 m/s, find the cylinder diameter for which a_{max} is 10 times the acceleration of gravity. Comment.

P8.60 One of the corner flow patterns of Fig. 8.19 is given by the cartesian stream function $\psi = A(3yx^2 - y^3)$. Which one? Can the correspondence be proved from Eq. (8.53)?

P8.61 Plot the streamlines of Eq. (8.53) in the upper right quadrant for $n = 4$. How does the velocity increase with x outward along the x axis from the origin? For what corner angle and value of n would this increase be linear in x? For what corner angle and n would the increase be as x^5?

P8.62 Combine stagnation flow, Fig. 8.19b, with a source at the origin:

$$f(z) = Az^2 + m \ln z$$

Plot the streamlines for $m = AL^2$, where L is a length scale. Interpret.

P8.63 The superposition in Prob. P8.62 leads to stagnation flow near a curved bump, in contrast to the flat wall of Fig. 8.19b. Determine the maximum height H of the bump as a function of the constants A and m.

P8.64 Consider the polar coordinate velocity potential $\phi = B\ r^{1.2}\cos(1.2\theta)$, where B is a constant. Determine whether $\nabla^2\phi = 0$. If so, find the associated stream function $\psi(r,\theta)$, plot the full streamline that includes the x axis ($\theta = 0$), and interpret.

P8.65 Potential flow past a wedge of half-angle θ leads to an important application of laminar boundary layer theory called the *Falkner-Skan flows* [15, pp. 239–245]. Let x denote distance along the wedge wall, as in Fig. P8.65, and let $\theta = 10°$. Use Eq. (8.53) to find the variation of surface velocity $U(x)$ along the wall. Is the pressure gradient adverse or favorable?

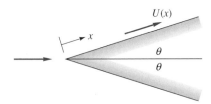

P8.65

*P8.66 The inviscid velocity along the wedge in Prob. P8.65 has the analytic form $U(x) = Cx^m$, where $m = n - 1$ and n is the exponent in Eq. (8.53). Show that, for any C and n, computation of the boundary layer by Thwaites's method, Eqs. (7.53) and (7.54), leads to a unique value of the Thwaites parameter λ. Thus wedge flows are called *similar* [15, p. 241].

P8.67 Investigate the complex potential function $f(z) = U_\infty(z + a^2/z)$ and interpret the flow pattern.

P8.68 Investigate the complex potential function $f(z) = U_\infty z + m\ln[(z + a)/(z-a)]$ and interpret the flow pattern.

P8.69 Investigate the complex potential $f(z) = A\cosh[\pi(z/a)]$, and plot the streamlines inside the region shown in Fig. P8.69. What hyphenated word (originally French) might describe such a flow pattern?

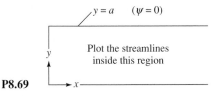

P8.69

P8.70 Show that the complex potential $f = U_\infty\{z + \tfrac{1}{4}a\coth[\pi(z/a)]\}$ represents flow past an oval shape placed midway between two parallel walls $y = \pm\tfrac{1}{2}a$. What is a practical application?

P8.71 Figure P8.71 shows the streamlines and potential lines of flow over a thin-plate weir as computed by the complex potential method. Compare qualitatively with Fig. 10.16a. State the proper boundary conditions at all boundaries. The velocity potential has equally spaced values. Why do the flow-net "squares" become smaller in the overflow jet?

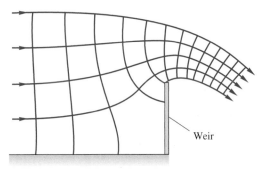

P8.71

P8.72 Use the method of images to construct the flow pattern for a source $+m$ near two walls, as shown in Fig. P8.72. Sketch the velocity distribution along the lower wall ($y = 0$). Is there any danger of flow separation along this wall?

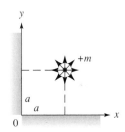

P8.72

P8.73 Set up an image system to compute the flow of a source at unequal distances from two walls, as in Fig. P8.73. Find the point of maximum velocity on the y axis.

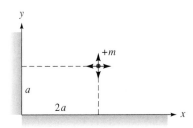

P8.73

P8.74 A positive line vortex K is trapped in a corner, as in Fig. P8.74. Compute the total induced velocity vector at point B, $(x, y) = (2a, a)$, and compare with the induced velocity when no walls are present.

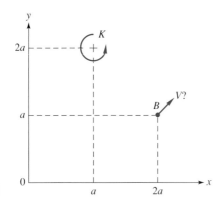

P8.74

P8.75 Using the four-source image pattern needed to construct the flow near a corner in Fig. P8.72, find the value of the source strength m that will induce a wall velocity of 4.0 m/s at the point $(x, y) = (a, 0)$ just below the source shown, if $a = 50$ cm.

P8.76 Use the method of images to approximate the flow pattern past a cylinder a distance $4a$ from a single wall, as in Fig. P8.76. To illustrate the effect of the wall, compute the velocities at corresponding points A, B, C, and D, comparing with a cylinder flow in an infinite expanse of fluid.

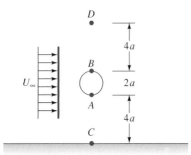

P8.76

P8.77 Discuss how the flow pattern of Prob. P8.58 might be interpreted to be an image system construction for circular walls. Why are there two images instead of one?

*P8.78 Indicate the system of images needed to construct the flow of a uniform stream past a Rankine half-body constrained between two parallel walls, as in Fig. P8.78. For the particular dimensions shown in this figure, estimate the position of the nose of the resulting half-body.

P8.79 Explain the system of images needed to simulate the flow of a line source placed unsymmetrically between two parallel walls as in Fig. P8.79. Compute the velocity on the lower wall at $x = a$. How many images are needed to estimate this velocity within 1 percent?

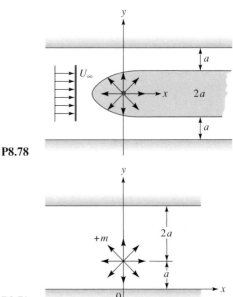

P8.78

P8.79

*P8.80 The beautiful expression for lift of a two-dimensional airfoil, Eq. (8.72), arose from applying the Joukowski transformation, $\zeta = z + a^2/z$, where $z = x + iy$ and $\zeta = \eta + i\beta$. The constant a is a length scale. The theory transforms a certain circle in the z plane into an airfoil in the ζ plane. Taking $a = 1$ unit for convenience, show that (a) a circle with center at the origin and radius > 1 will become an ellipse in the ζ plane and (b) a circle with center at $x = -\epsilon \ll 1$, $y = 0$, and radius $(1 + \epsilon)$ will become an airfoil shape in the ζ plane. *Hint:* The Excel spreadsheet is excellent for solving this problem.

*P8.81 Given an airplane of weight W, wing area A, aspect ratio AR, and flying at an altitude where the density is ρ. Assume all drag and lift is due to the wing, which has an infinite-span drag coefficient $C_{D\infty}$. Further assume sufficient thrust to balance whatever drag is calculated. (a) Find an algebraic expression for the *best cruise velocity* V_b, which occurs when the ratio of drag to speed is a minimum. (b) Apply your formula to the data in Prob. P7.119 for which a laborious graphing procedure gave an answer $V_b \approx 180$ m/s.

P8.82 The ultralight plane *Gossamer Condor* in 1977 was the first to complete the Kremer Prize figure-eight course under human power. Its wingspan was 29 m, with $C_{av} = 2.3$ m and a total mass of 95 kg. The drag coefficient was approximately 0.05. The pilot was able to deliver $\frac{1}{4}$ hp to propel the plane. Assuming two-dimensional flow at sea level, estimate (a) the cruise speed attained, (b) the lift coefficient, and (c) the horsepower required to achieve a speed of 15 kn.

P8.83 Two-dimensional lift–drag data for the NACA 2412 airfoil with 2 percent camber (from Ref. 12) may be curve-fitted accurately as follows:

$$C_L \approx 0.178 + 0.109\alpha - 0.00109\alpha^2$$
$$C_D \approx 0.0089 + 1.97\,\text{E-4}\,\alpha + 8.45\,\text{E-5}\,\alpha^2$$
$$- 1.35\,\text{E-5}\,\alpha^3 + 9.92\,\text{E-7}\,\alpha^4$$

with α in degrees in the range $-4° < \alpha < +10°$. Compare (a) the lift curve slope and (b) the angle of zero lift with theory, Eq. (8.72). (c) Prepare a polar lift–drag plot and compare with Fig. 7.26.

P8.84 Reference 12 contains inviscid theory calculations for the upper and lower surface velocity distributions $V(x)$ over an airfoil, where x is the chordwise coordinate. A typical result for small angle of attack is as follows:

x/c	V/U_∞(upper)	V/U_∞(lower)
0.0	0.0	0.0
0.025	0.97	0.82
0.05	1.23	0.98
0.1	1.28	1.05
0.2	1.29	1.13
0.3	1.29	1.16
0.4	1.24	1.16
0.6	1.14	1.08
0.8	0.99	0.95
1.0	0.82	0.82

Use these data, plus Bernoulli's equation, to estimate (a) the lift coefficient and (b) the angle of attack if the airfoil is symmetric.

P8.85 A wing of 2 percent camber, 5-in chord, and 30-in span is tested at a certain angle of attack in a wind tunnel with sea-level standard air at 200 ft/s and is found to have lift of 30 lbf and drag of 1.5 lbf. Estimate from wing theory (a) the angle of attack, (b) the minimum drag of the wing and the angle of attack at which it occurs, and (c) the maximum lift-to-drag ratio.

P8.86 An airplane has a mass of 20,000 kg and flies at 175 m/s at 5000-m standard altitude. Its rectangular wing has a 3-m chord and a symmetric airfoil at 2.5° angle of attack. Estimate (a) the wing span, (b) the aspect ratio, and (c) the induced drag.

P8.87 A freshwater boat of mass 400 kg is supported by a rectangular hydrofoil of aspect ratio 8, 2 percent camber, and 12 percent thickness. If the boat travels at 7 m/s and $\alpha = 2.5°$, estimate (a) the chord length, (b) the power required if $C_{D\infty} = 0.01$, and (c) the top speed if the boat is refitted with an engine that delivers 20 hp to the water.

P8.88 The Boeing 727 airplane has a gross weight of 125,000 lbf, a wing area of 1200 ft^2, and an aspect ratio of 6. It is fitted with two turbofan engines and cruises at 532 mi/h at 30,000-ft standard altitude. Assume for this problem that its airfoil is the NACA 2412 section described in Prob. P8.83. If we neglect all drag except the wing, what thrust is required from each engine for these conditions?

P8.89 The Beechcraft T-34C aircraft has a gross weight of 5500 lbf and a wing area of 60 ft^2 and flies at 322 mi/h at 10,000-ft standard altitude. It is driven by a propeller that delivers 300 hp to the air. Assume for this problem that its airfoil is the NACA 2412 section described in Prob. P8.83, and neglect all drag except the wing. What is the appropriate aspect ratio for the wing?

P8.90 NASA is developing a swing-wing airplane called the Bird of Prey [37]. As shown in Fig. P8.90, the wings pivot like a pocketknife blade: forward (a), straight (b), or backward (c). Discuss a possible advantage for each of these wing positions. If you can't think of one, read the article [37] and report to the class.

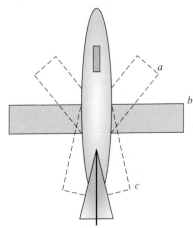

P8.90

P8.91 If $\phi(r, \theta)$ in axisymmetric flow is defined by Eq. (8.85) and the coordinates are given in Fig. 8.29, determine what partial differential equation is satisfied by ϕ.

P8.92 A point source with volume flow $Q = 30$ m^3/s is immersed in a uniform stream of speed 4 m/s. A Rankine half-body of revolution results. Compute (a) the distance from source to the stagnation point and (b) the two points (r, θ) on the body surface where the local velocity equals 4.5 m/s.

P8.93 The Rankine half-body of revolution (Fig. 8.31) could simulate the shape of a pitot-static tube (Fig. 6.30). According to inviscid theory, how far downstream from the nose should the static pressure holes be placed so that the local velocity is within ±0.5 percent of U_∞? Compare your answer with the recommendation $x \approx 8D$ in Fig. 6.30.

P8.94 Determine whether the Stokes streamlines from Eq. (8.86) are everywhere orthogonal to the Stokes potential lines from Eq. (8.87), as is the case for cartesian and plane polar coordinates.

P8.95 Show that the axisymmetric potential flow formed by superposition of a point source $+m$ at $(x, y) = (-a, 0)$, a point sink $-m$ at $(+a, 0)$, and a stream U_∞ in the x direction forms a Rankine body of revolution as in Fig. P8.95. Find analytic expressions for determining the length $2L$ and maximum diameter $2R$ of the body in terms of m, U_∞, and a.

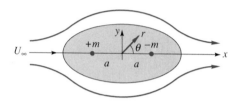

P8.95

P8.96 Consider inviscid flow along the streamline approaching the front stagnation point of a sphere, as in Fig. 8.32. Find (a) the maximum fluid deceleration along this streamline and (b) its position.

P8.97 The Rankine body of revolution in Fig. P8.97 is 60 cm long and 30 cm in diameter. When it is immersed in the low-pressure water tunnel as shown, cavitation may appear at point A. Compute the stream velocity U, neglecting surface wave formation, for which cavitation occurs.

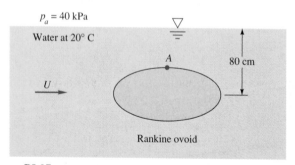

$p_a = 40$ kPa
Water at 20° C
A
80 cm
U

Rankine ovoid

P8.97

P8.98 We have studied the point source (sink) and the line source (sink) of infinite depth into the paper. Does it make any sense to define a finite-length line sink (source) as in Fig. P8.98? If so, how would you establish the mathematical properties of such a finite line sink? When combined with a uniform stream and a point source of equivalent strength as in Fig. P8.98, should a closed-body shape be formed? Make a guess and sketch some of these possible shapes for various values of the dimensionless parameter $m/(U_\infty L^2)$.

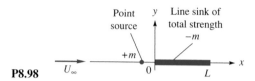

Point source y Line sink of total strength $-m$
$+m$
U_∞ 0 L

P8.98

*P8.99 Consider air flowing past a hemisphere resting on a flat surface, as in Fig. P8.99. If the internal pressure is p_i, find an expression for the pressure force on the hemisphere. By analogy with Prob. P8.49, at what point A on the hemisphere should a hole be cut so that the pressure force will be zero according to inviscid theory?

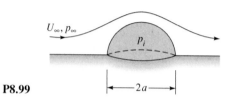

U_∞, p_∞
p_i
$2a$

P8.99

P8.100 A 1-m-diameter sphere is being towed at speed V in fresh water at 20°C as shown in Fig. P8.100. Assuming inviscid theory with an undistorted free surface, estimate the speed V in m/s at which cavitation will first appear on the sphere surface. Where will cavitation appear? For this condition, what will be the pressure at point A on the sphere, which is 45° up from the direction of travel?

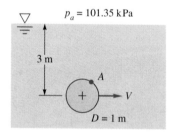

$p_a = 101.35$ kPa
3 m
A
V
$D = 1$ m

P8.100

P8.101 Consider a steel sphere (SG = 7.85) of diameter 2 cm, dropped from rest in water at 20°C. Assume a constant drag coefficient $C_D = 0.47$. Accounting for the sphere's hydrodynamic mass, estimate (a) its terminal velocity; and (b) the time to reach 99% of terminal velocity. Compare these to the results when hydrodynamic mass is neglected, $V_{terminal} \approx 1.95$ m/s and $t_{99\%} \approx 0.605$ s, and discuss.

P8.102 A golf ball weighs 0.102 lbf and has a diameter of 1.7 in. A professional golfer strikes the ball at an initial velocity of 250 ft/s, an upward angle of 20°, and a backspin (front of the ball rotating upward). Assume that the lift coefficient on the ball (based on frontal area) follows Fig. P7.108. If the ground is level and drag is neglected, make a simple analysis to predict the impact point (a) without spin and (b) with backspin of 7500 r/min.

P8.103 Consider inviscid flow past a sphere, as in Fig. 8.32. Find (a) the point on the front surface where the fluid acceleration a_{max} is maximum and (b) the magnitude of a_{max}. (c) If the stream velocity is 1 m/s, find the sphere diameter for which a_{max} is 10 times the acceleration of gravity. Comment.

P8.104 Consider a cylinder of radius a moving at speed U_∞ through a still fluid, as in Fig. P8.104. Plot the streamlines relative to the cylinder by modifying Eq. (8.32) to give the relative flow with $K = 0$. Integrate to find the total relative kinetic energy, and verify the hydrodynamic mass of a cylinder from Eq. (8.104).

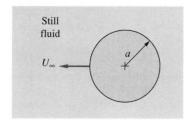

P8.104

*P8.105** In Table 7.2 the drag coefficient of a 4:1 elliptical cylinder in laminar boundary layer flow is 0.35. According to Patton [17], the hydrodynamic mass of this cylinder is $\pi\rho hb/4$, where b is width into the paper and h is the maximum thickness. Use these results to derive a formula from the time history $U(t)$ of the cylinder if it is accelerated from rest in a still fluid by the sudden application of a constant force F.

P8.106 Laplace's equation in plane polar coordinates, Eq. (8.11), is complicated by the variable radius. Consider the finite difference mesh in Fig. P8.106, with nodes (i, j) equally spaced $\Delta\theta$ and Δr apart. Derive a finite difference model for Eq. (8.11) similar to the cartesian expression (8.109).

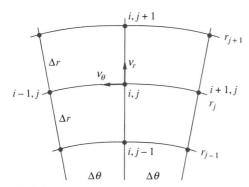

P8.106

P8.107 Set up the numerical problem of Fig. 8.35 for an expansion of 30°. A new grid system and a nonsquare mesh may be needed. Give the proper nodal equation and boundary conditions. If possible, program this 30° expansion and solve on a computer.

P8.108 Consider two-dimensional potential flow into a step contraction as in Fig. P8.108. The inlet velocity $U_1 = 7$ m/s, and the outlet velocity U_2 is uniform. The nodes (i, j) are labeled in the figure. Set up the complete finite difference algebraic relations for all nodes. Solve, if possible, on a computer and plot the streamlines in the flow.

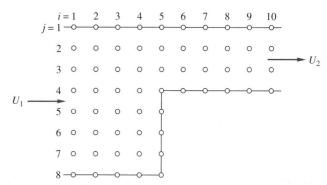

P8.108

P8.109 Consider inviscid flow through a two-dimensional 90° bend with a contraction, as in Fig. P8.109. Assume uniform flow at the entrance and exit. Make a finite difference computer analysis for small grid size (at least 150 nodes), determine the dimensionless pressure distribution along the walls, and sketch the streamlines. (You may use either square or rectangular grids.)

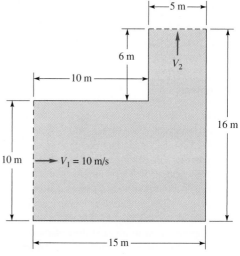

P8.109

P8.110   For fully developed laminar incompressible flow through a straight noncircular duct, as in Sec. 6.8, the Navier-Stokes equations (4.38) reduce to

$$\frac{\partial^2 u}{\partial y^2} + \frac{\partial^2 u}{\partial z^2} = \frac{1}{\mu}\frac{dp}{dx} = \text{const} < 0$$

where (y, z) is the plane of the duct cross section and x is along the duct axis. Gravity is neglected. Using a non-square rectangular grid $(\Delta x, \Delta y)$, develop a finite difference model for this equation, and indicate how it may be applied to solve for flow in a rectangular duct of side lengths a and b.

P8.111   Solve Prob. P8.110 numerically for a rectangular duct of side length b by $2b$, using at least 100 nodal points. Evaluate the volume flow rate and the friction factor, and compare with the results in Table 6.4:

$$Q \approx 0.1143\frac{b^4}{\mu}\left(-\frac{dp}{dx}\right) \qquad f\mathrm{Re}_{D_h} \approx 62.19$$

where $D_h = 4A/P = 4b/3$ for this case. Comment on the possible truncation errors of your model.

P8.112   In CFD textbooks [5, 23–27], one often replaces the left-hand sides of Eqs. (8.119b and c) with the following two expressions, respectively:

$$\frac{\partial}{\partial x}(u^2) + \frac{\partial}{\partial y}(vu) \qquad \text{and} \qquad \frac{\partial}{\partial x}(uv) + \frac{\partial}{\partial y}(v^2)$$

Are these equivalent expressions, or are they merely simplified approximations? Either way, why might these forms be better for finite difference purposes?

P8.113   Repeat Example 8.7 using the implicit method of Eq. (8.118). Take $\Delta t = 0.2$ s and $\Delta y = 0.01$ m, which ensures that an explicit model would diverge. Compare your accuracy with Example 8.7.

P8.114   If your institution has an online potential flow boundary element computer code, consider flow past a symmetric airfoil, as in Fig. P8.114. The basic shape of an NACA symmetric airfoil is defined by the function [12]

$$\frac{2y}{t_{max}} \approx 1.4845\zeta^{1/2} - 0.63\zeta - 1.758\zeta^2$$
$$+ 1.4215\zeta^3 - 0.5075\zeta^4$$

where $\zeta = x/C$ and the maximum thickness t_{max} occurs at $\zeta = 0.3$. Use this shape as part of the lower boundary for zero angle of attack. Let the thickness be fairly large, say, $t_{max} = 0.12, 0.15,$ or 0.18. Choose a generous number of nodes (≥ 60), and calculate and plot the velocity distribution V/U_∞ along the airfoil surface. Compare with the theoretical results in Ref. 12 for NACA 0012, 0015, or 0018 airfoils. If time permits, investigate the effect of the boundary lengths $L_1, L_2,$ and L_3, which can initially be set equal to the chord length C.

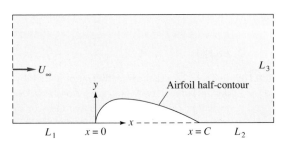

P8.114

P8.115   Use the explicit method of Eq. (8.115) to solve Prob. P4.85 numerically for SAE 30 oil at 20°C with $U_0 = 1$ m/s and $\omega = M$ rad/s, where M is the number of letters in your surname. (This author will solve the problem for $M = 5$.) When steady oscillation is reached, plot the oil velocity versus time at $y = 2$ cm.

Word Problems

W8.1   What simplifications have been made, in the potential flow theory of this chapter, which result in the elimination of the Reynolds number, Froude number, and Mach number as important parameters?

W8.2   In this chapter we superimpose many basic solutions, a concept associated with *linear* equations. Yet Bernoulli's equation (8.3) is *nonlinear*, being proportional to the square of the velocity. How, then, do we justify the use of superposition in inviscid flow analysis?

W8.3   Give a physical explanation of circulation Γ as it relates to the lift force on an immersed body. If the line integral defined by Eq. (8.23) is zero, it means that the integrand is a perfect differential—but of what variable?

W8.4   Give a simple proof of Eq. (8.46)—namely, that both the real and imaginary parts of a function $f(z)$ are laplacian if $z = x + iy$. What is the secret of this remarkable behavior?

W8.5   Figure 8.18 contains five body corners. Without carrying out any calculations, explain physically what the value of the inviscid fluid velocity must be at each of these five corners. Is any flow separation expected?

W8.6 Explain the Kutta condition physically. Why is it necessary?

W8.7 We have briefly outlined finite difference and boundary element methods for potential flow but have neglected the *finite element* technique. Do some reading and write a brief essay on the use of the finite element method for potential flow problems.

Comprehensive Problems

C8.1 Did you know that you can solve simple fluid mechanics problems with Microsoft Excel? The successive relaxation technique for solving the Laplace equation for potential flow problems is easily set up on a spreadsheet, since the stream function at each interior cell is simply the average of its four neighbors. As an example, solve for the irrotational potential flow through a contraction, as given in Fig. C8.1. *Note:* To avoid the "circular reference" error, you must turn on the iteration option. Use the help index for more information. For full credit, attach a printout of your spreadsheet, with stream function converged and the value of the stream function at each node displayed to four digits of accuracy.

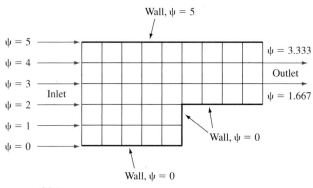

C8.1

C8.2 Use an explicit method, similar to but not identical to Eq. (8.115), to solve the case of SAE 30 oil at 20°C starting from rest near a *fixed* wall. Far from the wall, the oil accelerates linearly; that is, $u_\infty = u_N = at$, where $a = 9$ m/s^2. At $t = 1$ s, determine (*a*) the oil velocity at $y = 1$ cm and (*b*) the instantaneous boundary layer thickness (where $u \approx 0.99\, u_\infty$). *Hint:* There is a nonzero pressure gradient in the outer (nearly shear-free) stream, $n = N$, which must be included in Eq. (8.114) and your explicit model.

C8.3 Consider plane inviscid flow through a symmetric diffuser, as in Fig. C8.3. Only the upper half is shown. The flow is to expand from inlet half-width h to exit half-width $2h$, as shown. The expansion angle θ is 18.5° ($L \approx 3h$). Set up a nonsquare potential flow mesh for this problem, and calculate and plot (*a*) the velocity distribution and (*b*) the pressure coefficient along the centerline. Assume uniform inlet and exit flows.

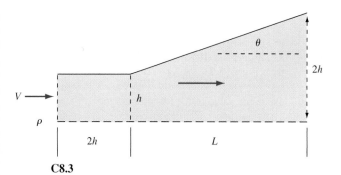

C8.3

C8.4 Use potential flow to approximate the flow of air being sucked up into a vacuum cleaner through a two-dimensional slit attachment, as in Fig. C8.4. In the xy plane through the centerline of the attachment, model the flow as a line sink of strength $(-m)$, with its axis in the z direction at height a above the floor. (*a*) Sketch the streamlines and locate any stagnation points in the flow. (*b*) Find the magnitude of velocity $V(x)$ along the floor in terms of the parameters a and m. (*c*) Let the pressure far away be p_∞, where velocity is zero. Define a velocity scale $U = m/a$. Determine the variation of dimensionless pressure coefficient, $C_p = (p - p_\infty)/(\rho U^2/2)$, along the floor. (*d*) The vacuum cleaner is most effective where C_p is a minimum—that is, where velocity is maximum. Find the locations of minimum pressure coefficient along the x axis. (*e*) At which points along the x axis do you expect the vacuum cleaner to work most effectively? Is it best at $x = 0$ directly beneath the slit, or at some other x location along the floor? Conduct a scientific experiment at home with a vacuum cleaner and

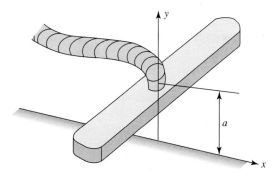

C8.4

some small pieces of dust or dirt to test your prediction. Report your results and discuss the agreement with prediction. Give reasons for any disagreements.

C8.5 Consider a three-dimensional, incompressible, irrotational flow. Use the following two methods to prove that the viscous term in the Navier-Stokes equation is identically zero: (*a*) using vector notation; and (*b*) expanding out the scalar terms and substituting terms from the definition of irrotationality.

C8.6 Reconsider the lift–drag data for the NACA 4412 airfoil from Prob. P8.83. (*a*) Again draw the polar lift–drag plot and compare qualitatively with Fig. 7.26. (*b*) Find the maximum value of the lift-to-drag ratio. (*c*) Demonstrate a straight-line construction on the polar plot that will immediately yield the maximum L/D in (*b*). (*d*) If an aircraft could use this two-dimensional wing in actual flight (no induced drag) and had a perfect pilot, estimate how far (in miles) this aircraft could glide to a sea-level runway if it lost power at 25,000 ft altitude.

C8.7 Find a formula for the stream function for flow of a doublet of strength λ a distance a from a wall, as in Fig. C8.7. (*a*) Sketch the streamlines. (*b*) Are there any stagnation points? (*c*) Find the maximum velocity along the wall and its position.

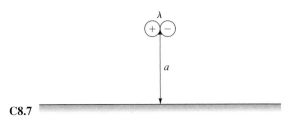

C8.7

Design Projects

D8.1 In 1927, Theodore von Kármán developed a scheme to use a uniform stream, plus a row of sources and sinks, to generate an arbitrary closed-body shape. A schematic of the idea is sketched in Fig. D8.1. The body is symmetric and at zero angle of attack. A total of N sources and sinks are distributed along the axis within the body, with strengths m_i at positions x_i, for $i = 1$ to N. The object is to find the correct distribution of strengths that approximates a given body shape $y(x)$ at a finite number of surface locations and then to compute the approximate surface velocity and pressure. The technique should work for either two-dimensional bodies (distributed line sources) or bodies of revolution (distributed point sources).

For our body shape let us select the NACA 0018 airfoil, given by the formula in Prob. P8.114 with $t_{max} = 0.18$. Develop the ideas stated here into N simultaneous algebraic equations that can be used to solve for the N unknown line source/sink strengths. Then program your equations for a computer, with $N \geq 20$; solve for m_i; compute the surface velocities; and compare with the theoretical velocities for this shape in Ref. 12. Your goal should be to achieve accuracy within ±1 percent of the classical results. If necessary, you should adjust N and the locations of the sources.

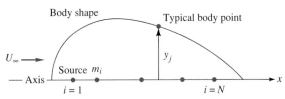

D8.1

D8.2 Modify Prob. D8.1 to solve for the point-source distribution that approximates an "0018" body-of-revolution shape. Since no theoretical results are published, simply make sure that your results converge to ±1 percent.

D8.3 Consider water at 20°C flowing at 12 m/s in a water channel. A Rankine oval cylinder, 40 cm long, is to be placed parallel to the flow, where the water static pressure is 120 kPa. The oval's thickness is a design parameter. Prepare a plot of the minimum pressure on the oval's surface as a function of body thickness. Especially note the thicknesses where (*a*) the local pressure is 50 kPa and (*b*) cavitation first occurs on the surface.

References

1. M. Rahman (ed.), *Potential Flow of Fluids*, WIT Press, Billerica, MA, 1995.
2. J. M. Robertson, *Hydrodynamics in Theory and Application*, Prentice-Hall, Englewood Cliffs, NJ, 1965.
3. L. M. Milne-Thomson, *Theoretical Hydrodynamics*, 4th ed., Dover, New York, 1996.
4. T. Ransford, *Potential Theory in the Complex Plane*, Cambridge Univ. Press, New York, 1995.
5. T. Cebeci, *Computational Fluid Dynamics for Engineers*, Springer-Verlag, New York, 2005.
6. R. W. Lewis, P. Nithiarasu, and K. N. Seetharamu, *Fundamentals of the Finite Element Method for Heat and Fluid Flow*, Wiley, New York, 2004.
7. L. C. Wrobel, *The Boundary Element Method: Applications in Thermo-Fluids and Acoustics*, vol. 1, Wiley, New York, 2002.

8. A. D. Moore, "Fields from Fluid Flow Mappers," *J. Appl. Phys.,* vol. 20, 1949, pp. 790–804.

9. H. J. S. Hele-Shaw, "Investigation of the Nature of the Surface Resistance of Water and of Streamline Motion under Certain Experimental Conditions," *Trans. Inst. Nav. Archit.,* vol. 40, 1898, p. 25.

10. S. W. Churchill, *Viscous Flows: The Practical Use of Theory,* Butterworth, Stoneham, MA, 1988.

11. J. D. Anderson, Jr., *Fundamentals of Aerodynamics,* 4th ed., McGraw-Hill, New York, 2007.

12. I. H. Abbott and A. E. von Doenhoff, *Theory of Wing Sections,* Dover, New York, 1981.

13. F. O. Smetana, *Introductory Aerodynamics and Hydrodynamics of Wings and Bodies: A Software-Based Approach,* AIAA, Reston, VA, 1997.

14. L. Prandtl, "Applications of Modern Hydrodynamics to Aeronautics," *NACA Rep. 116,* 1921.

15. F. M. White, *Viscous Fluid Flow,* 3d ed., McGraw-Hill, New York, 2005.

16. C. S. Yih, *Fluid Mechanics,* McGraw-Hill, New York, 1969.

17. K. T. Patton, "Tables of Hydrodynamic Mass Factors for Translational Motion," *ASME Winter Annual Meeting,* Paper 65-WA/UNT-2, 1965.

18. J. L. Hess and A. M. O. Smith, "Calculation of Nonlifting Potential Flow about Arbitrary Three-Dimensional Bodies," *J. Ship Res.,* vol. 8, 1964, pp. 22–44.

19. K. H. Huebner, *The Finite Element Method for Engineers,* 4th ed., Wiley, New York, 2001.

20. J. C. Tannehill, D. A. Anderson, and R. H. Pletcher, *Computational Fluid Mechanics and Heat Transfer,* 2d ed., Taylor and Francis, Bristol, PA, 1997.

21. J. N. Newman, *Marine Hydrodynamics,* M.I.T. Press, Cambridge, MA, 1977.

22. P. T. Tokumaru and P. E. Dimotakis, "The Lift of a Cylinder Executing Rotary Motions in a Uniform Flow," *J. Fluid Mechanics,* vol. 255, 1993, pp. 1–10.

23. J. H. Ferziger and M. Peric, *Computational Methods for Fluid Dynamics,* Springer-Verlag, New York, 1996.

24. P. J. Roache, *Fundamentals of Computational Fluid Dynamics,* Hermosa Pub., Albuquerque, NM, 1998.

25. P. Wesseling, *Principles of Computational Fluid Dynamics,* Springer-Verlag, New York, 2001.

26. J. D. Anderson, *Computational Fluid Dynamics: The Basics with Applications,* McGraw-Hill, New York, 1995.

27. T. P. Sengupta, *Fundamentals of Computational Fluid Dynamics,* Orient Longman, Hyderabad, India, 2004.

28. M. Deshpande, J. Feng, and C. L. Merkle, "Numerical Modeling of the Thermodynamic Effects of Cavitation," *J. Fluids Eng.,* June 1997, pp. 420–427.

29. P. A. Durbin and R. B. A. Pettersson, *Statistical Theory and Modeling for Turbulent Flows,* Wiley, New York, 2001.

30. C. J. Freitas, "Perspective: Selected Benchmarks from Commercial CFD Codes," *J. Fluids Eng.,* vol. 117, June 1995, pp. 208–218.

31. R. Martinuzzi and C. Tropea, "The Flow around Surface-Mounted, Prismatic Obstacles in a Fully Developed Channel Flow," *J. Fluids Eng.,* vol. 115, March 1993, pp. 85–92.

32. K. B. Shah and J. H. Ferzier, "Fluid Mechanicians View of Wind Engineering: Large Eddy Simulation of Flow Past a Cubic Obstacle," *J. Wind Engineering and Industrial Aerodynamics,* vol. 67–68, 1997, pp. 221–224.

33. P. Sagaut et al., *Large Eddy Simulations for Incompressible Flows: An Introduction,* Springer-Verlag, New York, 2001.

34. W. J. Palm, *Introduction to MATLAB 7 for Engineers,* McGraw-Hill, New York, 2003.

35. A. Gilat, *MATLAB: An Introduction with Applications,* 2d ed., Wiley, New York, 2004.

36. J. W. Hoyt and R. H. J. Sellin, "Flow over Tube Banks—A Visualization Study," *J. Fluids Eng.,* vol. 119, June 1997, pp. 480–483.

37. S. Douglass, "Switchblade Fighter Bomber," *Popular Science,* Nov. 2000, pp. 52–55.

38. G. Beer, *Programming the Boundary Element Method: An Introduction for Engineers,* Wiley, New York, 2001.

39. J. D. Anderson, *A History of Aerodynamics and Its Impact on Flying Machines,* Cambridge University Press, Cambridge, UK, 1997.

40. B. Robins, *Mathematical Tracts 1 & 2,* J. Nourse, London, 1761.

41. T. K. Sengupta, A. Kasliwal, S. De, and M. Nair, "Temporal Flow Instability for Magnus-Robins Effect at High Rotation Rates," *J. Fluids and Structures,* vol. 17, 2003, pp. 941–953.

42. G. F. Dargush and M. M. Grigoriev, "Fast and Accurate Solutions of Steady Stokes Flows Using Multilevel Boundary Element Methods," *J. Fluids Eng.,* vol. 127, July 2005, pp. 640–646.

43. R. H. Kirchhoff, *Potential Flows: Computer Graphic Solutions,* Marcel Dekker, New York, 2001.

44. H. Werle, "Hydrodynamic Visualization of the Flow around a Streamlined Cylinder with Suction: Cousteau-Malavard Turbine Sail Model," *Le Recherche Aerospatiale,* vol. 4, 1984, pp. 29–38.

45. T. K. Sengupta and S. R. Talla, "Robins-Magnus Effect: A Continuing Saga," *Current Science,* vol. 86, no. 7, 2004, pp. 1033–1036.

46. P. R. Spalart, "Airplane Trailing Vortices," *Annual Review Fluid Mechanics,* vol. 30, 1998, pp. 107–138.

This is a one-in-a-million shot of the F-18 Hornet fighter plane passing through the speed of sound. Ensign John Gay, a photographer for the U.S. Navy, caught the photo just as the aircraft approached sonic speed in wet air. The speed is slightly below Ma = 1, and visible condensation shocks form on the surfaces where local velocity is supersonic. In an instant, the F-18 will be fully supersonic, and these shocks will be replaced by sharp conical shocks from the nose and other leading edges of the airplane. *(Photo supplied by the U.S. Navy.)*

Chapter 9
Compressible Flow

Motivation. All eight of our previous chapters have been concerned with "low-speed" or "incompressible" flow, where the fluid velocity is much less than its speed of sound. In fact, we did not even develop an expression for the speed of sound of a fluid. That is done in this chapter.

When a fluid moves at speeds comparable to its speed of sound, density changes become significant and the flow is termed *compressible*. Such flows are difficult to obtain in liquids, since high pressures of order 1000 atm are needed to generate sonic velocities. In gases, however, a pressure ratio of only 2:1 will likely cause sonic flow. Thus compressible gas flow is quite common, and this subject is often called *gas dynamics*.

Probably the two most important and distinctive effects of compressibility on flow are (1) *choking,* wherein the duct flow rate is sharply limited by the sonic condition, and (2) *shock waves,* which are nearly discontinuous property changes in a supersonic flow. The purpose of this chapter is to explain such striking phenomena and to familiarize the reader with engineering calculations of compressible flow.

Speaking of calculations, the present chapter is made to order for the Engineering Equation Solver (EES) in App. E. Compressible flow analysis is filled with scores of complicated algebraic equations, most of which are very difficult to manipulate or invert. Consequently, for nearly a century, compressible flow textbooks have relied on extensive tables of Mach number relations (see App. B) for numerical work. With EES, however, any set of equations in this chapter can be typed out and solved for any variable—see part (*b*) of Example 9.13 for an especially intricate example. With such a tool, App. B serves only as a backup and indeed may soon vanish from textbooks.

9.1 Introduction: Review of Thermodynamics

We took a brief look in Chap. 4 [Eqs. (4.13) to (4.17)] to see when we might safely neglect the compressibility inherent in every real fluid. We found that the proper criterion for a nearly incompressible flow was a small Mach number

$$\text{Ma} = \frac{V}{a} \ll 1$$

where V is the flow velocity and a is the speed of sound of the fluid. Under small Mach number conditions, changes in fluid density are everywhere small in the flow field. The energy equation becomes uncoupled, and temperature effects can be either ignored or put aside for later study. The equation of state degenerates into the simple statement that density is nearly constant. This means that an incompressible flow requires only a momentum and continuity analysis, as we showed with many examples in Chaps. 7 and 8.

This chapter treats compressible flows, which have Mach numbers greater than about 0.3 and thus exhibit nonnegligible density changes. If the density change is significant, it follows from the equation of state that the temperature and pressure changes are also substantial. Large temperature changes imply that the energy equation can no longer be neglected. Therefore the work is doubled from two basic equations to four

1. Continuity equation
2. Momentum equation
3. Energy equation
4. Equation of state

to be solved simultaneously for four unknowns: pressure, density, temperature, and flow velocity (p, ρ, T, V). Thus the general theory of compressible flow is quite complicated, and we try here to make further simplifications, especially by assuming a reversible adiabatic or *isentropic* flow.

The Mach Number

The Mach number is the dominant parameter in compressible flow analysis, with different effects depending on its magnitude. Aerodynamicists especially make a distinction between the various ranges of Mach number, and the following rough classifications are commonly used:

Ma < 0.3:	*incompressible flow,* where density effects are negligible.
0.3 < Ma < 0.8:	*subsonic flow,* where density effects are important but no shock waves appear.
0.8 < Ma < 1.2:	*transonic flow,* where shock waves first appear, dividing subsonic and supersonic regions of the flow. Powered flight in the transonic region is difficult because of the mixed character of the flow field.
1.2 < Ma < 3.0:	*supersonic flow,* where shock waves are present but there are no subsonic regions.
3.0 < Ma:	*hypersonic flow* [11], where shock waves and other flow changes are especially strong.

The numerical values listed are only rough guides. These five categories of flow are appropriate to external high-speed aerodynamics. For internal (duct) flows, the most important question is simply whether the flow is subsonic (Ma < 1) or supersonic (Ma > 1), because the effect of area changes reverses, as we show in Sec. 9.4. Since supersonic flow effects may go against intuition, you should study these differences carefully.

The Specific-Heat Ratio

In addition to geometry and Mach number, compressible flow calculations also depend on a second dimensionless parameter, the *specific-heat ratio* of the gas:

$$k = \frac{c_p}{c_v} \tag{9.1}$$

Earlier, in Chaps. 1 and 4, we used the same symbol k to denote the thermal conductivity of a fluid. We apologize for the duplication; thermal conductivity does not appear in these later chapters of the text.

Recall from Fig. 1.5 that k for the common gases decreases slowly with temperature and lies between 1.0 and 1.7. Variations in k have only a slight effect on compressible flow computations, and air, $k \approx 1.40$, is the dominant fluid of interest. Therefore, although we assign some problems involving other gases like steam and CO_2 and helium, the compressible flow tables in App. B are based solely on the single value $k = 1.40$ for air.

This text contains only a single chapter on compressible flow, but, as usual, whole books have been written on the subject. The previous edition listed some 30 books, but let us trim that now to recent or classical texts. References 1 to 4 are introductory or intermediate treatments, while Refs. 5 to 10 are advanced books. One can also become specialized within this specialty of compressible flow. Reference 11 concerns *hypersonic flow*—that is, at very high Mach numbers. Reference 12 explains the exciting new technique of direct simulation of gas flows with a *molecular dynamics model*. Compressible flow is also well suited for computational fluid dynamics (CFD), as described in Ref. 13. Finally, a short, thoroughly readable (no calculus) Ref. 14 describes the principles and promise of high-speed (supersonic) flight. From time to time we shall defer some specialized topic to these other texts.

We note in passing that at least two flow patterns depend strongly on very small density differences, acoustics, and natural convection. Acoustics [7, 9] is the study of sound wave propagation, which is accompanied by extremely small changes in density, pressure, and temperature. Natural convection is the gentle circulating pattern set up by buoyancy forces in a fluid stratified by uneven heating or uneven concentration of dissolved materials. Here we are concerned only with steady compressible flow where the fluid velocity is of magnitude comparable to that of the speed of sound.

The Perfect Gas

In principle, compressible flow calculations can be made for any fluid equation of state, and we shall assign problems involving the steam tables [15], the gas tables [16], and liquids [Eq. (1.19)]. But in fact most elementary treatments are confined to the perfect gas with constant specific heats:

$$p = \rho R T \qquad R = c_p - c_v = \text{const} \qquad k = \frac{c_p}{c_v} = \text{const} \tag{9.2}$$

For all real gases, c_p, c_v, and k vary with temperature but only moderately; for example, c_p of air increases 30 percent as temperature increases from 0 to 5000°F. Since we rarely deal with such large temperature changes, it is quite reasonable to assume constant specific heats.

Recall from Sec. 1.8 that the gas constant is related to a universal constant Λ divided by the gas molecular weight:

$$R_{gas} = \frac{\Lambda}{M_{gas}} \tag{9.3}$$

where $\qquad \Lambda = 49{,}720 \text{ ft-lbf/(lbmol} \cdot {}^\circ\text{R)} = 8314 \text{ J/(kmol} \cdot \text{K)}$

For air, $M = 28.97$, and we shall adopt the following property values for air throughout this chapter:

$$R = 1716 \text{ ft}^2/(\text{s}^2 \cdot {}^\circ\text{R}) = 287 \text{ m}^2/(\text{s}^2 \cdot \text{K}) \qquad k = 1.400$$

$$c_v = \frac{R}{k-1} = 4293 \text{ ft}^2/(\text{s}^2 \cdot {}^\circ\text{R}) = 718 \text{ m}^2/(\text{s}^2 \cdot \text{K}) \tag{9.4}$$

$$c_p = \frac{kR}{k-1} = 6009 \text{ ft}^2/(\text{s}^2 \cdot {}^\circ\text{R}) = 1005 \text{ m}^2/(\text{s}^2 \cdot \text{K})$$

Experimental values of k for eight common gases were shown in Fig. 1.5. From this figure and the molecular weight, the other properties can be computed, as in Eqs. (9.4).

The changes in the internal energy $\hat{u}$ and enthalpy h of a perfect gas are computed for constant specific heats as

$$\hat{u}_2 - \hat{u}_1 = c_v(T_2 - T_1) \qquad h_2 - h_1 = c_p(T_2 - T_1) \tag{9.5}$$

For variable specific heats one must integrate $\hat{u} = \int c_v \, dT$ and $h = \int c_p \, dT$ or use the gas tables [16]. Most modern thermodynamics texts now contain software for evaluating properties of nonideal gases [17], as does EES.

Isentropic Process

The isentropic approximation is common in compressible flow theory. We compute the entropy change from the first and second laws of thermodynamics for a pure substance [17 or 18]:

$$T \, ds = dh - \frac{dp}{\rho} \tag{9.6}$$

Introducing $dh = c_p \, dT$ for a perfect gas and solving for ds, we substitute $\rho T = p/R$ from the perfect-gas law and obtain

$$\int_1^2 ds = \int_1^2 c_p \frac{dT}{T} - R \int_1^2 \frac{dp}{p} \tag{9.7}$$

If c_p is variable, the gas tables will be needed, but for constant c_p we obtain the analytic results

$$s_2 - s_1 = c_p \ln \frac{T_2}{T_1} - R \ln \frac{p_2}{p_1} = c_v \ln \frac{T_2}{T_1} - R \ln \frac{\rho_2}{\rho_1} \tag{9.8}$$

Equations (9.8) are used to compute the entropy change across a shock wave (Sec. 9.5), which is an irreversible process.

For isentropic flow, we set $s_2 = s_1$ and obtain these interesting power-law relations for an isentropic perfect gas:

$$\frac{p_2}{p_1} = \left(\frac{T_2}{T_1}\right)^{k/(k-1)} = \left(\frac{\rho_2}{\rho_1}\right)^k \qquad (9.9)$$

These relations are used in Sec. 9.3.

EXAMPLE 9.1

Argon flows through a tube such that its initial condition is $p_1 = 1.7$ MPa and $\rho_1 = 18$ kg/m^3 and its final condition is $p_2 = 248$ kPa and $T_2 = 400$ K. Estimate (a) the initial temperature, (b) the final density, (c) the change in enthalpy, and (d) the change in entropy of the gas.

Solution

From Table A.4 for argon, $R = 208$ m^2/(s$^2 \cdot$ K) and $k = 1.67$. Therefore estimate its specific heat at constant pressure from Eq. (9.4):

$$c_p = \frac{kR}{k-1} = \frac{1.67(208)}{1.67 - 1} \approx 519 \text{ m}^2/(\text{s}^2 \cdot \text{K})$$

The initial temperature and final density are estimated from the ideal-gas law, Eq. (9.2):

$$T_1 = \frac{p_1}{\rho_1 R} = \frac{1.7 \text{ E6 N/m}^2}{(18 \text{ kg/m}^3)[208 \text{ m}^2/(\text{s}^2 \cdot \text{K})]} = 454 \text{ K} \qquad \textit{Ans. (a)}$$

$$\rho_2 = \frac{p_2}{T_2 R} = \frac{248 \text{ E3 N/m}^2}{(400 \text{ K})[208 \text{ m}^2/(\text{s}^2 \cdot \text{K})]} = 2.98 \text{ kg/m}^3 \qquad \textit{Ans. (b)}$$

From Eq. (9.5) the enthalpy change is

$$h_2 - h_1 = c_p(T_2 - T_1) = 519(400 - 454) \approx -28{,}000 \text{ J/kg (or m}^2/\text{s}^2) \quad \textit{Ans. (c)}$$

The argon temperature and enthalpy decrease as we move down the tube. Actually, there may not be any external cooling; that is, the fluid enthalpy may be converted by friction to increased kinetic energy (Sec. 9.7).

Finally, the entropy change is computed from Eq. (9.8):

$$s_2 - s_1 = c_p \ln \frac{T_2}{T_1} - R \ln \frac{p_2}{p_1}$$

$$= 519 \ln \frac{400}{454} - 208 \ln \frac{0.248 \text{ E6}}{1.7 \text{ E6}}$$

$$= -66 + 400 \approx 334 \text{ m}^2/(\text{s}^2 \cdot \text{K}) \qquad \textit{Ans. (d)}$$

The fluid entropy has increased. If there is no heat transfer, this indicates an irreversible process. Note that entropy has the same units as the gas constant and specific heat.

This problem is not just arbitrary numbers. It correctly simulates the behavior of argon moving subsonically through a tube with large frictional effects (Sec. 9.7).

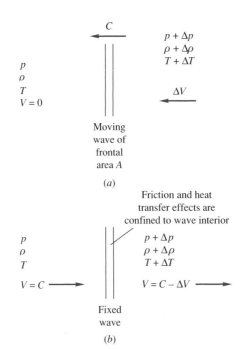

Fig. 9.1 Control volume analysis of a finite-strength pressure wave: (*a*) control volume fixed to still fluid at left; (*b*) control volume moving left at wave speed C.

9.2 The Speed of Sound

The so-called speed of sound is the rate of propagation of a pressure pulse of infinitesimal strength through a still fluid. It is a thermodynamic property of a fluid. Let us analyze it by first considering a pulse of finite strength, as in Fig. 9.1. In Fig. 9.1*a* the pulse, or pressure wave, moves at speed C toward the still fluid (p, ρ, T, $V = 0$) at the left, leaving behind at the right a fluid of increased properties ($p + \Delta p$, $\rho + \Delta \rho$, $T + \Delta T$) and a fluid velocity ΔV toward the left following the wave but much slower. We can determine these effects by making a control volume analysis across the wave. To avoid the unsteady terms that would be necessary in Fig. 9.1*a*, we adopt instead the control volume of Fig. 9.1*b*, which moves at wave speed C to the left. The wave appears fixed from this viewpoint, and the fluid appears to have velocity C on the left and $C - \Delta V$ on the right. The thermodynamic properties p, ρ, and T are not affected by this change of viewpoint.

The flow in Fig. 9.1*b* is steady and one-dimensional across the wave. The continuity equation is thus, from Eq. (3.24),

$$\rho A C = (\rho + \Delta \rho)(A)(C - \Delta V)$$

or
$$\Delta V = C \frac{\Delta \rho}{\rho + \Delta \rho} \tag{9.10}$$

This proves our contention that the induced fluid velocity on the right is much smaller than the wave speed C. In the limit of infinitesimal wave strength (sound wave) this speed is itself infinitesimal.

Notice that there are no velocity gradients on either side of the wave. Therefore, even if fluid viscosity is large, frictional effects are confined to the interior of the wave. Advanced texts [for example, 9] show that the thickness of pressure waves in gases is of order 10^{-6} ft at atmospheric pressure. Thus we can safely neglect friction and apply the one-dimensional momentum equation (3.40) across the wave:

$$\sum F_{\text{right}} = \dot{m}(V_{\text{out}} - V_{\text{in}})$$

or

$$pA - (p + \Delta p)A = (\rho A C)(C - \Delta V - C) \tag{9.11}$$

Again the area cancels, and we can solve for the pressure change:

$$\Delta p = \rho C\, \Delta V \tag{9.12}$$

If the wave strength is very small, the pressure change is small.

Finally, combine Eqs. (9.10) and (9.12) to give an expression for the wave speed:

$$C^2 = \frac{\Delta p}{\Delta \rho}\left(1 + \frac{\Delta \rho}{\rho}\right) \tag{9.13}$$

The larger the strength $\Delta \rho/\rho$ of the wave, the faster the wave speed; that is, powerful explosion waves move much more quickly than sound waves. In the limit of infinitesimal strength $\Delta \rho \to 0$, we have what is defined to be the speed of sound a of a fluid:

$$a^2 = \frac{\partial p}{\partial \rho} \tag{9.14}$$

But the evaluation of the derivative requires knowledge of the thermodynamic process undergone by the fluid as the wave passes. Sir Isaac Newton in 1686 made a famous error by deriving a formula for sound speed that was equivalent to assuming an isothermal process, the result being 20 percent too low for air, for example. He rationalized the discrepancy as being due to the "crassitude" (dust particles and so on) in the air; the error is certainly understandable when we reflect that it was made 180 years before the proper basis was laid for the second law of thermodynamics.

We now see that the correct process must be *adiabatic* because there are no temperature gradients except inside the wave itself. For vanishing-strength sound waves we therefore have an infinitesimal adiabatic or isentropic process. The correct expression for the sound speed is

$$\boxed{a = \left(\left.\frac{\partial p}{\partial \rho}\right|_s\right)^{1/2} = \left(k\left.\frac{\partial p}{\partial \rho}\right|_T\right)^{1/2}} \tag{9.15}$$

for any fluid, gas or liquid. Even a solid has a sound speed.

For a perfect gas, From Eq. (9.2) or (9.9), we deduce that the speed of sound is

$$a = \left(\frac{kp}{\rho}\right)^{1/2} = (kRT)^{1/2} \tag{9.16}$$

The speed of sound increases as the square root of the absolute temperature. For air, with $k = 1.4$, an easily memorized dimensional formula is

$$a(\text{ft/s}) \approx 49[T(°\text{R})]^{1/2}$$
$$a(\text{m/s}) \approx 20[T(\text{K})]^{1/2} \tag{9.17}$$

Table 9.1 Sound Speed of Various Materials at 60°F (15.5°C) and 1 atm

Material	a, ft/s	a, m/s
Gases:		
H_2	4,246	1,294
He	3,281	1,000
Air	1,117	340
Ar	1,040	317
CO_2	873	266
CH_4	607	185
$^{238}UF_6$	297	91
Liquids:		
Glycerin	6,100	1,860
Water	4,890	1,490
Mercury	4,760	1,450
Ethyl alcohol	3,940	1,200
Solids:*		
Aluminum	16,900	5,150
Steel	16,600	5,060
Hickory	13,200	4,020
Ice	10,500	3,200

*Plane waves. Solids also have a *shear-wave speed.*

At sea-level standard temperature, 60°F = 520°R, a = 1117 ft/s. This decreases in the upper atmosphere, which is cooler; at 50,000-ft standard altitude, $T = -69.7°F = 389.9°R$ and $a = 49(389.9)^{1/2} = 968$ ft/s, or 13 percent less.

Some representative values of sound speed in various materials are given in Table 9.1. For liquids and solids it is common to define the *bulk modulus K* of the material:

$$K = -\mathcal{V} \frac{\partial p}{\partial \mathcal{V}}\bigg|_s = \rho \frac{\partial p}{\partial \rho}\bigg|_s \tag{9.18}$$

In terms of bulk modulus, then, $a = (K/\rho)^{1/2}$. For example, at standard conditions, the bulk modulus of liquid carbon tetrachloride is 1.12 GPa absolute, and its density is 1590 kg/m³. Its speed of sound is therefore $a = (1.12E9 \text{ Pa}/1590 \text{ kg/m}^3)^{1/2} = 840$ m/s = 2750 ft/s. Steel has a bulk modulus of about 2.0E11 Pa and water about 2.2E9 Pa (see Table A.3), or 90 times less than steel.

For solids, it is sometimes assumed that the bulk modulus is approximately equivalent to Young's modulus of elasticity E, but in fact their ratio depends on Poisson's ratio σ:

$$\frac{E}{K} = 3(1 - 2\sigma) \tag{9.19}$$

The two are equal for $\sigma = \frac{1}{3}$, which is approximately the case for many common metals such as steel and aluminum.

EXAMPLE 9.2

Estimate the speed of sound of carbon monoxide at 200-kPa pressure and 300°C in m/s.

Solution

From Table A.4, for CO, the molecular weight is 28.01 and $k \approx 1.40$. Thus from Eq. (9.3) $R_{CO} = 8314/28.01 = 297$ m²/(s² · K), and the given temperature is 300°C + 273 = 573 K. Thus from Eq. (9.16) we estimate

$$a_{CO} = (kRT)^{1/2} = [1.40(297)(573)]^{1/2} = 488 \text{ m/s} \qquad Ans.$$

9.3 Adiabatic and Isentropic Steady Flow

As mentioned in Sec. 9.1, the isentropic approximation greatly simplifies a compressible flow calculation. So does the assumption of adiabatic flow, even if nonisentropic.

Consider high-speed flow of a gas past an insulated wall, as in Fig. 9.2. There is no shaft work delivered to any part of the fluid. Therefore every streamtube in the flow satisfies the steady flow energy equation in the form of Eq. (3.66):

$$h_1 + \tfrac{1}{2}V_1^2 + gz_1 = h_2 + \tfrac{1}{2}V_2^2 + gz_2 - q + w_v \tag{9.20}$$

where point 1 is upstream of point 2. You may wish to review the details of Eq. (3.66) and its development. We saw in Example 3.16 that potential energy changes of a gas

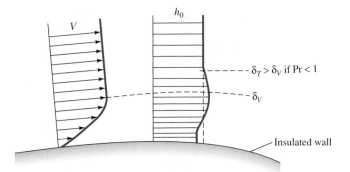

Fig. 9.2 Velocity and stagnation enthalpy distributions near an insulated wall in a typical high-speed gas flow.

are extremely small compared with kinetic energy and enthalpy terms. We shall neglect the terms gz_1 and gz_2 in all gas dynamic analyses.

Inside the thermal and velocity boundary layers in Fig. 9.2 the heat transfer and viscous work terms q and w_v are not zero. But outside the boundary layer q and w_v are zero by definition, so that the outer flow satisfies the simple relation

$$h_1 + \tfrac{1}{2}V_1^2 = h_2 + \tfrac{1}{2}V_2^2 = \text{const} \tag{9.21}$$

The constant in Eq. (9.21) is equal to the maximum enthalpy that the fluid would achieve if brought to rest adiabatically. We call this value h_0, the *stagnation enthalpy* of the flow. Thus we rewrite Eq. (9.21) in the form

$$h + \tfrac{1}{2}V^2 = h_0 = \text{const} \tag{9.22}$$

This should hold for steady adiabatic flow of any compressible fluid outside the boundary layer. The wall in Fig. 9.2 could be either the surface of an immersed body or the wall of a duct. We have shown the details of Fig. 9.2; typically the thermal layer thickness δ_T is greater than the velocity layer thickness δ_V because most gases have a dimensionless Prandtl number Pr less than unity (see, for example, Ref. 19, sec. 4-3.2). Note that the stagnation enthalpy varies inside the thermal boundary layer, but its average value is the same as that at the outer layer due to the insulated wall.

For nonperfect gases we may have to use EES or the steam tables [15] or the gas tables [16] to implement Eq. (9.22). But for a perfect gas $h = c_pT$, and Eq. (9.22) becomes

$$c_pT + \tfrac{1}{2}V^2 = c_pT_0 \tag{9.23}$$

This establishes the stagnation temperature T_0 of an adiabatic perfect-gas flow—that is, the temperature it achieves when decelerated to rest adiabatically.

An alternate interpretation of Eq. (9.22) occurs when the enthalpy and temperature drop to (absolute) zero, so that the velocity achieves a maximum value:

$$V_{\text{max}} = (2h_0)^{1/2} = (2c_pT_0)^{1/2} \tag{9.24}$$

No higher flow velocity can occur unless additional energy is added to the fluid through shaft work or heat transfer (Sec. 9.8).

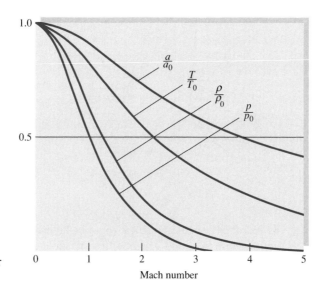

Fig. 9.3 Adiabatic (T/T_0 and a/a_0) and isentropic (p/p_0 and ρ/ρ_0) properties versus Mach number for $k = 1.4$.

Mach Number Relations

The dimensionless form of Eq. (9.23) brings in the Mach number Ma as a parameter, by using Eq. (9.16) for the speed of sound of a perfect gas. Divide through by $c_p T$ to obtain

$$1 + \frac{V^2}{2c_p T} = \frac{T_0}{T} \tag{9.25}$$

But, from the perfect-gas law, $c_p T = [kR/(k-1)]T = a^2/(k-1)$, so that Eq. (9.25) becomes

$$1 + \frac{(k-1)V^2}{2a^2} = \frac{T_0}{T}$$

or
$$\boxed{\frac{T_0}{T} = 1 + \frac{k-1}{2}\,\text{Ma}^2 \qquad \text{Ma} = \frac{V}{a}} \tag{9.26}$$

This relation is plotted in Fig. 9.3 versus the Mach number for $k = 1.4$. At Ma = 5 the temperature has dropped to $\frac{1}{6}T_0$.

Since $a \propto T^{1/2}$, the ratio a_0/a is the square root of (9.26):

$$\frac{a_0}{a} = \left(\frac{T_0}{T}\right)^{1/2} = \left[1 + \frac{1}{2}(k-1)\text{Ma}^2\right]^{1/2} \tag{9.27}$$

Equation (9.27) is also plotted in Fig. 9.3. At Ma = 5 the speed of sound has dropped to 41 percent of the stagnation value.

Isentropic Pressure and Density Relations

Note that Eqs. (9.26) and (9.27) require only adiabatic flow and hold even in the presence of irreversibilities such as friction losses or shock waves.

If the flow is also *isentropic,* then for a perfect gas the pressure and density ratios can be computed from Eq. (9.9) as a power of the temperature ratio:

$$\frac{p_0}{p} = \left(\frac{T_0}{T}\right)^{k/(k-1)} = \left[1 + \frac{1}{2}(k-1)\,\mathrm{Ma}^2\right]^{k/(k-1)} \qquad (9.28a)$$

$$\frac{\rho_0}{\rho} = \left(\frac{T_0}{T}\right)^{1/(k-1)} = \left[1 + \frac{1}{2}(k-1)\,\mathrm{Ma}^2\right]^{1/(k-1)} \qquad (9.28b)$$

These relations are also plotted in Fig. 9.3; at Ma = 5 the density is 1.13 percent of its stagnation value, and the pressure is only 0.19 percent of stagnation pressure.

The quantities p_0 and ρ_0 are the isentropic stagnation pressure and density, respectively that is, the pressure and density that the flow would achieve if brought isentropically to rest. In an adiabatic nonisentropic flow p_0 and ρ_0 retain their local meaning, but they vary throughout the flow as the entropy changes due to friction or shock waves. The quantities h_0, T_0, and a_0 are constant in an adiabatic nonisentropic flow (see Sec. 9.7 for further details).

Relationship to Bernoulli's Equation

The isentropic assumptions (9.28) are effective, but are they realistic? Yes. To see why, take the differential of Eq. (9.22):

Adiabatic: $dh + V\,dV = 0$ (9.29)

Meanwhile, from Eq. (9.6), if $ds = 0$ (isentropic process),

$$dh = \frac{dp}{\rho} \qquad (9.30)$$

Combining (9.29) and (9.30), we find that an isentropic streamtube flow must be

$$\frac{dp}{\rho} + V\,dV = 0 \qquad (9.31)$$

But this is exactly the Bernoulli relation, Eq. (3.75), for steady frictionless flow with negligible gravity terms. Thus we see that the isentropic flow assumption is equivalent to use of the Bernoulli or streamline form of the frictionless momentum equation.

Critical Values at the Sonic Point

The stagnation values (a_0, T_0, p_0, ρ_0) are useful reference conditions in a compressible flow, but of comparable usefulness are the conditions where the flow is sonic, Ma = 1.0. These sonic, or *critical,* properties are denoted by asterisks: p^*, ρ^*, a^*, and T^*. They are certain ratios of the stagnation properties as given by Eqs. (9.26) to (9.28) when Ma = 1.0; for $k = 1.4$

$$\frac{p^*}{p_0} = \left(\frac{2}{k+1}\right)^{k/(k-1)} = 0.5283 \qquad \frac{\rho^*}{\rho_0} = \left(\frac{2}{k+1}\right)^{1/(k-1)} = 0.6339$$

$$\frac{T^*}{T_0} = \frac{2}{k+1} = 0.8333 \qquad \frac{a^*}{a_0} = \left(\frac{2}{k+1}\right)^{1/2} = 0.9129$$

(9.32)

In all isentropic flow, all critical properties are constant; in adiabatic nonisentropic flow, $a*$ and $T*$ are constant, but $p*$ and $\rho*$ may vary.

The critical velocity $V*$ equals the sonic sound speed $a*$ by definition and is often used as a reference velocity in isentropic or adiabatic flow:

$$V* = a* = (kRT*)^{1/2} = \left(\frac{2k}{k+1}RT_0\right)^{1/2} \tag{9.33}$$

The usefulness of these critical values will become clearer as we study compressible duct flow with friction or heat transfer later in this chapter.

Some Useful Numbers for Air

Since the great bulk of our practical calculations are for air, $k = 1.4$, the stagnation property ratios p/p_0 and so on from Eqs. (9.26) to (9.28) are tabulated for this value in Table B.1. The increments in Mach number are rather coarse in this table because the values are meant as only a guide; these equations are now a trivial matter to manipulate on a hand calculator. Thirty years ago every text had extensive compressible flow tables with Mach number spacings of about 0.01, so that accurate values could be interpolated. Even today, reference books are available [20, 21, 29] with tables and charts and computer programs for a wide variety of compressible flow situations. Reference 22 contains formulas and charts applying to the thermodynamics of *real* (nonperfect) gas flows.

For $k = 1.4$, the following numerical versions of the isentropic and adiabatic flow formulas are obtained:

$$\frac{T_0}{T} = 1 + 0.2\,\text{Ma}^2 \qquad \frac{\rho_0}{\rho} = (1 + 0.2\,\text{Ma}^2)^{2.5}$$

$$\frac{p_0}{p} = (1 + 0.2\,\text{Ma}^2)^{3.5} \tag{9.34}$$

Or, if we are given the properties, it is equally easy to solve for the Mach number (again with $k = 1.4$):

$$\text{Ma}^2 = 5\left(\frac{T_0}{T} - 1\right) = 5\left[\left(\frac{\rho_0}{\rho}\right)^{2/5} - 1\right] = 5\left[\left(\frac{p_0}{p}\right)^{2/7} - 1\right] \tag{9.35}$$

Note that these isentropic flow formulas serve as the equivalent of the frictionless adiabatic momentum and energy equations. They relate velocity to physical properties for a perfect gas, but they are *not* the "solution" to a gas dynamics problem. The complete solution is not obtained until the continuity equation has also been satisfied, for either one-dimensional (Sec. 9.4) or multidimensional (Sec. 9.9) flow.

One final note: These isentropic-ratio–versus–Mach-number formulas are seductive, tempting one to solve all problems by jumping right into the tables. Actually, many problems involving (dimensional) velocity and temperature can be solved more easily from the original raw dimensional energy equation (9.23) plus the perfect-gas law (9.2), as the next example will illustrate.

EXAMPLE 9.3

Air flows adiabatically through a duct. At point 1 the velocity is 240 m/s, with $T_1 = 320$ K and $p_1 = 170$ kPa. Compute (a) T_0, (b) p_0, (c) ρ_0, (d) Ma, (e) V_{max}, and (f) V^*. At point 2 further downstream $V_2 = 290$ m/s and $p_2 = 135$ kPa. (g) What is the stagnation pressure p_{02}?

Solution

- *Assumptions:* Let air be approximated as an ideal gas with constant k. The flow is adiabatic but *not* isentropic. Isentropic formulas are used *only* to compute local p_0 and ρ_0, which vary.
- *Approach:* Use adiabatic and isentropic formulas to find the various properties.
- *Ideal gas parameters:* For air, $R = 287$ m²/(s² · K), $k = 1.40$, and $c_p = 1005$ m²/(s² · K).
- *Solution steps (a, b, c, d):* With T_1, p_1 and V_1 known, other properties at point 1 follow:

$$T_{01} = T_1 + \frac{V_1^2}{2c_p} = 320 + \frac{(240 \text{ m/s})^2}{2[1005 \text{ m}^2/(\text{s}^2 \cdot \text{K})]} = 320 + 29 = 349 \text{ K} \qquad Ans. \ (a)$$

Once the Mach number is found from Eq. (9.35), local stagnation pressure and density follow:

$$\text{Ma}_1 = \sqrt{5\left(\frac{T_{01}}{T_1} - 1\right)} = \sqrt{5\left(\frac{349 \text{ K}}{320 \text{ K}} - 1\right)} = \sqrt{0.448} = 0.67 \qquad Ans. \ (b)$$

$$p_{01} = p_1(1 + 0.2 \text{ Ma}_1^2)^{3.5} = (170 \text{ kPa})[1 + 0.2(0.67)^2]^{3.5} = 230 \text{ kPa} \qquad Ans. \ (d)$$

$$\rho_{01} = \frac{p_{01}}{RT_{01}} = \frac{230{,}000 \text{ N/m}^2}{[287 \text{ m}^2/(\text{s}^2 \cdot \text{K})](349 \text{ K})} = 2.29 \frac{\text{N} \cdot \text{s}^2/\text{m}}{\text{m}^3} = 2.29 \frac{\text{kg}}{\text{m}^3} \qquad Ans. \ (c)$$

- *Comment:* Note that we used dimensional (non-Mach-number) formulas where convenient.
- *Solution steps (e, f):* Both V_{max} and V^* are directly related to stagnation temperature from Eqs. (9.24) and (9.33):

$$V_{max} = \sqrt{2c_pT_0} = \sqrt{2[1005 \text{ m}^2/(\text{s}^2 \cdot \text{K})](349 \text{ K})} = 837 \frac{\text{m}}{\text{s}} \qquad Ans. \ (e)$$

$$V^* = \sqrt{\frac{2k}{k+1}RT_0} = \sqrt{\frac{2(1.4)}{(1.4+1)}\left(287 \frac{\text{m}^2}{\text{s}^2 \cdot \text{K}}\right)(349 \text{ K})} = 342 \frac{\text{m}}{\text{s}} \qquad Ans. \ (f)$$

- At point 2 downstream, the temperature is unknown, but since the flow is adiabatic, the stagnation temperature is constant: $T_{01} = T_{02} = 349$ K. Thus, from Eq. (9.23),

$$T_2 = T_{02} - \frac{V_2^2}{2c_p} = 349 - \frac{(290 \text{ m/s})^2}{2[1005 \text{ m}^2/(\text{s}^2 \cdot \text{K})]} = 307 \text{ K}$$

Hence, from Eq. (9.28a), the isentropic stagnation pressure at point 2 is

$$p_{02} = p_2\left(\frac{T_{02}}{T_2}\right)^{k/(k-1)} = (135 \text{ kPa})\left(\frac{349 \text{ K}}{307 \text{ K}}\right)^{1.4/0.4} = 211 \text{ kPa} \qquad Ans. \ (g)$$

- *Comments:* Part (g), a ratio-type ideal-gas formula, is more direct than finding the Mach number, which turns out to be $Ma_2 = 0.83$, and using the Mach number formula, Eq. (9.34) for p_{02}. Note that p_{02} is 8 percent less than p_{01}. The flow is nonisentropic: Entropy rises downstream, and stagnation pressure and density drop, due in this case to frictional losses.

9.4 Isentropic Flow with Area Changes

By combining the isentropic and/or adiabatic flow relations with the equation of continuity we can study practical compressible flow problems. This section treats the one-dimensional flow approximation.

Figure 9.4 illustrates the one-dimensional flow assumption. A real flow, Fig. 9.4a, has no slip at the walls and a velocity profile $V(x, y)$ that varies across the duct section (compare with Fig. 7.8). If, however, the area change is small and the wall radius of curvature large

$$\frac{dh}{dx} \ll 1 \qquad h(x) \ll R(x) \tag{9.36}$$

then the flow is approximately one-dimensional, as in Fig. 9.4b, with $V \approx V(x)$ reacting to area change $A(x)$. Compressible flow nozzles and diffusers do not always satisfy conditions (9.36), but we use the one-dimensional theory anyway because of its simplicity.

For steady one-dimensional flow the equation of continuity is, from Eq. (3.24),

$$\rho(x)V(x)A(x) = \dot{m} = \text{const} \tag{9.37}$$

Before applying this to duct theory, we can learn a lot from the differential form of Eq. (9.37):

$$\frac{d\rho}{\rho} + \frac{dV}{V} + \frac{dA}{A} = 0 \tag{9.38}$$

The differential forms of the frictionless momentum equation (9.31) and the sound-speed relation (9.15) are recalled here for convenience:

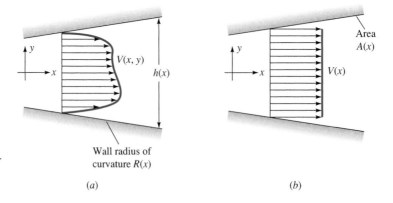

Fig. 9.4 Compressible flow through a duct: (a) real-fluid velocity profile; (b) one-dimensional approximation.

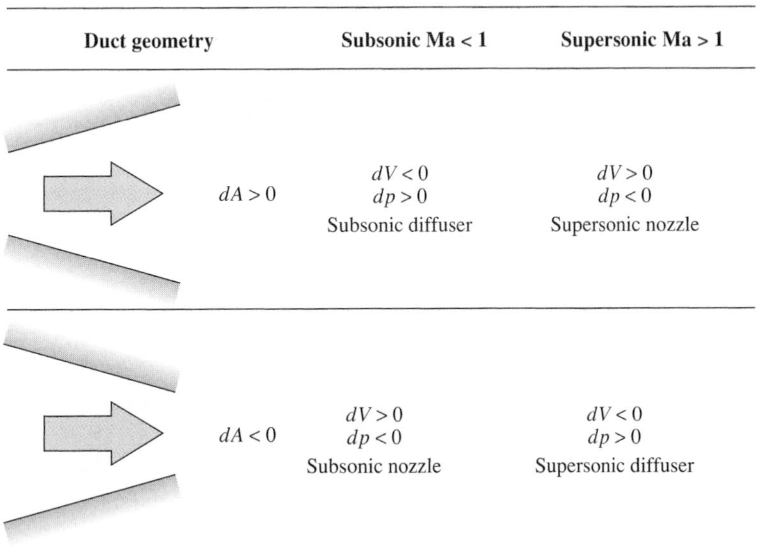

Duct geometry		Subsonic Ma < 1	Supersonic Ma > 1
	$dA > 0$	$dV < 0$ $dp > 0$ Subsonic diffuser	$dV > 0$ $dp < 0$ Supersonic nozzle
	$dA < 0$	$dV > 0$ $dp < 0$ Subsonic nozzle	$dV < 0$ $dp > 0$ Supersonic diffuser

Fig. 9.5 Effect of Mach number on property changes with area change in duct flow.

Momentum

$$\frac{dp}{\rho} + V\,dV = 0$$

Sound speed:

$$dp = a^2\,d\rho$$

(9.39)

Now eliminate dp and $d\rho$ between Eqs. (9.38) and (9.39) to obtain the following relation between velocity change and area change in isentropic duct flow:

$$\frac{dV}{V} = \frac{dA}{A}\frac{1}{\text{Ma}^2 - 1} = -\frac{dp}{\rho V^2}$$

(9.40)

Inspection of this equation, without actually solving it, reveals a fascinating aspect of compressible flow: Property changes are of opposite sign for subsonic and supersonic flow because of the term $\text{Ma}^2 - 1$. There are four combinations of area change and Mach number, summarized in Fig. 9.5.

From earlier chapters we are used to subsonic behavior (Ma < 1): When area increases, velocity decreases and pressure increases, which is denoted a subsonic diffuser. But in supersonic flow (Ma > 1), the velocity actually increases when the area increases, a supersonic nozzle. The same opposing behavior occurs for an area decrease, which speeds up a subsonic flow and slows down a supersonic flow.

What about the sonic point Ma = 1? Since infinite acceleration is physically impossible, Eq. (9.40) indicates that dV can be finite only when $dA = 0$—that is, a minimum area (throat) or a maximum area (bulge). In Fig. 9.6 we patch together a throat section and a bulge section, using the rules from Fig. 9.5. The throat or converging–diverging section can smoothly accelerate a subsonic flow through sonic to supersonic flow, as in Fig. 9.6a. This is the only way a supersonic flow can be created by expanding the gas from a stagnant reservoir. The bulge section fails; the bulge Mach number moves away from a sonic condition rather than toward it.

Although supersonic flow downstream of a nozzle requires a sonic throat, the opposite is not true: A compressible gas can pass through a throat section without becoming sonic.

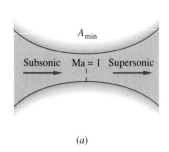

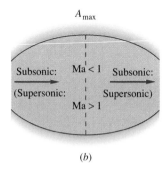

Fig. 9.6 From Eq. (9.40), in flow through a throat (*a*) the fluid can accelerate smoothly through sonic and supersonic flow. In flow through the bulge (*b*) the flow at the bulge cannot be sonic on physical grounds.

Perfect-Gas Area Change

We can use the perfect-gas and isentropic flow relations to convert the continuity relation (9.37) into an algebraic expression involving only area and Mach number, as follows. Equate the mass flow at any section to the mass flow under sonic conditions (which may not actually occur in the duct):

$$\rho V A = \rho^* V^* A^*$$

or

$$\frac{A}{A^*} = \frac{\rho^*}{\rho} \frac{V^*}{V} \tag{9.41}$$

Both the terms on the right are functions only of Mach number for isentropic flow. From Eqs. (9.28) and (9.32)

$$\frac{\rho^*}{\rho} = \frac{\rho^*}{\rho_0} \frac{\rho_0}{\rho} = \left\{ \frac{2}{k+1} \left[1 + \frac{1}{2}(k-1)\,\mathrm{Ma}^2 \right] \right\}^{1/(k-1)} \tag{9.42}$$

From Eqs. (9.26) and (9.32) we obtain

$$\frac{V^*}{V} = \frac{(kRT^*)^{1/2}}{V} = \frac{(kRT)^{1/2}}{V} \left(\frac{T^*}{T_0} \right)^{1/2} \left(\frac{T_0}{T} \right)^{1/2}$$

$$= \frac{1}{\mathrm{Ma}} \left\{ \frac{2}{k+1} \left[1 + \frac{1}{2}(k-1)\,\mathrm{Ma}^2 \right] \right\}^{1/2} \tag{9.43}$$

Combining Eqs. (9.41) to (9.43), we get the desired result:

$$\boxed{\frac{A}{A^*} = \frac{1}{\mathrm{Ma}} \left[\frac{1 + \frac{1}{2}(k-1)\,\mathrm{Ma}^2}{\frac{1}{2}(k+1)} \right]^{(1/2)(k+1)(k-1)}} \tag{9.44}$$

For $k = 1.4$, Eq. (9.44) takes the numerical form

$$\frac{A}{A^*} = \frac{1}{\mathrm{Ma}} \frac{(1 + 0.2\,\mathrm{Ma}^2)^3}{1.728} \tag{9.45}$$

which is plotted in Fig. 9.7. Equations (9.45) and (9.34) enable us to solve any one-dimensional isentropic airflow problem given, say, the shape of the duct $A(x)$ and the stagnation conditions and assuming that there are no shock waves in the duct.

Figure 9.7 shows that the minimum area that can occur in a given isentropic duct flow is the sonic, or critical, throat area. All other duct sections must have A greater

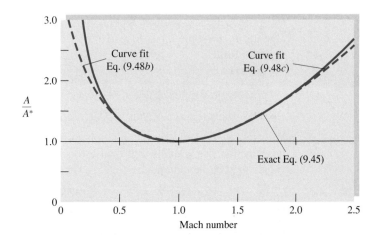

Fig. 9.7 Area ratio versus Mach number for isentropic flow of a perfect gas with $k = 1.4$.

than A^*. In many flows a critical sonic throat is not actually present, and the flow in the duct is either entirely subsonic or, more rarely, entirely supersonic.

Choking

From Eq. (9.41) the inverse ratio A^*/A equals $\rho V/(\rho^* V^*)$, the mass flow per unit area at any section compared with the critical mass flow per unit area. From Fig. 9.7 this inverse ratio rises from zero at Ma $= 0$ to unity at Ma $= 1$ and back down to zero at large Ma. Thus, for given stagnation conditions, the maximum possible mass flow passes through a duct when its throat is at the critical or sonic condition. The duct is then said to be *choked* and can carry no additional mass flow unless the throat is widened. If the throat is constricted further, the mass flow through the duct must decrease.

From Eqs. (9.32) and (9.33) the maximum mass flow is

$$\dot{m}_{max} = \rho^* A^* V^* = \rho_0 \left(\frac{2}{k+1}\right)^{1/(k-1)} A^* \left(\frac{2k}{k+1} RT_0\right)^{1/2}$$

$$= k^{1/2} \left(\frac{2}{k+1}\right)^{(1/2)(k+1)/(k-1)} A^* \rho_0 (RT_0)^{1/2} \tag{9.46a}$$

For $k = 1.4$ this reduces to

$$\boxed{\dot{m}_{max} = 0.6847 A^* \rho_0 (RT_0)^{1/2} = \frac{0.6847 p_0 A^*}{(RT_0)^{1/2}}} \tag{9.46b}$$

For isentropic flow through a duct, the maximum mass flow possible is proportional to the throat area and stagnation pressure and inversely proportional to the square root of the stagnation temperature. These are somewhat abstract facts, so let us illustrate with some examples.

The Local Mass Flow Function

Equations (9.46) give the *maximum* mass flow, which occurs at the choking condition (sonic exit). They can be modified to predict the actual (nonmaximum) mass flow at any section where local area A and pressure p are known.[1] The algebra is convoluted, so here we give only the final result, expressed in dimensionless form:

$$\text{Mass flow function} = \frac{\dot{m}}{A}\frac{\sqrt{RT_0}}{p_0} = \sqrt{\frac{2k}{k-1}\left(\frac{p}{p_0}\right)^{2/k}\left[1-\left(\frac{p}{p_0}\right)^{(k-1)/k}\right]} \quad (9.47)$$

We stress that p and A in this relation are the *local* values at position x. As p/p_0 falls, this function rises rapidly and then levels out at the maximum of Eqs. (9.46). A few values may be tabulated here for $k = 1.4$:

p/p_0	1.0	0.98	0.95	0.9	0.8	0.7	0.6	≤ 0.5283
Function	0.0	0.1978	0.3076	0.4226	0.5607	0.6383	0.6769	0.6847

Equation (9.47) is handy if stagnation conditions are known and the flow is not choked.

The only cumbersome algebra in these problems is the inversion of Eq. (9.45) to compute the Mach number when A/A^* is known. If available, EES is ideal for this situation and will yield Ma in a flash. In the absence of EES, the following curve-fitted formulas are suggested; given A/A^*, they estimate the Mach number within ± 2 percent for $k = 1.4$ if you stay within the ranges listed for each formula:

$$\text{Ma} \approx \begin{cases} \dfrac{1 + 0.27(A/A^*)^{-2}}{1.728A/A^*} & 1.34 < \dfrac{A}{A^*} < \infty \quad \left. \right\} \text{subsonic flow} & (9.48a) \\[2ex] 1 - 0.88\left(\ln\dfrac{A}{A^*}\right)^{0.45} & 1.0 < \dfrac{A}{A^*} < 1.34 & (9.48b) \\[2ex] 1 + 1.2\left(\dfrac{A}{A^*} - 1\right)^{1/2} & 1.0 < \dfrac{A}{A^*} < 2.9 \quad \left. \right\} \text{supersonic flow} & (9.48c) \\[2ex] \left[216\dfrac{A}{A^*} - 254\left(\dfrac{A}{A^*}\right)^{2/3}\right]^{1/5} & 2.9 < \dfrac{A}{A^*} < \infty & (9.48d) \end{cases}$$

Formulas (9.48a) and (9.48d) are asymptotically correct as $A/A^* \to \infty$, while (9.48b) and (9.48c) are just curve fits. However, formulas (9.48b) and (9.48c) are seen in Fig. 9.7 to be accurate within their recommended ranges.

Note that two solutions are possible for a given A/A^*, one subsonic and one supersonic. The proper solution cannot be selected without further information, such as known pressure or temperature at the given duct section.

EXAMPLE 9.4

Air flows isentropically through a duct. At section 1 the area is 0.05 m² and $V_1 = 180$ m/s, $p_1 = 500$ kPa, and $T_1 = 470$ K. Compute (a) T_0, (b) Ma_1, (c) p_0, and (d) both A^* and $\dot{m}$. If at section 2 the area is 0.036 m², compute Ma_2 and p_2 if the flow is (e) subsonic or (f) supersonic. Assume $k = 1.4$.

---

[1] The author is indebted to Georges Aigret, of Chimay, Belgium, for suggesting this useful function.

Solution

Part (a)

A general sketch of the problem is shown in Fig. E9.4. With V_1 and T_1 known, the energy equation (9.23) gives

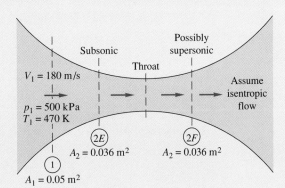

Subsonic

Possibly supersonic

Throat

$V_1 = 180$ m/s

Assume isentropic flow

$p_1 = 500$ kPa
$T_1 = 470$ K

(2E)

(2F)

$A_2 = 0.036$ m^2

$A_2 = 0.036$ m^2

(1)

E9.4 $A_1 = 0.05$ m^2

$$T_0 = T_1 + \frac{V_1^2}{2c_p} = 470 + \frac{(180)^2}{2(1005)} = 486 \text{ K} \qquad Ans. (a)$$

Part (b)

The local sound speed $a_1 = \sqrt{kRT_1} = [(1.4)(287)(470)]^{1/2} = 435$ m/s. Hence

$$\text{Ma}_1 = \frac{V_1}{a_1} = \frac{180}{435} = 0.414 \qquad Ans. (b)$$

Part (c)

With Ma_1 known, the stagnation pressure follows from Eq. (9.34):

$$p_0 = p_1(1 + 0.2\,\text{Ma}_1^2)^{3.5} = (500 \text{ kPa})[1 + 0.2(0.414)^2]^{3.5} = 563 \text{ kPa} \qquad Ans. (c)$$

Part (d)

Similarly, from Eq. (9.45), the critical sonic throat area is

$$\frac{A_1}{A^*} = \frac{(1 + 0.2\,\text{Ma}_1^2)^3}{1.728\,\text{Ma}_1} = \frac{[1 + 0.2(0.414)^2]^3}{1.728(0.414)} = 1.547$$

or

$$A^* = \frac{A_1}{1.547} = \frac{0.05 \text{ m}^2}{1.547} = 0.0323 \text{ m}^2 \qquad Ans. (d)$$

This throat must *actually be present* in the duct if the flow is to become supersonic.

We now know A^*. So to compute the mass flow we can use Eqs. (9.46), which remain valid, based on the numerical value of A^*, whether or not a throat actually exists:

$$\dot{m} = 0.6847 \frac{p_0 A^*}{\sqrt{RT_0}} = 0.6847 \frac{(563,000)(0.0323)}{\sqrt{(287)(486)}} = 33.4 \text{ kg/s} \qquad Ans. (d)$$

Or we could fare equally well with our new "local mass flow" formula, Eq. (9.47), using, say, the pressure and area at section 1. Given $p_1/p_0 = 500/563 = 0.889$, Eq. (9.47) yields

$$\dot{m}\frac{\sqrt{287(486)}}{563,000(0.05)} = \sqrt{\frac{2(1.4)}{0.4}(0.889)^{2/1.4}[1 - (0.889)^{0.4/1.4}]} = 0.444 \quad \dot{m} = 33.4\,\frac{\text{kg}}{\text{s}}\;Ans. (d)$$

Part (e)

Assume *subsonic* flow corresponds to section 2E in Fig. E9.4. The duct contracts to an area ratio $A_2/A^* = 0.036/0.0323 = 1.115$, which we find on the left side of Fig. 9.7 or the subsonic part of Table B.1. Neither the figure nor the table is that accurate. There are two accurate options. First, Eq. (9.48*b*) gives the estimate $Ma_2 \approx 1 - 0.88 \ln (1.115)^{0.45} \approx 0.676$ (error less than 0.5 percent). Second, EES (App. E) will give an arbitrarily accurate solution with only three statements (in SI units):

$$A2 = 0.036$$

$$Astar = 0.0323$$

$$A2/Astar = (1 + 0.2*Ma2\hat{}2)\hat{}3/1.2\hat{}3/Ma2$$

Specify that you want a *subsonic* solution (e.g., limit $Ma_2 < 1$), and EES reports

$$Ma_2 = 0.6758 \qquad\qquad\qquad\qquad \textit{Ans. (e)}$$

[Ask for a supersonic solution (require $Ma_2 > 1$) and you receive $Ma_2 = 1.4001$, which is the answer to part (*f*).] The pressure is given by the isentropic relation

$$p_2 = \frac{p_0}{[1 + 0.2(0.676)^2]^{3.5}} = \frac{563 \text{ kPa}}{1.358} \approx 415 \text{ kPa} \qquad \textit{Ans. (e)}$$

Part (*e*) does *not* require a throat, sonic or otherwise; the flow could simply be contracting subsonically from A_1 to A_2.

 Part (f)

This time assume *supersonic* flow, corresponding to section 2*F* in Fig. E9.4. Again the area ratio is $A_2/A^* = 0.036/0.0323 = 1.115$, and we look on the *right* side of Fig. 9.7 or the supersonic part of Table B.1—the latter can be read quite accurately as $Ma_2 \approx 1.40$. Again there are two other accurate options. First, Eq. (9.48*c*) gives the curve-fit estimate $Ma_2 \approx 1 + 1.2(1.115 - 1)^{1/2} \approx 1.407$, only 0.5 percent high. Second, EES will give a very accurate solution with the same three statements from part (*e*). Specify that you want a *supersonic* solution (e.g., limit $Ma_2 > 1$), and EES reports

$$Ma_2 = 1.4001 \qquad\qquad\qquad\qquad \textit{Ans. (f)}$$

Again the pressure is given by the isentropic relation at the new Mach number:

$$p_2 = \frac{p_0}{[1 + 0.2(1.4001)^2]^{3.5}} = \frac{563 \text{ kPa}}{3.183} = 177 \text{ kPa} \qquad \textit{Ans. (f)}$$

Note that the supersonic flow pressure level is much less than p_2 in part (*e*), and a sonic throat *must* have occurred between sections 1 and 2*F*.

EXAMPLE 9.5

It is desired to expand air from $p_0 = 200$ kPa and $T_0 = 500$ K through a throat to an exit Mach number of 2.5. If the desired mass flow is 3 kg/s, compute (*a*) the throat area and the exit (*b*) pressure, (*c*) temperature, (*d*) velocity, and (*e*) area, assuming isentropic flow, with $k = 1.4$.

Solution

The throat area follows from Eq. (9.47), because the throat flow must be sonic to produce a supersonic exit:

$$A^* = \frac{\dot{m}(RT_0)^{1/2}}{0.6847 p_0} = \frac{3.0[287(500)]^{1/2}}{0.6847(200,000)} = 0.00830 \text{ m}^2 = \frac{1}{4}\pi D^{*2}$$

or $\qquad\qquad\qquad\qquad\qquad\qquad D_{\text{throat}} = 10.3 \text{ cm} \qquad\qquad\qquad\qquad \textit{Ans. (a)}$

With the exit Mach number known, the isentropic flow relations give the pressure and temperature:

$$p_e = \frac{p_0}{[1 + 0.2(2.5)^2]^{3.5}} = \frac{200{,}000}{17.08} = 11{,}700 \text{ Pa} \qquad \textit{Ans. (b)}$$

$$T_e = \frac{T_0}{1 + 0.2(2.5)^2} = \frac{500}{2.25} = 222 \text{ K} \qquad \textit{Ans. (c)}$$

The exit velocity follows from the known Mach number and temperature:

$$V_e = \text{Ma}_e(kRT_e)^{1/2} = 2.5[1.4(287)(222)]^{1/2} = 2.5(299 \text{ m/s}) = 747 \text{ m/s} \quad \textit{Ans. (d)}$$

The exit area follows from the known throat area and exit Mach number and Eq. (9.45):

$$\frac{A_e}{A^*} = \frac{[1 + 0.2(2.5)^2]^3}{1.728(2.5)} = 2.64$$

or $$A_e = 2.64A^* = 2.64(0.0083 \text{ m}^2) = 0.0219 \text{ m}^2 = \tfrac{1}{4}\pi D_e^2$$

or $$D_e = 16.7 \text{ cm} \qquad \textit{Ans. (e)}$$

One point might be noted: The computation of the throat area A^* did not depend in any way on the numerical value of the exit Mach number. The exit was supersonic; therefore the throat is sonic and choked, and no further information is needed.

9.5 The Normal Shock Wave

A common irreversibility occurring in supersonic internal or external flows is the normal shock wave sketched in Fig. 9.8. Except at near-vacuum pressures such shock waves are very thin (a few micrometers thick) and approximate a discontinuous change in flow properties. We select a control volume just before and after the wave, as in Fig. 9.8.

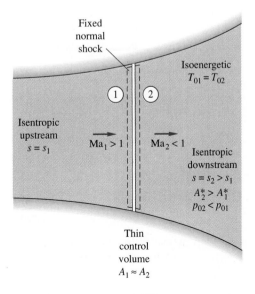

Fig. 9.8 Flow through a fixed normal shock wave.

The analysis is identical to that of Fig. 9.1; that is, a shock wave is a fixed strong pressure wave. To compute all property changes rather than just the wave speed, we use all our basic one-dimensional steady flow relations, letting section 1 be upstream and section 2 be downstream:

Continuity:
$$\rho_1 V_1 = \rho_2 V_2 = G = \text{const} \tag{9.49a}$$

Momentum:
$$p_1 - p_2 = \rho_2 V_2^2 - \rho_1 V_1^2 \tag{9.49b}$$

Energy:
$$h_1 + \tfrac{1}{2}V_1^2 = h_2 + \tfrac{1}{2}V_2^2 = h_0 = \text{const} \tag{9.49c}$$

Perfect gas:
$$\frac{p_1}{\rho_1 T_1} = \frac{p_2}{\rho_2 T_2} \tag{9.49d}$$

Constant c_p:
$$h = c_p T \qquad k = \text{const} \tag{9.49e}$$

Note that we have canceled out the areas $A_1 \approx A_2$, which is justified even in a variable duct section because of the thinness of the wave. The first successful analyses of these normal shock relations are credited to W. J. M. Rankine (1870) and A. Hugoniot (1887), hence the modern term *Rankine-Hugoniot relations*. If we assume that the upstream conditions (p_1, V_1, ρ_1, h_1, T_1) are known, Eqs. (9.49) are five algebraic relations in the five unknowns (p_2, V_2, ρ_2, h_2, T_2). Because of the velocity-squared term, two solutions are found, and the correct one is determined from the second law of thermodynamics, which requires that $s_2 > s_1$.

The velocities V_1 and V_2 can be eliminated from Eqs. (9.49a) to (9.49c) to obtain the Rankine-Hugoniot relation:

$$h_2 - h_1 = \frac{1}{2}(p_2 - p_1)\left(\frac{1}{\rho_2} + \frac{1}{\rho_1}\right) \tag{9.50}$$

This contains only thermodynamic properties and is independent of the equation of state. Introducing the perfect-gas law $h = c_p T = kp/[(k - 1)\rho]$, we can rewrite this as

$$\frac{\rho_2}{\rho_1} = \frac{1 + \beta p_2/p_1}{\beta + p_2/p_1} \qquad \beta = \frac{k + 1}{k - 1} \tag{9.51}$$

We can compare this with the isentropic flow relation for a very weak pressure wave in a perfect gas:

$$\frac{\rho_2}{\rho_1} = \left(\frac{p_2}{p_1}\right)^{1/k} \tag{9.52}$$

Also, the actual change in entropy across the shock can be computed from the perfect-gas relation:

$$\frac{s_2 - s_1}{c_v} = \ln\left[\frac{p_2}{p_1}\left(\frac{\rho_1}{\rho_2}\right)^k\right] \tag{9.53}$$

Assuming a given wave strength p_2/p_1, we can compute the density ratio and the entropy change and list them as follows for $k = 1.4$:

$\dfrac{p_2}{p_1}$	ρ_2/ρ_1		$\dfrac{s_2 - s_1}{c_v}$
	Eq. (9.51)	Isentropic	
0.5	0.6154	0.6095	−0.0134
0.9	0.9275	0.9275	−0.00005
1.0	1.0	1.0	0.0
1.1	1.00704	1.00705	0.00004
1.5	1.3333	1.3359	0.0027
2.0	1.6250	1.6407	0.0134

We see that the entropy change is negative if the pressure decreases across the shock, which violates the second law. Thus a rarefaction shock is impossible in a perfect gas.[2] We see also that weak shock waves ($p_2/p_1 \le 2.0$) are very nearly isentropic.

Mach Number Relations

For a perfect gas all the property ratios across the normal shock are unique functions of k and the upstream Mach number Ma_1. For example, if we eliminate p_2 and V_2 from Eqs. (9.49a) to (9.49c) and introduce $h = kp/[(k-1)\rho]$, we obtain

$$\frac{p_2}{p_1} = \frac{1}{k+1}\left[\frac{2\rho_1 V_1^2}{p_1} - (k-1)\right] \tag{9.54}$$

But for a perfect gas $\rho_1 V_1^2/p_1 = kV_1^2/(kRT_1) = k\,Ma_1^2$, so that Eq. (9.54) is equivalent to

$$\boxed{\frac{p_2}{p_1} = \frac{1}{k+1}\left[2k\,Ma_1^2 - (k-1)\right]} \tag{9.55}$$

From this equation we see that, for any k, $p_2 > p_1$ only if $Ma_1 > 1.0$. Thus for flow through a normal shock wave, the upstream Mach number must be supersonic to satisfy the second law of thermodynamics.

What about the downstream Mach number? From the perfect-gas identity $\rho V^2 = kp\,Ma^2$, we can rewrite Eq. (9.49b) as

$$\frac{p_2}{p_1} = \frac{1 + k\,Ma_1^2}{1 + k\,Ma_2^2} \tag{9.56}$$

which relates the pressure ratio to both Mach numbers. By equating Eqs. (9.55) and (9.56) we can solve for

$$\boxed{Ma_2^2 = \frac{(k-1)\,Ma_1^2 + 2}{2k\,Ma_1^2 - (k-1)}} \tag{9.57}$$

Since Ma_1 must be supersonic, this equation predicts for all $k > 1$ that Ma_2 must be subsonic. Thus a normal shock wave decelerates a flow almost discontinuously from supersonic to subsonic conditions.

---

[2]This is true also for most real gases; see Ref. 9, Sec. 7.3.

Further manipulation of the basic relations (9.49) for a perfect gas gives additional equations relating the change in properties across a normal shock wave in a perfect gas:

$$\frac{\rho_2}{\rho_1} = \frac{(k+1)\,\mathrm{Ma}_1^2}{(k-1)\,\mathrm{Ma}_1^2 + 2} = \frac{V_1}{V_2}$$

$$\frac{T_2}{T_1} = [2 + (k-1)\,\mathrm{Ma}_1^2]\frac{2k\,\mathrm{Ma}_1^2 - (k-1)}{(k+1)^2\,\mathrm{Ma}_1^2} \qquad (9.58)$$

$$T_{02} = T_{01}$$

$$\frac{p_{02}}{p_{01}} = \frac{\rho_{02}}{\rho_{01}} = \left[\frac{(k+1)\,\mathrm{Ma}_1^2}{2 + (k-1)\,\mathrm{Ma}_1^2}\right]^{k/(k-1)}\left[\frac{k+1}{2k\,\mathrm{Ma}_1^2 - (k-1)}\right]^{1/(k-1)}$$

Of additional interest is the fact that the critical, or sonic, throat area A^* in a duct increases across a normal shock:

$$\frac{A_2^*}{A_1^*} = \frac{\mathrm{Ma}_2}{\mathrm{Ma}_1}\left[\frac{2 + (k-1)\,\mathrm{Ma}_1^2}{2 + (k-1)\,\mathrm{Ma}_2^2}\right]^{(1/2)(k+1)(k-1)} \qquad (9.59)$$

All these relations are given in Table B.2 and plotted versus upstream Mach number Ma_1 in Fig. 9.9 for $k = 1.4$. We see that pressure increases greatly while temperature and density increase moderately. The effective throat area A^* increases slowly at first and then rapidly. The failure of students to account for this change in A^* is a common source of error in shock calculations.

The stagnation temperature remains the same, but the stagnation pressure and density decrease in the same ratio; in other words, the flow across the shock is adiabatic but nonisentropic. Other basic principles governing the behavior of shock waves can

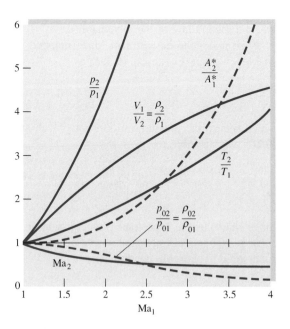

Fig. 9.9 Change in flow properties across a normal shock wave for $k = 1.4$.

be summarized as follows:

1. The upstream flow is supersonic, and the downstream flow is subsonic.
2. For perfect gases (and also for real fluids except under bizarre thermodynamic conditions) rarefaction shocks are impossible, and only a compression shock can exist.
3. The entropy increases across a shock with consequent decreases in stagnation pressure and stagnation density and an increase in the effective sonic throat area.
4. Weak shock waves are very nearly isentropic.

Normal shock waves form in ducts under transient conditions, such as in shock tubes, and in steady flow for certain ranges of the downstream pressure. Figure 9.10a shows a normal shock in a supersonic nozzle. Flow is from left to right. The oblique wave pattern to the left is formed by roughness elements on the nozzle walls and indicates

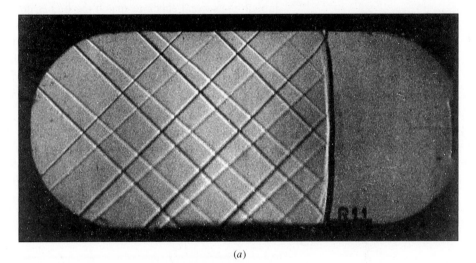

(a)

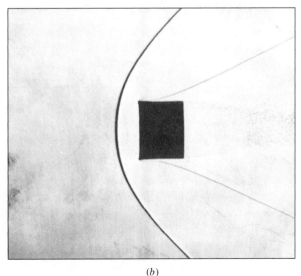

(b)

Fig. 9.10 Normal shocks form in both internal and external flows. (a) Normal shock in a duct; note the Mach wave pattern to the left (upstream), indicating supersonic flow. *(Courtesy of U.S. Air Force Arnold Engineering Development Center.)* (b) Supersonic flow past a blunt body creates a normal shock at the nose; the apparent shock thickness and body-corner curvature are optical distortions. *(Courtesy of U.S. Army Ballistic Research Laboratory, Aberdeen Proving Ground.)*

that the upstream flow is supersonic. Note the absence of these Mach waves (see Sec. 9.10) in the subsonic flow downstream.

Normal shock waves occur not only in supersonic duct flows but also in a variety of supersonic external flows. An example is the supersonic flow past a blunt body shown in Fig. 9.10b. The bow shock is curved, with a portion in front of the body that is essentially normal to the oncoming flow. This normal portion of the bow shock satisfies the property change conditions just as outlined in this section. The flow inside the shock near the body nose is thus subsonic and at relatively high temperature $T_2 > T_1$, and convective heat transfer is especially high in this region.

Each nonnormal portion of the bow shock in Fig. 9.10b satisfies the oblique shock relations to be outlined in Sec. 9.9. Note also the oblique recompression shock on the sides of the body. What has happened is that the subsonic nose flow has accelerated around the corners back to supersonic flow at low pressure, which must then pass through the second shock to match the higher downstream pressure conditions.

Note the fine-grained turbulent wake structure in the rear of the body in Fig. 9.10b. The turbulent boundary layer along the sides of the body is also clearly visible.

The analysis of a complex multidimensional supersonic flow such as in Fig. 9.10 is beyond the scope of this book. For further information see, e.g., Ref. 9, Chap. 9, or Ref. 5, Chap. 16.

Moving Normal Shocks

The preceding analysis of the fixed shock applies equally well to the moving shock if we reverse the transformation used in Fig. 9.1. To make the upstream conditions simulate a still fluid, we move the shock of Fig. 9.8 to the left at speed V_1; that is, we fix our coordinates to a control volume moving with the shock. The downstream flow then appears to move to the left at a slower speed $V_1 - V_2$ following the shock. The thermodynamic properties are not changed by this transformation, so that all our Eqs. (9.50) to (9.59) are still valid.

EXAMPLE 9.6

Air flows from a reservoir where $p = 300$ kPa and $T = 500$ K through a throat to section 1 in Fig. E9.6, where there is a normal shock wave. Compute (a) p_1, (b) p_2, (c) p_{02}, (d) A_2^*, (e) p_{03}, (f) A_3^*, (g) p_3, and (h) T_{03}.

Solution

- *System sketch:* This is shown in Fig. E9.6. Between sections 1 and 2 is a normal shock.
- *Assumptions:* Isentropic flow before and after the shock. Lower p_0 and ρ_0 after the shock.
- *Approach:* After first noting that the throat is *sonic*, work your way from 1 to 2 to 3.
- *Property values:* For air, $R = 287$ m^2/(s$^2 \cdot$ K), $k = 1.40$, and $c_p = 1005$ m^2/(s$^2 \cdot$ K). The inlet stagnation pressure of 300 kPa is constant up to point 1.
- *Solution step (a):* A shock wave cannot exist unless Ma$_1$ is supersonic. Therefore the throat is *sonic* and choked: $A_{\text{throat}} = A_1^* = 1$ m^2. The area ratio gives Ma$_1$ from Eq. (9.45) for $k = 1.4$:

$$\frac{A_1}{A_1^*} = \frac{2 \text{ m}^2}{1 \text{ m}^2} = 2.0 = \frac{1}{\text{Ma}_1} \frac{(1 + 0.2 \text{ Ma}_1^2)^3}{1.728} \qquad \text{solve for} \qquad \text{Ma}_1 = 2.1972$$

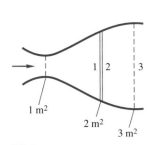

E9.6

1 m^2

2 m^2

3 m^2

1 | 2 | 3

Such four-decimal-place accuracy might require iteration or the use of EES. The curve-fit approximation (9.48c) would give $Ma_1 \approx 1 + 1.2(2.0 - 1)^{1/2} \approx 2.20$, an excellent estimate. Linear interpolation in Table B.1 would give $Ma_1 \approx 2.197$, quite good also. The pressure at section 1 then follows from the isentropic relation, Eq. (9.28):

$$p_1 = \frac{p_{01}}{(1 + 0.2Ma_1^2)^{3.5}} = \frac{300 \text{ kPa}}{[1 + 0.2(2.197)^2]^{3.5}} = 28.2 \text{ kPa} \qquad \text{Ans. } (a)$$

- *Steps (b, c, d):* The pressure p_2 is found from the normal shock Eq. (9.55) or Table B.2:

$$p_2 = \frac{p_1}{k + 1}\left[2k\,Ma_1^2 - (k - 1)\right] = \frac{28.2 \text{ kPa}}{(1.4 + 1)}\left[2(1.4)(2.197)^2 - (1.4 - 1)\right] = 154 \text{ kPa} \quad \text{Ans. } (b)$$

Similarly, for $Ma_1 \approx 2.20$, Table B.2 gives $p_{02}/p_{01} \approx 0.628$ (EES gives 0.6294) and $A_2^*/A_1^* \approx 1.592$ (EES gives 1.5888). Thus, to good accuracy,

$$p_{02} \approx 0.628 p_{01} = 0.628(300 \text{ kPa}) \approx 188 \text{ kPa} \qquad \text{Ans. } (c)$$

$$A_2^* = 1.59 A_1^* = 1.59(1.0 \text{ m}^2) \approx 1.59 \text{ m}^2 \qquad \text{Ans. } (d)$$

- *Comment:* To calculate A_2^* directly, without Table B.2, you would need to pause and calculate $Ma_2 \approx 0.547$ from Eq. (9.57), since Eq. (9.59) involves both Ma_1 and Ma_2.
- *Step (e, f):* The flow from 2 to 3 is isentropic (but at higher entropy than upstream of the shock); therefore

$$p_{03} = p_{02} \approx 188 \text{ kPa} \qquad \text{Ans. } (e)$$

$$A_3^* = A_2^* \approx 1.59 \text{ m}^2 \qquad \text{Ans. } (f)$$

- *Steps (g, h):* The flow is adiabatic throughout, so the stagnation temperature is constant:

$$T_{03} = T_{02} = T_{01} = 500 \text{ K} \qquad \text{Ans. } (h)$$

Next, the area ratio, using the *new* sonic area, gives the Mach number at section 3:

$$\frac{A_3}{A_3^*} = \frac{3 \text{ m}^2}{1.59 \text{ m}^2} = 1.89 = \frac{1}{Ma_3}\frac{(1 + 0.2\,Ma_3^2)^3}{1.728} \qquad \text{solve for} \qquad Ma_3 \approx 0.33$$

EES would yield $Ma_3 = 0.327$, and our curve fit (9.48a) would give $Ma_3 \approx 0.329$. Finally, with p_{02} known, Eq. (9.28) yields p_3:

$$p_3 = \frac{p_{02}}{(1 + 0.2\,Ma_3^2)^{3.5}} \approx \frac{188 \text{ kPa}}{[1 + 0.2(0.33)^2]^{3.5}} \approx 174 \text{ kPa} \qquad \text{Ans. } (g)$$

- *Comments:* EES would give $p_3 = 175$ kPa, so we see that Table B.2 and the curve fits are satisfactory for this type of problem. A duct flow with a normal shock wave requires straightforward application of algebraic perfect-gas relations, coupled with a little thought as to which formula is appropriate for the given property.

EXAMPLE 9.7

An explosion in air, $k = 1.4$, creates a spherical shock wave propagating radially into still air at standard conditions. At the instant shown in Fig. E9.7, the pressure just inside the shock is 200 lbf/in^2 absolute. Estimate (a) the shock speed C and (b) the air velocity V just inside the shock.

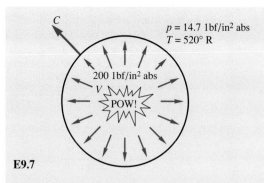

E9.7

Solution

Part (a)

In spite of the spherical geometry, the flow across the shock moves normal to the spherical wave front; hence the normal shock relations (9.50) to (9.59) apply. Fixing our control volume to the moving shock, we find that the proper conditions to use in Fig. 9.8 are

$$C = V_1 \qquad p_1 = 14.7 \text{ lbf/in}^2 \text{ absolute} \qquad T_1 = 520°\text{R}$$
$$V = V_1 - V_2 \qquad p_2 = 200 \text{ lbf/in}^2 \text{ absolute}$$

The speed of sound outside the shock is $a_1 \approx 49T_1^{1/2} = 1117$ ft/s. We can find Ma_1 from the known pressure ratio across the shock:

$$\frac{p_2}{p_1} = \frac{200 \text{ lbf/in}^2 \text{ absolute}}{14.7 \text{ lbf/in}^2 \text{ absolute}} = 13.61$$

From Eq. (9.55) or Table B.2

$$13.61 = \frac{1}{2.4}(2.8 \text{ Ma}_1^2 - 0.4) \qquad \text{or} \qquad \text{Ma}_1 = 3.436$$

Then, by definition of the Mach number,

$$C = V_1 = \text{Ma}_1 \, a_1 = 3.436(1117 \text{ ft/s}) = 3840 \text{ ft/s} \qquad\qquad \textit{Ans. (a)}$$

Part (b)

To find V_2, we need the temperature or sound speed inside the shock. Since Ma_1 is known, from Eq. (9.58) or Table B.2 for $\text{Ma}_1 = 3.436$ we compute $T_2/T_1 = 3.228$. Then

$$T_2 = 3.228T_1 = 3.228(520°\text{R}) = 1679°\text{R}$$

At such a high temperature we should account for non-perfect-gas effects or at least use the gas tables [16], but we won't. Here just estimate from the perfect-gas energy equation (9.23) that

$$V_2^2 = 2c_p(T_1 - T_2) + V_1^2 = 2(6010)(520 - 1679) + (3840)^2 = 815{,}000$$

or
$$V_2 \approx 903 \text{ ft/s}$$

Notice that we did this without bothering to compute Ma_2, which equals 0.454, or $a_2 \approx 49T_2^{1/2} = 2000$ ft/s.

Finally, the air velocity behind the shock is

$$V = V_1 - V_2 = 3840 - 903 \approx 2940 \text{ ft/s} \qquad \textit{Ans. (b)}$$

Thus a powerful explosion creates a brief but intense blast wind as it passes.[3]

9.6 Operation of Converging and Diverging Nozzles

By combining the isentropic flow and normal shock relations plus the concept of sonic throat choking, we can outline the characteristics of converging and diverging nozzles.

Converging Nozzle

First consider the converging nozzle sketched in Fig. 9.11a. There is an upstream reservoir at stagnation pressure p_0. The flow is induced by lowering the downstream outside, or *back,* pressure p_b below p_0, resulting in the sequence of states a to e shown in Fig. 9.11b and c.

For a moderate drop in p_b to states a and b, the throat pressure is higher than the critical value p^* that would make the throat sonic. The flow in the nozzle is subsonic throughout, and the jet exit pressure p_e equals the back pressure p_b. The mass flow is predicted by subsonic isentropic theory and is less than the critical value $\dot{m}_{max}$, as shown in Fig. 9.11c.

For condition c, the back pressure exactly equals the critical pressure p^* of the throat. The throat becomes sonic, the jet exit flow is sonic, $p_e = p_b$, and the mass flow equals its maximum value from Eqs. (9.46). The flow upstream of the throat is subsonic everywhere and predicted by isentropic theory based on the local area ratio $A(x)/A^*$ and Table B.1.

Finally, if p_b is lowered further to conditions d or e below p^*, the nozzle cannot respond further because it is choked at its maximum throat mass flow. The throat remains sonic with $p_e = p^*$, and the nozzle pressure distribution is the same as in state c, as sketched in Fig. 9.11b. The exit jet expands supersonically so that the jet pressure can be reduced from p^* down to p_b. The jet structure is complex and multidimensional and is not shown here. Being supersonic, the jet cannot send any signal upstream to influence the choked flow conditions in the nozzle.

If the stagnation plenum chamber is large or supplemented by a compressor, and if the discharge chamber is larger or supplemented by a vacuum pump, the converging nozzle flow will be steady or nearly so. Otherwise the nozzle will be blowing down, with p_0 decreasing and p_b increasing, and the flow states will be changing from, say, state e backward to state a. Blowdown calculations are usually made by a quasi-steady analysis based on isentropic steady flow theory for the instantaneous pressures $p_0(t)$ and $p_b(t)$.

---

[3]This is the principle of the *shock tube wind tunnel,* in which a controlled explosion creates a brief flow at very high Mach number, with data taken by fast-response instruments. See, e.g., Ref. 2, Sec. 4.5.

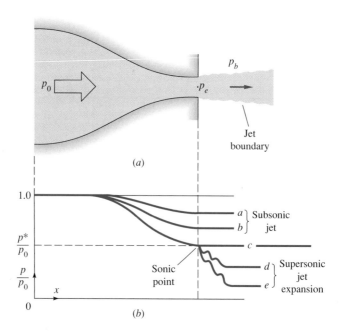

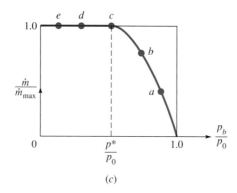

Fig. 9.11 Operation of a converging nozzle: (*a*) nozzle geometry showing characteristic pressures; (*b*) pressure distribution caused by various back pressures; (*c*) mass flow versus back pressure.

EXAMPLE 9.8

A converging nozzle has a throat area of 6 cm^2 and stagnation air conditions of 120 kPa and 400 K. Compute the exit pressure and mass flow if the back pressure is (*a*) 90 kPa and (*b*) 45 kPa. Assume $k = 1.4$.

Solution

From Eq. (9.32) for $k = 1.4$ the critical (sonic) throat pressure is

$$\frac{p^*}{p_0} = 0.5283 \qquad \text{or} \qquad p^* = (0.5283)(120 \text{ kPa}) = 63.4 \text{ kPa}$$

If the back pressure is less than this amount, the nozzle flow is choked.

Part (a)

For $p_b = 90$ kPa $> p*$, the flow is subsonic, not choked. The exit pressure is $p_e = p_b$. The throat Mach number is found from the isentropic relation (9.35) or Table B.1:

$$\text{Ma}_e^2 = 5\left[\left(\frac{p_0}{p_e}\right)^{2/7} - 1\right] = 5\left[\left(\frac{120}{90}\right)^{2/7} - 1\right] = 0.4283 \qquad \text{Ma}_e = 0.654$$

To find the mass flow, we could proceed with a serial attack on Ma_e, T_e, a_e, V_e, and ρ_e, hence to compute $\rho_e A_e V_e$. However, since the local pressure is known, this part is ideally suited for the dimensionless mass flow function in Eq. (9.47). With $p_e/p_0 = 90/120 = 0.75$, compute

$$\frac{\dot{m}\sqrt{RT_0}}{Ap_0} = \sqrt{\frac{2(1.4)}{0.4}(0.75)^{2/1.4}[1 - (0.75)^{0.4/1.4}]} = 0.6052$$

hence

$$\dot{m} = 0.6052\frac{(0.0006)(120{,}000)}{\sqrt{287(400)}} = 0.129 \text{ kg/s} \qquad \qquad Ans.\ (a)$$

for

$$p_e = p_b = 90 \text{ kPa} \qquad \qquad Ans.\ (a)$$

Part (b)

For $p_b = 45$ kPa $< p*$, the flow is choked, similar to condition d in Fig. 9.11b. The exit pressure is sonic:

$$p_e = p* = 63.4 \text{ kPa} \qquad \qquad Ans.\ (b)$$

The (choked) mass flow is a maximum from Eq. (9.46b):

$$\dot{m} = \dot{m}_{\text{max}} = \frac{0.6847p_0A_e}{(RT_0)^{1/2}} = \frac{0.6847(120{,}000)(0.0006)}{[287(400)]^{1/2}} = 0.145 \text{ kg/s} \qquad Ans.\ (b)$$

Any back pressure less than 63.4 kPa would cause this same choked mass flow. Note that the 50 percent increase in exit Mach number, from 0.654 to 1.0, has increased the mass flow only 12 percent, from 0.128 to 0.145 kg/s.

Converging–Diverging Nozzle

Now consider the converging–diverging nozzle sketched in Fig. 9.12a. If the back pressure p_b is low enough, there will be supersonic flow in the diverging portion and a variety of shock wave conditions may occur, which are sketched in Fig. 9.12b. Let the back pressure be gradually decreased.

For curves A and B in Fig. 9.12b the back pressure is not low enough to induce sonic flow in the throat, and the flow in the nozzle is subsonic throughout. The pressure distribution is computed from subsonic isentropic area-change relations, such as in Table B.1. The exit pressure $p_e = p_b$, and the jet is subsonic.

For curve C the area ratio A_e/A_t exactly equals the critical ratio $A_e/A*$ for a subsonic Ma_e in Table B.1. The throat becomes sonic, and the mass flux reaches a maximum in Fig. 9.12c. The remainder of the nozzle flow is subsonic, including the exit jet, and $p_e = p_b$.

Now jump for a moment to curve H. Here p_b is such that p_b/p_0 exactly corresponds to the critical area ratio $A_e/A*$ for a *supersonic* Ma_e in Table B.1. The diverging flow is entirely supersonic, including the jet flow, and $p_e = p_b$. This is called the *design pressure ratio* of the nozzle and is the back pressure suitable for operating a supersonic wind tunnel or an efficient rocket exhaust.

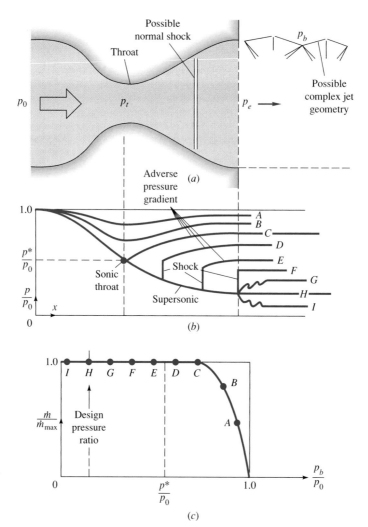

Fig. 9.12 Operation of a converging–diverging nozzle: (*a*) nozzle geometry with possible flow configurations; (*b*) pressure distribution caused by various back pressures; (*c*) mass flow versus back pressure.

Now back up and suppose that p_b lies between curves C and H, which is impossible according to purely isentropic flow calculations. Then back pressures D to F occur in Fig. 9.12*b*. The throat remains choked at the sonic value, and we can match $p_e = p_b$ by placing a normal shock at just the right place in the diverging section to cause a *subsonic diffuser* flow back to the back-pressure condition. The mass flow remains at maximum in Fig. 9.12*c*. At back pressure F the required normal shock stands in the duct exit. At back pressure G no single normal shock can do the job, and so the flow compresses outside the exit in a complex series of oblique shocks until it matches p_b.

Finally, at back pressure I, p_b is lower than the design pressure H, but the nozzle is choked and cannot respond. The exit flow expands in a complex series of supersonic wave motions until it matches the low back pressure. See, Ref. 7, Sec. 5.4, for further details of these off-design jet flow configurations.

Note that for p_b less than back pressure C, there is supersonic flow in the nozzle, and the throat can receive no signal from the exit behavior. The flow remains choked, and the throat has no idea what the exit conditions are.

Note also that the normal shock-patching idea is idealized. Downstream of the shock, the nozzle flow has an adverse pressure gradient, usually leading to wall boundary layer separation. Blockage by the greatly thickened separated layer interacts strongly with the core flow (recall Fig. 6.27) and usually induces a series of weak two-dimensional compression shocks rather than a single one-dimensional normal shock (see, Ref. 9, pp. 292 and 293, for further details).

EXAMPLE 9.9

A converging–diverging nozzle (Fig. 9.12a) has a throat area of 0.002 m^2 and an exit area of 0.008 m^2. Air stagnation conditions are $p_0 = 1000$ kPa and $T_0 = 500$ K. Compute the exit pressure and mass flow for (a) design condition and the exit pressure and mass flow if (b) $p_b \approx 300$ kPa and (c) $p_b \approx 900$ kPa. Assume $k = 1.4$.

Solution

Part (a)

The design condition corresponds to supersonic isentropic flow at the given area ratio $A_e/A_t = 0.008/0.002 = 4.0$. We can find the design Mach number either by iteration of the area ratio formula (9.45), using EES, or by the curve fit (9.48d):

$$\text{Ma}_{e,\text{design}} \approx [216(4.0) - 254(4.0)^{2/3}]^{1/5} \approx 2.95 \qquad (\text{exact} = 2.9402)$$

The accuracy of the curve fit is seen to be satisfactory. The design pressure ratio follows from Eq. (9.34):

$$\frac{p_0}{p_e} = [1 + 0.2(2.95)^2]^{3.5} = 34.1$$

or
$$p_{e,\text{design}} = \frac{1000 \text{ kPa}}{34.1} = 29.3 \text{ kPa} \qquad\qquad Ans. \text{ (}a\text{)}$$

Since the throat is clearly sonic at design conditions, Eq. (9.46b) applies:

$$\dot{m}_{\text{design}} = \dot{m}_{\text{max}} = \frac{0.6847 p_0 A_t}{(RT_0)^{1/2}} = \frac{0.6847(10^6 \text{ Pa})(0.002 \text{ m}^2)}{[287(500)]^{1/2}} \qquad Ans. \text{ (}a\text{)}$$

$$= 3.61 \text{ kg/s}$$

Part (b)

For $p_b = 300$ kPa we are definitely far below the subsonic isentropic condition C in Fig. 9.12b, but we may even be below condition F with a normal shock in the exit—that is, in condition G, where oblique shocks occur outside the exit plane. If it is condition G, then $p_e = p_{e,\text{design}} = 29.3$ kPa because no shock has yet occurred. To find out, compute condition F by assuming an exit normal shock with Ma$_1 = 2.95$—that is, the design Mach number just upstream of the shock. From Eq. (9.55)

$$\frac{p_2}{p_1} = \frac{1}{2.4} [2.8(2.95)^2 - 0.4] = 9.99$$

or
$$p_2 = 9.99 p_1 = 9.99 p_{e,\text{design}} = 293 \text{ kPa}$$

Since this is less than the given $p_b = 300$ kPa, there is a normal shock just upstream of the exit plane (condition E). The exit flow is subsonic and equals the back pressure:

$$p_e = p_b = 300 \text{ kPa} \qquad\qquad\qquad Ans.\ (b)$$

Also

$$\dot{m} = \dot{m}_{max} = 3.61 \text{ kg/s} \qquad\qquad\qquad Ans.\ (b)$$

The throat is still sonic and choked at its maximum mass flow.

Part (c)   Finally, for $p_b = 900$ kPa, which is up near condition C, we compute Ma_e and p_e for condition C as a comparison. Again $A_e/A_t = 4.0$ for this condition, with a subsonic Ma_e estimated from the curve-fitted Eq. (9.48a):

$$Ma_e(C) \approx \frac{1 + 0.27/(4.0)^2}{1.728(4.0)} = 0.147 \qquad (\text{exact} = 0.14655)$$

Then the isentropic exit pressure ratio for this condition is

$$\frac{p_0}{p_e} = [1 + 0.2(0.147)^2]^{3.5} = 1.0152$$

or

$$p_e = \frac{1000}{1.0152} = 985 \text{ kPa}$$

The given back pressure of 900 kPa is less than this value, corresponding roughly to condition D in Fig. 9.12b. Thus for this case there is a normal shock just downstream of the throat, and the throat is choked:

$$p_e = p_b = 900 \text{ kPa} \qquad \dot{m} = \dot{m}_{max} = 3.61 \text{ kg/s} \qquad\qquad Ans.\ (c)$$

For this large exit area ratio, the exit pressure would have to be larger than 985 kPa to cause a subsonic flow in the throat and a mass flow less than maximum.

9.7 Compressible Duct Flow with Friction[4]

Section 9.4 showed the effect of area change on a compressible flow while neglecting friction and heat transfer. We could now add friction and heat transfer to the area change and consider coupled effects, which is done in advanced texts [for example, 5, Chap. 8]. Instead, as an elementary introduction, this section treats only the effect of friction, neglecting area change and heat transfer. The basic assumptions are

1. Steady one-dimensional adiabatic flow.
2. Perfect gas with constant specific heats.
3. Constant-area straight duct.
4. Negligible shaft work and potential energy changes.
5. Wall shear stress correlated by a Darcy friction factor.

In effect, we are studying a Moody-type pipe friction problem but with large changes in kinetic energy, enthalpy, and pressure in the flow.

This type of duct flow—constant area, constant stagnation enthalpy, constant mass flow, but variable momentum (due to friction)—is often termed *Fanno flow,* after Gino

---

[4]This section may be omitted without loss of continuity.

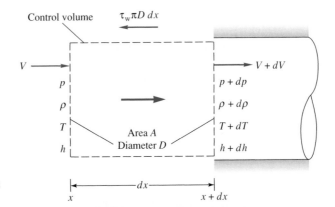

Fig. 9.13 Elemental control volume for flow in a constant-area duct with friction.

Fanno, an Italian engineer born in 1882, who first studied this flow. For a given mass flow and stagnation enthalpy, a plot of enthalpy versus entropy for all possible flow states, subsonic or supersonic, is called a *Fanno line*. See Probs. P9.94 and P9.111 for examples of a Fanno line.

Consider the elemental duct control volume of area A and length dx in Fig. 9.13. The area is constant, but other flow properties (p, ρ, T, h, V) may vary with x. Application of the three conservation laws to this control volume gives three differential equations:

Continuity:
$$\rho V = \frac{\dot{m}}{A} = G = \text{const}$$

or
$$\frac{d\rho}{\rho} + \frac{dV}{V} = 0 \qquad (9.60a)$$

x momentum:
$$pA - (p + dp)A - \tau_w \pi D \, dx = \dot{m}(V + dV - V)$$

or
$$dp + \frac{4\tau_w dx}{D} + \rho V \, dV = 0 \qquad (9.60b)$$

Energy:
$$h + \tfrac{1}{2}V^2 = h_0 = c_p T_0 = c_p T + \tfrac{1}{2}V^2$$

or
$$c_p \, dT + V \, dV = 0 \qquad (9.60c)$$

Since these three equations have five unknowns—p, ρ, T, V, and τ_w—we need two additional relations. One is the perfect-gas law:

$$p = \rho R T \qquad \text{or} \qquad \frac{dp}{p} = \frac{d\rho}{\rho} + \frac{dT}{T} \qquad (9.61)$$

To eliminate τ_w as an unknown, it is assumed that wall shear is correlated by a local Darcy friction factor f

$$\tau_w = \tfrac{1}{8} f \rho V^2 = \tfrac{1}{8} f k p \, \text{Ma}^2 \qquad (9.62)$$

where the last form follows from the perfect-gas speed-of-sound expression $a^2 = kp/\rho$. In practice, f can be related to the local Reynolds number and wall roughness from, say, the Moody chart, Fig. 6.13.

Equations (9.60) and (9.61) are first-order differential equations and can be integrated, by using friction factor data, from any inlet section 1, where p_1, T_1, V_1, and so on are known, to determine $p(x)$, $T(x)$, and the like along the duct. It is practically impossible to eliminate all but one variable to give, say, a single differential equation for $p(x)$, but all equations can be written in terms of the Mach number $\text{Ma}(x)$ and the friction factor, by using this definition of Mach number:

$$V^2 = \text{Ma}^2\, kRT$$

or

$$\frac{2\,dV}{V} = \frac{2\,d\,\text{Ma}}{\text{Ma}} + \frac{dT}{T} \tag{9.63}$$

Adiabatic Flow

By eliminating variables between Eqs. (9.60) to (9.63), we obtain the working relations

$$\frac{dp}{p} = -k\,\text{Ma}^2\,\frac{1 + (k-1)\,\text{Ma}^2}{2(1 - \text{Ma}^2)}\,f\,\frac{dx}{D} \tag{9.64a}$$

$$\frac{d\rho}{\rho} = -\frac{k\,\text{Ma}^2}{2(1 - \text{Ma}^2)}\,f\,\frac{dx}{D} = -\frac{dV}{V} \tag{9.64b}$$

$$\frac{dp_0}{p_0} = \frac{d\rho_0}{\rho_0} = -\frac{1}{2}\,k\,\text{Ma}^2\,f\,\frac{dx}{D} \tag{9.64c}$$

$$\frac{dT}{T} = -\frac{k(k-1)\,\text{Ma}^4}{2(1 - \text{Ma}^2)}\,f\,\frac{dx}{D} \tag{9.64d}$$

$$\frac{d\,\text{Ma}^2}{\text{Ma}^2} = k\,\text{Ma}^2\,\frac{1 + \frac{1}{2}(k-1)\,\text{Ma}^2}{1 - \text{Ma}^2}\,f\,\frac{dx}{D} \tag{9.64e}$$

All these except dp_0/p_0 have the factor $1 - \text{Ma}^2$ in the denominator, so that, like the area change formulas in Fig. 9.5, subsonic and supersonic flow have opposite effects:

Property	Subsonic	Supersonic
p	Decreases	Increases
ρ	Decreases	Increases
V	Increases	Decreases
p_0, ρ_0	Decreases	Decreases
T	Decreases	Increases
Ma	Increases	Decreases
Entropy	Increases	Increases

We have added to this list that entropy must increase along the duct for either subsonic or supersonic flow as a consequence of the second law for adiabatic flow. For the same reason, stagnation pressure and density must both decrease.

The key parameter in this discussion is the Mach number. Whether the inlet flow is subsonic or supersonic, the duct Mach number always tends downstream toward $\text{Ma} = 1$ because this is the path along which the entropy increases. If the pressure and density are computed from Eqs. (9.64a) and (9.64b) and the entropy from Eq. (9.53), the result can be plotted in Fig. 9.14 versus Mach number for $k = 1.4$.

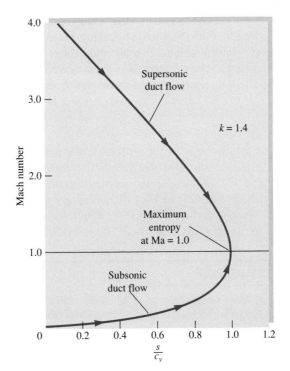

Fig. 9.14 Adiabatic frictional flow in a constant-area duct always approaches Ma = 1 to satisfy the second law of thermodynamics. The computed curve is independent of the value of the friction factor.

The maximum entropy occurs at Ma = 1, so that the second law requires that the duct flow properties continually approach the sonic point. Since p_0 and ρ_0 continually decrease along the duct due to the frictional (nonisentropic) losses, they are not useful as reference properties. Instead, the sonic properties p^*, ρ^*, T^*, p_0^*, and ρ_0^* are the appropriate constant reference quantities in adiabatic duct flow. The theory then computes the ratios p/p^*, T/T^*, and so forth as a function of local Mach number and the integrated friction effect.

To derive working formulas, we first attack Eq. (9.64e), which relates the Mach number to friction. Separate the variables and integrate:

$$\int_0^{L^*} f\,\frac{dx}{D} = \int_{\mathrm{Ma}^2}^{1.0} \frac{1 - \mathrm{Ma}^2}{k\,\mathrm{Ma}^4[1 + \tfrac{1}{2}(k-1)\,\mathrm{Ma}^2]}\,d\,\mathrm{Ma}^2 \tag{9.65}$$

The upper limit is the sonic point, whether or not it is actually reached in the duct flow. The lower limit is arbitrarily placed at the position $x = 0$, where the Mach number is Ma. The result of the integration is

$$\boxed{\frac{\bar{f}L^*}{D} = \frac{1 - \mathrm{Ma}^2}{k\,\mathrm{Ma}^2} + \frac{k+1}{2k}\ln\frac{(k+1)\,\mathrm{Ma}^2}{2 + (k-1)\,\mathrm{Ma}^2}} \tag{9.66}$$

where $\bar{f}$ is the average friction factor between 0 and L^*. In practice, an average f is always assumed, and no attempt is made to account for the slight changes in Reynolds number along the duct. For noncircular ducts, D is replaced by the hydraulic diameter $D_h = (4 \times \text{area})/\text{perimeter}$ as in Eq. 6.59.

Equation (9.66) is tabulated versus Mach number in Table B.3. The length L^* is the length of duct required to develop a duct flow from Mach number Ma to the sonic point. Many problems involve short ducts that never become sonic, for which the solution uses the differences in the tabulated "maximum," or sonic, length. For example, the length ΔL required to develop from Ma_1 to Ma_2 is given by

$$\bar{f}\frac{\Delta L}{D} = \left(\frac{\bar{f}L^*}{D}\right)_1 - \left(\frac{\bar{f}L^*}{D}\right)_2 \tag{9.67}$$

This avoids the need for separate tabulations for short ducts.

It is recommended that the friction factor $\bar{f}$ be estimated from the Moody chart (Fig. 6.13) for the average Reynolds number and wall roughness ratio of the duct. Available data [23] on duct friction for compressible flow show good agreement with the Moody chart for subsonic flow, but the measured data in supersonic duct flow are up to 50 percent less than the equivalent Moody friction factor.

EXAMPLE 9.10

Air flows subsonically in an adiabatic 2-cm-diameter duct. The average friction factor is 0.024. What length of duct is necessary to accelerate the flow from $\text{Ma}_1 = 0.1$ to $\text{Ma}_2 = 0.5$? What additional length will accelerate it to $\text{Ma}_3 = 1.0$? Assume $k = 1.4$.

Solution

Equation (9.67) applies, with values of $\bar{f}L^*/D$ computed from Eq. (9.66) or read from Table B.3:

$$\bar{f}\frac{\Delta L}{D} = \frac{0.024\,\Delta L}{0.02\text{ m}} = \left(\frac{\bar{f}L^*}{D}\right)_{\text{Ma}=0.1} - \left(\frac{\bar{f}L^*}{D}\right)_{\text{Ma}=0.5}$$

$$= 66.9216 - 1.0691 = 65.8525$$

Thus
$$\Delta L = \frac{65.8525(0.02\text{ m})}{0.024} = 55\text{ m} \qquad \textit{Ans. (a)}$$

The additional length $\Delta L'$ to go from Ma = 0.5 to Ma = 1.0 is taken directly from Table B.2:

$$f\frac{\Delta L'}{D} = \left(\frac{fL^*}{D}\right)_{\text{Ma}=0.5} = 1.0691$$

or
$$\Delta L' = L^*_{\text{Ma}=0.5} = \frac{1.0691(0.02\text{ m})}{0.024} = 0.9\text{ m} \qquad \textit{Ans. (b)}$$

This is typical of these calculations: It takes 55 m to accelerate up to Ma = 0.5 and then only 0.9 m more to get all the way up to the sonic point.

Formulas for other flow properties along the duct can be derived from Eqs. (9.64). Equation (9.64e) can be used to eliminate $f\,dx/D$ from each of the other relations, giving, for example, dp/p as a function only of Ma and $d\,\text{Ma}^2/\text{Ma}^2$. For convenience in tabulating the results, each expression is then integrated all the way from (p, Ma) to the sonic point (p^*, 1.0). The integrated results are

$$\frac{p}{p^*} = \frac{1}{\text{Ma}}\left[\frac{k+1}{2+(k-1)\text{Ma}^2}\right]^{1/2} \tag{9.68a}$$

$$\frac{\rho}{\rho^*} = \frac{V^*}{V} = \frac{1}{\text{Ma}}\left[\frac{2+(k-1)\text{Ma}^2}{k+1}\right]^{1/2} \tag{9.68b}$$

$$\frac{T}{T^*} = \frac{a^2}{a^{*2}} = \frac{k+1}{2+(k-1)\text{Ma}^2} \tag{9.68c}$$

$$\frac{p_0}{p_0^*} = \frac{\rho_0}{\rho_0^*} = \frac{1}{\text{Ma}}\left[\frac{2+(k-1)\text{Ma}^2}{k+1}\right]^{(1/2)(k+1)/(k-1)} \tag{9.68d}$$

All these ratios are also tabulated in Table B.3. For finding changes between points Ma_1 and Ma_2 that are not sonic, products of these ratios are used. For example,

$$\frac{p_2}{p_1} = \frac{p_2}{p^*}\frac{p^*}{p_1} \tag{9.69}$$

since p^* is a constant reference value for the flow.

EXAMPLE 9.11

For the duct flow of Example 9.10 assume that, at $\text{Ma}_1 = 0.1$, we have $p_1 = 600$ kPa and $T_1 = 450$ K. At section 2 farther downstream, $\text{Ma}_2 = 0.5$. Compute (a) p_2, (b) T_2, (c) V_2, and (d) p_{02}.

Solution

As preliminary information we can compute V_1 and p_{01} from the given data:

$$V_1 = \text{Ma}_1\, a_1 = 0.1[(1.4)(287)(450)]^{1/2} = 0.1(425 \text{ m/s}) = 42.5 \text{ m/s}$$
$$p_{01} = p_1(1 + 0.2\,\text{Ma}_1^2)^{3.5} = (600 \text{ kPa})[1 + 0.2(0.1)^2]^{3.5} = 604 \text{ kPa}$$

Now enter Table B.3 or Eqs. (9.68) to find the following property ratios:

Section	Ma	p/p^*	T/T^*	V/V^*	p_0/p_0^*
1	0.1	10.9435	1.1976	0.1094	5.8218
2	0.5	2.1381	1.1429	0.5345	1.3399

Use these ratios to compute all properties downstream:

$$p_2 = p_1\frac{p_2/p^*}{p_1/p^*} = (600 \text{ kPa})\frac{2.1381}{10.9435} = 117 \text{ kPa} \qquad Ans.\ (a)$$

$$T_2 = T_1\frac{T_2/T^*}{T_1/T^*} = (450 \text{ K})\frac{1.1429}{1.1976} = 429 \text{ K} \qquad Ans.\ (b)$$

$$V_2 = V_1\frac{V_2/V^*}{V_1/V^*} = (42.5 \text{ m/s})\frac{0.5345}{0.1094} = 208 \frac{\text{m}}{\text{s}} \qquad Ans.\ (c)$$

$$p_{02} = p_{01}\frac{p_{02}/p_0^*}{p_{01}/p_0^*} = (604 \text{ kPa})\frac{1.3399}{5.8218} = 139 \text{ kPa} \qquad Ans.\ (d)$$

Note the 77 percent reduction in stagnation pressure due to friction. The formulas are seductive, so check your work by other means. For example, check $p_{02} = p_2(1 + 0.2 \, Ma_2^2)^{3.5}$.

Software comment: EES is somewhat laborious for this type of problem because the basic pipe friction relations, Eqs. (9.68), have to be typed in twice, once for section 1 and once for section 2. Also, V_1, a_1, and p_{01} have to be computed as just shown. The beauty thereafter is that Mach number is no longer dominant. One could specify p_2 or T_2 or V_2 or p_{02} and EES would immediately give the full solution at section 2.

Choking due to Friction

The theory here predicts that for adiabatic frictional flow in a constant-area duct, no matter what the inlet Mach number Ma_1 is, the flow downstream tends toward the sonic point. There is a certain duct length $L^*(Ma_1)$ for which the exit Mach number will be exactly unity. The duct is then choked.

But what if the actual length L is greater than the predicted "maximum" length L^*? Then the flow conditions must change, and there are two classifications.

Subsonic Inlet. If $L > L^*(Ma_1)$, the flow slows down until an inlet Mach number Ma_2 is reached such that $L = L^*(Ma_2)$. The exit flow is sonic, and the mass flow has been reduced by *frictional choking*. Further increases in duct length will continue to decrease the inlet Ma and mass flow.

Supersonic Inlet. From Table B.3 we see that friction has a very large effect on supersonic duct flow. Even an infinite inlet Mach number will be reduced to sonic conditions in only 41 diameters for $\bar{f} = 0.02$. Some typical numerical values are

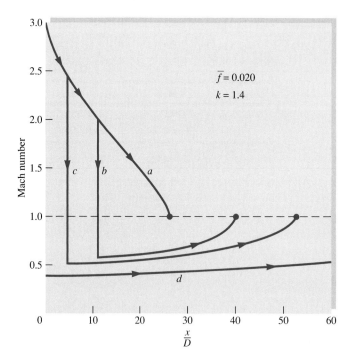

Fig. 9.15 Behavior of duct flow with a nominal supersonic inlet condition Ma = 3.0: (*a*) $L/D \le 26$, flow is supersonic throughout duct; (*b*) $L/D = 40 > L^*/D$, normal shock at Ma = 2.0 with subsonic flow then accelerating to sonic exit point; (*c*) $L/D = 53$, shock must now occur at Ma = 2.5; (*d*) $L/D > 63$, flow must be entirely subsonic and choked at exit.

shown in Fig. 9.15, assuming an inlet Ma = 3.0 and $\bar{f}$ = 0.02. For this condition L^* = 26 diameters. If L is increased beyond $26D$, the flow will not choke but a normal shock will form at just the right place for the subsequent subsonic frictional flow to become sonic exactly at the exit. Figure 9.15 shows two examples, for L/D = 40 and 53. As the length increases, the required normal shock moves upstream until, for Fig. 9.15, the shock is at the inlet for L/D = 63. Further increase in L causes the shock to move upstream of the inlet into the supersonic nozzle feeding the duct. Yet the mass flow is still the same as for the very short duct, because presumably the feed nozzle still has a sonic throat. Eventually, a very long duct will cause the feed-nozzle throat to become choked, thus reducing the duct mass flow. Thus supersonic friction changes the flow pattern if $L > L^*$ but does not choke the flow until L is much larger than L^*.

EXAMPLE 9.12

Air enters a 3-cm-diameter duct at p_0 = 200 kPa, T_0 = 500 K, and V_1 = 100 m/s. The friction factor is 0.02. Compute (a) the maximum duct length for these conditions, (b) the mass flow if the duct length is 15 m, and (c) the reduced mass flow if L = 30 m.

Solution

Part (a)

First compute

$$T_1 = T_0 - \frac{\frac{1}{2}V_1^2}{c_p} = 500 - \frac{\frac{1}{2}(100 \text{ m(s)}^2}{1005 \text{ m}^2/\text{s}^2 \cdot \text{K}} = 500 - 5 = 495 \text{ K}$$

$$a_1 = (kRT_1)^{1/2} \approx 20(495)^{1/2} = 445 \text{ m/s}$$

Thus

$$\text{Ma}_1 = \frac{V_1}{a_1} = \frac{100}{445} = 0.225$$

For this Ma_1, from Eq. (9.66) or interpolation in Table B.3,

$$\frac{\bar{f}L^*}{D} = 11.0$$

The maximum duct length possible for these inlet conditions is

$$L^* = \frac{(\bar{f}L^*/D)D}{\bar{f}} = \frac{11.0(0.03 \text{ m})}{0.02} = 16.5 \text{ m} \qquad \textit{Ans. (a)}$$

Part (b)

The given L = 15 m is less than L^*, and so the duct is not choked and the mass flow follows from the inlet conditions:

$$\rho_{01} = \frac{p_{01}}{RT_0} = \frac{200,000 \text{ Pa}}{287(500 \text{ K})} = 1.394 \text{ kg/m}^3$$

$$\rho_1 = \frac{\rho_{01}}{[1 + 0.2(0.225)^2]^{2.5}} = \frac{1.394}{1.0255} = 1.359 \text{ kg/m}^3$$

whence

$$\dot{m} = \rho_1 AV_1 = (1.359 \text{ kg/m}^3)\left[\frac{\pi}{4}(0.03 \text{ m})^2\right](100 \text{ m/s})$$

$$= 0.0961 \text{ kg/s} \qquad \textit{Ans. (b)}$$

Part (c)   Since $L = 30$ m is greater than L^*, the duct must choke back until $L = L^*$, corresponding to a lower inlet Ma_1:

$$L^* = L = 30 \text{ m}$$

$$\frac{\bar{f}L^*}{D} = \frac{0.02(30 \text{ m})}{0.03 \text{ m}} = 20.0$$

It is difficult to interpolate for $fL/D = 20$ in Table B.3 and impossible to invert Eq. (9.66) for the Mach number without laborious iteration. But it is a breeze for EES to solve Eq. (9.66) for the Mach number, using the following three statements:

$$k = 1.4$$

$$fLD = 20$$

$$fLD = (1 - Ma^2)/k/Ma^2 + (k + 1)/2/k*LN((k + 1)*Ma^2/(2 + (k - 1)*Ma^2))$$

Simply specify $Ma < 1$ in the Variable Information menu, and EES reports

$$Ma_{choked} = 0.174 \quad (23 \text{ percent less})$$

$$T_{1,new} = \frac{T_0}{1 + 0.2(0.174)^2} = 497 \text{ K}$$

$$a_{1,new} \approx 20(497 \text{ K})^{1/2} = 446 \text{ m/s}$$

$$V_{1,new} = Ma_1 \, a_1 = 0.174(446) = 77.6 \text{ m/s}$$

$$\rho_{1,new} = \frac{\rho_{01}}{[1 + 0.2(0.174)^2]^{2.5}} = 1.373 \text{ kg/m}^3$$

$$\dot{m}_{new} = \rho_1 AV_1 = 1.373 \left[\frac{\pi}{4}(0.03)^2 \right](77.6)$$

$$= 0.0753 \text{ kg/s} \quad (22 \text{ percent less}) \qquad Ans. (c)$$

Minor Losses in Compressible Flow

For incompressible pipe flow, as in Eq. (6.78), the loss coefficient K is the ratio of pressure head loss ($\Delta p/\rho g$) to the velocity head ($V^2/2g$) in the pipe. This is inappropriate in compressible pipe flow, where ρ and V are not constant. Benedict [24] suggests that the static pressure loss ($p_1 - p_2$) be related to downstream conditions and a *static loss coefficient* K_s:

$$K_s = \frac{2(p_1 - p_2)}{\rho_2 V_2^2} \tag{9.70}$$

Benedict [24] gives examples of compressible losses in sudden contractions and expansions. If data are unavailable, a first approximation would be to use $K_s \approx K$ from Section 6.9.

Isothermal Flow with Friction: Long Pipelines

The adiabatic frictional flow assumption is appropriate to high-speed flow in short ducts. For flow in long ducts, such as natural gas pipelines, the gas state more closely approximates an isothermal flow. The analysis is the same except that the isoenergetic energy equation (9.60c) is replaced by the simple relation

$$T = \text{const} \qquad dT = 0$$

Again it is possible to write all property changes in terms of the Mach number. Integration of the Mach number–friction relation yields

$$\frac{\bar{f} L_{\max}}{D} = \frac{1 - k\,\text{Ma}^2}{k\,\text{Ma}^2} + \ln\left(k\,\text{Ma}^2\right) \tag{9.71}$$

which is the isothermal analog of Eq. (9.66) for adiabatic flow.

This friction relation has the interesting result that $L_{\max}$ becomes zero not at the sonic point but at $\text{Ma}_{\text{crit}} = 1/k^{1/2} = 0.845$ if $k = 1.4$. The inlet flow, whether subsonic or supersonic, tends downstream toward this limiting Mach number $1/k^{1/2}$. If the tube length L is greater than $L_{\max}$ from Eq. (9.71), a subsonic flow will choke back to a smaller Ma_1 and mass flow and a supersonic flow will experience a normal shock adjustment similar to Fig. 9.15.

The exit isothermal choked flow is not sonic, and so the use of the asterisk is inappropriate. Let p', ρ', and V' represent properties at the choking point $L = L_{\max}$. Then the isothermal analysis leads to the following Mach number relations for the flow properties:

$$\frac{p}{p'} = \frac{1}{\text{Ma}\,k^{1/2}} \qquad \frac{V}{V'} = \frac{\rho'}{\rho} = \text{Ma}\,k^{1/2} \tag{9.72}$$

The complete analysis and some examples are given in advanced texts [for example, 5, Sec. 6.4].

Mass Flow for a Given Pressure Drop

An interesting by-product of the isothermal analysis is an explicit relation between the pressure drop and duct mass flow. This common problem requires numerical iteration for adiabatic flow, as outlined here. In isothermal flow, we may substitute $dV/V = -dp/p$ and $V^2 = G^2/[p/(RT)]^2$ in Eq. (9.63) to obtain

$$\frac{2p\,dp}{G^2 RT} + f\frac{dx}{D} - \frac{2\,dp}{p} = 0$$

Since $G^2 RT$ is constant for isothermal flow, this may be integrated in closed form between $(x, p) = (0, p_1)$ and (L, p_2):

$$G^2 = \left(\frac{\dot{m}}{A}\right)^2 = \frac{p_1^2 - p_2^2}{RT\left[\bar{f}L/D + 2\ln\left(p_1/p_2\right)\right]} \tag{9.73}$$

Thus mass flow follows directly from the known end pressures, without any use of Mach numbers or tables.

The writer does not know of any direct analogy to Eq. (9.73) for adiabatic flow. However, a useful adiabatic relation, involving velocities instead of pressures, is derived in several textbooks [2, p. 212]:

$$V_1^2 = \frac{a_0^2\left[1 - (V_1/V_2)^2\right]}{k\bar{f}L/D + (k + 1)\ln\left(V_2/V_1\right)} \tag{9.74}$$

where $a_0 = (kRT_0)^{1/2}$ is the stagnation speed of sound, constant for adiabatic flow. This may be combined with continuity for constant duct area $V_1/V_2 = \rho_2/\rho_1$, plus the following combination of adiabatic energy and the perfect-gas relation:

$$\frac{V_1}{V_2} = \frac{p_2}{p_1}\frac{T_1}{T_2} = \frac{p_2}{p_1}\left[\frac{2a_0^2 - (k-1)V_1^2}{2a_0^2 - (k-1)V_2^2}\right] \tag{9.75}$$

If we are given the end pressures, neither V_1 nor V_2 will likely be known in advance. Here, if EES is not available, we suggest only the following simple procedure. Begin with $a_0 \approx a_1$ and the bracketed term in Eq. (9.75) approximately equal to 1.0. Solve Eq. (9.75) for a first estimate of V_1/V_2, and use this value in Eq. (9.74) to get a better estimate of V_1. Use V_1 to improve your estimate of a_0, and repeat the procedure. The process should converge in a few iterations.

Equations (9.73) and (9.74) have one flaw: With the Mach number eliminated, the frictional choking phenomenon is not directly evident. Therefore, assuming a subsonic inlet flow, one should check the exit Mach number Ma_2 to ensure that it is not greater than $1/k^{1/2}$ for isothermal flow or greater than 1.0 for adiabatic flow. We illustrate both adiabatic and isothermal flow with the following example.

EXAMPLE 9.13

Air enters a pipe of 1-cm diameter and 1.2-m length at $p_1 = 220$ kPa and $T_1 = 300$ K. If $\bar{f} = 0.025$ and the exit pressure is $p_2 = 140$ kPa, estimate the mass flow for (a) isothermal flow and (b) adiabatic flow.

Solution

Part (a)

For isothermal flow Eq. (9.73) applies without iteration:

$$\frac{\bar{f}L}{D} + 2\ln\frac{p_1}{p_2} = \frac{(0.025)(1.2\text{ m})}{0.01\text{ m}} + 2\ln\frac{220}{140} = 3.904$$

$$G^2 = \frac{(220{,}000\text{ Pa})^2 - (140{,}000\text{ Pa})^2}{[287\text{ m}^2/(\text{s}^2\cdot\text{K})](300\text{ K})(3.904)} = 85{,}700 \qquad \text{or} \qquad G = 293\text{ kg/(s}\cdot\text{m}^2)$$

Since $A = (\pi/4)(0.01\text{ m})^2 = 7.85$ E-5 m^2, the isothermal mass flow estimate is

$$\dot{m} = GA = (293)(7.85\text{ E-5}) \approx 0.0230\text{ kg/s} \qquad\qquad Ans.\ (a)$$

Check that the exit Mach number is not choked:

$$\rho_2 = \frac{p_2}{RT} = \frac{140{,}000}{(287)(300)} = 1.626\text{ kg/m}^3 \qquad V_2 = \frac{G}{\rho_2} = \frac{293}{1.626} = 180\text{ m/s}$$

or

$$Ma_2 = \frac{V_2}{\sqrt{kRT}} = \frac{180}{[1.4(287)(300)]^{1/2}} = \frac{180}{347} \approx 0.52$$

This is well below choking, and the isothermal solution is accurate.

Part (b)

For adiabatic flow, we can iterate by hand, in the time-honored fashion, using Eqs. (9.74) and (9.75) plus the definition of stagnation speed of sound. A few years ago the author would have done just that, laboriously. However, EES makes handwork and manipulation of equations unnecessary, although careful programming and good guesses are required. If we ignore superfluous output such as T_2 and V_2, 13 statements are appropriate. First, spell out the given physical properties (in SI units):

$$k = 1.4$$

$$P1 = 220000$$

$$P2 = 140000$$

$$T1 = 300$$

Next, apply the adiabatic friction relations, Eqs. (9.66) and (9.67), to both points 1 and 2:

```
fLD1 = (1 − Ma1^2)/k/Ma1^2 + (k + 1)/2/k*LN((k + 1)*Ma1^2/(2 +
(k − 1)*Ma1^2))
fLD2 = (1 − Ma2^2)/k/Ma2^2 + (k + 1)/2/k*LN((k + 1)*Ma2^2/(2 +
(k − 1)*Ma2^2))
DeltafLD = 0.025*1.2/0.01
fLD1 = fLD2 + DeltafLD
```

Then apply the pressure ratio formula (9.68a) to both points 1 and 2:

```
P1/Pstar = ((k + 1)/(2 + (k − 1)*Ma1^2))^0.5/Ma1
P2/Pstar = ((k + 1)/(2 + (k − 1)*Ma2^2))^0.5/Ma2
```

These are *adiabatic* relations, so we need not further spell out quantities such as T_0 or a_0 unless we want them as additional output.

The above 10 statements are a closed algebraic system, and EES will solve them for Ma_1 and Ma_2. However, the problem asks for mass flow, so we complete the system:

```
V1 = Ma1*sqrt(1.4*287*T1)
Rho1 = P1/287/T1
Mdot = Rho1*(pi/4*0.01^2)*V1
```

If we apply no constraints, EES reports "cannot solve" because its default allows all variables to lie between $-\infty$ and $+\infty$. So we enter Variable Information and constrain Ma_1 and Ma_2 to lie between 0 and 1 (subsonic flow). EES still complains that it "cannot solve" but hints that "better guesses are needed." Indeed, the default guesses for EES variables are normally 1.0, too large for the Mach numbers. Guess the Mach numbers equal to 0.8 or even 0.5, and EES still complains, for a subtle reason: Since $f\Delta L/D = 0.025(1.2/0.01) = 3.0$, Ma_1 can be no larger than 0.36 (see Table B.3). Finally, then, we guess Ma_1 and $Ma_2 = 0.3$ or 0.4, and EES reports the solution:

$$Ma_1 = 0.3343 \quad Ma_2 = 0.5175 \quad \frac{fL}{D_1} = 3.935 \quad \frac{fL}{D_2} = 0.9348$$

$$p^* = 67,892 \text{ Pa} \quad \dot{m} = 0.0233 \text{ kg/s} \qquad \textit{Ans. (b)}$$

Though the programming is complicated, the EES approach is superior to hand iteration; and, of course, we can save this program for use again with new data.

9.8 Frictionless Duct Flow with Heat Transfer[5]

Heat addition or removal has an interesting effect on a compressible flow. Advanced texts [for example, 5, Chap. 8] consider the combined effect of heat transfer coupled with friction and area change in a duct. Here we confine the analysis to heat transfer with no friction in a constant-area duct.

This type of duct flow—constant area, constant momentum, constant mass flow, but variable stagnation enthalpy (due to heat transfer)—is often termed *Rayleigh flow* after John William Strutt, Lord Rayleigh (1842–1919), a famous physicist and engineer. For a given mass flow and momentum, a plot of enthalpy versus entropy for all possible flow states, subsonic or supersonic, forms a *Rayleigh line*. See Probs. P9.110 and P9.111 for examples of a Rayleigh line.

Consider the elemental duct control volume in Fig. 9.16. Between sections 1 and 2 an amount of heat δQ is added (or removed) to each incremental mass δm passing through. With no friction or area change, the control volume conservation relations are quite simple:

Continuity:
$$\rho_1 V_1 = \rho_2 V_2 = G = \text{const} \tag{9.76a}$$

x momentum:
$$p_1 - p_2 = G(V_2 - V_1) \tag{9.76b}$$

Energy:
$$\dot{Q} = \dot{m}(h_2 + \tfrac{1}{2}V_2^2 - h_1 - \tfrac{1}{2}V_1^2)$$

or
$$q = \frac{\dot{Q}}{\dot{m}} = \frac{\delta Q}{\delta m} = h_{02} - h_{01} \tag{9.76c}$$

The heat transfer results in a change in stagnation enthalpy of the flow. We shall not specify exactly how the heat is transferred—combustion, nuclear reaction, evaporation, condensation, or wall heat exchange—but simply that it happened in amount q between 1 and 2. We remark, however, that wall heat exchange is not a good candidate for the theory because wall convection is inevitably coupled with wall friction, which we neglected.

To complete the analysis, we use the perfect-gas and Mach number relations:

$$\frac{p_2}{\rho_2 T_2} = \frac{p_1}{\rho_1 T_1} \qquad h_{02} - h_{01} = c_p(T_{02} - T_{01})$$
$$\frac{V_2}{V_1} = \frac{\text{Ma}_2\, a_2}{\text{Ma}_1\, a_1} = \frac{\text{Ma}_2}{\text{Ma}_1}\left(\frac{T_2}{T_1}\right)^{1/2} \tag{9.77}$$

For a given heat transfer $q = \delta Q/\delta m$ or, equivalently, a given change $h_{02} - h_{01}$, Eqs. (9.76) and (9.77) can be solved algebraically for the property ratios p_2/p_1, Ma_2/Ma_1, and so on between inlet and outlet. Note that because the heat transfer allows the entropy to either increase or decrease, the second law imposes no restrictions on these solutions.

Before writing down these property ratio functions, we illustrate the effect of heat transfer in Fig. 9.17, which shows T_0 and T versus Mach number in the duct. Heating increases T_0, and cooling decreases it. The maximum possible T_0 occurs at $\text{Ma} = 1.0$, and we see that heating, whether the inlet is subsonic or supersonic, drives the duct Mach number toward unity. This is analogous to the effect of friction in the previous section. The temperature of a perfect gas increases from $\text{Ma} = 0$ up to $\text{Ma} = 1/k^{1/2}$ and then decreases. Thus there is a peculiar—or at least unexpected—region where heating (increasing T_0) actually decreases the gas temperature, the difference being reflected in

---

[5]This section may be omitted without loss of continuity.

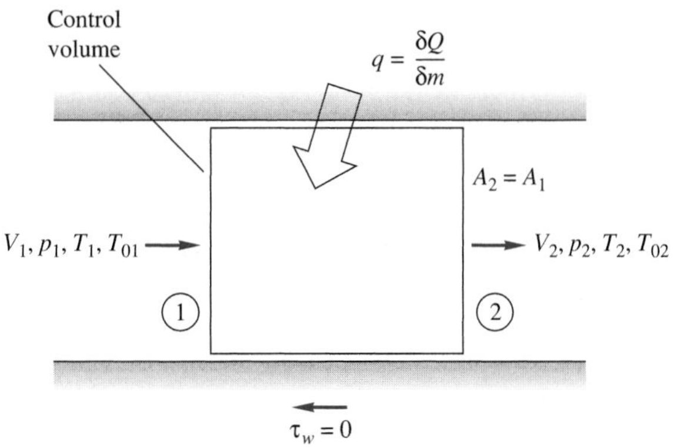

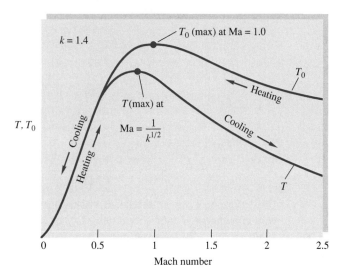

Fig. 9.16 Elemental control volume for frictionless flow in a constant-area duct with heat transfer. The length of the element is indeterminate in this simplified theory.

Fig. 9.17 Effect of heat transfer on Mach number.

a large increase of the gas kinetic energy. For $k = 1.4$ this peculiar area lies between $Ma = 0.845$ and $Ma = 1.0$ (interesting but not very useful information).

The complete list of the effects of simple T_0 change on duct flow properties is as follows:

	Heating		Cooling	
	Subsonic	**Supersonic**	**Subsonic**	**Supersonic**
T_0	Increases	Increases	Decreases	Decreases
Ma	Increases	Decreases	Decreases	Increases
p	Decreases	Increases	Increases	Decreases
ρ	Decreases	Increases	Increases	Decreases
V	Increases	Decreases	Decreases	Increases
p_0	Decreases	Decreases	Increases	Increases
s	Increases	Increases	Decreases	Decreases
T	*	Increases	†	Decreases

*Increases up to $Ma = 1/k^{1/2}$ and decreases thereafter.
†Decreases up to $Ma = 1/k^{1/2}$ and increases thereafter.

Probably the most significant item on this list is the stagnation pressure p_0, which always decreases during heating whether the flow is subsonic or supersonic. Thus heating does increase the Mach number of a flow but entails a loss in effective pressure recovery.

Mach Number Relations

Equations (9.76) and (9.77) can be rearranged in terms of the Mach number and the results tabulated. For convenience, we specify that the outlet section is sonic, Ma = 1, with reference properties T_0^*, T^*, p^*, ρ^*, V^*, and p_0^*. The inlet is assumed to be at arbitrary Mach number Ma. Equations (9.76) and (9.77) then take the following form:

$$\frac{T_0}{T_0^*} = \frac{(k+1)\,\text{Ma}^2\,[2+(k-1)\,\text{Ma}^2]}{(1+k\,\text{Ma}^2)^2} \tag{9.78a}$$

$$\frac{T}{T^*} = \frac{(k+1)^2\,\text{Ma}^2}{(1+k\,\text{Ma}^2)^2} \tag{9.78b}$$

$$\frac{p}{p^*} = \frac{k+1}{1+k\,\text{Ma}^2} \tag{9.78c}$$

$$\frac{V}{V^*} = \frac{\rho^*}{\rho} = \frac{(k+1)\,\text{Ma}^2}{1+k\,\text{Ma}^2} \tag{9.78d}$$

$$\frac{p_0}{p_0^*} = \frac{k+1}{1+k\,\text{Ma}^2}\left[\frac{2+(k-1)\,\text{Ma}^2}{k+1}\right]^{k/(k-1)} \tag{9.78e}$$

These formulas are all tabulated versus Mach number in Table B.4. The tables are very convenient if inlet properties Ma_1, V_1, and the like are given but are somewhat cumbersome if the given information centers on T_{01} and T_{02}. Let us illustrate with an example.

EXAMPLE 9.14

A fuel–air mixture, approximated as air with $k = 1.4$, enters a duct combustion chamber at $V_1 = 75$ m/s, $p_1 = 150$ kPa, and $T_1 = 300$ K. The heat addition by combustion is 900 kJ/kg of mixture. Compute (a) the exit properties V_2, p_2, and T_2 and (b) the total heat addition that would have caused a sonic exit flow.

Solution

Part (a)

First compute $T_{01} = T_1 + V_1^2/(2c_p) = 300 + (75)^2/[2(1005)] = 303$ K. Then compute the change in stagnation temperature of the gas:

$$q = c_p(T_{02} - T_{01})$$

or

$$T_{02} = T_{01} + \frac{q}{c_p} = 303\text{ K} + \frac{900{,}000\text{ J/kg}}{1005\text{ J/(kg}\cdot\text{K)}} = 1199\text{ K}$$

We have enough information to compute the initial Mach number:

$$a_1 = \sqrt{kRT_1} = [1.4(287)(300)]^{1/2} = 347 \text{ m/s} \qquad Ma_1 = \frac{V_1}{a_1} = \frac{75}{347} = 0.216$$

For this Mach number, use Eq. (9.78a) or Table B.4 to find the sonic value T^*_0:

At $Ma_1 = 0.216$: $\qquad \dfrac{T_{01}}{T^*_0} \approx 0.1992 \qquad$ or $\qquad T^*_0 = \dfrac{303 \text{ K}}{0.1992} \approx 1521 \text{ K}$

Then the stagnation temperature ratio at section 2 is $T_{02}/T^*_0 = 1199/1521 = 0.788$, which corresponds in Table B.4 to a Mach number $Ma_2 \approx 0.573$.

Now use Table B.4 at Ma_1 and Ma_2 to tabulate the desired property ratios.

Section	Ma	V/V^*	p/p^*	T/T^*
1	0.216	0.1051	2.2528	0.2368
2	0.573	0.5398	1.6442	0.8876

The exit properties are computed by using these ratios to find state 2 from state 1:

$$V_2 = V_1 \frac{V_2/V^*}{V_1/V^*} = (75 \text{ m/s}) \frac{0.5398}{0.1051} = 385 \text{ m/s} \qquad Ans. (a)$$

$$p_2 = p_1 \frac{p_2/p^*}{p_1/p^*} = (150 \text{ kPa}) \frac{1.6442}{2.2528} = 109 \text{ kPa} \qquad Ans. (a)$$

$$T_2 = T_1 \frac{T_2/T^*}{T_1/T^*} = (300 \text{ K}) \frac{0.8876}{0.2368} = 1124 \text{ K} \qquad Ans. (a)$$

Part (b)   The maximum allowable heat addition would drive the exit Mach number to unity:

$$T_{02} = T^*_0 = 1521 \text{ K}$$

$$q_{max} = c_p(T^*_0 - T_{01}) = [1005 \text{ J/(kg} \cdot \text{K)}](1521 - 303) \text{ K} \approx 1.22 \text{ E6 J/kg} \quad Ans. (b)$$

Choking Effects due to Simple Heating

Equation (9.78a) and Table B.4 indicate that the maximum possible stagnation temperature in simple heating corresponds to T^*_0, or the sonic exit Mach number. Thus, for given inlet conditions, only a certain maximum amount of heat can be added to the flow—for example, 1.22 MJ/kg in Example 9.14. For a subsonic inlet there is no theoretical limit on heat addition: The flow chokes more and more as we add more heat, with the inlet velocity approaching zero. For supersonic flow, even if Ma_1 is infinite, there is a finite ratio $T_{01}/T^*_0 = 0.4898$ for $k = 1.4$. Thus if heat is added without limit to a supersonic flow, a normal shock wave adjustment is required to accommodate the required property changes.

In subsonic flow there is no theoretical limit to the amount of cooling allowed: The exit flow just becomes slower and slower, and the temperature approaches zero. In supersonic flow only a finite amount of cooling can be allowed before the exit flow approaches infinite Mach number, with $T_{02}/T_0^* = 0.4898$ and the exit temperature equal to zero. There are very few practical applications for supersonic cooling.

EXAMPLE 9.15

What happens to the inlet flow in Example 9.14 if the heat addition is increased to 1400 kJ/kg and the inlet pressure and stagnation temperature are fixed? What will be the subsequent decrease in mass flow?

Solution

For $q = 1400$ kJ/kg, the exit will be choked at the stagnation temperature:

$$T_0^* = T_{01} + \frac{q}{c_p} = 303 + \frac{1.4 \text{ E6 J/kg}}{1005 \text{ J/(kg} \cdot \text{K)}} \approx 1696 \text{ K}$$

This is higher than the value $T_0^* = 1521$ K in Example 9.14, so we know that condition 1 will have to choke down to a lower Mach number. The proper value is found from the ratio $T_{01}/T_0^* = 303/1696 = 0.1787$. From Table B.4 or Eq. (9.78a) for this condition, we read the new, lowered entrance Mach number: $\text{Ma}_{1,\text{new}} \approx 0.203$. With T_{01} and p_1 known, the other inlet properties follow from this Mach number:

$$T_1 = \frac{T_{01}}{1 + 0.2 \text{ Ma}_1^2} = \frac{303}{1 + 0.2(0.203)^2} = 301 \text{ K}$$

$$a_1 = \sqrt{kRT_1} = [1.4(287)(301)]^{1/2} = 348 \text{ m/s}$$

$$V_1 = \text{Ma}_1 a_1 = (0.203)(348 \text{ m/s}) = 71 \text{ m/s}$$

$$\rho_1 = \frac{p_1}{RT_1} = \frac{150{,}000}{(287)(301)} = 1.74 \text{ kg/m}^3$$

Finally, the new lowered mass flow per unit area is

$$\frac{\dot{m}_{\text{new}}}{A} = \rho_1 V_1 = (1.74 \text{ kg/m}^3)(71 \text{ m/s}) = 123 \text{ kg/(s} \cdot \text{m}^2)$$

This is 7 percent less than in Example 9.14, due to choking by excess heat addition.

Relationship to the Normal Shock Wave

The normal shock wave relations of Sec. 9.5 actually lurk within the simple heating relations as a special case. From Table B.4 or Fig. 9.17 we see that for a given stagnation temperature less than T_0^* two flow states satisfy the simple heating relations, one subsonic and the other supersonic. These two states have (1) the same value of T_0, (2) the same mass flow per unit area, and (3) the same value of $p + \rho V^2$. Therefore these two states are exactly equivalent to the conditions on each side of a normal shock wave. The second law would again require that the upstream flow Ma_1 be supersonic.

To illustrate this point, take $Ma_1 = 3.0$ and from Table B.4 read $T_{01}/T_0^* = 0.6540$ and $p_1/p^* = 0.1765$. Now, for the same value $T_{02}/T_0^* = 0.6540$, use Table B.4 or Eq. (9.78a) to compute $Ma_2 = 0.4752$ and $p_2/p^* = 1.8235$. The value of Ma_2 is exactly what we read in the shock table, Table B.2, as the downstream Mach number when $Ma_1 = 3.0$. The pressure ratio for these two states is $p_2/p_1 = (p_2/p^*)/(p_1/p^*) = 1.8235/0.1765 = 10.33$, which again is just what we read in Table B.2 for $Ma_1 = 3.0$. This illustration is meant only to show the physical background of the simple heating relations; it would be silly to make a practice of computing normal shock waves in this manner.

9.9 Two-Dimensional Supersonic Flow

Up to this point we have considered only one-dimensional compressible flow theories. This illustrated many important effects, but a one-dimensional world completely loses sight of the wave motions that are so characteristic of supersonic flow. The only "wave motion" we could muster in a one-dimensional theory was the normal shock wave, which amounted only to a flow discontinuity in the duct.

Mach Waves

When we add a second dimension to the flow, wave motions immediately become apparent if the flow is supersonic. Figure 9.18 shows a celebrated graphical

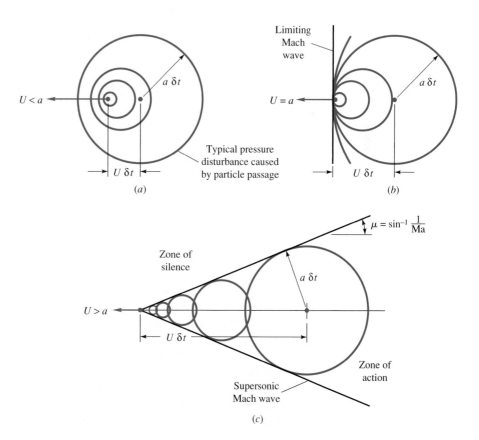

Fig. 9.18 Wave patterns set up by a particle moving at speed U into still fluid of sound velocity a: (a) subsonic, (b) sonic, and (c) supersonic motion.

construction that appears in every fluid mechanics textbook and was first presented by Ernst Mach in 1887. The figure shows the pattern of pressure disturbances (sound waves) sent out by a small particle moving at speed U through a still fluid whose sound velocity is a.

As the particle moves, it continually crashes against fluid particles and sends out spherical sound waves emanating from every point along its path. A few of these spherical disturbance fronts are shown in Fig. 9.18. The behavior of these fronts is quite different according to whether the particle speed is subsonic or supersonic.

In Fig. 9.18a, the particle moves subsonically, $U < a$, Ma $= U/a < 1$. The spherical disturbances move out in all directions and do not catch up with one another. They move well out in front of the particle also, because they travel a distance $a\,\delta t$ during the time interval δt in which the particle has moved only $U\,\delta t$. Therefore a subsonic body motion makes its presence felt everywhere in the flow field: You can "hear" or "feel" the pressure rise of an oncoming body before it reaches you. This is apparently why that pigeon in the road, without turning around to look at you, takes to the air and avoids being hit by your car.

At sonic speed, $U = a$, Fig. 9.18b, the pressure disturbances move at exactly the speed of the particle and thus pile up on the left at the position of the particle into a sort of "front locus," which is now called a *Mach wave,* after Ernst Mach. No disturbance reaches beyond the particle. If you are stationed to the left of the particle, you cannot "hear" the oncoming motion. If the particle blew its horn, you couldn't hear that either: A sonic car can sneak up on a pigeon.

In supersonic motion, $U > a$, the lack of advance warning is even more pronounced. The disturbance spheres cannot catch up with the fast-moving particle that created them. They all trail behind the particle and are tangent to a conical locus called the *Mach cone.* From the geometry of Fig. 9.18c the angle of the Mach cone is seen to be

$$\mu = \sin^{-1}\frac{a\,\delta t}{U\,\delta t} = \sin^{-1}\frac{a}{U} = \sin^{-1}\frac{1}{\text{Ma}} \tag{9.79}$$

The higher the particle Mach number, the more slender the Mach cone; for example, μ is 30° at Ma $= 2.0$ and 11.5° at Ma $= 5.0$. For the limiting case of sonic flow, Ma $= 1$, $\mu = 90°$; the Mach cone becomes a plane front moving with the particle, in agreement with Fig. 9.18b.

You cannot "hear" the disturbance caused by the supersonic particle in Fig. 9.18c until you are in the *zone of action* inside the Mach cone. No warning can reach your ears if you are in the *zone of silence* outside the cone. Thus an observer on the ground beneath a supersonic airplane does not hear the *sonic boom* of the passing cone until the plane is well past.

The Mach wave need not be a cone: Similar waves are formed by a small disturbance of any shape moving supersonically with respect to the ambient fluid. For example, the "particle" in Fig. 9.18c could be the leading edge of a sharp flat plate, which would form a Mach wedge of exactly the same angle μ. Mach waves are formed by small roughnesses or boundary layer irregularities in a supersonic wind tunnel or at the surface of a supersonic body. Look again at Fig. 9.10: Mach waves are clearly visible along the body surface downstream of

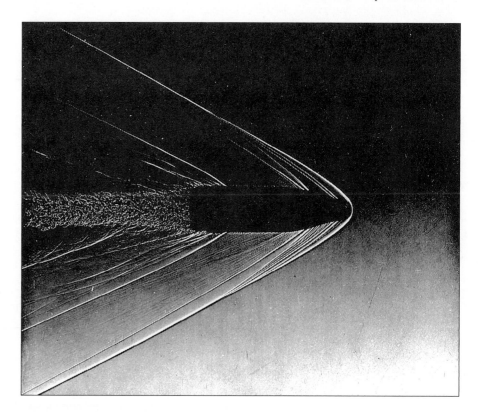

Fig. 9.19 Supersonic wave pattern emanating from a projectile moving at Ma ≈ 2.0. The heavy lines are oblique shock waves and the light lines Mach waves. *(Courtesy of U.S. Army Ballistic Research Laboratory, Aberdeen Proving Ground.)*

the recompression shock, especially at the rear corner. Their angle is about 30°, indicating a Mach number of about 2.0 along this surface. A more complicated system of Mach waves emanates from the supersonic projectile in Fig. 9.19. The Mach angles change, indicating a variable supersonic Mach number along the body surface. There are also several stronger oblique shock waves formed along the surface.

EXAMPLE 9.16

An observer on the ground does not hear the sonic boom caused by an airplane moving at 5-km altitude until it is 9 km past her. What is the approximate Mach number of the plane? Assume a small disturbance and neglect the variation of sound speed with altitude.

Solution

A finite disturbance like an airplane will create a finite-strength oblique shock wave whose angle will be somewhat larger than the Mach wave angle μ and will curve downward due to the variation in atmospheric sound speed. If we neglect these effects, the altitude and distance are a measure of μ, as seen in Fig. E9.16. Thus

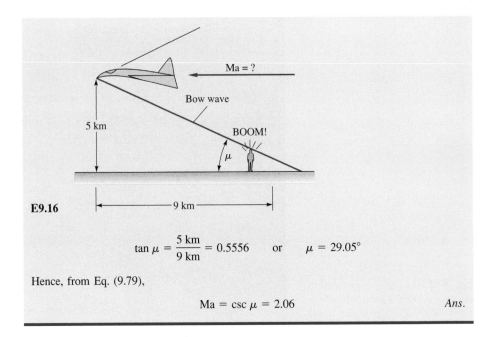

E9.16

$$\tan \mu = \frac{5 \text{ km}}{9 \text{ km}} = 0.5556 \quad \text{or} \quad \mu = 29.05°$$

Hence, from Eq. (9.79),

$$\text{Ma} = \csc \mu = 2.06 \qquad\qquad Ans.$$

The Oblique Shock Wave

Figures 9.10 and 9.19 and our earlier discussion all indicate that a shock wave can form at an oblique angle to the oncoming supersonic stream. Such a wave will deflect the stream through an angle θ, unlike the normal shock wave, for which the downstream flow is in the same direction. In essence, an oblique shock is caused by the necessity for a supersonic stream to turn through such an angle. Examples could be a finite wedge at the leading edge of a body and a ramp in the wall of a supersonic wind tunnel.

The flow geometry of an oblique shock is shown in Fig. 9.20. As for the normal shock of Fig. 9.8, state 1 denotes the upstream conditions and state 2 is downstream. The shock angle has an arbitrary value β, and the downstream flow V_2 turns at an angle θ which is a function of β and state 1 conditions. The upstream flow is always supersonic, but the downstream Mach number $\text{Ma}_2 = V_2/a_2$ may be subsonic, sonic, or supersonic, depending on the conditions.

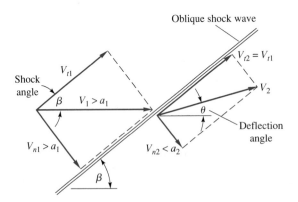

Fig. 9.20 Geometry of flow through an oblique shock wave.

It is convenient to analyze the flow by breaking it up into normal and tangential components with respect to the wave, as shown in Fig. 9.20. For a thin control volume just encompassing the wave, we can then derive the following integral relations, canceling out $A_1 = A_2$ on each side of the wave:

Continuity:
$$\rho_1 V_{n1} = \rho_2 V_{n2} \tag{9.80a}$$

Normal momentum:
$$p_1 - p_2 = \rho_2 V_{n2}^2 - \rho_1 V_{n1}^2 \tag{9.80b}$$

Tangential momentum:
$$0 = \rho_1 V_{n1}(V_{t2} - V_{t1}) \tag{9.80c}$$

Energy:
$$h_1 + \tfrac{1}{2}V_{n1}^2 + \tfrac{1}{2}V_{t1}^2 = h_2 + \tfrac{1}{2}V_{n2}^2 + \tfrac{1}{2}V_{t2}^2 = h_0 \tag{9.80d}$$

We see from Eq. (9.80c) that there is no change in tangential velocity across an oblique shock:

$$V_{t2} = V_{t1} = V_t = \text{const} \tag{9.81}$$

Thus tangential velocity has as its only effect the addition of a constant kinetic energy $\tfrac{1}{2}V_t^2$ to each side of the energy equation (9.80d). We conclude that Eqs. (9.80) are identical to the normal shock relations (9.49), with V_1 and V_2 replaced by the normal components V_{n1} and V_{n2}. All the various relations from Sec. 9.5 can be used to compute properties of an oblique shock wave. The trick is to use the "normal" Mach numbers in place of Ma_1 and Ma_2:

$$\boxed{\text{Ma}_{n1} = \frac{V_{n1}}{a_1} = \text{Ma}_1 \sin \beta}$$

$$\boxed{\text{Ma}_{n2} = \frac{V_{n2}}{a_2} = \text{Ma}_2 \sin (\beta - \theta)} \tag{9.82}$$

Then, for a perfect gas with constant specific heats, the property ratios across the oblique shock are the analogs of Eqs. (9.55) to (9.58) with Ma_1 replaced by Ma_{n1}:

$$\frac{p_2}{p_1} = \frac{1}{k+1}\left[2k\,\text{Ma}_1^2 \sin^2\beta - (k-1)\right] \tag{9.83a}$$

$$\frac{\rho_2}{\rho_1} = \frac{\tan\beta}{\tan(\beta-\theta)} = \frac{(k+1)\,\text{Ma}_1^2 \sin^2\beta}{(k-1)\,\text{Ma}_1^2 \sin^2\beta + 2} = \frac{V_{n1}}{V_{n2}} \tag{9.83b}$$

$$\frac{T_2}{T_1} = \left[2 + (k-1)\,\text{Ma}_1^2 \sin^2\beta\right]\frac{2k\,\text{Ma}_1^2 \sin^2\beta - (k-1)}{(k+1)^2\,\text{Ma}_1^2 \sin^2\beta} \tag{9.83c}$$

$$T_{02} = T_{01} \tag{9.83d}$$

$$\frac{p_{02}}{p_{01}} = \left[\frac{(k+1)\,\text{Ma}_1^2 \sin^2\beta}{2 + (k-1)\,\text{Ma}_1^2 \sin^2\beta}\right]^{k/(k-1)}\left[\frac{k+1}{2k\,\text{Ma}_1^2 \sin^2\beta - (k-1)}\right]^{1/(k-1)} \tag{9.83e}$$

$$\text{Ma}_{n2}^2 = \frac{(k-1)\,\text{Ma}_{n1}^2 + 2}{2k\,\text{Ma}_{n1}^2 - (k-1)} \tag{9.83f}$$

All these are tabulated in the normal shock Table B.2. If you wondered why that table listed the Mach numbers as Ma_{n1} and Ma_{n2}, it should be clear now that the table is also valid for the oblique shock wave.

Thinking all this over, we realize with hindsight that an oblique shock wave is the flow pattern one would observe by running along a normal shock wave (Fig. 9.8) at a constant tangential speed V_t. Thus the normal and oblique shocks are related by a galilean, or inertial, velocity transformation and therefore satisfy the same basic equations.

If we continue with this run-along-the-shock analogy, we find that the deflection angle θ increases with speed V_t up to a maximum and then decreases. From the geometry of Fig. 9.20, the deflection angle is given by

$$\theta = \tan^{-1}\frac{V_t}{V_{n2}} - \tan^{-1}\frac{V_t}{V_{n1}} \tag{9.84}$$

If we differentiate θ with respect to V_t and set the result equal to zero, we find that the maximum deflection occurs when $V_t/V_{n1} = (V_{n2}/V_{n1})^{1/2}$. We can substitute this back into Eq. (9.84) to compute

$$\theta_{max} = \tan^{-1} r^{1/2} - \tan^{-1} r^{-1/2} \qquad r = \frac{V_{n1}}{V_{n2}} \tag{9.85}$$

For example, if $Ma_{n1} = 3.0$, from Table B.2 we find that $V_{n1}/V_{n2} = 3.8571$, the square root of which is 1.9640. Then Eq. (9.85) predicts a maximum deflection of $\tan^{-1}$ 1.9640 $- \tan^{-1}$ (1/1.9640) $= 36.03°$. The deflection is quite limited even for infinite Ma_{n1}: From Table B.2 for this case $V_{n1}/V_{n2} = 6.0$, and we compute from Eq. (9.85) that $\theta_{max} = 45.58°$.

This limited-deflection idea and other facts become more evident if we plot some of the solutions of Eqs. (9.83). For given values of V_1 and a_1, assuming as usual that $k = 1.4$, we can plot all possible solutions for V_2 downstream of the shock. Figure 9.21 does this in velocity-component coordinates V_x and V_y, with x parallel to V_1. Such a plot is called a *hodograph*. The heavy dark line that looks like a fat airfoil is the locus, or *shock polar*, of all physically possible solutions for the given Ma_1. The two dashed-line fishtails are solutions that increase V_2; they are physically impossible because they violate the second law.

Examining the shock polar in Fig. 9.21, we see that a given deflection line of small angle θ crosses the polar at two possible solutions: the *strong* shock, which greatly decelerates the flow, and the *weak* shock, which causes a much milder deceleration.

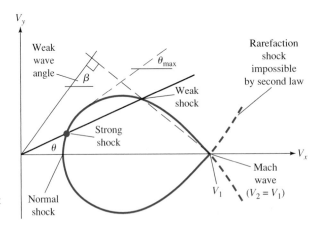

Fig. 9.21 The oblique shock polar hodograph, showing double solutions (strong and weak) for small deflection angle and no solutions at all for large deflection.

The flow downstream of the strong shock is always subsonic, while that of the weak shock is usually supersonic but occasionally subsonic if the deflection is large. Both types of shock occur in practice. The weak shock is more prevalent, but the strong shock will occur if there is a blockage or high-pressure condition downstream.

Since the shock polar is only of finite size, there is a maximum deflection θ_{max}, shown in Fig. 9.21, that just grazes the upper edge of the polar curve. This verifies the kinematic discussion that led to Eq. (9.85). What happens if a supersonic flow is forced to deflect through an angle greater than θ_{max}? The answer is illustrated in Fig. 9.22 for flow past a wedge-shaped body.

In Fig. 9.22a the wedge half-angle θ is less than θ_{max}, and thus an oblique shock forms at the nose of wave angle β just sufficient to cause the oncoming supersonic stream to deflect through the wedge angle θ. Except for the usually small effect of boundary layer growth (see, for example, Ref. 19, Sec. 7–5.2), the Mach number Ma_2 is constant along the wedge surface and is given by the solution of Eqs. (9.83). The pressure, density, and temperature along the surface are also nearly constant, as predicted by Eqs. (9.83). When the flow reaches the corner of the wedge, it expands to higher Mach number and forms a wake (not shown) similar to that in Fig. 9.10.

In Fig. 9.22b the wedge half-angle is greater than θ_{max}, and an attached oblique shock is impossible. The flow cannot deflect at once through the entire angle θ_{max}, yet somehow the flow must get around the wedge. A detached curve shock wave forms in front of the body, discontinuously deflecting the flow through angles smaller than θ_{max}. The flow then curves, expands, and deflects subsonically around the wedge, becoming sonic and then supersonic as it passes the corner region. The flow just inside each point on the curved shock exactly satisfies the oblique shock relations (9.83) for

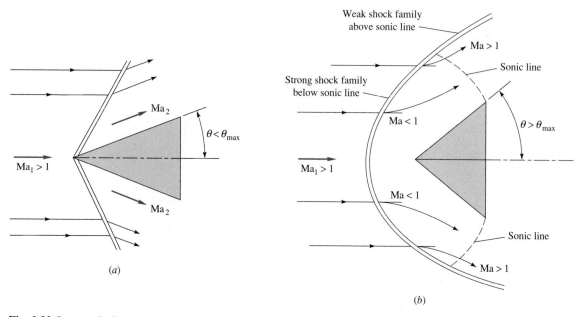

Fig. 9.22 Supersonic flow past a wedge: (a) small wedge angle, attached oblique shock forms; (b) large wedge angle, attached shock not possible, broad curved detached shock forms.

that particular value of β and the given Ma_1. Every condition along the curved shock is a point on the shock polar of Fig. 9.21. Points near the front of the wedge are in the strong shock family, and points aft of the sonic line are in the weak shock family. The analysis of detached shock waves is extremely complex [13], and experimentation is usually needed, such as the shadowgraph optical technique of Fig. 9.10.

The complete family of oblique shock solutions can be plotted or computed from Eqs. (9.83). For a given k, the wave angle β varies with Ma_1 and θ, from Eq. (9.83b). By using a trigonometric identity for $\tan(\beta - \theta)$ this can be rewritten in the more convenient form

$$\tan \theta = \frac{2 \cot \beta \, (Ma_1^2 \sin^2 \beta - 1)}{Ma_1^2 \, (k + \cos 2\beta) + 2} \tag{9.86}$$

All possible solutions of Eq. (9.86) for $k = 1.4$ are shown in Fig. 9.23. For deflections $\theta < \theta_{max}$ there are two solutions: a weak shock (small β) and a strong shock (large β), as expected. All points along the dash–dot line for θ_{max} satisfy Eq. (9.85). A dashed line has been added to show where Ma_2 is exactly sonic. We see that there is a narrow region near maximum deflection where the weak shock downstream flow is subsonic.

For zero deflections ($\theta = 0$) the weak shock family satisfies the wave angle relation

$$\beta = \mu = \sin^{-1} \frac{1}{Ma_1} \tag{9.87}$$

Thus weak shocks of vanishing deflection are equivalent to Mach waves. Meanwhile the strong shocks all converge at zero deflection to the normal shock condition $\beta = 90°$.

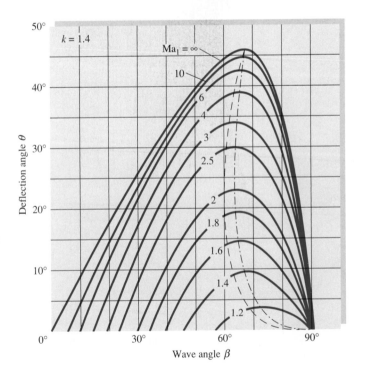

Fig. 9.23 Oblique shock deflection versus wave angle for various upstream Mach numbers, $k = 1.4$: dash–dot curve, locus of θ_{max}, divides strong (right) from weak (left) shocks; dashed curve, locus of sonic points, divides subsonic Ma_2 (right) from supersonic Ma_2 (left).

Two additional oblique shock charts are given in App. B for $k = 1.4$, where Fig. B.1 gives the downstream Mach number Ma_2 and Fig. B.2 the pressure ratio p_2/p_1, each plotted as a function of Ma_1 and θ. Additional graphs, tables, and computer programs are given in Refs. 20 and 21.

Very Weak Shock Waves

For any finite θ the wave angle β for a weak shock is greater than the Mach angle μ. For small θ Eq. (9.86) can be expanded in a power series in $\tan \theta$ with the following linearized result for the wave angle:

$$\sin \beta = \sin \mu + \frac{k+1}{4 \cos \mu} \tan \theta + \cdots + \mathbb{O}(\tan^2 \theta) + \cdots \qquad (9.88)$$

For Ma_1 between 1.4 and 20.0 and deflections less than $6°$ this relation predicts β to within $1°$ for a weak shock. For larger deflections it can be used as a useful initial guess for iterative solution of Eq. (9.86).

Other property changes across the oblique shock can also be expanded in a power series for small deflection angles. Of particular interest is the pressure change from Eq. (9.83a), for which the linearized result for a weak shock is

$$\frac{p_2 - p_1}{p_1} = \frac{k\,Ma_1^2}{(Ma_1^2 - 1)^{1/2}} \tan \theta + \cdots + \mathbb{O}(\tan^2 \theta) + \cdots \qquad (9.89)$$

The differential form of this relation is used in the next section to develop a theory for supersonic expansion turns. Figure 9.24 shows the exact weak shock pressure jump

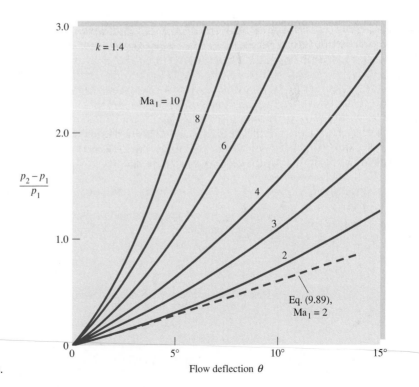

Fig. 9.24 Pressure jump across a weak oblique shock wave from Eq. (9.83a) for $k = 1.4$. For very small deflections Eq. (9.89) applies.

computed from Eq. (9.83a). At very small deflections the curves are linear with slopes given by Eq. (9.89).

Finally, it is educational to examine the entropy change across a very weak shock. Using the same power series expansion technique, we can obtain the following result for small flow deflections:

$$\frac{s_2 - s_1}{c_p} = \frac{(k^2 - 1)\mathrm{Ma}_1^6}{12(\mathrm{Ma}_1^2 - 1)^{3/2}} \tan^3 \theta + \cdots + \mathbb{O}(\tan^4 \theta) + \cdots \tag{9.90}$$

The entropy change is cubic in the deflection angle θ. Thus weak shock waves are very nearly isentropic, a fact that is also used in the next section.

EXAMPLE 9.17

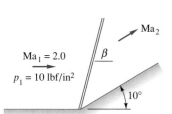

Ma$_2$

Ma$_1$ = 2.0

p_1 = 10 lbf/in^2

β

10°

E9.17

Air at Ma = 2.0 and $p = 10$ lbf/in^2 absolute is forced to turn through 10° by a ramp at the body surface. A weak oblique shock forms as in Fig. E9.17. For $k = 1.4$ compute from exact oblique shock theory (a) the wave angle β, (b) Ma$_2$, and (c) p_2. Also use the linearized theory to estimate (d) β and (e) p_2.

Solution

With Ma$_1$ = 2.0 and $\theta = 10°$ known, we can estimate $\beta \approx 40° \pm 2°$ from Fig. 9.23. For more (hand calculated) accuracy, we have to solve Eq. (9.86) by iteration. Or we can program Eq. (9.86) in EES with six statements (in SI units, with angles in degrees):

```
Ma = 2.0

k = 1.4

Theta = 10

Num = 2*(Ma^2*SIN(Beta)^2 − 1)/TAN(Beta)

Denom = Ma^2*(k + COS(2*Beta)) + 2

Theta = ARCTAN(Num/Denom)
```

Specify that Beta > 0 and EES promptly reports an accurate result:

$$\beta = 39.32° \qquad\qquad Ans. (a)$$

The normal Mach number upstream is thus

$$\mathrm{Ma}_{n1} = \mathrm{Ma}_1 \sin \beta = 2.0 \sin 39.32° = 1.267$$

With Ma$_{n1}$ we can use the normal shock relations (Table B.2) or Fig. 9.9 or Eqs. (9.56) to (9.58) to compute

$$\mathrm{Ma}_{n2} = 0.8031 \qquad \frac{p_2}{p_1} = 1.707$$

Thus the downstream Mach number and pressure are

$$\mathrm{Ma}_2 = \frac{\mathrm{Ma}_{n2}}{\sin (\beta - \theta)} = \frac{0.8031}{\sin (39.32° - 10°)} = 1.64 \qquad\qquad Ans. (b)$$

$$p_2 = (10 \text{ lbf/in}^2 \text{ absolute})(1.707) = 17.07 \text{ lbf/in}^2 \text{ absolute} \qquad\qquad Ans. (c)$$

Notice that the computed pressure ratio agrees with Figs. 9.24 and B.2.

For the linearized theory the Mach angle is $\mu = \sin^{-1}(1/2.0) = 30°$. Equation (9.88) then estimates that

$$\sin \beta \approx \sin 30° + \frac{2.4 \tan 10°}{4 \cos 30°} = 0.622$$

or
$$\beta \approx 38.5° \qquad \text{Ans. } (d)$$

Equation (9.89) estimates that

$$\frac{p_2}{p_1} \approx 1 + \frac{1.4(2)^2 \tan 10°}{(2^2 - 1)^{1/2}} = 1.57$$

or
$$p_2 \approx 1.57(10 \text{ lbf/in}^2 \text{ absolute}) \approx 15.7 \text{ lbf/in}^2 \text{ absolute} \qquad \text{Ans. } (e)$$

These are reasonable estimates in spite of the fact that 10° is really not a "small" flow deflection.

9.10 Prandtl-Meyer Expansion Waves

The oblique shock solution of Sec. 9.9 is for a finite compressive deflection θ that obstructs a supersonic flow and thus decreases its Mach number and velocity. The present section treats gradual changes in flow angle that is, are primarily *expansive;* they widen the flow area and increase the Mach number and velocity. The property changes accumulate in infinitesimal increments, and the linearized relations (9.88) and (9.89) are used. The local flow deflections are infinitesimal, so that the flow is nearly isentropic according to Eq. (9.90).

Figure 9.25 shows four examples, one of which (Fig. 9.25c) fails the test for gradual changes. The gradual compression of Fig. 9.25a is essentially isentropic, with a smooth increase in pressure along the surface, but the Mach angle increases along the surface and the waves tend to coalesce farther out into an oblique shock wave. The gradual expansion of Fig. 9.25b causes a smooth isentropic increase of Mach number and velocity along the surface, with diverging Mach waves formed.

The sudden compression of Fig. 9.25c cannot be accomplished by Mach waves: An oblique shock forms, and the flow is nonisentropic. This could be what you would see if you looked at Fig. 9.25a from far away. Finally, the sudden expansion of Fig. 9.25d is isentropic and forms a fan of centered Mach waves emanating from the corner. Note that the flow on any streamline passing through the fan changes smoothly to higher Mach number and velocity. In the limit as we near the corner the flow expands almost discontinuously at the surface. The cases in Fig. 9.25a, b, and d can all be handled by the Prandtl-Meyer supersonic wave theory of this section, first formulated by Ludwig Prandtl and his student Theodor Meyer in 1907 to 1908.

Note that none of this discussion makes sense if the upstream Mach number is subsonic, since Mach wave and shock wave patterns cannot exist in subsonic flow.

The Prandtl-Meyer Perfect-Gas Function

Consider a small, nearly infinitesimal flow deflection $d\theta$ such as occurs between the first two Mach waves in Fig. 9.25a. From Eqs. (9.88) and (9.89) we have, in

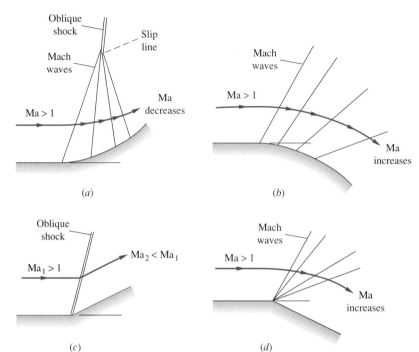

Fig. 9.25 Some examples of supersonic expansion and compression: (a) gradual isentropic compression on a concave surface, Mach waves coalesce farther out to form oblique shock; (b) gradual isentropic expansion on convex surface, Mach waves diverge; (c) sudden compression, nonisentropic shock forms; (d) sudden expansion, centered isentropic fan of Mach waves forms.

the limit,

$$\beta \approx \mu = \sin^{-1}\frac{1}{\text{Ma}} \tag{9.91a}$$

$$\frac{dp}{p} \approx \frac{k\,\text{Ma}^2}{(\text{Ma}^2 - 1)^{1/2}}\,d\theta \tag{9.91b}$$

Since the flow is nearly isentropic, we have the frictionless differential momentum equation for a perfect gas:

$$dp = -\rho V\,dV = -kp\,\text{Ma}^2\,\frac{dV}{V} \tag{9.92}$$

Combining Eqs. (9.91a) and (9.92) to eliminate dp, we obtain a relation between turning angle and velocity change:

$$d\theta = -(\text{Ma}^2 - 1)^{1/2}\,\frac{dV}{V} \tag{9.93}$$

This can be integrated into a functional relation for finite turning angles if we can relate V to Ma. We do this from the definition of Mach number:

$$V = \text{Ma}\,a$$

or

$$\frac{dV}{V} = \frac{d\,\text{Ma}}{\text{Ma}} + \frac{da}{a} \tag{9.94}$$

Finally, we can eliminate da/a because the flow is isentropic and hence a_0 is constant for a perfect gas:

$$a = a_0[1 + \tfrac{1}{2}(k - 1)\,\mathrm{Ma}^2]^{-1/2}$$

or

$$\frac{da}{a} = \frac{-\tfrac{1}{2}(k - 1)\,\mathrm{Ma}\,d\,\mathrm{Ma}}{1 + \tfrac{1}{2}(k - 1)\,\mathrm{Ma}^2} \tag{9.95}$$

Eliminating dV/V and da/a from Eqs. (9.93) to (9.95), we obtain a relation solely between turning angle and Mach number:

$$d\theta = -\frac{(\mathrm{Ma}^2 - 1)^{1/2}}{1 + \tfrac{1}{2}(k - 1)\,\mathrm{Ma}^2}\frac{d\,\mathrm{Ma}}{\mathrm{Ma}} \tag{9.96}$$

Before integrating this expression, we note that the primary application is to expansions: increasing Ma and decreasing θ. Therefore, for convenience, we define the Prandtl-Meyer angle $\omega(\mathrm{Ma})$, which increases when θ decreases and is zero at the sonic point:

$$d\omega = -d\theta \qquad \omega = 0 \quad \text{at} \quad \mathrm{Ma} = 1 \tag{9.97}$$

Thus we integrate Eq. (9.96) from the sonic point to any value of Ma:

$$\int_0^\omega d\omega = \int_1^{\mathrm{Ma}} \frac{(\mathrm{Ma}^2 - 1)^{1/2}}{1 + \tfrac{1}{2}(k - 1)\,\mathrm{Ma}^2}\frac{d\,\mathrm{Ma}}{\mathrm{Ma}} \tag{9.98}$$

The integrals are evaluated in closed form, with the result, in radians,

$$\boxed{\omega(\mathrm{Ma}) = K^{1/2}\tan^{-1}\left(\frac{\mathrm{Ma}^2 - 1}{K}\right)^{1/2} - \tan^{-1}(\mathrm{Ma}^2 - 1)^{1/2}} \tag{9.99}$$

where

$$\boxed{K = \frac{k + 1}{k - 1}}$$

This is the *Prandtl-Meyer supersonic expansion function,* which is plotted in Fig. 9.26 and tabulated in Table B.5 for $k = 1.4$, $K = 6$. The angle ω changes rapidly at first and then levels off at high Mach number to a limiting value as $\mathrm{Ma} \to \infty$:

$$\omega_{\max} = \frac{\pi}{2}(K^{1/2} - 1) = 130.45° \qquad \text{if} \quad k = 1.4 \tag{9.100}$$

Thus a supersonic flow can expand only through a finite turning angle before it reaches infinite Mach number, maximum velocity, and zero temperature.

Gradual expansion or compression between finite Mach numbers Ma_1 and Ma_2, neither of which is unity, is computed by relating the turning angle $\Delta\omega$ to the difference in Prandtl-Meyer angles for the two conditions

$$\Delta\omega_{1\to 2} = \omega(\mathrm{Ma}_2) - \omega(\mathrm{Ma}_1) \tag{9.101}$$

The change $\Delta\omega$ may be either positive (expansion) or negative (compression) as long as the end conditions lie in the supersonic range. Let us illustrate with an example.

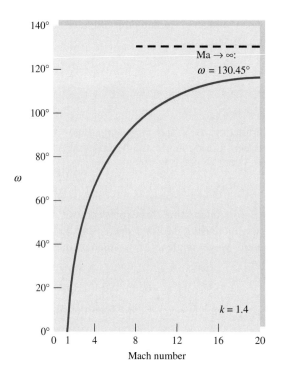

Fig. 9.26 The Prandtl-Meyer supersonic expansion from Eq. (9.99) for $k = 1.4$.

EXAMPLE 9.18

Air ($k = 1.4$) flows at $\text{Ma}_1 = 3.0$ and $p_1 = 200$ kPa. Compute the final downstream Mach number and pressure for (*a*) an expansion turn of 20° and (*b*) a gradual compression turn of 20°.

Solution

Part (a)

The isentropic stagnation pressure is

$$p_0 = p_1[1 + 0.2(3.0)^2]^{3.5} = 7347 \text{ kPa}$$

and this will be the same at the downstream point. For $\text{Ma}_1 = 3.0$ we find from Table B.5 or Eq. (9.99) that $\omega_1 = 49.757°$. The flow expands to a new condition such that

$$\omega_2 = \omega_1 + \Delta\omega = 49.757° + 20° = 69.757°$$

Linear interpolation in Table B.5 is quite accurate, yielding $\text{Ma}_2 \approx 4.32$. Inversion of Eq. (9.99), to find Ma when ω is given, is impossible without iteration. Once again, our friend EES easily handles Eq. (9.99) with four statements (angles specified in degrees):

```
k = 1.4
C = ((k + 1)/(k − 1))^0.5
Omega = 69.757
Omega = C*ARCTAN((Ma^2−1)^0.5/C) − ARCTAN((Ma^2−1)^0.5)
```

Specify that Ma > 1, and EES readily reports an accurate result:[6]

$$\text{Ma}_2 = 4.32 \qquad\qquad\qquad\qquad\qquad Ans.\ (a)$$

The isentropic pressure at this new condition is

$$p_2 = \frac{p_0}{[1 + 0.2(4.32)^2]^{3.5}} = \frac{7347}{230.1} = 31.9\ \text{kPa} \qquad Ans.\ (a)$$

Part (b)   The flow compresses to a lower Prandtl-Meyer angle:

$$\omega_2 = 49.757° - 20° = 29.757°$$

Again from Eq. (9.99), Table B.5, or EES we compute that

$$\text{Ma}_2 = 2.125 \qquad\qquad\qquad\qquad\qquad Ans.\ (b)$$

$$p_2 = \frac{p_0}{[1 + 0.2(2.125)^2]^{3.5}} = \frac{7347}{9.51} = 773\ \text{kPa} \qquad Ans.\ (b)$$

Similarly, we compute density and temperature changes by noticing that T_0 and ρ_0 are constant for isentropic flow.

Application to Supersonic Airfoils   The oblique shock and Prandtl-Meyer expansion theories can be used to patch together a number of interesting and practical supersonic flow fields. This marriage, called *shock expansion theory,* is limited by two conditions: (1) Except in rare instances the flow must be supersonic throughout, and (2) the wave pattern must not suffer interference from waves formed in other parts of the flow field.

A very successful application of shock expansion theory is to supersonic airfoils. Figure 9.27 shows two examples, a flat plate and a diamond-shaped foil. In contrast to subsonic flow designs (Fig. 8.21), these airfoils must have sharp leading edges, which form attached oblique shocks or expansion fans. Rounded supersonic leading edges would cause detached bow shocks, as in Fig. 9.19 or 9.22b, greatly increasing the drag and lowering the lift.

In applying shock expansion theory, one examines each surface turning angle to see whether it is an expansion ("opening up") or compression (obstruction) to the surface flow. Figure 9.27a shows a flat-plate foil at an angle of attack. There is a leading-edge shock on the lower edge with flow deflection $\theta = \alpha$, while the upper edge has an expansion fan with increasing Prandtl-Meyer angle $\Delta\omega = \alpha$. We compute p_3 with expansion theory and p_2 with oblique shock theory. The force on the plate is thus $F = (p_2 - p_3)Cb$, where C is the chord length and b the span width (assuming no wingtip effects). This force is normal to the plate, and thus the lift force normal to the stream is $L = F\cos\alpha$, and the drag parallel to the stream is $D = F\sin\alpha$. The dimensionless coefficients C_L and C_D have the same definitions as in low-speed flow, Eqs. (7.66), except that the perfect-gas law identity $\frac{1}{2}\rho V^2 \equiv \frac{1}{2}kp\,\text{Ma}^2$ is very useful here:

$$C_L = \frac{L}{\frac{1}{2}kp_\infty \text{Ma}_\infty^2\, bC} \qquad C_D = \frac{D}{\frac{1}{2}kp_\infty \text{Ma}_\infty^2\, bC} \qquad (9.102)$$

---

[6]The author saves these little programs for further use, giving them names such as *Prandtl-Meyer.*

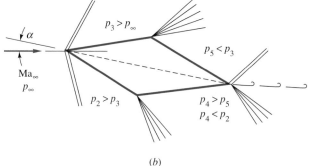

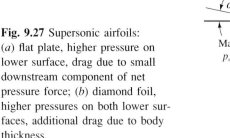

Fig. 9.27 Supersonic airfoils: (*a*) flat plate, higher pressure on lower surface, drag due to small downstream component of net pressure force; (*b*) diamond foil, higher pressures on both lower surfaces, additional drag due to body thickness.

The typical supersonic lift coefficient is much smaller than the subsonic value $C_L \approx 2\pi\alpha$, but the lift can be very large because of the large value of $\frac{1}{2}\rho V^2$ at supersonic speeds.

At the trailing edge in Fig. 9.27*a*, a shock and fan appear in reversed positions and bend the two flows back so that they are parallel in the wake and have the same pressure. They do not have quite the same velocity because of the unequal shock strengths on the upper and lower surfaces; hence a vortex sheet trails behind the wing. This is very interesting, but in the theory you ignore the trailing-edge pattern entirely, since it does not affect the surface pressures: The supersonic surface flow cannot "hear" the wake disturbances.

The diamond foil in Fig. 9.27*b* adds two more wave patterns to the flow. At this particular α less than the diamond half-angle, there are leading-edge shocks on both surfaces, the upper shock being much weaker. Then there are expansion fans on each shoulder of the diamond: The Prandtl-Meyer angle change $\Delta\omega$ equals the sum of the leading-edge and trailing-edge diamond half-angles. Finally, the trailing-edge pattern is similar to that of the flat plate (9.27*a*) and can be ignored in the calculation. Both lower-surface pressures p_2 and p_4 are greater than their upper counterparts, and the lift is nearly that of the flat plate. There is an additional drag due to thickness because p_4 and p_5 on the trailing surfaces are lower than their counterparts p_2 and p_3. The diamond drag is greater than the flat-plate drag, but this must be endured in practice to achieve a wing structure strong enough to hold these forces.

The theory sketched in Fig. 9.27 is in good agreement with measured supersonic lift and drag as long as the Reynolds number is not too small (thick boundary layers) and the Mach number not too large (hypersonic flow). It turns out that for large Re_C and moderate supersonic Ma_∞ the boundary layers are thin and separation seldom occurs, so that the shock expansion theory, although frictionless, is quite successful. Let us look now at an example.

EXAMPLE 9.19

A flat-plate airfoil with $C = 2$ m is immersed at $\alpha = 8°$ in a stream with $Ma_\infty = 2.5$ and $p_\infty = 100$ kPa. Compute (a) C_L and (b) C_D, and compare with low-speed airfoils. Compute (c) lift and (d) drag in newtons per unit span width.

Solution

Instead of using a lot of space outlining the detailed oblique shock and Prandtl-Meyer expansion computations, we list all pertinent results in Fig. E9.19 on the upper and lower surfaces. Using the theories of Secs. 9.9 and 9.10, you should verify every single one of the calculations in Fig. E9.19 to make sure that all details of shock expansion theory are well understood.

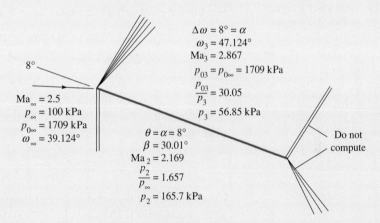

E9.19

The important final results are p_2 and p_3, from which the total force per unit width on the plate is

$$F = (p_2 - p_3)bC = (165.7 - 56.85)(\text{kPa})(1 \text{ m})(2 \text{ m}) = 218 \text{ kN}$$

The lift and drag per meter width are thus

$$L = F \cos 8° = 216 \text{ kN} \qquad\qquad \text{Ans. (c)}$$

$$D = F \sin 8° = 30 \text{ kN} \qquad\qquad \text{Ans. (d)}$$

These are very large forces for only 2 m^2 of wing area.

From Eq. (9.102) the lift coefficient is

$$C_L = \frac{216 \text{ kN}}{\frac{1}{2}(1.4)(100 \text{ kPa})(2.5)^2(2 \text{ m}^2)} = 0.246 \qquad\qquad \text{Ans. (a)}$$

The comparable low-speed coefficient from Eq. (8.67) is $C_L = 2\pi \sin 8° = 0.874$, which is 3.5 times larger.

From Eq. (9.102) the drag coefficient is

$$C_D = \frac{30 \text{ kN}}{\frac{1}{2}(1.4)(100 \text{ kPa})(2.5)^2(2 \text{ m}^2)} = 0.035 \qquad\qquad Ans. (b)$$

From Fig. 7.25 for the NACA 0009 airfoil C_D at $\alpha = 8°$ is about 0.009, or about 4 times smaller.

Notice that this supersonic theory predicts a finite drag in spite of assuming frictionless flow with infinite wing aspect ratio. This is called *wave drag*, and we see that the d'Alembert paradox of zero body drag does not occur in supersonic flow.

Thin-Airfoil Theory

In spite of the simplicity of the flat-plate geometry, the calculations in Example 9.19 were laborious. In 1925 Ackeret [28] developed simple yet effective expressions for the lift, drag, and center of pressure of supersonic airfoils, assuming small thickness and angle of attack.

The theory is based on the linearized expression (9.89), where $\tan \theta \approx$ surface deflection relative to the free stream and condition 1 is the free stream, $\text{Ma}_1 = \text{Ma}_\infty$. For the flat-plate airfoil, the total force F is based on

$$\frac{p_2 - p_3}{p_\infty} = \frac{p_2 - p_\infty}{p_\infty} - \frac{p_3 - p_\infty}{p_\infty}$$

$$= \frac{k \, \text{Ma}_\infty^2}{(\text{Ma}_\infty^2 - 1)^{1/2}} \left[\alpha - (-\alpha) \right] \qquad\qquad (9.103)$$

Substitution into Eq. (9.102) gives the linearized lift coefficient for a supersonic flat-plate airfoil:

$$C_L \approx \frac{(p_2 - p_3)bC}{\frac{1}{2}kp_\infty \, \text{Ma}_\infty^2 \, bC} \approx \frac{4\alpha}{(\text{Ma}_\infty^2 - 1)^{1/2}} \qquad\qquad (9.104)$$

Computations for diamond and other finite-thickness airfoils show no first-order effect of thickness on lift. Therefore Eq. (9.104) is valid for any sharp-edged supersonic thin airfoil at a small angle of attack.

The flat-plate drag coefficient is

$$C_D = C_L \tan \alpha \approx C_L \alpha \approx \frac{4\alpha^2}{(\text{Ma}_\infty^2 - 1)^{1/2}} \qquad\qquad (9.105)$$

However, the thicker airfoils have additional thickness drag. Let the chord line of the airfoil be the x axis, and let the upper-surface shape be denoted by $y_u(x)$ and the lower profile by $y_l(x)$. Then the complete Ackeret drag theory (see Ref. 5, Sec. 14.6, for details) shows that the additional drag depends on the mean square of the slopes of the upper and lower surfaces, defined by

$$\overline{y'^2} = \frac{1}{C} \int_0^C \left(\frac{dy}{dx} \right)^2 dx \qquad\qquad (9.106)$$

The final expression for drag [5, p. 442] is

$$C_D \approx \frac{4}{(\mathrm{Ma}_\infty^2 - 1)^{1/2}} \left[\alpha^2 + \frac{1}{2} (\overline{y_u'^2} + \overline{y_l'^2}) \right] \tag{9.107}$$

These are all in reasonable agreement with more exact computations, and their extreme simplicity makes them attractive alternatives to the laborious but accurate shock expansion theory. Consider the following example.

EXAMPLE 9.20

Repeat parts (a) and (b) of Example 9.19, using the linearized Ackeret theory.

Solution

From Eqs. (9.104) and (9.105) we have, for $\mathrm{Ma}_\infty = 2.5$ and $\alpha = 8° = 0.1396$ rad,

$$C_L \approx \frac{4(0.1396)}{(2.5^2 - 1)^{1/2}} = 0.244 \qquad C_D = \frac{4(0.1396)^2}{(2.5^2 - 1)^{1/2}} = 0.034 \qquad Ans.$$

These are less than 3 percent lower than the more exact computations of Example 9.19.

A further result of the Ackeret linearized theory is an expression for the position x_{CP} of the center of pressure (CP) of the force distribution on the wing:

$$\frac{x_{CP}}{C} = 0.5 + \frac{S_u - S_l}{2\alpha C^2} \tag{9.108}$$

where S_u is the cross-sectional area between the upper surface and the chord and S_l is the area between the chord and the lower surface. For a symmetric airfoil ($S_l = S_u$) we obtain x_{CP} at the half-chord point, in contrast with the low-speed airfoil result of Eq. (8.69), where x_{CP} is at the quarter-chord.

The difference in difficulty between the simple Ackeret theory and shock expansion theory is even greater for a thick airfoil, as the following example shows.

EXAMPLE 9.21

By analogy with Example 9.19 analyze a diamond, or double-wedge, airfoil of 2° half-angle and $C = 2$ m at $\alpha = 8°$ and $\mathrm{Ma}_\infty = 2.5$. Compute C_L and C_D by (a) shock expansion theory and (b) Ackeret theory. Pinpoint the difference from Example 9.19.

Solution

Part (a)

Again we omit the details of shock expansion theory and simply list the properties computed on each of the four airfoil surfaces in Fig. E9.21. Assume $p_\infty = 100$ kPa. There are both a force F normal to the chord line and a force P parallel to the chord. For the normal force the pressure difference on the front half is $p_2 - p_3 = 186.4 - 65.9 = 120.5$ kPa, and on the rear half it is $p_4 - p_5 = 146.9 - 48.8 = 98.1$ kPa. The average pressure difference is $\frac{1}{2}(120.5 + 98.1) = 109.3$ kPa, so that the normal force is

$$F = (109.3 \text{ kPa})(2 \text{ m}^2) = 218.6 \text{ kN}$$

For the chordwise force P the pressure difference on the top half is $p_3 - p_5 = 65.9 - 48.8 = 17.1$ kPa, and on the bottom half it is $p_2 - p_4 = 186.4 - 146.9 = 39.5$ kPa. The average difference is $\frac{1}{2}(17.1 + 39.5) = 28.3$ kPa, which when multiplied by the frontal area (maximum thickness times 1-m width) gives

$$P = (28.3 \text{ kPa})(0.07 \text{ m})(1 \text{ m}) = 2.0 \text{ kN}$$

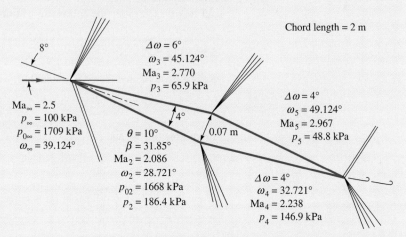

E9.21

Both F and P have components in the lift and drag directions. The lift force normal to the free stream is

$$L = F \cos 8° - P \sin 8° = 216.2 \text{ kN}$$

and

$$D = F \sin 8° + P \cos 8° = 32.4 \text{ kN}$$

For computing the coefficients, the denominator of Eq. (9.102) is the same as in Example 9.19: $\frac{1}{2} k p_\infty \text{Ma}_\infty^2 bC = \frac{1}{2}(1.4)(100 \text{ kPa})(2.5)^2(2 \text{ m}^2) = 875$ kN. Thus, finally, shock expansion theory predicts

$$C_L = \frac{216.2 \text{ kN}}{875 \text{ kN}} = 0.247 \qquad C_D = \frac{32.4 \text{ kN}}{875 \text{ kN}} = 0.0370 \qquad \textit{Ans. (a)}$$

Part (b)   Meanwhile, by Ackeret theory, C_L is the same as in Example 9.20:

$$C_L = \frac{4(0.1396)}{(2.5^2 - 1)^{1/2}} = 0.244 \qquad \textit{Ans. (b)}$$

This is 1 percent less than the shock expansion result above. For the drag we need the mean-square slopes from Eq. (9.106):

$$\overline{y_u'^2} = \overline{y_l'^2} = \tan^2 2° = 0.00122$$

Then Eq. (9.107) predicts this linearized result:

$$C_D = \frac{4}{(2.5^2 - 1)^{1/2}} \left[(0.1396)^2 + \frac{1}{2}(0.00122 + 0.00122)\right] = 0.0362 \qquad \textit{Ans. (b)}$$

This is 2 percent lower than shock expansion theory predicts. We could judge Ackeret theory to be "satisfactory." Ackeret theory predicts $p_2 = 167$ kPa $(-11$ percent), $p_3 = 60$ kPa $(-9$ percent), $p_4 = 140$ kPa $(-5$ percent), and $p_5 = 33$ kPa $(-6$ percent).

Three-Dimensional Supersonic Flow

We have gone about as far as we can go in an introductory treatment of compressible flow. Of course, there is much more, and you are invited to study further in the references at the end of the chapter.

Three-dimensional supersonic flows are highly complex, especially if they concern blunt bodies, which therefore contain embedded regions of subsonic and transonic flow, as in Fig. 9.10. Some flows, however, yield to accurate theoretical treatment such as flow past a cone at zero incidence, as shown in Fig. 9.28. The exact theory of cone flow is discussed in advanced texts [for example, 5, Chap. 17], and extensive tables of such solutions have been published [25]. There are similarities between cone flow and the wedge flows illustrated in Fig. 9.22: an attached oblique shock, a thin turbulent boundary layer, and an expansion fan at the rear corner. However, the conical shock deflects the flow through an angle less than the cone half-angle, unlike the wedge shock. As in the wedge flow, there is a maximum cone angle above which the shock must detach, as in Fig. 9.22b. For $k = 1.4$ and $Ma_\infty = \infty$, the maximum cone half-angle for an attached shock is about 57°, compared with the maximum wedge angle of 45.6° (see Ref. 25).

The use of computational fluid dynamics (CFD) is now very popular and successful in compressible flow studies [13]. For example, a supersonic cone flow such as Fig. 9.28, even at an angle of attack, can be solved by numerical simulation of the full three-dimensional (viscous) Navier-Stokes equations [26].

Fig. 9.28 Shadowgraph of flow past an 8° half-angle cone at $Ma_\infty = 2.0$. The turbulent boundary layer is clearly visible. The Mach lines curve slightly, and the Mach number varies from 1.98 just inside the shock to 1.90 at the surface. *(Courtesy of U.S. Army Ballistic Research Laboratory, Aberdeen Proving Ground.)*

Fig. 9.29 Wind tunnel test of the Cobra P-530 supersonic interceptor. The surface flow patterns are visualized by the smearing of oil droplets. *(Courtesy of Northrop Grumman.)*

For more complicated body shapes one usually resorts to experimentation in a supersonic wind tunnel. Figure 9.29 shows a wind tunnel study of supersonic flow past a model of an interceptor aircraft. The many junctions and wingtips and shape changes make theoretical analysis very difficult. Here the surface flow patterns, which indicate boundary layer development and regions of flow separation, have been visualized by the smearing of oil drops placed on the model surface before the test.

As we shall see in the next chapter, there is an interesting analogy between gas dynamic shock waves and the surface water waves that form in an open-channel flow. Chapter 11 of Ref. 9 explains how a water channel can be used in an inexpensive simulation of supersonic flow experiments.

New Trends in Aeronautics

The previous edition of this text discussed the Boeing sonic cruiser, the Airbus A380, and the Lockheed-Martin X-35 supersonic fighter. The Boeing plan has evolved slowly, perhaps due to post-9/11 considerations, into a "superefficient" 7E7 Dreamliner, designed to achieve markedly lower operating costs than typical modern commercial transports. The 7E7 will not be "nearly sonic" but instead will fly at a more typical Ma ≈ 0.85. If the airlines show interest, Boeing plans to put the 7E7 into service in 2008.

The Airbus 380 made its successful debut in January 2005 and already has at least 14 airline customers. With 555 seats and two decks, the A-380 is the world's largest airliner. It began service with Singapore Airlines in 2006. A later freight version will carry up to 150 tonnes (331,000 lbf) of cargo. The A-380 cruises at Ma ≈ 0.85 and has a range of 9200 miles.

Fig. 9.30 The Lockheed-Martin X-35 Joint Strike Fighter (JSF) will serve the U.S. Air Force, Navy, and Marine Corps and the UK Royal Air Force and Royal Navy. It has stealth capability, relatively low cost, and emphasizes advanced weapons concepts. *(Reproduced by permission of the Lockheed-Martin Company.)*

The Lockheed-Martin F-35 has been accepted as the new Joint Strike Fighter (JSF) to be used by both the U.S. and UK military. The first version, the F-35A, is a conventional takeoff and landing aircraft that cruises at about Ma = 1.5. As many as 3000 units may be purchased, at a cost of about $45 million each. A later version, the F-35B, will have Short Takeoff and Vertical Landing (STOVL) capability. Service should begin in 2008.

Summary

This chapter briefly introduced a very broad subject, compressible flow, sometimes called *gas dynamics*. The primary parameter is the Mach number $Ma = V/a$, which is large and causes the fluid density to vary significantly. This means that the continuity and momentum equations must be coupled to the energy relation and the equation of state to solve for the four unknowns (p, ρ, T, V).

The chapter reviewed the thermodynamic properties of an ideal gas and derived a formula for the speed of sound of a fluid. The analysis was then simplified to one-dimensional steady adiabatic flow without shaft work, for which the stagnation enthalpy of the gas is constant. A further simplification to isentropic flow enables formulas to be derived for high-speed gas flow in a variable-area duct. This reveals the phenomenon of sonic-flow *choking* (maximum mass flow) in the throat of a nozzle. At supersonic velocities there is the possibility of a normal shock wave, where the gas discontinuously reverts to subsonic conditions. The normal shock explains the effect of back pressure on the performance of converging–diverging nozzles.

To illustrate nonisentropic flow conditions, the chapter briefly focused on constant-area duct flow with friction and with heat transfer, both of which lead to choking of the exit flow.

The chapter ended with a discussion of two-dimensional supersonic flow, where oblique shock waves and Prandtl-Meyer (isentropic) expansion waves appear. With a proper combination of shocks and expansions one can analyze supersonic airfoils.

Problems

Most of the problems herein are fairly straightforward. More difficult or open-ended assignments are labeled with an asterisk. Problems labeled with an EES icon will benefit from the use of the Engineering Equation Solver (EES), while problems labeled with a computer icon may require the use of a computer. The standard end-of-chapter problems P9.1 to P9.157 (categorized in the problem list here) are followed by word problems W9.1 to W9.8, fundamentals of engineering exam problems FE9.1 to FE9.10, comprehensive problems C9.1 to C9.7, and design projects D9.1 and D9.2.

Problem Distribution

Section	Topic	Problems
9.1	Introduction	P9.1–P9.9
9.2	The speed of sound	P9.10–P9.18
9.3	Adiabatic and isentropic flow	P9.19–P9.33
9.4	Isentropic flow with area changes	P9.34–P9.53
9.5	The normal shock wave	P9.54–P9.62
9.6	Converging and diverging nozzles	P9.63–P9.85
9.7	Duct flow with friction	P9.86–P9.107
9.8	Frictionless duct flow with heat transfer	P9.108–P9.115
9.9	Mach waves	P9.116–P9.121
9.9	The oblique shock wave	P9.122–P9.139
9.10	Prandtl-Meyer expansion waves	P9.140–P9.148
9.10	Supersonic airfoils	P9.149–P9.157

P9.1 An ideal gas flows adiabatically through a duct. At section 1, $p_1 = 140$ kPa, $T_1 = 260°C$, and $V_1 = 75$ m/s. Farther downstream, $p_2 = 30$ kPa and $T_2 = 207°C$. Calculate V_2 in m/s and $s_2 - s_1$ in J/(kg · K) if the gas is (a) air, $k = 1.4$, and (b) argon, $k = 1.67$.

P9.2 Solve Prob. P9.1 if the gas is steam. Use two approaches: (a) an ideal gas from Table A.4 and (b) real gas data from the steam tables [15].

P9.3 If 8 kg of oxygen in a closed tank at 200°C and 300 kPa is heated until the pressure rises to 400 kPa, calculate (a) the new temperature, (b) the total heat transfer, and (c) the change in entropy.

P9.4 Compressibility effects become important when the Mach number exceeds approximately 0.3. How fast can a two-dimensional cylinder travel in sea-level standard air before compressibility becomes important *somewhere* in its vicinity?

P9.5 Steam enters a nozzle at 377°C, 1.6 MPa, and a steady speed of 200 m/s and accelerates isentropically until it exits at saturation conditions. Estimate the exit velocity and temperature.

P9.6 Use EES, other software, or the gas tables to estimate c_p and c_v, their ratio, and their difference, for CO_2 at 800 K and compare with ideal-gas estimates similar to Eqs. (9.4).

P9.7 Air flows through a variable-area duct. At section 1, $A_1 = 20$ cm², $p_1 = 300$ kPa, $\rho_1 = 1.75$ kg/m³, and $V_1 = 122.5$ m/s. At section 2, the area is exactly the same, but the density is much lower: $\rho_2 = 0.266$ kg/m³ and $T_2 = 281$ K. There is no transfer of work or heat. Assume one-dimensional steady flow. (a) How can you reconcile these differences? (b) Find the mass flow at section 2. Calculate (c) V_2, (d) p_2, and (e) $s_2 - s_1$. *Hint:* This problem requires the continuity equation.

P9.8 Atmospheric air at 20°C enters and fills an insulated tank that is initially evacuated. Using a control volume analysis from Eq. (3.63), compute the tank air temperature when it is full.

P9.9 Liquid hydrogen and oxygen are burned in a combustion chamber and fed through a rocket nozzle that exhausts at $V_{exit} = 1600$ m/s to an ambient pressure of 54 kPa. The nozzle exit diameter is 45 cm, and the jet exit density is 0.15 kg/m³. If the exhaust gas has a molecular weight of 18, estimate (a) the exit gas temperature, (b) the mass flow, and (c) the thrust developed by the rocket.

P9.10 A certain aircraft flies at the same Mach number regardless of its altitude. Compared to its speed at 12,000-m standard altitude, it flies 127 km/h faster at sea level. Determine its Mach number.

P9.11 At 300°C and 1 atm, estimate the speed of sound of (a) nitrogen, (b) hydrogen, (c) helium, (d) steam, and (e) $^{238}UF_6$ ($k \approx 1.06$).

P9.12 Assume that water follows Eq. (1.19) with $n \approx 7$ and $B \approx 3000$. Compute the bulk modulus (in kPa) and the speed of sound (in m/s) at (a) 1 atm and (b) 1100 atm (the deepest part of the ocean). (c) Compute the speed of sound at 20°C and 9000 atm and compare with the measured value of 2650 m/s (A. H. Smith and A. W. Lawson, *J. Chem. Phys.*, vol. 22, 1954, p. 351).

P9.13 Assume that the airfoil of Prob. P8.84 is flying at the same angle of attack at 6000 m standard altitude. Estimate the forward velocity, in mi/h, at which supersonic flow (and possible shock waves) will appear on the airfoil surface.

P9.14 Assume steady adiabatic flow of a perfect gas. Show that the energy equation (9.21), when plotted as speed of sound versus velocity, forms an ellipse. Sketch this ellipse; label the intercepts and the regions of subsonic, sonic, and supersonic flow; and determine the ratio of the major and minor axes.

P9.15 The pressure-density relation for ethanol is approximated by Eq. (1.19) with $B = 1600$. Use this relation to estimate the speed of sound of ethanol at a pressure of 2000 atmospheres.

P9.16 A weak pressure pulse Δp propagates through still air. Discuss the type of reflected pulse that occurs and the boundary conditions that must be satisfied when the wave strikes normal to, and is reflected from, (a) a solid wall and (b) a free liquid surface.

P9.17 A submarine at a depth of 800 m sends a sonar signal and receives the reflected wave back from a similar submerged object in 15 s. Using Prob. P9.12 as a guide, estimate the distance to the other object.

P9.18 Race cars at the Indianapolis Speedway average speeds of 185 mi/h. After determining the altitude of Indianapolis, find the Mach number of these cars and estimate whether compressibility might affect their aerodynamics.

P9.19 In 1976, the SR-71A, flying at 20 km standard altitude, set a jet-powered aircraft speed record of 3326 km/h. Estimate the temperature, in °C, at its front stagnation point. At what Mach number would it have a front stagnation-point temperature of 500°C?

P9.20 A gas flows at $V = 200$ m/s, $p = 125$ kPa, and $T = 200$°C. For (a) air and (b) helium, compute the maximum pressure and the maximum velocity attainable by expansion or compression.

P9.21 CO_2 expands isentropically through a duct from $p_1 = 125$ kPa and $T_1 = 100$°C to $p_2 = 80$ kPa and $V_2 = 325$ m/s. Compute (a) T_2, (b) Ma_2, (c) T_0, (d) p_0, (e) V_1, and (f) Ma_1.

P9.22 Given the pitot stagnation temperature and pressure and the static pressure measurements in Fig. P9.22, estimate the air velocity V, assuming (a) incompressible flow and (b) compressible flow.

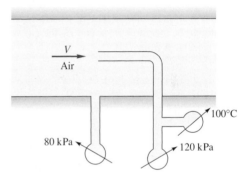

P9.22

P9.23 A gas, assumed ideal, flows isentropically from point 1, where the velocity is negligible, the pressure is 200 kPa, and the temperature is 300°C, to point 2, where the pressure is 40 kPa. What is the Mach number Ma_2 if the gas is (a) air, (b) argon, or (c) CH_4? (d) Can you tell, without calculating, which gas will be the coldest at point 2?

P9.24 For low-speed (nearly incompressible) gas flow, the stagnation pressure can be computed from Bernoulli's equation:

$$p_0 = p + \frac{1}{2}\rho V^2$$

(a) For higher subsonic speeds, show that the isentropic relation (9.28a) can be expanded in a power series as follows:

$$p_0 \approx p + \frac{1}{2}\rho V^2\left(1 + \frac{1}{4}Ma^2 + \frac{2-k}{24}Ma^4 + \cdots\right)$$

(b) Suppose that a pitot-static tube in air measures the pressure difference $p_0 - p$ and uses the Bernoulli relation, with stagnation density, to estimate the gas velocity. At what Mach number will the error be 4 percent?

P9.25 If it is known that the air velocity in the duct is 750 ft/s, use the mercury manometer measurement in Fig. P9.25 to estimate the static pressure in the duct in lbf/in² absolute.

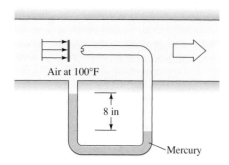

P9.25

P9.26 Show that for isentropic flow of a perfect gas if a pitot-static probe measures p_0, p, and T_0, the gas velocity can be calculated from

$$V^2 = 2c_p T_0 \left[1 - \left(\frac{p}{p_0} \right)^{(k-1)/k} \right]$$

What would be a source of error if a shock wave were formed in front of the probe?

P9.27 In many problems the sonic (*) properties are more useful reference values than the stagnation properties. For isentropic flow of a perfect gas, derive relations for p/p^*, T/T^*, and ρ/ρ^* as functions of the Mach number. Let us help by giving the density ratio formula:

$$\frac{\rho}{\rho^*} = \left[\frac{k+1}{2 + (k-1)\,\mathrm{Ma}^2} \right]^{1/(k-1)}$$

P9.28 A large vacuum tank, held at 60 kPa absolute, sucks sea-level standard air through a converging nozzle whose throat diameter is 3 cm. Estimate (a) the mass flow rate through the nozzle and (b) the Mach number at the throat.

P9.29 Steam from a large tank, where $T = 400°C$ and $p = 1$ MPa, expands isentropically through a nozzle until, at a section of 2-cm diameter, the pressure is 500 kPa. Using EES or the steam tables [15], estimate (a) the temperature, (b) the velocity, and (c) the mass flow at this section. Is the flow subsonic?

P9.30 When does the incompressible-flow assumption begin to fail for pressures? Construct a graph of p_0/p for incompressible flow of a perfect gas as compared to Eq. (9.28a). Plot both versus Mach number for $0 \le \mathrm{Ma} \le 0.6$ and decide for yourself where the deviation is too great.

P9.31 Air flows adiabatically through a duct. At one section $V_1 = 400$ ft/s, $T_1 = 200°F$, and $p_1 = 35$ lbf/in² absolute, while farther downstream $V_2 = 1100$ ft/s and $p_2 =$

18 lbf/in² absolute. Compute (a) Ma_2, (b) U_{max}, and (c) p_{02}/p_{01}.

P9.32 The large compressed-air tank in Fig. P9.32 exhausts from a nozzle at an exit velocity of 235 m/s. The mercury manometer reads $h = 30$ cm. Assuming isentropic flow, compute the pressure (a) in the tank and (b) in the atmosphere. (c) What is the exit Mach number?

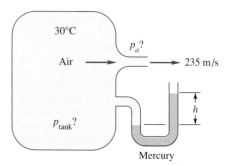

P9.32

P9.33 Air flows isentropically from a reservoir, where $p = 300$ kPa and $T = 500$ K, to section 1 in a duct, where $A_1 = 0.2$ m² and $V_1 = 550$ m/s. Compute (a) Ma_1, (b) T_1, (c) p_1, (d) $\dot{m}$, and (e) A^*. Is the flow choked?

P9.34 Carbon dioxide, in a large tank at 100°C and 151 kPa, exhausts through a converging nozzle whose throat area is 5 cm². Using isentropic ideal-gas theory, calculate (a) the exit temperature and (b) the mass flow.

P9.35 Helium, at $T_0 = 400$ K, enters a nozzle isentropically. At section 1, where $A_1 = 0.1$ m², a pitot-static arrangement (see Fig. P9.25) measures stagnation pressure of 150 kPa and static pressure of 123 kPa. Estimate (a) Ma_1, (b) mass flow $\dot{m}$, (c) T_1, and (d) A^*.

P9.36 An air tank of volume 1.5 m³ is initially at 800 kPa and 20°C. At $t = 0$, it begins exhausting through a converging nozzle to sea-level conditions. The throat area is 0.75 cm². Estimate (a) the initial mass flow in kg/s, (b) the time required to blow down to 500 kPa, and (c) the time at which the nozzle ceases being choked.

P9.37 Make an exact control volume analysis of the blowdown process in Fig. P9.37, assuming an insulated tank with negligible kinetic and potential energy within. Assume critical flow at the exit, and show that both p_0 and T_0 decrease during blowdown. Set up first-order differential equations for $p_0(t)$ and $T_0(t)$, and reduce and solve as far as you can.

P9.38 Prob. P9.37 makes an ideal senior project or combined laboratory and computer problem, as described in Ref. 27, Sec. 8.6. In Bober and Kenyon's lab experiment, the tank had a volume of 0.0352 ft³ and was initially filled with air at 50 lb/in² gage and 72°F. Atmospheric pressure was 14.5 lb/in² absolute, and the nozzle exit diameter was 0.05 in. After 2 s of blowdown, the measured tank pressure was 20 lb/in^2 gage and the tank temperature was $-5°F$. Compare these values with the theoretical analysis of Prob. P9.37.

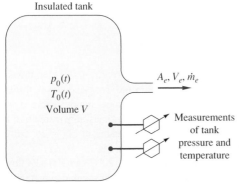

Insulated tank

$p_0(t)$
$T_0(t)$
Volume V

$A_e, V_e, \dot{m}_e$

Measurements of tank pressure and temperature

P9.37

P9.39 Consider isentropic flow in a channel of varying area, from section 1 to section 2. We know that $Ma_1 = 2.0$ and desire that the velocity ratio V_2/V_1 be 1.2. Estimate (a) Ma_2 and (b) A_2/A_1. (c) Sketch what this channel looks like. For example, does it converge or diverge? Is there a throat?

P9.40 Air, with stagnation conditions of 800 kPa and 100°C, expands isentropically to a section of a duct where $A_1 = 20$ cm^2 and $p_1 = 47$ kPa. Compute (a) Ma_1, (b) the throat area, and (c) $\dot{m}$. At section 2 between the throat and section 1, the area is 9 cm^2. (d) Estimate the Mach number at section 2.

P9.41 Air, with a stagnation pressure of 100 kPa, flows through the nozzle in Fig. P9.41, which is 2 m long and has an area variation approximated by

$$A \approx 20 - 20x + 10x^2$$

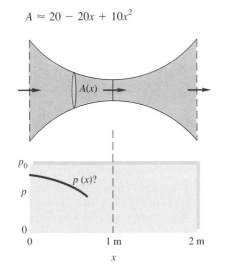

$A(x)$

p_0

$p(x)$?

p

0

0   1 m   2 m

P9.41 x

with A in cm^2 and x in m. It is desired to plot the complete family of isentropic pressures $p(x)$ in this nozzle, for the range of inlet pressures $1 < p(0) < 100$ kPa. Indicate which inlet pressures are not physically possible and discuss briefly. If your computer has an online graphics routine, plot at least 15 pressure profiles; otherwise just hit the highlights and explain.

P9.42 A bicycle tire is filled with air at an absolute pressure of 169.12 kPa, and the temperature inside is 30.0°C. Suppose the valve breaks, and air starts to exhaust out of the tire into the atmosphere ($p_a = 100$ kPa absolute and $T_a = 20.0°C$). The valve exit is 2.00 mm in diameter and is the smallest cross-sectional area of the entire system. Frictional losses can be ignored here; one-dimensional isentropic flow is a reasonable assumption. (a) Find the Mach number, velocity, and temperature at the exit plane of the valve (initially). (b) Find the initial mass flow rate out of the tire. (c) Estimate the velocity at the exit plane using the incompressible Bernoulli equation. How well does this estimate agree with the "exact" answer of part (a)? Explain.

P9.43 Air flows isentropically through a variable-area duct. At section 1, $A_1 = 20$ cm^2, $p_1 = 300$ kPa, $\rho_1 = 1.75$ kg/m^3, and $Ma_1 = 0.25$. At section 2, the area is exactly the same, but the flow is much faster. Compute (a) V_2, (b) Ma_2, (c) T_2, and (d) the mass flow. (e) Is there a sonic throat between sections 1 and 2? If so, find its area.

P9.44 In Prob. P3.34 we knew nothing about compressible flow at the time, so we merely assumed exit conditions p_2 and T_2 and computed V_2 as an application of the continuity equation. Suppose that the throat diameter is 3 in. For the given stagnation conditions in the rocket chamber in Fig. P3.34 and assuming $k = 1.4$ and a molecular weight of 26, compute the actual exit velocity, pressure, and temperature according to one-dimensional theory. If $p_a = 14.7$ lbf/in^2 absolute, compute the thrust from the analysis of Prob. P3.68. This thrust is entirely independent of the stagnation temperature (check this by changing T_0 to 2000°R if you like). Why?

P9.45 At a point upstream of the throat of a converging-diverging nozzle the properties are $V_1 = 200$ m/s, $T_1 = 300$ K, and $p_1 = 125$ kPa. If the exit flow is supersonic, compute, from isentropic theory, (a) $\dot{m}$ and (b) A_1. The throat area is 35 cm^2.

P9.46 A one-dimensional isentropic airflow has the following properties at one section where the area is 53 cm^2: $p = 12$ kPa, $\rho = 0.182$ kg/m^3, and $V = 760$ m/s. Determine (a) the throat area, (b) the stagnation temperature, and (c) the mass flow.

P9.47 In wind tunnel testing near Mach 1, a small area decrease caused by model blockage can be important.

Suppose the test section area is 1 m², with unblocked test conditions Ma = 1.10 and $T = 20°C$. What model area will first cause the test section to choke? If the model cross section is 0.004 m² (0.4 percent blockage), what percentage change in test section velocity results?

P9.48 A force $F = 1100$ N pushes a piston of diameter 12 cm through an insulated cylinder containing air at 20°C, as in Fig. P9.48. The exit diameter is 3 mm, and $p_a = 1$ atm. Estimate (a) V_e, (b) V_p, and (c) $\dot{m}_e$.

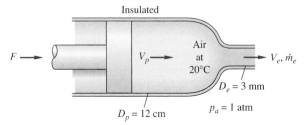

P9.48

P9.49 Consider the venturi nozzle of Fig. 6.40c, with $D = 5$ cm and $d = 3$ cm. Stagnation temperature is 300 K, and the upstream velocity $V_1 = 72$ m/s. If the throat pressure is 124 kPa, estimate, with isentropic flow theory, (a) p_1, (b) Ma₂, and (c) the mass flow.

P9.50 Argon expands isentropically in a converging nozzle whose entrance conditions are $D_1 = 10$ cm, $p_1 = 150$ kPa, $T_1 = 100°C$, and $\dot{m} = 1$ kg/s. The flow discharges smoothly to an ambient pressure of 101 kPa. (a) What is the exit diameter of the nozzle? (b) How much further can the ambient pressure be reduced before it affects the inlet mass flow?

P9.51 Air, at stagnation conditions of 500 K and 200 kPa, flows through a nozzle. At section 1, where the area is 12 cm², the density is 0.32 kg/m³. Assuming isentropic flow, (a) find the mass flow. (b) Is the flow choked? If so, estimate A^*. Also estimate (c) p_1 and (d) Ma₁.

P9.52 A converging–diverging nozzle exits smoothly to sea-level standard atmosphere. It is supplied by a 40-m³ tank initially at 800 kPa and 100°C. Assuming isentropic flow in the nozzle, estimate (a) the throat area and (b) the tank pressure after 10 s of operation. The exit area is 10 cm².

P9.53 Air flows steadily from a reservoir at 20°C through a nozzle of exit area 20 cm² and strikes a vertical plate as in Fig. P9.53. The flow is subsonic throughout. A force of 135 N is required to hold the plate stationary. Compute (a) V_e, (b) Ma$_e$, and (c) p_0 if $p_a = 101$ kPa.

P9.54 The airflow in Prob. P9.46 undergoes a normal shock just past the section where data was given. Determine the (a) Mach number, (b) pressure, and (c) velocity just downstream of the shock.

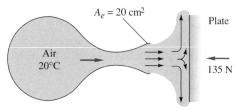

P9.53

P9.55 Air, supplied by a reservoir at 450 kPa, flows through a converging–diverging nozzle whose throat area is 12 cm². A normal shock stands where $A_1 = 20$ cm². (a) Compute the pressure just downstream of this shock. Still farther downstream, at $A_3 = 30$ cm², estimate (b) p_3, (c) A_3^*, and (d) Ma₃.

P9.56 Air from a reservoir at 20°C and 500 kPa flows through a duct and forms a normal shock downstream of a throat of area 10 cm². By an odd coincidence it is found that the stagnation pressure downstream of this shock exactly equals the throat pressure. What is the area where the shock wave stands?

P9.57 Air flows from a tank through a nozzle into the standard atmosphere, as in Fig. P9.57. A normal shock stands in the exit of the nozzle, as shown. Estimate (a) the pressure in the tank and (b) the mass flow.

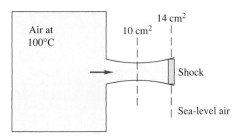

P9.57

P9.58 Argon (Table A.4) approaches a normal shock with $V_1 = 700$ m/s, $p_1 = 125$ kPa, and $T_1 = 350$ K. Estimate (a) V_2 and (b) p_2. (c) What pressure p_2 would result if the same velocity change V_1 to V_2 were accomplished isentropically?

P9.59 Air, at stagnation conditions of 450 K and 250 kPa, flows through a nozzle. At section 1, where the area is 15 cm², there is a normal shock wave. If the mass flow is 0.4 kg/s, estimate (a) the Mach number and (b) the stagnation pressure just downstream of the shock.

P9.60 When a pitot tube such as in Fig. 6.30 is placed in a supersonic flow, a normal shock will stand in front of the probe. Suppose the probe reads $p_0 = 190$ kPa and $p = 150$ kPa. If the stagnation temperature is 400 K, estimate the (supersonic) Mach number and velocity upstream of the shock.

P9.61 Air flows from a large tank, where $T = 376$ K and $p = 360$ kPa, to a design condition where the pressure is 9800 Pa. The mass flow is 0.9 kg/s. However, there is a normal shock in the exit plane just after this condition is reached. Estimate (*a*) the throat area and, just downstream of the shock, (*b*) the Mach number, (*c*) the temperature, and (*d*) the pressure.

P9.62 An atomic explosion propagates into still air at 14.7 lbf/in^2 absolute and 520°R. The pressure just inside the shock is 5000 lbf/in^2 absolute. Assuming $k = 1.4$, what are the speed C of the shock and the velocity V just inside the shock?

P9.63 Sea-level standard air is sucked into a vacuum tank through a nozzle, as in Fig. P9.63. A normal shock stands where the nozzle area is 2 cm^2, as shown. Estimate (*a*) the pressure in the tank and (*b*) the mass flow.

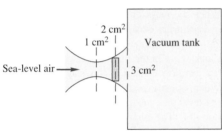

P9.63

P9.64 Air in a large tank at 100°C and 150 kPa exhausts to the atmosphere through a converging nozzle with a 5-cm^2 throat area. Compute the exit mass flow if the atmospheric pressure is (*a*) 100 kPa, (*b*) 60 kPa, and (*c*) 30 kPa.

P9.65 Air flows through a converging–diverging nozzle between two large reservoirs, as shown in Fig. P9.65. A mercury manometer between the throat and the downstream reservoir reads $h = 15$ cm. Estimate the downstream reservoir pressure. Is there a normal shock in the flow? If so, does it stand in the exit plane or farther upstream?

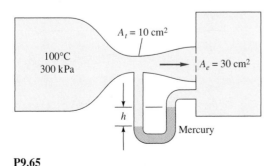

P9.65

P9.66 In Prob. P9.65 what would be the mercury manometer reading h if the nozzle were operating exactly at supersonic design conditions?

P9.67 A supply tank at 500 kPa and 400 K feeds air to a converging diverging nozzle whose throat area is 9 cm^2. The exit area is 46 cm^2. State the conditions in the nozzle if the pressure outside the exit plane is (*a*) 400 kPa, (*b*) 120 kPa, and (*c*) 9 kPa. (*d*) In each of these cases, find the mass flow.

P9.68 Air in a tank at 120 kPa and 300 K exhausts to the atmosphere through a 5-cm^2-throat converging nozzle at a rate of 0.12 kg/s. What is the atmospheric pressure? What is the maximum mass flow possible at low atmospheric pressure?

P9.69 With reference to Prob. P3.68, show that the thrust of a rocket engine exhausting into a vacuum is given by

$$F = \frac{p_0 A_e (1 + k \, \text{Ma}_e^2)}{\left(1 + \dfrac{k-1}{2} \text{Ma}_e^2\right)^{k/(k-1)}}$$

where A_e = exit area
Ma_e = exit Mach number
p_0 = stagnation pressure in combustion chamber

Note that stagnation temperature does not enter into the thrust.

P9.70 Air, at stagnation temperature 100°C, expands isentropically through a nozzle of 6-cm^2 throat area and 18-cm^2 exit area. The mass flow is at its maximum value of 0.5 kg/s. Estimate the exit pressure for (*a*) subsonic and (*b*) supersonic exit flow.

P9.71 A converging–diverging nozzle has a throat area of 10 cm^2 and an exit area of 20 cm^2. It is supplied by an air tank at 250 kPa and 350 K. (*a*) What is the design pressure at the exit? At one operating condition, the exit properties are $p_e = 183$ kPa, $T_e = 340$ K, and $V_e = 144$ m/s. (*b*) Can this condition be explained by a normal shock inside the nozzle? (*c*) If so, at what Mach number does the normal shock occur? [*Hint*: Use the change in A^* to locate, if necessary, this location.]

P9.72 A large tank at 500 K and 165 kPa feeds air to a converging nozzle. The back pressure outside the nozzle exit is sea-level standard. What is the appropriate exit diameter if the desired mass flow is 72 kg/h?

P9.73 Air flows isentropically in a converging–diverging nozzle with a throat area of 3 cm^2. At section 1, the pressure is 101 kPa, the temperature is 300 K, and the velocity is 868 m/s. (*a*) Is the nozzle choked? Determine (*b*) A_1 and (*c*) the mass flow. Suppose, without changing stagnation conditions or A_1, the (flexible) throat is reduced to 2 cm^2. Assuming shock-free flow, will there be any change in the gas properties at section 1? If so, compute new p_1, V_1, and T_1 and explain.

P9.74 The perfect-gas assumption leads smoothly to Mach number relations that are very convenient (and tabulated). This is not so for a real gas such as steam. To illustrate, let steam at $T_0 = 500°C$ and $p_0 = 2$ MPa expand isentropically through a converging nozzle whose exit area is 10 cm². Using the steam tables, find (a) the exit pressure and (b) the mass flow when the flow is sonic, or choked. What complicates the analysis?

*P9.75 A double-tank system in Fig. P9.75 has two identical converging nozzles of 1-in² throat area. Tank 1 is very large, and tank 2 is small enough to be in steady-flow equilibrium with the jet from tank 1. Nozzle flow is isentropic, but entropy changes between 1 and 3 due to jet dissipation in tank 2. Compute the mass flow. (If you give up, Ref. 9, pp. 288–290, has a good discussion.)

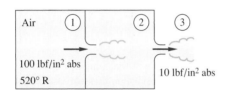

P9.75

P9.76 A large reservoir at 20°C and 800 kPa is used to fill a small insulated tank through a converging–diverging nozzle with 1-cm² throat area and 1.66-cm² exit area. The small tank has a volume of 1 m³ and is initially at 20°C and 100 kPa. Estimate the elapsed time when (a) shock waves begin to appear inside the nozzle and (b) the mass flow begins to drop below its maximum value.

P9.77 A perfect gas (not air) expands isentropically through a supersonic nozzle with an exit area 5 times its throat area. The exit Mach number is 3.8. What is the specific-heat ratio of the gas? What might this gas be? If $p_0 = 300$ kPa, what is the exit pressure of the gas?

P9.78 The orientation of a hole can make a difference. Consider holes A and B in Fig. P9.78, which are identical but reversed. For the given air properties on either side, compute the mass flow through each hole and explain why they are different.

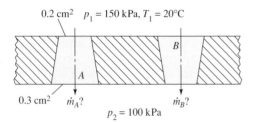

P9.78

P9.79 A large reservoir at 600 K supplies air flow through a converging–diverging nozzle with a throat area of 2 cm². A normal shock wave forms at a section of area 6 cm². Just downstream of this shock, the pressure is 150 kPa. Calculate (a) the pressure in the throat, (b) the mass flow, and (c) the pressure in the reservoir.

P9.80 A sea-level automobile tire is initially at 32 lbf/in² gage pressure and 75°F. When it is punctured with a hole that resembles a converging nozzle, its pressure drops to 15 lbf/in² gage in 12 min. Estimate the size of the hole, in thousandths of an inch. The tire volume is 2.5 ft².

P9.81 Helium, in a large tank at 100°C and 400 kPa, discharges to a receiver through a converging–diverging nozzle designed to exit at Ma = 2.5 with exit area 1.2 cm². Compute (a) the receiver pressure and (b) the mass flow at design conditions. (c) Also estimate the range of receiver pressures for which mass flow will be a maximum.

P9.82 Air at 500 K flows through a converging–diverging nozzle with throat area of 1 cm² and exit area of 2.7 cm². When the mass flow is 182.2 kg/h, a pitot-static probe placed in the exit plane reads $p_0 = 250.6$ kPa and $p = 240.1$ kPa. Estimate the exit velocity. Is there a normal shock wave in the duct? If so, compute the Mach number just downstream of this shock.

P9.83 When operating at design conditions (smooth exit to sea-level pressure), a rocket engine has a thrust of 1 million lbf. The chamber pressure and temperature are 600 lbf/in² absolute and 4000°R, respectively. The exhaust gases approximate $k = 1.38$ with a molecular weight of 26. Estimate (a) the exit Mach number and (b) the throat diameter.

P9.84 Air flows through a duct as in Fig. P9.84, where $A_1 = 24$ cm², $A_2 = 18$ cm², and $A_3 = 32$ cm². A normal shock stands at section 2. Compute (a) the mass flow, (b) the Mach number, and (c) the stagnation pressure at section 3.

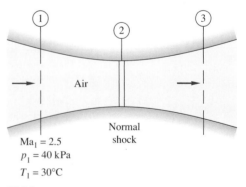

P9.84

P9.85 A typical carbon dioxide tank for a paintball gun holds about 12 oz of liquid CO_2. The tank is filled no more than one-third with liquid, which, at room temperature, maintains the gaseous phase at about 850 psia. (a) If a valve is opened that simulates a converging nozzle with an exit diameter of 0.050 in, what mass flow and exit velocity results? (b) Repeat the calculations for helium.

P9.86 Air enters a 3-cm-diameter pipe 15 m long at $V_1 = 73$ m/s, $p_1 = 550$ kPa, and $T_1 = 60°C$. The friction factor is 0.018. Compute V_2, p_2, T_2, and p_{02} at the end of the pipe. How much additional pipe length would cause the exit flow to be sonic?

P9.87 Air enters a duct of $L/D = 40$ at $V_1 = 170$ m/s and $T_1 = 300$ K. The flow at the exit is choked. What is the average friction factor in the duct for adiabatic flow?

P9.88 Air flows adiabatically through a 2-cm-diameter pipe. Conditions at section 2 are $p_2 = 100$ kPa, $T_2 = 15°C$, and $V_2 = 170$ m/s. The average friction factor is 0.024. At section 1, which is 55 meters upstream, find (a) the mass flow, (b) p_1, and (c) p_{01}.

P9.89 Carbon dioxide flows through an insulated pipe 25 m long and 8 cm in diameter. The friction factor is 0.025. At the entrance, $p = 300$ kPa and $T = 400$ K. The mass flow is 1.5 kg/s. Estimate the pressure drop by (a) compressible and (b) incompressible (Sec. 6.6) flow theory. (c) For what pipe length will the exit flow be choked?

P9.90 Air, supplied at $p_0 = 700$ kPa and $T_0 = 330$ K, flows through a converging nozzle into a pipe of 2.5-cm diameter that exits to a near vacuum. If $\bar{f} = 0.022$, what will be the mass flow through the pipe if its length is (a) 0 m, (b) 1 m, and (c) 10 m?

P9.91 Air flows steadily from a tank through the pipe in Fig. P9.91. There is a converging nozzle on the end. If the mass flow is 3 kg/s and the nozzle is choked, estimate (a) the Mach number at section 1 and (b) the pressure inside the tank.

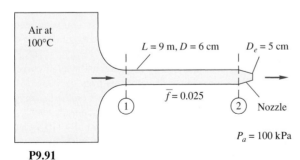

P9.91

P9.92 Air enters a 5-cm-diameter pipe at 380 kPa, 3.3 kg/m³, and 120 m/s. The friction factor is 0.017. Find the pipe length for which the velocity (a) doubles, (b) triples, and (d) quadruples.

P9.93 Air flows adiabatically in a 3-cm-diameter duct. The average friction factor is 0.015. If, at the entrance,

$V = 950$ m/s and $T = 250$ K, how far down the tube will (a) the Mach number be 1.8 or (b) the flow be choked?

P9.94 Compressible pipe flow with friction, Sec. 9.7, assumes constant stagnation enthalpy and mass flow but variable momentum. Such a flow is often called *Fanno flow,* and a line representing all possible property changes on a temperature–entropy chart is called a *Fanno line.* Assuming a perfect gas with $k = 1.4$ and the data of Prob. P9.86, draw a Fanno curve of the flow for a range of velocities from very low (Ma $\ll$ 1) to very high (Ma $\gg$ 1). Comment on the meaning of the maximum-entropy point on this curve.

P9.95 Helium (Table A.4) enters a 5-cm-diameter pipe at $p_1 = 550$ kPa, $V_1 = 312$ m/s, and $T_1 = 40°C$. The friction factor is 0.025. If the flow is choked, determine (a) the length of the duct and (b) the exit pressure.

P9.96 Methane (CH_4) flows through an insulated 15-cm-diameter pipe with $f = 0.023$. Entrance conditions are 600 kPa, 100°C, and a mass flow of 5 kg/s. What lengths of pipe will (a) choke the flow, (b) raise the velocity by 50 percent, or (c) decrease the pressure by 50 percent?

P9.97 By making a few algebraic substitutions, show that Eq. (9.74) may be written in the density form

$$\rho_1^2 = \rho_2^2 + \rho^{*2}\left(\frac{2k}{k+1}\frac{\bar{f}L}{D} + 2\ln\frac{\rho_1}{\rho_2}\right)$$

Why is this formula awkward if one is trying to solve for the mass flow when the pressures are given at sections 1 and 2?

P9.98 Compressible *laminar* flow, $f \approx 64/Re$, may occur in capillary tubes. Consider air, at stagnation conditions of 100°C and 200 kPa, entering a tube 3 cm long and 0.1 mm in diameter. If the receiver pressure is near vacuum, estimate (a) the average Reynolds number, (b) the Mach number at the entrance, and (c) the mass flow in kg/h.

P9.99 A compressor forces air through a smooth pipe 20 m long and 4 cm in diameter, as in Fig. P9.99. The air leaves at 101 kPa and 200°C. The compressor data for

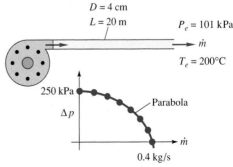

P9.99

pressure rise versus mass flow are shown in the figure. Using the Moody chart to estimate $\bar{f}$, compute the resulting mass flow.

P9.100 Air in a large tank, at 300 kPa and 200°C, discharges adiabatically through a smooth pipe 1 cm in diameter and 2.5 m long. The pipe exits to an atmosphere at 20°C and 100 kPa. Estimate the mass flow in the pipe. For convenience, assume $f \approx 0.020$.

P9.101 How do the compressible pipe flow formulas behave for small pressure drops? Let air at 20°C enter a tube of diameter 1 cm and length 3 m. If $\bar{f} = 0.028$ with $p_1 = 102$ kPa and $p_2 = 100$ kPa, estimate the mass flow in kg/h for (a) isothermal flow, (b) adiabatic flow, and (c) incompressible flow (Chap. 6) at the entrance density.

P9.102 Air at 550 kPa and 100°C enters a smooth 1-m-long pipe and then passes through a second smooth pipe to a 30-kPa reservoir, as in Fig. P9.102. Using the Moody chart to compute $\bar{f}$, estimate the mass flow through this system. Is the flow choked?

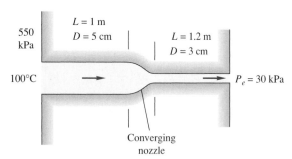

P9.102

P9.103 Natural gas, with $k \approx 1.3$ and a molecular weight of 16, is to be pumped through 100 km of 81-cm-diameter pipeline. The downstream pressure is 150 kPa. If the gas enters at 60°C, the mass flow is 20 kg/s, and $\bar{f} = 0.024$, estimate the required entrance pressure for (a) isothermal flow and (b) adiabatic flow.

P9.104 A tank of oxygen (Table A.4) at 20°C is to supply an astronaut through an umbilical tube 12 m long and 1.5 cm in diameter. The exit pressure in the tube is 40 kPa. If the desired mass flow is 90 kg/h and $\bar{f} = 0.025$, what should be the pressure in the tank?

P9.105 Air enters a 5-cm-diameter pipe at $p_1 = 200$ kPa and $T_1 = 350$ K. The downstream receiver pressure is 74 kPa. The friction factor is 0.02. If the exit is choked, what is (a) the length of the pipe and (b) the mass flow? (c) If p_1, T_1, and p_{receiver} stay the same, what pipe length will cause the mass flow to increase by 50 percent over (b)? *Hint:* In part (c) the exit pressure does not equal the receiver pressure.

P9.106 Air, from a 3 cubic meter tank initially at 300 kPa and 200°C, blows down adiabatically through a smooth pipe 1 cm in diameter and 2.5 m long. Estimate the time required to reduce the tank pressure to 200 kPa. For simplicity, assume constant tank temperature and $f \approx 0.020$.

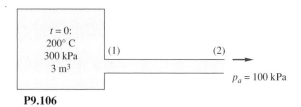

P9.106

P9.107 A fuel–air mixture, assumed equivalent to air, enters a duct combustion chamber at $V_1 = 104$ m/s and $T_1 = 300$ K. What amount of heat addition in kJ/kg will cause the exit flow to be choked? What will be the exit Mach number and temperature if 504 kJ/kg are added during combustion?

P9.108 What happens to the inlet flow of Prob. P9.107 if the combustion yields 1500 kJ/kg heat addition and p_{01} and T_{01} remain the same? How much is the mass flow reduced?

P9.109 A jet engine at 7000-m altitude takes in 45 kg/s of air and adds 550 kJ/kg in the combustion chamber. The chamber cross section is 0.5 m², and the air enters the chamber at 80 kPa and 5°C. After combustion the air expands through an isentropic converging nozzle to exit at atmospheric pressure. Estimate (a) the nozzle throat diameter, (b) the nozzle exit velocity, and (c) the thrust produced by the engine.

P9.110 Compressible pipe flow with heat addition, Sec. 9.8, assumes constant momentum $(p + \rho V^2)$ and constant mass flow but variable stagnation enthalpy. Such a flow is often called *Rayleigh flow*, and a line representing all possible property changes on a temperature–entropy chart is called a *Rayleigh line*. Assuming air passing through the flow state $p_1 = 548$ kPa, $T_1 = 588$ K, $V_1 = 266$ m/s, and $A = 1$ m², draw a Rayleigh curve of the flow for a range of velocities from very low (Ma ≪ 1) to very high (Ma ≫ 1). Comment on the meaning of the maximum-entropy point on this curve.

P9.111 Add to your Rayleigh line of Prob. P9.110 a Fanno line (see Prob. P9.94) for stagnation enthalpy equal to the value associated with state 1 in Prob. P9.110. The two curves will intersect at state 1, which is subsonic, and at a certain state 2, which is supersonic. Interpret these two states vis-à-vis Table B.2.

P9.112 Air enters a duct subsonically at section 1 at 1.2 kg/s. When 650 kW of heat are added, the flow chokes at the exit at $p_2 = 95$ kPa and $T_2 = 700$ K. Assuming frictionless heat addition, estimate (a) the velocity and (b) the stagnation pressure at section 1.

P9.113 Air enters a constant-area duct at $p_1 = 90$ kPa, $V_1 = 520$ m/s, and $T_1 = 558°C$. It is then cooled with negligible friction until it exits at $p_2 = 160$ kPa. Estimate (a) V_2, (b) T_2, and (c) the total amount of cooling in kJ/kg.

P9.114 We have simplified things here by separating friction (Sec. 9.7) from heat addition (Sec. 9.8). Actually, they often occur together, and their effects must be evaluated simultaneously. Show that, for flow with friction *and* heat transfer in a constant-diameter pipe, the continuity, momentum, and energy equations may be combined into the following differential equation for Mach number changes:

$$\frac{d\,\mathrm{Ma}^2}{\mathrm{Ma}^2} = \frac{1 + k\,\mathrm{Ma}^2}{1 - \mathrm{Ma}^2}\frac{dQ}{c_p T} + \frac{k\,\mathrm{Ma}^2[2 + (k-1)\,\mathrm{Ma}^2]}{2(1 - \mathrm{Ma}^2)}\frac{f\,dx}{D}$$

where dQ is the heat added. A complete derivation, including many additional combined effects such as area change and mass addition, is given in Chap. 8 of Ref. 5.

P9.115 Air enters a 5-cm-diameter pipe at 380 kPa, 3.3 kg/m³, and 120 m/s. Assume frictionless flow with heat addition. Find the amount of heat addition for which the velocity (a) doubles, (b) triples, and (d) quadruples.

P9.116 An observer at sea level does not hear an aircraft flying at 12,000-ft standard altitude until it is 5 (statute) mi past her. Estimate the aircraft speed in ft/s.

P9.117 A tiny scratch in the side of a supersonic wind tunnel creates a very weak wave of angle 17°, as shown in Fig. P9.117, after which a normal shock occurs. The air temperature in region (1) is 250 K. Estimate the temperature in region (2).

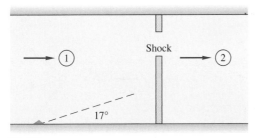

P9.117

P9.118 A particle moving at uniform velocity in sea-level standard air creates the two disturbance spheres shown in Fig. P9.118. Compute the particle velocity and Mach number.

P9.119 The particle in Fig. P9.119 is moving supersonically in sea-level standard air. From the two given disturbance spheres, compute the particle Mach number, velocity, and Mach angle.

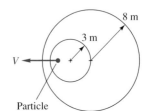

P9.118

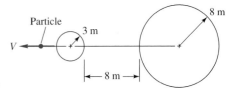

P9.119

P9.120 The particle in Fig. P9.120 is moving in sea-level standard air. From the two disturbance spheres shown, estimate (a) the position of the particle at this instant and (b) the temperature in °C at the front stagnation point of the particle.

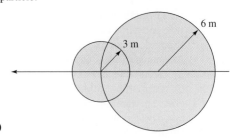

P9.120

P9.121 A thermistor probe, in the shape of a needle parallel to the flow, reads a static temperature of −25°C when inserted into a supersonic airstream. A conical disturbance cone of half-angle 17° is created. Estimate (a) the Mach number, (b) the velocity, and (c) the stagnation temperature of the stream.

P9.122 Supersonic air takes a 5° compression turn, as in Fig. P9.122. Compute the downstream pressure and Mach number and the wave angle, and compare with small-disturbance theory.

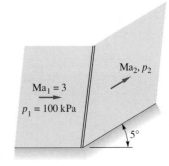

P9.122

P9.123 Modify Prob. P9.122 as follows. Let the 5° total turn be in the form of five separate compression turns of 1° each. Compute the final Mach number and pressure, and compare the pressure with an isentropic expansion to the same final Mach number.

P9.124 When a sea-level flow approaches a ramp of angle 20°, an oblique shock wave forms as in Figure P9.124. Calculate (a) Ma_1, (b) p_2, (c) T_2, and (d) V_2.

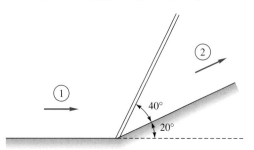

P9.124

P9.125 We saw in the text that, for $k = 1.40$, the maximum possible deflection caused by an oblique shock wave occurs at infinite approach Mach number and is $\theta_{max} = 45.58°$. Assuming an ideal gas, what is θ_{max} for (a) argon and (b) carbon dioxide?

P9.126 Consider airflow at $Ma_1 = 2.2$. Calculate, to two decimal places, (a) the deflection angle for which the downstream flow is sonic and (b) the maximum deflection angle.

P9.127 Do the Mach waves upstream of an oblique shock wave intersect with the shock? Assuming supersonic downstream flow, do the downstream Mach waves intersect the shock? Show that for small deflections the shock wave angle β lies halfway between μ_1 and $\mu_2 + \theta$ for any Mach number.

P9.128 Air flows past a two-dimensional wedge-nosed body as in Fig. P9.128. Determine the wedge half-angle δ for which the horizontal component of the total pressure force on the nose is 35 kN/m of depth into the paper.

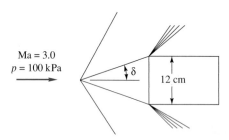

P9.128

P9.129 Air flows at supersonic speed toward a compression ramp, as in Fig. P9.129. A scratch on the wall at point

a creates a wave of 30° angle, while the oblique shock created has a 50° angle. What is (a) the ramp angle θ and (b) the wave angle ϕ caused by a scratch at *b*?

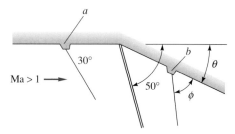

P9.129

P9.130 A supersonic airflow, at a temperature of 300K, strikes a wedge and is deflected 12°. If the resulting shock wave is attached, and the temperature after the shock is 450K, (a) estimate the approach Mach number and wave angle. (b) Why are there two solutions?

P9.131 The following formula has been suggested as an alternate to Eq. (9.86) to relate upstream Mach number to the oblique shock wave angle β and turning angle θ:

$$\sin^2 \beta = \frac{1}{Ma_1^2} + \frac{(k + 1) \sin \beta \sin \theta}{2 \cos (\beta - \theta)}$$

Can you prove or disprove this relation? If not, try a few numerical values and compare with the results from Eq. (9.86).

P9.132 Air flows at $Ma = 3$ and $p = 10$ lbf/in² absolute toward a wedge of 16° angle at zero incidence in Fig. P9.132. If the pointed edge is forward, what will be the pressure at point A? If the blunt edge is forward, what will be the pressure at point B?

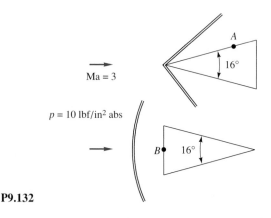

P9.132

P9.133 Air flows supersonically toward the double-wedge system in Fig. P9.133. The (x, y) coordinates of the tips are given.

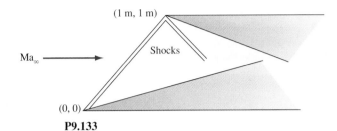

P9.133

The shock wave of the forward wedge strikes the tip of the aft wedge. Both wedges have 15° deflection angles. What is the free-stream Mach number?

P9.134 When an oblique shock strikes a solid wall, it reflects as a shock of sufficient strength to cause the exit flow Ma_3 to be parallel to the wall, as in Fig. P9.134. For airflow with $Ma_1 = 2.5$ and $p_1 = 100$ kPa, compute Ma_3, p_3, and the angle ϕ.

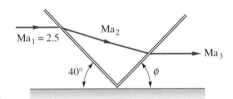

P9.134

P9.135 A bend in the bottom of a supersonic duct flow induces a shock wave that reflects from the upper wall, as in Fig. P9.135. Compute the Mach number and pressure in region 3.

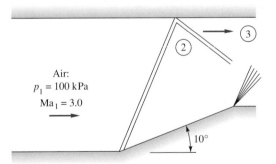

P9.135

P9.136 Figure P9.136 is a special application of Prob. P9.135. With careful design, one can orient the bend on the lower wall so that the reflected wave is exactly canceled by the return bend, as shown. This is a method of reducing the Mach number in a channel (a supersonic diffuser). If the bend angle is $\phi = 10°$, find (a) the downstream width h and (b) the downstream Mach number. Assume a weak shock wave.

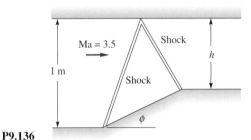

P9.136

P9.137 A 6° half-angle wedge creates the reflected shock system in Fig. P9.137. If $Ma_3 = 2.5$, find (a) Ma_1 and (b) the angle α.

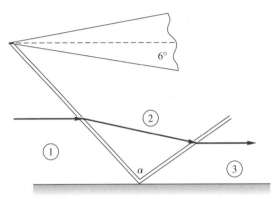

P9.137

P9.138 The supersonic nozzle of Fig. P9.138 is overexpanded (case G of Fig. 9.12b) with $A_e/A_t = 3.0$ and a stagnation pressure of 350 kPa. If the jet edge makes a 4° angle with the nozzle centerline, what is the back pressure p_r in kPa?

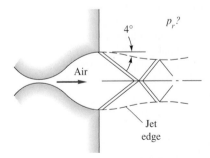

P9.138

P9.139 Airflow at Ma = 2.2 takes a compression turn of 12° and then another turn of angle θ in Fig. P9.139. What is the maximum value of θ for the second shock to be attached? Will the two shocks intersect for any θ less than θ_{max}?

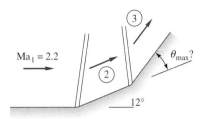

P9.139

P9.140 The solution to Prob. P9.122 is $Ma_2 = 2.750$ and $p_2 = 145.5$ kPa. Compare these results with an isentropic compression turn of 5°, using Prandtl-Meyer theory.

P9.141 Supersonic airflow takes a 5° expansion turn, as in Fig. P9.141. Compute the downstream Mach number and pressure, and compare with small-disturbance theory.

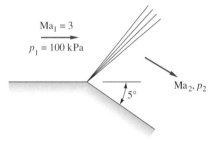

P9.141

P9.142 A supersonic airflow at $Ma_1 = 3.2$ and $p_1 = 50$ kPa undergoes a compression shock followed by an isentropic expansion turn. The flow deflection is 30° for each turn. Compute Ma_2 and p_2 if (a) the shock is followed by the expansion and (b) the expansion is followed by the shock.

P9.143 Airflow at $Ma_1 = 3.2$ passes through a 25° oblique shock deflection. What isentropic expansion turn is required to bring the flow back to (a) Ma_1 and (b) p_1?

P9.144 Consider a smooth isentropic compression turn of 20°, as shown in Fig. P9.144. The Mach waves thus generated will form a converging fan. Sketch this fan as accurately as possible, using at least five equally spaced waves, and demonstrate how the fan indicates the probable formation of an oblique shock wave.

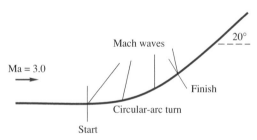

P9.144

P9.145 Air at $Ma_1 = 2.0$ and $p_1 = 100$ kPa undergoes an isentropic expansion to a downstream pressure of 50 kPa. What is the desired turn angle in degrees?

P9.146 Air flows supersonically over a surface that changes direction twice, as in Fig. P9.146. Calculate (a) Ma_2 and (b) p_3.

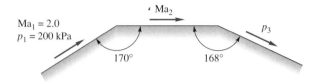

P9.146

P9.147 A converging–diverging nozzle with a 4:1 exit-area ratio and $p_0 = 500$ kPa operates in an underexpanded condition (case *I* of Fig. 9.12*b*) as in Fig. P9.147. The receiver pressure is $p_a = 10$ kPa, which is less than the exit pressure, so that expansion waves form outside the exit. For the given conditions, what will the Mach number Ma_2 and the angle ϕ of the edge of the jet be? Assume $k = 1.4$ as usual.

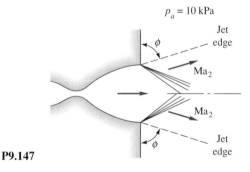

P9.147

P9.148 Air flows supersonically over a circular-arc surface as in Fig. P9.148. Estimate (a) the Mach number Ma_2 and (b) the pressure p_2 as the flow leaves the circular surface.

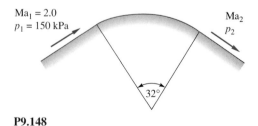

P9.148

P9.149 Air flows at $Ma_\infty = 3.0$ past a doubly symmetric diamond airfoil whose front and rear included angles are

both 24°. For zero angle of attack, compute the drag coefficient obtained using shock-expansion theory and compare with Ackeret theory.

P9.150 A flat-plate airfoil with $C = 1.2$ m is to have a lift of 30 kN/m when flying at 5000-m standard altitude with $U_\infty = 641$ m/s. Using Ackeret theory, estimate (a) the angle of attack and (b) the drag force in N/m.

P9.151 Air flows at Ma = 2.5 past a half-wedge airfoil whose angles are 4°, as in Fig. P9.151. Compute the lift and drag coefficient at α equal to (a) 0° and (b) 6°.

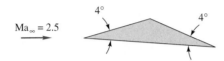

P9.151

P9.152 A supersonic airfoil has a parabolic symmetric shape for upper and lower surfaces

$$y_{u,l} = \pm 2t\left(\frac{x}{C} - \frac{x^2}{C^2}\right)$$

such that the maximum thickness is t at $x = \frac{1}{2}C$. Compute the drag coefficient at zero incidence by Ackeret theory, and compare with a symmetric double wedge of the same thickness.

P9.153 A supersonic transport has a mass of 65 Mg and cruises at 11-km standard altitude at a Mach number of 2.25. If the angle of attack is 2° and its wings can be approximated by flat plates, estimate (a) the required wing area in m² and (b) the thrust required in N.

P9.154 A symmetric supersonic airfoil has its upper and lower surfaces defined by a sine-wave shape:

$$y = \frac{t}{2}\sin\frac{\pi x}{C}$$

where t is the maximum thickness, which occurs at $x = C/2$. Use Ackeret theory to derive an expression for the drag coefficient at zero angle of attack. Compare your result with Ackeret theory for a symmetric double-wedge airfoil of the same thickness.

*P9.155 The F-35 airplane in Fig. 9.29 has a wingspan of 10 m and a wing area of 41.8 m². It cruises at about 10 km altitude with a gross weight of about 200 kN. At that altitude, the engine develops a thrust of about 50 kN. Assume the wing has a symmetric diamond airfoil with a thickness of 8 percent, and accounts for all lift and drag. Estimate the cruise Mach number of the airplane. For extra credit, explain why there are *two* solutions.

P9.156 A thin circular-arc airfoil is shown in Fig. P9.156. The leading edge is parallel to the free stream. Using linearized (small-turning-angle) supersonic flow theory, derive a formula for the lift and drag coefficient for this orientation, and compare with Ackeret-theory results for an angle of attack $\alpha = \tan^{-1}(h/L)$.

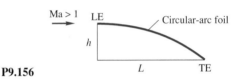

P9.156

P9.157 The Ackeret airfoil theory of Eq. (9.104) is meant for *moderate* supersonic speeds, $1.2 < \text{Ma} < 4$. How does it fare for *hypersonic* speeds? To illustrate, calculate (a) C_L and (b) C_D for a flat-plate airfoil at $a = 5°$ and $\text{Ma}_\infty = 8.0$, using shock-expansion theory, and compare with Ackeret theory. Comment.

Word Problems

W9.1 Notice from Table 9.1 that (a) water and mercury and (b) aluminum and steel have nearly the same speeds of sound, yet the second of each pair of materials is much denser. Can you account for this oddity? Can molecular theory explain it?

W9.2 When an object approaches you at Ma = 0.8, you can hear it, according to Fig. 9.18a. But would there be a Doppler shift? For example, would a musical tone seem to you to have a higher or a lower pitch?

W9.3 The subject of this chapter is commonly called *gas dynamics*. But can liquids not perform in this manner?

Using water as an example, make a rule-of-thumb estimate of the pressure level needed to drive a water flow at velocities comparable to the sound speed.

W9.4 Suppose a gas is driven at compressible subsonic speeds by a large pressure drop, p_1 to p_2. Describe its behavior on an appropriately labeled Mollier chart for (a) frictionless flow in a converging nozzle and (b) flow with friction in a long duct.

W9.5 Describe physically what the "speed of sound" represents. What kind of pressure changes occur in air sound waves during ordinary conversation?

W9.6 Give a physical description of the phenomenon of choking in a converging-nozzle gas flow. Could choking happen even if wall friction were not negligible?

W9.7 Shock waves are treated as discontinuities here, but they actually have a very small finite thickness. After giving it some thought, sketch your idea of the distribution of gas velocity, pressure, temperature, and entropy through the inside of a shock wave.

W9.8 Describe how an observer, running along a normal shock wave at finite speed V, will see what appears to be an oblique shock wave. Is there any limit to the running speed?

Fundamentals of Engineering Exam Problems

One-dimensional compressible flow problems have become quite popular on the FE Exam, especially in the afternoon sessions. In the following problems, assume one-dimensional flow of ideal air, $R = 287$ J/(kg·K) and $k = 1.4$.

FE9.1 For steady isentropic flow, if the absolute temperature increases 50 percent, by what ratio does the static pressure increase?
(a) 1.12, (b) 1.22, (c) 2.25, (d) 2.76, (e) 4.13

FE9.2 For steady isentropic flow, if the density doubles, by what ratio does the static pressure increase?
(a) 1.22, (b) 1.32, (c) 1.44, (d) 2.64, (e) 5.66

FE9.3 A large tank, at 500 K and 200 kPa, supplies isentropic airflow to a nozzle. At section 1, the pressure is only 120 kPa. What is the Mach number at this section?
(a) 0.63, (b) 0.78, (c) 0.89, (d) 1.00, (e) 1.83

FE9.4 In Prob. FE9.3 what is the temperature at section 1?
(a) 300 K, (b) 408 K, (c) 417 K, (d) 432 K, (e) 500 K

FE9.5 In Prob. FE9.3, if the area at section 1 is 0.15 m², what is the mass flow?
(a) 38.1 kg/s, (b) 53.6 kg/s, (c) 57.8 kg/s, (d) 67.8 kg/s, (e) 77.2 kg/s

FE9.6 For steady isentropic flow, what is the maximum possible mass flow through the duct in Fig. FE9.6?
(a) 9.5 kg/s, (b) 15.1 kg/s, (c) 26.2 kg/s, (d) 30.3 kg/s, (e) 52.4 kg/s

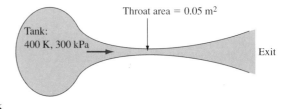

FE9.6

FE9.7 If the exit Mach number in Fig. FE9.6 is 2.2, what is the exit area?
(a) 0.10 m², (b) 0.12 m², (c) 0.15 m², (d) 0.18 m², (e) 0.22 m²

FE9.8 If there are no shock waves and the pressure at one duct section in Fig. FE9.6 is 55.5 kPa, what is the velocity at that section?
(a) 166 m/s, (b) 232 m/s, (c) 554 m/s, (d) 706 m/s, (e) 774 m/s

FE9.9 If, in Fig. FE9.6, there is a normal shock wave at a section where the area is 0.07 m², what is the air density just upstream of that shock?
(a) 0.48 kg/m³, (b) 0.78 kg/m³, (c) 1.35 kg/m³, (d) 1.61 kg/m³, (e) 2.61 kg/m³

FE9.10 In Prob. FE9.9, what is the Mach number just downstream of the shock wave?
(a) 0.42, (b) 0.55, (c) 0.63, (d) 1.00, (e) 1.76

Comprehensive Problems

C9.1 The converging–diverging nozzle sketched in Fig. C9.1 is designed to have a Mach number of 2.00 at the exit plane (assuming the flow remains nearly isentropic). The flow travels from tank a to tank b, where tank a is much larger than tank b. (a) Find the area at the exit A_e and the back pressure p_b that will allow the system to operate at design conditions. (b) As time goes on, the back pressure will grow, since the second tank slowly fills up with more air. Since tank a is huge, the flow in the nozzle will remain the same, however, until a normal shock wave appears at the exit plane. At what back pressure will this occur? (c) If tank b is held at constant temperature, $T = 20°C$, estimate how long it will take for the flow to go from design conditions to the condition of part (b)—that is, with a shock wave at the exit plane.

C9.2 Two large air tanks, one at 400 K and 300 kPa and the other at 300 K and 100 kPa, are connected by a straight tube 6 m long and 5 cm in diameter. The average friction factor is 0.0225. Assuming adiabatic flow, estimate the mass flow through the tube.

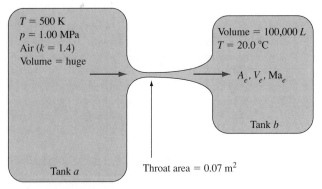

C9.1

*C9.3   Figure C9.3 shows the exit of a converging–diverging nozzle, where an oblique shock pattern is formed. In the exit plane, which has an area of 15 cm², the air pressure is 16 kPa and the temperature is 250 K. Just outside the exit shock, which makes an angle of 50° with the exit plane, the temperature is 430 K. Estimate (*a*) the mass flow, (*b*) the throat area, (*c*) the turning angle of the exit flow, and, in the tank supplying the air, (*d*) the pressure and (*e*) the temperature.

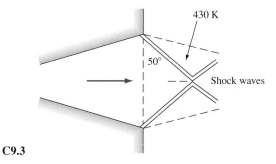

C9.3

C9.4   The properties of a dense gas (high pressure and low temperature) are often approximated by van der Waals's equation of state [17, 18]:

$$p = \frac{\rho RT}{1 - b_1 \rho} - a_1 \rho^2$$

where constants a_1 and b_1 can be found from the critical temperature and pressure

$$a_1 = \frac{27 R^2 T_c^2}{64 p_c} = 9.0 \times 10^5 \ \text{lbf} \cdot \text{ft}^4/\text{slug}^2$$

for air, and

$$b_1 = \frac{R T_c}{8 p_c} = 0.65 \ \text{ft}^3/\text{slug}$$

for air. Find an analytic expression for the speed of sound of a van der Waals gas. Assuming $k = 1.4$, compute the speed of sound of air in ft/s at $-100°F$ and 20 atm for (*a*) a perfect gas and (*b*) a van der Waals gas. What percentage higher density does the van der Waals relation predict?

C9.5   Consider one-dimensional steady flow of a nonideal gas, steam, in a converging nozzle. Stagnation conditions are $p_0 = 100$ kPa and $T_0 = 200°C$. The nozzle exit diameter is 2 cm. If the nozzle exit pressure is 70 kPa, calculate the mass flow and the exit temperature for real steam, either from the steam tables or using EES. (As a first estimate, assume steam to be an ideal gas from Table A.4.) Is the flow choked? Why is EES unable to estimate the exit Mach number? (*b*) Find the nozzle exit pressure and mass flow for which the steam flow *is* choked, using EES or the steam tables.

C9.6   Extend Prob. C9.5 as follows. Let the nozzle be converging–diverging, with an exit diameter of 3 cm. Assume isentropic flow. (*a*) Find the exit Mach number, pressure, and temperature for an ideal gas from Table A.4. Does the mass flow agree with the value of 0.0452 kg/s in Prob. C9.5? (*b*) Investigate, briefly, the use of EES for this problem and explain why part (*a*) is unrealistic and poor convergence of EES is obtained. [*Hint*: Study the pressure and temperature state predicted by part (*a*).]

C9.7   Professor Gordon Holloway and his student, Jason Bettle, of the University of New Brunswick obtained the following tabulated data for blow-down air flow through a converging–diverging nozzle similar in shape to Fig. P3.22. The supply tank pressure and temperature were 29 psig and 74°F, respectively. Atmospheric pressure was 14.7 psia. Wall pressures and centerline stagnation pressures were measured in the expansion section, which was a frustrum of a cone. The nozzle throat is at $x = 0$.

x(cm)	0	1.5	3	4.5	6	7.5	9
Diameter (cm)	1.00	1.098	1.195	1.293	1.390	1.488	1.585
p_{wall} (psig)	7.7	−2.6	−4.9	−7.3	−6.5	−10.4	−7.4
$p_{stagnation}$ (psig)	29	26.5	22.5	18	16.5	14	10

Use the stagnation pressure data to estimate the local Mach number. Compare the measured Mach numbers and wall pressures with the predictions of one-dimensional theory. For $x > 9$ cm, the stagnation pressure data was not thought by Holloway and Bettle to be a valid measure of Mach number. What is the probable reason?

Design Projects

D9.1 It is desired to select a rectangular wing for a fighter aircraft. The plane must be able (*a*) to take off and land on a 4500-ft-long sea-level runway and (*b*) to cruise supersonically at Ma = 2.3 at 28,000-ft altitude. For simplicity, assume a wing with zero sweepback. Let the aircraft maximum weight equal $(30 + n)(1000)$ lbf, where n is the number of letters in your surname. Let the available sea-level maximum thrust be one-third of the maximum weight, decreasing at altitude proportional to ambient density. Making suitable assumptions about the effect of finite aspect ratio on wing lift and drag for both subsonic and supersonic flight, select a wing of minimum area sufficient to perform these takeoff/landing and cruise requirements. Some thought should be given to analyzing the wingtips and wing roots in supersonic flight, where Mach cones form and the flow is not two-dimensional. If no satisfactory solution is possible, gradually increase the available thrust to converge to an acceptable design.

D9.2 Consider supersonic flow of air at sea-level conditions past a wedge of half-angle θ, as shown in Fig. D9.2. Assume that the pressure on the back of the wedge equals the fluid pressure as it exits the Prandtl-Meyer fan. (*a*) Suppose $Ma_\infty = 3.0$. For what angle θ will the supersonic wave drag coefficient C_D, based on frontal area, be exactly 0.5? (*b*) Suppose that $\theta = 20°$. Is there a free-stream Mach number for which the wave drag coefficient C_D, based on frontal area, will be exactly 0.5? (*c*) Investigate the percentage increase in C_D from (*a*) and (*b*) due to including boundary layer friction drag in the calculation.

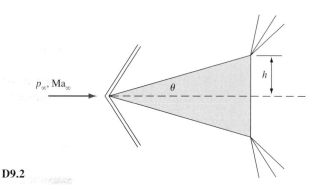

D9.2

References

1. J. E. A. John and T. G. Keith, *Gas Dynamics,* 3d ed., Pearson Education, Upper Saddle River, NJ, 2005.

2. A. J. Chapman and W. F. Walker, *Introductory Gas Dynamics,* Holt, New York, 1971.

3. R. D. Zucker and O. Biblarz, 2d ed., *Fundamentals of Gas Dynamics,* Wiley, New York, 2002.

4. J. D. Anderson, *Modern Compressible Flow: with Historical Perspective,* 3d ed., McGraw-Hill, New York, 2002.

5. A. H. Shapiro, *The Dynamics and Thermodynamics of Compressible Fluid Flow,* 2 vols., Wiley, New York, 1953.

6. R. Courant and K. O. Friedrichs, *Supersonic Flow and Shock Waves,* Interscience, New York, 1948; reprinted by Springer-Verlag, New York, 1992.

7. H. W. Liepmann and A. Roshko, *Elements of Gas Dynamics,* Dover, New York, 2001.

8. I. Straskraba, *Introduction to the Mathematical Theory of Compressible Flow,* Oxford University Press, New York, 2004.

9. P. A. Thompson, *Compressible Fluid Dynamics,* McGraw-Hill, New York, 1972.

10. P. H. Oosthuizen and W. E. Carscallen, *Compressible Fluid Flow,* McGraw-Hill, New York, 2003.

11. M. L. Rasmussen, *Hypersonic Flow,* Wiley, New York, 1994.

12. G. A. Bird, *Molecular Gas Dynamics and the Direct Simulation of Gas Flows,* Clarendon Press, Oxford, 1994.

13. C. B. Laney, *Computational Gas Dynamics,* Cambridge Univ, Press, New York, 1998.

14. L. W. Reithmaier, *Mach 1 and Beyond: The Illustrated Guide to High-Speed Flight,* McGraw-Hill, 1994.

15. W. T. Parry, *ASME International Steam Tables for Industrial Use,* ASME, New York, 2000.

16. J. H. Keenan et al., *Gas Tables: International Version,* Krieger Publishing, Melbourne, FL, 1992.

17. Y. A. Cengel and M. A. Boles, *Thermodynamics: An Engineering Approach,* 5th ed., McGraw-Hill, New York, 2005.

18. K. Wark, *Thermodynamics,* 6th ed., McGraw-Hill, New York, 1999.

19. F. M. White, *Viscous Fluid Flow,* 3d ed., McGraw-Hill, New York, 2005.

20. J. Palmer, K. Ramsden, and E. Goodger, *Compressible Flow Tables for Engineers: With Appropriate Computer Programs,* Scholium Intl., Port Washington, NY, 1989.

21. M. R. Lindeburg, *Consolidated Gas Dynamics Tables,* Professional Publications, Inc., Belmont, CA, 1994.

22. A. M. Shektman, *Gasdynamic Functions of Real Gases,* Taylor and Francis, New York, 1991.

23. J. H. Keenan and E. P. Neumann, "Measurements of Friction in a Pipe for Subsonic and Supersonic Flow of Air," *Journal of Applied Mechanics,* vol. 13, no. 2, 1946, p. A-91.

24. R. P. Benedict, *Fundamentals of Pipe Flow,* John Wiley, New York, 1980.

25. J. L. Sims, *Tables for Supersonic Flow around Right Circular Cones at Zero Angle of Attack,* NASA SP-3004, 1964 (see also NASA SP-3007).

26. J. L. Thomas, "Reynolds Number Effects on Supersonic Asymmetrical Flows over a Cone," *Journal of Aircraft,* vol. 30, no. 4, 1993, pp. 488–495.

27. W. Bober and R. A. Kenyon, *Fluid Mechanics,* Wiley, New York, 1980.

28. J. Ackeret, "Air Forces on Airfoils Moving Faster than Sound Velocity," *NACA Tech. Memo.* 317, 1925.

29. W. B. Brower, *Theory, Tables and Data for Compressible Flow,* Taylor & Francis, New York, 1990.

A giant water release from the Xiaolangdi reservoir in central China's Henan province. The release washes silt from the riverbed and provides water to drought areas downstream. This spectacular open channel flow can be analyzed by the methods of the present chapter, although it is somewhat beyond the imagination of this writer. (*Courtesy of the Associated Press*).

Chapter 10
Open-Channel Flow

Motivation. An *open-channel flow* denotes a flow with a free surface touching an atmosphere, like a river or a canal or a flume. Closed-duct flows (Chap. 6) are full of fluid, either liquid or gas, have no free surface within, and are driven by a pressure gradient along the duct axis. The open-channel flows here are driven by gravity alone, and the pressure gradient at the atmospheric interface is negligible. The basic force balance in an open channel is between gravity and friction.

Open-channel flows are an especially important mode of fluid mechanics for civil and environmental engineers. One needs to predict the flow rates and water depths that result from a given channel geometry, whether natural or artificial, and a given wet-surface roughness. Water is almost always the relevant fluid, and the channel size is usually large. Thus open-channel flows are generally turbulent, three-dimensional, sometimes unsteady, and often quite complex. This chapter presents some simple engineering theories and experimental correlations for steady flow in straight channels of regular geometry. We can borrow and use some concepts from duct flow analysis: hydraulic radius, friction factor, and head losses.

10.1 Introduction

Simply stated, open-channel flow is the flow of a liquid in a conduit with a free surface. There are many practical examples, both artificial (flumes, spillways, canals, weirs, drainage ditches, culverts) and natural (streams, rivers, estuaries, floodplains). This chapter introduces the elementary analysis of such flows, which are dominated by the effects of gravity.

The presence of the free surface, which is essentially at atmospheric pressure, both helps and hurts the analysis. It helps because the pressure can be taken as constant along the free surface, which therefore is equivalent to the *hydraulic grade line* (HGL) of the flow. Unlike flow in closed ducts, the pressure gradient is not a direct factor in open-channel flow, where the balance of forces is confined to gravity and friction.[1] But the free surface complicates the analysis because its shape is a priori unknown:

---

[1]Surface tension is rarely important because open channels are normally quite large and have a very large Weber number. Surface tension affects small models of large channels.

The depth profile changes with conditions and must be computed as part of the problem, especially in unsteady problems involving wave motion.

Before proceeding, we remark, as usual, that whole books have been written on open-channel hydraulics [1 to 7, 32]. There are also specialized texts devoted to wave motion [8 to 10] and to engineering aspects of coastal free-surface flows [11 to 13]. This chapter is only an introduction to broader and more detailed treatments. The writer recommends, as an occasional break from free-surface flow analysis, Ref. 31, which is an enchanting and spectacular gallery of ocean wave photographs.

The One-Dimensional Approximation

An open channel always has two sides and a bottom, where the flow satisfies the no-slip condition. Therefore even a straight channel has a three-dimensional velocity distribution. Some measurements of straight-channel velocity contours are shown in Fig. 10.1. The profiles are quite complex, with maximum velocity typically occurring in the midplane about 20 percent below the surface. In very broad shallow channels the maximum velocity is near the surface, and the velocity profile is nearly logarithmic from the bottom to the free surface, as in Eq. (6.65). In noncircular channels there are also secondary motions similar to Fig. 6.16 for closed-duct flows. If the channel curves or meanders, the secondary motion intensifies due to centrifugal effects, with high velocity occurring near the outer radius of the bend. Curved natural channels are subject to strong bottom erosion and deposition effects.

With the advent of the supercomputer, it is possible to make numerical simulations of complex flow patterns such as in Fig. 10.1 [27, 28]. However, the practical engineering approach, used here, is to make a one-dimensional flow approximation, as in Fig. 10.2. Since the liquid density is nearly constant, the steady flow continuity equation reduces to constant-volume flow Q along the channel

$$Q = V(x)A(x) = \text{const} \tag{10.1}$$

where V is average velocity and A the local cross-sectional area, as sketched in Fig. 10.2.

A second one-dimensional relation between velocity and channel geometry is the energy equation, including friction losses. If points 1 (upstream) and 2 (downstream) are on the free surface, $p_1 = p_2 = p_a$, and we have, for steady flow,

$$\frac{V_1^2}{2g} + z_1 = \frac{V_2^2}{2g} + z_2 + h_f \tag{10.2}$$

where z denotes the total elevation of the free surface, which includes the water depth y (see Fig. 10.2a) plus the height of the (sloping) bottom. The friction head loss h_f is analogous to head loss in duct flow from Eq. (6.10):

$$h_f \approx f \frac{x_2 - x_1}{D_h} \frac{V_{av}^2}{2g} \qquad D_h = \text{hydraulic diameter} = \frac{4A}{P} \tag{10.3}$$

where f is the average friction factor (Fig. 6.13) between sections 1 and 2. Since channels are irregular in shape, their "size" is taken to be the hydraulic *radius*:

$$\boxed{R_h = \frac{1}{4}D_h = \frac{A}{P}} \tag{10.4}$$

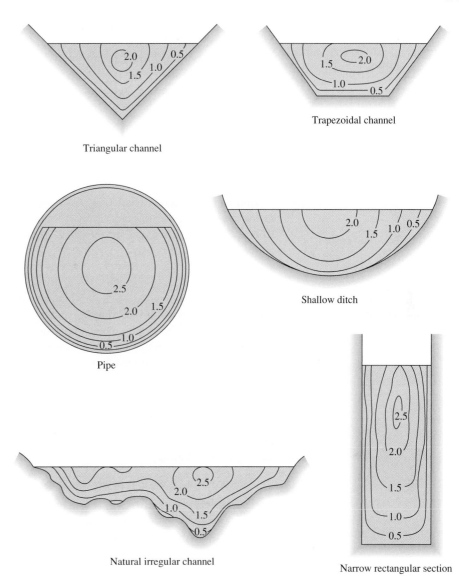

Fig. 10.1 Measured isovelocity contours in typical straight open-channel flows. *(From Ref. 2.)*

Triangular channel

Trapezoidal channel

Pipe

Shallow ditch

Natural irregular channel

Narrow rectangular section

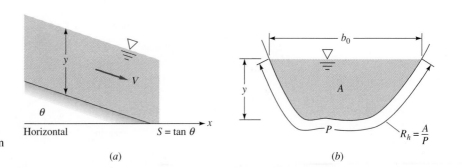

Fig. 10.2 Geometry and notation for open-channel flow: (*a*) side view; (*b*) cross section. All these parameters are constant in uniform flow.

$S = \tan \theta$

$R_h = \dfrac{A}{P}$

(*a*)

(*b*)

The local Reynolds number of the channel would be Re $= VR_h/\nu$, which is usually highly turbulent (>1 E5). The only commonly occurring laminar channel flows are the thin sheets that form as rainwater drains from crowned streets and airport runways.

The wetted perimeter P (see Fig. 10.2b) includes the sides and bottom of the channel but not the free surface and, of course, not the parts of the sides above the water level. For example, if a rectangular channel is b wide and h high and contains water to depth y, its wetted perimeter is

$$P = b + 2y$$

not $2b + 2h$.

Although the Moody chart (Fig. 6.13) would give a good estimate of the friction factor in channel flow, in practice it is seldom used. An alternative correlation due to Robert Manning, discussed in Sec. 10.2, is the formula of choice in open-channel hydraulics.

Flow Classification by Depth Variation

The most common method of classifying open-channel flows is by the rate of change of the free-surface depth. The simplest and most widely analyzed case is *uniform flow,* where the depth (hence the velocity in steady flow) remains constant. Uniform flow conditions are approximated by long straight runs of constant-slope and constant-area channel. A channel in uniform flow is said to be moving at its *normal depth* y_n, which is an important design parameter.

If the channel slope or cross section changes or there is an obstruction in the flow, then the depth changes and the flow is said to be *varied.* The flow is *gradually varying* if the one-dimensional approximation is valid and *rapidly varying* if not. Some examples of this method of classification are shown in Fig. 10.3. The classes can be summarized as follows:

1. Uniform flow (constant depth and slope).
2. Varied flow:
    a.   Gradually varied (one-dimensional).
    b.   Rapidly varied (multidimensional).

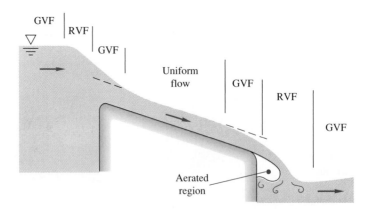

Fig. 10.3 Open-channel flow classified by regions of rapidly varying flow (RVF), gradually varying flow (GVF), and uniform flow depth profiles.

Typically uniform flow is separated from rapidly varying flow by a region of gradually varied flow. Gradually varied flow can be analyzed by a first-order differential equation (Sec. 10.6), but rapidly varying flow usually requires experimentation or three-dimensional computational fluid dynamics [14].

Flow Classification by Froude Number

A second and very useful classification of open-channel flow is by the dimensionless Froude number, Fr, which is the ratio of channel velocity to the speed of propagation of a small-disturbance wave in the channel. For a rectangular or very wide constant-depth channel, this takes the form

$$\text{Fr} = \frac{\text{flow velocity}}{\text{surface wave speed}} = \frac{V}{\sqrt{gy}} \tag{10.5}$$

where y is the water depth. The flow behaves differently depending on these three flow regimes:

$$\begin{aligned}
\text{Fr} &< 1.0 &\quad &\text{subcritical flow} \\
\text{Fr} &= 1.0 &\quad &\text{critical flow} \\
\text{Fr} &> 1.0 &\quad &\text{supercritical flow}
\end{aligned} \tag{10.6}$$

The Froude number for irregular channels is defined in Sec. 10.4. As mentioned in Sec. 9.10, there is a strong analogy here with the three compressible flow regimes of the Mach number: subsonic (Ma < 1), sonic (Ma = 1), and supersonic (Ma > 1). We shall pursue the analogy in Sec. 10.4.

Surface Wave Speed

The Froude number denominator $(gy)^{1/2}$ is the speed of an infinitesimal shallow-water surface wave. We can derive this with reference to Fig. 10.4a, which shows a wave of height δy propagating at speed c into still liquid. To achieve a steady flow inertial frame of reference, we fix the coordinates on the wave as in Fig. 10.4b, so that the still water moves to the right at velocity c. Figure 10.4 is exactly analogous to Fig. 9.1, which analyzed the speed of sound in a fluid.

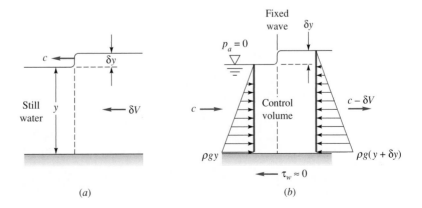

Fig. 10.4 Analysis of a small surface wave propagating into still shallow water; (a) moving wave, nonsteady frame; (b) fixed wave, inertial frame of reference.

For the control volume of Fig. 10.4b, the one-dimensional continuity relation is, for channel width b,

$$\rho c y b = \rho(c - \delta V)(y + \delta y)b$$

or

$$\delta V = c\frac{\delta y}{y + \delta y} \qquad (10.7)$$

This is analogous to Eq. (9.10); the velocity change δV induced by a surface wave is small if the wave is "weak," $\delta y \ll y$. If we neglect bottom friction in the short distance across the wave in Fig. 10.4b, the momentum relation is a balance between the net hydrostatic pressure force and momentum:

$$-\tfrac{1}{2}\rho g b[(y + \delta y)^2 - y^2] = \rho c b y(c - \delta V - c)$$

or

$$g\left(1 + \frac{\tfrac{1}{2}\delta y}{y}\right)\delta y = c\,\delta V \qquad (10.8)$$

This is analogous to Eq. (9.12). By eliminating δV between Eqs. (10.7) and (10.8) we obtain the desired expression for wave propagation speed:

$$c^2 = gy\left(1 + \frac{\delta y}{y}\right)\left(1 + \frac{\tfrac{1}{2}\delta y}{y}\right) \qquad (10.9)$$

The "stronger" the wave height δy, the faster the wave speed c, by analogy with Eq. (9.13). In the limit of an infinitesimal wave height $\delta y \to 0$, the speed becomes

$$\boxed{c_0^2 = gy} \qquad (10.10)$$

This is the surface-wave equivalent of fluid sound speed a, and thus the Froude number in channel flow $\mathrm{Fr} = V/c_0$ is the analog of the Mach number. For $y = 1$ m, $c_o = 3.1$ m/s.

As in gas dynamics, a channel flow can accelerate from subcritical to critical to supercritical flow and then return to subcritical flow through a sort of normal shock called a *hydraulic jump* (Sec. 10.5). This is illustrated in Fig. 10.5. The flow upstream of the sluice gate is subcritical. It then accelerates to critical and supercritical flow as it passes under the gate, which serves as a sort of "nozzle." Further downstream the

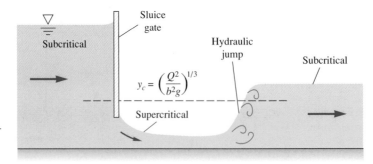

Fig. 10.5 Flow under a sluice gate accelerates from subcritical to critical to supercritical flow and then jumps back to subcritical flow.

flow "shocks" back to subcritical flow because the downstream "receiver" height is too high to maintain supercritical flow. Note the similarity with the nozzle gas flows of Fig. 9.12.

The critical depth $y_c = [Q^2/(b^2 g)]^{1/3}$ is sketched as a dashed line in Fig. 10.5 for reference. Like the normal depth y_n, y_c is an important parameter in characterizing open-channel flow (see Sec. 10.4).

An excellent discussion of the various regimes of open-channel flow is given in Ref. 15.

10.2 Uniform Flow; The Chézy Formula

Uniform flow can occur in long straight runs of constant slope and constant channel cross section. The water depth is constant at $y = y_n$, and the velocity is constant at $V = V_0$. Let the slope be $S_0 = \tan \theta$, where θ is the angle the bottom makes with the horizontal, considered positive for downhill flow. Then Eq. (10.2), with $V_1 = V_2 = V_0$, becomes

$$h_f = z_1 - z_2 = S_0 L \tag{10.11}$$

where L is the horizontal distance between sections 1 and 2. The head loss thus balances the loss in height of the channel. The flow is essentially fully developed, so that the Darcy-Weisbach relation, Eq. (6.10), holds

$$h_f = f \frac{L}{D_h} \frac{V_0^2}{2g} \qquad D_h = 4R_h \tag{10.12}$$

with $D_h = 4A/P$ used to accommodate noncircular channels. The geometry and notation for open-channel flow analysis are shown in Fig. 10.2.

By combining Eqs. (10.11) and (10.12) we obtain an expression for flow velocity in uniform channel flow:

$$V_0 = \left(\frac{8g}{f}\right)^{1/2} R_h^{1/2} S_0^{1/2} \tag{10.13}$$

For a given channel shape and bottom roughness, the quantity $(8g/f)^{1/2}$ is constant and can be denoted by C. Equation (10.13) becomes

$$V_0 = C(R_h S_0)^{1/2} \qquad Q = CA(R_h S_0)^{1/2} \tag{10.14}$$

These are called the *Chézy formulas*, first developed by the French engineer Antoine Chézy in conjunction with his experiments on the Seine River and the Courpalet Canal in 1769. The quantity C, called the *Chézy coefficient*, varies from about 60 ft$^{1/2}$/s for small rough channels to 160 ft$^{1/2}$/s for large smooth channels (30 to 90 m$^{1/2}$/s in SI units).

Over the past century a great deal of hydraulics research [16] has been devoted to the correlation of the Chézy coefficient with the roughness, shape, and slope of various open channels. Correlations are due to Ganguillet and Kutter in 1869, Manning in 1889, Bazin in 1897, and Powell in 1950 [16]. All these formulations are discussed in delicious detail in Ref. 2, Chap. 5. Here we confine our treatment to Manning's correlation, the most popular.

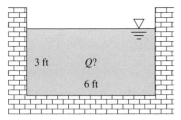

E10.1

EXAMPLE 10.1

A straight rectangular channel is 6 ft wide and 3 ft deep and laid on a slope of 2°. The friction factor is 0.022. Estimate the uniform flow rate in cubic feet per second.

Solution

- *System sketch:* The channel cross section is shown in Fig. E10.1.
- *Assumptions:* Steady, uniform channel flow with $\theta = 2°$.
- *Approach:* Evaluate the Chézy formula, Eq. (10.13) or (10.14).
- *Property values:* Please note that there are *no* fluid physical properties involved in the Chézy formula. Can you explain this?
- *Solution step:* Simply evaluate each term in the Chézy formula, Eq. (10.13):

$$C = \sqrt{\frac{8g}{f}} = \sqrt{\frac{8(32.2 \text{ ft/s}^2)}{0.022}} = 108 \frac{\text{ft}^{1/2}}{\text{s}} \qquad A = by = (6 \text{ ft})(3 \text{ ft}) = 18 \text{ ft}^2$$

$$R_h = \frac{A}{P_{\text{wet}}} = \frac{18 \text{ ft}^2}{(3 + 6 + 3 \text{ ft})} = 1.5 \text{ ft} \qquad S_0 = \tan(\theta) = \tan(2°)$$

Then $Q = CAR_h^{1/2}S_0^{1/2} = \left(108 \frac{\text{ft}^{1/2}}{\text{s}}\right)(18 \text{ ft}^2)(1.5 \text{ ft})^{1/2}(\tan 2°)^{1/2} \approx 450 \text{ ft}^3/\text{s}$ *Ans.*

- *Comments:* Uniform flow estimates are straightforward if the geometry is simple. Results are independent of water density and viscosity because the flow is fully rough and driven by gravity. Note the high flow rate, larger than some rivers. Two degrees is a substantial channel slope.

The Manning Roughness Correlation

The most fundamentally sound approach to the Chézy formula is to use Eq. (10.13) with f estimated from the Moody friction factor chart, Fig. 6.13. Indeed, the open-channel research establishment [18] strongly recommends use of the friction factor in all calculations. Since typical channels are large and rough, we would generally use the fully rough turbulent flow limit of Eq. (6.48)

$$f \approx \left(2.0 \log \frac{14.8R_h}{\epsilon}\right)^{-2} \tag{10.15}$$

where ϵ is the roughness height, with typical values listed in Table 10.1.

In spite of the attractiveness of this friction factor approach, most engineers prefer to use a simple (dimensional) correlation published in 1891 by Robert Manning [17], an Irish engineer. In tests with real channels, Manning found that the Chézy coefficient C increased approximately as the sixth root of the channel size. He proposed the simple formula

$$C = \left(\frac{8g}{f}\right)^{1/2} \approx \alpha \frac{R_h^{1/6}}{n} \tag{10.16}$$

where n is a roughness parameter. Since the formula is clearly not dimensionally consistent, it requires a conversion factor α that changes with the system of units used:

$$\alpha = 1.0 \quad \text{SI units} \qquad \alpha = 1.486 \quad \text{BG units} \tag{10.17}$$

Table 10.1 Experimental Values of Manning's n Factor*

	n	Average roughness height ϵ	
		ft	mm
Artificial lined channels:			
Glass	0.010 ± 0.002	0.0011	0.3
Brass	0.011 ± 0.002	0.0019	0.6
Steel, smooth	0.012 ± 0.002	0.0032	1.0
Painted	0.014 ± 0.003	0.0080	2.4
Riveted	0.015 ± 0.002	0.012	3.7
Cast iron	0.013 ± 0.003	0.0051	1.6
Concrete, finished	0.012 ± 0.002	0.0032	1.0
Unfinished	0.014 ± 0.002	0.0080	2.4
Planed wood	0.012 ± 0.002	0.0032	1.0
Clay tile	0.014 ± 0.003	0.0080	2.4
Brickwork	0.015 ± 0.002	0.012	3.7
Asphalt	0.016 ± 0.003	0.018	5.4
Corrugated metal	0.022 ± 0.005	0.12	37
Rubble masonry	0.025 ± 0.005	0.26	80
Excavated earth channels:			
Clean	0.022 ± 0.004	0.12	37
Gravelly	0.025 ± 0.005	0.26	80
Weedy	0.030 ± 0.005	0.8	240
Stony, cobbles	0.035 ± 0.010	1.5	500
Natural channels:			
Clean and straight	0.030 ± 0.005	0.8	240
Sluggish, deep pools	0.040 ± 0.010	3	900
Major rivers	0.035 ± 0.010	1.5	500
Floodplains:			
Pasture, farmland	0.035 ± 0.010	1.5	500
Light brush	0.05 ± 0.02	6	2000
Heavy brush	0.075 ± 0.025	15	5000
Trees	0.15 ± 0.05	?	?

*A more complete list is given in Ref. 2, pp. 110–113.

Recall that we warned about this awkwardness in Example 1.4. You may verify that α is the cube root of the conversion factor between the meter and your chosen length scale: In BG units, $\alpha = (3.2808 \text{ ft/m})^{1/3} = 1.486.$[2]

The Manning formula for uniform flow velocity is thus

$$
V_0 \text{ (m/s)} \approx \frac{1.0}{n} [R_h \text{ (m)}]^{2/3} S_0^{1/2}
$$
$$
V_0 \text{ (ft/s)} \approx \frac{1.486}{n} [R_h \text{ (ft)}]^{2/3} S_0^{1/2}
$$

(10.18)

[2]An interesting discussion of the history and "dimensionality" of Manning's formula is given in Ref. 2, pp. 98–99.

The channel slope S_0 is dimensionless, and n is taken to be the same in both systems. The volume flow rate simply multiplies this result by the area:

Uniform flow:

$$Q = V_0 A \approx \frac{\alpha}{n} A R_h^{2/3} S_0^{1/2} \qquad (10.19)$$

Experimental values of n (and the corresponding roughness height) are listed in Table 10.1 for various channel surfaces. There is a factor-of-15 variation from a smooth glass surface ($n \approx 0.01$) to a tree-lined floodplain ($n \approx 0.15$). Due to the irregularity of typical channel shapes and roughness, the scatter bands in Table 10.1 should be taken seriously. For routine calculations, always use the average roughness in Table 10.1.

Since Manning's sixth-root size variation is not exact, real channels can have a variable n depending on the water depth. The Mississippi River near Memphis, Tennessee, has $n \approx 0.032$ at 40-ft flood depth, 0.030 at normal 20-ft depth, and 0.040 at 5-ft low-stage depth. Seasonal vegetative growth and factors such as bottom erosion can also affect the value of n. Even nearly identical man-made channels can vary. Brater et al. [19] report that U.S. Bureau of Reclamation tests, on large concrete-lined canals, yielded values of n ranging from 0.012 to 0.017.

EXAMPLE 10.2

Engineers find that the most efficient rectangular channel (maximum uniform flow for a given area) flows at a depth equal to half the bottom width. Consider a rectangular brickwork channel laid on a slope of 0.006. What is the best bottom width for a flow rate of 100 ft³/s?

Solution

- *Assumptions:* Uniform flow in a straight channel of constant of slope $S = 0.006$.
- *Approach:* Use the Manning formula in English units, Eq. (10.19), to predict the flow rate.
- *Property values:* For brickwork, from Table 10.1, the roughness factor $n \approx 0.015$.
- *Solution:* For bottom width b, take the water depth to be $y = b/2$. Equation (10.19) becomes

$$A = by = b(b/2) = \frac{b^2}{2} \quad R_h = \frac{A}{P} = \frac{by}{b + 2y} = \frac{b^2/2}{b + 2(b/2)} = \frac{b}{4}$$

$$Q = \frac{\alpha}{n} A R_h^{2/3} S^{1/2} = \frac{1.486}{0.015}\left(\frac{b^2}{2}\right)\left(\frac{b}{4}\right)^{2/3}(0.006)^{1/2} = 100 \; \frac{\text{ft}^3}{\text{s}}$$

Clean this up: $b^{8/3} = 65.7$ solve for $b \approx 4.8$ ft   *Ans.*

- *Comments:* The Manning approach is simple and effective. The Moody friction factor method, Eq. (10.14), requires laborious iteration and leads to a result $b \approx 4.81$ ft.

Normal Depth Estimates

With water depth y known, the computation of Q is quite straightforward. However, if Q is given, the computation of the normal depth y_n may require iteration. Since the normal depth is a characteristic flow parameter, this is an important type of problem.

EXAMPLE 10.3

The asphalt-lined trapezoidal channel in Fig. E10.3 carries 300 ft^3/s of water under uniform flow conditions when $S = 0.0015$. What is the normal depth y_n?

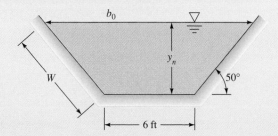

Note: See Fig. 10.7 for generalized trapezoid notation.

E10.3

Solution

From Table 10.1, for asphalt, $n \approx 0.016$. The area and hydraulic radius are functions of y_n, which is unknown:

$$b_0 = 6 \text{ ft} + 2y_n \cot 50° \qquad A = \tfrac{1}{2}(6 + b_0)y_n = 6y_n + y_n^2 \cot 50°$$

$$P = 6 + 2W = 6 + 2y_n \csc 50°$$

From Manning's formula (10.19) with a known $Q = 300$ ft^3/s, we have

$$300 = \frac{1.49}{0.016}(6y_n + y_n^2 \cot 50°)\left(\frac{6y_n + y_n^2 \cot 50°}{6 + 2y_n \csc 50°}\right)^{2/3}(0.0015)^{1/2}$$

or

$$(6y_n + y_n^2 \cot 50°)^{5/3} = 83.2(6 + 2y_n \csc 50°)^{2/3} \qquad (1)$$

One can iterate Eq. (1) laboriously and eventually find $y_n \approx 4.6$ ft. However, it is a perfect candidate for EES. Instead of manipulating and programming the final formula, one might simply evaluate each separate part of the Chézy equation (in English units, with angles in degrees):

```
P = 6 + 2*yn/sin(50)

A = 6*yn + yn^2/tan(50)

Rh = A/P

300 = 1.49/0.016*A*Rh^(2/3)*0.0015^0.5
```

Select Solve from the menu bar, and EES complains of "negative numbers to a power." Go back to Variable Information on the menu bar and make sure that y_n is positive. EES then immediately solves for

$P = 17.95 \qquad A = 45.04 \qquad R_h = 2.509 \qquad y_n = 4.577$ ft   *Ans.*

Generally, EES is ideal for open-channel flow problems where the depth is unknown.

Uniform Flow in a Partly Full Circular Pipe

Consider the partially full pipe of Fig. 10.6a in uniform flow. The maximum velocity and flow rate actually occur before the pipe is completely full. In terms of the pipe radius R and the angle θ up to the free surface, the geometric properties are

$$A = R^2\left(\theta - \frac{\sin 2\theta}{2}\right) \qquad P = 2R\theta \qquad R_h = \frac{R}{2}\left(1 - \frac{\sin 2\theta}{2\theta}\right)$$

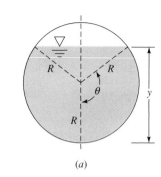

(a)

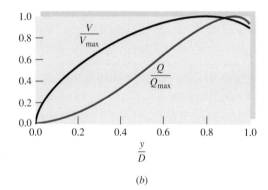

Fig. 10.6 Uniform flow in a partly full circular channel: (a) geometry; (b) velocity and flow rate versus depth.

(b)

The Manning formulas (10.19) predict a uniform flow as follows:

$$V_0 \approx \frac{\alpha}{n}\left[\frac{R}{2}\left(1 - \frac{\sin 2\theta}{2\theta}\right)\right]^{2/3} S_0^{1/2} \qquad Q = V_0 R^2\left(\theta - \frac{\sin 2\theta}{2}\right) \qquad (10.20)$$

For a given n and slope S_0, we may plot these two relations versus y/D in Fig. 10.6b. There are two different maxima, as follows:

$$V_{max} = 0.718\frac{\alpha}{n} R^{2/3} S_0^{1/2} \quad \text{at} \quad \theta = 128.73° \quad \text{and} \quad y = 0.813D$$

$$(10.21)$$

$$Q_{max} = 2.129\frac{\alpha}{n} R^{8/3} S_0^{1/2} \quad \text{at} \quad \theta = 151.21° \quad \text{and} \quad y = 0.938D$$

As shown in Fig. 10.6b, the maximum velocity is 14 percent more than the velocity when running full, and similarly the maximum discharge is 8 percent more. Since real pipes running nearly full tend to have somewhat unstable flow, these differences are not that significant.

10.3 Efficient Uniform-Flow Channels

The engineering design of an open channel has many parameters. If the channel surface can erode or scour, a low-velocity design might be sought. A dirt channel could be planted with grass to minimize erosion. For nonerodible surfaces, construction and lining costs might dominate, suggesting a cross section of minimum wetted perimeter. Nonerodible channels can be designed for maximum flow.

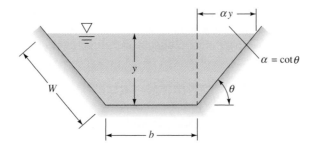

Fig. 10.7 Geometry of a trapezoidal channel section.

The simplicity of Manning's formulation (10.19) enables us to analyze channel flows to determine the most efficient low-resistance sections for given conditions. The most common problem is that of maximizing R_h for a given flow area and discharge. Since $R_h = A/P$, maximizing R_h for given A is the same as minimizing the wetted perimeter P. There is no general solution for arbitrary cross sections, but an analysis of the trapezoid section will show the basic results.

Consider the generalized trapezoid of angle θ in Fig. 10.7. For a given side angle θ, the flow area is

$$A = by + \alpha y^2 \qquad \alpha = \cot\theta \qquad (10.22)$$

The wetted perimeter is

$$P = b + 2W = b + 2y(1 + \alpha^2)^{1/2} \qquad (10.23)$$

Eliminating b between (10.22) and (10.23) gives

$$P = \frac{A}{y} - \alpha y + 2y(1 + \alpha^2)^{1/2} \qquad (10.24)$$

To minimize P, evaluate dP/dy for constant A and α and set equal to zero. The result is

$$A = y^2[2(1 + \alpha^2)^{1/2} - \alpha] \quad P = 4y(1 + \alpha^2)^{1/2} - 2\alpha y \quad R_h = \tfrac{1}{2}y \quad (10.25)$$

The last result is very interesting: For any angle θ, the most efficient cross section for uniform flow occurs when the hydraulic radius is half the depth.

Since a rectangle is a trapezoid with $\alpha = 0$, the most efficient rectangular section is such that

$$A = 2y^2 \qquad P = 4y \qquad R_h = \tfrac{1}{2}y \qquad b = 2y \qquad (10.26)$$

To find the correct depth y, these relations must be solved in conjunction with Manning's flow rate formula (10.19) for the given discharge Q.

Best Trapezoid Angle

Equations (10.25) are valid for any value of α. What is the best value of α for a given depth and area? To answer this question, evaluate $dP/d\alpha$ from Eq. (10.24) with A and y held constant. The result is

$$2\alpha = (1 + \alpha^2)^{1/2} \qquad \alpha = \cot\theta = \frac{1}{3^{1/2}}$$

or

$$\theta = 60° \qquad (10.27)$$

Thus the maximum-flow trapezoid section is half a hexagon.

Similar calculations with a circular channel section running partially full show best efficiency for a semicircle, $y = \frac{1}{2}D$. In fact, the semicircle is the best of all possible channel sections (minimum wetted perimeter for a given flow area). The percentage improvement over, say, half a hexagon is very slight, however.

EXAMPLE 10.4

(a) What are the best dimensions y and b for a rectangular brick channel designed to carry 5 m³/s of water in uniform flow with $S_0 = 0.001$? (b) Compare results with a half-hexagon and semicircle.

Solution

Part (a)

From Eq. (10.26), $A = 2y^2$ and $R_h = \frac{1}{2}y$. Manning's formula (10.19) in SI units gives, with $n \approx 0.015$ from Table 10.1,

$$Q = \frac{1.0}{n} AR_h^{2/3}S_0^{1/2} \quad \text{or} \quad 5 \text{ m}^3/\text{s} = \frac{1.0}{0.015}(2y^2)\left(\frac{1}{2}y\right)^{2/3}(0.001)^{1/2}$$

which can be solved for

$$y^{8/3} = 1.882 \text{ m}^{8/3}$$
$$y = 1.27 \text{ m} \qquad \qquad Ans.$$

The proper area and width are

$$A = 2y^2 = 3.21 \text{ m}^2 \qquad b = \frac{A}{y} = 2.53 \text{ m} \qquad Ans.$$

Part (b)

It is constructive to see what flow rate a half-hexagon and semicircle would carry for the same area of 3.214 m².

For the half-hexagon (HH), with $\alpha = 1/3^{1/2} = 0.577$, Eq. (10.25) predicts

$$A = y_{HH}^2[2(1 + 0.577^2)^{1/2} - 0.577] = 1.732y_{HH}^2 = 3.214$$

or $y_{HH} = 1.362$ m, whence $R_h = \frac{1}{2}y = 0.681$ m. The half-hexagon flow rate is thus

$$Q = \frac{1.0}{0.015}(3.214)(0.681)^{2/3}(0.001)^{1/2} = 5.25 \text{ m}^3/\text{s}$$

or about 5 percent more than that for the rectangle.

For a semicircle, $A = 3.214 \text{ m}^2 = \pi D^2/8$, or $D = 2.861$ m, whence $P = \frac{1}{2}\pi D = 4.494$ m and $R_h = A/P = 3.214/4.494 = 0.715$ m. The semicircle flow rate will thus be

$$Q = \frac{1.0}{0.015}(3.214)(0.715)^{2/3}(0.001)^{1/2} = 5.42 \text{ m}^3/\text{s}$$

or about 8 percent more than that of the rectangle and 3 percent more than that of the half-hexagon.

10.4 Specific Energy; Critical Depth

The total head of any incompressible flow is the sum of its velocity head $\alpha V^2/(2g)$, pressure head p/γ, and potential head z. For open-channel flow, surface pressure is everywhere atmospheric, so that channel energy is a balance between velocity and elevation head only. Since the flow is turbulent, we assume that $\alpha \approx 1$—recall Eq. (3.73). The final result is the quantity called *specific energy E*, as suggested by Bakhmeteff [1] in 1913:

$$E = y + \frac{V^2}{2g}$$ (10.28)

where y is the water depth. It is seen from Fig. 10.8 that E is the height of the *energy grade line* (EGL) above the channel bottom. For a given flow rate, there are usually two states possible, called *alternate states,* for the same specific energy. There is a minimum energy, E_{min}, which corresponds to a Froude number of unity.

Rectangular Channels

Consider the possible states at a given location. Let $q = Q/b = Vy$ be the discharge per unit width of a rectangular channel. Then, with q constant, Eq. (10.28) becomes

$$E = y + \frac{q^2}{2gy^2} \qquad q = \frac{Q}{b}$$ (10.29)

Figure 10.8 is a plot of y versus E for constant q from Eq. (10.29). There is a minimum value of E at a certain value of y called the *critical depth*. By setting $dE/dy = 0$ at constant q, we find that E_{min} occurs at

$$y = y_c = \left(\frac{q^2}{g}\right)^{1/3} = \left(\frac{Q^2}{b^2 g}\right)^{1/3}$$ (10.30)

The associated minimum energy is

$$E_{min} = E(y_c) = \tfrac{3}{2} y_c$$ (10.31)

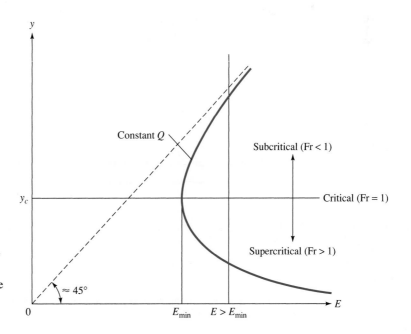

Fig. 10.8 Illustration of a specific energy curve. The curve for each flow rate Q has a minimum energy corresponding to critical flow. For energy greater than minimum, there are two *alternate* flow states, one subcritical and one supercritical.

The depth y_c corresponds to channel velocity equal to the shallow-water wave propagation speed C_0 from Eq. (10.10). To see this, rewrite Eq. (10.30) as

$$q^2 = gy_c^3 = (gy_c)y_c^2 = V_c^2y_c^2 \tag{10.32}$$

By comparison it follows that the critical channel velocity is

$$V_c = (gy_c)^{1/2} = C_0 \qquad \text{Fr} = 1 \tag{10.33}$$

For $E < E_{min}$ no solution exists in Fig. 10.8, and thus such a flow is impossible physically. For $E > E_{min}$ two solutions are possible: (1) large depth with $V < V_c$, called *subcritical*, and (2) small depth with $V > V_c$, called *supercritical*. In subcritical flow, disturbances can propagate upstream because wave speed $C_0 > V$. In supercritical flow, waves are swept downstream: Upstream is a zone of silence, and a small obstruction in the flow will create a wedge-shaped wave exactly analogous to the Mach waves in Fig. 9.18c.[3] The angle of these waves must be

$$\mu = \sin^{-1}\frac{C_0}{V} = \sin^{-1}\frac{(gy)^{1/2}}{V} \tag{10.34}$$

The wave angle and the depth can thus be used as a simple measurement of supercritical flow velocity.

Note from Fig. 10.8 that small changes in E near E_{min} cause a large change in the depth y, by analogy with small changes in duct area near the sonic point in Fig. 9.7. Thus critical flow is neutrally stable and is often accompanied by waves and undulations in the free surface. Channel designers should avoid long runs of near-critical flow.

EXAMPLE 10.5

A wide rectangular clean-earth channel has a flow rate $q = 50$ ft³/(s · ft). (a) What is the critical depth? (b) What type of flow exists if $y = 3$ ft?

Solution

Part (a)

The critical depth is independent of channel roughness and simply follows from Eq. (10.30):

$$y_c = \left(\frac{q^2}{g}\right)^{1/3} = \left(\frac{50^2}{32.2}\right)^{1/3} = 4.27 \text{ ft} \qquad \textit{Ans. (a)}$$

Part (b)

If the actual depth is 3 ft, which is less than y_c, the flow must be *supercritical*.   *Ans. (b)*

Nonrectangular Channels

If the channel width varies with y, the specific energy must be written in the form

$$E = y + \frac{Q^2}{2gA^2} \tag{10.35}$$

---

[3]This is the basis of the water channel analogy for supersonic gas dynamics experimentation [21, Chap. 11].

The critical point of minimum energy occurs where $dE/dy = 0$ at constant Q. Since $A = A(y)$, Eq. (10.35) yields, for $E = E_{min}$,

$$\frac{dA}{dy} = \frac{gA^3}{Q^2} \tag{10.36}$$

But $dA = b_0\,dy$, where b_0 is the channel width at the free surface. Therefore Eq. (10.36) is equivalent to

$$A_c = \left(\frac{b_0 Q^2}{g}\right)^{1/3} \tag{10.37a}$$

$$V_c = \frac{Q}{A_c} = \left(\frac{gA_c}{b_0}\right)^{1/2} \tag{10.37b}$$

For a given channel shape $A(y)$ and $b_0(y)$ and a given Q, Eqs. (10.37) have to be solved by trial and error or by EES to find the critical area A_c, from which V_c can be computed.

By comparing the actual depth and velocity with the critical values, we can determine the local flow condition.

$$y > y_c, V < V_c: \quad \text{subcritical flow (Fr} < 1)$$
$$y = y_c, V = V_c: \quad \text{critical flow (Fr} = 1)$$
$$y < y_c, V > V_c: \quad \text{supercritical flow (Fr} < 1)$$

Critical Uniform Flow: The Critical Slope

If a critical channel flow is also moving uniformly (at constant depth), it must correspond to a *critical slope* S_c, with $y_n = y_c$. This condition is analyzed by equating Eq. (10.37a) to the Chézy (or Manning) formula:

$$Q^2 = \frac{gA_c^3}{b_0} = C^2 A_c^2 R_h S_c = \frac{\alpha^2}{n^2} A_c^2 R_h^{4/3} S_c$$

or

$$S_c = \frac{n^2 g A_c}{\alpha^2 b_0 R_{hc}^{4/3}} = \frac{n^2 V_c^2}{\alpha^2 R_{hc}^{4/3}} = \frac{n^2 g}{\alpha^2 R_{hc}^{1/3}} \frac{P}{b_0} = \frac{f}{8} \frac{P}{b_0} \tag{10.38}$$

where α^2 equals 1.0 for SI units and 2.208 for BG units. Equation (10.38) is valid for any channel shape. For a wide rectangular channel, $b_0 \gg y_c$, the formula reduces to

Wide rectangular channel: $\qquad S_c \approx \dfrac{n^2 g}{\alpha^2 y_c^{1/3}} \approx \dfrac{f}{8}$

This is a special case, a reference point. In most channel flows $y_n \neq y_c$. For fully rough turbulent flow, the critical slope varies between 0.002 and 0.008.

E10.6

EXAMPLE 10.6

The 50° triangular channel in Fig. E10.6 has a flow rate $Q = 16$ m^3/s. Compute (a) y_c, (b) V_c, and (c) S_c if $n = 0.018$.

Solution

Part (a)

This is an easy cross section because all geometric quantities can be written directly in terms of depth y:

$$P = 2y \csc 50° \qquad A = y^2 \cot 50°$$
$$R_h = \tfrac{1}{2}y \cos 50° \qquad b_0 = 2y \cot 50° \tag{1}$$

The critical flow condition satisfies Eq. (10.37a):

$$gA_c^3 = b_0 Q^2$$

or

$$g(y_c^2 \cot 50°)^3 = (2y_c \cot 50°)Q^2$$

$$y_c = \left(\frac{2Q^2}{g \cot^2 50°}\right)^{1/5} = \left[\frac{2(16)^2}{9.81(0.839)^2}\right]^{1/5} = 2.37 \text{ m} \qquad Ans. \ (a)$$

Part (b)

With y_c known, from Eqs. (1) we compute $P_c = 6.18$ m, $R_{hc} = 0.760$ m, $A_c = 4.70$ m^2, and $b_{0c} = 3.97$ m. The critical velocity from Eq. (10.37b) is

$$V_c = \frac{Q}{A_c} = \frac{16 \text{ m}^3/\text{s}}{4.70 \text{ m}^2} = 3.41 \text{ m/s} \qquad Ans. \ (b)$$

Part (c)

With $n = 0.018$, we compute from Eq. (10.38) a critical slope:

$$S_c = \frac{gn^2P}{\alpha^2 R_h^{1/3} b_0} = \frac{9.81(0.018)^2(6.18)}{1.0(0.760)^{1/3}(3.97)} = 0.00542 \qquad Ans. \ (c)$$

Frictionless Flow over a Bump

A rough analogy to compressible gas flow in a nozzle (Fig. 9.12) is open-channel flow over a bump, as in Fig. 10.9a. The behavior of the free surface is sharply different according to whether the approach flow is subcritical or supercritical. The height of the bump also can change the character of the results. For frictionless two-dimensional flow, sections 1 and 2 in Fig. 10.9a are related by continuity and momentum:

$$V_1 y_1 = V_2 y_2 \qquad \frac{V_1^2}{2g} + y_1 = \frac{V_2^2}{2g} + y_2 + \Delta h$$

Eliminating V_2 between these two gives a cubic polynomial equation for the water depth y_2 over the bump:

$$y_2^3 - E_2 y_2^2 + \frac{V_1^2 y_1^2}{2g} = 0 \qquad \text{where } E_2 = \frac{V_1^2}{2g} + y_1 - \Delta h \tag{10.39}$$

This equation has one negative and two positive solutions if Δh is not too large. Its behavior is illustrated in Fig. 10.9b and depends on whether condition 1 is on the upper or lower leg of the energy curve. The specific energy E_2 is exactly Δh less than the approach energy E_1, and point 2 will lie on the same leg of the curve as E_1. A subcritical approach, $\text{Fr}_1 < 1$, will cause the water level to decrease at the bump. Supercritical approach flow, $\text{Fr}_1 > 1$, causes a water level increase over the bump.

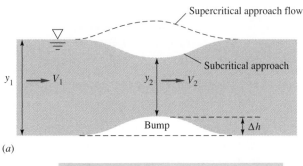

Fig. 10.9 Frictionless two-dimensional flow over a bump: (*a*) definition sketch showing Froude number dependence; (*b*) specific energy plot showing bump size and water depths.

If the bump height reaches $\Delta h_{max} = E_1 - E_c$, as illustrated in Fig. 10.9*b*, the flow at the crest will be exactly critical (Fr = 1). If $\Delta h > \Delta h_{max}$, there are no physically correct solutions to Eq. (10.39). That is, a bump too large will "choke" the channel and cause frictional effects, typically a hydraulic jump (Sec. 10.5).

These bump arguments are reversed if the channel has a *depression* ($\Delta h < 0$): Subcritical approach flow will cause a water level rise and supercritical flow a fall in depth. Point 2 will be $|\Delta h|$ to the right of point 1, and critical flow cannot occur.

EXAMPLE 10.7

Water flow in a wide channel approaches a 10-cm-high bump at 1.5 m/s and a depth of 1 m. Estimate (*a*) the water depth y_2 over the bump and (*b*) the bump height that will cause the crest flow to be critical.

Solution

Part (a)   First check the approach Froude number, assuming $C_0 = \sqrt{gy}$:

$$\text{Fr}_1 = \frac{V_1}{\sqrt{gy_1}} = \frac{1.5 \text{ m/s}}{\sqrt{(9.81 \text{ m/s}^2)(1.0 \text{ m})}} = 0.479 \quad \text{(subcritical)}$$

For subcritical approach flow, if Δh is not too large, we expect a depression in the water level over the bump and a higher subcritical Froude number at the crest. With $\Delta h = 0.1$ m, the specific energy levels must be

$$E_1 = \frac{V_1^2}{2g} + y_1 = \frac{(1.5)^2}{2(9.81)} + 1.0 = 1.115 \text{ m} \qquad E_2 = E_1 - \Delta h = 1.015 \text{ m}$$

This physical situation is shown on a specific energy plot in Fig. E10.7. With y_1 in meters, Eq. (10.39) takes on the numerical values

$$y_2^3 - 1.015y_2^2 + 0.115 = 0$$

There are three real roots: $y_2 = +0.859$ m, $+0.451$ m, and -0.296 m. The third (negative) solution is physically impossible. The second (smaller) solution is the *supercritical* condition for E_2 and is not possible for this subcritical bump. The first solution is correct:

$$y_2(\text{subcritical}) \approx 0.859 \text{ m} \qquad\qquad Ans. \ (a)$$

The surface level has dropped by $y_1 - y_2 - \Delta h = 1.0 - 0.859 - 0.1 = 0.041$ m. The crest velocity is $V_2 = V_1 y_1/y_2 = 1.745$ m/s. The Froude number at the crest is $\text{Fr}_2 = 0.601$. Flow downstream of the bump is subcritical. These flow conditions are shown in Fig. E10.7.

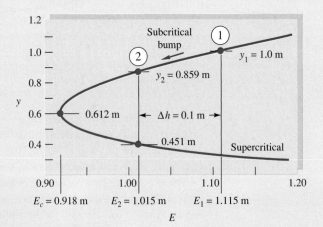

E10.7

Part (b)   For critical flow in a wide channel, with $q = Vy = 1.5$ m²/s, from Eq. (10.31),

$$E_{2,\text{min}} = E_c = \frac{3}{2}y_c = \frac{3}{2}\left(\frac{q^2}{g}\right)^{1/3} = \frac{3}{2}\left[\frac{(1.5 \text{ m}^2/\text{s})^2}{9.81 \text{ m/s}^2}\right]^{1/3} = 0.918 \text{ m}$$

Therefore the maximum height for frictionless flow over this particular bump is

$$\Delta h_{\text{max}} = E_1 - E_{2,\text{min}} = 1.115 - 0.918 = 0.197 \text{ m} \qquad\qquad Ans. \ (b)$$

For this bump, the solution of Eq. (10.39) is $y_2 = y_c = 0.612$ m, and the Froude number is unity at the crest. At critical flow the surface level has dropped by $y_1 - y_2 - \Delta h = 0.191$ m.

Flow under a Sluice Gate   A sluice gate is a bottom opening in a wall, as sketched in Fig. 10.10a, commonly used in control of rivers and channel flows. If the flow is allowed free discharge through the gap, as in Fig. 10.10a, the flow smoothly accelerates from subcritical

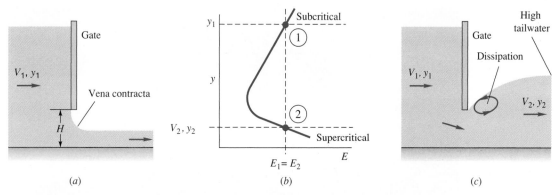

Fig. 10.10 Flow under a sluice gate passes through critical flow: (*a*) free discharge with vena contracta; (*b*) specific energy for free discharge; (*c*) dissipative flow under a drowned gate.

(upstream) to critical (near the gap) to supercritical (downstream). The gate is then analogous to a converging–diverging nozzle in gas dynamics, as in Fig. 9.12, operating at its *design condition* (similar to point H in Fig. 9.12*b*).

For free discharge, friction may be neglected, and since there is no bump ($\Delta h = 0$), Eq. (10.39) applies with $E_1 = E_2$:

$$y_2^3 - \left(\frac{V_1^2}{2g} + y_1\right)y_2^2 + \frac{V_1^2 y_1^2}{2g} = 0 \tag{10.40}$$

Given subcritical upstream flow (V_1, y_1), this cubic equation has only one positive real solution: supercritical flow at the same specific energy, as in Fig. 10.10*b*. The flow rate varies with the ratio y_2/y_1; we ask, as a problem exercise, to show that the flow rate is a maximum when $y_2/y_1 = \frac{2}{3}$.

The free discharge, Fig. 10.10*a*, contracts to a depth y_2 about 40 percent less than the gate's gap height, as shown. This is similar to a free *orifice* discharge, as in Fig. 6.39. If H is the height of the gate gap and b is the gap width into the paper, we can approximate the flow rate by orifice theory:

$$Q = C_d Hb \sqrt{2gy_1} \qquad \text{where} \qquad C_d \approx \frac{0.61}{\sqrt{1 + 0.61H/y_1}} \tag{10.41}$$

in the range $H/y_1 < 0.5$. Thus a continuous variation in flow rate is accomplished by raising the gate.

If the tailwater is high, as in Fig. 10.10*c*, free discharge is not possible. The sluice gate is said to be *drowned* or partially drowned. There will be energy dissipation in the exit flow, probably in the form of a drowned hydraulic jump, and the downstream flow will return to subcritical. Equations (10.40) and (10.41) do not apply to this situation, and experimental discharge correlations are necessary [3, 19]. See Prob. P10.77.

10.5 The Hydraulic Jump

In open-channel flow a supercritical flow can change quickly back to a subcritical flow by passing through a hydraulic jump, as in Fig. 10.5. The upstream flow is fast and shallow, and the downstream flow is slow and deep, analogous to the normal

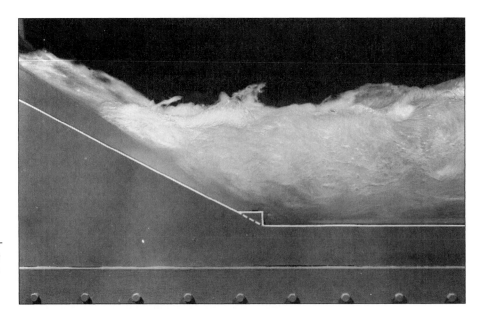

Fig. 10.11 Hydraulic jump formed on a spillway model for the Karnafuli Dam in Bangladesh. *(Courtesy of the St. Anthony Falls Hydraulic Laboratory, University of Minnesota.)*

shock wave of Fig. 9.8. Unlike the infinitesimally thin normal shock, the hydraulic jump is quite thick, ranging in length from 4 to 6 times the downstream depth y_2 [20].

Being extremely turbulent and agitated, the hydraulic jump is a very effective energy dissipator and is a feature of stilling-basin and spillway applications [20]. Figure 10.11 shows the jump formed at the bottom of a dam spillway in a model test. It is very important that such jumps be located on specially designed aprons; otherwise the channel bottom will be badly scoured by the agitation. Jumps also mix fluids very effectively and have application to sewage and water treatment designs.

Classification

The principal parameter affecting hydraulic jump performance is the upstream Froude number $Fr_1 = V_1/(gy_1)^{1/2}$. The Reynolds number and channel geometry have only a secondary effect. As detailed in Ref. 20, the following ranges of operation can be outlined, as illustrated in Fig. 10.12:

$Fr_1 < 1.0$:	Jump impossible, violates second law of thermodynamics.
$Fr_1 = 1.0$ to 1.7:	Standing-wave or *undular jump* about $4y_2$ long; low dissipation, less than 5 percent.
$Fr_1 = 1.7$ to 2.5:	Smooth surface rise with small rollers, known as a *weak jump;* dissipation 5 to 15 percent.
$Fr_1 = 2.5$ to 4.5:	Unstable, *oscillating jump;* each irregular pulsation creates a large wave that can travel downstream for miles, damaging earth banks and other structures. Not recommended for design conditions. Dissipation 15 to 45 percent.
$Fr_1 = 4.5$ to 9.0:	Stable, well-balanced, *steady jump;* best performance and action, insensitive to downstream conditions. Best design range. Dissipation 45 to 70 percent.

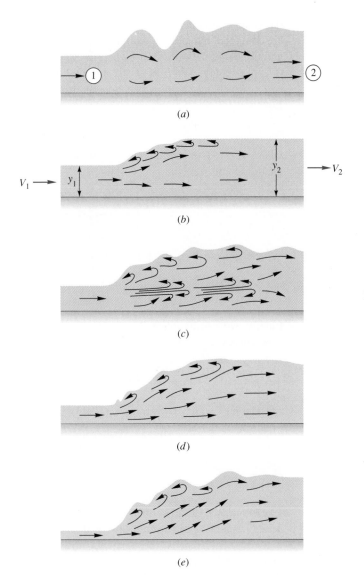

Fig. 10.12 Classification of hydraulic jumps: (*a*) Fr = 1.0 to 1.7: undular jump; (*b*) Fr = 1.7 to 2.5: weak jump; (*c*) Fr = 2.5 to 4.5: oscillating jump; (*d*) Fr = 4.5 to 9.0: steady jump; (*e*) Fr > 9.0: strong jump. *(Adapted from Ref. 20.)*

$Fr_1 > 9.0$:   Rough, somewhat intermittent *strong jump,* but good performance. Dissipation 70 to 85 percent.

Further details can be found in Ref. 20 and Ref. 2, Chap. 15.

Theory for a Horizontal Jump   A jump that occurs on a steep channel slope can be affected by the difference in water-weight components along the flow. The effect is small, however, so that the classic theory assumes that the jump occurs on a horizontal bottom.

You will be pleased to know that we have already analyzed this problem in Sec. 10.1. A hydraulic jump is exactly equivalent to the strong fixed wave in Fig. 10.4*b*, where the change in depth δy is not neglected. If V_1 and y_1 upstream are known, V_2 and y_2 are computed by applying continuity and momentum across the wave, as in Eqs. (10.7) and (10.8). Equation (10.9) is therefore the correct solution for a jump if we interpret C and y in Fig. 10.4b as upstream conditions V_1 and y_1, respectively, with $C - \delta V$ and $y + \delta y$ being the downstream conditions V_2 and y_2, respectively, as in Fig. 10.12b. Equation (10.9) becomes

$$V_1^2 = \tfrac{1}{2}gy_1\,\eta(\eta + 1) \tag{10.42}$$

where $\eta = y_2/y_1$. Introducing the Froude number $\mathrm{Fr}_1 = V_1/(gy_1)^{1/2}$ and solving this quadratic equation for η, we obtain

$$\boxed{\frac{2y_2}{y_1} = -1 + (1 + 8\,\mathrm{Fr}_1^2)^{1/2}} \tag{10.43}$$

With y_2 thus known, V_2 follows from the wide-channel continuity relation:

$$V_2 = \frac{V_1 y_1}{y_2} \tag{10.44}$$

Finally, we can evaluate the dissipation head loss across the jump from the steady flow energy equation:

$$h_f = E_1 - E_2 = \left(y_1 + \frac{V_1^2}{2g}\right) - \left(y_2 + \frac{V_2^2}{2g}\right)$$

Introducing y_2 and V_2 from Eqs. (10.43) and (10.44), we find after considerable algebraic manipulation that

$$\boxed{h_f = \frac{(y_2 - y_1)^3}{4y_1 y_2}} \tag{10.45}$$

Equation (10.45) shows that the dissipation loss is positive only if $y_2 > y_1$, which is a requirement of the second law of thermodynamics. Equation (10.43) then requires that $\mathrm{Fr}_1 > 1.0$; that is, the upstream flow must be supercritical. Finally, Eq. (10.44) shows that $V_2 < V_1$ and the downstream flow is subcritical. All these results agree with our previous experience analyzing the normal shock wave.

The present theory is for hydraulic jumps in wide horizontal channels. For the theory of prismatic or sloping channels see advanced texts [for example, 2, Chaps. 15 and 16].

EXAMPLE 10.8

Water flows in a wide channel at $q = 10$ m³/(s · m) and $y_1 = 1.25$ m. If the flow undergoes a hydraulic jump, compute (a) y_2, (b) V_2, (c) Fr_2, (d) h_f, (e) the percentage dissipation, (f) the power dissipated per unit width, and (g) the temperature rise due to dissipation if $c_p = 4200$ J/(kg · K).

Solution

Part (a)   The upstream velocity is

$$V_1 = \frac{q}{y_1} = \frac{10 \text{ m}^3/(\text{s} \cdot \text{m})}{1.25 \text{ m}} = 8.0 \text{ m/s}$$

The upstream Froude number is therefore

$$\text{Fr}_1 = \frac{V_1}{(gy_1)^{1/2}} = \frac{8.0}{[9.81(1.25)]^{1/2}} = 2.285$$

From Fig. 10.12 this is a weak jump. The depth y_2 is obtained from Eq. (10.43):

$$\frac{2y_2}{y_1} = -1 + [1 + 8(2.285)^2]^{1/2} = 5.54$$

or $y_2 = \frac{1}{2}y_1(5.54) = \frac{1}{2}(1.25)(5.54) = 3.46 \text{ m}$ *Ans.* (*a*)

Part (b)   From Eq. (10.44) the downstream velocity is

$$V_2 = \frac{V_1 y_1}{y_2} = \frac{8.0(1.25)}{3.46} = 2.89 \text{ m/s}$$ *Ans.* (*b*)

Part (c)   The downstream Froude number is

$$\text{Fr}_2 = \frac{V_2}{(gy_2)^{1/2}} = \frac{2.89}{[9.81(3.46)]^{1/2}} = 0.496$$ *Ans.* (*c*)

Part (d)   As expected, Fr_2 is subcritical. From Eq. (10.45) the dissipation loss is

$$h_f = \frac{(3.46 - 1.25)^3}{4(3.46)(1.25)} = 0.625 \text{ m}$$ *Ans.* (*d*)

Part (e)   The percentage dissipation relates h_f to upstream energy:

$$E_1 = y_1 + \frac{V_1^2}{2g} = 1.25 + \frac{(8.0)^2}{2(9.81)} = 4.51 \text{ m}$$

Hence $\text{Percentage loss} = (100)\dfrac{h_f}{E_1} = \dfrac{100(0.625)}{4.51} = 14 \text{ percent}$ *Ans.* (*e*)

Part (f)   The power dissipated per unit width is

$$\text{Power} = \rho g q h_f = (9800 \text{ N/m}^3)[10 \text{ m}^3/(\text{s} \cdot \text{m})](0.625 \text{ m})$$
$$= 61.3 \text{ kW/m}$$ *Ans.* (*f*)

Part (g)   Finally, the mass flow rate is $\dot{m} = \rho q = (1000 \text{ kg/m}^3)[10 \text{ m}^3/(\text{s} \cdot \text{m})] = 10{,}000 \text{ kg/(s} \cdot \text{m})$, and the temperature rise from the steady flow energy equation is

$$\text{Power dissipated} = \dot{m}c_p \, \Delta T$$

or $61{,}300 \text{ W/m} = [10{,}000 \text{ kg/(s} \cdot \text{m})][4200 \text{ J/(kg} \cdot \text{K)}]\Delta T$

from which

$$\Delta T = 0.0015 \text{ K} \qquad \qquad Ans. (g)$$

The dissipation is large, but the temperature rise is negligible.

10.6 Gradually Varied Flow[4]

In practical channel flows both the bottom slope and the water depth change with position, as in Fig. 10.3. An approximate analysis is possible if the flow is gradually varied, such as if the slopes are small and changes not too sudden. The basic assumptions are

1. Slowly changing bottom slope.
2. Slowly changing water depth (no hydraulic jumps).
3. Slowly changing cross section.
4. One-dimensional velocity distribution.
5. Pressure distribution approximately hydrostatic.

The flow then satisfies the continuity relation (10.1) plus the energy equation with bottom friction losses included. The two unknowns for steady flow are velocity $V(x)$ and water depth $y(x)$, where x is distance along the channel.

Basic Differential Equation

Consider the length of channel dx illustrated in Fig. 10.13. All the terms that enter the steady flow energy equation are shown, and the balance between x and $x + dx$ is

$$\frac{V^2}{2g} + y + S_0 \, dx = S \, dx + \frac{V^2}{2g} + d\left(\frac{V^2}{2g}\right) + y + dy$$

or

$$\frac{dy}{dx} + \frac{d}{dx}\left(\frac{V^2}{2g}\right) = S_0 - S \qquad (10.46)$$

where S_0 is the slope of the channel bottom (positive as shown in Fig. 10.13) and S is the slope of the EGL (which drops due to wall friction losses).

To eliminate the velocity derivative, differentiate the continuity relation:

$$\frac{dQ}{dx} = 0 = A\frac{dV}{dx} + V\frac{dA}{dx} \qquad (10.47)$$

But $dA = b_0 \, dy$, where b_0 is the channel width at the surface. Eliminating dV/dx between Eqs. (10.46) and (10.47), we obtain

$$\frac{dy}{dx}\left(1 - \frac{V^2 b_0}{gA}\right) = S_0 - S \qquad (10.48)$$

---

[4]This section may be omitted without loss of continuity.

Horizontal

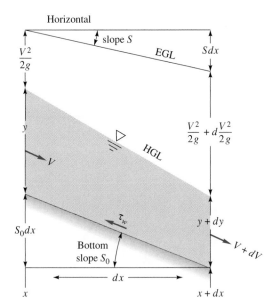

Fig. 10.13 Energy balance between two sections in a gradually varied open-channel flow.

Finally, recall from Eq. (10.37) that $V^2 b_0/(gA)$ is the square of the Froude number of the local channel flow. The final desired form of the gradually varied flow equation is

$$\frac{dy}{dx} = \frac{S_0 - S}{1 - \mathrm{Fr}^2} \qquad (10.49)$$

This equation changes sign according as the Froude number is subcritical or supercritical and is analogous to the one-dimensional gas dynamic area-change formula (9.40).

The numerator of Eq. (10.49) changes sign according as S_0 is greater or less than S, which is the slope equivalent to uniform flow at the same discharge Q:

$$S = S_{0n} = \frac{f}{D_h} \frac{V^2}{2g} = \frac{V^2}{R_h C^2} = \frac{n^2 V^2}{\alpha^2 R_h^{4/3}} \qquad (10.50)$$

where C is the Chézy coefficient. The behavior of Eq. (10.49) thus depends on the relative magnitude of the local bottom slope $S_0(x)$, compared with (1) uniform flow, $y = y_n$, and (2) critical flow, $y = y_c$. As in Eq. (10.38), the dimensional parameter α^2 equals 1.0 for SI units and 2.208 for BG units.

Classification of Solutions

It is customary to compare the actual channel slope S_0 with the critical slope S_c for the same Q from Eq. (10.38). There are five classes for S_0, giving rise to 12 distinct types of solution curves, all of which are illustrated in Fig. 10.14:

Slope class	Slope notation	Depth class	Solution curves
$S_0 > S_c$	Steep	$y_c > y_n$	S-1, S-2, S-3
$S_0 = S_c$	Critical	$y_c = y_n$	C-1, C-3
$S_0 < S_c$	Mild	$y_c < y_n$	M-1, M-2, M-3
$S_0 = 0$	Horizontal	$y_n = \infty$	H-2, H-3
$S_0 < 0$	Adverse	$y_n = $ imaginary	A-2, A-3

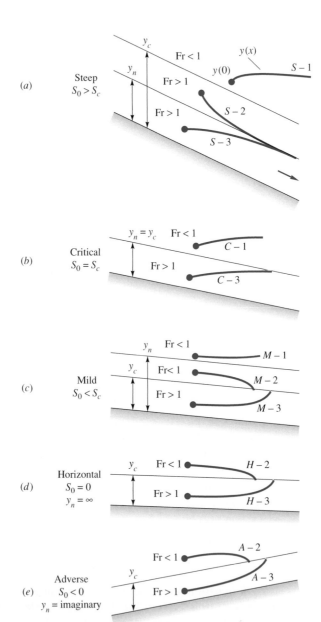

Fig. 10.14 Gradually varied flow for five classes of channel slope, showing the 12 basic solution curves.

The solution letters S, C, M, H, and A obviously denote the names of the five types of slope. The numbers 1, 2, 3 relate to the position of the initial point on the solution curve with respect to the normal depth y_n and the critical depth y_c. In type 1 solutions, the initial point is above both y_n and y_c, and in all cases the water depth solution $y(x)$ becomes even deeper and farther away from y_n and y_c. In type 2 solutions, the initial point lies between y_n and y_c, and if there is no change in S_0 or roughness, the solution tends asymptotically toward the lower of y_n or y_c. In type 3 cases, the initial point lies below both y_n and y_c, and the solution curve tends asymptotically toward the lower of these.

Figure 10.14 shows the basic character of the local solutions, but in practice, of course, S_0 varies with x, and the overall solution patches together the various cases to form a continuous depth profile $y(x)$ compatible with a given initial condition and a given discharge Q. There is a fine discussion of various composite solutions in Ref. 2, Chap. 9; see also Ref. 22, Sec. 12.7.

Numerical Solution

The basic relation for gradually varied flow, Eq. (10.49), is a first-order ordinary differential equation that can be easily solved numerically. For a given constant-volume flow rate Q, it may be written in the form

$$\frac{dy}{dx} = \frac{S_0 - n^2 Q^2/(\alpha^2 A^2 R_h^{4/3})}{1 - Q^2 b_0/(gA^3)} \tag{10.51}$$

subject to an initial condition $y = y_0$ at $x = x_0$. It is assumed that the bottom slope $S_0(x)$ and the cross-sectional shape parameters (b_0, P, A) are known everywhere along the channel. Then one may solve Eq. (10.51) for local water depth $y(x)$ by any standard numerical method. The author uses an Excel spreadsheet for a personal computer. Step sizes Δx may be selected so that each change Δy is limited to no greater than, say, 1 percent. The solution curves are generally well behaved unless there are discontinuous changes in channel parameters. Note that if one approaches the critical depth y_c, the denominator of Eq. (10.51) approaches zero, so small step sizes are required. It helps physically to know what type solution curve (M-1, S-2, or the like) you are proceeding along, but this is not mathematically necessary.

EXAMPLE 10.9

Let us extend the data of Example 10.5 to compute a portion of the profile shape. Given is a wide channel with $n = 0.022$, $S_0 = 0.0048$, and $q = 50$ ft³/(s · ft). If $y_0 = 3$ ft at $x = 0$, how far along the channel $x = L$ does it take the depth to rise to $y_L = 4$ ft? Is the 4-ft depth position upstream or downstream in Fig. E10.9a?

Solution

In Example 10.5 we computed $y_c = 4.27$ ft. Since our initial depth $y = 3$ ft is less than y_c, we know the flow is supercritical. Let us also compute the normal depth for the given slope S_0 by setting $q = 50$ ft³/(s · ft) in the Chézy formula (10.19) with $R_h = y_n$:

$$q = \frac{\alpha}{n} A R_h^{2/3} S_0^{1/2} = \frac{1.486}{0.022} [y_n(1 \text{ ft})] y_n^{2/3} (0.0048)^{1/2} = 50 \text{ ft}^3/(\text{s} \cdot \text{ft})$$

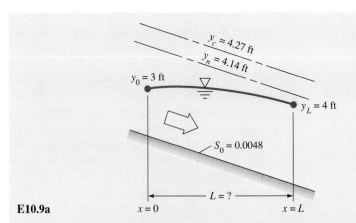

E10.9a

Solve for:
$$y_n \approx 4.14 \text{ ft}$$

Thus both $y(0) = 3$ ft and $y(L) = 4$ ft are less than y_n, which is less than y_c, so we *must* be on an S-3 curve, as in Fig. 10.14*a*. For a wide channel, Eq. (10.51) reduces to

$$\frac{dy}{dx} = \frac{S_0 - n^2 q^2/(\alpha^2 y^{10/3})}{1 - q^2/(gy^3)}$$

$$\approx \frac{0.0048 - (0.022)^2 (50)^2/(2.208 y^{10/3})}{1 - (50)^2/(32.2 y^3)} \qquad \text{with } y(0) = 3 \text{ ft}$$

The initial slope is $y'(0) \approx 0.00494$, and a step size $\Delta x = 5$ ft would cause a change $\Delta y \approx (0.00494)(5 \text{ ft}) \approx 0.025$ ft, less than 1 percent. We therefore integrate numerically with $\Delta x = 5$ ft to determine when the depth $y = 4$ ft is achieved. Tabulate some values:

x, ft	0	50	100	150	200	230
y, ft	3.00	3.25	3.48	3.70	3.90	4.00

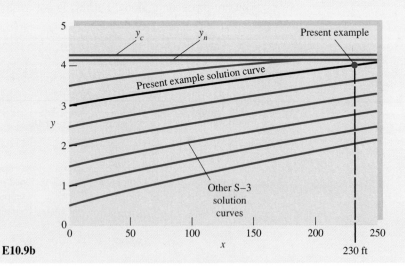

E10.9b

The water depth, still supercritical, reaches $y = 4$ ft at

$$x \approx 230 \text{ ft downstream} \qquad\qquad Ans.$$

We verify from Fig. 10.14a that water depth does increase downstream on an S-3 curve. The solution curve $y(x)$ is shown as the bold line in Fig. E10.9b.

For little extra effort we can investigate the entire family of S-3 solution curves for this problem. Figure E10.9b also shows what happens if the initial depth is varied from 0.5 to 3.5 ft in increments of 0.5 ft. All S-3 solutions smoothly rise and asymptotically approach the uniform flow condition $y = y_n = 4.14$ ft.

Approximate Solution for Irregular Channels

Direct numerical solution of Eq. (10.51) is appropriate when we have analytical formulas for the channel variations $A(x)$, $S_0(x)$, $n(x)$, $b_0(x)$, and $R_h(x)$. For natural channels, however, cross sections are often highly irregular, and data can be sparse and unevenly spaced. For such cases, civil engineers use an approximate method to estimate gradual flow changes. Write Eq. (10.46) in finite-difference form between two depths y and $y + \Delta y$:

$$\Delta x \approx \frac{E(y + \Delta y) - E(y)}{(S_0 - S)_{\text{avg}}} \qquad \text{where} \quad E = y + \frac{V^2}{2g} \qquad (10.52)$$

Average values of velocity, slope, and hydraulic radius are estimated between the two sections. For example,

$$V_{\text{avg}} \approx \frac{1}{2}[V(y) + V(y + \Delta y)]; \; R_{h,\,\text{avg}} \approx \frac{1}{2}[R_h(y) + R_h(y + \Delta y)]; \; S_{\text{avg}} \approx \frac{n^2 V_{\text{avg}}^2}{\alpha^2 R_{h,\,\text{avg}}^{4/3}}$$

Again, computation can proceed either upstream or downstream, using small values of Δy. Further details of such computations are given in Chap. 10 of Ref. 2.

EXAMPLE 10.10

Repeat Example 10.9 using the approximate method of Eq. (10.52) with a 0.25-foot increment in Δy. Find the distance required for y to rise from 3 ft to 4 ft.

Solution

Recall from Example 10.9 that $n = 0.022$, $S_0 = 0.0048$, and $q = 50$ ft³/(s-ft). Note that $R_h = y$ for a wide channel. Make a table with y varying from 3.0 to 4.0 ft in increments of 0.25 ft, computing $V = q/y$, $E = y + V^2/(2g)$, and $S_{\text{avg}} = [n^2V^2/(2.208y^{4/3})]_{\text{avg}}$:

y, ft	V (ft/s) = 50/y	$E = y + V^2/(2g)$	S	S_{avg}	$\Delta x = \Delta E/(S_0 - S)_{\text{avg}}$	$x = \Sigma \Delta x$
3.0	16.67	7.313	0.01407	—	—	0
3.25	15.38	6.925	0.01078	0.01243	51	51
3.5	14.29	6.669	0.00842	0.00960	53	104
3.75	13.33	6.511	0.00669	0.00756	57	161
4.0 ft	12.50 ft/s	6.426 ft	0.00539	0.00604	69 ft	230 ft

Comment: The accuracy is excellent, giving the same result, $x = 230$ ft, as the Excel spreadsheet numerical integration in Example 10.9. Much of this accuracy is due to the smooth, slowly varying nature of the profile. Less precision is expected when the channel is irregular and given as uneven cross sections.

Some Illustrative Composite-Flow Transitions

The solution curves in Fig. 10.14 are somewhat simplistic, since they postulate constant-bottom slopes. In practice, channel slopes can vary greatly, $S_0 = S_0(x)$, and the solution curves can cross between two regimes. Other parameter changes, such as $A(x)$, $b_0(x)$, and $n(x)$, can cause interesting composite-flow profiles. Some examples are shown in Fig. 10.15.[5]

Figure 10.15a shows a transition from a mild slope to a steep slope in a constant-width channel. The initial M-2 curve must change to an S-2 curve farther down the steep slope. The only way this can happen physically is for the solution curve to pass smoothly through the critical depth, as shown. The critical point is mathematically *singular* [2, Sec. 9.6], and the flow near this point is generally *rapidly,* not gradually, varied. The flow pattern, accelerating from subcritical to supercritical, is similar to a converging–diverging nozzle in gas dynamics. Other scenarios for Fig. 10.15a are impossible. For example, the upstream curve cannot be M-1, for the break in slope would cause an S-1 curve that would move away from uniform steep flow.

Figure 10.15b shows a mild slope that suddenly changes to an even milder slope. The approach flow is assumed uniform, and the break in slope makes its presence known upstream. The water depth moves smoothly along an M-1 curve until it exactly merges, at the break point, with a uniform flow at the new (milder) depth y_{n2}.

Figure 10.15c shows a steep slope that suddenly changes to a less steep slope. Note for both slopes that $y_n < y_c$. Because of the supercritical ($V > V_c$) approach flow, the break in slope cannot make its presence known upstream. Thus, not until the break point does an S-3 curve form, and then this profile proceeds smoothly to uniform flow at the new (higher) normal depth.

Figure 10.15d shows a steep slope that suddenly changes to a mild slope. Various cases may occur, possibly beyond the ability of this author to describe. The two cases shown depend on the relative magnitude of the mild slope. If the downstream depth y_{n2} is shallow, an M-3 curve will start at the break and develop until the local supercritical flow is just sufficient to form a hydraulic jump up to the new normal depth. As y_{n2} increases, the jump moves upstream until, for the "high" case shown, it forms on the steep side, followed by an S-1 curve that merges into normal depth y_{n2} at the break point.

Figure 10.15e illustrates a *free overfall* with a mild slope. This acts as a *control section* to the upstream flow, which then forms an M-2 curve and accelerates to critical flow near the overfall. The falling stream will be supercritical. The overfall "controls" the water depths upstream and can serve as an initial condition for computation of $y(x)$. This is the type of flow that occurs in a weir or waterfall, Sec. 10.7.

The examples in Fig. 10.15 show that changing conditions in open-channel flow can result in complex flow patterns. Many more examples of composite-flow profiles are given in Ref. 2, pp. 229–233.

---

[5]The author is indebted to Prof. Bruce Larock for clarification of these transition profiles.

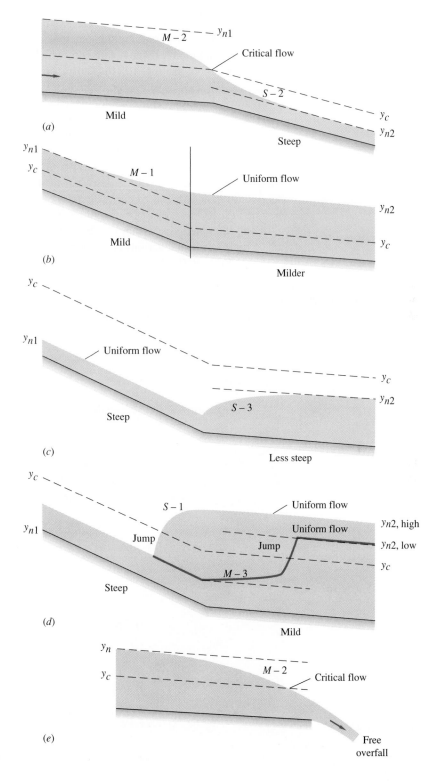

Fig. 10.15 Some examples of composite-flow transition profiles.

10.7 Flow Measurement and Control by Weirs

A *weir,* of which the ordinary dam is an example, is a channel obstruction over which the flow must deflect. For simple geometries the channel discharge Q correlates with gravity and with the blockage height H to which the upstream flow is backed up above the weir elevation (see Fig. 10.16). Thus a weir is a simple but effective open-channel flowmeter. We used a weir as an example of dimensional analysis in Prob. P5.32.

Figure 10.16 shows two common weirs, sharp-crested and broad-crested, assumed to be very wide. In both cases the flow upstream is subcritical, accelerates to critical near the top of the weir, and spills over into a supercritical *nappe.* For both weirs the discharge q per unit width is proportional to $g^{1/2}H^{3/2}$ but with somewhat different coefficients. The short-crested (or thin-plate) weir nappe should be *ventilated* to the atmosphere; that is, it should spring clear of the weir crest. Unventilated or drowned nappes are more difficult to correlate and depend on tailwater conditions. (The spillway of Fig. 10.11 is a sort of unventilated weir.)

A very complete discussion of weirs, including other designs such as the polygonal "Crump" weir and various contracting flumes, is given in the text by Ackers et al. [23]. See Prob. P10.122.

Analysis of Sharp-Crested Weirs

It is possible to analyze weir flow by inviscid potential theory with an unknown (but solvable) free surface, as in Fig. P8.71. Here, however, we simply use one-dimensional flow theory plus dimensional analysis to develop suitable weir flow rate correlations.

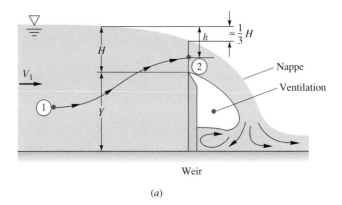

Weir

(*a*)

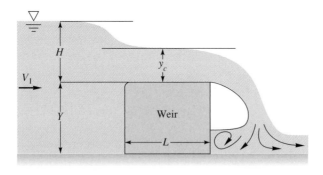

Fig. 10.16 Flow over wide, well-ventilated weirs: (*a*) sharp-crested; (*b*) broad-crested.

(*b*)

A very early theoretical approach is credited to J. Weisbach in 1855. The velocity head at any point 2 above the weir crest is assumed to equal the total head upstream; in other words, Bernoulli's equation is used with no losses:

$$\frac{V_2^2}{2g} + H - h \approx \frac{V_1^2}{2g} + H \qquad \text{or} \qquad V_2(h) \approx \sqrt{2gh + V_1^2}$$

where h is the vertical distance down to point 2, as shown in Fig. 10.16a. If we accept for the moment, without proof, that the flow over the crest draws down to $h_{min} \approx H/3$, the volume flow $q = Q/b$ over the crest is approximately

$$q = \int_{\text{crest}} V_2 \, dh \approx \int_{H/3}^{H} (2gh + V_1^2)^{1/2} \, dh$$

$$= \frac{2}{3} \sqrt{2g} \left[\left(H + \frac{V_1^2}{2g} \right)^{3/2} - \left(\frac{H}{3} + \frac{V_1^2}{2g} \right)^{3/2} \right]$$

Normally the upstream velocity head $V_1^2/(2g)$ is neglected, so this expression reduces to

Sharp-crested theory: $\qquad q \approx 0.81(\tfrac{2}{3})(2g)^{1/2}H^{3/2}$ (10.53)

This formula is functionally correct, but the coefficient 0.81 is too high and should be replaced by an experimentally determined discharge coefficient.

Analysis of Broad-Crested Weirs

The broad-crested weir of Fig. 10.16b can be analyzed more accurately because it creates a short run of nearly one-dimensional critical flow, as shown. Bernoulli's equation from upstream to the weir crest yields

$$\frac{V_1^2}{2g} + Y + H \approx \frac{V_c^2}{2g} + Y + y_c$$

If the crest is very wide into the paper, $V_c^2 = gy_c$ from Eq. (10.33). Thus we can solve for

$$y_c \approx \frac{2H}{3} + \frac{V_1^2}{3g} \approx \frac{2H}{3}$$

This result was used without proof to derive Eq. (10.53). Finally, the flow rate follows from wide-channel critical flow, Eq. (10.32):

Broad-crested theory: $q = \sqrt{gy_c^3} \approx \dfrac{1}{\sqrt{3}} \left(\dfrac{2}{3} \right) \sqrt{2g} \left(H + \dfrac{V_1^2}{2g} \right)^{3/2}$ (10.54)

Again we may usually neglect the upstream velocity head $V_1^2/(2g)$. The coefficient $1/\sqrt{3} \approx 0.577$ is about right, but experimental data are preferred.

Experimental Weir Discharge Coefficients

Theoretical weir flow formulas may be modified experimentally as follows. Eliminate the numerical coefficients $\frac{2}{3}$ and $\sqrt{2}$, for which there is much sentimental attachment in the literature, and reduce the formula to

$$Q_{\text{weir}} = C_d b \sqrt{g} \left(H + \frac{V_1^2}{2g} \right)^{3/2} \approx C_d b \sqrt{g} H^{3/2}$$ (10.55)

where b is the crest width and C_d is a dimensionless, experimentally determined *weir discharge coefficient,* which may vary with the weir geometry, Reynolds number, and Weber number. Many data for many different weirs have been reported in the literature, as detailed in Ref. 23.

An accurate (± 2 percent) composite correlation for wide ventilated sharp crests is recommended as follows [23]:

$$\text{Wide sharp-crested weir: } C_d \approx 0.564 + 0.0846 \frac{H}{Y} \quad \text{for} \quad \frac{H}{Y} \le 2 \tag{10.56}$$

The Reynolds numbers $V_1 H / \nu$ for these data vary from 1 E4 to 2 E6, but the formula should apply to higher Re, such as large dams on rivers.

The broad-crested weir of Fig. 10.16b is considerably more sensitive to geometric parameters, including the surface roughness ϵ of the crest. If the leading-edge nose is rounded, $R/L \ge 0.05$, available data [23, Chap. 7] may be correlated as follows:

$$\text{Round-nosed broad-crested weir: } C_d \approx 0.544\left(1 - \frac{\delta^*/L}{H/L}\right)^{3/2} \tag{10.57}$$

where

$$\frac{\delta^*}{L} \approx 0.001 + 0.2\sqrt{\epsilon/L}$$

The chief effect is due to turbulent boundary layer displacement-thickness growth δ^* on the crest as compared to upstream head H. The formula is limited to $H/L < 0.7$, $\epsilon/L \le 0.002$, and $V_1 H/\nu > 3$ E5. If the nose is round, there is no significant effect of weir height Y, at least if $H/Y < 2.4$.

If the broad-crested weir has a sharp leading edge, thus commonly called a *rectangular* weir, the discharge may depend on the weir height Y. However, in a certain range of weir height and length, C_d is nearly constant:

$$\begin{aligned} \text{Sharp-nosed} \quad & C_d \approx 0.462 \quad \text{for} \quad 0.08 < \frac{H}{L} < 0.33 \\ \text{broad-crested weir:} \quad & \qquad\qquad\qquad \text{and} \quad 0.22 < \frac{H}{Y} < 0.56 \end{aligned} \tag{10.58}$$

Surface roughness is not a significant factor here. For $H/L < 0.08$ there is large scatter (± 10 percent) in the data. For $H/L > 0.33$ and $H/Y > 0.56$, C_d increases up to 10 percent due to each parameter, and complex charts are needed for the discharge coefficient [19, Chap. 5].

EXAMPLE 10.11

A weir in a horizontal channel is 1 m high and 4 m wide. The water depth upstream is 1.6 m. Estimate the discharge if the weir is (a) sharp-crested and (b) round-nosed with an unfinished concrete broad crest 1.2 m long. Neglect $V_1^2/(2g)$.

Solution

Part (a)

We are given $Y = 1$ m and $H + Y \approx 1.6$ m, hence $H \approx 0.6$ m. Since $H \ll b$, we assume that the weir is "wide." For a sharp crest, Eq. (10.56) applies:

$$C_d \approx 0.564 + 0.0846 \frac{0.6 \text{ m}}{1 \text{ m}} \approx 0.615$$

Then the discharge is given by the basic correlation, Eq. (10.55):

$$Q = C_d b \sqrt{g} H^{3/2} = (0.615)(4 \text{ m}) \sqrt{(9.81 \text{ m/s}^2)}(0.6 \text{ m})^{3/2} \approx 3.58 \text{ m}^3/\text{s} \qquad Ans. (a)$$

We check that $H/Y = 0.6 < 2.0$ for Eq. (10.56) to be valid. From continuity, $V_1 = Q/(by_1) = 3.58/[(4.0)(1.6)] = 0.56$ m/s, giving a Reynolds number $V_1 H/\nu \approx 3.4$ E5.

Part (b)

For a round-nosed broad-crested weir, Eq. (10.57) applies. For an unfinished concrete surface, read $\epsilon \approx 2.4$ mm from Table 10.1. Then the displacement thickness is

$$\frac{\delta^*}{L} \approx 0.001 + 0.2\sqrt{\epsilon/L} = 0.001 + 0.2\left(\frac{0.0024 \text{ m}}{1.2 \text{ m}}\right)^{1/2} \approx 0.00994$$

Then Eq. (10.57) predicts the discharge coefficient:

$$C_d \approx 0.544\left(1 - \frac{0.00994}{0.6 \text{ m}/1.2 \text{ m}}\right)^{3/2} \approx 0.528$$

The estimated flow rate is thus

$$Q = C_d b \sqrt{g} H^{3/2} = 0.528(4 \text{ m})\sqrt{(9.81 \text{ m}^2/\text{s})}(0.6 \text{ m})^{3/2} \approx 3.07 \text{ m}^3/\text{s} \qquad Ans. (b)$$

Check that $H/L = 0.5 < 0.7$ as required. The approach Reynolds number is $V_1 H/\nu \approx 2.9$ E5, just barely below the recommended limit in Eq. (10.57).

Since $V_1 \approx 0.5$ m/s, $V_1^2/(2g) \approx 0.012$ m, so the error in taking total head equal to 0.6 m is about 2 percent. We could correct this for upstream velocity head if desired.

Other Thin-Plate Weir Designs

Weirs are often used for flow measurement and control of artificial channels. The two most common shapes are a rectangle and a V notch, as shown in Table 10.2. All should be fully ventilated and not drowned.

Table 10.2a shows a full-width rectangle, which will have slight end-boundary-layer effects but no end contractions. For a thin-plate design, the top is approximately sharp-crested, and Eq. (10.56) should give adequate accuracy, as shown in the table. Since the overfall spans the entire channel, artificial ventilation may be needed, such as holes in the channel walls.

Table 10.2b shows a partial-width rectangle, $b < L$, which will cause the sides of the overfall to contract inward and reduce the flow rate. An adequate contraction correction [23, 24] is to reduce the effective weir width by $0.1H$, as shown in the table. It seems, however, that this type of weir is rather sensitive to small effects, such as plate thickness and sidewall boundary layer growth. Small heads ($H < 75$ mm) and small slot widths ($b < 30$ cm) are not recommended. See Refs. 23 and 24 for further details.

The V notch, in Table 10.2c, is intrinsically interesting in that its overfall has only one length scale, H—there is no separate "width." The discharge will thus be

Table 10.2 Thin-Plate Weirs for Flow Measurement

Thin-plate weir	Flow-rate correlation
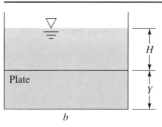 (a) Full-width rectangle.	$Q \approx \left(0.564 + 0.0846\dfrac{H}{Y}\right)bg^{1/2}H^{3/2}$
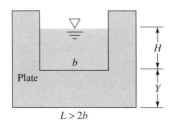 (b) Rectangle with side contractions.	$Q \approx 0.581(b - 0.1H)g^{1/2}H^{3/2} \qquad H < 0.5Y$
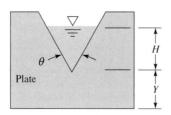 (c) V notch.	$Q \approx 0.44 \tan\dfrac{\theta}{2}g^{1/2}H^{5/2} \qquad 20° < \theta < 100°$

proportional to $H^{5/2}$, rather than a power of $\frac{3}{2}$. Application of Bernoulli's equation to the triangular opening, in the spirit of Eq. (10.52), leads to the following ideal flow rate for a V notch:

$$\text{V notch:} \qquad Q_{\text{ideal}} = \frac{8\sqrt{2}}{15}\tan\frac{\theta}{2}g^{1/2}H^{5/2} \tag{10.59}$$

where θ is the total included angle of the notch. The actual measured flow is about 40 percent less than this, due to contraction similar to a thin-plate orifice. In terms of an experimental discharge coefficient, the recommended formula is

$$Q_{\text{V notch}} \approx C_d \tan\frac{\theta}{2}g^{1/2}H^{5/2} \qquad C_d \approx 0.44 \qquad \text{for} \qquad 20° < \theta < 100° \tag{10.60}$$

for heads $H > 50$ mm. For smaller heads, both Reynolds number and Weber number effects may be important, and a recommended correction [23] is

$$\text{Low heads, } H < 50 \text{ mm:} \qquad C_{d,\text{ V notch}} \approx 0.44 + \frac{0.9}{(\text{Re We})^{1/6}} \tag{10.61}$$

where $Re = \rho g^{1/2} H^{3/2}/\mu$ and $We = \rho g H^2/Y$, with Y being the coefficient of surface tension. Liquids other than water may be used with this formula, as long as $Re > 300/\tan(\theta/2)^{3/4}$ and $We > 300$.

A number of other thin-plate weir designs—trapezoidal, parabolic, circular arc, and U-shaped—are discussed in Ref. 25, which also contains considerable data on broad-crested weirs. See also Refs. 29 and 30.

EXAMPLE 10.12

A V notch weir is to be designed to meter an irrigation channel flow. For ease in reading the upstream water-level gage, a reading $H \geq 30$ cm is desired for the design flow rate of 150 m^3/h. What is the appropriate angle θ for the V notch?

Solution

- *Assumptions:* Steady flow, negligible Weber number effect because $H > 50$ mm.
- *Approach:* Equation (10.60) applies with, we hope, a notch angle $20° < \theta < 100°$.
- *Property values:* If surface tension is neglected, no fluid properties are needed. Why?
- *Solution:* Apply Equation (10.60) to the known flow rate and solve for θ:

$$Q = \frac{150 \text{ m}^3/\text{h}}{3600 \text{ s/h}} = 0.0417 \frac{\text{m}^3}{\text{s}} \geq C_d \tan\left(\frac{\theta}{2}\right) g^{1/2} H^{5/2} = 0.44 \tan\left(\frac{\theta}{2}\right)\left(9.81 \frac{\text{m}}{\text{s}^2}\right)^{1/2} (0.3 \text{ m})^{5/2}$$

$$\text{Solve for } \tan\left(\frac{\theta}{2}\right) \leq 0.613 \quad \text{or} \quad \theta \leq 63° \qquad Ans.$$

- *Comments:* An angle of 63° will create an upstream head of 30 cm. Any angle less than that will create an even larger head. Weir formulas depend primarily on gravity and geometry. Fluid properties such as (ρ, μ, Y) enter only as slight modifications or as correction factors.

Backwater Curves

A weir is a flow barrier that not only alters the local flow over the weir but also modifies the flow depth distribution far upstream. Any strong barrier in an open-channel flow creates a *backwater curve*, which can be computed by the gradually varied flow theory of Sec. 10.6. If Q is known, the weir formula, Eq. (10.55), determines H and hence the water depth just upstream of the weir, $y = H + Y$, where Y is the weir height. We then compute $y(x)$ upstream of the weir from Eq. (10.51), following in this case an M-1 curve (Fig. 10.14c). Such a barrier, where the water depth correlates with the flow rate, is called a channel *control point*. These are the starting points for numerical analysis of floodwater profiles in rivers [26].

EXAMPLE 10.13

A rectangular channel 8 m wide, with a flow rate of 30 m^3/s, encounters a 4-m-high sharp-edged dam, as shown in Fig. E10.13a. Determine the water depth 2 km upstream if the channel slope is $S_0 = 0.0004$ and $n = 0.025$.

Solution

First determine the head H produced by the dam, using sharp-crested full-width weir theory, Eq. (10.56):

$$Q = 30 \text{ m}^3/\text{s} = C_d b g^{1/2} H^{3/2} = \left(0.564 + 0.0846\frac{H}{4 \text{ m}}\right)(8 \text{ m})(9.81 \text{ m/s}^2)^{1/2} H^{3/2}$$

Since the term $0.0846H/4$ in parentheses is small, we may proceed by iteration or EES to the solution $H \approx 1.59$ m. Then our initial condition at $x = 0$, just upstream of the dam, is $y(0) = Y + H = 4 + 1.59 = 5.59$ m. Compare this to the critical depth from Eq. (10.30):

$$y_c = \left(\frac{Q^2}{b^2 g}\right)^{1/3} = \left[\frac{(30 \text{ m}^3/\text{s})^2}{(8 \text{ m})^2 (9.81 \text{ m/s}^2)}\right]^{1/3} = 1.13 \text{ m}$$

Since $y(0)$ is greater than y_c, the flow upstream is subcritical. Finally, for reference purposes, estimate the normal depth from the Chézy equation (10.19):

$$Q = 30 \text{ m}^3/\text{s} = \frac{\alpha}{n} b y \, R_h^{2/3} S_0^{1/2} = \frac{1.0}{0.025}(8 \text{ m}) y_n \left(\frac{8 y_n}{8 + 2 y_n}\right)^{2/3} (0.0004)^{1/2}$$

By trial and error or EES solve for $y_n \approx 3.20$ m. If there are no changes in channel width or slope, the water depth far upstream of the dam will approach this value. All these reference values $y(0)$, y_c, and y_n are shown in Fig. E10.13b.

Since $y(0) > y_n > y_c$, the solution will be an M-1 curve as computed from gradually varied theory, Eq. (10.51), for a rectangular channel with the given input data:

$$\frac{dy}{dx} \approx \frac{S_0 - n^2 Q^2/(\alpha^2 A^2 R_h^{4/3})}{1 - Q^2 b_0/(g A^3)} \qquad \alpha = 1.0 \quad A = 8y \quad n = 0.025 \quad R_h = \frac{8y}{8 + 2y} \quad b_0 = 8$$

Beginning with $y = 5.59$ m at $x = 0$, we integrate backward to $x = -2000$ m. For the Runge-Kutta method, four-figure accuracy is achieved for $\Delta x = -100$ m. The complete solution curve is shown in Fig. E10.13b. The desired solution value is

At $x = -2000$ m: $\qquad\qquad\qquad\qquad y \approx 5.00$ m $\qquad\qquad\qquad\qquad$ *Ans.*

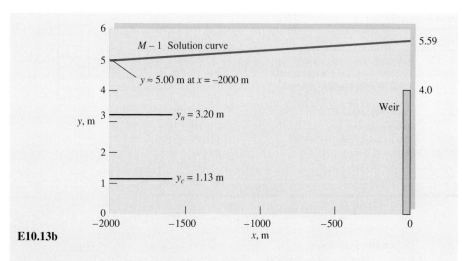

E10.13b

Thus, even 2 km upstream, the dam has produced a "backwater" that is 1.8 m above the normal depth that would occur without a dam. For this example, a near-normal depth of, say, 10 cm greater than y_n, or $y \approx 3.3$ m, would not be achieved until $x = -13{,}400$ m. Backwater curves are quite far-reaching upstream, especially in flood stages.

Summary

This chapter has introduced open-channel flow analysis, limited to steady, one-dimensional flow conditions. The basic analysis combines the continuity equation with the extended Bernoulli equation including friction losses.

Open-channel flows are classified either by depth variation or by Froude number, the latter being analogous to the Mach number in compressible duct flow (Chap. 9). Flow at constant slope and depth is called uniform flow and satisfies the classical Chézy equation (10.19). Straight prismatic channels can be optimized to find the cross section that gives maximum flow rate with minimum friction losses. As the slope and flow velocity increase, the channel reaches a *critical* condition of Froude number unity, where velocity equals the speed of a small-amplitude surface wave in the channel. Every channel has a critical slope that varies with the flow rate and roughness. If the flow becomes supercritical (Fr > 1), it may undergo a hydraulic jump to a greater depth and lower (subcritical) velocity, analogous to a normal shock wave.

The analysis of gradually varied flow leads to a differential equation (10.51) that can be solved by numerical methods. The chapter ends with a discussion of the flow over a dam or weir, where the total flow rate can be correlated with upstream water depth.

Problems

Most of the problems herein are fairly straightforward. More difficult or open-ended assignments are labeled with an asterisk. Problems labeled with an EES icon will benefit from the use of the Engineering Equation Solver (EES), while problems labeled with a computer icon may require the use of a computer. The standard end-of-chapter problems P10.1 to P10.128 (categorized in the problem list here) are followed by word problems W10.1 to W10.13, fundamentals of engineering exam problems FE10.1 to FE10.7, comprehensive problems C10.1 to C10.7, and design projects D10.1 and D10.2.

Problem Distribution

P10.1 The formula for shallow-water wave propagation speed, Eq. (10.9) or (10.10), is independent of the physical properties of the liquid, like density, viscosity, or surface tension. Does this mean that waves propagate at the same speed in water, mercury, gasoline, and glycerin? Explain.

P10.2 Water at 20°C flows in a 30-cm-wide rectangular channel at a depth of 10 cm and a flow rate of 80,000 cm³/s. Estimate (a) the Froude number and (b) the Reynolds number.

P10.3 Narragansett Bay is approximately 21 (statute) mi long and has an average depth of 42 ft. Tidal charts for the area indicate a time delay of 30 min between high tide at the mouth of the bay (Newport, Rhode Island) and its head (Providence, Rhode Island). Is this delay correlated with the propagation of a shallow-water tidal crest wave through the bay? Explain.

P10.4 The water channel flow in Fig. P10.4 has a free surface in three places. Does it qualify as an open-channel flow? Explain. What does the dashed line represent?

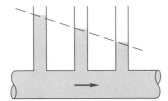

P10.4

P10.5 Water flows rapidly in a channel 25 cm deep. Piercing the surface with a pencil point creates a wedgelike downstream wave of included angle 38°. Estimate the velocity V of the water.

P10.6 Pebbles dropped successively at the same point, into a water channel flow of depth 42 cm, create two circular ripples, as in Fig. P10.6. From this information estimate (a) the Froude number and (b) the stream velocity.

P10.7 Pebbles dropped successively at the same point, into a water channel flow of depth 65 cm, create two circular ripples, as in Fig. P10.7. From this information

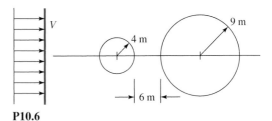

P10.6

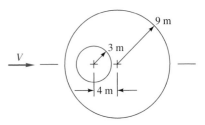

P10.7

estimate (a) the Froude number and (b) the stream velocity.

P10.8 An earthquake near the Kenai Peninsula, Alaska, creates a single "tidal" wave (called a *tsunami*) that propagates southward across the Pacific Ocean. If the average ocean depth is 4 km and seawater density is 1025 kg/m³, estimate the time of arrival of this tsunami in Hilo, Hawaii.

P10.9 Equation (10.10) is for a single disturbance wave. For *periodic* small-amplitude surface waves of wavelength λ and period T, inviscid theory [8 to 10] predicts a wave propagation speed

$$c_0^2 = \frac{g\lambda}{2\pi} \tanh \frac{2\pi y}{\lambda}$$

where y is the water depth and surface tension is neglected. (a) Determine if this expression is affected by the Reynolds number, Froude number, or Weber number. Derive the limiting values of this expression for (b) $y \ll \lambda$ and (c) $y \gg \lambda$. (d) For what ratio y/λ is the wave speed within 1 percent of limit (c)?

P10.10 If surface tension Υ is included in the analysis of Prob. P10.9, the resulting wave speed is [8 to 10]

$$c_0^2 = \left(\frac{g\lambda}{2\pi} + \frac{2\pi\Upsilon}{\rho\lambda} \right) \tanh \frac{2\pi y}{\lambda}$$

(a) Determine if this expression is affected by the Reynolds number, Froude number, or Weber number. Derive the limiting values of this expression for (b) $y \ll \lambda$ and (c) $y \gg \lambda$. (d) Finally determine the wavelength λ_{crit} for a minimum value of c_0, assuming that $y \gg \lambda$.

P10.11 A rectangular channel is 2 m wide and contains water 3 m deep. If the slope is 0.85° and the lining is corrugated metal, estimate the discharge for uniform flow.

P10.12 (*a*) For laminar draining of a wide thin sheet of water on pavement sloped at angle θ, as in Fig. P4.36, show that the flow rate is given by

$$Q = \frac{\rho g b h^3 \sin \theta}{3\mu}$$

where b is the sheet width and h its depth. (*b*) By (somewhat laborious) comparison with Eq. (10.13), show that this expression is compatible with a friction factor $f = 24/\text{Re}$, where $\text{Re} = V_{av}h/\nu$.

P10.13 The laminar-draining flow from Prob. P10.12 may undergo transition to turbulence if $\text{Re} > 500$. If the pavement slope is 0.0045, what is the maximum sheet thickness, in mm, for which laminar flow is ensured?

P10.14 The Chézy formula (10.18) is independent of fluid density and viscosity. Does this mean that water, mercury, alcohol, and SAE 30 oil will all flow down a given open channel at the same rate? Explain.

P10.15 The painted-steel channel of Fig. P10.15 is designed, without the barrier, for a flow rate of 6 m³/s at a normal depth of 1 m. Determine (*a*) the design slope of the channel and (*b*) the reduction in total flow rate if the proposed painted-steel central barrier is installed.

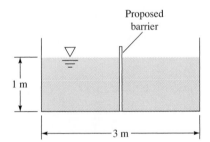

Proposed barrier

1 m

3 m

P10.15

P10.16 A brickwork rectangular channel is 120 cm wide and laid on a slope of 2 m per km. (*a*) Find the normal flow rate when the water depth is 40 cm. (*b*) For the same slope, find the water depth that will double the flow rate in part (*a*). Comment.

P10.17 The trapezoidal channel of Fig. P10.17 is made of brickwork and slopes at 1:500. Determine the flow rate if the normal depth is 80 cm.

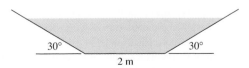

30°   30°

2 m

P10.17

P10.18 Modify Prob. P10.17 as follows. Determine the normal depth for which the flow rate will be 8 m³/s.

P10.19 Modify Prob. P10.17 as follows. Let the surface be clean earth, which erodes if V exceeds 1.5 m/s. What is the maximum depth to avoid erosion?

P10.20 A circular corrugated metal storm drain is flowing half full over a slope 4 ft/mi. Estimate the normal discharge if the drain diameter is 8 ft.

P10.21 An engineer makes careful measurements with a weir (see Sec. 10.7) that monitors a rectangular unfinished concrete channel laid on a slope of 1°. She finds, perhaps with surprise, that when the water depth doubles from 2 ft 2 inches to 4 ft 4 inches, the normal flow rate more than doubles, from 200 to 500 ft³/s. (*a*) Is this plausible? (*b*) If so, estimate the channel width.

P10.22 A trapezoidal aqueduct (Fig. 10.7) has $b = 5$ m and $\theta = 40°$ and carries a normal flow of 60 m³/s of water when $y = 3.2$ m. For clay tile surfaces, estimate the required elevation drop in m/km.

P10.23 It is desired to excavate a clean-earth channel as a trapezoidal cross section with $\theta = 60°$ (see Fig. 10.7). The expected flow rate is 500 ft³/s, and the slope is 8 ft per mile. The uniform flow depth is planned, for efficient performance, such that the flow cross section is half a hexagon. What is the appropriate bottom width of the channel?

P10.24 A riveted steel channel slopes at 1:500 and has a V shape with an included angle of 80°. Find the normal depth if the flow rate is 900 m³/h.

*P10.25 The equilateral-triangle channel in Fig. P10.25 has constant slope S_0 and constant Manning factor n. Find Q_{max} and V_{max}. Then, by analogy with Fig. 10.6*b*, plot the ratios Q/Q_{max} and V/V_{max} as a function of y/a for the complete range $0 < y/a < 0.866$.

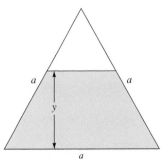

a a

y

P10.25 a

P10.26 In the spirit of Fig. 10.6*b*, analyze a rectangular channel in uniform flow with constant area $A = by$, constant slope, but varying width b and depth y. Plot the resulting flow rate Q, normalized by its maximum value Q_{max}, in the range $0.2 < b/y < 4.0$, and comment on whether it is crucial for discharge efficiency to have the channel flow at a depth exactly equal to half the channel width.

P10.27 A circular unfinished concrete water channel has a slope of 1:600 and a diameter of 5 ft. Estimate the normal discharge in gal/min for which the average wall shear stress is 0.15 lbf/ft^2, and compare your result to the maximum possible discharge for this channel.

P10.28 Show that, for any straight prismatic channel in uniform flow, the average wall shear stress is given by

$$\tau_{av} \approx \rho g R_h S_0$$

If you happen to spot this result early, you can use it in solving Prob. P10.27.

P10.29 Suppose that the trapezoidal channel of Fig. P10.17 contains sand and silt that we wish not to erode. According to an empirical correlation by A. Shields in 1936, the average wall shear stress τ_{crit} required to erode sand particles of diameter d_p is approximated by

$$\frac{\tau_{crit}}{(\rho_s - \rho)g\, d_p} \approx 0.5$$

where $\rho_s \approx 2400$ kg/m^3 is the density of sand. If the slope of the channel in Fig. P10.17 is 1:900 and $n \approx 0.014$, determine the maximum water depth to keep from eroding particles of 1-mm diameter.

P10.30 A clay tile V-shaped channel, with an included angle of 90°, is 1 km long and is laid out on a 1:400 slope. When running at a depth of 2 m, the upstream end is suddenly closed while the lower end continues to drain. Assuming quasi-steady normal discharge, find the time for the channel depth to drop to 20 cm.

P10.31 A storm drain has the cross section shown in Fig. P10.31 and is laid on a slope of 1.5 m/km. If it is constructed of brickwork, find the normal discharge when the water level passes through the center of the circle.

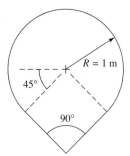

P10.31

P10.32 Does half a V-shaped channel perform as well as a full V-shaped channel? The answer to Prob. P10.24 is $y_n = 0.56$ m. (Do not reveal this to your friends still working on P10.24.) For the riveted-steel half-V in Fig. P10.32, find the normal depth when the flow is the same as in

P10.24, $Q = 900$ m^3/h. Compare the resulting flow area to Prob. P10.24.

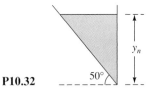

P10.32

P10.33 Five sewer pipes, each a 2-m-diameter clay tile pipe running half full on a slope of 0.25°, empty into a single asphalt pipe, also laid out at 0.25°. If the large pipe is also to run half full, what should be its diameter?

P10.34 A brick rectangular channel with $S_0 = 0.002$ is designed to carry 230 ft^3/s of water in uniform flow. There is an argument over whether the channel width should be 4 or 8 ft. Which design needs fewer bricks? By what percentage?

P10.35 In flood stage a natural channel often consists of a deep main channel plus two floodplains, as in Fig. P10.35. The floodplains are often shallow and rough. If the channel has the same slope everywhere, how would you analyze this situation for the discharge? Suppose that $y_1 = 20$ ft, $y_2 = 5$ ft, $b_1 = 40$ ft, $b_2 = 100$ ft, $n_1 = 0.020$, and $n_2 = 0.040$, with a slope of 0.0002. Estimate the discharge in ft^3/s.

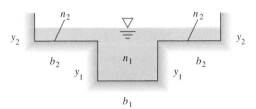

P10.35

P10.36 The Blackstone River in northern Rhode Island normally flows at about 25 m^3/s and resembles Fig. P10.35 with a clean-earth center channel, $b_1 \approx 20$ m and $y_1 \approx 3$ m. The bed slope is about 2 ft/mi. The sides are heavy brush with $b_2 \approx 150$ m. During hurricane Carol in 1955, a record flow rate of 1000 m^3/s was estimated. Use this information to estimate the maximum flood depth y_2 during this event.

P10.37 A triangular channel (see Fig. E10.6) is to be constructed of corrugated metal and will carry 8 m^3/s on a slope of 0.005. The supply of sheet metal is limited, so the engineers want to minimize the channel surface. What are (a) the best included angle θ for the channel, (b) the normal depth for part (a), and (c) the wetted perimeter for part (b)?

P10.38 An unfinished concrete conduit is running at one-quarter depth ($y = R/2$), as in Fig. P10.38. If the flow is uniform, at a rate of 2.03 m³/s, determine the slope of the conduit. The conduit diameter is 2 m. *Partial hint:* The flow area is $A = [2\theta - \sin(2\theta)]R^2/2$, with θ in radians.

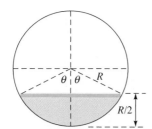

P10.38

P10.39 A trapezoidal channel has $n = 0.022$ and $S_0 = 0.0003$ and is made in the shape of a half-hexagon for maximum efficiency. What should the length of the side of the hexagon be if the channel is to carry 225 ft³/s of water? What is the discharge of a semicircular channel of the same cross-sectional area and the same S_0 and n?

P10.40 Using the geometry of Fig. 10.6a, prove that the most efficient circular open channel (maximum hydraulic radius for a given flow area) is a semicircle.

P10.41 Determine the most efficient value of θ for the V-shaped channel of Fig. P10.41.

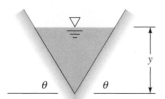

P10.41

P10.42 Suppose that the side angles of the trapezoidal channel in Prob. P10.39 are reduced to 15° to avoid earth slides. If the bottom flat width is 8 ft, (*a*) determine the normal depth and (*b*) compare the resulting wetted perimeter with the solution $P = 24.1$ ft from Prob. P10.39. (Do not reveal this answer to friends still struggling with Prob. P10.39.)

P10.43 What are the most efficient dimensions for a riveted steel rectangular channel to carry 4.8 m³/s at a slope of 1:900?

P10.44 What are the most efficient dimensions for a half-hexagon cast iron channel to carry 15,000 gal/min on a slope of 0.16°?

P10.45 Calculus tells us that the most efficient wall angle for a V-shaped channel (Fig. P10.41) is $\theta = 45°$. It yields the highest normal flow rate for a given area. But is this a sharp or a flat maximum? For a flow area of 1 m² and an unfinished-concrete channel with a slope of 0.004, plot the normal flow rate Q, in m³/s, versus angle for the range $30° \leq \theta \leq 60°$ and comment.

P10.46 It is suggested that a channel that reduces erosion has a parabolic shape, as in Fig. P10.46. Formulas for area and perimeter of the parabolic cross section are as follows [7, p. 36]:

$$A = \frac{2}{3}bh_0; \quad P = \frac{b}{2}\left[\sqrt{1 + \alpha^2} + \frac{1}{\alpha}\ln(\alpha + \sqrt{1 + \alpha^2})\right]$$

$$\text{where} \quad \alpha = \frac{4\,h_0}{b}$$

For uniform flow conditions, determine the most efficient ratio h_0/b for this channel (minimum perimeter for a given constant area).

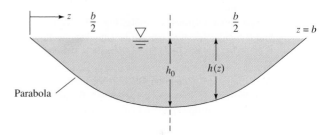

Parabola

P10.46

P10.47 Calculus tells us that the most efficient water depth for a rectangular channel (such as Fig. E10.1) is $y/b = 1/2$. It yields the highest normal flow rate for a given area. But is this a sharp or a flat maximum? For a flow area of 1 m² and a clay tile channel with a slope of 0.006, plot the normal flow rate Q, in m³/s, versus y/b for the range $0.3 \leq y/b \leq 0.7$ and comment.

P10.48 A wide, clean-earth river has a flow rate $q = 150$ ft³/(s · ft). What is the critical depth? If the actual depth is 12 ft, what is the Froude number of the river? Compute the critical slope by (*a*) Manning's formula and (*b*) the Moody chart.

P10.49 Find the critical depth of the brick channel in Prob. P10.34 for both the 4- and 8-ft widths. Are the normal flows subcritical or supercritical?

P10.50 A pencil point piercing the surface of a rectangular channel flow creates a wedgelike 25° half-angle wave, as in Fig. P10.50. If the channel surface is painted steel and the depth is 35 cm, determine (*a*) the Froude number, (*b*) the critical depth, and (*c*) the critical slope for uniform flow.

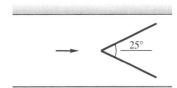

P10.50

P10.51   An asphalt circular channel, of diameter 75 cm, is flowing half-full at an average velocity of 3.4 m/s. Estimate (a) the volume flow rate, (b) the Froude number, and (c) the critical slope.

P10.52   Water flows full in an asphalt half-hexagon channel of bottom width W. The flow rate is 12 m³/s. Estimate W if the Froude number is exactly 0.60.

P10.53   For the river flow of Prob. P10.48, find the depth y_2 that has the same specific energy as the given depth $y_1 = 12$ ft. These are called *conjugate depths*. What is Fr_2?

P10.54   A clay tile V-shaped channel has an included angle of 70° and carries 8.5 m³/s. Compute (a) the critical depth, (b) the critical velocity, and (c) the critical slope for uniform flow.

P10.55   A trapezoidal channel resembles Fig. 10.7 with $b = 1$ m and $\theta = 50°$. The water depth is 2 m, and the flow rate is 32 m³/s. If you stick your fingernail in the surface, as in Fig. P10.50, what half-angle wave might appear?

P10.56   A riveted steel triangular duct flows partly full as in Fig. P10.56. If the critical depth is 50 cm, compute (a) the critical flow rate and (b) the critical slope.

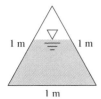

P10.56

P10.57   Consider the V-shaped channel of arbitrary angle in Fig. P10.41. If the depth is y, (a) find an analytic expression for the propagation speed c_0 of a small-disturbance wave along this channel. [*Hint:* Eliminate flow rate from the analyses in Sec. 10.4.] If $\theta = 45°$ and the depth is 1 m, determine (b) the propagation speed and (c) the flow rate if the channel is running at a Froude number of 1/3.

P10.58   A circular corrugated metal water channel is half-full and in uniform flow when laid on a slope of 0.0118. The average shear stress on the channel walls is 29 Pa. Estimate (a) the channel diameter, (b) the Froude number, and (c) the volume flow rate.

P10.59   Uniform water flow in a wide brick channel of slope 0.02° moves over a 10-cm bump as in Fig. P10.59. A slight depression in the water surface results. If the minimum water depth over the bump is 50 cm, compute (a) the velocity over the bump and (b) the flow rate per meter of width.

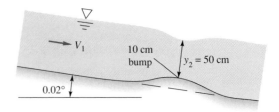

P10.59

P10.60   Modify Prob. P10.59 as follows. Again assuming uniform subcritical approach flow (V_1, y_1), find (a) the flow rate and (b) y_2 for which the Froude number Fr_2 at the crest of the bump is exactly 0.7.

P10.61   Modify Prob. P10.59 as follows. Again assuming uniform subcritical approach flow (V_1, y_1), find (a) the flow rate and (b) y_2 for which the flow at the crest of the bump is exactly critical ($Fr_2 = 1.0$).

P10.62   Consider the flow in a wide channel over a bump, as in Fig. P10.62. One can estimate the water depth change or *transition* with frictionless flow. Use continuity and the Bernoulli equation to show that

$$\frac{dy}{dx} = -\frac{dh/dx}{1 - V^2/(gy)}$$

Is the drawdown of the water surface realistic in Fig. P10.62? Explain under what conditions the surface might rise above its upstream position y_0.

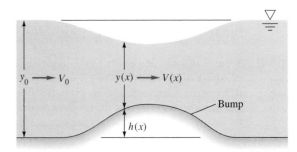

P10.62

P10.63   In Fig. P10.62 let $V_0 = 1$ m/s and $y_0 = 1$ m. If the maximum bump height is 15 cm, estimate (a) the Froude number over the top of the bump and (b) the maximum depression in the water surface.

P10.64 In Fig. P10.62 let $V_0 = 1$ m/s and $y_0 = 1$ m. If the flow over the top of the bump is exactly critical (Fr = 1.0), determine the bump height h_{max}.

P10.65 Program and solve the differential equation of "frictionless flow over a bump," from Prob. P10.62, for entrance conditions $V_0 = 1$ m/s and $y_0 = 1$ m. Let the bump have the convenient shape $h = 0.5h_{max}[1 - \cos (2\pi x/L)]$, which simulates Fig. P10.62. Let $L = 3$ m, and generate a numerical solution for $y(x)$ in the bump region $0 < x < L$. If you have time for only one case, use $h_{max} = 15$ cm (Prob. P10.63), for which the maximum Froude number is 0.425. If more time is available, it is instructive to examine a complete family of surface profiles for $h_{max} \approx 1$ cm up to 35 cm (which is the solution of Prob. P10.64).

P10.66 In Fig. P10.62 let $V_0 = 6$ m/s and $y_0 = 1$ m. If the maximum bump height is 35 cm, estimate (a) the Froude number over the top of the bump and (b) the maximum increase in the water surface level.

P10.67 In Fig. P10.62 let $V_0 = 5$ m/s and $y_0 = 1$ m. If the flow over the top of the bump is exactly critical (Fr = 1.0), determine the bump height h_{max}.

P10.68 Modify Prob. P10.65 to have a supercritical approach condition $V_0 = 6$ m/s and $y_0 = 1$ m. If you have time for only one case, use $h_{max} = 35$ cm (Prob. P10.66), for which the maximum Froude number is 1.47. If more time is available, it is instructive to examine a complete family of surface profiles for 1 cm $< h_{max} < 52$ cm (which is the solution to Prob. P10.67).

*P10.69 Given is the flow of a channel of large width b under a sluice gate, as in Fig. P10.69. Assuming frictionless steady flow with negligible upstream kinetic energy, derive a formula for the dimensionless flow ratio $Q^2/(y_1^3 b^2 g)$ as a function of the ratio y_2/y_1. Show by differentiation that the maximum flow rate occurs at $y_2 = 2y_1/3$.

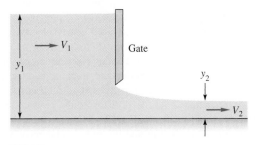

P10.69

P10.70 The spectacular water release in the chapter-opener photo flows through a giant sluice gate. Assume that the gate is 23 m wide, and its opening is 8 m high.

The water depth far upstream is 32 m. Assuming free discharge, estimate the volume flow rate through the gate.

P10.71 In Fig. P10.69 let $y_1 = 95$ cm and $y_2 = 50$ cm. Estimate the flow rate per unit width if the upstream kinetic energy is (a) neglected and (b) included.

*P10.72 Water approaches the wide sluice gate of Fig. P10.72 at $V_1 = 0.2$ m/s and $y_1 = 1$ m. Accounting for upstream kinetic energy, estimate at the outlet, section 2, the (a) depth, (b) velocity, and (c) Froude number.

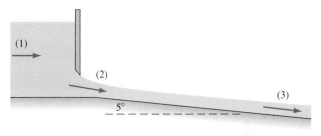

P10.72

P10.73 In Fig. P10.69 suppose that $y_1 = 1.4$ m and the gate is raised so that its gap is 15 cm. Estimate the resulting flow rate per unit width and the downstream depth.

P10.74 With respect to Fig. P10.69, show that, for frictionless flow, the upstream velocity may be related to the water levels by

$$V_1 = \sqrt{\frac{2g(y_1 - y_2)}{K^2 - 1}}$$

where $K = y_1/y_2$.

P10.75 A tank of water 1 m deep, 3 m long, and 4 m wide into the paper has a closed sluice gate on the right side, as in Fig. P10.75. At $t = 0$ the gate is opened to a gap of 10 cm. Assuming quasi-steady sluice gate theory, estimate the time required for the water level to drop to 50 cm. Assume free outflow.

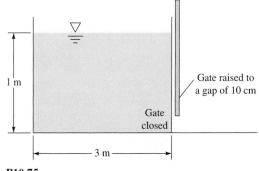

P10.75

P10.76   Figure P10.76 shows a horizontal flow of water through a sluice gate, a hydraulic jump, and over a 6-ft sharp-crested weir. Channel, gate, jump, and weir are all 8 ft wide unfinished concrete. Determine (a) the flow rate in ft³/s and (b) the normal depth.

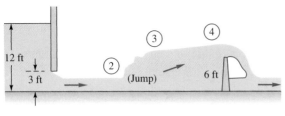

P10.76

*P10.77   Equation (10.41) for the discharge coefficient is for free (nearly frictionless) outflow. If the outlet is *drowned*, as in Fig. 10.10c, there is dissipation and C_d drops sharply. Figure P10.77 shows data from Ref. 2 on drowned vertical sluice gates. Use this chart to repeat Prob. P10.73, and plot the estimated flow rate versus y_2 in the range $0 < y_2 < 110$ cm.

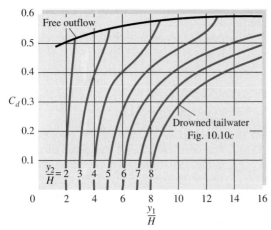

P10.77 *(From Ref. 2, p. 509.)*

P10.78   Repeat Prob. P10.75 if the gate is drowned at $y_2 = 40$ cm.

P10.79   Show that the Froude number downstream of a hydraulic jump will be given by

$$\text{Fr}_2 = 8^{1/2}\,\text{Fr}_1/[(1 + 8\,\text{Fr}_1^2)^{1/2} - 1]^{3/2}$$

Does the formula remain correct if we reverse subscripts 1 and 2? Why?

P10.80   Water, flowing horizontally in a wide channel of depth 30 cm, undergoes a hydraulic jump whose energy

dissipation is 71 percent. Estimate (a) the downstream depth and (b) the volume flow rate per meter of width.

P10.81   Water flows in a wide channel at $q = 25$ ft³/(s · ft), $y_1 = 1$ ft, and then undergoes a hydraulic jump. Compute y_2, V_2, Fr_2, h_f, the percentage of dissipation, and the horsepower dissipated per unit width. What is the critical depth?

P10.82   Downstream of a wide hydraulic jump the flow is 4 ft deep and has a Froude number of 0.5. Estimate (a) y_1, (b) V_1, (c) Fr_1, (d) the percentage of dissipation, and (e) y_c.

P10.83   A wide-channel flow undergoes a hydraulic jump from 40 to 140 cm. Estimate (a) V_1, (b) V_2, (c) the critical depth, in cm, and (d) the percentage of dissipation.

*P10.84   Consider the flow under the sluice gate of Fig. P10.84. If $y_1 = 10$ ft and all losses are neglected except the dissipation in the jump, calculate y_2 and y_3 and the percentage of dissipation, and sketch the flow to scale with the EGL included. The channel is horizontal and wide.

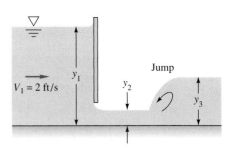

P10.84

P10.85   In Prob. P10.72 the exit velocity from the sluice gate is 4.33 m/s. If there is a hydraulic jump just downstream of section 2, determine the downstream (a) velocity, (b) depth, (c) Froude number, and (d) percentage of dissipation. Neglect the effect of the nonhorizontal bottom (see Prob. P10.91).

P10.86   A *bore* is a hydraulic jump that propagates upstream into a still or slower-moving fluid, as in Fig. 10.4a. Suppose that the still water is 2 m deep and the water behind the bore is 3 m deep. Estimate (a) the propagation speed of the bore and (b) the induced water velocity.

P10.87   A *tidal bore* may occur when the ocean tide enters an estuary against an oncoming river discharge, such as on the Severn River in England. Suppose that the tidal bore is 10 ft deep and propagates at 13 mi/h upstream into a river that is 7 ft deep. Estimate the river current in kn.

P10.88   For the situation in Fig. P10.84, suppose that at section 3 the depth is 2 m and the Froude number is 0.25. Estimate (a) the flow rate per meter of width, (b) y_c, (c) y_1, (d) the percentage of dissipation in the jump, and (e) the gap height H of the gate.

P10.89   Water 30 cm deep is in uniform flow down a 1° unfinished-concrete slope when a hydraulic jump occurs, as in Fig.

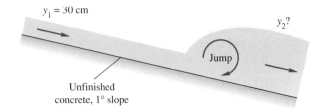

P10.89

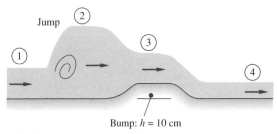

P10.95

P10.89. If the channel is very wide, estimate the water depth y_2 downstream of the jump.

P10.90 For the gate/jump/weir system sketched in Fig. P10.76, the flow rate was determined to be 379 ft³/s. Determine (a) the water depths y_2 and y_3, and (b) the Froude numbers Fr_2 and Fr_3 before and after the hydraulic jump.

*P10.91 No doubt you used the horizontal-jump formula (10.43) to solve Prob. P10.89, which is reasonable since the slope is so small. However, Chow [2, p. 425] points out that hydraulic jumps are *higher* on sloped channels, due to "the weight of the fluid in the jump." Make a control volume sketch of a sloping jump to show why this is so. The sloped-jump chart given in Chow's Fig. 15–20 may be approximated by the following curve fit:

$$\frac{2y_2}{y_1} \approx [(1 + 8\,Fr_1^2)^{1/2} - 1]e^{3.5S_0}$$

where $0 < S_0 < 0.3$ are the channel slopes for which data are available. Use this correlation to modify your solution to Prob. P10.89. If time permits, make a graph of y_2/y_1 (≤ 20) versus Fr_1 (≤ 15) for various S_0 (≤ 0.3).

P10.92 At the bottom of an 80-ft-wide spillway is a horizontal hydraulic jump with water depths 1 ft upstream and 10 ft downstream. Estimate (a) the flow rate and (b) the horsepower dissipated.

P10.93 Water in a horizontal channel accelerates smoothly over a bump and then undergoes a hydraulic jump, as in Fig. P10.93. If $y_1 = 1$ m and $y_3 = 40$ cm, estimate (a) V_1, (b) V_3, (c) y_4, and (d) the bump height h.

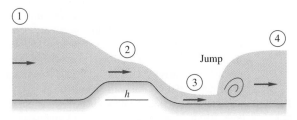

P10.93

P10.94 Water flows in a wide channel and undergoes a hydraulic jump. Before the jump, $V_1 = 6.0$ m/s. After the jump, $V_2 = 2.0$ m/s. Determine (a) y_1, (b) y_2, (c) Fr_1, and (d) Fr_2.

P10.95 A 10-cm-high bump in a wide horizontal water channel creates a hydraulic jump just upstream and the flow pattern in Fig. P10.95. Neglecting losses except in the jump, for the case $y_3 = 30$ cm, estimate (a) V_4, (b) y_4, (c) V_1, and (d) y_1.

P10.96 Show that the Froude numbers on either side of a wide hydraulic jump are related by the simple relation $Fr_2 = Fr_1(y_1/y_2)^{3/2}$.

P10.97 A brickwork rectangular channel 4 m wide is flowing at 8.0 m³/s on a slope of 0.1°. Is this a mild, critical, or steep slope? What type of gradually varied solution curve are we on if the local water depth is (a) 1 m, (b) 1.5 m, and (c) 2 m?

P10.98 A gravelly earth wide channel is flowing at 10 m³/s per meter of width on a slope of 0.75°. Is this a mild, critical, or steep slope? What type of gradually varied solution curve are we on if the local water depth is (a) 1 m, (b) 2 m, or (c) 3 m?

P10.99 A clay tile V-shaped channel of included angle 60° is flowing at 1.98 m³/s on a slope of 0.33°. Is this a mild, critical, or steep slope? What type of gradually varied solution curve are we on if the local water depth is (a) 1 m, (b) 2 m, or (c) 3 m?

P10.100 If bottom friction is included in the sluice gate flow of Prob. P10.84, the depths (y_1, y_2, y_3) will vary with x. Sketch the type and shape of gradually varied solution curve in each region (1, 2, 3), and show the regions of rapidly varied flow.

P10.101 Consider the gradual change from the profile beginning at point a in Fig. P10.101 on a mild slope S_{01} to a mild but steeper slope S_{02} downstream. Sketch and label the curve $y(x)$ expected.

*P10.102 The wide-channel flow in Fig. P10.102 changes from a steep slope to one even steeper. Beginning at points a and b, sketch and label the water surface profiles expected for gradually varied flow.

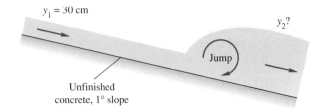

Within figure P10.89: $y_1 = 30$ cm; y_2?; Jump; Unfinished concrete, 1° slope

Within figure P10.95: Jump; (1) (2) (3) (4); Bump: $h = 10$ cm

Within figure P10.93: (1) (2) (3) (4); Jump; h

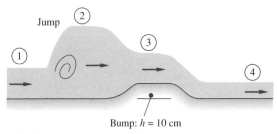

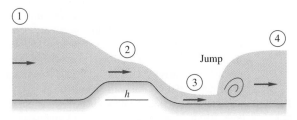

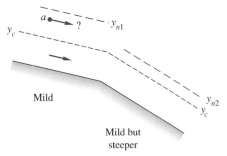

P10.101

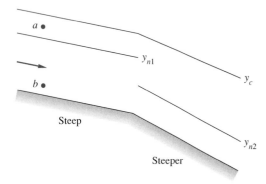

P10.102

P10.103 A circular painted-steel channel, of radius 50 cm, is running half-full at 1.2 m³/s on a slope of 5 m/km. Determine (a) whether the slope is mild or steep and (b) what type of gradually varied solution applies at this point. (c) Use the approximate method of Eq. (10.52), and a single depth increment $\Delta y = 5$ cm, to calculate the estimated Δx for this new y.

P10.104 The rectangular-channel flow in Fig. P10.104 expands to a cross section 50 percent wider. Beginning at points a and b, sketch and label the water surface profiles expected for gradually varied flow.

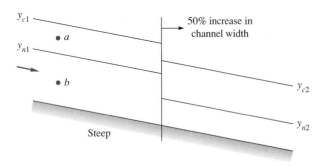

P10.104

P10.105 In Prob. P10.84 the frictionless solution is $y_2 = 0.82$ ft, which we denote as $x = 0$ just downstream of the gate. If the channel is horizontal with $n = 0.018$ and there is no hydraulic jump, compute from gradually varied theory the downstream distance where $y = 2.0$ ft.

P10.106 A rectangular channel with $n = 0.018$ and a constant slope of 0.0025 increases its width linearly from b to $2b$ over a distance L, as in Fig. P10.106. (a) Determine the variation $y(x)$ along the channel if $b = 4$ m, $L = 250$ m, the initial depth is $y(0) = 1.05$ m, and the flow rate is 7 m³/s. (b) Then, if your computer program is running well, determine the initial depth $y(0)$ for which the exit flow will be exactly critical.

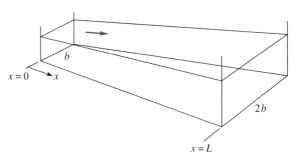

P10.106

P10.107 A clean-earth wide-channel flow is climbing an adverse slope with $S_0 = -0.002$. If the flow rate is $q = 4.5$ m³/(s · m), use gradually varied theory to compute the distance for the depth to drop from 3.0 to 2.0 m.

P10.108 Water flows at 1.5 m³/s along a straight, riveted-steel 90° V-shaped channel (see Fig. 10.41, $\theta = 45°$). At section 1, the water depth is 1.0 m. (a) As we proceed downstream, will the water depth rise or fall? Explain. (b) Depending upon your answer to part (a), calculate, in one numerical swoop, from gradually varied theory, the distance downstream for which the depth rises (or falls) 0.1 m.

P10.109 Figure P10.109 illustrates a free overfall or *dropdown* flow pattern, where a channel flow accelerates down a slope and falls freely over an abrupt edge. As shown, the flow reaches critical just before the overfall. Between y_c and the edge the flow is rapidly varied and does not satisfy gradually varied theory. Suppose that the flow rate is $q = 1.3$ m³/(s · m) and the surface is unfinished concrete. Use Eq. (10.51) to estimate the water depth 300 m upstream as shown.

P10.110 We assumed frictionless flow in solving the bump case, Prob. P10.65, for which $V_2 = 1.21$ m/s and $y_2 = 0.826$ m over the crest when $h_{max} = 15$ cm, $V_1 = 1$ m/s, and

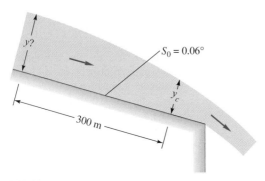

P10.109

find the depth y_2 for which the flow becomes critical in the throat.

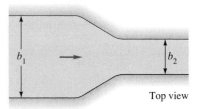

Top view

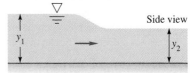

Side view

P10.113

$y_1 = 1$ m. However, if the bump is long and rough, friction may be important. Repeat Prob. P10.65 for the same bump shape, $h = 0.5h_{max}[1 - \cos(2\pi x/L)]$, to compute conditions (a) at the crest and (b) at the end of the bump, $x = L$. Let $h_{max} = 15$ cm and $L = 100$ m, and assume a clean-earth surface.

P10.111 Solve Prob. P10.105 (a horizontal variation along an H-3 curve) by the approximate method of Eq. (10.52), beginning at $(x, y) = (0, 0.82$ ft) and using a depth increment $\Delta y = 0.2$ ft. (The final increment should be $\Delta y = 0.18$ ft to bring us exactly to $y = 2.0$ ft.)

P10.112 The clean-earth channel in Fig. P10.112 is 6 m wide and slopes at 0.3°. Water flows at 30 m³/s in the channel and enters a reservoir so that the channel depth is 3 m just before the entry. Assuming gradually varied flow, how far is the distance L to a point in the channel where $y = 2$ m? What type of curve is the water surface?

P10.114 For the gate/jump/weir system sketched in Fig. P10.76, the flow rate was determined to be 379 ft³/s. Determine the water depth y_4 just upstream of the weir.

P10.115 Gradually varied theory, Eq. (10.49), neglects the effect of *width* changes, db/dx, assuming that they are small. But they are not small for a short, sharp contraction such as the venturi flume in Fig. P10.113. Show that, for a rectangular section with $b = b(x)$, Eq. (10.49) should be modified as follows:

$$\frac{dy}{dx} \approx \frac{S_0 - S + [V^2/(gb)](db/dx)}{1 - \text{Fr}^2}$$

Investigate a criterion for reducing this relation to Eq. (10.49).

P10.116 A Cipolletti weir, popular in irrigation systems, is trapezoidal, with sides sloped at 1:4 horizontal to vertical, as in Fig. P10.116. The following are flow-rate values, from the U.S. Dept. of Agriculture, for a few different system parameters:

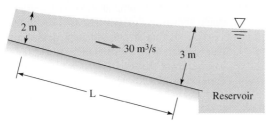

P10.112

P10.113 Figure P10.113 shows a channel contraction section often called a *venturi flume* [23, p. 167] because measurements of y_1 and y_2 can be used to meter the flow rate. Show that if losses are neglected and the flow is one-dimensional and subcritical, the flow rate is given by

$$Q = \left[\frac{2g(y_1 - y_2)}{1/(b_2^2 y_2^2) - 1/(b_1^2 y_1^2)}\right]^{1/2}$$

Apply this to the special case $b_1 = 3$ m, $b_2 = 2$ m, and $y_1 = 1.9$ m. (a) Find the flow rate if $y_2 = 1.5$ m. (b) Also

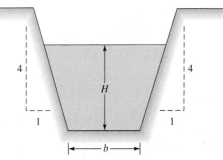

P10.116

H, ft	0.8	1.0	1.35	1.5
b, ft	1.5	2.0	2.5	3.5
Q, gal/min	1620	3030	5920	9740

Use this data to correlate a Cipolletti weir formula with a reasonably constant weir coefficient.

P10.117 A popular flow-measurement device in agriculture is the *Parshall flume* [33], Fig. P10.117, named after its inventor, Ralph L. Parshall, who developed it in 1922 for the U.S. Bureau of Reclamation. The subcritical approach flow is driven, by a steep constriction, to go critical ($y = y_c$) and then supercritical. It gives a constant reading H for a wide range of tailwaters. Derive a formula for estimating Q from measurement of H and knowledge of constriction width b. Neglect the entrance velocity head.

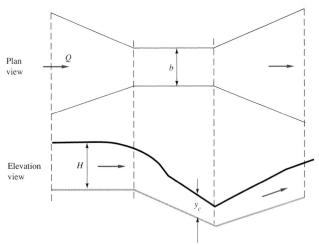

P10.117 The Parshall flume

***P10.118** Using a Bernoulli-type analysis similar to Fig. 10.16a, show that the theoretical discharge of the V-shaped weir in Fig. P10.118 is given by

$$Q = 0.7542g^{1/2} \tan \alpha \, H^{5/2}$$

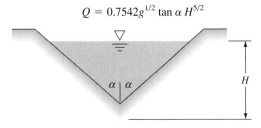

***P10.118**

P10.119 Data by A. T. Lenz for water at 20°C (reported in Ref. 23) show a significant increase of discharge coefficient of V-notch weirs (Fig. P10.118) at low heads. For $\alpha = 20°$, some measured values are as follows:

H, ft	0.2	0.4	0.6	0.8	1.0
C_d	0.499	0.470	0.461	0.456	0.452

Determine if these data can be correlated with the Reynolds and Weber numbers vis-à-vis Eq. (10.61). If not, suggest another correlation.

P10.120 The rectangular channel in Fig. P10.120 contains a V-notch weir as shown. The intent is to meter flow rates between 2.0 and 6.0 m³/s with an upstream hook gage set to measure water depths between 2.0 and 2.75 m. What are the most appropriate values for the notch height Y and the notch half-angle α?

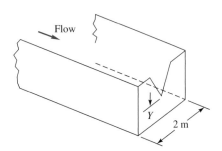

P10.120

P10.121 Water flow in a rectangular channel is to be metered by a thin-plate weir with side contractions, as in Table 10.2b, with $L = 6$ ft and $Y = 1$ ft. It is desired to measure flow rates between 1500 and 3000 gal/min with only a 6-in change in upstream water depth. What is the most appropriate length for the weir width b?

P10.122 In 1952 E. S. Crump developed the triangular weir shape shown in Fig. P10.122 [23, Chap. 4]. The front slope is 1:2 to avoid sediment deposition, and the rear slope is 1:5 to maintain a stable tailwater flow. The beauty of the design is that it has a unique discharge correlation up to near-drowning conditions, $H_2/H_1 \leq 0.75$:

$$Q = C_d bg^{1/2}\left(H_1 + \frac{V_1^2}{2g} - k_h\right)^{3/2}$$

where $C_d \approx 0.63$ and $k_h \approx 0.3$ mm

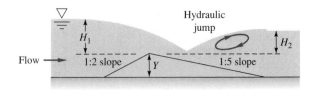

P10.122 The Crump weir [23, Chap. 4]

The term k_h is a low-head loss factor. Suppose that the weir is 3 m wide and has a crest height $Y = 50$ cm. If the water depth upstream is 65 cm, estimate the flow rate in gal/min.

*P10.123 The Crump weir calibration in Prob. P10.122 is for *modular* flow, which occurs when the flow rate is independent of downstream tailwater. When the weir becomes drowned, the flow rate decreases by the following factor:

$$Q = Q_{\text{mod}} f$$

where

$$f \approx 1.035 \left[0.817 - \left(\frac{H_2^*}{H_1^*} \right)^4 \right]^{0.0647}$$

for $0.70 \le H_2^*/H_1^* \le 0.93$, where H^* denotes $H_1 + V_1^2/(2g) - k_h$ for brevity. The weir is then *double-gaged* to measure both H_1 and H_2. Suppose that the weir crest is 1 m high and 2 m wide. If the measured upstream and downstream water depths are 2.0 and 1.9 m, respectively, estimate the flow rate in gal/min. Comment on the possible uncertainty of your estimate.

P10.124 Water flows at 600 ft³/s in a rectangular channel 22 ft wide with $n \approx 0.024$ and a slope of 0.1°. A dam increases the depth to 15 ft, as in Fig. P10.124. Using gradually varied theory, estimate the distance L upstream at which

the water depth will be 10 ft. What type of solution curve are we on? What should be the water depth asymptotically far upstream?

P10.125 The Tupperware dam on the Blackstone River is 12 ft high, 100 ft wide, and sharp-edged. It creates a backwater similar to Fig. P10.124. Assume that the river is a weedy-earth rectangular channel 100 ft wide with a flow rate of 800 ft³/s. Estimate the water depth 2 mi upstream of the dam if $S_0 = 0.001$.

P10.126 Suppose that the rectangular channel of Fig. P10.120 is made of riveted steel and carries a flow of 8 m³/s on a slope of 0.15°. If the V-notch weir has $\alpha = 30°$ and $Y = 50$ cm, estimate, from gradually varied theory, the water depth 100 m upstream.

P10.127 A horizontal gravelly earth channel 2 m wide contains a full-width Crump weir (Fig. P10.122) 1 m high. If the weir is not drowned, estimate, from gradually varied theory, the flow rate for which the water depth 100 m upstream will be 2 m.

P10.128 A rectangular channel 4 m wide is blocked by a broad-crested weir 2 m high, as in Fig. P10.128. The channel is horizontal for 200 m upstream and then slopes at 0.7° as shown. The flow rate is 12 m³/s, and $n = 0.03$. Compute the water depth y at 300 m upstream from gradually varied theory.

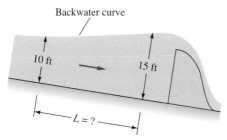

P10.124

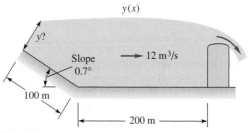

P10.128

Word Problems

W10.1 Free-surface problems are driven by gravity. Why do so many of the formulas in this chapter contain the *square root* of the acceleration of gravity?

W10.2 Explain why the flow under a sluice gate, Fig. 10.10, either is or is not analogous to compressible gas flow through a converging–diverging nozzle, Fig. 9.12.

W10.3 In uniform open-channel flow, what is the balance of forces? Can you use such a force balance to derive the Chézy equation (10.13)?

W10.4 A shallow-water wave propagates at the speed $c_0 \approx (gy)^{1/2}$. What makes it propagate? That is, what is the

balance of forces in such wave motion? In which direction does such a wave propagate?

W10.5 Why is the Manning friction correlation, Eq. (10.16), used almost universally by hydraulics engineers, instead of the Moody friction factor?

W10.6 During horizontal channel flow over a bump, is the specific energy constant? Explain.

W10.7 Cite some similarities, and perhaps some dissimilarities, between a hydraulic jump and a gas dynamic normal shock wave.

W10.8 Give three examples of rapidly varied flow. For each case, cite reasons why it does not satisfy one or more of the five basic assumptions of gradually varied flow theory.

W10.9   Is a free overfall, Fig. 10.15e, similar to a weir? Could it be calibrated versus flow rate in the same manner as a weir? Explain.

W10.10 Cite some similarities, and perhaps some dissimilarities, between a weir and a Bernoulli obstruction flowmeter from Sec. 6.12.

W10.11 Is a bump, Fig. 10.9a, similar to a weir? If not, when does a bump become large enough, or sharp enough, to be a weir?

W10.12 After doing some reading and/or thinking, explain the design and operation of a *long-throated flume*.

W10.13 Describe the design and operation of a *critical-depth flume*. What are its advantages compared to the venturi flume of Prob. P10.113?

Fundamentals of Engineering Exam Problems

The FE Exam is fairly light on open-channel problems in the general (morning) session, but this subject plays a big part in the specialized civil engineering (afternoon) exam.

FE10.1   Consider a rectangular channel 3 m wide laid on a $1°$ slope. If the water depth is 2 m, the hydraulic radius is
(a) 0.43 m, (b) 0.6 m, (c) 0.86 m, (d) 1.0 m, (e) 1.2 m

FE10.2   For the channel of Prob. FE10.1, the most efficient water depth (best flow for a given slope and resistance) is (a) 1 m, (b) 1.5 m, (c) 2 m, (d) 2.5 m, (e) 3 m

FE10.3   If the channel of Prob. FE10.1 is built of rubble cement (Manning's $n \approx 0.020$), what is the uniform flow rate when the water depth is 2 m?
(a) 6 m^3/s, (b) 18 m^3/s, (c) 36 m^3/s, (d) 40 m^3/s, (e) 53 m^3/s

FE10.4   For the channel of Prob. FE10.1, if the water depth is 2 m and the uniform flow rate is 24 m^3/s, what is the approximate value of Manning's roughness factor n?
(a) 0.015, (b) 0.020, (c) 0.025, (d) 0.030, (e) 0.035

FE10.5   For the channel of Prob. FE10.1, if Manning's roughness factor $n \approx 0.020$ and $Q \approx 29$ m^3/s, what is the normal depth y_n?
(a) 1 m, (b) 1.5 m, (c) 2 m, (d) 2.5 m, (e) 3 m

FE10.6   For the channel of Prob. FE10.1, if $Q \approx 24$ m^3/s, what is the critical depth y_c?
(a) 1.0 m, (b) 1.26 m, (c) 1.5 m, (d) 1.87 m, (e) 2.0 m

FE10.7   For the channel of Prob. FE10.1, if $Q \approx 24$ m^3/s and the depth is 2 m, what is the Froude number of the flow?
(a) 0.50, (b) 0.77, (c) 0.90, (d) 1.00, (e) 1.11

Comprehensive Problems

C10.1   February 1998 saw the failure of the earthen dam impounding California Jim's Pond in southern Rhode Island. The resulting flood raised temporary havoc in the nearby village of Peace Dale. The pond is 17 acres in area and 15 ft deep and was full from heavy rains. The breach in the dam was 22 ft wide and 15 ft deep. Estimate the time required for the pond to drain to a depth of 2 ft.

C10.2   A circular, unfinished concrete drainpipe is laid on a slope of 0.0025 and is planned to carry from 50 to 300 ft^3/s of runoff water. Design constraints are that (1) the water depth should be no more than three-fourths of the diameter and (2) the flow should always be subcritical. What is the appropriate pipe diameter to satisfy these requirements? If no commercial pipe is exactly this calculated size, should you buy the next smallest or the next largest pipe?

C10.3   Extend Prob. P10.72, whose solution was $V_2 \approx 4.33$ m/s. (a) Use gradually varied theory to estimate the water depth 10 m downstream at section (3) for the $5°$ unfin-

ished concrete slope shown in Fig. P10.72. (b) Repeat your calculation for an *upward* (adverse) slope of $5°$. (c) When you find that part (b) is impossible with gradually varied theory, explain why and repeat for an adverse slope of $1°$.

C10.4   It is desired to meter an asphalt rectangular channel of width 1.5 m, which is designed for uniform flow at a depth of 70 cm and a slope of 0.0036. The vertical sides of the channel are 1.2 m high. Consider using a thin-plate rectangular weir, either full or partial width (Table 10.2a,b) for this purpose. Sturm [7, p. 51] recommends, for accurate correlation, that such a weir have $Y \geq 9$ cm and $H/Y \leq 2.0$. Determine the feasibility of installing such a weir that will be accurate and yet not cause the water to overflow the sides of the channel.

C10.5   Figure C10.5 shows a hydraulic model of a *compound weir,* one that combines two different shapes. (a) Other than measurement, for which it might be poor, what could be the engineering reason for such a weir? (b) For the prototype river, assume that both sections have sides at a

C10.5 *(Courtesy of the U.S. Army Corps of Engineers Waterways Experiment Station.)*

70° angle to the vertical, with the bottom section having a base width of 2 m and the upper section having a base width of 4.5 m, including the cut-out portion. The heights of lower and upper horizontal sections are 1 m and 2 m, respectively. Use engineering estimates and make a plot of upstream water depth versus Petaluma River flow rate in the range 0 to 4 m^3/s. (c) For what river flow rate will the water overflow the top of the dam?

C10.6 Figure C10.6 shows a horizontal flow of water through a sluice gate, a hydraulic jump, and over a 6-ft sharp-crested weir. Channel, gate, jump, and weir are all 8 ft wide unfinished concrete. Determine (a) the flow rate, (b) the normal depth, (c) y_2, (d) y_3, and (e) y_4.

C10.7 Consider the V-shaped channel in Fig. C10.7, with an arbitrary angle θ. Make a continuity and momentum analysis of a small disturbance $\delta y \ll y$, as in Fig. 10.4. Show that the wave propagation speed in this channel is independent of θ and does *not* equal the wide-channel result $c_0 = (gy)^{1/2}$.

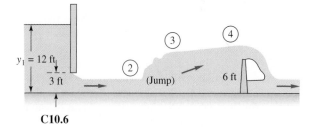

C10.6

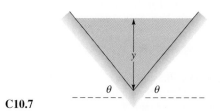

C10.7

Design Projects

D10.1 A straight weedy-earth channel has the trapezoidal shape of Fig. 10.7, with $b = 4$ m and $\theta = 35°$. The channel has a constant bottom slope of 0.001. The flow rate varies seasonally from 5 up to 10 m³/s. It is desired to place a sharp-edged weir across the channel so that the water depth 1 km upstream remains at 2.0 m ± 10 percent throughout the year. Investigate the possibility of accomplishing this with a full-width weir; if successful, determine the proper weir height Y. If unsuccessful, try other alternatives, such as (*a*) a full-width broad-crested weir or (*b*) a weir with side contractions or (*c*) a V-notch weir. Whatever your final design, cite the seasonal variation of normal depths and critical depths for comparison with the desired year-round depth of 2 m.

D10.2 The Caroselli Dam on the Pawcatuck River is 10 ft high, 90 ft wide, and sharp edged. The Coakley Company uses this head to generate hydropower electricity and wants *more* head. They ask the town for permission to raise the dam higher. The river above the dam may be approximated as rectangular, 90 ft wide, sloping upstream at 12 ft per statute mile, and with a stony, cobbled bed. The average flow rate is 400 ft³/s, with a 30-year predicted flood rate of 1200 ft³/s. The river sides are steep until 1 mi upstream, where there are low-lying residences. The town council agrees the dam may be heightened if the new river level near these houses, during the 30-year flood, is no more than 3 ft higher than the present level during average flow conditions. You, as project engineer, have to predict how high the dam crest can be raised and still meet this requirement.

References

1. B. A. Bakhmeteff, *Hydraulics of Open Channels,* McGraw-Hill, New York, 1932.
2. V. T. Chow, *Open Channel Hydraulics,* Prentice-Hall, Upper Saddle River, NJ, 1959.
3. M. H. Chaudhry, *Open Channel Flow,* Prentice-Hall, Upper Saddle River, NJ, 1993.
4. S. Montes, *Hydraulics of Open Channel Flow,* ASCE, Reston, VA, 1998.
5. H. Chanson, *The Hydraulics of Open Channel Flow,* 2d ed., Elsevier, New York, 2004.
6. S. C. Jain, *Open-Channel Flow,* Wiley, New York, 2000.
7. T. W. Sturm, *Open Channel Hydraulics,* McGraw-Hill, New York, 2001.
8. C. C. Mei, *Applied Dynamics of Ocean Surface Waves,* World Scientific Pub., Hackensack, NJ, 1994.
9. R. G. Dean and R. A. Dalrymple, *Water Wave Mechanics for Engineers and Scientists,* 2 vols., World Scientific Pub. Co., River Edge, NJ, 1991.
10. M. J. Lighthill, *Waves in Fluids,* Cambridge University Press, London, 2002.
11. A. T. Ippen, *Estuary and Coastline Hydrodynamics,* McGraw-Hill, New York, 1966.
12. M. B. Abbott and W. A. Price, *Coastal, Estuarial, and Harbor Engineers Reference Book,* Taylor & Francis, New York, 1994.
13. P. D. Komar, *Beach Processes and Sedimentation,* Pearson Education, Upper Saddle River, NJ, 1998.
14. W. Yue, C.-L. Lin, and V. C. Patel, "Large Eddy Simulation of Turbulent Open Channel Flow with Free Surface Simulated by Level Set Method," *Physics of Fluids,* vol. 17, no. 2, Feb. 2005, p. 025108.
15. J. M. Robertson and H. Rouse, "The Four Regimes of Open Channel Flow," *Civ. Eng.,* vol. 11, no. 3, March 1941, pp. 169–171.
16. R. W. Powell, "Resistance to Flow in Rough Channels," *Trans. Am. Geophys. Union,* vol. 31, no. 4, August 1950, pp. 575–582.
17. R. Manning, "On the Flow of Water in Open Channels and Pipes," *Trans. I.C.E. Ireland,* vol. 20, 1891, pp. 161–207.
18. "Friction Factors in Open Channels, Report of the Committee on Hydromechanics," *ASCE J. Hydraul. Div.,* March 1963, pp. 97–143.
19. E. F. Brater, H. W. King, J. E. Lindell, and C. Y. Wei, *Handbook of Hydraulics,* 7th ed., McGraw-Hill, New York, 1996.
20. U.S. Bureau of Reclamation, "Research Studies on Stilling Basins, Energy Dissipators, and Associated Appurtenances," *Hydraulic Lab. Rep.* Hyd-399, June 1, 1955.
21. P. A. Thompson, *Compressible-Fluid Dynamics,* McGraw-Hill, New York, 1972.
22. R. M. Olson and S. J. Wright, *Essentials of Engineering Fluid Mechanics,* 5th ed., Harper & Row, New York, 1990.
23. P. Ackers et al., *Weirs and Flumes for Flow Measurement,* Wiley, New York, 1978.
24. M. G. Bos, J. A. Replogle, and A. J. Clemmens, *Flow Measuring Flumes for Open Channel Systems,* American Soc. Agricultural and Biological Engineers, St. Joseph, MI, 1991.
25. M. G. Bos, *Long-Throated Plumes and Broad-Crested Weirs,* Springer-Verlag, New York, 1984.

26. D. H. Hoggan, *Computer-Assisted Floodplain Hydrology and Hydraulics,* 2d ed., McGraw-Hill, New York, 1996.

27. W. Rodi, *Turbulence Models and Their Application in Hydraulics,* A. A. Balkema, Leiden, The Netherlands, 1993.

28. P. D. Durbin, B. A. Pettersson, and B. A. Reif, *Theory and Modeling of Turbulent Flows,* Wiley, New York, 2001.

29. R. Baban, *Design of Diversion Weirs: Small Scale Irrigation in Hot Climates,* Wiley, New York, 1995.

30. H. Chanson, *Hydraulic Design of Stepped Cascades, Channels, Weirs, and Spillways,* Pergamon Press, New York, 1994.

31. D. Kampion and A. Brewer, *The Book of Waves: Form and Beauty on the Ocean,* 3d ed., Rowman and Littlefield, Lanham, MD, 1997.

32. L. Mays, *Water Resources Engineering,* Wiley, New York, 2005.

33. D. M. Grant and B. D. Dawson, *ISCO Open Channel Flow Measurement Handbook,* 5th ed., ISCO Inc., Lincoln, Nebraska, 1997.

With demand increasing for finite fossil fuel resources, renewable energy projects are crucial. As of October 2004, this 5M model, constructed by REpower Systems AG of Hamburg, Germany, became the world's largest wind turbine. The rotor is 126 m in diameter and, at peak conditions, delivers 5 MW of power. It rotates from 7 to 12 r/min, cuts in at winds of 3.5 m/s, and cuts out at 30 m/s. Subsequent models will be installed in the North Sea. [*Photo courtesy of REpower Systems AG.*]

Chapter 11
Turbomachinery

Motivation. The most common practical engineering application for fluid mechanics is the design of fluid machinery. The most numerous types are machines that *add* energy to the fluid (the pump family), but also important are those that *extract* energy (turbines). Both types are usually connected to a rotating shaft, hence the name *turbomachinery.*

The purpose of this chapter is to make elementary engineering estimates of the performance of fluid machines. The emphasis will be on nearly incompressible flow: liquids or low-velocity gases. Basic flow principles are discussed, but not the detailed construction of the machines.

11.1 Introduction and Classification

Turbomachines divide naturally into those that add energy (pumps) and those that extract energy (turbines). The prefix *turbo-* is a Latin word meaning "spin" or "whirl," appropriate for rotating devices.

The pump is the oldest fluid energy transfer device known. At least two designs date before Christ: (1) the undershot-bucket waterwheels, or *norias,* used in Asia and Africa (1000 B.C.) and (2) Archimedes' screw pump (250 B.C.), still being manufactured today to handle solid–liquid mixtures. Paddlewheel turbines were used by the Romans in 70 B.C., and Babylonian windmills date back to 700 B.C. [1].

Machines that deliver liquids are simply called *pumps,* but if gases are involved, three different terms are in use, depending on the pressure rise achieved. If the pressure rise is very small (a few inches of water), a gas pump is called a *fan;* up to 1 atm, it is usually called a *blower;* and above 1 atm it is commonly termed a *compressor.*

Classification of Pumps

There are two basic types of pumps: positive-displacement and dynamic or momentum-change pumps. There are several billion of each type in use in the world today.

Positive-displacement pumps (PDPs) force the fluid along by volume changes. A cavity opens, and the fluid is admitted through an inlet. The cavity then closes, and the fluid is squeezed through an outlet. The mammalian heart is a good example, and many mechanical designs are in wide use. References 35–38 give a summary of PDPs.

A brief classification of PDP designs is as follows:

A. Reciprocating
    1. Piston or plunger
    2. Diaphragm
B. Rotary
    1. Single rotor
        a. Sliding vane
        b. Flexible tube or lining
        c. Screw
        d. Peristaltic (wave contraction)
    2. Multiple rotors
        a. Gear
        b. Lobe
        c. Screw
        d. Circumferential piston

All PDPs deliver a pulsating or periodic flow as the cavity volume opens, traps, and squeezes the fluid. Their great advantage is the delivery of any fluid regardless of its viscosity.

Figure 11.1 shows schematics of the operating principles of seven of these PDPs. It is rare for such devices to be run backward, so to speak, as turbines or energy extractors, the steam engine (reciprocating piston) being a classic exception.

Since PDPs compress mechanically against a cavity filled with liquid, a common feature is that they develop immense pressures if the outlet is shut down for any reason. Sturdy construction is required, and complete shutoff would cause damage if pressure relief valves were not used.

Dynamic pumps simply add momentum to the fluid by means of fast-moving blades or vanes or certain special designs. There is no closed volume: The fluid increases momentum while moving through open passages and then converts its high velocity to a pressure increase by exiting into a diffuser section. Dynamic pumps can be classified as follows:

A. Rotary
    1. Centrifugal or radial exit flow
    2. Axial flow
    3. Mixed flow (between radial and axial)
B. Special designs
    1. Jet pump or ejector (see Fig. P3.36)
    2. Electromagnetic pumps for liquid metals
    3. Fluid-actuated: gas lift or hydraulic ram

We shall concentrate in this chapter on the rotary designs, sometimes called *rotodynamic pumps*. Other designs of both PDP and dynamic pumps are discussed in specialized texts [for example, 3, 31].

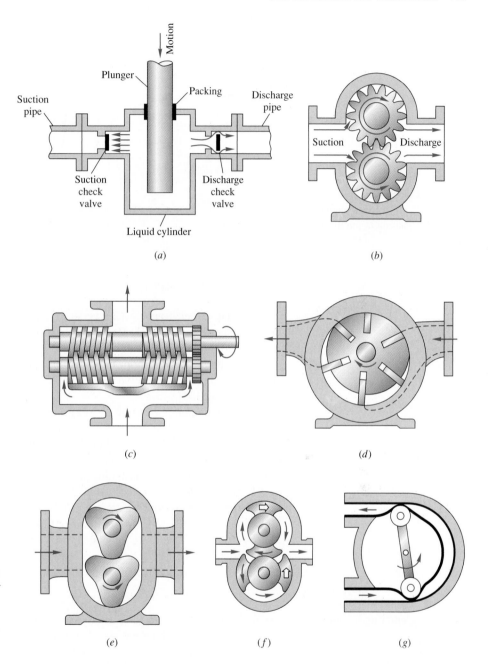

Fig. 11.1 Schematic design of positive-displacement pumps: (*a*) reciprocating piston or plunger, (*b*) external gear pump, (*c*) double-screw pump, (*d*) sliding vane, (*e*) three-lobe pump, (*f*) double circumferential piston, (*g*) flexible-tube squeegee.

Dynamic pumps generally provide a higher flow rate than PDPs and a much steadier discharge but are ineffective in handling high-viscosity liquids. Dynamic pumps also generally need *priming;* if they are filled with gas, they cannot suck up a liquid from below into their inlet. The PDP, on the other hand, is self-priming for most applications. A dynamic pump can provide very high flow rates (up to 300,000 gal/min) but usually with moderate pressure rises (a few atmospheres). In contrast, a PDP can

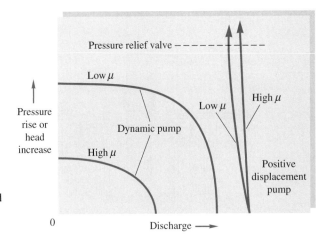

Fig. 11.2 Comparison of performance curves of typical dynamic and
positive-displacement pumps at
constant speed.

operate up to very high pressures (300 atm) but typically produces low flow rates
(100 gal/min).

The relative *performance* (Δp versus Q) is quite different for the two types of
pump, as shown in Fig. 11.2. At constant shaft rotation speed, the PDP produces nearly
constant flow rate and virtually unlimited pressure rise, with little effect of viscosity.
The flow rate of a PDP cannot be varied except by changing the displacement or the
speed. The reliable constant-speed discharge from PDPs has led to their wide use in
metering flows [35].

The dynamic pump, by contrast in Fig. 11.2, provides a continuous constant-speed
variation of performance, from near-maximum Δp at zero flow (shutoff conditions) to
zero Δp at maximum flow rate. High-viscosity fluids sharply degrade the performance
of a dynamic pump.

As usual—and for the last time in this text—we remind the reader that this is
merely an introductory chapter. Many books are devoted solely to turbomachines: generalized treatments [2 to 7], texts specializing in pumps [8 to 16, 30, 31], fans [17 to
20], compressors [21 to 23], gas turbines [24 to 26], hydropower [27, 28, 32], and
PDPs [35 to 38]. There are several useful handbooks [29 to 32], and at least two
undergraduate textbooks [33, 34] have a comprehensive discussion of turbomachines.
The reader is referred to these sources for further details.

11.2 The Centrifugal Pump

Let us begin our brief look at rotodynamic machines by examining the characteristics of the centrifugal pump. As sketched in Fig. 11.3, this pump consists of an
impeller rotating within a casing. Fluid enters axially through the *eye* of the casing,
is caught up in the impeller blades, and is whirled tangentially and radially outward
until it leaves through all circumferential parts of the impeller into the diffuser part
of the casing. The fluid gains both velocity and pressure while passing through the
impeller. The doughnut-shaped diffuser, or *scroll,* section of the casing decelerates the
flow and further increases the pressure.

The impeller blades are usually *backward-curved,* as in Fig. 11.3, but there are also
radial and forward-curved blade designs, which slightly change the output pressure.

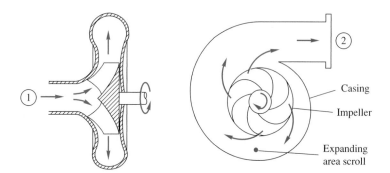

Fig. 11.3 Cutaway schematic of a typical centrifugal pump.

The blades may be *open* (separated from the front casing only by a narrow clearance) or *closed* (shrouded from the casing on both sides by an impeller wall). The diffuser may be *vaneless,* as in Fig. 11.3, or fitted with fixed vanes to help guide the flow toward the exit.

Basic Output Parameters

Assuming steady flow, the pump basically increases the Bernoulli head of the flow between point 1, the eye, and point 2, the exit. From Eq. (3.67), neglecting viscous work and heat transfer, this change is denoted by H:

$$H = \left(\frac{p}{\rho g} + \frac{V^2}{2g} + z \right)_2 - \left(\frac{p}{\rho g} + \frac{V^2}{2g} + z \right)_1 = h_s - h_f \tag{11.1}$$

where h_s is the pump head supplied and h_f the losses. The net head H is a primary output parameter for any turbomachine. Since Eq. (11.1) is for incompressible flow, it must be modified for gas compressors with large density changes.

Usually V_2 and V_1 are about the same, $z_2 - z_1$ is no more than a meter or so, and the net pump head is essentially equal to the change in pressure head:

$$H \approx \frac{p_2 - p_1}{\rho g} = \frac{\Delta p}{\rho g} \tag{11.2}$$

The power delivered to the fluid simply equals the specific weight times the discharge times the net head change:

$$P_w = \rho g Q H \tag{11.3}$$

This is traditionally called the *water horsepower*. The power required to drive the pump is the *brake horsepower*[1]

$$\text{bhp} = \omega T \tag{11.4}$$

where ω is the shaft angular velocity and T the shaft torque. If there were no losses, P_w and brake horsepower would be equal, but of course P_w is actually less, and the *efficiency* η of the pump is defined as

$$\eta = \frac{P_w}{\text{bhp}} = \frac{\rho g Q H}{\omega T} \tag{11.5}$$

[1]Conversion factors may be needed: 1 hp = 550 ft · lbf/s = 746 W.

The chief aim of the pump designer is to make η as high as possible over as broad a range of discharge Q as possible.

The efficiency is basically composed of three parts: volumetric, hydraulic, and mechanical. The *volumetric efficiency* is

$$\eta_v = \frac{Q}{Q + Q_L} \tag{11.6}$$

where Q_L is the loss of fluid due to leakage in the impeller casing clearances. The *hydraulic efficiency* is

$$\eta_h = 1 - \frac{h_f}{h_s} \tag{11.7}$$

where h_f has three parts: (1) *shock* loss at the eye due to imperfect match between inlet flow and the blade entrances, (2) *friction* losses in the blade passages, and (3) *circulation* loss due to imperfect match at the exit side of the blades.

Finally, the *mechanical efficiency* is

$$\eta_m = 1 - \frac{P_f}{\text{bhp}} \tag{11.8}$$

where P_f is the power loss due to mechanical friction in the bearings, packing glands, and other contact points in the machine.

By definition, the total efficiency is simply the product of its three parts:

$$\eta \equiv \eta_v \eta_h \eta_m \tag{11.9}$$

The designer has to work in all three areas to improve the pump.

Elementary Pump Theory

You may have thought that Eqs. (11.1) to (11.9) were formulas from pump *theory*. Not so; they are merely definitions of performance parameters and cannot be used in any predictive mode. To actually *predict* the head, power, efficiency, and flow rate of a pump, two theoretical approaches are possible: (1) simple one-dimensional flow formulas and (2) complex computer models that account for viscosity and three-dimensionality. Many of the best design improvements still come from testing and experience, and pump research remains a very active field [39]. The last 10 years have seen considerable advances in *computational fluid dynamics* (CFD) modeling of flow in turbomachines [42], and at least eight commercial turbulent flow three-dimensional CFD codes are now available.

To construct an elementary theory of pump performance, we assume one-dimensional flow and combine idealized fluid velocity vectors through the impeller with the angular momentum theorem for a control volume, Eq. (3.55).

The idealized velocity diagrams are shown in Fig. 11.4. The fluid is assumed to enter the impeller at $r = r_1$ with velocity component w_1 tangent to the blade angle β_1 plus circumferential speed $u_1 = \omega r_1$ matching the tip speed of the impeller. Its absolute entrance velocity is thus the vector sum of w_1 and u_1, shown as V_1. Similarly, the flow

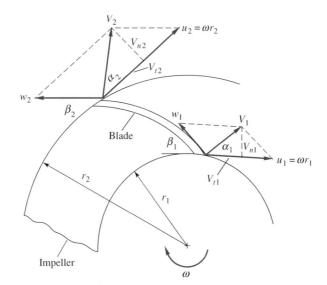

Fig. 11.4 Inlet and exit velocity diagrams for an idealized pump impeller.

exits at $r = r_2$ with component w_2 parallel to the blade angle β_2 plus tip speed $u_2 = \omega r_2$, with resultant velocity V_2.

We applied the angular momentum theorem to a turbomachine in Example 3.14 (Fig. 3.13) and arrived at a result for the applied torque T:

$$T = \rho Q(r_2 V_{t2} - r_1 V_{t1}) \tag{11.10}$$

where V_{t1} and V_{t2} are the absolute circumferential velocity components of the flow. The power delivered to the fluid is thus

$$P_w = \omega T = \rho Q(u_2 V_{t2} - u_1 V_{t1})$$

or

$$H = \frac{P_w}{\rho g Q} = \frac{1}{g}(u_2 V_{t2} - u_1 V_{t1}) \tag{11.11}$$

These are the *Euler turbomachine equations,* showing that the torque, power, and ideal head are functions only of the rotor-tip velocities $u_{1,2}$ and the absolute fluid tangential velocities $V_{t1,2}$, independent of the axial velocities (if any) through the machine.

Additional insight is gained by rewriting these relations in another form. From the geometry of Fig. 11.4

$$V^2 = u^2 + w^2 - 2uw \cos \beta \qquad w \cos \beta = u - V_t$$

or

$$uV_t = \tfrac{1}{2}(V^2 + u^2 - w^2) \tag{11.12}$$

Substituting this into Eq. (11.11) gives

$$H = \frac{1}{2g}[(V_2^2 - V_1^2) + (u_2^2 - u_1^2) - (w_2^2 - w_1^2)] \tag{11.13}$$

Thus the ideal head relates to the absolute plus the relative kinetic energy change of the fluid minus the rotor-tip kinetic energy change. Finally, substituting for H from its definition in Eq. (11.1) and rearranging, we obtain the classic relation

$$\frac{p}{\rho g} + z + \frac{w^2}{2g} - \frac{r^2 \omega^2}{2g} = \text{const} \qquad (11.14)$$

This is the *Bernoulli equation in rotating coordinates* and applies to either two- or three-dimensional ideal incompressible flow.

For a centrifugal pump, the power can be related to the radial velocity $V_n = V_t \tan \alpha$ and the continuity relation

$$P_w = \rho Q(u_2 V_{n2} \cot \alpha_2 - u_1 V_{n1} \cot \alpha_1) \qquad (11.15)$$

where

$$V_{n2} = \frac{Q}{2\pi r_2 b_2} \qquad \text{and} \qquad V_{n1} = \frac{Q}{2\pi r_1 b_1}$$

and where b_1 and b_2 are the blade widths at inlet and exit. With the pump parameters r_1, r_2, β_1, β_2, and ω known, Eq. (11.11) or Eq. (11.15) is used to compute idealized power and head versus discharge. The "design" flow rate Q^* is commonly estimated by assuming that the flow enters exactly normal to the impeller:

$$\alpha_1 = 90° \qquad V_{n1} = V_1 \qquad (11.16)$$

We can expect this simple analysis to yield estimates within ± 25 percent for the head, water horsepower, and discharge of a pump. Let us illustrate with an example.

EXAMPLE 11.1

Given are the following data for a commercial centrifugal water pump: $r_1 = 4$ in, $r_2 = 7$ in, $\beta_1 = 30°$, $\beta_2 = 20°$, speed $= 1440$ r/min. Estimate (*a*) the design point discharge, (*b*) the water horsepower, and (*c*) the head if $b_1 = b_2 = 1.75$ in.

Solution

Part (a)

The angular velocity is $\omega = 2\pi$ r/s $= 2\pi(1440/60) = 150.8$ rad/s. Thus the tip speeds are $u_1 = \omega r_1 = 150.8(4/12) = 50.3$ ft/s and $u_2 = \omega r_2 = 150.8(7/12) = 88.0$ ft/s. From the inlet velocity diagram, Fig. E11.1*a*, with $\alpha_1 = 90°$ for design point, we compute

$$V_{n1} = u_1 \tan 30° = 29.0 \text{ ft/s}$$

whence the discharge is

$$Q = 2\pi r_1 b_1 V_{n1} = (2\pi)\left(\frac{4}{12} \text{ ft}\right)\left(\frac{1.75}{12} \text{ ft}\right)\left(29.0 \frac{\text{ft}}{\text{s}}\right)$$

$$= (8.87 \text{ ft}^3/\text{s})(60 \text{ s/min})\left(\frac{1728}{231} \text{ gal/ft}^3\right)$$

$$= 3980 \text{ gal/min} \qquad \qquad \textit{Ans. (a)}$$

(The actual pump produces about 3500 gal/min.)

V_1

$90°$ $\qquad$ $30°$

$u_1 = 50.3$ ft/s

E11.1a

Part (b)

The outlet radial velocity follows from Q:

$$V_{n2} = \frac{Q}{2\pi r_2 b_2} = \frac{8.87 \text{ ft}^3/\text{s}}{2\pi(\frac{7}{12}\text{ft})(\frac{1.75}{12}\text{ft})} = 16.6 \text{ ft/s}$$

This enables us to construct the outlet velocity diagram as in Fig. E11.1b, given $\beta_2 = 20°$. The tangential component is

$$V_{t2} = u_2 - V_{n2} \cot \beta_2 = 88.0 - 16.6 \cot 20° = 42.4 \text{ ft/s}$$

$$\alpha_2 = \tan^{-1}\frac{16.6}{42.4} = 21.4°$$

E11.1b

The power is then computed from Eq. (11.11) with $V_{t1} = 0$ at the design point:

$$P_w = \rho Q u_2 V_{t2} = (1.94 \text{ slugs/ft}^3)(8.87 \text{ ft}^3/\text{s})(88.0 \text{ ft/s})(42.4 \text{ ft/s})$$

$$= \frac{64,100 \text{ ft} \cdot \text{lbf/s}}{550 \text{ ft}-\text{lbf/(s}-\text{hp})} = 117 \text{ hp} \qquad \qquad Ans. (b)$$

(The actual pump delivers about 125 water horsepower, requiring 147 bhp at 85 percent efficiency.)

Part (c)

Finally, the head is estimated from Eq. (11.11):

$$H \approx \frac{P_w}{\rho g Q} = \frac{64,100 \text{ ft} \cdot \text{lbf/s}}{(62.4 \text{ lbf/ft}^3)(8.87 \text{ ft}^3/\text{s})} = 116 \text{ ft} \qquad \qquad Ans. (c)$$

(The actual pump develops about 140-ft head.) Improved methods for obtaining closer estimates are given in advanced references [for example, 7, 8, and 31].

Effect of Blade Angle on Pump Head

The simple theory just discussed can be used to predict an important blade-angle effect. If we neglect inlet angular momentum, the theoretical water horsepower is

$$P_w = \rho Q u_2 V_{t2} \qquad \qquad (11.17)$$

where

$$V_{t2} = u_2 - V_{n2} \cot \beta_2 \qquad V_{n2} = \frac{Q}{2\pi r_2 b_2}$$

Then the theoretical head from Eq. (11.11) becomes

$$H \approx \frac{u_2^2}{g} - \frac{u_2 \cot \beta_2}{2\pi r_2 b_2 g} Q \qquad \qquad (11.18)$$

The head varies linearly with discharge Q, having a shutoff value u_2^2/g, where u_2 is the exit blade-tip speed. The slope is negative if $\beta_2 < 90°$ (backward-curved blades) and positive for $\beta_2 > 90°$ (forward-curved blades). This effect is shown in Fig. 11.5 and is accurate only at low flow rates.

The measured shutoff head of centrifugal pumps is only about 60 percent of the theoretical value $H_0 = \omega^2 r_2^2/g$. With the advent of the laser-doppler anemometer, researchers can now make detailed three-dimensional flow measurements inside pumps and can even animate the data into a movie [40].

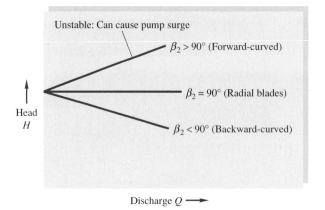

Fig. 11.5 Theoretical effect of blade exit angle on pump head versus discharge.

The positive slope condition in Fig. 11.5 can be unstable and can cause pump *surge,* an oscillatory condition where the pump "hunts" for the proper operating point. Surge may cause only rough operation in a liquid pump, but it can be a major problem in gas compressor operation. For this reason a backward-curved or radial blade design is generally preferred. A survey of the problem of pump stability is given by Greitzer [41].

11.3 Pump Performance Curves and Similarity Rules

Since the theory of the previous section is rather qualitative, the only solid indicator of a pump's performance lies in extensive testing. For the moment let us discuss the centrifugal pump in particular. The general principles and the presentation of data are exactly the same for mixed flow and axial flow pumps and compressors.

Performance charts are almost always plotted for constant shaft rotation speed n (in r/min usually). The basic independent variable is taken to be discharge Q (in gal/min usually for liquids and ft^3/min for gases). The dependent variables, or "output," are taken to be head H (pressure rise Δp for gases), brake horsepower (bhp), and efficiency η.

Figure 11.6 shows typical performance curves for a centrifugal pump. The head is approximately constant at low discharge and then drops to zero at $Q = Q_{max}$. At this speed and impeller size, the pump cannot deliver any more fluid than Q_{max}. The positive slope part of the head is shown dashed; as mentioned earlier, this region can be unstable and can cause hunting for the operating point.

The efficiency η is always zero at no flow and at Q_{max}, and it reaches a maximum, perhaps 80 to 90 percent, at about $0.6Q_{max}$. This is the *design flow rate $Q*$* or *best efficiency point* (BEP), $\eta = \eta_{max}$. The head and horsepower at BEP will be termed $H*$ and $P*$ (or bhp*), respectively. It is desirable that the efficiency curve be flat near η_{max}, so that a wide range of efficient operation is achieved. However, some designs simply do not achieve flat efficiency curves. Note that η is not independent of H and P but rather is calculated from the relation in Eq. (11.5), $\eta = \rho g Q H / P$.

As shown in Fig. 11.6, the horsepower required to drive the pump typically rises monotonically with the flow rate. Sometimes there is a large power rise beyond

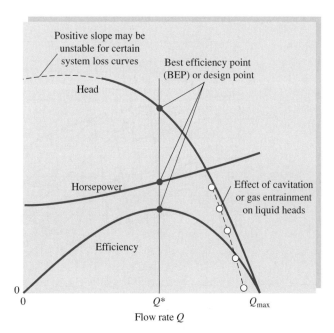

Positive slope may be unstable for certain system loss curves

Best efficiency point (BEP) or design point

Head

Horsepower

Effect of cavitation or gas entrainment on liquid heads

Efficiency

Fig. 11.6 Typical centrifugal pump performance curves at constant impeller rotation speed. The units are arbitrary.

0

0   Q^* Q_{max}

Flow rate Q

the BEP, especially for radial-tipped and forward-curved blades. This is considered undesirable because a much larger motor is then needed to provide high flow rates. Backward-curved blades typically have their horsepower level off above BEP ("nonoverloading" type of curve).

Measured Performance Curves

Figure 11.7 shows actual performance data for a commercial centrifugal pump. Figure 11.7a is for a basic casing size with three different impeller diameters. The head curves $H(Q)$ are shown, but the horsepower and efficiency curves have to be inferred from the contour plots. Maximum discharges are not shown, being far outside the normal operating range near the BEP. Everything is plotted raw, of course [feet, horsepower, gallons per minute (1 U.S. gal = 231 in^3)] since it is to be used directly by designers. Figure 11.7b is the same pump design with a 20 percent larger casing, a lower speed, and three larger impeller diameters. Comparing the two pumps may be a little confusing: The larger pump produces exactly the same discharge but only half the horsepower and half the head. This will be readily understood from the scaling or similarity laws we are about to formulate.

A point often overlooked is that raw curves like Fig. 11.7 are strictly applicable to a fluid of a certain density and viscosity, in this case water. If the pump were used to deliver, say, mercury, the brake horsepower would be about 13 times higher while Q, H, and η would be about the same. But in that case H should be interpreted as feet of *mercury*, not feet of water. If the pump were used for SAE 30 oil, *all* data would change (brake horsepower, Q, H, and η) due to the large change in viscosity (Reynolds number). Again this should become clear with the similarity rules.

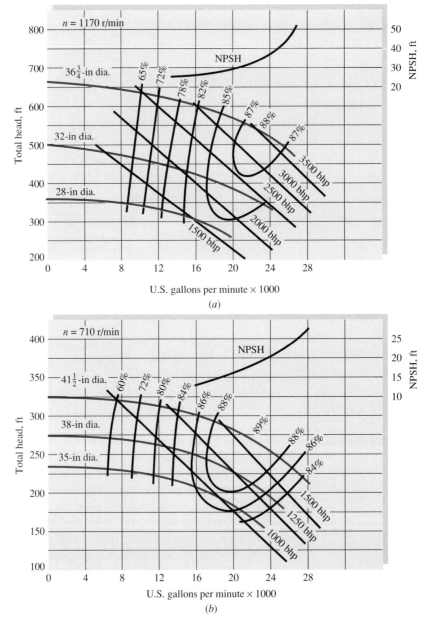

Fig. 11.7 Measured-performance curves for two models of a centrifugal water pump: (*a*) basic casing with three impeller sizes; (*b*) 20 percent larger casing with three larger impellers at slower speed. *(Courtesy of Ingersoll-Rand Corporation, Cameron Pump Division.)*

Net Positive-Suction Head

In the top of Fig. 11.7 is plotted the *net positive-suction head* (NPSH), which is the head required at the pump inlet to keep the liquid from cavitating or boiling. The pump inlet or suction side is the low-pressure point where cavitation will first occur. The NPSH is defined as

$$\text{NPSH} = \frac{p_i}{\rho g} + \frac{V_i^2}{2g} - \frac{p_v}{\rho g} \tag{11.19}$$

where p_i and V_i are the pressure and velocity at the pump inlet and p_v is the vapor pressure of the liquid. Given the left-hand side, NPSH, from the pump performance curve, we must ensure that the right-hand side is equal or greater in the actual system to avoid cavitation.

If the pump inlet is placed at a height Z_i above a reservoir whose free surface is at pressure p_a, we can use Bernoulli's equation to rewrite NPSH as

$$\text{NPSH} = \frac{p_a}{\rho g} - Z_i - h_{fi} - \frac{p_v}{\rho g} \tag{11.20}$$

where h_{fi} is the friction head loss between the reservoir and the pump inlet. Knowing p_a and h_{fi}, we can set the pump at a height Z_i that will keep the right-hand side greater than the "required" NPSH plotted in Fig. 11.7.

If cavitation does occur, there will be pump noise and vibration, pitting damage to the impeller, and a sharp dropoff in pump head and discharge. In some liquids this deterioration starts before actual boiling, as dissolved gases and light hydrocarbons are liberated.

Deviations from Ideal Pump Theory

The actual pump head data in Fig. 11.7 differ considerably from ideal theory, Eq. (11.18). Take, for example, the 36.75-in-diameter pump at 1170 r/min in Fig. 11.7a. The theoretical shutoff head is

$$H_0(\text{ideal}) = \frac{\omega^2 r_2^2}{g} = \frac{[1170(2\pi/60)\ \text{rad/s}]^2[(36.75/2)/(12)\ \text{ft}]^2}{32.2\ \text{ft/s}^2} = 1093\ \text{ft}$$

From Fig. 11.7a, at $Q = 0$, we read the actual shutoff head to be only 670 ft, or 61 percent of the theoretical value (see Prob. P11.24). This is a sharp dropoff and is indicative of nonrecoverable losses of three types:

1. *Impeller recirculation loss,* significant only at low flow rates.
2. *Friction losses* on the blade and passage surfaces, which increase monotonically with the flow rate.
3. *"Shock" loss* due to mismatch between the blade angles and the inlet flow direction, especially significant at high flow rates.

These are complicated three-dimensional flow effects and hence are difficult to predict. Although, as mentioned, numerical (CFD) techniques are becoming more important [42], modern performance prediction is still a blend of experience, empirical correlations, idealized theory, and CFD modifications [45].

EXAMPLE 11.2

The 32-in pump of Fig. 11.7a is to pump 24,000 gal/min of water at 1170 r/min from a reservoir whose surface is at 14.7 lbf/in^2 absolute. If head loss from reservoir to pump inlet is 6 ft, where should the pump inlet be placed to avoid cavitation for water at (a) 60°F, $p_v = 0.26$ lbf/in^2 absolute, SG = 1.0 and (b) 200°F, $p_v = 11.52$ lbf/in^2 absolute, SG = 0.9635?

| | Solution |

Part (a)   For either case read from Fig. 11.7a at 24,000 gal/min that the required NPSH is 40 ft. For this case $\rho g = 62.4$ lbf/ft³. From Eq. (11.20) it is necessary that

$$\text{NPSH} \le \frac{p_a - p_v}{\rho g} - Z_i - h_{fi}$$

or

$$40 \text{ ft} \le \frac{(14.7 - 0.26 \text{ lbf/in}^2)(144 \text{ in}^2/\text{ft}^2)}{62.4 \text{ lbf/ft}^3} - Z_i - 6.0$$

or

$$Z_i \le 27.3 - 40 = -12.7 \text{ ft} \qquad\qquad Ans. (a)$$

The pump must be placed at least 12.7 ft below the reservoir surface to avoid cavitation.

Part (b)   For this case $\rho g = 62.4(0.9635) = 60.1$ lbf/ft³. Equation (11.20) applies again with the higher p_v:

$$40 \text{ ft} \le \frac{(14.7 - 11.52 \text{ lbf/in}^2)(144 \text{ in}^2/\text{ft}^2)}{60.1 \text{ lbf/ft}^3} - Z_i - 6.0$$

or

$$Z_i \le 1.6 - 40 = -38.4 \text{ ft} \qquad\qquad Ans. (b)$$

The pump must now be placed at least 38.4 ft below the reservoir surface. These are unusually stringent conditions because a large, high-discharge pump requires a large NPSH.

Dimensionless Pump Performance   For a given pump design, the output variables H and brake horsepower should be dependent on discharge Q, impeller diameter D, and shaft speed n, at least. Other possible parameters are the fluid density ρ, viscosity μ, and surface roughness ϵ. Thus the performance curves in Fig. 11.7 are equivalent to the following assumed functional relations:[2]

$$gH = f_1(Q, D, n, \rho, \mu, \epsilon) \qquad \text{bhp} = f_2(Q, D, n, \rho, \mu, \epsilon) \qquad (11.21)$$

This is a straightforward application of dimensional analysis principles from Chap. 5. As a matter of fact, it was given as an exercise (Ex. 5.3). For each function in Eq. (11.21) there are seven variables and three primary dimensions (M, L, and T); hence we expect $7 - 3 = 4$ dimensionless pi groups, and that is what we get. You can verify as an exercise that appropriate dimensionless forms for Eqs. (11.21) are

$$\frac{gH}{n^2 D^2} = g_1\left(\frac{Q}{nD^3}, \frac{\rho n D^2}{\mu}, \frac{\epsilon}{D}\right)$$

$$\frac{\text{bhp}}{\rho n^3 D^5} = g_2\left(\frac{Q}{nD^3}, \frac{\rho n D^2}{\mu}, \frac{\epsilon}{D}\right)$$

$$(11.22)$$

[2]We adopt gH as a variable instead of H for dimensional reasons.

The quantities $\rho n D^2/\mu$ and ϵ/D are recognized as the Reynolds number and roughness ratio, respectively. Three new pump parameters have arisen:

$$
\begin{aligned}
\text{Capacity coefficient } C_Q &= \frac{Q}{nD^3} \\[2mm]
\text{Head coefficient } C_H &= \frac{gH}{n^2D^2} \\[2mm]
\text{Power coefficient } C_P &= \frac{\text{bhp}}{\rho n^3 D^5}
\end{aligned}
\tag{11.23}
$$

Note that only the power coefficient contains fluid density, the parameters C_Q and C_H being kinematic types.

Figure 11.7 gives no warning of viscous or roughness effects. The Reynolds numbers are from 0.8 to 1.5×10^7, or fully turbulent flow in all passages probably. The roughness is not given and varies greatly among commercial pumps. But at such high Reynolds numbers we expect more or less the same percentage effect on all these pumps. Therefore it is common to assume that the Reynolds number and the roughness ratio have a constant effect, so that Eqs. (11.23) reduce to, approximately,

$$
C_H \approx C_H(C_Q) \qquad C_P \approx C_P(C_Q)
\tag{11.24}
$$

For geometrically similar pumps, we expect head and power coefficients to be (nearly) unique functions of the capacity coefficient. We have to watch out that the pumps are geometrically similar or nearly so because (1) manufacturers put different-sized impellers in the same casing, thus violating geometric similarity, and (2) large pumps have smaller ratios of roughness and clearances to impeller diameter than small pumps. In addition, the more viscous liquids will have significant Reynolds number effects; for example, a factor-of-3 or more viscosity increase causes a clearly visible effect on C_H and C_P.

The efficiency η is already dimensionless and is uniquely related to the other three. It varies with C_Q also:

$$
\eta \equiv \frac{C_H C_Q}{C_P} = \eta(C_Q)
\tag{11.25}
$$

We can test Eqs. (11.24) and (11.25) from the data of Fig. 11.7. The impeller diameters of 32 and 38 in are approximately 20 percent different in size, and so their ratio of impeller to casing size is the same. The parameters C_Q, C_H, and C_P are computed with n in r/s, Q in ft^3/s (gal/min $\times$ 2.23 $\times$ 10^{-3}), H and D in ft, $g =$ 32.2 ft/s^2, and brake horsepower in horsepower times 550 ft $\cdot$ lbf/(s $\cdot$ hp). The nondimensional data are then plotted in Fig. 11.8. A dimensionless suction head coefficient is also defined:

$$
C_{HS} = \frac{g(\text{NPSH})}{n^2D^2} = C_{HS}(C_Q)
\tag{11.26}
$$

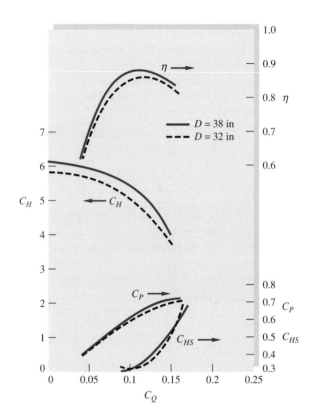

Fig. 11.8 Nondimensional plot of the pump performance data from Fig. 11.7. These numbers are not representative of other pump designs.

The coefficients C_P and C_{HS} are seen to correlate almost perfectly into a single function of C_Q, while η and C_H data deviate by a few percent. The last two parameters are more sensitive to slight discrepancies in model similarity; since the larger pump has smaller roughness and clearance ratios and a 40 percent larger Reynolds number, it develops slightly more head and is more efficient. The overall effect is a resounding victory for dimensional analysis.

The best efficiency point in Fig. 11.8 is approximately

$$C_{Q*} \approx 0.115 \qquad C_{P*} \approx 0.65$$

$$\eta_{\text{max}} \approx 0.88 \tag{11.27}$$

$$C_{H*} \approx 5.0 \qquad C_{HS*} \approx 0.37$$

These values can be used to estimate the BEP performance of any size pump in this geometrically similar family. In like manner, the shutoff head is $C_H(0) \approx 6.0$, and by extrapolation the shutoff power is $C_P(0) \approx 0.25$ and the maximum discharge is $C_{Q,\text{max}} \approx 0.23$. Note, however, that Fig. 11.8 gives no reliable information about, say, the 28- or 35-in impellers in Fig. 11.7, which have a different impeller-to-casing-size ratio and thus must be correlated separately.

By comparing values of n^2D^2, nD^3, and n^3D^5 for two pumps in Fig. 11.7, we can see readily why the large pump had the same discharge but less power and head:

	D, ft	n, r/s	Discharge nD^3, ft³/s	Head n^2D^2/g, ft	Power $\rho n^3 D^5/550$, hp
Fig. 11.7a	32/12	1170/60	370	84	3527
Fig. 11.7b	38/12	710/60	376	44	1861
Ratio	—	—	1.02	0.52	0.53

Discharge goes as nD^3, which is about the same for both pumps. Head goes as n^2D^2 and power as n^3D^5 for the same ρ (water), and these are about half as much for the larger pump. The NPSH goes as n^2D^2 and is also half as much for the 38-in pump.

EXAMPLE 11.3

A pump from the family of Fig. 11.8 has $D = 21$ in and $n = 1500$ r/min. Estimate (a) discharge, (b) head, (c) pressure rise, and (d) brake horsepower of this pump for water at 60°F and best efficiency.

Solution

Part (a)

In BG units take $D = 21/12 = 1.75$ ft and $n = 1500/60 = 25$ r/s. At 60°F, ρ of water is 1.94 slugs/ft³. The BEP parameters are known from Fig. 11.8 or Eqs. (11.27). The BEP discharge is thus

$$Q* = C_{Q*}nD^3 = 0.115(25 \text{ r/s})(1.75 \text{ ft})^3 = (15.4 \text{ ft}^3/\text{s})\left(448.8 \frac{\text{gal/min}}{\text{ft}^3/\text{s}}\right) = 6900 \text{ gal/min}$$

Ans. (a)

Part (b)

Similarly, the BEP head is

$$H* = \frac{C_{H*}n^2D^2}{g} = \frac{5.0(25)^2(1.75)^2}{32.2} = 300\text{-ft water} \qquad Ans. (b)$$

Part (c)

Since we are not given elevation or velocity head changes across the pump, we neglect them and estimate

$$\Delta p \approx \rho g H = 1.94(32.2)(300) = 18,600 \text{ lbf/ft}^2 = 129 \text{ lbf/in}^2 \qquad Ans. (c)$$

Part (d)

Finally, the BEP power is

$$P^* = C_{P*}\rho n^3 D^5 = 0.65(1.94)(25)^3(1.75)^5$$

$$= \frac{323,000 \text{ ft} \cdot \text{lbf/s}}{550} = 590 \text{ hp} \qquad Ans. (d)$$

EXAMPLE 11.4

We want to build a pump from the family of Fig. 11.8, which delivers 3000 gal/min water at 1200 r/min at best efficiency. Estimate (a) the impeller diameter, (b) the maximum discharge, (c) the shutoff head, and (d) the NPSH at best efficiency.

Solution

Part (a)   3000 gal/min = 6.68 ft³/s   and   1200 r/min = 20 r/s. At BEP we have

$$Q^* = C_{Q*}nD^3 = 6.68 \text{ ft}^3/\text{s} = (0.115)(20)D^3$$

$$D = \left[\frac{6.68}{0.115(20)}\right]^{1/3} = 1.43 \text{ ft} = 17.1 \text{ in} \qquad \qquad Ans. (a)$$

Part (b)   The maximum Q is related to Q^* by a ratio of capacity coefficients:

$$Q_{max} = \frac{Q^* C_{Q,max}}{C_{Q^*}} \approx \frac{3000(0.23)}{0.115} = 6000 \text{ gal/min} \qquad \qquad Ans. (b)$$

Part (c)   From Fig. 11.8 we estimated the shutoff head coefficient to be 6.0. Thus

$$H(0) \approx \frac{C_H(0)n^2D^2}{g} = \frac{6.0(20)^2(1.43)^2}{32.2} = 152 \text{ ft} \qquad \qquad Ans. (c)$$

Part (d)   Finally, from Eq. (11.27), the NPSH at BEP is approximately

$$\text{NPSH}^* = \frac{C_{HS*}n^2D^2}{g} = \frac{0.37(20)^2(1.43)^2}{32.2} = 9.4 \text{ ft} \qquad \qquad Ans. (d)$$

Since this is a small pump, it will be less efficient than the pumps in Fig. 11.8, probably about 85 percent maximum.

Similarity Rules

The success of Fig. 11.8 in correlating pump data leads to simple rules for comparing pump performance. If pump 1 and pump 2 are from the same geometric family and are operating at homologous points (the same dimensionless position on a chart such as Fig. 11.8), their flow rates, heads, and powers will be related as follows:

$$\frac{Q_2}{Q_1} = \frac{n_2}{n_1}\left(\frac{D_2}{D_1}\right)^3 \qquad \frac{H_2}{H_1} = \left(\frac{n_2}{n_1}\right)^2\left(\frac{D_2}{D_1}\right)^2$$

$$\frac{P_2}{P_1} = \frac{\rho_2}{\rho_1}\left(\frac{n_2}{n_1}\right)^3\left(\frac{D_2}{D_1}\right)^5 \qquad \qquad (11.28)$$

These are the *similarity rules,* which can be used to estimate the effect of changing the fluid, speed, or size on any dynamic turbomachine—pump or turbine—within a geometrically similar family. A graphic display of these rules is given in Fig. 11.9, showing the effect of speed and diameter changes on pump performance. In Fig. 11.9a the size is held constant and the speed is varied 20 percent, while Fig. 11.9b shows a 20 percent size change at constant speed. The curves are plotted to scale but with arbitrary units. The speed effect (Fig. 11.9a) is substantial, but the size effect (Fig. 11.9b) is even more dramatic, especially for power, which varies as D^5. Generally we see that a given pump family can be adjusted in size and speed to fit a variety of system characteristics.

Strictly speaking, we would expect for perfect similarity that $\eta_1 = \eta_2$, but we have seen that larger pumps are more efficient, having a higher Reynolds number and lower roughness and clearance ratios. Two empirical correlations are recommended for

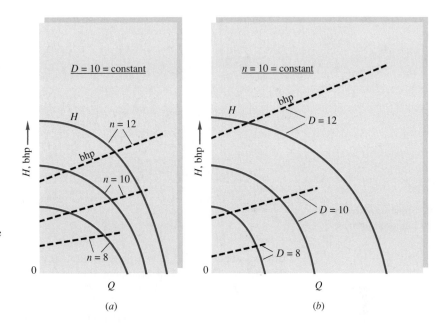

Fig. 11.9 Effect of changes in size and speed on homologous pump performance: (*a*) 20 percent change in speed at constant size; (*b*) 20 percent change in size at constant speed.

maximum efficiency. One, developed by Moody [43] for turbines but also used for pumps, is a size effect. The other, suggested by Anderson [44] from thousands of pump tests, is a flow rate effect:

Size changes [43]:
$$\frac{1 - \eta_2}{1 - \eta_1} \approx \left(\frac{D_1}{D_2}\right)^{1/4} \tag{11.29a}$$

Flow rate changes [44]:
$$\frac{0.94 - \eta_2}{0.94 - \eta_1} \approx \left(\frac{Q_1}{Q_2}\right)^{0.32} \tag{11.29b}$$

Anderson's formula (11.29*b*) makes the practical observation that even an infinitely large pump will have losses. He thus proposes a maximum possible efficiency of 94 percent, rather than 100 percent. Anderson recommends that the same formula be used for turbines if the constant 0.94 is replaced by 0.95. The formulas in Eq. (11.29) assume the same value of surface roughness for both machines—one could micropolish a small pump and achieve the efficiency of a larger machine.

Effect of Viscosity

Centrifugal pumps are often used to pump oils and other viscous liquids up to 1000 times the viscosity of water. But the Reynolds numbers become low turbulent or even laminar, with a strong effect on performance. Figure 11.10 shows typical test curves of head and brake horsepower versus discharge. High viscosity causes a dramatic drop in head and discharge and increases in power requirements. The efficiency also drops substantially according to the following typical results:

μ/μ_{water}	1.0	10.0	100	1000
η_{max}, %	85	76	52	11

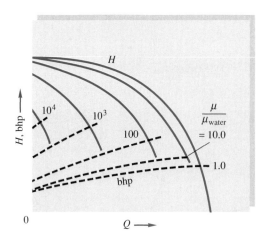

Fig. 11.10 Effect of viscosity on centrifugal pump performance.

Beyond about $300\mu_{\text{water}}$ the deterioration in performance is so great that a positive-displacement pump is recommended.

11.4 Mixed- and Axial-Flow Pumps: The Specific Speed

We have seen from the previous section that the modern centrifugal pump is a formidable device, able to deliver very high heads and reasonable flow rates with excellent efficiency. It can match many system requirements. But basically the centrifugal pump is a high-head, low-flow machine, whereas there are many applications requiring low head and high discharge. To see that the centrifugal design is not convenient for such systems, consider the following example.

EXAMPLE 11.5

We want to use a centrifugal pump from the family of Fig. 11.8 to deliver 100,000 gal/min of water at 60°F with a head of 25 ft. What should be (a) the pump size and speed and (b) brake horsepower, assuming operation at best efficiency?

Solution

Part (a)

Enter the known head and discharge into the BEP parameters from Eq. (11.27):

$$H^* = 25 \text{ ft} = \frac{C_{H*}n^2D^2}{g} = \frac{5.0n^2D^2}{32.2}$$

$$Q^* = 100,000 \text{ gal/min} = 222.8 \text{ ft}^3/\text{s} = C_{Q*}nD^3 = 0.115nD^3$$

The two unknowns are n and D. Solve simultaneously for

$$D = 12.4 \text{ ft} \qquad n = 1.03 \text{ r/s} = 62 \text{ r/min} \qquad \textit{Ans. (a)}$$

If you wish to avoid algebraic manipulation, simply program the two simultaneous equations from Part (a) in EES, using English units:

```
25 = 5.0*n^2*D^2/32.2
222.8 = 0.115*n*D^3
```

Specify in Variable Information that n and D are positive, and EES promptly returns the correct solution: D = 12.36 ft and n = 1.027 r/s.

Part (b) The most efficient horsepower is then, from Eq. (11.27),

$$\text{bhp*} \approx C_{P*}\rho n^3 D^5 = \frac{0.65(1.94)(1.03)^3(12.4)^5}{550} = 720 \text{ hp} \qquad \textit{Ans. (b)}$$

The solution to Example 11.5 is mathematically correct but results in a grotesque pump: an impeller more than 12 ft in diameter, rotating so slowly one can visualize oxen walking in a circle turning the shaft.

Other dynamic pump designs provide low head and high discharge. For example, there is a type of 38-in, 710 r/min pump, with the same input parameters as Fig. 11.7b, which will deliver the 25-ft head and 100,000 gal/min flow rate called for in Example 11.5. This is done by allowing the flow to pass through the impeller with an axial-flow component and less centrifugal component. The passages can be opened up to the increased flow rate with very little size increase, but the drop in radial outlet velocity decreases the head produced. These are the mixed-flow (part radial, part axial) and axial-flow (propeller-type) families of dynamic pump. Some vane designs are sketched in Fig. 11.11, which introduces an interesting new "design" parameter, the specific speed N_s or N_s'.

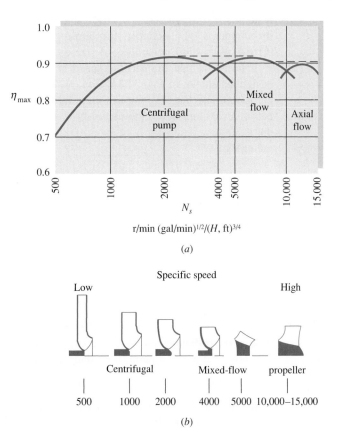

Fig. 11.11 (*a*) Optimum efficiency and (*b*) vane design of dynamic pump families as a function of specific speed.

The Specific Speed

Most pump applications involve a known head and discharge for the particular system, plus a speed range dictated by electric motor speeds or cavitation requirements. The designer then selects the best size and shape (centrifugal, mixed, axial) for the pump. To help this selection, we need a dimensionless parameter involving speed, discharge, and head but not size. This is accomplished by eliminating the diameter between C_Q and C_H, applying the result only to the BEP. This ratio is called the *specific speed* and has both a dimensionless form and a somewhat lazy, practical form:

Rigorous form:
$$N_s' = \frac{C_{Q*}^{1/2}}{C_{H*}^{3/4}} = \frac{n(Q^*)^{1/2}}{(gH^*)^{3/4}} \tag{11.30a}$$

Lazy but common:
$$N_s = \frac{(\text{r/min})(\text{gal/min})^{1/2}}{[H\,(\text{ft})]^{3/4}} \tag{11.30b}$$

In other words, practicing engineers do not bother to change n to revolutions per second or Q^* to cubic feet per second or to include gravity with head, although the latter would be necessary for, say, a pump on the moon. The conversion factor is

$$N_s = 17{,}182 N_s'$$

Note that N_s is applied only to BEP; thus a single number characterizes an entire family of pumps. For example, the family of Fig. 11.8 has $N_s' \approx (0.115)^{1/2}/(5.0)^{3/4} = 0.1014$, $N_s = 1740$, regardless of size or speed.

It turns out that the specific speed is directly related to the most efficient pump design, as shown in Fig. 11.11. Low N_s means low Q and high H, hence a centrifugal pump, and large N_s implies an axial pump. The centrifugal pump is best for N_s between 500 and 4000, the mixed-flow pump for N_s between 4000 and 10,000, and the axial-flow pump for N_s above 10,000. Note the changes in impeller shape as N_s increases.

Suction Specific Speed

If we use NPSH rather than H in Eq. (11.30), the result is called *suction-specific speed:*

Rigorous:
$$N_{ss}' = \frac{nQ^{1/2}}{(g\,\text{NPSH})^{3/4}} \tag{11.31a}$$

Lazy:
$$N_{ss} = \frac{(\text{r/min})(\text{gal/min})^{1/2}}{[\text{NPSH}\,(\text{ft})]^{3/4}} \tag{11.31b}$$

where NPSH denotes the available suction head of the system. Data from Wislicenus [4] show that a given pump is in danger of inlet cavitation if

$$N_{ss}' \geq 0.47 \qquad N_{ss} \geq 8100$$

In the absence of test data, this relation can be used, given n and Q, to estimate the minimum required NPSH.

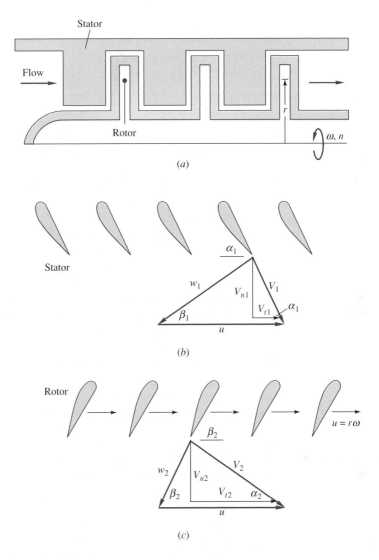

Fig. 11.12 Analysis of an axial-flow pump: (*a*) basic geometry; (*b*) stator blades and exit velocity diagram; (*c*) rotor blades and exit velocity diagram.

Axial-Flow Pump Theory

A multistage axial-flow geometry is shown in Fig. 11.12*a*. The fluid essentially passes almost axially through alternate rows of fixed *stator* blades and moving *rotor* blades. The incompressible flow assumption is frequently used even for gases because the pressure rise per stage is usually small.

The simplified vector diagram analysis assumes that the flow is one-dimensional and leaves each blade row at a relative velocity exactly parallel to the exit blade angle. Figure 11.12*b* shows the stator blades and their exit velocity diagram. Since the stator is fixed, ideally the absolute velocity V_1 is parallel to the trailing edge of the blade. After vectorially subtracting the rotor tangential velocity u from V_1, we obtain the velocity w_1 relative to the rotor, which ideally should be parallel to the rotor leading edge.

Figure 11.12*c* shows the rotor blades and their exit velocity diagram. Here the relative velocity w_2 is parallel to the blade trailing edge, while the absolute velocity V_2 should be designed to smoothly enter the next row of stator blades.

The theoretical power and head are given by Euler's turbine relation (11.11). Since there is no radial flow, the inlet and exit rotor speeds are equal, $u_1 = u_2$, and one-dimensional continuity requires that the axial-velocity component remain constant:

$$V_{n1} = V_{n2} = V_n = \frac{Q}{A} = \text{const}$$

From the geometry of the velocity diagrams, the normal velocity (or volume flow) can be directly related to the blade rotational speed u:

$$u = \omega r_{av} = V_{n1}(\cot \alpha_1 + \cot \beta_1) = V_{n2}(\cot \alpha_2 + \cot \beta_2) \qquad (11.32)$$

Thus the flow rate can be predicted from the rotational speed and the blade angles. Meanwhile, since $V_{t1} = V_{n1} \cot \alpha_1$ and $V_{t2} = u - V_{n2} \cot \beta_2$, Euler's relation (11.11) for the pump head becomes

$$gH = uV_n(\cot \alpha_2 - \cot \alpha_1)$$
$$= u^2 - uV_n(\cot \alpha_1 + \cot \beta_2) \qquad (11.33)$$

the preferred form because it relates to the blade angles α_1 and β_2. The shutoff or no-flow head is seen to be $H_0 = u^2/g$, just as in Eq. (11.18) for a centrifugal pump. The blade-angle parameter $\cot \alpha_1 + \cot \beta_2$ can be designed to be negative, zero, or positive, corresponding to a rising, flat, or falling head curve, as in Fig. 11.5.

Strictly speaking, Eq. (11.33) applies only to a single streamtube of radius r, but it is a good approximation for very short blades if r denotes the average radius. For long blades it is customary to sum Eq. (11.33) in radial strips over the blade area. Such complexity may not be warranted since theory, being idealized, neglects losses and usually predicts the head and power larger than those in actual pump performance.

Performance of an Axial-Flow Pump

At high specific speeds, the most efficient choice is an axial-flow, or propeller, pump, which develops high flow rate and low head. A typical dimensionless chart for a propeller pump is shown in Fig. 11.13. Note, as expected, the higher C_Q and lower C_H compared with Fig. 11.8. The head curve drops sharply with discharge, so that a large system head change will cause a mild flow change. The power curve drops with head also, which means a possible overloading condition if the system discharge should suddenly decrease. Finally, the efficiency curve is rather narrow and triangular, as opposed to the broad, parabolic-shaped centrifugal pump efficiency (Fig. 11.8).

By inspection of Fig. 11.13, $C_{Q*} \approx 0.55$, $C_{H*} \approx 1.07$, $C_{P*} \approx 0.70$, and $\eta_{max} \approx 0.84$. From this we compute $N_s' \approx (0.55)^{1/2}/(1.07)^{3/4} = 0.705$, $N_s = 12,000$. The relatively low efficiency is due to small pump size: $d = 14$ in, $n = 690$ r/min, $Q* = 4400$ gal/min.

A repetition of Example 11.5 using Fig. 11.13 would show that this propeller pump family can provide a 25-ft head and 100,000 gal/min discharge if $D = 46$ in and $n = 430$ r/min, with bhp $= 750$; this is a much more reasonable design solution, with improvements still possible at larger-N_s conditions.

Pump Performance versus Specific Speed

Specific speed is such an effective parameter that it is used as an indicator of both performance and efficiency. Figure 11.14 shows a correlation of the optimum efficiency of a pump as a function of the specific speed and capacity. Because the

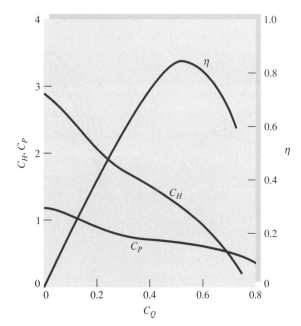

Fig. 11.13 Dimensionless performance curves for a typical axial-flow pump, $N_s = 12{,}000$. Constructed from data given by Stepanoff [8] for a 14-in pump at 690 r/min.

dimensional parameter Q is a rough measure of both size and Reynolds number, η increases with Q. When this type of correlation was first published by Wislicenus [4] in 1947, it became known as *the* pump curve, a challenge to all manufacturers. We can check that the pumps of Figs. 11.7 and 11.13 fit the correlation very well.

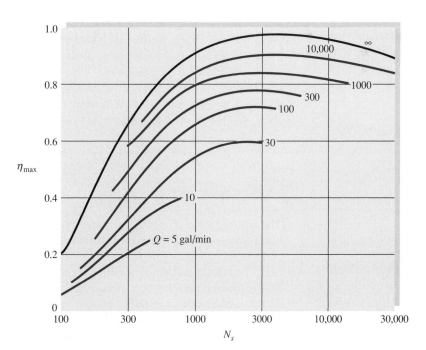

Fig. 11.14 Optimum efficiency of pumps versus capacity and specific speed. *(Adapted from Refs. 4 and 31.)*

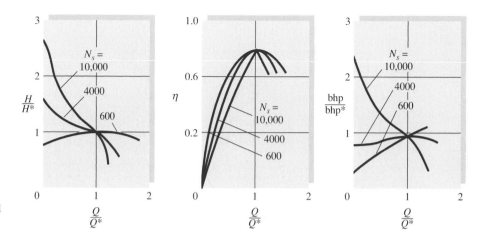

Fig. 11.15 Effect of specific speed on pump performance curves.

Figure 11.15 shows the effect of specific speed on the shape of the pump performance curves, normalized with respect to the BEP point. The numerical values shown are representative but somewhat qualitative. The high-specific-speed pumps ($N_s \approx 10{,}000$) have head and power curves that drop sharply with discharge, implying overload or start-up problems at low flow. Their efficiency curve is very narrow.

A low-specific-speed pump ($N_s = 600$) has a broad efficiency curve, a rising power curve, and a head curve that "droops" at shutoff, implying possible surge or hunting problems.

The Free Propeller

The propeller-style pump of Fig. 11.12 is enclosed in a duct and captures all the approach flow. In contrast, the *free propeller,* for either aircraft or marine applications, acts in an unbounded fluid and thus is much less effective. The analog of propeller-pump pressure rise is the free propeller *thrust* per unit area ($\pi D^2/4$) swept out by the blades. In a customary dimensional analysis, thrust T and power required P are functions of fluid density ρ, rotation rate n (rev/s), forward velocity V, and propeller diameter D. Viscosity effects are small and neglected. You might enjoy analyzing this as a Chap. 5 assignment. The NACA (now the NASA) chose (ρ, n, D) as repeating variables, and the results are the accepted parameters:

$$C_T = \text{thrust coefficient} = \frac{T}{\rho n^2 D^4} = \text{fcn}(J), \ J = \text{advance ratio} = \frac{V}{nD}$$

$$C_P = \text{power coefficient} = \frac{P}{\rho n^3 D^5} = \text{fcn}(J), \ \eta = \text{efficiency} = \frac{VT}{P} = \frac{JC_T}{C_P} \quad (11.34)$$

The advance ratio, J, which compares forward velocity to a measure proportional to blade tip speed, has a strong effect upon thrust and power.

Figure 11.16 shows performance data for a propeller used on the Cessna 172 aircraft. The thrust and power coefficients are small, of $\Theta(0.05)$, and are multiplied by 10 for plotting convenience. Maximum efficiency is 83 percent at $J = 0.7$, where $C_T^* \approx 0.040$ and $C_P^* \approx 0.034$.

There are several engineering methods for designing propellers. These theories are described in specialized texts, both for marine [60] and aircraft [61] propellers.

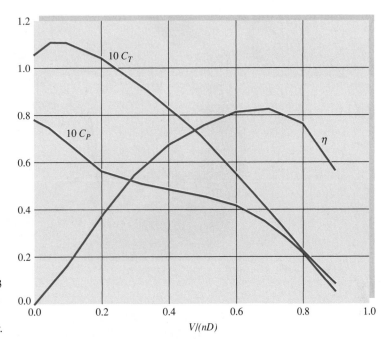

Fig. 11.16 Performance data for a free propeller used on the Cessna 172 aircraft. Compare to Fig. 11.13 for a (ducted) propeller pump. The thrust and power coefficients are much smaller for the free propeller.

Computational Fluid Dynamics

The design of turbomachinery has traditionally been highly experimental, with simple theories, such as in Sec. 11.2, only able to predict trends. Dimensionless correlations, such as Fig. 11.15, are useful but require extensive experimentation. Consider that flow in a pump is three-dimensional; unsteady (both periodic and turbulent); and involves flow separation, recirculation in the impeller, unsteady blade wakes passing through the diffuser, and blade roots, tips, and clearances. It is no wonder that one-dimensional theory cannot give firm quantitative predictions.

Modern computer analysis can give realistic results and is becoming a useful tool for turbomachinery designers. A good example is Ref. 56, reporting combined experimental and computational results for a centrifugal pump diffuser. A photograph of the device is shown in Fig. 11.17a. It is made of clear Perspex, so that laser measurements of particle tracking velocimetry (LPTV) and doppler anemometry (LDA) could be taken throughout the system. The data were compared with a CFD simulation of the impeller and diffuser, using the grids shown in Fig. 11.17b. The computations used a turbulence formulation called the k-ϵ model, popular in commercial CFD codes (see Sec. 8.9). Results were good but not excellent. The CFD model predicted velocity and pressure data adequately up until flow separation, after which it was only qualitative. Clearly, CFD is developing a significant role in turbomachinery design [42, 45].

11.5 Matching Pumps to System Characteristics

The ultimate test of a pump is its match with the operating system characteristics. Physically, the system head must match the head produced by the pump, and this intersection should occur in the region of best efficiency.

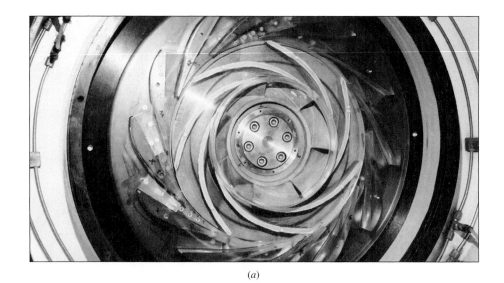

(a)

Impeller

Fig. 11.17 Turbomachinery design now involves both experimentation and computational fluid dynamics (CFD): (a) a centrifugal impeller and diffuser *(courtesy of K. Eisele et al., "Flow Analysis in a Pump Diffuser: Part 1, Measurements: Part 2, CFD," Journal of Fluids Eng. Vol. 119, December 1997, pp. 968–984/American Society of Mechanical Engineers)*; (b) a three-dimensional CFD model grid for this system *(from Ref. 56 by permission of the American Society of Mechanical Engineers).*

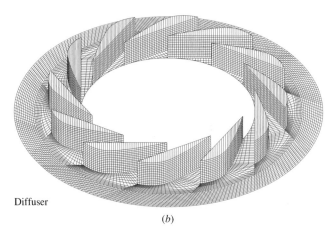

Diffuser

(b)

The system head will probably contain a static elevation change $z_2 - z_1$ plus friction losses in pipes and fittings:

$$H_{\text{sys}} = (z_2 - z_1) + \frac{V^2}{2g}\left(\sum \frac{fL}{D} + \sum K \right)$$

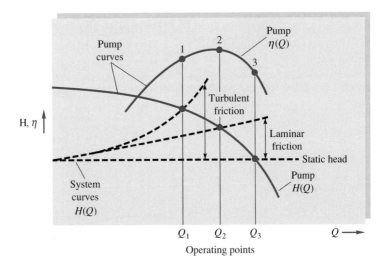

Fig. 11.18 Illustration of pump operating points for three types of system head curves.

where ΣK denotes minor losses and V is the flow velocity in the principal pipe. Since V is proportional to the pump discharge Q, the equation represents a system head curve $H_s(Q)$. Three examples are shown in Fig. 11.18: a static head $H_s = a$, static head plus laminar friction $H_s = a + bQ$, and static head plus turbulent friction $H_s = a + cQ^2$. The intersection of the system curve with the pump performance curve $H(Q)$ defines the operating point. In Fig. 11.18 the laminar friction operating point is at maximum efficiency while the turbulent and static curves are off design. This may be unavoidable if system variables change, but the pump should be changed in size or speed if its operating point is consistently off design. Of course, a perfect match may not be possible because commercial pumps have only certain discrete sizes and speeds. Let us illustrate these concepts with an example.

EXAMPLE 11.6

We want to use the 32-in pump of Fig. 11.7a at 1170 r/min to pump water at 60°F from one reservoir to another 120 ft higher through 1500 ft of 16-in-ID pipe with friction factor $f = 0.030$. (*a*) What will the operating point and efficiency be? (*b*) To what speed should the pump be changed to operate at the BEP?

Solution

Part (a)

For reservoirs the initial and final velocities are zero; thus the system head is

$$H_s = z_2 - z_1 + \frac{V^2}{2g}\frac{fL}{D} = 120 \text{ ft} + \frac{V^2}{2g}\frac{0.030(1500 \text{ ft})}{\frac{16}{12} \text{ ft}}$$

From continuity in the pipe, $V = Q/A = Q/[\frac{1}{4}\pi(\frac{16}{12} \text{ ft})^2]$, and so we substitute for V to get

$$H_s = 120 + 0.269Q^2 \qquad Q \text{ in ft}^3\text{/s} \tag{1}$$

Since Fig. 11.7a uses thousands of gallons per minute for the abscissa, we convert Q in Eq. (1) to this unit:

$$H_s = 120 + 1.335Q^2 \qquad Q \text{ in } 10^3 \text{ gal/min} \qquad (2)$$

We can plot Eq. (2) on Fig. 11.7a and see where it intersects the 32-in pump head curve, as in Fig. E11.6. A graphical solution gives approximately

$$H \approx 430 \text{ ft} \qquad Q \approx 15{,}000 \text{ gal/min}$$

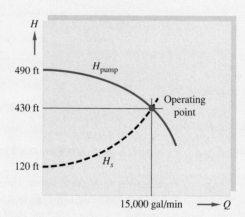

E11.6

The efficiency is about 82 percent, slightly off design.

An analytic solution is possible if we fit the pump head curve to a parabola, which is very accurate:

$$H_{\text{pump}} \approx 490 - 0.26Q^2 \qquad Q \text{ in } 10^3 \text{ gal/min} \qquad (3)$$

Equations (2) and (3) must match at the operating point:

$$490 - 0.26Q^2 = 120 + 1.335Q^2$$

or $\qquad\qquad\qquad Q^2 = \dfrac{490 - 120}{0.26 + 1.335} = 232$

$$Q = 15.2 \times 10^3 \text{ gal/min} = 15{,}200 \text{ gal/min} \qquad\qquad \textit{Ans. (a)}$$

$$H = 490 - 0.26(15.2)^2 = 430 \text{ ft} \qquad\qquad\qquad \textit{Ans. (a)}$$

Part (b)

To move the operating point to BEP, we change n, which changes both $Q \propto n$ and $H \propto n^2$. From Fig. 11.7a, at BEP, $H^* \approx 386$ ft; thus for any n, $H^* = 386(n/1170)^2$. Also read $Q^* \approx 20 \times 10^3$ gal/min; thus for any n, $Q^* = 20(n/1170)$. Match H^* to the system characteristics, Eq. (2):

$$H^* = 386\left(\frac{n}{1170}\right)^2 \approx 120 + 1.335\left(20\frac{n}{1170}\right)^2 \qquad\qquad \textit{Ans. (b)}$$

which gives $n^2 < 0$. Thus it is impossible to operate at maximum efficiency with this particular system and pump.

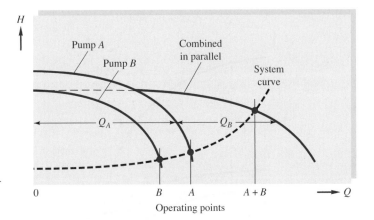

Fig. 11.19 Performance and operating points of two pumps operating singly and combined in parallel.

Pumps Combined in Parallel

If a pump provides the right head but too little discharge, a possible remedy is to combine two similar pumps in parallel, sharing the same suction and inlet conditions. A parallel arrangement is also used if delivery demand varies, so that one pump is used at low flow and the second pump is started up for higher discharges. Both pumps should have check valves to avoid backflow when one is shut down.

The two pumps in parallel need not be identical. Physically, their flow rates will sum for the same head, as illustrated in Fig. 11.19. If pump A has more head than pump B, pump B cannot be added in until the operating head is below the shutoff head of pump B. Since the system curve rises with Q, the combined delivery Q_{A+B} will be less than the separate operating discharges $Q_A + Q_B$ but certainly greater than either one. For a very flat (static) curve two similar pumps in parallel will deliver nearly twice the flow. The combined brake horsepower is found by adding brake horsepower for each of pumps A and B at the same head as the operating point. The combined efficiency equals $\rho g(Q_{A+B})(H_{A+B})/(550 \text{ bhp}_{A+B})$.

If pumps A and B are not identical, as in Fig. 11.19, pump B should not be run and cannot even be started up if the operating point is above its shutoff head.

Pumps Combined in Series

If a pump provides the right discharge but too little head, consider adding a similar pump in series, with the output of pump B fed directly into the suction side of pump A. As sketched in Fig. 11.20, the physical principle for summing in series is that the two heads add at the same flow rate to give the combined performance curve. The two need not be identical at all, since they merely handle the same discharge; they may even have different speeds, although normally both are driven by the same shaft.

The need for a series arrangement implies that the system curve is steep that is, it requires higher head than either pump A or B can provide. The combined operating point head will be more than either A or B separately but not as great as their sum. The combined power is the sum of brake horsepower for A and B at the operating point flow rate. The combined efficiency is

$$\frac{\rho g(Q_{A+B})(H_{A+B})}{550 \text{ bhp}_{A+B}}$$

similar to parallel pumps.

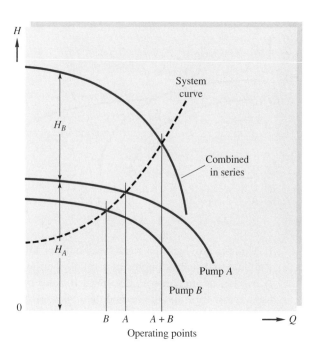

Fig. 11.20 Performance of two pumps combined in series.

Whether pumps are used in series or in parallel, the arrangement will be uneconomical unless both pumps are operating near their best efficiency.

Multistage Pumps

For very high heads in continuous operation, the solution is a multistage pump, with the exit of one impeller feeding directly into the eye of the next. Centrifugal, mixed-flow, and axial-flow pumps have all been grouped in as many as 50 stages, with heads up to 8000 ft of water and pressure rises up to 5000 lbf/in^2 absolute. Figure 11.21 shows a section of a seven-stage centrifugal propane compressor that develops 300 lbf/in^2 rise at 40,000 ft^3/min and 35,000 bhp.

Compressors

Most of the discussion in this chapter concerns incompressible flow—that is, negligible change in fluid density. Even the pump of Fig. 11.7, which can produce 600 ft of head at 1170 r/min, will increase standard air pressure only by 46 lbf/ft^2, about a 2 percent change in density. The picture changes at higher speeds, $\Delta p \propto n^2$, and multiple stages, where very large changes in pressure and density are achieved. Such devices are called *compressors*, as in Fig. 11.21. The concept of static head, $H = \Delta p/\rho g$, becomes inappropriate, since ρ varies. Compressor performance is measured by (1) the pressure ratio across the stage p_2/p_1 and (2) the change in stagnation enthalpy ($h_{02} - h_{01}$), where $h_0 = h + \frac{1}{2}V^2$ (see Sec. 9.3). Combining m stages in series results in $p_{\text{final}}/p_{\text{initial}} \approx (p_2/p_1)^m$. As density increases, less area is needed: note the decrease in impeller size from right to left in Fig. 11.21. Compressors may be either of the centrifugal or axial-flow type [21 to 23].

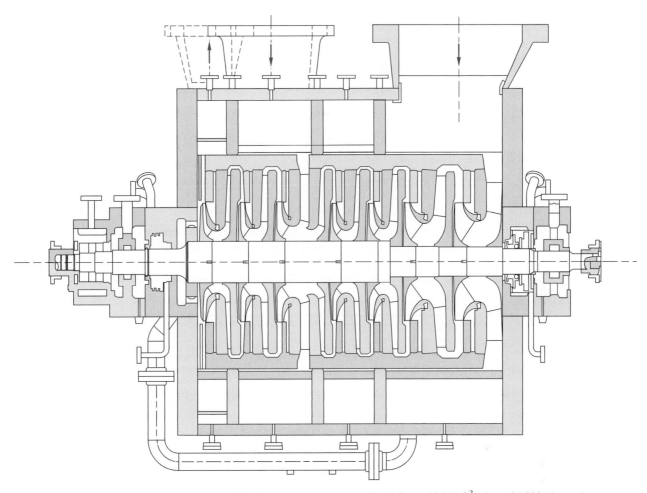

Fig. 11.21 Cross section of a seven-stage centrifugal propane compressor that delivers 40,000 ft³/min at 35,000 bhp and a pressure rise of 300 lbf/in². Note the second inlet at stage 5 and the varying impeller designs. *(Courtesy of DeLaval-Stork V.O.F., Centrifugal Compressor Division.)*

Compressor efficiency, from inlet condition 1 to final outlet f, is defined by the change in gas enthalpy, assuming an adiabatic process:

$$\eta_{comp} = \frac{h_f - h_{01}}{h_{0f} - h_{01}} \approx \frac{T_f - T_{01}}{T_{0f} - T_{01}}$$

Compressor efficiencies are similar to hydraulic machines ($\eta_{max} \approx 70$ to 80 percent), but the mass flow range is more limited: on the low side by compressor *surge,* where blade stall and vibration occur, and on the high side by *choking* (Sec. 9.4), where the Mach number reaches 1.0 somewhere in the system. Compressor mass flow is normally plotted using the same type of dimensionless function formulated in Eq. (9.47): $\dot{m}(RT_0)^{1/2}/(D^2 p_0)$, which will reach a maximum when choking occurs. For further details, see Refs. 21 to 23.

EXAMPLE 11.7

Investigate extending Example 11.6 by using two 32-in pumps in parallel to deliver more flow. Is this efficient?

Solution

Since the pumps are identical, each delivers $\frac{1}{2}Q$ at the same 1170 r/min speed. The system curve is the same, and the balance-of-head relation becomes

$$H = 490 - 0.26(\tfrac{1}{2}Q)^2 = 120 + 1.335Q^2$$

or
$$Q^2 = \frac{490 - 120}{1.335 + 0.065} \qquad Q = 16{,}300 \text{ gal/min} \qquad\qquad Ans.$$

This is only 7 percent more than a single pump. Each pump delivers $\frac{1}{2}Q = 8130$ gal/min, for which the efficiency is only 60 percent. The total brake horsepower required is 3200, whereas a single pump used only 2000 bhp. This is a poor design.

EXAMPLE 11.8

Suppose the elevation change in Example 11.6 is raised from 120 to 500 ft, greater than a single 32-in pump can supply. Investigate using 32-in pumps in series at 1170 r/min.

Solution

Since the pumps are identical, the total head is twice as much and the constant 120 in the system head curve is replaced by 500. The balance of heads becomes

$$H = 2(490 - 0.26Q^2) = 500 + 1.335Q^2$$

or
$$Q^2 = \frac{980 - 500}{1.335 + 0.52} \qquad Q = 16.1 \times 10^3 \text{ gal/min} \qquad\qquad Ans.$$

The operating head is $500 + 1.335(16.1)^2 = 845$ ft, or 97 percent more than that for a single pump in Example 11.5. Each pump is operating at 16.1×10^3 gal/min, which from Fig. 11.7a is 83 percent efficient, a pretty good match to the system. To pump at this operating point requires 4100 bhp, or about 2050 bhp for each pump.

11.6 Turbines

A turbine extracts energy from a fluid that possesses high head, but it is fatuous to say a turbine is a pump run backward. Basically there are two types, reaction and impulse, the difference lying in the manner of head conversion. In the *reaction turbine,* the fluid fills the blade passages, and the head change or pressure drop occurs within the impeller. Reaction designs are of the radial-flow, mixed-flow, and axial-flow types and are essentially dynamic devices designed to admit the high-energy fluid and extract its momentum. An *impulse turbine* first converts the high head through a nozzle into a high-velocity jet, which then strikes the blades at one position as they

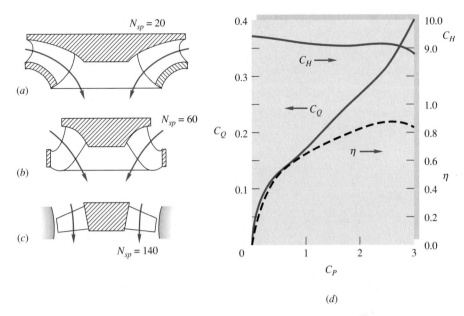

Fig. 11.22 Reaction turbines:
(*a*) Francis, radial type; (*b*) Francis
mixed-flow; (*c*) propeller axial-
flow; (*d*) performance curves for a
Francis turbine, $n = 600$ r/min,
$D = 2.25$ ft, $N_{sp} = 29$.

pass by. The impeller passages are not fluid-filled, and the jet flow past the blades is essentially at constant pressure. Reaction turbines are smaller because fluid fills all the blades at one time.

Reaction Turbines

Reaction turbines are low-head, high-flow devices. The flow is opposite that in a pump, entering at the larger-diameter section and discharging through the eye after giving up most of its energy to the impeller. Early designs were very inefficient because they lacked stationary guide vanes at the entrance to direct the flow smoothly into the impeller passages. The first efficient inward-flow turbine was built in 1849 by James B. Francis, a U.S. engineer, and all radial- or mixed-flow designs are now called *Francis turbines*. At still lower heads, a turbine can be designed more compactly with purely axial flow and is termed a *propeller turbine* [52]. The propeller may be either fixed-blade or adjustable (Kaplan type), the latter being complicated mechanically but much more efficient at low-power settings. Figure 11.22 shows sketches of runner designs for Francis radial, Francis mixed-flow, and propeller-type turbines.

Idealized Radial Turbine Theory

The Euler turbomachine formulas (11.11) also apply to energy-extracting machines if we reverse the flow direction and reshape the blades. Figure 11.23 shows a radial turbine runner. Again assume one-dimensional frictionless flow through the blades. Adjustable inlet guide vanes are absolutely necessary for good efficiency. They bring the inlet flow to the blades at angle α_2 and absolute velocity V_2 for minimum "shock" or directional-mismatch loss. After vectorially adding in the runner tip speed $u_2 = \omega r_2$, the outer blade angle should be set at angle β_2 to accommodate the relative velocity w_2, as shown in the figure. (See Fig. 11.4 for the analogous radial pump velocity diagrams.)

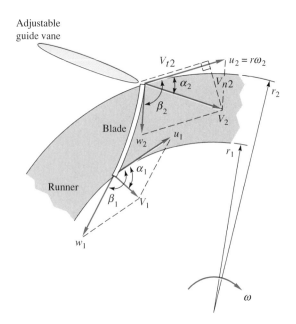

Fig. 11.23 Inlet and outlet velocity diagrams for an idealized radial-flow reaction turbine runner.

Application of the angular momentum control volume theorem, Eq. (3.55), to Fig. 11.23 (see Example 3.14 for a similar case) yields an idealized formula for the power P extracted by the runner:

$$P = \omega T = \rho \omega Q (r_2 V_{t2} - r_1 V_{t1}) = \rho Q (u_2 V_2 \cos \alpha_2 - u_1 V_1 \cos \alpha_1) \quad (11.35)$$

where V_{t2} and V_{t1} are the absolute inlet and outlet circumferential velocity components of the flow. Note that Eq. (11.35) is identical to Eq. (11.11) for a radial pump, except that the blade shapes are different.

The absolute inlet normal velocity $V_{n2} = V_2 \sin \alpha_2$ is proportional to the flow rate Q. If the flow rate changes and the runner speed u_2 is constant, the vanes must be adjusted to a new angle α_2 so that w_2 still follows the blade surface. Thus adjustable inlet vanes are very important to avoid shock loss.

Power Specific Speed

Turbine parameters are similar to those of a pump, but the dependent variable is the output brake horsepower, which depends on the inlet flow rate Q, available head H, impeller speed n, and diameter D. The efficiency is the output brake horsepower divided by the available water horsepower $\rho g Q H$. The dimensionless forms are C_Q, C_H, and C_P, defined just as for a pump, Eqs. (11.23). If we neglect Reynolds number and roughness effects, the functional relationships are written with C_P as the independent variable:

$$C_H = \frac{gH}{n^2 D^2} = C_H(C_P) \qquad C_Q = \frac{Q}{nD^3} = C_Q(C_P) \qquad \eta = \frac{bhp}{\rho g Q H} = \eta(C_P) \quad (11.36)$$

where

$$C_P = \frac{bhp}{\rho n^3 D^5}$$

Figure 11.22d shows typical performance curves for a small Francis radial turbine. The maximum efficiency point is called the *normal power,* and the values for this particular turbine are

$$\eta_{max} = 0.89 \qquad C_{P*} = 2.70 \qquad C_{Q*} = 0.34 \qquad C_{H*} = 9.03$$

A parameter that compares the output power with the available head, independent of size, is found by eliminating the diameter between C_H and C_P. It is called the *power specific speed:*

Rigorous form:
$$N'_{sp} = \frac{C_P^{*1/2}}{C_H^{*5/4}} = \frac{n(\text{bhp})^{1/2}}{\rho^{1/2}(gH)^{5/4}} \qquad (11.37a)$$

Lazy but common:
$$N_{sp} = \frac{(\text{r/min})(\text{bhp})^{1/2}}{[H\,(\text{ft})]^{5/4}} \qquad (11.37b)$$

For water, $\rho = 1.94$ slugs/ft^3 and $N_{sp} = 273.3 N'_{sp}$. The various turbine designs divide up nicely according to the range of power specific speed, as follows:

Turbine type	N_{sp} range	C_H range
Impulse	1–10	15–50
Francis	10–110	5–25
Propeller:		
Water	100–250	1–4
Gas, steam	25–300	10–80

Note that N_{sp}, like N_s for pumps, is defined only with respect to the BEP and has a single value for a given turbine family. In Fig. 11.22d, $N_{sp} = 273.3(2.70)^{1/2}/(9.03)^{5/4} = 29$, regardless of size.

Like pumps, turbines of large size are generally more efficient, and Eqs. (11.29) can be used as an estimate when data are lacking.

The design of a complete large-scale power-generating turbine system is a major engineering project, involving inlet and outlet ducts, trash racks, guide vanes, wicket gates, spiral cases, generator with cooling coils, bearings and transmission gears, runner blades, draft tubes, and automatic controls. Some typical large-scale reaction turbine designs are shown in Fig. 11.24. The reversible pump-and-turbine design of Fig. 11.24d requires special care for adjustable guide vanes to be efficient for flow in either direction.

The largest (1000-MW) hydropower designs are awesome when viewed on a human scale, as shown in Fig. 11.25. The economic advantages of small-scale model testing are evident from this photograph of the Francis turbine units at Grand Coulee Dam.

Impulse Turbines

For high head and relatively low power (that is, low N_{sp}) not only would a reaction turbine require too high a speed but also the high pressure in the runner would require a massive casing thickness. The impulse turbine of Fig. 11.26 is ideal for this situation. Since N_{sp} is low, n will be low and the high pressure is confined to the small nozzle, which converts the head to an atmospheric pressure jet of high velocity V_j.

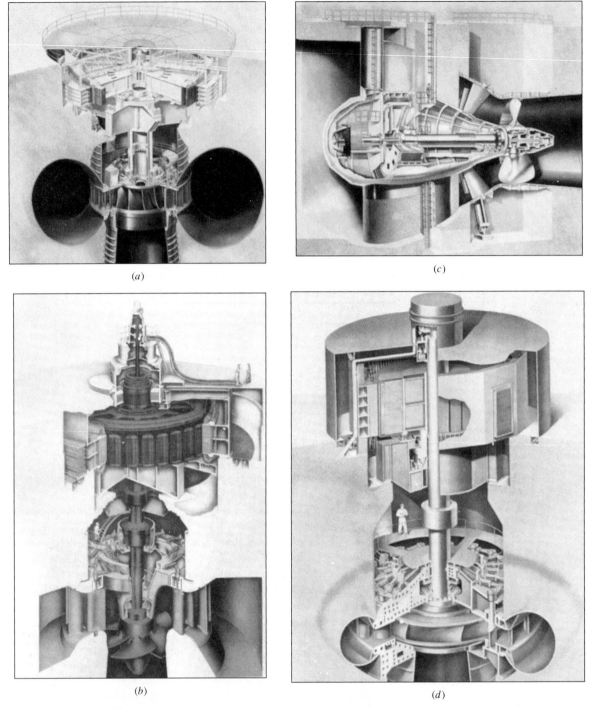

Fig. 11.24 Large-scale turbine designs depend on available head and flow rate and operating conditions: (*a*) Francis (radial); (*b*) Kaplan (propeller); (*c*) bulb mounting with propeller runner; (*d*) reversible pump turbine with radial runner. *(Courtesy of Voith Siemens Hydro Power.)*

Fig. 11.25 Interior view of the 1.1-million hp (820-MW) turbine units on the Grand Coulee Dam of the Columbia River, showing the spiral case, the outer fixed vanes ("stay ring"), and the inner adjustable vanes ("wicket gates"). *(Courtesy of Voith Siemens Hydro Power.)*

The jet strikes the buckets and imparts a momentum change similar to that in our control volume analysis for a moving vane in Example 3.10 or Prob. P3.51. The buckets have an elliptical split-cup shape, as in Fig. 11.26b. They are named *Pelton wheels,* after Lester A. Pelton (1829–1908), who produced the first efficient design.

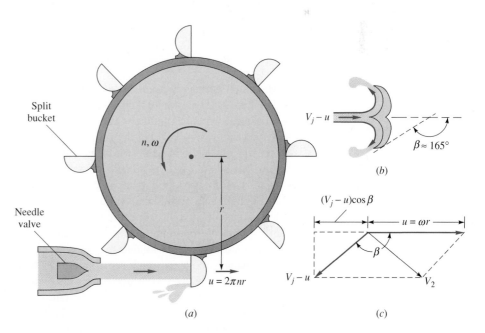

Fig. 11.26 Impulse turbine: (*a*) side view of wheel and jet; (*b*) top view of bucket; (*c*) typical velocity diagram.

In Example 3.10 we found that the force per unit mass flow on a single moving vane, or in this case a single Pelton bucket, was $(V_j - u)(1 - \cos\beta)$, where u is the vane velocity and β is the exit angle of the jet. For a single vane, as in Example 3.10, the mass flow would be $\rho A_j(V_j - u)$, but for a Pelton wheel, where buckets keep entering the jet and capture all the flow, the mass flow would be $\rho Q = \rho A_j V_j$. An alternative analysis uses the Euler turbomachine equation (11.11) and the velocity diagram of Fig. 11.26c. Noting that $u_1 = u_2 = u$, we substitute the absolute exit and inlet tangential velocities into the turbine power relation:

$$P = \rho Q(u_1 V_{t1} - u_2 V_{t2}) = \rho Q\{uV_j - u[u + (V_j - u)\cos\beta]\}$$

or
$$P = \rho Q u(V_j - u)(1 - \cos\beta) \tag{11.38}$$

where $u = 2\pi nr$ is the bucket linear velocity and r is the *pitch radius*, or distance to the jet centerline. A bucket angle $\beta = 180°$ gives maximum power but is physically impractical. In practice, $\beta \approx 165°$, or $1 - \cos\beta \approx 1.966$ or only 2 percent less than maximum power.

From Eq. (11.38) the theoretical power of an impulse turbine is parabolic in bucket speed u and is maximum when $dP/du = 0$, or

$$u^* = 2\pi n^* r = \tfrac{1}{2}V_j \tag{11.39}$$

For a perfect nozzle, the entire available head would be converted to jet velocity $V_j = (2gH)^{1/2}$. Actually, since there are 2 to 8 percent nozzle losses, a velocity coefficient C_v is used:

$$V_j = C_v(2gH)^{1/2} \qquad 0.92 \leq C_v \leq 0.98 \tag{11.40}$$

By combining Eqs. (11.36) and (11.40), the theoretical impulse turbine efficiency becomes

$$\boxed{\eta = 2(1 - \cos\beta)\phi(C_v - \phi)} \tag{11.41}$$

where
$$\boxed{\phi = \frac{u}{(2gH)^{1/2}} = \text{peripheral velocity factor}}$$

Maximum efficiency occurs at $\phi = \tfrac{1}{2}C_v \approx 0.47$.

Figure 11.27 shows Eq. (11.41) plotted for an ideal turbine ($\beta = 180°$, $C_v = 1.0$) and for typical working conditions ($\beta = 160°$, $C_v = 0.94$). The latter case predicts $\eta_{max} = 85$ percent at $\phi = 0.47$, but the actual data for a 24-in Pelton wheel test are somewhat less efficient due to windage, mechanical friction, backsplashing, and nonuniform bucket flow. For this test $\eta_{max} = 80$ percent, and, generally speaking, an impulse turbine is not quite as efficient as the Francis or propeller turbines at their BEPs.

Figure 11.28 shows the optimum efficiency of the three turbine types, and the importance of the power specific speed N_{sp} as a selection tool for the designer. These efficiencies are optimum and are obtained in careful design of large machines.

The water power available to a turbine may vary due to either net head or flow rate changes, both of which are common in field installations such as hydroelectric plants. The demand for turbine power also varies from light to heavy, and the operating response is a change in the flow rate by adjustment of a gate valve or needle valve (Fig. 11.26a). As shown in Fig. 11.29, all three turbine types achieve fairly uniform efficiency as a function of the level of power being extracted. Especially

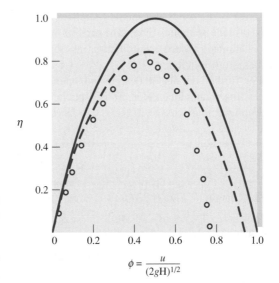

Fig. 11.27 Efficiency of an impulse turbine calculated from Eq. (11.41): solid curve = ideal, $\beta = 180°$, $C_v = 1.0$; dashed curve = actual, $\beta = 160°$, $C_v = 0.94$; open circles = data, Pelton wheel, diameter = 2 ft.

$$\phi = \frac{u}{(2gH)^{1/2}}$$

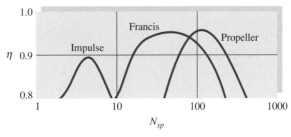

Fig. 11.28 Optimum efficiency of turbine designs.

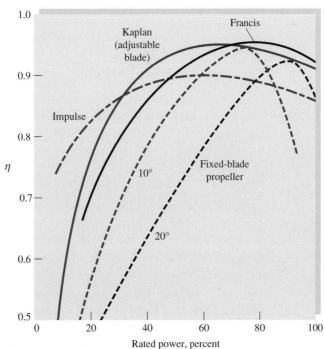

Fig. 11.29 Efficiency versus power level for various turbine designs at constant speed and head.

Rated power, percent

effective is the adjustable-blade (Kaplan-type) propeller turbine, while the poorest is a fixed-blade propeller. The term *rated power* in Fig. 11.29 is the largest power delivery guaranteed by the manufacturer, as opposed to *normal power,* which is delivered at maximum efficiency.

For further details of design and operation of turbomachinery, the readable and interesting treatment in Ref. 33 is especially recommended. The feasibility of micro-hydropower is discussed in Refs. 27, 28, and 46.

EXAMPLE 11.9

Investigate the possibility of using (*a*) a Pelton wheel similar to Fig. 11.27 or (*b*) the Francis turbine family of Fig. 11.22*d* to deliver 30,000 bhp from a net head of 1200 ft.

Solution

Part (a)

From Fig. 11.28, the most efficient Pelton wheel occurs at about

$$N_{sp} \approx 4.5 = \frac{(\text{r/min})(30{,}000 \text{ bhp})^{1/2}}{(1200 \text{ ft})^{1.25}}$$

or

$$n = 183 \text{ r/min} = 3.06 \text{ r/s}$$

From Fig. 11.27 the best operating point is

$$\phi \approx 0.47 = \frac{\pi D(3.06 \text{ r/s})}{[2(32.2)(1200)]^{1/2}}$$

or

$$D = 13.6 \text{ ft} \qquad\qquad\qquad Ans. (a)$$

This Pelton wheel is perhaps a little slow and a trifle large. You could reduce D and increase n by increasing N_{sp} to, say, 6 or 7 and accepting the slight reduction in efficiency. Or you could use a double-hung, two-wheel configuration, each delivering 15,000 bhp, which changes D and n by the factor $2^{1/2}$:

Double wheel: $n = (183)2^{1/2} = 260 \text{ r/min}$ $D = \dfrac{13.6}{2^{1/2}} = 9.6 \text{ ft}$ *Ans. (a)*

Part (b)

The Francis wheel of Fig. 11.22*d* must have

$$N_{sp} = 29 = \frac{(\text{r/min})(30{,}000 \text{ bhp})^{1/2}}{(1200 \text{ ft})^{1.25}}$$

or

$$n = 1183 \text{ r/min} = 19.7 \text{ r/s}$$

Then the optimum power coefficient is

$$C_{P*} = 2.70 = \frac{P}{\rho n^3 D^5} = \frac{30{,}000(550)}{(1.94)(19.7)^3 D^5}$$

or

$$D^5 = 412 \qquad D = 3.33 \text{ ft} = 40 \text{ in} \qquad\qquad Ans. (b)$$

This is a faster speed than normal practice, and the casing would have to withstand 1200 ft of water or about 520 lbf/in² internal pressure, but the 40-in size is extremely attractive. Francis turbines are now being operated at heads up to 1500 ft.

Wind Turbines

Wind energy has long been used as a source of mechanical power. The familiar four-bladed windmills of Holland, England, and the Greek islands have been used for centuries to pump water, grind grain, and saw wood. Modern research concentrates on the ability of wind turbines to generate electric power. Koeppl [47] stresses the potential for propeller-type machines. Spera [49] gives a detailed discussion of the technical and economic feasibility of large-scale electric power generation by wind. See also Refs. 47, 48, 50, and 51.

Some examples of wind turbine designs are shown in Fig. 11.30. The familiar American multiblade farm windmill (Fig. 11.30a) is of low efficiency, but thousands are in use as a rugged, reliable, and inexpensive way to pump water. A more efficient design is the propeller mill in Fig. 11.30b, similar to the pioneering Smith-Putnam 1250-kW two-bladed system that operated on Grampa's Knob, 12 mi west of Rutland, Vermont, from 1941 to 1945. The Smith-Putnam design broke because of inadequate blade strength, but it withstood winds up to 115 mi/h and its efficiency was amply demonstrated [47].

The Dutch, American multiblade, and propeller mills are examples of *horizontal-axis wind turbines* (HAWTs), which are efficient but somewhat awkward in that they require extensive bracing and gear systems when combined with an electric generator. Thus a competing family of *vertical-axis wind turbines* (VAWTs) has been proposed to simplify gearing and strength requirements. Figure 11.30c shows the "eggbeater" VAWT invented by G. J. M. Darrieus in 1925. To minimize centrifugal stresses, the twisted blades of the Darrieus turbine follow a *troposkien* curve formed by a chain anchored at two points on a spinning vertical rod. The Darrieus design has the advantage that the generator and gearbox may be placed on the ground for easy access. It is not as efficient, though, as a HAWT, and, further more, it is not self-starting. The largest Darrieus device known to the writer is a 4.2 MW turbine, 100 m in diameter, at Cap Chat, Québec, Canada.

An alternative VAWT, simpler to construct than the troposkien, is the straight-bladed Darrieus-type turbine in Fig. 11.30d. This design, proposed by Reading University in England, has blades that pivot due to centrifugal action as wind speeds increase, thus limiting bending stresses.

Idealized Wind Turbine Theory

The ideal, frictionless efficiency of a propeller windmill was predicted by A. Betz in 1920, using the simulation shown in Fig. 11.31. The propeller is represented by an *actuator disk,* which creates across the propeller plane a pressure discontinuity of area A and local velocity V. The wind is represented by a streamtube of approach velocity V_1 and a slower downstream wake velocity V_2. The pressure rises to p_b just before the disk and drops to p_a just after, returning to free-stream pressure in the far wake. To hold the propeller rigid when it is extracting energy from the wind, there must be a leftward force F on its support, as shown.

A control-volume–horizontal-momentum relation applied between sections 1 and 2 gives

$$\sum F_x = -F = \dot{m}(V_2 - V_1)$$

A similar relation for a control volume just before and after the disk gives

$$\sum F_x = -F + (p_b - p_a)A = \dot{m}(V_a - V_b) = 0$$

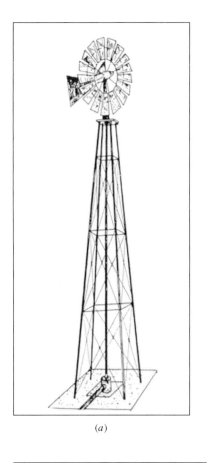

(a)

(c)

(b)

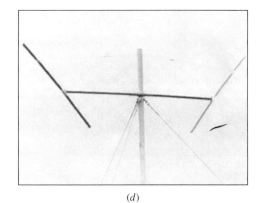

(d)

Fig. 11.30 Wind turbine designs: (a) the American multiblade farm HAWT; (b) propeller HAWT *(courtesy of Northrop Grumman);* (c) the Darrieus VAWT *(courtesy of National Research Council Canada);* (d) modified straight-blade Darrieus VAWT *(courtesy of Dr. Peter Musgrove).*

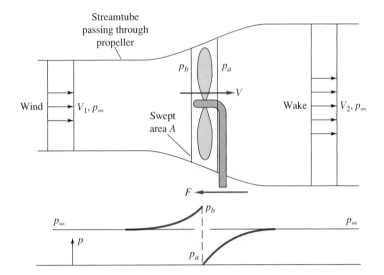

Fig. 11.31 Idealized actuator disk and streamtube analysis of flow through a windmill.

Equating these two yields the propeller force:

$$F = (p_b - p_a)A = \dot{m}(V_1 - V_2) \qquad (11.42)$$

Assuming ideal flow, the pressures can be found by applying the incompressible Bernoulli relation up to the disk:

From 1 to b:
$$p_\infty + \tfrac{1}{2}\rho V_1^2 = p_b + \tfrac{1}{2}\rho V^2$$

From a to 2:
$$p_a + \tfrac{1}{2}\rho V^2 = p_\infty + \tfrac{1}{2}\rho V_2^2$$

Subtracting these and noting that $\dot{m} = \rho A V$ through the propeller, we can substitute for $p_b - p_a$ in Eq. (11.42) to obtain

$$p_b - p_a = \tfrac{1}{2}\rho(V_1^2 - V_2^2) = \rho V(V_1 - V_2)$$

or
$$V = \tfrac{1}{2}(V_1 + V_2) \qquad (11.43)$$

Continuity and momentum thus require that the velocity V through the disk equal the average of the wind and far-wake speeds.

Finally, the power extracted by the disk can be written in terms of V_1 and V_2 by combining Eqs. (11.42) and (11.43):

$$P = FV = \rho A V^2(V_1 - V_2) = \tfrac{1}{4}\rho A(V_1^2 - V_2^2)(V_1 + V_2) \qquad (11.44)$$

For a given wind speed V_1, we can find the maximum possible power by differentiating P with respect to V_2 and setting equal to zero. The result is

$$P = P_{\text{max}} = \tfrac{8}{27}\rho A V_1^3 \qquad \text{at } V_2 = \tfrac{1}{3}V_1 \qquad (11.45)$$

which corresponds to $V = 2V_1/3$ through the disk.

The maximum available power to the propeller is the mass flow through the propeller times the total kinetic energy of the wind:

$$P_{\text{avail}} = \tfrac{1}{2}\dot{m}V_1^2 = \tfrac{1}{2}\rho A V_1^3$$

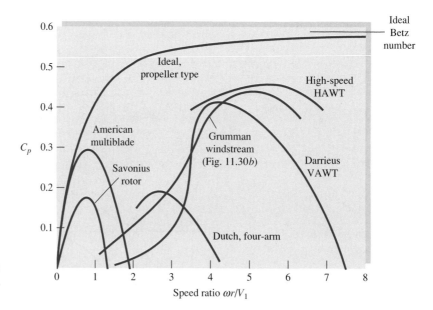

Fig. 11.32 Estimated performance of various wind turbine designs as a function of blade-tip speed ratio. *(From Ref. 53.)*

Thus the maximum possible efficiency of an ideal frictionless wind turbine is usually stated in terms of the *power coefficient:*

$$C_P = \frac{P}{\frac{1}{2}\rho A V_1^3} \tag{11.46}$$

Equation (11.45) states that the total power coefficient is

$$C_{p,\text{max}} = \tfrac{16}{27} = 0.593 \tag{11.47}$$

This is called the *Betz number* and serves as an ideal with which to compare the actual performance of real windmills.

Figure 11.32 shows the measured power coefficients of various wind turbine designs. The independent variable is not V_2/V_1 (which is artificial and convenient only in the ideal theory) but the ratio of blade-tip speed ωr to wind speed. Note that the tip can move much faster than the wind, a fact disturbing to the laity but familiar to engineers in the behavior of iceboats and sailing vessels. The Darrieus has the many advantages of a vertical axis but has little torque at low speeds (see Fig. 11.32) and also rotates more slowly at maximum power than a propeller, thus requiring a higher gear ratio for the generator. The Savonius rotor (Fig. 6.29b) has been suggested as a VAWT design because it produces power at very low wind speeds, but it is inefficient and susceptible to storm damage because it cannot be feathered in high winds.

As shown in Fig. 11.33, there are many areas of the world where wind energy is an attractive alternative, such as Ireland, Greenland, Iceland, Argentina, Chile, New Zealand, and Newfoundland. Robinson [53] points out that Australia, with only moderate winds, could generate half its electricity with wind turbines. Inexhaustible and available, the winds, coupled with low-cost turbine designs, promise a bright future for this alternative.

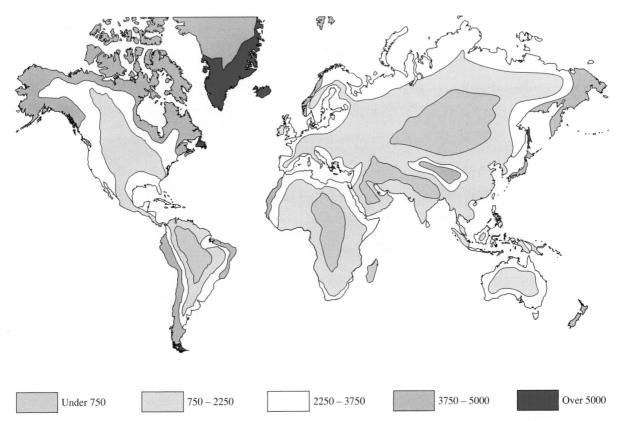

| Under 750 | 750 – 2250 | 2250 – 3750 | 3750 – 5000 | Over 5000 |

Fig. 11.33 World availability of land-based wind energy: estimated annual electric output in kWh/kW of a wind turbine rated at 11.2 m/s (25 mi/h). *(From Ref. 54.)*

With both fossil fuel demand and oil prices increasing alarmingly, the future for wind power seems bright. New global wind data by Archer and Jacobson [62] show that harnessing a modest fraction of available wind energy could, in fact, supply *all* of the earth's electricity needs. We are far from that, of course, but making progress. The very informative Web site for the American Wind Energy Association [59] cites the following wind energy totals as of 2005:

Country	2005 capacity, MW	Percent of world total
Germany	18,430	31
Spain	10,030	17
United States	9150	15
India	4430	8
Denmark	3120	5
Rest of the world	13,600	24
TOTALS	**58,760**	**100**

Wind already supplies a lot of power, 59 GW, but it is dwarfed by world demand. Of the total number of installed wind turbines, Europe has 70 percent, or 40 GW, but this is only 3 percent of Europe's present electricity usage. The cost will be high, and the political debate will be fractious, especially about site locations, but the future of wind energy seems assured.

Summary

Turbomachinery design is perhaps the most practical and most active application of the principles of fluid mechanics. There are billions of pumps and turbines in use in the world, and thousands of companies are seeking improvements. This chapter has discussed both positive-displacement devices and, more extensively, rotodynamic machines. With the centrifugal pump as an example, the basic concepts of torque, power, head, flow rate, and efficiency were developed for a turbomachine. Nondimensionalization leads to the pump similarity rules and some typical dimensionless performance curves for axial and centrifugal machines. The single most useful pump parameter is the specific speed, which delineates the type of design needed. An interesting design application is the theory of pumps combined in series and in parallel.

Turbines extract energy from flowing fluids and are of two types: impulse turbines, which convert the momentum of a high-speed stream, and reaction turbines, where the pressure drop occurs within the blade passages in an internal flow. By analogy with pumps, the power specific speed is important for turbines and is used to classify them into impulse, Francis, and propeller-type designs. A special case of reaction turbine with unconfined flow is the wind turbine. Several types of windmills were discussed and their relative performances compared.

Problems

Most of the problems herein are fairly straightforward. More difficult or open-ended assignments are labeled with an asterisk. Problems labeled with an EES icon will benefit from the use of the Engineering Equation Solver (EES), while problems labeled with a computer icon may require the use of a computer. The standard end-of-chapter problems P11.1 to P11.104 (categorized in the problem list here) are followed by word problems W11.1 to W11.10, comprehensive problems C11.1 to C11.8, and design project D11.1.

Problem Distribution

Section	Topic	Problems
11.1	Introduction and classification	P11.1–P11.14
11.2	Centrifugal pump theory	P11.15–P11.21
11.3	Pump performance and similarity rules	P11.22–P11.41
11.3	Net positive-suction head	P11.42–P11.44
11.4	Specific speed: mixed- and axial-flow pumps	P11.45–P11.62
11.5	Matching pumps to system characteristics	P11.63–P11.73
11.5	Pumps in parallel or series	P11.74–P11.81
11.5	Pump instability	P11.82–P11.83
11.6	Reaction and impulse turbines	P11.84–P11.99
11.6	Wind turbines	P11.100–P11.104

P11.1 Describe the geometry and operation of a human peristaltic PDP that is cherished by every romantic person on earth. How do the two ventricles differ?

P11.2 What would be the technical classification of the following turbomachines: (*a*) a household fan, (*b*) a windmill, (*c*) an aircraft propeller, (*d*) a fuel pump in a car, (*e*) an eductor, (*f*) a fluid-coupling transmission, and (*g*) a power plant steam turbine?

P11.3 A PDP can deliver almost any fluid, but there is always a limiting very high viscosity for which performance will deteriorate. Can you explain the probable reason?

P11.4 An interesting turbomachine is the *torque converter,* which combines both a pump and a turbine to change torque between two shafts. Do some research on this concept and describe it, with a report, sketches, and performance data, to the class.

P11.5 What type of pump is shown in Fig. P11.5? How does it operate?

P11.5

P11.6   Figure P11.6 shows two points a half-period apart in the operation of a pump. What type of pump is this [13]? How does it work? Sketch your best guess of flow rate versus time for a few cycles.

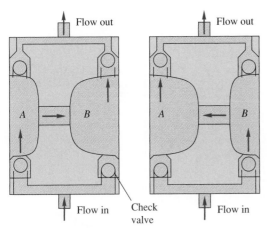

Flow out   Flow out

A → B   A ← B

Flow in   Check valve   Flow in

P11.6

P11.7   A piston PDP has a 5-in diameter and a 2-in stroke and operates at 750 r/min with 92 percent volumetric efficiency. (*a*) What is its delivery, in gal/min? (*b*) If the pump delivers SAE 10W oil at 20°C against a head of 50 ft, what horsepower is required when the overall efficiency is 84 percent?

P11.8   A centrifugal pump delivers 550 gal/min of water at 20°C when the brake horsepower is 22 and the efficiency is 71 percent. (*a*) Estimate the head rise in ft and the pressure rise in lbf/in^2. (*b*) Also estimate the head rise and horsepower if instead the delivery is 550 gal/min of gasoline at 20°C.

P11.9   Figure P11.9 shows the measured performance of the Vickers model PVQ40 piston pump when delivering SAE 10W oil at 180°F ($\rho \approx 910$ kg/m^3). Make some general observations about these data vis-à-vis Fig. 11.2 and your intuition about the behavior of piston pumps.

P11.10   Suppose that the pump of Fig. P11.9 is run at 1100 r/min against a pressure rise of 210 bar. (*a*) Using the measured displacement, estimate the theoretical delivery in gal/min. From the chart, estimate (*b*) the actual delivery and (*c*) the overall efficiency.

P11.11   A pump delivers 1500 L/min of water at 20°C against a pressure rise of 270 kPa. Kinetic and potential energy changes are negligible. If the driving motor supplies 9 kW, what is the overall efficiency?

P11.12   In a test of the centrifugal pump shown in Fig. P11.12, the following data are taken: $p_1 = 100$ mmHg (vacuum)

and $p_2 = 500$ mmHg (gage). The pipe diameters are $D_1 = 12$ cm and $D_2 = 5$ cm. The flow rate is 180 gal/min of light oil (SG = 0.91). Estimate (*a*) the head developed, in meters, and (*b*) the input power required at 75 percent efficiency.

P11.13   A 1.25-hp pump delivers 30 gal/min of kerosene at 20°C with 72-percent efficiency. What head and pressure rise, in BG units, result across the pump?

P11.14   A pump delivers gasoline at 20°C and 12 m^3/h. At the inlet $p_1 = 100$ kPa, $z_1 = 1$ m, and $V_1 = 2$ m/s. At the exit $p_2 = 500$ kPa, $z_2 = 4$ m, and $V_2 = 3$ m/s. How much power is required if the motor efficiency is 75 percent?

P11.15   A lawn sprinkler can be used as a simple turbine. As shown in Fig. P11.15, flow enters normal to the paper in the center and splits evenly into $Q/2$ and V_{rel} leaving each nozzle. The arms rotate at angular velocity ω and do work on a shaft. Draw the velocity diagram for this turbine. Neglecting friction, find an expression for the power delivered to the shaft. Find the rotation rate for which the power is a maximum.

P11.16   For the "sprinkler turbine" of Fig. P11.15, let $R = 18$ cm, with total flow rate of 14 m^3/h of water at 20°C. If the nozzle exit diameter is 8 mm, estimate (*a*) the maximum power delivered in W and (*b*) the appropriate rotation rate in r/min.

P11.17   A centrifugal pump has $d_1 = 7$ in, $d_2 = 13$ in, $b_1 = 4$ in, $b_2 = 3$ in, $\beta_1 = 25°$, and $\beta_2 = 40°$ and rotates at 1160 r/min. If the fluid is gasoline at 20°C and the flow enters the blades radially, estimate the theoretical (*a*) flow rate in gal/min, (*b*) horsepower, and (*c*) head in ft.

P11.18   A jet of velocity V strikes a vane that moves to the right at speed V_c, as in Fig. P11.18. The vane has a turning angle θ. Derive an expression for the power delivered to the vane by the jet. For what vane speed is the power maximum?

P11.19   A centrifugal pump has $r_2 = 9$ in, $b_2 = 2$ in, and $\beta_2 = 35°$ and rotates at 1060 r/min. If it generates a head of 180 ft, determine the theoretical (*a*) flow rate in gal/min and (*b*) horsepower. Assume near-radial entry flow.

P11.20   Suppose that Prob. P11.19 is reversed into a statement of the theoretical power $P_w \approx 153$ hp. Can you then compute the theoretical (*a*) flow rate and (*b*) head? Explain and resolve the difficulty that arises.

P11.21   The centrifugal pump of Fig. P11.21 develops a flow rate of 4200 gal/min of gasoline at 20°C with near-radial absolute inflow. Estimate the theoretical (*a*) horsepower, (*b*) head rise, and (*c*) appropriate blade angle at the inner radius.

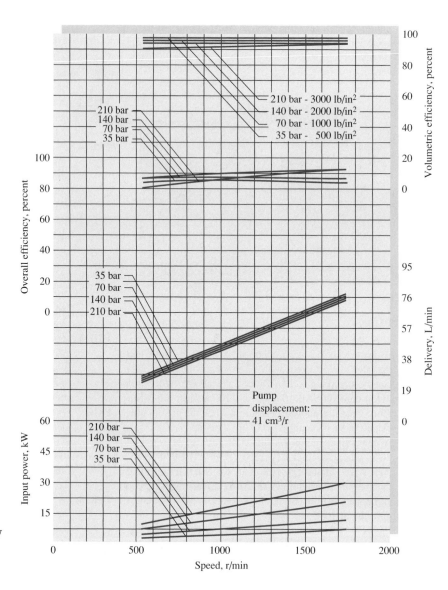

Fig. P11.9 Performance of the model PVQ40 piston pump delivering SAE 10W oil at 180°F. *(Courtesy of Vickers Inc., PDN/PACE Division.)*

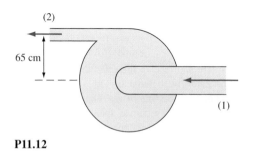

P11.12

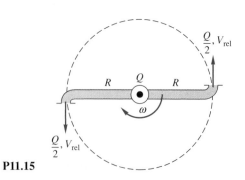

P11.15

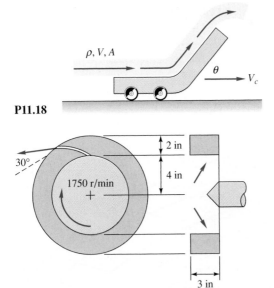

P11.18

P11.21

P11.22 A 37-cm-diameter centrifugal pump, running at 2140 r/min with water at 20°C, produces the following performance data:

Q, m³/s	0.0	0.05	0.10	0.15	0.20	0.25	0.30
H, m	105	104	102	100	95	85	67
P, kW	100	115	135	171	202	228	249

(a) Determine the best efficiency point. (b) Plot C_H versus C_Q. (c) If we desire to use this same pump family to deliver 7000 gal/min of kerosene at 20°C at an input power of 400 kW, what pump speed (in r/min) and impeller size (in cm) are needed? What head will be developed?

P11.23 If the 38-in-diameter pump of Fig. 11.7b is used to deliver 20°C kerosene at 850 r/min and 22,000 gal/min, what (a) head and (b) brake horsepower will result?

P11.24 Figure P11.24 shows performance data for the Taco, Inc., model 4013 pump. Compute the ratios of measured shutoff head to the ideal value U^2/g for all seven impeller sizes. Determine the average and standard deviation of this ratio and compare it to the average for the six impellers in Fig. 11.7.

P11.25 At what speed in r/min should the 35-in-diameter pump of Fig. 11.7b be run to produce a head of 400 ft at a discharge of 20,000 gal/min? What brake horsepower will be required? *Hint:* Fit $H(Q)$ to a formula.

P11.26 Would the smallest, or the largest, of the seven Taco Inc. pumps in Fig. P11.24 be better (a) for producing, near best efficiency, a water flow rate of 600 gal/min and a

head of 95 ft? (b) At what speed, in r/min, should this pump be run? (c) What input power is required?

P11.27 The 12-in pump of Fig. P11.24 is to be scaled up in size to provide a head of 90 ft and a flow rate of 1000 gal/min at BEP. Determine the correct (a) impeller diameter, (b) speed in r/min, and (c) horsepower required.

P11.28 Tests by the Byron Jackson Co. of a 14.62-in-diameter centrifugal water pump at 2134 r/min yield the following data:

Q, ft³/s	0	2	4	6	8	10
H, ft	340	340	340	330	300	220
bhp	135	160	205	255	330	330

What is the BEP? What is the specific speed? Estimate the maximum discharge possible.

P11.29 If the scaling laws are applied to the pump of Prob. P11.28 for the same impeller diameter, determine (a) the speed for which the shutoff head will be 280 ft, (b) the speed for which the BEP flow rate will be 8.0 ft³/s, and (c) the speed for which the BEP conditions will require 80 hp.

P11.30 A pump from the same family as in Prob. P11.28 is built with a BEP power of 100 bhp and an impeller diameter of 1 ft for *methanol* (not water). Using the scaling laws, estimate the resulting (a) speed in r/min, (b) the BEP head and, (c) the BEP flow rate.

P11.31 A centrifugal pump with backward-curved blades has the following measured performance when tested with water at 20°C:

Q, gal/min	0	400	800	1200	1600	2000	2400
H, ft	123	115	108	101	93	81	62
P, hp	30	36	40	44	47	48	46

(a) Estimate the best efficiency point and the maximum efficiency. (b) Estimate the most efficient flow rate, and the resulting head and brake horsepower, if the diameter is doubled and the rotation speed increased by 50 percent.

P11.32 The data of Prob. P11.31 correspond to a pump speed of 1200 r/min. (Were you able to solve Prob. P11.31 without this knowledge?) (a) Estimate the diameter of the impeller. (*Hint:* See Prob. P11.24 for a clue.) (b) Using your estimate from part (a), calculate the BEP parameters C_Q^*, C_H^*, and C_P^* and compare with Eqs. (11.27). (c) For what speed of this pump would the BEP head be 280 ft?

P11.33 In Prob. P11.31, the pump BEP flow rate is 2000 gal/min, the impeller diameter is 16 in, and the speed is 1200 r/min. Scale this pump with the similarity rules to find (a) the diameter and (b) the speed that will deliver a BEP water flow rate of 4000 gal/min and a head of

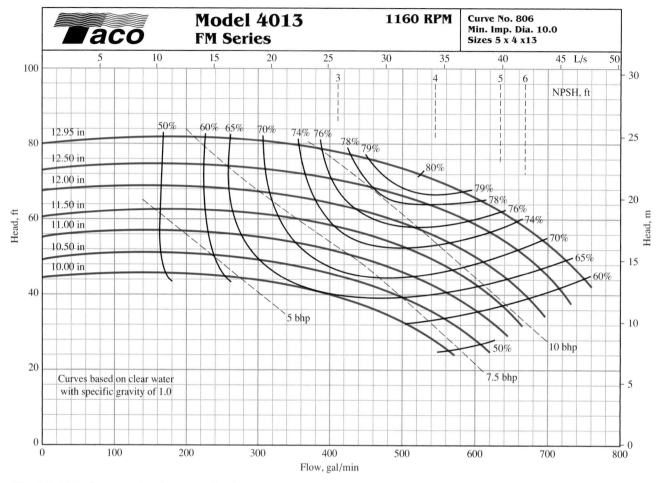

Fig. P11.24 Performance data for a centrifugal pump. *(Courtesy of Taco, Inc., Cranston, Rhode Island.)*

180 ft. (*c*) What brake horsepower will be required for this new condition?

P11.34 You are asked to consider a pump geometrically similar to the 9-in-diameter Taco pump of Fig. P11.34 to deliver 1200 gal/min at 1500 r/min. Determine the appropriate (*a*) impeller diameter, (*b*) BEP horsepower, (*c*) shutoff head, and (*d*) maximum efficiency. The fluid is kerosene, not water.

P11.35 An 18-in-diameter centrifugal pump, running at 880 r/min with water at 20°C, generates the following performance data:

Q, gal/min	0.0	2000	4000	6000	8000	10,000
H, ft	92	89	84	78	68	50
P, hp	100	112	130	143	156	163

Determine (*a*) the BEP, (*b*) the maximum efficiency, and (*c*) the specific speed. (*d*) Plot the required input power versus the flow rate.

P11.36 Plot the dimensionless performance curves for the pump of Prob. P11.35 and compare with Fig. 11.8. Find the appropriate diameter in inches and speed in r/min for a geometrically similar pump to deliver 400 gal/min against a head of 200 ft. What brake horsepower would be required?

P11.37 Consider the two pumps of Problems P11.28 and P11.35. If the diameters are not changed, which is better for delivering water at 3000 gal/min and a head of 400 ft? What is the appropriate rotation speed for the better pump?

P11.38 A 6.85-in pump, running at 3500 r/min, has the following measured performance for water at 20°C:

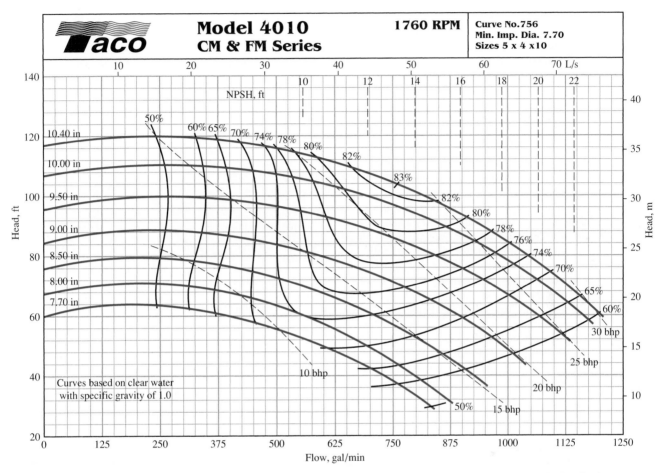

Fig. P11.34 Performance data for a family of centrifugal pump impellers. *(Courtesy of Taco, Inc., Cranston, Rhode Island.)*

Q, gal/min	50	100	150	200	250	300	350	400	450
H, ft	201	200	198	194	189	181	169	156	139
η, %	29	50	64	72	77	80	81	79	74

(*a*) Estimate the horsepower at BEP. If this pump is rescaled in water to provide 20 bhp at 3000 r/min, determine the appropriate (*b*) impeller diameter, (*c*) flow rate, and (*d*) efficiency for this new condition.

P11.39 The Allis-Chalmers D30LR centrifugal compressor delivers 33,000 ft³/min of SO₂ with a pressure change from 14.0 to 18.0 lbf/in² absolute using an 800-hp motor at 3550 r/min. What is the overall efficiency? What will the flow rate and Δp be at 3000 r/min? Estimate the diameter of the impeller.

P11.40 The specific speed N_s, as defined by Eqs. (11.30), does not contain the impeller diameter. How then should we size the pump for a given N_s? Logan [7] suggests a parameter called the *specific diameter* D_s, which is a dimensionless combination of Q, gH, and D. (*a*) If D_s is proportional to D, determine its form. (*b*) What is the relationship, if any, of D_s to C_{Q*}, C_{H*}, and C_{P*}? (*c*) Estimate D_s for the two pumps of Figs. 11.8 and 11.13.

P11.41 It is desired to build a centrifugal pump geometrically similar to that of Prob. P11.28 to deliver 6500 gal/min of gasoline at 20°C at 1060 r/min. Estimate the resulting (*a*) impeller diameter, (*b*) head, (*c*) brake horsepower, and (*d*) maximum efficiency.

P11.42 An 8-in model pump delivering 180°F water at 800 gal/min and 2400 r/min begins to cavitate when the inlet pressure and velocity are 12 lbf/in² absolute and 20 ft/s, respectively. Find the required NPSH of a prototype that is 4 times larger and runs at 1000 r/min.

P11.43 The 28-in-diameter pump in Fig. 11.7a at 1170 r/min is used to pump water at 20°C through a piping system at 14,000 gal/min. (a) Determine the required brake horsepower. The average friction factor is 0.018. (b) If there is 65 ft of 12-in-diameter pipe upstream of the pump, how far below the surface should the pump inlet be placed to avoid cavitation?

P11.44 The pump of Prob. P11.28 is scaled up to an 18-in diameter, operating in water at best efficiency at 1760 r/min. The measured NPSH is 16 ft, and the friction loss between the inlet and the pump is 22 ft. Will it be sufficient to avoid cavitation if the pump inlet is placed 9 ft below the surface of a sea-level reservoir?

P11.45 Determine the specific speeds of the seven Taco, Inc., pump impellers in Fig. P11.24. Are they appropriate for centrifugal designs? Are they approximately equal within experimental uncertainty? If not, why not?

P11.46 The answer to Prob. P11.40 is that the dimensionless "specific diameter" takes the form $D_s = D(gH^*)^{1/4}/Q^{*1/2}$, evaluated at the BEP. Data collected by the author for 30 different pumps indicate, in Fig. P11.46, that D_s correlates well with specific speed N_s. Use this figure to estimate the appropriate impeller diameter for a pump that delivers 20,000 gal/min of water and a head of 400 ft when running at 1200 r/min. Suggest a curve-fitted formula to the data. *Hint:* Use a hyperbolic formula.

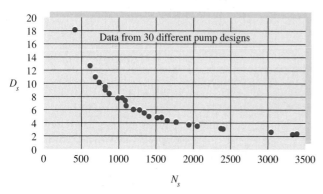

Fig. P11.46 Specific diameter at BEP for 30 commercial pumps.

P11.47 A typical household basement sump pump provides a discharge of 5 gal/min against a head of 15 ft. Estimate (a) the maximum efficiency and (b) the minimum horsepower required to drive such a pump at 1750 r/min.

P11.48 A commercial pump runs at 1750 r/min and delivers, near BEP, a flow of 2300 gal/min and a head of 40 m. (a) What type of pump is this? (b) Estimate the impeller diameter using the data of Prob. P11.46. (c) Estimate C_Q^* and add another data point to Fig. P11.49.

P11.49 Data collected by the author for flow coefficient at BEP for 30 different pumps are plotted versus specific speed in Fig. P11.49. Determine if the values of C_Q^* for the three pumps in Probs. P11.28, P11.35, and P11.38 also fit on this correlation. If so, suggest a curve-fitted formula for the data.

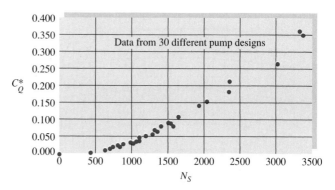

Fig. P11.49 Flow coefficient at BEP for 30 commercial pumps.

P11.50 Data collected by the author for power coefficient at BEP for 30 different pumps are plotted versus specific speed in Fig. P11.50. Determine if the values of C_P^* for the three pumps in Prob. P11.49 also fit on this correlation. If so, suggest a curve-fitted formula for the data.

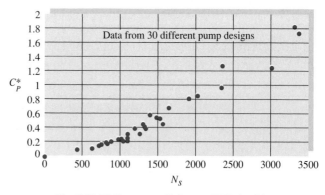

Fig. P11.50 Power coefficient at BEP for 30 commercial pumps.

P11.51 An axial-flow blower delivers 40 ft³/s of air that enters at 20°C and 1 atm. The flow passage has a 10-in outer radius and an 8-in inner radius. Blade angles are $\alpha_1 = 60°$ and $\beta_2 = 70°$, and the rotor runs at 1800 r/min. For the first stage compute (a) the head rise and (b) the power required.

P11.52 An axial-flow fan operates in sea-level air at 1200 r/min and has a blade-tip diameter of 1 m and a root diameter of 80 cm. The inlet angles are $\alpha_1 = 55°$ and $\beta_1 = 30°$, while at the outlet $\beta_2 = 60°$. Estimate the theoretical values of the (a) flow rate, (b) horsepower, and (c) outlet angle α_2.

P11.53 Figure P11.46 is an example of a centrifugal pump correlation, where D_s is defined in the problem. Logan and Roy [3] suggest the following correlation for axial-flow pumps and fans:

$$D_s \approx \frac{130}{N_s^{0.485}} \qquad \text{for } N_s > 8000$$

where N_s is the *dimensional* specific speed, Eq. (11.30b). Use this correlation to find the appropriate size for a fan that delivers 24,000 ft^3/min of air at sea-level conditions when running at 1620 r/min with a pressure rise of 2 inches of water. [*Hint:* Express the fan head in feet of *air,* not feet of water.]

P11.54 The Colorado River Aqueduct uses Worthington Corp. pumps that deliver 200 ft^3/s at 450 r/min against a head of 440 ft. What types of pump are these? Estimate the impeller diameter.

P11.55 We want to pump 70°C water at 20,000 gal/min and 1800 r/min. Estimate the type of pump, the horsepower required, and the impeller diameter if the required pressure rise for one stage is (a) 170 kPa and (b) 1350 kPa.

P11.56 A pump is needed to deliver 40,000 gal/min of gasoline at 20°C against a head of 90 ft. Find the impeller size, speed, and brake horsepower needed to use the pump families of (a) Fig. 11.8 and (b) Fig. 11.13. Which is the better design?

P11.57 Performance data for a 21-in-diameter air blower running at 3550 r/min are as follows:

Δp, in H_2O	29	30	28	21	10
Q, ft^3/min	500	1000	2000	3000	4000
bhp	6	8	12	18	25

Note the fictitious expression of pressure rise in terms of water rather than air. What is the specific speed? How does the performance compare with Fig. 11.8? What are C_Q^*, C_H^*, and C_P^*?

P11.58 The Worthington Corp. model A-12251 water pump, operating at maximum efficiency, produces 53 ft of head at 3500 r/min, 1.1 bhp at 3200 r/min, and 60 gal/min at 2940 r/min. What type of pump is this? What is its efficiency, and how does this compare with Fig. 11.14? Estimate the impeller diameter.

P11.59 Suppose it is desired to deliver 700 ft^3/min of propane gas (molecular weight = 44.06) at 1 atm and 20°C with

a single-stage pressure rise of 8.0 in H_2O. Determine the appropriate size and speed for using the pump families of (a) Prob. P11.57 and (b) Fig. 11.13. Which is the better design?

P11.60 A 45-hp pump is desired to generate a head of 200 ft when running at BEP with 20°C gasoline at 1200 r/min. Using the correlations in Figs. P11.49 and P11.50, determine the appropriate (a) specific speed, (b) flow rate, and (c) impeller diameter.

P11.61 A mine ventilation fan, running at 295 r/min, delivers 500 m^3/s of sea-level air with a pressure rise of 1100 Pa. Is this fan axial, centrifugal, or mixed? Estimate its diameter in ft. If the flow rate is increased 50 percent for the same diameter, by what percentage will the pressure rise change?

P11.62 The actual mine ventilation fan discussed in Prob. P11.61 had a diameter of 20 ft [Ref. 20, p. 339]. What would be the proper diameter for the pump family of Fig. 11.14 to provide 500 m^3/s at 295 r/min and BEP? What would be the resulting pressure rise in Pa?

P11.63 The 36.75-in pump in Fig. 11.7a at 1170 r/min is used to pump water at 60°F from a reservoir through 1000 ft of 12-in-ID galvanized iron pipe to a point 200 ft above the reservoir surface. What flow rate and brake horsepower will result? If there are 40 ft of pipe upstream of the pump, how far below the surface should the pump inlet be placed to avoid cavitation?

P11.64 A leaf blower is essentially a centrifugal impeller exiting to a tube. Suppose that the tube is smooth PVC pipe, 4 ft long, with a diameter of 2.5 in. The desired exit velocity is 73 mi/h in sea-level standard air. If we use the pump family of Eqs. (11.27) to drive the blower, what approximate (a) diameter and (b) rotation speed are appropriate? (c) Is this a good design?

*P11.65 An 11.5-in diameter centrifugal pump, running at 1750 rev/min, delivers 850 gal/min and a head of 105 ft at best efficiency (82 percent). (a) Can this pump operate efficiently when delivering water at 20°C through 200 m of 10-cm-diameter smooth pipe? Neglect minor losses. (b) If your answer to (a) is negative, can the speed n be changed to operate efficiently? (c) If your answer to (b) is also negative, can the impeller diameter be changed to operate efficiently and still run at 1750 rev/min?

P11.66 It is proposed to run the pump of Prob. P11.35 at 880 r/min to pump water at 20°C through the system in Fig. P11.66. The pipe is 20-cm-diameter commercial steel. What flow rate in ft^3/min will result? Is this an efficient application?

P11.67 The pump of Prob. P11.35, running at 880 r/min, is to pump water at 20°C through 75 m of horizontal galvanized iron pipe. All other system losses are neglected.

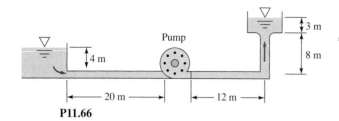

P11.66

Determine the flow rate and input power for (a) pipe diameter = 20 cm and (b) the pipe diameter found to yield maximum pump efficiency.

P11.68 Suppose that we use the axial-flow pump of Fig. 11.13 to drive the leaf blower of Prob. P11.64. What approximate (a) diameter and (b) rotation speed are appropriate? (c) Is this a good design?

P11.69 The pump of Prob. P11.38, running at 3500 r/min, is used to deliver water at 20°C through 600 ft of cast iron pipe to an elevation 100 ft higher. Determine (a) the proper pipe diameter for BEP operation and (b) the flow rate that results if the pipe diameter is 3 in.

P11.70 The pump of Prob. P11.28, operating at 2134 r/min, is used with 20°C water in the system of Fig. P11.70. (a) If it is operating at BEP, what is the proper elevation z_2? (b) If $z_2 = 225$ ft, what is the flow rate if $d = 8$ in.?

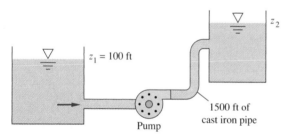

P11.70

P11.71 The pump of Prob. P11.38, running at 3500 r/min, delivers water at 20°C through 7200 ft of horizontal 5-in-diameter commercial steel pipe. There are a sharp entrance, sharp exit, four 90° elbows, and a gate valve. Estimate (a) the flow rate if the valve is wide open and (b) the valve closing percentage that causes the pump to operate at BEP. (c) If the latter condition holds continuously for 1 year, estimate the energy cost at 10 ¢/kWh.

P11.72 Performance data for a small commercial pump are as follows:

Q, gal/min	0	10	20	30	40	50	60	70
H, ft	75	75	74	72	68	62	47	24

This pump supplies 20°C water to a horizontal $\frac{5}{8}$-in-diameter garden hose ($\epsilon \approx 0.01$ in) that is 50 ft long.

Estimate (a) the flow rate and (b) the hose diameter that would cause the pump to operate at BEP.

*P11.73 The Cessna 172 aircraft has a wing area of 174 ft², an aspect ratio of 7.38, and a basic drag coefficient $C_{D\infty} = 0.037$. Its propeller, whose data is shown in Fig. 11.16, has a diameter of 6.25 ft. If the plane weighs 2300 lbf and flies at 180 ft/s at 1500 m standard altitude, estimate (a) the appropriate propeller speed, in rev/min and (b) the power required. Is the propeller efficient? [*Hint:* The efficiency is good, but not best.]

P11.74 The 32-in pump in Fig. 11.7a is used at 1170 r/min in a system whose head curve is H_s (ft) = $100 + 1.5Q^2$, with Q in thousands of gallons of water per minute. Find the discharge and brake horsepower required for (a) one pump, (b) two pumps in parallel, and (c) two pumps in series. Which configuration is best?

P11.75 Two 35-in pumps from Fig. 11.7b are installed in parallel for the system of Fig. P11.75. Neglect minor losses. For water at 20°C, estimate the flow rate and power required if (a) both pumps are running and (b) one pump is shut off and isolated.

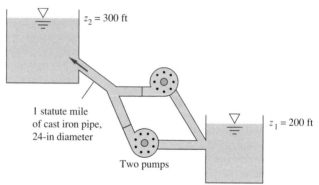

P11.75

P11.76 Two 32-in pumps from Fig. 11.7a are combined in parallel to deliver water at 60°F through 1500 ft of horizontal pipe. If $f = 0.025$, what pipe diameter will ensure a flow rate of 35,000 gal/min for $n = 1170$ r/min?

P11.77 Two pumps of the type tested in Prob. P11.22 are to be used at 2140 r/min to pump water at 20°C vertically upward through 100 m of commercial steel pipe. Should they be in series or in parallel? What is the proper pipe diameter for most efficient operation?

P11.78 Suppose that the two pumps in Fig. P11.75 are modified to be in series, still at 710 r/min. What pipe diameter is required for BEP operation?

P11.79 Two 32-in pumps from Fig. 11.7a are to be used in series at 1170 r/min to lift water through 500 ft of vertical cast iron pipe. What should the pipe diameter be for most efficient operation? Neglect minor losses.

P11.80 Determine if either (a) the smallest or (b) the largest of the seven Taco pumps in Fig. P11.24, running in series at 1160 r/min, can efficiently pump water at 20°C through 1 km of horizontal 12-cm-diameter commercial steel pipe.

P11.81 Reconsider the system of Fig. P6.62. Use the Byron Jackson pump of Prob. P11.28 running at 2134 r/min, no scaling, to drive the flow. Determine the resulting flow rate between the reservoirs. What is the pump efficiency?

P11.82 The S-shaped head-versus-flow curve in Fig. P11.82 occurs in some axial-flow pumps. Explain how a fairly flat system loss curve might cause instabilities in the operation of the pump. How might we avoid instability?

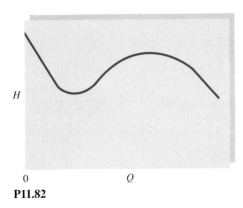

P11.82

P11.83 The low-shutoff head-versus-flow curve in Fig. P11.83 occurs in some centrifugal pumps. Explain how a fairly flat system loss curve might cause instabilities in the operation of the pump. What additional vexation occurs when two of these pumps are in parallel? How might we avoid instability?

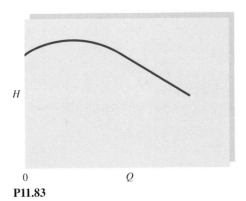

P11.83

P11.84 Turbines are to be installed where the net head is 400 ft and the flow rate 250,000 gal/min. Discuss the type, number, and size of turbine that might be selected if the generator selected is (a) 48-pole, 60-cycle ($n = 150$ r/min) and (b) 8-pole ($n = 900$ r/min). Why are at least two turbines desirable from a planning point of view?

P11.85 For a high-flow site with a head of 45 ft, it is desired to design a single 7-ft-diameter turbine that develops 4000 bhp at a speed of 360 r/min and 88-percent efficiency. It is decided first to test a geometrically similar model of diameter 1 ft, running at 1180 r/min. (a) What likely type of turbine is in the prototype? What are the appropriate (b) head and (c) flow rate for the model test? (d) Estimate the power expected to be delivered by the model turbine.

P11.86 The Tupperware hydroelectric plant on the Blackstone River has four 36-in-diameter turbines, each providing 447 kW at 200 r/min and 205 ft³/s for a head of 30 ft. What type of turbines are these? How does their performance compare with Fig. 11.22?

P11.87 An idealized radial turbine is shown in Fig. P11.87. The absolute flow enters at 30° and leaves radially inward. The flow rate is 3.5 m³/s of water at 20°C. The blade thickness is constant at 10 cm. Compute the theoretical power developed.

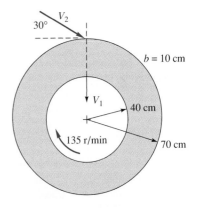

P11.87

P11.88 Performance data for a very small ($D = 8.25$ cm) model water turbine, operating with an available head of 49 ft, are as follows:

Q, m³/h	18.7	18.7	18.5	18.3	17.6	16.7	15.1	11.5
RPM	0	500	1000	1500	2000	2500	3000	3500
η	0	14%	27%	38%	50%	65%	61%	11%

(a) What type of turbine is this likely to be? (b) What is so different about these data compared to the dimensionless performance plot in Fig. 11.22d? Suppose it is desired to use a geometrically similar turbine to serve where the available head and flow are 150 ft and 6.7 ft^3/s, respectively. Estimate the most efficient (c) turbine diameter, (d) rotation speed, and (e) horsepower.

P11.89 A Pelton wheel of 12-ft pitch diameter operates under a net head of 2000 ft. Estimate the speed, power output, and flow rate for best efficiency if the nozzle exit diameter is 4 in.

P11.90 An idealized radial turbine is shown in Fig. P11.90. The absolute flow enters at 25° with the blade angles as shown. The flow rate is 8 m^3/s of water at 20°C. The blade thickness is constant at 20 cm. Compute the theoretical power developed.

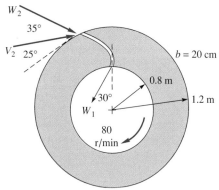

P11.90

P11.91 The flow through an axial-flow *turbine* can be idealized by modifying the stator–rotor diagrams of Fig. 11.12 for energy absorption. Sketch a suitable blade and flow arrangement and the associated velocity vector diagrams.

P11.92 A dam on a river is being sited for a hydraulic turbine. The flow rate is 1500 m^3/h, the available head is 24 m, and the turbine speed is to be 480 r/min. Discuss the estimated turbine size and feasibility for (a) a Francis turbine and (b) a Pelton wheel.

P11.93 Figure P11.93 shows a cutaway of a *cross-flow* or "Banki" turbine [55], which resembles a squirrel cage with slotted curved blades. The flow enters at about 2 o'clock and passes through the center and then again through the blades, leaving at about 8 o'clock. Report to the class on the operation and advantages of this design, including idealized velocity vector diagrams.

P11.94 A simple cross-flow turbine, Fig. P11.93, was constructed and tested at the University of Rhode Island. The blades were made of PVC pipe cut lengthwise into three 120°-arc pieces. When it was tested in water at a

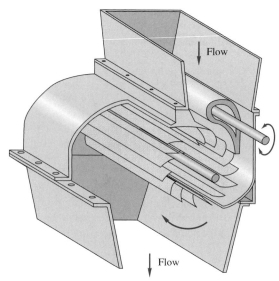

P11.93

head of 5.3 ft and a flow rate of 630 gal/min, the measured power output was 0.6 hp. Estimate (a) the efficiency and (b) the power specific speed if $n = 200$ rev/min.

P11.95 One can make a theoretical estimate of the proper diameter for a penstock in an impulse turbine installation, as in Fig. P11.95. Let L and H be known, and let the turbine performance be idealized by Eqs. (11.38) and (11.39). Account for friction loss h_f in the penstock, but neglect minor losses. Show that (a) the maximum power is generated when $h_f = H/3$, (b) the optimum jet velocity is $(4gH/3)^{1/2}$, and (c) the best nozzle diameter is $D_j = [D^5/(2\,fL)]^{1/4}$, where f is the pipe friction factor.

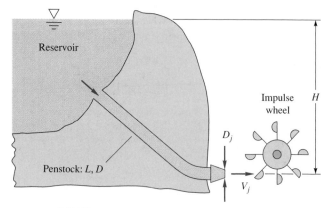

P11.95

P11.96 Apply the results of Prob. P11.95 to determining the optimum (a) penstock diameter and (b) nozzle diameter for a head of 800 ft and a flow rate of 40,000 gal/min with a commercial steel penstock of length 1500 ft.

P11.97 Consider the following nonoptimum version of Prob. P11.95: $H = 450$ m, $L = 5$ km, $D = 1.2$ m, $D_j = 20$ cm. The penstock is concrete, $\epsilon = 1$ mm. The impulse wheel diameter is 3.2 m. Estimate (a) the power generated by the wheel at 80 percent efficiency and (b) the best speed of the wheel in r/min. Neglect minor losses.

P11.98 Francis and Kaplan turbines are often provided with *draft tubes,* which lead the exit flow into the tailwater region, as in Fig. P11.98. Explain at least two advantages in using a draft tube.

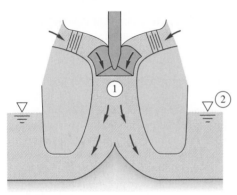

P11.98

P11.99 Turbines can also cavitate when the pressure at point 1 in Fig. P11.98 drops too low. With NPSH defined by Eq. (11.20), the empirical criterion given by Wislicenus [4] for cavitation is

$$N_{ss} = \frac{(r/min)(gal/min)^{1/2}}{[NPSH\ (ft)]^{3/4}} \geq 11,000$$

Use this criterion to compute how high $z_1 - z_2$, the impeller eye in Fig. P11.98, can be placed for a Francis turbine with a head of 300 ft, $N_{sp} = 40$, and $p_a = 14$ lbf/in^2 absolute before cavitation occurs in 60°F water.

P11.100 Consider the large wind turbine in the chapter-opener photo. Pretend that you do not know that 5M means five megawatts. Use the meager photo-caption data plus your knowledge of wind turbine theory and experiment to estimate the power delivered by the turbine for a rated wind speed of 12 m/s and a rotation rate of 10 rev/min.

P11.101 A Darrieus VAWT in operation in Lumsden, Saskatchewan, that is 32 ft high and 20 ft in diameter sweeps out an area of 432 ft^2. Estimate (a) the maximum power and (b) the rotor speed if it is operating in 16 mi/h winds.

P11.102 An American 6-ft diameter multiblade HAWT is used to pump water to a height of 10 ft through 3-in-diameter cast iron pipe. If the winds are 12 mi/h, estimate the rate of water flow in gal/min.

P11.103 A very large Darrieus VAWT was constructed by the U.S. Department of Energy near Sandia, New Mexico. It is 60 ft high and 30 ft in diameter, with a swept area of 1200 ft^2. If the turbine is constrained to rotate at 90 r/min, use Fig. 11.32 to plot the predicted power output in kW versus wind speed in the range $V = 5$ to 40 mi/h.

P11.104 The controversial Cape Cod Wind project proposes 130 large wind turbines in Nantucket Sound, intended to provide 75 percent of the electric power needs of Cape Cod and the Islands. The turbine diameter is 328 ft. For an average wind velocity of 14 mi/h, what are the best rotation rate and total power output estimates for (a) a HAWT and (b) a VAWT?

Word Problems

W11.1 We know that an enclosed rotating bladed impeller will impart energy to a fluid, usually in the form of a pressure rise, but how does it actually happen? Discuss, with sketches, the physical mechanisms through which an impeller actually transfers energy to a fluid.

W11.2 Dynamic pumps (as opposed to PDPs) have difficulty moving highly viscous fluids. Lobanoff and Ross [15] suggest the following rule of thumb: D (in) > $0.015\nu/\nu_{water}$, where D is the diameter of the discharge pipe. For example, SAE 30W oil ($\approx 300\nu_{water}$) should require at least a 4.5-in outlet. Can you explain some reasons for this limitation?

W11.3 The concept of NPSH dictates that liquid dynamic pumps should generally be immersed below the surface. Can you explain this? What is the effect of increasing the liquid temperature?

W11.4 For nondimensional fan performance, Wallis [20] suggests that the head coefficient should be replaced by FTP/$(\rho n^2 D^2)$, where FTP is the fan total pressure change. Explain the usefulness of this modification.

W11.5 Performance data for centrifugal pumps, even if well scaled geometrically, show a decrease in efficiency with decreasing impeller size. Discuss some physical reasons why this is so.

W11.6   Consider a dimensionless pump performance chart such as Fig. 11.8. What additional dimensionless parameters might modify or even destroy the similarity indicated in such data?

W11.7   One parameter not discussed in this text is the *number of blades* on an impeller. Do some reading on this subject, and report to the class about its effect on pump performance.

W11.8   Explain why some pump performance curves may lead to unstable operating conditions.

W11.9   Why are Francis and Kaplan turbines generally considered unsuitable for hydropower sites where the available head exceeds 1000 ft?

W11.10   Do some reading on the performance of the *free propeller* that is used on small, low-speed aircraft. What dimensionless parameters are typically reported for the data? How do the performance and efficiency compare with those for the axial-flow pump?

Comprehensive Problems

C11.1   The net head of a little aquarium pump is given by the manufacturer as a function of volume flow rate as listed below:

Q, m³/s	H, mH₂O
0	1.10
1.0 E-6	1.00
2.0 E-6	0.80
3.0 E-6	0.60
4.0 E-6	0.35
5.0 E-6	0.0

What is the maximum achievable flow rate if you use this pump to move water from the lower reservoir to the upper reservoir as shown in Fig. C11.1? *Note:* The tubing is smooth with an inner diameter of 5.0 mm and a total length of 29.8 m. The water is at room temperature and pressure. Minor losses in the system can be neglected.

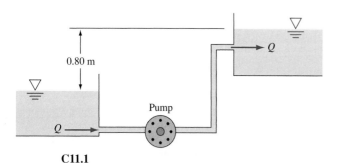

C11.1

C11.2   Reconsider Prob. P6.62 as an exercise in pump selection. Select an impeller size and rotational speed from the Byron Jackson pump family of Prob. P11.28 to deliver a flow rate of 3 ft³/s to the system of Fig. P6.68 at minimum input power. Calculate the horsepower required.

C11.3   Reconsider Prob. P6.77 as an exercise in turbine selection. Select an impeller size and rotational speed from the Francis turbine family of Fig. 11.22d to deliver maximum power generated by the turbine. Calculate the turbine power output and remark on the practicality of your design.

C11.4   The system of Fig. C11.4 is designed to deliver water at 20°C from a sea-level reservoir to another through new cast iron pipe of diameter 38 cm. Minor losses are $\Sigma K_1 = 0.5$ before the pump entrance and $\Sigma K_2 = 7.2$ after the pump exit. (*a*) Select a pump from either Fig. 11.7a or 11.7b, running at the given speeds, that can perform this task at maximum efficiency. Determine (*b*) the resulting flow rate, (*c*) the brake horsepower, and (*d*) whether the pump as presently situated is safe from cavitation.

C11.5   In Prob. P11.23, estimate the efficiency of the pump in two ways. (*a*) Read it directly from Fig. 11.7b (for the dynamically similar water pump); and (*b*) calculate it from Eq. (11.5) for the actual kerosene flow. Compare your results and discuss any discrepancies.

C11.6   An interesting turbomachine [58] is the *fluid coupling* of Fig. C11.6, which circulates fluid from a primary pump rotor and thus turns a secondary turbine on a separate shaft. Both rotors have radial blades. Couplings are common in all types of vehicle and machine transmissions and drives. The *slip* of the coupling is defined as the dimensionless difference between shaft rotation rates, $s = 1 - \omega_s/\omega_p$. For a given volume of fluid, the torque T transmitted is a function of s, ρ, ω_p, and impeller diameter D. (*a*) Nondimensionalize this function into two pi groups, with one pi proportional to T. Tests on a 1-ft-diameter coupling at 2500 r/min, filled with hydraulic fluid of density 56 lbm/ft³, yield the following torque versus slip data:

Slip, s	0%	5%	10%	15%	20%	25%
Torque T, ft-lbf	0	90	275	440	580	680

(*b*) If this coupling is run at 3600 r/min, at what slip value will it transmit a torque of 900 ft-lbf? (*c*) What is

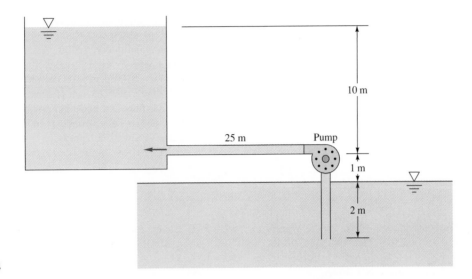

C11.4

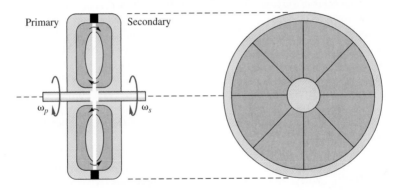

Primary   Secondary

ω_p ω_s

C11.6

the proper diameter for a geometrically similar coupling to run at 3000 r/min and 5 percent slip and transmit 600 ft-lbf of torque?

C11.7 Report to the class on the *Cordier method* [63] for optimal design of turbomachinery. The method is related to, and greatly expanded from, Prob. P11.46 and uses both software and charts to develop an efficient design for any given pump or compressor application.

C11.8 A *pump-turbine* is a reversible device that uses a reservoir to generate power in the daytime and then pumps water back up to the reservoir at night. Let us reset Prob. P6.62 as a pump-turbine. Recall that $\Delta z = 120$ ft, and the water flows through 2000 ft of 6-in-diameter cast iron pipe. For simplicity, assume that the pump operates at BEP (92%) with $H^*_p = 200$ ft and the turbine operates at BEP (89%) with $H^*_t = 100$ ft. Neglect minor losses. Estimate (*a*) the input power, in watts, required by the pump; and (*b*) the power, in watts, generated by the turbine. For further technical reading, consult the URL www.usbr.gov/pmts/hydraulics_lab/pubs/EM/EM39.pdf.

Design Project

D11.1 To save on electricity costs, a town water supply system uses gravity-driven flow from five large storage tanks during the day and then refills these tanks from 10 P.M. to 6 A.M. at a cheaper night rate of 7 ¢/kWh. The total resupply needed each night varies from 5 E5 to 2 E6 gal, with no more than 5 E5 gallons to any one tank. Tank elevations vary from 40 to 100 ft. A single constant-speed pump, drawing from a large groundwater aquifer and valved into five different cast iron tank supply lines, does this job. Distances from the pump to the five tanks vary more or less evenly from 1 to 3 mi. Each line averages one elbow every 100 ft and has four butterfly valves that can be controlled at any desirable angle. Select a suitable pump family from one of the six data sets in this chapter: Figs. 11.8, P11.24, and P11.34 plus Probs. P11.28, P11.35, and P11.38. Assume ideal similarity (no Reynolds number or pump roughness effects). The goal is to determine pump and pipeline sizes that achieve minimum total cost over a 5-year period. Some suggested cost data are

(a) Pump and motor: $2500 plus $1500 per inch of pipe size
(b) Valves: $100 plus $100 per inch of pipe size
(c) Pipelines: 50¢ per inch of diameter per foot of length

Since the flow and elevation parameters vary considerably, a random daily variation within the specified ranges might give a realistic approach.

References

1. D. G. Wilson, "Turbomachinery—From Paddle Wheels to Turbojets," *Mech. Eng.,* vol. 104, Oct. 1982, pp. 28–40.
2. D. Japikse and N. C. Baines, *Introduction to Turbomachinery,* Concepts ETI Inc., Hanover, NH, 1997.
3. E. S. Logan and R. Roy (Eds.), *Handbook of Turbomachinery,* 2d ed., Marcel Dekker, New York, 2003.
4. G. F. Wislicenus, *Fluid Mechanics of Turbomachinery,* 2d ed., McGraw-Hill, New York, 1965.
5. S. L. Dixon, *Fluid Mechanics and Thermodynamics of Turbomachinery*, 5th ed., Elsevier, New York, 2005.
6. J. A. Schetz and A. E. Fuhs, *Handbook of Fluid Dynamics and Fluid Machinery,* Wiley, New York, 1996.
7. E. S. Logan, Jr., *Turbomachinery: Basic Theory and Applications,* 2d ed., Marcel Dekker, New York, 1993.
8. A. J. Stepanoff, *Centrifugal and Axial Flow Pumps,* 2d ed., Wiley, New York, 1957.
9. J. Tuzson, *Centrifugal Pump Design,* Wiley, New York, 2000.
10. P. Girdhar and O. Moniz, *Practical Centrifugal Pumps,* Elsevier, New York, 2004.
11. L. Nelik, *Centrifugal and Rotary Pumps,* CRC Press, Boca Raton, FL, 1999.
12. D. Japikse, R. Furst, and W. D. Marscher, *Centrifugal Pump Design and Performance,* Concepts ETI Inc., Hanover, NH, 1997.
13. R. K. Turton, *Rotodynamic Pump Design,* Cambridge University Press, Cambridge, UK, 2005.
14. I. J. Karassik and T. McGuire, *Centrifugal Pumps,* 2d ed., Springer-Verlag, New York, 1996.
15. V. L. Lobanoff and R. R. Ross, *Centrifugal Pumps: Design and Application,* 2d ed., Elsevier, New York, 1992.
16. H. L. Stewart, *Pumps,* 5th ed. Macmillan, New York, 1991.
17. A. B. McKenzie, *Axial Flow Fans and Compressors: Aerodynamic Design and Performance,* Ashgate Publishing, Brookfield, VT, 1997.
18. A. J. Wennerstrom, *Design of Highly Loaded Axial-Flow Fans and Compressors,* Concepts ETI Inc., Hanover, NH, 1997.
19. F. P. Bleier, *Fan Handbook: Selection, Application, and Design,* McGraw-Hill, New York, 1997.
20. R. A. Wallis, *Axial Flow Fans and Ducts,* Wiley, New York, 1983.
21. H. P. Bloch, *A Practical Guide to Compressor Technology,* McGraw-Hill, New York, 1996.
22. M. T. Gresh, *Compressor Performance: Aerodynamics for the User,* Elsevier, New York, 2001.
23. Ronald H. Aungier, *Axial-Flow Compressors: A Strategy for Aerodynamic Design and Analysis,* ASME Press, New York, 2003.
24. H. I. H. Saravanamuttoo, H. Cohen, and G. F. C. Rogers, *Gas Turbine Theory,* Prentice Hall, Upper Saddle River, NJ, 2001.
25. P. P. Walsh and P. Fletcher, *Gas Turbine Performance,* ASME Press, New York, 2004.
26. M. P. Boyce, *Gas Turbine Engineering Handbook,* Butterworth-Heinemann, Woburn, MA, 2001.
27. Fluid Machinery Group, Institution of Mechanical Engineers, *Hydropower,* Wiley, New York, 2005.
28. Jeremy Thake, *The Micro-Hydro Pelton Turbine Manual,* Intermediate Technology Pub., Colchester, Essex, UK, 2000.
29. P. C. Hanlon (ed.), *Compressor Handbook,* McGraw-Hill, New York, 2001.
30. Hydraulic Institute, *Hydraulic Institute Pump Standards Complete,* 4th ed. New York, 1994.
31. P. Cooper, J. Messina, C. Heald, and I. J. Karassik (ed.), *Pump Handbook,* 3d ed., McGraw-Hill, New York, 2000.
32. J. S. Gulliver and R. E. A. Arndt, *Hydropower Engineering Handbook,* McGraw-Hill, New York, 1990.
33. R. L. Daugherty, J. B. Franzini, and E. J. Finnemore, *Fluid Mechanics and Engineering Applications,* 9th ed., McGraw-Hill, New York, 1997.
34. R. H. Sabersky, E. M. Gates, A. J. Acosta, and E. G. Hauptmann, *Fluid Flow: A First Course in Fluid Mechanics,* 4th ed., Pearson Education, Upper Saddle River, NJ, 1994.

35. J. P. Poynton, *Metering Pumps,* Marcel Dekker, New York, 1983.

36. Hydraulic Institute, *Reciprocating Pump Test Standard,* New York, 1994.

37. T. L. Henshaw, *Reciprocating Pumps,* Wiley, New York, 1987.

38. J. E. Miller, *The Reciprocating Pump: Theory, Design and Use,* Wiley, NewYork, 1987.

39. D. G. Wilson and T. Korakianitis, *The Design of High Efficiency Turbomachinery and Gas Turbines,* 2d ed., Pearson Education, Upper Saddle River, NJ, 1998.

40. S. O. Kraus et al., "Periodic Velocity Measurements in a Wide and Large Radius Ratio Automotive Torque Converter at the Pump/Turbine Interface," *J. Fluids Engineering*, vol. 127, no. 2, 2005, pp. 308–316.

41. E. M. Greitzer, "The Stability of Pumping Systems: The 1980 Freeman Scholar Lecture," *J. Fluids Eng.,* vol. 103, June 1981, pp. 193–242.

42. R. Elder et al. (eds.), *Advances of CFD in Fluid Machinery Design,* Wiley, New York, 2003.

43. L. F. Moody, "The Propeller Type Turbine," *ASCE Trans.,* vol. 89, 1926, p. 628.

44. H. H. Anderson, "Prediction of Head, Quantity, and Efficiency in Pumps—The Area-Ratio Principle," in *Performance Prediction of Centrifugal Pumps and Compressors,* vol. I00127, ASME Symp., New York, 1980, pp. 201–211.

45. Y. T. Lee and C. Hah (eds.), *Symposium on Numerical Modeling of Aerodynamics and Hydrodynamics in Turbomachinery,* Joint ASME–European Fluids Engineering Conference, Montreal, Canada, July, 2002.

46. D. J. Mahoney (ed.), *Proceedings of the 1997 International Conference on Hydropower,* ASCE, Reston, VA, 1997.

47. G. W. Koeppl, *Putnam's Power from the Wind,* 2d ed., Van Nostrand Reinhold, New York, 1982.

48. P. Gipe, *Wind Energy Basics: A Guide to Small and Micro Wind Systems,* Chelsea Green Publishing, White River Junction, VT, 1999.

49. D. A. Spera, *Wind Turbine Technology: Fundamental Concepts of Wind Turbine Engineering,* ASME Press, New York, 1994.

50. E. Hau, *Wind Turbines: Fundamentals, Technologies, Application, Economics,* 2d ed., Springer-Verlag, New York, 2005.

51. R. Harrison, E. Hau, and H. Snel, *Large Wind Turbines,* Wiley, New York, 2000.

52. L. Fielding, *Turbine Design,* ASME Press, New York, 2000.

53. M. L. Robinson, "The Darrieus Wind Turbine for Electrical Power Generation," *Aeronaut. J.,* June 1981, pp. 244–255.

54. D. F. Warne and P. G. Calnan, "Generation of Electricity from the Wind," *IEE Rev.,* vol. 124, no. 11R, November 1977, pp. 963–985.

55. L. A. Haimerl, "The Crossflow Turbine," *Waterpower,* January 1960, pp. 5–13; see also *ASME Symp. Small Hydropower Fluid Mach.,* vol. 1, 1980, and vol. 2, 1982.

56. K. Eisele et al., "Flow Analysis in a Pump Diffuser: Part 1, Measurements; Part 2, CFD," *J. Fluids Eng.,* vol. 119, December 1997, pp. 968–984.

57. D. Japikse and N. C. Baines, *Turbomachinery Diffuser Design Technology,* Concepts ETI Inc., Hanover, NH, 1998.

58. B. Massey and J. Ward-Smith, *Mechanics of Fluids,* 7th ed., Nelson Thornes Publishing, Cheltenham, UK, 1998.

59. American Wind Energy Association, "Global Wind Energy Market Report," URL: <http://www.awea.org/pubs/documents/globalmarket2004.pdf>.

60. J. P. Breslin and P. Andersen, *Hydrodynamics of Ship Propellers,* Cambridge University Press, Cambridge, UK, 2003.

61. M. Hollmann, *Modern Propeller and Duct Design,* Aircraft Designs, Inc., Monterey, CA, 1993.

62. C. L. Archer and M. Z. Jacobson, "Evaluation of Global Wind Power," *J. Geophys. Res.-Atm.,* vol. 110, 2005, doi:10.1029/2004JD005462.

63. M. Farinas and A. Garon, "Application of DOE for Optimal Turbomachinery Design," Paper AIAA-2004-2139, AIAA Fluid Dynamics Conference, Portland, OR, June 2004.

Physical Properties of Fluids

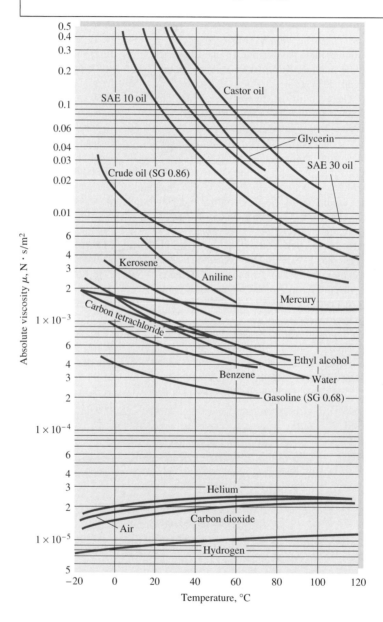

Fig. A.1 Absolute viscosity of common fluids at 1 atm.

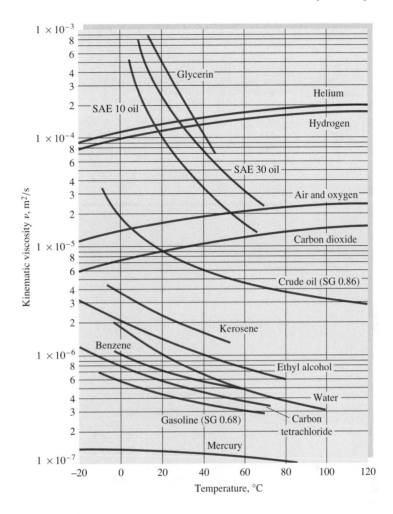

Fig. A.2 Kinematic viscosity of common fluids at 1 atm.

Table A.1 Viscosity and Density of Water at 1 atm

T, °C	ρ, kg/m³	μ, N · s/m²	ν, m²/s	T, °F	ρ, slug/ft³	μ, lb · s/ft²	ν, ft²/s
0	1000	1.788 E−3	1.788 E−6	32	1.940	3.73 E−5	1.925 E−5
10	1000	1.307 E−3	1.307 E−6	50	1.940	2.73 E−5	1.407 E−5
20	998	1.003 E−3	1.005 E−6	68	1.937	2.09 E−5	1.082 E−5
30	996	0.799 E−3	0.802 E−6	86	1.932	1.67 E−5	0.864 E−5
40	992	0.657 E−3	0.662 E−6	104	1.925	1.37 E−5	0.713 E−5
50	988	0.548 E−3	0.555 E−6	122	1.917	1.14 E−5	0.597 E−5
60	983	0.467 E−3	0.475 E−6	140	1.908	0.975 E−5	0.511 E−5
70	978	0.405 E−3	0.414 E−6	158	1.897	0.846 E−5	0.446 E−5
80	972	0.355 E−3	0.365 E−6	176	1.886	0.741 E−5	0.393 E−5
90	965	0.316 E−3	0.327 E−6	194	1.873	0.660 E−5	0.352 E−5
100	958	0.283 E−3	0.295 E−6	212	1.859	0.591 E−5	0.318 E−5

Suggested curve fits for water in the range $0 \leq T \leq 100°C$:

$$\rho(kg/m^3) \approx 1000 - 0.0178 \left| T°C - 4°C \right|^{1.7} \pm 0.2\%$$

$$\ln \frac{\mu}{\mu_0} \approx -1.704 - 5.306z + 7.003z^2$$

$$z = \frac{273 \text{ K}}{T \text{ K}} \qquad \mu_0 = 1.788 \text{ E}-3 \text{ kg/(m} \cdot \text{s)}$$

Table A.2 Viscosity and Density of Air at 1 atm

T, °C	ρ, kg/m³	μ, N · s/m²	ν, m²/s	T, °F	ρ, slug/ft³	μ, lb · s/ft²	ν, ft²/s
−40	1.52	1.51 E−5	0.99 E−5	−40	2.94 E−3	3.16 E−7	1.07 E−4
0	1.29	1.71 E−5	1.33 E−5	32	2.51 E−3	3.58 E−7	1.43 E−4
20	1.20	1.80 E−5	1.50 E−5	68	2.34 E−3	3.76 E−7	1.61 E−4
50	1.09	1.95 E−5	1.79 E−5	122	2.12 E−3	4.08 E−7	1.93 E−4
100	0.946	2.17 E−5	2.30 E−5	212	1.84 E−3	4.54 E−7	2.47 E−4
150	0.835	2.38 E−5	2.85 E−5	302	1.62 E−3	4.97 E−7	3.07 E−4
200	0.746	2.57 E−5	3.45 E−5	392	1.45 E−3	5.37 E−7	3.71 E−4
250	0.675	2.75 E−5	4.08 E−5	482	1.31 E−3	5.75 E−7	4.39 E−4
300	0.616	2.93 E−5	4.75 E−5	572	1.20 E−3	6.11 E−7	5.12 E−4
400	0.525	3.25 E−5	6.20 E−5	752	1.02 E−3	6.79 E−7	6.67 E−4
500	0.457	3.55 E−5	7.77 E−5	932	0.89 E−3	7.41 E−7	8.37 E−4

Suggested curve fits for air:

$$\rho = \frac{p}{RT} \qquad R_{air} \approx 287 \text{ J/(kg} \cdot \text{K)}$$

Power law:
$$\frac{\mu}{\mu_0} \approx \left(\frac{T}{T_0} \right)^{0.7}$$

Sutherland law:
$$\frac{\mu}{\mu_0} \approx \left(\frac{T}{T_0} \right)^{3/2} \left(\frac{T_0 + S}{T + S} \right) \qquad S_{air} \approx 110.4 \text{ K}$$

with $T_0 = 273$ K, $\mu_0 = 1.71$ E−5 kg/(m · s), and T in kelvins.

Table A.3 Properties of Common Liquids at 1 atm and 20°C (68°F)

Liquid	ρ, kg/m³	μ, kg/(m · s)	Y, N/m*	p_v, N/m²	Bulk modulus, N/m²	Viscosity parameter $C^\dagger$
Ammonia	608	2.20 E−4	2.13 E−2	9.10 E+5	—	1.05
Benzene	881	6.51 E−4	2.88 E−2	1.01 E+4	1.4 E+9	4.34
Carbon tetrachloride	1590	9.67 E−4	2.70 E−2	1.20 E+4	9.65 E+8	4.45
Ethanol	789	1.20 E−3	2.28 E−2	5.7 E+3	9.0 E+8	5.72
Ethylene glycol	1117	2.14 E−2	4.84 E−2	1.2 E+1	—	11.7
Freon 12	1327	2.62 E−4	—	—	—	1.76
Gasoline	680	2.92 E−4	2.16 E−2	5.51 E+4	9.58 E+8	3.68
Glycerin	1260	1.49	6.33 E−2	1.4 E−2	4.34 E+9	28.0
Kerosene	804	1.92 E−3	2.8 E−2	3.11 E+3	1.6 E+9	5.56
Mercury	13,550	1.56 E−3	4.84 E−1	1.1 E−3	2.55 E+10	1.07
Methanol	791	5.98 E−4	2.25 E−2	1.34 E+4	8.3 E+8	4.63
SAE 10W oil	870	1.04 E−1‡	3.6 E−2	—	1.31 E+9	15.7
SAE 10W30 oil	876	1.7 E−1‡	—	—	—	14.0
SAE 30W oil	891	2.9 E−1‡	3.5 E−2	—	1.38 E+9	18.3
SAE 50W oil	902	8.6 E−1‡	—	—	—	20.2
Water	998	1.00 E−3	7.28 E−2	2.34 E+3	2.19 E+9	Table A.1
Seawater (30%)	1025	1.07 E−3	7.28 E−2	2.34 E+3	2.33 E+9	7.28

*In contact with air.

†The viscosity–temperature variation of these liquids may be fitted to the empirical expression

$$\frac{\mu}{\mu_{20°C}} \approx \exp\left[C\left(\frac{293\ \text{K}}{T\ \text{K}} - 1\right)\right]$$

with accuracy of ±6 percent in the range $0 \leq T \leq 100°C$.

‡Representative values. The SAE oil classifications allow a viscosity variation of up to ±50 percent, especially at lower temperatures.

Table A.4 Properties of Common Gases at 1 atm and 20°C (68°F)

Gas	Molecular weight	R, m²/(s² · K)	ρg, N/m³	μ, N · s/m²	Specific-heat ratio	Power-law exponent n^*
H_2	2.016	4124	0.822	9.05 E−6	1.41	0.68
He	4.003	2077	1.63	1.97 E−5	1.66	0.67
H_2O	18.02	461	7.35	1.02 E−5	1.33	1.15
Ar	39.944	208	16.3	2.24 E−5	1.67	0.72
Dry air	28.96	287	11.8	1.80 E−5	1.40	0.67
CO_2	44.01	189	17.9	1.48 E−5	1.30	0.79
CO	28.01	297	11.4	1.82 E−5	1.40	0.71
N_2	28.02	297	11.4	1.76 E−5	1.40	0.67
O_2	32.00	260	13.1	2.00 E−5	1.40	0.69
NO	30.01	277	12.1	1.90 E−5	1.40	0.78
N_2O	44.02	189	17.9	1.45 E−5	1.31	0.89
Cl_2	70.91	117	28.9	1.03 E−5	1.34	1.00
CH_4	16.04	518	6.54	1.34 E−5	1.32	0.87

*The power-law curve fit, Eq. (1.27), $\mu/\mu_{293K} \approx (T/293)^n$, fits these gases to within ±4 percent in the range $250 \leq T \leq 1000$ K. The temperature must be in kelvins.

Table A.5 Surface Tension, Vapor Pressure, and Sound Speed of Water

T, °C	Y, N/m	p_v, kPa	a, m/s
0	0.0756	0.611	1402
10	0.0742	1.227	1447
20	0.0728	2.337	1482
30	0.0712	4.242	1509
40	0.0696	7.375	1529
50	0.0679	12.34	1542
60	0.0662	19.92	1551
70	0.0644	31.16	1553
80	0.0626	47.35	1554
90	0.0608	70.11	1550
100	0.0589	101.3	1543
120	0.0550	198.5	1518
140	0.0509	361.3	1483
160	0.0466	617.8	1440
180	0.0422	1002	1389
200	0.0377	1554	1334
220	0.0331	2318	1268
240	0.0284	3344	1192
260	0.0237	4688	1110
280	0.0190	6412	1022
300	0.0144	8581	920
320	0.0099	11,274	800
340	0.0056	14,586	630
360	0.0019	18,651	370
374*	0.0*	22,090*	0*

*Critical point.

Table A.6 Properties of the Standard Atmosphere

z, m	T, K	p, Pa	ρ, kg/m^3	a, m/s
−500	291.41	107,508	1.2854	342.2
0	288.16	101,350	1.2255	340.3
500	284.91	95,480	1.1677	338.4
1000	281.66	89,889	1.1120	336.5
1500	278.41	84,565	1.0583	334.5
2000	275.16	79,500	1.0067	332.6
2500	271.91	74,684	0.9570	330.6
3000	268.66	70,107	0.9092	328.6
3500	265.41	65,759	0.8633	326.6
4000	262.16	61,633	0.8191	324.6
4500	258.91	57,718	0.7768	322.6
5000	255.66	54,008	0.7361	320.6
5500	252.41	50,493	0.6970	318.5
6000	249.16	47,166	0.6596	316.5
6500	245.91	44,018	0.6237	314.4
7000	242.66	41,043	0.5893	312.3
7500	239.41	38,233	0.5564	310.2
8000	236.16	35,581	0.5250	308.1
8500	232.91	33,080	0.4949	306.0
9000	229.66	30,723	0.4661	303.8
9500	226.41	28,504	0.4387	301.7
10,000	223.16	26,416	0.4125	299.5
10,500	219.91	24,455	0.3875	297.3
11,000	216.66	22,612	0.3637	295.1
11,500	216.66	20,897	0.3361	295.1
12,000	216.66	19,312	0.3106	295.1
12,500	216.66	17,847	0.2870	295.1
13,000	216.66	16,494	0.2652	295.1
13,500	216.66	15,243	0.2451	295.1
14,000	216.66	14,087	0.2265	295.1
14,500	216.66	13,018	0.2094	295.1
15,000	216.66	12,031	0.1935	295.1
15,500	216.66	11,118	0.1788	295.1
16,000	216.66	10,275	0.1652	295.1
16,500	216.66	9496	0.1527	295.1
17,000	216.66	8775	0.1411	295.1
17,500	216.66	8110	0.1304	295.1
18,000	216.66	7495	0.1205	295.1
18,500	216.66	6926	0.1114	295.1
19,000	216.66	6401	0.1029	295.1
19,500	216.66	5915	0.0951	295.1
20,000	216.66	5467	0.0879	295.1
22,000	218.6	4048	0.0645	296.4
24,000	220.6	2972	0.0469	297.8
26,000	222.5	2189	0.0343	299.1
28,000	224.5	1616	0.0251	300.4
30,000	226.5	1197	0.0184	301.7
40,000	250.4	287	0.0040	317.2
50,000	270.7	80	0.0010	329.9
60,000	255.7	22	0.0003	320.6
70,000	219.7	6	0.0001	297.2

Appendix B
Compressible Flow Tables

Table B.1
Isentropic Flow of a Perfect Gas, $k = 1.4$

Ma	p/p_0	ρ/ρ_0	T/T_0	A/A^*	Ma	p/p_0	ρ/ρ_0	T/T_0	A/A^*
0.00	1.0000	1.0000	1.0000	∞	1.75	0.1878	0.3029	0.6202	1.3865
0.05	0.9983	0.9988	0.9995	11.5914	1.80	0.1740	0.2868	0.6068	1.4390
0.10	0.9930	0.9950	0.9980	5.8218	1.85	0.1612	0.2715	0.5936	1.4952
0.15	0.9844	0.9888	0.9955	3.9103	1.90	0.1492	0.2570	0.5807	1.5553
0.20	0.9725	0.9803	0.9921	2.9635	1.95	0.1381	0.2432	0.5680	1.6193
0.25	0.9575	0.9694	0.9877	2.4027	2.00	0.1278	0.2300	0.5556	1.6875
0.30	0.9395	0.9564	0.9823	2.0351	2.05	0.1182	0.2176	0.5433	1.7600
0.35	0.9188	0.9413	0.9761	1.7780	2.10	0.1094	0.2058	0.5313	1.8369
0.40	0.8956	0.9243	0.9690	1.5901	2.15	0.1011	0.1946	0.5196	1.9185
0.45	0.8703	0.9055	0.9611	1.4487	2.20	0.0935	0.1841	0.5081	2.0050
0.50	0.8430	0.8852	0.9524	1.3398	2.25	0.0865	0.1740	0.4969	2.0964
0.55	0.8142	0.8634	0.9430	1.2549	2.30	0.0800	0.1646	0.4859	2.1931
0.60	0.7840	0.8405	0.9328	1.1882	2.35	0.0740	0.1556	0.4752	2.2953
0.65	0.7528	0.8164	0.9221	1.1356	2.40	0.0684	0.1472	0.4647	2.4031
0.70	0.7209	0.7916	0.9107	1.0944	2.45	0.0633	0.1392	0.4544	2.5168
0.75	0.6886	0.7660	0.8989	1.0624	2.50	0.0585	0.1317	0.4444	2.6367
0.80	0.6560	0.7400	0.8865	1.0382	2.55	0.0542	0.1246	0.4347	2.7630
0.85	0.6235	0.7136	0.8737	1.0207	2.60	0.0501	0.1179	0.4252	2.8960
0.90	0.5913	0.6870	0.8606	1.0089	2.65	0.0464	0.1115	0.4159	3.0359
0.95	0.5595	0.6604	0.8471	1.0021	2.70	0.0430	0.1056	0.4068	3.1830
1.00	0.5283	0.6339	0.8333	1.0000	2.75	0.0398	0.0999	0.3980	3.3377
1.05	0.4979	0.6077	0.8193	1.0020	2.80	0.0368	0.0946	0.3894	3.5001
1.10	0.4684	0.5817	0.8052	1.0079	2.85	0.0341	0.0896	0.3810	3.6707
1.15	0.4398	0.5562	0.7908	1.0175	2.90	0.0317	0.0849	0.3729	3.8498
1.20	0.4124	0.5311	0.7764	1.0304	2.95	0.0293	0.0804	0.3649	4.0376
1.25	0.3861	0.5067	0.7619	1.0468	3.00	0.0272	0.0762	0.3571	4.2346
1.30	0.3609	0.4829	0.7474	1.0663	3.05	0.0253	0.0723	0.3496	4.4410
1.35	0.3370	0.4598	0.7329	1.0890	3.10	0.0234	0.0685	0.3422	4.6573
1.40	0.3142	0.4374	0.7184	1.1149	3.15	0.0218	0.0650	0.3351	4.8838
1.45	0.2927	0.4158	0.7040	1.1440	3.20	0.0202	0.0617	0.3281	5.1210
1.50	0.2724	0.3950	0.6897	1.1762	3.25	0.0188	0.0585	0.3213	5.3691
1.55	0.2533	0.3750	0.6754	1.2116	3.30	0.0175	0.0555	0.3147	5.6286
1.60	0.2353	0.3557	0.6614	1.2502	3.35	0.0163	0.0527	0.3082	5.9000
1.65	0.2184	0.3373	0.6475	1.2922	3.40	0.0151	0.0501	0.3019	6.1837
1.70	0.2026	0.3197	0.6337	1.3376	3.45	0.0141	0.0476	0.2958	6.4801

Table B.1
(*Concluded*)
Isentropic Flow of
a Perfect Gas,
$k = 1.4$

Ma	p/p_0	ρ/ρ_0	T/T_0	A/A^*	Ma	p/p_0	ρ/ρ_0	T/T_0	A/A^*
3.45	0.0141	0.0476	0.2958	6.4801	3.75	0.0092	0.0352	0.2623	8.5517
3.50	0.0131	0.0452	0.2899	6.7896	3.80	0.0086	0.0335	0.2572	8.9506
3.55	0.0122	0.0430	0.2841	7.1128	3.85	0.0081	0.0320	0.2522	9.3661
3.60	0.0114	0.0409	0.2784	7.4501	3.90	0.0075	0.0304	0.2474	9.7990
3.65	0.0106	0.0389	0.2729	7.8020	3.95	0.0070	0.0290	0.2427	10.2496
3.70	0.0099	0.0370	0.2675	8.1691	4.00	0.0066	0.0277	0.2381	10.7188

Table B.2 Normal Shock Relations
for a Perfect Gas, $k = 1.4$

Ma_{n1}	Ma_{n2}	p_2/p_1	$V_1/V_2 = \rho_2/\rho_1$	T_2/T_1	p_{02}/p_{01}	A_2^*/A_1^*
1.00	1.0000	1.0000	1.0000	1.0000	1.0000	1.0000
1.05	0.9531	1.1196	1.0840	1.0328	0.9999	1.0001
1.10	0.9118	1.2450	1.1691	1.0649	0.9989	1.0011
1.15	0.8750	1.3763	1.2550	1.0966	0.9967	1.0033
1.20	0.8422	1.5133	1.3416	1.1280	0.9928	1.0073
1.25	0.8126	1.6563	1.4286	1.1594	0.9871	1.0131
1.30	0.7860	1.8050	1.5157	1.1909	0.9794	1.0211
1.35	0.7618	1.9596	1.6028	1.2226	0.9697	1.0312
1.40	0.7397	2.1200	1.6897	1.2547	0.9582	1.0436
1.45	0.7196	2.2863	1.7761	1.2872	0.9448	1.0584
1.50	0.7011	2.4583	1.8621	1.3202	0.9298	1.0755
1.55	0.6841	2.6363	1.9473	1.3538	0.9132	1.0951
1.60	0.6684	2.8200	2.0317	1.3880	0.8952	1.1171
1.65	0.6540	3.0096	2.1152	1.4228	0.8760	1.1416
1.70	0.6405	3.2050	2.1977	1.4583	0.8557	1.1686
1.75	0.6281	3.4063	2.2791	1.4946	0.8346	1.1982
1.80	0.6165	3.6133	2.3592	1.5316	0.8127	1.2305
1.85	0.6057	3.8263	2.4381	1.5693	0.7902	1.2655
1.90	0.5956	4.0450	2.5157	1.6079	0.7674	1.3032
1.95	0.5862	4.2696	2.5919	1.6473	0.7442	1.3437
2.00	0.5774	4.5000	2.6667	1.6875	0.7209	1.3872
2.05	0.5691	4.7363	2.7400	1.7285	0.6975	1.4337
2.10	0.5613	4.9783	2.8119	1.7705	0.6742	1.4832
2.15	0.5540	5.2263	2.8823	1.8132	0.6511	1.5360
2.20	0.5471	5.4800	2.9512	1.8569	0.6281	1.5920
2.25	0.5406	5.7396	3.0186	1.9014	0.6055	1.6514
2.30	0.5344	6.0050	3.0845	1.9468	0.5833	1.7144
2.35	0.5286	6.2763	3.1490	1.9931	0.5615	1.7810
2.40	0.5231	6.5533	3.2119	2.0403	0.5401	1.8514
2.45	0.5179	6.8363	3.2733	2.0885	0.5193	1.9256
2.50	0.5130	7.1250	3.3333	2.1375	0.4990	2.0039
2.55	0.5083	7.4196	3.3919	2.1875	0.4793	2.0865
2.60	0.5039	7.7200	3.4490	2.2383	0.4601	2.1733
2.65	0.4996	8.0262	3.5047	2.2902	0.4416	2.2647
2.70	0.4956	8.3383	3.5590	2.3429	0.4236	2.3608
2.75	0.4918	8.6562	3.6119	2.3966	0.4062	2.4617
2.80	0.4882	8.9800	3.6636	2.4512	0.3895	2.5676
2.85	0.4847	9.3096	3.7139	2.5067	0.3733	2.6788
2.90	0.4814	9.6450	3.7629	2.5632	0.3577	2.7954
2.95	0.4782	9.9862	3.8106	2.6206	0.3428	2.9176
3.00	0.4752	10.3333	3.8571	2.6790	0.3283	3.0456

Table B.2 (*Concluded*) Normal Shock Relations for a Perfect Gas, $k = 1.4$

Ma_{n1}	Ma_{n2}	p_2/p_1	$V_1/V_2 = \rho_2/\rho_1$	T_2/T_1	p_{02}/p_{01}	A_2^*/A_1^*
3.00	0.4752	10.3333	3.8571	2.6790	0.3283	3.0456
3.05	0.4723	10.6863	3.9025	2.7383	0.3145	3.1796
3.10	0.4695	11.0450	3.9466	2.7986	0.3012	3.3199
3.15	0.4669	11.4096	3.9896	2.8598	0.2885	3.4667
3.20	0.4643	11.7800	4.0315	2.9220	0.2762	3.6202
3.25	0.4619	12.1563	4.0723	2.9851	0.2645	3.7806
3.30	0.4596	12.5383	4.1120	3.0492	0.2533	3.9483
3.35	0.4573	12.9263	4.1507	3.1142	0.2425	4.1234
3.40	0.4552	13.3200	4.1884	3.1802	0.2322	4.3062
3.45	0.4531	13.7196	4.2251	3.2472	0.2224	4.4969
3.50	0.4512	14.1250	4.2609	3.3151	0.2129	4.6960
3.55	0.4492	14.5363	4.2957	3.3839	0.2039	4.9036
3.60	0.4474	14.9533	4.3296	3.4537	0.1953	5.1200
3.65	0.4456	15.3763	4.3627	3.5245	0.1871	5.3456
3.70	0.4439	15.8050	4.3949	3.5962	0.1792	5.5806
3.75	0.4423	16.2396	4.4262	3.6689	0.1717	5.8253
3.80	0.4407	16.6800	4.4568	3.7426	0.1645	6.0801
3.85	0.4392	17.1263	4.4866	3.8172	0.1576	6.3454
3.90	0.4377	17.5783	4.5156	3.8928	0.1510	6.6213
3.95	0.4363	18.0363	4.5439	3.9694	0.1448	6.9084
4.00	0.4350	18.5000	4.5714	4.0469	0.1388	7.2069
4.05	0.4336	18.9696	4.5983	4.1254	0.1330	7.5172
4.10	0.4324	19.4450	4.6245	4.2048	0.1276	7.8397
4.15	0.4311	19.9263	4.6500	4.2852	0.1223	8.1747
4.20	0.4299	20.4133	4.6749	4.3666	0.1173	8.5227
4.25	0.4288	20.9063	4.6992	4.4489	0.1126	8.8840
4.30	0.4277	21.4050	4.7229	4.5322	0.1080	9.2591
4.35	0.4266	21.9096	4.7460	4.6165	0.1036	9.6484
4.40	0.4255	22.4200	4.7685	4.7017	0.0995	10.0522
4.45	0.4245	22.9362	4.7904	4.7879	0.0955	10.4711
4.50	0.4236	23.4583	4.8119	4.8751	0.0917	10.9054
4.55	0.4226	23.9862	4.8328	4.9632	0.0881	11.3556
4.60	0.4217	24.5200	4.8532	5.0523	0.0846	11.8222
4.65	0.4208	25.0596	4.8731	5.1424	0.0813	12.3057
4.70	0.4199	25.6050	4.8926	5.2334	0.0781	12.8065
4.75	0.4191	26.1562	4.9116	5.3254	0.0750	13.3251
4.80	0.4183	26.7133	4.9301	5.4184	0.0721	13.8620
4.85	0.4175	27.2762	4.9482	5.5124	0.0694	14.4177
4.90	0.4167	27.8450	4.9659	5.6073	0.0667	14.9928
4.95	0.4160	28.4196	4.9831	5.7032	0.0642	15.5878
5.00	0.4152	29.0000	5.0000	5.8000	0.0617	16.2032

Table B.3 Adiabatic Frictional Flow in a Constant-Area Duct for $k = 1.4$

Ma	$\bar{f}L/D$	p/p^*	T/T^*	$\rho^*/\rho = V/V^*$	p_0/p_0^*
0.00	∞	∞	1.2000	0.0000	∞
0.05	280.0203	21.9034	1.1994	0.0548	11.5914
0.10	66.9216	10.9435	1.1976	0.1094	5.8218
0.15	27.9320	7.2866	1.1946	0.1639	3.9103
0.20	14.5333	5.4554	1.1905	0.2182	2.9635
0.25	8.4834	4.3546	1.1852	0.2722	2.4027
0.30	5.2993	3.6191	1.1788	0.3257	2.0351
0.35	3.4525	3.0922	1.1713	0.3788	1.7780
0.40	2.3085	2.6958	1.1628	0.4313	1.5901
0.45	1.5664	2.3865	1.1533	0.4833	1.4487
0.50	1.0691	2.1381	1.1429	0.5345	1.3398
0.55	0.7281	1.9341	1.1315	0.5851	1.2549
0.60	0.4908	1.7634	1.1194	0.6348	1.1882
0.65	0.3246	1.6183	1.1065	0.6837	1.1356
0.70	0.2081	1.4935	1.0929	0.7318	1.0944
0.75	0.1273	1.3848	1.0787	0.7789	1.0624
0.80	0.0723	1.2893	1.0638	0.8251	1.0382
0.85	0.0363	1.2047	1.0485	0.8704	1.0207
0.90	0.0145	1.1291	1.0327	0.9146	1.0089
0.95	0.0033	1.0613	1.0165	0.9578	1.0021
1.00	0.0000	1.0000	1.0000	1.0000	1.0000
1.05	0.0027	0.9443	0.9832	1.0411	1.0020
1.10	0.0099	0.8936	0.9662	1.0812	1.0079
1.15	0.0205	0.8471	0.9490	1.1203	1.0175
1.20	0.0336	0.8044	0.9317	1.1583	1.0304
1.25	0.0486	0.7649	0.9143	1.1952	1.0468
1.30	0.0648	0.7285	0.8969	1.2311	1.0663
1.35	0.0820	0.6947	0.8794	1.2660	1.0890
1.40	0.0997	0.6632	0.8621	1.2999	1.1149
1.45	0.1178	0.6339	0.8448	1.3327	1.1440
1.50	0.1361	0.6065	0.8276	1.3646	1.1762
1.55	0.1543	0.5808	0.8105	1.3955	1.2116
1.60	0.1724	0.5568	0.7937	1.4254	1.2502
1.65	0.1902	0.5342	0.7770	1.4544	1.2922
1.70	0.2078	0.5130	0.7605	1.4825	1.3376
1.75	0.2250	0.4929	0.7442	1.5097	1.3865
1.80	0.2419	0.4741	0.7282	1.5360	1.4390
1.85	0.2583	0.4562	0.7124	1.5614	1.4952
1.90	0.2743	0.4394	0.6969	1.5861	1.5553
1.95	0.2899	0.4234	0.6816	1.6099	1.6193
2.00	0.3050	0.4082	0.6667	1.6330	1.6875
2.05	0.3197	0.3939	0.6520	1.6553	1.7600
2.10	0.3339	0.3802	0.6376	1.6769	1.8369
2.15	0.3476	0.3673	0.6235	1.6977	1.9185
2.20	0.3609	0.3549	0.6098	1.7179	2.0050
2.25	0.3738	0.3432	0.5963	1.7374	2.0964
2.30	0.3862	0.3320	0.5831	1.7563	2.1931
2.35	0.3983	0.3213	0.5702	1.7745	2.2953
2.40	0.4099	0.3111	0.5576	1.7922	2.4031
2.45	0.4211	0.3014	0.5453	1.8092	2.5168
2.50	0.4320	0.2921	0.5333	1.8257	2.6367
2.55	0.4425	0.2832	0.5216	1.8417	2.7630
2.60	0.4526	0.2747	0.5102	1.8571	2.8960
2.65	0.4624	0.2666	0.4991	1.8721	3.0359

Table B.3 (*Concluded*) Adiabatic Frictional Flow in a Constant-Area Duct for $k = 1.4$

Ma	$\bar{f}L/D$	p/p^*	T/T^*	$\rho^*/\rho = V/V^*$	p_0/p_0^*
2.65	0.4624	0.2666	0.4991	1.8721	3.0359
2.70	0.4718	0.2588	0.4882	1.8865	3.1830
2.75	0.4809	0.2513	0.4776	1.9005	3.3377
2.80	0.4898	0.2441	0.4673	1.9140	3.5001
2.85	0.4983	0.2373	0.4572	1.9271	3.6707
2.90	0.5065	0.2307	0.4474	1.9398	3.8498
2.95	0.5145	0.2243	0.4379	1.9521	4.0376
3.00	0.5222	0.2182	0.4286	1.9640	4.2346
3.05	0.5296	0.2124	0.4195	1.9755	4.4410
3.10	0.5368	0.2067	0.4107	1.9866	4.6573
3.15	0.5437	0.2013	0.4021	1.9974	4.8838
3.20	0.5504	0.1961	0.3937	2.0079	5.1210
3.25	0.5569	0.1911	0.3855	2.0180	5.3691
3.30	0.5632	0.1862	0.3776	2.0278	5.6286
3.35	0.5693	0.1815	0.3699	2.0373	5.9000
3.40	0.5752	0.1770	0.3623	2.0466	6.1837
3.45	0.5809	0.1727	0.3550	2.0555	6.4801
3.50	0.5864	0.1685	0.3478	2.0642	6.7896
3.55	0.5918	0.1645	0.3409	2.0726	7.1128
3.60	0.5970	0.1606	0.3341	2.0808	7.4501
3.65	0.6020	0.1568	0.3275	2.0887	7.8020
3.70	0.6068	0.1531	0.3210	2.0964	8.1691
3.75	0.6115	0.1496	0.3148	2.1039	8.5517
3.80	0.6161	0.1462	0.3086	2.1111	8.9506
3.85	0.6206	0.1429	0.3027	2.1182	9.3661
3.90	0.6248	0.1397	0.2969	2.1250	9.7990
3.95	0.6290	0.1366	0.2912	2.1316	10.2496
4.00	0.6331	0.1336	0.2857	2.1381	10.7187

Table B.4 Frictionless Duct Flow with Heat Transfer for $k = 1.4$

Ma	T_0/T_0^*	p/p^*	T/T^*	$\rho^*/\rho = V/V^*$	p_0/p_0^*
0.00	0.0000	2.4000	0.0000	0.0000	1.2679
0.05	0.0119	2.3916	0.0143	0.0060	1.2657
0.10	0.0468	2.3669	0.0560	0.0237	1.2591
0.15	0.1020	2.3267	0.1218	0.0524	1.2486
0.20	0.1736	2.2727	0.2066	0.0909	1.2346
0.25	0.2568	2.2069	0.3044	0.1379	1.2177
0.30	0.3469	2.1314	0.4089	0.1918	1.1985
0.35	0.4389	2.0487	0.5141	0.2510	1.1779
0.40	0.5290	1.9608	0.6151	0.3137	1.1566
0.45	0.6139	1.8699	0.7080	0.3787	1.1351
0.50	0.6914	1.7778	0.7901	0.4444	1.1141
0.55	0.7599	1.6860	0.8599	0.5100	1.0940
0.60	0.8189	1.5957	0.9167	0.5745	1.0753
0.65	0.8683	1.5080	0.9608	0.6371	1.0582
0.70	0.9085	1.4235	0.9929	0.6975	1.0431
0.75	0.9401	1.3427	1.0140	0.7552	1.0301
0.80	0.9639	1.2658	1.0255	0.8101	1.0193
0.85	0.9810	1.1931	1.0285	0.8620	1.0109
0.90	0.9921	1.1246	1.0245	0.9110	1.0049
0.95	0.9981	1.0603	1.0146	0.9569	1.0012
1.00	1.0000	1.0000	1.0000	1.0000	1.0000

Table B.4 (*Cont.*) Frictionless Duct Flow with Heat Transfer for $k = 1.4$

Ma	T_0/T_0^*	p/p^*	T/T^*	$\rho^*/\rho = V/V^*$	p_0/p_0^*
1.00	1.0000	1.0000	1.0000	1.0000	1.0000
1.05	0.9984	0.9436	0.9816	1.0403	1.0012
1.10	0.9939	0.8909	0.9603	1.0780	1.0049
1.15	0.9872	0.8417	0.9369	1.1131	1.0109
1.20	0.9787	0.7958	0.9118	1.1459	1.0194
1.25	0.9689	0.7529	0.8858	1.1765	1.0303
1.30	0.9580	0.7130	0.8592	1.2050	1.0437
1.35	0.9464	0.6758	0.8323	1.2316	1.0594
1.40	0.9343	0.6410	0.8054	1.2564	1.0777
1.45	0.9218	0.6086	0.7787	1.2796	1.0983
1.50	0.9093	0.5783	0.7525	1.3012	1.1215
1.55	0.8967	0.5500	0.7268	1.3214	1.1473
1.60	0.8842	0.5236	0.7017	1.3403	1.1756
1.65	0.8718	0.4988	0.6774	1.3580	1.2066
1.70	0.8597	0.4756	0.6538	1.3746	1.2402
1.75	0.8478	0.4539	0.6310	1.3901	1.2767
1.80	0.8363	0.4335	0.6089	1.4046	1.3159
1.85	0.8250	0.4144	0.5877	1.4183	1.3581
1.90	0.8141	0.3964	0.5673	1.4311	1.4033
1.95	0.8036	0.3795	0.5477	1.4432	1.4516
2.00	0.7934	0.3636	0.5289	1.4545	1.5031
2.05	0.7835	0.3487	0.5109	1.4652	1.5579
2.10	0.7741	0.3345	0.4936	1.4753	1.6162
2.15	0.7649	0.3212	0.4770	1.4848	1.6780
2.20	0.7561	0.3086	0.4611	1.4938	1.7434
2.25	0.7477	0.2968	0.4458	1.5023	1.8128
2.30	0.7395	0.2855	0.4312	1.5103	1.8860
2.35	0.7317	0.2749	0.4172	1.5180	1.9634
2.40	0.7242	0.2648	0.4038	1.5252	2.0451
2.45	0.7170	0.2552	0.3910	1.5320	2.1311
2.50	0.7101	0.2462	0.3787	1.5385	2.2218
2.55	0.7034	0.2375	0.3669	1.5446	2.3173
2.60	0.6970	0.2294	0.3556	1.5505	2.4177
2.65	0.6908	0.2216	0.3448	1.5560	2.5233
2.70	0.6849	0.2142	0.3344	1.5613	2.6343
2.75	0.6793	0.2071	0.3244	1.5663	2.7508
2.80	0.6738	0.2004	0.3149	1.5711	2.8731
2.85	0.6685	0.1940	0.3057	1.5757	3.0014
2.90	0.6635	0.1879	0.2969	1.5801	3.1359
2.95	0.6586	0.1820	0.2884	1.5843	3.2768
3.00	0.6540	0.1765	0.2803	1.5882	3.4245
3.05	0.6495	0.1711	0.2725	1.5920	3.5790
3.10	0.6452	0.1660	0.2650	1.5957	3.7408
3.15	0.6410	0.1612	0.2577	1.5992	3.9101
3.20	0.6370	0.1565	0.2508	1.6025	4.0871
3.25	0.6331	0.1520	0.2441	1.6057	4.2721
3.30	0.6294	0.1477	0.2377	1.6088	4.4655
3.35	0.6258	0.1436	0.2315	1.6117	4.6674
3.40	0.6224	0.1397	0.2255	1.6145	4.8783
3.45	0.6190	0.1359	0.2197	1.6172	5.0984
3.50	0.6158	0.1322	0.2142	1.6198	5.3280
3.55	0.6127	0.1287	0.2088	1.6223	5.5676
3.60	0.6097	0.1254	0.2037	1.6247	5.8173
3.65	0.6068	0.1221	0.1987	1.6271	6.0776

Table B.4 (*Concluded*)
Frictionless Duct Flow with Heat Transfer for $k = 1.4$

Ma	T_0/T_0^*	p/p^*	T/T^*	$\rho^*/\rho = V/V^*$	p_0/p_0^*
3.65	0.6068	0.1221	0.1987	1.6271	6.0776
3.70	0.6040	0.1190	0.1939	1.6293	6.3488
3.75	0.6013	0.1160	0.1893	1.6314	6.6314
3.80	0.5987	0.1131	0.1848	1.6335	6.9256
3.85	0.5962	0.1103	0.1805	1.6355	7.2318
3.90	0.5937	0.1077	0.1763	1.6374	7.5505
3.95	0.5914	0.1051	0.1722	1.6392	7.8820
4.00	0.5891	0.1026	0.1683	1.6410	8.2268

Table B.5 Prandtl-Meyer Supersonic Expansion Function for $k = 1.4$

Ma	ω, deg	Ma	ω, deg	Ma	ω, deg	Ma	ω, deg
1.00	0.00						
1.05	0.49	3.05	50.71	5.05	77.38	7.05	91.23
1.10	1.34	3.10	51.65	5.10	77.84	7.10	91.49
1.15	2.38	3.15	52.57	5.15	78.29	7.15	91.75
1.20	3.56	3.20	53.47	5.20	78.73	7.20	92.00
1.25	4.83	3.25	54.35	5.25	79.17	7.25	92.24
1.30	6.17	3.30	55.22	5.30	79.60	7.30	92.49
1.35	7.56	3.35	56.07	5.35	80.02	7.35	92.73
1.40	8.99	3.40	56.91	5.40	80.43	7.40	92.97
1.45	10.44	3.45	57.73	5.45	80.84	7.45	93.21
1.50	11.91	3.50	58.53	5.50	81.24	7.50	93.44
1.55	13.38	3.55	59.32	5.55	81.64	7.55	93.67
1.60	14.86	3.60	60.09	5.60	82.03	7.60	93.90
1.65	16.34	3.65	60.85	5.65	82.42	7.65	94.12
1.70	17.81	3.70	61.60	5.70	82.80	7.70	94.34
1.75	19.27	3.75	62.33	5.75	83.17	7.75	94.56
1.80	20.73	3.80	63.04	5.80	83.54	7.80	94.78
1.85	22.16	3.85	63.75	5.85	83.90	7.85	95.00
1.90	23.59	3.90	64.44	5.90	84.26	7.90	95.21
1.95	24.99	3.95	65.12	5.95	84.61	7.95	95.42
2.00	26.38	4.00	65.78	6.00	84.96	8.00	95.62
2.05	27.75	4.05	66.44	6.05	85.30	8.05	95.83
2.10	29.10	4.10	67.08	6.10	85.63	8.10	96.03
2.15	30.43	4.15	67.71	6.15	85.97	8.15	96.23
2.20	31.73	4.20	68.33	6.20	86.29	8.20	96.43
2.25	33.02	4.25	68.94	6.25	86.62	8.25	96.63
2.30	34.28	4.30	69.54	6.30	86.94	8.30	96.82
2.35	35.53	4.35	70.13	6.35	87.25	8.35	97.01
2.40	36.75	4.40	70.71	6.40	87.56	8.40	97.20
2.45	37.95	4.45	71.27	6.45	87.87	8.45	97.39
2.50	39.12	4.50	71.83	6.50	88.17	8.50	97.57
2.55	40.28	4.55	72.38	6.55	88.47	8.55	97.76
2.60	41.41	4.60	72.92	6.60	88.76	8.60	97.94
2.65	42.53	4.65	73.45	6.65	89.05	8.65	98.12
2.70	43.62	4.70	73.97	6.70	89.33	8.70	98.29
2.75	44.69	4.75	74.48	6.75	89.62	8.75	98.47
2.80	45.75	4.80	74.99	6.80	89.89	8.80	98.64
2.85	46.78	4.85	75.48	6.85	90.17	8.85	98.81
2.90	47.79	4.90	75.97	6.90	90.44	8.90	98.98
2.95	48.78	4.95	76.45	6.95	90.71	8.95	99.15
3.00	49.76	5.00	76.92	7.00	90.97	9.00	99.32

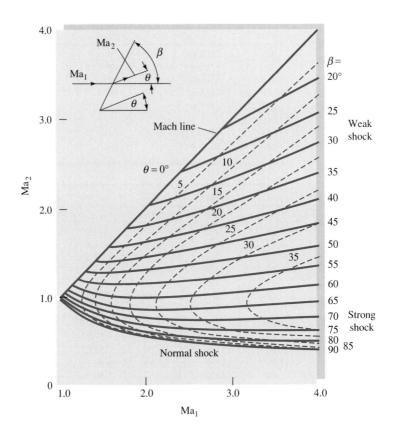

Mach number
downstream of an oblique shock
for $k = 1.4$.

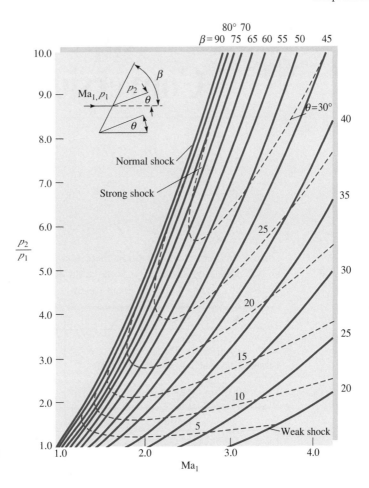

Fig. B.2 Pressure ratio downstream of an oblique shock for $k = 1.4$.

Appendix C
Conversion Factors

During this period of transition there is a constant need for conversions between BG and SI units (see Table 1.2). Some additional conversions are given here. Conversion factors are given inside the front cover.

Length	Volume
1 ft = 12 in = 0.3048 m	1 ft^3 = 0.028317 m^3
1 mi = 5280 ft = 1609.344 m	1 U.S. gal = 231 in^3 = 0.0037854 m^3
1 nautical mile (nmi) = 6076 ft = 1852 m	1 L = 0.001 m^3 = 0.035315 ft^3
1 yd = 3 ft = 0.9144 m	1 U.S. fluid ounce = 2.9574 E−5 m^3
1 angstrom (Å) = 1.0 E−10 m	1 U.S. quart (qt) = 9.4635 E−4 m^3

Mass	Area
1 slug = 32.174 lbm = 14.594 kg	1 ft^2 = 0.092903 m^2
1 lbm = 0.4536 kg	1 mi^2 = 2.78784 E7 ft^2 = 2.59 E6 m^2
1 short ton = 2000 lbm = 907.185 kg	1 acre = 43,560 ft^2 = 4046.9 m^2
1 tonne = 1000 kg	1 hectare (ha) = 10,000 m^2

Velocity	Acceleration
1 ft/s = 0.3048 m/s	1 ft/s^2 = 0.3048 m/s^2
1 mi/h = 1.466666 ft/s = 0.44704 m/s	
1 kn = 1 nmi/h = 1.6878 ft/s = 0.5144 m/s	

Mass flow	Volume flow
1 slug/s = 14.594 kg/s	1 gal/min = 0.002228 ft^3/s = 0.06309 L/s
1 lbm/s = 0.4536 kg/s	1 × 10^6 gal/day = 1.5472 ft^3/s = 0.04381 m^3/s

Pressure	Force
1 lbf/ft^2 = 47.88 Pa	1 lbf = 4.448222 N = 16 oz
1 lbf/in^2 = 144 lbf/ft^2 = 6895 Pa	1 kgf = 2.2046 lbf = 9.80665 N
1 atm = 2116.2 lbf/ft^2 = 14.696 lbf/in^2 = 101,325 Pa	1 U.S. (short) ton = 2000 lbf
1 inHg (at 20°C) = 3375 Pa	1 dyne = 1.0 E−5 N
1 bar = 1.0 E5 Pa	1 ounce (avoirdupois) (oz) = 0.27801 N

Energy	Power
1 ft · lbf = 1.35582 J 1 Btu = 252 cal = 1055.056 J = 778.17 ft · lbf 1 kilowatt hour (kWh) = 3.6 E6 J	1 hp = 550 ft · lbf/s = 745.7 W 1 ft · lbf/s = 1.3558 W

Specific weight	Density
1 lbf/ft^3 = 157.09 N/m^3	1 slug/ft^3 = 515.38 kg/m^3 1 lbm/ft^3 = 16.0185 kg/m^3 1 g/cm^3 = 1000 kg/m^3

Viscosity	Kinematic viscosity
1 slug/(ft · s) = 47.88 kg/(m · s) 1 poise (P) = 1 g/(cm · s) = 0.1 kg/(m · s)	1 ft^2/h = 0.000025806 m^2/s 1 stokes (St) = 1 cm^2/s = 0.0001 m^2/s

Temperature scale readings

$$T_F = \tfrac{9}{5}T_C + 32 \qquad T_C = \tfrac{5}{9}(T_F - 32) \qquad T_R = T_F + 459.69 \qquad T_K = T_C + 273.16$$

where subscripts F, C, R, and K refer to readings on the Fahrenheit, Celsius, Kelvin, and Rankine scales, respectively.

Specific heat or gas constant*	Thermal conductivity*
1 ft · lbf/(slug · °R) = 0.16723 N · m/(kg · K) 1 Btu/(lbm · °R) = 4186.8 J/(kg · K)	1 Btu/(h · ft · °R) = 1.7307 W/(m · K)

*Although the absolute (Kelvin) and Celsius temperature scales have different starting points, the intervals are the same size: 1 kelvin = 1 Celsius degree. The same holds true for the nonmetric absolute (Rankine) and Fahrenheit scales: 1 Rankine degree = 1 Fahrenheit degree. It is customary to express temperature differences in absolute temperature units.

Equations of Motion in Cylindrical Coordinates

The equations of motion of an incompressible newtonian fluid with constant μ, k, and c_p are given here in cylindrical coordinates (r, θ, z), which are related to cartesian coordinates (x, y, z) as in Fig. 4.2:

$$x = r \cos \theta \qquad y = r \sin \theta \qquad z = z \tag{D.1}$$

The velocity components are v_r, v_θ, and v_z. Here are the equations:

Continuity:

$$\frac{1}{r}\frac{\partial}{\partial r}(rv_r) + \frac{1}{r}\frac{\partial}{\partial \theta}(v_\theta) + \frac{\partial}{\partial z}(v_z) = 0 \tag{D.2}$$

Convective time derivative:

$$\mathbf{V} \cdot \nabla = v_r\frac{\partial}{\partial r} + \frac{1}{r}v_\theta\frac{\partial}{\partial \theta} + v_z\frac{\partial}{\partial z} \tag{D.3}$$

Laplacian operator:

$$\nabla^2 = \frac{1}{r}\frac{\partial}{\partial r}\left(r\frac{\partial}{\partial r}\right) + \frac{1}{r^2}\frac{\partial^2}{\partial \theta^2} + \frac{\partial^2}{\partial z^2} \tag{D.4}$$

The r-momentum equation:

$$\frac{\partial v_r}{\partial t} + (\mathbf{V} \cdot \nabla)v_r - \frac{1}{r}v_\theta^2 = -\frac{1}{\rho}\frac{\partial p}{\partial r} + g_r + \nu\left(\nabla^2 v_r - \frac{v_r}{r^2} - \frac{2}{r^2}\frac{\partial v_\theta}{\partial \theta}\right) \tag{D.5}$$

The θ-momentum equation:

$$\frac{\partial v_\theta}{\partial t} + (\mathbf{V} \cdot \nabla)v_\theta + \frac{1}{r}v_r v_\theta = -\frac{1}{\rho r}\frac{\partial p}{\partial \theta} + g_\theta + \nu\left(\nabla^2 v_\theta - \frac{v_\theta}{r^2} + \frac{2}{r^2}\frac{\partial v_r}{\partial \theta}\right) \tag{D.6}$$

The z-momentum equation:

$$\frac{\partial v_z}{\partial t} + (\mathbf{V} \cdot \nabla) v_z = -\frac{1}{\rho}\frac{\partial p}{\partial z} + g_z + \nu \nabla^2 v_z \tag{D.7}$$

The energy equation:

$$\rho c_p \left[\frac{\partial T}{\partial t} + (\mathbf{V} \cdot \nabla)T \right] = k\nabla^2 T + \mu[2(\epsilon_{rr}^2 + \epsilon_{\theta\theta}^2 + \epsilon_{zz}^2) + \epsilon_{\theta z}^2 + \epsilon_{rz}^2 + \epsilon_{r\theta}^2] \tag{D.8}$$

where

$$\epsilon_{rr} = \frac{\partial v_r}{\partial r} \qquad\qquad \epsilon_{\theta\theta} = \frac{1}{r}\left(\frac{\partial v_\theta}{\partial \theta} + v_r\right)$$

$$\epsilon_{zz} = \frac{\partial v_z}{\partial z} \qquad\qquad \epsilon_{\theta z} = \frac{1}{r}\frac{\partial v_z}{\partial \theta} + \frac{\partial v_\theta}{\partial z} \tag{D.9}$$

$$\epsilon_{rz} = \frac{\partial v_r}{\partial z} + \frac{\partial v_z}{\partial r} \qquad \epsilon_{r\theta} = \frac{1}{r}\left(\frac{\partial v_r}{\partial \theta} - v_\theta\right) + \frac{\partial v_\theta}{\partial r}$$

Viscous stress components:

$$\tau_{rr} = 2\mu\epsilon_{rr} \qquad \tau_{\theta\theta} = 2\mu\epsilon_{\theta\theta} \qquad \tau_{zz} = 2\mu\epsilon_{zz}$$

$$\tau_{r\theta} = \mu\epsilon_{r\theta} \qquad \tau_{\theta z} = \mu\epsilon_{\theta z} \qquad \tau_{rz} = \mu\epsilon_{rz} \tag{D.10}$$

Angular velocity components:

$$2\omega_r = \frac{1}{r}\frac{\partial v_z}{\partial \theta} - \frac{\partial v_\theta}{\partial z}$$

$$2\omega_\theta = \frac{\partial v_r}{\partial z} - \frac{\partial v_z}{\partial r} \tag{D.11}$$

$$2\omega_z = \frac{1}{r}\frac{\partial}{\partial r}(rv_\theta) - \frac{1}{r}\frac{\partial v_r}{\partial \theta}$$

Appendix E
Introduction to EES

Overview

EES (pronounced "ease") is an acronym for Engineering Equation Solver. The basic function provided by EES is the numerical solution of nonlinear algebraic and differential equations. In addition, EES provides built-in thermodynamic and transport property functions for many fluids, including water, dry and moist air, refrigerants, combustion gases, and others. Additional property data can be added by the user. The combination of equation-solving capability and engineering property data makes EES a very powerful tool.

A license for EES is provided to departments of educational institutions that adopt this text by McGraw-Hill. If you need more information, contact your local McGraw-Hill representative, call 1-800-338-3987, or visit the McGraw-Hill website at http://www.mhhe.com. A commercial or professional version of EES can be obtained from

F-Chart Software
Box 444042
Madison, WI 53744
Phone: (608) 836-8531
Fax: (608) 836-8536
http://fchart.com e-mail: info@fchart.com

Background Information

The EES program is probably installed on your departmental computer network. In addition, the license agreement for use of EES allows students and faculty in a participating educational department to copy the program for educational use onto their personal computer systems. Ask your instructor for details.

To start EES, double-click on the **EES** program icon shown at left or on any file created by EES having the .EES filename extension. You can also start EES from the Windows **Run** command in the **Start** menu by entering EES and clicking the **OK** button. EES begins by displaying a dialog window, which shows registration information, the version number, and other information. Click the **OK** button to dismiss the dialog window.

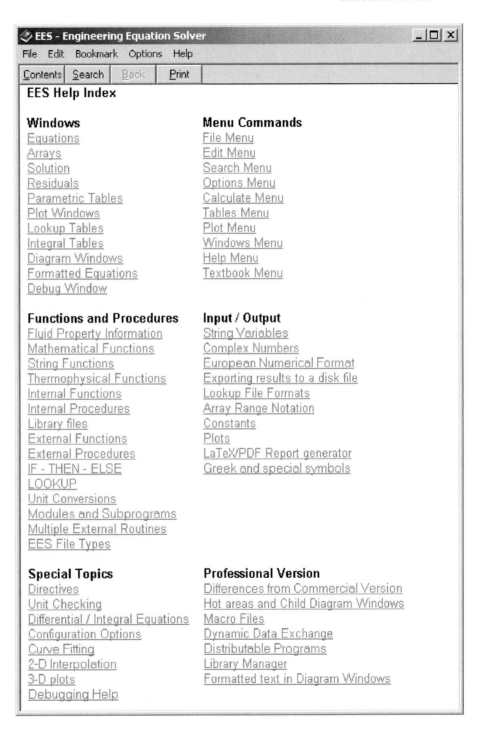

Fig. E.1 EES Help Index.

Detailed help is available at any point in EES. Pressing the **F1** key will bring up a Help window relating to the foremost window. Clicking the **Contents** button will present the **Help Index** shown in Fig. E.1. Clicking on an underlined word (shown in green on color monitors) will provide help relating to that subject.

EES commands are distributed among 11 pull-down menus. A brief summary of their functions follows:

EES Academic Version:

File   Edit   Search   Options   Calculate   Tables   Plots   Windows   Help   Fluid Mechanics

The **System** menu is accessible by clicking on the **EES** icon above the file menu. The **System** menu is not part of EES, but rather a feature of the Windows operating system. It holds commands that allow window moving, resizing, and switching to other applications.

The **File** menu provides commands for loading, merging, and saving work files and libraries, and printing. The **Load Textbook** command in this menu reads the problem disk developed for this text and creates a new menu to the right of the **Help** menu for easy access to EES problems accompanying this text.

The **Edit** menu provides the editing commands to cut, copy, and paste information.

The **Search** menu provides **Find** and **Replace** commands for use in the **Equations** window.

The **Options** menu provides commands for setting the guess values and bounds of variables, the unit system, default information, and program preferences. A command is also provided for displaying information on built-in and user-supplied functions.

The **Calculate** menu contains the commands to check, format, and solve the equation set. A command to check the units of the equations is also provided.

The **Tables** menu contains commands to set up and alter the contents of the **Parametric Table** and **Lookup Table** and to do linear regression on the data in these tables. The **Parametric Table,** which is similar to a spreadsheet, allows the equation set to be solved repeatedly while varying the values of one or more variables. The **Lookup Table** holds user-supplied data, which can be interpolated and used in the solution of the equation set.

The **Plots** menu provides commands to prepare a new plot of data in the **Parametric, Lookup, Array,** or **Integral Tables,** or to modify an existing plot. Curve-fitting capability and thermodynamic property plots are also provided.

The **Windows** menu provides a convenient method of bringing any of the EES windows to the front or to organize the windows.

The **Help** menu provides commands for accessing the online help documentation.

The **Fluid Mechanics** menu provides access to EES solutions to problems in this text.

A basic capability provided by EES is the solution of a set of nonlinear algebraic equations. To demonstrate this capability, start EES and enter this simple example problem in the **Equations** window:

Text is entered in the same manner as for any word processor. Formatting rules are as follows:

1. Uppercase and lowercase letters are not distinguished. EES will (optionally) change the case of all variables to match the manner in which they first appear.
2. Blank lines and spaces may be entered as desired since they are ignored.
3. Comments must be enclosed within braces { } or within quote marks " ". Comments may span as many lines as needed. Comments within braces may be nested, in which case only the outermost set of { } is recognized. Comments within quotes will also be displayed in the **Formatted Equations** window.
4. Variable names must start with a letter and consist of any keyboard characters except ( ) ' | */ + − ^ { } : " or ;. Array variables are identified with square braces around the array index or indices (for example, X[5,3]). The maximum variable length is 30 characters.
5. Multiple equations may be entered on one line if they are separated by a semi-colon (;). The maximum line length is 255 characters.
6. The caret symbol (^) or ** is used to indicate raising to a power.
7. The order in which the equations are entered does not matter.
8. The position of knowns and unknowns in the equation does not matter.

If you wish, you may view the equations in mathematical notation by selecting the **Formatted Equations** command from the **Windows** menu or from the **Formatted Equations** speedbutton located below the menu bar.

Select the **Solve** command from the **Calculate** menu or press **F2.** A dialog window will appear indicating the progress of the solution. When the calculations are completed, the button will change from **Abort** to **Continue.**

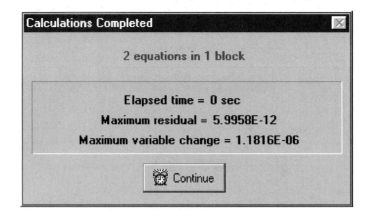

Click the **Continue** button. The solution to this equation set will then be displayed.

A Pipe Friction Example Problem

Let us now solve Prob. P6.55 from the text, for a cast iron pipe, to illustrate the capabilities of the EES program. This problem, without EES, would require iteration for Reynolds number, velocity, and friction factor, a daunting task. State the problem:

> **P6.55** As shown in Fig. E.2, reservoirs 1 and 2 contain water at 20°C. The pipe is cast iron, with $L = 4500$ m and $D = 4$ cm. What will be the flow rate in m³/hr if $\Delta z = 100$ m?

This is a representative problem in pipe flow; and with water in a reasonably large (noncapillary) pipe, it will probably be turbulent (Re > 4000). The steady flow energy

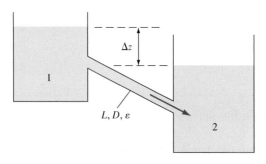

Fig. E.2 Sketch of the flow system.

equation (3.71) may be written between the surfaces of reservoirs 1 and 2:

$$\frac{p_1}{\rho g} + \frac{V_1^2}{2g} + z_1 = \frac{p_2}{\rho g} + \frac{V_2^2}{2g} + z_2 + h_f \qquad \text{where} \qquad h_f = f\frac{L}{D}\frac{V_{pipe}^2}{2g}$$

Since $p_1 = p_2 = p_{atm}$ and $V_1 \approx V_2 \approx 0$, this relation simplifies to

$$\Delta z = f\frac{L}{D}\frac{V^2}{2g} \tag{E.1}$$

where $V = Q/A$ is the velocity in the pipe. The friction factor f is a function of Reynolds number and pipe roughness ratio, if the flow is turbulent, from Eq. (6.48):

$$\frac{1}{f^{1/2}} = -2.0 \log_{10}\left(\frac{\epsilon/D}{3.7} + \frac{2.51}{\text{Re}\,f^{1/2}}\right) \qquad \text{if Re} > 4000 \tag{E.2}$$

Finally, we need the definitions of Reynolds number and volume flow rate

$$\text{Re} = \rho V D/\mu \qquad \text{(E.3)} \qquad \text{and} \qquad Q = V\frac{\pi}{4}D^2 \tag{E.4}$$

where ρ and μ are the fluid density and viscosity, respectively.

There are a total of 11 variables involved in this problem: (L, D, Δz, ε, g, μ, ρ, V, Re, f, Q). Of these, seven can be specified at the start (L, D, Δz, ε, g, μ, ρ), while four (V, Re, f, Q) must be calculated from Equations (E.1–4). These four equations in four unknowns are well-posed and solvable, but only by laborious iteration—exactly what EES is designed to do.

Start EES or select the **New** command from the **File** menu if you have already been using the program. A blank **Equations** window will appear. Our recommendation; Always set the unit system immediately: Select **Unit System** from the **Options** menu (Fig. E.3). Select *SI* and *Mass* units and trig *Degrees*, although we do not actually have trig functions this time. We select *kPa* for pressure and *Celsius* for temperature, *kJ* energy units that will be handy for using the EES built-in physical properties of water.

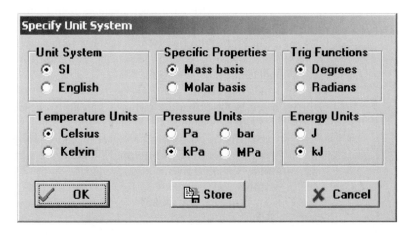

Fig. E.3 Unit selection dialog window.

Now, on the blank screen, enter the equations for this problem (Fig. E.4), five of which are known input values, two are property evaluations, and four are Equations (E.1–4).

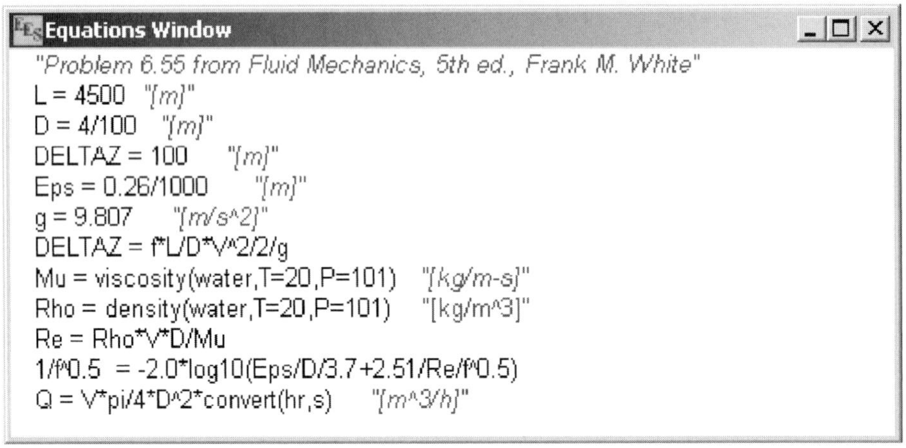

Fig. E.4 Equations window.

```
"Problem 6.55 from Fluid Mechanics, 5th ed., Frank M. White"
L = 4500   "[m]"
D = 4/100   "[m]"
DELTAZ = 100      "[m]"
Eps = 0.26/1000     "[m]"
g = 9.807    "[m/s^2]"
DELTAZ = f*L/D*V^2/2/g
Mu = viscosity(water,T=20,P=101)    "[kg/m-s]"
Rho = density(water,T=20,P=101)    "[kg/m^3]"
Re = Rho*V*D/Mu
1/f^0.5  = -2.0*log10(Eps/D/3.7+2.51/Re/f^0.5)
Q = V*pi/4*D^2*convert(hr,s)     "[m^3/h]"
```

Notice several things in Fig. E.4. First, quantities in quotes, such as "[m]", indicate the units of the variable on the left of the equal sign. There are other ways to enter the units of variables. Unit specifications do not affect the numerical results, but they are used in the unit checking that EES provides. You do not have to enter units in EES; but it is a good idea to do so since EES cannot check the units if you do not enter them, and unit conversions are a likely source of errors. Second, we changed *Eps* and *D* to meters right away to keep the SI units consistent. We could have used the **Convert** function to convert units as used in the last equation. Third, we called on EES to input the viscosity and density of water at 20°C and 1 atm, a procedure well explained in the **Help** menu. For example, **viscosity(water, T = 20, P = 101)** meets the EES requirement that temperature (T) and pressure (P) should be input in °C and kPa—EES will then evaluate μ in kg/m-s. Finally, note that EES recognizes **pi** to be 3.141593.

In Fig. E.4 we used only one built-in function, **log10.** There are many such functions, found by scrolling down the **Function Information** command in the **Options** menu.

Having entered the equations, check the syntax by using the **Check/Format** command in the **Calculate** menu. If you did well, EES will report that the 11 equations in 11 unknowns look OK. If not, EES will guess at what might be wrong. If OK, go for it: Choose the **Solve** command in the **Options** menu. EES reports "logarithm of a negative number—try setting limits on the variables". We might have known. Go to the **Variable Information** command in the **Options** menu. A box, listing the 11 variables, will appear (Fig. E.5). All default EES "guesses" are unity; all default limits are $-\infty$ to $+\infty$, too broad. Enter (as already shown in Fig. E.6) guesses for $f = 0.02$

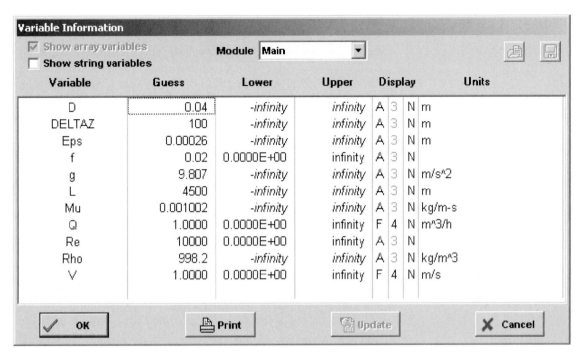

Fig. E.5 Variable Information window with units and guess values entered.

and Re = 10,000, while $V = 1$ and $Q = 1$ seem adequate, and other variables are fixed. Make sure that f, Re, V, and Q cannot be negative. The "display" columns normally say "A", automatic, satisfactory for most variables. We have changed "A" to "F" (fixed decimal) for Q and V to make sure they are displayed to four decimal

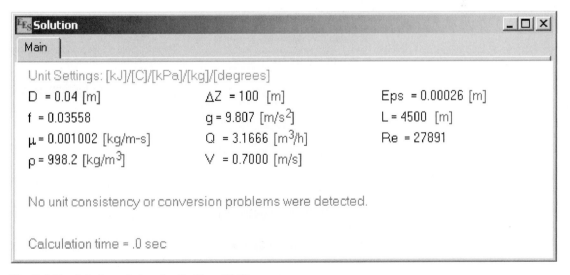

Fig. E.6 The Solution window for Problem P6.55.

places. The "units" column shows the units that were set within comments and square braces in the **Equations** window. Units can also be set in this dialog window.

Our guesses and limits are excellent, and the **Solve** command now iterates and reports success: "max residual = 2E−10", a negligible error. (The default runs for 100 iterations, which can be modified by the **Stop Criteria** command in the **Options** menu.) Hit **Continue** and the complete solution is displayed for all variables (Fig. E.6). Note that EES also checked the unit consistency of all the equations and found no problems.

This is the correct solution to Prob. P6.55: this cast iron pipe, when subjected to a 100 m elevation difference, will deliver $Q = 3.17$ m³/hr of water. EES did all the iteration.

Parametric Studies with Tabular Input

One of the most useful features of EES is its ability to provide parametric studies. For example, suppose we wish to know how varying Δz changed the flow rate Q. First comment out the equation that reads **DELTAZ = 100** by enclosing it within braces {}. If you select the equation and press the right mouse button, a menu will appear with **Comment** as its first item. If you select that menu item, EES will automatically enter the braces. Select the **New Parametric Table** command in the **Options** menu. A dialog will be displayed (Fig. E.7) listing all the variables in the problem. Highlight what you wish to vary: Δz. Also highlight variables to be calculated and tabulated: V, Q, Re, and f.

Click the **Add** button and then the **OK** button and the new table will be displayed (Fig. E.8). Enter 10 values of Δz that cover the range of interest; we have selected the linear range 10 m < Δz < 500 m. Note that it is not necessary to type these values in, although you can if you wish. Clicking the triangular icon at the upper right of each column header cell brings up a dialog that allows values to be automatically entered into the table.

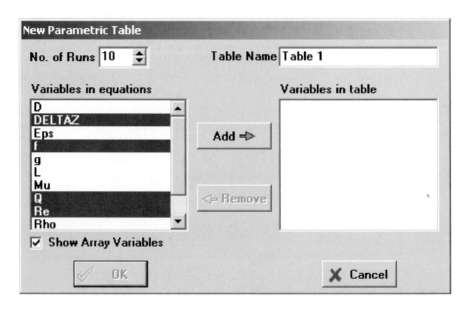

Fig. E.7 New Parametric Table showing selected variables (V is not shown).

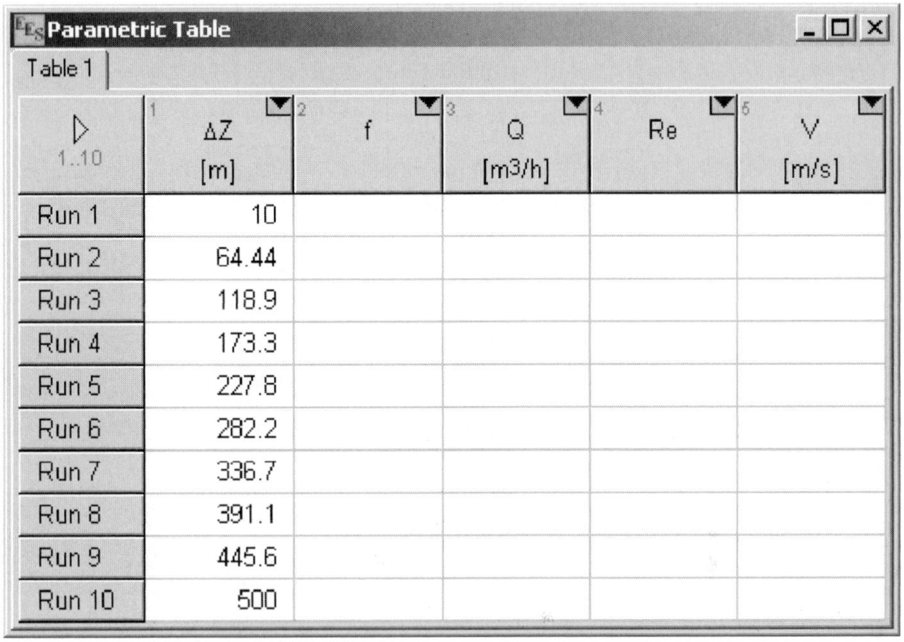

1..10	ΔZ [m]	f	Q [m3/h]	Re	V [m/s]
Run 1	10				
Run 2	64.44				
Run 3	118.9				
Run 4	173.3				
Run 5	227.8				
Run 6	282.2				
Run 7	336.7				
Run 8	391.1				
Run 9	445.6				
Run 10	500				

Fig. E.8 Parametric Table window.

Clearly the **Parametric Table** operates much like a spreadsheet. Select **Solve Table** from the **Calculate** menu and the **Solve Table** dialog window will appear (Fig. E.9). These are satisfactory default values; the writer has changed nothing. Hit the **OK** button, and the calculations will be made and the entire **Parametric Table** filled out, as in Fig. E.10.

The flow rates are there to see in Fig. E.10, but as always, in the writer's experience, a plot is more illuminating. Select **New Plot Window** from the **Plot** menu. The **New Plot Setup** dialog box (Fig. E.11) will appear. Choose Δz as the x-axis and Q as the y-axis.

We added grid lines. Click the **OK** button and the desired plot will appear in the **Plot** window (Fig. E.12). We see a nonlinear relationship, roughly a square root type, and learn that flow rate Q is not linearly proportional to head difference Δz.

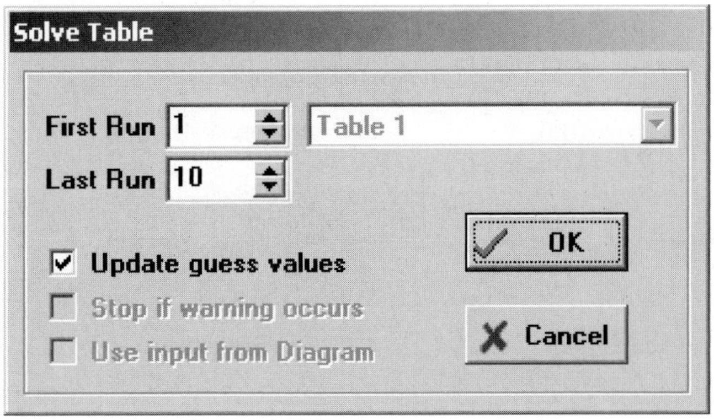

Fig. E.9 Solve Table dialog.

Parametric Table — Table 1

1..10	ΔZ [m]	f	Q [m3/h]	Re	V [m/s]
Run 1	10	0.04048	0.9389	8270	0.2075
Run 2	64.44	0.03619	2.5207	22202	0.5572
Run 3	118.9	0.03538	3.4629	30500	0.7655
Run 4	173.3	0.03498	4.2051	37038	0.9295
Run 5	227.8	0.03473	4.8379	42611	1.0694
Run 6	282.2	0.03455	5.3987	47552	1.1934
Run 7	336.7	0.03442	5.9078	52036	1.3059
Run 8	391.1	0.03431	6.3773	56170	1.4097
Run 9	445.6	0.03423	6.8151	60026	1.5065
Run 10	500	0.03416	7.2269	63654	1.5975

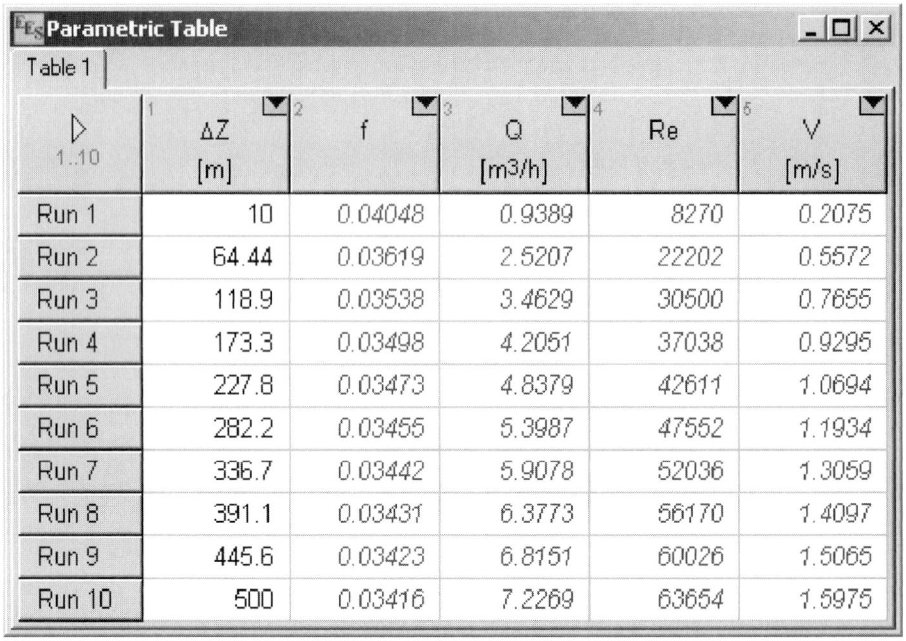

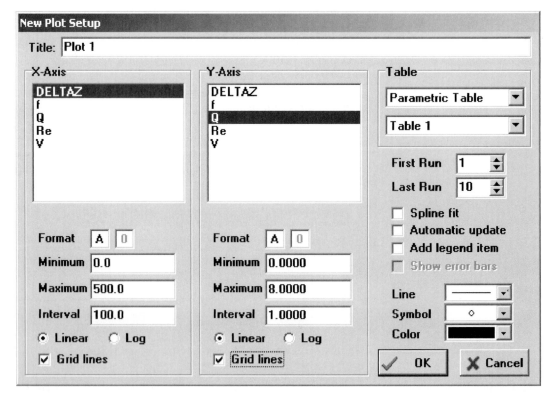

Fig. E.10 Parametric Table window after calculations are completed.

Fig. E.11 New Plot Setup dialog.

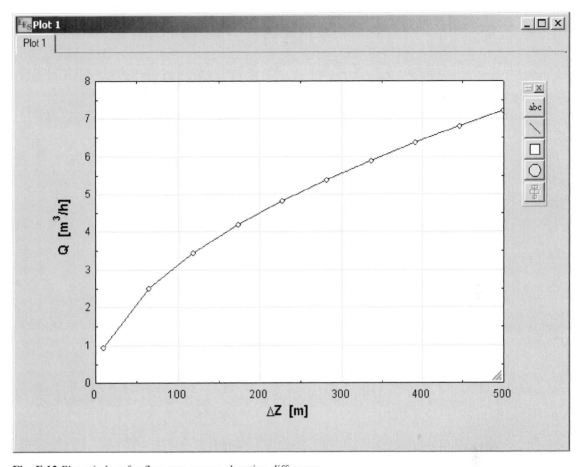

Fig. E.12 Plot window for flow rate versus elevation difference.

The plot's appearance in Fig. E.12 can be modified in several ways. Double-click the mouse in the plot rectangle to see some of these options. The toolbar at the right of the **Plot** window can be used to place text or graphics on the plot.

Fluid Dynamics Example Problems

A number of fluid dynamics problems developed for EES have been included with this textbook. In the menu bar at the top of the screen, you should see a menu called **Fluid Mechanics** to the right of the **Help** menu. This menu will provide access to all of the EES problem solutions developed for this book, organized by chapter. As an example, select **Chapter 6** from the **White Fluid Mechanics** menu. A dialog window will appear listing the problems in Chapter 6. Select **Problem P6.55–Flow Between Reservoirs.** This problem is a smooth wall alternative to the problem you just entered. It provides a **Diagram** window in which you can enter the Δz and other information. Enter values, and then select the **Solve** command in the **Calculate** menu to see their effect.

At this point, you should explore. Try whatever you wish. You can't hurt anything. The online help (invoked by pressing **F1**) will provide details for the EES commands. EES is a powerful tool that you will find very useful in your studies.

Answers to Selected Problems

Chapter 1

P1.2 6.1E18 kg; 1.3E44 molecules
P1.4 (a) inconsistent; (d) Yes, they are the same unit.
P1.6 (a) 4400 Pa; (b) 101,350 Pa
P1.8 $\sigma \approx 1.00\ My/I$
P1.10 Yes, all terms are $\{ML/T^2\}$
P1.12 $\{B\} = \{L^{-1}\}$
P1.14 $Q = \text{Const } B\ g^{1/2}H^{3/2}$
P1.16 All terms are $\{ML^{-2}T^{-2}\}$
P1.18 $V = V_0 e^{-mt/K}$
P1.20 $z_{\max} = 64.2$ m at $t = 3.36$ s
P1.24 (a) $\rho = 7.97$ kg/m^3; (b) $c_p = 819$
       J/(kg-K); $p = 79$ kPa (ideal gas)
P1.26 $W_{\text{air}} = 0.71$ lbf
P1.28 $\rho_{\text{wet}} = 1.10$ kg/m^3, $\rho_{\text{dry}} = 1.13$ kg/m^3
P1.30 $W_{1-2} = 21$ ft · lbf
P1.32 (a) 76 kN; (b) 501 kN
P1.34 (a) $\rho_1 = 5.05$ kg/m^3; (b) $\rho_2 = 2.12$ kg/m^3 (ideal gas)
P1.36 (a) $B_{\text{N}_2\text{O}} = 1.33$ E5 Pa; (b) $B_{\text{water}} = 2.13$ E9 Pa
P1.38 $\tau = 1380$ Pa, $\text{Re}_L = 28$
P1.40 $A = 0.0016$ kg/(m · s), $B = 1903$ K
P1.42 $\mu/\mu_{200\text{K}} \approx (T\ \text{K}/200\ \text{K})^{0.68}$
P1.44 Data 50 percent higher; Andrade fit varies ±50 percent
P1.46 (d) 3.0 m/s; (e) 0.79 m/s; (f) 22 m/s
P1.48 $F \approx (\mu_1/h_1 + \mu_2/h_2)AV$
P1.50 (a) Yes; (b) $\mu \approx 0.40$ kg/(m-s)
P1.52 $P \approx 73$ W
P1.54 $M \approx \pi\mu\Omega R^4/h$
P1.56 $\mu = 3M \sin\theta/(2\pi\Omega R^3)$
P1.58 $\mu = 0.040$ kg/(m · s), last 2 points are *turbulent* flow
P1.60 39,500 Pa
P1.62 28,500 Pa
P1.64 $D = 0.73$ mm
P1.66 $F = 0.014$ N

P1.68 $h = (Y/\rho g)^{1/2} \cot\theta$
P1.70 $h = 2Y \cos\theta/(\rho gW)$
P1.72 $z \approx 4800$ m
P1.74 Cavitation occurs for both (a) and (b)
P1.76 (a) 539 m/s; (b) 529 m/s
P1.78 (a) 25°C; (b) 4°C
P1.80 $x^2y - y^3/3 = $ constant
P1.82 $y = x \tan\theta + $ constant
P1.84 $x = x_0 \exp[\ln(y/y_0) + \ln^2(y/y_0)]$
P1.86 Approximately 5.0 percent
P1.88 (a) 0.29 kg/(m-s); (b) 4.4 percent
P1.90 5.6%

Chapter 2

P2.2 $\sigma_{xy} = -289$ lb/ft^2, $\tau_{AA} = -577$ lb/ft^2
P2.6 (a) 30.3 ft; (b) 30.0 in; (c) 10.35 m; (d) 13,100 mm
P2.8 (a) 140 kPa; (b) ±10 m
P2.10 10,500 Pa
P2.12 8.0 cm
P2.14 92,700 Pa
P2.16 (a) 4.58 m; (b) 0.627 m
P2.18 1.56
P2.20 14 lbf
P2.22 0.94 cm
P2.24 $p_{\text{sealevel}} \approx 117$ kPa, $m_{\text{exact}} = 5.3$ E18 kg
P2.28 (a) 421 ft; (b) 452 ft
P2.30 $p_1 - p_2 = 43.1$ kPa
P2.32 22.6 cm
P2.34 $\Delta p = \Delta h[\gamma_{\text{water}}(1 + d^2/D^2) - \gamma_{\text{oil}}(1 - d^2/D^2)]$
P2.36 25°
P2.38 (a) $p_{1,\text{gage}} = (\rho_m - \rho_a)gh - (\rho_t - \rho_a)gH$
P2.40 Left leg drops 19.3 cm, right leg rises 5 cm vertically
P2.42 $p_A - p_B = (\rho_2 - \rho_1)gh$
P2.44 (a) 171 lb/ft^2; (b) 392 lb/ft^2; manometer reads friction loss

P2.46	1.45
P2.48	$F = 39700$ N
P2.50	(a) 524 kN; (b) 350 kN; (c) 100 kN
P2.52	(a) 38,400 lbf; (b) 5.42 ft from A
P2.56	16.08 ft
P2.58	0.40 m
P2.60	(a) 3.28E6 lbf; (b) 3.2E-6 ft below
P2.62	10.6 ft
P2.64	1.35 m
P2.66	$F = 1.18$ E9 N, $M_C = 3.13$ E9 N · m counterclockwise, no tipping
P2.68	18,040 N
P2.70	3.28 m
P2.72	$h \approx 1.12$ m
P2.74	$H = R[\pi/4 + \{(\pi/4)^2 + 2/3\}^{1/2}]$
P2.76	(a) 239 kN; (c) 388 kN · m
P2.78	$P = \pi\gamma R^3/4$
P2.80	$\theta > 77.4°$
P2.82	$F_H = 97.9$ MN, $F_V = 153.8$ MN
P2.84	$F_H = 4895$ N, $F_V = 7343$ N
P2.86	$P = 59$ kN
P2.88	$F_H = 176$ kN, $F_V = 31.9$ kN, yes
P2.90	$F_V = 22,600$ N; $F_H = 16,500$ N
P2.92	$F_{one\ bolt} \approx 11,300$ N
P2.94	$C_x = 2996$ lb, $C_z = 313$ lbf
P2.96	$F_H = 336$ kN; $F_V = 162$ kN
P2.98	$F_H = 7987$ lbf, $F_V = 2280$ lbf
P2.100	$F_H = 0, F_V = 297$ kN
P2.102	(a) 238 kN; (b) 125 kN
P2.104	5.0 N
P2.106	$z \approx 4000$ m
P2.108	(a) 0.0427 m; (b) 1592 kg/m^3
P2.110	$h \approx$ (a) 7.05 mm; (b) 7.00 mm
P2.112	(a) 39 N; (b) 0.64
P2.114	0.636
P2.116	6.4 mm
P2.118	6.14 ft
P2.120	34.3°
P2.122	$a/b \approx 0.834$
P2.124	6850 m
P2.126	3130 Pa (vacuum)
P2.128	Yes, stable if $S > 0.789$
P2.130	Slightly unstable, MG $= -0.007$ m
P2.132	Stable if $R/h > 3.31$
P2.134	(a) unstable; (b) stable
P2.136	$MG = L^2/(3\pi R) - 4R/(3\pi) > 0$ if $L > 2R$
P2.138	2.77 in deep; volume $= 10.8$ fluid ounces
P2.140	$a_x =$ (a) -1.96 m/s^2 (deceleration); (b) -5.69 m/s^2 (deceleration)
P2.142	(a) 16.3 cm; (b) 15.7 N
P2.144	(a) $a_x \approx 319$ m/s^2; (b) no effect, $p_A = p_B$

P2.146	Leans to the right at $\theta = 27°$
P2.148	Leans to the left at $\theta = 27°$
P2.150	5.5 cm; linear scale OK
P2.152	(a) 224 r/min; (b) 275 r/min
P2.154	552 r/min
P2.156	420 r/min
P2.157	77 r/min, minimum pressure halfway between B and C
P2.158	10.57 r/min

Chapter 3

P3.2	$\mathbf{r} =$ position vector from point O
P3.4	1.73 in
P3.6	$Q = (2b/3)(2g)^{1/2}[(h + L)^{3/2} - (h - L)^{3/2}]$
P3.8	(a) 5.45 m/s; (b) 5.89 m/s; (c) 5.24 m/s
P3.10	(a) 3 m/s; (b) 6 m/s; (c) 5 cm/s out
P3.12	$\Delta t = 46$ s
P3.14	$dh/dt = (Q_1 + Q_2 - Q_3)/(\pi d^2/4)$
P3.16	$Q_{top} = 3U_0 b\delta/8$
P3.18	(b) $Q = 16bhu_{max}/9$
P3.20	(a) 7.8 mL/s; (b) 1.24 cm/s
P3.22	(a) 0.06 kg/s; (b) 1060 m/s; (c) 3.4
P3.24	$h = [3Kt^2d^2/(8 \tan^2 \theta)]^{1/3}$
P3.26	$Q = 2U_0 bh/3$
P3.28	$t_{drain} = (A_b/A_0)(h_0/2g)^{1/2}$
P3.30	1100 N per meter of width
P3.32	$V_{hole} = 6.1$ m/s
P3.34	$V_2 = 4660$ ft/s
P3.36	$U_3 = 6.33$ m/s
P3.38	$V = V_0 r/(2h)$
P3.40	500 N to the left
P3.42	$F = (p_1 - p_a)A_1 - \rho_1 A_1 V_1^2[(D_1/D_2)^2 - 1]$
P3.44	$F = \rho U^2 Lb/3$
P3.46	$\alpha = (1 + \cos \theta)/2$
P3.48	$V_0 \approx 2.27$ m/s
P3.50	102 kN
P3.52	$F = \rho WhV_1^2[1/(1 - \sin \theta) - 1]$ to the left
P3.54	163 N
P3.56	(a) 18.5 N to left; (b) 7.1 N up
P3.58	40 N
P3.60	2100 N
P3.62	3100 N
P3.64	980 N
P3.66	8800 N
P3.70	91 lbf
P3.72	Drag ≈ 4260 N
P3.74	$F_x = 0, F_y = -17$ N, $F_z = 126$ N
P3.76	(a) 1670 N/m; (b) 3.0 cm; (c) 9.4 cm
P3.80	$F = (\rho/2)gb(h_1^2 - h_2^2) - \rho h_1 bV_1^2(h_1/h_2 - 1)$
P3.82	25 m/s
P3.84	23 N
P3.86	274 kPa

P3.88 $V = \zeta + [\zeta^2 + 2\zeta V_j]^{1/2}, \zeta = \rho Q/2k$

P3.90 $dV/dt = g$

P3.92 $dV/dt = gh/(L + h)$

P3.94 $h = 0$ at $t \approx 70$ s

P3.96 $d^2Z/dt^2 + 2gZ/L = 0$

P3.100 (a) 507 m/s and 1393 m; (b) 14.5 km

P3.102 $h_2/h_1 = -\frac{1}{2} + \frac{1}{2}[1 + 8V_1^2/(gh_1)]^{1/2}$

P3.104 $\Omega = (-V_e/R) \ln (1 - \dot{m}t/M_0)$

P3.106 $\Omega_{final} = 75$ rad/s

P3.108 (a) $V = V_0/(1 + CV_0t/M), C = \rho bh(1 - \cos \theta)$

P3.110 (a) 0.113 ft · lbf; (b) 250 r/min

P3.112 $T = \dot{m}R_0^2\Omega$

P3.114 (a) 414 r/min; (b) 317 r/min

P3.116 $P = \rho Qr_2\omega[r_2\omega - Q \cot \theta_2/(2\pi r_2 b_2)]$

P3.118 $P = \rho Q^2\omega \cot \theta_2/(2\pi b_2)$

P3.120 (a) 22 ft/s; (b) 110 ft/s; (c) 710 hp

P3.122 $L = -h_1 (\cot \theta)/2$

P3.124 41 r/min

P3.126 -15.5 kW (work done *on* the fluid)

P3.128 1.07 m³/s

P3.130 34 kW

P3.134 5060 hp

P3.136 $z_1 = 115$ m

P3.138 $\mu = \pi\rho gd^4(H + L)/(128LQ) - \alpha_2\rho Q/(16\pi L)$

P3.140 1640 hp

P3.142 (a) 1150 gal/min; (b) 67 hp

P3.144 26 kW

P3.146 $h = 3.6$ ft

P3.148 8.2 m

P3.150 Lift = 119 kN

P3.152 (a) 85.9°; (b) 55.4°

P3.154 8.51 m

P3.156 (a) 52 cm; (b) 3.12 s

P3.158 (a) 169.4 kPa; (b) 209 m³/h

P3.160 (a) 31 m³/s; (b) 54 kW

P3.162 $Q = 166$ ft³/min, $\Delta p = 0.0204$ lbf/in²

P3.164 (a) 5.25 kg/s; (b) 0.91 m

P3.166 (a) 60 mi/h; (b) 1 atm

P3.168 $h = 1.08$ ft

P3.170 $h = 1.76$ m

P3.172 $D = 0.132$ ft

P3.174 (a) 5.61 ft/s; (b) further constriction *reduces* V_2

P3.176 (a) 9.3 m/s; (b) 68 kN/m

P3.178 $h_2 = 2.03$ ft (subcritical) or 0.74 ft (supercritical)

P3.180 $V = V_f \tanh (V_f t/2L), V_f = (2gh)^{1/2}$

P3.182 $kp/[(k - 1)\rho] + V^2/2 + gz = $ constant

P3.184 0.37 hp

Chapter 4

P4.2   (a) $du/dt = (2V_0^2/L)(1 + 2x/L)$

P4.4   (b) $a_x = (U_0^2/L)(1 + x/L); a_y = (U_0^2/L)(y/L)$

P4.6   (a) $6V_0^2/L$; (b) $L \ln 3/(2V_0)$

P4.8   (a) $0.0196 V^2/L$; (b) at $t = 1.05 L/U$

P4.10 (a) $v = -xy^2$; (b) $u = -x^3/3$

P4.12 If $v_\theta = v_\phi = 0, v_r = r^{-2}$ fcn (θ, ϕ)

P4.14 $v_\theta = $ fcn(r) only

P4.16 (a) Yes, continuity is satisfied.

P4.18 $\rho = \rho_0 L_0/(L_0 - Vt)$

P4.20 $v = v_0 = $ const, $\{K\} = \{L/T\}, \{a\} = \{L^{-1}\}$

P4.22 $v_r = U_0\cos\theta + V_0\sin\theta;$
       $v_\theta = -U_0\sin\theta + V_0\cos\theta$

P4.28 Exact solution for any a or b

P4.30 (a, b) Yes, continuity and Navier-Stokes are satisfied.

P4.32 $f_1 = C_1 r; f_2 = C_2/r$

P4.36 $C = \rho g \sin \theta/(2\mu)$

P4.38 $C_z = \tau_{yx} - \tau_{xy}$

P4.48 $\psi = U_0 r \sin\theta - V_0 r \cos\theta + $ const

P4.50 Inviscid flow around a 180° turn

P4.52 $\psi = -4Q\theta/(\pi b)$

P4.54 $Q = ULb$

P4.60 Irrotational, $z_0 = H - \omega^2R^2/(2g)$

P4.62 $\psi = Vy^2/(2h) + $ const

P4.66 $\psi = -K \sin \theta/r$

P4.68 (a) Yes, a velocity potential exists.

P4.70 $\phi = \lambda \cos \theta/r^2, \lambda = 2am$

P4.72 (a) $\psi = -0.0008 \theta$; (b) $\phi = -0.0008 \ln(r)$

P4.74 $\psi = B r \sin \theta + B L \ln r + $ const

P4.76 Yes, ψ exists.

P4.78 (a) $V_{wall,max} = m/L$; (b) p_{min} at $x = L$

P4.80 (a) $w = (\rho g/2\mu)(2\delta x - x^2)$

P4.82 Obsessive result: $v_\theta = \Omega R^2/r$

P4.84 $v_z = (\rho gb^2/2\mu) \ln (r/a) - (\rho g/4\mu)(r^2 - a^2)$

P4.86 $Q = 0.0031$ m³/(s · m)

P4.88 $v_z = U \ln (r/b)/[\ln (a/b)]$

P4.90 (a) $D = 10$ cm; (b) $Q = 34$ m³/h

P4.92 $h = h_0 \exp[-\pi D^4\rho gt/(128\mu LA_0)]$

P4.94 $v_\theta = \Omega R^2/r$

Chapter 5

P5.2   Prototype $V = 22.8$ mi/h

P5.4 $V = 1.55$ m/s, $F = 1.3$ N

P5.6 $F \approx 450$ N

P5.8 $Mo = g\mu^4/(\rho Y^3)$

P5.10 (a) $\{ML^{-2}T^{-2}\}$; (b) $\{MLT^{-2}\}$

P5.12 $St = \mu U/(\rho gD^2)$

P5.14 There are 3 pi groups, not just 2.

P5.16 Stanton number = $h/(\rho Vc_p)$

P5.18 $Q\mu/[(\Delta p/L)b^4] = $ const

P5.20 (a) $\{C\} = \{ML^{-1}T^{n-2}\}$

P5.22 $\Omega D/V = $ fcn(N, H/L)

P5.24 $F/(\rho V^2L^2) = $ fcn$(\alpha, \rho VL/\mu, L/D, V/a)$

P5.26 (a) Indeterminate; (b) $T = 2.75$ s

P5.28 $\delta/L = \text{fcn}[L/D, \rho VD/\mu, E/(\rho V^2)]$

P5.30 $\tau_w/(\rho\Omega^2 R^2) = \text{fcn}(\rho\Omega R^2/\mu, \Delta r/R)$

P5.32 $Q/(bg^{1/2}H^{3/2}) = \text{const}$

P5.34 $k_{\text{hydrogen}} \approx 0.182 \; W/(m \cdot K)$

P5.36   (a) $Q_{\text{loss}}R/(A\Delta T) = \text{constant}$

P5.38 $d/D = \text{fcn}(\rho UD/\mu, \rho U^2D/Y)$

P5.40 $h/L = \text{fcn}(\rho g L^2/Y, \alpha, \theta)$

P5.42   Halving m increases f by about 41 percent.

P5.44   (a) $\{\sigma\} = \{L^2\}$

P5.48 $F \approx 0.17 \; N$; (doubling U quadruples F)

P5.50   (a) $F/(\mu UL) = \text{constant}$

P5.52   (b) 180 m/s^2

P5.54   Power $\approx$ 7 hp

P5.56 $F_{\text{air}} \approx 25 \; N/m$

P5.58 $V \approx 2.8$ m/s

P5.60   (b) 4300 N

P5.62 $\Omega_{\text{max}} \approx 26.5$ r/s; $\Delta p \approx 22{,}300$ Pa

P5.64 $\omega_{\text{aluminum}} = 0.77$ Hz

P5.66   (a) $V = 27$ m/s; (b) $z = 27$ m

P5.68   (a) $F/(\mu U) = \text{constant}$; (b) no, not plausible

P5.70 $F = 87$ lbf (extrapolated)

P5.72 $V = 25$ ft/s

P5.74   Prototype moment = 88 kN $\cdot$ m

P5.76   Drag = 107,000 lbf

P5.78   Weber no. ≈ 100 if $L_m/L_p = 0.0090$

P5.80   (a) 1.86 m/s; (b) 42,900; (c) 254,000

P5.82   Speeds: 19.6, 30.2, and 40.8 ft/s;
        Drags: 14,600; 31,800; and 54,600 lbf

P5.84 $V_m = 39$ cm/s; $T_m = 3.1$ s; $H_m = 0.20$ m

P5.88   At 340 W, $D = 0.109$ m

P5.90 $\Delta p D/(\rho V^2 L) = 0.155(\rho VD/\mu)^{-1/4}$

Chapter 6

P6.4   (a) 106 m^3/h; (b) 3.6 m^3/h

P6.6   (a) hydrogen, $x = 43$ m

P6.8   (a) −3600 Pa/m; (b) −13,400 Pa/m

P6.10   (a) From A to B; (b) $h_f = 7.8$ m

P6.12 $\mu = 0.29$ kg/m-s

P6.14 $Q = 0.0067$ m^3/h if $H = 50$ cm

P6.16 $d_{\text{min}} = 1.67$ m

P6.18 $\mu = 0.0026$ kg/m-s (laminar flow)

P6.20   4500 cc/h

P6.22 $F = 4.0$ N

P6.24   (a) 0.019 m^3/h, laminar; (b) $d = 2.67$ mm

P6.26   (a) $D_2 = 5.95$ cm

P6.28 $\Delta p = 65$ Pa

P6.30   (a) 19.3 m^3/h; (b) flow is *up*

P6.32   (a) flow is *up*; (b) 1.86 m^3/h

P6.36   (a) 0.029 lbf/ft^2; (b) 70 ft/s

P6.38   5.72 m/s

P6.44 $h_f = 10.4$ m, $\Delta p = 1.4$ MPa

P6.46   Input power $\approx$ 11.2 MW

P6.48 $r/R = 1 - e^{-3/2}$

P6.50   (a) −4000 Pa/m; (b) 50 Pa; (c) 46 percent

P6.52 $p_1 = 2.38$ MPa

P6.54 $t_{\text{drain}} = [4WY/(\pi D^2)][2h_0(1 + f_{\text{av}}L/D)/g]^{1/2}$

P6.56   (a) 188 km; (b) 27 MW

P6.58   Power $\approx$ 870 kW

P6.60   (a) Not identical to Haaland

P6.62   204 hp

P6.64 $Q = 19.6$ m^3/h (laminar, Re = 1450)

P6.66   (a) 56 kPa; (b) 85 m^3/h; (c) $u = 3.3$ m/s at $r = 1$ cm

P6.70 $V = 6.05$ ft/s, $Q = 1.19$ ft^3/s

P6.72 $D \approx 9.2$ cm

P6.74 $D = 350$ m

P6.76 $Q = 15$ m^3/h

P6.78 $Q = 25$ m^3/h (to the left)

P6.80 $Q = 0.905$ m^3/s

P6.82   0.384 m

P6.84 $D \approx 0.104$ m

P6.86   (a) 3.0 m/s; (b) 0.325 m/m; (c) 2770 Pa/m

P6.88   About 17 passages

P6.90 $Q = 19.6$ ft^3/s

P6.92   (a) 1530 m^3/h; (b) 6.5 Pa (vacuum)

P6.94   260 Pa/m

P6.96 $h \leq 4$ mm

P6.98   Approximately 128 squares

P6.102   (a) 5.55 hp; (b) 5.31 hp with 6° cone

P6.104 $\Delta p = 0.0305$ lbf/in^2

P6.106 $Q = 0.0296$ ft^3/s

P6.108   (a) $K \approx 9.7$; (b) $Q \approx 0.48$ ft^3/s

P6.110   840 W

P6.112 $Q = 0.0151$ ft^3/s

P6.114   Short duct: $Q = 6.92$ ft^3/s

P6.116 $Q = 0.027$ m^3/s

P6.118 $\Delta p = 131$ lbf/in^2

P6.120 $Q_1 = 0.0281$ m^3/s, $Q_2 = 0.0111$ m^3/s, $Q_3 = 0.0164$ m^3/s

P6.122   Increased ϵ/d and L/d are the causes

P6.124 $Q_1 = -2.09$ ft^3/s, $Q_2 = 1.61$ ft^3/s, $Q_3 = 0.49$ ft^3/s

P6.126 $\theta_{\text{opening}} = 35°$

P6.128 $Q_{AB} = 3.47$, $Q_{BC} = 2.90$, $Q_{BD} = 0.58$, $Q_{CD} = 5.28$,
         $Q_{AC} = 2.38$ ft^3/s (all)

P6.130 $Q_{AB} = 0.95$, $\;\; Q_{BC} = 0.24$, $\;\; Q_{BD} = 0.19$, $\;\; Q_{CD} = 0.31$,
         $Q_{AC} = 1.05$ ft^3/s (all)

P6.132 $2\theta = 6°$, $D_e = 2.0$ m, $p_e = 224$ kPa

P6.134 $2\theta = 10°$, $W_e = 8.4$ ft, $p_e = 2180$ lbf/ft^2

P6.136   (a) 25.5 m/s, (b) 0.109 m^3/s, (c) 1.23 Pa

P6.138   46.7 m/s

P6.140 $\Delta p = 273$ kPa

P6.142 $Q = 18.6$ gal/min, $d_{\text{reducer}} = 0.84$ cm

P6.144 $Q = 54$ m^3/h

P6.146   (a) 0.00653 m^3/s; (b) 100 kPa

P6.148   (*a*) 1.58 m; (*b*) 1.7 m

P6.150 $\Delta p = 27$ kPa

P6.152 $D = 4.12$ cm

P6.154 $h = 58$ cm

P6.156 $Q = 0.924$ ft^3/s

P6.158   (*a*) 49 m^3/h; (*b*) 6200 Pa

Chapter 7

P7.2 $\text{Re}_c = 1.5$ E7

P7.4   (*a*) 4 μm; (*b*) 1 m

P7.6 $H = 2.5$ (versus 2.59 for Blasius)

P7.8   Approximately 0.08 N

P7.12   Does not satisfy $\partial^2 u/\partial y^2 = 0$ at $y = 0$

P7.14 $C = \rho v_0/\mu = \text{const} < 0$ (wall suction)

P7.16   (*a*) F = 181 N; (*b*) 256 N

P7.18   (*a*) 1.54 m/s; (*b*) 0.0040 Pa

P7.20 $x \approx 0.91$ m

P7.22 $\psi = (\nu x U)^{1/2} f(\eta)$

P7.24 $h_1 = 9.2$ mm; $h_2 = 5.5$ mm

P7.26 $F_a = 2.83\,F_1$, $F_b = 2.0\,F_1$

P7.28   (a) $F_{\text{drag}} = 2.66\,N^2(\rho\mu L)^{1/2}U^{3/2}a$

P7.30   Predicted thickness is about 10 percent higher

P7.32 $F = 0.0245\,\rho\nu^{1/7}\,L^{6/7}\,U_0^{13/7}\,\delta$

P7.34 $F = 725$ N

P7.36   7.2 m/s = 14 kn

P7.38   (*a*) 7.6 m/s; (*b*) 6.2 m/s

P7.40 $L = 3.51$ m, $b = 1.14$ m

P7.42   (*a*) 63 mi/h

P7.44   Accurate to about ± 6 percent

P7.46 $\varepsilon \approx 9$ mm, $U = 11.1$ m/s = 22 kn

P7.48   Separation at $x/L = 0.158$ (1 percent error)

P7.50   Separation at $x/R = 1.80$ rad = 103.1°

P7.52   (*a*) $\text{Re}_b = 0.84 < 1$; (*b*) $2a = 30$ mm

P7.54   Moment $\approx$ 200,000 N · m

P7.56   (*a*) 14 N; (*b*) crosswind creates a very large side force

P7.58   (*a*) 3200 N/m; (*b*) 2300 N/m

P7.60   Tow power = 140 hp

P7.62   Square side length $\approx$ 0.83 m

P7.64 $\Delta t_{1000-2000\text{m}} = 202$ s

P7.68   (*a*) 34 m/s; (*b*) no, only 67 percent of terminal velocity at impact

P7.70   40 ft

P7.72   (*a*) $L = 6.3$ m; (*b*) 120 m

P7.74   About 130 mi/h

P7.76   (*a*) 343 hp

P7.78 $\Delta p = 100$ Pa

P7.80 $\theta = 72°$

P7.82   (*a*) 46 s

P7.84 $V = 9$ m/s

P7.86   Approximately 2.9 m by 5.8 m

P7.88   (*a*) 62 hp; (*b*) 86 hp

P7.90 $V_{\text{overturn}} \approx 145$ ft/s = 99 mi/h

P7.94   Torque $\approx (C_D/4)\rho\Omega^2 DR^4$, $\Omega_{\max} = 85$ r/min

P7.96 $\Omega_{\text{avg}} \approx 0.21\,U/D$

P7.98   (*b*) $h \approx 0.18$ m

P7.100   (*a*) 73 mi/h; (*b*) 79 mi/h

P7.104   29.5 knots

P7.106   (*a*) 300 m; (*b*) 380 m

P7.108 $\Delta x_{\text{ball}} \approx 13$ m

P7.110 $\Delta y \approx 1.9$ ft

P7.114 $V_{\text{final}} \approx 18.3$ m/s = 66 km/h

P7.116   (*a*) 87 mi/h; (*b*) 680 hp

P7.118   (*a*) 21 m/s; (*b*) 360 m

P7.120 $(L/D)_{\max} = 21$; $\alpha = 4.8°$

P7.122   (*a*) 6.7 m/s; (*b*) 13.5 m/s = 26 kn

P7.124 $\Omega_{\text{crude theory}} \approx 340$ r/s

Chapter 8

P8.2 $\Gamma = \pi\Omega(R_2^2 - R_1^2)$

P8.4   No, $1/r$ is not a proper two-dimensional potential

P8.6 $\psi = B\,r^2\sin(2\theta)$

P8.8 $\Gamma = 4B$

P8.10   (*a*) 1.27 cm

P8.12 $\Gamma = 0$

P8.14   Irrotational outer, rotational inner; minimum $p = p_\infty - \rho\omega^2 R^2$ at $r = 0$

P8.16   (*a*) 0.106 m to the left of *A*

P8.18   From afar: a single source $4m$

P8.20   Vortex near a wall (see Fig. 8.17*b*)

P8.22   Same as Fig. 8.6 except upside down

P8.24 $C_p = -\{2(x/a)/[1 + (x/a)^2]\}^2$, $C_{p,\min} = -1.0$ at $x = a$

P8.26   (*a*) 8.75 m; (*b*) 27.5 m on each side

P8.28   Creates a source in a square corner

P8.30 $r = 25$ m

P8.34   Two stagnation points, at $x = \pm a/\sqrt{3}$

P8.36 $U_\infty = 12.9$ m/s, $2L = 53$ cm, $V_{\max} = 22.5$ m/s

P8.40   1.47 m

P8.42 $K/(U_\infty a) = 0.396$, $h/a = 1.124$

P8.44 $K = 3.44$ m^2/s; (*a*) 218 kPa; (*b*) 205 kPa *upper*, 40 kPa *lower*

P8.46 $F_{\text{1-bolt}} = 5060$ N

P8.50 $h = 3a/2$, $U_{\max} = 5U/4$

P8.52 $V_{\text{boat}} = 9.4$ ft/s with wind at 50°

P8.54 $F_{\text{parallel}} = 6700$ lbf, $F_{\text{normal}} = 2700$ lbf, power $\approx$ 560 hp (very approximate)

P8.60   This is Fig. 8.15*a*, flow in a 60° corner

P8.62   Stagnation flow near a "bump"

P8.64   (*a*) Yes; (*b*) $\psi = Br^{1.2}\sin(1.2\theta)$

P8.66 $\lambda = 0.45m/(5m + 1)$ if $U = Cx^m$

P8.68   Flow past a Rankine oval

P8.70   Applied to wind tunnel "blockage"

P8.72   Adverse gradient for $x > a$

P8.74	$V_{B,\text{total}} = (8K\mathbf{i} + 4K\mathbf{j})/(15a)$
P8.78	Need an infinite array of images
P8.82	(a) 4.5 m/s; (b) 1.13; (c) 1.26 hp
P8.84	(a) 0.21; (b) 1.9°
P8.86	(a) 26 m; (b) 8.7; (c) 1600 N
P8.88	$\text{Thrust}_{1\text{-engine}} \approx 2900$ lbf
P8.92	(a) 0.77 m; (b) $V = 4.5$ m/s at $(r, \theta) = (1.81, 51°)$ and (1.11, 88°)
P8.94	Yes, they are orthogonal
P8.96	(a) $0.61\, U_\infty^2/a$
P8.98	Yes, a closed teardrop shape appears
P8.100	$V = 14.1$ m/s, $p_A = 115$ kPa
P8.102	(a) 1250 ft; (b) 1570 ft (crudely)

Chapter 9

P9.2	(a) $V_2 = 450$ m/s, $\Delta s = 515$ J/(kg · K); (b) $V_2 = 453$ m/s, $\Delta s = 512$ J/(kg · K)
P9.4	About 50 m/s
P9.6	$c_p = 1169$ J/(kg · K); $c_v = 980$ J/(kg · K)
P9.8	410 K
P9.10	Ma = 0.78
P9.12	(a) 2.13 E9 Pa and 1460 m/s; (b) 2.91 E9 Pa and 1670 m/s; (c) 2645 m/s
P9.18	Ma ≈ 0.24
P9.20	(a) Air: 144 kPa and 995 m/s; (b) helium: 128 kPa and 2230 m/s
P9.22	(a) 267 m/s; (b) 286 m/s
P9.24	(b) at Ma ≈ 0.576
P9.28	(a) 0.17 kg/s; (b) 0.90
P9.30	Deviation less than 1 percent at Ma = 0.3
P9.32	(a) 141 kPa; (b) 101 kPa; (c) 0.706
P9.34	(a) 340 K
P9.40	(a) 2.50; (b) 7.6 cm²; (c) 1.27 kg/s; (d) $\text{Ma}_2 = 1.50$
P9.42	(a) Ma = 0.90, $T = 260$ K, $V = 291$ m/s
P9.44	$V_e = 5680$ ft/s, $p_e = 15.7$ psia, $T_e = 1587°$R, thrust = 4000 lbf
P9.46	(a) 0.0020 m²
P9.48	(a) 313 m/s; (b) 0.124 m/s; (c) 0.00331 kg/s
P9.50	(a) $D_{\text{exit}} = 5.6$ cm; (b) can reduce to 75 kPa
P9.52	(a) 5.9 cm²; (b) 773 kPa
P9.54	$\text{Ma}_2 = 0.513$
P9.56	At about $A_1 \approx 24.7$ cm²
P9.58	(a) 306 m/s; (b) 599 kPa; (c) 498 kPa
P9.60	Upstream: Ma = 1.92, $V = 585$ m/s
P9.62	$C = 19,100$ ft/s, $V_{\text{inside}} = 15,900$ ft/s
P9.64	(a) 0.150 kg/s; (b, c) 0.157 kg/s
P9.66	$h = 1.09$ m
P9.68	$p_{\text{atm}} = 92.6$ kPa; max flow = 0.140 kg/s
P9.70	(a) 388 kPa; (b) 19 kPa
P9.72	$D \approx 9.3$ mm
P9.74	(a) 1.09 MPa; (b) 2.24 kg/s

P9.76	$\Delta t_{\text{shocks}} \approx 23$ s; $\Delta t_{\text{choking-stops}} \approx 39$ s
P9.78	Case A: 0.0071 kg/s; B: 0.0068 kg/s
P9.80	$A^* = 2.4$ E-6 ft² or $D_{\text{hole}} = 0.021$ in
P9.82	$V_e = 110$ m/s, $\text{Ma}_e = 0.67$ (yes)
P9.84	(a) 0.96 kg/s; (b) 0.27; (c) 435 kPa
P9.86	$V_2 = 107$ m/s, $p_2 = 371$ kPa, $T_2 = 330$ K, $p_{02} = 394$ kPa
P9.88	(a) 0.0646 kg/s
P9.90	(a) 0.764 kg/s; (b) 0.590 kg/s; (c) 0.314 kg/s
P9.92	(a) 14.46 m
P9.96	(a) 128 m; (b) 80 m; (c) 105 m
P9.98	(a) 430; (b) 0.12; (c) 0.00243 kg/h
P9.100	0.0219 kg/s
P9.102	Flow is choked at 0.56 kg/s
P9.104	$p_{\text{tank}} = 190$ kPa
P9.106	about 91 s
P9.108	Mass flow drops by about 32 percent
P9.112	(a) 105 m/s; (b) 215 kPa
P9.116	$V_{\text{plane}} \approx 2640$ ft/s
P9.118	$V = 204$ m/s, Ma = 0.6
P9.120	P is 3 m ahead of the small circle, Ma = 2.0, $T_{\text{stag}} = 518$ K
P9.122	$\beta = 23.13°$, $\text{Ma}_2 = 2.75$, $p_2 = 145$ kPa
P9.124	(a) 1.87; (b) 293 kPa; (c) 404 K; (d) 415 m/s
P9.126	(a) 25.9°; (b) 26.1°
P9.128	$\delta_{\text{wedge}} \approx 15.5°$
P9.132	(a) $p_A = 18.0$ psia; (b) $p_B = 121$ psia
P9.134	$\text{Ma}_3 = 1.02$, $p_3 = 727$ kPa, $\phi = 42.8°$
P9.136	(a) $h = 0.40$ m; (b) $\text{Ma}_3 = 2.43$
P9.138	$p_r = 21.7$ kPa
P9.140	$\text{Ma}_2 = 2.75$, $p_2 = 145$ kPa
P9.142	(a) $\text{Ma}_2 = 2.641$, $p_2 = 60.3$ kPa; (b) $\text{Ma}_2 = 2.299$, $p_2 = 24.1$ kPa
P9.146	(a) 2.385; (b) 47 kPa
P9.148	(a) 4.44; (b) 9.6 kPa
P9.150	(a) $\alpha = 4.10°$; (b) drag = 2150 N/m
P9.152	Parabolic shape has 33 percent more drag

Chapter 10

P10.2	(a) Fr = 2.69
P10.4	These are piezometer tubes (no flow)
P10.6	(a) Fr = 3.8; (b) $V_{\text{current}} = 7.7$ m/s
P10.8	$\Delta t_{\text{travel}} = 6.3$ h
P10.10	$\lambda_{\text{crit}} = 2\pi(Y/\rho g)^{1/2}$
P10.14	Flow must be fully rough turbulent (high Re) for Chézy to be valid
P10.16	(a) 0.553 m³/s
P10.18	$y_n = 0.993$ m
P10.20	$Q = 74$ ft³/s
P10.22	$S_0 = 0.00038$ (or 0.38 m/km)
P10.24	$y_n = 0.56$ m
P10.30	$\Delta t \approx 32$ min
P10.32	$y_n = 0.837$ m

P10.34   If $b = 4$ ft, $y = 9.31$ ft, $P = 22.62$ ft; if $b = 8$ ft, $y = 4.07$ ft, $P = 16.14$ ft

P10.36 $y_2 = 3.6$ m

P10.38 $S_0 = 0.011$

P10.42 $P = 41.3$ ft (71 percent more than Prob. P10.39)

P10.44   Hexagon side length $b = 2.12$ ft

P10.46 $h_0/b \approx 0.49$

P10.48   (a) 0.00634; (b) 0.00637

P10.50   (a) 2.37; (b) 0.62 m; (c) 0.0023

P10.52 $W = 2.06$ m

P10.54   (a) 1.98 m; (b) 3.11 m/s; (c) 0.00405

P10.56   (a) 1.02 m^3/s; (b) 0.0205

P10.58   (a) 1.0 m; (b) 1.0; (c) 0.77 m^3/s

P10.60   (a) 0.055 m^3/s/m; (b) 0.086 m

P10.64 $h_{max} \approx 0.35$ m

P10.66   (a) 1.47; (b) $y_2 = 1.19$ m

P10.70   2600 m^3/s

P10.72   (a) 0.046 m; (b) 4.33 m/s; (c) 6.43

P10.76   (a) 379 ft^3/s

P10.78 $\Delta t \approx 8.6$ s (crude analysis)

P10.80   (a) 3.83 m; (b) 4.83 m^3/(s · m)

P10.82   (a) 1.46 ft; (b) 15.5 ft/s; (c) 2.26; (d) 13%; (e) 2.52 ft

P10.84 $y_2 = 0.82$ ft; $y_3 = 5.11$ ft; 47 percent

P10.86   (a) 6.07 m/s; (b) $\Delta V = 2.03$ m/s

P10.88   (a) 2.22 m^3/s/m; (b) 0.79 m; (c) 5.17 m; (d) 60%; (e) 0.37 m

P10.90   (a) $y_2 = 1.83$ ft; $y_3 = 7.86$ ft

P10.92   (a) 3370 ft^3/s; (b) 7000 hp

P10.94   (a) 0.612 m

P10.98   (a) steep S-3; (b) S-2; (c) S-1

P10.106   No entry depth leads to critical flow

P10.108   Approximately 6.6 m

P10.110   (a) $y_{crest} \approx 0.782$ m; (b) $y(L) \approx 0.909$ m

P10.112   M-1 curve, with $y = 2$ m at $L \approx 214$ m

P10.114   11.5 ft

P10.116 $Q \approx 9.51$ m^3/s

P10.120 $Y = 0.64$ m, $\alpha = 34°$

P10.122   5500 gal/min

P10.124   M-1 curve, $y = 10$ ft at $x = -3040$ ft

P10.126   At $x = -100$ m, $y = 2.81$ m

P10.128   At 300 m upstream, $y = 2.37$ m

Chapter 11

P11.6   This is a diaphragm pump

P11.8   (a) $H = 112$ ft and $\Delta p = 49$ lb/in^2; (b) $H = 112$ ft (of gasoline); $P = 15$ hp

P11.10   (a) 12 gal/min; (b) 12 gal/min; (c) 87%

P11.12   (a) 11.3 m; (b) 1520 W

P11.14   1870 W

P11.16   (a) 1450 W; (b) 1030 r/min

P11.18 $V_{vane} = (1/3)V_{jet}$ for max power

P11.20   (a) 2 roots: $Q = 7.5$ and 38.3 ft^3/s; (b) 2 roots; $H = 180$ ft and 35 ft

P11.22   (a) BEP = 92 percent at $Q = 0.20$ m^3/s

P11.26   Both are fine, the largest is more efficient.

P11.28   BEP at about 6 ft^3/s; $N_s \approx 1430$, $Q_{max} \approx 12$ ft^3/s

P11.30   (a) 2350 r/min; (b) 270 ft

P11.32   (a) $D \approx 15.5$ in; (c) $n \approx 2230$ r/min

P11.34   (a) 11.5 in; (b) 28 hp; (c) 100 ft; (d) 78 percent

P11.36 $D = 3.1$ in, $n = 8800$ r/min, $P = 25$ hp

P11.38   (a) 18.5 hp; (b) 7.64 in; (c) 415 gal/min; (d) 81 percent

P11.40   (a) $D_s = D(gH^*)^{1/4}/Q^{*1/2}$

P11.42   NPSH$_{proto} \approx 23$ ft

P11.44   No cavitation, required depth is only 5 ft

P11.46 $D_s \approx C/N_s$, $C = 7800 \pm 7$ percent

P11.48   (b) about 12 in

P11.52   (a) 6.56 m^3/s; (b) 12.0 kW; (c) 28.3°

P11.54   Centrifugal pumps, $D \approx 7.2$ ft

P11.56   (a) $D = 5.67$ ft, $n = 255$ r/min, $P = 700$ hp; (b) $D = 1.76$ ft, $n = 1770$ r/min, $P = 740$ hp

P11.58   Centrifugal pump, $\eta = 67$ percent, $D = 0.32$ ft

P11.60   (a) 623; (b) 762 gal/min; (c) 1.77 ft

P11.62 $D = 18.7$ ft, $\Delta p = 1160$ Pa

P11.64   (a) 15.4 in; (b) 900 r/min

P11.66   720 ft^3/min, non-BEP efficiency 78 percent

P11.68   (a) 4.8 in; (b) 6250 r/min

P11.70   (a) 212 ft; (b) 5.8 ft^3/s

P11.72   (a) 10 gal/min; (b) 1.3 in

P11.74   (a) 14.9; (b) 15.9; (c) 20.7 kgal/min

P11.76 $D_{pipe} \approx 1.70$ ft

P11.78 $D_{pipe} \approx 1.67$ ft, $P \approx 2000$ hp

P11.80   Both pumps work with three each in series, the largest being more efficient.

P11.84   Two turbines: (a) $D \approx 9.6$ ft; (b) $D \approx 3.3$ ft

P11.86 $N_{sp} \approx 70$, hence Francis turbines

P11.88   (a) Francis; (c) 16 in; (d) 900 r/min; (e) 87 hp

P11.90 $P \approx 800$ kW

P11.94   (a) 71 percent; (b) $N_{sp} \approx 19$

P11.96   (a) 1.68 ft; (b) 0.78 ft

P11.100   About 5.7 MW

P11.102 $Q \approx 29$ gal/min

P11.104   (a) 69 MW

Index

D

W

Z